THE DICTIONARY OF MODERN WAR

THE DICTIONARY
OF MODERN
WAR

Edward Luttwak
and Stuart Koehl

HarperCollins*Publishers*

Library of Congress Cataloging-in-Publication Data

Luttwak, Edward.
  The dictionary of modern war / Edward Luttwak and Stuart Koehl.
    p. cm.
  ISBN 0–06–270021–9
  1. Military art and science—Dictionaries. I. Koehl, Stuart L.,
  1956– . II. Title.
U24.L93 1991
355′.03—dc20   90–55998

91 92 93 94 95 CC/HC 10 9 8 7 6 5 4 3 2 1

PREFACE

This book on the ideas, institutions and weapons of modern military power and its ancillaries (from ARMS CONTROL to INTELLIGENCE) is meant to serve both as a reference source and as a readable guide. To that end, no abbreviations or acronyms are used in the text, unless preceded by a full rendition within the same entry. Every effort has also been made to avoid obscure technical language; the exceptions are each explained in the separate entries on LASER, SONAR, NUCLEAR and other current technologies of military interest.

In the same vein, the meretricious language of many published weapon descriptions has been deconstructed whenever possible ("advanced" for weapons not yet properly tried; "semi-automatic" for key functions that remain entirely manual; and "highly successful" for weapons perhaps notoriously inadequate but successfully sold all the same).

Too short to be an encyclopedia, this book is nevertheless more than a dictionary because it explains and does not merely describe its contents. In addition to the many entries devoted to single weapons (e.g., MIG-29), weapon categories (e.g., BALLISTIC MISSILE), and the national inventories of leading military powers (e.g., DESTROYERS, FRENCH), this book includes: i. Concepts, both very general (e.g., *strategy*) and very specific (e.g., NUCLEAR FREEZE); ii. Organizations, notably the major forces of the major powers, arranged in a hierarchy of entries (e.g., AIR FORCE, US, cross-referenced to STRATEGIC AIR COMMAND); iii. Technological applications, both in general (e.g., *radar*) and in detail (e.g., MTI— *moving target indicator*); iv. Methods of war, from the very broad (e.g., ELECTRONIC WARFARE) to the very specific (e.g., SPOT JAMMING); and, v.

Arms control treaties (e.g., the 1979 IMF TREATY) and long-term negotiations (e.g., START).

To assist those who want to read the text extensively, cross-references link discursive accounts of broader subjects (e.g., COMBAT AIRCRAFT), with more specific intermediate entries (e.g., FIGHTER), which are cross-referenced in turn to detailed descriptions of single items (e.g., EAGLE F-15). Readers may follow such a sequence from the general to the particular (or vice versa) within a given subject, or proceed laterally via cross-references that lead to different subjects (e.g., from *combat aircraft* to RADAR), which may be pursued in turn more specifically (e.g., from *radar* to DOPPLER SHIFT) or more broadly (e.g., from *radar* to ELECTRONIC COUNTERMEASURES).

Many military matters are controversial, and the authors have not tried to conceal their views behind the mask of an impossible objectivity. On the contrary, they have not hesitated to infuse this book with their own views; but they have also tried to provide enough information on each side of issues under debate to allow readers to independently form their own opinions.

Finally, a word about weights and measures: both US and metric measures are given in all cases, as are weights except for ship tonnages; "tons" should be read as metric tonnes in the case of European and Soviet vessels.

A

A-: U.S. designation for ATTACK AIRCRAFT, including: A-4 Macdonnell Douglas SKYHAWK; A-6 Grumman INTRUDER; A-7 LTV CORSAIR II; A-10 Fairchild THUNDERBOLT II; and A-37 Cessna DRAGONFLY. Additional prefix letters designate variants for diverse missions (e.g., R for RECONNAISSANCE). In some cases a different name is given to the aircraft, e.g., the EA-6B PROWLER.

AA: ANTI-AIRCRAFT (weapons). See also AIR DEFENSE.

AA-: NATO designation for Soviet air-to-air missiles. Models in service include: AA-2 ATOLL; AA-3 ANAB; AA-5 ASH; AA-6 ACRID; AA-7 APEX; AA-8 APHID; AA-9 AMOS; AA-10 ALAMO; and AA-11 Archer.

AAA: Anti-Aircraft Artillery. See ANTI-AIRCRAFT (Weapons).

AAC: Army Air Corps, British army branch whose inventory is limited to light HELICOPTERS (mainly the Westland LYNX and Aerospatiale GAZELLE) for observation and RECONNAISSANCE, anti-tank, artillery SPOTTING, command and control, liaison, casualty evacuation, and light transport roles.

AAFCE: Allied Air Forces, Central Europe, the NATO command responsible for the operational control of all air units allocated to the Central Front.

AAM: Air-to-Air Missile. See MISSILES, GUIDED.

AAW: See ANTI-AIR WARFARE.

ABBOT (FV433): British light self-propelled gun, in service since 1965. A tracked vehicle with a turret mounting a 105-mm. GUN-HOWITZER, Abbot has a crew of four: commander, gunner, loader, and driver. Forty rounds of ammunition are carried in the vehicle. Range with HE shells in the INDIRECT FIRE mode is 17,000 m. In both range and destructive effect, Abbot is badly outclassed by standard 155-mm. gun-howitzers. It does, however, have a high surge rate of fire (eight rounds per minute).

Abbot has a maximum armor thickness of 12 mm., sufficient only against splinters and small arms. A positive overpressure COLLECTIVE FILTRATION unit is provided for NBC defense.

A member of the FV430 family, which also includes the FV432 TROJAN armored personnel carrier (APC) and the FV438 SWINGFIRE anti-tank guided missile carrier, Abbot is now being replaced by the U.S.-made M109 155-mm. SP gun.

Specifications Length: 18.8 ft. (5.73 m.). Width: 8.67 ft. (2.64 m.). Height: 8.16 ft. (2.48 m.). Weight, combat: 16.5 tons. Powerplant: 240-hp. Rolls Royce 6-cylinder diesel. Speed, road: 30 mph (50 kph). Range, max.: 240 mi. (400 km.).

ABC: Atomic, Biological, and Chemical weapons (or warfare). Synonyms include CBR (Chemical, Biological, and Radiological) and NBC (Nuclear, Biological, and Chemical), the U.S.-preferred term. See also BIOLOGICAL WARFARE; CHEMICAL WARFARE; NUCLEAR WEAPONS.

ABM: See ANTI-BALLISTIC MISSILE (SYSTEM).

ABM-1: NATO designation for the Soviet anti-ballistic missile system. See GALOSH.

ABM TREATY: See ANTI-BALLISTIC MISSILE TREATY.

ABRAMS (M1): U.S. MAIN BATTLE TANK (MBT). After the rejection of the highly innovative but costly MBT-70, the M1 was developed during the 1970s as a more conservative alternative. Of all modern tanks, only the West German LEOPARD 2, British CHALLENGER, and Israeli MERKAVA are in its class.

The M1 was originally produced with British-developed CHOBHAM armor, a composite of steel, ceramics, and exotic alloys which provides twice the protection of an equivalent thickness of steel (the well-sloped glacis plate is equivalent to more than 650 mm. of homogeneous steel armor). Later M1s have an improved, U.S.-modified armor, and from 1988 production switched to an entirely new composite armor containing depleted uranium (STABALLOY).

The turret has the angular front and sides typical of Chobham construction. Of conventional layout, the hull is divided into a driving compartment in front, a fighting compartment in the middle, and an engine compartment in the rear. The driver has a reclining seat to reduce vehicle height, three observation periscopes, and a passive night-vision scope. The commander sits in the turret on the right, behind and above the gunner, and the loader is seated on the left. The commander has a low-profile cupola with six periscopes for 360° observation, plus a 3x machine-gun sight and an optical extension of the gunner's primary sight. The latter, with daylight magnification settings of 10x, 3x, and 1x, plus a 10x THERMAL IMAGING System (TIS) night sight, is stabilized in elevation and incorporates a Hughes LASER rangefinder. The entire sight unit is linked to a digital FIRE CONTROL computer, which incorporates range, target relative motion, ammunition type, barrel warp, barrel tilt, and meteorological factors into the ballistic solution, to provide a very high first-round hit probability.

Main armament was originally the Patton's M68E1 105-mm. rifled gun (the U.S. version of the British L7, which also arms the CENTURION, LEOPARD 1, Merkava, S-TANK, and many other Western MBTs), which can fire APFSDS, APDS, and HEAT ammunition. The APFSDS round weighs some 41 lb. (18.64 kg.) with its depleted uranium (Staballoy) penetrator, and has a muzzle velocity of roughly 1600 m./sec. and an effective range of more than 2000 m.; it can still penetrate the armor of all known Soviet tanks. A total of 55 rounds are carried, including 11 ready rounds in the fighting compartment and 44 rounds in a turret bustle behind a blast-resistant sliding door; the roof of the ammunition compartment is fitted with blowout panels to vent explosions upward. The gun is fully stabilized, allowing fairly accurate fire on the move; its elevation limits are +20° and −8°.

Secondary armament consists of a coaxial 7.62-mm. machine gun, a pintle-mounted 7.62-mm. machine gun by the loader's hatch, and a remotely operated 12.7-mm M2 HB BROWNING HEAVY MACHINE GUN over the commander's hatch. Six smoke grenade launchers are fitted on each side of the turret. The vehicle has a sophisticated fire detection system and an automatic Halon fire extinguisher. There is a COLLECTIVE FILTRATION unit to which each crewman can connect his personal gas mask.

The M1's most radical innovation is its 1500-shp. Avco-Lycoming AGT-1500 gas turbine engine, which provides a very high power-to-weight ratio (25 to 1) for excellent acceleration (0–20 mph/33.4 kph in less than 7 seconds) and good cross-country mobility. The gas turbine weighs and costs less than an equivalent diesel, but also has a higher fuel consumption. A reinforced torsion-bar suspension cushions the crew in some degree during high-speed runs over rough terrain.

In 1985 production switched to the M1A1, which has additional armor (in the form of applique plates on the turret face), a positive overpressure collective filtration system for NBC defense, and a 120-mm. Rheinmetall smoothbore gun (as on the Leopard 2), which fires APFSDS and HEAT rounds. Its APFSDS "arrow" round has a total weight of 41.8 lb. (19 kg.), a penetrator weight of 16 lb. (7.27 kg.), a muzzle velocity of more than 1600 m./sec., and an effective range in excess of 2000 m. against the frontal armor of all known Soviet tanks. The less accurate HEAT round weighs 37.4 lb. (17 kg.), including the 12.33-lb. (5.6-kg.) projectile, and is effective beyond 3000 m. Because of the ammunition's larger size, only 40 rounds can be carried. These changes increased the M1A1's combat weight by six tons, reducing its road speed and range.

The M1's shortcomings include its reduced range as compared to the diesel-powered PATTON and its manual-reloading layout. It has also been said that the M1 is too complex for maneuver-oriented operational methods; it is nevertheless superior to all other tanks.

Specifications Length: 25.98 ft. (7.92 m.). **Width:** 12 ft. (3.26 m.) **Height:** 9.46 ft. (2.88 m.).

Weight, combat: (M1) 54.5 tons, (M1A1) 57 tons. Powerplant: one AVCO-Lycoming AGT-1580 gas turbine, 1500 shp. Speed, road: (M1) 45 mph (75 kph), (M1A1) 40 mph (67 kph). Speed, cross-country: 29 mph (48.5 kph). Fuel: 475 gal. (2090 lit.). Range, max.: (M1) 280 mi. (468 km), (M1A1) 279 mi. (466 km).

AC: Hydrogen Cyanide, a nonpersistent, lethal BLOOD AGENT with CHEMICAL WARFARE applications. A highly volatile liquid with a faint odor of peaches, AC vaporizes at room temperature; it acts by binding with the enzyme cytochrome oxidase, thereby preventing the blood from absorbing oxygen. Highly effective, it has an LD-50 (the dose which will kill 50 percent of exposed personnel) of only 5000 mg./min./m.^{3}, with death by asphyxiation within 15 minutes of exposure. Lighter than air, AC dissipates within a few minutes, and is thus suitable for use immediately before a ground attack. AC can be delivered by artillery shells, missile warheads, aerial bombs, and aerosol dispensers. AC quickly clogs gas mask filters, which—most impractically—would have to be changed frequently. First aid consists of immediate doses of amyl nitrate and artificial respiration.

An easy chemical to manufacture and dispense, AC forms a major portion of the Soviet chemical arsenal, and was used by Iraq in 1988 to attack Kurdish villages.

Cyanogen chloride (CK) is very similar; it causes severe irritations of the eyes and throat, and can clog mask filters within ten minutes. When inhaled, CK is metabolized into AC.

ACDA: Arms Control and Disarmament Agency, an independent agency of the U.S. government which, in conjunction with the State Department, is responsible for the planning and conduct of arms control negotiations. At times, ACDA has been criticized for its bureaucratic tendency to value "negotiability" over strategy; equally, it has been criticized for passivity vis à vis the Defense Department. See also ARMS CONTROL; DISARMAMENT; INF TREATY; START.

ACE: Allied Command, Europe, in practice NATO's highest command, with control of AFCENT, AFNORTH, and AFSOUTH, plus the United Kingdom Air Command Region and the ACE MOBILE FORCE (AMF). The commander of ACE, always so far a U.S. Army general, is known as Supreme Allied Commander, Europe (SACEUR). See also ACHAN and ACLANT, its nominal equals.

ACE MOBILE FORCE: A small multinational NATO force under the direct control of SACEUR, and generally based on LIGHT INFANTRY. Its mission is rapid intervention in crises, mainly to introduce a multinational "presence" in exposed NATO sectors otherwise manned only by national forces.

ACHAN: Allied Command (English) Channel, hierarchically (but not substantively) a major NATO command responsible for the security of the channel and part of the North Sea; for the Eastern Atlantic Region, it is subordinate to ACLANT.

ACLANT: Allied Command, Atlantic, one of the three highest NATO commands, responsible for the security of the Atlantic from the North Pole to the Tropic of Cancer, including Portuguese coastal waters but excluding the English Channel.

ACM: 1. See AIR COMBAT MANEUVERING.
2. Advanced Cruise Missile. See ALCM.

ACQUISITION: See TARGET ACQUISITION.

ACR: Armored Cavalry Regiment, a U.S. Army formation. See CAVALRY, ARMORED.

ACRID (AA-6): NATO code name for a Soviet medium-range air-to-air missile (AAM), successor to the earlier AA-3 ANAB and AA-5 ASH missiles. Developed since the late 1950s to counter the abortive U.S. XB-70 Valkyrie supersonic bomber, Acrid entered service in the mid-1960s. The world's largest AAM, Acrid has four cropped delta wings with ailerons for roll control, and four smaller canards for steering. Like most Soviet AAMs, Acrid has both SEMI-ACTIVE RADAR HOMING (SARH) and INFRARED HOMING (IR) variants; the latter has limited ALL-ASPECT capability (i.e., it can attack from ahead of the target, but not directly head-on). Powered by a solid-fuel rocket engine, Acrid is known to be equipped with a 200-lb. (90-kg.) HE warhead, but a nuclear-armed version may also exist. Designed to attack large bombers, it cannot cope with highly maneuverable fighters. The MiG-25 FOXBAT is generally armed with 4 Acrids (2 of each variant), while the Su-15 FLAGON carries only 2. Acrid is currently being replaced by the AA-9 AMOS.

Specifications Length: 20 ft. (6.29 m.). Diameter: 15.7 in. (400 mm.). Weight, launch: 1650 lb. (700 kg.). Span: 7.4 ft. (2.25 m.). Speed: Mach 4 (2600 mph/4342 kph). Ceiling: 70,000 ft. (21,340 m.). Range, max.: (SARH) 50 mi. (80 km.); (IR) 15.5 mi. (25 km.).

ACTIVE DEFENSE: A defensive OPERATIONAL METHOD based upon the combination of a flexible resistance (to slow, weaken, and eventually halt enemy attacks) with localized counterattacks meant to throw the enemy off balance, and eventually force his retreat.

The U.S. Army version, the AIR-LAND BATTLE doctrine introduced in the 1980s, emphasizes the use of strongpoints held by infantry and anti-tank forces to channel enemy attacks into MAIN BATTLE AREAS. Ideally, counterattacks would be aimed against the flanks of the enemy force.

In the West German *Bundeswehr*, defensive units are assigned to "rooms." Units can move freely within their rooms, but "key terrain" upon which the defense is to be based, is identified for every room. Should key terrain be lost to the enemy, German doctrine prescribes an immediate counterattack with whatever forces are at hand. Commanders on the spot determine the direction and objective of the enemy attack, and maneuver their units to create a SCHWERPUNKT (center of gravity) at that point. A commander may thus accept the temporary encirclement of his forces in order to hold key terrain at the *schwerpunkt*, with encircled units then forming a fortified HEDGEHOG, to hold up disproportionate enemy forces. Operational reserves from the rear could then be committed against the enemy forces besieging the hedgehog, in a HAMMER AND ANVIL attack.

Israeli operational methods closely resemble the German, but multiple, alternative firing positions are prepared to reduce the vulnerability of defensive forces to enemy fire. Fortified positions held by infantry and small tank detachments cover key terrain (e.g., dominating ridge lines), minefields, and anti-tank ditches, and serve as OBSERVATION POSTS. The British and Soviet armies both favor a POSITIONAL DEFENSE. See also MOBILE DEFENSE.

ACTIVE HOMING: A form of missile (including torpedo) guidance in which the weapon has its own sensor (radar or sonar) with both a transmitter and a receiver. The launch platform may feed positional data to the weapon until its sensor can acquire it ("lock on"). Once locked onto the target and released, the weapon steers on a collision course on the basis of its own transmitted signals reflected back off the target to its own receiver, and processed by an on-board computer programmed according to one of several guidance laws (the most common being PROPORTIONAL GUIDANCE). Some active-homing weapons need not be locked on before launch; instead, they follow a preprogrammed course in the general direction of the target, until the transmitter is switched on to acquire and lock onto it. Some can be programmed to execute a search pattern until a target is acquired by its own sensors.

The advantage of active-homing weapons is

their FIRE AND FORGET capability; having launched the weapon, the ship or aircraft can maneuver without constraint to engage another target or evade the enemy. The major disadvantage is the high cost and complexity as compared to passive or SEMI-ACTIVE HOMING weapons. Moreover, the emissions of the active sensor can warn the target of the weapon's approach. See also MISSILE, GUIDED; TORPEDO.

ACTIVE JAMMING: The use of powerful transmissions in radar, sonar, or communications frequencies to blind or confuse enemy systems; sometimes also includes the use of flares or heat-generating devices against infrared sensors. See also BARRAGE JAMMING; DECEPTION JAMMING; ELECTRONIC COUNTERMEASURES; INFRARED COUNTERMEASURES; TORPEDO COUNTERMEASURES.

ACTIVE MEASURES: Soviet term (*Aktivnyye meropriyatiya*) for a variety of overt and covert methods meant to influence the policy of other governments, disrupt their relations with third countries, or discredit opponents. The most common covert method is the dissemination of false or misleading information (*dezinformatsia*, or DISINFORMATION). Overt active measures include official PROPAGANDA and the manipulation of diplomatic relations and cultural exchanges. Covert active measures include disguised propaganda disseminated by non-Soviet sources; forged documents; clandestine ("black") radio broadcasts; and the use of agents of influence and front organizations (such as the World Peace Council). Active measures are primarily political, but military maneuvers and terrorism may also be used.

Conversely, active measures are integrated into Soviet military operations as one form of MASKIROVKA (deception), to mislead others as to Soviet military intentions, influence world opinion on Soviet military activities, and dissuade resistance to Soviet military actions.

Covert active measures are conducted by Department A of the First Chief Directorate of the KGB, in charge of covert propaganda, agents of influence, forgery, disinformation, the manipulation of foreign media, and the provision of assistance to terrorists.

ACV: 1. Armored Command Vehicle. Any ARMORED FIGHTING VEHICLE (AFV) with additional COMMUNICATIONS equipment and room for a small command staff. See also COMMAND AND CONTROL.

2. Air Cushion Vehicle. See HOVERCRAFT.

ADAM: Area Denial Artillery Munition, a mine-dispensing 155-mm. ARTILLERY round in service with the U.S. Army. ADAM comes in two variants: the M692 Long Active Life and M731 Short Active Life. Both are thin-walled canisters containing 36 M74 anti-personnel MINES, which are scattered by a time-fuzed burster charge. The mines can be activated both by a self-deploying tripwire or a motion-sensing device (to prevent removal). Upon detonation, a fragmentation charge is blown into the air by a liquid propellant, exploding some 3 feet above the ground to produce more than 600 potentially lethal fragments with an initial velocity of 100 m./sec. If not detonated after a predetermined time, the mines self-destruct. ADAM rounds have a maximum range of 17,400 m. when fired from M109A2/3 self-propelled, or M198 towed, GUN-HOWITZERS. Six rounds can deny an area of 400 by 400 m. ADAM is often used in conjunction with RAAMS (Remote Anti-Armor Mine System). See also IMPROVED CONVENTIONAL MUNITIONS.

ADAMS: A class of 29 guided-missile DE-STROYERS, of which 23 (DDG-2 through -24) serve in the U.S. Navy, 3 in the Australian navy (PERTH class), and 3 in the West Germany navy (LUTJENS class). Commissioned between 1960 and 1964, these relatively small but highly capable destroyers were the most numerous class of U.S. missile-armed warships prior to the PERRY-class frigates. Flush-decked, they have a graceful shear, a sharply raked bow, twin stacks, and a compact superstructure.

Intended mainly for ANTI-AIR WARFARE, the primary armament is the STANDARD MR-2 surface-to-air missile, launched from either a twin-arm Mk.11 Mod 0 launcher (on early ships), or a single-arm Mk.13 Mod 0 launcher; both are mounted aft, with 40-round magazines. Two illuminator radars allow two separate targets to be engaged simultaneously. The same launchers can also accommodate HARPOON anti-ship missiles, eight of which are normally carried. The main ANTI-SUBMARINE WARFARE (ASW) weapon is an 8-round ASROC pepper-box launcher with no reloads. Two sets of Mk.32 triple tubes for MK.46 homing torpedoes are also fitted for close-in ASW. Two 5-in. 54-caliber DUAL PURPOSE guns (one forward, one aft) supplement both Standard and Harpoon. See also DESTROYERS, UNITED STATES.

Specifications Length: 420 ft. (128.05 m.). **Beam:** 47 ft. (14.33 m.). **Draft:** 22 ft. (6.7 m.). **Displacement:** 3380 tons standard/4500 tons full load. **Propulsion:** 4 oil-fired boilers, 2 sets of geared steam turbines, 2 shafts, 70,000 shp. **Speed,** max.: 31.5 kt. **Range:** 1600 n.mi. at 30 kt./4500 n.mi. at 20 kt. **Crew:** 360. **Sensors:** 1 SPS-10F surface-search radar, 1 SPS-52B 3-dimensional air-search radar, 1 SPS-40 2-dimensional air-search radar, 1 SPG-53 fire control radar, 2 SPG-51C missile-guidance radars, 1 SQS-23 bow-mounted low-frequency sonar. **Electronic warfare equipment:** 1 SLQ-32(V)2 electronic countermeasures array.

ADAPTIVE JAMMING: A form of ACTIVE JAMMING (a.k.a. "responsive jamming") in which the equipment can respond almost immediately to changes in enemy RADAR parameters, including frequency, band width, and pulse-repetition frequency. Against radars with alternative modes, adaptive jamming is a key ELECTRONIC COUNTERMEASURES capability. See also ELECTRONIC WARFARE.

ADAPTIVE OPTICS: A technique meant to overcome the distortion of laser beams caused by the refractive effects of the atmosphere. A series of mirrors are mounted on computer-controlled actuators, to change the focal point of the laser beam in response to feedback on beam distortion from the laser radiation reflected off the target. Also known as "rubber mirrors," adaptive optics are essential for earth-based laser weapons. The concept has been tested successfully (with low-powered lasers). See also BALLISTIC MISSILE DEFENSE; DIRECTED ENERGY WEAPON; LASER.

ADATS: Air-Defense Anti-Tank System, a DUAL PURPOSE missile system developed by Oerlikon-Buhle from 1979 as a private venture intended specifically for export. Testing began in 1983, and in 1986, ADATS was selected by the Canadian Forces; in 1988, it was also selected by the U.S. Army as the Line-of-Sight, Forward, Heavy (LOS-F-H) component of the Forward Area Air-Defense System (FAADS). ADATS is to enter service in the early 1990s.

The system consists of the missile itself in a sealed launch/storage canister, and an 8-round, power-operated, turreted launcher. The latter has a centrally mounted FIRE CONTROL system with ELECTRO-OPTICAL (TV) and forward-looking infrared (FLIR) sensors, a LASER rangefinder, and a missile-guidance laser; a surveillance RADAR at the rear of the turret; and two elevating launcher arms, each of which can hold four missile canisters. With 360° traverse and elevation limits of −5° and +85°, the turret is compatible with many armored vehicles. The Canadian version is mounted on an M113 armored personnel carrier, while the U.S. version is mounted on a modified M2 Bradley infantry fighting vehicle. Eight reload missiles inside the vehicle can be loaded automatically through hatches in the vehicle's roof.

The ADATS missile has a cylindrical body, a long, pointed nose, and four small folding tailfins for steering. Powered by a smokeless solid-fuel rocket, it has a 26-lb. (12-kg.) shaped-charge (HEAT) warhead capable of penetrating up to 900 mm. (35.4 in.) of homogeneous steel armor, whose fragmentation jacket is effective against aircraft.

ADATS is controlled by laser BEAM-RIDING guidance. In the surface-to-air role, ADATS first detects targets with its surveillance radar (out to a maximum range of 12.4 mi./20 km.); if the target is confirmed as hostile by IFF interrogation, the turret is trained to its bearing, and the operator acquires the target with the FLIR or TV sensors. The sensor then locks onto the target, and the range is determined by the laser rangefinder. When the target is in range, a missile can be launched to fly down the laser guidance beam to the target.

In ADATS's anti-tank role, targets are detected with the TV or FLIR, range is determined by the laser rangefinder, and the missile is launched down the guidance beam.

The main disadvantages of ADATS are its cost, its reliance on an easily detected surveillance radar, and its lack of true all-weather guidance (laser beams cannot penetrate clouds, fog, heavy precipitation, or smoke).

Specifications Length: 6.7 ft. (2.05 m.). Diameter: 5.98 in. (150 mm.). Weight, launch: 112 lb. (51 kg.). Speed: Mach 3 (2100 mph/3500 kph). Range, min.: (anti-tank) 500 m. Range, max.: (surface-to-air) 8000 m. (5 mi.); (anti-tank) 5000 m. Ceiling: 5000 m. (16,400 ft.).

ADCAP: Advanced Capability, the latest version of the U.S. MK.48 homing torpedo.

ADD: *Aviatsia Dal'Nevo Deistviya*, Long-Range Aviation; the Soviet Air Force counterpart of the U.S. STRATEGIC AIR COMMAND, but responsible only for strategic BOMBERS and long-range RECONNAISSANCE aircraft, and not for ballistic missiles. See also VVS.

ADEN: A British copy of the Mauser MG213C 30-mm. aircraft CANNON of World War II. The Aden arms several aircraft, including the HARRIER, HAWK, and HUNTER. Highly accurate and effective against both air and ground targets, the Aden generally fires HE ammunition, but there are also special ARMOR PIERCING rounds for ground attack. The new Aden 25 fires the more powerful NATO 25-mm. round; it can be fitted in a twin pod installation, with one pod holding the gun, and the other 100 rounds of ammunition.

Specifications Length (Aden) 62.6 in. (1.59 m.); (Aden 25) 90 in. (2.285 m.). Weight: (Aden) 191.8 lb. (87 kg.); (Aden 25) 202 lb. (92 kg.). Muzzle velocity: (Aden) 790 m./sec. (2590 ft./sec.); (Alden 25) 1050 m./sec. (3445 ft./sec.). Cyclic rate: (Alden) 1400 rds./min.; (Aden 25) 1650–1850 rds./min.

ADM: Atomic Demolition Mine, a small nuclear weapon designed to be emplaced by hand. When the use of ADMs was still seriously contemplated, typical targets included bridges, tunnels, and dams too large to be destroyed by nonnuclear demolition charges. In addition, barriers can be formed by detonating ADMs underground, to leave large craters. In the U.S. Army and Marine Corps, two types of ADM remain in service. The Medium Atomic Demolition Mine (MADM), first deployed in 1965, weighs approximately 400 lb. (181 kg.), and has a variable yield between 1 and 15 kilotons (KT); it is designed to form barriers ahead of advancing enemy forces. The smaller Special Atomic Demolition Mine (SADM) weighs only 150 lb. (68 kg.), and has a low yield (0.1–1 kT). Assigned to RANGERS and SPECIAL FORCES units for emplacement behind enemy lines, SADM has been called a "suitcase nuke." Soviet SPETSNAZ reportedly also have a SADM-type device. See also NUCLEAR WEAPONS.

ADTAC: Air Defense, Tactical Air Command; the U.S. Air Force organization responsible for the defense of the continental United States against enemy air attack—formerly the Air Defense Command. In October 1986, ADTAC was redesignated the First Air Force (TAC); its commander is also the commander of the CONUS (Continental U.S.) NORAD region, responsible to the Commander-in-Chief NORAD for the air sovereignty surveillance and air defense of CONUS. Compared to the Soviet Air Defense Troops (PVO), U.S. air defenses are very weak. See also AIR DEFENSE; BUIC; JOINT SURVEILLANCE SYSTEM; SAGE.

AEGIS: A shipboard combat system for ANTI-AIR WARFARE (AAW), installed in the U.S. Navy's TICONDEROGA-class cruisers and BURKE-class destroyers. Intended to defeat massed attacks by ANTI-SHIP MISSILES (the key Soviet anti-carrier tactic), Aegis combines an advanced form of radar with FIRE CONTROL computers to detect and track hundreds of fast-moving targets simultaneously, and direct weapons against them in a prioritized sequence. Conventional, mechanically scanned RADARS could not generate a sufficient number of concurrent tracking beams, but Aegis relies on SPY-1 PHASED ARRAY radars, each with 4,100 radiating elements whose interaction is controlled by a UYK-1 digital computer to produce multiple beams for a TRACK-WHILE-SCAN capability. The fire controls linked to the SPY-1 automatically evaluate, prioritize, and engage targets with all suitable shipboard weapons, including the STANDARD 2-MR surface-to-air missile, SEA SPARROW point defense missile, 5-inch DUAL PURPOSE gun, and PHALANX anti-missile cannon.

While conventional shipboard AAW systems can engage a maximum of four targets at one time because of the limited number of illuminator radars, Aegis employs a unique procedure whereby a Standard missile is first launched on a ballistic trajectory towards the target, and the fire control computer directs a radar to illuminate it only when the missile is near enough to need reflected signals to home on. Hence the available illumination beams can cope with many more targets; moreover, when operated in this "kinematic" mode, Standard actually flies a more efficient trajectory, extending its range by almost 50 percent.

By employing the intership Naval Tactical Data System (NTDS), an Aegis ship can direct the weapons of other ships in its BATTLE GROUP.

Aegis ships are the cornerstone of the U.S. defense-in-depth of CARRIER BATTLE GROUPS against air and missile attack. While much more effective for AAW than escorts with conventional radar, even Aegis ships can be overwhelmed, and because of their great cost there are too few for much backup. It is not clear, moreover, if Aegis can cope with the true surface-skimmers among current anti-ship missiles. See also BATTLE MANAGEMENT; COMBAT INFORMATION CENTER.

AERIAL REFUELING: The replenishment of fuel in flight by pipe transfer from one aircraft to another. Pioneered by the U.S. Air Force, aerial refueling is now in general use. The two established methods are "probe-and-drogue" and "flying boom."

In the more common probe-and-drogue method, a tanker aircraft unreels a flexible hose with a winged "drogue" at the end, which stabilizes the hose in the airflow and contains a receptacle for the rigid probe of the receiving aircraft. The pilot of the receiving aircraft flies the probe into the drogue; when a solid connection is achieved, fuel is transferred. Probe-and-drogue is used by most

air forces (except the USAF), and by all naval air services. Its advantage is that almost any aircraft can be modified to carry a drogue assembly, while probes can also be added to most aircraft. A "buddy store" is a self-contained drogue unit fitted inside an external fuel tank; with it, any aircraft that can carry external tanks can also act as a tanker.

Only the USAF, faced with the problem of refueling large bombers with poor maneuverability, uses the flying-boom technique. The boom, a rigid telescopic tube equipped with small, movable wings, is extended from the tanker into a flush-mounted receptacle on the upper fuselage of the receiving aircraft; an operator "flies the boom" from a cabin in the tail of the tanker. The receiving aircraft is thus not required to maneuver, and need only maintain formation. Flying booms can be equipped with adaptors to refuel probe-equipped aircraft, but aircraft which have only flying-boom receptacles cannot be refueled by drogue-equipped tankers. See also EXTENDER; INTRUDER; MIDAS; SKYWARRIOR; STRATOTANKER; VICTOR.

AEROSPACE DEFENSE: An inclusive term for all activities meant to intercept or otherwise neutralize enemy aircraft, missiles, and space-based systems. See also AIR DEFENSE; ANTI-BALLISTIC MISSILE; ANTI-SATELLITE SYSTEM; BALLISTIC MISSILE DEFENSE; SPACE, MILITARY USES OF.

AEROSTAT: An unmanned, tethered AIRSHIP (or balloon), used as a platform for sensors, including RADAR. While very vulnerable, aerostats extend the RADAR HORIZON by virtue of their height, are relatively cheap, and offer almost unlimited time-on-station compared to powered platforms. Aerostats are in operational use in various countries, including Israel, and have also been used by the U.S. Drug Enforcement Agency to detect low-flying aircraft and small boats. The U.S. Navy is now considering the employment of aerostats with naval BATTLE GROUPS. See also AIRBORNE EARLY WARNING.

AEW: See AIRBORNE EARLY WARNING.

AFCENT: Allied Forces, Central Europe. See SHAPE.

AFLC: Air Force Logistics Command, the organization responsible for the logistic support of all equipment employed by the U.S. Air Force. It consists of nine logistics centers in the United States and at U.S. bases overseas, which provide supply and depot maintenance services. In addition, AFLC carries out research on logistic support, sets calibration standards for all electronic equipment, and supervises all USAF maintenance contracts. Also responsible for the Air Force Band, the USAF Medical Center, and the Air Force Museum, AFLC currently has some 12,000 active-duty Air Force personnel and more than 90,000 civilians—numbers many think excessive.

AFNORTH: Allied Forces, Northern Europe. See SHAPE.

AFSC: United States Air Force Systems Command, the umbrella organization responsible for the research, development, and procurement of almost all equipment employed by the U.S. Air Force, including aircraft, weapons, sensors, and ground support equipment. Headquartered at Andrews AFB, Maryland, its subordinate commands include: the Armament Division (AD), responsible for the development of aerial ordnance; the Electronic Systems Division (ESD), responsible for radar, computers, and electronic warfare systems; the Space Division (SD), responsible for satellites, ground tracking stations, space boosters, and space weapons systems; and the Aeronautical Systems Division (ASD), responsible for the development of new aircraft. There are also several independent offices within AFSC, of which the most important are the Ballistic Missile Office, responsible for development of ICBMS; and the Foreign Technologies Division (FTD), which evaluates foreign (mainly Soviet) weapons and technology for the intelligence community. In addition, AFSC is responsible for the management of Air Force equipment contracts through its Contract Management Division. At present, AFSC has over 27,000 military personnel (mostly officers), and nearly 30,000 civilians. The sheer size of this bureaucracy has been identified by some as one of the primary causes of the high costs and slow development of USAF weapons.

AFSOUTH: Allied Forces, Southern Europe. See SHAPE.

AFV: See ARMORED FIGHTING VEHICLE.

AGM: U.S. acronym for air-to-ground missile. Models currently in service or in development include: AGM-12 BULLPUP; AGM-45 SHRIKE; AGM-65 MAVERICK; AGM-62 WALLEYE; AGM-69 SRAM; AGM-78 STANDARD ARM; AGM-84 HARPOON; AGM-86 Air-Launched Cruise Missile (ALCM); AGM-88 HARM; AGM-109 TOMAHAWK; AGM-114 HELLFIRE; AGM-119 PENGUIN; AGM-122 SIDEARM; AGM-123 SKIPPER; AGM-129 Advanced Cruise Missile (see ALCM); AGM-130; and AGM-136 TACIT RAINBOW.

AGM-130: A rocket-boosted, modular GLIDE BOMB built by Rockwell for the USAF. Assembled from a kit, with either a 2000-lb. (909-kg.) Mk.84 LDGP bomb or an SUU-54 CLUSTER BOMB as the payload, the AGM-130 has alternative ELECTRO-OPTICAL (TV) or IMAGING INFRARED (IIR) guidance units. With both, a DATA LINK permits release of the weapon from beyond the visual range of the target. The view from the seeker head is transmitted back to the aircraft and displayed on a cockpit video monitor; when the target is acquired by the seeker, the operator can lock onto it, after which the missile will home autonomously. Alternatively, the missile may be guided manually with a joystick, with the steering commands transmitted via the data link. The AGM-130 is based on the earlier GBU-15, but has a rocket motor under the bomb to extend its standoff range, allowing the launch aircraft to avoid close-in AIR DEFENSES.

When assembled, the AGM-130 has four short-chord cruciform wings at the rear, and four smaller stabilizing canards at the nose. Flight control is accomplished by pneumatically actuated control surfaces on the wings. The nose houses the seeker head and a radar altimeter (for terrain avoidance). Modules bolted behind the bomb house a guidance computer, an autopilot, control actuators, and the data link. The solid rocket motor, attached by exploding bolts to the nose and guidance sections, can be activated either immediately after release to loft the weapon into a higher trajectory, or after an initial glide period, for low-level attack. After burnout, the booster is jettisoned to reduce weight. Maximum range, dependent on altitude and speed at release, is roughly triple that of the GBU-15.

The AGM-130 entered limited service with F-4E PHANTOM and F-111F squadrons, and has been used successfully in the 1991 Operation Desert Storm. However the range of the AGM-130 is now considered inadequate to cope with the latest Soviet air defense weapons. Hence the USAF plans to supplant both the AGM-130 and GBU-15 with a "Modular Standoff Weapon."

Specifications Length: 16.33 ft. (4.98 m.). Diameter: 18 in. (457 mm.). **Weight, launch:** 3086 lb. (1403 kg.). **Span:** 55 in. (1.4 m.). **Range, max.:** 15 mi. (25 km.).

AGOSTA: A class of ten French diesel-electric attack SUBMARINES (SSs), commissioned between 1977 and 1985. Successors to the DAPHNE class (which proved too small), the Agostas have a longer range, higher speed, and improved sensors and fire controls. Four Agostas were delivered to the French navy in 1977–78, with an additional four built under license for the Spanish navy between 1977 and 1985. Two more ordered by South Africa in 1977 were embargoed in 1979, and subsequently sold to Pakistan.

The Agostas have a modified teardrop hull in a double-hull configuration, with fuel, ballast and fresh water tanks between the pressure hull and the outer casing. A tall, streamlined sail (conning tower) of glass-reinforced plastic is located well forward. Control surfaces consist of retractable diving planes at the bow, and cruciform rudders and stern planes ahead of the propeller.

Armament consists of four 550-mm. (21.7-in.) tubes in the bow, for a variety of 21-in. (533-mm.) and 21.7-in. homing TORPEDOES, as well as EXOCET SM.39 anti-ship missiles. The basic load is 19 torpedoes and 4 missiles, but TSM 3510 MINES can replace other weapons on a 3-for-1 basis. The torpedo tubes are fitted with a very quiet pneumatic loader for high-speed reloading, and a positive-impulse ejector allows the torpedoes to be launched at any depth.

The Agostas have single-shaft diesel-electric propulsion. A retractable snorkel allows diesel operation from shallow submergence, and a 23-hp. "creep" motor can propel the submarine at 3.5 knots for "silent running." When the ship is fully submerged, power is supplied to the electric motor by two 160-cell Type N batteries, with twice the capacity of earlier French submarines. See also SUBMARINES, FRANCE.

Specifications Length: 222 ft. (67.68 m.). Beam: 22 ft. (6.7 m.). Displacement: 1490 tons surfaced/1740 tons submerged. Max. operating depth: 1150 ft. (350 m.). Collapse depth: 2300 ft. (700 m.). Propulsion: 2 1270-hp. SEMT-Pielstick 16PA4 185 diesel-generators, 1 4725-hp. Jeumont-Schneider electric motor. Speed: 12.5 kt. surfaced/20.5 kt. submerged (17.5 kt. sustained). Range: 8500 n.mi. at 9 kt. (snorkeling). Crew: 54. Sensors: 1 DUUA 1 and 1 DUUA 2 medium-frequency active/passive sonar array, 1 DSUV 22 low-frequency passive sonar array, 1 DUUX 2 passive ranging sonar, 1 DUUG/AUUG sonar intercept array, 1 DRUA 33 surface-search radar, 1 ARUR and 1 ARUD electronic signal monitoring array, 2 periscopes. Fire controls: 1 DLT D-3 automated system based on an Iris 35M digital computer.

AH-: U.S. designation for Attack Helicopter. Models in service include the AH-1 COBRA and AH-64 APACHE. See also HELICOPTER.

AI: 1. Airborne Intercept, a term used in certain air forces (but not the USAF) for airborne RADARS optimized for aerial interception.

2. Artificial Intelligence, a computer design and programming concept whose objective is the development of computers capable of emulating human thought processes—e.g., reacting to situations, learning through experience, etc. Although now rudimentary, AI could become very important in future BATTLE MANAGEMENT systems for BALLISTIC MISSILE DEFENSE, and in future combat aircraft as a "pilot's associate," i.e., a super-autopilot.

AIM-: U.S. acronym for Air Intercept Missile, i.e., an air-to-air missile. Models currently in service or under development include: AIM-4 FALCON; AIM-7 SPARROW; AIM-9 SIDEWINDER; AIM-26 FALCON; AIM-54 PHOENIX; AIM-120 AMRAAM; and AIM-132 ASRAAM.

AIR ASSAULT: An operational concept developed by the U.S. Army in Vietnam, based on the use of helicopters to surprise the enemy by landing troops directly on, or very near, his positions. Attack helicopters accompany the transports as aerial artillery. When ANTI-TANK GUIDED MISSILES (ATGMs) such as TOW, HOT, or HELLFIRE are mounted on its attack helicopters, an air assault force can also act as a highly mobile ANTI-TANK reserve to block enemy penetrations.

The U.S. Army has one AIR ASSAULT DIVISION, the 101st (formerly Airborne). From the early 1970s the Soviet Union formed AIR ASSAULT BRIGADES.

Air assault was originally developed to pin down elusive guerillas; it has never been practiced in a high-threat environment.

AIR ASSAULT BRIGADE: *Desantnii Shturmovaia Brigada*, a Soviet army formation trained and equipped for AIR ASSAULT operations behind enemy lines, first formed in the early 1970s. At present there are some ten Air Assault Brigades. Each consists of 4 rifle battalions—1 or 2 "heavy" battalions mounted in BMD airborne assault vehicles, and 2 or 3 "light" (dismounted) battalions—an artillery battalion of 18 122-mm. howitzers and 6 122-mm. multiple rocket launchers, a reconnaissance company, an anti-tank company, an air defense company, an engineer company, and support troops, for a total of some 2000 to 2600 men and about 80 vehicles.

Transport and attack helicopters are not organic to the brigades; instead, they are assigned by FRONTOVAYA AVIATSIYA (Frontal Aviation, or FA) as directed by the FRONT command. A full brigade without BMDs would require 75 Mil-8 HIP medium and 35 Mi-6 HOOK heavy transport helicopters for its lift, while with BMDs it would require 41 Hips and 125 Hooks.

In war, Air Assault Brigades would be assigned to Fronts to execute tactical DESANTS, i.e., seizures of key terrain, bridges, defiles, etc., the blocking of enemy reinforcements, attacks on headquarters and communications, and the neutralization of enemy nuclear weapons. These missions differ from those of the Airborne Troops (VDV) only in their depth: while VDV units would operate under Theater (TVD) command up to several hundred kilometers behind the lines, Air Assault Brigades would operate only 25 to 100 kilometers behind the lines, under Front command. Though not part of the VDV elite, Air Assault troops are better than Soviet line infantry, and were much used in Afghanistan to fight elusive *Mujahideen* guerillas. See also AIRBORNE FORCES, SOVIET UNION.

AIR ASSAULT DIVISION: A U.S. Army formation. The only U.S. Air Assault Division is the 101st ("Screaming Eagles"), which was converted from an airborne division during the Vietnam War. At present, it consists of nine rifle BATTALIONS, each of 733 men; three light artillery battalions, each with 18 105-mm. towed HOWITZERS; an ENGINEER battalion; and an air defense battalion. Its organic helicopter regiment consists of a scout helicopter company; an attack helicopter squadron with 63 AH-1 COBRAS or AH-64 APACHES; a medium transport squadron with 48 CH-47 CHINOOKS; an air cavalry squadron with 30 OH-58 Kiowa scout helicopters, 22 UH-60A BLACKHAWK utility helicopters, and 27 Cobras; and two transport squadrons with 45 Blackhawks. The division is thus self-sufficient for both transport and attack helicopter support. It includes 15,000 men, 54 howitzers, and 373 helicopters. The 101st now forms part of the Rapid Deployment Force of the U.S. CENTRAL COMMAND.

AIRBORNE EARLY WARNING (AEW): A means of detecting enemy aircraft at long range by placing air-search RADARS on fixed-wing aircraft, helicopters, or AEROSTATS. The concept originated in the U.S. Navy's experience with Japanese kamikaze attacks. Because the RADAR HORIZON of even the largest ships with the tallest masts is limited to some 30 mi. (50 km.), low-flying kamikazes could penetrate close to the fleet before being detected. To place destroyers in an outer PICKET SCREEN would expose them to the brunt of the enemy at-

tack. By the early 1950s, the U.S. Navy succeeded in placing large search radars on carrier aircraft, eliminating the need for picket ships.

The advent of Soviet nuclear weapons and long-range bombers prompted the development of more powerful AEW platforms by the U.S. Air Force. Early AEW systems were semirigid AIRSHIPS (blimps), with large radar antennas inside the gasbag. Air operations off North Vietnam after 1965 revealed that AEW platforms could also direct fighters, leading to the development of AWACS.

The increased emphasis on very low level flight paths, and the particular threat of sea-skimming ANTI-SHIP MISSILES, now makes AEW essential on land and at sea. During the 1982 Falklands War, the British were severely handicapped by their lack of AEW: the use of destroyers as radar pickets resulted in the loss of HMS *Sheffield* and *Coventry*. By contrast, the success of the Israeli air force against Syrian aircraft in the 1982 Lebanon War owed much to the early detection of Syrian fighters by U.S.-built E-2 Hawkeye AEW aircraft in Israeli service. Current AEW aircraft include the E-2 Grumman HAWKEYE, E-3 Boeing SENTRY (AWACS), Westland SEA KING (with Searchwater radar), Il-76 MAINSTAY, and Tu-126 MOSS. (Britain was forced to cancel its NIMROD AEW program in 1986 because of technical problems.) See also AIR DEFENSE.

AIRBORNE FIGHTING VEHICLE: A vehicle designed specifically to be parachute-dropped or air-landed. Since their inception, their lack of mobility on the ground has been recognized as the major shortcoming of parachute- and air-landed forces (strategic mobility = tactical immobility). Weight and volume constraints have stimulated many expedients. Soviet experiments have included the dropping of amphibious tanks into lakes without parachutes, and the remarkable Antonov AT-1 tank-glider (a tank with wings attached); neither met with success. During World War II, both the Allies and the Germans used glider aircraft to bring light tanks into landing zones.

The Soviet army was the first to acquire a parachute-droppable armored vehicle, the ASU-57. Lightly armored, it mounted a 57-mm. anti-tank gun, and could also transport troops. The subsequent ASU-85 assault gun, armed with an 85-mm. anti-tank gun, is too heavy to be parachuted. The Soviet BMD, introduced in the 1970s, can be dropped by parachute, and has a 73-mm. anti-tank gun or a 30-mm. cannon, an ANTI-TANK GUIDED MISSILE launcher, and room for six riflemen behind its (thin) armor. The BMD thus combines mobility, some protection, firepower, and troop transport in one vehicle; with it, Soviet airborne forces have been mechanized.

The only U.S. airborne fighting vehicle is the M551 SHERIDAN, too heavy to be dropped by parachute, armed with the unreliable SHILLELEIGH gun/missile system, and without space for troops. The Sheridan can be delivered by extreme low-altitude parachute extraction (whereby the vehicle is pulled out of the tail of a transport by parachute to skid across the ground), but that technique is more spectacular than practical. The French army has VAB and other ARMORED CARS in its airborne forces. See also AIRBORNE FORCES; AIRBORNE FORCES, SOVIET UNION; AIRBORNE FORCES, UNITED STATES.

AIRBORNE FORCES: More or less elite troops trained for air insertion (generally into the enemy rear), by parachute drop, helicopter assault (AIR ASSAULT in U.S. terminology), helicopter landing, or fixed-wing aircraft landing. In parachute drops, airborne forces jump from transport aircraft into a drop zone (DZ). Because their equipment must also land by parachute, they can be only lightly equipped. In helicopter assaults, the troops are delivered in close proximity to the enemy, and sometimes directly upon enemy positions. In helicopter landings, troops are delivered at a safe distance from the enemy and then move overland to attack. In fixed-wing aircraft landings, the troops are disembarked from transport aircraft behind enemy lines after the capture of airfields; thus advance elements must first be inserted by one of the other three methods to seize a landing ground. Glider landings do not require airfields, and have allowed the insertion of heavier equipment, but the risk of interception is extreme, and long-range towing is impractical.

Airborne forces were hypothesized by J. F. C. Fuller in his famous Plan 1919, but the first actual air drops were carried out by the Soviet army in the 1920s. During World War II, German airborne troops assisted BLITZKRIEG operations with great success. The U.S. and Britain imitated the German model on a much larger scale, and eventually employed airborne forces in Sicily, Normandy, Holland, and Germany.

Britain, France, Germany, the U.S., Israel, the Soviet Union, and other countries still maintain airborne forces, even though AIR DEFENSES now generally rule out massed parachute assaults in World War II style. Remaining airborne roles include RAIDS, the disruption of COMMUNICATIONS, and the VERTICAL ENVELOPMENT of enemy forces. Land mobility has always been a problem for airborne troops, but only the U.S.S.R. has developed satisfactory, purpose-built AIRBORNE FIGHTING VEHICLES.

Airborne forces acquired a new mission in the 1980s: the suppression of terrorist acts, notably hostage-taking—with the Israeli Entebbe rescue of 3 July 1976 setting the example. But their primary mission is the conduct of interventions in Third World settings, either autonomously (where defenses are weak), or as the spearhead of a larger invasion force, as in the Soviet air-landed assault on Kabul, Afghanistan in December 1979, the U.S. seizure of Grenada in 1983, and the attack on Panama of December 1989. See also separate AIRBORNE FORCES entries for BRITAIN; FEDERAL REPUBLIC OF GERMANY; FRANCE; ISRAEL; SOVIET UNION; UNITED STATES.

AIRBORNE FORCES, BRITAIN: Formed in 1940, at their peak in 1944, British airborne forces consisted of two divisions, the 1st Airborne (destroyed at Arnhem) and 6th Airborne, employed with mixed results in the North African, Normandy, Holland, and Rhine campaigns.

Current British airborne forces consist of the Parachute Regiment (of three regular BATTALIONS, plus three battalions of the TERRITORIAL ARMY), which forms the 5th Airborne Brigade together with units of GURKHAS, artillery, engineers, and reconnaissance troops. Each parachute battalion consists of 590 men organized into three rifle COMPANIES and one heavy weapons company. But the 5th Brigade's engineers, artillery, and anti-aircraft and anti-tank support units are not parachute-qualified at this time, and would have to be air landed.

British airborne troops have a reputation for skill, esprit de corps, and aggressiveness, and are certainly among the best infantry in the world. The shortage of transport aircraft in the Royal Air Force, however, would make it difficult to employ the Brigade en masse at any distance from Great Britain.

AIRBORNE FORCES, FEDERAL REPUBLIC OF GERMANY: German paratroops, or *Fallschirmjager* (FJ), constitute three brigades (25th, 26th, and 27th), each of four 575-man battalions, including anti-tank troops with light vehicles. A headquarters company, engineer company, supply company, and transport company raise the brigade total to 2500 men. One *Fallschirmjager* brigade is attached to each West German corps, to serve as its rapid-reaction force. Although untested in battle, the *Fallschirmjager* are considered elite troops.

AIRBORNE FORCES, FRANCE: Constituted after World War II, French airborne forces have served with distinction in Algeria, Indochina, and several small-scale interventions. The Foreign Legion's *Regiments Etrangere Parachutiste* (REPs) were originally three in number, but the 1ere REP and 3eme REP were disbanded during the Algerian War. Only the 2eme REP remains, and it is now part of the 11th Airborne Division, which in turn belongs to the *Force d'Action Rapide* (FAR), the French rapid deployment force.

The 11th Division, which has 12,600 men and some 570 vehicles, contains 1 reconnaissance battalion with light armored cars and jeeps; 7 line battalions, including 2 parachute regiments and 4 marine parachute regiments (RPIMa); an artillery battalion with 18 105-mm. howitzers; an engineer battalion; and a support battalion. The 2eme REP, 3eme RPIMa, and 8eme RPIMa, which contain no conscripts, constitute the *Groupement Aeroporté*, a rapid-intervention force which can be deployed overseas without parliamentary approval.

The *Paras* are excellent troops, but France lacks suitable long-range transport aircraft for them; past operations in Africa have relied on U.S. aircraft and commandeered Air France airliners.

In addition to the 11th Division, the 4th air mobile division, formed in 1982 to serve as a mobile anti-tank reserve, has 3 combat helicopter squadrons, 1 infantry battalion, and 2 engineer companies.

AIRBORNE FORCES, ISRAEL: Israel formed a parachute unit during the 1947–48 War of Independence, with scant success, and a new battalion formed in 1949 on British lines soon degenerated into a show unit. Only after being amalgamated with Unit 101 (a very unorthodox commando unit formed by then-Major Ariel Sharon in 1952) did the Israeli paratroops become effective, using tactics based on stealth, shock, and surprise. Unlike their counterparts in other armies (which tend to remain an isolated elite), the paratroops are the very core of the Israeli Defense Force (IDF); all officers in all combat arms must undergo jump training, and many senior officers come from the paratroops, including most IDF chiefs of staff.

At present the IDF has four parachute brigades,

including the 35th, the active-duty "school brigade," always maintained at full strength; the other three reserve brigades are capable of mobilization within hours. Each brigade has three rifle BATTALIONS plus an artillery battalion (with lightweight 120-mm. MORTARS). Each rifle battalion consists in turn of some 400 men. ARMORED PERSONNEL CARRIERS are attached to the brigades, which are also trained as mechanized infantry.

The Israeli paratroops have fought successfully in the 1956 Sinai Campaign, the 1967 Six Day War, the 1973 Yom Kippur War, and the 1982 Lebanon War, as well as in hundreds of small operations; their combat experience is unequaled at this time. The paratroops also provide recruits for the elite General Staff Reconnaissance unit (SAYARET MATKAL).

AIRBORNE FORCES, SOVIET UNION: The Soviet army was the first to field parachute and airborne infantry from the 1920s, forming ten "corps" (actually small divisions) by the outbreak of World War II. The lack of suitable transports afflicted the airborne troops until late in the war, and they served mostly on foot as shock infantry. The 1945 Manchurian Campaign, however, included a successful, large-scale airborne assault. After 1945, the Airborne Troops (*Vozdushno-desantnyye voyska*, VDV) were reorganized into divisions assigned directly to the High Command (GKO) as a strategic reserve. There are now six airborne divisions: the 7th, 76th, 103rd, 104th, and 106th, all "Guard" divisions (the 105th was disbanded in 1987). All are maintained at full readiness, except for the 106th, the training division at the Ryazan Airborne School.

By the mid-1950s, VDV units acquired air-portable tracked assault guns (ASU-57 and ASU-85) as well as anti-tank and anti-aircraft weapons; by the mid-1960s, they were conducting division-sized air drops as part of major military exercises. But during the 1970s the VDV was transformed into a light MECHANIZED INFANTRY force, equipped with BMD air-portable infantry fighting vehicles. In a corresponding change of tactics, instead of full-scale divisional drops, company and battalion-sized VDV forces would execute raids, or concentrate on specific objectives with their BMDs.

Soviet doctrine identifies three types of airborne operation, or DESANTS: tactical in support of divisions or "Armies" (= corps), at a depth of up to 100 km. behind enemy lines; operational in support of FRONTS, at depths of 100–300 km.; and strategic in support of Theaters of Operations

(TVDS), which may take place at very long ranges.

Since 1967, the Soviet army has formed AIR AS-SAULT BRIGADES specifically for tactical *desants,* thus releasing VDV units for operational and strategic *desants.* Each division consists of three airborne regiments and an artillery regiment equipped with towed 122-mm. howitzers and 140-mm. multiple rocket launchers, as well as ASU-85 self-propelled anti-tank guns. Each airborne regiment has three battalions, plus an ANTI-TANK GUIDED MISSILE (ATGM) company, a 120-mm. mortar battery, an anti-aircraft battery, and an anti-tank battery with towed 85-mm. guns. Each airborne battalion has 335 men. Total divisional manpower is now 800 officers and 7,673 enlisted men.

Soviet Airborne Troops are mostly conscripts, but the VDV receives the highest quality manpower, on an equal basis with the Strategic Rocket Forces (RVSN). Troops are enlisted for three years rather than the usual 18 months, and receive higher pay, better clothing, and better rations. Many have had preliminary parachute training in DOSAAF, the paramilitary youth training organization.

Training is rigorous: it includes demolitions, martial arts, foreign languages, and weapons, as well as small-unit tactics. Every effort is made to instill individual initiative, a quality lacking in the rest of the Soviet army. Since 1945 the VDV has participated in the Soviet army's most visible operations, including the 1956 suppression of Hungary, the 1968 invasion of Czechoslovakia, and the 1979 invasion of Afghanistan.

The VDV depends on the Soviet military air-transport command (VTA), which can also draw upon Aeroflot as needed. See also AIR ASSAULT; AIRBORNE FIGHTING VEHICLES; AIRBORNE FORCES.

AIRBORNE FORCES, UNITED STATES:
The U.S. Army established its first parachute units in 1940, immediately after the success of German airborne troops in the invasion of Belgium and the Netherlands. By 1945 there were four U.S. airborne divisions and several independent regiments, plus detachments of Marine Corps paratroopers. The airborne establishment was then cut back to two divisions, the 82nd ("All Americans") and the 101st ("Screaming Eagles"), plus the 173rd independent airborne brigade, an independent parachute battalion (1/509), and several Reserve and National Guard parachute units. This establishment persists, except that the 101st, now redesignated an AIR ASSAULT DIVISION, is no longer

parachute qualified, the 173rd Brigade has been disbanded, and the 1/509 has been redesignated 4/325 Parachute Infantry.

The 82nd Airborne, part of XVIII Airborne CORPS (which also includes the 101st division, an airborne ENGINEER brigade, three SPECIAL FORCES groups, a field artillery brigade, a MECHANIZED INFANTRY division, a mechanized brigade, and an armored brigade), consists of 12,700 men, organized in three airborne brigades and an airborne artillery brigade (of three BATTALIONS, each with 18 105-mm. towed howitzers). Each airborne brigade in turn consists of three 687-man parachute infantry battalions and an ANTI-TANK battalion armed with 18 TOW anti-tank guided missiles (ATGMs). In addition to its four brigades, the division also has an engineer battalion, an AIR DEFENSE battalion, an AIR CAVALRY battalion with 27 AH-1 COBRA attack helicopters, and an aviation squadron with OH-58 KIOWAS, UH-1 HUEYS, UH-60 BLACK-HAWKS, and additional Cobra helicopters. A battalion of 54 M551 SHERIDAN light tanks is still retained.

The 82nd Airborne is very much the U.S. Army's fire brigade. One battalion is always kept at 18 hours' readiness, with one duty company available within 2 hours. The division's ready brigade can follow the lead battalion within 24 hours, with the rest of the division quickly deployable thereafter. Once on the ground, however, the 82nd Airborne has limited tactical mobility because of its lack of suitable AIRBORNE FIGHTING VEHICLES; this proved a serious deficiency even in the 1983 Grenada invasion. In a large-scale war, the basic mission of the 82nd Airborne would still be to seize and hold, as in World War II. After the initial assault, the division would form mutually supporting "islands," or HEDGEHOGS, supported by its artillery and air cavalry. Obstacles emplaced by airborne engineers would be used to channel the enemy into preselected "kill zones" between the hedgehogs, for attack by ATGMs and attack helicopters. Supplies for three days of intensive combat can be inserted with the division; after that, it must either be relieved or resupplied by air.

U.S. Airborne troops pass through a training course at the Airborne School at Fort Bragg, North Carolina; all divisional personnel, including female riggers, must remain jump-qualified. In the 1983 invasion of Grenada, and the 1989 attack on Panama, Airborne troops demonstrated high morale but not always the skill levels appropriate for an elite force. See also AIR ASSAULT; AIRBORNE FIGHTING VEHICLES: AIRBORNE FORCES.

AIRBURST: The detonation of a bomb, artillery shell, or missile warhead at some height above the surface of the earth. Nonnuclear airbursts are meant to destroy "soft targets" (trucks, guns, and troops) without overhead cover. A nuclear detonation that occurs at a height greater than the radius of the fireball it generates is defined as an airburst, whereby the principal destructive effects would be blast, heat, and prompt radiation, while NUCLEAR FALLOUT would be minimized. To achieve an airburst, special fuzes must be used, either mechanical time fuzes (which require an accurate estimate of the weapon's time of flight), barometric fuzes (which determine height from atmospheric pressure), or radar proximity (VT) fuzes, the most accurate. See also ARTILLERY; NUCLEAR WEAPONS.

AIR CAVALRY: See CAVALRY, AIR.

AIR COMBAT MANEUVERING: Tactics employed by fighter aircraft, individually or in groups, to gain positional advantages in firing their weapons; colloquially, "dogfighting." The basic principles of ACM, already established during World War I, are dictated by the ability of aircraft to maneuver reciprocally in three dimensions; by aerodynamic limitations; and by the prevalence of weapons that fire frontally.

With gun armament, the ideal attack position is directly astern of the target: the chances of detection are reduced, the enemy's weapons generally cannot bear, and the relative differences in the motion of the two aircraft (and thus the need to "lead") are minimized. With guns, air tactics are simple: get behind the enemy, close to effective range, and open fire. Because the enemy strives to do the same, air combat often begins as a stalking match, with each side attempting to surprise the other from the rear ("bouncing"), often by diving out of the sun (historically, 80 percent of all fighter pilots shot down never saw the aircraft attacking them). If both sides see each other, air combat tends to degenerate into the dogfight proper, as each attempts to gain a stern position on the other. The greater the number of aircraft involved in such dogfights, the more difficult it becomes to achieve a firing position without being attacked in turn; survival, not victory, then becomes the predominant aim.

After World War II, as aircraft speeds increased, and air-to-air missiles were introduced, less emphasis was placed on ACM. By the 1960s, most fighters were optimized for long-range air-to-

air missile attacks, and pilots were trained accordingly. The need for good cockpit visibility and agility in fighters, and dogfight training for pilots, had been forgotten, and guns had been neglected. Hence over North Vietnam, highly sophisticated, missile-armed F-4 PHANTOMS could barely cope with small, agile, cannon-armed Vietnamese MIGS. U.S. pilots had not been trained in ACM, while missiles proved to be unreliable and mostly ineffective against agile fighters. Guns, maneuvering slats, and better training were all provided by the U.S. services, with eventual success. (The Israeli air force, by contrast, remained well prepared for air combat during the 1960s "missile era"—because it could not afford many missiles, and thus stressed ACM with guns.)

Air combat maneuvering has not been eliminated, but only complicated by today's more effective air-to-air missiles and modern sensors such as RADAR, RADAR WARNING RECEIVERS, and INFRARED detectors; the advent of airborne radar and control aircraft such as AWACS makes surprise much more difficult to achieve, but it is still the decisive factor. It is now generally agreed that fighter pilots must be specifically trained in ACM, preferably against aircraft similar to those of potential enemies. The U.S. has two training programs, TOP GUN for the navy and RED FLAG for the air force. After instituting Top Gun in 1970, the U.S. Navy's kill ratio in Vietnam increased from 2.5 to 1 to a respectable 8 to 1. See also AIR SUPERIORITY; FLUID FOUR; LOOSE DEUCE; WELDED WING.

AIRCRAFT CARRIER: A warship designed to launch, retrieve, and maintain fixed-wing aircraft and helicopters. Aircraft carriers are characterized by a large flight deck, the hangar deck below it (which provides a service and storage area), and the elevators that connect them, the latter being either on the ship's centerline or along the edge of the flight deck. The ship is conned (and flight operations controlled) from a superstructure or "island," which is usually cantilevered outboard (to starboard) in order to maximize the width of the flight deck. The island also serves as a platform for radar and other sensors, and contains the exhaust stacks of conventionally powered carriers. Though carriers usually have some weapons for close-in defense, their primary weapons are their aircraft, the "air wing" or "air group" with its FIGHTERS, ATTACK AIRCRAFT, and supporting AIRBORNE EARLY WARNING (AEW), ANTI-SUBMARINE WARFARE (ASW), and ELECTRONIC COUNTERMEASURES (ECM) fixed-wing aircraft and helicopters.

Conventional fixed-wing aircraft are launched by steam-powered or hydraulic catapults that can accelerate them to a flying speed of 150–170 mph in less than three seconds; on landing, they are halted by arresting wires snagged by a hook mounted on their underside. (The arresting hook, the catapult bridle, and the structural reinforcements that go with them are the distinctive features of carrier aircraft.) When launching or retrieving aircraft, carriers must turn into the prevailing wind to reduce relative takeoff and landing speeds; that is a major constraint on carrier tactics.

Originally, carriers had straight ("axial") flight decks, which precluded launching and retrieving aircraft simultaneously. Axial decks became more problematic when jets were introduced, because of their reduced endurance, increased weight, and higher landing speeds as compared to propeller aircraft. If they missed their approach, early jets could not accelerate fast enough to abort, and would crash into the barriers stretched across the flight deck to protect aircraft parked at the bow. The Royal Navy invented the angled deck after World War II: a portion of the flight deck is cantilevered outboard to port (up to 15°) to serve for landings. With it, aircraft that miss their approach or fail to snag the arresting wire ("bolter") can take off again without hitting parked aircraft. Angled decks also allow carriers to launch and retrieve aircraft simultaneously, greatly increasing the flexibility of their flight operations.

Carrier hulls contain aviation fuel tanks and magazines for aerial ordnance, both highly explosive. Thus fire-detection, fire-suppression and DAMAGE CONTROL provisions are essential elements of carrier design and key capabilities in combat.

Carriers are categorized by size, mission, and the type of aircraft they operate. In World War II, there were two basic classes: Fleet Carriers (CVs) and Light Carriers (CVLs), the former with between 70 and 90 aircraft, the latter with about half that number. In addition, many tanker and merchant hulls were converted to Escort Carriers (CVEs), with 10–20 aircraft, mainly for anti-submarine warfare (ASW).

The very large U.S. "super carriers" built after 1945 were eventually designated Attack Carriers (CVAs), while earlier CVs converted for ASW duties were reclassified as Support Carriers (CVSs)—a designation now generally applied to any smaller aircraft or helicopter carrier (carriers which can only operate V/STOL aircraft are sometimes designated CVVs).

In the 1970s, the CVSs of the U.S. Navy were retired, and their ASW aircraft added to the air wings of CVAs, which were then redesignated CVs (or CVNs for nuclear-powered carriers). NUCLEAR PROPULSION allows nearly unlimited endurance at high speed, and also leaves more volume for aviation fuel, but reactors are very costly.

Originally developed at the end of World War I as auxiliaries to the battle fleet for SCOUTING, gunnery SPOTTING, and defense against enemy aircraft, aircraft carriers became the decisive naval arm early in World War II. Modern jet aircraft give a long reach and great firepower to carriers, but their cost is such that only a few navies can afford them, and then only in small numbers.

The resulting concentration of power in a limited number of hulls makes carriers prime targets. Thus they must operate in CARRIER BATTLE GROUPS (CVBGs) with several costly escort vessels for ASW and ANTI-AIR WARFARE (AAW) defense; moreover, a large portion of their aircraft must also be devoted to fighter, AEW, and ASW self-protection, at the expense of attack capabilities. On the other hand, aircraft carriers allow the projection of air power into remote areas. Historically, reliance upon land-based aircraft as a substitute for carrier air power has not been successful, though increasing ranges may invalidate that lesson. Carrier advocates also claim that the defense-in-depth of the CVBG can defeat most missile threats, while the high speed of carriers makes them difficult targets for submarines.

V/STOL aircraft (e.g., the HARRIER and Soviet FORGER), which do not need catapults, arresting gear, or long flight decks, make smaller, less expensive carriers possible. Operations by the British V/STOL carriers HMS *Hermes* and INVINCIBLE in the 1982 Falklands War were impressive, but their lack of long-range interceptors and AEW aircraft was a serious handicap even against weak opposition. Nevertheless, as performance differences between V/STOL and conventional aircraft continue to diminish, V/STOL carriers become more attractive, as does the conversion of tankers and container ships to escort carriers by using prefabricated kits (e.g., ARAPAHO and SCADS). See also separate AIRCRAFT CARRIERS entries for various nations.

AIRCRAFT CARRIERS, ARGENTINA: See VEINTICINCO DE MAYO.

AIRCRAFT CARRIERS, BRAZIL: See MINAS GERAIS.

AIRCRAFT CARRIERS, BRITAIN: The Royal Navy disposed of its last large aircraft carriers in 1979, and now operates three smaller V/STOL carriers of the INVINCIBLE class, equipped with Sea Harrier fighters and Sea King helicopters (plans to sell the HMS *Invincible* to Australia were canceled after the 1982 Falklands War). In addition, four World War II light carriers remain in service with Third World navies: the VEINTICINCO DE MAYO of Argentina, the MINAS GERALS of Brazil, and VIKRANT and VIRAAT of India).

AIRCRAFT CARRIERS, FRANCE: The two 32,000-ton French aircraft carriers of the CLEMENCEAU class were built to a French design between 1955 and 1960. They are to be replaced by two nuclear-powered aircraft carriers of the 38,000-ton CHARLES DE GAULLE class between 1995 and 1998.

AIRCRAFT CARRIERS, INDIA: The Indian Navy operates two ex-British light aircraft carriers: VIKRANT (formerly HMS *Hercules*), purchased in 1957; and VIRAAT (formerly HMS *Hermes*), purchased in 1986. Now outfitted as V/STOL carriers, both operate air groups of British Sea Harrier fighters and anti-submarine helicopters.

AIRCRAFT CARRIERS, ITALY: Italy operates two specialized air-capable ships, the 13,240-ton GARIBALDI and the 9500-ton VITTORIO VENETO. The former, in service since 1985, has a full-length flight deck and HARRIER V/STOL fighters and anti-submarine helicopters. The latter, commissioned in 1969, is a hybrid helicopter-carrier/cruiser, with the superstructure and guided-missile armament of a cruiser forward, and a flight deck for anti-submarine helicopters aft.

AIRCRAFT CARRIERS, SOVIET UNION: The Soviet navy's first aviation ships were two 20,000-ton MOSKVA-class helicopter carriers, launched in 1967–68. Designated anti-submarine cruisers, they in fact have a cruiser forward half (with surface-to-air, anti-ship, and anti-submarine missile launchers), and a flight deck aft for 18 Ka-25 HORMONE anti-submarine helicopters. They were followed by four 30,000-ton KIEV-class V/STOL carriers commissioned between 1975 and 1985, which also have heavy cruiser armament forward, but an angled flight deck cantilevered to port allows short rolling takeoffs by Yak-38 FORGER V/STOL fighters. Both the Moskva and Kiev classes are primarily ANTI-SUBMARINE WARFARE (ASW) vessels, and their aircraft are meant to provide ASW support for surface action groups; their primary weapons are ANTI-SHIP MISSILES.

The first Soviet conventional-takeoff carrier, the 65,000-ton TBILISI, is powered by a uniquely Soviet combination of nuclear and steam (CONAS) propulsion. Completed in 1988, it began sea trials in 1989, and will enter service in the early 1990s. Two additional Tbilisis are now under construction. See also AV-MF.

AIRCRAFT CARRIERS, SPAIN: Spain acquired the 16,416-ton *Dedalo*, a surplus U.S. light fleet carrier (ex-USS *Cabot*, CVL-28) in 1967. Originally operated as a helicopter carrier for ANTI-SUBMARINE WARFARE (ASW), in 1973 it was modified to operate nine AV-8 Harrier V/STOL fighters. *Dedalo* is more than 40 years old, and Spain has built a new carrier, the PRINCIPE DE ASTURIAS, as a replacement. Based on the design of the U.S. Navy's Sea Control Ship (which was never built), with the addition of a ski-jump bow to enhance Harrier performance, the ship was launched in 1982, and entered service in 1986. Twelve AV-8B Harrier IIs were acquired in 1987–88.

AIRCRAFT CARRIERS, UNITED STATES: The U.S. Navy has by far the largest force of aircraft carriers, and most of its operational methods are focused around CARRIER BATTLE GROUPS and the employment of their air power.

At present, the U.S. has 15 operational carriers, but this number is expected to fall to 13 or fewer by the mid-1990s. At any given time, one carrier is undergoing a Service Life Extension Program (SLEP) overhaul, which can last up to 24 months, while several more are "working up" their air wings; hence, the number of fully operational carriers rarely exceeds 8.

The oldest active carriers are the 51,000-ton MIDWAY and *Coral Sea*, commissioned in 1945–46 and subsequently modernized; both will be retired in the early 1990s (the third Midway, the USS *Franklin D. Roosevelt*, was retired in the 1970s).

The first post-1945 carriers were the four 62,000-ton FORRESTALS commissioned from 1958. They were followed by three similar but improved KITTY HAWKS. In 1961 the U.S. Navy commissioned the 90,000-ton ENTERPRISE (CVAN-65), the first nuclear-powered aircraft carrier. Its unlimited range at high speed, and its ability to stow much more aviation fuel and ordnance in place of bunker oil, were offset by the huge initial cost of nuclear power. As a result, the next carrier launched was the conventionally powered USS *John F. Kennedy*, a modified Kitty Hawk commissioned in 1968.

In 1966, Congress authorized three new nuclear

carriers of the 96,000-ton NIMITZ class, with 90 percent more aviation fuel and 50 percent more ordnance storage than the Forrestals. After 1977, the Carter administration canceled a fourth Nimitz, and proposed instead a 50,000-ton "Sea Control Ship" (SCS) to operate smaller air groups of helicopters and V/STOL aircraft. The navy, however, successfully blocked the SCS, and in 1979 funding was restored for a fourth Nimitz, the USS *Theodore Roosevelt* (an SCS was eventually built by Spain as the PRINCIPE DE ASTURIAS). Under the Reagan administration, two more improved Nimitz carriers were authorized in 1983. Though the desirability of large carriers was increasingly questioned, the Falklands War, operations off Libya and the Persian Gulf, and the construction of the first large Soviet aircraft carrier TBILISI have all militated in their favor, and two more improved Nimitzes were ordered in 1988.

In addition to its large aircraft carriers, the navy has large amphibious assault ships which also have flight decks, including one WASP-class LHD, five TARAWA-class LHAS, and seven Iwo Jima-class LPHS. Normally, only transport helicopters are based on those ships, but the marine corps can operate AV-8 HARRIERS from them to provide close air support for its landing forces. A promising program to convert container ships into V/STOL carriers (ARAPAHO) was canceled in 1983 despite the success of a similar British system in the 1982 Falklands War.

AIR CUSHION VEHICLE: See HOVERCRAFT.

AIR DEFENSE (SYSTEM/FORCE):   A combination of weapons, sensors, and COMMAND AND CONTROL facilities designed to defend areas or specific sites ("points") against air attack. An air defense system consists of surveillance elements to detect enemy aircraft, usually RADARS but sometimes ELECTRO-OPTICAL devices; acquisition and FIRE CONTROL elements to vector weapons to their targets, usually based on radar or electro-optical devices; command and control elements, facilities where sensor reports can be correlated and actions coordinated; and ANTI-AIRCRAFT weapons, ground-based guns and missiles as well as manned fighter/interceptors.

TACTICAL air-defense forces accompany ground forces into battle to protect them from CLOSE AIR SUPPORT or INTERDICTION strikes. They are armed with self-propelled or towed surface-to-air missiles and light-to-medium anti-aircraft guns. Some of these weapons are mounted on ARMORED FIGHTING VEHICLE (AFV) chassis to move with armored for-

mations, generally as self-contained units with their own surveillance, acquisition, and fire control equipment (see, e.g., *Flakpanzer* GEPARD and ZSU-23-4). The primary effect of tactical air defenses is deterrence: their mere presence, once known, can seriously degrade the accuracy of air attacks as pilots concentrate on evasion and self-protection.

STRATEGIC air defenses are designed to defend rear areas and home territories; thus, their sensors and command facilities can be larger and more sophisticated, because they are generally static. Radar data can be fed into central plotting and direction facilities to track incoming aircraft and coordinate the response.

In both tactical and strategic air defenses, combinations of diverse weapons and sensors are preferred, in spite of the resulting inefficiencies, in order to complicate enemy attempts to neutralize the system. In addition, weapons are differentiated in range and ceiling to achieve a layered defense, with guns and short-range missiles for the lowest altitude bands, medium-range missiles for medium altitudes, and long-range missiles for the high-altitude band. Thus aircraft attempting to evade weapons in one band are driven into another. That was the effect achieved by Soviet-supplied Arab air defenses against the Israeli Air Force in 1973: aircraft attempting to evade the SA-6 GAINFUL by diving to low altitude were attacked by ZSU-23-4 guns and SA-7 GRAIL missiles. Given time, resources, and imagination, air defenses can be wholly neutralized (as, e.g., by Israel in the 1982 Lebanon War and by U.S.-led coalition forces in the 1991 Persian Gulf conflict.), but even so, the attacker is thereby forced to divert resources from his primary offensive purposes. See also COUNTERAIR; NORAD: PVO: ROLLBACK; SEAD.

AIR FORCE: Independent air service, autonomous arm, or air branch (of unified military services). A numbered or geographic "air force" is a higher formation responsible for air activities within a specific THEATER OF OPERATIONS. Thus there are U.S., Soviet, British (RAF Germany), and Chinese "air forces" in different geographic regions, while NATO has two "Allied Tactical Air Forces" (ATAFS). In the Soviet air force, the same echelon is called an "air army," and its subordinate formations are designated "air divisions" and "air regiments."

AIR FORCE, SOVIET UNION: See VVS.

AIR FORCE, UNITED STATES: The USAF is organized into three major commands, two overseas commands, and several support commands.

The three major commands are the STRATEGIC AIR COMMAND (SAC), the TACTICAL AIR COMMAND (TAC), and the MILITARY AIRLIFT COMMAND (MAC). The major overseas commands are U.S. Air Force in Europe (USAFE) and the Pacific Air Force (PACAF).

The USAF's support commands include the Air Force Communications Command, Air Force Logistics Command (AFLC), and Air Force Systems Command (AFSC).

AIR INTERDICTION: See INTERDICTION.

AIR-LAND BATTLE: The prescribed U.S. Army OPERATIONAL METHOD since 1982, which embodies an attempt to shift from firepower-based ATTRITION tactics to MANEUVER. While it also contains a long-range attritional element ("Follow-On Forces Attack," FOFA), the maneuver dimension of Air-Land Battle doctrine owes much to the classic lessons of armored warfare, as practiced by the Germans and Israelis. Its goal is to disrupt the enemy's ability to fight rather than to destroy his forces cumulatively. Surprise blows against forces and facilities in critical sectors are meant to dislocate the enemy's plans and dispositions, and ideally the entire functioning of his chain of command. The Air-Land Battle concept of maneuver is summarized in FM 100-5, *Operations*, the U.S. Army's basic doctrinal manual, which emphasizes initiative, depth, agility, synchronization, maneuver, firepower, protection, and leadership.

Initiative is defined as the ability to take advantage of fluid situations; subordinates must swiftly act independently at the tactical level within the scope of the operational plan but without waiting for orders.

Depth refers to the need for action throughout the entire operational zone, to prevent the enemy from freely concentrating his firepower or maneuvering his forces.

Agility is the ability to change plans, operational direction and tactics quickly and with minimal disruption, and is achieved by a flexible organizational structure, as well as by commanders who can decide and act quickly, improvising as needed.

Synchronization requires a unity of purpose in the action of all elements involved in an operation.

Maneuver is movement designed to achieve concentration, surprise, SHOCK, and momentum.

Firepower implies the concentration of fire against critical points in the enemy's deployment, to suppress his own fire, neutralize his TACTICS, and generally destroy his ability to fight.

Protection is the converse of firepower: U.S.

forces must be shielded from enemy fire to preserve their own freedom of maneuver. It includes the use of cover, deception, concealment, suppression, and mobility.

Finally, *leadership* is described as the crucial element; to implement Air-Land Battle methods, leaders must plan and act on their own initiative within the general concept of the operation, while inspiring their troops.

These definitions of initiative, agility, synchronization, and leadership show the influence of the German AUFSTRAGSTAKTIK concept.

The adoption of a maneuver-based doctrine by the U.S. Army reflected its sense of material weakness vis à vis the Soviet army during the 1970s and early 1980s. But the army's continuing centralization of COMMAND AND CONTROL and ingrained habits of tactical micromanagement by higher echelons are serious obstacles to a full application of Air-Land Battle doctrine. The collision between the new doctrine and traditional preferences is reflected in the divisional organization that was adopted at the same time as Air-Land Battle (DIVISION 86); instead of smaller, leaner, and more numerous units, the new divisions are still very large and unwieldy. Moreover, the U.S. Army's equipment bureaucracy remains focused on weapons designed to fight an attritional war. See also ACTIVE DEFENSE; BLITZKRIEG.

AIR-LANDING: A method of inserting ground forces behind enemy lines, whereby a landing ground is seized by parachuted AIRBORNE FORCES or helicoptered AIR ASSAULT forces, before other troops and heavier equipment are flown in by fixed-wing transports. See also AIR MOBILE FORCES.

AIR MOBILE (FORCES): Forces that can be transported by helicopter or fixed-wing aircraft. Unlike AIRBORNE FORCES, air mobile forces cannot be parachuted, but must instead be AIR-LANDED. In general, only lightly equipped forces can be truly air mobile. The subcategory of AIR ASSAULT forces defines units specifically trained and equipped for forced entry by helicopter.

AIR POWER: The ability to project military power from the air, whether "strategically"—i.e., autonomously—or in support of ground or naval forces. Air power missions include CLOSE AIR SUPPORT, battlefield and deep INTERDICTION, AIR SUPERIORITY, and strategic BOMBING.

More specifically, "air power" connotes the theories developed between the world wars by Gen. Giulio Douhet of Italy, Gen. William Mitchell and Alexander de Seversky in the U.S., and Air Chief Marshal Trenchard in Britain, all of whom advocated the primacy of "strategic" (independent) air forces, arguing that future wars would be decided by the bombardment of industry (and cities, to break civilian morale), thus saving both sides from the stalemate of trench warfare. The results of the strategic bombing campaigns of World War II discredited those theories (civilian morale did not break; industry proved resilient) by 1945, when they were immediately rehabilitated by the advent of NUCLEAR WEAPONS.

AIRSHIP: A self-propelled, lighter-than-air vehicle. Rigid airships, or "dirigibles," have not been built since the late 1930s, but semi-rigid airships (blimps) are still produced. Consisting of a balloon filled with a light gas (usually helium) to generate lift, and a suspended gondola that contains the cockpit, payload, fuel tanks, and engines, semi-rigid airships are controlled in flight by rudders and elevators on the tail surfaces, and/or by gimballed propellers. Airships have much greater endurance than any other aircraft (days, not hours), albeit at speeds not much above 100 mph (167 kph), which limit their maneuverability in high winds.

Semi-rigid airships have been used for ANTI-SUBMARINE WARFARE (ASW) patrol, AIRBORNE EARLY WARNING (AEW), and as electronic intelligence (ELINT) platforms. The U.S. Navy is currently reassessing the airship as a platform for fleet AEW and ASW support partly because of its economy and low radar signature. See also AEROSTAT; ANTI-AIR WARFARE; AWACS.

AIR SUPERIORITY: A condition realized (in war) when one side is so preponderant that its aircraft can carry out attack and air-transport operations with little or no interference from enemy fighters. Air superiority may derive from an overwhelming initial advantage (if the enemy does not even contest the airspace), or it may be won by successful COUNTERAIR operations; in either case, it may extend throughout a THEATER OF OPERATIONS, or only in certain sectors (local air superiority).

Air superiority is usually a prerequisite for success on the ground (except in a guerrilla war) and at sea, unless there is an inordinate disparity in ground or naval strength, or exceptionally effective AIR DEFENSES are present. Local air superiority may suffice, if it coincides at each remove with the critical sector (SCHWERPUNKT) of ground operations.

"Air-superiority fighter" defines aircraft designed primarily for air combat with other fighters (e.g., the F-15 EAGLE and MiG-29 FULCRUM), as opposed to (anti-bomber) interceptors and (ground) attack aircraft.

AIR SUPREMACY: Another term for AIR SUPERIORITY but generally implying a decisive advantage that extends throughout a THEATER OF OPERATIONS.

AIST: A class of some 17 Soviet air-cushion landing craft (HOVERCRAFT) completed since 1971. The largest hovercraft worldwide until the Soviet navy introduced the POMORNIK in 1986, the Aists have a boat-shaped hull surrounded by an inflatable rubber skirt to trap the air cushion on which the vehicle rides. There is a small pilothouse at the bow, and a folding ramp door at each end of the vehicle for the roll-on/roll-off loading of cargo, including two MAIN BATTLE TANKS or four PT-76 light tanks, and up to 220 fully equipped troops—a total payload of some 270 tons.

Defensive armament consists of two twin 30-mm. ANTI-AIRCRAFT guns in the bow; later versions also have two quadruple pedestal launchers for SA-N-5 GRAIL short-range surface-to-air missiles.

The Aists have two pairs of shrouded lift fans and four aircraft-type propellers (two tractor and two pusher). Steering is facilitated by rudders on twin vertical fins at the stern, and by varying propeller pitch. See also AMPHIBIOUS ASSAULT/WARFARE; LCAC.

Specifications **Length:** 155.16 ft. (47.3 m.). **Beam:** 57.08 ft. (17.4 m.). **Draft:** 1 ft. (0.3 m.). **Displacement:** 150 tons full load. **Propulsion:** 2 14,795-shp. NK-12MV gas turbines. **Speed, max:** 80 kt. **Range:** 100 n.mi. at 65 kt./350 n.mi. at 60 kt. **Sensors:** 1 "Spin Trough" surface-search radar, 1 "Drum Tilt" fire control radar. **Electronic warfare equipment:** 1 "High Pole B" IFF, 1 "Square Head" IFF interrogator, 2 chaff launchers.

AK: *Avtomat Kalashnikova*, a ubiquitous and highly successful family of Soviet automatic weapons, known for their ruggedness and ease of operation.

Introduced in 1946, the basic model, the famous AK-47, is a gas-operated ASSAULT RIFLE which fires reduced-power 7.62- × 39-mm. ammunition from a 30-round "banana" clip. Made largely of stamped steel, and with a wooden stock, the AK-47 is capable of semi- or full-automatic fire; it has sight settings of up to 800 m., but is only accurate out to a range of 300 m. These characteristics suit Soviet tactics, which emphasize suppression rather than deliberate, aimed fire. Widely used in Vietnam and in almost all other wars since 1950, the

AK-47 is extremely popular with troops because of its excellent "feel," absolute reliability, and ease of maintenance.

The AK-47 was superseded in the late 1950s by the AKM and AKMS, which are lighter and even easier to manufacture, but otherwise identical to the AK-47. The folding-stock AKMS was issued to Soviet Airborne Forces.

From 1974, the AK-47 and AKM were superseded by the AK-74, chambered for a smaller, higher-velocity 5.45- × 39-mm. round. The AK-74 is almost identical to the AKM, except that the barrel and receiver mechanism are modified to accept the smaller round; a folding-stock version is issued to the Airborne Forces. The lighter 5.45-mm. round allows soldiers to carry more ammunition.

The AK design is also the basis of a number of LIGHT MACHINE GUNS, most notably the RPK (an AKM with a bipod and a heavier barrel) and the RPK-74 (adapted from the AK-74). The RPK fires the same 7.62- × 39-mm. round as the AK-47 at a cyclic rate of 660 rounds per minute from a 75-round drum magazine, but because the barrel cannot be quick-changed, the practical rate of fire is limited to 80 rounds per minute to avoid overheating. The RPK-74 is essentially similar, but chambered for the 5.45-mm. round. One RPK or RPK-74 is issued to each Motorized Rifle or airborne infantry squad in the Soviet and Warsaw Pact armies.

The AK family is manufactured throughout the Warsaw Pact, and in China, Yugoslavia, Finland, and elsewhere. Handmade copies are produced in the arms bazaars of Pakistan. The Afghan MUJAHIDEEN used both copies and captured originals.

Specifications **Length OA:** (AK-47, AKM, AK-74) 35.27 in. (896 mm.); (AKMS, AKS-74) 25.4 in. (655 mm.); (RPK, RPK-74) 40.75 in. (1.035 m.). **Length, barrel:** (AK-47, AKM, AKMS, AK-74) 16.3 in. (414 mm.); (RPK, RPK-74) 23.27 in. (591 mm.). **Weight, loaded:** (AK-47) 11.31 lb. (5.14 kg.); (AKM, AKMS) 8.77 lb. (3.98 kg.); (AK-74) 8.5 lb. (3.86 kg.); (RPK, RPK-74) 15.65 lbs. (7.11 kg.). **Muzzle velocity:** (AK-47, AKM, AKMS) 710 m./sec.; (AK-74) 900 m./sec. **Cyclic rate:** (AK-47, AKM, AKMS, AK-74) 600 rds./min.; (RPK, RPK-74) 660 rds./min. **Practical rate:** (AK-47, AKM, AKMS, AK-74) 100 rds./min.; (RPK, RPK-74) 80 rds./min. **Effective range:** (AK-47, AKM, AKMS, RPK) 300 m.; (AK-74, RPK-74) 300–400 m.

AKULA: NATO code name for a class of three Soviet nuclear-powered attack SUBMARINES,

the first of which was completed in 1984. Believed to be an incremental development of the VICTOR III class, the Akula may also be a low-risk backup to the SIERRA submarines.

Like most recent Soviet submarines, the Akula has a modified teardrop hullform in a double-hull configuration. The pressure hull is probably fabricated of high-yield steel, and the outer hull is covered with CLUSTER GUARD anechoic tiles to reduce radiated noise and impede active sonars. A very long and low sail (conning tower) amidships is partially blended into the outer hull. Control surfaces have the standard Soviet configuration of retractable bow diving planes, and cruciform rudders and stern planes ahead of the propeller.

Armament consists of 6 bow-mounted torpedo tubes, 4 of 21-in. (533-mm.) and 2 of 25.6-in. (650-mm.) diameter. The 21-inch tubes can launch a variety of free-running and acoustic homing TORPEDOES, plus the SS-N-15 STARFISH nuclear anti-submarine missile, while the 25.6-inch tubes can launch the SS-N-21 SAMPSON long-range cruise missile, the Type 65 wake-homing anti-ship torpedo, and the SS-N-16 STALLION anti-submarine missile. The basic load is estimated at 24 weapons of all types.

The Akulas have single-shaft NUCLEAR PROPULSION, but unlike earlier Soviet submarines, they are extremely quiet (perhaps as quiet as early units of the U.S. LOS ANGELES class), and that has generated much concern among Western specialists. It is believed that the Akula's entire powerplant is mounted on resilient sound-isolation rafts, with its components individually soundproofed as well. Further, it seems that smuggled Western computer-controlled precision milling machines have allowed the Soviet navy to manufacture large, low-speed, non-cavitating propellers, further reducing noise levels. Overall, with the Akulas, the silencing advantage of Western submarines is much reduced. See also SUBMARINES, SOVIET UNION.

Specifications **Length:** 352.7 ft. (107.53 m.). **Beam:** 37 ft. (11.3 m.). **Displacement:** 7500 tons surfaced/10,000 tons submerged. **Max. operating depth:** 1800 ft. (550 m.). **Collapse depth:** 3000 ft. (915 m.). **Propulsion:** 2 pressurized-water reactors with two sets of geared steam turbines, 45,000 shp. **Speed:** 20 kt. surfaced/35 kt. submerged. **Crew:** 120. **Sensors:** 1 bow-mounted low-frequency active/passive sonar; 1 low-frequency passive towed array sonar; 1 "Snoop Pair" surface-search radar; 1 electronic signal monitoring array; 1 "Park Lamp" radio direction-finder (D/F); 2 periscopes;

UHF, VHF, and SHF antennas; one VLF communications buoy; 1 "Pert Spring" satellite navigation system.

AL ABBAS: Iraqi short-range ballistic missile derived from the Soviet 55-lb Scud-b. See AL HUSAYN.

ALAMO (AA-10): NATO code name for the Soviet R-72 medium-range air-to-air missile introduced in the mid-1980s as armament for the MiG-29 FULCRUM and Su-27 FLANKER. A large missile, Alamo has four small nose strakes, mid-mounted cruciform wings for steering, and four fixed broad-chord tail fins.

Three different versions have been identified to date. The original AA-10A Alamo-A had a fast-burning solid-rocket engine and SEMI-ACTIVE RADAR HOMING (SARH); the AA-10B Alamo-B has the same engine but is guided by INFRARED HOMING. The current AA-10C Alamo-C combines SARH with a long-burning rocket engine.

The infrared version is believed to have ALL-ASPECT capability, while the SARH versions supposedly have full LOOK-DOWN/SHOOT-DOWN capability (HOME-ON-JAM guidance has also been reported). All versions are armed with a high-explosive warhead fitted with proximity fuzes. Flanker and Fulcrum can both carry up to six Alamos, and usually fly with a mix of SARH and infrared models. Given the AIR SUPERIORITY mission of these aircraft, the AA-10 should be highly maneuverable.

Specifications Length: 10.5 ft. (3.2 m.) to 13.1 ft. (4 m.), depending on version. **Diameter:** 7.3 in. (185 mm.). **Span:** 27.5 in. (698 mm.). **Weight, launch:** 350–450 lb. (160–182 kg.). **Speed, max.:** Mach 3 (2100 mph/3507 kph). **Range, max.:** (A, B) 5 mi. (8.35 km.); (C) 18.5 mi. (31 km.).

ALARM: Air-Launched Anti-Radiation Missile, an ANTI-RADIATION MISSILE (ARM) under development by British Aerospace for the RAF and other NATO air forces. Similar in function to the U.S. AGM-88 HARM, but smaller and lighter, ALARM has four highly swept cruciform wings set well back, and four movable tail fins of equal span for steering. Powered by a solid rocket motor, ALARM is armed with a high-explosive fragmentation warhead fitted with an advanced laser-proximity fuze.

ALARM's broad-spectrum scanner can be reprogrammed in the field to attack newly identified hostile radars. Like most third-generation ARMs, ALARM has an INERTIAL GUIDANCE platform which can continue to guide it to the target even if the target radar shuts down. But ALARM is unique in having a "loiter mode" whereby the missile is fired inertially to a height of 40,000 ft. (21,200 m.), at which point the motor shuts down, a parachute deploys, and the missile descends, scanning for enemy radars. Should one be detected, the missile locks on, jettisons the parachute, and executes a diving glide attack. Alarm was used by the RAF during the 1991 Persian Gulf conflict. See also SHRIKE; STANDARD ARM; TACIT RAINBOW.

Specifications Length: 13.15 ft. (4 m.). **Diameter:** 9 in. (229 mm.). **Span:** 28.75 in. (730 mm.). **Weight, launch:** 661 lb. (300 kg.). **Speed, max.:** Mach 2+. **Range, max.:** 15 mi. (25 km.).

ALAT: *Aviation Legère de l'Armée de Terre*, the air arm of the French army, which operates helicopters and light fixed-wing aircraft for transport, casualty evacuation (CASEVAC), SCOUTING, and light attack. See also AAC; CAVALRY, AIR.

ALCM: 1. Air-Launched Cruise Missile.
2. The Boeing AGM-86B air-to-surface missile, in service with the U.S. STRATEGIC AIR COMMAND (SAC). Intended for the long-range nuclear attack of strategic targets, ALCM is essentially a very small, pilotless aircraft that can be dropped from a bomber (or any other large aircraft) well outside enemy-controlled airspace to fly a preprogrammed course to the target at very low altitude. Developed from the early 1970s in response to the increasing vulnerability of the B-52 STRATOFORTRESS to modern Soviet AIR DEFENSES, the original AGM-86A began test flights in 1976, but was found to have inadequate range. It was therefore superseded by the AGM-86B ALCM-B, with a longer fuselage, larger wing span, and greater internal fuel capacity. In 1982, ALCM-B entered service aboard specially converted B-52Gs, each of which can carry up to 12 missiles on external pylons; B-52Hs are also to be be converted to carry up to 20 ALCMs (12 externally, 8 internally). The B-1B can carry up to 22 ALCMs, but will not normally be used for that purpose unless and until it is replaced as a penetrating bomber by the B-2 "Stealth Bomber" in the late 1990s.

ALCM-B has two low-mounted, swept wings that fold inside the missile until launch, and three small tail fins for steering. It is powered by a small Williams Research F107-WR-100 turbofan engine developing 600 lb. (272 kg.) of thrust, and is armed with a 250-kT W80 nuclear warhead weighing some 270 lb. (122.72 kg.).

The missile is guided by a combination of INER-

TIAL GUIDANCE and TERRAIN COMPARISON AND MATCHING GUIDANCE (TERCOM): its computer compares the returns from an on-board ground-mapping RADAR with a digitized radar map of the programmed flight path. By maneuvering to match the radar returns with the stored map, the missile can achieve very high accuracy; the median error radius (CEP) on test flights is generally less than 100 ft. (30.5 m.).

ALCM-B thus has a definite HARD target capability (against missile silos, command bunkers, etc.), but would still be unsuitable as a FIRST STRIKE weapon because of its long flight time (more than three hours at maximum range). The missile's very small RADAR CROSS SECTION, combined with its low-level flight path, makes its detection difficult in spite of its low speed. Since the 1980s, however, the Soviet Union has deployed weapons intended specifically to intercept cruise missiles, notably the SA-10 GRUMBLE and SA-12A GLADIATOR surface-to-air missiles, and the MiG-31 FOXHOUND interceptor.

Large numbers of ALCMs could still swamp such defenses, but only 1547 ALCM-Bs were procured because the U.S. Air Force is now developing the AGM-129 Advanced Cruise Missile (ACM), with "low observables" (STEALTH) characteristics, a higher cruising speed, and longer range. Believed to be a wingless lifting body with a triangular cross-section and powered by a Williams F112 turbofan, the ACM is expected to arm both the B-1 and B-2 from the mid-1990s. See also KENT; SAMPSON; TOMAHAWK.

Specifications Length: (ALCM-B) 20.75 ft. (6.325 m.); (ACM) 10.83 ft. (3.3 m.). **Diameter:** (ALCM-B) 24.5 in. (620 mm.); (ACM) NA. **Span:** (ALCM-B) 12 ft. (3.66 m.); (ACM) NA. **Weight, launch:** (ALCM-B) 3200 lb. (1452 kg.); (ACM) 3645 lb. (1657 kg.). **Speed, cruise:** (ALCM-B) 500 mph (805 kph); (ACM) Mach 0.85 (600 mph/1000 kph). **Range, max.:** (ALCM-B) 1550 mi. (2500 km.); (ACM) 3583 mi. (5780 km.).

ALE-: U.S. designation for airborne countermeasures, dispensers for packages of CHAFF (used against tracking and weapon guidance RADARS) and high-intensity flares (used against INFRARED-HOMING missiles). The most widely issued ALE-40 can be carried internally, or externally, in a pod. Miniaturized expendable ACTIVE JAMMERS, now under development, may also be dispensed by ALE-40s. See also ELECTRONIC COUNTERMEASURES; INFRARED COUNTERMEASURES.

ALFA: NATO code name for a class of six Soviet nuclear-powered attack SUBMARINES completed between 1970 and 1980. When first detected, the Alfa caused much concern among Western navies because of its unprecedented speed and deep-diving capacity, the results of highly innovative technology.

A small submarine by current standards, the Alfa has a modified teardrop hullform in a double-hull configuration, with ballast tanks between the outer casing and the pressure hull. Because it is fabricated of titanium alloy (a major innovation), which is both lighter and stronger than high-yield (HY) steel, the Alfa's maximum operating depth of 2297 ft. (700 m.) and collapse depth of 3805 ft. (1160 m.) greatly exceed the diving abilities of Western submarines. A long, low sail amidships is blended into the outer casing to reduce drag (another innovative feature) and the outer casing is covered with CLUSTER GUARD anechoic tiles to reduce radiated noise and impede detection by active sonars. The control surfaces are arranged in standard Soviet fashion, with retractable bow diving planes, and cruciform rudders and stern planes ahead of the propeller.

Alfa has six 21-in. (533-mm.) tubes in the bow for a normal load of 8 21-inch anti-ship TORPEDOES, 10 16-in. (406-mm.) anti-submarine torpedoes, 2 nuclear torpedoes, and 2 SS-N-15 STARFISH anti-submarine missiles.

The Alfa has single-shaft NUCLEAR PROPULSION with a very compact, closed-cycle reactor cooled by liquid sodium (vs. pressurized water) that operates at very high temperatures. Liquid metal reactors have a much greater power density than the pressurized water reactors of most (and all U.S.) submarines, but are much less reliable, and cannot be shut down without causing the coolant to solidify. Extensively automated, the Alfa has a crew of only 45 men (mostly officers), as compared to 127 on the Los Angeles.

The total result is a small, streamlined submarine which achieves its high speed by a very high power-to-displacement ratio. The reduction of internal volume made possible by automation also allows more weight for a stronger pressure hull, to achieve very great depths. But innovative technology has a high price: the first Alfa suffered from severe mechanical problems, including a possible reactor accident, and was scrapped in 1974; the second did not enter service until 1979. Each subsequent Alfa was completed with slight variations, suggestive of the experimental nature of the entire program.

Too small to incorporate much acoustical silencing, the Alfa is extremely noisy at higher speeds, and would have to rely on its speed, maneuverability, and active sonar to evade attack—as opposed to the reliance of all U.S. and most Soviet submarines on stealth and passive sensors. Much of the technology developed for the Alfas was incorporated (unsuccessfully) into the later MIKE SSN. See also SUBMARINES, SOVIET UNION.

Specifications Length: 267 ft. (81.4 m.). **Beam:** 31.16 ft. (9.5 m.). **Displacement:** 2900 tons surfaced/3700 tones submerged. **Max. operating depth:** 2297 ft. (700 m.). **Collapse depth:** 3805 ft. (1160 m.). **Powerplant:** 1 liquid sodium reactor, 2 sets of geared steam turbines, 45,000 shp. **Speed:** 20 kt. surfaced/45 kt. submerged. **Crew:** 45. **Sensors:** 1 bow-mounted low-frequency sonar array, 1 medium-frequency active attack sonar, 1 "Snoop Tray" surface-search radar, 1 "Park Lamp" radio direction finder (D/F), 1 electronic signal monitoring array, 2 periscopes.

AL HUSAYN: Iraqi short-range BALLISTIC MISSILE (SRBM) project. After the outbreak of the Iran-Iraq War, in which both sides used Soviet-supplied FROG unguided rockets and SS-1 SCUD short-range ballistic missiles, Iraq began the development of an extended-range Scud, probably with assistance from West German companies for the production facilities, and East German technicians for the design modifications.

In March 1985, Iran launched a Libyan-supplied Scud against Kirkuk, then 13 more against Baghdad; after that, Iran periodically launched small numbers of Scuds against Baghdad. Iraq, however, could not retaliate against Teheran because the distance exceeded the Scud's range. But by August 1987, Iraq's modified Scud was ready: in it, the original 2200-lb. (1000-kg.) warhead was reduced to 792 lb. (360 kg.), including 352 lb. (160 kg.) of high explosives. The resulting "Al Husayn" missile had a range of just over 360 mi. (600 km.); more than 180 were launched against Teheran and other Iranian cities in March–April 1988. Several dozen of these missiles were launched against targets in Saudi Arabia and Israel during the 1991 Persian Gulf conflict.

The first Al Husayns were built from cannibalized Scuds, but under the designation "Project 124," Iraq also established facilities for the complete production of the SRBM, with the covert assistance of West German companies. Under the

same project, Iraq next attempted to produce a longer-range missile, the "Al Abbas," with a 558-mi. (900-km.) range. Test-launched in April 1988, the Al Abbas is seemingly in production, and the project appears to have evolved into a still larger space-launch booster/medium-range ballistic missile (MRBM). In December 1989, Iraq tested a 48-ton "Al Abid" booster capable of generating some 154,000 lb. (70,000 kg.) of thrust, which could be capable of placing a small payload into low earth orbit. At the same time, Iraq announced that it had an MRBM with a range of 1240 mi. (2000 km.).

By then, Iraq's solid-fuel-propulsion "Project 395," which was started in 1985, had made sufficient progress to reduce Iraq's interest in the Al Husayn/Al Abbas liquid-fuel approach. Several versions of the "Fahd" SRBM produced under Project 395 have been reported, with ranges between 155 and 372 mi. (250 and 600 km.), apparently forming a family of weapons, rather than an experimental succession. Large production facilities have been established under Project 395 with concealed assistance from West German companies, including a chemical plant for the production of solid rocket fuel south of Baghdad, a missile component factory and assembly line near Faluja west of Baghdad, and a rocket test stand, also south of Baghdad.

Iraq's "Sa'ad 16" research facility includes laboratories, machine shops, a wind tunnel, and missile test stands in a large complex near Mosul; because Sa'ad 16 was presented as a scientific project, it originally had U.S. industrial as well as West German and other European participation. All of these facilities were severely damaged by allied bombing in 1991. See also CONDOR.

ALIZÉ: Dassault-Breguet BR.1150 carrier-based, turboprop ANTI-SUBMARINE WARFARE (ASW) aircraft, in service with the French and Indian navies. Developed beginning in the late 1940s as a carrier-based attack aircraft similar to the U.S. A-1 SKYRAIDER, the Alizé had evolved into an ASW patrol aircraft by the time of its first flight in 1956. Of a total of 89 delivered between 1957 and 1962, some 39 remain in service with France (aboard the two CLEMENCEAU carriers) and 20 with India (aboard the carrier VIKRANT), which used them in the 1971 Indo-Pakistan War in an anti-shipping role.

The Alizé's engine is mounted in the nose, just ahead of a rather cramped cockpit for the pilot, radar operator, and ASW systems operator. Below the cabin, an enclosed weapons bay normally houses either one ASW TORPEDO or three 350-lb. (160-kg.) DEPTH CHARGES. A surface search RADAR is housed in a retractable "dustbin" radome behind the weapons bay.

The low-mounted wings fold upward for deck parking. Each has a pylon under the inner section for an ASW torpedo or a 386-lb. (175-kg.) depth charge, and each outer section has five more pylons for 5-inch (127-mm.) rockets or light missiles. SONOBUOYS can be carried in two large wing fairings behind the main landing gear.

The Alizé's sensors are obsolescent, but under a modernization program 28 French aircraft are being retrofitted with new a Thomson-CFS Iguane radar, a SERCEL-Crouzet INERTIAL NAVIGATION system, ELECTRONIC SIGNAL MONITORING (ESM) equipment, new communications gear, and more advanced AVIONICS. Even so, the Alizé is simply too small for modern ASW. The French navy, however, continues to rely on it because the two Clemenceaus cannot operate larger aircraft. See also AIRCRAFT CARRIERS, FRANCE; AIRCRAFT CARRIERS, INDIA.

Specifications **Length:** 45.5 ft. (13.86 m.). **Span:** 51.16 ft. (15.6 m.). **Powerplant:** 1 1975-shp. Rolls Royce Dart RDa.21 turboprop. **Weight, empty:** 12,566 lb. (5700 kg.) **Weight, max. takeoff:** 18,078 lb. (8200 kg.) **Speed, max.:** 322 mph (518 kph) at 10,000 ft. (3050 m.). **Speed, patrol:** 150–230 mph (240–370 kph). **Range:** 1553 mi. (2500 km.). **Endurance, max.:** 7 hrs. 40 min.

ALL-ASPECT (MISSILE): An air-to-air or surface-to-air missile that is capable of attacking targets head-on and laterally, as opposed to rear-quarter-only missiles. With all-aspect missiles, fighters need not achieve a position behind the target aircraft before firing the missile, a major tactical advantage. Until recently only ACTIVE HOMING or SEMI-ACTIVE RADAR HOMING (SARH) missiles such as SPARROW, PHOENIX, ASPIDE, etc., had all-aspect capability, while INFRARED-HOMING missiles such as SIDEWINDER were limited to rear-quarter attacks for which they could home on the very intense infrared signatures of jet exhausts. But late-model IR missiles, such as the AIM-9L/M versions of Sidewinder, the Rafael PYTHON, and Soviet AA-8B APHID, have homing sensors of greater sensitivity which allow them to detect and track aircraft by the less intense heat radiated by (or reflected) from their airframes. See also AIR COMBAT MANEUVERING.

ALLIGATOR: NATO code name for a class of 16 Soviet tank landing ships (LSTs) commissioned between 1966 and 1974. The largest Soviet AMPHIBIOUS ASSAULT ships until the advent of the IVAN ROGOV class in the 1980s, the Alligators have a short forecastle and a small superstructure aft, with large ramp doors at the bow and stern for the roll-on/roll-off loading of vehicles. Their total capacity is 1500 tons of cargo, or 12 main battle tanks, or up to 50 smaller vehicles, as well as some 120 troops.

There are four distinct versions, with 1, 2, or 3 cargo cranes amidships; the later, single-crane type predominates. For self-defense, all are armed with 1 twin 57-mm. and 2 twin 25-mm. anti-aircraft guns. The last two subclasses also have 2 quadruple pedestal launchers for SA-N-5 GRAIL short-range surface-to-air missiles, and the last subclass has a 40-round 122-mm. multiple rocket launcher for shore bombardment. The Alligators have been supplemented by the newer but smaller POLNOCNY LSTs built in Poland. See also NAVAL INFANTRY.

Specifications **Length:** 374 ft. (114 m.). **Beam:** 50.95 ft. (15.5 m.). **Draft:** 12.05 ft. (3.7 m.). **Displacement:** 3400 tons standard/4800 tons full load. **Powerplant:** 2 9000-hp. diesels, 2 shafts. **Speed:** 18 kt. **Crew:** 100. **Sensors:** 2 "Don Kay" navigation radars, one "Muff Cob" 57-mm. fire control radar, one "High Pole B" IFF.

ALOUETTE: French light helicopter built by Aerospatiale, in service with some 27 countries. First flown in 1955, the initial SA.313 Alouette II was a three-seat light helicopter originally intended for crop dusting and other agricultural uses, but was soon adopted by the French army's aviation branch (ALAT) and by other armed forces for light attack, liaison, RECONNAISSANCE, light transport, casualty evacuation (CASEVAC), and training. More than 1300 were built in France through 1975, and the type is now produced under license in India and Brazil (as the "Lama").

Successful because of its simplicity, ease of operation, and low costs, the Alouette II has a fuselage consisting of a plexiglass pod and an open lattice tail boom which supports the anti-torque rotor; avionics are limited to basic blind-flying instruments and tactical radios. A fuel tank is located immediately behind the cabin, and the engine, mounted directly over the fuel tank, drives a fully articulated three-bladed rotor. Most Alouette IIs are unarmed, but ASW versions carry a lightweight homing TORPEDO, and light attack versions can be armed with two ANTI-TANK GUIDED MISSILES (ATGMs) or two rocket pods.

The SA.316B Alouette III, a larger, more powerful development of the Alouette II, entered service in 1959 and was quickly adopted by the France and many other countries as a light multi-role helicopter. More than 1450 have been built so far, and licensed production continues in India, Romania, and Switzerland.

The Alouette III retains the engine, transmission, and rotors of the II, but has an enclosed cabin for the pilot and up to six passengers or two stretcher cases; the seats can be removed to provide an unobstructed hold for up to 1200 lb. (545 kg.) of cargo. Alternatively, up to 1653 lb. (750 kg.) of cargo can be carried externally. Armament includes a variety of machine guns and cannon, rocket pods, ATGMs, and (on ASW versions) homing torpedoes.

While lacking the speed, maneuverability, armor protection, and crashworthiness of heavier and more modern helicopters, both Alouettes remain attractive to users because of their economy.

Specifications **Length:** (II) 33.6 ft. (10.24 m.); (III) 32.95 ft. (1.05 m.). **Rotor diameter:** 36.15 ft. (11.02 m.). **Powerplant:** 1 870-shp. Turbomeca Artouste IIIB turboshaft. **Weight, empty:** (II) 2251 lb. (1021 kg.); (III) 2520 lb. (1143 kg.) **Weight, max. takeoff:** (II) 5071 lb. (2300 kg.); (III) 4850 lb. (2200 kg.) **Speed, max. cruise:** (II) 119 mph (192 kph); (III) 115 mph (185 kph). **Initial climb:** (II) 1083 ft./min (330 m./min.); (III) 853 ft./min. (260 m./min.). **Range, max. fuel:** (II) 350 mi. (565 km.). **Range w/860-lb. (390-kg.) payload:** (II) 186 mi. (300 km.). **Operators:** (both) Alg, Ang, Arg, Bang, Bel, Ben, Bra, Cam, CAR, Chad, Chi, Col, Con, Dji, Dom Rep, Ecu, Fin, Fra, FRG, Gui-Bis, Ind, Indo, Isr, Iv Cst, Por, Peru, Leb, Lib, Mali, Mex, S Af, Swe, Swi, Tun, Tur, UK.

ALPHA JET: A Franco-German twin-jet trainer/light attack aircraft built by a consortium of Dassault-Breguet and Dornier. Developed from 1969 in response to a French and West German requirement for an advanced jet trainer with secondary attack capabilities, the Alpha Jet was selected over the more powerful Sepecat JAGUAR in 1970, but technical problems (an inevitable fact of multinational programs) caused a delay of more than two years beyond the intended delivery date. Series production for the French air force did not begin until 1977, while the German air force did not receive its first Alpha Jet until 1978. Although designed for economy, design changes intended to enhance its attack capabilities raised the Alpha Jet's unit cost (approximately $4.5 million) to more than that of any other trainer. On the other hand, the Alpha Jet has proven to be capable in both its roles, and also rugged and simple to maintain. The French air force now operates some 175 Alpha Jets, and the West German *Luftwaffe* 126. It is also in service with the air forces of Belgium (33 built under license), Egypt (30 built under license), Morocco (24), the Ivory Coast (12), and Togo (5). The Franco-German consortium, however, failed to sell the Alpha Jet to the U.S. Navy, which chose the British Aerospace HAWK.

The pilot and instructor sit in a tandem cockpit, with the rear seat elevated to improve the instructor's forward vision for landing. French *(Ecole)* Alpha Jets, intended mainly for training, have a lead-computing optical gunsight and simple AVIONICS; German *(Appui)* versions, on the other hand, have a head-up display (HUD), a DOPPLER navigation radar, a navigation/attack computer, and Elletronica RADAR WARNING RECEIVERS, while Egyptian-made aircraft have a LASER rangefinder in the nose, a HUD, an INERTIAL NAVIGATION unit, and a radar altimeter. The space behind the fuselage contains six fuel tanks and the main landing gear. The moderately swept, shoulder-mounted wing has a fixed leading edge with a prominent dogtooth, outboard ailerons, and inboard flaps. The engines are mounted in pods blended into the fuselage just behind the cockpit.

The Alpha Jet has one fuselage pylon and four wing pylons, all rated at 1100 lb. (500 kg.). The fuselage pylon is used exclusively for a Mauser 27-mm. or DEFA 30-mm. cannon pod with 150 rounds of ammunition, while the wing pylons can accommodate a range of low-drag general-purpose (LDGP) bombs, CLUSTER BOMBS, NAPALM canisters, and rocket pods; German and Egyptian versions can also launch AGM-65 MAVERICK air-to-ground missiles.

While its high-speed maneuverability is mediocre compared to that of the Hawk, its low-speed handling makes the Alpha Jet a potential anti-helicopter aircraft of value. But the Alpha Jet lacks the performance and avionics needed to survive against strong air defenses; if employed for CLOSE AIR SUPPORT, it would probably suffer very heavy casualties.

Specifications **Length:** 40.66 ft. (12.29 m.). **Span:** 29.85 ft. (9.10 m.). **Powerplant:** 2 SNECMA/ Turbomeca Lazarc-04 turbofans, 2976 lb. (1350 kg.) of thrust each. **Weight, empty:** (trainer) 7374 lb. (3345 kg.); (attack) 7716 lb. (3500 kg.). **Weight, max. takeoff:** (attack) 16,535 lb. (7500 kg.). **Speed,**

max.: Mach 0.85 (553 mph/923 kph) at 30,000 ft. (9145 m); 576 mph (927 kph) at sea level. **Time to 30,000 ft. (9145 m.):** 7 min. **Service ceiling:** 48,000 ft. (14,630 m.). **Combat radius, max. payload:** 255 mi. (410 km.). **Ferry range:** 1727 mi. (2780 km.).

ALPINE TROOPS: LIGHT INFANTRY forces specifically trained and equipped for combat in high mountains, but not normally trained in mountaineering skills such as rock climbing and rappelling, except for small special units (e.g., the Italian *Alpieri*). Even in the age of the helicopter, the tactical and operational modalities of mountain warfare remain distinct from those of flatland operations, because much greater emphasis must be placed on small-unit action and dispersal. Hence COMMAND AND CONTROL must be decentralized, and units must often perform missions that would be assigned to the next higher echelon in flatland operations (i.e., PLATOONS must perform COMPANY missions, while companies must perform BATTALION missions, etc.).

Defensive mountain operations are normally aimed at blocking vehicular movement through passes by controlling the dominating ridgelines, while offensive operations are normally meant to seize terrain overlooking enemy positions or vehicular corridors. For both offense and defense, INFILTRATION and DEMOLITIONS are much used. Even with helicopter lift, ordinary infantry forces are generally unable to cope with the requirements of mountain warfare, because they are structured for operations in large, centrally controlled formations which rely heavily on artillery support. In addition, the troops themselves may lack the physical conditioning required for mountainous terrain. Hence attempts to use regular infantry in mountainous terrain have usually failed (e.g., the Allied operations in Italy, 1943–45, and Soviet operations in Afghanistan).

Not all armies that have mountain warfare forces (which now includes both the Soviet and U.S. armies) can emulate the European practice of recruiting natives of mountain regions, but all attempt to specifically select and train mountain troops. Generally, mountain units are lightly equipped, their ANTI-TANK elements are of course small, and they tend to emphasize MORTARS (more portable and effective than other artillery in mountains), and also portable anti-aircraft weapons, essential in the helicopter era. Battalions are generally organized as COMBINED-ARMS units capable of independent action, and thus tend to be larger than equivalent flatland forces.

At present, the West German army has one alpine brigade of *Gebirgsjagers*, the Italian army has several brigades of *Alpini*, the French army has the 27th Alpine Division, the Spanish army has a "high-mountain" brigade, while the Austrian and Swiss infantry is mostly Alpine. The U.S. Army has reactivated the 10th Light (Mountain) Infantry Division, and the Soviet army has several mountain-trained divisions of MOTORIZED-RIFLE TROOPS.

ALQ-: U.S. designation for airborne active ELECTRONIC COUNTERMEASURES systems, both internal and podded. The earliest ALQs were simple BARRAGE JAMMERS, but later versions added DECEPTION JAMMING capabilities. In early ALQs jamming transmitters had to be preset on the ground to the presumed frequencies of enemy radars, but the latest ALQs have ADAPTIVE JAMMING capabilities to counter a wide variety of radars and radios. Current models include the ALQ-161, an internal system for the B-1B, the ALQ-119 pod for ATTACK and WILD WEASEL aircraft, and the ALQ-131 pod, often used with the F-15 EAGLE and F-16 FALCON. See also ELECTRONIC WARFARE.

ALR-: U.S. designation for airborne ELECTRONIC SIGNAL MONITORING (ESM) equipment. See also RADAR HOMING AND WARNING RECEIVER.

ALWT: U.S. Mk.50 Advanced Lightweight Torpedo. See BARRACUDA.

AMAZON (TYPE 21): A class of six (originally eight) British FRIGATES commissioned between 1974 and 1978. Jointly produced by Vosper-Thorneycraft and Yarrow on the basis of a low-cost export design (similar to that of the Brazilian NITEROI class), the Amazons were criticized from the start as structurally weak, vulnerable, overloaded, and unstable. Elegant ships with yachtlike lines, they have a sharply raked bow, a long, low superstructure, a tower mast forward, and a low stack amidships. They were fitted with fixed internal ballast while under construction, to raise their metacentric height; despite that, none could carry the full design load of weapons and sensors. Serious cracks in frames and hull plates which developed during the 1982 Falklands War required structural reinforcement with additional hull plating amidships.

Armament consists of a 4.5-inch (114-mm.) DUAL PURPOSE gun on the bow, 4 EXOCET MM38 anti-ship missiles in canisters before the bridge, a quadruple launcher with 20 (obsolescent) SEA CAT short-range surface-to-air missiles on the superstructure aft, 2 sets of triple tubes for lightweight anti-submarine TORPEDOES, and 2 or 4 manually operated

20-mm. cannons for close-in defense. The primary ANTI-SUBMARINE WARFARE weapon is a multi-purpose LYNX helicopter operated from a hangar and a fantail landing deck. The Amazons have a SCOT satellite communications terminal, but ELECTRONIC WARFARE equipment is inadequate by modern standards.

The Amazons have a highly automated, twin-shaft COGOG powerplant that can be operated by remote control from the bridge.

All eight Amazons participated in the Falklands War, and two (HMS *Antelope* and *Ardent*) were lost to air attacks. During the fighting, it was determined that the ships were inadequately armed, even against unsophisticated forms of air attack; that they lacked effective sensors and electronic warfare equipment; and that their DAMAGE CONTROL arrangements were poor. While some structural reinforcement was possible, and two 20-mm. cannons were added, the Amazons had insufficient stability margins for more effective weapons such as PHALANX radar-controlled anti-missile guns. Because of their severe shortcomings, the Amazons will probably be retired early, with older but more solid LEANDER-class frigates retained instead. See also FRIGATES, BRITAIN.

Specifications **Length:** 384 ft. (117 mi.). **Beam:** 41.7 ft. (12.7 m.). **Draft:** 19.5 ft. (5.9 m.). **Displacement:** 2850 tons standard/3350 tons full load. **Powerplant:** 2 4250-shp. Rolls Royce Tyne gas turbines (cruising); 2 25,000-shp. Rolls Royce Olympus TM3B gas turbines (high-speed sprints). **Speed:** 32 kt. **Range:** 1200 n.mi. at 30 kt./4500 n.mi. at 18 kt. **Crew:** 177. **Sensors:** 1 Type 1006 navigation radar, 1 Type 992Q long-range surveillance radar, 2 Type 912 fire-control radars with electro-optical backup directors, 1 Type 184M hull-mounted medium-frequency sonar, 1 Type 162M side-looking classification sonar. **Electronic warfare equipment:** 1 UAA-1 "Abbey Hill" electronic signal monitoring array; 2 Knebworth Corvus chaff rocket launchers.

AMF: See ACE MOBILE FORCE.

AML: *Automitrailleuse Legere,* a small French 4×4 ARMORED CAR similar to the British FERRET, but with much heavier armament. Developed in the late 1950s, the AML entered service with the French army in 1960, and more than 4000 have been produced by Panhard since then. Widely exported, it was also produced under license in South Africa as the "Eland." Though superseded by newer armored cars in French active

units, the AML still equips French reserve infantry units.

The all-welded steel hull is divided into a driver's compartment up front, a turreted fighting compartment in the middle, and an engine compartment in the rear. Maximum armor protection is 12 mm., sufficient only against small arms and splinters. The driver, up front on the left, has an armored window/hatch with three observation periscopes. Inside the turret, the commander sits on the left and the gunner on the right; the former has several observation periscopes around his hatch, while the latter has an optical gunsight with a coincidence rangefinder (a LASER rangefinder, night vision sights, and an NBC filtration unit are offered in export versions).

In the most common AML-90 version, the main armament is a GIAT 90-mm. smoothbore gun that can fire HE, HEAT, APFSDS, CANISTER, and smoke rounds; a total of 20 rounds are stowed in the hull. The APFSDS round has a muzzle velocity of 1350 m./sec. and can penetrate most tank armor out to 1000 m., while the HEAT round can penetrate up to 320 mm. (12.6 in.) of homogeneous steel armor; the AML-90 is thus the smallest TANK DESTROYER in service worldwide. Secondary armament consists of a 7.62-mm. coaxial machine gun, and a second 7.62-mm. machine gun on a pintle mount by the commander's hatch. Four smoke grenade launchers are fitted to each side of the turret. Though not normally amphibious, the AML can be fitted with a flotation screen and propelled through still waters by tire action at a speed of 4 mph (6 kph).

Variants include the AML 60-7, with a breech-loading 60-mm. gun-mortar and two 7.62-mm. machine guns; the AML 60-20, with a 60-mm. gun-mortar and a 20-mm. cannon; an anti-aircraft vehicle with twin 20-mm. cannon; and scout cars armed only with machine guns.

Specifications **Length:** 12.45 ft. (5.11 m.). **Width:** 6.45 ft. (1.97 m.). **Height:** 6.8 ft. (2.07 m.). **Weight, combat:** 5.5 tons. **Powerplant:** 1 90-hp. Panhard 4-cylinder diesel. **Speed, road:** 56 mph (90 kph). **Range, max.:** 375 mi. (600 km.).

AMOS (AA-9): NATO code name for a Soviet long-range, radar-guided air-to-air missile which is the standard armament of the MiG-31 FOXHOUND interceptor. First reported in 1986, Amos is a large missile superficially similar to the U.S. AIM-54 PHOENIX, with an ogival nose, a cylindrical body, four cropped-delta tail fins, and separate rectangular control surfaces. Powered by a solid-fuel rocket engine, Amos is armed with a

176-lb. (80-kg.) high-explosive warhead fitted with radar proximity fuzes.

Unlike the active-homing Phoenix, Amos is controlled by SEMI-ACTIVE RADAR HOMING (SARH) with target illumination provided by the MiG-31's large pulse-Doppler radar. Thought to have a true LOOK-DOWN/SHOOT-DOWN capability, the missile has reportedly been successfully tested against CRUISE MISSILES flying as low as 200 ft. (50 m.) above the ground. Given its size, however, Amos is probably optimized for the interception of less maneuverable aircraft, specifically strategic bombers.

Specifications Length: 13.75 ft. (4.2 m.). **Diameter:** 16.1 in. (410 mm.). **Span:** (wings) 51 in. (1.3 m.); (control surfaces) 71 in. (1.8 m.). **Weight, launch:** 1280 lb. (580 kg.). **Speed, max.:** Mach 3.5 (2275 mph/3800 kph). **Range, max.:** (est). 46.5–92 mi. (75–150 km.).

AMPHIBIOUS ASSAULT/WARFARE:

The landing of troops onto a hostile shore. The modern form of amphibious warfare was developed by the U.S. Navy and Marine Corps before World War II, and its basic principles have changed little, except that HELICOPTER lift can now assist, or replace, beach landings. Beach landings presuppose a suitable beach: flat and hard for vehicular mobility, not subject to extreme tides, with good inland access, and located near the operational objective. Optimally, the beach should also be within range of friendly land-based aircraft, but not of hostile aircraft; alternatively, superior carrier air power must be available until airfields can be prepared ashore.

Once the beach is selected, it must be thoroughly reconnoitered by aerial surveillance and/or frogmen (e.g., SEALS, SBS, and naval SPETSNAZ) to determine the location of enemy forces, fortified areas, and any underwater obstacles and minefields. For amphibious assaults upon heavily defended shores, naval and aerial bombardment is also a prerequisite. During World War II, BATTLESHIPS, CRUISERS, and DESTROYERS bombarded plotted (or suspected) enemy defenses on or near the beach, while targets farther inland were bombarded from the air. Today, few navies have ships with heavy gun batteries, and tactical air support is accordingly even more important (a major reason for reactivating the U.S. Navy's four IOWA-class battleships was their 16-inch guns).

Under World War II procedures still in effect, landing areas are divided into distinct beaches, each under its own separate command. In a multi-wave assault, the first wave of INFANTRY would normally be reinforced with strong ENGINEER elements to clear minefields and other obstacles. Fire support may be provided by tanks in the first wave, before follow-on waves bring in ARTILLERY and other heavy equipment.

The normal objective of amphibious assaults is not the beach itself but an inland beachhead, which should be deep enough to prevent enemy observation of the beach and direct-fire attacks upon it (at least 20,000 m., given modern artillery ranges).

Assault craft capable of being beached to disembark men and equipment include tank landing ships (LSTS), smaller tank landing craft (LCTS), and the basic vehicle and personnel landing craft (LCVPS); there are also amphibious combat vehicles, notably the US LVTP-7 and the Soviet PT-76 light tank.

Larger amphibious ships have troop accommodations and vehicle decks; some (LHAS, LHDS, LPDS, and the Soviet IVAN ROGOVS) also have floodable docking wells, from which landing craft can float out. Special portable breakwaters, piers, and prefabricated roadways have also been developed to facilitate the unloading of supplies over the beach by creating, in effect, temporary ports.

The use of helicopters for transport *over* the beach from ships, pioneered by the U.S. MARINE CORPS, is a form of AIR ASSAULT. Aside from specialized helicopter carriers (LPHS), most larger amphibious ships have helicopter decks, including LSDS, LHAS, and LHDS. The advent of air cushion vehicles, or HOVERCRAFT, that can move over the water, across beaches, and as deep inland as desired, is a major innovation (long resisted by the U.S. Navy), because hovercraft are: (1) far more seaworthy than most landing craft; (2) capable of speeds in excess of 50 mph/80 kph (allowing them to be launched far from the beach under the cover of darkness, to arrive at first light); and (3) immune to most MINES. The Soviet navy has pioneered the use of hovercraft, and now operates the large AIST and POMORNIK classes, each capable of carrying several hundred troops or several MAIN BATTLE TANKS. The U.S. Navy's deployment of the much smaller (100-ton) LCAC (Landing Craft, Air Cushion) began only in 1989.

The largest and best-equipped amphibious force is the U.S. MARINE CORPS. The much smaller British Royal MARINES are very well trained, but not much more than a raiding force. The Soviet Union's NAVAL INFANTRY is also an amphibious force, as are the French navy's marine battalions, but many other "marine" units (including the French army's *Infanterie de Marine*) are not trained or equipped for amphibious operations.

AMRAAM: The Hughes AIM-120 Advanced Medium Range Air-to-Air Missile, under development for the U.S. Air Force (USAF) and Navy. Developed from the late 1970s, AMRAAM is intended as a replacement for the AIM-7 SPARROW, but while the latter relies on SEMI-ACTIVE RADAR HOMING (SARH), AMRAAM has ACTIVE HOMING radar guidance; as such, it is a true "fire-and-forget" missile which allows the pilot to disengage immediately after launch. SARH missiles, by contrast, require continuous radar illumination of the target, forcing the pilot to fly a set course until intercept. In addition, AMRAAM is intended to be smaller, lighter, and more maneuverable than Sparrow, to better engage maneuvering fighters. Flight tests began in 1981, but technical problems and rising costs delayed its full-scale production.

Compatible with most Sparrow launchers, AMRAAM has four fixed delta wings at midbody for lift, and four movable tail fins of equal span for steering. The nose houses a small active radar antenna and the guidance electronics, the 50-lb. (27-kg.) high-explosive fragmentation warhead fitted with both radar-proximity and contact fuzes is mounted at midbody, and the low-smoke, dual-impulse (booster/sustainer) solid-rocket engine is located at the rear.

In action, the pilot would first lock onto a target with his aircraft's radar, and the fire control computer would then transfer the target's course and speed to the missile. After launch, AMRAAM would fly an inertially guided course towards the target's estimated position; near that point, the missile's guidance computer would activate the small K-band radar in its nose, to automatically acquire the target, and home in for the kill.

AMRAAM has been criticized because of its high cost, which reflects the inherent complexity of active homing and also very ambitious ELECTRONIC COUNTER-COUNTERMEASURE specifications; its cost may yet interrupt the program. AMRAAM has also been criticized conceptually, on the grounds that beyond visual range (BVR) missiles cannot be used in many air combat situations, because pilots must visually identify their targets to avoid attacking friendly aircraft; but advances in IFF technology and fighter direction by AWACS should assuage such concerns, while the risk of "own goals" would be dwarfed by the tactical advantage of BVR engagements in any larger combat.

Specifications **Length:** 11.75 ft. (3.58 m.). **Diameter:** 7 in. (178 mm.). **Span:** 25 in. (635 mm.). **Weight, launch:** 335 lb. (152 kg.). **Speed, max.:** Mach 4 (2600 mph/4342 kph). **Range, Max.:** 30 mi. (48 km.).

AMSA: Advanced Manned Strategic Aircraft. See B-1.

AMX: Aeritalia/Macchi Experimental, a single-seat, single-engine ATTACK AIRCRAFT built by the Italo-Brazilian consortium of Aeritalia, Aeromacchi, and EMBRAER. It was originally developed from 1977 to an Italian Air Force requirement for a replacement to the Fiat G.91 and F-104 STARFIGHTER in the attack, anti-shipping, and RECONNAISSANCE roles. Aeromacchi and Aeritalia were later joined by EMBRAER when the Brazilian air force issued a parallel requirement (though the Brazilian focus is on exports). The AMX first flew in 1986, and the first production aircraft were delivered in 1989; Italy presently plans to buy 187 aircraft and Brazil 79.

Of solid, conventional design, the AMX has an area-ruled ("coke-bottle") fuselage with a simple range-only RADAR in the nose (though space is reserved for a larger set), behind which is an avionics bay and (in Italian aircraft) an M61 VULCAN 20-mm. cannon with 350 rounds of ammunition, or (in Brazilian versions) two 30-mm. DEFA cannons with 125 rounds per gun. The pilot sits in a raised cockpit under a clamshell canopy faired into the fuselage spine that provides excellent forward and lateral visibility, but restricts rearward visibility. Cockpit AVIONICS include a head-up display (HUD), a head-down radar display, UHF and VHF radio, a TACAN receiver, RADAR WARNING RECEIVERS, an air-data computer, and navigation/attack computers.

Mounted in the tail, the engine is fed by rectangular cheek inlets behind the cockpit. The landing gear (which retract into the fuselage) are reinforced and fitted with low-pressure tires for rough-field operations. The moderately swept, mid-mounted wings have full-span leading-edge slats, small outboard ailerons, large double-slotted Fowler flaps, and spoilers to augment the ailerons at low speeds. The slab tailplane is fabricated mainly of composites, while the vertical stabilizer has a computer-controlled FLY-BY-WIRE rudder to provide greater stability at high angles of attack.

The AMX has 1 centerline pylon rated at 2000 lb. (909 kg.), 2 inboard wing pylons rated at 2000 lb. (909 kg.) each, 2 outboard pylons rated at 1000 lb. (454 kg.) each, and 2 wingtip pylons (for AIM-9 SIDEWINDER or similar INFRARED-HOMING air-to-air missiles) rated at 220 lb. (100 kg.) each; total payload capacity is thus 8376 lb. (3800 kg.). Weapons could include low-drag general purpose (LDGP) bombs, NAPALM, and CLUSTER BOMBS, DURANDAL runway-cratering bombs, AGM-65 MAVERICK air-to-ground missiles, KORMORAN anti-ship missiles, rocket pods, and drop tanks. In addition, the AMX can carry a reconnaissance pallet on the centerline pylon, as well as ELECTRONIC COUNTERMEASURE pods on the outboard wing pylons. In many respects, the AMX is competitive with the British HAWK and the Franco-German ALPHA JET, which are cheaper but also have smaller payloads.

Specifications **Length:** 44.52 ft. (13.57 m.). **Span:** 29.1 ft. (8.88 m.). **Powerplant:** 1 Rolls Royce Spey 807 turbofan, 11,030 lb. (50003 kg.) of thrust. **Weight, empty:** 14,330 lb. (6500 kg.). **Weight, max. takeoff:** 26,455 lb. (12,000 kg.). **Speed, max.:** Mach 0.92 (722 mph/1162 kph) at sea level. **Speed, cruise:** Mach 0.75 (525 mph/876 kph). **Combat radius:** (lo-lo-lo) 230 mi. (385 km.) w/6000-lb. (2727-kg.) payload.

AMX-10P: A French tracked INFANTRY FIGHTING VEHICLE (IFV) similar to the German MARDER, but less well protected. Designed from 1965 to replace the AMX-VCI armored personnel carrier, the AMX-10P entered service in 1973, and more than 2000 have been produced for the French army and exported to Greece, Indonesia, Qatar, Saudi Arabia, and the United Arab Emirates.

The all-welded aluminum hull, divided into a driver/engine compartment up front, a turreted fighting compartment in the middle, and a troop compartment in the rear, has a maximum armor thickness of some 13 mm., sufficient only against small arms and splinters. The driver, up front on the left with the engine to his right, is provided with three observation periscopes, one of which can be replaced with a night-vision scope. The commander and gunner, in the power-operated turret, are provided with observation periscopes and optical gunsights; the latter can be replaced by INFRARED night sight.

The turret, armed with a 20-mm. CANNON (with separate ammunition feeds for HE and AP rounds), has 360° traverse and elevation limits of −8° and +50°; a total of 760 rounds can be stowed in the vehicle. Secondary armament consists of a coaxial 7.62-mm. machine gun. Two smoke grenade launchers are attached to each side of the turret.

The troop compartment in the rear can hold up to eight infantrymen, with access through a rear ramp door and two roof hatches. Though an IFV, the AMX-10P has only two firing ports in the ramp-door through which the troops can fire from under armor; this reflects French doctrine, whereby the AMX-10P would transport troops to the battle zone to fight dismounted, with fire support from the vehicle. A MILAN anti-tank guided missile can also be mounted by the roof hatches, but the operator must remain exposed to guide the missile. A COLLECTIVE FILTRATION unit is provided for NBC defense. Amphibious without extensive preparation, the vehicle can be propelled through still waters by track action at a speed of 4.33 mph (7 kph).

Variants include the AMX-10 PAC-90 fire support vehicle, with a GIAT two-man turret and a 90-mm. low-velocity gun but only four infantrymen; the AMX-10 ambulance; the AMX-10ECH repair vehicle (with a crane for engine replacements); the AMX-10PC armored command vehicle (ACV), with extra communications equipment; the AMX-10SAO for artillery SPOTTING; the AMX-10TM for towing MORTARS and other ARTILLERY pieces; and the AMX-10RC wheeled TANK DESTROYER.

Specifications **Length:** 18.95 ft. (5.78 m.). **Width:** 9.05 ft. (2.78 m.). **Height:** 6.33 ft. (1.92 m.). **Weight, combat:** 14.2 tons. **Powerplant:** 1 280-hp. Hispano-Suiza HS-115 V-8 diesel. **Fuel:** 120 gal. (528 lit.). **Speed, road:** 40 mph (65 kph). **Range, Max.:** 373 mi. (600 km.).

AMX-10RC: French wheeled TANK DESTROYER derived from the AMX-10P infantry fighting vehicle. Developed from the early 1970s to replace the EBR heavy armored car, the AMX-10RC is a 6×6 vehicle with the same basic hull and chassis as the AMX-10P. Very sophisticated, it entered service in 1979 with the French and Moroccan armies, but because of its cost, it had to be supplemented in the French army by the cheaper ERC Sagaie. Production ended in 1987.

As compared to the AMX-10P, the internal arrangements are completely revised, with the driver on the left, a turreted fighting compartment in the middle, and the engine in the rear. The driver has three observation periscopes and an INFRARED night scope. The rest of the crew are in the large, welded turret, with the commander on the right, above and behind the gunner, and the loader to his left. The commander has a cupola with three observation periscopes and a large, periscopic gunsight. The gunner has an optical sight integrated with a LASER rangefinder and a ballistic computer.

Both the gunner and the commander also have a low-light television (LLTV) system for night operations.

The turret is armed with a low-pressure MECA 105-mm. gun that can fire APFSDS, HEAT, and HE rounds; the APFSDS round can penetrate most main battle tanks. Thirty-eight 105-mm. rounds can be stowed in the turret and in racks to the right of the driver. Elevation limits are −8° and +20°. Secondary armament consists of a coaxial 7.62-mm. machine gun, and two smoke grenade launchers are attached on each side of the turret. The AMX-10RC has a positive overpressure COLLECTIVE FILTRATION unit for NBC defense. Amphibious without extensive preparation, it can be propelled through the water by two rear-mounted pumpjets at 4.5 mph (7.2 kph).

Specifications Length: 20.83 ft. (6.35 m.). Width: 9.67 ft. (2.95 m.). Height: 8.8 ft. (2.68 m.). Weight, combat: 15.8 tons. Powerplant: 260-hp. Hispano-Suiza HS-115 supercharged V-8 diesel. Fuel: 120 gal. (528 lit.). Speed, road: 53 mph (85 kph). Range, Max.: 500 mi. (800 km.).

AMX-13: French LIGHT TANK, now obsolescent, but still in service with the French army and the forces of Argentina, Chile, the Dominican Republic, Ecuador, El Salvador, Indonesia, the Ivory Coast, Djibouti, Lebanon, Morocco, Nepal, Peru, Singapore, Tunisia, and Venezuela. The Israeli army used the AMX-13 in the 1956 and 1967 wars, but found it unsatisfactory. It was developed from 1946, and more than 3000 were completed between 1952 and the 1960s.

The all-welded steel hull, divided into a driver/engine compartment up front and a turreted fighting compartment in the rear, has a maximum armor thickness of only 1 in. (25 mm.), sufficient only against light cannon, small arms, and splinters. The driver, on the left with the engine to his right, has three observation periscopes; most users have retrofitted a night-vision scope. Most AMX-13s have gasoline engines, but France, Singapore, Argentine, Peru, and Venzuela have retrofitted diesel engines for increased range and to reduce the risk of fire.

The most unusual feature of the AMX-13 is its oscillating turret: the main gun is mounted rigidly in the upper half of the turret, which pivots on the lower half to change the elevation; an automatic loader replaces the fourth crewman. The commander sits on the left under a large, dome-shaped cupola equipped with periscopes for 360° observation. In most versions, the gunner has a simple stadiametric gunsight, but some users have retrofitted improved sights with LASER rangefinders, night vision, and a ballistic computer. The original FL-10 turret had a high-velocity 75-mm. gun, but this was replaced in the 1960s by the FL-12 turret with a 90-mm. low-velocity gun, and a third version with a low-velocity 105-mm. gun was also built for export. The 90-mm. gun can fire smoke, HE, and HEAT rounds, the latter capable of penetrating up to 450 mm. of homogeneous steel armor; an APFSDS round introduced in the 1980s is said to be capable of pentrating most MAIN BATTLE TANKS out to a range of 2000 m.

Specifications Length: 16 ft. (4.88 m.). Width: 8.2 ft. (2.5 m.). Height: 7.55 ft. (2.3 m.). Weight, combat: 15 tons. Powerplant: 250-hp. 8-cylinder gasoline. Speed, road: 36 mph (60 kph). Range, max.: 210 mi. (350 km.).

AMX-30: French MAIN BATTLE TANK (MBT), in service with the French army and the armies of Chile, Greece, Qatar, Saudi Arabia, Spain, the United Arab Emirates, and Venezuela. Developed from 1956 (together with the West German LEOPARD 1) as a potential NATO standard tank, the first prototypes were delivered in 1960, and completed trials in 1963, but because of financial reasons, the AMX-30 entered service with the French army only in 1966. More than 2000 have been delivered to date, and the vehicle has also been produced under license in Spain.

The AMX-30 is the lightest of all Western MBTs. Its design emphasized speed, a low silhouette, and good ballistic shaping (the same priorities as for Soviet tanks), while armor protection is only 80 mm. on the frontal glacis and turret face (as opposed to 120 mm. on the contemporary U.S. M60 Patton).

The all-welded steel hull has a conventional configuration, with the driver's compartment up front, a turreted fighting compartment in the middle, and the engine in the rear. The driver, on the left with ammunition racks to his right, has three observation periscopes, one of which can be replaced by an INFRARED night-vision scope. The remaining crew are in the long, cast turret, with the commander on the right above and behind the gunner, and the loader on the left. The commander sits under a rotating cupola with periscopes for 360° observation and a periscopic gunsight, while the gunner has an optical gunsight with a coincidence rangefinder. Both the commander and gunner have infrared night sights.

Main armament, a medium-velocity F1 105-

mm. rifled gun, is much lighter and has less recoil than the more powerful high-velocity 105-mm. L7 rifled gun of most Western tanks. Originally, it could fire only special *Obus* G HEAT rounds, in which the warhead is mounted on ball bearings inside the shell (to minimize the loss of shaped-charge performance caused by the spin that rifled guns impart). More recently, HE, illuminating, and APFSDS rounds have been introduced; the latter *(Obus Fleché)* has a muzzle velocity of 1525 m./sec. and is said to be capable of penetrating up to 300 mm. of armor out to a remarkable 5000 m. A total of 47 rounds are carried, 28 next to the driver and 19 in the turret. The gun has elevation limits of −8° and +20°. Secondary armament consists of an unusual 20-mm. coaxial CANNON which can be elevated independently of the main gun (to a maximum of +40°) in order to engage attack helicopters, and a pintle-mounted 7.62-mm. machine gun by the commander's cupola. Two smoke grenade launchers are attached to each side of the turret bustle, and a white light/infared searchlight is mounted on the right side of the turret mantlet. A positive overpressure COLLECTIVE FILTRATION unit is fitted for NBC defense.

In 1979 the French army introduced the AMX-30B2, with a LASER rangefinder and a ballistic computer, an improved NBC system, a new transmission, a low-light television (LLTV) system, and additional frontal armor. The French army ordered 231 new production vehicles and retrofitted 730 AMX-30s to B2 standard; production of the B2 continues for export.

Major variants include the AMX-30D armored recovery vehicle, the AMX-30 bridgelayer, and the AMX-30EBG combat engineer vehicle. The chassis has also been used for the ROLAND self-propelled surface-to-air missile launcher and GCT self-propelled 155-mm. gun. The AMX-30 was also the basis of the AMX-40, developed by GIAT for export. The AMX-30 is to be replaced in the French army by the 120-mm.-gunned Leclerc tank.

Specifications **Length:** 21.6 ft. (6.59 m.). **Width:** 10.16 ft. (3.1 m.). **Height:** 9.33 ft. (2.86 m.). **Weight, combat:** 36 tons. **Powerplant:** Hispano-Suiza 720-hp. multi-fuel diesel. **Fuel:** 220 gal. (970 lit.). **Speed, road:** 40 mph (65 kph). **Speed, cross-country:** 25 mph (41 kph). **Range, max.:** 311 mi. (500 km.).

AN-: Designation of aircraft developed by the Soviet design bureau founded by O. K. Antonov, best known for its transport aircraft, including the An-2 COLT, An-12 CUB, An-22 COCK, and An-124 CONDOR (the largest military aircraft in the world).

ANAB (AA-3): NATO code name for a large Soviet air-to-air missile that armed Su-11 Fishpot, Yak-28P FITTER, and early Su-15 FLAGON interceptors of the IA-PVO.

First observed in 1961, Anab has four small delta canards to control pitch and yaw, and four broad tailfins with ailerons for roll control. Powered by a solid-fuel rocket engine, the missile is armed with a proximity-fuzed high-explosive warhead in the 100-lb. (45-kg.) class. Anab was produced in two versions, one with SEMI-ACTIVE RADAR HOMING (SARH) and one with passive INFRARED-HOMING guidance; normally fighters would carry a mix of both. Optimized for the interception of large strategic bombers, Anab is relatively ineffective against more maneuverable fighters. Even the later AA-3-2 Advanced Anab is obsolescent by modern standards, and has been replaced by the AA-6 ACRID.

Specifications **Length:** 11.83 ft. (3.6 m.). **Diameter:** 11 in. (280 mm.). **Span:** 51 in. (1.3 m.) **Weight, launch:** 600 lb. (275 kg.). **Speed, max.:** Mach 2.5 (1625 mph/2714 kph). **Range, max.:** (SARH) 50 mi. (80 km.); (IR) 15 mi. (25 km.).

ANDREA DORIA: A class of two Italian air-capable, guided missile CRUISERS *(Andrea Doria* and *Caio Duilio)* commissioned in 1964. Though rather small by cruisers standards, these ships have a heavy and balanced armament, and also a large fantail flight deck for ANTI-SUBMARINE WARFARE (ASW) helicopters.

The flight deck extends from the stern to a large hangar in the aft superstructure. Freeboard is rather low (a typical Italian design feature, but no handicap for Mediterranean operations), and fin stabilizers are fitted to improve seakeeping.

ANTI-AIR WARFARE (AAW) armament consists of a twin-arm Mk. 10 launcher for 40 long-range STANDARD-ER surface to air missiles on the bow, supplemented by 8 OTO-Melara 76.2-mm. DUAL PURPOSE guns (6 in *Caio Duilio*). Primary ASW armament consists of 3 Agusta Bell AB-212 HUEY helicopters. For close-in ASW, the ships have 2 sets of Mk. 32 triple tubes for lightweight homing TORPEDOES.

In 1979–80, *Caio Duilio* was refitted as a school ship. The aft pair of 76.2-mm. guns and their associated fire control radar were removed, the hangar was converted into classrooms and accommodations, and a smaller hangar was built out on the flight deck for two helicopters. The Andrea Dorias served as the design basis for the larger, more capable VITTORIO VENETO, and probably influenced the design of the Soviet MOSKVA-class helicopter carriers. See also CRUISERS, ITALY.

Specifications Length: 489.7 ft. (149. 3 m.). **Beam:** 56.6 ft. (17.25 m.). **Draft:** 24.6 ft. (7.5 m.). **Displacement:** 5000 tons standard/6500 tons full load. **Flight deck:** 98.4 × 52.46 ft. (30 × 16 m.). **Powerplant:** 4 oil-fired boilers, 2 sets of geared turbines, 60,000 shp., 2 shafts. **Speed:** 31 kt. **Fuel:** 1100 tons. **Range:** 5000 n.mi. at 17 kt. **Crew:** 484. **Sensors:** 1 RAN-20S air-search radar, 1 SPQ-2D surveillance radar, 1 SPS-52B 3-dimensional air-search radar, 2 SPG-55C missile guidance radars, 4 Orion RTN-10X fire control radars, 1 SQS-23F (*Andrea Doria*) or SQS-29 (*Caio Duilio*) hull-mounted medium-frequency sonar. **Electronic warfare equipment:** 1 British UAA-1 "Abbey Hill" electronic signal monitoring array, 2 20-barrel SCLAR chaff rocket launchers.

ANECHOIC (MATERIAL): Material that impedes the reflection of electromagnetic radiation (especially sound or radio waves), and which is increasingly used to suppress the radar and sonar signatures of submarines, ships, and aircraft. Anechoic materials for radio frequency bands, usually called RADAR ABSORBANT MATERIALS (RAM), are a critical element of STEALTH technology. RAM or other anechoic materials can be used to build aircraft structures to reduce their RADAR CROSS SECTIONS, while submarine hulls can now be covered with anechoic tiles. The Soviet and British navies in particular make much use of such tiles, which may reduce the effective detection range of active sonar by 33–50 percent. See also ANTI-SUBMARINE WARFARE; CLUSTER GUARD.

ANIMOSO: A class of two Italian DESTROYERS, improved versions of the AUDACE class. See also DESTROYERS, ITALY.

ANTARCTIC TREATY: A multinational agreement on the demilitarization of the continent of Antarctica and the surrounding seas below the 60° parallel. The treaty was signed in 1959 by representatives of Argentina, Australia, Belgium, Chile, France, Great Britain, Japan, New Zealand, Norway, South Africa, the Soviet Union, and the United States, and came into effect in 1961 for a renewable term of 30 years. The original signatories all had territorial claims and/or ongoing activities in Antarctica.

The treaty prohibits all military activities (including the storage of nuclear devices) in Antarc-

tica, and the unrestricted on-site inspection of all facilities was accepted by all signatories. The Antarctic Treaty is a successful example of international ARMS CONTROL, even if the military interests it negates are weak (it is the Arctic that is of great military importance). The discovery of oil or other important natural resources in Antarctica could place a greater strain on the treaty. See also DEMILITARIZED ZONE.

ANTI-AIRCRAFT (WEAPONS): Ground- or ship-based guns and missiles designed to attack aircraft, helicopters, and tactical missiles. Introduced in World War I, the earliest anti-aircraft (AA) weapons were high-velocity guns with high-elevation mounts (Anti-Aircraft Artillery, or AAA), firing normal high explosive (HE) ammunition. To aim AAA accurately, the course and speed of the target, the elevation and deflection required to compensate for the shell's time of flight, and the effects of gravity, drag, and crosswinds must all be calculated precisely; hence, for early AAA the chances of a direct hit were minimal. Before the advent of radar, AAA effectiveness could be increased by using mechanical ballistic computers (predictors); by firing many guns in present pattern BARRAGES; and by time fuzes set to detonate the shells at the estimated target altitude. With enough guns, if the target's altitude is correctly estimated, and if it maintains a steady course, a reasonable probability of kill (P_K) can be achieved. With AAA ranging in size from 75 mm. to 120 mm. that was the the aim of the classic form of "flak" in World War II, but even against large formations of slow, nonmaneuvering bombers, the primary benefit of AAA was to degrade bombing accuracy by inducing premature release (ahead of the flak over the target). By the end of World War II, the effectiveness of heavy AAA had been increased by the introduction of radar directors (which eliminated guesswork from the determination of target altitude, course, and speed), and of shells with radar proximity (VT) fuzes—small radar rangefinders set to detonate the shell if it passed within lethal radius of an aircraft. VT fuzes increased the lethality of heavy AAA from roughly 2 percent to 20 percent per round (based on U.S. Navy experience against kamikazes).

But these improvements in heavy AAA were overtaken by the advent of jet aircraft with higher speeds and low-level attack profiles; in the short firing time available, heavy guns could not fire enough rounds to achieve reasonable P_ks. Only the Soviet Union continued to field heavy AAA, deploying many 85-mm. and 100-mm. guns for its strategic air defenses (PVO)—a policy justified by North Vietnam's experience: the combination of heavy AAA, light AAA, and missiles was effective against U.S. aircraft by denying a "free ride" at any altitude. In addition, the highly visible shell bursts again had a significant deterrent effect, degrading bombing accuracy as in World War II.

Light AAA, i.e., automatic CANNON and HEAVY MACHINE GUNS ranging from 12.7 mm. to 57 mm., lack the ceiling and lethal radius of heavy AAA, but have much higher rates of fire (120–3000 rounds per minute), which make it possible to aim them with tracer rounds: gunners can instinctively correct their aim to send up a "steel curtain" in front of enemy aircraft. The classic light AAA weapons of World War II were the 20-mm. Oerlikon and 40-mm. BOFORS GUNS used by all sides; in modified form they are still in standard issue. By 1945, light AAA was also equipped with radar gunsights, but VT ammunition was (and is) confined to shells of 40-mm. and larger diameter. Light AAA remains effective below 10,000 ft. (3050 m.), within ranges of under 2 mi. (3.4 km.). Because they are light, these weapons can be mounted on ARMORED FIGHTING VEHICLES, with machine guns the universal secondary armament for self-defense, and 20-mm. or larger weapons used to arm dedicated AA vehicles to provide mobile air defense for ground units. Some completely self-contained mobile AAA systems have been produced, with search and tracking radars, fire control computers, and/or sophisticated optical sights (e.g., the Soviet ZSU-23-4 and West German GEPARD). Because large numbers of light AAA weapons can often be deployed, the sheer volume of their fire is a major impediment to aircraft operations.

The development of guided anti-aircraft (surface-to-air) missiles (SAMs) began with German projects during World War II; after 1945, U.S. and Soviet efforts focused on the interception of nuclear-armed bombers at long ranges from their targets. The first practical U.S. systems were the navy's "3-T" missiles (TALOS, TERRIER, and TARTAR), and the army's Nike Ajax and NIKE HERCULES, with SEMI-ACTIVE RADAR HOMING (SARH), BEAM RIDING, or COMMAND GUIDANCE; the first Soviet SAMs, the SA-1 Guild and SA-2 GUIDELINE, were similar, as was the later British BLOODHOUND—all were optimized for use against large, high-altitude bombers, and none could cope with agile fighters. In 1960, however, an SA-2 intercepted a U.S. U-2

reconnaissance aircraft, and that episode induced a widespread overestimation of all SAMs, until combat experience showed their limitations. Much used in Vietnam, the SA-2 achieved some initial success until the U.S. developed effective ELECTRONIC COUNTERMEASURES (ECM) to jam and deceive its guidance radars (some 120 were fired for each kill in the Vietnam War). Israeli experience in 1967–70 against the SA-2 and the subsequent SA-3 GOA was similar, but in the 1973 Yom Kippur War the Israelis encountered the SA-6 GAINFUL, for which they had no ECM. Gainfuls exacted a heavy toll at first, but once specific countermeasures were developed, its effectiveness dropped rapidly; in the 1982 Lebanon War, Gainful achieved no kills.

All missiles can be negated through evasive tactics, but to do so aircraft must often jettison their bomb loads and/or dive to low altitudes where they are vulnerable to light AAA. As for ECM, they displace offensive payloads, and are often very costly. Specialized aircraft (such as U.S. WILD WEASELS) developed to suppress air defenses also divert resources from offensive purposes.

By the 1970s, moreover, small, portable surface-to-air missiles became analogous to light AAA: they too came to be effective at low altitudes, cheap, and available in large numbers. The smallest shoulder-fired, man-portable SAMs (notably the Soviet SA-7 GRAIL and U.S. REDEYE and STINGER) are guided by passive INFRARED HOMING; hence, they are effective only in clear weather, at altitudes up to 10,000 ft. (3050 m.), and within ranges of 5 mi. (8.5 km.) or less. Somewhat larger missiles such as the Franco-German ROLAND, British RAPIER, and Soviet SA-8 GECKO, which are mounted severally on wheeled or tracked vehicles, can have radar illuminators for semi-active radar guidance, or at least surveillance radars for warning and target acquisition. With such medium-small SAMs, completely self-contained, mobile systems can engage even small maneuvering targets effectively at low-to-medium altitudes, complementing both light AAA and large, long-range SAMs to achieve overlapping air defense coverage from ultra-low to very high altitudes. See also AIR DEFENSE; ANTI-AIR WARFARE; NORAD; SEAD.

ANTI-AIR WARFARE (AAW): U.S. Navy and Marine Corps term for the weapons and tactics of naval defense against air and missile attack. U.S. AAW doctrine is based on DEFENSE-IN-DEPTH, to detect and engage air threats as far as possible from high-value ships such as aircraft carriers, battleships, etc. To that end, AIRBORNE EARLY WARN-ING (AEW) aircraft, notably the E-2C HAWKEYE or E-3 SENTRY (AWACS), are to provide warning as much as 400 mi. (670 km.) out, and if no AEW is available, picket ships must be stationed as far out as possible (though their detection range is limited to the radar horizon). Long range INTERCEPTORS are to begin engagements some 200 mi. (334 km.) out, to try to destroy enemy missile-carrying aircraft before they can launch their weapons. Surviving enemy aircraft or missiles are next to be engaged by long-range missiles from the fleet's outer screen of guided missile DESTROYERS and CRUISERS. Ships equipped with the AEGIS system are especially valuable for that phase of the defense because of their ability to engage several dozen targets simultaneously; in any case, that phase of AAW can be coordinated by the NTDS data link, which can transfer target data from ship to ship. Finally, remaining air threats are to be engaged by the self-defense weapons of each ship, i.e., DUAL PURPOSE guns, short-range missiles such as SEA SPARROW, and, finally, 20-mm. PHALANX radar-controlled guns. Such active defense is to be supplemented by ELECTRONIC COUNTERMEASURES, including ACTIVE JAMMING and the projection of CHAFF, as well as by evasive maneuvers.

Various means can be used to avoid detection in the first place, including emissions control (EMCON) and the use of passive sensors (RADAR WARNING RECEIVERS and other ELECTRONIC SIGNAL MONITOR-ING) to evade enemy surveillance. See also AIR DE-FENSE; ANTI-SHIP MISSILE; BATTLE MANAGEMENT.

ANTI-BALLISTIC MISSILE (SYSTEM): A form of BALLISTIC MISSILE defense designed to destroy the REENTRY VEHICLES (RVs) of intercontinental ballistic missiles during the final (terminal) phase of their trajectory. The 1972 ANTI-BALLISTIC MISSILE TREATY defines ABM systems as consisting of ground-launched interceptor missiles, their launchers, and associated surveillance, tracking, and missile-guidance radars. Spaced-based defenses and other exotic systems studied under the U.S. STRATEGIC DEFENSE INITIATIVE are usually described as BALLISTIC MISSILE DEFENSE (BMD) systems to distinguish them from pre-1972 projects.

ABMs can be categorized by their intended coverage: HARDPOINT ABMs would defend only a small area around a single, small, and fortified high-value target such as a command post or missile silo, because they would intercept RVs at low altitude (endoatmospherically), with possible nuclear warhead effects that only HARD targets could

withstand; area-defense ABMs, on the other hand, would destroy RVs outside the atmosphere (exoatmospherically), and could thus protect SOFT targets such as cities—being also more efficient because a single system could defend all potential targets within its "footprint." Obviously, ABM systems combining area and hardpoint defenses could have the former as the main defense, with the latter kept in reserve to engage RVs that leak through.

Any ABM system requires the following elements:

1. Large PHASED ARRAY RADARS to detect and track incoming RVs. For exoatmospheric area defense, such radars require high resolutions to discriminate between RVs and decoys (PENAIDS). Such fine discrimination is not required endoatmospherically, because the low-mass PENAIDS decelerate quickly and break up ("atmospheric sifting"), revealing the real RVs. Because all large radars would be attractive targets in their own right, they must be defended individually and hardened against weapon effects.

2. Missile guidance radars. Interceptor missiles controlled by SEMI-ACTIVE RADAR HOMING require target illumination by radars which can function in the presence of ELECTRONIC COUNTERMEASURES (ECM) and nuclear effects such as BLACKOUT and ELECTROMAGNETIC PULSE (EMP). Missiles controlled by COMMAND GUIDANCE require appropriate DATA LINKS. ACTIVE HOMING missiles are deemed unsuitable for ABM because of their inherent vulnerability to ECM.

3. A BATTLE MANAGEMENT system to integrate the various defensive sites and allocate targets among them, as well as to set instantaneous priorities.

4. Interceptor missiles. Area defense requires large, multi-stage missiles, such as the Soviet GA-LOSH or the abortive U.S. SPARTAN, with ranges of 150–250 mi. (250–400 km.) to engage targets at altitudes up to 100–120 mi. (150–200 km.). Terminal defense requires short-range missiles with very high acceleration to intercept RVs between reentry and impact. The pre-1972 U.S. SPRINT missile was capable of maximum speeds above Mach 5 within seconds of launch, and the Soviet GAZELLE is similar. Both could be armed with low yield (1-10 kT) ENHANCED RADIATION (ER) warheads to destroy RVs with X-RAY and NEUTRON effects; ER warheads minimize blast, and therefore "collateral damage" on the ground (fallout would be minimal).

With MULTIPLE INDEPENDENTLY TARGETED REEN-TRY VEHICLES (MIRVs), ABM systems can be saturated, albeit not cheaply; moreover, in action the nuclear warheads of ABM interceptors would seriously degrade the effectiveness of radar sensors and also cause widespread EMP damage to communications and electronics (though EMP can be resisted by shielding). For these reasons, traditional ABM systems of pre-1972 configuration could defend only against lesser ballistic missile threats (accidental, unauthorized, or third-party attacks).

Arms control advocates have always insisted that ABM defenses are inherently destabilizing, because they would be ineffective against a planned PREEMPTIVE STRIKE, but not against a "ragged" post-attack "second strike," thereby creating an incentive to shoot first. Further, the relative ineffectiveness of ABM-type defenses also encourages increases in offensive forces (to saturate them), undermining the U.S. doctrine of AS-SURED DESTRUCTION.

From the 1950s to the 1970s, both the U.S. and U.S.S.R. developed ABM systems. The U.S. eventually completed the SAFEGUARD system at Grand Forks, North Dakota, consisting of 100 Spartan long-range missiles and Sprint terminal interceptors, and a large phased-array battle management radar. Though allowed under the terms of the ABM Treaty, the U.S. system was dismantled unilaterally. The Soviet Union built its own allowed 100-launcher Galosh system by 1972, and began to modernize it in 1982 with a new generation of interceptors, including the high-acceleration Gazelle missile, analogous to Sprint. In addition, in violation of the ABM Treaty, the Soviet Union added a large battle management radar at Krasnoyarsk (the violation was admitted in 1989, and the radar left incomplete). Notwithstanding its relative inefficiency, the Soviet system is quite adequate to defeat lesser third-party threats, including Chinese ICBMs.

ANTI-BALLISTIC MISSILE TREATY: An ARMS CONTROL agreement between the United States and the Soviet Union to limit the deployment and development of ANTI-BALLISTIC MISSILE (ABM) systems, signed in Moscow on 26 May 1972. Under the treaty, ABM systems are defined as consisting of: (1) ground-based anti-ballistic missile interceptors; (2) launchers for those missiles; and (3) ballistic missile tracking and ABM-guidance radars. The treaty permitted each signatory to build two ABM sites (one for the national

capital, and one around ICBM silos), limited to a total of 100 interceptor missiles each—with rapid-reloading launchers also prohibited. Each side was further limited to a total of six ABM radars within 150 mi. of its ABM site. Any other ballistic missile early-warning radars had to be sited along the periphery of each country with outward-facing arrays (to limit their utility for BATTLE MANAGE-MENT). A 1974 protocol revised the treaty to limit each side to only one ABM site with 100 launchers. The Soviet Union chose the Moscow GALOSH system, while the U.S., which had constructed its own SAFEGUARD system at Grand Forks, North Dakota, chose unilaterally to dismantle the site and has no deployed ABM system.

The treaty also bans the deployment or development of air-, sea-, space-, or land-based mobile ABM systems, and prohibits the testing of surface-to-air missiles in an ABM mode. But a key clause in Article III can be interpreted to mean that the treaty does not prohibit the development or deployment of ABM systems or components based on "other physical principles," thus allowing space-based kinetic energy weapons such as ELEC-TROMAGNETIC LAUNCHERS (EMLs) or DIRECTED EN-ERGY WEAPONS (DEWs); that interpretation, however, is highly controversial.

The treaty provides for a Consultative Commission to resolve disputes over interpretation, violations, and verification; it also calls for reviews at five-year intervals. The treaty may be abrogated by either side six months after a warning notice. The U.S. signing was predicated on the assumptions of the doctrine of ASSURED DESTRUCTION: that in the absence of effective defenses, the two superpowers can maintain a relatively stable offensive-only equilibrium with relatively few weapons. In the memoranda of understanding that accompanied the treaty, the U.S. side stated its expectation that the signing of the treaty would be followed within six months by substantial reductions of both sides' offensive nuclear forces. The U.S. further stated that if such reductions were not forthcoming, it would feel free to abrogate the treaty. But the number of Soviet offensive nuclear weapons was roughly tripled in the aftermath (1972–79), and the U.S. did not abrogate. Nor was abrogation seriously considered thereafter, despite several unambiguous Soviet violations, including the development of missile launchers with rapid-reload capability, the testing of the SA-12B GIANT surface-to-air missiles in an ABM mode, and the construction of a large ABM radar at Krasnoyarsk in the interior of the country facing inward—a violation eventually admitted by the U.S.S.R. in 1989.

The ABM Treaty became a major obstacle to the U.S. STRATEGIC DEFENSE INITIATIVE launched by the Reagan administration in 1983. The administration's call for a loose interpretation of Article III on "other physical principles" was rejected by the U.S. Senate, which insisted on a strict interpretation (more strict, in fact, than the Soviet interpretation). That greatly constrained the development and testing of new ballistic missile defenses even before the Gorbachev revolution in Soviet policy undermined support for a Soviet-directed ballistic missile defense. See also ANTI-SATELLITE SYSTEM; BALLISTIC MISSILE DEFENSE; SALT II; SPACE, MILITARY USES OF; START.

ANTI-RADIATION MISSILE (ARM): A missile that homes on radar or radio line-of-sight emissions. Developed in the 1960s by the U.S. for use against North Vietnamese AIR DEFENSES, the first ARM (AGM-45 SHRIKE) had to be pretuned to the frequency of a given target radar, had a relatively short range and low speed, and could not hit the target radar if it shut down even very briefly (an obvious countermeasure). Developed as an interim conversion of the STANDARD surface-to-air missile, the subsequent AGM-78 Standard ARM had a longer range, covered a wider frequency range, and had an autopilot to keep it on its attack course even if the target radar ceased transmitting. Carried by F-4 PHANTOM, F-105G WILD WEASEL, and A-6B INTRUDER aircraft, these two missiles were the chief weapons for SEAD (suppression of enemy air defenses) missions in the Vietnam War. The first Soviet ARMs tended to be versions of larger radar-homing air-to-surface missiles such as the AS-5 KELT (used with success in the 1973 Yom Kippur War).

The latest ARMs, notably the U.S. AGM-88 HARM and British ALARM, have longer ranges and higher speeds, to allow SEAD attacks from outside the envelope of enemy air defenses. In addition, they no longer require pretuning, because they can scan automatically through the entire range of known radar frequencies; their programmable signal processors also allow new frequencies to be added as soon as new target radars emerge. Finally, their inertial platforms and autopilots can guide these missiles to the target's position even if the radar shuts down shortly after its acquisition. ALARM also has a loiter mode, to stay aloft for several minutes while scanning for enemy radars. The much larger U.S. AGM-136 TACIT RAINBOW

(derived from a REMOTELY PILOTED VEHICLE) can loiter over suspected air defense sites for up to 30 minutes, thus causing many radars to shut down, even if it can only attack one (a ground-launched version is also in development). Recent Soviet ARMs include the AS-9 KYLE and AS-12 KEGLER, comparable to Shrike and Standard ARM, respectively.

Alongside these complex weapons have been produced simpler, cheaper short-range ARMs, such as the AGM-122 SIDEARM, an anti-radiation version of the SIDEWINDER air-to-air missile, and Stinger-Post, a variant of the STINGER surface-to-air missile. No anti-radiation missiles have yet been developed for air-to-air combat, but several radar-guided air-to-air missiles have HOME-ON-JAM capabilities, a specialized type of anti-radiation guidance. See also ELECTRONIC COUNTERMEASURES; ELECTRONIC SIGNAL MONITORING; ELECTRONIC WARFARE.

ANTI-SUBMARINE WARFARE (ASW): Weapons designed to destroy or otherwise neutralize man-made space platforms, categorized into four basic types: direct ascent ASATs, co-orbital ASATs, space-based interceptors, and DIRECTED ENERGY WEAPONS (DEWs).

Direct ascent ASATs, the simplest in concept, would consist of a satellite tracking system to determine the ephemeris of the target, and a sounding rocket-type missile armed with a nuclear, high-explosive, or fragmentation warhead, with or without terminal guidance. Even without terminal guidance, missiles of this type could achieve high kill probabilities (P_K) by placing many small fragments in the path of the target; the combined velocity of the fragments and the satellite would ensure destruction. Missiles with terminal guidance (either passive INFRARED or radar ACTIVE HOMING) could actually collide with the target and would not need a warhead at all. Long-range surface-to-air missiles (SAMs), such as the Soviet SA-5 GAMMON and SA-12B GIANT and the U.S. NIKE HERCULES, as well as the Soviet ABM-1 GALOSH anti-ballistic missile, could function as direct ascent ASATs. The U.S. has also developed an air-launched direct ascent interceptor, designed to be launched at high altitude by the F-15 EAGLE fighter. In action, the pilot would fly the aircraft to the release position under ground control; the ASAT would then fly under INERTIAL GUIDANCE towards the predicted intercept point. When its engine burned out, the interceptor would release a Miniature Homing Vehicle (MHV), equipped with small steering rockets and infrared terminal guidance to achieve a collision with the target. An air-launched ASAT has much greater flexibility than fixed ground-based missiles because it could be positioned rapidly to intercept satellites in a variety of orbital inclinations. In addition, by having an aircraft boost the ASAT to high altitude, the missile can be made that much smaller and cheaper than ground-launched missiles. Direct ascent ASATs, however, are only efficient for the attack of satellites in low orbits (i.e., under 150 mi./250 km.). To attack satellites in higher orbits, larger, more costly missiles would be needed, and the time required to reach higher altitudes could allow the target to detect the attack and initiate evasive maneuvers.

Co-orbital ASATs overcome this problem by launching a "killer" satellite into orbit. That orbit is then matched to the target orbit, and when the two satellites pass in close proximity, the killer satellite explodes to create a lethal cloud of high-velocity debris. The U.S. investigated co-orbital ASATs in the late 1950s under Project SAINT (Satellite Interceptor), without completing an operational weapon. The Soviet Union began the development of a co-orbital ASAT in the early 1970s, and has conducted actual intercepts in space with a high rate of success. In 1985, however, the Soviet Union announced a unilateral moratorium on ASAT development and halted further testing; if the Soviet goal was to discourage U.S. ASAT development, that goal was accomplished, because the U.S. Congress banned further U.S. tests. Reportedly operational, Soviet co-orbital ASATs are based at the TYURATAM and PLESETSK Missile Test Centers; up to 12 ASAT missions per day could be launched for three or four days. Even with only a 30 percent P_k, the U.S.S.R. could thus cripple U.S. reconnaissance and other satellite systems in low-medium orbits (up to 600 mi./1000 km.). The major drawback of all co-orbital ASATs is the need for accurate target tracking; if the target's orbit changes even slightly, it will not arrive at the pre-computed intercept point. Future U.S. military satellites may therefore be provided with the capability to perform random or pseudo-random maneuvers.

That countermeasure, however, can be negated by placing ASATs in permanent orbit to make direct intercepts without regard to target altitude. With several interceptor missiles per satellite (see SPACE-BASED INTERCEPTOR), a radar or infrared sensor to detect and track targets, and a fire control system to launch interceptors as appropriate, such systems would be essentially identical to space-based BALLISTIC MISSILE DEFENSES; in fact, any space-based BMD would have an inherent ASAT capability as well (which could in turn be countered by a parallel ASAT system).

Finally, DEWs such as high energy LASERS (HELs) or neutral particle beams (NPBs) could be used to destroy or at least neutralize satellites by overheating their surfaces or overloading their electronics. DEWs may be based either in space or on the ground. Large Soviet ground-based lasers (at SARY SHAGAN and elsewhere) appear already to have the capacity to damage the optics of U.S. reconnaissance satellites. So far, however, no DEWs powerful enough to destroy satellites from the ground have been developed, in the U.S. at least.

Although an ASAT ban may be negotiable, the high military utility of surveillance and communications satellites makes it highly unlikely that they would be allowed to operate with impunity during major hostilities. The proliferation of space launch technology also increases the probability that ASATs, at least simple, direct-ascent weapons, will be developed by lesser powers. See also SATELLITES, MILITARY. SPACE, MILITARY USES OF.

ANTI-SHIP MISSILE (ASM): A guided missile designed to attack surface ships. Now the principal weapon of ANTI-SURFACE WARFARE (ASUW), ASMs can be launched from ships, submarines, aircraft (including helicopters), and shore bases. Developed by Germany in World War II, the first anti-ship missiles, air-launched and controlled by radio command guidance, were used with much success in 1943, but were soon neutralized by ACTIVE JAMMING. After 1945, the U.S. developed short-range air-launched missiles (e.g., the AGM-12 BULLPUP and AGM-62 WALLEYE), which could also be used as ASMs, but it was only the Soviet Union that seriously pursued the development of specialized, long-range, air- and ship-launched anti-ship missiles as substitutes for carrier air power. Early Soviet ASMs such as the ship-launched SS-N-2 STYX and SS-N-3 SHADDOCK, and the air-launched AS-2 KIPPER and AS-4 KITCHEN, are very large (up to 11,000 lbs./5000 kg.), resemble small jet aircraft, and fly parabolic trajectories to their target; ranges vary from 70 to 400 mi. (117 to 668 km.). Longer-range systems such as Kipper and Shaddock are controlled by a combination of INERTIAL GUIDANCE with radar ACTIVE HOMING for terminal guidance; the shorter-ranged Styx has a combination of radio COMMAND GUIDANCE with terminal radar homing.

Despite the proliferation of Soviet ASMs from 1959, they were not considered a serious threat by the U.S. Navy until the sinking of the Israeli destroyer *Eilat* by three Styx missiles in 1967. By then, U.S., European, and Israeli development programs were already far advanced, and the resulting second generation of ASMs are all "surface skimmers": instead of flying a high parabolic trajectory, they evade detection by flying most of their course at only 30–50 ft. (10–15 m) above the waves with the aid of radar altimetry. The U.S. AGM-84 HARPOON, French EXOCET, and Italian OTOMAT are controlled by inertial guidance until they reach the general vicinity of the target; they then switch to terminal guidance by either radar active homing or passive INFRARED HOMING (and can also fly preset search patterns, if they do not immediately acquire a target when their terminal guidance is activated). The Israeli GABRIEL is controlled by SEMI-ACTIVE RADAR HOMING, which requires target illumination from the launch platform (but there is now an active-homing version as well), while the Norwegian PENGUIN relies on inertial guidance and passive IR homing.

These second-generation Western missiles are much smaller than their Soviet predecessors because of miniaturized electronics and more efficient engines. FAST ATTACK CRAFT (FACs) as small as 80 tons, and most tactical aircraft and helicopters, can thus be armed with ASMs, while Harpoon and Exocet also have submarine-launched versions that fit standard 21-inch (533-mm.) torpedo tubes. The Soviet Union did not develop small ASMs until the 1980s, while Soviet missiles of the 1970s, such as the AS-5 KELT, AS-6 KINGISH, and SS-N-12 SANDBOX were even larger than their 1960s predecessors, for the sake of longer ranges, supersonic speed, and better terminal guidance. To attack heavily armored U.S. AIRCRAFT CARRIERS, large warheads and thus large missiles are necessary, and the Soviet navy could not therefore pursue miniaturization; instead, it built specialized platforms for large ASMs, such as the ECHO, CHARLIE, and OSCAR submarines, and relied on large, land-based bombers (Bison and BACKFIRE) instead of fighter-type aircraft to deliver long-range ASMs. The first Soviet surface-skimmers of the 1980s were short-range tactical missiles, but these were followed by the powerful SS-N-22 SUNBURN, and by small yet long-range cruise missiles similar to the U.S. TOMAHAWK: the air-launched

AS-15 KENT and the submarine-launched SS-N-21 SAMPSON. In addition, the Soviet navy developed the large SS-N-19 SHIPWRECK, a much improved member of the Shaddock/Sandbox family, and the submarine-launched SS-N-24, a very large Mach 3 missile expected to fly at least a portion of its trajectory at low altitude. Most Soviet ASMs have nuclear as well as high-explosive warheads.

Several other countries have produced anti-ship missiles, mostly small, relatively short-range surface-skimmers, including the British SEA SKUA and SEA EAGLE, South African and Taiwanese developments of Gabriel, the West German KORMORAN and the Swedish RBS-14.

The most notable response to the ASM threat has been the development of terminal defenses such as the SEA SPARROW, SEA WOLF, and RAM point-defense missiles, and radar-controlled guns such as the U.S. PHALANX, the Dutch GOALKEEPER, and the Spanish MEROKA. Because surface-skimmers are small and fast, and fly so close to the water, they are difficult to detect by radar except at close range. By then, the time to impact is so short that anti-missile defenses must be automated to be effective at all; i.e., once switched on, they engage targets as they acquire them, without human intervention. In a second response, ELECTRONIC COUNTERMEASURES, including CHAFF launchers and ACTIVE JAMMERS, have become crucial naval capabilities, which platform-fixated navies ignore at their peril—as the Royal Navy discovered in the 1982 Falklands War, and the U.S. Navy discovered in the Persian Gulf from 1986. Anti-ship missiles have had a specific impact on international politics, because almost any state can now purchase the capability to sink even large surface vessels. See also AEGIS; AIRBORNE EARLY WARNING; ANTI-AIR WARFARE; AV-MF.

ANTI-SUBMARINE WARFARE (ASW): Measures intended to prevent SUBMARINES from accomplishing their missions (defensive ASW) or to locate and destroy them (offensive ASW). The continuing ability of submarines to attack undetected with particularly devastating weapons makes them singularly effective in an era of global surveillance by remote sensing. Though few in number (some 600 worldwide), modern submarines—both nuclear-powered and diesel-electric—are far more more capable than their predecessors of World War II, which could already disrupt SEA LINES OF COMMUNICATION and affect the conduct of most naval operations; in addition, nuclear-powered ballistic-missile submarines (SSBNs) are now the most important category of strategic-nuclear weapons.

Defensive ASW focuses on avoidance and deterrence through the grouping of ships into CONVOYS, which: (1) concentrates potential targets into a limited ocean area, making it difficult for submarines to find them; (2) dissuades attack by concentrating ASW escort vessels and air cover around the ships, increasing the risk to submarines; and (3) if intelligence on enemy submarine movements is available, allows the evasive routing of convoys to preclude attack. In both world wars, convoys were very successful—but, of course, submarines did not then have long-range weapons.

Offensive ASW entails four phases: search and acquisition, classification, attack, and assessment. Search and acquisition are conducted with a variety of sensors on surface ships, helicopters, land- and carrier-based aircraft, or other ("hunter-killer") submarines, or fixed on the seabed. The primary ASW sensor is SONAR, both passive and active. Passive sonar (hydrophones), essentially an underwater microphone, can detect propeller, machinery, and other noises generated by submerged submarines, at ranges up to several hundred miles, if the noise of marine life, waves and currents, and friendly surface ships can be filtered out by computerized signal processing. Active sonar is analogous to RADAR, but functions in a far less favorable medium because the propagation of sound waves is seriously affected by variations in water temperature, density, and salinity (see THERMOCLINE).

Open-ocean ASW escorts and larger submarines can incorporate large, low-frequency active/passive sonar arrays, and use bottom-bounce, convergence, and other techniques to function at long range. In coastal waters with much more difficult sonar conditions, smaller medium- or high-frequency sonars of shorter range but better resolution are more effective. Passive TOWED ARRAY sonars, such as the U.S. TACTAS and SURTASS, can be trailed behind ships and submarines to provide a long baseline for triangulation, and VARIABLE DEPTH SONARS (VDS) operated from ship transoms can penetrate below sonar-opaque temperature, salinity, and density layers. Helicopters such as the SH-3 SEA KING can be equipped with small VDS "dipping sonars," which can be lowered to the chosen depth while the helicopter hovers at low altitude. Both helicopters and fixed-wing ASW patrol aircraft can drop patterns of expendable sonars (SONOBUOYS), both active and passive, in the vicinity of a suspected submarine; the relative

strength of the signal received by each sonobuoy indicates the submarine's position.

Fixed seabed sonars, notably the U.S. SOSUS, consist of large, very sensitive passive arrays emplaced on the ocean floor and linked by cable to shore stations, where powerful signal processing computers can plot the position of detected submarines. The North Atlantic SOSUS barrier across the GIUK GAP can direct ASW units to (and convoys away from) enemy submarines.

In addition, radar can detect the SNORKEL masts of diesel-electric submarines recharging their batteries (as well as any surfaced submarines): the emissions of diesel engines can be detected by ionization sensors ("sniffers"), such as AUTOLYCUS; and, at very close ranges, the MAGNETIC ANOMALY DETECTORS (MAD) probes of ASW aircraft can detect fluctuations in the earth's magnetic field caused by the mass of ferrous materials in submarine hulls. Radio direction finding (D/F) was very useful in World War II because submarines had to report to headquarters on a regular basis and also pass target contacts to one another. High frequency direction finding (HF/DF, or "Huff Duff") is still used, but submarines can circumvent radio interception by burst transmissions, and the use of message buoys which float to the surface to transmit a recorded message, but only after a delay that allows the submarine to get away. Finally, recent experiments indicate the possibility of detecting submerged submarines from space: sensitive radar and INFRARED sensors can detect the wakes generated by submarines, or even the submarines themselves if they are near the surface.

Because all ASW sensors are unreliable compared to radar, successful ASW operations require the coordinated action of many ships and aircraft for each target submarine.

Once classified as hostile, submarines can be attacked with a variety of weapons. The most basic is the DEPTH CHARGE, an explosive-filled canister fitted with a hydrostatic fuze to detonate at a selected depth, so as to cause an overpressure sufficient to damage submarine pressure hulls, either to sink the submarine outright or force it to surface. Initially, depth charges were simply rolled over ship fantails; later, side-throwing projectors (K-guns and Y-guns) were developed to increase the area that could be covered by depth charge barrages. Late in World War II, the U.S. and Britain developed forward-throwing depth charge mortars (HEDGEHOG and Squid), whose derivatives (see LIMBO) are still in use.

Depth charges are relatively ineffective because the depth of the submarine must be determined with some accuracy to set the fuze; because their detonation deafens all sonars in the vicinity, allowing submarines to escape if they survive the first attack; and because the "dead time" between the dropping of the charge and its explosion can allow submarines to maneuver away. Thus, the probability of kill (P_K) of individual depth charges is very low; in fact, the latest submarines, which can dive deep and move fast, are virtually immune to depth charge attack. Nevertheless, forward firing depth charge mortars and rocket launchers are still in service for ASW operations in shallow waters. Nuclear depth charges, in service with both the U.S. and Soviet navies, have a lethal radius of several hundred meters (vs. several feet), but their use would disrupt sonar operations over a wide area for extended periods (BLUEOUT).

Developed late in World War II, the acoustical homing TORPEDO is now the primary ASW weapon, both in lightweight, short-range versions (such as the ubiquitous U.S. MK.46) and in heavy, long-range versions, notably the British TIGERFISH and SPEARFISH, and the U.S. MK.48. Lightweight torpedoes can be launched from ships, helicopters, and aircraft in the immediate vicinity of the target; they are, in effect, mobile depth charges. Long-range "heavy" torpedoes now mostly arm submarines, though the Soviet and Swedish navies still retain them aboard surface ships; their ranges can exceed 20 mi. (34 km.).

To extend the reach of lightweight torpedoes and depth charges, they can be incorporated into standoff missiles such as the ASROC, SUBROC, and SEA LANCE, the Australian IKARA, the French MALAFON, and the Soviet SILEX, STARFISH, and STALLION, all meant to be launched from the first convergence zone (some 10 mi./16 km.).

MINES can be used to block submarine base approaches and narrow straits. In addition to moored mines, there are bottom-laid influence mines and there is even a "mobile" mine, U.S. Mk.60 CAPTOR, which incorporates a torpedo, a passive sonar, and FIRE CONTROLS on a moored platform: if a submarine passes within its range, CAPTOR launches the torpedo.

By far the most efficient ASW method is to destroy submarines in their bases, but those bases are usually well defended. The next most efficient method is to intercept submarines as they enter or leave their ports; submarines are the best platforms for this tactic, but again, the waters adjacent to submarine bases are usually well patrolled by ASW forces. The least efficient method is the use of "hunter-killer" (HUK) groups of DESTROYERS, FRIGATES, and an aircraft carrier to search for submarines with the aid of generic intelligence, SOSUS-type data, and convoy reports; by cruising at low speed, submarines can minimize their acoustical signature (and hence the ranges at which they can be detected), while HUK groups cannot avoid detection and can thus be evaded. HUK groups could still be useful, however, to clear the way for convoys.

ANTI-SURFACE WARFARE (ASUW): U.S. Navy term for weapons and tactics meant to be employed against enemy warships and merchant vessels by land-based and carrier aircraft, SUBMARINES, and surface combatants. The dominating factor in ASUW is now the very long range and high lethality of aircraft-delivered ordnance and ship-launched ANTI-SHIP MISSILES. Hence, the overriding goal in ASUW is to detect and attack targets at the longest possible range, without being detected and attacked in turn. That has greatly increased the importance of SCOUTING by the use of all available SURVEILLANCE sensors (and their integration), and of counterscouting by screening forces meant to foil enemy scouting. Because RADAR, SONAR, ELECTRONIC SIGNAL MONITORING (ESM), satellite surveillance, and signals intelligence (SIGINT) all have their limitations, visual searches by aircraft, submarines, PICKET vessels, and REMOTELY PILOTED VEHICLES (RPVs) remain important; and because the use of radar can be detected by enemy ESM, counterscouting requires a heavy emphasis on emissions control (EMCON), as well as screening by surface vessels and by COMBAT AIR PATROLS (CAPs), to impede enemy air reconnaissance.

ASUW activities must therefore strike a delicate balance between the need to find the enemy and the need to avoid detection. That makes passive sensors and long-range surveillance aircraft more useful.

Once detected and located, enemy ships are to be attacked at the longest possible range, both to preserve the initiative and to make it possible to reengage if necessary. The longest-range ASUW weapons are land- or carrier-based aircraft, whose high speed minimizes the "dead time" between acquisition and attack, thereby limiting the enemy's ability to strike first or evade the attack. Land-based aircraft are more economical, but their combat radii are limited to 300–500 mi. (500–

1000 km.) from their bases, except for large maritime BOMBERS such as those used by Soviet Naval Aviation (AV-MF); U.S. B-52 STRATOFORTRESSES are now adapted for maritime strike as well. Bombers, however, are vulnerable to interception by carrier-based fighters. Carrier-based aircraft, on the other hand, are few in number, and may not be present when needed.

Free-fall and GLIDE BOMBS, short-range anti-ship missiles (ASMs), and unguided rockets can all be used for airborne ASUW, but ASMs predominate because of the increasing range and lethality of shipboard AIR DEFENSES. Submarines can also attack enemy ships independently of the main body of the fleet (if targets are reported to them) with long-range TORPEDOES as well as anti-ship missiles. Within ranges of 200–300 mi. (350–500 km.), on the other hand, ship-launched ASM strikes are feasible.

The proliferation of shipboard ASMs has revolutionized Western ASUW because surface ships are no longer confined to the role of escorts for AIRCRAFT CARRIERS and merchant ships; CRUISERS, DESTROYERS, and FRIGATES can now form SURFACE ACTION GROUPS (SAGs) which can use their own ASMs against enemy ships. To reach launch positions without suffering losses from enemy ASMs, modern surface warships are now equipped with surface-to-air missiles, DUAL PURPOSE guns, and radar-controlled guns (e.g., PHALANX, GOALKEEPER, and MEROKA) designed specifically to destroy surface-skimming ASMs, as well as ELECTRONIC COUNTERMEASURES (ECM) including ACTIVE JAMMERS and CHAFF launchers. Because the range of anti-ship missiles greatly exceeds the range of shipboard sensors (essentially limited to the masthead radar horizon), some form of OVER-THE-HORIZON TARGETING (OTH-T), from fixed-wing aircraft, helicopters, RPVs, satellites, or submarines, is required to provide both precise targeting data (including classification) and midcourse guidance corrections; that, however, entails coordination problems of its own.

Finally, should the range between opposing forces become close enough, gunnery actions may still be feasible, especially at night or in confined waters such as the Baltic or the Eastern Mediterranean, where ships can hide close to the shore or behind islands (In 1973, Israeli missile boats actually sunk more Arab craft with guns than with missiles). Gunnery has not changed much since World War II, in spite of the introduction of computerized FIRE CONTROLS and rapid-fire guns rang-

ing in caliber from 76 mm. to 127 mm. (3 to 5 in.); but the recent development of LASER- guided projectiles could yield high first-round hit probabilities. See also ANTI-AIR WARFARE; BATTLESHIP; CARRIER BATTLE GROUP.

ANTI-TANK (AT): Weapons and tactics designed to destroy or neutralize MAIN BATTLE TANKS (MBTs) and other ARMORED FIGHTING VEHICLES (AFVs).

WEAPONS

The earliest anti-tank weapons were ordinary field ARTILLERY pieces firing solid shot (an early form of AP, or armor piercing, ammunition) or high-explosive shells over open sights.

As tanks became more heavily armored, specialized anti-tank guns firing AP shot at high velocities from low-profile carriages were developed, at first in small calibers (in 1940 they ranged from 20 to 45 mm.). As tanks added more armor, calibers increased to 76 mm., 85 mm., 88 mm., 90 mm., and finally 100 mm.; concurrently, higher-performance ammunition was also developed, first composite and capped (APC), then shaped-charge (HEAT) and rigid composite or "Hyper-Velocity, Armor Piercing" (HVAP), and later discarding sabot (APDS) rounds. As AT guns grew larger and needed heavy tractors to tow them, a logical progression led to the development of tanklike, self-propelled anti-tank guns, or TANK DESTROYERS. Some were turreted but lightly armored, while others were heavily armored but turretless, both being cheaper than tanks.

At the start of World War II, tanks were not regarded as anti-tank weapons; they were designed rather either for exploitation in the enemy rear or for infantry support. Hence, their guns were low-velocity weapons, suitable against infantry, artillery, and unarmored vehicles. As the war progressed and tanks were increasingly engaged in tank-to-tank combat, high-velocity guns became indispensable. After 1945, most armies focused on tanks as the best weapons against other tanks, so that both tank destroyers and towed anti-tank guns greatly declined in importance.

The search for AT weapons that could be used by foot infantry began as soon as tanks emerged in 1915. The first result was the anti-tank rifle, essentially an oversized "elephant gun" that fired very heavy bullets; with poor armor-piercing performance even at very short range, AT rifles became completely worthless as tanks added armor. Until the middle of World War II, the only alternatives were anti-tank hand GRENADES, or improvised incendiary devices ("Molotov cocktails"). By 1943, however, the SHAPED-CHARGE principle was exploited in various portable weapons, including the British spring-driven PIAT, the recoilless German *Panzerfaust,* and the U.S. rocket-propelled BAZOOKA (imitated in the German *Panzerschreck* and early Soviet RPGS). Hardly changed since 1945, these remain the most common infantry anti-tank weapons. Though their range is 300 m. or less, they can still be effective in close terrain, e.g., forests and urban areas.

MINES have been important anti-tank weapons from the start; in World War II, many millions of mines were laid, especially on the Eastern Front and in the Western Desert, with later types encased in wood or plastic to avoid magnetic detection. Nowadays, small shaped-charge mines can be delivered in large numbers by artillery shells, rocket warheads, and aircraft dispensers.

Free-fall bombs are too inaccurate to hit tanks (though their concussion can injure crews and derange fire control equipment). In World War II, the principal AT weapons for aircraft were thus unguided rockets and automatic CANNON. Cannon are still used (as, e.g., the 30-mm. GAU-8 of the A-10 THUNDERBOLT), but the main aircraft AT weapons are guided missiles (e.g., the AGM-65 MAVERICK) and CLUSTER BOMBS. The U.S. Mk.20 ROCKEYE cluster bomb, for example, contains several hundred small, shaped-charge bomblets that can penetrate the thin top armor of tanks.

The most innovative anti-tank weapon, however, is the ANTI-TANK GUIDED MISSILE (ATGM), light enough to be carried on foot, or at least to be mounted on jeeps and even the smallest helicopters, yet with long range and very high lethality; on the other hand, ATGMs require uninterrupted intervisibility, and their relatively long engagement times make them vulnerable to countermeasures. Moreover, all current ATGMs have shaped-charge HEAT warheads, and the development of advanced spaced, composite (CHOBHAM), and REACTIVE ARMORS has thus much reduced their effectiveness. To protect their operators from enemy counterfire, larger ATGMs are increasingly mounted in AFVs, which thus become tank destroyers. Finally, precision-guided AT artillery shells (e.g., COPPERHEAD) and mortar bombs are now in development.

METHODS

The mass and momentum of armored attacks must be absorbed or blocked before any eventual counterattack. MOBILE DEFENSE methods rely on forces operating throughout the depth of a defensive area to absorb the offensive impact, while static defense methods rely on obstacles covered by anti-tank weapons and arrayed in several lines, or which form "islands" (HEDGEHOGS) in the depth of the defended area. The aim is to draw the attacker into combat against successive lines or islands, until his momentum is exhausted, while armored forces are kept in reserve behind the last line of defense, to deal with any breakthroughs and then counterattack once the enemy is weakened. Thus both mobile and static methods require adequate operational RESERVES and geographic depth—though minefields and other obstacles can replace depth by reducing the speed of enemy movement. The third essential ingredient is an adequate number of effective anti-tank weapons. See also ARMORED FORCES.

ANTI-TANK GUIDED MISSILE (ATGM):

A small missile designed to attack tanks and other armored fighting vehicles (AFVs), launched from the ground, soft vehicles, AFVs, or helicopters. An ATGM consists of a small, solid-fuel rocket with a shaped charge (HEAT) warhead, miniature wings and aerodynamic controls, and some form of launcher and guidance unit. Most ATGMs are COMMAND GUIDED by wire: the operator transmits steering signals via fine wires which pay out from spools in the missile as it flies downrange; radio command-guided and SEMI-ACTIVE LASER HOMING, and BEAM-RIDING ATGMs have also been developed.

The earliest ATGMs, such as the Soviet AT-3 SAGGER, the French SS-11, and the German COBRA, have manual command to line-of-sight (MCLOS) guidance: with a small joystick, the operator flies the missile like a model airplane. Much skill and practice are required. Immediately after launch, the operator must first bring the ATGM onto his line of sight to the target ("gathering") and then direct it against a target which may be moving or firing back; if smoke or dust interrupt intervisibility even briefly, that will usually result in the loss of the ATGM. With MCLOS, the need to gather the missile imposes a rather long minimum range (on the order of 500 m.), while the pace of manual control imposes a speed limit on the missile, increasing engagement times and, hence, the time available for the target to evade or take countermeasures. On the other hand, MCLOS is cheap, and it does not require the operator to be collocated with the launcher, allowing him to remain under cover away from the missile's boost "signature."

Most current ATGMs have semi-automatic command to line-of-sight (SACLOS) guidance. With SACLOS, the operator need only keep the target fixed in his sights, because an autopilot generates the required steering signals for the missile, whose deviations from the line of sight are detected by an infrared sensor built into the launcher. Because the sight, the missile, and the target are all in line, "gathering" is eliminated, and minimum range (75–100 m.) is constrained only by the warhead's safety fuzing. Because guidance is no longer manual, the size of the control surfaces can be reduced and the maximum speed of the missile increased. (The British SWINGFIRE has no control surfaces at all, being steered by the vectoring of the rocket exhaust.) Because the launcher and operator must be collocated, SACLOS missiles are more safely operated from armored vehicles or helicopters. Current SACLOS missiles include the U.S. TOW and DRAGON, the Euromissile HOT and MILAN, and the Soviet SPIGOT, SPANDREL, and SPIRAL.

The U.S. AGM-114 HELLFIRE, on the other hand, has laser homing guidance, and thus can be launched by one operator at a target outside his line-of sight but laser-designated by another operator.

Tactically, the great advantage of all ATGMs is their high accuracy and long range, up to 4000 m. in some cases (most tank guns are inaccurate beyond 3000 m.). Moreover, ATGMs are cheap, rarely costing more than $10,000 each, as compared to $2 million to $3 million for tanks. In theory, therefore, ATGMs are extremely cost-effective; in practice, however, several factors degrade their effectiveness. First, their putative range advantage requires intervisibility, but in Germany, for example, the mean line of sight is under 1000 m. (the range at which tank guns are most effective). Second, ATGMs are slow (400–600 mph/680–1000 kph) and may take as long as 12 seconds to reach their target; tanks can therefore spot the launch of ATGMs and take countermeasures, including evasive maneuvers, the generation of smoke screens, and counterfire. Third, unless under armor themselves, ATGMs are highly vulnerable to ARTILLERY fire, whereas tanks are not.

As a result, ATGMs do not dominate the modern battlefield, though they do inhibit the offensive dash of tank forces by imposing anti-ATGM precautions (see OVERWATCH).

Even if man-portable (and some are not), ATGMs are increasingly mounted on vehicles for the sake of better fire controls as well as protection and mobility; the M901 IMPROVED TOW VEHICLE, for example, has sophisticated computerized sights, as does the M2/M3 BRADLEY fighting vehicle. The ATGM has thus become the primary weapon of modern TANK DESTROYERS (including the lightest ones, without sophisticated fire controls, e.g., STRIKER, BRDM).

Helicopter-launched ATGMs were first employed in Vietnam in 1972, with much success, but against enemy forces very weak in anti-aircraft weapons. Helicopters can launch ATGMs from maximum range, and they can move rapidly to stage successive ambushes or form concentrations; but they are also vulnerable, and expensive to acquire and maintain. At present, helicopters armed with ATGMs are mostly fielded as mobile fire brigades, to deal with enemy spearheads that penetrate main defensive lines.

The advent of tanks with composite (CHOBHAM) armor, and of applique REACTIVE ARMOR boxes, has now reduced the effectivness of shaped charges, and thus of ATGMs, though some have double charges to defeat reactive armor (e.g., TOW-2A) and others (TOW-2B and the Bofors BILL) fly top-attack profiles to circumvent the thickest frontal armor. See also ANTI-TANK; ARMOR.

ANZUS: Australia–New Zealand–United States, the Pacific Security Treaty of September 1951, offered by the U.S. in the wake of the unsuccessful application of the other two signatories to join NATO. Under its terms, each party undertakes to assist any other party in the event of an armed attack on its territory or military forces anywhere in the Pacific. (As usual, the U.S. commitment is subject to congressional approval.) Australian and New Zealand forces participated in the Vietnam War, as in the Korean War before it. The United Kingdom at one time expressed an interest in joining the treaty, but did not do so. The continued survival of the treaty was jeopardized in the late 1980s by the rise of anti-nuclear sentiment in New Zealand. See also NUCLEAR FREE ZONE.

AP: ARMOR PIERCING, a solid (monobloc) shot of hardened steel, designed to penetrate armor plate. AP rounds inflict damage either by hitting vulnerable components (engines, fuel, ammunition, crewmen) after they penetrate, or by fragmenting; sometimes a small burster charge enhances their fragmentation. AP was the most common type of anti-tank ammunition in World War II, but has been supplanted by APDS and APFSDS rounds, which can penetrate more armor for any given gun. See also APC; APHE; API.

APACHE: The McDonnell Douglas (formerly Hughes) AH-64 attack helicopter. Developed from 1972 to replace the U.S. Army's AH-1 COBRA, the Apache's design drew heavily on the abortive AH-56 Cheyenne program to serve as a specialized ANTI-TANK helicopter. The prototype flew in September 1975, and the first Apaches entered service in late 1983. Originally, the army planned to acquire some 1300 Apaches, but the purchase was cut to 675 because of rising costs. In 1989, Israel became the first export user, leasing 9.

The Apache's two-man crew sits in a heavily armored tandem cockpit, with the pilot in an elevated seat behind the weapon system operator (WSO) to enhance his forward view. The nose houses a Martin Marietta Target Acquisition and Designation Sight/Pilots Night Vision System (TADS/PNVS), which combines a LASER DESIGNATOR and a Forward-Looking Infrared (FLIR) sensor in a trainable turret controlled by the WSO. Both crewmen have integrated helmet-mounted sights, enabling them to acquire and engage targets merely by looking at them. A Hughes M230 CHAIN GUN (with 1200 rounds of ammunition) in a flexible turret under the cockpit can be sighted with the TADS/PNVS or with the helmet sights. The space behind the cockpit houses two self-sealing fuel tanks.

Designed for all-weather, day/night operations, the Apache's AVIONICS include a Litton ASN-143 strap-down INERTIAL NAVIGATION unit, a digital autopilot/autostabilization system, a Singer-Kearfott lightweight DOPPLER navigation radar, UHF and VHF radios with secure communications links, an omnidirection air-data probe for precise low-speed handling, a RADAR WARNING RECEIVER, CHAFF/flare dispensers, and an ALQ-144 INFRARED COUNTERMEASURES beacon.

The fuel tanks and critical components are armored, and the tricycle landing gear are reinforced for crash landings. The engines, mounted in nacelles on each side of the fuselage behind the cockpit, are fitted with "Black Hole" infrared suppressors as protection against INFRARED-HOMING missiles. The rotor hub is fully articulated, and both the hub and the composite rotor blades can withstand hits from anti-aircraft guns up to 23 mm. in caliber (i.e., the Soviet ZSU-23-4 *Shilka*).

Two large stub wings behind the cockpit have four underwing pylons for up to 16 AGM-114 HELLFIRE laser-homing anti-tank guided missiles; or 8 Hellfires and 2 7- or 19-round pods for "Hydra-70" 2.75-in. (70-mm.) folding-fin aerial rockets (FFARS); or up to 16 BGM-71 TOW ATGMs, or 4 FIM-92 STINGER IR-homing air-to-air missiles. There are also two wingtip pylons for AIM-9 SIDEWINDER air-to-air missiles or AGM-122 SIDEARM anti-radiation missiles.

Though a formidable anti-tank helicopter, the Apache's nose-mounted TADS/PNVS forces it to rise above the horizon to spot targets, thus exposing itself to enemy gun and missile fire. Other helicopters, by contrast, have mast-mounted sights above their rotors, allowing them to track targets while remaining "hull down." (Beginning in the early 1990s, Apaches are to be retrofitted with the Airborne Adverse Weather Weapon System, a mast-mounted millimeter wave [MMW] surveillance radar linked to a radar-homing version of Hellfire.) Moreover, because of the weight of its armor protection, the Apache is somewhat underpowered (it cannot lift a full load of Hellfires on a hot day) and surprisingly slow. See also CAVALRY, AIR; HAVOC; HIND.

Specifications Length: 48.16 ft. (14.68 m.). **Rotor diameter:** 48 ft. (14.63 m.). **Powerplant:** 2 1696-shp. General Electric T-700-GE-701 turboshafts. **Internal fuel:** 4550 lb. (2068 kg.). **Weight, empty:** 11,015 lb. (4996 kg.). **Weight, normal mission:** 14,694 lb. (6665 kg.). **Weight, max. takeoff:** 17,650 lb. (8006 kg.). **Speed, max.:** (clean) 186 mph (300 kph). **Speed, cruising:** (external stores) 150 mph (250 kph). **Initial climb:** 2500 ft./min. (762 m./min.). **Hover ceiling:** 13,400 ft. (4084 m.). **Range, max.:** (internal fuel) 380 mi. (611 km.); (external tanks) 1121 mi. (1804 km.).

APC: 1. Armor Piercing, Capped; a type of tank and anti-tank ammunition. APC is similar to AP solid shot, but has a superhardened nose cap (usually of tungsten) to enhance penetration. Armor Piercing, Capped, Ballistic Cap (APCBC) is similar to APC, but has an additional streamlined "ballistic cap," or nose cone, over the superhardened cap to improve the aerodynamics of the round and thus its ballistic performance. Both APC and APCBC have largely been replaced by APDS and APFSDS, though they remain the principal ammunition of Soviet-made T-54/55 tanks.

2. See ARMORED PERSONNEL CARRIER.

APCBC: Armor Piercing, Capped, Ballistic Cap; a type of tank and anti-tank ammunition. See APC.

APDS: Armor Piercing Discarding Sabot, an ARMOR-PIERCING round consisting of a dense penetrator of tungsten alloy or depleted uranium (STABALLOY) with a lightweight sleeve (sabot). Much smaller in caliber than the gun, the penetrator is surrounded by the sabot to seal the barrel; the sabot breaks up and falls away after leaving the muzzle. With that, the frontal area of the round, and thus its drag coefficient, are greatly reduced, and the penetrator therefore decelerates more slowly than full-bore AP, retaining more kinetic energy over greater ranges. Its dense material also concentrates mass behind a small frontal area, further improving armor penetration. APDS can be fired from most rifled tank (and anti-tank) guns, and was the standard (105-mm.) armor-piercing round in NATO armies during the late 1960s and 1970s; it has now been superseded by APFSDS.

APDS rounds have also been developed for the small-caliber (25–30-mm.) CANNON of light armored vehicles (e.g., the 25-mm. M249 Bushmaster of the M2/3 BRADLEY), and for shipboard antiaircraft guns (e.g., the 20-mm. PHALANX), but APDS cannot be fired by jet aircraft because the engines could ingest the discarded sabots.

APEX (AA-7): NATO code name for the Soviet R-23 air-to-air missile, first reported in the West in 1976. The standard medium-range missile of MiG-23 FLOGGER fighters in the Soviet air force, Apex has a unique aerodynamic configuration with three sets of cruciform aerodynamic surfaces arranged in line: small canards at the nose, delta wings at midbody, and small, movable rectangular tailfins. Steering is accomplished by the tailfins, with roll control provided by ailerons on the wings; the canards are apparently fixed aerials for radar interferometric guidance.

In keeping with standard Soviet design practice, there are separate SEMI-ACTIVE RADAR HOMING (SARH) and INFRARED-HOMING versions, and Floggers usually carry one of each type on their wing glove hardpoints (AA-7s have also been seen on the outer wing pylons of MiG-25 FOXBATS as an alternative to the AA-6 ACRID). The SARH (R-23R) version relies on target illumination by the Flogger's I-J band "High Lark" radar, which does not have a LOOK-DOWN/SHOOT-DOWN capability. The IR-homing (R-23T) version appears to have a limited all-aspect capability; i.e., it may be capable of head-on attacks under favorable circumstances.

Both versions are powered by a dual-impulse (boost/sustainer) solid-rocket engine, and armed with a 66-lb. (30-kg.) high-explosive fragmentation warhead fitted with both contact and proximity fuzes.

Specifications **Length:** 15.08 ft. (4.6 m.). **Diameter:** 8.8 in. mm.). **Span:** 41.4 in. (1.05 m.). **Weight, launch:** 705 lb. (320 kg.). **Speed, max.:** Mach 3 (2000 mph/3400 kph). **Range, max.:** (SARH) 21.75 mi. (36.33 km.); (IR) 9.3 mi. (15.5 km.).

APFSDS: Armor Piercing, Fin-Stabilized, Discarding Sabot ("arrow round"), an armor piercing round that consists of a subcaliber penetrator of dense alloy stabilized by fins, which is inserted into a lightweight sleeve (sabot) to seal the barrel; the sabot breaks up and falls away as the round leaves the muzzle. APFSDS is normally fired from smooth-bore cannon, which are easier to manufacture and do not wear out as quickly as rifled guns, but can also be fired from the latter if the sabot is designed to serve as a slip-ring. The main advantage of fin-stabilized penetrators is that they can be made much longer than spin-stabilized penetrators of the same caliber (the latter become unstable once their length/diameter ratio exceeds 5:1), improving their aerodynamics, and allowing that much more mass to be concentrated behind their frontal area, thereby increasing armor penetration as compared to APDS rounds of the same caliber. On the other hand, APFSDS rounds have a tendency to yaw into the wind because of the weathervane effect of their fins, and tend to decelerate more rapidly than APDS because of fin drag.

The Soviet army was the first to introduce APFSDS rounds (for the 115-mm. smoothbore gun of the T-62 main battle tank of 1962); when the concept was copied in the West, the sabot was made with a slip-ring to fit 105-mm. rifled guns, but by 1979 the West German army introduced a 120-mm. smoothbore gun on the LEOPARD 2 MBT, and that gun now also arms the M1A1 version of the U.S. ABRAMS tank. U.S.-made APFSDS rounds have depleted uranium (STABALLOY) penetrators; Israeli 105-mm. APFSDS rounds, also used by several NATO armies, have tungsten penetrators.

APHE: An ARMOR-PIERCING round that consists of a hardened body containing a high-explosive burster charge fitted with a base-detonating fuze. APHE rounds are designed to penetrate before exploding so as to damage the target by both blast and fragmentation. They have relatively low kinetic energy (as compared to AP, APDS, or

APFSDS) because the lower density of the explosive filler reduces the ratio of mass to frontal area. Hence, APHE rounds have lower muzzle velocities and tend to decelerate more quickly (reducing their effectiveness still further). No longer in use for tank and anti-tank guns, APHE rounds remain optimal for anti-aircraft CANNON.

APHID (AA-8): NATO code name for the Soviet R-60 air-to-air missile. First observed in 1976, Aphid is a highly maneuverable, short-range dogfight missile that arms the MiG-21 FISHBED, MiG-23 FLOGGER, MiG-25M FOXBAT, MiG-31 FOXHOUND, MiG-29 FULCRUM, Su-27 FLANKER, and Yak-38 FORGER. In service with the Soviet air force (VVS) and Air Defense Troops (PVO), it has also been widely exported.

One of the smallest air-to-air missiles, Aphid has a round nose, four small, fixed rectangular fins ahead of four movable delta canards for steering, and four long-chord cropped-delta wings at the tail for lift.

Guided by passive INFRARED HOMING, Aphid's seeker may have limited all-aspect capability; i.e., it is capable of beam attacks, and may also be capable of head-on attacks under favorable conditions. Powered by a single-pulse solid-rocket engine, Aphid is armed with a 20-lb. (9-kg.) high-explosive fragmentation warhead fitted with both contact and proximity fuzes. Aphid is now being replaced by the all-aspect AA-11 Archer.

Specifications Length: 6.85 ft. (2.15 m.). Diameter: 4.72 in. (120 mm.). Span: 15.75 in. (400 mm.). Weight, launch: 121 lb. (55 kg.). Speed, max.: Mach 2.5 (1625 mph/2713 kph). Range envelope: 0.3–3.5 mi. (0.5–5.0 km.).

API: Armor Piercing Incendiary ammunition. Mainly for heavy MACHINE GUNS or CANNON for use against light armored vehicles and aircraft, API rounds consist of a hardened steel or tungsten body designed to pierce ARMOR, with a cavity filled with phosphorus, thermite, or other substances that burn quickly at high temperatures. API is unsuitable for anti-tank use because of its low kinetic energy as compared to AP, APDS, and APFSDS rounds.

ARAPAHO: An experimental U.S. Navy system of prefabricated modules for the quick conversion of merchant ships into V/STOL auxiliary AIRCRAFT CARRIERS. Standard shipping containers prepackaged with fuel, munitions, and support equipment were to be bolted to the deck of a container ship, to form a superstructure on which a prefabricated flight deck could be secured. The

ship could then operate ANTI-SUBMARINE WARFARE (ASW) helicopters and V/STOL AV-6 HARRIERS. Developed in the 1970s, Arapaho could have provided ASW carrier escorts for convoys very economically, but the navy abandoned the concept, *after* the success of the similar British SCADS (Shipboard Containerized Air Defense System) during the 1982 Falklands War. See also AIRCRAFT CARRIERS, BRITAIN; AIRCRAFT CARRIERS, UNITED STATES.

ARM: See ANTI-RADIATION MISSILE.

ARMBRUST: West German disposable, man-portable, RECOILLESS anti-tank weapon. Developed as a private venture by MBB, Armbrust is now in service with the West German army and the U.S. DELTA FORCE.

Armbrust consists of a fin-stabilized, shaped-charge HEAT round in a sealed barrel/storage container that is issued as a "certified" round. The projectile is reportedly able to penetrate up to 11.8 in. (300 mm.) of homogeneous steel armor—though it would be only marginally effective against tanks fitted with composite (CHOBHAM) armor or add-on REACTIVE ARMOR. The launcher has a simple optical sight, but a costly computerized sight can also be fitted for special operations.

In action, the operator simply removes protective caps from the ends of the tube, pulls down two hand grips, aims the weapon, and pulls the trigger. As in any gun, the expansion of propellant gases in the cartridge ejects the round from the barrel, but Armbrust simultaneously ejects 5000 light plastic flakes out the rear of the barrel as a recoil-cancelling countermass. In contrast to most recoilless weapons, there is no backblast, allowing Armbrust to be fired from enclosed spaces; moreover, the noise it generates is reportedly no louder than a pistol shot, to further conceal firing positions. See also BAZOOKA; LAW; RPG-7.

Specifications Length, launcher: 2.75 ft. (835 mm.). Diameter, launcher: 3.1 in. (78 mm.). Weight, loaded: 13.86 lb. (6.3 kg.). Weight, warhead: 2.178 lb. (0.99 kg.) Muzzle velocity: 220 m./sec. (720 ft./sec.). Range, max. effective: 300 m.

ARMOR: The most common form of armor is still homogeneous steel plate of uniform hardness. Because very hard plate can resist small, high-velocity ARMOR-PIERCING (AP) rounds, but will be shattered by larger, low-velocity rounds (while softer, more resilient plate is too easily penetrated), attempts were made before World War II to develop laminated "face hardened" plate. Consisting of a thin, very hard plate cemented over a thicker, softer plate, face-hardened armor could resist high-velocity rounds, while having the elasticity to cope with lower-velocity rounds. But because face-hardened armor was difficult to manufacture and tended to delaminate when hit, homogeneous steel plate, rolled or cast, remained predominant for armored vehicles and warships until the late 1970s. For aircraft, however, steel plate was too heavy, and various aluminum alloys were used instead. Aluminum armor requires greater thickness than steel for a given degree of protection, but is still much lighter, and is widely used in light armored vehicles and warships; the excessive thickness that would be required to stop large-caliber AP rounds prevented its use in MAIN BATTLE TANKS. Though it softens and loses its rigidity when heated, contrary to popular belief aluminum armor does not burn.

From the 1970s, a new composite armor, known as CHOBHAM after the original British version, has increasingly been used in tanks especially. Consisting of layers of steel and ceramic embedded in a manner that prevents debonding, Chobham-type armor spreads the kinetic energy of anti-tank projectiles over a wide area, thereby dissipating their penetrative effect, and is even more effective against SHAPED CHARGES because it disrupts their hypervelocity jets. Chobham-type modules protect the U.S. M1 ABRAMS, British CHALLENGER, West German LEOPARD 2, and Israeli MERKAVA III tanks, and also the magazines of the latest U.S. AIRCRAFT CARRIERS.

The shaped charges of lighter anti-tank weapons, including most ANTI-TANK GUIDED MISSILES (ATGMs), can penetrate homogenous plate several times as thick as their diameter. SPACED ARMOR, thin outer plates (or even wire mesh), can detonate shaped charges ahead of the main armor, thereby degrading their effect, but the spacing must be at least 3 ft. (1 m.) deep to be effective, as in Israeli MERKAVA tanks.

A West German invention, explosive REACTIVE ARMOR, was ignored by the West German and U.S. armies only to be adopted by the Israeli army for its tanks and marketed worldwide (as BLAZER). It consists of metal boxes that contain an explosive compound; when hit by the jet of a shaped charge, they detonate, creating a shock wave that breaks up the jet. Blazer was very successful in the 1982 Lebanon War against RPG-7 rockets and SAGGER missiles, and a copy of it has been widely adopted by the Soviet army for its T-80, T-72, and T-62 tanks, negating many Western anti-tank weapons.

Plastic, or "soft," armor of bonded layers of fiberglass, nylon, or KEVLAR, can defeat small-caliber bullets and splinters by absorbing and dissipating their energy; it is widely used in body armor, and to protect light armored vehicles, aircraft, and combat vessels.

ARMOR, BODY: Full body plate armor was made obsolete by firearms, and the breastplates retained by some heavy cavalry units into the 20th century were mainly decorative. But the trench fighting of World War I caused the revival of body armor: helmets were issued universally to protect soldiers from splinters and concussion, while body armor was issued to SAPPERS and assault troops in the form of rigid, overlapping plates for the upper and lower torso. Between the world wars, body armor again disappeared except for helmets, but it was reintroduced in World War II for bomber aircrew, whose "flak jackets" usually consisted of steel plates inserted into an upper-body garment. That in turn became the classic bulletproof vest, sometimes issued to assault troops though heavy and of scant effectivness.

The use of body armor increased with the introduction of "soft" armor made of fiberglass or ballistic nylon, which absorbs the momentum of projectiles by spreading their impact over many layers of fabric. Soft armored vests issued by the U.S. Army in Vietnam were quite effective against grenade or artillery splinters, but not against high-powered bullets. The introduction of KEVLAR, a lightweight artificial (aramid) fiber of very high tensile strength, has greatly improved the ballistic performance of body armor, which is now widely issued to combat troops and aircrew.

ARMORED CAR: A wheeled ARMORED FIGHTING VEHICLE normally designed for SCOUTING and rear-area security functions, but also widely used as weapon platforms. Generally with all-wheel drive in 4×4, 6×6, or even 8×8 configurations, armored cars are commonly protected only against small arms fire and splinters, and must therefore rely on speed and stealth for protection. Capable of at least 40 mph (64 kph) on roads—and often more—most armored cars are not as mobile as most tracked vehicles in sand or mud, but terrain impassable to armored cars is rather rare, while they are much easier to maintain and consume much less fuel than tracked vehicles of the same weight.

Armament ranges from light MACHINE GUNS and 20 to 30-mm. CANNON, to 90-mm. low-velocity guns; with the latter, even small 4×4 armored cars can serve as TANK DESTROYERS. British armored cars, intended mainly for scouting, tend to be lightly armed: the FOX has a 30-mm. cannon, while the FERRET has only machine guns, though the older SALADIN has a 76-mm. gun. French armored cars, by contrast, are meant to serve as tank destroyers, and even the 4×4 AML, as well as the 6×6 ERC, EBR, and AMX-10RC have 90- or even 105-mm. low-velocity guns. The West German *Spahpanzer* LUCHS (Lynx) is a very large 8×8 armored car, but is armed only with a 20-mm. cannon because it was designed for (long-range) scouting. By contrast, the Italian Centauro, also a large 8×8 vehicle, is armed with a *high*-velocity 105-mm. gun to serve as a powerful tank destroyer with tank-like firepower. The Soviet army uses wheeled armored vehicles for troop transport as well as for reconnaissance (see BTR-60); its BRDM armored cars are mainly meant for scouting (their mission being often to discover the enemy by drawing fire), but also serve as platforms for ANTI-TANK GUIDED MISSILES. The U.S. Army has hardly used armored cars since World War II, though the U.S.-made Cadillac Gage COMMANDO series has been widely exported. The main wheeled armored vehicles in U.S. service are the Swiss-designed, Canadian-built 6×6 Light Armored Vehicles (LAVs) of the Marine Corps. A dozen more countries produce armored cars, mostly unambitious 4×4 or 6×6 truck conversions. See also ARMORED FORCES; ARMORED PERSONNEL CARRIER.

ARMORED FIGHTING VEHICLE (AFV): A generic term for all armor-protected military vehicles with a direct combat role, including MAIN BATTLE TANKS, LIGHT TANKS, ARMORED CARS, INFANTRY FIGHTING VEHICLES, ARMORED PERSONNEL CARRIERS, armored command vehicles (ACVs), and COMBAT ENGINEER VEHICLES. Self-propelled ARTILLERY vehicles, though also (thinly) armored, are not usually considered AFVs. See also ARMORED FORCES.

ARMORED FORCES: Ground forces equipped with ARMORED FIGHTING VEHICLES (AFVs). Since their introduction in World War I, AFVs have variously been used as: (1) supporting weapons for the INFANTRY, attached to foot infantry units, and meant to advance at a walking pace; (2) mechanized CAVALRY with the same roles as horse cavalry, in support of the main (infantry and ARTILLERY) forces; and (3) the basis of autonomous COMBINED-ARMS formations of AFVs, with infantry, artillery, and other supporting forces, and organized to conduct all phases of battle.

Until quite late in World War II, British practice

favored the first two uses, and the British army therefore developed two different types of tanks, slow "I" tanks and fast "C" tanks, both lightly armed. French practice until 1940 favored infantry support roles, while the third, autonomous use of AFVs was pioneered by the German army, was quickly emulated by the U.S. and Soviet armies during World War II, and has since been accepted as the norm by almost all mechanized armies, though only a few have mastered the demanding COMMAND AND CONTROL requirements of mobile armored warfare. Israeli practice in 1967 included cavalry-style all-tank forces operating in depth, modern-style combined-arms armored formations, and also the use of tanks for infantry support. Since then the Israeli army has emphasized the combined-arms method.

In modern armies, all but LIGHT INFANTRY troops are now carried in ARMORED PERSONNEL CARRIERS (APCs), or more heavily armed INFANTRY FIGHTING VEHICLES (IFVs), and much of the artillery (and most combat-support units) are also mechanized, so that traditional distinctions have become blurred; mechanized "infantry" formations merely have fewer tanks and more APCs or IFVs than "armored" or "tank" formations. In the Soviet army, MOTORIZED RIFLE divisions and regiments differ from tank divisions and regiments only in their ratios of tanks to APCs; in the U.S. Army, armored divisions have six tank and five MECHANIZED INFANTRY battalions, while in mechanized divisions the proportions are reversed.

Apart from internal security operations, or combat in jungle, mountain, or urban terrain, almost any major ground operation would now be conducted by combined-arms formations of tanks with infantry, artillery, ENGINEERS, and other support units, all mounted in armored fighting vehicles. See also ACTIVE DEFENSE; BLITZKRIEG; DEFENSE-IN-DEPTH; INDIRECT APPROACH; MAIN BATTLE TANK.

ARMORED PERSONNEL CARRIER (APC):
An ARMORED FIGHTING VEHICLE designed to transport infantry troops and protect them from artillery and small-arms fire. APCs may be wheeled or tracked; half-tracked vehicles, much used in World War II, have become very rare. Armor protection, generally very light, may not include overhead cover, and is often provided by aluminum alloy to reduce weight and thereby allow some amphibious capability. INFANTRY FIGHTING VEHICLES (IFVs) resemble APCs, but are generally more heavily armored, and are invariably armed with a turret-mounted CANNON and/or

missiles; most also have firing ports for mounted combat. Unlike IFVs, which are designed to move with tanks *on* the battlefield, APCs are only "battle taxis" designed to transport infantry to the close-combat battle zone, where the troops are to dismount to fight on foot.

The most common Western APC is the U.S. M113, which also serves as the basis for a variety of specialized vehicles; it is a fully enclosed tracked vehicle with aluminum armor. The British FV432 TROJAN is similar in most respects, but the British army also has wheeled, 4 × 4 SAXON APCs for economy reasons. The French army likewise has both the tracked AMX-10P and the wheeled 6 × 6 VAB. The Soviet army has wheeled APCs of the BTR-60 series in MOTORIZED RIFLE divisions, while the tracked BTR-50 has been replaced in tank divisions by the BMP, which is an IFV. The Soviet army also has some MT-LB tracked APCs suitable for soft terrain. The Israeli army had large numbers of U.S. M3 half-tracks of World War II vintage until the 1970s, but has since relied on the M113. Elsewhere, civilian trucks with added armor often serve as APCs, especially for internal security roles. Combat ambulances, armored command vehicles (ACVS), mortar and other weapon carriers, scout vehicles, communications vehicles, and other specialized vehicles are all usually derived from APCs. See also ARMORED FORCES.

ARMORED RECONNAISSANCE VEHICLE:

A tracked ARMORED FIGHTING VEHICLE designed for SCOUTING, SCREENING, and RECONNAISSANCE; wheeled armored reconnaissance vehicles are classified as ARMORED CARS. Once assigned to LIGHT TANKS, reconnaissance functions are now generally carried out by ARMORED PERSONNEL CARRIERS (APCs). Specialized (tracked) armored reconnaissance vehicles include the purpose-built British SCORPION, the U.S. M3 BRADLEY cavalry fighting vehicle variant, the LYNX derivative of the M113 APC, and the Soviet BMP-R derivative of the BMP infantry fighting vehicle. See also ARMORED FORCES.

ARMOR PIERCING (ROUNDS, WARHEADS):

Munitions designed specifically to defeat ARMOR, which can function either chemically or kinetically. In the former case, penetration is achieved by the shaped charges of high explosive anti-tank (HEAT) rounds, or by the spalling effect of high explosive squash head (HESH) rounds (a.k.a. high explosive plastic, HEP). Kinetic energy rounds include basic armor piercing (AP) monobloc shot and its variations, API, APC, and APCBC, as well as rigid composite or high velocity armor piercing

(HVAP) rounds, but discarding-sabot APDS and APFSDS rounds are far more effective, especially the latter, which have long-rod "arrow" penetrators.

ARMS CONTROL: A general term for the pursuit of strategic goals by the subtraction rather than the addition of armaments, often used in contradistinction to humanitarian DISARMAMENT, and often pursued by negotiated ARMS LIMITATION. Arms control agreements between two or more states to limit or reduce the number and/or performance characteristics of weapons (and related support equipment), or to prohibit the development or deployment of given weapons altogether, or to reduce the number, deployment, or use of armed forces, need not be negotiated explicitly; any mutual restraint which is recognized as intentional falls within the scope of arms control, which may also be unilateral, insofar as strategic goals are still being pursued thereby. Since 1946, arms contol efforts have focused on the spread and deployment of NUCLEAR WEAPONS, but nonnuclear arms control became important in the late 1980s. See also ACDA; ANTARCTIC TREATY; CAFE; MBFR; NUCLEAR FREEZE; START; TEST BAN TREATIES.

ARMS LIMITATION: A form of ARMS CONTROL meant to prevent quantitative or qualitative increases in specified weapons and/or military forces, or to prevent specified activities, usually weapon tests. The U.S.-British-Japanese Naval Treaty of 1922 and the Anglo-German Naval Treaty of 1935, which established mutually agreed tonnage ratios for battleships and other categories of warships, were early examples of negotiated arms limitation. The 1963 Nuclear Test Ban Treaty was also specific in what it limited, but instead of encompassing only specific signatories it was intended as a universal treaty. Arms control and disarmament are broader concepts: arms limitation does not include either reductions or tacit agreements. Arms limitation agreements covering particular weapons are a natural concomitant of ARMS RACES, because the parties can usually pursue their own strategic goals more easily if certain avenues of competition are selectively limited or prohibited.

ARMS RACE: An increase in the quantity and/or quality of weapons and/or military forces of two or more states which is seen as mutually competitive. An arms race begins when one party is perceived as increasing its military strength, and that evokes a competitive reaction. That initial perception may accurately reflect a deliberate attempt to improve relative capabilities; or it may misconstrue bureaucratic initiatives that have no strategic purpose at all, or normal equipment-replacement cycles, or changes in recruiting policies induced only by demographic factors. Regardless of its origins, if the parties to an arms race seek to maintain specific relative force ratios through their (inevitably) lagged responses, no stable equilibrium may be possible. Even if the principle of parity is accepted, the long lead times of weapon development can deny stability, unless the delivery dates of future weapons, and their numbers and performance, are all accurately predicted on both sides, removing incentives for more-than-competitive anticipatory adjustments. NUCLEAR FREEZE proposals are meant to contend with these dynamics. There is little or no historical correlation between arms races and war. See also ARMS CONTROL; ARMS LIMITATION; DISARMAMENT.

ARMY: In general, the ground service within national armed forces. More specifically, as a "numbered" army, a higher formation of ground forces consisting of two or more CORPS, with or without organic ARTILLERY, AIR DEFENSE, ENGINEER, or other supporting forces. Numbered armies can be independent, or subordinated to an ARMY GROUP. In the Soviet army, the corps echelon is new and still experimental; a Soviet "army" consists of three or more divisions, and is directly subordinated to a FRONT.

ARMY GROUP: A higher formation of ground forces force consisting of two or more "numbered" ARMIES, with or without organic ARTILLERY, AIR DEFENSE, ENGINEER, or other supporting forces. Independent DIVISIONS or CORPS can be directly subordinated to an army group echelon, to serve in its operational RESERVE or for special missions such as RAIDS. An army group may also have operational control of air or naval units for specific purposes. The Soviet equivalent is the FRONT.

ARTILLERY: (1) A general term for weapons that fire explosive projectiles (vs. inert bullets) which exceed small-arms caliber (20 mm. and above) and also for battlefield ROCKETS and short-range BALLISTIC MISSILES. (2) The branch of the armed forces equipped with those weapons.

Artillery is the most lethal branch of the ground forces, and has inflicted the majority of all casualties in most wars since 1914. Its destructive power was multiplied by the advent of nuclear munitions during the 1950s, and more recently it has also been enhanced by the introduction of nonnuclear cluster shells and warheads (see IMPROVED CONVENTIONAL MUNITIONS). Hence, for high-intensity combat, the relative importance of the artillery, as compared to infantry and armor, has greatly increased since 1945.

Artillery weapons fall into five basic categories: GUNS, HOWITZERS, GUN-HOWITZERS, MORTARS, and ROCKETS and missiles. Except for the latter, artillery weapons are categorized by their bore diameter and by their ratio of barrel length to bore diameter, or "CALIBER" (confusingly, the same term also denotes the bore diameter of small arms).

Guns, with length-to-bore ratios (calibers) in excess of 40:1, fire high-velocity projectiles with relatively flat trajectories, and are meant for accurate, long-range fire (up to 30,000 m. and more). ANTI-TANK guns are a specialized subtype that fire ARMOR-PIERCING ammunition from low-elevation carriages. ANTI-AIRCRAFT guns, in contrast, have carriages capable of elevations up to 90°.

Howitzers do not exceed 30 calibers, and fire lower-velocity projectiles in parabolic trajectories over ranges that rarely exceed 20,000 m. Howitzer shells do not have to withstand the barrel pressures of high-velocity guns, and can therefore have thinner walls to contain larger explosive charges than shells fired by guns of the same bore diameter. Howitzers are meant for use against area targets such as troop concentrations, built-up areas, supply routes, etc., and their lofted trajectories are effective in attacking forces sheltered on reverse slopes.

Gun-howitzers are an intermediate type between 30 and 40 calibers; they combine the longer range and accuracy of guns with the high trajectory and larger explosive charges of howitzers. While not optimized for either role, gun-howitzers are good all-purpose weapons, and appear to be supplanting both guns and howitzers in modern artillery forces.

Guns, howitzers, and gun-howitzers (collectively known as cannon) all operate on the same general principles. A projectile, the shell, is loaded through a sliding or doorlike breech, to be followed into the chamber by the propellant charge in flammable bags (for large-bore guns of 155 mm. or more) or in a metal cartridge. The latter is attached to the shell (fixed ammunition) in small-bore weapons (20 to 105 mm.), and separate (semi-fixed ammunition) in medium-bore weapons (105 to 155 mm.); large-bore weapons have bag charges. When the propellant is ignited, the expansion of combustion gases ejects the shell from the barrel. The shell is then stabilized in flight by a spin imparted by spiral bands within the barrel (rifling). The recoil energy generated by firing is absorbed by often complex hydraulic mechanisms (sometimes supplemented by muzzle brakes) as well as by the inertia of the mount (RECOILLESS rifles, sometimes classified as artillery, eliminate recoil by venting gases through an opening in their breech, or sometimes by ejecting a countermass, in either case at the cost of dissipating much of their propellant energy; they serve mainly as anti-tank weapons).

Mortars, the simplest form of artillery, consist of a steel tube open at one end, which is placed on a heavy baseplate meant to be planted firmly in the ground. The tube is supported by a telescopic bipod, and elevation is adjusted with a calibrated screw that extends or retracts the bipod. Most mortars are muzzle-loaded by simply dropping a round (a "bomb," with a variable number of ring-shaped propellant charges) down the tube; when the round reaches the bottom of the tube, a fixed firing pin ignites the propellant by impact. Most mortars are smoothbore (with no rifling), and their bombs are stabilized in flight by fins attached to their bases. Mortars have very low muzzle velocities and high, parabolic trajectories, which make them ideal against targets protected by frontal cover; and their bombs contain a much higher proportion of explosive than either gun or howitzer shells, because they do not have to withstand high pressures within the tube. Cheap and easy to manufacture, mortars have bore diameters ranging from 50 mm. to 240 mm., but 60-, 81/82-, and 120-mm. weapons are the most common. In general, mortars below 120-mm. are issued to infantry and not artillery forces.

ROCKET ARTILLERY was first introduced by the Soviet army during World War II as the famous *Katyusha*, or "Stalin Organ." Rockets are easy to manufacture, and because they have no recoil can be launched from lightweight frames, often mounted on trucks. Unguided, and stabilized by spinning, aerodynamic fins, or both, rockets are relatively inaccurate as compared to tube artillery, but can still be highly effective when launched in large numbers to saturate an area; even if they inflict few casualties, they can demoralize attacking troops and suppress defensive fire. The Soviet army has emphasized rocket artillery much more than most, but in recent years U.S. and NATO armies have also acquired their own weapons of this type (e.g., LARS, MULTIPLE LAUNCH ROCKET SYSTEM). Rockets such as the Soviet FROG series, and short-range ballistic missiles such as the U.S.

LANCE and the Soviet SS-21 SPIDER, which can deliver high-explosive, nuclear, or chemical warheads over ranges out to 100 km., are also operated by artillery forces.

Artillery is further categorized functionally; it was traditionally divided into field artillery, fortress and coastal artillery, and siege artillery. Fortress and coastal artillery, with heavy, fixed weapons in protected emplacements, greatly declined after the introduction of aircraft and nuclear weapons, although coastal artillery has survived in Scandinavia and has even undergone a resurgence in the form of shore-based ANTI-SHIP MISSILES. Air power also caused the virtual disappearance of siege artillery, although the Soviet army retains some very heavy howitzers and huge 240-mm. mortars suitable for use against fortifications. Field artillery, now predominant, consists entirely of mobile weapons classified as light, medium, and heavy, although light artillery (with bore diameter of 105 mm. or less), is now mostly confined to AIRBORNE, AIR MOBILE, and ALPINE infantry. Medium artillery, between 105 mm. and 155 mm., now the most numerous, is deployed to support BRIGADES, REGIMENTS, and DIVISIONS. Heavy artillery, usually assigned to CORPS, ARMY, and ARMY GROUP echelons and their Soviet counterparts (army FRONTS), is meant to support the main effort by concentrated fires.

During World War II, artillery pieces on wheeled carriages pulled by horses, or even towed by motor vehicles, could not keep up with the advance of tracked armored vehicles. To accompany tanks and mechanized infantry, guns and howitzers began to be mounted on tracked chassis as "self-propelled" artillery. When the post-1945 development of RADARS capable of detecting firing positions by extrapolating shell trajectories made COUNTERBATTERY fire a major threat, the shift from towed to self-propelled artillery accelerated, because the latter can rapidly move to new positions after firing. In addition, self-propelled weapons can be enclosed in (lightly) armored turrets, to protect their crews from small arms and artillery fire. The Soviet army was the last major one to shift from towed to self-propelled artillery, beginning this process only in 1972; the delay reflected the magnitude of the undertaking—by then the Soviet army had more artillery pieces than did all other armies combined.

Artillery of all types is organized in BATTERIES of four to eight weapons, with the associated ammunition supply and FIRE CONTROL elements. Three or four batteries form a BATTALION, while three or four battalions form an artillery regiment or brigade; only in the Soviet army are several brigades or regiments organized into artillery divisions attached at the army or Front level to form an artillery reserve. In Western armies, artillery brigades are usually incorporated into divisions, or attached to corps, armies, and army groups.

When artillery is used for DIRECT FIRE, i.e., with a direct line of sight from weapon to target ("over open sights"), it can be highly effective, because each shot can be aimed. On the other hand, the weapon will then be in plain sight of the enemy, and its range will be limited by visibility, often well below its maximum range. Only Soviet doctrine envisages the nonemergency use of direct fire; it is estimated that artillery thus used can be 3 to 5 times as effective as with indirect fire, but casualties will be 2 or 3 times as great. In most other armies, direct fire is only a last resort in the face of imminent attack.

INDIRECT FIRE is aimed at map coordinates, nowadays often provided via radio from a FORWARD OBSERVER on the ground or in the air who can see the target and transmit fire corrections. Indirect fire allows artillery to: (1) use its full range and to intervene over a corresponding portion of the battlefield; (2) avoid enemy observation and counterfire; and (3) attack enemy forces sheltered from direct fire by intervening terrain. Indirect fire can also be aimed (without forward observers) at fixed map references or grid coordinates in a set pattern, to lay down a BARRAGE. Barrage fire can be used to weaken enemy positions prior to a direct assault, or (defensively) to impede enemy movements. A "rolling barrage" moves forward along the line of advance just ahead of the assaulting forces to disrupt the enemy response. FINAL PROTECTIVE FIRE (FPF) is a barrage laid down close to defensive positions, as a final line of defense.

Counterbattery fire is aimed at enemy artillery detected by varied SURVEILLANCE and RECONNAISSANCE methods. It is more effective against towed artillery because the crews are unprotected, but can also disrupt the operations of self-propelled artillery, by forcing it to move to new positions. Hence modern artillery must normally "shoot and scoot," firing only a few rounds and then moving before the enemy can locate the battery.

The most common artillery ammunition is still high explosive (HE) shell, which inflicts damage by blast and fragmentation, and is most effective against troops in the open and other "soft" targets,

including transport and towed artillery. For self-defense, artillery units normally have shaped-charge HEAT rounds for use against armored vehicles, and CANISTER or BEEHIVE rounds for use against infantry. The great innovation has been the introduction of IMPROVED CONVENTIONAL MUNITIONS (ICM), cargo rounds for scatterable mines or bomblets whose dimensions can be sized to varied purposes; against most types of targets, ICM rounds can *multiply* the effectiveness of artillery units. Guided projectiles are now being developed for guns, howitzers, and mortars, while ROCKET-ASSISTED PROJECTILES (RAP rounds) can increase the range of tube artillery by up to 50 percent, albeit at the cost of a reduced explosive charge.

During World War I, the principal mode of delivery for CHEMICAL weapons was tube artillery; currently, rocket artillery would fulfill this role, because it could saturate large areas in a single salvo. Artillery has also been the principal mode of delivery for battlefield NUCLEAR WEAPONS, with energy yields ranging from less than 0.1 to some 5 kilotons (kT). The early nuclear-capable artillery of the 1950s was of huge dimensions (the first U.S. "nuclear cannon" was mounted on a railway car) because the warheads were very large; current nuclear shells can be fired from ordinary guns and howitzers, as well as delivered by short-range ballistic missiles. See also ARTILLERY, BRITAIN; ARTILLERY, FEDERAL REPUBLIC OF GERMANY; ARTILLERY, FRANCE; ARTILLERY, SOVIET UNION; ARTILLERY, UNITED STATES.

ARTILLERY, BRITAIN: The British army's only light artillery weapon is the L118 105-mm. Light Gun (actually a GUN-HOWITZER), which is also fielded in a self-propelled version as the FV433 ABBOT. Capable of high rates of fire, the L118 has a maximum range of 17,000 m. Abbot is being phased out, while the L118 is being otherwise replaced by the 155-mm. FH-70 (British designation L121) except in light forces. A towed 155-mm. gun-howitzer with a range of 24,000–30,000 m., the FH-70 has an automatic loader which allows a sustained rate of fire of six rounds per minute.

The principal medium artillery of the British army is now the U.S. M109A2 self-propelled 155-mm. gun-howitzer; its planned replacement, the multinational SP-70, has been canceled.

British heavy artillery consists of U.S. M110A1 203-mm. self-propelled howitzers, and M107 175-mm. self-propelled guns, fielded primarily for nuclear delivery along with U.S.-supplied LANCE short-range ballistic missiles. The British army plans to acquire U.S. MULTIPLE LAUNCH ROCKET SYSTEMS for suppression fire.

The British artillery is organized into battalion-sized (12 to 18-tube) REGIMENTS. The British Army of the Rhine includes an artillery "division" of two regiments with M107 SP guns, one regiment with M110A2 203-mm. SP howitzers, and one regiment with Lance missiles.

Technically and tactically excellent, it was the British artillery that developed the ABCA (American-British-Canadian-Australian) system of indirect artillery control, which has unparalleled flexibility (it enables the artillery forces of an entire corps or army to support any unit within their range); it was also among the first to adopt digital fire control equipment. Its current FACE system is being replaced by the Battlefield Artillery Target Engagement System (BATES), which allows more batteries to be concentrated against a single target.

British artillery pieces of World War II vintage still found in Third World armies include the excellent 25-pounder (87.5-mm.) gun-howitzer, and 17-pounder (76-mm.) anti-tank gun.

ARTILLERY, FEDERAL REPUBLIC OF GERMANY: The West German artillery has both U.S.- and German-made equipment. The main towed weapon is the FH-70 155-mm. GUN-HOWITZER; old U.S.-built M114 155-mm. HOWITZERS remain in the reserve units along with some U.S. M101 105-mm. howitzers. The M109G, which accounts for the bulk of the self-propelled (SP) artillery, is a variant of the U.S.-built M109 155-mm. gun-howitzer, while the U.S.-made M110 203-mm. SP howitzer is the only heavy piece. (The M109Gs are to be refitted with FH-70 barrels to improve their performance.) ROCKET ARTILLERY consists of the German-made LARS (Light Artillery Rocket System) 110-mm. BARRAGE rocket, while the U.S. LANCE tactical ballistic missile is reserved for nuclear delivery. The U.S. MULTIPLE LAUNCH ROCKET SYSTEM (MLRS) is now being coproduced to supplement LARS. The German army also greatly emphasizes MORTARS, and has many 120-mm. mortars, both towed and SP, organic to most maneuver battalions to provide highly responsive DIRECT SUPPORT fires.

German artillery is organized into BATTALIONS of three 6-tube batteries. Each *Panzer* and *Panzergrenadier* brigade includes one battalion of M109G SP gun-howitzers; the only *Gebirgsjager* (mountain) brigade has one battalion of Italian-made 105-mm. pack howitzers, which can be broken down into mule-portable components. Each German division includes an artillery REGIMENT with one battalion of tube artillery and one battalion of rocket artillery. Lance missiles are held at the CORPS level, with one battalion of four launchers for each corps. German Territorial Army units have older U.S.-built M114 and M101 towed howitzers.

German artillery tactics differ significantly from their U.S. and British counterparts. Because it is the brigade rather than the division that is the primary maneuver echelon, DIRECT SUPPORT artillery is concentrated at brigade, rather than divisional, level. Divisional artillery, with its heavy pieces and rocket units, is responsible for GENERAL SUPPORT as in the U.S. and British armies, but also has the COUNTERBATTERY role. FIRE CONTROL is also focused on the brigade level: fire support officers coordinate the artillery fire plan, calling on divisional artillery as needed. Because German procedures do not normally permit one brigade to control the artillery fire of another, they lack the flexibility of Anglo-American ABCA control procedures, but German doctrine holds that responsiveness at lower command levels is more important (see AUFSTRAGSTAKTIK).

ARTILLERY, FRANCE: The French army deploys a full range of modern, French-produced weapons, some of excellent technical quality, but older weapons continue to predominate in reserve and second-line units. The basic field artillery weapons are the AUF-1 155-mm. self-propelled (SP) GUN and the TR 155-mm. towed gun, which share the same barrel. Mounted on the chassis of the AMX-30 tank, the AUF-1 has a mechanical loader which allows burst fires of six rounds in 45 seconds. Each vehicle has a digital FIRE CONTROL system, an on-board ELECTRO-OPTICAL sight for direct fire, and an INERTIAL NAVIGATION unit. The TR also has a mechanical loader, and can fire three rounds in 15 seconds. Both guns have a maximum range of 30,500 m., and can fire a full range of ammunition.

The old U.S. M101 105-mm. howitzer is still issued to reserve forces, and also to the 9th Marine, 11th Airborne, and 27th Alpine divisions. Older French weapons still in service include the F3 155-mm. gun and the 105-mm. AU-50 SP howitzer, both mounted on modified AMX-13 light tank chassis. The standard medium towed gun of second-line units is the BF-50 155-mm. howitzer.

In addition to tube artillery, the French army

also has Pluton tactical ballistic missiles exclusively for nuclear delivery (French tube artillery is not nuclear-capable). Controlled directly by the army high command (though only the French president can authorize its use), Pluton is described as a "prestrategic weapon" whose use would be meant to demonstrate that further hostile action would result in full-scale nuclear retaliation. Pluton has a maximum range of 120 kilometers, and carries a warhead with a nominal yield of between 15 and 25 T. HADES, a new missile of longer range, is to replace Pluton in the 1990s.

French artillery forces are organized in battalion-sized REGIMENTS of 12–18 tubes. A Pluton regiment consists of three firing batteries, each with two missiles.

Each French armored division has one or two SP artillery regiments, with either AUF-1s or F3s, while infantry divisions have one regiment with 155-mm. towed howitzers, except for the air-portable 9th Marines, 11th Airborne, and 27th Alpine divisions, which have one regiment of lighter 105-mm. howitzers. Each French CORPS has one or two Pluton regiments, and several 155-mm. regiments, towed or SP.

The French army believes that COUNTERBATTERY fires now dominate artillery tactics, and its new 155-mm. weapons, the Atila fire control system, and various new sensors were all designed accordingly.

ARTILLERY, SOVIET UNION: From the time of Peter the Great, the Russian army has been lavishly equipped with artillery (the "King of Battles"), and has based its operational methods on massed artillery fires. Under the Soviet regime, the importance of artillery declined in the 1920s and 1930s (when mobility was emphasized), only to rise again in World War II; it declined again in the 1950s (because of a great reliance on nuclear weapons) and rose again with the denuclearization that began in the late 1960s. At present, on the eve of prospective reductions, the Soviet artillery has more than 40,000 GUNS, GUN-HOWITZERS, MORTARS, rocket launchers, and TACTICAL BALLISTIC MISSILES. Much of its equipment is modern and highly effective. As late as the 1960s, however, the bulk of Soviet artillery was still towed, and thus vulnerable to counterbattery fire and also unable to accompany rapidly advancing armored forces. From 1972, however, several types of self-propelled (SP) artillery were introduced, and much of the Soviet artillery is now self-propelled. Its major weapons are described below.

Towed Artillery

- 122-mm. howitzer D-30: The standard regimental and divisional howitzer, the D-30 was introduced in 1967 to replace World War II weapons of the same caliber. Though simple, rugged, and effective, the D-30 has a maximum range of 15,300 m., inferior to that of more recent Western 155-mm. gun-howitzers (notably the FH-70), but in the Soviet Army, the D-30 is only the lightest piece, used mainly for direct support of leading combat units.

- 122-mm. gun D-74: The D-74 is an older weapon in service with second-line reserves and Soviet-supplied armies. It is heavier than the D-30, but it has a greater maximum range of 23,900 m.

- 152-mm. gun-howitzer D-20: The standard medium howitzer of army-level artillery brigades and divisions, the D-20 was introduced in 1955, and has been much used in combat by Egypt, India, and Vietnam. It has a maximum range of 17,000 m., which can be extended to 30,000 m. with ROCKET-ASSISTED PROJECTILES (RAP); it is nuclear capable.

- 130-mm. gun M-46: An accurate long-range weapon derived from a naval gun and meant for COUNTERBATTERY fire, the M46 first appeared in 1954 and has a maximum range of 27,490 m. An M-46 regiment is organic to every Soviet artillery division, and each independent artillery brigade has an M-46 battalion. Much used in combat in Vietnam and the Middle East, the M46 has been widely exported.

- 152-mm. howitzer D-1: An older gun designed during World War II, the D-1 still serves in Soviet reserve formations and has been widely exported. It has a maximum range of 12,400 m.

- 152-mm. gun M1976: The nuclear-capable M1976 is replacing the M-46 in the towed-gun battalions of artillery regiments and brigades at army and FRONT level. Its maximum range is in excess of 27,000 m. (The same tube is used in the self-propelled 2S5.)

- 180-mm. gun S-23: The primary Soviet heavy artillery piece, introduced in 1954, the S-23 is the largest towed gun in service worldwide. Derived from a naval gun, it is issued to army and Front-level artillery formations for counterbattery fire, and for use against fortified positions. It can fire standard high-explosive shells, concrete-piercing shells, and tactical nuclear rounds. The S-23 was used successfully by the Egyptian army against Israeli Suez Canal fortifications in 1973.

- 76.2-mm. mountain gun M-1969: A light weapon issued to some eight divisions specifically trained to operate in mountain terrain, the M-1969 can be disassembled into man-portable components. It serves as a regimental artillery piece in place of

122-mm. howitzers. The M-1969 has a maximum range of 11,000 m. and can fire high-explosive, smoke, and anti-tank rounds.

Towed Anti-Tank Guns

The Soviet army is now unique in keeping large numbers of towed ANTI-TANK guns in service to provide divisions with a last line of defense against enemy armor. Of the three types in service, the newest (1965) is the T-12M, a 100-mm. smoothbore gun which fires fin-stabilized, discarding sabot (APFSDS) and high explosive anti-tank (HEAT) rounds. Accurate out to 2000 m., the T-12M can still defeat the armor of most MAIN BATTLE TANKS (MBTs). The M-1955, a rifled 100-mm. anti-tank gun introduced in 1954, fires a variety of ARMOR-PIERCING ammunition out to an accurate range of 1000 m. It is only marginally effective against the very latest MBTs. The much smaller SD-85, first introduced in 1954, is an 85-mm. rifled gun which equips the anti-tank companies of airborne regiments and air assault brigades. Though it has an effective range of at least 1000 m. against LIGHT TANKS, INFANTRY FIGHTING VEHICLES, and other light ARMORED FIGHTING VEHICLES, the SD-85 is only marginally effective against MBTs.

Self-Propelled Artillery

The Soviet army fielded both ASSAULT GUNS and TANK DESTROYERS during World War II, but did not deploy SP field artillery until the early 1970s. Since then, the following SP weapons have been introduced:

- 122-mm. SP howitzer 2S1: A D-30 howitzer in a lightly armored turret mounted on a chassis derived from the MTLB artillery tractor, the 2S1 equips the howitzer battalions of first-line Motorized Rifle regiments, and of some tank regiments. The 2S1 is amphibious, and has enough mobility to accompany armored formations across country. Although it has a maximum range of 15,300 m., the 2S1 would often be used for direct fires; it was first seen in 1974 and was much used in Afghanistan.
- 152-mm. SP howitzer 2S3: A D-20 howitzer in a lightly armored turret mounted on chassis derived from the PT-76 light tank, the 2S3 has replaced the D-20 in the artillery regiment of Motorized Rifle divisions, one of the D-30 battalions of each tank division, and the D-20s of artillery regiments at

army and Front level. Maximum range is 17,300 m. (30,000 with RAP rounds).
- 152-mm. SP gun 2S5: Introduced in 1981, the 2S5 is mounted (without an enclosed turret) on a tracked chassis also derived from the PT-76; the tube itself is identical to that of the towed M1976. The 2S5 is replacing the M-46 in artillery regiments and brigades at army and Front levels. It is nuclear capable, and its range exceeds 27,000 m.
- 203-mm. SP gun 2S7: The nuclear-capable 2S7 is mounted (without an enclosed turret) on a heavy tracked chassis. With a maximum range of more than 30,000 m. it equips heavy artillery units at the army and Front level.

Mortars

The Soviet army has a wide variety of medium and heavy mortars for direct support at the battalion level, and also for ultraheavy fire support at army and Front level. The basic types in service include the 120-mm. mortar M1943, the 160-mm. mortar M-160, the 240-mm. mortar M-240, and the 240-mm. SP mortar M1975. The latter is nuclear capable and has a maximum range of 9700 m. All Soviet heavy mortars can also fire chemical ammunition. Their high rates of fire make them ideal for suppression and for the direct support of assault echelons. Each Soviet Motorized Rifle battalion has a company of six 120-mm. mortars.

Rocket Artillery

The Soviet Union pioneered the use of barrage rockets in World War II (the famous *Katyusha*). Soviet multiple rocket launchers include:

- 122-mm. BM-21: The standard multiple rocket launcher (MRL) in tank and Motorized Rifle divisions, the BM-21 consists of 40 launch tubes mounted on the back of an unarmored 6 × 6 truck. Its high-explosive, chemical, or smoke rockets, with a range of 20,000 m., are unguided and spin stabilized. They are thus relatively inaccurate, but they are when fired in salvo, a single BM-21 can saturate an area of roughly one square kilometer. Widely exported, the BM-21 has been much used in combat, in Southeast Asia, the Middle East, and Afghanistan.
- 220-mm. BM-27: First seen in 1977, the BM-27 consists of 16 launch tubes mounted on a special-purpose 8 × 8 transporter. With an extraordinary

range of 35,000–40,000 m., the BM-27 is slowly replacing BM-21s in divisional MRL battalions and Soviet artillery divisions.

- 240-mm. BM-24: No longer in service with first-line Soviet units, but still widely fielded by Soviet-supplied armies (and once produced without license in Israel), the BM-24 consists of 12 open frame launchers mounted on a 6 × 6 truck. Its high-explosive, smoke, or chemical rockets have a maximum range of 11,000 m.
- 140-mm. BM-14: No longer in service with the Soviet army, the BM-14 consists of 16 launch tubes mounted on a 6 × 6 truck. The rockets have a maximum range of 9800 m.
- 140-mm. RPU-14: A towed, lightweight version of the BM-14, the RPU-14 is issued to the Soviet Airborne Troops and the Airborne Assault Brigades. The Polish 6th Airborne Division has an even lighter, 8-tube version.

Tactical Ballistic Missiles

The Soviet army's long-range FROG rockets and SS-21 SPIDER short-range tactical BALLISTIC MISSILES are mainly fielded to deliver chemical and nuclear warheads deep into the enemy rear.

ORGANIZATION

The Soviet artillery is generally organized in BATTALIONS of three 6-tube (or -launcher) BATTERIES. Every Motorized Rifle REGIMENT has a battalion of D-30 122-mm. howitzers, except for regiments equipped with the BMP infantry fighting vehicle, each of which has a battalion of 2S1 122-mm. SP howitzers. Likewise, every Soviet DIVISION has an organic artillery regiment of two battalions of 122-mm. howitzers; one battalion of 2S3 152-mm. SP howitzers; and a battalion of BM-21 or BM-27 multiple rocket launchers. In addition, each division has an anti-tank gun battalion equipped with 100-mm. T-12M guns. The artillery organization of Soviet numbered armies is flexible but usually includes at least one artillery regiment or brigade, a rocket launcher regiment, and a tactical ballistic missile brigade. Army artillery regiments consist in turn of two gun battalions with 130-mm. or 152-mm. guns, and two battalions of 152-mm. howitzers (either towed or SP). Rocket launcher regiments have three BM-21 or BM-27 battalions, while a tactical ballistic missile brigade

consists of three battalions, each with six launchers.

The organization of Front artillery is also variable, but generally includes an artillery division, a heavy artillery brigade, and several tactical ballistic missile brigades. An artillery division consists in turn of two regiments with 130-mm. or 152-mm. guns, two regiments with 152-mm. howitzers, two anti-tank regiments, each with three battalions of T-12Ms, and a rocket launcher brigade of three BM-21/27 battalions. The heavy artillery brigade has two battalions of 203-mm. howitzers and two battalions of 240-mm. SP mortars. Each tactical ballistic missile brigade has three battalions, each with six launchers for SS-21 missiles.

TACTICS

Soviet artillery tactics used to emphasize massed, preplanned barrage fires with densities as high as 400 tubes per kilometer of front. The introduction of self-propelled artillery during the 1970s was associated with the increased fluidity of Soviet operational methods, particularly the use of OPERATIONAL MANEUVER GROUPS (OMGs) deep in the enemy rear—a method now presumably abandoned under the new Gorbachev-era defensive doctrine.

ARTILLERY, UNITED STATES: Gen. George Patton once said, "The worse the infantry, the more artillery it needs. American infantry needs a lot of artillery." In fact, the U.S. ended World War II with artillery technically more advanced and tactically more proficient than that of any other belligerent. Today, the U.S. Army still relies very heavily on its artillery.

WEAPONS

The army's infantry, airborne, and light divisions, as well as the Marine Corps, are equipped with towed artillery; armored and mechanized divisions, and corps and army artillery groups are equipped with self-propelled (SP) artillery. Until fairly recently, the U.S. Army was uninterested in multiple rocket launchers (MRLs), but has now acquired the advanced MULTIPLE LAUNCH ROCKET SYSTEM (MLRS). The army also has LANCE short-range ballistic missiles for nuclear delivery. Otherwise modern and effective, U.S. artillery has lacked the automatic loaders, inertial navigation systems, and digital fire control displays of the latest European weapons; "product improvement" programs are now being implemented to retrofit those devices.

Towed Artillery

- 105-mm. HOWITZER M102: A light, air-portable weapon, the M102 equips the artillery battalions of AIRBORNE, AIR ASSAULT, and LIGHT INFANTRY divisions, as well as U.S. Marine Corps forces. Over a maximum range of 15,000 m., it can fire high-explosive (HE), illuminating, chemical, HEAT, and anti-personnel BEEHIVE rounds. It is mainly a DIRECT SUPPORT weapon for use in the absence of heavier artillery pieces. Much used during the Vietnam War, it has also been acquired by the Brazilian army.
- 155-mm. GUN-HOWITZER M198: A relatively light weapon, the M198 also equips the artillery of infantry, airborne, air mobile, and light divisions, as well as marine forces. First deployed in 1979 to replace the M114 howitzer, the M198 has a maximum range of 18,000 m., which can be extended to 24,000 m. with a special "supercharge" cartridge, or to 30,000 m. with ROCKET-ASSISTED PROJECTILES (RAP). It can fire HE, chemical, illuminating, anti-personnel, smoke, and guided projectiles, and IMPROVED CONVENTIONAL MUNITIONS (ICMs); it is also nuclear capable.
- 155-mm. howitzer M114A1: The M114A1 is a World War II–vintage weapon still found in U.S. reserve formations and widely used by U.S.-supplied armies. It is heavier than the M198 and has a shorter range (14,600 m.); in most combat conditions, however, it remains an effective weapon.
- Older U.S. towed artillery pieces still in service around the world include the 105-mm. howitzer M101; the 203-mm. howitzer M115; and the "Long Tom" 155-mm. gun M2.

Self-Propelled Artillery

The U.S. Army has fielded SP artillery since early in World War II, when "Armored Artillery" provided direct support to armored divisions. SP artillery in service includes:

- 155-mm. SP gun-howitzer M109A2/A3: One of the most widely fielded artillery weapons worldwide, the M109 originally had the short-barreled howitzer of the M114. From 1966, however, the U.S. Army retrofitted an extended barrel to the M109, converting it to a gun-howitzer which remains the standard medium artillery piece of armored and mechanized divisions, and of independent artillery battalions assigned to corps and army artillery groups. Enclosed in a lightly armored turret mounted on a fully tracked vehicle, the gun has a maximum range of 18,000 m., which can be extended to 24,000 m. with supercharge cartridges, or 30,000. with RAP; it can fire HE, chemical, smoke, illuminating, anti-personnel, ICMs, and guided projectiles, but there is no HEAT round for use against tanks. The M109 is nuclear capable.
- 203-mm. gun-howitzer M110A2: Originally, the U.S. Army fielded two separate heavy artillery pieces, the M110 howitzer and the 175-mm. M107 gun, mounted on identical tracked vehicles. The M107 could fire 66-kg. shells at ranges of more than 37,000 m., but was inaccurate and subject to rapid barrel wear. The army therefore decided to replace M107 tubes with a long-barrel version of the M110, which entered service in 1977 as the M110A1, later adding muzzle brakes and other improvements under the designation M110A2. The weapon is mounted on a fully tracked vehicle but is not enclosed in a turret, so that the crew is unprotected, though a KEVLAR splinter cover can be erected. Maximum range is more than 30,000 m. (or 40,000+ with RAP rounds). With HE, antipersonnel, ICM, chemical, and nuclear rounds, the M110 equips the heavy artillery battalions of corps and army artillery groups, and also serves with Belgium, Britain, South Korea, Iran, Israel, and West Germany.
- Older U.S.-made SP weapons still in service include the 175-mm. gun M107, the 105-mm. SP howitzer M108, the 155-mm. SP howitzer M44, and the 105-mm. howitzer M52.

Rocket Artillery

The U.S. Army developed a family of 4.5-in. barrage rockets in World War II. Although short-ranged and inaccurate, these "Screaming Meemies" mounted on tanks, half-tracks, and other vehicles, proved valuable for area suppressive fires. After 1945, U.S. interest in such weapons waned, and no U.S. rocket artillery (other than large nuclear-armed "free rockets" such as the HONEST JOHN) was fielded until the early 1980s, when the M993 MULTIPLE LAUNCH ROCKET SYSTEM (MLRS) was introduced. With 12 227-mm. launcher tubes in an armored box mounted on top of a lightly armored, fully tracked derivative of the BRADLEY fighting vehicle, the MLRS is completely self-contained, with its own fire control and communications equipment; as such, it is ideally suited for "shoot and scoot" tactics. The rockets are both spin and fin stabilized, and have ranges in excess of 40,000 m., with a wide variety of warheads including smoke, chemical, HE, and ICM. Each armored and mechanized division has a battery of nine MLRS, with additional weapons at corps and army level.

Mortars

The U.S. Army now has no heavy ("artillery") mortars. Its heaviest mortar is the old 4.2-in. (107-mm.) M30, but a 120-mm. mortar is now being acquired as a battalion direct support weapon by both the army and Marine Corps (See MORTARS, UNITED STATES).

Tactical Ballistic Missiles

The U.S. Army now has only the Lance tactical ballistic missile; with a maximum range of 120 km., it can deliver a 10 kT nuclear warhead or a 1000-lb. (454-kg.) cluster or chemical warhead. It serves in corps artillery groups, while the longer-ranged PERSHING 1A and Pershing 2 missiles have been withdrawn under the terms of the 1987 INF TREATY. The Army Tactical Missile System (ATACMS), a short-range ballistic missile for the delivery of ICM warheads, is now under development, as is a nuclear-armed "Follow-On To Lance" (FOTL) missile.

ORGANIZATION

U.S. tube artillery is organized into BATTALIONS of 3 (6- or 8-tube) BATTERIES; MLRS batteries have 9 launchers. Each U.S. armored and mechanized DIVISION has an artillery BRIGADE of 3 medium battalions with M109s, and a heavy battalion with 2 batteries of M110A2s and 1 of MLRS. Infantry and light divisions have a brigade of 3 M102 battalions and one M198 battalion, while the airborne division only has 3 M102 battalions. In practice, additional artillery is frequently attached to divisions from CORPS or ARMY echelons, to allow a battalion of artillery to be assigned to each maneuver brigade for direct support. At the corps and army level, U.S. artillery organization is quite flexible, but usually includes 2 or more artillery brigades. Each corps level brigade includes several battalions of M110s and M109s, and one or more battalions with Lance.

TACTICS

U.S. artillery tactics are based on the ABCA (America-Britain-Canada-Australia) method of artillery control, which permits the fire of many batteries to be concentrated against a single target. Thus the batteries can remain dispersed (and less vulnerable to counterbattery fire) while still being able to concentrate their fire. Artillery fire is controlled by FIRE DIRECTION CENTERS (FDCs) which function down to the battalion level, and receive their fire missions from FORWARD OBSERVERS (FOs) assigned down to the company level. Battalion FDCs can call upon their own battalion mortars and on the direct support batteries of their parent brigade. If additional fire support is required, the brigade FDC can request it from divisional artillery. If even more is required, divisional FDCs can call on corps and army artillery. This process can be very rapid: most fire calls take no more than two minutes from request to delivery. This is made possible by the prior establishment of integrated fire support plans from corps down to battalion FDCs and by the TACFIRE digital FIRE CONTROL system, which can plot the fire of dozens of batteries against any one of more than 1000 individual targets. To avoid COUNTERBATTERY fires, the U.S. has adopted "shoot-and-scoot" tactics, whereby a battery fires one or two salvos at a target, and then displaces to a new firing position before it can be located. But the ABCA system, and the essential communications between FOs and FDCs, could be completely disrupted by ACTIVE JAMMING and other ELECTRONIC COUNTERMEASURES. The U.S. Army has thus devoted considerable efforts to develop jam-resistant communications (the centrally controlled Soviet method, while less flexible than ABCA, is much less vulnerable to jamming). The U.S. artillery has also tried to increase its ANTI-TANK capabilities by developing the laser guided COPPERHEAD projectile and improved conventional munitions, notably RAAMS (Remote Anti-Armor Mine System) and SADARM (Search-and-Destroy Armor).

Nuclear artillery rounds range in yield from 0.5 to 10 kT, and include ENHANCED RADIATION (ER) or "neutron" weapons for use against armored forces. The release of nuclear weapons can only be authorized by the U.S. NATIONAL COMMAND AUTHORITY (NCA); a request for the use of nuclear artillery would have to pass up from division, through corps and army, to the NATO supreme command and finally to the NCA, in a process that might take as long as twelve hours. See also AIR-LAND BATTLE; ARTILLERY; ASSAULT BREAKER.

AS-: NATO designation for Soviet air-to-surface missiles (ASM), including long-range CRUISE MISSILES, tactical missiles, and ANTI-RADIATION MISSILES. Models in service include: AS-2 KIPPER; AS-4 KITCHEN; AS-5 KELT; AS-6 KINGFISH; AS-7 KERRY; AS-9 KYLE; AS-10 KAREN; AS-11 KILTER; AS-12 KEGLER; AS-14 KEDGE; and AS-15 KENT.

ASAGIRI: A class of eight Japanese DESTROYERS commissioned between 1988 and 1991. Improved versions of the HATSUYUKI class, they are optimized for ANTI-SUBMARINE WARFARE (ASW), but also have a considerable ANTI-SURFACE WARFARE (ASUW) capabilities. Flush-decked, the Asagiris have a sharply raked bow, a graceful shear line, a blocky superstructure forward, and a large, separate deckhouse/hangar aft. A U.S. Prairie Masker bubble generator is fitted to the hull under the engine rooms, to minimize radiated noise.

Armament consists of an OTO-Melara 76.2-mm. DUAL PURPOSE gun at the bow, an 8-round ASROC launcher behind the gun, 2 20-mm. PHALANX radar-controlled guns for anti-missile defense on the forward superstructure, 2 quadruple launchers for HARPOON anti-ship missiles amidships, 2 sets of Mk.32 triple tubes for lightweight ASW homing torpedoes (also amidships), and an 8-round launcher for NATO SEA SPARROW short-range surface-to-air missiles on the fantail. An HSS-2B SEA KING ASW helicopter is operated from a raised landing deck just forward of the Sea Sparrow launcher. See also DESTROYERS, JAPAN.

Specifications Length: 447.72 ft. (136.5 m.). **Beam:** 47.9 ft. (14.6 m.). **Draft:** 14.6 ft. (4.45 m.). **Displacement:** 3400 tons standard/4200 tons full load. **Powerplant:** twin-shaft COGAG, 4 13,500-shp. Kawasaki-Rolls Royce Spey SM1A gas turbines. **Speed:** 30 kt. **Crew:** 230. **Sensors:** 1 OPS-28C air and surface-search radar, 1 OPS-14C air-search radar, 2 FCS-2 fire control radars, 1 OQS-4 hull-mounted low-frequency sonar, 1 SQR-18 TACTASS passive towed array sonar. **Electronic warfare equipment:** 1 OLR-9C electronic signal monitoring array, 1 OLR-6C active jamming array, 1 OLT-3 radio direction-finder (D/F), 2 Mk.36 SRBOC chaff launchers.

ASALM: Advanced Strategic Air Launched Missile, a supersonic, long-range CRUISE MISSILE developed in the late 1970s by Martin Marietta for the U.S. Air Force (USAF). Intended to replace both the AGM-69 SRAM and AGM-86B ALCM on strategic bombers in the late 1980s (when Soviet AIR DEFENSES were expected to become more effective against subsonic cruise missiles), ASALM was a wingless supersonic lifting body with small cruciform tailfins for stability and control.

Powered by a revolutionary integral rocket-ramjet engine, ASALM would have been armed with a 200-kT nuclear warhead. Several guidance systems were proposed, including passively updated INERTIAL GUIDANCE, ANTI-RADIATION homing, and active radar terminal homing.

After ASALM's 1981 cancellation, much of the engineering research was carried over to the SRAM II and Advanced Cruise Missile programs. Martin Marietta continues to develop the airframe as the BQM-127A Supersonic Air-Launched Target (SLAT), as a proposed replacement for the HARPOON anti-ship missiles, and as a long-range air-to-air missile. See also KENT; TOMAHAWK.

Specifications Length: 14 ft. (4.27 m.). **Diameter:** 25 in. (635 mm.). **Weight, launch:** 2700 lb. (1227 kg.). **Speed, max.:** Mach 4 (2600 mph/4342 kph). **Range, max.:** 1500 mi. (2500 km.).

ASAT: See ANTI-SATELLITE SYSTEMS.

ASDIC: British term for SONAR.

ASH (AA-5): NATO code name for a large Soviet air-to-air missile. First seen in 1961, Ash was developed specifically for the Tu-28P FIDDLER long-range, all-weather interceptor, though it was also carried by early versions of the MiG-25 FOXBAT.

Ash has four cruciform delta wings fitted with ailerons for roll control, and smaller cruciform tailfins for steering. Powered by a solid-fuel rocket engine, Ash is available in both SEMI-ACTIVE RADAR HOMING (SARH) and INFRARED-HOMING versions; Fiddlers usually carry two of each type on underwing pylons. The SARH versions require target illumination from the Fiddler's powerful "Big Nose" I-band radar, while the IR version has a relatively simple seeker, and is thus limited to rear-quarter attacks. Designed to attack large intercontinental bombers, Ash is ineffective against agile fighters. It was replaced on Foxbats by the AA-6 ACRID.

Specifications Length: 18 ft. (5.5 m.). **Diameter:** 12 in. (305 mm.). **Span:** 51 in. (1.3 m.). **Weight, launch:** 860 lb. (390 kg.). **Speed, max.:** Mach 3 (2000 mph/3400 kph). **Range, max.:** (SARH) 35 mi. (55 km.); (IR) 13 mi. (21 km.).

ASM: 1. ANTI-SHIP MISSILE.
2. Air-to-Surface Missile.

ASMP: *Air-Sol Moyenne Portée* (Air-to-Surface, Medium-Range), a French air-launched nuclear standoff missile deployed on MIRAGE IV and MIRAGE 2000 bombers of the French air force, and Super ETENDARD attack aircraft of French naval aviation *(Aeronavale)*. Developed from 1971 by Aerospatiale, ASMP was flight-tested in June 1983 and entered service in 1986. Roughly equivalent to the U.S. AGM-68 SRAM, ASMP is meant to allow French bombers to strike heavily defended targets without actually approaching the air defenses around them.

Relying entirely on body lift, ASMP has only four small tail fins for steering. The missile incorporates some "low observable" (STEALTH) features, including radar-absorbant materials, and is hardened against ELECTRO-MAGNETIC PULSE (EMP). Armed with a 300-kT nuclear warhead, ASMP is controlled by INERTIAL GUIDANCE and a radar-altimeter, and can be programmed to fly one of three alternative flight profiles: high-altitude with a terminal dive; low-altitude CONTOUR FLYING; or low altitude with a terminal "pop-up" maneuver for airburst detonation.

ASMP is powered by a unique integral rocket-ramjet: at launch, a solid-fuel rocket burns for 5 seconds to accelerate the missile to ramjet-ignition speed (Mach 2); the rocket nozzle is then jettisoned, two wedge-shaped air intakes are opened, and the empty rocket casing serves as the combustion chamber for a kerosene-fueled ramjet. That eliminates the need for heavy tandem or strap-on boosters.

Specifications Length: 17.67 ft. (5.38 m.). **Diameter:** 32.2 in. (820 mm.). **Weight, launch:** 1851 lb. (840 kg.). **Speed, max.:** Mach 4 (2600 mph/ 4342 kph). **Range, max.:** (low altitude) 50 mi. (80 km.); (high altitude) 155 mi. (250 km.).

ASPIDE: Italian medium-range, radar-guided air-to-air missile, also available in a ship-launched variant. Developed from 1969, Aspide is nearly identical to the U.S.-built AIM-7 SPARROW, but has an Italian-made rocket motor and guidance electronics, and various detailed differences. Flight tests began in 1974, full-scale production began in 1978, and Aspide entered service with the Italian air force in 1981 as the primary armament of F-104S STARFIGHTERS.

Like Sparrow, Aspide has four delta wings at midbody for lift and steering, and four smaller fixed tailfins. The nose houses a radar receiver antenna, the guidance electronics, and a 77.2-lb. (35-kg.) high-explosive fragmentation warhead

with both proximity and contact fuzes, and the missile is powered by an advanced, single-pulse solid-rocket motor.

Aspide guided by SEMI-ACTIVE RADAR HOMING (SARH), whereby the missile homes on radiation reflected off the target by the launch aircraft's I-band monopulse radar; it is claimed that this system has better resistance to jamming and other ELECTRONIC COUNTERMEASURES than the latest AIM-7M version of Sparrow. Aspide also has a backup HOME-ON-JAM guidance mode.

Aspide can also be launched from 8-round *Albatros* Mk.2 shipboard box launchers, generally equivalent to the NATO SEA SPARROW launchers, but with 16-round below-decks magazines (a lighter 4-round launcher is under development). The Naval Aspide is essentially unchanged, except for cropped wings. In service with the Italian navy from 1979, the *Albatros* system has also been exported to Argentina, Columbia, Ecuador, Egypt, Greece, Iraq, Libya, Morocco, Nigeria, Peru, Spain, Thailand, and Venezuela. From 1984, an ACTIVE-HOMING variant *(Idra)* has been under development as a lower-cost alternative to the U.S.-built AIM-120 AMRAAM. For shipboard use, *Idra* will be launched from vertical launchers.

Specifications Length: 12.14 ft. (3.7 m.). Diameter: 8 in. (203 mm.). **Span:** (Aspide) 39.37 in. (1 m.); (Naval Aspide) 31.5 in. (800 mm.). **Weight,** launch: 485 lb. (220 kg.). **Speed, max.:** (Aspide) Mach 4 (2600 mph/4342 kph); (Naval Aspide) Mach 2.5 (1900 mph/3173 kph). **Range, max.:** (Aspide) 62 mi. (100 km.); (Naval Aspide) 11.5 mi. (18.5 km.). **Height envelope:** (Naval Aspide) 50–16,405 ft. (15–5000 m.).

ASRAAM: The AIM-132 Advanced Short Range Air-to-Air Missile. Under development by the U.S.-U.K.-German BBG consortium of *Bodenseewerke Geratetechnik*, British Aerospace, and General Dynamics to replace the AIM-9 SIDEWINDER, ASRAAM's progress has been slow because of the continued upgrading of the Sidewinder, and the higher priority of the long-range, radar-guided AIM-120 AMRAAM. West Germany withdrew from the program in 1989, leaving its future uncertain.

Shorter and lighter than Sidewinder, ASRAAM is a wingless lifting body configuration with only four folding tailfins for steering. Speed and maneuverability are reported to be better than in late-model Sidewinders. ASRAAM would reportedly be controlled by INERTIAL GUIDANCE in the early phases of flight, with an advanced IMAGING

INFRARED seeker for terminal homing—a combination that allows the missile to lock on after launch, and could also reduce the missile's vulnerability to flares and other INFRARED COUNTERMEASURES. Should it actually reach the production stage and operate according to specification, it would supplement and eventually replace the AIM-9.

Specifications Length: 8.2 ft. (2.5 m.). Diameter: 5.9 in. (150 mm.). **Span:** 17.7 in. (450 mm.). **Weight, launch:** 187 lb. (85 kg.). **Range, max.:** 6 mi. (10 km.).

ASROC: The U.S. Navy's RUR-5A Anti-Submarine Rocket, a ship-launched, solid-fuel rocket which delivers either a MK.46 homing torpedo, or a Mk.17 nuclear DEPTH CHARGE. Developed from the late 1950s as an all-weather backup for the unsuccessful DASH (Drone Anti-Submarine Helicopter), ASROC entered service in 1962 and has since become the U.S. Navy's standard medium-range ANTI-SUBMARINE WARFARE (ASW) weapon; it has also been widely exported. Some 12,000 AS-ROCs were produced through 1970.

ASROC consists of a blunt nose section which houses the payload, and a solid-rocket booster; two sets of folding cruciform fins provide stabilization. ASROC is launched from Mk.112 8-round pepperbox launchers, Mk.10 TERRIER launchers, or Mk.26 STANDARD launchers.

In action, the ship's FIRE CONTROL system receives target data from sonar, calculates the time of flight to the vicinity of the target, and controls the range by elevating the launcher and setting the cutoff time for the rocket motor. After launch, ASROC flies a parabolic trajectory; when the rocket shuts down, the torpedo or depth bomb separates, and a braking parachute opens to slow its descent. Once in the water, the torpedo would initiate an autonomous search pattern, while the nuclear depth charge would be detonated by a hydrostatic or time fuze.

The principal drawback of ASROC is "dead time": because the rocket cannot change course, the target can maneuver in the time between launch and impact (as much as 30 seconds), to move beyond the search range of the torpedo, or the lethal radius of the depth bomb. ASROC was to be replaced by the SEA LANCE ASW Standoff Weapon, a missile with longer range and mid-course guidance updates from the ship to minimize dead time. But Sea Lance encountered technical problems, and it was decided instead to develop Vertical Launch ASROC (VLA) as an interim weapon. VLA is simply a standard ASROC with a

vectored-thrust nozzle, which can be launched from the Mk.41 VERTICAL LAUNCH SYSTEMS of SPRU-ANCE- and BURKE-class destroyers and TICON-DEROGA-class cruisers. Length and weight are increased slightly, but performance is otherwise unchanged. VLA is now several years behind schedule, and the capability of ASROC against the newest Soviet submarines is dubious. See also IKARA; MALAFON; SILEX.

Specifications Length: (ASROC) 14.79 ft. (4.51 m.); (VLA) 16.66 ft. (5.08 m.). **Diameter:** 13.25 in. (325 mm.). **Span:** 2.77 ft. (845 mm.). **Weight, launch:** (ASROC) 1073 lb. (487 kg.); (VLA) 1650 lb. (750 kg.). **Speed, max.:** Mach 0.8 (560 mph/960 kph). **Range, min.:** 900 m. **Range, max.:** 10,000 m.

ASSAULT BREAKER: A proposed U.S. battlefield INTERDICTION concept of the 1970s which was meant to enhance NATO's strength by fielding advanced, nonnuclear systems capable of attacks deep behind the front. As in the later "Follow-on Forces Attack" (FOFA), the aim was to hold off the second and third echelons of a Soviet offensive before they could reach the front lines. To do this, Assault Breaker was to combine three elements: airborne sensors to locate and classify targets as small as tank platoons several hundred kilometers behind the front; a command, control, and communications (c^3) system to process this information in "real time"; and delivery systems with the range and accuracy to attack those targets with a variety of IMPROVED CONVENTIONAL MUNITIONS (ICMs).

The proposed sensor, a large, airborne SYN-THETIC APERTURE RADAR originally known as "Pave Mover," has now been developed as J-STARS. Orbiting at a safe distance behind the battlefield, J-STARS can detect both moving and stationary man-made objects even (it is claimed) when hidden beneath camouflage or trees. For the weapon element, the T-22 tactical ballistic missile (based on LANCE) was developed to deliver a variety of ICMs, including SKEET and TERMINALLY GUIDED SUBMISSILES both capable of autonomously homing on armored vehicles and other targets. Other potential ICMs for the T-22 included ERAM (Extended Range Anti-armor Mine) and a variety of air-delivered MINES and CLUSTER BOMBS. Targets for Assault Breaker would have included bridges, tunnels, and other chokepoints; fuel and munition storage areas; airfields; and ARMORED FORCES. A product of the U.S. Army's ATTRITION-based operational methods of the 1970s, and now strategically obsolete as well, Assault Breaker nevertheless lives on in the equipment it has spawned.

ASSAULT GUN: An ARTILLERY piece mounted within an armored casemate on a tank chassis for DIRECT FIRE. Much used by the German and Soviet armies in World War II (partly to utilize obsolete light tanks, by removing their turrets to fit much heavier weapons instead), assault guns were armed with low-velocity howitzers to provide mobile fire support to infantry forces. Anti-tank guns were also mounted on up-armored tank chassis to create TANK DESTROYERS, the most powerful of which were the German *Jagdpanzer* (Hunting Panther), and the Soviet SU (*Samokhodnaya ustanovka*, self-propelled mount) series.

Although they lacked 360° traverse, assault guns had a lower silhouette, and often heavier armor, than equivalent tanks, and were also cheaper to produce. The Western Allies never developed this type of weapon, while the Soviet Union continued to produce assault guns (as well as tank destroyers) after 1945, notably the the ISU-130 and ISU-152. Since the 1960s, however, Soviet assault guns have been phased out except for the ASU-85 of Soviet airborne divisions.

ASSAULT RIFLE (AR): Now the prevalent form of military rifle, assault rifles fire intermediate cartridges smaller than traditional rifle rounds but much more powerful than the pistol rounds of SUBMACHINE GUNS. The assault rifle was first developed by the German army in World War II to obtain semi- and full-automatic infantry fire at ranges longer than the 100 m. or so of submachine guns. Battle experience had long since shown that: (1) the typical infantryman was incapable of aimed fire at ranges much over 200 m.; (2) large volumes of automatic fire could suppress enemy fire and movement out to 400 m.; and (3) infantrymen with automatic weapons were much more likely to fire in the first place. But marksmen-dominated bureaucracies remained attached to full-power rifles, and only the Soviet army generally adopted assault rifles soon after 1945. Western armies did not abandon traditional, full-power (7.62-mm.) rifles until the 1960s, when assault rifles firing smaller, higher-velocity 5.56-mm. rounds were adopted by the U.S. and later its allies. The Soviet army in turn switched from "short" 7.62-mm. to 5.45-mm. rounds in the 1980s.

Almost all assault rifles have selectable rates of fire, from single shot to full automatic, and some, such as the U.S.-made M16A2, can fire three-round bursts. Most assault rifles replicate conventional forms, but the foreshortened BULLPUP configuration (with the receiver behind the trigger) is gaining ground. Automatic rifles currently in service include the U.S. M16-series, the Soviet AK series, the German G3, the Israeli GALIL, and the British ENFIELD L85A1.

ASSURED DESTRUCTION: A concept associated with the 1960s policies of former U.S. Secretary of Defense Robert S. McNamara, officially repudiated since the 1970s, but still in fact authoritative, at least for the U.S. Congress. The goal of Assured Destruction was to limit the rapid growth of U.S. strategic-nuclear offensive forces by establishing a criterion of sufficiency: any deliberate nuclear attack on the United States and its allies was to be deterred by "maintaining, continuously, a highly reliable ability to inflict an unacceptable degree of damage upon any single aggressor, or combination of aggressors, at any time during the course of a strategic nuclear exchange, even after absorbing a surprise first strike." The required degree of damage was officially assessed in the mid-1960s as "about one-fifth to one-third of the population and one-half to two-thirds of the industrial capacity" of the Soviet Union. Translated into nuclear warheads deliverable on target, that was in turn said to equate to some 400 "equivalent megatons" (EMTs) to meet the population requirement (the industrial capacity requirement was also thereby met). But 400 EMTs did not mean 400 warheads of one-megaton yield each, because two or more smaller MULTIPLE INDEPENDENTLY TARGETED REENTRY VEHICLES (MIRVs) can inflict more than one megaton's worth of damage; while on the other hand, the required number of weapons must allow for maintenance and reliability detractions, the effect of a prior FIRST STRIKE attack, and any losses to BALLISTIC MISSILE DEFENSES (BMD) as well as AIR DEFENSES in place. Considering only intercontinental ballistic missiles (ICBMS), "degradation factors" listed in open sources imply that 400 EMTs on target would require a force of 1081–2222 ICBMs with one-megaton warheads (see table A1). Even if interpreted conservatively, Assured Destruction thus provides a logical framework to set limits on weapon requirements. It is also implicitly an offense-only formula, because less than totally reliable defenses cannot be "assured," while any form of BMD interferes with the stabilization of strategic inventories.

On the basis of this logic, it became U.S. policy after 1963 to essentially freeze the U.S. strategic-nuclear inventory, and virtually encourage the So-

TABLE A1

Assured Destruction Strike Degradation Factors

Degradation Factor	Estimated Values (expressed as demicals)	
	Nominal	Conservative
Missiles available on station	0.9	0.8
Missiles ready to fire	0.9	0.8
Missiles that launch correctly	0.9	0.8
Missiles that release RVs correctly	0.9	0.8
Fraction surviving all engineering hazards	0.66	0.41
Fraction surviving a Soviet first strike	0.49	0.27
Fraction penetrating Soviet BMD	0.37	0.18

viet Union to catch up; at the same time, U.S. policymakers loudly deplored the then-continued development of Soviet ballistic missile defenses. The stability of a situation in which both sides would have assured destructive capabilities (Mutual Assured Destruction, or MAD), clearly fascinated U.S. policymakers; but in fact the concept rested on a combination of transient technological conditions, specifically relatively inaccurate ICBMs without MIRVs. The development of highly accurate ICBMs with MIRVs (now the SS-18 SATAN, SS-19 STILETTO, and MX PEACEKEEPER), which can have high kill probabilities even against hardened missile silos, made the offense-only MAD posture questionable, because only some form of ballistic missile defense can absorb a substantial proportion of "first strike" capabilities, thus averting an ARMS RACE by way of numbers or mobility, or both. Moreover, the logic of Assured Destruction assumed symmetrical goals and values on both sides, while in fact Soviet planners continued to reject "sufficiency" and thus Assured Destruction until the Gorbachev revolution; instead, deterrence was seen by Soviet planners as a derivative of war-fighting/war-winning capabilities. Hence, while the number of U.S. strategic delivery systems remained fairly constant from 1967 (though their quality did not), the number of Soviet nuclear delivery systems continued to increase (while also improving qualitatively), to far exceed U.S. levels by the late 1970s. See also ABM TREATY; BALANCE OF TERROR; DETERRENCE; ICBMS, UNITED STATES; PREEMPTIVE STRIKE; RVSN

ASU-: *Aviadesantnaya samokhodnaya ustanovka,* literally, airborne self-propelled (gun)

mount, the designator of two airborne TANK DESTROYERS, the ASU-57 (superseded by the BMD) and ASU-85, in service with the Soviet Airborne Troops. See also AIRBORNE FIGHTING VEHICLES.

ASU-85: A Soviet Airborne ASSAULT GUN, in service with the anti-tank units of Soviet (and one Polish) airborne divisions. First spotted in 1962, the ASU-85 is derived from the chassis and drive train of the ubiquitous PT-76 light tank. The all-welded steel hull is fully enclosed and has a maximum armor thickness of 40 mm.—sufficient against light cannon as well as small arms and splinters. The four-man crew of commander, driver, gunner, and loader all ride inside the armored casemate, with the driver up front on the right, the commander behind him, the gunner on the left, and the loader in the rear. The driver, commander, and gunner all have INFRARED night sights, but FIRE CONTROLS are limited to a simple stadiametric sight.

Main armament is an 85-mm. D-5-S85 anti-tank gun in a ball mounting on the casemate's front, which can fire APHE and HVAP ammunition. With a maximum armor penetration of 130 mm. at 1000 m., the gun is reasonably effective against most targets except modern MAIN BATTLE TANKS. It has traverse limits of 12° left and right, and elevation limits of −4° to +15°. Forty rounds of ammunition are stored in the fighting compartment. Secondary armament is usually a pintle-mounted 7.62-mm. PKT light MACHINE GUN, but some ASU-85s have 12.7-mm. DSHK anti-aircraft machine guns. Fully loaded, the ASU-85 is too heavy to be dropped by parachute; it must therefore be AIRLANDED by fixed-wing transports or heavy lift helicopters (such as the Mil-26 HALO). See also AIRBORNE FIGHTING VEHICLES; AIRBORNE FORCES, SOVIET UNION.

Specifications Length: 20 ft. (6 m.). **Width:** 9 ft. (2.8 m.). **Height:** 7 ft. (2.1 m.). **Weight, combat:** 14 tons. **Powerplant:** 240-hp. V-6 diesel. **Speed, road:** 27 mph (44 kph). **Range, max.:** 161 mi. (260 km.).

ASUW: See ANTI-SURFACE WARFARE.

ASW: See ANTI-SUBMARINE WARFARE.

ASW-SOW: Anti-Submarine Warfare–Standoff Weapon, designation formerly applied to the U.S. SEA LANCE anti-submarine missile.

AT-: NATO designation for Soviet ANTI-TANK GUIDED MISSILES. Models in service include: AT-3 SAGGER; AT-4 SPIGOT; AT-5 SPANDREL; AT-6 SPIRAL; AT-7 SAXHORN; and AT-8 SONGSTER.

ATA: Advanced Tactical Aircraft, the General Dynamics/McDonnell Douglas A-12 all-weather, carrier-based ATTACK AIRCRAFT, under development for the U.S. Navy to replace the A-6 INTRUDER. Developed from the early 1980s in extreme secrecy as a "black" program outside of normal budget scrutiny, the ATA was to incorporate state-of-the-art "low-observables" (STEALTH) technology, including extensive use of composites, RADAR ABSORBANT MATERIALS, and contouring to minimize its RADAR CROSS SECTION. The aircraft would have had a two-man crew, and would penetrate air defenses by flying at high-subsonic speeds at low altitudes, mainly at night. The aircraft would have had two nonafterburning turbofan engines, a maximum takeoff weight of some 55,000 lb., and a combat radius of 1125 mi. (1850 km.). Recent artist renderings depicted a pure flying wing design with a small bubble canopy set at the leading edge. The program was terminated in Jan. 1991 because of cost overruns and technical problems.

ATAF: Allied Tactical Air Force, a NATO command organization responsible for all air operations within an Army Group sector. Each NATO member earmarks national air units for service under an ATAF for peacetime exercises and wartime operations. The two most important ATAFs, 2 ATAF and 4 ATAF, are subordinate to Allied Air Forces, Central Europe (AAFCE), which is the air counterpart to NATO's Central Army Group (CENTAG). Their contingents include the totality of RAF Germany, the German air force, and the U.S. Air Force in Europe (USAFE). In Southern Europe, 5 ATAF and 6 ATAF provide air support for the Southern Army Group (SOUTHAG) with Portuguese, Spanish, Italian, Greek, and Turkish units. In Northern Europe, there is no ATAF command; the Northern Army Group (NORTAG) is supported by the Danish and Norwegian air forces and RAF Norway under the direct control of Allied Forces Northern Europe (AFNORTH).

ATB: Advanced Technology Bomber. See B-2.

ATF: Advanced Technology Fighter, a U.S. Air Force program to develop a successor to the F-15 EAGLE in the late 1990s. Developed from the early 1980s in extreme secrecy as a "black" program free from normal budget scrutiny, the ATF will incorporate state-of-the-art "low observables" (STEALTH) technology, including extensive use of RADAR-ABSORBANT MATERIALS and contouring to reduce its RADAR CROSS SECTION; advanced aerodynamics for sustained supersonic cruising speed and

radical maneuvering capabilities; vectored-thrust engines for short takeoff and landing; and advanced cockpit displays to reduce pilot workload. (One of the more interesting systems associated with the ATF is the "pilot's associate," an artificial intelligence [AL] program to monitor aircraft systems and serve as a super-autopilot.) Maximum takeoff weight is to be limited to 50,000 lb. (22,680 kg.).

Two competitive prototypes, the Lockheed/General Dynamics/Boeing YF-22A and the McDonnell-Douglas/Northrop YF-23A, began flight tests in 1990. Both prototype designs have angular, wedge-shaped fuselages, trapezoidal wings, and contoured tail planes. These features serve to reflect radar signals away from their source, thereby attenuating the aircraft's radar signature. Engine intakes are similarly fitted with baffles to prevent radar signals being reflected off engine turbines. Production aircraft will have 2-dimensional vectored exhausts to enhance maneuverability and reduce infrared emissions.

Pratt & Whitney and General Electric are developing alternative engine for the ATF. Both are afterburning turbo fans in the 30,000 lb. (13,636 kg.) class. They will permit the twin-engine ATF to fly faster than Mach 1 without resorting to fuel-guzzling afterburners (a characteristic called "supercruise"). See also NIGHTHAWK.

ATGM: See ANTI-TANK GUIDED MISSILE.

ATLANTIC AND *ATLANTIQUE*: A French LONG-RANGE MARITIME PATROL (LRMP) and ANTI-SUBMARINE WARFARE (ASW) aircraft. Developed from 1957 by Breguet in response to a NATO requirement, the first prototype flew in 1961, and the aircraft has been in intermittent production since then. In 1965, the French *Aeronavale* acquired forty aircraft, and the West German *Marineflieger* twenty, with five of the German Atlantics modified for electronic intelligence (ELINT) duties. Other users included the Netherlands navy, which ordered 9 between 1969 and 1972 (replaced by P-3C ORIONS in 1984), the Italian navy, which acquired 18 between 1972 and 1974, and Pakistan, which purchased 3 ex-French Atlantics in 1975.

Of conventional construction throughout, the fuselage consists of two superimposed pressurized cabins in a "double bubble" configuration. The nose houses a small weather RADAR, and the flight crew sit in an airline-type flight deck in the upper cabin. Behind the cockpit, the main cabin is configured as a tactical center for five ASW system operators, with various displays and computers (the original ASW system was based on that of the U.S. P-2 NEPTUNE). A CSF surface-search radar is housed in a retractable "dustbin" radome below the cockpit, and a MAGNETIC ANOMALY DETECTOR boom is mounted in a retractable tail "stinger." The lower cabin includes a 30-ft. (9.15-m.) unpressurized weapons bay for DEPTH CHARGES and homing TORPEDOES; a smaller bay in the rear can hold up to 78 SONOBUOYS.

The low-mounted wing has outboard ailerons and inboard Fowler flaps. The vertical stabilizer is surmounted by a small pod for an ELECTRONIC SIGNAL MONITORING (ESM) array. There are four underwing pylons with a total capacity of 7716 lb. (3500 kg.), including additional depth charges and torpedoes, rocket pods, bombs, and EXOCET antiship missiles.

German Atlantics, except for the ELINT versions, were extensively modified in the 1970s with a new Texas Instruments search radar, a Loral ESM array in wingtip pods, and airframe improvements to extend the service life by 10,000 hours. In 1978 the French navy decided to replace its Atlantic with an improved model initially designated Atlantic NG (*Nouvelle Generation*) but renamed *Atlantique* in 1984. Externally identical to its predecessor, it is manufactured of corrosion-resistant materials with new construction techniques to attain a planned service life of more than 15,000 hours; its completely new avionics include a Thomson-CFS *Iguane* radar, a forward-looking infrared (FLIR) sensor mounted in a chin turret, ESM arrays in wingtip pods, and a high-precision INERTIAL NAVIGATION system. The weapons bay has also been modified to hold two Exocets. The prototype flew in 1981, and the first of some forty production aircraft were delivered to the *Aeronavale* in 1988

Although the *Atlantique* is a good ASW platform, whose low operating costs make it attractive for smaller navies, its sensors remain inferior to those of the U.S. ORION and British NIMROD.

Specification **Length:** 107 ft. (32.62 m.). **Span:** 123.7 ft. (37.7m.). **Powerplant:** two 6220-shp. Rolls Royce Dart turboprops. **Weight, empty:** 56,659 lb. (25,700 kg.). **Weight, max. takeoff:** 101,-854 lb. (46,200 kg.). **Speed, max.:** 408 mph (657 kph). **Speed, patrol:** 196 mph (315 kph). **Patrol radius:** (8-hr. mission) 690 mi. (1110 km.). **Endurance, max.:** 18 hr.

ATOLL (AA-2): NATO code name for the Soviet K-13 air-to-air missile, in service with the Soviet Air Force (VVS) and Air Defense Troops (PVO), and widely exported; China produces a copy as the PL-2.

The original AA-2A Atoll-A (K-13A) version was a direct copy of early AIM-9B SIDEWINDERS. Its INFRARED-HOMING guidance relied on an uncooled lead sulfide (PbS) seeker with a narrow field of view, which was limited to rear-quarter attacks and easily distracted by the sun, cloud reflections, and ground clutter; as with early Sidewinders, it was relatively easy for maneuverable fighters to break lock-on. Atoll-A had the same configuration as Sidewinder, with four nose canards for steering and four fixed tailfins, a solid-fuel rocket engine, and a 13.2-lb. (6-kg.) high-explosive fragmentation warhead. Much used in the Vietnam and Middle East wars, early Atolls proved to be susceptible to relatively simple countermeasures. Despite its limitations, the introduction of Atoll in the early 1960s finally gave Soviet fighters a reasonably effective air-to-air missiles. Early Atolls armed MiG-17 FRESCOS, MiG-19 FARMERS, MiG-21 FISHBEDS, and the interceptors of the IA-PVO.

Soon after deploying the initial model, the Soviet Union developed the SEMI-ACTIVE RADAR HOMING (SARH) Atoll-B (generally similar to the U.S. AIM-9C version of Sidewinder) to give its aircraft some all-weather and ALL-ASPECT capability, but the effectiveness of Atoll-B was limited by its short range and narrow sensor field of view. Never widely deployed, Atoll-B partially replaced the earlier AA-1 Alkalai in the IA-PVO.

In 1967 the Soviets began deploying the SARH AA-2C "Advanced Atoll," on late-model MiG-21 Fishbeds; though better than its predecessors, it is inferior even to second-generation Sidewinders. The infrared-homing AA-2D version of 1972 almost certainly incorporates an improved seeker head that has some form of cryogenic cooling, a wider field of view, and limited all-aspect capability; it is generally comparable to the AIM-9J version of Sidewinder. Atoll was superseded from the late 1970s by the AA-8 APHID, but because of the vast numbers supplied worldwide, Atolls will remain in service for many years.

Specifications **Length:** (A, B) 9.16 ft. (2.8 m.); (C, D) 11.5 ft. (3.5 m.). **Diameter:** 4.72 in. (120 mm.). **Span:** 20.9 in. (530 mm.). **Weight, launch:** (A, B) 154 lb. (70 kg.); (C, D) 243 lb. (110 kg.). **Speed, max.:** Mach 2.5 (1635 mph/2714 kph). **Range, max.:** (A, B) 4 mi. (6.5 km.); (C, D) 5 mi. (8.5 km.).

ATTACK AIRCRAFT: Tactical combat aircraft optimized for the delivery of air-to-ground ordnance, including guided and unguided bombs, rockets, air-to-surface missiles, and cannon fire. Attack aircraft developed out of, and have mostly replaced, light BOMBERS. Though they are superficially similar to FIGHTERS, the design of attack aircraft emphasizes fuel economy at high subsonic cruise speeds (for long range with a heavy payload) as opposed to supersonic speed and maneuverability. Moreover, their FIRE CONTROL systems are optimized for air-to-ground delivery (although this has become less of a differentiating factor, because of the introduction of software-driven, multimode systems).

Attack aircraft fall into several subcategories. INTERDICTION aircraft require higher speeds, longer ranges, and sophisticated navigation, attack, and ELECTRONIC COUNTERMEASURE systems to penetrate air defenses; current examples include the U.S. F-111, European TORNADO, and Soviet Su-24 FENCER. At the opposite extreme, CLOSE AIR SUPPORT aircraft are slow, very maneuverable at low altitudes, and heavily armored, and usually lack all-weather avionics; the A-10 THUNDERBOLT and Su-25 FROGFOOT epitomize the type. In between, there are general-purpose attack aircraft which can be used in either role, such as the U.S. A-4 SKYHAWK and A-7 CORSAIR, the Anglo-French JAGUAR, the Italo-Brazilian AMX, and the Soviet Su-17 FITTER.

Multi-role fighter/attack aircraft replaced the earlier fighter/fighter-bomber combination during the 1960s; in the latter, the fighter-bomber was usually an older fighter adopted for ground attack functions, such as the F-100 SUPER SABRE. The F-4 PHANTOM was the first multi-role tactical aircraft, and its successors include the French MIRAGE series, the U.S. F-16 FALCON and FA-18 HORNET, and the Soviet MiG-23/27 FLOGGER.

Unlike the U.S. Air Force (USAF), the U.S. Navy sharply differentiates its fighter and attack aircraft; the Navy's A-6 INTRUDER, for example, has virtually no air combat capability. The Soviet air force relied on older fighters (e.g., the MiG-17 FRESCO and MiG-21 FISHBED) for the attack role until the 1970s, when it developed the very specialized Su-17 Fitter and Su-24 Fencer.

During the 1980s the advent of software-based systems reversed the trend towards specialization. The U.S. Navy replaced the A-7 Corsair with the multi-role FA-18 Hornet, while both the U.S. and Israeli air forces deploy the F-16 FALCON for attack as well; USAF is also supplementing the F-111 with the F-15 Strike Eagle variant of the F-15 EAGLE air superiority fighter. At present, however, the U.S. is developing the "Advanced Technology Aircraft" (ATA), a specialized attack aircraft with "low observables" (STEALTH) characteristics.

ATTRITION: A STRATEGY or OPERATIONAL METHOD whereby the enemy is to be defeated by the cumulative destruction of his forces rather than by disruption and demoralization. Although attrition does not require a favorable exchange ratio for victory (only for a less costly victory), armed forces that follow an attritional approach tend to emphasize the improvement of the technical performance of their weapons, necessarily with diminishing marginal returns. In fact, in attritional combat exchange rates seldom exceed unity, except when very considerable differences in the quality of both manpower and equipment exist.

For the side with materiel superiority, the crucial advantage of attritional methods is that they minimize uncertainties; with them, war can be reduced to an industrial process, subject to normal criteria of efficiency. Success cannot be assured, but failure (i.e., a less than expected cumulation of destruction) need not be catastrophic. On the other hand, the materially inferior force can only avert assured defeat by attrition if it adopts a relational MANEUVER approach based on high-risk/high-payoff methods that seek to exploit specific weaknesses in the operational methods, force structure, morale, tactics, or equipment of the enemy. Historically, inferior forces that applied relational-maneuver methods have often defeated attrition-oriented forces, but the larger the scale of combat, the less likely are such reversals.

Until recently, the U.S. was the major practitioner of attritional warfare. The preferred weapons, tactics, and force structures of all U.S. forces reflected a striving for high exchange ratios in attritional engagements, with units treated as industrial plants: manpower and munitions were the inputs, firepower the output. The influence of the systems analysis techniques introduced under Robert McNamara further reinforced attritional proclivities, because such techniques are based on quantitative measures of success such as munitions expended, targets "serviced," enemy soldiers killed, etc. The results of relational maneuver, by contrast, are not subject to "objective" measurement on an incremental basis.

Nevertheless, by the late 1970s attrition had generally fallen into disfavor in U.S. military circles; relational-maneuver warfare was not only more attractive intellectually, but also offered the only possibility of victory in the face of Soviet material superiority. U.S. doctrine, equipment, and training are still marked by attritional antecedents, but their evolution has been fairly rapid. See also AIR-LAND BATTLE; AUFSTRAGSTAKTIK.

AUDACE: Class of two Italian guided-missile DESTROYERS commissioned in 1972–73. Based on the earlier Impavido class, the Audaces are general-purpose escorts optimized for Mediterranean conditions. Flush-decked, they have a tall, blocky superstructure forward, a long, detached deckhouse/hangar aft, and two stacks. Freeboard is rather low (a typical Italian design feature, but no handicap in the Mediterranean), and fin stabilizers are fitted to improve seakeeping.

Armament consists of two single OTO-Melara 5-in. 54-caliber (127-mm.) DUAL-PURPOSE guns forward, four OTO-Melara 76.2-mm. dual-purpose guns amidships, two sets of Mk.32 triple tubes for lightweight anti-submarine TORPEDOES aft, two 21-inch (533-mm.) tubes for heavy WIRE-GUIDED torpedoes in the transom, and a Mk. 13 launcher with forty STANDARD-MR surface-to-air missiles on the aft deckhouse. Two AB-212 HUEY anti-submarine helicopters, or one SH-3D SEA KING, can be operated from a fantail landing deck.

Two "Improved Audace" or Animoso-class destroyers are now under construction. With a modified hullform to improve seakeeping, they will have superstructures made of steel (vs. aluminum), and KEVLAR armor will protect critical systems. Freeboard will be increased forward, and two sets of fin stabilizers will be fitted.

Armament will consist of one 5-inch dual-purpose gun on the bow, an 8-round *Albatros* launcher for ASPIDE short-range surface-to-air missiles before the bridge, three OTO-Melara 76.2-mm. "Super Rapid" guns (two abreast the bridge and one on the aft deckhouse), four Otomat anti-ship missiles amidships, two sets of ILAS-3 triple torpedo tubes, and a Mk.13 launcher. The helicopter facilities have been enlarged to accommodate two SH-3D Sea King or EH-101 helicopters, and the landing deck is fitted with a haul-down system to allow safer operations in heavy seas. See also DESTROYERS, ITALY.

Specifications Length: 446.4 ft. (136.6 m.). Beam: 47.1 ft. (14.5 m.). Draft: 15 ft. (4.6 m.). Displacement: (Audace) 3600 tons standard/4400 tons full load; (Animoso) 4500 tons full load. Powerplant: (Audace) twin-shaft steam, four oil-fired boilers, two sets of geared turbines, 73,000 shp.; (Animoso) twin-shaft CODOG, two 6300-hp.

GMT diesels (cruising), two 27,500-shp Fiat/GE LM2500 gas turbines (sprint). **Speed:** (Audace) 34 kt; (Animoso) 31.5 kt. **Range:** (Audace) 4000 n.mi. at 25 kt.; (Animoso) 7000 n.mi. at 18 kt. (on diesels). **Crew:** (Audace) 380; (Animoso) 400. **Sensors:** (Audace) one SPS-25A surveillance radar, one RAN-3L air-search radar, one SPQ-2D surface-search radar, one 3RM-7 navigation radar, three Orion RTN-10X fire control radars, two SPG-51 missile guidance radars, one CWE-610A bow-mounted medium-frequency sonar; (Animoso) one RAN-10S 2-dimensional air-search radar, one SPS-52C 3-dimensional air-search radar, one RAN-11S surface-search radar, 1 RAN-3L surveillance radar, four RTN-30X fire control radars, two SPG-51C missile guidance radars, one Raytheon DE1164 bow-mounted medium-frequency sonar. **Electronic warfare equipment:** (Audace) passive electronic signal monitoring arrays, one 20-barrel SCLAR chaff rocket launcher.

AUFSTRAGSTAKTIK: A form of decentralized COMMAND AND CONTROL developed by the German army in the nineteenth century, much imitated since then, and still the basis of OPERATIONAL METHODS and TACTICS of the West German army. Under the *Aufstragstaktik*, commanders are given brief "mission orders" which define the objective to be achieved, the resources available, and a time limit, but methods and tactics are left to the commander's discretion.

The principle applies from the highest levels of command down to the platoon or even the squad level. That both requires, and allows scope for, competence and initiative at all levels, but especially at the brigade and battalion levels, wherein commanders are expected to react (perhaps drastically) to the evolving situation, without awaiting orders from above. That shortens reaction times, and forces trained in the *Aufstragstaktik* are more likely to exploit fleeting opportunities and evade sudden dangers.

Under the *Aufstragstaktik*, higher commanders do not so much direct as "choreograph" the battle, constantly monitoring the progress of subordinate units but intervening only rarely, notably to feed reinforcements into the battle at points deemed critical (in German military doctrine, the SCHWERPUNKT). Higher commanders are supposed to issue orders only when they are warranted because of external factors beyond the horizon of their subordinates. Such a form of command requires mutual trust and a shared professional culture, so that officers can anticipate the reactions of their colleagues to particular situations, and accomplish their own missions even if isolated from their superiors. Each commander must know his mission, that of the units on his flanks, and that of the next higher formation, so that he can improvise within the framework of the operational concept when the original plan breaks down. *Aufstragstaktik* is an essential ingredient of armored warfare practiced in the BLITZKRIEG style, and is also an intrinsic part of German ACTIVE DEFENSE doctrine: the officer on the spot is responsible for deciding when to hold and when to abandon terrain, and, most importantly, when to launch counterattacks.

The *Aufstragstaktik* has been adopted by other armies under different names. Although it does not use the term, the Israeli army employs the *Aufstragstaktik* to a much greater extent than any other army, perhaps more than the *Bundeswehr*. The U.S. Army now employs the same method in the form of "Mission-Type Orders," as defined by the AIR-LAND BATTLE doctrine.

AUTOLYCUS: U.S. Navy sensor designed to detect the exhaust emissions of diesel submarines while surfaced or snorkelling; its range is very short. Called a "sniffer" by the U.S. Navy, it is installed on a number of platforms, including the P-3 ORION. See also ANTI-SUBMARINE WARFARE.

AUTOMATIC WEAPON: A weapon in which the entire cycle of loading, firing, cartridge extraction, and reloading is performed continuously as long as the trigger is depressed and ammunition remains in the magazine. The rate of fire is usually selectable, enabling the user to fire single shots, multi-round bursts, or continuously. Automatic weapons include CANNON, MACHINE GUNS, ASSAULT RIFLES, and SUBMACHINE GUNS.

AVIONICS: Aviation electronics; any electronic system used to control the working of aircraft, as well as airborne RADAR, ELECTRONIC COUNTERMEASURES (ECM), ELECTRONIC SIGNAL MONITORING (ESM), and FIRE CONTROL equipment. In combat against fighters or AIR DEFENSE systems, the quality of avionics nowadays contributes more to combat effectiveness than speed or maneuverability. In modern combat aircraft, avionics may constitute more than 50 percent of the total acquisition cost. The avionics categories are:

1. COMMUNICATIONS: radios, radio navigation aids, and digital DATA LINKS. The greater the selection of frequencies and devices available, the greater the operational flexibility of the aircraft.

2. Radar: air-to-air, air-to-ground, and multi-mode radars for target detection, tracking, and weapon guidance. The simplest type of radar is range-only; the most sophisticated are multi-mode radars which can perform a variety of tasks by changing their frequency or waveform. The latest types have digital signal processors to suppress unwanted "noise," resist ECM, and present their data in symbolic formats for enhanced comprehension. A transponder (IFF) is usually incorporated into aircraft radars to enable friendly radars to identify them as friendly.

3. INERTIAL NAVIGATION systems, which plot the aircraft's course and position independently of all external aids or active sensors, thereby enabling an aircraft to reach its target without generating detectable electronic emissions.

4. ELECTRONIC WARFARE equipment, including ESM to detect and classify threats; and ECM to jam, deceive, or suppress enemy sensors and communications, including tracking radars, missile or gun fire-control radars, and communication nets. As modern missiles have become more difficult to evade by maneuver, ESM and ECM have increased in importance. ELECTRONIC COUNTER-COUNTERMEASURES (ECCM) are capabilities built into both radars and radios to resist enemy ECM.

5. Digital computers: used increasingly to monitor flight systems and to control aircraft operations. In FLY-BY-WIRE systems, the pilot's control inputs are actually fed to a computer, which interprets them and in turn feeds appropriate commands to high-speed electronic control actuators. Computers are also the essential component in weapon delivery systems, including gunsights and head up displays (HUDS), automatic bomb release systems, and missile guidance systems.

AV-MF: *Aviatisia Voyenno-Morskoy Flot*, Soviet Naval Aviation, a sizable force of over 1500 aircraft that provides maritime patrol, anti-shipping, ANTI-SUBMARINE WARFARE, SURVEILLANCE, and TARGET ACQUISITION support for the four fleets of the Soviet navy. Of some 1500 AV-MF aircraft, roughly 275 are now assigned to the Baltic Fleet, 400 to the Black Sea Fleet, 425 to the Northern Fleet, and 440 to the Pacific Fleet.

Equipped mainly with land-based aircraft, the AV-MF now has 120 Tu-26 BACKFIRE, 25 Tu-22 BLINDER, and 150 Tu-16 BADGER bombers armed with AS-2 KIPPER, AS-4 KITCHEN, AS-5 KELT, and AS-6 KINGFISH anti-ship missiles for attacks against U.S. CARRIER BATTLE GROUPS and merchant convoys. Target acquisition for the bombers is performed by some 95 Tu-142 BEAR-D electronic intelligence (ELINT) and maritime surveillance aircraft.

For maritime surveillance and ASW duties, the AV-MF has some 60 II-38 MAY and 60 Tu-142 Bear-F land-based patrol aircraft and 90 Be-12 MAIL amphibious flying boats, as well as some 95 Mi-14 HAZE ASW helicopters. Short-range maritime strike (mainly in the Baltic) would be carried out by 70 Su-17 FITTERS armed with AS-7 KERRY tactical missiles.

AV-MF shipboard aircraft include more than 100 Ka-25 HORMONE and 70 Ka-27 HELIX ASW helicopters operated from Soviet DESTROYERS, CRUISERS, and FRIGATES and the MOSKVA-class helicopter carriers. The four KIEV-class aircraft carriers can each operate a dozen Yak-38 FORGER V/STOL fighter/attack aircraft, in addition to Hormones and Hazes. Although technically inferior to the U.S.-British HARRIER, Forger provided important experience that will undoubtedly be exploited in the new TBILISI-class aircraft carrier if and when it enters service. The Tbilisi is to operate some two dozen high-performance fighters, probably variants of the Su-27 FLANKER, as well as Forgers and ASW helicopters. See also AIRCRAFT CARRIERS, SOVIET UNION; AIR FORCE, SOVIET UNION.

AWACS: Airborne Warning and Control System, an advanced form of AIRBORNE EARLY WARNING (AEW) aircraft that combines a large surveillance radar capable of detecting, classifying, and tracking large numbers of aircraft at ranges exceeding several hundred miles; with command, control, and communications (c^3) facilities for the

tactical direction of fighters and other friendly aircraft. The advantage of AWACS over traditional GROUND-CONTROLLED INTERCEPT systems is the greatly extended radar horizon of its airborne radar, and also its ability to evade attack, by electronic means as well. By placing the controllers directly on the airborne radar platform, AWACS reduces reaction times and eliminates dependence on ground facilities, an important consideration for expeditionary operations far from friendly bases.

Tactically, AWACS aircraft can confer a great advantage: during operations over Lebanon in 1982, Israeli AWACS aircraft were able to detect Syrian aircraft as soon as they took off, and directed Israeli fighters into optimal intercept positions.

"AWACS" frequently refers to the Boeing E-3A SENTRY, but other AWACS aircraft include the Grumman E-2C HAWKEYE (also in service with Egypt, Japan, and Israel) and the Soviet Tu-126 MOSS and II-76 MAINSTAY. As the importance of AWACS in modern air combat becomes evident, more countries are attempting to acquire them despite their very high costs. More economical alternatives to the E-3 include an AWACS variant of the Lockheed P-3 ORION, equipped with the radar system of the HAWKEYE; several of those aircraft have been ordered by the U.S. Customs Service to combat airborne drug smuggling. AIRSHIPS are also being studied by the U.S. Navy as long-endurance AWACS platforms.

B

B-: U.S. designation for BOMBER. Additional prefix letters indicate aircraft modified for specific missions: EB- for ELECTRONIC WARFARE, RB- for RECONNAISSANCE, and KB- for tanker variants.

The U.S. STRATEGIC AIR COMMAND now operates three types of bombers: the Rockwell B-1B, the Boeing B-52 STRATOFORTRESS, and the General Dynamics FB-111A strategic bomber variant of the F-111 attack aircraft. The Northrop B-2 "Stealth" bomber began test flights in 1989.

B-1B: Supersonic strategic BOMBER (official name "Lancer") built by Rockwell International for the U.S. Air Force STRATEGIC AIR COMMAND (SAC). Developed from the early 1970 under the AMSA (Advanced Manned Strategic Aircraft) program, the B-1 was originally meant as a replacement for the abortive XB-70 Valkyrie high-altitude Mach 3 bomber, itself intended as the successor to the Boeing B-52 STRATOFORTRESS. The 1965 AMSA requirement called for an aircraft capable of Mach 2 at high altitude and Mach 1 at sea level. The initial contract was placed in 1970, and the prototype B-1 first flew in 1974. By then, however, Mach 2 speed at altitude was no longer considered as important as low radar observability (STEALTH), and the mission profile was changed from "hi-lo-hi" to extended low-altitude flight. At the same time, a new debate had begun over the utility of manned penetrating bombers, and President Carter, committed to both arms limitation and reduced military expenditures, canceled the B-1 in 1977. USAF, however, was allowed to continue flight-testing the four prototype aircraft already built, and between 1977 and 1981 numerous modifications were made to reduce the B-1's RADAR CROSS SECTION (RCS), and increase its range and payload. The B-1's RCS was in fact reduced from 10 square m. to only 1 square m. for the modified B-1B version (as compared to 100 square m. for the B-52).

During the late 1970s, the Air-Launched Cruise Missile (ALCM) program encountered technical problems, and concurrently the Soviet Union introduced new surface-to-air missiles optimized against cruise missiles, making sole reliance on ALCM-equipped B-52s problematic. An advanced "Stealth" bomber (later the B-2) was already being developed, but it could not be ready before the mid-1990s at the earliest. Upon entering office in 1981, President Reagan therefore reinstated the B-1B as an interim bomber to supplement B-52s, and Rockwell received a contract for 100 B-1Bs at a fixed cost of $20.5 billion. The first production aircraft flew in March 1985, an INITIAL OPERATING CAPABILITY (IOC) was achieved in 1986, and the last B-1B was delivered in 1989. The B-1B now equips four wings of the SAC: the 28th, 96th, 319th, and 384th.

The fuselage is smoothly contoured to reduce both drag and RCS. The slim nose houses an APQ-164 multi-mode pulse-Doppler RADAR whose PHASED-ARRAY antenna is canted downward to reduce the RCS. Capable of high-resolution ground mapping, the APQ-164 also has TERRAIN-FOLLOWING and TERRAIN-AVOIDANCE modes, a ground moving target indicator, Doppler velocity update, and several attack modes. An AERIAL REFUELING receptacle (for the USAF "flying boom" system) is mounted in the nose, just ahead of the windscreen. The crew of pilot, copilot, offensive systems operator, and defensive system operator are all seated in an airline-type cockpit in the nose. Behind the cockpit, the central electronics compartment contains most of the AVIONICS, including an INERTIAL NAVIGATION system with GLOBAL POSITIONING SYSTEM (GPS) updates, a satellite communications link to the U.S. National Command Authorities, VHF, UHF, and HF radios, a TACAN receiver, a flight-data computer, and an integrated navigation/attack computer. Three weapons bays behind the avionics compartment each contain an 8-round rotary launcher sized for the AGM-69 SRAM missile, free-fall bombs, and the ALCM-A cruise missile. But ALCM-A was canceled in favor of the longer ALCM-B, which cannot fit into a single bay. Hence, a removable bulkhead is installed between the first and second bays to allow eight ALCM-Bs to be carried on a 14-ft. (4.26-m.) "common strategic launcher" (CSL), together with an auxiliary fuel tank. More fuel, or eight SRAMs, or other weapons can be accommodated in the aft (no. 3) weapons bay, as well as in the two forward bays, instead of ALCMs. Aft of the no. 3 weapons bay is the main defensive avionics compartment, which contains the core of the ALQ-161 ELECTRONIC COUNTERMEASURES (ECM) suite. The highly elaborate ALQ-161 incorporates a variety of RADAR WARNING RECEIVERS, an ELECTRONIC SIGNAL MONITORING array, a tail warning radar, a number of ADAPTIVE JAMMING and BARRAGE JAMMING devices, and a countermeasures dispenser for CHAFF and flares. Modular and programmable (so as to evolve with Soviet AIR DEFENSE systems), the ALQ-161 is supposed to automatically detect, classify, and prioritize radar and missile threats and to initiate countermeasures (though its jamming equipment would be activated only if the ALQ-161 determines that an enemy radar has already detected the aircraft). At present, however, the system is deemed ineffective because of "integration" difficulties.

The low-mounted, variable-geometry (swing) wing is pivoted inboard within small fixed "gloves" which are blended smoothly into the fuselage to generate additional lift, reduce platform drag, and eliminate radar-reflective angles at the wing-fuselage joint. Wing sweep can be adjusted from 15° for takeoff and landing to 67.5° for high-speed dashes. The outer wing sections are fitted with leading-edge slats and double-slotted Fowler flaps to reduce takeoff and landing runs, and with spoilers for lateral control; the outer wing boxes also form integral fuel tanks. The sharply swept tailplanes, midway up the vertical fin, move in unison for pitch control, and move differentially for roll control when the wings are fully swept. The B-1B also has two small movable vanes in the nose below the cockpit; controlled by the flight-data computer, these act to damp out vertical accelerations caused by wind gusts when the aircraft is flying at high speeds at low altitudes.

The engines are mounted in pairs under each inboard wing section; the carefully contoured air inlets are fitted with radar absorbant baffles to prevent the reflection of radar signals off the turbine blades, thus eliminating a major radar signature.

For strategic-nuclear delivery, the B-1B can carry a maximum of 24 SRAM internally, plus 14 on fuselage and wing-glove pylons; or 8 ALCM internally plus 14 externally; or up to 24 B-83 free-fall thermonuclear bombs; or a mix of all three. External loads would normally be avoided due to their increased drag and radar signature. As a nonnuclear bomber, the B-1B could carry up to 128 Mk.82 500-lb. (227-kg.) bombs internally, plus 44 more externally, for a maximum load of 86,000 lb. (39,090 kg.).

Since entering service the B-1B has suffered from a variety of teething problems, including fuel leaks from the integral wing tanks, software errors in the terrain-following mode of the APQ-164, and, most critically, electromagnetic interference problems with the ALQ-161. While the other problems were quickly resolved, the ECM suite remains defective.

Specifications Length: 147 ft. (44.81 m.). **Span:** (15°) 136.7 ft. (41.67 m.); (67.5°) 78.2 ft. (23.84 m.) **Powerplant:** 4 General Electric F101-GE-102 afterburning turbofans, 30,000 lb. (13608 kg.) of thrust each. **Fuel, internal:** 202,254 lb. (91,-740 kg.). **Weight, empty:** 192,000 lb. (87,090 kg.). **Weight, max. takeoff:** 477,000 lb. (216,364 kg.). **Speed, max.:** Mach 1.25 (825 mph/1328 kph) at 36,000 ft. (11,000 m.). **Speed, cruise:** Mach 0.92 (699 mph/1125 kph) at 500 ft. **Range, max.:** 7450 mi. (11,990 km.).

B-2: The Northrop "STEALTH" bomber under development for the U.S. STRATEGIC AIR COMMAND (SAC). Developed from the late 1970s as the "Advanced Technology Bomber" (ATB), the B-2 was originally intended as an alternative to the B-1 as a replacement for the aging B-52 STRATOFORTRESS. Incorporating the most advanced "low observables" (stealth) technology, the B-2 has a frontal RADAR CROSS SECTION (RCS) of only 0.1 square m., one-tenth that of the B-1B (whose RCS only is one percent of the B-52's), and a greatly reduced INFRARED signature as well, to allow it to penetrate air defenses undetected even at medium or high altitudes. While the B-1B was procured as an interim weapon, mid-1980s plans envisaged the acquisition of 132 B-2s from the mid-1990s.

Developed in extreme secrecy as a "black" program omitted from published budgets, the prototype B-2 was publicly displayed on 12 November 1988, and flight tests began on 17 July 1989. Extremely controversial, mainly because of its costs, the B-2 has aroused much congressional opposition. Nonetheless, funding for the first four aircraft was authorized under the 1991 defense budget. If acquired in quantity, the B-2 would replace the B-1 in the penetrating bomber role from 1996.

A true "flying wing," without a tail or a distinct fuselage, the B-2 is fabricated mainly of composites and incorporates RADAR ABSORBANT MATERIALS (RAM) to reduce RCS. The latter is further reduced by absence of right-angle surfaces that could act as resonant "corner reflectors."

The two-man crew is accommodated in the bulged center section of the wing, which also houses two large weapon bays. AVIONICS are believed to include a "low probability of intercept" ground-mapping radar with conformal antennas under the nose, a terrain-following laser radar (LADAR), a forward-looking infrared (FLIR) sensor, an advanced INERTIAL NAVIGATION unit with ring-laser gyros and GLOBAL POSITIONING SYSTEM (GPS) update, a satellite DATA LINK to the U.S. NATIONAL COMMAND AUTHORITIES, an advanced weapon-delivery computer, and comprehensive ELECTRONIC WARFARE equipment, including passive ELECTRONIC SIGNAL MONITORING arrays and several ACTIVE JAMMING arrays.

The two weapon bays have a total capacity of some 40,000 lb. (18,000 kg.). Planned weapons include the AGM-131 SRAM-II standoff missile and the AGM-129 Advanced Cruise Missile (both with stealth characteristics), as well as free-fall nuclear and conventional bombs.

The wing has straight, moderately swept leading edges intended to deflect radar signals away from their source, and a sawtooth trailing edge composed of ten separate surfaces, which is also intended to "dump" radar signals away from the rear of the aircraft; RAM is used extensively in both the leading and trailing edges. Because the B-2 has no vertical tail, its outboard trailing edges are fitted with "drag rudders" to control both pitch and yaw; two sets of "elevons" on the inboard sawtooth edges control pitch and roll, while another movable control surface behind the wing center section apparently acts as a speed brake and landing flap. Inherently unstable, the B-2 relies on computer-controlled, quadruples FLY-BY-WIRE (possibly FLY-BY-LIGHT) flight controls.

The engines are housed in twin nacelles built into the wing on each side of the cockpit. The engine intakes are on the upper wing surface to impede radar observation from below, and are fitted with RAM baffles to prevent the reflection of radar signals off the turbine blades. The exhausts, also on the upper wing surface to prevent observation from below, are fitted with suppressors to reduce infrared emissions. The engines are also equipped with a contrail-suppression system, to impede visual acquisition at high altitudes. Intended to penetrate hostile airspace at low-to-medium altitudes at subsonic speed, the B-2 would rely mainly on stealth for protection.

Specifications Length: 69 ft. (21 m.). **Span:** 172 ft. (52.43 m.). **Powerplant:** four General Electric F118-GE-100 turbofans, 19,000 lb. (8620 kg.) of thrust each. **Weight, empty:** (est.) 100,000 lb. (45,400 kg.). **Weight, max. takeoff:** (est.) 330,000 lb. (149,700 kg.). **Speed, cruising:** Mach 0.85 (595 mph/995 kph). **Ceiling:** 50,000 ft. (15,000 m.). **Range, max.:** 5165 mi. (8330 km.).

B-52: See STRATOFORTRESS.

BACKFIRE: NATO code name for the Soviet Tupolev Tu-26 long-range, twin engine supersonic BOMBER, in service with the long-range aviation (ADD) branch of the Soviet air force (VVS) and with maritime strike regiments of Soviet Naval Aviation (AV-MF). Backfire originated in an attempt to improve the range of the Tu-22 BLINDER medium bomber by fitting it with variable geometry ("swing") wings with pivoting outboard sections; prototype conversions started in 1966, and by 1973 at least twelve were flying as Tu-22Ms (Backfire-A); some may still be in service with the ADD. They were followed by a much more extensive redesign, the Tu-26 (Backfire-B), which was produced at the Kazan aircraft-manufacturing complex from 1973. The Backfire-B retains much of the Tu-22M's wing, but has improved high-lift devices on the outboard (pivoting) wing; the landing gear was also moved from wing pods to wells inside the fuselage. The engines, moved from the Tu-22M's external pods, are fed by large chin inlets. From 1980, Backfire B was superseded in production by the definitive Backfire C, with vertical wedge inlets similar to those of the MiG-25 FOXBAT fighter, improved avionics, and numerous detail improvements.

The Tu-26's fuselage cross section is unchanged from that of the original Tu-22 Blinder, but the crew compartment is completely new, with the pilot, copilot, and two system operators seated "2-by-2" in ejection seats. The nose glazing of the Tu-22 has been replaced by a large navigation-and-bombing RADAR (NATO code name "Down Beat"). Other AVIONICS include a "Fan Tail" tail-mounted gunnery radar, a Sirena-3 RADAR-WARNING RECEIVER, and an ACTIVE JAMMING array. An AERIAL REFUELING probe can be fitted over the radar, but to comply with the terms of the SALT II Treaty it is normally absent. The space behind the cockpit contains a large avionics bay, the midfuselage houses several large fuel tanks, the main landing gear, and an internal weapons bay with a capacity of 26,455 lb. (12,000 kg.), and the engines are mounted in the rear. Defensive armament consists of two radar-directed 23-mm. NR-23 tail cannons in a remotely controlled barbette at the base of the vertical stabilizer.

The low-mounted wings have large fixed inboard sections ("gloves") swept at 60°. The outer sections, which can be swept between 20° and 65°, are fitted with full-span leading-edge slats and slotted flaps, as well as spoilers for low-speed control and short-field capability; they also house integral fuel tanks.

Up to three long-range CRUISE MISSILES or ANTI-SHIP MISSILES (including the AS-4 KITCHEN and AS-6 KINGFISH) can be carried in a recessed belly well and on two wing-glove pylons. Alternatively, up to 27,000 lb. (12,272 kg.) of free-fall bombs can be carried internally and on external pylons under the engine intakes and wing gloves.

When the Backfire-B first appeared, there was disagreement in the U.S. about its classification—its range could be "strategic" (i.e., intercontinental), but its size was not quite in that class. Under the 1979 SALT II Treaty, the aircraft was defined as a "theater weapon" provided that it is not fitted with an aerial refueling probe—but not more than fifteen minutes are required to attach the probe. However, the Backfire is certainly not a dedicated strategic bomber; it has been observed training over Eastern Europe, the Pacific, and the Atlantic

in a variety of roles, including level bombing, low-level attack, strategic RECONNAISSANCE, electronic intelligence gathering (ELINT), anti-shipping attack, and over-the-horizon missile guidance. Some 180 Backfire-B/Cs are in service with the ADD, with 180 more in service with the AV-MF.

Specifications Length: 139.5 ft. (42.5 m.). **Span:** (20°) 113 ft. (34.45 m.); (65°) 86 ft. (26.21 m.). **Powerplant:** two Kuznetsov NK-144 afterburning turbojets, 44,092 lb. (20,000 kg.) of thrust each. **Fuel, internal:** 125,663 lb. (57,000 kg.). **Weight, empty:** 119,050 lb. (54,000 kg.). **Weight, max. takeoff:** 270,066 lb. (122,500 kg.). **Speed, max.:** Mach 1.92 (1276 mph/2039 kph) at 36,000 ft. (11,000 m.)/Mach 0.9 (685 mph/1102 kph) at sea level. **Service ceiling:** 60,000 ft. (18,290 m.). **Combat radius:** 3399 mi. (5470 km.). **Range, max.:** 8699 mi. (14,000 km.).

BADGER: NATO code name for the Soviet Tupolev Tu-16 twin-engine, subsonic medium BOMBER. A contemporary of the U.S. B-47 Stratojet, the Tu-16 flew as a prototype in 1952 and entered service with Soviet Long-range Aviation (ADD) in 1955. Production was terminated in 1959, after some 2000 had been built; two decades later, China produced a copy (Xian H-6). The Tu-16 also served as the basis for the Tu-104, the first Soviet jet airliner.

The Badger's design is a mix of conservative and innovative features. The fuselage, taken from an unsuccessful propellor-driven bomber, is combined with sharply swept wings and wing-root turbojet engines. The pilot and copilot sit side-by-side in an airliner-style flight deck, with the radio operator and flight engineer behind the cockpit, and the bombardier-navigator in a glazed nose position. In the initial Badger-A, bomb aiming was strictly visual, although a small navigation radar was mounted in a ventral chin blister. The heavy defensive armament of Badger-A replicated that of the Tu-4 Bull (the Soviet copy of the Boeing B-29 Superfortress), with a pair of 23-mm. cannon in a manned tail turret, another two pairs in remotely controlled dorsal and ventral barbettes, and one more, fixed and firing forward under the nose. The midfuselage houses four large fuel tanks and a weapons bay with a capacity of some 19,842 lb. (9000 kg.) for free-fall nuclear or high-explosive bombs.

The midmounted wing, swept 42° inboard and 35° over most of the span, is fitted with outboard ailerons and inboard Fowler flaps, and houses integral fuel tanks. The main landing gear retracts into large trailing-edge fairings, a typical Tupolev design feature. Most Badgers are also fitted with a probe for AERIAL REFUELING via an unusual wing-tip-to-wingtip connection.

By the mid-1960s, the Badger was outdated as a bomber, but it received a new lease on life in Soviet Naval Aviation (AV-MF) as maritime patrol aircraft and ANTI-SHIP MISSILE carrier. The first missile-armed variant (Badger-B) carried one AS-1 KENNEL missile under each wing. Later, these aircraft were upgraded to Badger-Gs with the more advanced AS-5 KELT; a number supplied to the Egyptian air force were used in combat against Israeli radar stations in 1973 (armed with anti-radiation versions of Kelt). In the 1970s some Badger-Gs were further upgraded (as "Badger-G Modified") with two supersonic AS-6 KINGFISH missiles. The Badger-C version, on the other hand, carried one very large AS-2 KIPPER anti-ship missile guided by a powerful "Puff Ball" radar which replaced the standard glazed nose. A similar radar equips the Badger-D maritime RECONNIASSANCE version, which also has a variety of specialized electronic intelligence (ELINT) antennas and radomes. Many Badger-Cs were further upgraded (as Badger-C Modified) to launch both Kipper and Kingfish. Badger-E is a multi-spectral reconnais-sance conversion of Badger-A, and Badger-F is a similar ELINT conversion. Badger-H is a special-ized standoff ELECTRONIC COUNTERMEASURES (ECM) aircraft equipped with powerful ACTIVE JAMMERS and more than *nine tons* of CHAFF. Badger-J is a more advanced ECM aircraft with jammers mounted in a large "canoe" fairing in the belly. Finally, Badger-K is an upgraded ELINT version with large antenna blisters in the belly.

Some 300 Badgers still serve with ADD, includ-ing bomber, tanker, and ELINT versions. In 1986 they carried out high-level conventional bombing in Afghanistan. About 250 Badgers serving with AV-MF in a variety of roles are being replaced (slowly) by Tu-26 BACKFIRES.

Specifications **Length:** 114 ft. (34.75 m.). **Span:** 108 ft. (32.92 m.). **Powerplant:** two Mikulin AM-3 turbojets, 20,943 lb. (9520 kg.). of thrust each. **Fuel, internal:** 72,386 lb. (32,902 kg.). **Weight, empty:** 80,000 lb. (36,000 kg.). **Weight, max. takeoff:** 158,733 lb. (72,151 kg.). **Speed, max.:** 620 mph (1000 kph). **Speed, cruising:** 530 mph (850 kph). **Service ceiling:** 46,000 ft. (14,000 m.). **Range, max.:** 4000 mi. (6400 km.).

BAINBRIDGE: A U.S. nuclear-powered guided missile CRUISER (CGN-25) laid down in 1959 and commissioned in 1962. Designed as an ANTI-AIR WARFARE (AAW) escort for the nuclear carrier ENTERPRISE, Bainbridge was the third nu-clear-powered surface warship in the world (after the *Enterprise* and the cruiser LONG BEACH). Based upon the conventionally powered LEAHY design, the *Bainbridge* was the first attempt to fit a nuclear powerplant into a DESTROYER-type hull. It is slightly longer than the Leahys, but it displaced nearly 2000 tons more, largely due to the shielding of the nuclear reactor.

Like the Leahys, the *Bainbridge* was originally armed exclusively with guided missiles, with no guns at all. The main battery consisted of a twin-arm Mk.10 missile launcher at each end of the ship, each with eighty TERRIER surface-to-air mis-siles. The primary ANTI-SUBMARINE WARFARE weapon was an eight-round launcher for ASROC, supplemented by two sets of Mk.32 triple tubes for MK.46 ASW torpedoes; there is also a landing deck aft for ASW helicopters, but no hangar. Due to the poor reliability of Terrier, two twin 3-in. DUAL PUR-POSE guns were installed soon after the ship was launched to provide some close-in protection. In 1977, those guns were replaced with manually op-erated 20-mm. cannon, which in 1979 were re-placed in turn by two quadruple launch canisters for HARPOON anti-ship missiles. In 1983, the *Bain-bridge* was finally equipped with effective self-defense weapons: two PHALANX radar-controlled guns for anti-missile defense. The Mk.10 launch-ers and fire controls have also been modified to launch the more effective STANDARD SM2-ER mis-siles in place of Terrier. In 1979 the ship was equipped with the NTDS data link system, which allows its weapons to be integrated into a network with other NTDS-equipped ships in a battle group. See also CRUISERS, UNITED STATES.

Specifications **Length:** 560 ft. (170.73 m.). **Beam:** 57.1 ft. (17.4 m.). **Draft:** 31 ft. (9.45 m.). **Displacement:** 8000 tons standard/9100 tons full load. **Powerplant:** twin-shaft nuclear, two General Electric D2G pressurized-water reactors, 2 sets of geared turbines, 70,000 shp. **Speed:** 30 kt. **Crew:** 542 men plus 18 staff. **Sensors:** one SPS-10 surface-search radar, one SPS-49 2-dimensional air-search radar, one SPS-48 3-dimensional air-search radar, one LN-66 navigational radar, four SPG-55 mis-sile-guidance radars, one bow-mounted SQS-23 low-frequency active/passive sonar. **Electronic warfare equipment:** one SLQ-32(V)3 electronic countermeasures array, four Mk.36 SRBOC chaff launchers.

BALANCE OF TERROR: An equilibrium between powers based only on the possession by each of particular (i.e., nuclear) weapons which allow each party to inflict unacceptable levels of damage upon the other(s). In contrast to the classic "balance of power," wherein all sources of strength and weakness produce a combined equilibrium independent of narrowly specific strengths—which is therefore operative at all levels of interaction—the balance of terror can only be relevant in extreme situations. In theory, the "balance of terror" is also delicate because the introduction of new weapons, or the expectation that they might be introduced, could upset the equilibrium; in practice, with nuclear weapons on each side, the balance is insensitive to all but the most drastic changes in relative capabilities. The U.S. strategy of MASSIVE RETALIATION (circa 1954–60) was an attempt to equate the "balance of power" with "balance of terror," so as to freeze the global status quo with nuclear weapons alone (thus avoiding the costs of a global military presence with nonnuclear forces). See also ASSURED DESTRUCTION; DETERRENCE.

BALEARES: A class of five Spanish guided-missile FRIGATES derived from the U.S. Navy's KNOX class (FF-1052). Commissioned between 1973 and 1976, their hulls, superstructure, and propulsion are essentially identical to those of the Knoxes, with the addition of a Mk.22 single-arm launcher for STANDARD-MR surface-to-air missiles in place of the helicopter pad and hangar. With that, a single-channel SPG-51C missile guidance radar was added to provide target illumination. The ANTI-SUBMARINE WARFARE (ASW) weapons suite, also changed, consists of two twin tubes for MK.46 lightweight ASW TORPEDOES and two fixed 21-in. (533-mm.) tubes for Mk.37 homing torpedoes launched over the fantail. Four HARPOON anti-ship missiles are also added, in single launch canisters. Other armament is essentially identical to that of the Knox class, but each Baleares is equipped with an SQS-23 bow-mounted low-frequency SONAR, rather than the more powerful SQS-26 of the Knox class. Under an ongoing modernization program, a Meroka 20-mm. radar-controlled gun is being fitted on each ship as a defense against anti-ship missiles. See also FRIGATES, SPANISH.

BALLISTIC MISSILE: A rocket-propelled missile that follows an elliptical (ballistic) trajectory to deliver nuclear or nonnuclear warheads over long distances. The flight of a ballistic missile begins with the powered ascent, or boost phase, during which the missile climbs through the atmosphere, rapidly gaining velocity. In most cases, the missile is then steered by INERTIAL GUIDANCE towards a "correlated velocity vector," the set of initial dynamic conditions (velocity, azimuth, and elevation) required to carry the missile to its desired impact point (within a desired time of flight) after booster burnout. The boost phase lasts several minutes in most cases; during that time intercontinental ballistic missiles (ICBMS) can reach an altitude of between 50 and 100 km., and a velocity of 5–7 km/sec. To save energy, such long-range missiles invariably have two or three tandem stages (each with its own engines and control systems), which are jettisoned in turn as each exhausts its fuel to offload the mass of empty fuel tanks and engines.

When the final stage burns out, the payload, usually a REENTRY VEHICLE (RV), is released to fly a ballistic trajectory, determined only by its momentum and the force of gravity. During this "midcourse" phase, which can last between ten and thirty minutes depending on the range and the desired time of flight, the RV can reach an apogee (maximum height) of up to 1500 km.

Any number of trajectories can connect any given launch and impact points; the most fuel-efficient is described as the "minimum energy" trajectory. By depressing the trajectory below the minimum-energy path, the time of flight can be shortened, while lofting the trajectory can extend the time of flight (both can be useful to coordinate multiple-point attacks).

Eventually, the force of gravity pulls the RV into the upper atmosphere. During the reentry phase, which begins at a height of approximately 250,000–300,000 ft. (76,000–91,000 m.), and lasts some two minutes, the RV is subjected to atmospheric heating, rapid deceleration, and lateral displacement due to wind shear. Hence, RVs must be covered with an ablative coating to dissipate heat, and more recent RVs are shaped as slightly rounded cones to minimize aerodynamic perturbations. Most RVs are also spin stabilized to ensure a proper nose-down orientation.

Most larger ballistic missiles deployed since the late 1960s have more than one reentry vehicle. At first, these were MULTIPLE REENTRY VEHICLES (MRVs) released in a fixed shotgun pattern around the centroid of the impact point for the booster's correlated velocity vector. From 1970, however, first the U.S. and then the U.S.S.R. deployed MUL-

TIPLE INDEPENDENTLY TARGETED REENTRY VEHICLES (MIRVs), which can be directed towards widely spaced targets on different correlated velocity vectors. This is achieved by a POST-BOOST VEHICLE (PBV), or "bus," an additional substage equipped with a small motor, an attitude-control system, and a guidance system.

Ballistic missiles are conventionally classified as intercontinental ballistic missiles (ICBMs) when their maximum range exceeds 6000 mi. (10,000 km.); as intermediate-range ballistic missiles (IRBMs) between 2000 and 6000 mi. (3340 and 10,000 km.); as medium-range ballistic missiles (MRBMS) between 1000 and 2000 mi. (1670 and 3340 km.); and as short-range ballistic missiles (SRBMs) when the range is under 1000 mi. (1670 km.).

In addition, ballistic missiles are also classified by their payload capacity as light, medium, or heavy; and by their method of launch, if not land-based, as submarine-launched ballistic missiles (SLBMS) or air-launched ballistic missiles (ALBMs).

A further classification is by fuel type. The first ballistic missiles, such as the German V-2, Soviet SS-6 Sapwood, and U.S. Atlas, all burned liquid fuels, composed of alcohol or kerosene mixed with liquid oxygen (LOX). Because LOX is highly volatile, and cannot be stored for long periods, early liquid-fuel missiles had to be fueled just before launch—a process lasting hours. All modern ballistic missiles use either solid fuel or storable liquid fuels which do not require refrigeration, most commonly some form of hydrazine and oxides of nitrogen. Stored in separate tanks, these chemicals are hypergolic, i.e., they burn spontaneously when mixed in a combustion chamber. Most Soviet ballistic missiles use storable liquids, which have a greater specific impulse (a measure of energy per unit of mass) than solid fuels, but which require complex pumps and fuel lines, reducing mechanical reliability. The U.S. has therefore preferred solid-fuel missiles, whose rocket engines are no more than steel or fiberglass tubes open at one end, into which a combustible, rubberlike compound is poured. An igniter inside the tube initiates combustion, and the hot gas exists through a nozzle. Because of the reliability advantages of solid-fuel ICBMs, the Soviet Union now also produces two, the SS-24 STILETTO and SS-25 SICKLE.

The proliferation of ballistic missile technology is now a significant military trend. In addition to the U.S., Britain, France, China, and the U.S.S.R., countries that have fielded or are developing ballistic missiles (SRBMs and MRBMs) include Brazil (see CONDOR), India, Iran, Iraq (see AL HUSAYN), Israel, Pakistan, and Saudi Arabia. See also ANTI-BALLISTIC MISSILE SYSTEMS; ANTI-BALLISTIC MISSILE TREATY; BALLISTIC MISSILE DEFENSE; SPACE, MILITARY USES OF.

BALLISTIC MISSILE DEFENSE: An array meant to intercept or neutralize BALLISTIC MISSILES, either with ground-launched, nuclear-armed interceptor missiles (as envisaged by the 1972 ABM TREATY), or by nonnuclear means, including space-based weapons such as those studied under the STRATEGIC DEFENSE INITIATIVE (SDI).

A BMD can be layered to intercept ballistic missiles during more than one of the three phases (boost, freefall or midcourse, and reentry or terminal) of their trajectory to increase the cumulative probability of kill (P_K).

But BOOST-PHASE INTERCEPT (BPI) is potentially the most cost-effective form of BMD, because it could destroy ballistic missiles before they can deploy their MULTIPLE INDEPENDENTLY TARGETED REENTRY VEHICLES (MIRVs). Ground-based weapons cannot be used for BPI, which requires weapons already in orbits from which ascending missiles can be observed. BPI concepts were first developed in the U.S. under Project Defender, which began in 1958. It was found that boosters are easy to detect with satellite-based INFRARED (IR) sensors because of their exhaust plumes; that missiles are constrained to narrow trajectory envelopes ("threat tubes") by their correlated velocity vectors; and that boosters are very vulnerable when still loaded with their highly volatile fuels. In addition, decoys and other penetration aids (PENAIDS) cannot be deployed during the boost phase, so the firepower of the defense is not diluted by false targets.

The main technical difficulty of BPI is its compressed timing. The entire boost phase lasts only 180–360 sec., during which time target missiles would have to be detected, tracked, designated to a weapon, and intercepted. Under Project Defender, it was proposed to intercept boosters with guided missiles, and to destroy them with the kinetic energy of their collision (at velocities approaching 8–10 km./sec.). The BAMBI (Ballistic Missile Boost Interceptor) concept of c. 1964 envisaged up to twelve small interceptors mounted on larger satellite "carrier vehicles" (CV), equipped with a FIRE CONTROL computer and a target-aquisition sensor. The boosters were to be detected by orbiting infrared sensors, which would report their positions to the nearest CV. The fire control computer would then program an interceptor for a collision course, and launch it at the appropriate moment to fly a ballistic trajectory to the vicinity of the target, where its own INFRARED-HOMING system would guide it against the booster.

The SPACE-BASED INTERCEPTOR (SBI) proposed for the first phase of SDI is in fact very similar to the BAMBI scheme; the high velocity required to achieve intercepts is now feasible because of advances in miniaturization and propulsion. SBIs, which rely on mechanical energy to destroy the target, are defined as Kinetic Energy Weapons (KEWS) in SDI jargon.

Under the SDI program, a number of other KEWs have been studied, including ELECTROMAGNETIC LAUNCHERS (EMLs), also called "rail guns." In theory, rail guns could launch hundreds of small guided interceptors at velocities up to 15 km./sec. (much faster than missiles), thereby increasing the number of targets which could be engaged by a single platform, and improving P_ks. The major technical problem with rail guns is the difficulty of generating (and storing) the electrical energy they require.

Another approach relies on the direct transmission of energy; proposed DIRECTED ENERGY WEAPONS (DEWs) include high-energy LASERS (HELs), charged particle beams (CPBs), and neutral particle beams (NPBs). Lasers can transfer thermal energy to the missile's skin, causing melting and rupture; they can also destroy targets by generating shock waves in the missile's structure. Charged particle beams would incapacitate missiles primarily by overloading electronic circuits, while neutral particle beams would do so by generating shock waves and by affecting the molecular composition of the missile.

Lasers appear to be the most feasible DEWs at present; a laser with a 10-megawatt (MW) output could suffice for BPI, and 2.5-MW chemical lasers have already been built in the U.S., while the U.S.S.R. is believed to have lasers in the 10-MW range. The main drawbacks of lasers are the need to optically focus the beam, and the vast amount of energy required to "pump" them; the latter is not a problem on earth, but is hard to do in space; the former is a trivial problem in space, but difficult to do on earth because of atmospheric distortion. Ground-based lasers could be used for BPI by aiming their beams at orbital relay mirrors, which would in turn direct the beams against their targets; distortion would be minimized by ADAPTIVE OPTICS ("rubber mirrors"), which sense distortions and change shape to compensate.

The main problem of NPBs is the difficulty of providing sufficient energy in space. CPBs, on the other hand, are distorted by the earth's magnetic field; various esoteric techniques have been proposed to overcome that problem, but none appear to be technically feasible at present. Hence chemically propelled SBIs, or possibly ground-based lasers, are the most feasible schemes at present, followed by rail guns and space-based lasers.

The second BMD is a "midcourse defense" against REENTRY VEHICLES (RVs) or POST-BOOST VEHICLES (PBVs) for missiles with several MIRVs. In the first 180–600 sec. of the midcourse phase, the PBV maneuvers to release the RVs; they could also release a large number of low-mass decoys (PENAIDS) designed to simulate RVs (PBVs can carry between 100 and 1000 PENAIDS for each RV).

Thus, at the start of its operation the PBV is a high value target, because it still carries all or most of its RVs, and has released few PENAIDS. As PBV operation continues, its value decreases until, with the release of the last RV, the value of the PBV as a target becomes nil. RVs are difficult to detect, because they have small radar and infrared signatures, and may be masked by PENAIDS. Discriminating RVs from PENAIDS is the most difficult task for a midcourse defense, and requires very high resolution radar and infrared sensors, as well as sophisticated signal processing.

If discrimination can be highly effective (85–95 percent), the engagement of midcourse targets is relatively easy. Current RVs fly purely ballistic trajectories, and their tracks can therefore be predicted with great accuracy to compute intercept trajectories. In addition, the time available is long—up to 20–30 minutes with ICBMs. But this advantage is offset by the proliferation of targets: each surviving PBV can release as many as ten RVs, each a relatively hard target compared to a booster because they are built to absorb the shock of reentry, dissipate heat, and protect the warhead inside from nuclear effects. Hence, DEWs are not well suited for midcourse intercept, whereas KEWs could be very effective (SBIs or rail guns would be equally useful for midcourse and BPI).

RVs that survive to reenter the atmosphere could be engaged by the final BMD layer, the "Terminal Defense" of long-range, exoatmospheric interceptor missiles that could defend ground "footprints" some 300–400 mi. (500–667 km.) in diameter; and/or short-range endoatmospheric interceptors, which could provide a last-ditch defense for point targets such as missile silos. The Soviet GALOSH and GAZELLE missiles, and the proposed long-range ERIS and short-range HEDI of current SDI schemes, correspond to the two types; both missiles would be guided on the basis of data provided by large phased array radars (TERMINAL IMAGING RADAR) and by airborne infrared sensors. Ultra-short-range unguided rockets ("Swarmjets") or PHALANX automatic radar-controlled guns have also been proposed to defend silos and command bunkers, which only a direct hit could reliably destroy.

PENAIDS are not a problem for terminal defenses, because light objects decelerate quickly and burn up within the atmosphere ("atmospheric sifting"). Reentry decoys could simulate RVs, but because their mass would be high, only a few could be carried for each RV. The main problem of a terminal defense would be saturation; if many RVs are concentrated against a single target, the localized defenses would be swamped. In theory, boost-phase and midcourse BMD layers could raise the saturation threshold to unacceptably high levels, but otherwise, only Phalanx-type HARD-POINT defenses could match the target numbers.

To integrate the different layers of a BMD system, vast amounts of data would have to be transferred between and within the layers; only automated BATTLE MANAGEMENT systems with very high-speed processors and (still theoretical) artificial intelligence (AI) capabilities could cope with that task; humans would be left in the decision loop mainly to activate fail-safe overrides.

In the U.S. debate, some critics have continued to argue that an effective BMD is not technically feasible, in spite of the technical successes of the SDI program. Other critics hold that a BMD would merely trigger an offensive ARMS RACE (or even that the Soviet Union would be forced into a PREEMPTIVE STRIKE by U.S. deployment of a BMD system). A final objection is that no BMD can prevent some warheads from "leaking" through, making the entire defense worthless. Proponents argue that even an imperfect BMD could deter a preemptive counterforce attack, because the attacker would have no way of determining beforehand which of his warheads would hit its (hard-point) target. Moreover, they note that even a thin BMD could provide an effective defense against accidental, unauthorized, or "third party" attacks. The recent spread of ballistic missiles has strength-

ened that claim. See also ARMS CONTROL; ASSURED DESTRUCTION; SPACE, MILITARY USES OF.

BALLISTIC MISSILE EARLY WARNING SYSTEM (BMEWS): A long-range RADAR and communications network designed to detect BALLISTIC MISSILES launched from the Soviet Union towards North America over the northern polar region. It consists of three large PHASED ARRAY radars at Thule, Greenland; Clear, Alaska; and Flyingdales Moor, United Kingdom. The FPS-50 phased array radars at Clear and Thule have been updated to improve their ability to track and discriminate REENTRY VEHICLES in the presence of Soviet PENAIDS. All three BMEWS radars are connected by secure DATA LINKS and land lines ("Rag Mop," "Blue Grass," and "White Alice") to the North American Air Defense Command (NORAD) headquarters under Cheyenne Mountain, Colorado. See also BALLISTIC MISSILE DEFENSE.

BANZAI JAMMING: A form of DECEPTIVE JAMMING provided by escort vessels to protect more valuable ships such as AIRCRAFT CARRIERS from attack by radar-guided missiles. One or two escorts, usually FRIGATES, are positioned between the probable threat vector and the high-value ship; when enemy missiles are detected, they activate BLLP ENHANCEMENT devices, to simulate a single large target. Once enemy missiles are locked onto the deceptive return, the blip enhancers are turned off. In theory, this should cause the missiles to "break lock" and fly out of control. At worst, the missiles will remain locked onto an escort, sparing the more valuable ship. This duty is unpopular with crews. See also ANTI-SHIP MISSILES; ANTI-SURFACE WARFARE; ELECTRONIC WARFARE.

BARBEL: A class of three U.S. diesel-electric attack SUBMARINES commissioned in 1959. The last conventionally powered submarines built for the U.S. Navy, their design was based on that of the revolutionary experimental submarine *Albacore* (AGSS-569), whose teardrop-shaped hull allowed a then-unprecedented submerged speed (30+ kt.) and maneuverability. The same hullform was later applied in the nuclear-powered SKIPJACK class, and the Barbels also served as the design basis for the Dutch ZWAARDVIS class and Japanese YUUSHIO class. All three Barbels were retired by the early 1990s.

As built, the class had its forward diving planes mounted in the bow; these were later moved up to the sail, as in most U.S. nuclear-powered submarines. The Barbels are armed with six bow-mounted 21-in. (533-mm.) tubes for eighteen

MK.48 homing TORPEDOES. A retractable snorkel allows diesel operation from shallow submergence. See also SUBMARINES, UNITED STATES.

Specifications Length: 219.5 ft. (66.92 m.). **Beam:** 29 ft. (8.84 m.). **Displacement:** 2145 tons surfaced/2895 tons submerged. **Max. operating depth:** 656 ft. (200 m.). **Collapse depth:** 984 ft. (300 m.). **Powerplant:** single-shaft diesel-electric, three 1600-hp. Fairbanks-Morse diesel generators, one 3150-hp. Westinghouse electric motor. **Speed:** 15 kt. surfaced/25 kt. submerged. **Crew:** 88. **Sensors:** one BQS-4 active/passive medium-frequency sonar, one BQR-2 passive ranging sonar; one BPS-4 surface-search radar, one WLR-1 radar warning receiver, two periscopes, one Mk.101 analog torpedo fire control system.

BARCAP: Barrier Combat Air Patrol, a form of COMBAT AIR PATROL (CAP) meant to prevent enemy aircraft from infiltrating friendly airspace by attaching themselves to formations of returning friendly aircraft. The BARCAP orbits well ahead of the friendly base (an airfield or an AIRCRAFT CARRIER) to interrogate each approaching aircraft by IFF (Identification Friend or Foe) radar. Aircraft which do not respond to IFF must usually be identified visually, a potentially dangerous procedure. BARCAP is especially important when aircraft routinely return from missions over enemy territory, because battle damage can disable IFF sets. See also ANTI-AIR WARFARE; COUNTERAIR.

BARRACUDA: The U.S. Navy's Mk.50 Advanced Lightweight Torpedo (ALWT). Developed from 1972 by Honeywell, the Barracuda is intended to replace the Mk.46 ANTI-SUBMARINE WARFARE (ASW) homing torpedo (now deemed inadequate against late-model Soviet submarines because of its low speed, short range, shallow diving depth, and small explosive warhead). While roughly the same size as its predecessor, the Mk.50 is reportedly much faster and has a much more effective warhead. Although it was originally intended to enter service in the late 1980s, technical problems have delayed its introduction until the early 1990s.

Of conventional configuration, the Barracuda has a blunt nose housing an active/passive SONAR unit; a cylindrical centerbody containing the warhead, guidance electronics, and fuel; and a tapered tail for the propulsion unit, terminating in four control fins and a shrouded pumpjet propulsor instead of a conventional propeller (a low-speed turbine mounted in a duct between two sets of stator vanes, a pump jet is much quieter than a

propeller, especially at higher speeds). Barracuda also has an on-board computer programmed to recognize hostile countermeasures; it can alter the sonar mode as required and maneuver the torpedo to ensure direct contact with the target. The 95-lb. (43.1-kg.) warhead is reported to operate on the SHAPED-CHARGE principle to better penetrate Soviet double-hull submarines.

Barracuda has a unique closed-cycle Stored Chemical Energy Propulsion System (SCEPS), whereby a solid block of lithium is melted by an electrical charge and mixed with sulfur hexafluoride to produce heat, which in turn generates steam for a turbine engine. Because the product of the reaction is a solid ash of less volume than the original reactants, there is no need to discharge combustion products against the backpressure of the sea (as in the Otto-propellant Mk.46); thus the Barracuda's performance is independent of depth.

In action, Barracuda could be launched from standard Mk.32 tubes on surface ships, dropped from helicopters, or carried as the payload of ASROC and SEA LANCE anti-submarine missiles. Once the torpedo entered the water, it would begin searching for the target with its sonar in the passive mode, switching to active sonar for the final run to the target. If no target is detected, the torpedo can execute several preprogrammed spiral search patterns until a target is detected or its fuel is exhausted. See also ANTI-SUBMARINE WARFARE; TORPEDO.

Specification Length: 9.5 ft. (2.9 m.). **Diameter:** 12.75 in. (324 mm.). **Weight:** 800 lb. (362.9 kg.). **Speed, max.:** 55 kt. **Range, max.:** 8.5 mi. (13.7 km.). **Max. operating depth:** 1970 ft. (600 m.).

BARRAGE: A heavy ARTILLERY bombardment of large areas. Barrages are employed against enemy fortifications, to demoralize enemy troops, disrupt enemy plans of attack or defense, suppress enemy fire during an attack, or isolate portions of the battlefield. All known barrage techniques were developed during World War I, including the box barrage and the rolling (or creeping) barrage. The former is laid behind and on both flanks of a given sector to prevent the enemy from reinforcing it before or during an attack; on the defensive, a box barrage can be laid on three or even all four sides of a friendly position to impede enemy attacks. The latter, intended to suppress enemy defenses, is laid down directly ahead of advancing friendly troops.

Barrage fire can shatter the morale of inexperienced troops, even if physical casualties remain low; in addition, it can destroy defensive positions, disrupt attacking formations, and cut communications. Prolonged barrages, however, can lose their shock effect, and in wet conditions, they can churn the terrain into a quagmire that favors the defense. Barrage fire was much used in the Iran-Iraq War of 1980–1989. See also ARTILLERY; ARTILLERY, SOVIET UNION; INDIRECT FIRE.

BARRAGE JAMMING: An ELECTRONIC COUNTERMEASURE (ECM) technique meant to neutralize enemy RADARS and radios by broadcasting powerful signals across a broad range of frequencies, to simply overwhelm enemy signals. Unlike ADAPTIVE JAMMING or DECEPTIVE JAMMING, which respond to a specific enemy emitter, barrage jamming neutralizes all emitters, including friendly ones. Barrage jammers also need much power, and that tends to limit their use to larger aircraft, ships, and fixed ground installations. ELECTRONIC COUNTER-COUNTERMEASURES (ECCM) to overcome barrage jamming rely on signal processing techniques which suppress electronic "noise" and pick out coherent radar and radio signals from the background. On the other hand, barrage jammers can be very effective against emitters that operate in narrow, well-defined wavelengths, and they do not require much technical skill to build or operate; the Soviet army relies on them to disrupt enemy radio communications. See also ELECTRONIC WARFARE.

BATTALION: An army unit, often the basic unit of tactical maneuver, and usually part of a REGIMENT or BRIGADE. Specialized support battalions (ANTI-TANK, ANTI-AIRCRAFT, ENGINEER, etc.), however, can be directly subordinated to divisions or higher formations (CORPS, ARMIES, and ARMY GROUPS).

Infantry (including MECHANIZED infantry) battalions usually consist of three or more infantry COMPANIES (often called "rifle" companies), as well as a heavy-weapons company usually armed with MORTARS and anti-tank and anti-aircraft weapons, and a headquarters company, for a total of 400 to 1000 men. Tank battalions also normally consist of three or more companies, each with 10 to 17 tanks, so that tank battalions generally include between 30 and 54 tanks. Artillery battalions are normally organized into three BATTERIES of 6 to 8 tubes. In general, NATO armies have larger battalions, while Soviet-style armies have smaller ones.

BATTERY: The basic unit of the ARTILLERY, usually having between 4 and 10 tube weapons or multiple rocket launchers, with 6 to 8 being the norm. Between 2 and 4 batteries, plus headquarters and service elements, form an artillery BATTALION in most armies.

BATTLE: An engagement between two or more sizable forces conducted within defined geographic and time limits. Until the twentieth century, those limits rarely exceeded an area of several hundred square miles and a duration of a few days. The increasing lethality of modern weapons, and the resulting dispersal of forces, have expanded the geographic scope of battles, and extended their duration to weeks, or even months, thus blurring the distinction previously made between battles and campaigns (a prolonged combination of battles and maneuvers).

BATTLE CRUISER: A large, heavily armed, but lightly armored fast naval vessel, superficially similar to a BATTLESHIP. The battle cruiser concept, developed just before World War I, called for ships powerful enough to quickly defeat enemy CRUISERS, yet fast enough to evade enemy battleships. Though armed with battleship-sized guns, most battle cruisers had only 4–6 in. of ARMOR (as compared to 8–14 in. for battleships). After 1918, the type gave way to the fast battleship, which combined battle-cruiser speed with battleship armor. This obsolete term has been revived to describe the large Soviet nuclear-powered KIROV class cruisers.

BATTLE DRILL: Rehearsed tactical evolutions meant to assure rapid reactions in battle and allow commanders to communicate their intentions quickly and unambiguously. As the maneuver commands that comprise an operational concept, battle drills are essential in fluid forms of armored warfare, which require quick responses to evolving situations; they were a key component of the German AUFSTRAGSTAKTIK, and are much used by the Israeli army. See also BLITZKRIEG.

BATTLE GROUP: 1. An ad hoc force of BATTALION size or larger, organized for a specific operation. German World War II *Kampfgruppen* combined armor, artillery, infantry, engineers, etc., to form mission-tailored combined-arms forces. Highly successful, the technique was emulated by British army task forces and U.S. Army COMBAT TEAMS. In the Soviet army, battalions, regiments, and divisions reinforced by army and Front support units are to form OPERATIONAL MANEUVER GROUPS for deep penetrations. The technique is also used by the Israeli army, which specializes in improvising mission-tailored forces.

2. In naval warfare, the grouping of ships of different types to accomplish a particular mission, or perform a specific type of operation, as, e.g., the CARRIER BATTLE GROUP (CVBG).

BATTLE MANAGEMENT (SYSTEM): An automated or semi-automated COMMAND AND CONTROL (C²) system meant to process data from sensors and other sources to identify threats, formulate and prioritize responses, and allocate weapons accordingly. So far, battle management systems have been used for air defense, including naval air defense for CARRIER BATTLE GROUPS (CVBGs). Some form of battle management system would be the critical element of BALLISTIC MISSILE DEFENSES (BMD), notably under the STRATEGIC DEFENSE INITIATIVE (SDI). Because of the possible intensity of BALLISTIC MISSILE attacks (up to tens of thousands of targets, always in a span of less than 30 minutes), BMD battle management systems would have to be almost completely autonomous, with humans in the loop only to observe and override. SDI battle management requirements are now driving the U.S. development of very high speed data-processing and artificial intelligence (AI) techniques.

BATTLESHIP: A large, heavily armored naval vessel armed with very large caliber guns. The first modern battleship was the British *Dreadnought* of 1906. By 1914, the battleship was the primary weapon of modern navies; it retained its perceived primacy after 1918, despite the development of the SUBMARINE and AIRCRAFT CARRIER. Even after the post-1941 recognition of the aircraft carrier as *the* capital ship, battleships were still considered useful for shore bombardment, anti-aircraft defense, and close-range surface actions, especially at night and in confined waters. By 1945, it was recognized that the residual utility of battleships in modern war did not justify their high operating (let alone new construction) costs, and most were scrapped or placed in reserve. The U.S. Navy, however, now has four reactivated IOWA-class battleships of World War II in service. On the grounds that their big guns can provide fire support in amphibious operations, that their armor makes them practically immune to most modern naval weapons (designed for use against unarmored ships), and that they can carry many ANTI-SHIP MISSILES to form the nucleus of SURFACE ACTION GROUPS (SAGs), from 1982, the U.S. Navy reactivated the four Iowas, adding, among other things, HARPOON and TOMAHAWK missiles. See also ANTI-SURFACE WARFARE; BATTLE CRUISER.

BAZOOKA: Specifically, an obsolete U.S. anti-tank rocket launcher; loosely, any weapon of that type. The original 1942 bazooka was an open tube with a sight and a trigger mechanism for the launch of 2.36-in. (60-mm.), fin-stabilized rockets fitted with a shaped charge (HEAT) warhead which could penetrate up to seven in. (178 mm.) of armor. Because of its low velocity, the effective range was only about 100 m.; that and the very visible "backblast" of the rocket made the bazooka a desperation weapon when used against tanks in the open. Nonetheless, as the first effective, man-portable, anti-tank weapon (especially in forests and built-up areas), and because of its more reliable effect against fortifications, the bazooka inaugurated a new class of weapons. By 1943, the German army introduced its own copy of captured bazookas, the *Racketenpanzerbuchse* (anti-tank rocket gun). An enlarged, 3.5-in. (89-mm.) version (M20B1) of the U.S. bazooka was developed by 1945 but not issued until the Korean War (the 2.36-in. rockets having proved useless against North Korean T-34 tanks); copies include the Belgian *Blindecide*, the Canadian *Heller*, and the Czech *Tarasnice*. The lighter M72A1 LAW (Light Anti-Tank Weapon) replaced the M20B1 in U.S. service during the 1960s, but both are still widely issued to non-U.S. forces. See also ANTI-TANK; ANTI-TANK GUIDED MISSILES.

BE-12: Soviet amphibious patrol aircraft. See MALL.

BEAM RIDING: A form of missile guidance whereby the missile guides itself along an electromagnetic beam generated by a RADAR on a separate platform. The technique is used mainly for surface-to-air missiles, as, e.g., the U.S. Navy's TALOS and TERRIER. Once launched, the missile flies a ballistic trajectory until it enters the cone of a wide "capture beam," which turns the missile towards the target; next the radar generates a very narrow guidance beam, which is detected by antennas in the rear of the missile to guide it to the target. Beam riding has the advantage of relative simplicity, and also provides POSITIVE CONTROL—essential for nuclear-armed weapons under U.S. doctrine. Its disadvantages include the need to keep the beam pointed at the target throughout the missile's flight, thereby limiting the number of missiles which can be launched simultaneously to the number of available radars; the technique also makes the radar (and its platform) vulnerable to attack by ANTI-RADIATION MISSILES. Hence, beam riding is seldom used today. See also AIR DEFENSE; ANTI-AIRCRAFT WEAPONS; ANTI-AIR WARFARE.

BEAR: NATO code name for the Soviet Tupolev Tu-95/Tu-142 four-engine turboprop heavy BOMBER/strategic RECONNAISSANCE aircraft. Still the largest Soviet bomber in service, with its cavernous fuselage and massive, shoulder-mounted swept wings, the Bear resembles a propeller-driven B-52 STRATOFORTRESS.

Developed from 1952, the original Tu-95 (Bear-A) flew in 1954 and entered service with Soviet Long Range Aviation (ADD) as a strategic bomber the following year. The fuselage incorporated an internal bomb bay with a capacity of (only) 22,000 lb. (10,000 kg.), and six remote-controlled turrets armed with twin 23-mm. cannon. The seven man crew of pilot, copilot, bombardier, navigator, radio operator, flight engineer, and tail gunner is housed in two pressurized cabins connected by a tunnel. The original weapons load consisted of free-fall thermonuclear bombs. The nose is glazed for optical bomb-aiming, a small navigation radar is located in a chin fairing, and an aerial refueling probe is mounted over the nose.

The midmounted wings, swept at 35°, have outboard ailerons and massive inboard Fowler flaps. The main landing gear are housed in trailing-edge fairings, a typical Tupolev design feature. The tail surfaces are also swept, with radar warning receivers and active jamming antennas in tip fairings on the tailplane. The engines, housed in nacelles along the wing leading edges, each drive a pair of bladed, counterrotating propellers with a diameter of 18.35 ft. (5.6 m.), making the Bear the fastest turboprop aircraft in the world.

Bear-A accounted for the bulk of the first production run of some 300 aircraft, which was completed by 1961; few remain in service as bombers. Bear-B, first seen in 1961, was modified to carry the large AS-3 KANGAROO nuclear CRUISE MISSILE, by the replacement of the glazed nose with a "duckbill" radome for a large "Crown Drum" surveillance radar. Some 110 are still in ADD service, having been rebuilt as Bear-Gs (see below). Bear-C, in service with Soviet Naval Aviation (AV-MF), is similar to Bear-B, but has additional avionics in large blisters on each side of the fuselage. Bear-D, the variant most frequently observed because it is used by AV-MF for maritime reconnaissance and OVER-THE-HORIZON TARGETING, has a large chin radome housing an I-band "Mushroom" mapping and navigation radar, and an even larger "canoe" radome in the weapons bay for a "Big Bulge" surveillance radar. ELECTRONIC SIGNAL MONITORING (ESM) blisters are attached to both sides of the rear fuselage and on the outer wing panels; some have additional ELECTRONIC WARFARE equipment in place of the tail turret. Bear-E, also used by the AV-MF, is a rebuilt Bear-A configured for maritime reconnaissance, with additional fuel tanks and a pallet for reconnaissance cameras in the weapons bay.

Production of the Bear was restarted in 1970 under the designation Tu-142. Bear-F, first noticed in 1973, is substantially different from earlier Tu-95s. Designed for long-range ANTI-SUBMARINE WARFARE (ASW), it has a strengthened wing for higher gross weights, a greater fuel capacity, and the more resistance to stresses of low-altitude ASW patrols. It is equipped with a SONOBUOY dispenser, a MAGNETIC ANOMALY DETECTOR, an ASW tactical control system, and an I/J-band surveillance radar housed in the forward weapons bay; all defensive weapons except for the tail guns have been removed. About 70 were built between 1972 and 1985.

Bear-G is a rebuild of earlier Bear-B and -C missile carriers, reconfigured to carry up to three supersonic AS-4 KITCHENS, and more AVIONICS. Bear-H, still in production, is a Tu-142 variant designed specifically to carry the new AS-15 KENT cruise missile, which, unlike earlier Soviet cruise missiles, is quite compact (it is similar to the U.S. TOMAHAWK). Two Kents are carried in recessed pylons on the inner wing sections, with perhaps eight more carried internally. Bear-H has a large surveillance radar in the nose, probably for target designation. As of 1989, there were some 150 Bears of various types in service with ADD, and at least 95 more in service with AV-MF.

Specifications **Length:** 162.4 ft. (49.51 m.). **Span:** 167.67 ft. (51.11 m.). **Powerplant:** four 15,000-shp. Kuznetsov NK12MV turboprops. **Weight, empty:** 165,000 lb. (75,000 kg.). **Weight max. takeoff:** 375,000 lb. (117,000 kg.). **Speed, max.:** 540 mph (870 kph) at 36,000 ft. (11,000 m.). **Speed, cruising:** 465 mph (750 kph). **Service ceiling:** 41,000 ft. (12,500 m.). **Combat radius:** 5150 mi. (8600 km.). **Range, max.:** (max. weapons) 7800 mi. (13,026 km.); (max. fuel) 11,000 mi. (17,500 km.). **Mission endurance:** 25 hr.

BEEHIVE: Close-range anti-personnel ammunition for tube ARTILLERY and tank guns. It consists of a thin metal canister filled with hundreds, even thousands, of small darts ("fléchettes"), and bursts upon emerging from the barrel, spreading fléchettes in a fan-shaped pattern out to ranges of several hundred meters. Alternatively, a time fuze can delay the bursting of the canister for several seconds to extend the effective range.

BELKNAP: A class of nine U.S. guided-missile CRUISERS (CG-26 through -34) commissioned between 1964 and 1967, intended as high-speed escorts for CARRIER BATTLE GROUPS (CVBGs). Their design reflected the experience gained by the earlier LEAHY-class cruisers (CG-16 through -24), and served as the basis for the nuclear-powered cruiser TRUXTON (CGN-35). Unlike the Leahys, the Belknaps are general-purpose escorts, with ANTI-SUBMARINE WARFARE (ASW) as well as ANTI-AIR WARFARE capabilities. In November 1975, the *Belknap* was severely damaged in a collision with the aircraft carrier *John F. Kennedy* (CV-67) and extensively rebuilt with updated electronics.

The Belknaps' original armament consisted of a twin-arm Mk.10 Mod 7 missile launcher on the bow with sixty TERRIER (and later STANDARD-ER) surface-to-air missiles and ASROC ASW missiles (thus eliminating the need for a separate ASROC launcher); the normal load is now forty Standard Missiles and twenty ASROC. The missiles were supplemented by a Mk.42 5-in. 54-caliber DUAL PURPOSE gun on the fantail and by two twin 3-in. anti-aircraft guns amidships for close-in protection. From 1976, the latter were replaced by two quadruple launch canisters for HARPOON anti-ship missiles. In addition to ASROC, the Belknaps also have two sets of Mk.32 triple tubes for MK.46 lightweight ASW TORPEDOES and a landing deck for one SH-2 SEA SPRITE ASW helicopter. By 1986 two PHALANX radar-controlled guns were added for defense against anti-ship missiles. See also CRUISERS, UNITED STATES.

Specifications Length: 542 ft. (165.25 m.). **Beam:** 54.5 ft. (15.62 m.). **Draft:** 27 ft. (8.23 m.). **Displacement:** 6570 tons standard/8065 tons full load; (*Belknap*) 8575 tons. **Powerplant:** twin-shaft steam, four oil-fired boilers, two sets of geared turbines, 85,000 shp. **Speed:** 33 kt. **Range:** 2500 n.mi. at 30 kt./8000 n.mi. at 14 kt. **Crew:** 492 men plus 18 staff. **Sensors:** one LN-66 navigation radar, one SPS-10F surface-search radar, 1 SPS-49(V)3 2-dimensional air-search radar, 1 SPS-48A 3-dimensional air-search radar, 2 SPG-55D missile guidance radars, 1 SPG-53A fire control radar, 1 SQS-26 (*Belknap* SQS-53) bow-mounted active/passive low-frequency sonar. **Electronic warfare equipment:** one SLQ-32(V)3 electronic counter-measures array, four Mk.36 SRBOC chaff launchers, 1 SLQ-25 Nixie anti-torpedo decoy

BIG BIRD: The U.S. KH-9 strategic RECONNAISSANCE satellite, the first of which was launched in June 1971. Approximately 50 ft. (15.24 m.) long and 10 ft. (3.05 m.) in diameter, the KH-9 weighs some 30,000 lb. (9146 kg.). Carried aloft by a Titan IIID booster, it was typically placed in a polar, sun-synchronous orbit of 155 by 100 mi. (258 by 167 km.), to enable it to photograph any spot on the surface of the earth every few days.

Big Bird, a milestone in satellite reconnaissance, combined wide area surveillance with the high-resolution photography of specifically selected targets. Film from the "search and find" cameras was developed on-board, and then transmitted via DATA LINK to a ground station. If a target of interest was discovered, it could be rephotographed by a high-resolution camera when the satellite passed overhead a few days later. The high-resolution Perkin-Elmer camera, which accounted for the bulk of the satellite payload, reportedly had a ground resolution of less than one ft. (0.3 m.), allowing the identification of targets as small as single military vehicles. High-resolution film was returned to earth in reentry pods recovered in mid-air by HC-130 HERCULES transport planes equipped with capture hooks. At least four reentry capsules were available on each satellite; at predetermined intervals, one was loaded with film and returned to earth. Big Bird was also reportedly equipped with multispectral cameras for INFRARED and ultraviolet photogrammetry to detect camouflaged facilities. The Big Bird satellites had a life span of 6 to 12 months, limited by the number of film canisters.

In 1976 the KH-9 was supplemented by the more advanced KH-11, which provides near–real time reconnaissance, albeit with lower resolutions than the KH-9's close-look system. The KH-9's replacement by satellites of the KH-12 type was delayed by the loss of the Space Shuttle *Challenger*. See also SATELLITES, MILITARY; SPACE, MILITARY USES OF.

BIGEYE: U.S. BLU-80/B air-delivered, chemical BINARY WEAPON, designed to produce the VX nerve agent through the chemical reaction of two less toxic compounds. Bigeye was developed to replace unitary chemical bombs, whose handling is much more dangerous. The bomb has a blunt nose with an FMU-140 PROXIMITY FUZE, and a cylindrical body with two ports through which the nerve agent is dispensed; four canted, pop-out tail fins make the weapon spin, thereby mixing the binary compounds. See also CHEMICAL WARFARE.

Specifications Length: 7.5 ft. (2.28 m.). **Diameter:** 13 in. (330 mm.). **Weight:** 416 lb. (189 kg.). **Filler:** 216 lb. (65.85 kg.) of binary VX.

BILL: The Bofors Rbs-56 medium-range ANTITANK GUIDED MISSILE (ATGM), in service with the Swedish army and under evaluation in the U.S. and elsewhere. Developed from 1979, Bill was the first ATGM designed specifically to defeat CHOBHAM-type and REACTIVE ARMOR with a "top attack" profile; i.e., the missile would attack the thin turret roof and engine deck armor of MAIN BATTLE TANKS (MBTs) rather than their very thick frontal glacis and turret mantlets. The system consists of the missile itself in a sealed KEVLAR launch/storage canister, a sight/tracker unit, and a tripod mount. Vehicular and helicopter installations are also in development.

The Bill missile has four thin, pop-out wings at midbody, and four small pop-out tail fins for steering. The missile body is also fabricated of Kevlar, to reduce weight. In its unique configuration, the solid-fuel rocket motor is mounted in the nose (with four radial nozzles), the warhead is in the midsection, and the guidance electronics are in the rear.

In another unique feature, the shaped-charge (HEAT) warhead, fitted with a PROXIMITY FUZE, is angled to fire downward at 30°, and the guidance electronics have a built-in offset to allow the missile to pass some 3.3 ft. (1 m.) over the target. When detonated by the proximity fuze, the warhead projects a high-velocity jet of molten metal through the thin top armor of the target. Though warhead diameter is only some 4 in. (100 mm.), this attack profile makes Bill far more effective against the latest MBTs than larger missiles that attack tank frontal armor directly. In addition to bypassing the target's most formidable protection, the oblique warhead and proximity fuze facilitates attacks against vehicles deployed in hull-down positions.

Bill is controlled in flight by WIRE GUIDANCE with semi-automatic command to line-of sight (SACLOS). In action, the missile in its canister is attached to the sight unit on its tripod. The operator must first acquire the target in his sights, determine that it is in range, and then pull the trigger. The missile is expelled from the canister by a cold-gas generator, and the main engine ignites at a safe distance from the launcher. As the missile flies downrange, a sensor in the sight/tracker unit detects a coded LASER beacon in the base of the missile, and compares the missile's position to the operator's line of sight to the target. It then automatically generates steering commands to bring the missile onto this line, which are transmitted via two thin wires which pay out from an inertialess reel in the base of the missile. All the operator needs to do is keep the target centered in his sight until warhead detonation.

While Bill is undoubtedly more effective than more conventional ATGMs against tanks fitted with composite and reactive armor, much of its lethality can be neutralized by extending reactive armor protection over the turret roof (as in the latest versions of the Soviet T-72 and T-80). Its other shortcomings are common to all wire-guided ATGMs: a long time of flight during which the target can evade or use smoke and terrain masking to disrupt tracking; and the need for the operator to remain exposed to counterfire throughout the engagement.

Specifications Length: 35.4 in. (900 mm.). **Diameter:** 5.9 in. (100 mm.). **Fin span:** 16.1 in. (410 mm.). **Weight, missile:** 35.3 lb. (16 kg.). **Weight, system:** 59.4 lb. (27 kg.). **Speed, max.:** 656 ft./sec. (200 m./sec.). **Range, min.:** 150 m. (164 yd.). **Range, max.:** 2000 m. (2154 yd.). **Time of flight:** 10 sec.

BINARY WEAPON: A chemical-warfare bomb, missile, or artillery shell in which the lethal agent is formed by two relatively harmless compounds which are combined only after the weapon is fired or launched; when detonated the mixture is dispensed as a gas or aerosol. See also BIGEYE; CHEMICAL WARFARE.

BIOLOGICAL WARFARE (BW): The use of disease-producing viruses or bacteria as weapons—in violation of the 1925 GENEVA CONVENTION, as reaffirmed by the United Nations General Assembly in 1966. Several states maintain BW establishments to develop suitable strains of disease-causing microorganisms, distribution systems, and protective vaccines. Bacteria, viruses, and rickettsiae produce about 160 known infectious diseases, but each exists in many different strains, each in turn a possible BW agent; e.g., bubonic plague (*Yersinia pestis*) has more than 140 different strains. Seven criteria determine military utility: (1) virulence, the damage inflicted by the infection, which must be severe although not necessarily fatal; (2) infectivity, the size of the dose required to initiate an infection, which should be low, to ensure economy in distribution; (3) stability, or the ability of the organism to survive until it reaches the host/victim; (4) the extent of natural immunity, which must be low to ensure a significant impact against the target population; (5) the availability of vaccines to the user (to prevent backlash infection), but not to the enemy; (6) the availability and ease of therapy to mitigate the effects of the disease (the organism should not be susceptible to commonly available remedies); and (7) transmissibility, or the ability of the disease to spread from person to person, which should be low because the goal is to neutralize a target group, and not to start a worldwide pandemic. Characteristics 1, 2, and 3 can be changed in the laboratory, by producing "artificial strains" through the use of recombinant DNA techniques; that is a major focus of BW research.

BW agents are cheap to produce, and the weight of effective amounts is low. BW is thus especially attractive to poorly equipped states and terrorist groups. While BW agents can be delivered by missile, artillery shell, aerial bomb, or spray canister, maximum accuracy and maximum infectivity are hard to combine: the very small particles (1–5 microns) that are best for high infectivity (because they can penetrate the walls of the lung) descend so slowly that they can easily be dispersed by wind.

Lethal agents with high BW potential include smallpox *(Poxvirus variolae)*, bubonic plague *(Yersinia pestis)*, melioidosis *(Whitmorella pseudomallei)*, Glanders *(Malleomyces)*, and Anthrax *(Bacillus anthracis)*. Also highly lethal, but not strictly speaking biological (they are not microorganisms), are the biotoxins produced by the metabolic action of bacteria and molds, notably botulism and mycotoxins ("YELLOW RAIN").

For most military uses, nonlethal agents are quite sufficient. Such incapacitating agents with high BW potential include tularemia *(Pasturella tularensis)*, brucellosis *(Brucella melitensis)*, dengue fever, and Q fever *(Coxiella burnetti)*. Against populations heavily dependent on agriculture or animal transport, livestock diseases can be devastating, even if they have no effect on humans. The U.S. does not maintain stockpiles of military BW agents, but conducts research on BW vaccines, for which small quantities of BW agents are required. Countries with BW capabilities include Libya, Iraq, and Syria. See also CHEMICAL WARFARE.

BLACKBIRD: The Lockheed SR-71 long-range, supersonic strategic RECONNAISSANCE aircraft, in service with the U.S. Air Force (USAF). Developed to a 1959 requirement for a high-speed, high-altitude replacement for the U-2, which had become vulnerable to Soviet surface-to-air missiles, the SR-71 was designed at the famous "Skunkworks" by the brilliant Kelly Johnson. The prototype (A-12) first flew by 1963, and it was followed by the definitive SR (strategic reconnaissance)-71, which first flew in 1964. The most advanced aircraft of its time, the SR-71 remains the fastest operational aircraft worldwide, capable of sustained cruise at Mach 3 at altitudes greater than 80,000 ft. (24,400 m.).

To achieve this, the SR-71 incorporated a number of revolutionary design innovations, including a lifting-body fuselage, a blended supersonic delta wing with blended engine nacelles, an active cooling system which circulates fuel beneath the aircraft skin to dissipate heat generated by atmospheric friction, and much use of heat-resistant titanium alloy. It was also the first plane to make large-scale use of RADAR ABSORBANT MATERIALS (RAM) to reduce the aircraft's RADAR CROSS SECTION (RCS).

The pilot and system operator sit in a tandem cockpit in the nose, under a heavily braced canopy that restricts visibility; though the cockpit is pressurized, both crewmen must also wear astronaut-type pressure suits. The fuselage has a flattened-oval cross section with prominent "chines" from the nose to the wing roots; these generate much lift at cruising speeds, improve stability, and provide space for a palletized sensors. Another equipment bay in the nose houses additional palleted payloads, including ground-mapping radar and multi-spectral cameras. The space behind the cockpit contains an avionics bay and seven cells for the special low-volatility JP-7 fuel, which does not evaporate at high altitudes. An AERIAL REFUELING receptacle is located on the fuselage spine near the tail.

The wings blend smoothly into the fuselage, contain inserts of RAM to reduce the Blackbird's radar signature, are covered with a corrugated titanium skin to radiate heat from atmospheric friction (which ranges from 450° to more than 1000° C), and house integral fuel tanks. The outboard sections have small elevons to control pitch and roll, while the inboard sections have two large elevons which also act as landing flaps.

The engines, mounted in nacelles blended into the wings at midspan, are of a unique design: more than one-third of total thrust at Mach 3 is generated by intake action (i.e., the aircraft is literally "sucked" along); in addition, the positioning of the large intake shock cones is automatically controlled by computers, to prevent engine stalls caused by shock wave ingestion.

Sensors include panoramic and side-looking cameras, infrared and multispectral line scanners, SIDE-LOOKING AERIAL RADAR (SLAR), and electronic intelligence (ELINT) receivers. One of the most spectacular payloads was the GTD-21, a Mach-3 drone carried on the back of the SR-71 for the penetration of especially dangerous air space. GTD-21s were used to overfly the Soviet Union, while the SR-71s remained outside Soviet airspace to receive the relayed data. (The drones landed at air bases in Europe or Asia if they survived.) Each SR-71 can reportedly survey 100,000 square mi. (278,500 sq. km.) in one hour.

Unarmed, the SR-71 relies for protection on its speed, altitude, ELECTRONIC COUNTERMEASURES, low-observability (STEALTH) features, and discretion. It is believed that the air force purchased 32 SR-71s between 1964 and 1968, including a number of SR-71B two-seat trainers. When production was terminated in 1968, the government-owned jigs and other manufacturing tools were destroyed (in a preemptive maneuver to resist lobbying for a fighter version). Several SR-71s have been lost in accidents (not one has ever been intercepted), and the 20-odd survivors continued to serve in two squadrons of the 9th Strategic Reconnaissance Wing of the U.S. Strategic Air Command (at Beale AFB, with detachments on Okinawa and at Mildenhall in the U.K.) until 1990, when they were retired. A replacement aircraft with much better stealth characteristics (code-named Aurora) is believed to be under development by Lockheed.

Specifications **Length:** 107.4 ft. (32.74 m.). **Span:** 55.62 ft. (16.94 m.). **Powerplant:** two Pratt and Whitney J58-P-1 continuous turbo-ramjets, 32,500 lb. (14,700 kg.) of thrust each. **Fuel:** 100,000 lb. (45,454 kg.). **Weight, empty:** 60,000 lb. (27,216 kg.). **Weight, max. takeoff:** 170,000 lb. (77,000 kg.). **Speed, max.:** Mach 3.35 (2177 mph/3636 kph) at 80,000 ft. (24,400 m.). **Speed, cruising:** Mach 3 (1950 mph/3256 kph). **Service ceiling:** 86,000 ft. (26,000 m.). **Range, max.:** 3250 mi. (5427 km.). **Mission endurance:** 10 hr.

BLACKHAWK: The Sikorsky S-70/UH-60 utility helicopter, in service with the U.S. Army and Navy. The design originated as the U.S. Army's Utility Tactical Transport Aircraft System (UTTAS), a replacement for the ubiquitous Bell

UH-1 HUEY. The prototype flew in 1976, and the first of some 1200 Blackhawks entered army service in 1978. The UH-60 is faster than the Huey, has a longer range, can lift a greater load, and is less vulnerable, but its maintenance costs remain much higher, a major disappointment.

The pilot and copilot sit side-by-side in the nose, ahead of a main cabin, which has room for eleven equipped troops or eight stretcher cases; alternatively, up to 8000 lb. (3600 kg.) of cargo can be slung from a belly hook. The fuselage and key components are armored against 7.62-mm. armor-piercing rounds, and the tricycle landing gear is reinforced to withstand crash landings. The engines are housed in a streamlined fairing over the cabin. The four-bladed main rotor has swept tips to reduce noise, and each blade is stressed to survive a direct hit from a 23-mm. shell. The rotor hub, of advanced design, requires no lubrication, reducing both maintenance needs and vulnerability.

Specialized variants include the EH-60A ELECTRONIC COUNTERMEASURES platform, equipped with 1800 lb. (820 kg.) of stand-off jamming transmitters; and the HH-60A Night Hawk, equipped with night vision devices, external fuel tanks, and an AERIAL REFUELING probe for long-range SPECIAL OPERATIONS and rescue missions.

The most numerous variant is the SH-60B Seahawk ANTI-SUBMARINE WARFARE helicopter, selected as the U.S. Navy's LAMPS III for U.S. destroyers and frigates. It has a chin-mounted APS-124 RADAR, a dispenser for 25 SONOBUOYS in the port side, a MAGNETIC ANOMALY DETECTOR attached to the starboard side, an ELECTRONIC SIGNAL MONITORING array, and two hardpoints for lightweight ASW TORPEDOES or ANTI-SHIP MISSILES. Other changes include a folding rotor and tail, and revised landing gear with a smaller "footprint." Another navy version, the SH-60F, with an active/passive dipping SONAR, is scheduled to replace the SH-3 SEA KING aboard U.S. aircraft carriers in the 1990s. Seahawks are also in service with the Australian, Spanish, and Italian navies; China purchased S-70 civil variants, but used them in military operations in Tibet in 1987.

Specifications **Length:** 50.1 ft. (15.26 m.). **Rotor diameter:** 53.67 ft. (13.36 m.). **Powerplant:** two 1560-shp. General Electric T700-GE-700 turboshafts. **Fuel:** 2520 lb. (1145 kg.). **Weight, empty:** 10,624 lb. (4819 kg.). **Weight, normal loaded:** 16,260 lb. (7375 kg.). **Weight, max. takeoff:** 20,250 lb. (9185 kg.). **Speed, max.:** 184 mph (296 kph) at sea level. **Speed, cruising:** 167 mph (269 kph). **Range, max.:** 373 mi. (600 km.) at max. weight.

BLACKJACK: NATO code name for the Soviet Tupolev Tu-160 supersonic BOMBER. Developed from the early 1970s, the prototype flew in 1981, but development has been slow: by 1990 only some twelve aircraft had been delivered.

Superficially similar to the U.S. B-1B but larger, the Blackjack has a very long, pointed nose which houses a multi-mode navigation/attack RADAR and an AERIAL REFUELING probe. The flight crew of pilot, copilot, and two system operators sit in a well-equipped two-by-two cockpit. The space behind the cockpit contains a large bay for a ELEC-TRONIC WARFARE equipment, including RADAR WARNING RECEIVERS, and ELECTRONIC SIGNAL MONITORING and ACTIVE JAMMING systems. Other AVIONICS are believed to include an INERTIAL NAVIGATION system, DOPPLER navigation radar, a satellite DATA LINK, and a digital weapon-delivery computer.

Two large weapon bays amidships have a total capacity of 36,000 lb. (16,300 kg.). Primary armament consists of 12 AS-15 KENT nuclear-armed cruise missiles carried in two rotary launchers, or 24 smaller missiles similar to the U.S. AGM-69 SRAM. There are apparently no provisions for external weapons.

The low-mounted, variable-geometry ("swing") wings are blended into the fuselage as in the B-1, but the wing pivot point is much farther outboard. The wings can be swept between 20° and 65°, and the pivoting sections have outboard ailerons and inboard Fowler flaps. The tail surfaces are sharply swept, with the tailplane mounted midway up the vertical stabilizer.

The engines are mounted in twin nacelles under the fixed portion of the wing. In contrast to the B-1, the engine inlets have the complex variable ramps that allow speeds greater than Mach 2, but they lack radar-absorbent baffles to prevent the reflection of radar signals off the turbine blades. It is believed, therefore, that the Blackjack is inferior to the B-1 in the use of low observables (STEALTH) technology, but this could be incorporated into later versions of the aircraft.

Specifications **Length:** 177 ft. (53.9 m.). **Span:** (20°) 182 ft. (55.5 m.); (65°) 110 ft. (33.53 m.). **Powerplant:** four afterburning turbofans, 50,-000 lb. (22,700 kg.) of thrust each. **Weight, empty:** 250,000 lb. (113,636 kg.). **Weight, max. takeoff:** 590,000 lb. (267,600 kg.). **Speed, max.:** Mach 2.3 (1495 mph/2496 kph) at 36,000 ft. (11,000

m.)/Mach 1 (760 mph/1270 kph) at sea level. **Combat radius:** (est.) 4526 mi. (7300 km.).

BLACKOUT, NUCLEAR: The effect on radar and radios of a nuclear explosion in the atmosphere. Such an explosion causes intense heat, which ionizes air molecules, as do beta rays from radioactive debris. Ionized air absorbs and distorts electromagnetic radiation, thereby degrading the performance of RADAR and radios; such "blackout" effects are similar to those of ELECTROMAGNETIC PULSE (EMP). Deliberate blackout detonations have been proposed as a countermeasure against ANTI-BALLISTIC MISSILE systems based on radar; this tactic (LADDER-DOWN) envisages the use of precursor ICBMS to black out detection and tracking radars. The intensity of the blackout depends on the yield and altitude of the detonation, the relative position of the radar, and its wavelength. The effect on a system of several radars would depend on the number of precursors, and the precision with which their detonations are coordinated. Allowing for mechanical unreliability, timing errors, and defensive interceptions, the tactic might require many precursor warheads. See also BALLISTIC MISSILE DEFENSE.

BLAZER: Trade name for Israeli-made REACTIVE ARMOR, consisting of boxlike modules which can be bolted onto the armor of heavier ARMORED FIGHTING VEHICLES. Each "box" contains a small charge of low-volatility explosive between two thin steel plates; when hit by a shaped charge (HEAT) round, ANTI-TANK GUIDED MISSILE, or BAZOOKA-type rocket warhead, the explosive detonates, breaking up the jet of high-velocity gas with which shaped charges penetrate armor. Much used by the Israeli army and also exported, Blazer is estimated to degrade the effectiveness of HEAT rounds and warheads by up to 70 percent. Similar reactive armor is now deployed on Soviet tanks. See also ARMOR.

BLINDER: NATO code name for the Soviet Tupolev Tu-22 twin-engine, supersonic medium BOMBER. Developed from 1955–56, Blinder was an attempt to overcome increasingly lethal Western AIR DEFENSES by greatly improving on the speed and altitude of the earlier Tu-16 BADGER. The prototype Tu-105 (Blinder-A) first flew in 1959, and the production Tu-22 entered service in 1961–62. By that time, however, it was realized that higher altitude and supersonic speed were not adequate safeguards against modern surface-to-air missiles and radar-directed interceptors; thus the entire program was something of a white elephant. Only some 160 Blinders were supplied to Soviet Long-

Range Aviation (ADD) from 1964 to 1967; many more would have been needed to replace all the Badgers.

Blinder's area-ruled ("coke bottle") fuselage has a long, pointed nose that houses a small navigation/attack RADAR and an AERIAL REFUELING probe. The pilot and defensive systems operator sit in a small tandem cockpit with very poor visibility, while the bombardier/navigator sits below them in a small, glazed bomb-aiming position; all three crewmen have *downward*-firing ejection seats. In Blinder-A, there is a small weapons bay below the wings, with a capacity of 17,500 lb. (8000 kg) for free-fall nuclear and high explosive bombs. Defensive armament consists of a remote-controlled tail barbette with two 23-mm cannons. The low-mounted wing, swept at 45°, is identical to that of the Tu-28P FIDDLER interceptor (both were derived from the abortive Tu-108 bomber). The main landing gear are housed in large trailing-edge fairings, a typical Tupolev design feature. The engines are mounted in pods at the base of the vertical stabilizer.

Blinder-A was soon superseded in the ADD by Blinder-B, with one AS-4 KITCHEN missile in a recessed belly well and a larger target acquisition radar in a bulged radome. All Blinders were withdrawn from ADD service in the late 1970s, but fifty or so are still in service with Soviet Naval Aviation (AV-MF). The main naval variant, Blinder-C, is used for RECONNAISSANCE and electronic intelligence (ELINT) gathering, but it can also carry the Kitchen anti-ship missile.

A few Blinder-Bs were supplied to Libya and Iraq. Iraqi Blinders bombed Kurdish villages as well as Iranian targets, and Libyan Blinders attacked Tanzania in support of Idi Amin's Ugandan troops, as well as targets in Chad, where at least one was lost to STINGER missiles. In the late 1960s the Tupolev design bureau set out to improve the Blinder's speed, payload, and range by fitting it with variable geometry wings. The final result was an entirely new aircraft, the Tu-26 BACKFIRE, originally designated Tu-22M.

Specifications **Length:** 133 ft. (40.53 m.). **Span:** 90.83 ft. (27.7 m.). **Powerplant:** two Kolesov VD-7 turbojets, 33,069 lb. (15,000 kg.) of thrust each. **Fuel:** 79,366 lb. (36075 kg.). **Weight, empty:** 88,185 lb. (40,000 kg.). **Weight, max. takeoff:** 187,-393 lb. (85,000 kg.). **Speed, max.:** Mach 1.5 (994 mph/1600 kph) at 36,000 ft. (11,000 m.). **Speed, Cruising:** Mach 0.85 (559 mph/900 kph). **Service ceiling:** 60,040 ft. (18,300 m.). **Combat radius:** 1740 mi. (2800 km.) incl. 250-mi. (418-km.) supersonic dash. **Max. range:** 4039 mi. (6500 km.).

BLIP ENHANCEMENT: A form of DECEPTIVE JAMMING whereby enemy radar signals are detected, duplicated, and rebroadcast at amplified levels to make the object being scanned seem much larger than it really is. Blip enhancement allows small decoy missiles or drones to simulate the radar signature of a large aircraft. The technique is also much used by naval forces; aside from floating decoys with blip enhancers, all the ships in a CARRIER BATTLE GROUP can be given the same radar signature, preventing enemy missiles from homing onto the aircraft carrier by size discrimination. A specialized form of blip enhancement called BANZAI JAMMING is sometimes exercised by the U.S. Navy. See also ELECTRONIC COUNTERMEASURES; ELECTRONIC WARFARE.

BLISTER AGENT: A chemical agent that causes skin burns, blisters, and inflammation of the eyes and mucous membranes. If aspirated, blister agents can cause fatal chemical pneumonia, and in large doses they can act as systemic toxins.

The basic blister agent, "mustard" or "mustard gas" (thus named for its smell), was developed during the First World War, and is easily produced; its large-scale use in the Iran-Iraq War has been confirmed. Mustard ("H" in the U.S. classification) has several varieties, including distilled mustard (HD), nitrogen mustard (HN), lewisite (L), and mustard lewisite (HL). All mustard agents are thick, viscous liquids dispensed as aerosols. Heavier than air, they can saturate the soil and pool in low-lying terrain, remaining effective for up to a month, depending upon weather conditions. Personnel in contaminated areas must wear protective masks and clothing, but mustard tends to stick to such equipment, affecting anyone who touches it without protection. The classic persistent agent, mustard is most effective to deny passage, or at least slow down the tempo of enemy operations by forcing the use of protective clothing and decontamination procedures. See also CHEMICAL WARFARE.

BLITZKRIEG: German, lit. "lightning war," both Hitler's own psychological-overthrow strategy and an OPERATIONAL METHOD developed by the German army before World War II. In the latter, tank-spearheaded motorized forces achieve deep penetrations of the enemy front to disrupt communications, cut supply lines, and sow panic, thus setting the stage for the envelopment of the (disorganized) defenses. The initial breakthrough is achieved with closely coordinated tactical air attacks, concentrated artillery fire, and foot infantry. But once the penetrations begin, engagements with any strong enemy forces behind the front are avoided in order to maintain momentum. Obstacles are either bypassed (hence the need for tanks and other cross-country vehicles) or neutralized by focused air attacks. As the columns penetrate deeper into the enemy rear, they sever communications, cut supply lines, and above all, "create data" in massive amounts: lags, leads, and contradictions in reported tank sightings can confuse the enemy command on the strength and direction of the penetrating column, dislocating its response or even inducing a decisional paralysis that prevents purposeful counterattacks against the very vulnerable flanks of the penetrations. Eventually, the different columns converge (in "pincer attacks"), and the enemy forces they cut off may collapse from a lack of supplies or a general breakdown of coordination and morale. The *Blitzkrieg* can still be effective even against a materially superior enemy, but requires several prerequisites: air superiority, a mobility advantage, and a command structure that allows even rather junior officers to act on their own to rapidly exploit weakly held passages along the general direction of the advance. The last truly successful *Blitzkrieg* was conducted by the Israeli army against the Egyptian, Jordanian, and Syrian armies in 1967. See also ARMORED FORCES; AUFSTRAGSTAKTIK; INDIRECT APPROACH.

BLOOD AGENT: A lethal chemical agent that attacks the respiratory and circulatory systems by interfering with the ability of blood to absorb and transport oxygen. Three blood agents are known to be in the chemical arsenals of the Soviet Union and the United States as well as other countries: cyanogen chloride (CK), arsine (SA), and hydrogen cyanide (AC). Each can be dispensed as a gas or aerosol from artillery shells, missile warheads, or aerial bombs. All are nonpersistent, and will dissipate in hours. The action of blood agents is extremely rapid, and there is no practical form of first aid, but they cannot penetrate skin, and effective protection can be provided by standard respiratory masks. See also CHEMICAL WARFARE.

BLOODHOUND: British surface-to-air missile deployed by the Royal Air Force (RAF) for home defense. Developed from the early 1950s, the initial Bloodhound I entered service in 1957; it was replaced in 1964 by Bloodhound II, with bet-

ter performance against small, low-flying targets. Bloodhound is an airplane-configured missile, with a long, cylindrical body, an ogival nose cap housing a radar receiver antenna, midmounted trapezoidal wings, and a square, midmounted horizontal stabilizer. For steering, the wings move differentially to control roll, and together to control pitch. The main engines, two Thor ramjets, are mounted above and below the wings. Four strap-on solid fuel boosters accelerate the missile to ramjet-ignition speed, and are released after burnout.

Bloodhound is guided by SEMI-ACTIVE RADAR HOMING on signals reflected off the target by a continuous wave (CW) DOPPLER illuminating radar. It is armed with a large, CONTINUOUS ROD high-explosive warhead fitted with a radar proximity fuze.

Bloodhound is launched from a fixed single-rail launcher with adjustable elevation and 360° traverse. A firing unit consists of four sections, each with 16 launchers, plus several AEI "Scorpion" CW illuminator radars, search radars, and FIRE CONTROL equipment. A mobile version has a lighter Ferranti "Firelight" illuminator, but still requires several hours to set up. Bloodhound is also deployed by Sweden (as the Rb.68), Switzerland (as the BL-84), and Singapore. The RAF still operates two squadrons (nos. 25 and 85) in Great Britain.

Specifications Length: 25.4 ft. (7.75 m.). Diameter: 21.5 in. (546 mm.). Span: 9.8 ft. (2.83 m.). Weight, launch: 5070 lb. (2300 kg.). Speed, max.: Mach 3 (2100 mph/3507 kph). Range, max.: 50 mi. (80 km.). Height envelope: 325–75,500 ft. (100–23,010 m.).

BLOWPIPE: A British man-portable, shoulder-fired, short-range surface-to-air missile, in service with the British army and the forces of Argentina, Canada, Chile, Ecuador, Malawi, Nigeria, Oman, Portugal, and Thailand.

Developed by Shorts from the early 1960s, Blowpipe is cheap, relatively simple, and of dubious effectiveness. Blowpipe consists of the missile itself, inside a sealed launch/storage canister, and an aiming/guidance unit. The missile has four delta canards for steering, and four fixed tailfins. Powered by a dual-impulse solid rocket motor, the missile is armed with a small high-explosive warhead fitted with both proximity and contact fuzes. The aiming/guidance unit consists of an optical sight, a gripstock, a guidance transmitter, and a trigger/controller assembly.

In action, the operator acquires the target in the monocular sight, and squeezes the trigger to initiate the launch sequence. A small booster pulse ejects the missile from the launch canister to a safe distance before the sustainer pulse ignites. After launch, flares in the base of the missile are detected by a tracking unit in the sight, which automatically "gathers" the missile into the operator's field of view; thereafter, the operator must continue to track both target and missile visually (assisted by the flares), to steer the missile via radio signals by manipulating the control mechanism on the aiming/transmission unit. Thus, Blowpipe is fundamentally different from most other shoulder-fired anti-aircraft missiles (e.g., REDEYE, STINGER, and GRAIL) which are guided by "fire-and-forget" INFARED HOMING. Blowpipe's radio COMMAND GUIDANCE allows the all-aspect engagement of enemy aircraft, a capability lacking in all IR missiles except Stinger, but also requires a highly skilled operator who must remain in position until intercept, exposed to enemy fire. That disadvantage is compounded by the very large backblast at launch. Blowpipe is now being replaced in the British army by the Shorts Javelin, a similar missile with automatic laser beam-riding guidance.

Specifications **Length:** 4.55 ft. (1.39 m.). **Diameter:** 3 in. (76.2 mm.). **Span:** 10.8 in. (275 mm.). **Weight, missile:** 24.5 lb. (7.47 kg.). **Weight, launcher:** 19.6 lb. (8.9 kg.). **Weight, system:** 48.3 lb. (21.9 kg.). **Speed, max.:** Mach 1.5 (1050 mph/1753 kph). **Range, max.:** 2.5 mi. (4000 m.). **Height envelope:** 100–6560 ft. (30–2000 m.).

BLUEOUT, NUCLEAR: The widespread disruption of SONAR that would be caused by the turbulence of a subsurface nuclear explosion; it is analogous to the effect of nuclear BLACKOUT on radar.

BMD: 1. *Boyevaya mashina desantnaya* (Airborne Fighting Vehicle), a light armored fighting vehicle which equips the Soviet Airborne Troops and Air Assault Brigades. Introduced in 1970, the BMD resembles a scaled-down version of the Soviet BMP infantry fighting vehicle, and has the same turret and armament, but on a completely different chassis. Constructed mainly of aluminum, the boat-shaped hull has armor protection sufficient only against small arms and splinters. The driver, up front in the center of the glacis, has several observation periscopes. The commander sits to the driver's left, and a bow machine gunner sits to his right; the latter can fire 7.62-mm. machine guns through fixed ports.

The turret is armed with a PG-9 smoothbore 73-mm. ANTI-TANK gun; in addition, an AT-3 SAG-GER or AT-4 SPIGOT anti-tank guided missile mounted over the gun barrel can be controlled from inside the turret by the vehicle commander/gunner. Secondary armament consists of a coaxial 7.62-mm. machine gun. An open well in the rear of the vehicle can accommodate up to six (cramped) troops, but without overhead protection. The engine is located under the troop well, and a hydraulic suspension allows the ground clearance to be lowered from 18 in. (400 mm.). to 4 in. (100 mm.). for loading onto smaller transport aircraft. Amphibious without preparation, the BMD can be propelled through still waters by track action at a speed of 3 mph (5 kph). The BMD can be dropped by parachute from fixed-wing aircraft, or carried as a slung load by large helicopters. Soviet airborne rifle companies of 90 men are equipped with 14 BMDs, which allow them to act as highly mobile raiding parties.

Variants include the BMD-U command vehicle, with a longer chassis and additional radios; an 82-mm. mortar carrier; the 2S9 self-propelled 120-mm. breech-loading mortar; and the BMD-2, with the longer chassis of the BMD-U and a revised turret with a 30-mm. cannon. There is no Western counterpart to the BMD. See also AIR ASSAULT BRIGADE; AIRBORNE FIGHTING VEHICLE; AIRBORNE FORCES, SOVIET UNION; AIR MOBILE.

Specifications **Length:** 17.75 ft. (5.4 m.). **Width:** 8.67 ft. (2.63 m.). **Height:** 6.5 ft. (1.97 m.). **Weight, combat:** 6.7 tons. **Powerplant:** 280-hp. V-6 diesel. **Speed, road:** 43 mph (70 kph). **Range, max.:** 200 mi. (343 km.).

2. See BALLISTIC MISSILE DEFENSE.

BMEWS: See **Ballistic Missile Early Warning System.**

BMP: *Boyevaya mashina pekhoty* (Armored Vehicle, Infantry), a Soviet INFANTRY FIGHTING VEHICLE (IFV) in service with Soviet and other Warsaw Pact forces, and also widely exported. When first reported in 1967, the BMP was a revolutionary innovation, as the first true IFV armed with both cannon and ANTI-TANK GUIDED MISSILES (ATGMs). Developed from the early 1960s, originally for high-tempo armored warfare on the nuclear battlefield, the BMP combined good mobility, a COLLECTIVE FILTRATION unit for NBC defense, and sufficient armament to deal with remnants of enemy armored forces. It was assumed that the infantry would mostly fight mounted, because the BMP's armor would suffice for postnuclear combat.

A sleek, low-slung, fully tracked vehicle with a

boat-shaped hull and a small, midmounted turret, the BMP-1 is constructed mainly of aluminum alloy. The hull divided into an engine compartment in the bow, a fighting compartment including the driver and commander's positions as well as the turret, and a troop compartment in the rear. The maximum armor thickness of 20 mm. is sufficient only against small arms and splinters. The driver, in the bow on the left with the engine mounted to his right, has three observation periscopes, one replaceable by a semi-active INFRARED night scope. The commander sits behind him under a separate hatch, also with three observation periscopes. The gunner sits behind them inside the turret; behind it, the troop compartment can accomodate eight infantrymen seated back to back on benches, with access via four roof hatches and two rear doors. The troop compartment has eight gunports (three on each side, and one in each rear door), sealed against the entry of chemical or nuclear contaminants.

The one-man turret is armed with a 73-mm. PG-9 smoothbore gun whose shaped-charge HEAT rounds can penetrate up to 18 in. (400 mm.) of steel armor out to an effective range of 1500 m. in ideal conditions. Fed by an automatic loader (which has suffered from reliability problems), the PG-9 can fire up to eight rounds per minute; a total of 40 rounds are carried in the loader and in racks throughout the hull. The effectiveness of the gun is degraded by its low muzzle velocity and the lack of a sophisticated sighting system. A launch rail for AT-3 SAGGER ATGMs is mounted above the gun barrel; the missile can be controlled from inside the turret, but the launcher can be reloaded only from outside the vehicle; five reload missiles are carried. Amphibious without preparation, the BMP is propelled through still water by track action at a speed of 3 mph (5 kph).

When the BMP was first used in combat during the 1973 Yom Kippur War, the 73-mm. gun proved hard to operate, armor protection was found to be insufficient to allow BMPs to accompany tanks in the face of enemy fire, and the mass of fuel and ammunition crammed into a small hull led to catastrophic explosions even from minor hits. Moreover, by then Soviet strategy had changed to emphasize nonnuclear combat. As a result, the Soviet army developed new tactics under which BMPs are to provide fire support for their infantry— which must dismount for combat. Because mounted combat was to be limited to exploitation and pursuit only, most Soviet Motorized Rifle divi-

sions could retain their BTR-60 wheeled ARMORED PERSONNEL CARRIERS previously scheduled for replacement by BMPs.

In 1980 the Soviet army introduced the BMP-2, in which the 73-mm. gun is replaced a 30-mm. automatic cannon in a two-man turret, with an AT-5 SPANDREL ATGM in place of Sagger. With selectable rates of fire of 300 and 500 rounds per minute, and a maximum effective range of 1200 m., the cannon is effective against lighter armored vehicles and can be elevated to 74° for use against helicopters. The BMP-2's commander sits in the turret with a better field of view, but only seven troops can be carried.

Specialized variants include the BMP-SON artillery radar vehicle, the BMP-U ARMORED COMMAND VEHICLE, and the BMP-R reconnaissance vehicle. All Soviet tank and Motorized Rifle divisions have at least one BMP regiment, and the Motorized Rifle battalions of tank regiments are fully equipped with BMPs. See also BRADLEY; MARDER; WARRIOR.

Specifications **Length:** 22.1 ft. (6.74 m.). **Width:** 9.67 ft. (2.94 m.). **Height:** 6.25 ft. (1.9 m.). **Weight, combat:** 13.5 tons. **Powerplant:** 300-hp. V-6 diesel. **Speed, road:** 50 mph (80 kph). **Range, max.:** 310 mi. (500 km.).

BO-105: A West German utility and light attack HELICOPTER produced by Messerschmitt-Bolkow-Blohm (MBB). Developed from 1964, the first prototype flew in 1967, and the first production models were delivered in 1971. Though costly, the BO-105 has been widely sold because of its good speed, range, and maneuverability, because it was the first small helicopter to offer twin engines and all-weather AVIONICS, and because its rigid main rotor and hingeless rotor hub stressed to withstand negative acceleration make it fully aerobatic and capable of NAP-OF-THE-EARTH (NOE) flight at high speed.

The fuselage consists of an oval pod with a tubular tail boom which supports the small anti-torque rotor. The pilot and copilot/observer sit side-by-side in the glazed nose, next to the main cabin which has room for three passengers, or two stretcher cases; a small cargo compartment behind the main cabin is seldom used. The tubular skid landing gear is crash-resistant, and can be fitted with flotation bags for water landings. Standard avionics include complete blind-flying instruments, VHF and UHF radios, an autopilot, and VOR radio-navigation equipment. The engines, mounted over the main cabin, drive a four-bladed fiberglass main rotor with titanium anti-erosion strips on the leading edges.

The West German army's Aviation (*Heeresflieger*) has both BO-105M SCOUTING and SURVEILLANCE variants equipped with night vision devices, and BO-105P (or PAH-1) ANTI-TANK variants armed with six HOT guided missiles directed via a stabilized, roof-mounted, all-weather sight which allows the helicopter to remain "hull-down" while engaging enemy forces. BP-105Ps are also equipped with a DOPPLER navigation system and additional communications. The West German army has 212 PAH-1s in its anti-tank helicopter regiments, one of which is attached to each army corps; 227 BO-105P scouts are also in service. The BO-105 is also produced under license in Canada, Indonesia, and Spain.

Specifications **Length:** 28.05 ft. (8.55 m.). **Rotor diameter:** 32.3 ft. (9.85 m.). **Powerplant:** 2 420-shp. Allison 250-C20B turboshafts. **Weight, empty:** 2813 lb. (1276 kg.). **Weight, normal loaded:** 5291 lb. (2400 kg.). **Weight, max. takeoff:** 5511 lb. (2500 kg.). **Speed, max.:** 150 mph (242 kph). **Speed, cruising:** 137 mph (220 kph). **Initial climb:** 1770 ft./min. (540 m./min.). **Ceiling:** 5298 ft. (1615 m.). **Range, max. payload:** 408 mi. (657 km.). **Mission endurance:** 90 min. **Operators:** Bah, Bru, Col, Indo, Iraq, Les, Mex, Neth, Nig, Phi, S Leo, Sp, Su.

BOFORS GUN: Swedish-designed 40-mm. CANNON used as a light anti-aircraft weapon by both ground and naval forces worldwide. The original 1930s cannon was 56 CALIBERS long, and was produced in single, twin, and quadruple mounts. Single mounts were generally fielded by ground forces in a towed configuration, with manual elevation and traverse. Twin and quad mounts, which armed warships of DESTROYER size and larger, had powered traverse and elevation, and water cooling for the barrels. The post-1945 L/60 (60 calibers long) was generally similar. The current L/70 fires a 2-lb. (1-kg.) shell with a muzzle velocity of 2900 ft. (884 m.) per second at a cyclic rate of fire of 160 rounds per minute. The maximum effective range is 6000 m. (3.72 mi.), and the effective ceiling is 4000 m. (13,123 ft.). Ground versions are mounted on a light four-wheeled carriage; in most cases they are combined with RADAR directors and/or ELECTRO-OPTICAL sights. Naval versions for FAST ATTACK CRAFT and naval auxiliaries are similar.

Bofors also manufactures a 155-mm. self-propelled gun (whose purchase was the object of the greatest of Indian political scandals) and a family

of 375-mm. anti-submarine rocket launchers, among other weapons.

BOMBER: A large, multi-engine combat aircraft optimized for the delivery of large loads of air-to-surface munitions over long ranges. In World War II, bombers were divided into light, medium, and heavy classes on the basis of both range and payload. Most frequently used in tactical roles (notably CLOSE AIR SUPPORT, BATTLEFIELD INTERDICTION, and maritime strike), light bombers have largely been supplanted by ATTACK AIRCRAFT, which resemble FIGHTERS and are much more maneuverable than bombers. Only the MIRAGE IV and some old light bombers are still in service, notably the Soviet Il-28 Beagle and the British CANBERRA. Medium bombers have partly given way to interdiction aircraft, notably the TORNADO, F-111, and Su-24 FENCER, but three supersonic medium bombers (the FB-111, Tu-26 BACKFIRE, and Tu-22 BLINDER) remain in service, along with Chinese copies of the old-style Soviet medium Tu-16 BADGER, retained in Soviet service as a cruise missile carrier, as an electronic intelligence (ELINT) gatherer, and for maritime patrol.

Only the U.S. and U.S.S.R. operate heavy bombers. The U.S. retains B-52 STRATOFORTRESSES developed in the 1950s, and nowadays armed with Short Range Attack Missiles (SRAMS) or air-launched cruise missiles (ALCMS), or configured as conventional bombers and maritime strike aircraft (with HARPOON anti-ship missiles). B-1Bs built in the 1980s supplement the B-52s as penetrators pending the planned introduction of the B-2 "Stealth Bomber" in the mid-1990s. The Soviet Union developed two heavy bombers in during the 1950s, the Tu-95/142 BEAR turborpop, and the Mya-4 Bison jet; the former, first flown in 1954, has proven remarkably versatile and durable, and now serves as a maritime patrol and ANTI-SUBMARINE WARFARE aircraft, and as a strategic cruise missile carrier. In fact, the Bear was put back into production, nearly 30 years after its first flight. After a long hiatus, the Soviet air force developed the Tupolev Tu-160 BLACKJACK, a large supersonic bomber, during the 1980s.

The continued viability of the heavy bomber has long been questioned because of the increasing lethality of AIR DEFENSES. Large aircraft are easily detected, and even STEALTH technology is only a partial remedy, while low-altitude flight profiles require power and maneuverability generally lacking in heavy bombers. On the other hand, bombers are far more versatile than strategic missiles, and also more flexible as instruments of policy (they can be alerted, placed on airborne alert, redeployed, etc.). See also ADD; AV-MF; BOMBING TECHNIQUES; SAC.

BOMBING TECHNIQUES: To drop a free-fall bomb from a moving aircraft accurately enough to hit a target on the ground requires precise calculations of the motion and altitude of the aircraft, of atmospheric effects (including surface winds, barometric pressure, temperature, and transient wind shear), and of the aerodynamic characteristics of the bomb itself, to determine the release parameters. For level bombing, the most elementary technique, if the aircraft flies a steady course at a set altitude, computing bombsights can determine the needed offsets for altitude and drift (because once released, bombs follow a ballistic trajectory). If dropped directly over their targets, the forward velocity of the aircraft would carry bombs far beyond them, while any crosswinds would cause lateral diversions. Computing bombsights can also be linked to an autopilot, to fly the aircraft along the proper course until the computed moment of release. Because it requires precise data, this method is quite inaccurate even with sophisticated radar sights and digital computers because of the many transient variables involved (e.g., wind shear).

Dive bombing, pioneered by the U.S. during the 1930s for the precision bombing of tactical positions and moving ships, simplifies the problem because the aircraft aims itself at the target. By diving at angles in excess of 70°, the aircraft cancels out most of the relative motion between aircraft and target, allowing great accuracy. But most modern jet aircraft accelerate too quickly in a dive to allow use of this technique. The feasible compromise is glide bombing from a shallow dive, which enables the aircraft to aim the bomb with some precision without a requiring high-speed pullout from a steep dive.

Another current method is low-level bombing, whereby the pilot flies towards the target at high speeds only a few hundred feet off the ground until the computed release point. To prevent aircraft from being damaged by the blast of their own weapons, retarded bombs (e.g., the U.S. SNAKEYE) have air brakes which deploy to slow them, allowing the aircraft to fly clear before the detonation. Modern weapon-delivery ("navigation-attack") systems allow a variety of aiming methods: automatic, whereby the pilot flies towards a steering dot projected onto his gunsight until the computer

releases the bombs; continually computed release point (CCRP), whereby the pilot steers towards the point at which the bombs must be released manually; and continuously computed impact point (CCIP), whereby the sight displays the point at which the bombs would fall at that moment, so that bombs can be released once the point is positioned over the target. All three methods can be remarkably accurate, with bombs dropped within 30 ft. (10 m.) of their target.

For the delivery of nuclear weapons, accuracy is less necessary, but the ability to fly clear is essential. Level bombing from high altitude being unacceptable, two techniques are used instead: (1) toss bombing, whereby the aircraft flies towards the target at high speed and low altitude until the weapon-delivery computer prompts the pilot to executes a sharp climb during which the bomb is released to fly on a tangent towards the target, while the pilot extends his climb into a loop, to escape in the opposite direction; and (2) the "over-the-shoulder" delivery, whereby the climb is executed directly over the target, with the bomb released while the aircraft is still climbing vertically to fly straight up until its momentum is exhausted, and then come straight down. Again, the aircraft executes a vertical reversal and exits. See also AIR DEFENSE; ATTACK AIRCRAFT; BOMBERS.

BOOST-PHASE INTERCEPT (BPI): One layer of projected multi-layered BALLISTIC MISSILE DEFENSES, meant to destroy BALLISTIC MISSILES during their initial powered ascent. BPI provides great leverage, because with each booster, all its MIRV warheads and PENAIDS are also destroyed. In theory, BPI could also be relatively easy, because boosters can easily be spotted and tracked by INFRARED sensing of their exhaust plumes; because they fly a fairly predictable trajectory; and because, loaded with highly volatile fuel, they are highly vulnerable. What would make BPI difficult in practice is the short duration of the boost phase: 3–5 minutes in all to detect and track the missile, compute a FIRE CONTROL solution, assign a weapon, and decide to engage. Both directed energy weapons (DEWS) such as lasers and particle beams, and kinetic energy weapons (KEWS) such as interceptor missiles and RAIL GUNS, have been suggested for BPI. See also BATTLE MANAGEMENT; STRATEGIC DEFENSE INITIATIVE.

BRADLEY: The U.S. M2 INFANTRY FIGHTING VEHICLE (IFV) and the very similar M3 CAVALRY FIGHTING VEHICLE (CFV), officially named Devens; both are in fact generally known as the Bradley fighting Vehicle (BFV). The Bradley originated in a 1965 U.S. Army requirement for a replacement for the M113 armored personnel carrier, which is only a "battle taxi" that can move troops under thin protection, but which lacks the armor, firepower, and mobility needed to maneuver with tanks on the battlefield. Instead of a better M113, the army wanted an infantry fighting vehicle with tanklike mobility and armed with a gun capable of defeating its Soviet counterpart, the BMP. Under the MICV-65 (Mechanized Infantry Combat Vehicle) program, several candidate vehicles were built, but development problems and the escalating costs of Vietnam delayed action until 1972. After the Vietnam withdrawal, the army's attention was refocused on armored warfare in Europe, the MICV was revived, and a new program was also started to replace the unsuccessful M551 SHERIDAN and the M114 scout vehicle with a wheeled ASRV (Armored Reconnaissance Scout Vehicle). Under congressional pressure, the army later merged the MICV and ASRV programs, and the resulting M2/M3 entered production in 1981.

The Bradley is constructed largely of aluminium alloys to minimize weight; the boxlike hull is divided into a driver/engine compartment in front, a turreted fighting compartment in the middle, and a troop compartment in the rear. Maximum armor thickness, on the order of 25 mm. is sufficient to defeat armor-piercing rounds up to 23 mm. in caliber; additional laminate armor is fitted to the frontal glacis and side plates, but no special protection (e.g., REACTIVE ARMOR) is provided against shaped-charge (HEAT) rounds.

The driver, on the left with the engine to his right, has five observation periscopes and a passive night-vision scope. Immediately behind the driver, the turret is manned by the vehicle commander/squad leader and the gunner, while the troop compartment has seven seats, with access via a rear ramp door and a small roof hatch, and six firing ports: two on each side and two in the ramp door. In spite of the vehicle's size, the troop compartment is rather cramped, with two of the infantrymen sitting awkwardly abreast of the turret. Because the standard M16 rifle will not fit into the firing ports, special M321 variants are built in; though the ports have vision blocks, the chances of hitting a target from a moving vehicle are very small, but suppression fires could be useful. The M3 cavalry version is generally similar, but lacks firing ports, carries only two scouts in the troop

compartment, and has more main gun ammunition.

The T-BAT-II (TOW-Bushmaster Armored Turret-Two Man) turret is armed with a Hughes M249 Bushmaster 25-mm. CHAIN GUN on a costly, fully stabilized mount. The gun can fire both high-explosive (HE) and armor piercing discarding sabot (APDS) rounds with depleted uranium (STABALLOY) penetrators, effective against BMPs and other light armored vehicles out to 1000 m.; secondary armament is a coaxial 7.62-mm. machine gun. All fire controls are duplicated so that weapons can be fired by either the commander or the gunner. Long-range anti-tank capability is provided by a twin box launcher for TOW anti-tank guided missiles on the left side of the turret, which can be erected from inside the vehicle for firing. A computerized Integrated Sight Unit (ISU) serves both missiles and the Bushmaster cannon; fully stabilized, the ISU also incorporates THERMAL-IMAGING night sights. The M2 IFV has room for 900 rounds of 25-mm. ammunition and 5 reload TOWs; the CFV carries 1500 rounds and 12 TOWs.

The Bradley has a power-to-weight ratio of 22 hp./ton for good cross-country mobility. The vehicle can swim (on track action alone) with a flotation screen and inflatable flotation bags. Equipped with a COLLECTIVE FILTRATION system to which gas masks can be connected, it lacks a positive overpressure system for a shirt-sleeve environment inside the vehicle.

The upgraded M2A1 and M3A1, which entered service in 1986, have an improved NBC system and can launch the heavier TOW-2 missile; detailed improvements have also been made to the fuel and fire-suppression systems to reduce the risk of explosions. Future upgrades are intended to increase the level of protection by adding composite ceramic armor.

Though undoubtedly the most elaborate infantry fighting vehicle in the world, and greatly superior to the BMP by virtue of its armor, Bushmaster gun, TOW missiles, and ISU, the Bradley has been much criticized for its vulnerability. As visible and almost as costly as a tank, it lacks the protection of tanks, while the modern battlefield is filled with anti-armor weapons. The Bradley's main role is to carry infantry close to the enemy, where it can dismount to attack on foot with the Bradley's fire support. Thus even if all deliberate confrontation with enemy tanks is avoided, use of the costly T-BAT-II turret places the vehicle in the direct fire zone, for which its protection is insufficient. An operational criticism is that the rifle strength of Bradley units is too low: the M113 has a crew of 2 and can carry 11 riflemen; the Bradley has a crew of 3, but can dismount only 7 riflemen—a 54 percent reduction in the dismounted strength of a mechanized infantry units. On the other hand, the Bradley is reliable and well liked by its crews, and can move well in COMBINED ARMS exercises with M1 ABRAMS tanks.

The Bradley chassis has also been adapted for the MULTIPLE LAUNCH ROCKET SYSTEM (MLRS) and the ADATS surface-to-air missile system.

Specifications Length: 21.16 ft. (6.45 m.). Width: 10.5 ft. (3.2 m.). Height: 9.75 ft. (2.97 m.). Weight, combat: 22 tons. Powerplant: 500-hp. Cummins VTA-903T 8-cylinder diesel. Speed, road: 41 mph (66 kph). Range, max.: 300 mi. (483 km.).

BRDM: *Bronirovannaya Razvedyvatyelno Dozornaya Maschina* (Armored Reconnaissance Patrol Vehicle), family of Soviet 4 × 4 ARMORED CARS derived from the World War II BTR-40 scout car. Unlike the BTR-40, the BRDM (BTR-40P) is amphibious (with water jet propulsion), and has central tire-pressure control and two pairs of small retractable wheels between the main wheels for use in soft terrain. Introduced in 1959, the basic BRDM has no turret, and also served as the basis for a variety of specialized vehicles, including ANTI-TANK GUIDED MISSILE (ATGM) launchers (with SAGGER and SPIGOT ATGMs) and AIR DEFENSE vehicles (with SA-9 GASKIN surface-to-air missiles). The BRDM-2 (BTR-40PB) of 1966 is armed with a 14.5-mm. KPVT heavy machine gun and a 7.62-mm. machine gun in a one-man turret, and has better vision for the crew, a slightly more powerful engine, and a COLLECTIVE FILTRATION system for NBC defense.

Both versions have a boat-shaped hull with minimal (10-mm.) armor protection—sufficient only against small arms. The engine is at the front of the hull. With its four-wheel drive (and auxiliary mid-wheels) the BRDM is highly mobile; it serves in the RECONNAISSANCE and light air defense units of Soviet tank and Motorized Rifle formations, and has been widely exported.

Specifications Length: 18.7 ft. (5.7 m.). Width: 7.4 ft. (2.25 m.). Height: 6.2 ft. (1.9 m.). Weight, combat: 6 tons. Powerplant: 140-hp. GAZ-41 V-8 gasoline engine. Speed, road: 50 mph (80 kph). Range, max.: 310 mi. (500 km.).

BREMEN: A class of eight West German FRIGATES commissioned between 1982 and 1990. Modified versions of the Dutch KORTENAER frigates, with different weapons, sensors, and propulsion, the Bremens are flush-decked, with a small, blocky superstructure forward, a large stack amidships, and a separate deckhouse/hangar aft. The hull is fitted with a U.S. PRAIRIE MASKER bubble generator to reduce radiated noise, as well as fin stabilizers to improve seakeeping. A COLLECTIVE FILTRATION system is provided for a central citadel inside the ship as protection against nuclear or chemical attack.

Armament consists of an OTO-Melara 76.2-mm. DUAL PURPOSE gun on the bow, an 8-round NATO SEA SPARROW short-range surface-to-air missile launcher (with 24 missiles) before the bridge, two quadruple launchers for HARPOON anti-ship missiles amidships, and four Mk.32 tubes for lightweight anti-submarine TORPEDOES. Two launchers for RIM-116 RAM point-defense missiles are to be added in the 1990s. There is a hangar for two LYNX anti-submarine helicopters operated from a landing deck on the fantail; a Canadian "Bear Claw" haul-down apparatus allows safe recoveries even in heavy seas. See also FRIGATES, FEDERAL REPUBLIC OF GERMANY.

Specifications Length: 426.4 ft. (130 m.). Beam: 47.23 ft. (14.4 m.). Draft: 14 ft. (4.26 m.). Displacement: 2950 tons standard/3800 tons full load. Powerplant: twin-shaft CODOG, 2 5200-hp. MTU diesels (crusing), 2 25,000-shp. GE-Fiat LM2500 gas turbines (sprint). Speed: 18 kt. (diesels)/30 kt. (turbines). Range: (on diesels) 5700 n.mi. at 17 kt. Crew: 181. Sensors: 1 3RM20 surface-search radar, 1 DA-08 air-search radar, 1 WM-25 fire control radar, 1 STIR (Separate Tracking and Illuminating Radar), 1 DSQS-21B(Z) bow-mounted medium-frequency sonar. Electronic warfare equipment: 1 FL1800S electronic signal monitoring array, 4 Mk.36 SRBOC chaff launchers, 1 SLQ-25 Nixie torpedo countermeasures sled.

BREN GUN: British LIGHT MACHINE GUN, first introduced in 1938 and still in service with British and other Commonwealth forces. Derived from a Czech design (hence Bren, from Brno-Enfield) and originally chambered for .303-caliber (7.7-mm.) ammunition, the Bren was the standard British squad automatic weapon during World War II. After 1945, it was redesigned for NATO-standard 7.62 × 51-mm. rounds, and many older Brens were actually rechambered for the new round (as the L4A1). The present L4A4 model is now issued only to reserve and support units in the British army, having been supplanted from 1985 by the

5.56-mm. L86 Light Support Weapon, a variant of the ENFIELD L85A1 assault rifle.

The Bren has a conventional layout, with a straight wooden stock, a wooden pistol grip, and a bipod; the quick-detachable barrel is fitted with a carrying handle. A gas-operated weapon, the Bren is fed from 30-round box magazines fitted over the receiver, reflecting the pre-1945 British army preference for deliberate aimed fire, as opposed to high-volume (but relatively inaccurate) suppressive fire.

Specifications Length OA: 44.6 in. (1.13 m.). **Length, barrel:** 21.1 in. (536 mm.). **Weight, empty:** 21 lb. (9.53 kg.). **Muzzle velocity:** 2700 ft./sec. (823 m./sec.). **Cyclic rate:** 500 rds./min. **Practical rate:** 80–100 rds./min. **Effective Range:** 640 m.

BREWER: NATO code name for the Soviet Yakolev Yak-28 twin-jet, multi-purpose combat aircraft. With no direct equivalent in the West, fast, rugged, and with substantial internal volume, Yak-28s have been used as light BOMBERS, INTERCEPTORS, and RECONNAISSANCE and ELECTRONIC COUNTERMEASURES (ECM) platforms. A development of the earlier Yak-25/27, with more powerful engines, improved aerodynamics, and revised internal systems, Brewer first flew in 1961, and entered service in 1963; production terminated in 1963.

The basic Yak-281 Brewer-B two-seat light bomber has a slender, area-ruled ("coke bottle") fuselage with a fighter-type clamshell canopy for the pilot, a glazed bomb-aiming position for the bombardier/navigator in the pointed nose, and a small navigation radar in a ventral fairing. The midfuselage houses fuel tanks, an internal weapons bay with a capacity of 6600 lb. (3000 kg.), and bicycle landing gear. The sharply swept, mid-mounted wings have small outboard ailerons and larger inboard flaps, with small outrigger landing gear in the tips to stabilize the aircraft on the ground. The engines are mounted in underwing nacelles at midspan.

Brewer-B formed the backbone of FRONTOVAYA AVIATSIYA's long-range strike force until the introduction of the Su-24 FENCER in the 1970s; the Brewer-C light bomber was essentially similar. The Yak-28R Brewer-D reconnaissance aircraft, first spotted in 1969, is externally similar to the Brewer-B, but has a ground mapping and surveillance radar in the forward half of the weapons bay, with provisions for cameras and other sensors behind. Brewer-E is an ECM conversion of

Brewer-B. Designed to escort strike formations, it has a powerful ACTIVE JAMMING set in the weapons bay. A few Brewer-Ds and -Es are still in service with the Soviet Air Force (VVS).

Brewer was also developed into a long-range interceptor for the Soviet Air Defense Troops (PVO) as the Yak-28P (NATO code name Firebar), which entered service in small numbers in 1964; the definitive version appeared in 1967. In Firebar, the glazed nose is replaced with a long, pointed radome for a "Skip Spin" radar (also on the Su-15 FLAGON). While not as fast as the latter, Firebar has a superior combat radius, especially useful for patrolling the Soviet Union's Arctic frontier. The pilot and a systems operator are seated in tandem under an extended canopy. The standard armament consists of two AA-3 ANAB radar guided missiles and two AA-2 ATOLL infrared guided missiles. The performance of Brewer and Firebar are essentially similar.

Specifications Length: 76 ft. (21.17 m.). **Span:** 42.5 ft. (12.95 m.). **Powerplant:** 2 Turmansky R-11 turbojets, 13,120 lb. (5960 kg.) of thrust each. **Weight, empty:** 30,000 lb. (13,600 kg.). **Weight, max. Takeoff:** 50,000 lb. (22,000 kg.). **Speed, max.:** Mach 1.87 (1249 mph/2000 kph) at 36,000 ft. (11,000 m.). **Initial climb:** 28,000 ft./min. (8536 m./min.). **Service ceiling:** 55,000 ft. (16,500 m.). **Combat radius:** (hi-hi-hi) 575 mi. (957 km.).

BRIGADE: An army formation which usually includes three BATTALIONS, and may be independent or else subordinate to a DIVISION (with three brigades to a division). The smallest formations capable of independent operations, brigades are normally organized on COMBINED-ARMS principles: an armored brigade may include two or more tank battalions, a battalion or two of mechanized infantry, and a battery or more of artillery. In mechanized brigades, the ratio of tank and mechanized infantry battalions is reversed, while infantry brigades are often independent, and may include organic ENGINEER, ANTI-TANK and ANTI-AIRCRAFT units, to act as minidivisions. In some armies, brigade organizations are fixed. In others, such as the U.S. Army, divisions have a pool of battalions which are attached as required to a brigade headquarters. ARTILLERY brigades consist of three or more artillery battalions. In well-equipped armies, most divisions have at least one organic artillery brigade, with additional artillery brigades at CORPS, ARMY, and ARMY GROUP levels. Soviet-style divisions are generally composed of REGIMENTS, roughly equivalent to brigades, but smaller. In Soviet terminology, "brigade" *(brigada)* refers to any ad hoc team, but SPETSNAZ, AIR ASSAULT, and NAVAL INFANTRY brigades are in fact sizable units.

BRINKSMANSHIP: A diplomatic technique associated with the policies of John Foster Dulles, U.S. secretary of state from 1952 to 1958, which was ancillary to the strategy of MASSIVE RETALIATION. In the event of any Soviet aggression, even if minor or carried out by proxy, the prescribed brinksmanship response was to threaten a nuclear attack against the U.S.S.R. The threat was supposed to initiate a diplomatic exchange which would lead to the renunciation of aggression. Both brinkmanship and massive retaliation never represented more than a segment of U.S. policy; both were formally renounced by the end of the 1950s. See also ASSURED DESTRUCTION.

BRISTOL (TYPE 82): A large British guided-missile DESTROYER commissioned in 1973. Originally intended as the first of four ANTI-AIR WARFARE escorts for a planned class of large aircraft carriers that was canceled in 1966, only the *Bristol* was completed. Cheaper, less capable versions of its weapons and sensors were later incorporated into the smaller SHEFFIELD (Type 42) destroyers.

Armament, rather light for a ship of this size, originally consisted of a 4.5-in. (114-mm.) DUAL-PURPOSE gun at the bow, a LIMBO depth charge mortar behind the gun, a twin-arm SEA DART surface-to-air missile launcher (with 40 missiles) aft, and 2 manually operated 20-mm. cannons. The Limbo was removed in 1978, and 2 twin 30-mm. cannons and 2 additional 20-mm. cannons were added after the 1982 Falklands War. There is a helicopter landing deck on the fantail, but no hangar. Outfitted as a flagship, *Bristol* has an NTDS data link and 2 SCOT satellite communications terminals. See also DESTROYERS, BRITAIN.

Specifications **Length:** 507 ft. (154.5 m.). **Beam:** 55 feet (16.8 m.). **Draft:** 23 feet (7 m.). **Displacement:** 6100 tons standard/7100 tons full load. **Powerplant:** twin-shaft COSAG, 2 oil-fired boilers, 2 1500-shp. steam turbines (cruising); 2 22,300-shp. gas turbines (sprint). **Fuel:** 900 tons. **Speed:** 28 kt. **Range:** 5000 n.mi. at 18 kt. **Crew:** 397 (+ 100 cadets). **Sensors:** 1 Type 1022 air-search radar, 1 Type 1006 navigation radar, 1 Type 992 long-range surveillance radar, 2 Type 909 Sea Dart guidance radars, 1 hull-mounted Type 184 medium-frequency sonar, 1 Type 162 side-scanning sonar. **Electronic warfare equipment:** 1 UAA-1 "Abbey Hill" electronic signal monitoring array, 1 Type 970 active jamming array, 4 Mk.36

SRBOC chaff mortars, 2 Corvus chaff rocket launchers.

BROADSWORD (TYPE 22): A class of 14 British FRIGATES commissioned between 1979 and 1990. Originally intended as replacements for the widely sold LEANDER anti-submarine warfare (ASW) frigates, the Broadswords were built in three distinct subclasses ("batches"). The 4 Batch 1 ships are essentially ASW escorts with only a close-in air defense capability; the 6 Batch 2 ships have a longer hull for better speed, range, and seakeeping, and also to accommodate additional ASW systems; the 4 Batch 3 ships, which incorporate some lessons of the 1982 Falklands War, combine the long hull of Batch 2 ships with a stronger general-purpose armament, to act in effect as DESTROYERS.

Commissioned between 1979 and 1982, the Batch 1s are flush-decked, with a long, low superstructure forward, a large stack amidships, and a detached deckhouse/hangar aft. Armament consists of 4 EXOCET MM 38 anti-ship missiles mounted in single canisters on the foredeck, 2 6-round SEA WOLF point defense missile launchers, one before the bridge, the other on top of the hangar (each with 30 missiles), 2 40-mm. BOFORS GUNS amidships, and 2 sets of STWS-1 triple tubes for light ASW TORPEDOES; 2 20-mm. cannons were added in 1982–83 (post-Falklands) for close-in defense. The main ASW armament consists of 2 Westland LYNX helicopters operated from a hangar and a landing deck on the fantail. These ships also have 2 SCOT satellite communications terminals and a satellite navigation unit.

Commissioned between 1984 and 1988, the Batch 2s ("Boxer class") are longer, and have a flared clipper bow to improve seakeeping. Armament and sensors are slightly revised, with improved Sea Wolf GWS-25 Mod 3 launchers.

Commissioned between 1987 and 1990, the Batch 3s ("Cornwall" class) combine the long hull with a heavier, general-purpose armament. A 4.5-in. (114-mm.) DUAL PURPOSE gun on the foredeck and 2 quadruple HARPOON launchers on the superstructure amidships replace the Exocets; and a 30-mm. GOALKEEPER radar-controlled gun, for defense against sea-skimming missiles, and 2 power-driven Oerlikon twin 20-mm. cannons replace the Bofors guns in supplementing the 2 Sea Wolf launchers. Displacement is now greater than that of some destroyers.

The Batch 1 HMS *Broadsword* and *Battleaxe* served in the Falklands War, where their Sea Wolf missiles proved effective against low-flying aircraft (unlike other British missiles), and even provided some defense against Argentinian air-launched Exocets. Early ships of the class are to be upgraded with Harpoon instead of Exocet, Type 2031 towed arrays, vertical launchers for Sea Wolf, and landing deck and hangar modifications to accommodate larger SEA KING helicopters. See also FRIGATES, BRITAIN.

Specifications Length: (Batch 1) 430 ft. (131.1 m.); (Batch 2/3) 475.7–480.6 ft. (145–146.5 m.). Beam: 48.5 ft. (14.8 m.). Draft: 19.9 ft. (6.1 m.). Displacement: (Batch 1) 3500 tons standard/ 4400 tons full load; (Batch 2) 4100 tons standard/ 4800 tons full load; (Batch 3) 4900 tons full load. Powerplant: (Batch 1 and first 2 Batch 2) twin-shaft COGOG, 2 4100-shp. Rolls Royce Tyne RM1C gas turbines (cruising), 2 25,000-shp. Rolls Royce Olympus TM3B gas turbines (sprints); (third Batch 2) twin-shaft COGOG, 2 Tynes, 2 18,770-shp. Rolls Royce Spey SM1A gas turbines; (last 3 Batch 2 and Batch 3) twin-shaft COSAG, 2 Speys and 2 Tynes, 48,220 shp. Speed: (Batch 1) 30 kt. *(Olympus)*/18 kt. *(Tynes)*; (Batch 2/3) 32 kt. Range: (Batch 1) 4500 n.mi. at 18 kt./1200 n.mi. at 29 kt.; (Batch 2/3) 7000 n.mi. Crew: (Batch 1); 223 (Batch 2) 273; (Batch 3) 250. Sensors: (Batch 1) 1 Type 1006 navigational radar, 1 Type 967/968 surveillance radar, 2 Type 910 Sea Wolf radar/electro-optical missile directors, 1 Type 2016 hull-mounted medium-frequency sonar; (Batch 2) same as Batch 1, but with Sea Wolf Type 911 directors and a Type 203 towed array sonar; (Batch 3) same as Batch 2, but with 1 Type 2050 hull-mounted sonar. Electronic warfare equipment: 1 UAA-1 "Abbey Hill" electronic signal monitoring array, 2 Mk.36 SRBOC chaff mortars, 2 i-round Corvus chaff rocket launchers.

BRONSTEIN: A class of two U.S. FRIGATES (FF-1037 and -1038) commissioned in 1962. The prototypes for all subsequent U.S. deep ocean escorts until the introduction of the Perry class (FFG-7) in the late 1970s, the Bronsteins were optimized for ANTI-SUBMARINE WARFARE (ASW).

As built, armament consisted of a twin 3-in. 50-caliber anti-aircraft gun at each end of the ship, an 8-round ASROC launcher before the bridge, two sets of Mk.32 triple tubes for lightweight ASW torpedoes, and a DASH (Drone Anti-Submarine Helicopter) system. DASH was removed in the early 1970s, and the aft 3-in. gun was replaced by an SQR-15 TASS towed array sonar shortly thereafter.

In many ways "proof-of-concept" experiments, the Bronsteins were the first U.S. frigates with large low-frequency sonar, but they proved too slow for modern ASW operations, and also lacked range and seakeeping. Too small to retrofit with modern weapons and sensors, they were retired in 1989–90. See also FRIGATES, UNITED STATES.

Specifications Length: 350 ft. (106.7 m.). Beam: 40.5 ft. (12.3 m.). Draft: 23 ft. (7 m.). Displacement: 2360 tons standard/2650 tons full load. Powerplant: single-shaft steam, 2 oil-fired boilers, 2 sets of geared turbines, 20,000 shp. Speed: 24 kt. Range: 4000 n.mi. at 15 kt./3200 n.mi. at 20 kt. Crew: 218. Sensors: 1 SQR-15 TASS, 1 bow-mounted SQS-26 low-frequency active/passive sonar, 1 SPS-40 air-search radar, 1 Mk. 35 fire control radar. Electronic warfare equipment: WLR-1 and WLR-2 radar warning receivers, 1 ULQ-6 active jamming array.

BROOKE: A class of six U.S. guided-missile frigates commissioned in 1966–67. Essentially ANTI-AIR WARFARE versions of earlier GARCIA class ANTI-SUBMARINE WARFARE (ASW) frigates, they combine the ASW capability of the Garcias with enhanced air defense capability of a surface-to-air missile system. Flush-decked, they have a short, blocky superstructure amidships with a single "mack" (combination of mast and stack). Gyro-controlled fin stabilizers are fitted to improve seakeeping.

Armament consists of a 5-in. 38-caliber DUAL-PURPOSE gun on the bow, an 8-round ASROC launcher behind the gun, two sets of Mk.32 triple tubes for lightweight ASW TORPEDOES amidships, and a Mk.22 launcher for 16 STANDARD MR surface-to-air missiles amidships (in place of the second 5-in. gun of the Garcias). As completed, the Brookes had a fantail landing deck and hangar for DASH (Drone Anti-Submarine Helicopter), but this was enlarged in the 1970s to allow them to operate an SH-2 SEA SPRITE ASW helicopter.

With their limited magazine capacity and their inability to deal with massed attacks, the Brookes are poor anti-air escorts, and their lack of a PHALANX radar-controlled gun also leaves them vulnerable to surface-skimming ANTI-SHIP MISSILES. On the other hand, with their large sonar and helicopter, they remain valuable ASW escorts. All were retired by the U.S. Navy, but two have been leased to Pakistan and two more to Turkey. See also FRIGATES, UNITED STATES.

Specifications Length: 414.5 ft. (118.9 m.). Beam: 44.16 ft. (13.5 m.). Draft: 24 ft. (7.3 m.).

Displacement: 2640 tons standard/3600 tons full load. **Powerplant:** single-shaft steam, 2 oil-fired boilers, 2 sets geared turbines, 35,000 shp. **Speed:** 27 kt. **Range:** 4000 n.mi. at 20 kt. **Crew:** 276. **Sensors:** 1 SPS-10F surface-search radar, 1 SPS-52A/D 3-dimensional air-search radar, 1 SPG-51 missile guidance radar, 1 Mk. 35 fire control radar, 1 SQS-26 bow-mounted low-frequency active/passive sonar. **Electronic warfare equipment:** 1 SLQ-32(V)2 electronic countermeasures suite, 2 Mk.36 SRBOC chaff launchers.

BTR-: *Bronetransporter* (Armored Transporter), Soviet designation for ARMORED PERSONNEL CARRIERS (APCs). During World War II, cooperation between Soviet tank and infantry units was hampered by the lack of protected, cross-country vehicles for the infantry; troops were often carried on top of tanks (*tank* DESANTS), but that was a costly stopgap. During the war, the Soviet army received some 3340 M3A1 scout cars and 804 M2/M5 half-tracks from the U.S. under Lend-Lease, and also made some use of captured German half-track carriers, as well as Soviet ARMORED CARS. It was not until its 1946 reorganization that the Soviet army introduced a successful troop carrier, the BTR-152, a converted 6 × 6 truck with an armored hull, a two-man crew, and room for up to 19 infantrymen, but originally without overhead protection (added only in 1961), and 13-mm. armor, sufficient only against small arms fire. Built from 1956 with central tire-pressure controls, the BTR-152 serves in some Third World armies. In Soviet service, it was slowly replaced during the 1960s by the BTR-60, an 8 × 8 wheeled APC with improved armor and mobility, and overhead protection. In various models, it remains the standard APC in Soviet and Warsaw Pact Motorized Rifle divisions and has been widely exported. In Soviet tank divisions, on the other hand, the BTR-152 was replaced by the BTR-50, a fully tracked APC derived from the PT-76 light tank, which was in turn replaced by the BMP infantry fighting vehicle in the early 1970s. A light 4 × 4 armroed car, the BTR-40, served in Soviet reconnaissance units from 1948 until it was replaced in the 1960s by the BRDM.

BTR-40: A Soviet light ARMORED CAR, introduced in 1946, which was in fact a light armored truck with room for a driver and seven troops. A 4 × 4 vehicle weighing some 5 tons, it is lightly armored (10 mm.), has no overhead protection, and is normally armed only with a 7.62-mm. light machine gun. Powered by 90-hp. GAZ V-6 gasoline engine, it lacks range and cross-country mobil-

ity. The BTR-40 has been replaced in the Soviet army by the BRDM, but is still much used in the Third World.

BTR-50: A Soviet tracked ARMORED PERSONNEL CARRIER, formerly standard equipment for the MOTORIZED RIFLE battalions of Soviet tank divisions. It has now mostly been replaced in Soviet service by the BMP infantry fighting vehicle, but remains in service with Warsaw Pact and Third World armies. Introduced from 1955 (before the wheeled BTR-60 destined for Motorized Rifle divisions), the BTR-50 is a conversion of the PT-76 light tank (with the turret replaced with an armored box structure), and bench seats for 12 (or more) infantrymen. Like the PT-76, it is amphibious and is propelled through the water by pump jets. The first model was open-topped, but the subsequent BTR-50PK of 1960 introduced an armored roof with hatches for entry and exit. The maximum armor thickness of 15 mm. is sufficient only against small-arms fire and splinters. Armament is limited to a pintle-mounted 7.62-mm. machine gun, but the infantry on board can fire from the roof hatches. A COLLECTIVE FILTRATION unit for NBC protection is provided in the PK version.

Specialized variants include the BTR-50PU ARMORED COMMAND VEHICLE, ENGINEER vehicles, MORTAR and RECOILLESS rifle carriers, and armored repair vehicles. Czechoslovakia has produced its own version, the OT-62 TOPAS (*Transporter Obojzivelay Pasovy Stredni*, or Medium Tracked Amphibious Vehicle), in many ways superior to the original. It has a more powerful engine, 20 mm. of armor plate, an NBC system, and side exit doors. The improved TOPAS-2A also has a small turret on the right front equipped with a 7.62-mm. machine gun internally and a T-21 82-mm. recoilless rifle externally, while the subsequent TOPAS-2AP has the 14.5 mm KPV heavy machine gun turret of the SKOT 2AP, a Czech version of the BTR-60. TOPAS is made in a number of specialized variants, including an armored recovery vehicle and a mortar carrier. It equips Czech tank and motor rifle divisions and the Polish 7th Marine Division, and has also been exported.

Specifications **Length:** 22 ft. (6.7 m.). **Width:** 10.5 ft. (3.2 m.). **Height:** 6 ft. (1.83 m.). **Weight, combat:** 14 tons. **Powerplant:** (BTR-50) 240-hp. V-6 diesel; (OT-62) 300-hp. diesel. **Speed, road:** 28 mph (47 kph.). **Speed, water:** 8 mph (13.4 kph.). **Range, Max.:** 161 mi. (270 km.).

BTR-60: A Soviet 8 × 8 wheeled ARMORED PERSONNEL CARRIER (APC), the standard APC of

Soviet MOTORIZED RIFLE divisions, and widely exported. Introduced from 1961 to replace the BTR-152 (a converted truck), the BTR-60 has a boat-shaped armored hull with seating for a driver and up to 13 infantrymen. The initial version had an open top, but in 1964 production switched to the BTR-60PK, with an armored roof fitted with hatches for troop egress. That was followed in 1965 by the definitive BTR-60PB, with a small turret armed with a 14.5-mm KPV heavy machine gun and a 7.62-mm. machine gun, and room for a driver and twelve infantrymen.

The all-welded steel hull is divided into a driver's compartment up front, a troop compartment in the middle, and an engine compartment in the rear. Maximum armor thickness is 14 mm., sufficient only against small arms and splinters; top armor is even thinner, and bullets can enter through the unarmored front wheel wells to disable the driver. The BTR-60PB was not built with an NBC filter system, but one may be retrofitted. Amphibious without preparation, it is propelled through the water by two pump jets at 6 mph (10 kph). Specialized variants include the BTR-60PU ARMORED COMMAND VEHICLE and a FORWARD AIR CONTROL (FAC) version.

The BTR-60's shortcomings include an awkward, twin-engine powerplant of explosion-prone gasoline engines, and the lack of side or rear doors for tactical crew exits. The Czech army attempted to remedy these deficiencies in their OT-64 SKOT (*Stredny Kolovy Obrneny Transporter*, or Medium Armored Wheeled Transporter) based on a Tatra truck chassis. In service from 1964, it has two large rear exit doors, a turret armed with a 14.5-mm. KPV heavy machine gun capable of anti-aircraft use, increased frontal armor protection, propellers in place of pump jets, and better mechanical reliability. Specialized variants include an armored command vehicle, a radio vehicle, and an armored repair vehicle. The Czechs have also produced a version armed with the same turret fitted to the OT-65 tracked troop carrier (see BTR-50), equipped with a 7.62-mm. machine gun and an 82-mm. RECOILLESS rifle. The SKOT is used by motor rifle formations in the Czech and Polish armies, and has been exported to India, Uganda, and elsewhere.

The Soviet army introduced the BTR-70 from the early 1970s, with side exit doors, upgraded armor protection, two 100-hp engines, and a positive overpressure NBC protection system. It is now being supplanted in turn by the BTR-80.

Specifications Length: 25 ft. (7.62 m.). Width: 7 ft. (1.13 m.). Height: 7 ft. (1.13 m.). Weight, combat: 10.3 tons. Powerplant: 2 rear-mounted 90-hp. GAZ-49B 6-cylinder gasoline engines. Speed, road: 50 mph (80 kph). Range, Max.: 311 mi. (500 km.).

BTR-70: An improved version of the Soviet BTR-60PB wheeled armored personnel carrier.

BTR-80: A further development of the Soviet BTR-60 and BTR-70 8 × 8 wheeled ARMORED PERSONNEL CARRIERS. Externally similar to its predecessors, the BTR-80 has more armor protection (also for the wheel wells), and a large rear exit door. The most notable change, however, is the installation of a single 260-hp. diesel engine in place of the two gasoline engines of the earlier models; with it, off-road mobility is considerably improved, maintenance is simplified, and the risk of fire is reduced. Road speed and other characteristics are similar to those of the BTR-60.

BUCCANEER: British twin-jet, long-range ATTACK AIRCRAFT, originally built for the FLEET AIR ARM, and still in service with the Royal Air Force and the South African air force. Developed by Blackburn (now British Aerospace) in the late 1950s as a high-speed, low-altitude carrier strike aircraft, the Buccaneer reconciled the requirement for a high-speed design with the low-speed stability needed for carrier landings through "boundary layer controls" achieved by blowing engine bleed air over the wings and the horizontal stabilizer. Other innovative features include an internal rotary weapons bay (to eliminate the drag of externally carried stores), and, more remarkably, double curvatures on the fuselage and wings to reduce radar reflectivity (an early if unintended example of STEALTH). Because supersonic speed was not required at low altitudes, fuel-efficient nonafterburning engines were installed to achieve long range. The first of 40 naval Buccaneer S.1s flew in 1963, but their two Rolls Royce Gyron Junior turbojets did not provide adequate power, and were replaced by more powerful Rolls Royce Spey turbofans in the 87 subsequent S.2s. In 1968, the RAF bought an additional 46 S.2A/Bs, identical to the naval S.2s. When the Royal Navy decommissioned its last conventional aircraft carrier in 1978, its surviving S.2s were transferred to the RAF and redesignated S.2C/D. The RAF originally intended the Buccaneer to serve as an interim aircraft until delivery of the TORNADO, but it is now scheduled to remain in service at least until 1995 in three squadrons (12, 208, and 237 OCU) specialized for maritime strike. South Africa purchased 16 Mk.50 versions in 1965–66, which are identical to the S.2 except for the installation of a BS.605 rocket motor rated at 8000 lb. (3636 kg.) of thrust for hot-and-high takeoffs; six surviving Mk.50s still serve with No. 24 Squadron in a ground attack role.

The Buccanneer has an area-ruled ("coke bottle") fuselage for low transsonic drag, with a Ferranti "Blue Parrot" maritime search radar in the nose, and an AERIAL REFUELING probe attached to the right side. The pilot and systems operator sit in a tandem cockpit under a single-piece sliding canopy. The outdated AVIONICS include a optical reflecting gunsight and head-down radar display for the pilot, a radar screen, threat-warning display and video terminal for the system operator, plus "Sky Guardian" RADAR-WARNING RECEIVERS, UHF and HF radios, a TACAN beacon, a radar altimeter, an INERTIAL NAVIGATION system, and DOPPLER navigation radar.

The space behind the cockpit contains three large fuel tanks. The rotary weapons bay amidships can carry up to four 1000-lb. (454-kg.) bombs or a variety of other stores, including up to 3900 lb. (1772 kg.) of additional fuel. The tail ends in four petal-type airbrakes. The engines are in pods at the wing roots, with simple semicircular cheek inlets. The midmounted wings have a compound sweep, and incorporate large leading edge slats, blown flaps, and boundary layer control for low-speed handling. There are four hardpoints under the wing for additional ordnance or external fuel tanks. The maximum external load of 12,000 lb. (5455 kg.) can include free-fall bombs, MARTEL or SEA EAGLE anti-ship missiles, PAVEWAY laser guided bombs, AIM-9 SIDEWINDER air-to-air missiles for self-defense, or ALQ-101 ELECTRONIC COUNTERMEASURES (ECM) pods.

While Buccaneer is not supersonic, its low-altitude performance is outstanding. In U.S. RED FLAG duel against much faster aircraft, Buccanneers can evade interception by flying at high speeds just above the ground. On several occasions, Buccaneers have conducted strike exercises involving an unrefueled round trip between the Irish Sea and Gibraltar. The amazingly strong airframe shows few signs of fatigue after nearly 25 years of hard service, and a squadron of Buccaneers performed well in the 1991 Persian Gulf conflict.

Specifications Length: 63.5 ft. (19.35 m.). Span: 44 ft. (13.41 m.). **Powerplant:** 2 Rolls Royce Spey turbofans, 11,200 lb. of thrust each. **Fuel:** 15,612 lb. (7096 kg.). **Weight, empty:** 29,980 lb. (13,599 kg.). **Weight, max. takeoff:** 62,000 lb. (28,123 kg.). **Speed, max.:** Mach 0.91 (691 mph/1112 kph) at sea level. **Speed, Cruise:** 621 mph (1000 kph) at sea level. **Combat radius:** (hi-lo-lo-hi, max. payload) 600 mi. (966 km.). **Range, max.:** (external fuel tanks, full internal bomb load) 2300 mi. (3701 km.).

BUIC: Back-Up Interceptor Control, an emergency computer and communication system meant to supplement the primary SAGE ground-control intercept system within the North American Aerospace Defense Command (NORAD). BUIC was intended to operate if enemy attack neutralized the larger SAGE facilities. Both SAGE and BUIC were replaced in the 1980s by the JOINT SURVEILLANCE SYSTEM. See also ADTAC; AIR DEFENSE.

BULLPUP: The U.S. AGM-12 air-to-surface missile. Developed from 1954 by the U.S. Naval Weapons Center at China Lake, Bullpup was intended as a standoff weapon for attacks against well-defended point targets. The initial AGM-12A of 1959 combined a standard 250-lb. (114-kg.) high-explosive bomb with a navy-designed solid rocket motor. It had four cruciform wings in the tail for lift, and four small, movable canards in the nose for steering. The missile was controlled by radio COMMAND GUIDANCE, whereby the operator flies the missile manually with a small joystick in his cockpit; flares in the base of the missile facilitated visual tracking.

The AGM-12A was soon superseded by the AGM-12B Bullpup-A, with a storable liquid-fueled Thiokol LR-58RM4 motor rated at 12,000 lb. (5455 kg.) of thrust, and armed with a 250-lb. (114-kg.) bomb of improved design. In 1961, Bullpup-A was adopted by the U.S. Air Force, which modified the guidance system to allow the pilot to control the missile while flying off to one side of its trajectory. From 1963, the missile was also produced under license by Kongsberg Wapenfabrik of Norway. Bullpup-A production totaled 22,000 missiles in the U.S. and 8000 in Norway.

In 1964, the U.S. Navy introduced the AGM-12C Bullpup-B, with a 1000-lb. (454-kg.) warhead and a 33,000-lb. (15,000-kg.) Thiokol LR-62RM-2 liquid-fueled motor. The guidance system was unchanged, but the maximum range was extended. Some 4600 were produced by 1969; of these, some 840 were refitted with fragmentation warheads and designated AGM-12E. A proposed nuclear variant (AGM-12D) was also designed but did not enter service, and a proposed ELECTRO-OPTICALLY guided version (AGM-79A Bulleye) was rejected in favor of the AGM-65A MAVERICK. A laser-guided version, the AGM-83 Bulldog, was produced in small numbers for the U.S. Marine Corps, but was soon displaced by laser-guided Mavericks. A small, subcaliber training missile (ATM-12A) was produced in quantity, but later training was conducted with surplus Bullpup-As. All Bullpups have the tactical disadvantage of a command guidance system that requires the launching aircraft to have a direct line of sight to the target throughout the missile's flight, precluding evasive maneuvers.

Bullpup can be carried by most U.S. and allied fighters and attack aircraft, including the SKY-HAWK, PHANTOM, CORSAIR, STARFIGHTER, SUPER SABRE, and INTRUDER. Though no longer in U.S. service, Bullpups are still in service elsewhere.

Specifications Length: (AGM-12B) 10.5 ft. (3.2 m.); (AGM-12C) 13.33 ft. (4.06 m.). **Diameter:** (AGM-12B) 12 in. (305 mm.); (AGM-12C) 17.3 in. (440 mm.). **Span:** (AGM-12B) 37 in. (940 mm.); (AGM-12C) 46.5 in. (1.18 m.). **Weight, launch:** (AGM-12B) 571 lb. (259.5 kg.); (AGM-12C) 1785 lb. (811 kg.). **Speed, max.:** Mach 2.5. **Range, max.:** (AGM-12B) 7.5 mi. (12.5 km.); (AGM-12C) 10 mi. (16.7 km.) **Operators:** Arg, Aus, Bra, Chi, Den, Gre, Isr, Nor, NZ, Phi, ROK, Tai, Tur, Ven.

BURKE: A new class of U.S. Navy guided-missile DESTROYERS (DDG-51), the first of which was commissioned in 1990. Intended to replace ADAMS- and COONTZ-class guided-missile destroyers as ANTI-AIR WARFARE (AAW) escorts for CARRIER BATTLE GROUPS (CVBGs) or SURFACE ACTION GROUPS (SAGs), the Burkes are equipped with a lighter version of the AEGIS air defense system but also have considerable capabilities for ANTI-SUBMARINE WARFARE (ASW) and ANTI-SURFACE WARFARE (ASUW). Navy plans call for more than 30 Burkes; three are now under construction.

The Burke has a shorter, broader hullform than earlier U.S. destroyers, for better seakeeping and maneuverability, as well as increased internal volume. Also unlike its predecessors, the ship is constructed almost exclusively of steel, with aluminum only for the funnels. In a further reaction to British experiences in the 1982 Falklands War, the Burke also incorporates some 130 tones of KEVLAR armor around machinery and magazine spaces.

Armament consists of 2 Mk.41 vertical launch systems (VLS), for STANDARD surface-to-air missiles, TOMAHAWK cruise missiles, and, eventually, vertically launched ASROC or SEA LANCE anti-submarine missiles. The Mk.41 Mod 0 launcher on the

foredeck with 32 cells is intended primarily for Standard, while the 64-cell Mk.41 Mod 0 launcher on the quarterdeck is shared by Tomahawk, ASROC, and more Standard. Other armament consists of 8 HARPOON anti-ship missiles in 2 quadruple launch canisters just ahead of the aft Mk.41, a 5-in., 54-caliber DUAL PURPOSE gun on the foredeck ahead of the forward Mk.41 launcher, 2 20-mm. PHALANX radar-controlled guns for anti-missile defense (one before the bridge, the other just ahead of the Harpoon launchers), and 2 sets of Mk.32 triple tubes for the Mk.50 BARRACUDA Advanced Lightweight Torpedo (ALWT). An SH-60B SEAHAWK anti-submarine helicopter can be operated from a landing deck on the fantail, but there is no hangar. The Burke has a distinctly uncluttered appearance because the vertical launchers do not protrude above the deck; while the ship may seem underarmed, the Mi. 41s have a greater magazine capacity, a higher rate of fire, and much better mechanical reliability than more visible twin-arm launchers.

The true strength of the DDG-51 is in its sensors and fire controls, primarily Aegis with its associated SPY-1D PHASED ARRAY RADAR. Four arrays mounted on the four faces of the forward superstructure allow 360° coverage out to a range of 200 mi. (334 km.) against very high-flying aircraft. The SPY-1 radar can track hundreds of targets simultaneously, while Aegis can prioritize targets according to threat estimates, and control up to 12 Standard missiles simultaneously to cope with mass attacks by anti-ship missiles. The Burke also has an NTDS data link, which allows it to control the sensors and weapons of other NTDS-equipped ships in the area.

Congress has mandated experiments with a Rankin Closed Cycle Energy Recovery (RACER) system, which uses heat from engine exhaust gases to produce steam used in turn to drive another turbine; it has been estimated that RACER could add 1000 n.mi. to the ship's range, but cost and space factors make its addition unlikely. A proposed "Improved Burke" would be lengthened to include a hangar, a second 5-in. gun, and larger missile magazines. See also DESTROYERS, UNITED STATES.

Specifications Length: 466 ft. (142.1 m.). **Beam:** 59 ft. (18 m.). **Draft:** 30.6 ft. (9.3 m.). **Displacement:** 6624 tons standard/8300 tons full load. **Powerplant:** twin-shaft COGAG, 4 25,000-shp. General Electric LM2500 gas turbines. **Speed:** 30 kt. **Range:** 5000 n.mi. at 20 kt. **Crew:** 303. **Sensors:**

4 SPY-1D phased array radars, 1 SPS-67(V) surface search radar, 3 SPG-62 missile guidance radars, 1 SQS-53 bow-mounted low-frequency active/passive sonar, 1 SQR-19 TACTASS towed array. **Electronic warfare equipment:** 1 SLQ-32(V)2 electronic countermeasure system, 4 Mk.36 SRBOC chaff launchers.

BUS: See POST-BOOST VEHICLE.

BVR: Beyond Visual Range—air engagements or long-range air-to-air missiles. BVR missiles are invariably radar-guided by either SEMI-ACTIVE RADAR HOMING (SARH) or ACTIVE HOMING, or both in sequence. BVR missiles can attack enemy aircraft before the launching aircraft can be spotted, thus allowing a force to attrit another (or at least disrupt its formation) at long range, and then disengage or close in for a shorter-range engagement at will. But BVR missiles also have disadvantages. In a crowded airspace, it may be difficult to distinguish between friendly and enemy aircraft, even with IFF (Identification Friend or Foe), creating the possibility of "own goals" and inhibiting BVR engagement. (In the Vietnam air war, U.S. RULES OF ENGAGEMENT required the visual identification of all targets, sacrificing the tactical potential of the BVR SPARROW missiles of U.S. PHANTOM fighters.) BVR missiles are also relatively heavy and unmaneuverable as compred to short-range "dogfight" missiles such as the Soviet AA-8 APHID or U.S. AIM-9 SIDEWINDER; thus, if spotted, they can be evaded by agile fighters. Moreover, BVR missiles with SARH guidance (e.g., Sparrow, the British SKYFLASH, the Italian ASPIDE, and the Soviet AA-9 AMOS) require the launching aircraft to illuminate the target for the entire duration of the engagement, thus flying a relatively straight course towards the enemy, at the end of which the launching aircraft may be vulnerable to attack. On the other hand, BVR missiles with fire-and-forget active homing guidance are as costly as (small) fighters used to be, as e.g., the U.S. AIM-54 PHOENIX and AIM-120 AMRAAM. BVR capability is nevertheless deemed desirable in spite of its cost and complexity. See also AIR COMBAT MANEUVERING.

BZ: An incapacitating psychochemical agent whose effects are similar to those of LSD. Inhalation causes disorientation and degrades mental abilities. It has been used (rarely) for riot control and in special operations. See also CHEMICAL WARFARE.

C

C-: U.S. designation for transport aircraft. Additional letters indicate special-purpose modifications, e.g., AC- for attack gunships, EC- for electronic warfare, HC- and MC- for special operations, KC- for tankers, and RC- for reconnaissance. Models in service include: C-5 GALAXY; C-10 EXTENDER; the C-47 Dakota; C-130 HERCULES; C-135 STRATOTANKER; and C-141 STARLIFTER. The new C-17 has no name as yet.

C-17: New tactical transport aircraft developed by McDonnell Douglas for the U.S. Air Force MILITARY AIRLIFT COMMAND (MAC). A heavy-lift, air refuelable cargo plane for both inter- and intra-theater airlift, the C-17 can accommodate outsize loads including heavy engineering equipment and MAIN BATTLE TANKS (MBTs). Designed for short takeoff and landing, and the low-altitude parachute extraction of outsize cargo, the C-17 fills the gap between the C-130 HERCULES, capable of flying from short fields but without intercontinental range, and the C-5 GALAXY, with long range and outsize cargo capacity, but too large for tactical use. The first contract was awarded in 1981 and full-scale development was approved in 1985, but because of technical and budget problems, deliveries will not begin until 1992.

The first U.S. military transport with an automated cockpit equipped with multi-functional display screens instead of analog instrument dials, the C-17 has a three-man crew of pilot, copilot, and loadmaster. The cargo hold can accommodate a maximum payload of some 172,200 lb., including 102 fully equipped paratroops, or 48 stretcher cases, or a single M1 Abrams MBT. Intended for operations from short, unpaved fields, it can take off and land in less than 3000 ft. (915 m.).

Specifications Length: 175.2 ft. (53.41 m.). Span: 165 ft. (50.3 m.). **Cargo Hold:** 88.0 ft. (26.82 m.). **Powerplant:** 4 Pratt and Whitney F-117-PW-100 turbofans, 37,600 lb. (17,090 kg.) of thrust each. **Weight, max. takeoff:** 570,000 lb. (259,090 kg.). **Speed, Cruising:** 518 mph (865 kph). **Range, max.:** 2765 mi. (4617 km.).

C²: See COMMAND AND CONTROL ("C-Squared").

C³: See COMMAND AND CONTROL and COMMUNICATIONS ("C-Cubed").

C³I: See COMMAND AND CONTROL, COMMUNICATIONS, and INTELLIGENCE ("C-Cubed I").

CAFE: Conventional Armed Forces in Europe (sometimes Conventional Forces in Europe, CFE), ongoing negotiations between NATO and the members of the dissolving WARSAW PACT for the reduction of their nonnuclear forces. As of early 1990, a tentative agreement called for the reduction of U.S. and Soviet forces in central Europe to 195,000 men each, with additional limits on main battle tanks, artillery, armored personnel carriers, attack aircraft, and helicopters. The collapse of the Communist regimes of Eastern Europe in 1989 led to unilateral Soviet withdrawals, which largely made the negotiations irrelevant. Nonetheless, a preliminary CFE Treaty between NATO and the Warsaw Pact was signed in December 1990. See also MBFR.

CALIBER: In loose usage, the diameter of a gun muzzle, measured either in millimeters or inches; more technically, the length of the barrel is divided by its diameter to derive the length of the gun in "calibers." For example, a gun with a 5-in. bore and a 270-in. barrel is defined as a "5-in. 54-caliber" gun. Some older British artillery is defined instead by the weight of the solid shot it can fire, as e.g. the 17-pounder and 25-pounder. See also ARTILLERY.

CALIFORNIA: A class of two U.S. nuclear-powered guided-missile CRUISERS (CGN-36 and CGN-37), commissioned in 1974 and 1975. The Californias are essentially enlarged, nuclear-powered versions of 1960s-vintage guided-missile DESTROYERS. The design was finalized in June 1968, but completion was delayed until 1972–73 by the Vietnam War. Originally the class was intended for series production, but after two ships the navy preferred to build instead the larger, more capable VIRGINIA cruisers.

Intended primarily as ANTI-AIR WARFARE (AAW) escorts for CARRIER BATTLE GROUPS, the Californias also have considerable ANTI-SUBMARINE WARFARE (ASW) capability. Primary armament consists of a Mk.13 Mod 3 single-arm launcher at each end of the ship for RIM-66 STANDARD MR surface-to-air missiles, each with a magazine capacity of 40 missiles. The Californias also have two Mk.45 5-in. 54-caliber DUAL-PURPOSE guns, fore and aft. An 8-round ASROC pepperbox launcher with 8 reloads is mounted behind the forward gun; other ASW weapons include four fixed tubes for MK.46 lightweight ASW TORPEDOES. A helicopter pad is provided on the fantail, for vertical replenishment (VERTREP) from other ships, but there is no hangar for an ASW helicopter. Two quadruple launchers for HARPOON anti-ship missiles amidships provide an ANTI-SURFACE WARFARE (ASUW) capability. In the 1980s, two PHALANX 20-mm. radar-controlled guns were added for anti-missile defense, and two quadruple box launchers for TOMAHAWK cruise missiles were fitted on the helicopter pad. The ships are equipped with WSC-3 satellite communications terminals and the NTDS data link.

While not as capable as the later Virginias, the Californias are still valuable for escort nuclear-powered carriers. Very well armed in other respects (especially since the addition of Tomahawk), these ships are, however, too small for the Mk.41 VERTICAL LAUNCH SYSTEM or AEGIS radar and fire control system. See also CRUISERS, UNITED STATES.

Specifications **Length:** 596 ft. (181.7 m.). **Beam:** 61 ft. (18.6 m.). **Draft:** 31.5 ft. (9.6 m.). **Displacement:** 9561 tons standard/10,473 tons full load. **Powerplant:** twin-shaft nuclear, 2 D2G pressurized-water reactors, 2 sets of geared steam turbines, 60,000 shp. **Speed:** 30+ kt. **Crew:** 549. **Sensors:** 1 SPS-40B 2-dimensional air-search radar, 1 SPS-48C 3-dimensional air-search radar, 1 SPS-67 surface-search radar, 1 SPG-60 and 1 SPG-9A fire control radar, 4 SPG-51D missile guidance radars, 1 bow-mounted SQS-26CX low-frequency active/passive sonar. **Fire controls:** 1 Mk.13 weapon direction system, 1 Mk.114 ASW system, 2 Mk.74 missile direction systems, 1 Mk. 86 gun control system. **Electronic warfare equipment:** 1 SLQ-32(V)3 electronic signal monitoring and active jamming array, 4 6-barrel Mk. 36 SRBOC chaff rocket launchers.

CAMOUFLAGE, CONCEALMENT, AND DECEPTION (CC&D): The U.S. military term for three interrelated counterintelligence methods meant to preserve surprise.

Camouflage describes the use of natural and man-made materials to disguise military forces and facilities. Complete disguise is rarely achievable, but by breaking up the profile and distorting the surface pattern of objects, their recognition can be impeded. In the past, camouflage was aimed solely against visual recognition, but it is now often designed to impede sensors operating in the INFRARED (IR), ultraviolet, microwave (RADAR), and millimeter wave (MMW) bands as well. Special materials and design techniques can absorb and distort radar signals (see STEALTH), and others can mask thermal (IR) "signatures," but it is inevitably more difficult to overcome multi-spectral observation.

Concealment has the same aim but its means are entirely passive: the use of terrain to mask forces from enemy observation is basic. Camouflage can enhance concealment by disguising unavoidably exposed signatures, but concealment itself is becoming progressively more difficult because of the ever-increasing use of airborne and space-based sensors.

Deception is the all-encompassing term for both active and passive measures intended to mislead the enemy. Normally, successful deception combines misdirection (diversionary actions meant to distract attention); simulation (the use of physical or electronic DECOYS, double agents, and rumors) to create false impressions; and security, to prevent the enemy from penetrating the deceptive scheme. Camouflage and concealment are thus the natural complements of deception.

Standard Western CC&D and Soviet MAS-KIROVKA are analogous at the tactical and operational levels, but *maskirovka* continues up to the highest level of grand strategic level (e.g., with ACTIVE MEASURES) as a routine instrument of Soviet policy, while Western, and particularly U.S., attempts at strategic deception have been desultory. Another major difference is in priority: CC&D is usually an afterthought, while *maskirovka* has been integral to Soviet political and military action.

CANBERRA: British-designed light BOMBER, of 1950s vintage, also produced in the U.S. as the Martin B-57. A few remain in service in the Third World. **Operators:** Arg, Chi, Ecu, Ind, Per, S Af, Ven, Zim.

CANDID: NATO code name for the Soviet Ilyushin Il-76 transport aircraft. A successor to the Antonov An-12 CUB strategic airlifter in the VTA (Soviet Military Transport Aviation), the Il-76 was first flown in 1971, and has proved to be rugged, reliable, and highly capable; there are also tanker and airborne early warning variants. The Il-76 is superficially similar to the Lockheed C-141 STAR-LIFTER, but incorporates a number of advanced features which improve short takeoff and landing (STOL) and rough field performance.

The pilot, copilot, and flight engineer are seated in an airliner-type flight deck above the cargo hold, while the navigator is in a glazed compartment ahead of and below the flight deck. There is small meteorological radar in the nose, and a larger ground-mapping radar is housed in a belly radome behind the navigator. The cargo hold, some 11.16 ft. (3.4 m.) wide, has a tail door/ramp for the roll-on/roll-off loading of vehicles and palletized cargo. Fully pressurized, the hold is equipped with a computer-controlled loading and positioning system. Most military versions are armed with two 23-mm. cannons mounted in a manned tail turret at the base of the vertical stabilizer.

The main landing gear, housed in bulges along the base of the fuselage, has 18 low pressure tires which allow the aircraft to land on snow, mud, or other soft surfaces; tire pressure can be controlled from the cockpit. The slightly swept wings have high-lift leading-edge slats and double slotted flaps, as well as 16 spoilers used for speed brakes, roll control, and lift dumping, all features that allow excellent short-field performance. The Il-76 is powered by four turbofans mounted in underw-

ing nacelles. Considerably more powerful than the engines of the Starlifter, they give the Il-76 a typical takeoff run of only 2800 ft. (850 m.); thrust reversers are used to reduce landing rolls.

Some 300 Il-76s are in service with the VTA. The Il-78 tanker variant (NATO code name MIDAS) has three drogue hoses for AERIAL REFUELING. There is also an AWACS derivative (NATO code name MAINSTAY), with a large rotary radome above the rear fuselage.

Specifications Length: 152.85 ft. (46.6 m.). Span: 165.85 ft. (50.56 m.). Powerplant: 4 Soloviev D-30KP turbofans, 26,455 lb. (12,000 kg.) of thrust each. Fuel: 151,300 lb. (6877 kg.). Weight, empty: 165,347 lb. (75,000 kg.). Weight, max. takeoff: 374,786 lb. (170,000 kg.). Speed, max.: 528 mph (850 kph). Speed, cruising: 497 mph (800 kph). Range, max.: (w/88,200-lb. [40,000-kg.] payload) 3,107 mi. (5000 km.). Operators: Cze, Ind, Iraq, NG, Pol, Swaz, Syr, USSR.

CANISTER (ROUND): A short-range, DI-RECT-FIRE anti-personnel round for tube ARTIL-LERY and tank guns. Canister rounds, essentially gigantic shotgun shells, consist of thin sheet-metal cylinders filled with large numbers of spherical shot. The cylinder bursts upon exiting the barrel, dispersing the shot at high velocity in a conical pattern. Effective at ranges up to several hundred meters, canister rounds are issued to artillery units for self-defense against infantry assaults, and to tanks for use against troops, trucks, and other soft targets. Fléchette-filled BEEHIVE rounds are more versatile.

CANNON, AUTOMATIC: A heavy AUTO-MATIC WEAPON, ranging in caliber from 20 mm. to 76.2 mm. Because of their high rates of fire, automatic cannon are suitable as aircraft and anti-aircraft weapons. Most modern fighters are armed with high-velocity cannon such as the ADEN, DEFA, GAU-8, and VULCAN; 20, 23, 27, and 30 mm. are the most common calibers. Anti-aircraft guns are similar, but with calibers up to 57 mm. ARMORED FIGHT-ING VEHICLES, including ARMORED CARS, INFANTRY FIGHTING VEHICLES, light AIR DEFENSE vehicles, and ARMORED RECONNAISSANCE VEHICLES, are also armed with 20 to 30-mm. cannon. Shipboard DUAL PURPOSE weapons, such as the OTO-MELARA 76.2-mm. gun, are also classified as cannon. Cannon actions in use include gas blowback, multiple barrel (Gatling), and rotary breech.

CAP: See COMBAT AIR PATROL.

CAPTOR: The en*cap*sulated *tor*pedo, U.S. Mk.60 ANTI-SUBMARINE WARFARE (ASW) mine. Anchored above the seabed by a cable, CAPTOR contains a passive SONAR array, FIRE CONTROLS programmed to detect enemy submarines, and a MK.46 lightweight ASW torpedo. When the sonar detects a submarine, the fire controls determine if it is within range and, if so, launch the torpedo, which homes autonomously on the target. Operational since 1970, CAPTOR would be used primarily in deep water, to deny passage to submerged submarines. Like other mines, it can be laid by ships, submarines, or aircraft. CAPTOR has reportedly suffered from a number of operational and technical problems which slowed its procurement; it is nonetheless a very effective weapon. See also MINES, NAVAL.

Specifications Length: 12.08 ft. (3.68 m.). Diameter: 21 in. (533 mm.). Weight: 2610 lb. (1184 kg.).

CARL GUSTAV: Swedish shoulder-fired 84-mm. RECOILLESS rifle, primarily an ANTI-TANK weapon, but also capable of engaging a variety of other targets. The original M2 version consists of a rifled tube with an upward-hinged venturi breech (through which rounds are loaded and exhaust gases are vented to neutralize the recoil), a spring-loaded bipod, a pistol-grip trigger assembly, a forward handgrip, and either open or telescopic sights. In action, a round is loaded into the breech with the venturi hinged open; the firing mechanism is hand-cocked; and the venturi must then be closed and locked to unlock the trigger for firing.

The primary round is a High Explosive Anti-Tank (HEAT) shell, which can penetrate up to 15 in. (381 mm.) of steel armor plate. Other types of ammunition include high-explosive (HE), smoke, and illumination. All rounds have a light alloy cartridge with a plastic blow-out disc in its base, to allow the backblast which produces the recoilless effect.

In the improved M2-550, the basic design is unchanged, but the barrel and breech have been strengthened to fire rocket-assisted HEAT projectiles for greater range and accuracy. The round is fin stabilized, and fitted with a plastic slip ring to eliminate the rotation imparted by the rifled barrel, thereby improving the performance of the shaped charge.

Specifications Length: 44.5 in. (1.13 m.). Caliber: 84 mm. Weight, loaded: 33 lb. (15 kg.). Muzzle velocity: 350 m./sec. Range, effective: (M2) 450 m. HEAT/2000 m. HE; (M2-550) 700 m. HEAT. Time of flight, max.: (M2-550) 2.2 sec. Op-

erators: Aut, Can, Den, Ire, FRG, Gha, Neth, Nor, Swe, UK, US.

CARRIER BATTLE GROUP (CVBG): U.S. Navy term for a force organized around one or more AIRCRAFT CARRIERS. A CVBG consists of 1 or 2 carriers, a number of escort vessels to screen them against enemy air and submarine attack, and UNDERWAY REPLENISHMENT ships. Escorts typically include 1 or 2 guided-missile CRUISERS and 2 or 3 guided-missile DESTROYERS, which together with the carriers' air groups provide ANTI-AIR WARFARE (AAW) defense; and 1 or more ANTI-SUBMARINE WARFARE (ASW) destroyers and ASW frigates, which provide ASW defense together with the carriers' own patrol aircraft and ASW helicopters. One or 2 nuclear attack SUBMARINES often operate in support of CVBGs, patrolling ahead to provide early warning and intercept any submarines. In addition, CVBGs are followed by logistic support ships with fuel, ordnance, and victuals; without resupply, battle groups cannot sustain operations for more than two or three days. Logistic ships must also be escorted by destroyers or frigates, raising the total number of ships in a typical battle group to 20–25. Critics have pointed out the magnitude of the effort required to defend the one or two carriers of a CVBG, and the high ratio of self-defense aircraft onboard. Proponents stress that carrier firepower is obtainable worldwide, and that the addition of ANTI-SHIP and CRUISE MISSILES such as HARPOON and TOMAHAWK to escorts has done much to redress the offense/defense imbalance. See also AEGIS; AIRCRAFT CARRIERS, UNITED STATES.

CAS: See CLOSE AIR SUPPORT.

CASEVAC: Casualty Evacuation; specifically, the use of helicopters to retrieve wounded men from the battlefield for prompt treatment in the rear. In Vietnam, CASEVAC significantly reduced mortality rates for the seriously wounded. Whether CASEVAC could be used so extensively in more conventional wars against better-armed enemies is in question, because of denser air defenses, and greater demands on limited helicopter resources. U.S. Army slang for CASEVAC is "dustoff." See also HELICOPTER.

CASSARD II: French ANTI-AIR WARFARE (AAW) destroyers, a version of the GEORGES LEYGUES class. Four ships have been ordered to date; the last is to be commissioned in 1997. See also DESTROYERS, FRANCE.

CATALYTIC WAR: A hypothetical nuclear war provoked by a third party; as, e.g., if China were to simulate a Soviet attack on the United States in order to induce an U.S. retaliatory strike on the Soviet Union. Catalytic war was a concern in the era of "first generation" BALLISTIC MISSILES which were not protected by hardened SILOS, and which thus allowed no time for deliberation and communication in the face of an attack, because any hesitation could result in the loss of the retaliatory force. With the advent of U.S./Soviet ASSURED DESTRUCTION capabilities, catalytic war has ceased to be even a theoretical concern.

CAT HOUSE: NATO code name for a very large Soviet BATTLE MANAGEMENT radar associated with the GALOSH anti-ballistic missile system. Cat House is a huge PHASED ARRAY radar located south of Moscow, which operates in conjunction with the similar DOG HOUSE radar nearby. Both are meant to track incoming REENTRY VEHICLES and pass target data to Galosh firing sites; both are to be replaced by a new "Pill Box" 360° phased array under construction at Pushkino, north of Moscow. See also ANTI-BALLISTIC MISSILE TREATY; BALLISTIC MISSILE DEFENSE; HEN HOUSE.

CAVALRY: There are still troops mounted on horses in China, Mongolia, and elsewhere, but the term mainly defines the (armored) RECONNAISSANCE troops of some armies. Classical horse cavalry was divided into three classes: light cavalry, dragoons, and heavy cavalry. Light cavalry, which included hussars, "light dragoons," *chasseurs a cheval,* and lancers, were trained principally for SCOUTING and PURSUIT. Dragoons were originally mounted infantry, with horses for transportation, not for fighting, but they evolved into a medium form of cavalry trained both for scouting and shock action. Heavy cavalry, which included cuirassiers, carabiniers, "dragoon guards," and horse grenadiers, were trained exclusively for shock action. By the time of the American Civil War, rifled muskets had rendered (heavy) cavalry charges suicidal; the cavalry of both sides quickly evolved into mounted riflemen who fought mostly on foot, while still serving as scouts and as raiders against enemy supply lines. The Boer Wars confirmed the effectiveness of mounted riflemen in mobile, GUERRILLA-style operations. During World War I, the MACHINE GUN, heavy ARTILLERY, and, above all, continuous-front trench warfare mostly precluded the use of cavalry, but the British used a large force of mounted riflemen against the Turks in Palestine, and entire cavalry armies were used successfully on the Russian Front, as in the subsequent civil war.

Between the world wars, horse cavalry was mostly replaced by armored car units with the same scouting, raiding, and pursuit missions; such units were often designated "armored cavalry." But during World War II the Soviet army still deployed 13 cavalry divisions, which were so effective that the Germans formed a cavalry division of their own. The vast distances of the Russian Front, which made continuous front lines impossible, allowed scope for raiding, while Russian terrain mostly confined trucks, ARMORED CARS, and even TANKS to the few main roads. Because of this, the Soviet army retained some cavalry units until 1956. The last notable success of cavalry troops came in the Rhodesian insurgency, when mounted scouts were extremely successful in tracking down guerrilla infiltrators.

For U.S. Army helicopter-equipped forces trained in traditional cavalry roles, see CAVALRY, AIR.

CAVALRY, AIR: U.S. Army term for forces equipped with scout, attack, and transport HELICOPTERS and trained for SCOUTING, SCREENING, delay, PURSUIT, and other traditional CAVALRY missions. In the air cavalry developed during the Vietnam War, scout helicopters were used to locate enemy forces, which were then pinned down and engaged by air-mobile infantry supported by attack helicopters. Helicopters greatly extended the range at which forces could operate from their bases, and increased the speed with which they could react when the enemy was detected. The concept was taken to its logical conclusion by the establishment of the AIR ASSAULT DIVISION, in which all combat elements were transported by helicopter. The U.S. Army now operates several different types of air cavalry unit under the DIVISION 86 organization. Every heavy (i.e., armored or mechanized) division includes a combat aviation BRIGADE, consisting of one cavalry squadron (i.e., BATTALION), one general support squadron, and two attack helicopter squadrons. Each U.S. Army CORPS has an attached combat aviation brigade, which consists of a combat aviation squadron and three attack helicopter squadrons.

The Air Assault Division, no longer all helicopter-borne, has an air cavalry squadron with 30 scout, 22 Blackhawk, and 27 AH-1S COBRA attack helicopters.

Each U.S. Armored Cavalry REGIMENT contains an organic air cavalry squadron of 565 men, organized into a headquarters company, one general support troop, two anti-tank troops, and three air cavalry troops.

The army is presently forming a new type of unit, the Air Cavalry Brigade (Air Combat), with four attack helicopter squadrons trained specifically for anti-tank operations. One such unit, the 6th Air Cavalry Brigade, now forms part of U.S. CENTRAL COMMAND. See also CAVALRY, ARMORED.

CAVALRY, ARMORED: U.S. Army term for armored RECONNAISSANCE units, now of two basic types: divisional cavalry squadrons (i.e., BATTALIONS), and Armored Cavalry REGIMENTS (ACRs).

Under the DIVISION 86 reorganization, the divisional cavalry squadron is part of the Combat Aviation Brigade of each armored and mechanized DIVISION. It consists of a headquarters troop (i.e., COMPANY), two armored cavalry troops, and two air cavalry troops.

The Armored Cavalry Regiment is an independent, BRIGADE-size, COMBINED-ARMS formation, usually attached to a CORPS or numbered ARMY. The ACR is to provide the army or corps commander with his own reconnaissance and target acquisition force, but ACRs are mostly used as covering forces. Each ACR has a total of 5000 men divided into three cavalry squadrons, an air cavalry squadron, an ARTILLERY battalion, an ENGINEER company, an AIR DEFENSE company, and assorted logistic support units. See also CAVALRY, AIR.

CBU: See Cluster Bomb.

CBW: Chemical and Biological Warfare, with reference to both weapons and protective measures. See BIOLOGICAL WARFARE; CHEMICAL WARFARE.

CC&D: See CAMOUFLAGE, CONCEALMENT, AND DECEPTION.

CCV: Control Configured Vehicle, an aircraft deliberately designed to be unstable ("relaxed static stability"), which requires active, computer-controlled stability augmentation systems in order to fly (usually, but not always, via FLY-BY-WIRE control systems). CCV design enhances maneuverability because inherently unstable aircraft respond more readily to flight controls; CCV also relaxes the design constraints imposed by the need for positive stability in conventional aircraft designs. Almost all the latest combat aircraft incorporate CCV technology.

CENTAG: Central Army Group, a NATO ground forces command subordinate to the AFCENT multi-service command. CENTAG is responsible for all Allied ground forces within a zone that ex-

tends from Bonn and Kassel down to the Alps and the Austrian border. Forces currently assigned to CENTAG include II and III West German Corps, and V and VII U.S. Corps, with a total of $13^2/_3$ divisions. These troops could be reinforced in an emergency by U.S. REFORGER units from CONUS, and (probably) by French army units. Other NATO army groups include NORTAG and SOUTHAG.

CENTCOM: See CENTRAL COMMAND.

CENTRAL COMMAND: U.S. UNIFIED COMMAND responsible for all operations in the Persian Gulf and Southwest Asia. CENTCOM originated in the RAPID DEPLOYMENT JOINT TASK FORCE (RDJTF), formed in March 1980 in response to the Iranian revolution and the Soviet invasion of Afghanistan. The RDJTF was controlled by the U.S. Army Readiness Command, but that caused interservice friction which induced the establishment of CENTCOM in 1983. With headquarters at Mac-Dill Air Force Base, Florida, CENTCOM includes army, navy, marine corps, and air force units earmarked for operations in the Persian Gulf/West Asia region, but mostly also committed to NATO or other theaters of operations as well.

CENTCOM now has some 325,000 men in its earmarked forces, but except for a few ships, the U.S. has no combat forces deployed in the region. In an emergency, CENTCOM ground- and land-based air forces would have to deploy from the United States. Studies have shown that during the first 20 days of a Soviet invasion of Iran, the U.S.S.R. could deploy at least 100,000 men into the theater, as compared to only 20,000 men from CENTCOM. In addition, most Soviet forces would consist of TANK or MOTORIZED RIFLE divisions, while most U.S. forces would consist of light armor, LIGHT INFANTRY, or airborne units.

The 2 August 1990 Iraqi invasion of Kuwait posed an entirely different challenge to CENTCOM. Rather than serving as a tripwire for nuclear deterrence of Soviet aggression, the Command was opposed to a large regional power with some 5000 tanks. Against such a force CENTCOM's light ground forces could only deploy defensively, and even then only with massive air support from land-based and carrier aircraft. During the first six months after the Iraqi invasion, CENTCOM was heavily reinforced by additional Marine and Army units, including about half the armored and mechanized units assigned to the Seventh Army in Germany, for a total of some 450,000 men.

CENTRAL INTELLIGENCE AGENCY (CIA): The primary U.S. government organization responsible for the collection, collation, analysis, and dissemination of foreign political, economic, and military INTELLIGENCE; for overseas counterintelligence; and for COVERT OPERATIONS.

Created under the 1947 National Security Act, the CIA is headed by a director of central intelligence (DCI), who is also responsible for coordinating the activities of all other U.S. intelligence services, and for advising the president on all intelligence matters. Under the 1947 act, the CIA has no mandate for internal (domestic) intelligence or security operations, which are the province of the Federal Bureau of Investigation (FBI).

At present, the CIA is divided into four directorates, each headed by a deputy director: the Directorate of Operations (DO); the Directorate of Intelligence (DI); the Directorate of Science and Technology (DS&T); and the Directorate of Administration (DA).

DO, also called Clandestine Services, comes closest to the popular conception of the CIA. It is responsible for clandestine intelligence collection from both human intelligence (HUMINT) and technical intelligence (TECHINT) sources; for counterespionage; and for covert operations. DO is divided into several staff bureaus for espionage, counterespionage, covert action, operational support, and technical support. The staff bureaus manage the activities of several area divisions organized on geographical lines, which operate overseas offices and process the intelligence data they collect. DO is also responsible for all paramilitary operations (a minor, subordinate function).

DI is responsible for the analysis of intelligence data and the production of "finished" national intelligence reports for the president, the defense establishment, and other users. It is organized into five geographic offices, an Office of Scientific and Weapons Research, an Office of Global Issues, an Office of Imagery Analysis (responsible for analysis of photographic intelligence), an Office of Current Production and Analytic Support (responsible for the publication of intelligence reports), and an Office of Central Reference.

DS&T researches and develops technical collection systems. It consists of an Office of Research and Development, an Office of Development and Engineering, an Office of Signals Intelligence (SIGINT) Operations, an Office of Technical Services, the National Photographic Interpretation Center (responsible for analysis of aerial and satellite surveillance imagery), and the Foreign Broadcast Information Service (FBIS), which monitors foreign press and broadcast media.

DA is responsible for internal housekeeping, including security, medical services, training, and finance.

The CIA's covert operations capability remains minimal. See also DEFENSE INTELLIGENCE AGENCY; GRU; KGB; NATIONAL SECURITY AGENCY.

CENTRAL WAR: A hypothetical war between the United States and the Soviet Union, or, more generally, between any nuclear powers, with or without the use of nuclear weapons. The term has been used to define direct combat between the United States and the Soviet Union, as opposed to fighting done by allies, clients, or other proxies. GENERAL WAR defines a central war that escalates into an all-out, all-weapons war.

CENTURION: British MAIN BATTLE TANK (MBT), originally designed to a 1943 requirement for a heavy "cruiser" tank armed with a 17-pounder (76.2-mm.) gun. The initial Mk.1 entered service in 1945, just too late to be used in World War II. Since then, Centurions have been much used in the Korean War, several Middle East wars, and in Southeast Asia. A total of 4423 Centurions were built between 1945 and 1962, and although the Centurion was replaced by the CHIEFTAIN MBT in the British army itself, at least 2400 are still in service. The Centurion's longevity is a tribute to the soundness of the basic design, including its ability to incorporate a host of modifications. By the end of production in 1962, armament had increased from the 17-pounder to the 105-mm. L7A2 rifled gun, and armor thickness had increased from a maximum of 76 mm. in the Mk. 1 to 120 mm. Other improvements included new engines, transmissions, and suspensions; increased fuel and ammunition stowage; and various detailed changes. No fewer than 12 distinct models (marks) of the Centurion were produced by the Royal Ordnance Factory.

The hull, constructed from welded steel ARMOR plates with a maximum thickness of 120 mm., is divided into a driver's compartment up front, a fighting compartment in the middle, and an engine compartment in the rear. The driver, on the right side of the bow, has two observation periscopes, and ammunition is stored to his left. The turret, over the fighting compartment, has cast steel sides and a welded roof plate; turret frontal armor is 152 mm. thick. There is an ammunition resupply hatch on the left side of the turret, and external stowage boxes are attached to both sides. The commander sits on the right side of the turret under a rotating cupola, behind and above the gunner, with the loader on his left. Both gunner and commander are provided with mechanically linked periscopic rangefinding sights. Semi-active INFRARED night sights can be fitted, with illumination provided by an infrared searchlight.

The 105-mm. L7A2 rifled gun (the standard NATO tank gun until 1982) can fire a variety of ARMOR-PIERCING ammunition, including APDS, APFSDS, HEAT, and HESH. The APFSDS round can penetrate all known Soviet tanks out to 3000 m., the APDS round has an effective range of 2000 m., and the HEAT round has a range of 4000 m. Up to 70 rounds can be carried in the hull and turret. The gun has elevation limits of −10° and +15°, and a maximum rate of fire of some 8 rounds per minute. Secondary armament includes a 7.62-mm. coaxial machine gun and a 12.7-mm. M2 Browning heavy machine gun by the commander's cupola. With the ranging machine gun, mounted over and ballistically matched to the main gun, the gunner fires bursts at the target, and when the RMG hits it, the main gun is fired in turn. Most Centurions have a 5- or 6-barrel smoke GRENADE launcher on each side of the turret.

Low speed, mediocre off-road mobility, and short range were the major deficiencies of the Centurion, as of later British tanks; several operators have therefore upgraded their tanks. The Israeli army, the largest single user with more than 1100 in service, has made many modifications, finally producing a rebuilt Improved Centurion, with a more powerful diesel engine for better speed, range, and cross-country mobility. In addition, the Improved Centurion has greater fuel and ammunition stowage, an automatic fire suppression system, modern FIRE CONTROLS (with a digital computer and LASER rangefinder), and fittings for BLAZER reactive armor for protection against ANTI-TANK GUIDED MISSILES (ATGMs) and other SHAPED CHARGE weapons. Sweden and Switzerland have implemented similar upgrade programs. Centurion will probably remain in service into the next century.

Specialized vehicles derived from the Centurion chassis include bridgelaying tanks, armored recovery vehicles, engineer vehicles, and mine-clearing tanks.

Specifications Length: 25.67 ft. (7.82 m.). Width: 11.1 ft. (3.39 m.). Height: 9.85 ft. (3.01 m.). Weight, combat: 52 tons. Powerplant: (Mk.13) 650-hp. Rolls Royce Mk. IVB 12-cylinder, liquid-cooled gasoline engine; (Imp. Cent.) 900-hp. Teledyne Continental AVDS-1790-2R diesel. **Hp./Wt. ratio:** (Mk. 13) 12.5:1; (Imp. Cent.) 17.3:1. **Fuel:** (Mk. 13) 260 gal. (1145 lit.). **Speed, road:** (Mk.13) 21 mph (35 kph); (Imp. Cent.) 26 mph (43.4 kph). **Range, max.:** (Mk.13) 115 mi. (192 km.); (Imp. Cent.) 230 mi. (384 km.). **Operators:** Den, Isr, Jor, Km, Neth, S Af, Sing, Som, Swe, Swi.

CEP: Circular Error Probable, a measure of weapon accuracy, defined as the radius of a circle centered on the target, within which half of all weapons of a given type are expected to fall (i.e., the median inaccuracy). CEP is the usual measure of accuracy for ballistic missiles, usually stated in meters.

CEV: See COMBAT ENGINEER VEHICLE.

CFV: Cavalry Fighting Vehicle, the M3 variant of the BRADLEY fighting vehicle. See also ARMORED RECONNAISSANCE VEHICLE; CAVALRY, ARMORED.

CG: Chemical warfare choking agent. See PHOSGENE.

CH-46: U.S. transport helicopter. See SEA KNIGHT.

CH-47: U.S. medium transport helicopter. See CHINOOK.

CH-53: U.S. heavy transport helicopter. See SEA STALLION.

CHAFF: A passive ELECTRONIC COUNTERMEASURE (ECM), used against enemy RADARS and radar-guided weapons; "window" is the analogous British term. Chaff consists of thin aluminium or metallized plastic strips cut to half the wavelength of the radar to be neutralized. Acting as half-wavelength dipole antennas, such strips resonate harmonically when illuminated by radar signals of the appropriate frequency, producing return signals of greatly amplified intensity. Bundles of chaff scattered in the air clutter enemy radar displays with many false targets, hiding the true target. Because a particular length of chaff will only neutralize radars of a particular wavelength, different types of chaff are needed against different radars. These can either be packaged together, for broad-band masking; or separately, provided in ECM systems designed to deal with radar threats specifically.

Aircraft can be equipped with special dispensers for chaff canisters, such as the U.S. ALE-40, which may be mounted internally or on wing pylons. Without any special equipment, chaff bundles can be dispensed from weapon pylons or even air-brake housings.

On ships, chaff canisters are usually dispensed

by specialized small mortars or rocket launchers, which boost the canisters to a height of several hundred feet before detonating to dispense the chaff. Typical systems include the British CORVUS and Shield, and the U.S. Mk.36 SRBOC (Super Rapid-Blooming Offboard Chaff). Chaff can also be dispensed from special shells fired from naval guns.

To suppress air defenses, chaff can be dispensed in bulk by dedicated "chaff bomber" aircraft (this technique was used in Vietnam to blind surveillance and tracking radars to the approach of strike forces); more commonly, strike aircraft dispense their own chaff more selectively, for self-defense against particular radars. An aircraft being illuminated by an enemy radar can respond immediately by releasing several chaff packages, to prevent that radar from "locking on." Likewise, if attacked by a radar homing missile, aircraft can jettison chaff bundles while executing evasive maneuvers; that may cause the missile to home on the chaff return instead of the aircraft. Similarly, ships fire chaff rockets on the approach of radar-guided ANTI-SHIP MISSILES, to create a large, intense (ship-like) target for the missile (while, if possible, simultaneously maneuvering out of its path).

The first limitation of chaff is that it is not persistent. The chaff cloud dissipates quickly, decreasing in effectiveness in a matter of minutes at best. Second, chaff can be negated by ELECTRONIC COUNTER-COUNTERMEASURES (ECCM), such as digital signal processing, moving target indicator (MTI) techniques, and frequency agility. The latest missiles combine radar homing with terminal INFRARED guidance, to resist both chaff and INFRARED COUNTERMEASURES (IRCM). In response, "thermal chaff" has been developed; this combines chaff strips with incendiary chemicals to create infrared "hot spots." See also ELECTRONIC WARFARE.

CHAIN GUN: A type of lightweight automatic CANNON whose simple rotating breech mechanism is actuated by an electrically powered chain drive. The simplified loading cycle makes chain guns extremely reliable.

The U.S. Army's M230 30-mm. cannon, which arms the AH-64A APACHE attack helicopter, is a chain gun mounted in a rotating turret under the helicopter's nose. The M230 and its mount weigh only 123 lb. (55.9 kg.); the maximum rate of fire is 1200 rounds per minute, but the gun is usually limited to 625–750 rounds per minute to conserve ammunition. Muzzle velocity varies with ammunition types; API rounds have a muzzle velocity of 2600 ft. (792 m.) per second. Other rounds include HEI, shrapnel, and APDS.

The other chain gun in service is the M242 Bushmaster, a 25-mm. weapon which arms the M2/M3 BRADLEY fighting vehicle and the LAV-25 light armored vehicle. Its low rate of fire (100–200 rounds per minute) is consistent with the M242's primary role as a ground weapon. It also fires API and HEI rounds, but the primary round is APDS with a STABALLOY (depleted uranium) penetrator, which can penetrate light armor from all angles, and the side and rear armor of MAIN BATTLE TANKS. The M242 with its mount weighs only 230 lb. (104.5 kg.).

Because of their light weight and reliability, chain guns are especially suitable for FAST ATTACK CRAFT, light armored vehicles, and aircraft, all of which can provide the electrical power they require.

CHALLENGER: MAIN BATTLE TANK (MBT) in service with the British army. Derived from the Shir II (a radically modified version of the CHIEFTAIN developed for the Shah of Iran), Challenger was ordered for the British army as an "interim" MBT after the collapse of the Anglo-German MBT-90 program in 1980.

Challenger's hull and turret are covered with composite CHOBHAM armor (more than twice as effective as steel armor of the same thickness); the glacis armor is sloped approximately 70° from the vertical (more than doubling its effective thickness), and maximum armor protection is equivalent to some 24 in. (610 mm.) of homogeneous steel plate. The layout is conventional, with the hull divided into a driver's compartment in the front, a fighting compartment in the middle, and the engine compartment in the rear. Seated on the center line in a reclining seat to reduce vehicle height, the driver is provided with both day and night periscopes. The commander is seated on the right side of the turret under a rotating cupola, with the gunner seated below and ahead of him, and the loader to his left. The commander's cupola is equipped with a THERMAL IMAGING periscopic sight and several periscopic vision blocks. The gunner has a similar thermal imaging sight and a LASER rangefinder, linked to a Marconi Improved Fire Control System (IFCS), whose ballistic sensors and digital computer can lay the gun, while compensating for target and vehicle movement and weather factors; but the IFCS is both less complete in its sensors and much harder to use than the FIRE CONTROLS of other recent MBTs. Both commander and gunner can train and fire the main gun with the IFCS.

The main armament is the Chieftain's L11A5 rifled gun, which has separate-loading bagged propellant charges instead of brass cartridge cases. Up to 52 rounds of APDS, APFSDS, HESH, or smoke ammunition can be loaded in individual containers below the turret ring, each of which is jacketed with a fire suppressing liquid (the British army plans to replace the L11A5 with either a new high-technology rifled 120-mm. gun or the Rheinmetall 120-mm. smoothbore gun of the LEOPARD II and M1A1 ABRAMS). Secondary armament includes a 7.62-mm. coaxial machine gun, and a second 7.62-mm. machine gun by the commander's cupola. A 5-barrel smoke GRENADE launcher is attached to each side of the turret. The Challenger has a positive overpressure COLLECTIVE FILTRATION system for NBC protection.

Challenger has a horsepower-to-weight ratio of 19.35 to 1, better than the Chieftain but much worse than the ratios of other recent MBTs. Maximum speed is some 10 mph (16 kph) less than the speed of Leopard II or Abrams, but more important, cross-country mobility is also inferior. Fuel is stowed in conformal cells in the hull, and the engine and transmission form a single power "egg" which can be replaced in less than 45 minutes under field conditions.

Planned improvements include a better transmission (another weak point), automatic engine controls, and improved tracks. An armored recovery vehicle based on the Challenger chassis went into production in 1985.

In the British army, 300 Challengers equip five armored regiments. A vast improvement over the Chieftain, especially in mobility and protection, Challenger still reflects the British concept of ARMORED WARFARE in which tanks are mainly used as ANTI-TANK guns; hence the design emphasizes firepower and (above all) protection, at the expense of mobility.

Specifications **Length:** 29 ft. (8.84 m.). **Width:** 11.4 ft. (3.47 m.). **Height:** 9.45 ft. (2.89 m.). **Weight, combat:** 62 tons. **Powerplant:** 1200-hp. Rolls Royce Condor V-12 diesel. **Fuel:** 450 gal. (1980 lit.). **Speed, road:** 36 mph (60 kph). **Range, max.:** 300 mi. (500 km.).

CHAPARRAL: The U.S. MIM-72 surface-to-air missile (SAM) system, derived from the AIM-9 SIDEWINDER air-to-air missile. Developed by Philco-Ford from 1965 as an "interim" weapon to replace the canceled Mauler mobile SAM, Chap-

arral adapted Sidewinder to ground launch in lieu of a new missile, to reduce costs, technical risks, and development time. Tested in 1965, Chaparral entered service the following year and has since been exported to Israel (which has used it successfully in combat), Morocco, Taiwan, and Tunisia.

The original MIM-72 missile, almost identical to the AIM-9D Sidewinder, had a passive INFRARED seeker which could be used only for rear-quarter attacks, limiting its tactical utility. From 1970, the army initiated improvements which resulted in the MIM-72C missile of 1977, with an improved motor for longer range, a larger warhead with an M817 directional Doppler fuze, and a DAW-1 ALL-ASPECT infrared seeker. It was superseded in the mid-1980s by the MIM-72G, with a low-smoke motor (making it more difficult to spot), an Identification, Friend or Foe (IFF) system derived from the FIM-92 STINGER missile, and a more sensitive "Rosette-Scan" seeker (RSS). Over 9000 Chaparrals have been produced to date.

Generally similar to Sidewinder in both appearance and operation, the MIM-72G has a minimum range of 500 m. and a maximum range of 9300 m. In action, the operator slews the launcher onto the target bearing, and fires the missile when the seeker head locks onto the target, as indicated by an audio tone received in the operator's headset. Within its range envelope, the MIM-72G has an estimated kill probability (P_K) of more than 50 percent against fighter-type targets.

The M730 launch vehicle, derived from the M113 armored personnel carrier, has a trainable 4-rail launcher which can be retracted for traveling (the missile cannot be launched while the vehicle is moving). Eight reload rounds are carried in the vehicle. Originally, the M730 had no on-board radar, and thus had to be netted into a surveillance and warning system based on a separate FAAR (Forward Area Alert Radar). In the late 1970s, an all-weather surveillance radar was added to the M730.

The U.S. Army planned to replace Chaparral with a tracked version of the Euromissile ROLAND, but this program was canceled, and Chaparral will remain the U.S. Army's primary mobile battlefield SAM well into the 1990s. It is to be replaced by the Martin-Oerlikon Buehle ADATS.

CHARLES DE GAULLE: A class of two French nuclear AIRCRAFT CARRIERS (*Porte Avions Nuclear*, PAN), the first of which was laid down in 1989, and is now scheduled to replace the conventional carrier CLEMENCEAU in 1996. The second carrier is to be laid down in the mid-1990s, for completion in 2001.

The de Gaulle's hull is a more robust version of the Clemenceau's, with additional armor and better damage control provisions. Four gyro-controlled stabilizers will improve seakeeping, and a COLLECTIVE FILTRATION system will provide NBC protection. The flight deck is 857.72 ft. (261.5 m.) long and 211.1 ft. (64.36 m.) wide, with a 633-ft. (195-m.) angled landing deck canted 8.3° to port, and two U.S.-built steam catapults (one on the bow, the other on the angled deck). Two deck-edge elevators to starboard, behind the island, provide access to the hangar deck. The hangar, 450 ft. (137.19 m.) long, 94 ft. (28.65 m.) wide, and 20 ft. (6.1 m.) high, is lower than that of the Clemenceau, but is otherwise much larger and has better armor protection.

Defensive weapons include seven 8-round vertical launchers for SAAM surface-to-air missiles, and two 6-round launchers for Sadrale short-range SAMs. The air group of 35–40 aircraft will initially be based on the Dassault Super ETENDARD, but they may be replaced by the proposed ACM (Avion de Combat Maritime), or the U.S.-built FA-18 HORNET. See also AIRCRAFT CARRIERS, FRANCE.

Specifications **Length:** 845 ft. (257.62 m.). **Beam:** 103 ft. (31.4 m.). **Draft:** 27.7 ft. (8.45 m.). **Displacement:** 36,000 tons full load. **Powerplant:** twin-shaft nuclear, 2 type K-15 pressurized-water reactors, two sets of geared steam turbines, 83,000 shp. **Speed:** 28 kt. **Range:** unlimited **Crew:** 1150, plus 550 air group. **Sensors:** 1 DRBJ11B 3-dimensional air-search radar, 1 DRBV27 2-dimensional air-search radar, 1 DRBV15 surface/low-altitude search radar. **Electronic warfare equipment:** 1 ARBR33 electronic signal monitoring array, 4 Sagaie 8-round chaff rocket launchers, 1 rocket-boosted torpedo countermeasures system.

CHARLIE: NATO code name for a class of 19 Soviet nuclear-powered guided-missile SUBMARINES (SSGNs) built in two subclasses between 1967 and 1980. Successors to the ECHO-class SSGNs, the Charlies were the first Soviet submarines able to launch cruise missiles while submerged.

The 12 Charlie Is completed by 1972 are similar to the VICTOR-class attack submarines, with a modified teardrop hullform in a double hull configuration, and a long streamlined sail amidships. Control surfaces are arranged in standard Soviet fashion, with retractable bow diving planes and cruciform rudders and stem planes ahead of the propeller.

The main armament consists of eight SS-N-7 SIREN anti-ship missiles, launched from individual tubes in the bow, inserted between the pressure hull and the outer casing. Angled forward and upward, the tubes are loaded in port through 14-ft. (4.26-m.) hatches in the outer hull. Siren has a range of approximately 35 mi. (64 km.) and carries a 1000-lb. (454-kg.) high-explosive warhead. In addition to missiles, Charlie Is have six 21-in. (533-mm.) torpedo tubes in the bow for a total of 12 TORPEDOES, or SS-N-15 STARFISH anti-submarine missiles.

The six Charlie IIs completed between 1973 and 1982 were lengthened by the addition of a 23-ft. (7-m.) insert in the forward hull, and the forward end of the sail was extended to accommodate additional sensors. The missile tubes are also enlarged for the SS-N-9 STARBRIGHT, which is 6 ft. longer than Siren and has a range of 68 mi. (113 km.). Sensors, torpedo tubes, and propulsion are unchanged, but because of the Charlie II's increased size, speed is reduced.

Though potent anti-ship platforms, the Charlies are also rather noisy, and hampered by the short range of their missiles. They are slowly being supplanted by OSCAR-class SSGNs. See also SUBMARINES; SUBMARINES, SOVIET UNION.

Specifications **Length:** (I) 312 ft. (95.12 m.); (II) 340 ft. (103.65 m.). **Beam:** 32.8 ft. (10 m.). **Displacement:** (I) 4000 tons surfaced/4800 tons submerged; (II) 4500 tons surfaced/5500 tons submerged. **Powerplant:** single-shaft nuclear, 1 pressurized-water reactor, 1 set of geared steam turbines, 15,000 shp. **Speed:** (I) 27 kt; (II) 26 kt. **Max. Operating Depth:** 1315 ft. (400 m.). **Collapse depth:** 1970 ft. (600 m.). **Sensors:** 1 large bow-mounted low-frequency active/passive sonar; 1 passive conformal array sonar, 1 "Snoop Tray" surface-search radar, 1 "Brick Pulp" electronic signal monitoring array, 2 periscopes.

CHEMICAL WARFARE (CW): The military use of toxic chemicals (usually gases or aerosols) to kill or incapacitate. Chemical warfare is sometimes defined to include the military use of smoke, pyrotechnics, and defoliants, because that too is usually the responsibility of specialist chemical troops.

Offensive chemical warfare comprises the development and delivery of toxic chemicals against the enemy. Such "agents" are classified by their effects. BLOOD AGENTS, which act by disrupting the ability of blood to absorb oxygen, include hydrogen cyanide (AC) and cyanogen chloride (CK); both are very fast-acting and lethal.

CHOKING AGENTS (lung irritants or asphyxiants) cause inflammation of the mucous membranes in the lungs, nose, and eyes. Slower-acting than blood agents but still lethal (or secretly incapacitating), choking agents include chlorine (CL), PHOSGENE (CG), and Di-Phosgene (DP).

BLISTER AGENTS, or vesicants, cause skin burns and inflammation of the mucous membranes. Their primary symptom is the formation of large liquid-filled blisters on exposed skin. Blister agents are not immediately lethal, but death frequently results from infection, chemical pneumonia (if inhaled), or systemic poisoning (if ingested). Common blister agents include MUSTARD (H), distilled mustard (HD), lewisite (L), and mustard lewisite (HL). Blister agents are effective primarily for harassment and area denial.

NERVE AGENTS act by disrupting the central nervous system and thus vital involuntary muscular activities such as respiration and circulation. The three most common nerve agents, Tabun (GA), Sarin (GB), and Soman (GD), were developed in Germany before World War II; highly lethal, they are very fast acting and skin permeable. VX, developed by Britain in the 1950s, is even more lethal than the "G" agents.

Tear agents, or lachrymators, are nonlethal chemicals often used for riot control, but also in military use for training, harassment, and to induce enemy troops to exit from tunnels or bunkers. BZ, a psychotropic agent similar to LSD in its effects, can also be used for harassment.

The lethality of chemical agents can be stated as their LD-50 (Lethal Dose-50 percent), the dosage (in milligrams per minute per cubic meter) required to inflict 50 percent fatalities on all exposed personnel; it varies from only 36 for VX to 19,000 for chlorine.

Chemical agents are also classified by the duration of their effects. "Persistent" agents, which may require days or even weeks to dissipate, are effective for area denial, because enemy troops must wear protective equipment to cross contaminated zones, at least slowing down their advance. In addition, persistent agents tend to adhere to clothing and equipment, forcing troops to decontaminate every time they leave a contaminated zone. Thus persistent agents act as effective barriers to movement, either absolute (for unprotected

troops) or relative (for protected troops). Mustard and VX are the most common persistent agents.

Nonpersistent agents can be used in direct support of attacks, because they dissipate rapidly. The attacker can release nonpersistent agents on enemy defenses a few minutes before advancing to disrupt and panic the defenders; by the time his troops arrive, the agent will have dissipated.

Chemical agents have been delivered in the past by ARTILLERY shells, mortar bombs, and aerial bombs, but multiple rocket launchers would be ideal for chemical warfare, because they can saturate a large area in a short time. Aerial spray dispensers can also be effective, especially for persistent agents. Short-range ballistic missiles (SRBMs) could also serve as chemical delivery systems for attacks on high-value area targets such as airfields.

Chemical agents are often highly corrosive as well as toxic, and thus require special handling. Agents stored in bombs, shells, or rocket warheads will eventually begin to leak, posing a hazard to friendly forces and civilians. As a result, some chemical agents, and especially nerve agents and mustard, are normally stored in special bulk tanks, to be loaded into weapons a short time before their planned use; this can be inconvenient if fast reaction is required. To cope with the extreme dangers of storing and handling nerve agents, both the United States and Soviet Union have developed BINARY WEAPONS, in which the chemical agent is divided into two less toxic compounds that are kept separate inside the warhead, and mixed together only after the weapon is launched or fired.

Defensive chemical warfare comprises the protection of troops, vehicles, and facilities; the treatment of casualties; and the decontamination of clothing, weapons, and equipment. Respirators (gas masks) can provide adequate individual protection against all chemical agents except blister and nerve agents, which are skin permeable. To defend against them, troops must wear complete protective clothing. Soviet forces issue rubberized canvas suits, while the U.S. and NATO armies favor fabric suits impregnated with chemicals meant to neutralize nerve agents. Those protective chemicals, however, soon deteriorate, and must be replaced after a few days of use. Both rubber and impregnated fabric suits are very restrictive, and extremely hot; unless protected troops are constantly trained to perform their military tasks while wearing them, their performance is severely degraded, with heat exhaustion a major problem.

To control the use of protective clothing, the

U.S. Army has adopted Mission-Oriented Protective Posture (MOPP) procedures, under which the extent of chemical protection is matched to the estimated threat. Under MOPP-1, the lowest state of alert, equipment is not worn, but is kept ready; under MOPP-5, the highest state of alert, full protective equipment is worn.

Ship, aircraft, and combat vehicle crews can be protected by a COLLECTIVE FILTRATION system. The platform, or at least its crew compartment, is hermetically sealed, and air is drawn through a filter system to allow the crew to operate in a shirtsleeve environment. Individual protective equipment must still be kept handy, however, because the protective environment can be ruptured by battle damage, and because quick exits might be necessary.

Treatment is specific for each agent. In the case of nerve agent casualties, specific antidotes exist but must be administered quickly, and artificial respiration is required. Blood agents can be countered with amyl nitrate if it is administered quickly. Blister, choking, and tear agents can be treated only symptomatically.

Decontamination is not needed for nonpersistent agents. Equipment exposed to persistent agents, particularly mustard, VX, and thickened nerve agents, can be washed with a neutralizing solution. Manual decontamination is dangerous, tiring, and time consuming. The Soviet army, however, has automatic decontamination equipment, which consists of jet engine sprayers mounted on trucks. Decontamination is dangerous because persistent agents can lodge in inaccessible parts of vehicles and equipment, posing a hidden threat.

The military utility of chemical warfare depends on the relative degree of preparation by each side. With total asymmetry (i.e., when one side has both offensive and defensive capability, while the other has neither), CW can be decisive, by inflicting casualties and shattering morale. If one side has both offensive and defensive capability, while the other only has defensive capabilities, CW can have a serious impact, because troops on the defense-only side are forced to wear protective equipment at all times, thereby degrading their combat effectiveness, while troops on the other side need to wear protective gear only when and where it chooses to launch chemical attacks. If both sides have similar offensive and defensive capabilities, the primary effect of CW is to slow the tempo of operations, since troops forced to wear protective

equipment move more slowly and tire more easily. That condition generally helps the defender.

The Soviet Union still maintains formidable offensive and defensive chemical warfare capabilities. At present, the Chemical Troops of the Soviet army number more than 138,000 (the U.S. Army Chemical Corps has some 7000), and roughly 20 percent of all Soviet artillery shells are filled with chemical agents. NATO forces, by contrast, have unequal defensive capabilities, and maintain a very small offensive arsenal. The Soviet army trains intensively for chemical warfare, which is ingrained in Soviet OPERATIONAL METHODS.

Historically, chemical weapons have inflicted relatively few casualties on properly trained and properly equipped troops. Their primary effects are rather harassment, low-level ATTRITION, and a general lowering of MORALE and combat effectiveness. But it may be impossible to avoid massive civilian casualties if CW is waged in a densely populated area. Even a serious CIVIL DEFENSE program is unlikely to prevent panic and heavy fatalities. Not only civilian personnel but entire economies can be devastated by CW, especially in less developed countries which depend heavily on draft animals for transportation and agriculture. Crops can be rendered inedible by contamination with persistent agents, especially mustard. Herbicides, used in Vietnam to clear jungle foliage, can just as easily be used against crops, leading to widespread famine in subsistence economies.

Chemical warfare has been outlawed by a variety of international agreements, including the Geneva Protocol of 1925 and the Biological and Toxin Weapons Convention of 1972. This has not prevented several countries, including signatories, from developing, maintaining, and using chemical weapons. Since World War II, documented uses of CW have been made by Egypt against Yemen (1965–67), by Vietnam against Laotian and Cambodian insurgents (1975 to the present), by the Soviet Union in Afghanistan (1979 to 1988), and by Iraq against Iran and its own Kurdish population (1985 to 1988). An effort is under way to prohibit chemical weapons, but the ease with which chemical agents can be manufactured, the inability of the major powers to control the raw materials used to produce them (most of which have civilian applications), and the ease with which chemical weapons can be concealed make effective verification impossible. The precedent set in World War II, when all parties had chemical arsenals but did not use them, seems to indicate that CW can be prevented by a combination of narrowly military self-interest and DETERRENCE. See also ARMS CONTROL; BIOLOGICAL WARFARE.

CHEVALINE: British-developed penetration aid (PENAID) system for the POLARIS A3 submarine-launched ballistic missiles (SLBMs) of the Royal Navy. Chevaline is based on the principle of dissimulation: the REENTRY VEHICLES (RVs) themselves were modified to make them resemble the low-mass inflatable decoys distributed in large numbers by the system, to present Soviet ANTI-BALLISTIC MISSILE (ABM) radars with a homogeneous cloud of objects. The very costly Chevaline system was developed in response to the Soviet GALOSH Anti-Ballistic Missile System around Moscow, which could invalidate the small British nuclear force even if inadequate against a U.S. attack. See also BALLISTIC MISSILE; BALLISTIC MISSILE DEFENSE; SLBMS, BRITAIN.

CHIEFTAIN: A British MAIN BATTLE TANK (MBT), in service with Iran, Iraq, Kuwait, and Oman as well as with the British army. Developed to a 1958 specification for a CENTURION replacement, the first prototypes were delivered in 1961, and the Chieftain entered service with the Royal Army in 1963. The design reflects British ARMORED WARFARE doctrine under which tanks are mostly employed as mobile anti-tank guns: Chieftain is heavily armed with a 120-mm. L11A5 rifled gun, and carries heavier armor than any other tank of its generation, at the expense of both road speed and cross-country mobility. Though repeatedly upgraded, Chieftain remains underpowered.

The British army received some 900 Chieftains between 1963 and 1971; in that year the Shah of Iran placed one order for 707 Chieftains, and one for 187 modified "Improved Chieftains." In response to Iranian requirements, Royal Ordnance developed two variants, the Shir I and Shir II. An order for 1400 was canceled after the Iranian revolution in 1979, but the Shir II was eventually purchased by the British army as the CHALLENGER MBT.

Chieftain has been repeatedly upgraded during its service life through 12 variants ("marks"). The representative MK.5 is constructed of several cast steel sections welded together, with the hull divided into driver's compartment up front, a fighting compartment in the middle, and an engine compartment in the rear. The frontal glacis is some 8 in. (203 mm.) thick, and sloped more than 70° from the vertical, more than doubling the effective thickness of the armor. The driver is provided with

both daylight and INFRARED night-vision periscopes (the latter are being replaced by passive IMAGE-INTENSIFIER periscopes). Cast in two pieces, the turret has a very thick and well-sloped frontal plate. The commander sits on the right under a rotating cupola, behind and above the gunner, with the loader on the left. The commander's cupola is equipped with nine vision observation periscopes and a telescopic sight. The gunner has a similar sight, plus a LASER rangefinder. Both commander and gunner can replace their day sights with 3x semi-active infrared sights for night vision; illumination is suppled by a large white light/infrared searchlight mounted in an armored box on the left side of the turret and boresighted to the main gun.

The main armament is a 120-mm. L11A5 rifled gun, still one of the most powerful tank guns in service. Its unique, separate-loading, bagged propellant charges (rather than brass cartridge cases) make for easier ammunition handling, but reduce the rate of fire. The gun has a fume extractor, and is usually fitted with a thermal sleeve to reduce barrel warping. ARMOR PIERCING ammunition includes APDS, APFSDS, and HESH; 64 rounds are stowed in the hull and turret. Secondary armament includes a coaxial 7.62-mm. machine gun, and a second 7.62-mm. gun by the commander's cupola.

Originally, Chieftain fire controls relied on a 12.7-mm. ranging machine gun (RMG) ballistically matched to the main gun: in action, the gunner fires several RMG bursts at the target; when shots are observed hitting the target the main gun is fired in turn. Accurate out to about 2000 m., the RMG can hardly be used at night or in poor visibility, and it has since been replaced by a laser rangefinder, accurate from 500 to 10,000 m., linked to a FIRE CONTROL computer. The current fire control system is the Marconi IFCS, whose digital computer is linked to the laser rangefinder, as well as ballistic sensors, allowing the gun to be fired out to the limits of its range; however, the IFCS is both less complete in its sensors and much harder to use than the fire controls of other recent MBTs. Both gunner and commander have identical controls to lay and fire the main gun. The IFCS is not installed in export versions of Chieftain.

The Chieftain Mk.5 has a horsepower-to-weight ratio of only 13.8 hp./ton, the lowest of any modern MBT. As a result, maximum speed is low and cross-country mobility mediocre.

A radically modified Chieftain 900 has been developed for export, with composite CHOBHAM armor, a laser rangefinder, digital fire controls, and a 900-hp. Rolls Royce Condor engine for a maximum speed of 31 mph (52 kph) and much improved off-road mobility; but none have been ordered to date. Jordan purchased a total of 278 Shir I variants originally intended for Iran. In service as the Khalid, it is essentially a Chieftain Mk.5 with the 1200-hp. engine of the Challenger, and the Marconi IFCS.

The Chieftain has also served as the basis for a family of specialized vehicles, including a bridge layer, armored recovery vehicle, and engineer assault vehicle.

Specifications **Length:** 24.75 ft. (7.52 m.). **Width:** 11.85 ft. (3.66 m.). **Height:** 9.4 ft. (2.89 m.). **Weight, combat:** 55 tongs. **Powerplant:** 750-hp. Leyland L60 (No. 4 Mk.8) 12-cylinder diesel. **Fuel:** 230 gal. (1012 lit.). **Speed, road:** 28.8 mph (48 kph). **Range, max.:** 300 mi. (500 km.) on roads/ 180 mi. (300 km.) cross-country.

CHIKUGO: A class of 11 small Japanese ANTI-SUBMARINE WARFARE (ASW) FRIGATES commissioned between 1970 and 1977; they are the smallest ships ever equipped with ASROC anti-submarine missiles.

Their primary armament is an 8-round ASROC pepperbox launcher amidships (without reloads), supplemented by two sets of Mk.32 triple tubes for lightweight ASW TORPEDOES. Anti-air warfare armament is limited to one twin 3-in. 50-caliber DUAL-PURPOSE gun on the foredeck, and one twin 40-mm. BOFORS GUN aft. See also FRIGATES, JAPAN.

Specifications **Length:** 305 ft. (93 m.). **Beam:** 35.5 ft. (10.82 m.). **Draft:** 11.5 ft. (3.5 m.). **Displacement:** 1470 tons standard/1700 tons full load. **Powerplant:** 4 4000-hp. Mitsubishi UEV 30/40 or Mitsui 28VBC-38 diesels, 2 shafts. **Speed:** 25 kt. **Range:** 10,700 n.mi. at 12 kt./12,000 n.mi. at 9 kt. **Crew:** 165. **Sensors:** 1 hull-mounted OQS-3 medium-frequency sonar, 1 SPS-35(J) variable depth sonar, 1 OPS-14 air-search radar, 1 OPS-16 surface-search radar, 1 OPS-19 navigation radar, 1 GCFS-1B fire control radar. **Electronic warfare equipment:** 1 NORL-5 electronic signal monitoring array.

CHINOOK: The Boeing Vertol CH-47 medium-lift transport HELICOPTER, also produced under license in Italy and Japan. Developed to a U.S. Army requirement for an all-weather medium transport helicopter, the prototype flew in 1961, and the first Chinooks entered service with the army in 1962. Instead of the more usual "penny-farthing" configuration (with a large main rotor and a smaller anti-torque tail rotor) the Chinook has a twin-tandem rotor configuration, with main rotors located at each end of the fuselage, powered by a common drive shaft. The two rotors interleave and rotate in opposite directions, thereby neutralizing torque and eliminating the need for a small tail rotor.

A door/ramp built into the rear fuselage allows the roll-on loading of vehicles and palletized cargo. The fuselage is watertight for amphibious landings, and a dam is built into the cargo bay to permit water landings with the ramp door open. The Chinook has four-poster landing gear, with large, low-pressure tires to allow operations from soft surfaces. Two large sponsons attached to the lower sides of the fuselage contain self-sealing, crash-resistant fuel cells.

The original CH-47A, much used in Vietnam to resupply fire bases and recover crashed aircraft, was powered by two 2200-shp. Avco Lycoming T55-L-5 turboshaft engines and could carry 16,000 lb. (7272 kg.) of cargo or 40 fully equipped troops (on one occasion a CH-47A lifted 147 Vietnamese refugees). The CH-47A was quickly superseded by the CH-47B, with 2850-shp. T55-L-7C engines, and finally, from 1968, by the CH-47C, with two 3750-shp. T55-L-11 engines, increased internal fuel capacity, and a stronger transmission allowing higher operating weights, including a maximum payload of 26,000 lb. (11,818 kg.). Over 550 were deployed to Vietnam, and many others were exported; the CH-47C was also built under license in Italy and Japan.

In 1969 the U.S. Army began upgrading its Chinooks to a new CH-47D standard. The CH-47D has two 4500-shp. T55-L-712 engines (more than twice the power of the CH-47A), a strengthened transmission, and more internal fuel. More than 15 percent of the structure is manufactured from composite materials to reduce weight, and the rotor blades, made from glassfiber and Nomex, are stressed to survive a direct hit from a 23-mm. shell. The CH-47D has night and all-weather instrumentation, and its ELECTRONIC COUNTERMEASURES (ECM) include an ALE-40 CHAFF dispenser and IN-FRARED COUNTERMEASURES. The CH-47D can carry up to 44 troops or 24 stretcher cases and 4 attendants, or up to 28,000 lb. (12,700 kg.) of cargo. The fuselage has three cargo hooks, two rated at 20,000 lb. (12,700 kg.) and one at 28,000 lb. (9072 kg.). Some 436 U.S. Army Chinooks will be rebuilt to

D-standard by 1992, and Kawasaki initiated license production of new CH-47Ds in 1986.

Specifications **Length:** 51 ft. (15.55 m.). **Rotor diameter:** 60 ft. (18.29 m.). **Cargo hold:** 30.16 × 8.25 × 6.5 ft. (9.2 × 2.51 × 1.98 m.). **Powerplant:** 2 Avco-Lycoming T-55 turboshafts. **Fuel:** (D) 55,770 lb. (25,350 kg.). **Weight, empty:** (C) 20,378 lb. (9243 kg.); (D) 23,149 lb. (10,500 kg.). **Weight, max. takeoff:** (C) 46,000 lb. (20,866 kg.); (D) 50,000 lb. (22,680 kg.). **Speed, max.:** (C) 190 mph (302 kph) at sea level; (D) 183 mph (295 kph) at sea level. **Mission radius:** (C) 115 mi. (185 km.) w/7300-lb. (3294-kg.) payload; (D) 35 mi. (56 km.) w/23,030-lb. (10,446-kg.) payload. **Ferry range:** (D) 1279 mi. (2058 km.). **Operators:** Arg, Aus, Can, Egy, Gre, Iran, Jap, Lib, Mor, Nig, ROK, Sp, Tan, Thai.

CHOBHAM (ARMOR): A form of composite ARMOR developed by the Royal Ordnance Laboratory at Chobham, England. The exact composition is secret, but informed speculation has it that blocks of Chobham armor consist of alternating layers of steel, ceramic, and titanium plates, all laminated between layers of ballistic nylon or Kevlar mesh, to resist both kinetic and chemical ARMOR PIERCING ammunition. The mesh serves to spread the impact of kinetic energy rounds (such as APC, APDS, or APFSDS) over the entire face of the plate, thereby dissipating its effect, while the ceramic and titanium layers are believed to absorb the effects of chemical energy rounds such as HEAT or HESH. It is claimed that Chobham armor is at least twice as effective as the same weight of conventional (homogeneous steel) armor against kinetic rounds, and more than that against HEAT. Chobham armor seems to have overcome the problem of delamination, which defeated pre–World War II attempts to produce composite armor. High manufacturing costs, however, limit its application to MAIN BATTLE TANKS. Chobham or similar composites protect the U.S. ABRAMS, British CHALLENGER, West German LEOPARD II, Israeli MERKAVA III, and probably the Soviet T-80.

CHOKEPOINT: A passage through which military forces and/or war-critical matériel must transit because of geographic or infrastructure constraints, such as straits or deep-water channels at sea; and bridges, tunnels, mountain passes, or roads though nontrafficable terrain on land. Chokepoints are critical to the success of INTERDICTION campaigns aimed at reducing overall enemy capabilities by attacking supply lines. On land, the importance of any one chokepoint depends on the density of the transportation network and the availability of alternate sources of supply. Normally, interdiction will only cause rerouting until the network is saturated. Similarly, even if normal sources of key materials are fully denied by interdiction, alternate sources or substitutes may circumvent the effect (e.g., German synthetic oil and rubber production in World War II). Finally, if logistic requirements are both light and simple, even the interdiction of most chokepoints most of the time may have little effect, as in the case of U.S. interdiction efforts against the Vietcong (U.S. interdiction was more effective against the more heavily armed North Vietnamese regular forces). See also LOGISTICS.

CHOKING AGENT: A lethal CHEMICAL WARFARE agent (also called an asphyxiant) that damages the respiratory tract by causing the mucous membranes to swell and fill with liquid. Symptoms include coughing, choking, difficulty in swallowing, chest tightness, shallow breathing, and pulmonary edema. Fatalities tend to occur only some hours after exposure. Most choking agents, notably PHOSGENE (CG) and di-phosgene (DP) gas, are nonpersistent, and often disperse within 12 hours; hence they can kill or incapacitate troops, but not deny terrain.

CIA: See CENTRAL INTELLIGENCE AGENCY.

CIC: See COMBAT INFORMATION CENTER.

CIVIL DEFENSE: All measures (other than active interception) meant to limit the effects of enemy attack upon civilians and civilian facilities, including evacuation, dispersal, concealment, the provision of protective shelters, the duplication of essential services, resilient or redundant procedures for production and distribution, and the political, educational, and psychological preparation of the population.

Sweden, Switzerland, Vietnam, and Israel have the most comprehensive programs, including shelters for the entire urban population and highly trained personnel. In the Soviet Union, evacuation as well as the costly dispersal of critical industries has been emphasized instead, as opposed to in-place disaster relief (hence the abysmal Soviet performance in the wake of earthquakes, etc.). Extremely hardened, very deep protective shelters are provided for the top Communist party and military leadership. China has a similar but cheaper program. Most advanced countries protect only key government personnel by sheltering and evacuation.

Civil defense is an integral part of overall military power, especially in relation to NUCLEAR, BIO-LOGICAL, or CHEMICAL WARFARE, because it can increase the credibility of threats by degrading the presumptive effects of enemy retaliation. Thus Swedish and Swiss civil defense programs reinforce neutrality, by rendering the two countries less vulnerable to nuclear blackmail, while the Israeli program is rather more focused on threats that may be immune to DETERRENCE (e.g., Iraqi Scud missile attacks).

Both anti-nuclear "peace activists" and many strategic analysts committed to the policy of ASSURED DESTRUCTION argue that any civil defense activities by a nuclear power could be perceived by another power as part of attack preparations; in any case, the object of assured destruction is deterrence through mutual vulnerability, and civil defense thus can only undermine such deterrence. That argument having prevailed, the U.S. civil defense effort is largely symbolic; current budgets allow no hope of practical results. See also DAMAGE LIMITATION.

CIWS: See CLOSE-IN WEAPON SYSTEM.

CLANDESTINE OPERATIONS: Properly, actions hidden by their nature rather than disguised as in covert actions; more loosely, any secret operation. Clandestine operations of military interest include sabotage and secret long-range SCOUTING (as, e.g., the British SAS and SBS explorations of Argentine dispositions prior to the Falklands landings). Such operations can be carried out by highly trained COMMANDO units such as the U.S. DELTA FORCE, the Soviet SPETSNAZ, and the Israeli *Sayaret Matkal*, as well as the SAS and SBS. Clandestine operations thus differ fundamentally from COVERT OPERATIONS, which are normally carried out by civilian personnel.

CLEMENCEAU: A class of two French AIRCRAFT CARRIERS (*Clemenceau* and *Foch*) built between 1955 and 1963. The Clemenceaus are the smallest conventional takeoff and landing carriers built since World War II. Designed to replace warsurplus British and U.S. light carriers, they incorporate most of the features of modern carrier design, including an enclosed bow, an angled and armored flight deck, steam catapults, and a mirror landing system.

The angled deck, canted 8° to port, is 543 ft. (156.55 m.) long and 97 ft. (29.57 m.) wide. The two 20-ton steam catapults are located on the bow and on the angled deck. Two elevators (one on the bow behind the catapult, the other on the starboard deck edge behind the island) provide access to the hangar deck, which is 499 ft. (152.13 m.) long, 73 ft. (22.25 m.) wide, and 23 ft. (7.01 m.) high. Defensive armament consists of four 100-mm. DUAL PURPOSE guns in single mounts located along the edge of the flight deck, and two 8-round CROTALE surface-to-air missile launchers with 36 missiles.

The air group normally consists of 16 Super ETENDARD attack aircraft, 3 ETENDARD IVP reconnaissance aircraft, 10 F-8E(FN) CRUSADER fighters, 7 ALIZE anti-submarine warfare aircraft, 2 ALOUETTE III light helicopters, and 2 SUPER FRELON heavy helicopters. Alternatively, 30 to 40 helicopters can be carried.

The Clemenceaus have been effective instruments of French foreign policy in spite of their operational limitations, which include aircraft weight limits imposed by the catapults and the lack of effective airborne early warning aircraft. The *Clemenceau* and *Foch* are to be replaced by the two 36,000-ton nuclear-powered carriers of the CHARLES DE GAULLE class in the mid-1990s. See also AIRCRAFT CARRIERS, FRANCE.

Specifications Length: 870 ft. (262.25 m.). **Beam:** 104 ft. (31.7 m.). **Draft:** 28.2 ft. (8.6 m.). **Displacement:** 27,307 tons standard/32,780 tons full load. **Powerplant:** twin-shaft steam, 6 oil-fired boilers, 2 sets of geared turbines, 126,000 shp. **Speed:** 32 kt. **Fuel:** 3720 tons. **Range:** 4800 n.mi. at 24 kt./7500 n.mi. at 18 kt. **Crew:** 1338. **Sensors:** 1 DRBV15 surface-search radar, 1 DRBV23B air-search radar, 1 DRBV23C warning radar, 2 DRBV31 fire control radars, 2 DRBI10 height-finding radars, 1 NRBA navigational radar, 1 SQS-503 hull-mounted medium-frequency sonar. **Electronic warfare equipment:** 1 ABRX10, 1 ABRX16, 1 ABRC17 electronic signal monitoring array, 2 Dagaie chaff rocket launchers.

CLGP: Cannon-Launched Guided Projectile, an ARTILLERY shell with a LASER, INFRARED, or millimeter wave (MMW) seeker in the nose and movable control surfaces which deploy after the shell leaves the gun barrel. Once such shells are fired into the general vicinity of the target, the seeker head can lock onto and guide the shell to the target. The only CLGP now in service is the U.S. 155-mm. M712 Copperhead. The U.S. Navy has been developing a similar 5-in. (127-mm.) projectile ("Deadeye"), and the U.S., West Germany, Britain, and several other countries are jointly developing an "Autonomous Precision-Guided Munition" (APGM). See also IMPROVED CONVENTIONAL MUNITIONS.

CLOSE AIR SUPPORT (CAS): The use of tactical aircraft as flying ARTILLERY over the battlefield in close coordination with ground forces. The concept was pioneered by the U.S. Marine Corps in the 1930s and further developed by the *Luftwaffe:* in the opening phases of World War II, hundreds of Ju-87 Stuka dive bombers and other aircraft would be massed in support of the momentary SCHWERPUNKT of armored offensives to obtain concentrations of firepower against enemy resistance that was slowing the armored columns of the BLITZKRIEG.

The German version of CAS was powerful but inflexible: once air units were committed to an attack in a particular sector, it was difficult to divert them to meet other contingencies, because of the *Luftwaffe*'s centralized direction. In addition, air-ground coordination at the front was poor. In 1944–45, the Allies rectified these deficiencies with two innovations: the FORWARD AIR CONTROLLER (FAC), and the "Cab Rank" system.

Initially, FACs were pilots seconded to ground forces, who communicated with CAS aircraft via radio to direct them as required. Later, FACs operated from spotter aircraft, to acquire and designate targets over wider areas. In Vietnam, AIR DEFENSES were sometimes too dangerous for slow moving FAC aircraft; in response, some two-seat jet fighters were converted to high speed "Misty FACs." Both airborne and ground-based FACs can now coordinate CAS operations, which are planned and controlled as low as the BATTALION echelon in some Western armies.

Under the Cab Rank system of 1944–45, several flights of CAS aircraft would be kept orbiting behind friendly lines until a FAC designated a target; when a flight left the Cab Rank to attack, a new flight would replace it, to keep a continuous supply of CAS firepower on call.

The early jets that replaced propeller aircraft in the CAS role lacked the endurance needed for Cab Rank operations. Current jets have ample endurance, but are usually too few in number to warrant the system. The attack HELICOPTER, armed with cannon, rockets, and ANTI-TANK GUIDED MISSILES (ATGMs), now competes with fixed-wing aircraft in the CAS role. Helicopters can be organic, or attached directly to DIVISIONS or BRIGADES for a very close air-ground coordination. V/STOL aircraft such as the AV-8 HARRIER, on the other hand, allow a variation of the Cab Rank: ground loitering. Instead of orbiting behind the battlefield (and burn-

ing fuel), Harriers can wait for calls in small clearings, with their engines shut down.

Targets for CAS are normally within a few thousand meters of friendly troops, and often within a few hundred. Hence, accurate weapons delivery is essential. Moreover, enemy forces are likely to be camouflaged or concealed. Because of these factors, CAS aircraft should fly at speeds under 400 mph (670 kph) at altitudes under 250 ft. (75 m.); maneuverability is critical, both for terrain avoidance and in order to bring weapons on to bear once a target is detected. But low/slow CAS flight profiles are of course highly vulnerable to air defenses, as the Yom Kippur War of 1973 demonstrated once again. In response, specialized CAS aircraft, such as the A-10 THUNDERBOLT or Su-25 FROGFOOT, are heavily armored to resist light antiaircraft guns, and have individually protected engines and redundant control systems as protection against hits from larger weapons. In addition, both aircraft are highly maneuverable at low speed, so that they can exploit terrain masking. Both aircraft have large fuel capacities for extended loitering in Cab Rank fashion, and carry heavy ordnance loads, to be able to repond to several requests for support in a single sortie. Other air forces, including the Israeli, now believe that CAS is simply too costly for the results obtainable.

CLOSE-IN WEAPON SYSTEM (CIWS): A naval gun or missile system meant for terminal (last-ditch) defense against attacking ANTI-SHIP MISSILES. Surface-skimming missiles such as HARPOON or EXOCET, flying less than 50 ft. (15 m.) above the sea at speeds in excess of 600 mph (1000 kph), are difficult to detect except at very short range, due to their small size and the clutter of spurious radar returns from waves. A CIWS therefore requires very fast reaction times and extreme accuracy to be able to destroy incoming missiles in the few seconds between detection and impact; thus most of them are completely automatic with manual override. Gun-type CWIS, typified by the U.S. PHALANX, Dutch GOALKEEPER, and Spanish MEROKA, usually consist of a 20-mm. or 30-mm. cannon with a very high rate of fire (2000–4000 rounds per minute), controlled by its own radar (optimized for target detection and tracking in severe ground clutter) and fire control computer.

Missile-type CWIS, such as the British SEA WOLF, NATO SEA SPARROW, and multi-national RAM (ROLLING AIRFRAME MISSILE) are highly accurate short-range missiles which destroy their target by impact (they are sometimes called "hit-tiles").

CLUSTER BOMB: An air-delivered, free-fall weapon which consists of a thin-walled, stream-lined, fin-stabilized canister that contains numerous small submunitions, or bomblets. A time or proximity fuze in the nose of the cluster bomb activates a burster charge, which splits open the canister after release to disperse the bomblets over a wide area. That compensates for the inherent inaccuracy of low-altitude bombing, and increases the probability of multiple target kills with a single weapon. Current cluster bombs carry a variety of submunitions, optimized against armored vehicles, or against personnel and soft-skinned vehicles, or else designed for runway cratering against airfields. Some submunitions are fitted with delay or pressure fuzes, to act as mines. An entirely new type is the intelligent submunition, which has its own sensor and guidance system to seek out tanks or other point targets after its release.

Cluster bombs were first used in Vietnam as anti-personnel weapons, and have since been acquired by every major air force. Common types include the US Mk.20 ROCKEYE anti-tank and SADEYE anti-personnel cluster bombs, and the British BL.755. The latter is 8 ft. (2.43 m.) long, weighs 582 lb. (265 kg.), and contains 147 SHAPED-CHARGE anti-armor bomblets that are scattered at different velocities to achieve uniform distribution on the ground; the shaped charge is sufficient to penetrate the thin roof armor of tanks, and the bomblets are also fitted with fragmentation jackets, to generate splinters more effective against soft vehicles and personnel.

A variation on the cluster bomb is the munitions dispenser, which releases submunitions while remaining attached to the aircraft. Munitions dispensers do not require good free-fall ballistics, and can therefore be designed to carry and dispense submunitions more efficiently while minimizing aerodynamic drag. Their primary disadvantage is that the aircraft must actually overfly its target, thus exposing itself to enemy air defenses. Free-flying munitions dispensers such has the Brunswick Defense LAD (Low Altitude Dispenser) overcome that problem at commensurate cost.

CLUSTER GUARD: NATO code name for tiles manufactured of ANECHOIC MATERIAL which are applied to the hulls of Soviet nuclear submarines to impede hostile active SONAR, and insulate noises generated within the submarine. Tile separation due to wave action has been a chronic problem.

CN: Chloroacetophene, a tear agent or lachrymator, widely used by military and police forces for crowd control. CN causes coughing, choking, and uncontrolled discharges from the tear ducts and mucous membranes of the nose and eyes, but is generally not lethal. Chloroacetophene in chloroform (CNC) and chloroacetophene in benzene and carbon tetrachloride (CNB) are similar. See also CHEMICAL WARFARE.

COBRA: 1. Bell AH-1 attack helicopter, in service with the U.S. Army and Marine Corps and several other countries. The Cobra was developed quickly in response to an urgent U.S. Army requirement for a fast, heavily armed helicopter to escort troop-carrying helicopters, and provide fire support for AIR MOBILE and AIR ASSAULT operations in Vietnam. Bell developed the Cobra in 1965 as a private venture, on the basis of the well-proven engine, transmission, and rotor system of the ubiquitous UH-1B HUEY utility helicopter (hence the AH-1 is sometimes called HueyCobra). Service tests began in December 1965, and an order was placed for 110 Cobras in April 1966. Initial production batches reached Vietnam in June 1967. The U.S. Army originally acquired the Cobra as an interim type pending the introduction of the more advanced AH-56 Cheyenne, but when that costly helicopter was canceled, the Cobra was placed in large-scale production.

Production of the initial AH-1G version totalled 1119, of which 39 were transferred to the marine corps for evaluation, and a few others were converted to dual-control trainers. In keeping with its counterinsurgency mission in Vietnam, the primary armament of the AH-1G consisted of 7.62-mm. GAU-2 miniguns and 40-mm. GRENADE LAUNCHERS in a chin turret, and several pods for 2.75-in. Folding Fin Aerial Rockets (FFARS) on stub wing pylons. In 1972, 92 were upgraded to AH-1Q standard and equipped with TOW Anti-Tank Guided Missiles (ATGMs); others were converted to AH-1Rs with more powerful engines but without TOW. A dozen AH-1Gs were transferred to Israel in 1973.

The U.S. Army planned to replace all of its Cobras with the more advanced AH-64 APACHE anti-tank helicopter during the 1980s, but production delays and cost overruns reduced the number of Apaches it could afford. Hence the Cobra remains the army's primary attack helicopter, supplemented by some 700 Apaches. As a substitute, the army funded improvements to upgrade the Cobra's performance for anti-armor missions on a modern battlefield. The TOW system of the AH-1Q and the powerplant of the AH-1R were retrofitted to 315 AH-1G models subsequently redesignated the "Modified AH-1S," which also has an armored, nonreflective (low glint) canopy consisting of optically flat plates, and armor around the cockpit and engines. These were followed by 100 new production AH-1Ss, with new avionics, a universal gun turret for 20-mm. or 30-mm. cannon, and detailed improvements. The definitive army version is the "Modernized AH-1S," equipped with a new air data computer, DOPPLER navigation, a digital FIRE CONTROL systems, and INFRARED COUNTERMEASURES. This version is still in production for the U.S. Army, Spain, Israel, Jordan, South Korea, Pakistan, and Turkey, and is manufactured under license in Japan.

The basic layout of the Cobra remains essentially unchanged from the AH-1G, but the data that follows applies to the Modernized AH-1S. The fuselage is only 3.16 ft. (963 mm.) wide, to minimize drag, visibility, and vulnerability. In the tandem cockpit the weapons operator sits in the front seat, with the pilot above and behind him, an arrangement, now standard on attack helicopters worldwide, that maximizes the weapons operator's field of view, while still affording the pilot good visibility. The cockpit is covered by a heavily braced canopy which incorporates nonglint armored glass (proof against 12.7-mm. rounds) in the windscreen and side windows. An M65 TOW THERMAL IMAGING sight and a gunsight are mounted in a small, trainable nose turret, which is aimed by a hand yoke at the weapons operator's station, or slaved to a Helmet Mounted Sight System (HMSS). Below the gunner is a universal gun turret for a variety of alternative weapons, including a 4-barrel M197 VULCAN 20-mm. cannon, a GAU-12/U 25-mm. rotary cannon, an M230 30mm CHAIN GUN, or a GAU-2A 7.62-mm. minigun. The turret, with a traverse of 110° left or right, and elevation limits of +10° and −60°, can be aimed with either the M65 TOW sight or the HMSS, and fired by either crewman. Additionally, each of the two stub wings behind the cockpit has two hardpoints for a variety of ordnance, including 7- and 19-round FFAR pods, 3-round ZUNI 5-in. rocket pods, up to 8 TOW missiles, 4 STINGER anti-aircraft missiles, additional pod-mounted guns, or free-fall weapons. The landing gear consist of two sturdy tubular skids.

The engine and transmission, mounted immediately behind the pilot, drive a 2-bladed main rotor with tapered tips to reduce noise. To protect against IR-homing missiles, the AH-1S has a passive infrared suppressor for the engine exhaust pipe, an ALQ-144 infrared jammer, and two ALE-44 dispensers for flares and CHAFF. The Cobra is highly maneuverable, but its rotor system is not stressed for negative-g maneuvers. Although slower than the Soviet Mil-24 HIND-D attack helicopter, the Cobra is much more agile, and being much smaller, is a more difficult target to detect.

After the U.S. Marine Corps received a number of AH-1Gs for evaluation, in May 1968 it placed an order for 49 AH-1J Sea Cobras, equipped with twin engines for increased reliability on overwater flights, but otherwise comparable to the AH-1G. Iran ordered 202 AH01Js modified to fire TOW missiles, and a further 20 AH-1Js were ordered by the marines in 1974–75, after which production switched to the AH-1T Improved Sea Cobra. The fuselage was lengthened to accommodate more fuel, and a new transmission and rotor system were provided to handle the output of two cross-coupled engines. Fifty-seven were delivered in 1976–77. The AH-1T is generally comparable to the AH-1S, except for retention of the original (curved) AH-1G canopy. In 1986 the U.S. Marines began taking delivery of 44 AH-1W Super Cobras, with uprated engines, the nonglint canopy, and provisions for up to eight AGM-114A HELLFIRE laser-guided ATGMs, two AIM-9L SIDEWINDER air-to-air missiles for self-defense, or two AGM-122 SIDEARM anti-radiation missiles for use against enemy radar. Maximum takeoff weight is increased, but the powerful engines raised the maximum speed with eight TOWs, and hot-weather performance is also much improved.

Specification **Length:** (S) 44.6 ft. (3.6 m.); (T, W) 58 ft. (17.68 m.). **Rotor diam.:** (S) 44 ft. (13.45 m.); (T) 48 ft (14.63 m.) **Powerplant:** (S) 1 1800-shp. Lycoming T53-L-703 turboshaft derated to 1100-shp.; (T) 2 Pratt and Whitney Canada T400-PW-402 turboshafts, 1970 shp.; (W) 2 General Electric T700-GE-4015, 3250 shp. **Weight, empty:** (S) 5479 lb. (2939 kg.); (T) 8030 lb. (3642 kg.). **Weight, max. takeoff:** (S) 10,000 lb. (4536 kg.); (T) 14,000 lb. (6350 kg.). **Speed, max.:** (S, T) 172 mph (287 kph) clean/141 mph (235 kph) w/8 TOWs; (W) 187 mph (312 kph). **Initial climb:** (S) 1620 ft./min. (494 m./min. **Range, max.:** (S) 315 mi. (526 km.); (T) 261 mi. (420 km.).

2. (BO-810) and Mamba: First-generation ANTI-

TANK GUIDED MISSILES (ATGMs) produced by Messerschmidt-Bolkow-Blohm. A lightweight WIRE-GUIDED missile designed for one-man operation, Cobra has a cylindrical body with a conical nose which can house one of two interchangeable 5.5-lb. (2.5-kg.) warheads: a shaped-charge HEAT warhead capable of penetrating up to 18.7 in. (475 mm.) of steel plate; and an anti-tank fragmentation warhead capable of penetrating 13.7 in. (348 mm.) of steel plate while generating lethal fragments over a radius of 10–15 m. Four square cruciform tail fins have movable control surfaces for steering.

Unlike most ATGMs, Cobra does not have a launch rail or tube; instead, its fins rest directly on the ground with the small booster rocket pointed downwards. At launch, the booster drives the missile up and forward until a sustainer motor ignites, accelerating the missile to its maximum speed. The operator steers the missile with a joystick through a control box weighing 9 lb. (4.09 kg.), with the steering commands transmitted to the missile via fine wires unreeled from spools inside the missile. With Cobra's Manual Command to Line-of-Sight (MCLOS) guidance, the operator must first manually bring the missile onto his line of sight to the target ("gathering"), then track both missile and target while steering via the joystick; that imposes a minimum range of 400 m. An operator usually carries two missiles, but can be linked by cables to as many as eight Cobras up to 240 m. from the control box.

Cobra is light, simple, and cheap. On the other hand, its MCLOS guidance requires intensive training; the missile's low speed results in a long time-of-flight (25 sec. out to 2000 m.) during which the operator is exposed to enemy fire; and the warhead is too small to be effective against MAIN BATTLE TANKS protected by SPACED, CHOBHAM, or REACTIVE ARMOR. Cobra is in service with 18 countries; a number have been used by the PLO.

The improved Cobra, now called Mamba, has an improved guidance system that facilitates gathering and steering, and a new, two-pulse rocket motor with a lower velocity for gathering, followed by a rapid acceleration to maximum speed, which significantly reduces the time of flight. A 7x optical telescope built into the controller also facilitates tracking at long range. Minimum range is reduced to 100 m. Mamba is now in service with the West German army.

Specification Length: (Cobra) 37.4 in. (950 mm.); (Mamba) 37.5 in. (952 mm.). **Diameter:** (Cobra) 3.9 in. (100 mm.); (Mamba) 4.75 in. (120 mm.). **Span:** (both) 18.9 in. (480 mm.). **Weight, launch:** (Cobra) 22.66 lb. (10.3 kg.); (Mamba) 26.2 lb. (11.9 kg.). **Speed, max.:** (Cobra) 187 mph (86.75 m./sec.); (Mamba) 311 mph (144 m./sec.). **Range, max.:** (both) 2000 m. **Operators:** Arg, Bra, Den, FRG, It, Pak, Tur, 11 others.

COBRA DANE AND COBRA JUDY: Code names of two large PHASED ARRAY RADAR systems used by the United States to track Soviet BALLISTIC MISSILES test-fired into the Kamchatka Peninsula and Pacific Broad Ocean Area (BOA). Located at Shemya Air Force Base, Alaska, Cobra Dane has a array approximately 90 m. in diameter, containing more than 35,000 transmitting and receiving elements. It can survey a corridor 2000 mi. long, which is perpendicular to the flight path of Soviet missile tests. With a fixed field of view of 120°, Cobra Dane can provide tracking and high-resolution signature data on Soviet POST-BOOST VEHICLES, REENTRY VEHICLES, booster debris, and PENAIDS; it also has some space tracking and early warning capabilities.

Cobra Judy is a similar but smaller array installed aboard a ship which can be positioned to best view Soviet test shots into the BOA.

COCK: NATO code name for the Soviet Antonov An-22 *Antei* long-range, heavy-lift transport aircraft, which was by far the largest aircraft in the world at the time of its introduction in 1965 (and even the Lockheed C-5A GALAXY of 1969, though heavier, had smaller dimensions). Some 100 An-22s were built between 1965 and 1974, of which 55 were allocated to the VTA (Soviet Military Transport Aviation), and the rest to Aeroflot. A number have been lost, mostly operating in the Siberian Arctic, and Cock is now slowly being replaced by the An-124 CONDOR.

In its general configuration, Cock is virtually a scaled-up version of the An-12 CUB, except for the tail section, which has twin fins and rudders extending above and below the horizontal stabilizer. The unpressurized cargo hold has a ramp/door in the upswept tail for the roll-on loading of vehicles and cargo, up to a maximum payload of 176,370 lb. (80,168 kg.). A pressurized passenger cabin seating 28 is located forward of the hold. Of the six-man crew, the pilot, copilot, engineer, radio operator, and loadmaster are seated in an airliner-type flight deck above the passenger cabin, while the navigator sits in an extensively glazed nose compartment, behind which is a ground-mapping radar in a chin radome.

The shoulder-mounted wings have a very high aspect ratio for economical cruising, resulting in a very high wing loading of 148.4 lb. per square ft. (724.6 kg./m.²) at maximum weight, vs. 135 lb. per square ft. (661.6 kg./m.²) for the smaller, heavier C-5A. The propeller slipstream is directed over the wings' large, double-slotted flaps to enhance short field performance and low-speed handling. Twelve large, low-pressure main tires allow rough-field operations. Despite its great weight and high wing loading, the An-22 has very good performance: at maximum weight the takeoff run is only 4300 ft. (1310 m.).

Specification   **Length:** 189.95 ft. (57.91 m.). **Span:** 211.28 ft. (64.41 m.). **Length, cargo hold:** 108.33 ft. (33 m.). **Powerplant:** 4 15,000-shp. Kuznetsov NK-12MA turboprops. **Fuel:** 94,800 lb. (43,000 kg.). **Weight, empty:** 251,327 lb. (114,240 kg.). **Weight, max. takeoff:** 551,156 lb. (250,525 kg.). **Speed, max.:** 460 mph (740 kph). **Speed, cruising:** 373 mph (600 kph). **Range, max. payload:** 3107 mi. (5000 km.). **Range, max. fuel:** 6804 mi. (10,950 km.).

COD: Carrier On-board Delivery (aircraft), small transports used by the U.S. Navy to transfer personnel and high-value equipment to and from aircraft carriers at sea. The current COD aircraft is the Grumman C-2 Greyhound, derived from the E-2 HAWKEYE airborne early warning plane.

CODAD: Combination Of Diesel And Diesel, a naval propulsion arrangement in which two diesel engines drive a single propeller shaft. A small diesel (the base plant) is used for sustained, fuel-efficient cruising at relatively low speeds; when higher speeds are required, the second, larger diesel (the boost plant) is started.

CODAG: Combination Of Diesel And Gas, a naval propulsion arrangement in which a low-power diesel (the base plant) and a high-power gas turbine (the boost plant) are connected to the same propeller shaft through a common gearbox. The diesel is used for sustained, fuel-efficient cruising at low speeds, while the turbine is started only when high speeds are required.

CODLAG: Combination Of Diesel-Electric And Gas, a naval propulsion arrangement in which the base plant consists of diesel-powered electric generators which power electric motors for fuel-efficient cruising at speeds up to 18 kt. or so. A boost plant of high-power gas turbines is started only when higher speeds are required. Both plants are connected to the same propeller shaft through a common gearbox.

CODOG: Combination Of Diesel Or Gas, a naval propulsion arrangement in which a low-power diesel base plant and a high-power gas turbine boost plant are connected to the same propeller shaft through a common gearbox. The diesel is used for sustained, fuel-efficient cruising at low speeds. When high speed is required, the diesel is disconnected from the propeller shaft, and the gas turbine is started. CODOG's primary disadvantage is the time required for the gas turbine to "rev up" to speed.

COGAG: Combination Of Gas And Gas, a naval propulsion arrangement in which two gas turbine engines are connected to the same propeller shaft through a common gearbox. Gas turbines are fuel efficient only when operating at or near their rated output; to achieve fuel efficiency, a low-power turbine base plant for cruising is combined with a high-power turbine boost plant, which supplements the smaller turbine when high speeds are required.

COGOG: Combination Of Gas Or Gas, a naval propulsion arrangement in which two gas turbines are connected to the same propeller shaft through a common gearbox. A small gas turbine base plant is used for sustained, fuel-efficient cruising at low speeds. When high speeds are required, the base plant is disconnected from the propeller shaft and a high-power gas turbine boost plant is connected instead. The primary disadvantage is the time needed for the high power turbine to "rev up" to speed.

COHESION: The moral force that enables fighting units to withstand the shock of combat as well as the privations of war. Cohesion is an essential component of combat effectiveness (i.e., fighting power), because it can overcome the natural instinct to flee, or at least take cover rather than fight. Until the late 19th century, armies normally fought in close-order formations upon which cohesion could be imposed by incessant drill and ferocious discipline (Frederick the Great: ". . . an army should fear its officers more than the enemy"). When automatic weapons forced armies to fight dispersed and under cover, cohesion could no longer be imposed externally, because the individual soldier no longer fought in drilled ranks under the close supervision of his officers. Hence the soldier could hide under cover during the fighting, or even run away, with little chance of being caught. In response, armies began to promote an internalized cohesion based upon the sense of mutual obligation and personal loyalty that naturally develops among soldiers who remain in stable, small units. Patriotism or other idealisms can have some effect (often ephemeral), but soldiers mainly fight because of their loyalty to their primary group, the stable small unit; because of the personal leadership of respected immediate commanders; and because of a sense of identity with a larger, but still exclusive, secondary group such as the regiment, an elite corps, etc.

The emotional ties that develop among soldiers in a "primary group," usually the SQUAD, section, or PLATOON engaged in combat, reflect their mutual reliance for mutual survival. Modern infantry tactics depend on these ties because they are based on the concept of "distributed risk." In elementary FIRE AND MOVEMENT tactics, for example, one group of soldiers advances, assuming the greater risk, while a second group also exposes itself to fire at the enemy, thereby suppressing the enemy's fire, and drawing his attention away from the advancing element. If the soldiers distrust each other, the movement group will refuse to advance for fear that the fire group will not accept the lesser risk, and the fire group will refuse to expose itself for fear that the movement group will not advance. When, by contrast, intimate association with shared experiences and shared suffering result in unit bonding (the German army's *Kameradschraft*), its members will fight because they do not wish to let down others in the unit, and to avoid being identified as cowards. They become "afraid to flee."

Leadership contributes to cohesion when soldiers identify with their commander, and because they feel that he in turn identifies with them, by showing sympathy and by fully sharing their risks and hardships. Combat leaders must normally lead by example, but that is not enough: soldiers must also believe that the leader is technically competent and that he will not expose them to unnecessary danger.

Loyalty to a larger secondary group can also contribute to cohesion. In the British army, for example, identification with the history and traditions of the REGIMENT is important; in other cases, such as the U.S. Marine Corps and the French Foreign Legion, it is an entire military branch that evokes such a wider loyalty. But except for long-service soldiers in long-established formations, secondary group loyalty is usually insignificant compared to primary group loyalty.

Some armies constantly subordinate until cohesion to administrative convenience. The U.S.

Army, for instance, has a long history of treating its soldiers as interchangeable units in an impersonal machine. Soldiers are moved about as individuals, and not kept together long enough to form strong primary groups. In addition, officers tend to be rotated too quickly to gain the trust and confidence of their soldiers. An attempt was made to correct these deficiencies through the short-lived COHORT system.

The Soviet army has not officially recognized the importance of unit cohesion. Instead, it has relied on political indoctrination and harsh discipline. Unofficially, many Soviet officers have recognized the need to build cohesive primary groups, but are hindered by the short term of conscription and the ethic diversity within units; neither applies to the elite Airborne Troops and SPETSNAZ units, which are highly cohesive.

Extreme examples of the weight of strong unit cohesion in overall fighting power include the Israeli victories over Arab armies since 1948 and the British victory over Argentina in 1982. See also MORALE.

COHORT: Cohesion, Operational Readiness, and Training, a U.S. Army personnel program in which COMPANY-size units of recruits, with their officers and NCOs, are kept together for their entire initial term of service, as opposed to traditional U.S. Army policy of training and assigning recruits as individuals. COHORT is designed to promote unit COHESION, MORALE, and motivation at the PLATOON and company level, to obtain greater fighting power and combat proficiency. COHORT was introduced in 1982 as part of the New Manning System (NMS), which also includes the adoption of regimental affiliations for all battalions. The first complete COHORT battalions were formed in 1984. By 1988, with COHORT still only partially implemented, the system was substantially abandoned, because of complaints by officers and NCOs that it (sometimes) delayed (by a few months) their promotions—and, more generally, because cohesion was too abstract a value compared to administrative notions of efficiency.

COIN: See COUNTERINSURGENCY WARFARE.

COLBERT: The only active French guided-missile CRUISER. Built between 1953 and 1956 but not commissioned until 1959, *Colbert* was originally designed as an all-gun anti-aircraft cruiser (CLAA), but was converted to a guided-missile cruiser between 1970 and 1972. Unlike most cruisers in service today, Colbert has substantial armor protection: 50 mm. of deck armor, and a side belt of 50–80 mm.

Intended mainly for ANTI-AIR WARFARE (AAW), the main armament is a twin-arm MASCURA surface-to-air missile launcher with 48 missiles, supplemented by 2 100-mm. DUAL-PURPOSE guns and 6 twin 57-mm. ANTI-AIRCRAFT guns. In 1980, 4 EXOCET MM38 anti-ship missiles in individual launch canisters were added to provide some ANTI-SURFACE WARFARE capability.

Colbert is often used as a flagship for overseas operations, and is fitted with a variety of long-range radios and satellite communications equipment. It is scheduled for disposal in 1997.

Specifications **Length:** 585 ft. (178.35 m.). **Beam:** 65.65 ft. (20 m.). **Draft:** 25.67 ft. (7.83 m.). **Displacement:** 8500 tons standard/11,300 tons full load. **Powerplant:** twin-shaft steam, 86,000 shp. **Speed:** 31.5 kt. **Range:** 4000 n.mi. at 25 kt. **Crew:** 562. **Sensors:** 1 DRBVC20 and 1 DRBV23C air-search radar, 1 DRBI10D height-finding radar, 1 DRBV50 surface-search radar, 1 DRBN32 navigation radar, 1 DRBR32C 57-mm. fire control radar, 2 DRBC31 100-mm. fire control radars, 2 DRBR51 missile guidance radars. **Electronic warfare equipment:** 1 ARBR10 and ARBR31 electronic signal monitoring array, 1 ARBB32 active jamming unit, 2 Syllex chaff rocket launchers.

COLD LAUNCH: A technique for launching BALLISTIC MISSILES in which the missile is loaded into its SILO or TEL (Transporter-Erector-Launcher) within a tight-fitting launch canister equipped with slip ring gaskets (to form a gas-tight seal with the missile), and with a cold gas generator in its base. On launch, the generator produces high-pressure gas which ejects the missile from the silo to a height of some 100 ft. (30 m.), at which point the missile's main engine ignites.

Cold launch does not cause much damage to the silo, unlike "hot launch," whereby the missile main engines ignite inside the silo. Thus a cold-launch silo can be reconditioned and reloaded relatively quickly. In addition, with cold launch a given silo can acept a missile of greater diameter than with hot launch, which must allow room inside the silo for the venting of the rocket exhaust. The Soviet SS-16 SINNER, SS-17 SPANKER, SS-18 SATAN, SS-24 SCALPEL, and SS-25 SICKLE ICBMs and the SS-20 SABER IRBM all have cold-launch systems. Of U.S. land-based ballistic missiles, only the PEACEKEEPER ICBM is cold-launched. All submarine-launched ballistic missiles (SLBMs) also use cold-launch techniques, but rely on high-pressure steam or water to eject the missile. See also ICBMS, SOVIET UNION; ICBMS, UNITED STATES.

COLD WAR: An international conflict in which all means other than overt military force are used; the term describes the state of East-West relations from the late 1940s to the early 1960s, by the narrowest definition. Combinations of economic warfare, propaganda, subversion, proxy war, and COVERT OPERATIONS are also sometimes described as CONFRONTATIONS.

COLLATERAL DAMAGE: Damage inflicted upon civilian or other unintended targets caused by the spillover of weapons effects aimed at a nearby military target (vs. the damage caused by aiming errors). The term is usually employed to describe the civilian casualties inflicted by a nuclear attack aimed exclusively at missile SILOS or other military installations. See also COUNTERFORCE.

COLLECTIVE FILTRATION: A form of NBC (Nuclear-Chemical-Biological) protection for ships and ARMORED FIGHTING VEHICLES, whereby all or part of the ship or vehicle is hermetically sealed, with air provided through appropriate filters. The crew can therefore operate within the ship or vehicle without having to wear cumbersome individual protective equipment. The most common form of collective filtration system works by positive overpressure: the pressure within the ship or vehicle is maintained at levels slightly higher than atmospheric pressure. Hence, even if the sealed environment is imperfect or penetrated by small projectiles, contaminants can be excluded by the pressure differential. See also BIOLOGICAL WARFARE; CHEMICAL WARFARE.

COLLECTIVE SECURITY: A strategy based on the proposition that potential aggressors can be dissuaded by a multilateral alliance of sufficient strength. The assumption is that all parties will be willing to defend any party threatened or under attack, regardless of their own interests at the time. The idea was applied (theoretically) in the Covenant of the League of Nations, and later in the Charter of the United Nations, whose Security Council is supposed to deal with threats to peace by collective action. But the five "permanent members" of the council can veto any action, and it is those five (Britain, China, France, the Soviet Union, and the United States) that are either directly or indirectly responsible for most threats to peace.

COLT: NATO code name for the Soviet Antonov An-2 single-engine light transport biplane aircraft. For all its antiquated appearance, the An-2 has been produced in greater numbers than any other aircraft since World War II; it first flew in 1947, and is still being produced in Poland and China.

The all-metal fuselage is a shortened version of the Li-2 (the Soviet copy of the DC-3/C-47 Dakota); it can carry up to 12 passengers or 2900 lb. (1318 kg.) of cargo, loaded through a large door on the left. The pilot and copilot sit side-by-side at the front of the cabin. The An-2 has rugged fixed landing gear with two large, low-pressure main wheels mounted under the lower wing and a small fixed tail wheel, which allows operations from rough or muddy surfaces. The main wheels can be replaced by skis, or by pontoons for water landings.

The biplane configuration was chosen for short takeoffs and landings, and good low-speed handling. Mounted above and below the fuselage, the two wings are connected by a single pair of struts. The upper wing has full-span leading edge slats, while slotted flaps are fitted on both upper and lower wings. Takeoff runs vary from 492 to 558 ft. (150 to 170 m.), and landing rolls from 558 to 607 ft. (160 to 185 m.). Fuel is carried in six fuel tanks in the upper wing. Both the wings and the tailplane are fabric-covered (!) over a metal frame.

More than 5000 Colts were produced in the Soviet Union between 1948 and 1960, when production was transferred to the Polish WSK-Mielec Factory, which has since delivered at least 9500 aircraft. Since 1967, more than 5000 AN-2s have been produced in China as the Fong Chou 2. Production was restarted in the Soviet Union between 1964 and 1970, when several hundred An-2M versions, with large, square-cut tails, were delivered. In 1972 the Soviet Union began producing the An-3, an An-2 with a turboprop engine. Some new production An-3s were built, but most were conversions of existing An-2s. Total An-2/3 production exceeds 18,000.

The Colt's amazing longevity is due largely to its simplicity, ruggedness, reliability, and versatility. It serves both as a civil and military transport, and as a search and rescue, crop duster, ambulance, scout, and liaison aircraft. It is used by both the Soviet Union and North Korea to deliver secret agents and SPETSNAZ (commando) units, because its low speed allows it to fly at very low altitudes (in the terrain) to avoid detection and evade modern high-speed fighters (though head-on the propeller has a large radar signature). In the Soviet Union the An-2/3 continues to serve with VTA (Military Transport Aviation), Aeroflot, and DOSAAF, the Soviet paramilitary training organization.

Specifications **Length:** 41.8 ft. (12.75 m.). **Span:** (upper) 59.64 ft. (18.18 m.); (lower) 46.71 ft. (14.24 m.). **Cargo hold:** 13.5 ft. (4.11 m.). **Powerplant:** (2) 1000-hp. Shvetsov ASh-62IR 9-cylinder radial; (3) 1081-shp. Glushenkov TVD-20 turboprop. **Fuel:** 1870 lb. (850 kg.). **Weight, empty:** 7605 lb. (3457 kg.). **Weight, max. takeoff:** 12,125 lb. (5511 kg.). **Speed, max.:** 160 mph (267 kph). **Speed, Cruising:** 115 mph (192 kph). **Range, 1100-lb. (500-kg.) payload:** 560 mi. (900 km.). **Operators:** Afg, Alg, Bul, Cuba, DDR, Egy, Eth, Hun, Iraq, Mali, Mon, N Kor, Pol, PRC, Som, Sud, Syr, Tan, Tun, Tg, USSR.

COMBAT AIRCRAFT: Fixed-wing aircraft and HELICOPTERS employed for air-to-air combat, air-to-surface attack, reconnaissance, troop and logistic transport, and Airborne Early Warning (AEW).

MILITARY REQUIREMENTS

Manned aircraft are still dominant in all dimensions of modern, nonnuclear warfare, with the exception of REVOLUTIONARY WARS (in which their role is, or ought to be, marginal). Because of technological change, and the evolution of military requirements, the many types of aircraft in service at any given time reflect both generational and functional differences. The two sometimes interact. For example, OBSOLESCENT fighters are often converted to attack aircraft (fighter-bombers) when their performance is judged insufficient for aerial combat, but still adequate for air-to-surface attack. A basic classification by primary mission follows.

Air-to-Air Combat

Aircraft designed to seek out and destroy enemy aircraft in aerial combat are termed FIGHTERS. Fighters deemed to have the speed and maneuverability required to destroy other fighters are classified as AIR SUPERIORITY fighters, while INTERCEPTORS are fighters optimized to detect and destroy intruding bombers over friendly territory, often with the assistance of ground control systems. Interceptors tend to be less maneuverable than air superiority fighters, but require high rates of climb and are normally equipped with long-range RADAR and air-to-air missile systems.

Air-to-Surface Attack

Aircraft designed for the long-range delivery of large weapon loads are defined as BOMBERS. They tend to be large and unmaneuverable, and carry a heavy bomb load. Bombers have tended to become less important in nonnuclear warfare, because the capabilities of both air defense systems and attack aircraft have increased (STEALTH may change that). The U.S. and U.S.S.R. retain several hundred bombers each, mainly for strategic nuclear delivery and maritime attack.

ATTACK AIRCRAFT may be older fighters employed in that role, or specialized aircraft superficially similar to fighters, but with configurations that emphasize ordnance loads and air-to-ground accuracy at the expense of maneuverability and air-combat capabilities. Attack aircraft diverge radically between CLOSE AIR SUPPORT (CAS) and INTERDICTION aircraft. CAS is performed over the battlefield in close coordination with ground forces, while interdiction is aimed at targets deep behind the lines (often in the attempt to sever enemy communications). Specialized attack helicopters, such as the US AH-1 COBRA and AH-64 APACHE and the Soviet Mil-24 HIND, are taking over much of the CAS mission from fixed-wing aircraft. The only specialized CAS aircraft in service, the US A-10 THUNDERBOLT and Soviet Su-25 FROGFOOT, are both slow, heavily armored, and highly maneuverable. Interdiction or "strike" aircraft, by contrast, are faster than many fighters but less maneuverable, and have greater range and payload; e.g., the U.S. F-111, European TORNADO, and Soviet Su-24 FENCER. Armed trainers such as the ALPHA JET, AMX, and HAWK are classified as light attack aircraft; they are of marginal value for CAS because they lack protection and redundant systems, but can be economical for shallow or "battlefield" interdiction.

For maritime attack against surface vessels, both ordinary attack aircraft armed with bombs or ANTI-SHIP MISSILES and specialized maritime patrol aircraft are in service. The latter are large, long-range aircraft, often converted civilian airliners equipped with surface search radar, ELECTRONIC SIGNAL MONITORING (ESM), and other specialized sensors as well as weapons.

ANTI-SUBMARINE WARFARE (ASW) aircraft are maritime patrol aircraft equipped with underwater sensors and armed with torpedoes and depth charges. Helicopters supplement fixed-wing aircraft in this role, often being operated from aboard FRIGATES, DESTROYERS, and other ASW vessels.

Reconnaissance

For long-range, or strategic, RECONNAISSANCE, two different types of aircraft are in service. Older bombers, transports, or adapted airliners equipped with ESM are used to gather electronic intelligence (ELINT) without actually overflying hostile airspace. On the other hand, photographic and photogrammetric strategic reconnaissance, which requires the penetration of hostile airspace, is performed by specialized high-altitude aircraft, such as the Lockheed U-2, TR-1, and SR-71 BLACKBIRD. The use of this type of aircraft has diminished because of the political risks of peacetime overflights, the increasing effectiveness of AIR DEFENSES even against high-altitude, high-speed aircraft, and the availability of reconnaissance satellites.

Short-range, or tactical, reconnaissance is performed at high speeds and low altitudes by fighter aircraft equipped with cameras and multi-spectral sensors. Some air forces have special variants of fighter aircraft, while others use regular fighters equipped with sensor pods.

Troop and Logistic Transport

For long range, or strategic, transport, both civilian airliners (as in the CRAF program) and specially designed transports are in use; only the latter can deliver MAIN BATTLE TANKS and other such cargoes, normally with roll-on landing through rear ramp doors. Typical of these long-range transports are the U.S. C-5 GALAXY, C-141 STARLIFTER, and C-17; and the Soviet An-22 COCK and An-124 CONDOR.

Short-range, or tactical, transports are smaller aircraft with short takeoff and landing (STOL) capability, to allow the use of fields closer to the front lines. Examples include the C-130 HERCULES, the An-12 CUB, and TRANSALL C-160. Helicopters have taken over most tactical transport duties in the battle area itself, because of their ability to land in small clearings.

Airborne Early Warning

AIRBORNE EARLY WARNING aircraft are usually derived from transports or airliners, and carry large SURVEILLANCE radars, often in rotary "saucer" antennas over the fuselage. When fighter-direction equipment and personnel are carried in the main cabin, the aircraft becomes an Airborne Warning and Control System, or AWACS, as in the E-2 HAWKEYE, E-3 SENTRY, and IL-76 MAINSTAY. Airborne command posts, a closely related type, are adpated long-range civilian airlines equipped with command, control, and communications (c^3) equipment.

TECHNOLOGY

Combat aircraft consist of five major elements: the airframe, powerplant, AVIONICS, weapons, and aircrew. The first four have developed much since World War II, while the last (possibly the most important) has not.

Diverse operating requirements that may be needed in a single combat mission are often technologically incompatible; e.g., for interdiction, a combination of long range and heavy payload at subsonic cruising speeds (for penetration to the target) require high-bypass-ratio engines, high wing loading, modest wing sweep, and high-aspect-ratio wings; while high speed and rapid acceleration (for the attack itself) require low-bypass-ratio engines, low wing loading, and high-sweep, low-aspect-ratio wings. More generally, long runways allow economical takeoffs and heavy payload, but are themselves a major vulnerability. Two relatively new techniques, variable-geometry (swing) wings and vertical takeoff and landing (VTOL), resolve some design conflicts while creating others (complexity, weight). Variable-geometry aircraft can combine fuel-efficient long-range cruise and short takeoff performance (at low sweep angles) with high speed, acceleration, and maneuverability (with the wings fully swept). VTOL aircraft, notably the British HARRIER, can hover or land in small clearings. The most primitive VTOL technique, two separate sets of engines employed for vertical and horizontal flight, is very inefficient from a weight and fuel standpoint, because the vertical lift engines are dead weight once horizontal flight begins. In the more complex Harrier, pivoted exhaust nozzles for vectored thrust allow one engine to be used for both vertical and horizontal flight. Vectoring can also be used during forward flight (VIFF) to radically improve maneuverability. Even so, VTOL still exacts a substantial penalty in range, payload, and speed. Short takeoff and landing (STOL) offers many of the benefits of VTOL with few of the penalties.

STOL aircraft combine high-lift devices, such as slats and flaps, with techniques such as boundary layer control (whereby engine bleed air is diverted over wing control surfaces to create the effect of higher speeds).

Airframes have been constructed mainly of aluminium and steel, but costly titanium alloys are required in high-performance aircraft because of their light weight, strength, and resistance to heat. In the 1980s, composite materials, such as embedded graphite fibers, began to supplant metals because of their light weight, strength, stiffness, and ease of manufacture.

Jet engines have greatly improved in reliability and fuel efficiency since the 1960s. Most military aircraft now have turbofans instead of less efficient turbojets. The number of moving parts has been greatly reduced, with significant improvement in reliability and maintainability.

But the greatest progress has been registered in all aspects of avionics, the on-board electronics for target acquisition, navigation, weapon direction, and flight control. In various applications, radar is used in all four functions, and is now increasingly supplemented by passive sensors such as FLIR (Forward-Looking Infrared) and, more tentatively, by laser radar (LADAR). INERTIAL NAVIGATION, supplemented by satellite navigation systems such as NAVSTAR, have led to a quantum improvement in navigational accuracy. Computers are also essential in all four functions. Radar data is now processed and presented in symbolic displays by computers, which also process other sensor inputs for FIRE CONTROL systems, now allowing median bombing accuracies of less than 10 m. Computers are also increasingly important in flight control, to process pilot inputs so as to optimally position control surfaces (FLY-BY-WIRE). A fifth branch of avionics, ELECTRONIC COUNTERMEASURES, has become critical to combat survival in the presence of air defenses.

Aerial ordnance has also evolved. For air-to-air combat, the CANNON is complemented in fighters, and supplanted in interceptors, by a wide variety of air-to-air missiles with radar or infrared guidance. The first missiles (especially radar-guided) were very unreliable, but in the most recent combat in the Middle East, the Falkland Islands, and the Persian Gulf, missiles scored the majority of kills.

Air-to-surface ordnance is still dominated by free-fall weapons such as low-drag general purpose (LDGP), NAPALM, fragmentation, and CLUSTER

BOMBS, but now includes LASER and ELECTRO-OPTI-CALLY (TV) guided GLIDE BOMBS and intelligent submunitions such as SKEET, as well as air-to-surface missiles. Originally, most air-to-ground missiles were controlled by some form of manual COMMAND GUIDANCE (i.e., the operator "flew" the weapon via a joystick and DATA LINK); a typical example is BULLPUP. The latest missiles, however, are launch-and-leave types with terminal guidance: TV, imaging infrared, and laser-guided point-target missiles such as MAVERICK; anti-radiation missiles such as HARM, ALARM, and SHRIKE; ANTI-SHIP MISSILES such as HARPOON and EXOCET; and long-range CRUISE MISSILES such as ALCM and TOMAHAWK, which are controlled by complex computer-correlated TERRAIN COMPARISON AND MATCHING GUIDANCE.

COMBAT AIR PATROL (CAP):   A standing patrol of fighters or interceptors assigned to the defense of a particular area on land and sea. In U.S. terminology specifically, a BARCAP, or Barrier Combat Air Patrol, is meant to prevent enemy aircraft from reaching friendly bases by mixing in with returning groups of friendly aircraft; a RESCAP, or Rescue Combat Air Patrol, is staged over the area of a search and rescue (SAR) mission, to prevent the enemy from intercepting rescue helicopters and close air support flights; a MIGCAP is a free-ranging fighter sweep along the flight path of a strike mission, to engage enemy aircraft before they can intecept the strike group; a FORCECAP is a combat air patrol staged over or near a carrier battle group; and a TARCAP, or Target Combat Air Patrol, is mounted over strike targets to protect attack aircraft during their bomb runs. See also AIR COMBAT MANEUVERING; ANTI-AIR WARFARE; COUNTERAIR.

COMBAT ENGINEER VEHICLE (CEV):   A modified MAIN BATTLE TANK (MBT), armed with a large-caliber, low-velocity short-barreled demolition gun, to reduce obstacles (e.g., roadblocks), bunkers, and fortified buildings. Most CEVs also have a bulldozer blade (for the building of defensive positions or clearing debris), and a boom and winch assembly. A typical CEV is the U.S. M728, derived from the M60 PATTON; it is armed with a 165-mm. demolition gun, and carries a dozer blade, an A-frame derrick, and a winch. See also ENGINEERS.

COMBAT INFORMATION CENTER (CIC):   In modern COMBAT VESSELS, the location where all primary sensor displays and FIRE CONTROLS are consolidated. Sensor inputs are combined to generate and maintain battle plots which show the positions of all friendly and hostile forces in the area, to enable CIC officers to keep track of the action, allocate weapons, and coordinate maneuvers. In most warships, the CIC is divided into ANTI-AIR WARFARE (AAW), ANTI-SURFACE WARFARE (ASUW), and ANTI-SUBMARINE WARFARE (ASW) sections, each commanded by an officer with specialized training, under a tactical coordinator. The captain traditionally commands from the bridge, but because naval warfare relies increasingly on remote sensing, captains are increasingly commanding from the CIC, leaving the executive officer to con the ship from the bridge. DATA LINKS such as the U.S. Naval Tactical Data System (NTDS), which gather sensor data from all ships in a battle group, have further increased the importance of the CIC.

COMBAT RADIUS: The distance at which a ship or aircraft can attack targets with a useful payload and then return to its base. Combat radius is usually less than half of maximum range for aircraft, and is greatly affected by mission profiles. Maximum range requires some optimal combination of (relatively low) speed and (relatively high) altitude. In combat, however, missions might require very high speeds or very low altitudes, or both, increasing fuel consumption accordingly. ATTACK AIRCRAFT, for instance, usually fly "hi-lo-hi" profiles; i.e., the penetration to the target is made at cruising speed and high altitude, the attack is made at high speed and low altitude, and the return to base is again at high altitude. For attack aircraft, combat radius also depends on the weapon load and its carriage: external loads impose additional drag penalties.

FIGHTERS usually cruise at high altitude, but burn fuel in prodigious amounts when using afterburners in AIR COMBAT MANEUVERING (ACM). Usually, their combat radius is calculated on the basis of a high-altitude, cruising-speed approach to and from the battle area, plus a certain number of minutes of ACM.

Ship combat radius depends very little on the weapons load, but is highly sensitive to speed. In general, warship ranges are stated on the basis of cruising at half to two-thirds power, normally equivalent to 18–20 kt. in modern warships. At combat speeds of 27–32 kt., range can be halved. Large warships, when operating at sea for protracted missions, rely on underway refueling. Small combat vessels, such as FAST ATTACK CRAFT (FACs), must usually return to base after two or three days, being limited by their combat radius in the same manner as aircraft.

COMBAT TALON: U.S. Air Force MC-130E/H special operations variants of the Lockheed HERCULES tactical transport aircraft. Combat Talons have special avionics for the low-level penetration of hostile airspace in support of CLANDESTINE and other SPECIAL OPERATIONS. Fourteen C-130Es were modified to MC-130E Combat Talon Is, which equip the 1st, 7th, and 8th Special Operations Squadrons, based in the Philippines, West Germany, and Florida, respectively. These are being supplemented by 24 MC-130H Combat Talon IIs, which began entering service in 1987. The Combat Talon II is equipped with a TERRAIN-FOLLOWING RADAR, precision navigation and air-drop equipment, AERIAL REFUELING equipment, a variety of ELECTRONIC COUNTERMEASURES, and the Fulton STAR midair recovery system (which can pick up men off the ground by catching a tethered balloon with a special hook). Performance characteristics are almost identical to those of the basic C-130E/H.

COMBATTANTE: A series of French-built fast guided-missile boats, in service with several navies. The original *Combattante,* completed in 1964, was an experimental 202-ton wood-and-fiberglass FAST ATTACK CRAFT based on a German Lurssen design. The Combattante II, larger and more capable, has a steel hull with round bilges for stability in rough seas, and a superstructure of light alloy to reduce weight.

The Combattante II is in service with West Germany, Greece, Iran, Libya, and Malaysia. The West German navy operates a total of 20 as the Type 148. Commissioned between 1972 and 1975, they are armed with four EXOCET MM.38 anti-ship missiles amidships, an OTO-Melara 76.2-mm. DUAL PURPOSE gun on the bow, and a 40-mm. BOFORS GUN on the stern; the latter can be replaced by eight MINES. Sensors include a 3RM20 navigational RADAR, a Thomson-CSF Triton surveillance radar, and a Pollux tracking radar, all linked to a Vega weapons control system. They have a crew of 22.

Greece operates four Combattante IIs completed in 1971–72, armed with 4 Exocet missiles, 2 stern-mounted 21-in. (533-mm.) tubes for wire-guided TORPEDOES, and 2 twin 35-mm. OERLIKON anti-aircraft guns. Sensors include a Decca 1226 navigation radar, a Triton surveillance radar, and a Castor tracking radar. They have a crew of 46.

Ten Combattante IIs were delivered to Libya in 1982–83; one was sunk by U.S. aircraft in 1986. Armament consists of four OTOmat anti-ship missiles, an OTO-Melara 76.2-mm. gun on the bow, and a twin 40-mm. gun on the stern. Sensors include a Triton surveillance radar and a Castor tracking radar, both linked to a Vega weapons control system.

Malaysia operates four boats completed in 1972–73, armed with 2 Exocet, 1 57-mm. dual-purpose gun, and one 40-mm. Bofors gun. Sensors are identical to those of the Greek boats. Iran took delivery of 13 Combattante IIs between 1977 and 1981. At least 2 have been lost in combat, and the condition of the others is unknown. They were to be armed with four HARPOON anti-ship missiles, but the U.S. arms embargo has forced Iran to equip them with Exocet, which is available on the open market. Other armament consists of one 76.2-mm. gun and one 40-mm. Bofors gun. Sensors include a Decca 1226 navigation radar and an HSA WM-25 search and tracking radar. ELECTRONIC WARFARE equipment includes a TMV-433 ELECTRONIC SIGNAL MONITORING array and an Alligator-5A ACTIVE JAMMING unit.

Developed in 1975 for the Greek navy, the enlarged Combattante III retains four-shaft diesel propulsion, but with increased power. Greece now has 10 Combattante IIIs completed between 1977 and 1982, of which 6 were built under license. The first 4 are armed with 4 Exocet, 2 OTO-Melara 76.2-mm. guns, 2 Emerson twin 30-mm. anti-aircraft guns, and 2 21-in. torpedo tubes. Sensors are similar to those in the German Type 148s, with the addition of an integrated optical sight system and a COMBAT INFORMATION CENTER (CIC). The boats have a crew of 42 men. The last 6 have 6 Penguin anti-ship missiles instead of Exocet, and simplified sensors.

Nigeria has three Combattante IIIs delivered in 1981, armed with 4 Exocet, 1 OTO-Melara 76.2-mm. gun, 1 twin 40-mm. gun, and 2 twin 30-mm. guns. Sensors include a Decca 1226 navigation radar, a Triton surveillance radar, and a Castor II tracking radar. Electronic warfare equipment is limited to a Decca RDL RADAR WARNING RECEIVER.

The French-built Israeli SA'AR missile boats are similar to the Combattante IIs. See also FAST ATTACK CRAFT, FRANCE.

Specifications Length: (II) 154.2 ft. (47 m.); (III) 184.2 ft. (56.15 m.). **Beam:** (II) 23.3 ft. (7.1 m.); (III) 26.2 ft. (8 m.). **Draft:** (II) 6.2 ft. (1.9 m.); (III) 8.2 ft. (2.5 m.). **Displacement:** (II) 234 tons standard/260 tons full load; (III) 385 tons standard/447 tons full load. **Powerplant:** (II) 4 3600-hp. MTU diesels; (III) 4 5000-hp. MTU diesels. **Speed:** (II) 35–40 kt.; (III) 40 kt. **Range:** (II) 2000 n.mi. at 15 kt.; (III) 700 mi. at 32.5 kt./2000 mi. at 15 kt. **Crew:** See text. **Sensors:** See text. **Electronic warfare equipment:** See text.

COMBAT TEAM: U.S. Army term for a COM-PANY-size COMBINED-ARMS force formed for a specific mission by the attachment or cross-attachment of PLATOONS from other companies. A combat team may have as few as two or as many as five platoons. A tank-heavy team may have two tank platoons and one infantry platoon, while an infantry-heavy team has an inverse ratio. A balanced team has the same number of tank and infantry platoons. Additional support, including ANTI-TANK, ANTI-AIRCRAFT, or ENGINEER units may be attached to a combat team as required. See also ARMORED WARFARE; BATTLE GROUP.

COMBAT VESSELS: Naval issues are not always amenable to the analytical techniques applied to land, tactical air, and strategic issues. This is due in part to the lack of symmetry among naval forces: most notably, the U.S. Navy maintains a fleet of large aircraft carriers, while the Soviet navy is only now building its first. Such asymmetries are caused largely by geographic circumstances, broadly defined. The geography that matters is that of narrow sea lanes, trade routes, shipping access, trade dependence, and the quantity and quality of ports; in other words, details. It is perhaps that mass of detail which enables many navies to resist the imposition of quantitative methods of analysis.

The U.S. Navy, for example, finds it necessary to maintain aircraft carriers in the Mediterranean even though the U.S. has air bases in Spain, Greece, Italy, and Turkey; the Soviet navy, by contrast, has relied on destroyer-cruiser surface action groups (SAGs) in the Mediterranean, without any effective air cover at all. These paradoxes illustrate the complexities of air power/sea power interactions and the role of visual symbolism in naval matters. A squadron of fighters at 30,000 ft. may have a greater attack capability than a squadron of destroyers, but the latter usually have greater political impact. And while air bases can be more economical than carriers in any one location, vulnerability considerations, the combat radii of preferred aircraft, and the political costs of using land bases for actual combat operations may justify the stationing of carriers in zones already covered by land-based aircraft.

To trace a logical path through the maze of naval equipment, certain modes of naval action are selected for appraisal in what follows. First, it is assumed that a certain kind of "primary" naval power is required. The opposition can be expected to deploy "threat" forces intended to neutralize the effectiveness of those primary forces. Threat forces are opposed in turn by "support" forces to protect the value of the primary forces. The following interactions then emerge:

1. Sea-Based Air Power (primarily tactical)
   Primary: Aircraft carriers
   Threat: Aircraft carriers and attack submarines
   Support: Anti-air warfare (AAW) escorts (e.g., guided-missile cruisers), anti-submarine warfare (ASW) carriers, ASW escorts (destroyers and frigates), and hunter-killer (ASW) submarines
   Secondary threat: Surface action groups (with anti-ship missiles), more submarines, and land-based air power
   Secondary support: More air power, including ASW aircraft
2. Submarine-Launched Ballistic Missiles (SLBMs)
   Primary: Nuclear-powered ballistic missile submarines (SSBNs)
   Threat: Attack and hunter-killer submarines
   Support: Attack submarines, ASW surface vessels, and ASW aircraft
   Secondary threat: ASW surface vessels and other SSBNs (deterrence)
   Secondary support: Aircraft carriers and surface action groups
3. Amphibious Warfare
   Primary: Landing ships, tank landing craft, helicopter carriers, and gunfire support vessels
   Threat: Land-based air power
   Support: Aircraft carriers (see 1, above) and AAW vessels
   Secondary threat: Attack submarines
   Secondary support: Hunter-killer submarines, ASW aircraft, and ASW vessels
4. Interdiction of Sea Lines of Communication (SLOCs)
   Primary: Attack submarines
   Threat: Hunter-killer submarines and ASW vessels

Support: Air power and submarines

Secondary threat: Aircraft carriers (see 1, above)

Secondary support: Surface action groups and fast attack craft (FACs)

These chains of action and reaction reveal that the pivotal combat vessels are submarines and aircraft carriers. The Soviet navy, with a primary sea-denial mission, relies most heavily on submarines, supported by surface action groups and land-based air power. The recent development of Soviet carrier air power indicates a potential switch to sea control missions. The U.S. Navy, intent on controlling and maintaining SLOCs, bases its sea power on aircraft carriers organized into CARRIER BATTLE GROUPS (CVBGs) with powerful AAW and ASW escort. Attack submarines are used both in support of CVBGs and against Soviet SSBNs and SAGs. Since 1980, the U.S. has begun formation of its own SAGs, armed with TOMAHAWK and HARPOON anti-ship missiles, to support CVBGs and provide gunfire support to amphibious forces.

Navies with purely regional interests, or those operating in constricted waters, are making increasing use of fast attack craft (FACs), especially missile-armed patrol boats and HYDROFOILS, which can sortie from a protected base, strike, and retire. Such vessels have much firepower for their size, and are far more cost-effective for small navies than traditional combat vessels.

COMBINED ARMS (OPERATIONS): The tactical integration of INFANTRY, ARMOR, and ARTILLERY down to the lowest possible echelon (usually the company), so as to balance the weaknesses of each individual arm with the strengths of the other two. For instance, tanks have both mobility and firepower, but are vulnerable in close terrain unless escorted by infantry. Infantry, in turn, needs the firepower and shock provided by tanks. Artillery can devastate infantry unless suppressed by COUNTERBATTERY fire, but cannot take or even hold terrain by itself. Supporting arms, such as ENGINEER, ANTI-TANK, RECONNAISSANCE, and AIR DEFENSE units can also contribute to the synergism. Combined-arms operations have been the key ingredient of success in armored warfare since the German BLITZKRIEGS early in World War II. To succeed, however, all units in a combined arms force must conform to a common operational and tactical DOCTRINE. Otherwise, the combination of dissimilar units in ad hoc formations is a prescrip-

tion for disaster. See also ARMORED FORCES; BATTLE GROUP; COMBAT TEAM.

COMMAND AND CONTROL: The direction of forces in the accomplishment of their missions, personnel, equipment, procedures, and facilities. Some command and control (C^2) systems are very informal.

The command and control *organization* consists of the headquarters and its staff. To manage the details of any sizable military organization is beyond the capacity of any one man, and the staff is supposed to act as a filter, to present the commander with organized data for his decision making, and to supervise the execution of his decisions. Staffs are usually organized on functional lines, with operations, intelligence, supply, and personnel sections, each with its own chief, who reports to the commander's chief of staff, who supervises all staff activities. As warfare has grown more complex, the size of staffs has increased; those of major formations number in the hundreds or thousands, though staff-to-soldier ratios vary sharply. In war, larger ground formations maintain two separate staffs: a forward or battle staff, and a main staff. The battle staff, with the heads of the operations and intelligence sections, and a small number of assistants, often has armored command vehicles (ACVs) for mobility, and to accompany the leading echelons. In some armies, the purpose is to allow the commander to be at the transient decisive point (SCHWERPUNKT) of the battle. The main staff remains in the rear to support the battle staff by planning and analysis, as well as routine "housekeeping" chores, notably logistics.

The command and control *process* varies with the personality of individual commanders. Some rely heavily on their staffs, and limit their role to the setting of objectives, intervening only to cope with difficulties at critical points in the battle. Others, unable to delegate, insist on making many decisions personally, often paralyzing the C^2 system in the process. Most commanders fall somewhere in between these extremes.

The exercise of command and control has become increasingly complex and fast-paced because of modern sensor and communications systems, which can overload manually operated C^2 processes. In response, many armies are increasingly using computer-assisted "decision aids" to correlate and display data in comprehensible patterns. In a more advanced application, computers at diverse command locations are networked through DATA LINKS to enable them to update and interrogate each other's data bases. Such systems can display instantaneously, in near-real time, the position of all units in a command, their current strength and logistic status, as well as the location of neighboring friendly units and known enemy forces. Artificial intelligence (AI) and complex simulation programs are being studied to see if they can allow commanders to test operational plans and otherwise assist their decisions.

Before the advent of long-range artillery and rifled weapons forced armies to disperse for survival, commanders could personally observe the battlefield, and pass orders personally or through messengers. With units now spread out over hundreds or thousands of square miles, COMMUNICATIONS have become critical for command and control. The commander receives information from land lines or radio networks, and issues his orders in the same way; because communications are now integrated into C² systems, they have in fact become command, control, and communication (C³) systems. And because INTELLIGENCE collection, analysis, and distribution are also critical for command and control, C³I, for Command, Control, Communications, and Intelligence, is now increasingly the term of choice.

COMMANDANTE RIVIERE: A class of ten French general-purpose FRIGATES, commissioned between 1962 and 1970. Only eight remain in service: one ship has been stricken and another rerated as a training vessel. The others are to be retired by the mid-1990s. Intended for escort duties and detached service in Middle Eastern, African, and Pacific waters, the Commandante Rivieres are air conditioned for tropical operations and have accommodations for an 80-man COMMANDO unit.

Armament consists of a 100-mm. DUAL PURPOSE gun at each end of the ship, a 305-mm. quadruple (ASW) mortar on the forward superstructure, 2 single 40-mm. anti-aircraft guns amidships, 2 sets of triple tubes for lightweight anti-submarine TORPEDOES, and 4 EXOCET MM38 anti-ship missiles in individual launch canisters on the after superstructure.

One ship, the *Balny*, was modified in 1964 as a testbed for the first French CODAG plant. Unlike its sister ships, the *Balny* has only one large propeller. Because the CODAG plant is so compact, more fuel can be carried to extend the maximum range. The *Balny* does not carry Exocet, but is otherwise identical to the other eight ships. See also FRIGATES, FRANCE.

Specifications **Length:** 333.75 ft. (102.7 m.). **Beam:** 38.35 ft. (11.8 m.). **Draft:** 14.13 ft. (4.35 m.). **Displacement:** 1750 tons standard/2230 tons full load. **Powerplant:** 4 4000-hp. SEMT-Pielstick 12PC diesels, 2 shafts; (*Balny*) single-shaft CODAG, 2 3600-hp. diesels (cruise)/1 11,500-shp. gas turbine engine (spring). **Speed:** 26 kt. **Range:** 2300 n.mi. at 26 kt./7500 n.mi. at 16.5 kt.; (*Balny*) 13,100 n.mi. at 10 kt. **Crew:** 166. **Sensors:** 1 DRBN32 navigational radar, 1 DRBV22A air-search radar, 1 DRBC32C fire control radar, 1 SQS-17 hull-mounted midfrequency sonar, 1 DUBA3 hull-mounted high-frequency attack sonar. **Electronic warfare equipment:** 1 ARBR16 radar warning receiver, 2 Dagaie chaff rocket launchers.

COMMAND GUIDANCE: A form of missile guidance in which the missile is under the continuous control of the operator or launch station throughout its flight. The most common forms are radio command and wire guidance. In the former, the operator or launch station tracks both the missile and its target (either optically or with radar), and steers the missile to intercept by commands transmitted over a radio DATA LINK. With wire guidance, steering commands are transmitted via two fine wires unreeled from spools inside the missile as it flies downrange; the technique is used primarily with ANTI-TANK GUIDED MISSILES, but with some homing TORPEDOES for midcourse updates. In both radio command and wire guidance, steering commands can be generated either manually (e.g., with a joystick), or automatically by a computerized fire control system. See also BULLPUP; MCLOS; MILAN; SACLOS; SAGGER; TOW.

COMMANDO: 1. Originally, Boer irregular CAVALRY units in the Anglo-Boer Wars. The term is now applied officially to units of the British Royal marines, and unofficially to many other elite units trained for SPECIAL OPERATIONS. Because the term has acquired connotations of bravery, efficiency, and ruthlessness, it is often used to flatter assorted inferior soldiery. See also COMMANDO, ROYAL MARINE.

2. A family of ARMORED CARS built by Cadillac Gage, in service with U.S. and several other forces. Originally developed as simple, amphibious armored cars for export, the Commando series incorporates commercial automotive components to reduce costs and simplify logistics.

The initial V100 4 × 4 scout car was armed with 7.62-mm. or 12.7-mm. MACHINE GUNS on pintle mounts or in power turrets. Intended primarily for troop transport and police riot control, the V100 is armored only against small arms fire. The U.S. Army acquired a few V100s (as the M706), for convoy escort and base security in Vietnam, and more than 100 were also supplied to South Vietnamese forces. Armed with twin 7.62-mm. machine guns in a power-operated turret, the M706 has a crew of three and can carry seven troops. The vehicle is amphibious and propelled through water by the rotation of the tires. Commercially successful, the V100 is in service with more than 11 countries.

To meet foreign demand for a more heavily armed vehicle, Cadillac Gage introduced the V200 in 1969. Longer, wider, and much heavier than the V100, the V200 is also slightly slower, and off-road mobility is somewhat degraded due to the increased weight and the higher ground pressure on only four tires. Not as successful as its predecessor, the V200 is in service only with Singapore.

The V150 4 × 4, sized between the V100 and the V200, closely resembles the V100, but has an enlarged turret ring that can accommodate turrets for weapons as large as the Cockerill 90-mm. ANTI-TANK gun; other weapon options include an Oerlikon 20-mm. cannon, twin light machine guns, an 81-mm. MORTAR, a 40-mm. automatic GRENADE LAUNCHER, a 76.2-mm. cannon, or an M61 VULCAN air defense gun, or up to 12 troops. Operators can therefore field a fleet of scout, troop carrier, and tank destroyer vehicles, all based on a common chassis and drive train, thereby simplifying logistics. The V150 has been very successful, with more than 3600 produced to date.

In 1979 Cadillac Gage introduced the enlarged V300, with 6 × 6 chassis. It can accommodate a wide range of armaments, including the Cockerill 90-mm. gun; as a troop carrier, it can hold 12 infantrymen.

Specifications Length: (V100) 18.5 ft. (5.64 m.). **Width:** (V100) 7.5 ft. (2.28 m.). **Height:** (V100) 7.8 ft. (2.38 m.). **Weight, combat:** (V100) 7.5 tons; (V150) 10 tons; (V200) 11 tons. **Powerplant:** (V100) 191-hp. Chrysler 361 gasoline; (V150) 202-hp. V8 gasoline; (V200) 275-hp. 440 CID gasoline; (V300) 250-hp V8. diesel. **Speed, road:** (V100) 60 mph (96 kph); (V150) 53 mph (88.5 kph); (V200 and V300) 58 mph (97 kph). **Range, max.:** (V100) 400 mi. (668 km.); (V150) 385 mi. (643 km.); (V200) 290 mi. (485 km.); (V300) 435 mi. (726 km.).

COMMANDO, ROYAL MARINE: A BAT-TALION-size formation of the British Royal Marines.

A Commando consists of 800 officers and men, organized into a headquarters COMPANY of 200 men, 3 rifle companies ("troops") of 128 men each, and a heavy weapons company (troop) of 156 men. Each rifle company is subdivided in turn into a headquarters platoon of 24 men and 3 rifle platoons of 34 men. Each platoon is equipped with 1 51-mm. MORTAR, 4 7.62-mm. machine guns, 3 BREN guns, 1 CARL GUSTAV 84-mm. recoilless rifle, and 21 ASSAULT RIFLES. The weapons company is divided into a mortar platoon with 6 81-mm. mortars, an ANTI-TANK platoon with 18 truck-mounted MILAN anti-tank guided missiles, a scout platoon with six machine-gun-armed jeeps, and an assault pioneer platoon equipped with MINES, DEMOLITION charges, and mine clearing equipment. See also MARINES, ROYAL.

COMMUNICATIONS (MILITARY): Until the invention of the telegraph and telephone in the 19th century, messengers (or "runners"), visual signals (with flags and hand signs), and aural signals (with drums, trumpets, etc.) remained the only forms of military communications. Those new devices (collectively, "wire communications") were not practical for tactical use on the battlefield under mobile conditions; hence runner, visual, and aural signals remained the primary forms of tactical communication through World War I, while the telephone and telegraph were much used at the operational and strategic levels. Radio telegraphy and telephony (collectively, "wireless communications") were used extensively by naval forces from their inception, and also by land forces at the operational level and above by World War I, but radio telephony (voice radio) became reliable only after the 1920s. Voice radio was the key to the BLITZKRIEG style of armored warfare, because it allowed German commanders to remain in contact with rapidly moving and widely separated armored columns. It also revolutionized air combat by allowing ground controllers to direct fighters to intercept enemy aircraft, by allowing tactical coordination between aircraft, and by allowing air and ground units to coordinate close air support.

Reliance on voice radio today persists at the tactical and operational levels, but wireless communications are not secure. Since their first use, radio signals have been intercepted by the enemy, and encryption is cumbersome and of uncertain effectiveness. Moreover, enemy ELECTRONIC SIGNAL MONITORING (ESM) can locate the source of radio signals with direction finding (D/F) equipment, allowing the enemy to target headquarters units and communications. In addition, radio is vulnerable to ELECTRONIC COUNTERMEASURES, notably jamming and deception. If military forces rely too heavily on radio for tactical coordination, jamming can totally disrupt their operations. Accordingly, it has been the Soviet practice to minimize the use of radio, and rely instead on wire communications, messengers, and thorough planning. A third drawback of radio communications is more insidious: as radio has become more reliable, the temptation to direct subordinates by detached "remote control" has grown apace. In Vietnam, U.S. battalion commanders in helicopters would direct the tactics of squad leaders 5000 ft. below. That stifles initiative in junior officers, and miseducates them to rely on orders from afar.

Even wire communications are not entirely secure. Switchboards and cable junctions can be attacked, disrupting entire networks. Commandos can infiltrate and clandestinely tap phone lines, and it is even possible to intercept field telephone messages, by detecting the ground currents generated by the wires. Thus, when ground units seek absolute communications security (COMSEC), they tend to fall back on the ancient methods of visual and aural signals, and runners, especially in small units.

In the field of communications, the most significant innovation since World War II has been the development of satellite communications (SATCOM), which have made fast, reliable, and global telecommunications commonplace; but of course, SATCOM has most of the drawbacks of radio, plus the added risk of ASAT attack on communications satellites.

Communications with ballistic missile submarines present a particular problem, because the submarines must remain deeply submerged to avoid detection, but must also remain in contact with the NATIONAL COMMAND AUTHORITIES. Radio waves cannot penetrate more than a few dozen feet beneath the surface of the water, except for VLF (Very Low Frequency) and ELF (Extremely Low Frequency) radio, which can penetrate to a depth of several hundred feet. But data rates (the rate at which information can be transmitted) are proportional to the frequency, so that ELF and VLF messages can only be transmitted slowly; in most cases, they are therefore used only as "bell ringers," to alert submarines to come up near the surface to receive conventional radio messages, or to perform some preplanned task. Attempts are now being made to develop satellite LASER communications for submarines by using blue-green lasers which can penetrate water easily; data rates would be very high.

Digitalized communications have become increasingly important. Radio DATA LINKS allow military computers to communicate (interface) directly; they are particularly important for fire control, weapon direction, and missile guidance.

Advances in COMSEC and ELECTRONIC COUNTER-COUNTERMEASURES include built-in voice scramblers for tactical radios, the use of packet switching devices, and frequency-agile radios.

COMPANY: A tactical subunit of ground combat formations. A company is usually subordinate to a BATTALION, and consists of several PLATOONS. Infantry battalions usually contain 3 or 4 rifle companies of 100–200 men, each of which consists of 3 or 4 rifle platoons and a heavy weapons platoon. In MECHANIZED INFANTRY formations the rifle company is similar, but the heavy platoon is often deleted because the fighting vehicles have their own heavy weapons. TANK battalions usually contain 3 or 4 tank companies, each with 3 or 4 tank platoons of 3–5 tanks, so that tank companies have between 10 and 17 tanks, including the commanders' tanks. In artillery units the BATTERY is equivalent to the company; it usually consists of 4–8 weapons and their crews. See also ARMORED FORCES; ARTILLERY; INFANTRY.

COMZ: Communications Zone, a U.S. Army term for the rear areas of overseas theaters of operations. Generally, the COMZ extends from the port of entry to the line at which field forces take over responsibility.

CONAS: Combination Of Nuclear And Steam, a naval propulsion arrangement in which a nuclear reactor with its steam turbine and an oil-fired steam turbine are connected to the same single propeller shaft through a common gearbox. The nuclear reactor base plant is used for sustained cruising, while the oil-fired turbine boost plant is started only when higher speeds are required. This highly efficient combination, applied in the Soviet KIROV battle cruisers, allows the use of a much smaller and cheaper nuclear reactor than would be needed if the reactor had to power high-speed surges as well (power requirements rise exponentially at speeds in excess of 20 kt.).

CONDOR: 1. NATO code name for the Soviet Antonov An-124 *Rusian* long-range, heavy-lift transport aircraft. A successor to the turboprop An-22 COCK (till then the world's largest aircraft), the An-124 first flew in 1982 and entered service in 1986.

Superficially similar to the Lockheed C-5A/B GALAXY, but in some respects more capable and technically advanced, the An-124's huge cargo hold has both a door/ramp in the tail and a swing-up door in the nose, to allow the roll-on/roll-off handling of vehicles and palletized cargo. Maximum payload is reported as 330,688 lb. (150,312 kg.), vs. 236,000 lb. (107,272 kg.) for the C-5 Galaxy. Two 10-ton traveling electric cranes are installed in the roof of the hold for cargo handling. An upper level above the cargo hold contains the flight deck and a cabin with 88 seats. The pilot, copilot, engineer, and navigator sit in an airliner-type flight deck, while a loadmaster rides in the hold. A cabin behind the cockpit has a galley and bunks for a relief crew. An important feature of the An-124, designed as a relaxed static stability aircraft, is its digital, quadruplex FLY-BY-WIRE flight control system, with a separate fifth channel for emergency control.

The multiple-bogie landing gear has 20 low-pressure main tires and 4 nose tires; the pressure of all 24 tires can be controlled from the flight deck, to facilitate landings on rough or soft airfields. The height of the landing gear is also adjustable, so that the aircraft can "kneel" for the loading and unloading of cargo directly to and from trucks.

The shoulder-mounted wing incorporates several high-lift devices, including leading-edge slats and Fowler flaps, to allow maximum-weight take-offs from fields only 4000 ft. (1220 m.) long. The engines are equipped with thrust reversers, which (along with powerful brakes) reduce the landing roll to only 2625 ft. (800 m.) at maximum landing weight.

It is anticipated that Condors will be acquired by both the VTA (Soviet Military Transport Aviation) and Aeroflot. The An-124 can carry almost every type of Soviet military equipment, including ICBMS, the T-80 tank, and HIND attack helicopters.

Specifications Length: 228 ft. (69.5 m.). **Span:** 240.49 ft. (73.3 m.). **Cargo hold:** 118.1 × 21 × 14.4 ft. (36 × 6.4 × 4.4 m.). **Powerplant:** 4 Lotarev D-18T high-bypass-ratio turbofans, 51,-650 lb. (23,430 kg.) of thrust each. **Fuel:** 440,000 lb. (200,000 kg.). **Weight empty:** 390,000 lb. (177,000 kg.). **Weight, max. takeoff:** 892,857 lb. (405,000 kg.). **Speed, max. cruising:** 537 mph (865 kph). **Range, max. payload:** 3100 mi. (5000 km.).

2. Transnational, medium-range ballistic missile (MRBM) program. Condor originated in 1979 as the Condor-1 sounding rocket, developed by the Argentinian air force with technical assistance from the West German MBB and associated European companies, and as such was a seemingly innocent scientific project. After the 1982 Falklands War, the program evolved with the goal of producing an MRBM.

In 1984, Argentina and Egypt secretly agreed to jointly produce a two-stage MRBM with an 1100-lb. (500-kg.) payload, with funding provided by Iraq (then at war with Iran). Concurrently, MBB and and other European technology suppliers formed front companies (notably the Monte Carlo–based CONSEN) to conceal their involvement. Development and production facilities for up to 400 Condors (with solid-fuel boosters) were established in Argentina and Egypt by 1988; the total cost of the project was then expected to exceed $3.2 billion, including nearly $1 billion for initial research and development.

During 1989, the Argentine-Egypt/Iraq agreement broke down because MBB and its partners had been induced by third parties to stop their participation (though the former front companies remain active, independently); Egypt was dissuaded by the U.S.; and Iraq had made sufficient progress with its own MRBM and kindred projects (see AL HUSAYN) to lose interest in Condor. The latter now survives as an exclusively Argentine project whose funding is now interrupted (though Libyan interest in funding the project has been reported). See also BALLISTIC MISSILE.

CONFORMAL ARRAY: RADAR or SONAR transmitting and receiving elements built into the surface structure of ships, submarines, or aircraft, in conformance with their contours. Conformal arrays have several advantages: they do not require special structural members to support them, and they minimize platform impact because they can be streamlined and distributed over almost any platform surface, allowing a larger total array to be carried. A conformal-array AWACS aircraft, for instance, would not need a rotating dish antennna. Conformal arrays could be incorporated in the rotors of helicopters and other unusual places, where conventional arrays would not fit. Conformal arrays are made possible by PHASED ARRAY and electronic beam steering technology.

CONFRONTATION: A protracted conflict in which all means other than overt warfare are used, as in the Malaysia-Indonesia conflict (1963–66) and the U.S.-Nicaragua conflict (1979–88). The term describes any combination of ECONOMIC WAR-FARE, POLITICAL WARFARE, SUBVERSION, TERRORISM, COVERT OPERATIONS, and CLANDESTINE OPERATIONS. The COLD WAR was in effect a confrontation, as is the Arab-Israeli conflict. See also COUNTERINSUR-GENCY WARFARE; GUERRILLA WARFARE; TERRORISM.

CONSCIENTIOUS   OBJECTION: The denial of the state's right to compel participation in the activities of its armed forces, on the basis of stated beliefs or values and consequent rules of behavior, which are claimed to transcend military obligations. Such an objection can occur at differ-ent levels of generality, and the state's recognition of the validity of the objection is often related to its generality.

The broadest form is an objection to any military service in any contingency. This is usually moti-vated by a religious faith or by the ideology of Pacifism. If comprehensive adherence can be doc-umented, an exemption is usually granted by lib-eral-democratic states, which generally recognize the prerogative of individuals in matters of con-science, though some form of alternative nonmili-tary service may be imposed instead. In states with developed economies, however, any economic ac-tivity is ultimately related to the war effort in a general war; thus conscientious objection would logically require an abstention from all economic activities except subsistence farming.

Somewhat less broad are objections, in any con-tingency, to activities which imply the possible ne-cessity of killing. Such objections, being also gen-eral with respect to the contingency, but less general with regard to the exemption claimed, are also frequently allowed in liberal-democratic soci-eties.

The most specific form of conscientious objec-tion applies only to military service in one stated situation. Such an objection is thus political, rather than religious or philosophical, and amounts to an individual claim of the right to formulate policy at variance with that of the state. Because overt ac-ceptance of such political objections by the author-ities would erode state prerogatives, exemptions are not normally granted, unless the individual can show that his objection is not political, despite being specific. Such nonreligious, nonpolitical ob-jections usually arise from membership in a partic-ular religious or ethnic group. For example, British authorities removed Jewish troops from the forces sent to Suez in 1956, because it seemed possible that they might engage Israeli troops in combat.

Similarly, Israel exempts its Arab citizens from military service. See also CONSCRIPTION.

CONSCRIPTION: The process whereby citi-zens or subjects are selected and compelled to serve in the armed forces, or to render other ser-vices. Conscription enables the modern state to mobilize its entire able-bodied population for mili-tary or supporting activities; without it, mass ar-mies are not possible. In many countries, it is widely believed that compulsory military service has a unifying effect on the population, and is thus maintained even if no immediate military need warrants the system. Most developed countries maintain some form of conscription, despite its in-termittent political unpopularity; the most notable exceptions are the United States and Great Brit-ain, which share a strong cultural bias against standing armies. See also CONSCIENTIOUS OBJEC-TION.

CONSTELLATION   (SATELLITE): The grouping of several satellites in orbit to carry out a specific function. In the case of communications satellites, the objective is to provide a continuous relay capability; the most common constellation of this type consists of three satellites placed in geo-synchronous equatorial orbits, spaced 120° apart, so as to maintain continuous lines of sight between each, and between at least one and any point on the ground. RECONNAISSANCE satellites, in low-alti-tude polar orbits, can observe any part of the world within several days, but if continuous cover-age of a particular point is desired, a constellation of several satellites in polar orbits must be formed. Constellations are defined by the number of orbi-tal planes, the number of satellites in each orbital plane, and the spacing between the satellites in each plane. See also SATELLITES, MILITARY; SPACE, MILITARY USES OF.

CONTINUOUS ROD WARHEAD: A type of blast-fragmentation warhead used most fre-quently in air-to-air and surface-to-air missiles. The warhead consists of a bundle of metal rods placed around a high-explosive burster charge. When the charge is detonated, the rods are thrust out radially to form a ring of metal of greatly in-creased diameter. The (connected) rods form two semicircles as they expand, which are intended to hit the target during their expansion and to cause damage by a cutting action. The damage inflicted by continuous rod warheads is usually more severe than that inflicted by conventional fragmentation warheads of equal weight, because of the greater mass of the rods, as compared to splinters.

CONTOUR FLYING: A mode of flight meant to take advantage of the poor performance of RADAR at very low altitudes. Because radars cannot generally detect targets below the horizon, aircraft can avoid detection by flying low enough to be shielded by the folds in the terrain. To do this at high speed, the aircraft requires an autopilot directed by a TERRAIN FOLLOWING RADAR (TFR), because human reflexes are too slow to detect obstacles and correct the flight path accordingly. Helicopters, which fly at very low speeds, can be flown manually in NAP-OF-THE-EARTH (NOE) profiles at extremely low altitudes. See also B-1; F-111; FENCER; TORNADO.

CONTROLLED RESPONSE: A policy under which responses to enemy attacks are deliberately kept within well-defined limits, on the assumption that the enemy will recognize the existence of those limits, and also the deliberate nature of their observance. Specifically, the term describes the policy enunciated by the Kennedy administration, which was supposed to apply even if nuclear weapons were used in the conflict. This policy implicitly assumed the existence of mutually recognized thresholds, of symmetrical capabilities, and of symmetrical targets. In the absence of such symmetries, the stabilizing effect is impaired: if A destroys a "medium value" target, but is itself devoid of medium value targets, then B would be forced either to show weakness by attacking a "low value" target, or to escalate the conflict by attacking a "high value" target.

The adoption of Controlled Response by the U.S. implied the need for a full range of military capabilities; for the earlier policy of MASSIVE RETALIATION, by contrast, only light forces and nuclear weapons were required, at least in theory. Controlled Response, which remains the basis of U.S. strategy, is sometimes called Flexible Response.

CONUS: U.S. term for the continental United States, a definition which includes the 48 contiguous states, but excludes Alaska, Hawaii, and overseas dependencies.

CONVENTIONAL WAR: An unsatisfactory term for a war in which nuclear weapons are not used. Because neither chemical weapons nor guerrilla tactics are "conventional," the term *nonnuclear* is more precise.

CONVOY: A concentration of merchant vessels headed to a common destination under naval escort. Convoy were much used from the 16th to 19th centuries to defend merchant vessels from pirates, privateers, and naval raiders; after the long peace of the late 19th century, convoys were reinstated in both world wars as a defense against surface raiders and SUBMARINES. As an ANTI-SUBMARINE WARFARE (ASW) tactic, convoys have three effects. First, they concentrate merchant vessels within a small area, making it more difficult for the enemy to find targets in the vastness of the ocean; and if the tracks of enemy submarines are known, convoys can be routed to avoid them. Second, convoys utilize available escorts more effectively, because once they exceed a certain size, the number of escorts required to form a screen increases only as the cube root of the number of merchant ships added. Third, by concentrating escorts in the vicinity of merchant vessels, the convoy acts as a deterrent to submarine attack, because even a successful attack is likely to result in a dangerous if not lethal counterattack by the escorts.

Convoys also proved to be effective against air attack. The concentration of ships again makes the detection of any more difficult, while the concentration of anti-aircraft weapons increases the chances of successfully repelling an attack.

A modern convoy would consist of as many as 40 merchant vessels of similar speed, under the command of a civilian convoy commodore (usually a retired naval officer) answerable to the commander of the escort force. The latter could include an air-capable ship with ASW HELICOPTERS and perhaps a few V/STOL fighters, as well as one or two ASW DESTROYERS (DD), one or two guided-missile FRIGATES (FFG) for AIR DEFENSE, and two or three ASW frigates (FF). The convoy could also receive indirect support from other ASW forces operating in the vicinity.

It is now often argued that convoys would not be effective against forces equipped with modern sensors and weapons. First, convoys would have large acoustical (SONAR) and RADAR signatures that could be detected at long range. Second, running convoys requires radio traffic which could be located by enemy direction finding (D/F) equipment. Finally, convoys are difficult to organize and control, because merchant crews are not trained to steam in close formation.

On the other hand, it can be argued that an unprotected single ship is as easy to detect as a convoy, and much easier to sink. The concentrated defenses of a convoy could still deter attack, and resist attacks that do take place. At any rate, U.S. and other Western naval forces include a large escort element that is still justified by putative convoy requirements.

COONTZ: A class of ten U.S. guided-missile DESTROYERS (DDG-37 through DDG-46), sometimes known as the "Farragut" class. Commissioned between 1959 and 1961 as "Destroyer Leaders" (i.e., large destroyers intended as flagships of destroyer squadrons), the Coontz design was essentially a scaled-up version of the ADAMS class, enlarged to carry a TERRIER missile system and a greater volume of electronics. Flush-decked, with a graceful shear, twin stacks, and a small superstructure concentrated forward, the Coontz class is to continue in service through the 1990s, when they are to be replaced by the BURKE class (DDG-51).

Armament, optimized for ANTI-AIR WARFARE (AAW) for the protection of CARRIER BATTLE GROUPS, includes a rapid-fire (Mk.42) 5-in. 54-caliber DUAL-PURPOSE gun on the foredeck, and an 8-round ASROC pepperbox launcher on the forward superstructure (no reloads were carried, except by the *Farragut*, which has 8 additional ASROCs stowed in the bridge face behind the launcher). The only other ANTI-SUBMARINE WARFARE (ASW) weapons are 2 sets of Mk.32 triple tubes for MK.46 lightweight homing TORPEDOES. The main armament, however, is a Mk.10 Mod 0 twin-arm Terrier surface-to-air missile launcher on the fantail, for either Terrier or STANDARD-ER missiles. A total of 40 missiles are stored in a magazine forward of the launcher. As built, the ships also had 2 twin 3-in. 50-caliber ANTI-AIRCRAFT guns amidships, but during the 1980s, these were replaced by 2 quadruple launch canisters for HARPOON anti-ship missiles. Two ships, USS *King* and *Mahan*, served in 1974 as trials ships for the PHALANX 20-mm. radar-controlled anti-missile gun, but these weapons were subsequently removed; the class is not scheduled to receive Phalanx due to their limited stability margins.

All ships of the class are fitted with the Naval Tactical Data System (NTDS) data link. The Coontzes are receiving minor modifications as part of the New Threat Upgrade Program, including provisions for the Standard-2ER missile, Mk.14 gun fire control system, and SPS-48E and SPS-49(V)5 radars. See also DESTROYERS, UNITED STATES.

Specifications Length: 512.5 ft. (156.2 m.). Beam: 52.5 ft. (16 m.). Draft: 23.4 ft. (7.1 m.). Displacement: 4700 tons standard/6150 tons full load. Powerplant: twin-shaft steam, 4 oil-fired boilers, 2 sets of geared turbines, 85,000 shp. Fuel: 900 tons. Speed: 33 kt. Range: 1500 n.mi. at 30 kt./4500 n.mi. at 20 kt./6000 n.mi. at 15 kt. Crew: 403. Sensors: 1 SPS-10B surface-search radar, 1 SPS-29E (in the 2 oldest ships) or SPS-49 (in the other 8) air-search radar, 1 SPS-48 3-dimensional air-search radar, 1 SPG-53A gun fire control radar, 2 SPG-55B missile guidance radars, 1 hull-mounted SPS-23 medium-frequency sonar. Fire Controls: Mk.11 weapons direction system (Mk. 14 in. *Mahan*, DDG-42), Mk.68 gun fire control system, Mk.111 ASW fire control system, 2 Mk.76 missile fire control systems. Electronic warfare equipment: 1 SQS-32(V)2 electronic countermeasures array, 4 6-barrel Mk.36 SBROC chaff rocket launchers.

COPPERHEAD: The Martin-Marietta M712 Cannon-Launched Guided Projectile, in service with the U.S. and the Egyptian armies. A "smart" guided munition meant to destroy tanks and other point targets, Copperhead can be fired by standard U.S. M109A2/3 self-propelled and M198 towed 155-mm. gun-howitzers. The round is 54 in. (1.37 m.) long, with a diameter of 155 mm., and a firing weight of 140 lb. (63.64 kg.). The nose houses a SEMI-ACTIVE LASER HOMING seeker, which can lock onto LASER energy reflected off a target by a ground-, vehicle-, or aircraft-mounted LASER DESIGNATOR. A guidance computer and a gyroscope are located behind the seeker, and the 50-lb. (22.7-kg.) HESH warhead (which can defeat most tank armor, as well as concrete bunkers) is fitted in the rear. Four midmounted, pop-out cruciform wings provide steering, and four smaller, nonmovable spring-loaded tail fins act as stabilizers. Copperhead has a minimum range of 3000 m. and maximum range of 17,000 m.

In action, a forward observer designates a target for the gun, and illuminates it with his laser. The gun crew need only place the round within a 1000-m. cone above the target. When the round is fired, the acceleration of launch activates its systems; as it emerges from the gun barrel, the four spring-loaded tailfins deploy to provide stabilization. When the round reaches the apogee of its trajectory and begins to descend in a nose-down attitude, it should pick up laser radiation from the target designated by the forward observer. If so, the four movable wings deploy as well, and the guidance system begins to provide steering inputs for them to bring Copperhead onto the target. It can be fired in a depressed trajectory mode when a low cloud ceiling would prevent the round from acquiring its target from a normal ballistic trajectory.

Copperhead went into low-volume production in 1981, but the technical difficulties of hardening the guidance electronics to resist 9000-g acceleration at launch, combined with large cost overruns, prevented large-scale procurement. An attempt to develop a common guided projectile for both the army and navy was made, but the services could not agree on its design. Copperhead production was terminated in 1990. See also IMPROVED CONVENTIONAL MUNITIONS.

CORPS: A higher formation of ground forces above a DIVISION and below a numbered ARMY. A corps generally consists of two or more divisions and ENGINEER, RECONNAISSANCE, ARTILLERY, ANTI-TANK, ANTI-AIRCRAFT, and maintenance supports. The corps echelon does not exist in the Soviet army; divisions are directly subordinate to numbered armies (which are equivalent to Western corps in composition).

CORRELATION OF FORCES: A Soviet term for a method of strategic analysis based on Marxist-Leninist doctrine, whereby every element of the power balance, including economic, political, and social factors, in addition to purely military factors, is evaluated to estimate the cost-risk-benefit ratios of any particular course of action. See also NET ASSESSMENT.

CORSAIR: The Vought A-7 single-seat light ATTACK AIRCRAFT, in service with the United States, Greece, and Portugal. Developed to meet a 1963 U.S. Navy requirement for a carrier-based aircraft to replace the McDonnell-Douglas A-4 SKYHAWK, the Corsair was derived from the F-8 CRUSADER carrier-based fighter; it has the same general configuration, but is shorter, and has a nonafterburning engine and a simplified wing. The initial A-7A version entered service in 1966. Powered by a Pratt and Whitney TF-30-PW-6 turbofan, it had relatively simple avionics optimized for CLOSE AIR SUPPORT (CAS), and was armed with two Mk. 12 20-mm. cannon, as well as a 12,000-lb. (5455-kg.) bomb load. After 199 had been built, production switched to the A-7B with a more powerful engine, of which 196 were built.

In 1966, the U.S. Air Force decided to acquire the Corsair as well, to replace the F-100 SUPER SABRE. Its A-7D variant had improved avionics for all-weather precision strike, an M61 VULCAN 20-mm. cannon, an Allison TF41-A-1 turbofan rated at 14,500 lb. (6590 kg.) of thrust, and an AERIAL REFUELING receptacle for the USAF "flying boom"

system; a total of 459 were built between 1968 and 1981. These aircraft demonstrated outstanding bombing accuracy in Vietnam, and proved rugged and reliable. In 1976, all surviving A-7Ds were transferred to the Air National Guard.

The navy next developed the A-7C, with even better avionics than the A-7D and also the Vulcan cannon, but with the TF-30 engine. Only 67 were built before they were superseded by the definitive A-7E, which eventually became the primary navy light attack aircraft until the advent of the FA-18 HORNET in 1982. The A-7E has further improved avionics and an uprated TF41-A-2 turbofan; a total of 596 were built between 1969 and 1979. Surviving A-7Es now serve only with the Naval Reserve squadrons, but have continued to receive upgrades, and notably the provision of a FLIR (Forward-Looking Infrared) pod for low-level night attack.

The boxy, slab-sided fuselage has an APQ-126 multi-mode air-to-ground radar in a nose radome above a chin intake for the engine. The pilot sits just behind the radome in a well-equipped cockpit faired into the rear fuselage. Cockpit AVIONICS include several CRT display screens and an AVQ-7(V) Head-Up Display (HUD). The Vulcan cannon is mounted on the left below the cockpit, and there is a retractable aerial refueling probe on the right. Two fuselage rails for AIM-9 SIDEWINDER missiles are mounted immediately behind the cockpit. The rear fuselage houses fuel cells, the main landing gear, and the engine.

The shoulder-mounted wing has full-span leading edge slats, single slotted flaps, and conventional ailerons. The outer wing sections fold up for carrier deck parking, and house integral fuel tanks. The Corsair has six wing pylons with a combined payload capacity of more than 15,000 lb. (6818 kg.) for LDGP and CLUSTER BOMBS, PAVEWAY laser-guided bombs, AGM-62 WALLEYE glide bombs, AGM-65 MAVERICK air-to-surface missiles, AGM-45 SHRIKE and AGM-88 HARM anti-radiation missiles, AGM-84 HARPOON anti-ship missiles, MINES, NAPALM bombs, rocket pods, or 30-mm. cannon pods.

In addition to the single-seat variants, the U.S. Navy acquired some 60 TA-7C 2-seat trainers, while the Air National Guard has 31 A-7K 2-seaters, with full combat capabilities. Greece has 60 A-7H variants, plus 5 2-seat TA-7H; Portugal has 44 A-7Ps (rebuilt A-7As), plus 6 TA-7P 2-seat trainers.

In 1986, Vought proposed a radically modified

Corsair (A-7F), to meet USAF's CAS requirement for the 1990s. This conversion of existing A-7D airframes has a lengthened fuselage, an afterburning F100 turbofan rated at 25,000 lb. (11,363 kg.) of thrust to provide supersonic dash speed, improved avionics, and enhanced maneuverability; two prototypes began flight tests in late 1989.

Specifications (A-7E) Length: 46.1 ft. (14.05 m.). **Span:** 38.73 ft. (11.08 m.). **Powerplant:** TF41-A-2 turbofan, 15,000 lb. (6818 kg.) of thrust. **Fuel:** 9500 lb. (4318 kg.). **Weight, empty:** 19,781 lb. (9032 kg.). **Weight, max. takeoff:** 42,000 lb. (19,-051 kg.). **Speed, max.:** 691 mph (1154 kph) at sea level. **Combat radius:** (hi-lo-hi) 600 mi. (1000 km.). **Range, max.:** 2861 mi. (4604 km.).

CORVETTE: A small, heavily armed warship usually displacing between 500 and 1000 tons. Corvettes are an intermediate type between smaller FAST ATTACK CRAFT (FACs) and larger FRIGATES. In general, corvettes resemble scaled-down frigates, with commensurately less firepower and endurance. Originally developed during World War II as very economical ANTI-SUBMARINE WARFARE (ASW) escorts, corvettes went out of fashion after 1945, when the fast attack submarine rendered them obsolete. But corvettes are now regaining importance because of the development of lightweight sensors, guns, ASW weapons, and especially ANTI-SHIP MISSILES, which can give them very heavy firepower for their displacement. The Soviet NANUCHKA corvette for example, has six SS-N-9 STARBRIGHT anti-ship missiles, one twin 57-mm. DUAL-PURPOSE gun, a twin SA-N-4 GECKO surface-to-air missile launcher, and a 30-mm. radar-controlled anti-missile gun—all on a displacement of only 900 tons. Even smaller corvettes can still carry a dual-purpose gun, several light cannon, ASW TORPEDOES, and anti-ship missiles. Some even carry a light ASW helicopter. Small navies tend to use corvettes as substitutes for larger, more expensive frigates, while large navies use corvettes primarily for inshore and narrow water patrols. See also ANTI-AIR WARFARE; ANTI-SUBMARINE WARFARE; ANTI-SURFACE WARFARE.

CORVUS: A British shipboard 8-tube, 4-in. (100-mm.) rocket launcher used to dispense CHAFF and flares against ANTI-SHIP MISSILES. Corvus entered service in 1982, but had to be supplemented during the 1982 Falklands War by the greatly superior U.S. Mk.36 SRBOC multi-barrel chaff mortar, and is now being replaced by the Plessey Shield system. See also ANTI-AIR WARFARE; ANTI-SURFACE WARFARE; ELECTRONIC COUNTERMEASURES.

COSAG: Combination of Steam and Gas, a naval propulsion arrangement in which a steam turbine and a gas turbine engine are connected to the same propeller shaft through a common gearbox. The steam turbine is used for fuel-efficient cruising at low speeds, while the gas turbine is started only when higher speeds are required. COSAG is not widely used.

COUNTERAIR: Weapons and tactics designed to neutralize enemy AIR POWER to achieve AIR SUPERIORITY. Defensive counterair includes all means of neutralizing enemy air power over friendly territory, notably ground-based AIR DEFENSE systems and INTERCEPTOR aircraft. Offensive counterair includes all means of neutralizing enemy air power in or over his own territory, and notably airfield attack, offensive fighter sweeps, and attacks against GROUND-CONTROLED INTERCEPT (GCI) facilities.

Historically, attacking enemy air forces on the ground has been the fastest method of winning the counterair battle, especially if surprise is achieved at the start of a war (e.g., the opening of Operation Barbarossa in 1941, and the Israeli preemptive strike of 5 June 1967). But many air forces now protect aircraft parking areas with revetments and hardened aircraft shelters, and have surrounded entire air bases with dense air defense belts. On the other hand, air-delivered guided bombs and munitions dispensers, special runway-cratering bombs (e.g., DURANDAL) and tactical BALLISTIC MISSILES armed with IMPROVED CONVENTIONAL MUNITIONS (ICM) have all been developed for airfield attack. During the opening phases of a war, air forces would normally be commited to counterair missions, and would not be available for CLOSE AIR SUPPORT (CAS) or other forms of air-to-ground attack. Hence any enhancement of NATO's counterair capabilities with missiles can accelerate the release of aircraft to support the ground war. See also ANTI-AIR WARFARE.

COUNTERBATTERY: The use of artillery to neutralize enemy artillery. The primary difficulty of this mission is TARGET ACQUISITION, because enemy batteries are normally far behind the lines, and not under direct observation. Among the methods used to locate enemy artillery, the oldest are sound ranging, whereby sensitive directional microphones detect artillery weapons by the sound of their firing, and flash ranging (mainly at night), whereby the weapons are located by their muzzle flashes. Both methods are inaccurate. The modern methods of target acquisition are aerial reconnaissance, radio direction finding (D/F), and counter-battery RADAR. Aerial reconnaissance can be greatly impeded by camouflage and air defenses. D/F can locate batteries from their radio emissions, which are generated in large volumes because of the need to make requests for INDIRECT FIRE (the Soviet army relies heavily on this method, and has D/F companies attached to all larger artillery formations); the separation of antennas and the use of land lines can impede this method. Counterbattery radar tracks enemy shells in flight, and an associated ballistic computer can extrapolate their point of origin. This method is fast and highly accurate, but the radar generates characteristic electronic emissions which can reveal its position in turn. The U.S. and NATO armies rely mainly on counterbattery radar.

Once a battery is detected, towed artillery is extremely vulnerable to counterbattery fire, because gun crews and ready ammunition are completely unprotected. Self-propelled (SP) artillery with armored protection is less vulnerable to fire, but its effectiveness can still be severely reduced by accurate counterbattery fire, if only because it compels frequent chages of position. Moreover, IMPROVED CONVENTIONAL MUNITIONS with anti-armor capability are increasing the vulnerability of SP guns (see SADARM)

As artillery has become more effective against both infantry and armor, more resources have been assigned to the counterbattery function. In combat between well-equipped armies, more than half of all artillery fire missions might be dedicated to counterbattery fire. In response, modern artillery is becoming more mobile. Modern artillery tactics call for the firing of only one or two salvoes before displacement to a new position ("shoot-and-scoot"). That requirement has dictated the progressive replacement of towed artillery by far more costly SP mounts, even in the Soviet army, which long resisted this trend. Efforts are now being made to increase maximum rates of fire, in order to generate more firepower during the brief periods between "scoots." See also ARTILLERY.

COUNTERFORCE: A nuclear targeting policy whereby attacks would be directed against the enemy's military forces in general, and nuclear forces in particular, rather than against enemy population centers. The policy was initially enunciated by U.S. Secretary of Defense Robert S. McNamara in June 1962 and repudiated within a year, because it requires nuclear delivery systems in sufficient numbers, and of sufficient accuracy, to destroy enemy missile SILOS, bomber bases, and ballistic missile submarines in port—a capability which is costly and highly threatening (neither consideration deterred Soviet planners). A counterforce capability must overcome any BALLISTIC MISSILE DEFENSES (BMD) as well as AIR DEFENSES. Operationally, moreover, a counterforce capability requires several warheads per target to ensure high kill probabilities, plus accurate INTELLIGENCE on the location and vulnerability of enemy delivery systems. Soon after the policy was repudiated, however, the improvement of surveillance satellites and the introduction of MULTIPLE INDEPENDENTLY TARGETED REENTRY VEHICLES (MIRVs) made the acquisition of a counterforce capability much easier than before, except in regard to submerged ballistic-missile submarines.

A full counterforce capability would imply the ability to mount a successful FIRST STRIKE. That in turn might tempt a potential victim to initiate a PREEMPTIVE STRIKE of its own in a crisis. Because of this, the U.S. repudiated counterforce and adopted the ASSURED DESTRUCTION policy in the 1960s and '70s. The Soviet Union, by contrast, has placed a high priority on eliminating enemy "weapons of mass destruction" at the very outset of a war, and developed its forces accordingly, including very large, powerful, and accurate ICBMs, the SS-18 SATAN and SS-19 STILETTO. It was Soviet doctrine that DETERRENCE is enhanced by a demonstrable "war-fighting capability," and that a first (counterforce) strike would be inherently defensive: by eliminating (or at least reducing) U.S. nuclear forces, the homeland is protected, at least partly. In response, the U.S. began in the 1970s to develop counterforce weapons of its own, namely the PEACEKEEPER ICBM and the TRIDENT D-5 submarine-launched ballistic missile. In addition, U.S. BMD research (the STRATEGIC DEFENSE INITIATIVE) was ultimately meant to negate at least a portion of Soviet counterforce capabilities. See also COUNTERVALUE.

COUNTERINSURGENCY (WARFARE): Action taken by, or on behalf of, a constituted government against forces waging REVOLUTIONARY WAR, or conducting a localized armed rebellion (insurgency). In the wake of various colonial conflicts and the Vietnam War, counterinsurgency warfare has taken a definite form; various methods and techniques have been developed, though none is universally applicable.

The key method is "civil action" at the village level, including technical and medical assistance, economic aid and propaganda to win support for the government, plus the formation of local militias to protect the larger villages; in addition, small, independent military forces can conduct small-scale actions against GUERRILLA forces. At the national level, economic aid and technical assistance have tended to be combined with large-scale operations by conventional forces against such guerrilla forces as can be found; these operations have rarely been productive, and their impact tends to negate the results of civil action. Experience in Vietnam clearly indicated that village-level operations are more likely to succeed, but conventional forces resist such dispersion. Moreover, main arm of revolutionary war is SUBVERSION, which is covert rather than clandestine, and is thus better resisted by intelligence than by military methods.

COUNTERVALUE: A nuclear targeting policy whereby attacks would be directed against the enemy's cities and industrial area; its purpose is to hold the enemy's population hostage so as to influence its government, usually to dissuade it from offensive actions. Countervalue targeting may reflect mere necessity: with a limited number of nuclear weapons, or with weapons that lack the accuracy required for COUNTERFORCE attacks, countervalue targeting is the only possible policy. But the U.S. chose to adopt a declared countervalue policy from 1964, because it was deemed less threatening to the Soviet Union than a counterforce policy would have been. That policy, ASSURED DESTRUCTION, emphasized SECOND STRIKE (or "Strike Back") countervalue capabilities, even though military planning continued to inject a counterforce element as well. With their small nuclear forces, Britain, France, and China have had to adopt countervalue targeting of necessity, simply because they lacked counterforce capabilities. In any case, if a countervalue policy is to succeed, it must threaten "values" actually valued by the ruling authorities, who in some cases may value the survival of the power structure, the armed forces, or certain key industries more than the population at large.

COUNTY: A class of eight British guided-missile DESTROYERS (DDG) commissioned between 1962 and 1970. Phased out of the Royal Navy in the 1980s, three have been scrapped, while four have been sold to Chile and one to Pakistan. The Counties were designed as general-purpose escorts, with a heavy and balanced armament for

ANTI-AIR WARFARE (AAW), ANTI-SURFACE WARFARE (ASUW), and ANTI-SUBMARINE WARFARE (ASW).

The primary AAW weapon was Sea Slug, a very large, unreliable surface-to-air missile (SAM) of early-1960s vintage, launched from a twin launcher aft, with a long below-decks magazine for 30 missiles. Relatively ineffective against low-flying targets (and useless against ANTI-SHIP MISSILES), Sea Slug could also be used against surface targets. The launcher and its magazine have been removed from the remaining ships (the Chilean navy has converted the space occupied by the magazine into a large landing deck and hangar for several anti-submarine helicopters). In addition, the Counties have 2 SEA CAT short-range SAM launchers amidships; Sea Cat is also ineffective, but efforts are under way to modify its launchers for the modern SEA WOLF. Two 40-mm. BOFORS GUNS and 2 manually operated twin 20-mm. anti-aircraft guns were added after the 1982 Falklands War, as stopgap weapons for close-in defense. ASUW armament consists of 4 EXOCET anti-ship missiles in individual launch canisters forward of the bridge, and a twin 4.5-in. DUAL-PURPOSE gun on the foredeck. The primary ASW weapon is a LYNX helicopter operated from a landing deck just forward of the Sea Slug launcher, but there is no hangar. In addition, there are 2 sets of Mk.32 triple tubes for lightweight ASW TORPEDOES.

The Counties have four gyro-controlled fin stabilizers and twin rudders for maneuverability. All are air conditioned, and the machinery can be run by remote control from an NBC-protected citadel equipped with a COLLECTIVE FILTRATION system. The Counties also have two SCOT satellite communications terminals, and one ADAWS combat data processing system.

Large, potentially powerful warships that remain with obsolete weapons and sensors, the Counties could become effective if retrofitted with modern equipment. See also DESTROYERS, BRITAIN.

Specifications Length: 515 ft. (157.01 m.). **Beam:** 53.5 ft. (16.31 m.). **Draft:** 20.5 ft. (6.25 m.). **Displacement:** 5440 tons standard/6200 full load. **Powerplant:** twin-shaft COSAG, 2 oil-fired boilers, 2 sets of geared steam turbines, 15,000 shp. (cruise)/4 7500-shp. G6 gas turbines (sprint). **Speed:** 32.5 kt. (30 kt. sustained). **Range:** 3500 n.mi. at 28 kt. **Crew:** 471. **Sensors:** 1 Type 798 navigational radar, 1 Type 965M early warning radar, 1 Type 992 search radar, 1 type Type 277 height-finding radar, 1 Type 901 missile guidance radar (now removed), 2 Type 904 Sea Cat fire control radars, 1 Type 903 fire control radar, 1 Type 184M hull-mounted medium-frequency sonar, 1 Type 162M side-looking classification sonar. **Electronic warfare equipment:** 1 UA-9 active jamming unit, 1 Knebworth Corvus chaff rocket launcher, 1 Mk.36 SRBOC chaff launcher.

COUP DE MAIN: A rapid military operation meant to seize an objective by surprise with a minimum of fighting. *Coups de main* are generally small-scale SPECIAL OPERATIONS performed by COMMANDOS or other elite troops; exceptions include the Soviet seizure of Kabul on 27 December 1979.

COUP D'ÉTAT: The overthrow of a constituted government by a small clique, usually formed within the country's armed forces. Coups d'état are rarely bloody, because one side or the other capitulates without fighting; operationally they consist of simultaneous arrests, seizures of key facilities, and blocking actions against armed forces loyal to the regime. Coups d'état differ politically from other violent methods of overthrowing governments because they do not involve the population at large. Coups are now the usual method of changing governments in Africa.

COVER: 1. Any form of protection against observation provided by natural or man-made means.

2. Any form of protection against the effects of fire, such as foxholes, slit trenches, bunkers, etc.

3. The use of false or misleading identity to hide true origin, as in COVERT OPERATIONS.

COVERING FORCE: U.S. term for a force deployed ahead of a larger (main) force in order to intercept, engage, delay, deceive, and disorganize enemy attacks before they reach the force being covered. U.S. covering forces generally consist of armored CAVALRY units, sometimes supplemented by LIGHT INFANTRY and/or supported by detachments of tanks and MECHANIZED INFANTRY. The "Covering Force Zone" extends from the FLOT (Forward Line of Own Troops) to the FEBA (Forward Edge of the Battle Area), so that it includes all friendly forces deployed between the enemy and the MAIN BATTLE AREA. See also ACTIVE DEFENSE; AIR-LAND BATTLE.

COVERT OPERATIONS: U.S. term for secret operations, i.e., government activities which are meant to remain secret, or at least whose perpetrators are meant to remain secret, or, at least whose sponsorship is meant to remain plausibly deniable. Covert operations include SUBVERSION, sabotage, TERRORISM, assassination, and espio-

nage. Covert operations, unlike CLANDESTINE OPERATIONS (commando, scouting, etc.), are not hidden by their nature but rather disguised as innocent activities. When total secrecy is impractical, the measures that can be taken to conceal the identity of the sponsor include the use of proxies, perhaps isolated from the sponsor by "cut-out" contacts; the use of nationals disguised as third-party citizens ("false flag" operations); and all forms of deceptive "cover" (diplomatic, commercial, academic, journalistic, etc.).

CRAF: Civil Reserve Air Fleet, a program whereby U.S. civilian airliners and cargo aircraft are prepared and registered for eventual mobilization by the U.S. Air Force in emergencies to supplement the MILITARY AIRLIFT COMMAND (MAC). CRAF aircraft now include some 135 long-range cargo jets and 94 passenger airliners. U.S. plans for the rapid reinforcement of NATO are predicated on the ability of CRAF to transport personnel and general military cargo to Europe, freeing MAC transports for specialized and outsize cargo shipments. Many CRAF aircraft are modified with strengthened decks and otherwise for military use, and the owners are compensated accordingly. Because their crews need not be reserve military personnel, and because no CRAF aircraft are capable of roll-on/roll-off operations from unprepared fields, CRAF is as fully integrated into USAF airlift capability as Aeroflot is integrated the Soviet air force. See also REFORGER.

CREDIBILITY: The presumed level of expectation on the part of others that a threat or promise related to a specific contingency will be implemented. Credibility therefore depends on an external estimate of the resolve to act as announced in a particular way in a particular situation. Because the magnitude of a threat is capability multiplied by credibility, in a nuclear context even CIVIL DEFENSE along with other defensive capabilities can increase the perceived magnitude of a threat by reducing the cost of its implementation. In general, the credibility of threats depends on their automaticity; thus a computer-controlled, totally automatic retaliation system would be most credible, while a highly vulnerable strike system can also be very credible, because hesitation in using it might be seen as unlikely. A completely invulnerable retaliatory system with a recall mechanism would be the least credible, because it would allow ample time for post-attack negotiations or simple blackmail. In spite of suggestions that perceptions of "resolve" can be manipulated, the most important factor in credibility is the perceived importance of the interests meant to be protected by the threat or promise.

CROTALE: The French Matra R.440 short-range, low-altitude surface-to-air missile system. Fully automatic, and capable of all-weather engagements against fast, maneuverable targets, the complete system can be mounted on a self-propelled wheeled vehicle or on trailers, and is airportable. Crotale is still in production for the French air force and has been ordered by at least ten other countries, including Egypt, Kuwait, Libya, Morocco, Pakistan, Saudi Arabia, and Spain. A shipboard version is in service with the French navy.

The Crotale missile has four small, movable canards for steering, and four cruciform delta tail fins for lift. The 33-lb. (15-kg.) high-explosive fragmentation warhead is fitted with an infrared PROXIMITY FUZE. Powered by a single-pulse solid-rocket engine, Crotale can achieve its maximum speed is only 2.3 sec.

Crotale is controlled in flight by radar COMMAND GUIDANCE; steering commands are transmitted to the missile via the monopulse fire control RADAR itself, with tracking facilitated by a transponder in the base of the missile. The claimed kill probability (P_K) against a fighter-type target is 82 percent.

When mounted on a wheeled vehicle, a complete firing unit consists of three launcher/command guidance vehicles and one surveillance radar vehicle. Each launcher vehicle carries four missiles in sealed storage/launch canisters, and houses the monopulse fire control radar, which is capable of guiding two missiles simultaneously. Acquisition of the missile after launch is assisted by an INFRARED sensor linked to the radar antenna, and backed up by an optical tracker. The surveillance radar is an E/F-band pulse-Doppler set with multiple TRACK-WHILE-SCAN capability and an automatic target evaluation system to reduce reaction time. The surveillance radar vehicle can provide command and control for up to three launch vehicles.

Two export variants have also been developed: Cactus, supplied to South Africa from 1969 to 1972; and Sica, slightly longer and heavier than Crotale, with the range extended to 6 mi., a more powerful acquisition radar, and a six-round launcher. Sica is in service with Saudi Arabia as the "Shahine."

Specifications Length: 9.5 ft. (2.9 m.). Diameter: 5.9 in. (150 mm.). Span: 21.25 in. (540 mm.). **Weight, launch:** 176 lb. (80 kg.). **Speed, max.:** Mach 2.3 (1750 mph/2922 kph). **Range, max.:** 7.5 mi. (12 km.). **Range, min.:** 1600 ft. (488 m.). **Ceiling:** 9842 ft. (3000 m.).

CRUISE MISSILE: A small, pilotless, jet-propelled aircraft designed to deliver a conventional or nuclear warhead over long ranges. The first cruise missile was the German V-1 of 1944–45. With a range of several hundred miles, it was guided only by an autopilot and was therefore inaccurate. The V-1 was copied by the United States after World War II as the "Loon" missile, which served as the developmental basis for a number of improved cruise missiles, notably the nuclear-armed Regulus I and II and, more distantly, the intercontinental-range, Mach 3 Navaho of 1957–60. These missiles were controlled in flight by INERTIAL and/or radio COMMAND GUIDANCE, combined in some cases with active radar TERMINAL GUIDANCE. Because it was believed in the U.S. that BALLISTIC MISSILES were more effective for nuclear delivery, while the U.S. Air Force preferred manned aircraft in any case, cruise missile technology languished until the mid-1970s.

The Soviet armed forces, by contrast, saw the cruise missile as an effective substitute for manned aircraft, and from the late 1940s they developed several cruise missiles for both anti-ship and surface attack applications, armed with conventional or nuclear warheads, and with ranges between 35 and 1000 mi. (see KANGAROO, KENNEL, KITCHEN, STYX, SHADDOCK).

The U.S. began developing a new generation of cruise missiles during the 1970s, in response to improvements in Soviet air defenses against manned bombers. These new weapons, the Air Launched Cruise Missile (ALCM) and sea-launched TOMAHAWK, were no mere pilotless aircraft; with miniaturized jet engines and advanced electronics, they combined a heavy payload, high accuracy, and 1500-mi. (2500-km.) ranges in missiles weighing less than 4000 lb. (earlier U.S. and Soviet missiles had weighed in excess of 10,000 lb.); the small size of the new cruise missiles made their interception extremely difficult.

ALCM and Tomahawk introduced a revolutionary guidance system, which combines midcourse inertial guidance with TERRAIN COMPARISON AND MATCHING GUIDANCE; the latter compares radar returns from an on-board ground mapping radar with a digitized terrain map stored in the guidance computer in order to compensate for inertial guidance errors, thus achieving an unprecedented CEP of under 10 m. ALCM is used exclusively for nuclear delivery by U.S. strategic bombers, but Tomahawk has been produced in nuclear, high-explosive, and IMPROVED CONVENTIONAL MUNITIONS versions; it can be launched from ships, submarines, aircraft, and ground launchers.

The Soviet Union has followed in turn with a new generation of cruise missiles, notably the AS-15 KENT and SS-N-21 SAMPSON, which closely resemble Tomahawk. The latest U.S. cruise missiles incorporate STEALTH technology. See also ANTI-SHIP MISSILES; ANTI-SURFACE WARFARE.

CRUISER: Originally a fast, heavily armed, lightly armored warship, smaller than a BATTLESHIP but larger than a DESTROYER, designed for long-range, independent operations, including commerce raiding, commerce protection, and "showing the flag."

By World War I, cruisers had evolved into distinct light and heavy categories. Light cruisers (CL), armed with guns of 5- to 6-in. caliber, were intended primarily as scouts for the battle fleet, and as destroyer flotilla leaders, while heavy cruisers, armed with guns up to 8-in. caliber, were intended mainly for commerce raiding and to counter enemy light cruisers.

During World War II, heavy cruisers sometimes acted as ersatz battleships, but both heavy and light cruisers mostly acted as anti-aircraft escorts for CARRIER BATTLE GROUPS and CONVOYS, and also served as shore bombardment ships during AMPHIBIOUS OPERATIONS.

After World War II, very few new cruisers were built; instead, during the 1950s and early 1960s, many remaining wartime cruisers were converted into guided-missile cruisers (CG). Most were retired in the 1970s. A few unconverted, gun-armed cruisers sold by the U.S. and Great Britain remain active in Latin American navies.

Current guided-missile cruisers do not conform to the traditional definition. They are not armored, and they are not designed for detached service; in fact, they differ little from destroyers except for their larger size—up to 12,000 tons vs. 4000 to 7000 tons for destroyers. The main roles of cruisers today are convoy and battlegroup escort, as well as surface action with anti-ship missiles. Because of their size they frequently act as flagships.

A recent development is the air-capable or "through-deck" cruiser, which has a large flight deck for helicopters and V/STOL aircraft, while retaining the heavy defensive armament of a cruiser (as, e.g., the Soviet MOSKVA and KIEV

classes). For some navies, cruiser designation is actually a political subterfuge, meant to avoid the AIRCRAFT CARRIER designation, with its offensive undertones. See also CRUISERS, BRITAIN; CRUISERS, FRANCE; CRUISERS, ITALY; CRUISERS, SOVIET UNION; CRUISERS, UNITED STATES.

CRUISERS, BRITAIN: The Royal Navy discarded the last of its cruisers in the early 1980s, and there are no plans to build new ones. The V/STOL carriers of the INVINCIBLE class were originally ordered as "through-deck cruisers," to avoid using the politically loaded term "aircraft carrier"; but the subterfuge was soon abandoned.

CRUSIERS, FRANCE: France retains only one guided-missile cruiser, the COLBERT, commissioned in 1959. There are no plans to build new cruisers.

CRUISERS, ITALY: The Italian navy operates three guided-missile cruisers: the VITTORIO VENETO, commissioned in 1969, and the two ships of the ANDREA DORIA class (*Andrea Doria* and *Caio Duilio*), commissioned in 1964. All three are primarily ANTI-AIR WARFARE ships, but are also equipped with large flight decks aft for ANTI-SUBMARINE WARFARE helicopters. The *Vittorio Veneto* was modernized in 1983–84, with OTOMAT anti-ship missiles and 40-mm. radar-controlled guns for anti-missile defense. The *Caio Duilio* was modified to serve as a school ship in 1979–80, but retains most of its combat capability.

CRUISERS, SOVIET UNION: From 1952 to 1956, the Soviet navy built 14 17,000-ton SVERDLOV-class light cruisers armed with 12 6-in. guns.

By the late 1950s, they began building missile cruisers (*Raketnyy Kreyser*) designed to engage U.S. CARRIER BATTLE GROUPS (CVBGs) and operated against SEA LINES OF COMMUNICATION (SLOCs). Because the primary mission of these ships was sea denial, their design emphasized speed and heavy ANTI-SURFACE WARFARE (ASUW) armament based on ANTI-SHIP MISSILES. Very heavily armed for their size, these ships lacked survivability and sustainability. The first four missile cruisers of the 5500-ton KYNDA class, commissioned between 1962 and 1965, are still in service.

The subsequent KRESTA class of 14 ships is divided into two subclasses: 4 Kresta Is completed in 1966–67, and 10 Kresta IIs commissioned between 1970 and 1978, designated Large Anti-Submarine Ships (*Bol'shoy Protivolodochnyy Korabl*, BPKs); they have the same hull as the Kresta Is, but are equipped specifically for ASW, with the Shaddocks replaced by eight SS-N-14 SILEX anti-submarine missiles.

In a clear departure from the sea-denial format, the Soviet navy introduced the seven KARA-class BPKs between 1973 and 1980. Much larger than the Krestas at 9700 tons, the Karas are more seaworthy and have much more internal volume for sensors and fire control systems.

Commissioned from 1982, latest Soviet cruisers of the 12,400-ton SLAVA class are optimized for ASUW. Believed to be a low-risk backup for the nuclear-powered KIROV-class battle cruisers, they are the largest conventionally powered cruisers yet built by the Soviet navy, and are excellent sea boats. The four Kirovs are so large, at 28,000 tons, that most Western naval analysts describe them as battle cruisers, although the Soviet navy modestly calls them *Raketnyy Kreysera*.

The KIEV-class V/STOL AIRCRAFT CARRIERS are designated *Taktycheskoye Avionosnyy Kreysera* (tactical aircraft cruisers), while the MOSKVA-class helicopter carriers are listed as BPKs—but that is only a subterfuge to circumvent international agreements that bar passage of capital ships through the Dardanelles.

CRUISERS, UNITED STATES: The introduction of jet aircraft and guided missiles made missile armament essential for all modern warships. During the 1950s and 1960s, the U.S. Navy converted a number of World War II cruisers into guided-missile cruisers (CGs) with TALOS, TERRIER, and/or TARTAR surface-to-air missiles. All have been decommissioned, and most have been scrapped.

The first U.S. guided-missile cruiser built from the keel up was the 17,100-ton nuclear-powered LONG BEACH (CGN-9), commissioned in 1961. Now armed with HARPOON and TOMAHAWK missiles and PHALANX radar-controlled guns for anti-missile defense, the *Long Beach* remains useful as a flagship.

The U.S. Navy did not acquire any more ships with conventional "cruiser" hulls after the *Long Beach*; instead, Guided Missile Destroyer Leaders (DLGs) with scaled-up DESTROYER hulls were built to serve as flagships for destroyer flotillas and escort groups. In 1975, all these ships were reclassified as guided-missile cruisers. In the absence of armored belts, only size differentiates U.S. cruisers and destroyers—and even new destroyers of the BURKE and SPRUANCE classes are larger than some older CGs.

The first of these DLG/CGs were the nine 8200-ton LEAHY class (CG-16 through CG-24) commis-

sioned between 1962 and 1964, noteworthy for their all-missile armament. The nuclear-powered BAINBRIDGE (CGN-25), commissioned in 1962, is essentially a Leahy with a slightly larger hull to accommodate the reactor.

The next large class of cruisers were the nine 8500-ton BELKNAPS (CG-26 through CG-34) commissioned between 1964 and 1967. Unlike the Leahys, they have a 5-in. dual-purpose gun and a landing deck for LAMPS ASW helicopters. The TRUXTON (CGN-35) is a nuclear-powered counterpart of the Belknaps; although slightly larger, it has the same hullform and armament.

After the Belknaps, the navy decided not to build any more conventionally powered cruisers. Hence its next cruisers were the two 10,500-ton CALIFORNIAS (CGN-36 and CGN-37) commissioned in 1974–75. They were followed from 1976 to 1980 by the four larger and more capable VIRGINIAS (CGN-38 through CGN-41); with a displacement of 11,300 tons, they have the sensors and weapons of the AAW variant of the Spruance-type destroyers (KIDD class). The Virginias were to be followed by the Aegis-equipped CGN-42 "Strike Cruiser," with two VERTICAL LAUNCH SYSTEMS (VLS) for Standard, Harpoon, and Tomahawk, but they were canceled because of inordinate costs. The navy was thus forced to adapt Aegis to the much smaller Spruance hull; the resulting TICONDEROGA class (CG-47 through CG-74) nonnuclear cruisers are still in production. Although they are excellent ships, the navy considers the 9600-ton Ticonderogas too small for their mission.

Six pre–World War II light cruisers of the Brooklyn class were sold to Latin American navies; two are still in service with Chile, and one, the *Phoenix*, sold to Argentina and renamed *General Belgrano*, was sunk by the British nuclear submarine HMS *Conqueror* during the 1982 Falklands War.

CRUSADER: Vought F-8 carrier-based, single-engine, single-seat *air superiority* fighter, originally built for the U.S. Navy and Marine Corps, and still in service with the French navy and Philippine air force. Developed in response to a 1951 requirement for a shipboard supersonic air superiority FIGHTER, the prototype flew in 1955 and the first Crusaders entered service in 1957. Originally a clear-weather "day" fighter armed with four 20-mm. cannons and two AIM-9 SIDEWINDER infrared-homing air-to-air missiles, the Crusader was faster and more maneuverable than most land-based fighters of its time. During its long service, the

Crusader was repeatedly improved; among other changes, RADAR and FIRE CONTROL systems were added to give it limited all-weather and air-to-ground capabilities. A total of 1281 were produced between 1956 and 1965, with the last new production variant designated F-8E. Outstandingly successful in Vietnam, the Crusader shot down 19 MiGs against 3 losses in air-to-air combat.

A rebuilding program initiated in the late 1960s kept the Crusader in service through the 1970s. Rebuilt models included the F-8H (rebuilt F-Ds), F-8J (rebuilt F-8Es), F-8K (rebuilt F-8Cs), and F-8L (rebuilt F-8Bs). A photographic RECONNAISSANCE version, the RF-8A, was rebuilt as the RF-8G, with a strengthened wing and fuselage, ventral fins, a more powerful TF-30 engine, and new avionics. Flying with the Naval Air Reserve until 1987, it was the last version of the Crusader in U.S. service.

The F-8E has an APQ-94 search and tracking radar in a pointed radome above a chin inlet for the engine. The pilot sits immediately behind the radome under a clamshell canopy faired into the fuselage (obstructed rear-quarter visibility is one of the Crusader's few shortcomings). Of 1960s vintage, the avionics include a lead-computing optical gunsight, a head-down radar display, and an infrared search-and-track (IRST) sensor built into the base of the windscreen. Four Colt Mk. 12 20-mm. cannons are mounted in the sides of the fuselage behind the cockpit, and up to four Sidewinders can be carried on fuselage rails. On most versions, a retractable AERIAL REFUELING probe is built into the right side of the fuselage. The rear fuselage contains fuel tanks, the main landing gear, and the engine.

The shoulder-mounted wing has a unique variable incidence system, whereby the wing is hinged at the trailing edge, allowing the leading edge to be raised during landings to improve pilot visibility without degrading low-speed handling. Its large wing area, necessary for low-speed carrier landings, also gives the Crusader excellent maneuverability. With leading-edge slats and large flaps for low-speed stability, the wing's outboard sections fold up for deck parking. The F-8E has four wing pylons with a total capacity of 4000 lb. (1816 kg.) of bombs, fuel tanks, or rocket pods.

France purchased 42 F-8E(FN) models which operate from the small CLEMENCEAU-class aircraft carriers. Modifications include drooping flaps and ailerons, and an engine bleed-air boundary layer control system, for improved takeoff and landing

performance. The French Crusaders have no air-to-ground capability, but their fire controls were modified to be compatibile with French Matra R.530 and R.550 MAGIC air-to-air missiles.

The Philippine air force acquired 25 surplus F-8H models at bargain prices in 1977. They are operational with the 7th Fighter Squadron at Basa Air Base.

Specifications Length: 54.5 ft. (16.61 m.). **Span:** 35.16 ft. (10.72 m.). **Powerplant:** Pratt and Whitney J57-PW-420 afterburning turbojet, 18,000 lb. (8165 kg.) of thrust. **Weight, empty:** 19,925 lb. (9038 kg.). **Weight, normal loaded:** 28,000 lb. (12,701 kg.). **Weight, max. takeoff:** 34,000 lb. (15,422 kg.). **Speed, max.:** 1120 mph (10,975 kph) at 36,000 ft. (11,000 m.). **Initial Climb:** 21,000 ft./min. (6400 m./min.). **Service ceiling:** 58,000 ft. (17,680 m.). **Range, max.:** (internal fuel) 1,100 mi. (1770 km.).

CS: O-chlorobenzylmalonontrile, a tear agent used by military and police forces for crowd control. CS induces choking and profuse discharges from the tear ducts and mucous membranes of the nose and eyes; bromobenzylcyanide (CA) is very similar. CS is not usually lethal. See also CHEMICAL WARFARE.

CUB: NATO code name for the Soviet Antonov An-12 four-engine turboprop tactical transport aircraft widely used in a variety of military and civil roles. The An-12 flew in prototype form in 1958, and entered Soviet service in 1959. For many years it equipped the majority of VTA (Soviet Military Transport Aviation) squadrons, and more than 900 were manufactured in the Soviet Union through 1973. Often regarded as the Soviet equivalent of the C-130 HERCULES (though the two aircraft differ considerably), the An-12 has been widely exported, and is also produced in China as the Y-8, in both transport and maritime patrol versions. The An-12 was much used in Afghanistan, where many were shot down by U.S. STINGER missiles.

The cargo hold is loaded through a large door in the tail, which is split longitudinally into two sections that fold up into the fuselage. No ramp is provided for roll-on/roll-off handling of vehicles and cargo, but the aircraft does have a built-in crane which can be used to load cargo directly from trucks. With a maximum capacity of 44,100 lb. (20,045 kg.), the hold can accommodate all Soviet ARMORED PERSONNEL CARRIERS, light and medium trucks, light ANTI-AIRCRAFT vehicles such as the ZSU-23-4 and SA-6 GUILD missile launcher, and the ASU-57 and ASU-85 airborne ASSAULT GUNS. When used as a troop transport, the Cub can carry up to 90 infantrymen or 60 paratroops; via the rear doors, a full load of paratroops can be dropped in under one minute.

The An-12 has a crew of six. The pilot and copilot sit side-by-side in an airliner-type flight deck above the cargo hold, the flight engineer sits behind the copilot, the radio operator is in a well behind and below the pilot, and the navigator sits in a fully glazed nose position. Finally, a tail gunner in a powered turret under the vertical stabilizer operates two 23-mm. cannon.

A small chin radome houses a ground mapping/navigational radar. The landing gear, optimized for operations from rough fields, consists of eight large low-pressure tires housed in blisters along the lower sides of the fuselage, and two nose tires housed under the cockpit.

The shoulder-mounted wing has a high aspect ratio for economical long-range cruising, and is fitted with a variety of high-lift devices, including Fowler flaps, to enhance short-field performance. Typically, the takeoff run is 2300 ft. (701 m.), while the landing roll is only 1640 ft. (500 m.).

Some 150 An-12s remain in service with the VTA, but most are assigned to the air armies of the 16 Military Districts, and many more are operated by Aeroflot. In the VTA the An-12 is being replaced by the larger Il-76 CANDID at a rate of some 30 per year. An-12 airframes have also been modified for use as ELECTRONIC WARFARE (EW) platforms, with four different EW variants identified: Cub-A, essentially identical to the transport version, has several large blade antennas behind the flight deck for electronic intelligence (ELINT) gathering; Cub-B, a more extensive ELINT conversion, has two small ventral radomes and a variety of antennas mounted on the fuselage; Cub-C is an ELECTRONIC COUNTERMEASURES (ECM) aircraft with several tons of electronic equipment in the cargo hold, and at least five antennas for ACTIVE JAMMING in belly fairings, plus several dispensers for CHAFF, and the tail turret replaced by an ogival fairing for a variety of receivers and transmitters; and Cub-D, yet another ECM variant equipped to deal with different radars than the Cub-C. Some 30–40 ECM Cubs are in service with the Soviet air force (VVS) and AV-MF.

Specifications Length: 108.6 ft. (33.1 m.). **Span:** 124.67 ft. (38 m.). **Cargo hold:** 44.4 × 11.5 × 8.5 ft. (13.54 × 3.5 × 2.6 m.). **Powerplant:** 4 4000-shp. Ivchenko Al-20K turboprops. **Fuel:** 30,000 lb. (13,636 kg.). **Weight empty:** 61,730 lb. (28,000 kg.). **Weight, max. takeoff:** 134,480 lb. (61,000 kg.). **Speed, max.:** 482 mph (777 kph). **Speed, cruising:** 416 mph (670 kph). **Range, max.:** 2236 mi. (3600 km.). w/max. payload/3540 mi. (5700 km.) w/max. fuel. **Operators:** Alg, Bang, Egy, Eth, Gui, Ind, Iraq, Jor, Mad, Pol, PRC, Syr, USSR, Yem, Yug.

CV: U.S. abbreviation for AIRCRAFT CARRIER.

CVBG: See CARRIER BATTLE GROUP.

CVR(T): Combat Vehicle, Reconnaissance (Tracked), the official British army designation for the SCORPION and SCIMITAR armored reconnaissance vehicles.

CVR(W): Combat Vehicle, Reconnaissance (Wheeled), the official British army designation for the FERRET and FOX armored cars.

D

DAMAGE CONTROL: Measures meant to limit or counteract the effects of battle damage. Naval damage control measures include firefighting, counterflooding to reduce listing, the shoring up of damaged bulkheads or hull frames, and ad hoc repairs. The effectiveness of such measures depends on crew training, as well as on design features, such as compartmentation, duplication of key facilities, the number and location of fire mains, emergency electrical generators, and pumps; such provisions account for a significant portion of warship costs. In World War II, *industrial* damage control measures were the key countermeasure against strategic bombing.

DAMAGE LIMITATION: Active and passive measures meant to restrict the level and/or extent of war losses; nuclear damage-limitation measures include COUNTERFORCE targeting, active defenses (mainly AIR DEFENSE and BALLISTIC MISSILE DEFENSE), and CIVIL DEFENSE.

DAPHNE: A class of 25 French diesel-electric attack SUBMARINES (SSs), commissioned between 1964 and 1975, in service with France, Portugal, Pakistan, and South Africa. When the Daphnes were designed in 1952 as successors to the Arethuse-class coastal submarines, low radiated noise levels, maneuverability, a small crew, and low maintenance were emphasized more than safety or endurance. The Daphnes are highly automated, and have modularized equipment for easy replacement; on the other hand, of the 11 Daphnes commissioned by the French navy between 1964 and 1967, 2 have been lost in accidents. Portugal

bought 4 French-built boats between 1967 and 1969, and had 4 more built under license in Spain between 1973 and 1975. Pakistan bought 3 boats in 1975, and South Africa 3 in 1970–71. (One Portuguese boat was sold to Pakistan in 1975.) In 1971, the Pakistani submarine *Hangor* sank the Indian frigate *Khukri* in the first submarine attack since World War II.

Based on the German Type XXI U-boats of 1944–45, the Daphnes have a cigar-shaped double hull, with fuel and ballast tanks installed between the pressure hull and outer casing. A short, streamlined sail is located amidships, and control surfaces are conventional, with retractable bow diving planes, a single rudder, and stern planes ahead of the propellers.

The Daphnes have eight externally mounted 21.7-in. (550-mm.) bow torpedo tubes, and four externally mounted stern tubes for variety of unguided and acoustic-homing TORPEDOES. Externally mounted tubes reduce crew size and workload, but cannot be reloaded at sea. All sensors are tied to a DLT D-3 torpedo FIRE CONTROL system. See also SUBMARINES, FRANCE.

Specifications Length: 189.6 ft. (57.8 m.). Beam: 22.3 ft. (6.8 m.). Displacement: 870 tons surfaced/1045 tons submerged. Powerplant: twin shaft diesel-electric, 2 615-hp. SEMT-Pielstick diesel generators, 2 790-hp. Jeumont Schneider electric motors Speed: 13.5 kt. surfaced/16 kt. submerged. Range: 4300 n.mi. at 7.5 kt. Max. operating depth: 984 ft. (300 m.). Collapse depth: 1886 ft. (575 m.). Crew: 45. Sensors: 1 DUUA1 or

DUUA2 active/passive scanning sonar, 1 DSUV2 passive circular sonar, 1 DUUX2 passive ranging sonar, 1 AUUD/DUUG1 sonar intercept system, 1 Calypso-II surveillance radar, 1 electronic signal monitoring (ESM) array, 2 periscopes.

DARPA: Defense Advanced Research Projects Agency, part of the U.S. Department of Defense (DOD). Originally established in 1958 as the Advanced Research Projects Agency (ARPA), and placed under the director of Defense Research and Engineering (DDR&E), it was renamed DARPA in 1972, and placed under the undersecretary of Defense for Research and Engineering (USDR&E). DARPA is responsible for the management of high-risk/high (potential) payoff basic research and applied technology projects; these are conducted under its direction by the armed services, other government agencies, private industry, universities, and national laboratories. DARPA manages projects only through feasibility demonstrations; approved programs are then transferred to one of the military services for further development.

DASH: The Gyrodyne QH-50 Drone Anti-Submarine Helicopter, a REMOTELY PILOTED VEHICLE of the early 1960s, formerly in service with the U.S. Navy and the Japanese Maritime Self-Defense Force (JMSFDF).

Developed from 1958 as an alternative to the RUR-5A ASROC rocket-assisted torpedo, DASH was intended to deliver a lightweight homing TORPEDO out to a range of some 20,000 m. A miniature helicopter with a simple skid landing gear and a tandem-rotor configuration, DASH had a cruising speed of 92 mph (154 kph), and an endurance of 25 min. while carrying two Mk.44 or Mk.46 torpedoes. Operated from a small landing deck and hangar, and controlled manually via a radio DATA LINK, DASH would be launched against submarines detected by sonar, and tracked in flight by the launch ship's radars to the target location, where the torpedoes would be released.

In practice, it was extremely difficult to correlate the drone's position and that of the target because of the variability of sonar performance. Moreover, the drone was extremely difficult to fly because of a lack of feedback in the control system; even worse, the radio command link was unreliable at longer ranges. Of 746 built between 1962 and 1968, more than half were lost in accidents. Thus DASH was replaced in the U.S. Navy by manned LAMPS helicopters. On the other hand, the JMSDF bought 17 DASHs and lost none, and a

television-equipped spotting variant ("Snoopy DASH") was used successfully in the Vietnam War. All surviving DASHs were acquired by Israel in 1980.

DATA ENCRYPTION: The deliberate distortion of telemetry data by an "encryption algorithm" so as to preserve the security of information transmitted over DATA LINKS by SURVEILLANCE satellites, RECONNAISSANCE aircraft, and missiles undergoing flight tests (Soviet encryption of BALLISTIC MISSILE test data was once a direct violation of the verification clauses of the SALT II treaty).

DATA LINK: A radio used to transmit digital or analog information in a machine-readable format. Data links are used to transmit RADAR, SONAR, imagery, and other sensor data, as well as weapons guidance commands, FIRE CONTROL information, and missile telemetry. See also JTIDS; NTDS.

DEALEY: A class of 13 U.S. ANTI-SUBMARINE WARFARE (ASW) escorts commissioned between 1952 and 1957. The first U.S. FRIGATES completed after 1945, the Dealeys were intended as economical replacements for World War II DESTROYER ESCORTS. However, they were soon judged to be both too slow and too lightly armed to cope with post-1945 Soviet submarines, and too expensive for mass production. All were retired from the U.S. Navy in the early 1970s. One was transferred to Colombia and another to Uruguay; only the latter remains in service. Three more Dealeys were built in Portugal between 1966 and 1968, and five of a modified design were built in Norway as the OSLO class.

The Dealeys are flush-decked, with a small superstructure forward and a single stack amidships. As built, they were armed with a twin 3-in. 50-caliber anti-aircraft gun at either end of the ship, a Mk. 108 "Weapon Alpha" automatic DEPTH CHARGE mortar before the bridge, and a depth charge rack on the fantail. In the survivors, Weapon Alpha has been replaced by two Bofors 4-barreled, 375-mm. ASW rocket launchers and two sets of Mk.32 triple tubes for lightweight ASW homing TORPEDOES. See also FRIGATES, UNITED STATES.

Specifications Length: 315 ft. (95.87 m.). Beam: 36.67 ft. (11.18 m.). Draft: 13.25 ft. (4.04 m.). Displacement: 1450 tons standard/1950 tons fully loaded. Powerplant: single-shaft steam, 2 oil-fired boilers, 1 set of geared turbines, 20,000 shp. Fuel: 360 tons. Speed: 26 kt. Range: 1600 n.mi. at 25 kt./4400 n.mi. at 11 kt. Crew: 165. Sensors: 1 SPS-6 air-search radar, 2 SPG-34 fire control radars, 1 hull-mounted SQS-4 medium-frequency sonar; local upgrades include variable-depth sonar or passive towed array sonar. **Electronic warfare equipment**: WLR-1 radar warning receiver.

DECAPITATION (STRIKE): A nuclear FIRST STRIKE directed against command, control, and communications (c³) facilities and NATIONAL COMMAND AUTHORITIES (NCA) in order to paralyze decision making and impede retaliation. A decapitation strike may be a precursor to a COUNTERFORCE strike meant to prevent the launch of missiles on warning; only shorter-range, high-speed weapons, notably submarine-launched ballistic missiles (SLBMS), would be suitable for the purpose. Against this contingency, the U.S. and U.S.S.R. have several alternative national command posts and hardened facilities, as well as airborne command posts. (Until 1990, one U.S. "Looking Glass" aircraft was always flying. This practice was discontinued because of improved U.S.-Soviet relations and to cut costs.)

DECEPTION: The use of misdirection and distortion to mask capabilities, intentions, and specific plans. See also CAMOUFLAGE, CONCEALMENT, AND DECEPTION; *Maskirovka*.

DECEPTION JAMMING: A form of active ELECTRONIC COUNTERMEASURE (ECM), in which signals from enemy RADAR, SONAR, or weapons-guidance transmitters are detected, analyzed, and deliberately reradiated in subtly altered form, in order to generate a misleading return signal (usually to simulate a false target position). Because radar range estimation is accomplished by timing the return signal, fractional delays in simulated return signals result in false ranges. In the case of COMMAND-GUIDED weapons, the jammer radiates spurious steering commands meant to divert the weapon from its course. Deception jamming requires less power than NOISE JAMMING (see BARRAGE JAMMING and SPOT JAMMING), and is more selective; i.e., it does not jam an entire waveband. The most modern types of deceptive jammers can detect enemy emissions over a broad frequency band and automatically select the proper imitation signal.

DECOY: A device used to attract enemy attention and divert or mislead weapons and sensors. Passive decoys imitate the shape, size, and material composition of weapons or platforms; active decoys also simulate their acoustical, thermal, radio, or radar emissions. CHAFF is a common passive RADAR decoy, while flares are one form of active INFRARED decoy. More sophisticated active decoys include submarine decoys (torpedolike vehicles equipped with noisemakers and DECEPTION JAMMERS to imitate submarines); anti-torpedo decoys, such as NIXIE, which imitate the acoustical signature of surface ships; and decoy missiles (such as Quail), equipped with BLIP ENHANCERS and deception jammers that simulate bombers. PENAIDS are decoys that simulate REENTRY VEHICLES, in order to confuse and dilute BALLISTIC MISSILE DEFENSES. See also ELECTRONIC COUNTER-COUNTERMEASURES; ELECTRONIC COUNTERMEASURES; INFRARED COUNTERMEASURES.

DEEP STRIKE: A proposed array of advanced weapons for the long-range, nonnuclear attack of Soviet forces in Eastern Europe; the concept was later absorbed by "Follow-On Forces Attack" (FOFA). Weapons studied under Deep Strike include ASSAULT BREAKER, and the Army Tactical Missile System (ATACMS). See also AIR-LAND BATTLE.

DEFA: French 30-mm. aircraft CANNON, fitted to a variety of aircraft, including the MIRAGE III/5, MIRAGE F.1, MIRAGE 2000, KFIR C2, and Israeli-operated A-4E/H/N SKYHAWKS. Copied from the German Mauser cannon of World War II, the DEFA operates on the revolver principle; it is quite similar to the British ADEN Mk.4 (also a Mauser copy), which uses the same ammunition, including AP, API, and HEI rounds.

The initial DEFA 552A, of which more than 10,000 were manufactured for the Mirage III/5, JAGUAR, Fiat G.91, and Super ETENDARD, was superseded by the DEFA 553, with a longer service life and simplified mountings, as well as electric controls for 0.5- and 1.0-sec. bursts. Both guns weigh 178 lb. (80.9 kg.), and have a cyclic rate of fire of 1300 rounds per minute and a muzzle velocity of 2690 ft. 1 sec. (820 m./sec.).

The new, lightweight DEFA 554 fitted to the Mirage 2000 fighter weighs 176 lb., and has a (very high) rate of fire of 1800 rounds per minute. A selector switch can reduce that to 1100 rounds per minute for ground strafing. There is also a pod-mounted version.

DEFENDER: The McDonnell-Douglas (formerly Hughes) 500-MD scout/light attack helicopter, a more powerful development of the OH-6A scout helicopter (much used in the Vietnam War and still in service with the U.S. Army Reserve and National Guard). After the U.S. Army chose to replace the OH-6A with the Bell OH-58 Kiowa, Hughes developed the 500 for export. The prototype flew in 1968, and the Defender is now in service with some 15 countries; it is also produced under license in South Korea and Italy. There are three main variants: the 500-MD for scouting, liaison, casualty evacuation (CASEVAC), and general support; the 500-MD/TOW for anti-tank missions; and the 500-MD/ASW for ANTI-SUBMARINE WARFARE.

Small, maneuverable, and economical, the 500-MD has an egg-shaped fuselage, with the pilot and copilot seated side-by-side in the well-glazed nose. There is seating for five (or two stretcher cases) in the main cabin. A tubular tail boom supports the small anti-torque rotor and a T-tail stabilizer. Located in the rear of the fuselage, the engine drives a five-bladed main rotor with noise-reducing "feathered" tips. The engine exhausts are fitted with "Black Hole" INFRARED suppressors to reduce vulnerability to heat-seeking missiles.

A wide range of avionics are available, including all-digital instrumentation integrated by a RAMS 3000 flight-data computer and a 1553B data bus. Basic equipment includes blind-flying instruments, a gyrocompass, an INERTIAL NAVIGATION unit, DOPPLER navigation radar, an autopilot, IFF, and an APR-39 RADAR WARNING RECEIVER. The 500-MD/TOW also has a sight and guidance unit for TOW anti-tank guided missiles in the left side of the nose, or in an optional mast-mounted sight with a FLIR (Forward-Looking Infrared) sensor and a laser rangefinder/designator. The 500-MD/ASW has a lightweight surface-search radar and an ASQ-81 MAGNETIC ANOMALY DETECTOR (MAD) in a towed "bird." CHAFF and flare dispensers are standard on all models.

Armament varies. The Defender Scout can carry a number of gun and rocket pods (including a 30-mm. CHAIN GUN), 40-mm. GRENADE LAUNCHERS, and STINGER anti-aircraft missiles (for self-defense). The anti-tank version can also carry up to four TOW missiles in two twin launcher pods attached to outrigger pylons, while the 500-MD/ASW can carry two Mk.46 homing TORPEDOES or two DEPTH CHARGES.

Specifications Length: 25 ft. (7.62 m.). **Rotor diameter:** 27.33 ft. (8.33 m.). **Powerplant:** 420-shp. Allison 250-C20B turboshaft derated to 375 shp. **Weight, empty:** 1976 lb. (896 kg.). **Weight, max. takeoff:** 3000 lb. (1361 kg.). **Speed, max.:** 152 mph (217 kph). **Initial climb:** 1650 ft./min. (503 m./min.). **Range, max.:** 242 mi. (389 km.). **Operators:** Bah, Den, El Sal, Fin, Hai, Iran, Iraq, Isr, Jap, Jor, Ken, Mor, ROK, Sp, US.

DEFENSE COUNCIL *(SOVIET OBORONY)*:
The supreme military command of the Soviet Union, responsible in peacetime for all aspects of war preparations. Chaired by the general secretary of the Communist Party of the Soviet Union (CPSU), the council includes selected members of the Politburo, in addition to the minister of defense and the chairman of the KGB; other civil and military officials participate when summoned.

The council is the U.S.S.R.'s paramount military-economic planning body that sets priorities for resource allocation between the civil and military sectors, and establishes policies for the mobilization of industry, transport, and manpower for war.

The Main Military Council *(Glavnyy voyennyy sovet)*, directly subordinate to the Defense Council, is responsible for the overall leadership and war readiness in the armed forces; it is chaired by the minister of defense, and the chairman of the Defense Council (i.e., the general secretary of the CPSU) is also a member. Other members include the chief of the General Staff, the commander of the WARSAW PACT Forces, the commanders of the five military services, the chief of the Main Political Administration (responsible for the political loyalty of the armed forces), the chief of the Rear Services *(Tyl)*, and the chief of CIVIL DEFENSE.

In war, the Defense Council would be reorganized as the State Committee of Defense *(Gosudartsvennyy komitet oborony*, or GKO), essentially a war cabinet in charge of all political, economic, and diplomatic matters relevant to the grand strategy for the conduct of war. Similarly, the Main Military Council would be transformed into the Supreme High Command *(Verkhovnoye glavnokomandovaniye)*, responsible for the planning and the conduct of war operations through the various Theaters of Military Operations (TVDS), to which it allocates resources in accordance with the broad strategic plan established by the GKO.

The Soviet General Staff is the primary executive arm of the Main Military Council in peacetime, as of the *Stavka* in war. It is charged with basic military planning and the control of the five military services, the sixteen MILITARY DISTRICTS, Groups of Forces abroad, and, in wartime, the TVDs and associated FRONTS.

DEFENSE DEPARTMENT: See DOD.
DEFENSE-IN-DEPTH: A defensive method based upon successive lines or perimeters (physical or notional), as opposed to a linear defense, in which all available strength is concentrated in a

single perimeter. Current defense-in-depth methods in ground warfare were developed to counter the BLITZKRIEG and, more generally, deep penetration offensives. Instead of a "thick" front with "soft" supply lines behind it, defensive forces are deployed in successive (and mutually supporting) fortified lines or chains of strong points which are intended to resist direct assaults with their concentrated ANTI-TANK and ANTI-AIRCRAFT fire, while employing their artillery to prevent the enemy from passing forces around or between them. A linear defense is disrupted once it is penetrated, if only because its supply lines are cut; the strong points or perimeters of a defense-in-depth, by contrast, are supposed to withstand envelopment through at least a temporary self-sufficiency in supplies and ancillary services. Instead, it is the attacker who is to be disrupted by the defense-in-depth, because his armor can penetrate the artillery fire zones, while his (unarmored) supporting infantry, artillery, and supplies cannot do so. Once the attack has thus been weakened, mobile forces held in reserve by the defense can emerge from the rear to counterattack.

The Soviet version relies upon successive lines, rather than on chains of strongpoints. Flanks are anchored on impassable terrain (e.g., swamps), or covered by minefields and other man-made obstacles. Minefields and obstacles are also used to partition the areas between the lines, so as to impede the enemy from moving laterally to enlarge his penetrations. The enemy may penetrate one or two of the lines, but is supposed to exhaust his momentum in the effort. Concentrated anti-tank and artillery fires are to be used against enemy forces trapped between the lines, while mobile forces are held behind the lines for decisive counterattacks.

In any version, if armored/mobile forces cannot be held in reserve, but must instead be sent into action to hold the lines of strongpoints, the defense loses its sword, becoming all shield. That degrades the method into a linear defense reproduced many times, and highly vulnerable to destruction in detail. The essence of defense-in-depth methods is thus the use of static forces to wear down enemy mobile forces, so as to set the stage for counterattacks by the mobile forces of the defense. See also ACTIVE DEFENSE; ARMORED WARFARE; COMBINED ARMS.

DEFENSE INTELLIGENCE AGENCY (DIA):
An agency of the U.S. Department of Defense (DOD), established in 1961 under the control of the Office of the Secretary of Defense (OSD), to combine intelligence activities previously conducted separately by each military service. DIA is responsible for the organization, management, and control of all DoD INTELLIGENCE resources. It also reviews and coordinates the activities of the army, navy and air force intelligence branches, and supervises their operations.

The director of DIA is the principal intelligence advisor to the secretary of defense, reports directly to the chairman of the JOINT CHIEFS OF STAFF (JCS), and acts as the chief intelligence officer of the JCS. See also CENTRAL INTELLIGENCE AGENCY.

DEFENSE SUPPORT PROGRAM (DSP):
Code name for U.S. surveillance satellites intended to provide early warning of a BALLISTIC MISSILE attack. Developed from 1966 as a successor to the earlier Midas satellites, the first DSP satellite was launched in 1971, and the system was declared operational in 1973. Since then, the system has been successively upgraded by the deployment of newer satellites equipped with better sensors. The latest "Block 14" satellites, built by TRW, are some 33 ft. (10 m.) long, and weigh 2.5 tons; they are placed into geosynchronous orbits by TITAN 34D boosters. The system is believed to consist of three active satellites spaced 120° apart to provide global coverage, and three in-orbit spares.

The DSP satellites are equipped with a sensitive INFRARED telescope operating in the 2.7- to 2.95-micron band, which can detect missile launches within one minute; it can also detect surface or atmospheric nuclear explosions. In addition to providing attack warning, the DSP satellites are also used for monitoring Soviet missile tests; tracking accuracy is roughly one mile.

These satellites are a vital part of the U.S. early warning system, providing up to 30 min. warning of an impending attack (twice the time available from ground sensors such as the BALLISTIC MISSILE EARLY WARNING SYSTEM, or (BMEWS); DSP data are transmitted to the NORAD headquarters at Cheyenne Mountain, Colorado. The DSP satellites will be replaced in the mid-1990s by the Boost Surveillance and Tracking System (BSTS), which has greater accuracy and even shorter reaction time. See also SATELLITES, MILITARY.

DELIVERY SYSTEM: A vehicle (or a system of vehicles) used to transport one or more warheads to their target(s), including associated guidance, communication, and field maintenance equipment. The term refers most often to BALLIS-

TIC MISSILE systems, with their ground controls, launch mechanisms, boosters, guidance or homing devices, and warheads.

The number and quality of nuclear-capable delivery systems is the most common arithmetical (and static) measure of offensive nuclear capabilities, because delivery systems rather than warheads are the constraining factor. The production of nuclear devices is less demanding than that of the vehicles required to carry them, unless they are simple transport aircraft, vulnerable to AIR DEFENSES. Thus nuclear warheads usually account for only a small fraction of the total cost of nuclear delivery systems.

DELTA: A class of 40 Soviet nuclear-powered ballistic-missile SUBMARINES (SSBNs), the first of which was commissioned in 1973. Successors to the earlier YANKEE class, the Deltas retain the same basic configuration: a modified teardrop hull with a round nose and elongated, cylindrical midsection, and the sail (conning tower) mounted well forward, behind which are two rows of missile tubes housed beneath an under "whaleback" superstructure. Like most Soviet nuclear submarines, the Deltas have a double-hull configuration, with ballast tanks, batteries, and fresh water stored between the outer casing and pressure hull. Control surfaces consist of sail-mounted "fairwater" planes, and a cruciform arrangement of rudders and stern planes mounted ahead of the propellers. A VLF buoy and an ELF trailing wire antenna are used for submerged communications with shore bases.

The Deltas have been built in four successively enlarged subclasses. The initial Delta I (18 of which were competed through 1977) are armed with 12 SS-N-8 submarine-launched ballistic missiles (SLBMS), arranged in 2 rows of 6. A large, 2-stage, liquid-fuel missile, the SS-N-8 has a range of 5655 mi. (9100 km.) and carries a single 800-kT nuclear warhead. Six torpedo tubes in the bow with 18 homing TORPEDOES are for self-defense. Four Delta IIs completed between 1973 and 1975 are essentially stretched Delta is with 16 SS-N-8s; weapons, sensors, and propulsion are otherwise unchanged.

The 14 Delta IIIs completed between 1976 and 1984 are armed with larger, more accurate SS-N-18 STINGRAY missiles, whose greater length required an increase in the height of the whaleback casing behind the sail. Hydrodynamic matching of the sail and the superstructure is poor, making the Delta IIIs rather more noisy than earlier Deltas.

Length was again increased, further reducing maximum speed.

The first of the Delta IVs was completed in 1984, and four have been delivered to date. Their SS-N-23 SKIFF missiles, with a range of 5190 mi. (8300 km.) and a payload of ten 100-kT MIRVs, are slightly smaller than the Stingrays, hence no changes were needed in the superstructure, but length was increased yet again. The torpedo armament is also unchanged, but a TOWED ARRAY SONAR, stowed in the whaleback, has been added to the sensor suite.

Of the 40 Deltas in service, 23 now serve with the Northern Fleet, and 17 with the Pacific Fleet. They are the backbone of the Soviet SSBN force, as the even larger, more powerful TYPHOONS enter service. The long range of their SS-N-18 and SS-N-23 missiles allow Deltas to remain close to Soviet coasts, where their acoustic signatures are not so serious handicaps. See also SLBMS, SOVIET UNION; SUBMARINES, SOVIET UNION.

Specifications Length: (I) 447.8 ft. (136.6 m.); (II) 501 ft. (152.7 m.); (III) 509 ft. (155.2 m.); (IV) 516.7 ft. (156.5 m.). Beam: 39.4 ft. (12 m.). Displacement: (I) 8600 tons surfaced/10,000 tons submerged; (II) 9600 tons surfaced/11,400 tons submerged; (III) 11,000 tons surfaced/13,250 tons submerged; (IV) 12,000 tons surfaced/14,250 tons submerged. Powerplant: 2 pressurized-water reactors w/geared steam turbines, 80,000 shp. Speed: (I) 20 kt. surfaced/26.5 kt. submerged; (II) 25 kt. submerged; (III) 24 kt. submerged; (IV) 23.5 kt. submerged. Max. operating depth: 970 ft. (600 m.). Collapse depth: 1315 ft. (400 m.). Crew: 130. Sensors: 1 large, bow-mounted low-frequency sonar array, 1 "Snoop Tray" surface-search radar, 1 "Pert Spring" electronic countermeasures (ECM) array, 1 "Brick Pup" electronic signal monitoring (ESM) array, 2 periscopes; (IV) 1 towed array sonar.

DELTA FORCE: U.S. Army hostage rescue unit, established in November 1977 as the "First Special Forces Operational Detachment-Delta." Its organization, training, and operations are all highly classified, but is known that the unit is based at Fort Bragg, North Carolina, that it is under the operational control of the U.S. CENTRAL COMMAND, and that it is very small. Organized along the same lines as the British SAS, Delta consists of three or four "squadrons" of 60–90 men, each divided into four "troops" of 16 men. The unit has no organic air or sea transport, but can call upon the 1st Special Operations Wing of 23rd Air

Force, which has specialized MC-130 E/H COMBAT TALON transports, AC-130A/H Spectre gunships, and HH-53H PAVE LOW helicopters.

Delta personnel are mostly volunteers from RANGER and SPECIAL FORCES units, selected with techniques copied from the SAS. Each man undergoes physical conditioning and individualized training in unarmed combat, DEMOLITIONS, marksmanship, parachuting, mountain climbing, scuba techniques, and night combat. Team training rotates through water, mountain, airport, and other assault situations. Marksmanship is greatly stressed (Delta snipers must score 100 percent at 600 yards and 90 percent at 1000 yards), and more realistic training is conducted in an elaborate "shooting house" which simulates typical hostage situations. Nonstandard equipment includes frame charges for blasting doors and windows, flash and stun grenades, infrared and light amplification telescopic sights, personal radios, and man-portable satellite ground terminals.

Delta became known because of Operation Eagle Claw, the abortive attempt to rescue American diplomats held hostage in Teheran on 24 April 1980. Delta's classified missions in the 1983 Grenada invasion were also unsuccessful (because of inept planning). Mobilized often but hardly employed, Delta Force is as well trained and equipped as the SAS, GSG-9, or SAYAR MATKAL, but its utility is undermined by inconsistent U.S. policies on counterterrorist operations and by the incomprehension of higher military echelons. See also CLANDESTINE OPERATIONS; COVERT OPERATIONS; SPECIAL OPERATIONS; TERRORISM.

DEMILITARIZED ZONE (DMZ): An area within which military forces and/or specified military facilities are prohibited. The nature of prohibited installations, and arrangements for internal security forces, if any, are defined by each particular DMZ agreement. Because such zones are established between states that are openly hostile, they tend to be contested in detail, with firefights and small raids. A NUCLEAR-FREE ZONE would be a highly selective DMZ, and likely to be advocated by the party with conventional superiority.

DEMOLITIONS: The deliberate destruction of structures or natural features with emplaced explosives. Demolitions have been used to clear obstacles to movement (roadblocks, minefields, rubble, etc.); to create obstacles (by knocking down buildings, trees, power pylons, etc., or by cratering road surfaces); to blast out defensive positions; and to destroy or disrupt enemy forces

directly. The most common explosives issued for demolitions are of the plastic type such as C4. Specialized demolition devices include the "Bangalore torpedo" and rocket-propelled line charges (used to slight wire entanglements and clear minefields), shaped charges (used for cratering and against steel and concrete structures), and incendiary devices such as thermite or phosphorus bombs (used to burn forests and buildings). Atomic demolitions (ADMS) are man-portable nuclear devices. Demolitions are one of the primary responsibilities of SAPPERS and ENGINEERS. See also EXPLOSIVES, MILITARY.

DENSE PACK: A proposed basing scheme for the U.S. PEACEKEEPER (MX) ICBM, with 100 closely spaced, hardened SILOS. In theory, it would be impossible to destroy all the silos in a single strike because hardening would preclude multiple kills, while FRATRICIDE (the destruction or deflection of incoming warheads by the fireball, radiation, and debris of earlier detonations) would preclude concurrent strikes. On the other hand, it was envisaged that surviving ICBMs could be launched through the debris clouds shortly after the first attack wave, while the attacker would have to wait several hours before attacking the Dense Pack again (i.e., ascending ICBMs were deemed to be less fragile than descending warheads). The scheme was rejected because of basic uncertainties about the nuclear effects on which it was based. Dense Pack might be a desirable basing mode if some form of terminal BALLISTIC MISSILE DEFENSE is deployed around the silos. See also COUNTERFORCE.

DEPRESSED TRAJECTORY: A BALLISTIC MISSILE flight path in which the apogee is kept lower than with standard (minimum-energy) trajectories, in order to achieve a shorter time of flight, and longer masking from line-of-sight BALLISTIC MISSILE EARLY WARNING SYSTEMS (BMEWS). Depressed trajectories are usually envisaged in connection with the employment of submarine-launched ballistic missiles (SLBMS) for DECAPITATION STRIKES.

DEPTH CHARGE: An ANTI-SUBMARINE WARFARE (ASW) weapon, essentially a container of high explosives with a hydrostatic fuze set to detonate at a selected depth (or with a proximity fuze). When triggered near a SUBMARINE, depth charges can cause localized peak overpressures on the pressure hull, which may result in its rupture. Depth charges were first developed in World War I as cylinders filled with 200–300 lb. (100–150 kg.)

of high explosive. Known as "ash cans," they were simply rolled over the stern of ASW vessels; later, side-throwing projectors (K-guns and Y-guns) were provided to increase the area that could be covered by a single ship.

Depth charging is complex, for it amounts to the three-dimensional bombing of a moving target. The speed, depth, and course of the submarine, and the rate at which the depth charge sinks, must all be calculated to "lead" the target accurately. To reduce the interval between the release of a charge and its detonation ("dead time"), during which the submarine can take evasive action, World War II depth charges were made denser and more streamlined to increase their sink rate.

Moreover, since SONAR becomes ineffective at close range, and cannot function to the rear because of propeller noises, each time a ship attacks with depth charges rolled off the stern, it loses contact with the submarine. As a remedy, World War II escorts would operate in pairs, with one ship to track the submarine and vector the other into an attack position, while maintaining continuous sonar contact throughout. Later in the war, forward-throwing multiple projectors were introduced. Essentially large MORTARS (e.g., the British Squid or LIMBO), by firing depth charges a few thousand yards ahead of the ship they allowed continuous sonar coverage to be maintained throughout an attack. Each depth charge explosion, however, still created underwater turbulence ("blueout"), which temporarily blinded sonar, giving submarines a time window in which to evade. To avoid this, depth charges are now fitted with magnetic or acoustical PROXIMITY FUZES, so that they will not detonate unless they pass close to a submarine. Proximity fuzes also eliminate the problem of estimating the submarine's depth, which is always difficult because of the uncertainties of sonar performance.

Depth charges have largely been supplanted by more costly lightweight homing TORPEDOES, but are still used from HELICOPTERS and fixed-wing patrol aircraft, and smaller vessels for inshore ASW are still equipped with forward-throwing mortars or low-recoil rocket launchers; in shallow waters, which impede sonar, depth charges are more reliable than homing torpedoes (and also intimidate submarine crews).

Nuclear depth charges with a yield of two or three kilotons (kT) have lethal radii of several hundred yards (vs. several dozen feet for high-explosive depth charges). But aside from the obvious political constraints on their use, underwater nuclear explosions would create extensive blueout conditions that might be a greater hazard than the target submarine. Britain and the U.S. still stock nuclear depth charges for use by patrol aircraft, and as a warhead option for ASROC. The Soviet navy undoubtedly deploys nuclear depth charges aboard its larger warships.

DESANT: Soviet military term for the insertion of troops behind enemy lines, usually AIRBORNE insertion but also AMPHIBIOUS landings, and even the carriage of infantrymen by tanks (*"Tank Desants"*). Soviet texts categorize desants as follows: (1) tactical *desants*, executed to depths of 40–60 km. behind enemy lines in support of divisional operations, normally by AIR ASSAULT BRIGADES; (2) operational *desants*, to depths of 60–150 km. in support of ARMY or FRONT echelons; (3) operational-strategic *desants*, executed more than 150 km. behind the lines in support of a Theater of Military Operations (TVD); and (4) strategic *desants*, independent operations (not meant as precursors to a ground advance), e.g., an airborne intervention in support of a pro-Soviet faction. See also AIRBORNE FORCES, SOVIET UNION.

DESCUBIERTA: A class of 9 Spanish-built FRIGATES commissioned between 1978 and 1982, in service with Spain (6), Egypt (2), and Morocco (1). Derived from the Portuguese JOAO COUTINHO class (also constructed at the Spanish Bazan shipyard), the Descubiertas are somewhat larger and carry more advanced sensors and weapons. The ships have gyro-controlled fin stabilizers, and may eventually receive the U.S. PRAIRIE MASKER bubble generator system to reduce radiated noise.

Meant as general-purpose escorts, they have a balanced armament of 1 OTO-MELARA 76.2-mm. DUAL PURPOSE gun forward, 2 40-mm. BOFORS GUNS aft, a 2-barrel Bofors DEPTH CHARGE mortar behind the 76.2-mm. gun, 2 sets of MK.32 triple tubes for lightweight ANTI-SUBMARINE WARFARE (ASW) TORPEDOES, and an 8-round NATO SEA SPARROW short-range surface-to-air missile (SAM) launcher on the fantail. In 1982–83, the Spanish ships also received 2 quadruple launch canisters for HARPOON missiles, which are mounted on the deckhouse behind the bridge. One 40-mm. gun is to be replaced by a MEROKA radar-controlled gun for anti-missile defense. The Moroccan ship has EXOCET missiles instead of Harpoon, and Italian ASPIDE SAMs instead of Sea Sparrow.

Specifications Length: 291.3 ft. (88.81 m.). Beam: 34.4 ft. (10.5 m.). Draft: 11.5 ft. (3.5 m.).

Displacement: 1200 tons standard/1575 tons full load. **Powerplant:** 4 4500-hp. MTU-Bazan 16MA956 TB91 diesels, 2 shafts w/variable-pitch propellers. **Speed:** 26 kt. **Fuel:** 250 tons. **Range:** 6100 n.mi. at 18 kt. **Crew:** 116. **Sensors:** surface-search, air-search, and fire control radars; 1 Raytheon 1160B hull-mounted medium-frequency sonar; (last 4 Spanish ships) 1 Raytheon 1167 passive towed array sonar. **Electronic warfare equipment:** 1 Electronica Beta passive warning system, 2 chaff launchers.

DESIGNATOR: See LASER DESIGNATOR.

D'ESTIENNE D'ORVES: A class of 23 French light FRIGATES *(avisos)* built between 1972 and 1984; 17 were commissioned by the French navy between 1976 and 1984, and 3 more, built for South Africa but embargoed, were later sold to Argentina. Meant as low-cost successors to the highly successful COMMANDANTE RIVIERE frigates, these *avisos* are considerably smaller and less capable overall, being well suited for ASW patrol only if inshore or alongside more capable units; their endurance is also quite adequate for a "gunboat" role, for which their simple sensors and weapons may suffice.

Barely larger than the Soviet NANUCHKA- and PAUK- class CORVETTES, the d'Estienne d'Orves are armed with a single 100-mm. DUAL PURPOSE gun forward, two manually operated 20-mm. ANTI-AIRCRAFT guns amidships, a 6-barrel Bofors 375-mm. Mk.54 DEPTH CHARGE mortar aft, and four 21.7-in. (550-mm.) tubes for heavy homing TORPEDOES. When the ships are deployed overseas, two EXOCET anti-ship missiles are installed behind the funnel. See also FRIGATES, FRANCE.

Specifications **Length:** 262.5 ft. (80 m.). **Beam:** 33.8 ft. (10.3 m.). **Draft:** 17.4 ft. (5.3 m.). **Displacement:** 980 tons standard/1170 tons full load. **Powerplant:** 2 6000-hp. SEMT-Pielstick 12PC2-V400 diesels, 2 shafts. **Speed:** 26 kt. **Range:** 4500 n.mi. at 15 kt. **Crew:** 79. **Sensors:** 1 DRBV51 surface-search radar, 1 DRBC32E gun fire control radar, 1 DRBN32 navigational radar, 1 hull-mounted Thomson-CSF DUBA medium-frequency sonar. **Electronic warfare equipment:** 1 ARBR16 radar warning receiver, 2 Dagaie chaff launchers.

DESTROYER: A fast, unarmored, heavily armed warship larger than a FRIGATE but smaller than a CRUISER. Destroyers originated in the 1880s as "torpedo-boat destroyers," designed to defend battle fleets against small, fast torpedo boats; but destroyers were themselves soon armed with

TORPEDOES. By 1914, destroyers were intended for SCOUTING, for torpedo attacks against the enemy battle fleet, and for SCREENING the friendly battle fleet from enemy destroyers and torpedo boats. By the end of World War I, a fourth mission had been added: ANTI-SUBMARINE WARFARE (ASW). Destroyers were fast and maneuverable enough to run down SUBMARINES speeding on the surface, and were soon equipped with hydrophones and DEPTH CHARGES to attack submerged submarines as well.

Between the world wars, destroyer configurations changed in response to the threat of air attack; DUAL PURPOSE main gun batteries supplemented by a few heavy MACHINE GUNS were often thought sufficient until 1939, but during World War II heavy secondary batteries of automatic ANTI-AIRCRAFT weapons (e.g., the 40-mm. BOFORS GUN and 20-mm. Oerlikon) were added by all navies. Early in World War II, destroyers were engaged in many surface actions, especially in the confined waters of the Mediterranean and South Pacific; but as the war progressed, ANTI-AIR WARFARE (AAW) as well as ASW became more important than ANTI-SURFACE WARFARE (ASUW), and many destroyers replaced all or some of their torpedo battery with additional light anti-aircraft guns. The advent of RADAR, which enhanced the effectiveness of destroyers in night surface actions, also led to their use as PICKET ships to direct fighter aircraft against enemy bombers (another role in which destroyers are still employed).

Long before World War II, the British and U.S. navies had equipped destroyers with active SONAR, greatly increasing the range at which submarines could be detected. ASW weapons, at first only simple depth charges, were supplanted from 1943 by more effective forward-throwing depth-charge mortars such as HEDGEHOG and SQUID, and now include lightweight homing torpedoes, ASW rockets, and guided missiles.

One result of the broadening role of destroyers was their increasing size and complexity. Most World War I destroyers displaced less than 1000 tons; by the mid-1930s 1500 tons was common; at the start of World War II, the largest destroyers in service were the 2100-ton U.S. FLETCHER class; by 1945 the 2600-ton GEARING class set the standard because of the need to accommodate heavy anti-aircraft batteries, new radar and sonar systems, and also the men to operate them. Despite repeated attempts to control or reverse the escalation of their size (and cost), current destroyers now displace 5000–7000 tons (more than many World War II cruisers).

After World War II most navies deemphasized ASUW capabilities, and destroyer designs evolved into two specialized types: the ASW destroyer and the guided-missile AAW destroyer. The former carries powerful sonars and ASW standoff weapons such as ASROC, ASW HELICOPTERS to extend engagement ranges out to more than 70 mi., and ASW torpedoes for close-in attacks. AIR DEFENSE weapons are limited to dual-purpose guns, short-range surface-to-air missiles (e.g., SEA SPARROW, SEA WOLF, ASPIDE), and radar-controlled anti-missile guns such as PHALANX, GOALKEEPER, and MEROKA.

Guided-missile destroyers, on the other hand, have long-range air defense SAMs such as TARTAR, TERRIER, SEA DART, or STANDARD, with the associated search, tracking, and guidance radars and FIRE CONTROL systems. Most also retain a secondary ASW capability.

The Soviet navy was, until recently, unique in having powerful ASUW capabilities in its destroyers armed with large ANTI-SHIP MISSILES. The development of lightweight anti-ship missiles such as HARPOON and EXOCET (which can be mounted on any available deck space) has allowed Western navies to follow suit. In the 1980s the general-purpose destroyer (combining AAW, ASW, and ASUW) has thus reemerged, a development much facilitated by the introduction of VERTICAL LAUNCH SYSTEMS which can accommodate any mix of anti-aircraft, anti-ship, and anti-submarine missiles, enabling the same ship to function in varied roles at short notice. See also separate DESTROYERS entries for BRITAIN; FEDERAL REPUBLIC OF GERMANY; FRANCE; ITALY; JAPAN; SOVIET UNION; UNITED STATES.

DESTROYER ESCORT (DE): Originally, a small, austere DESTROYER optimized for ANTI-SUBMARINE WARFARE (ASW)—a type first built in response to a World War II British requirement for cheap and abundant convoy escorts. DEs were smaller, slower, and otherwise more lightly armed except for their heavier ASW armament. Of more than 400 built between 1943 and 1945, a few are still in service with Third World navies.

After 1945, the U.S. Navy continued to classify smaller, more austere ASW vessels as DEs, while most other navies designated such ships as CORVETTES, FRIGATES (FFs), or guided-missile frigates (FFGs); in 1975 the U.S. Navy finally conformed to international practice by redesignating all of its DEs and DEGs as FFs and FFGs. Of the major naval powers, only Japan continues to designate frigate-sized vessels as destroyer escorts.

Under the Soviet navy's classification system, based on function as well as on size, frigates/destroyer escorts fall into three separate categories: Patrol Ships (*Storozhevoy Korabl*, or SKR); Small ASW Ships (*Maly Protivolodochnyy Korabl*, MPK); and Border Patrol Ships (*Pogranichniyy Storozhevoy Korabl*, PSKR). See also separate FRIGATES entries for BRITAIN; CANADA; FEDERAL REPUBLIC OF GERMANY; FRANCE; ITALY; JAPAN; NETHERLANDS; SOVIET UNION; UNITED STATES.

DESTROYERS, BRITAIN: The Royal Navy ended World War II with a large force of destroyers, mostly optimized for ANTI-SUBMARINE WARFARE (ASW) in the North Atlantic or ANTI-SURFACE WARFARE (ASUW) in the Mediterranean. Most displaced about 1500 tons, and only a few were large enough to remain useful in the postwar era. As Britain's imperial role diminished during the 1950s, wartime destroyers were replaced by a much smaller number of new ships.

By the late 1950s the Royal Navy decided that its destroyers would be primarily ANTI-AIR WARFARE (AAW) vessels armed with long-range surface-to-air missiles, while the ASW role would be left to FRIGATES. The first purpose-built guided-missile destroyers were eight COUNTY-class ships built in the early 1960s and phased out in the early 1980s (the last two were sold to Chile in 1987).

The Royal Navy now has only 13 destroyers in two classes: the 7100-ton BRISTOL (Type 82), built between 1967 and 1969 as a high-speed carrier escort; and 12 SHEFFIELD class (Type 42).

Commissioned between 1976 and 1985, the Sheffields reflect an attempt to build smaller, more economical destroyers (the original Batch 1 design displaces 4100 tons).

Although classified as frigates because they lack a long-range SAM, the BROADSWORD (Type 22) Batch 3 ships are larger (at 4200 tons) than the Batch 1 Sheffields, and are, in fact, powerful general-purpose warships with good ASW and ASUW capabilities.

DESTROYERS, FEDERAL REPUBLIC OF GERMANY: The first postwar German destroyers were 6 ex-U.S. FLETCHERS, transferred between 1958 and 1960 and retired in the early 1980s. The only German destroyers built since 1945 are the 4 3340-ton HAMBURGS, commissioned between 1964 and 1968. Optimized for operations in the Baltic and North Sea, they have a heavy gun and anti-ship missile armament; all 4 will continue in service through the 1990s. The West German navy also purchased 4 new, modified ADAMS-class guided-missile destroyers from the U.S., commissioned in 1969–70 as the LUTJENS class. The West German navy has no plans to build or purchase new destroyers, preferring instead to base its future strength on smaller FRIGATES, SUBMARINES, and FAST ATTACK CRAFT.

DESTROYERS, FRANCE: The first post-1945 French destroyers were the 2800-ton Cassard class, 18 ships commissioned between 1946 and 1958. Four were converted to guided-missile destroyers (DDGs) between 1961 and 1965, 5 more were converted to specialized ANTI-SUBMARINE WARFARE (ASW) vessels between 1968 and 1971, and 1 was converted to a special trials vessel between 1967 and 1971. The Cassards were followed by 4 2750-ton La Bourdonnaise destroyers, commissioned in 1958. Originally built as gun-armed, general-purpose escorts, the class has only one survivor, which been converted to an ASW vessel. The La Bourdonnaises were followed by a single ship, the 2750-ton *La Galissionere*, built as a flotilla leader with DUAL PURPOSE guns, Malafon, and an ASW helicopter.

Commissioned in 1967 and 1970, the two 5090-ton SUFFRENS were the first purpose-built DDGs of the French navy. The Suffrens were followed in 1973 by the 3500-ton *Aconit*, an attempt to produce a smaller, cheaper destroyer; less than successful, the design was not repeated. Instead, three 4580-ton TOURVILLE-class destroyers were commissioned in 1973–74.

The next French destroyers were the 3830-ton GEORGES LEYGUES class, specialized ASW vessels armed with one 100-mm. gun, four Exocet, Sea Crotale, ASW torpedoes, and two Lynx helicopters. Six were built between 1979 and 1988, and one more is under construction. Anti-air warfare (AAW) versions of the Georges Leygues, the two CASSARD II guided-missile destroyers entered service in 1988 and 1990.

DESTROYERS, ITALY: The first post-1945 Italian-designed destroyers were the two 2755-ton Impetuosos, ordered in 1950 and commissioned in 1958. Optimized for Mediterranean operations, with high speed at the expense of seakeeping and range, they were heavily influenced by contemporary U.S. design practices; both Impetuosos were stricken by 1983. The two 3200-ton IMPAVIDO-class destroyers of 1958–59 were enlarged and improved Impetuosos, with a 40-round Mk.13

launcher for TARTAR or STANDARD surface-to-air missiles, and a landing deck aft for light ASW HELICOPTERS. Both remain in service.

The Impavidos were followed by two 3600-ton AUDACE-class guided-missile destroyers, commissioned in 1972–73. Essentially improved Impavidos, they have a landing deck and a hangar for two ASW helicopters. The Audaces were to be followed immediately by two Improved Audaces, but these ships were delayed by several years. To fill the gap, Italy acquired three ex-U.S. FLETCHER destroyers in 1969–70, which were stricken in the early 1980s. Finally authorized in 1983, the Improved Audaces, now the 5000-ton ANIMOSO class, should enter service in late 1992.

DESTROYERS, JAPAN: The Japanese navy, officially the Maritime Self-Defense Force, or MSDF, is limited by the Japanese constitution to a purely defensive role, a limitation interpreted as precluding the acquisition of "offensive" vessels such as AIRCRAFT CARRIERS (even small ones) or CRUISERS. Moreover, the submarines of the Soviet Pacific Fleet have posed the salient threat to Japanese SEA LINES OF COMMUNICATION; accordingly Japan has built up an ANTI-SUBMARINE WARFARE (ASW) force of more than 30 DESTROYERS and 18 FRIGATES, all of indigenous design; additional ships are always under construction.

The first Japanese destroyers built after 1945 reflected U.S. influence in their hullforms, propulsion, armament, and electronics. Since the mid-1960s, however, Japanese destroyer designs have increasingly reflected the specific preferences of the MSDF for very seaworthy ships, heavily armed for their size.

The first postwar Japanese destroyers were 2 1700-ton Harukazes, commissioned in 1956; both were converted to training ships in 1981. They were followed by 7 1700-ton, Ayanami-class ASW destroyers commissioned between 1957 and 1960; 4 of these ships were converted to training vessels in 1982. Three 1800-ton Murasame class destroyers commissioned in 1959 have the same hull and machinery as the Ayanamis, but are general-purpose vessels. The 2 2300-ton Akizukis ships, commissioned in 1960, are essentially enlarged Murasames.

The MINEGUMO and YAMAGUMO classes, specialized ASW vessels designated "Hunter-Killer" destroyers (DDK) commissioned between 1966 and 1978, already reflect the "Japanization" of MSDF designs. Both types displace 2100 tons and have the same hull and machinery, but their armament differs. The four 3200-ton TAKATSUKIS, commissioned between 1967 and 1970, marked a sharp increase in the size and capability of Japanese destroyers, with SEA SPARROW short-range surface-to-air missiles (SAMs), HARPOON anti-ship missiles, and two PHALANX radar-controlled guns for anti-missile defense.

Because of its focus on the Soviet submarine threat, the MSDF built only one guided-missile destroyer (DDG) before the mid-1970s, the 3500-ton *Amatsukaze*. Essentially experimental, the ship is armed with a 40-round Mk.13 single-arm launcher for TARTAR or STANDARD SAMs. By the late-1970s the Soviet Naval Air Force (AVMF) was equipped with supersonic BACKFIRE bombers armed with anti-ship missiles. In response, the MSDF built three 3850-ton TACHIKAZE-class guided-missile destroyers optimized for ANTI-AIR WARFARE (AAW). In the early 1980s they were also fitted with Phalanx.

In a major expansion of the MSDF, the Tachikazes were followed by 12 general-purpose 3050-ton HATSUYUKI-class DDGs, the first general-purpose Japanese destroyers to carry a manned helicopter. The two newly commissioned, 4450-ton HATAKAZE-class DDGs have a landing pad on the fantail, but no hangar.

With aircraft carriers forbidden, the MSDF developed a new type of small, air-capable ship, the "helicopter destroyer" (DDH), which has a large landing deck and hangar aft to accommodate up to four large ASW helicopters, while retaining traditional destroyer armament forward. The first two 4700-ton HARUNAS, commissioned in 1973–74, were followed by the two 5200-ton SHIRANES, commissioned in 1980–81. The MSDF uses the DDHs as flagships for its four escort groups, each with one DDH, 1–2 DDGs, and several DDs or DDKs.

In 1987 Japan announced its intention to build four new 8500-ton DDGs equipped with AEGIS radar and fire control systems, and Mk.41 VERTICAL LAUNCH SYSTEMS. U.S. hesitations on the transfer of the advanced Aegis technology may delay the introduction of these ships. See also FRIGATES, JAPAN.

DESTROYERS, SOVIET UNION: Soviet destroyers are variously classified as "Fleet Minelayers" (*Eskadrenny minonosets*, EM), which are in fact general-purpose destroyers; "Large Missile Ships" (*Bol'shoy raketnyy korabl'*, BRK), guided-missile destroyers with a heavy ANTI-SURFACE WARFARE (ASUW) armament; and "Large Anti-Submarine Ships" (*Bol'shoy protivolodochnyy korabl'*,

BPK), guided-missile destroyers with heavy ANTI-SUBMARINE WARFARE (ASW) armament. Until the mid-1970s, all Soviet destroyers were characterized by high speed and a multiplicity of weapons (but with few reloads) on relatively small hulls, reflecting an emphasis on short-endurance sea-denial missions, rather than on long-endurance sea control. Since then, however, Soviet destroyers have been increasing in size as well as complexity, revealing a greater emphasis on oceanic power projection.

The first post-1945 Soviet destroyers were the 2600-ton SKORYY EMs, classic World War II–style destroyers, of which 72 were built between 1949 and 1954 (the largest destroyer program of any navy since World War II). Only a few Skoryys remain in Soviet service, but some 15 have been transferred to other navies.

The Skoryys were followed by the 2600-ton KOTLIN class, 27 ships completed between 1954 and 1958. Nine were subsequently converted to "Kotlin SAM" guided-missile destroyers (DDGs), and four incomplete Kotlin hulls were converted on the ways to Kildin-class DDGs. Eight similar hulls were built as the Krupnyy class, but were later converted to KANIN-class DDGs.

The 18 basic Kotlin EMs were the last Soviet destroyers built on classic World War II lines with heavy gun and (anti-ship) torpedo armament. Eleven Kotlins were modernized between 1960 and 1962, but are now phasing out.

The Kotlin SAM class, converted between 1966 and 1972, are slightly larger at 2700 tons, and retain the hull and machinery, but are armed with one twin-arm SA-N-1 GOA surface-to-air missile (SAM) launcher with 22 missiles; eight remain in Soviet service, and one was transferred to Poland in 1970.

The four 2800-ton KILDINS completed in 1959–60 are classified as BRKs. Originally armed with four SS-N-1 SCRUBBER anti-ship missiles, they were re-armed between 1973 and 1975 with four SS-N-2 STYX anti-ship missiles.

The eight 3700-ton Kanins were originally completed between 1960 and 1962 as Krupnyy-class BRKs, essentially enlarged Kotlins with SS-N-1 Scrubber missiles. All eight were converted to Kanin BPKs between 1968 and 1977, and rearmed with one twin SA-N-1 Goa launcher.

The subsequent 20 3750-ton KASHINS completed between 1963 and 1971 were the first gas turbine warships in the world. Originally BRKs, they were reclassified as BPKs in the early 1960s to reflect their priority role. Four more were built to a modified design (Kashin II) for the Indian navy in 1980–82. One Kashin was lost in the Black Sea in 1974, after a catastrophic fire and explosion. Six extensively modified Kashins are considered to form a separate class. Converted between 1973 and 1980, these "Modified Kashins" have a displacement of 3950 tons and carry four SS-N-2 Styx missiles as well as a landing pad for a Kamov Ka-25 HORMONE ASW helicopter, and a VARIABLE DEPTH SONAR.

In 1980–81 the Soviets introduced two fundamentally new destroyer designs: the SOVREMENYY and UDALOY classes; both much larger than the Kashins, they reflect a novel emphasis on endurance for oceanic operations.

Designated EMs, the 6200-ton Sovremenyys are specialized ASUW ships with only a minimal ASW capability. Armed with 8 advanced SS-N-22 SUNBURN anti-ship missiles, 2 single-arm launchers for the new SA-N-7 GADFLY SAM, and 4 130-mm. rapid-fire dual-purpose guns, they also have a landing deck and hangar for a Ka-25 Hormone helicopter, used mainly for OVER-THE HORIZON TARGETING (OTH-T) of the SS-N-22s.

Designated BPKs, the 6700-ton Udaloys are specialized ASW vessels similar in concept to the U.S. SPRUANCE-class destroyers. Powered by gas turbine, they have 8 vertical launchers for SA-N-9 point defense surface-to-air missiles, 8 SS-N-14 SILEX ASW missiles, as well as a landing deck and hangar for 2 Kamov Ka-27 HELIX ASW helicopters. See also CRUISERS, SOVIET UNION; FRIGATES, SOVIET UNION.

DESTROYERS, UNITED STATES:

By the end of World War II, the U.S. Navy had some 350 modern destroyers, mainly of three classes: the 2100-ton FLETCHER (181 built), the 2400-ton SUMNER (70 built), and the 2600-ton GEARING (98 built); they continued to account for the bulk of the U.S. Navy's destroyer force until the mid-1970s. After 1945, with no large and hostile surface fleet in existence, the role of destroyers remained defensive, i.e., to provide air defense and anti-submarine escort to CONVOYS and CARRIER BATTLE GROUPS (CVBGs); accordingly, many of the guns and torpedo tubes of wartime destroyers were replaced by additional ASW weapons.

By the late 1950s the Fletchers, Sumners, and Gearings, though not old, were becoming outdated; with replacement financially impossible, in 1958 the navy initiated the FRAM (Fleet Rehabilitation and Modernization) program to install new sensors and weapons (such as ASROC) and completely overhaul all machinery. The FRAM program enabled the navy to operate its wartime destroyers into the mid-1970s—ten years later than expected. More than 100 other Fletchers, Gearings, and Sumners were transferred to allied navies during the 1950s and '60s, and such was the soundness of their construction that some still remain in service.

Given the inventory of wartime destroyers, and the lack of a serious threat, the navy attempted to design an "ultimate destroyer," which would incorporate the lessons of World War II. The resulting 5000-ton Mitscher class was far too large and expensive for mass production; only 4 were commissioned in 1953–54, and all were retired by 1978. The navy next focused on the design of a more austere "mobilization" destroyer, which could form the low end of a "high-low mix" (with the Mitschers at the high end). Eighteen of these 2800-ton Forrest Shermans were commissioned between 1955 and 1959. Originally armed with rapid-fire guns, 4 were converted to guided-missile destroyers (DDGs) in 1967–68, and 8 others were converted to specialized ASW escorts; all retired in the early 1980s.

As the problem of fleet air defense became salient in the late 1950s, the navy built two groups of DDGs (by definition specialized for AAW): the COONTZ and ADAMS classes. Commissioned in 1960–61, the ten 4700-ton Coontzes represented the high end of another high-low mix. Originally designated destroyer-leaders (DLs), i.e., flagships for destroyer squadrons, they have a twin-arm TERRIER SAM launcher with 40 missiles, and were also fitted with eight HARPOON anti-ship missiles from the late 1970s.

The 3380-ton Adams class, derived from the Forrest Sherman hull, is armed with two 5-in. guns, ASROC, and a 40-round Mk.13 or 42-round Mk.22 Tartar launcher. Twenty-three were commissioned by the U.S. Navy between 1960 and 1964; three more were delivered to the Australian navy (as the PERTH class) between 1965 and 1967; and another three to the West German navy (as the LUTJENS class) between 1969 and 1970.

During the 1960s, the navy built a number of large guided-missile ships as specialized AAW escorts for aircraft carriers. Originally classified as guided-missile destroyer leaders (DLGs), they were all reclassified in 1975 as guided-missile cruisers (CGs). These ex-DLGs include the LEAHY and BELKNAP classes, and the nuclear-powered BAINBRIDGE and TRUXTON.

With the Vietnam War under way, no funds were allocated for the construction of new destroyers between 1965 and 1972, just when all the pre-1945 destroyers were reaching the end of their service lives. This "block obsolescence" caused a precipitous decline in the number of destroyers at a time when the Soviet fleet was expanding rapidly. Nevertheless, the U.S. Navy chose to build huge, 7000-ton SPRUANCE-class destroyers, inevitably in small numbers. Very large, gas-turbine ships, they were designed for weapon modularity, so that a common hull could serve for both a specialized ASW and a guided-missile destroyer. While it ordered none of the latter, the navy commissioned 31 of the ASW Spruances between 1975 and 1983, with varied armament and a landing deck and hangar for two LAMPS ASW helicopters. The Imperial Iranian Navy ordered four DDG variants in 1979–80; after the fall of the shah, they were purchased by the U.S. Navy in 1982 as the KIDD class. With two Mk.26 twin-arm missile launchers for any mix of Standard, ASROC, and Harpoon, in place of separate ASROC and Sea Sparrow launchers, they were by far the most powerful Spruances until in 1986 the navy began retrofitting its ASW versions with two Mk.41 VERTICAL LAUNCH SYSTEMS (VLSs), for up to 120 SAM, ASW, cruise, and anti-ship missiles. The TICONDEROGA-class (CG-47) AEGIS guided-missile cruisers also have Spruance hulls.

The navy is currently building the 8200-ton BURKE class (DDG-51) to replace the aging Coontz and Adams classes. Armed with two Mk.41 VLSs with a total of 90 missiles, and a helicopter deck (but no hangar), these ships are larger than many World War II heavy cruisers; they are intended as the low (!) end of a high-low mix with the Ticonderogas. The navy plans to acquire 29 of these ships, to be followed by 31 of an improved version. The lead ship was laid down in 1986 and launched in 1990. See also CRUISERS, UNITED STATES; FRIGATES, UNITED STATES.

DESTRUCTOR: U.S. term for an aerial bomb fitted with a special fuze, to convert it into a land or naval MINE. Usually derived from the Mk.80-series low-drag, general-purpose (LDGP) bombs, destructors are fitted with acoustic, magnetic, or pressure fuzes. Much used during the Vietnam War, on land and in shallow coastal waters, destructors are not as effective as purpose-built mines, but they are economical and quickly available from existing bomb inventories. The current destructors are known as QUICKSTRIKE mines.

DETERRENCE: Measures designed to narrow an opponent's freedom of choice among possible actions by raising the cost of some of them to levels thought to be unacceptable to that opponent. The term is often employed in a narrower sense to describe the dissuasion of nuclear attack by the threat of nuclear retaliation. Deterrence in the more general sense has of course always been central to human relations at all levels of organization.

To deter a nuclear attack, retaliation must be perceived as likely even after such an attack has actually taken place. Such deterrence therefore depends on the perceived likelihood of the survival of the means of retaliation, as well as on perceptions of their destructive capabilities once unleashed. ASSURED DESTRUCTION is an explicit formulation of the force requirements of nuclear deterrence from the U.S. point of view—that being (implicitly) the deterrence of *nuclear* attack upon the U.S. itself, rather than "extended" deterrence offered to allies.

Active deterrence describes a threat specifically intended to prevent a specific move on the part of an opponent; i.e., latent deterrence is the norm.

Extended deterrence applies to a particular third party or parties. It has long been asserted, for example, that any nuclear attack upon NATO territory would lead to retaliation in kind by the U.S.; in a double extension of nuclear deterrence, it is further asserted that the U.S. might retaliate with nuclear weapons even for a nonnuclear attack on NATO which could not otherwise be defeated. The limits of such "twice-extended deterrence" are set by the CREDIBILITY of the threat: will the U.S. risk nuclear use for Korea? Spain? West Germany?

Minimum deterrence, politically plausible but technically dubious, is a concept based on the recognition that even a small number of nuclear weapons can be sufficiently destructive to inflict damage deemed unacceptable by almost all opponents in almost all circumstances. This prescription for nuclear deterrence at low cost is historically associated with the advent of the first broadly secure nuclear delivery system, the Polaris submarine-launched ballistic missile. Enthusiasts argued that a few Polaris submarines would suffice to achieve strategic stability, security, and economy, so that the upkeep of other delivery systems (land-based ICBMs and bombers) was simply wasteful. Opponents argued then and later that the deployment of "excess" weapons, beyond those needed to inflict the desired damage, is necessary to assure that enough would survive an attack, after allowing for technical failures (likely to be widespread) and defensive barriers, if any. They further noted that reliance on a single system is vulnerable to a single technological breakthrough by the other side—and of course, if it faces a single system the other side has even more reason to invest in research and development to achieve just such a breakthrough.

Finite deterrence is a variant of minimum deterrence, in which a small number of weapons is deployed against a stated number of targets, as, e.g., in the early British targeting policy, whose goal was the capability to destroy four Soviet cities, including Moscow.

DEW: See DIRECTED ENERGY WEAPON.

DEW LINE: Distant Early Warning Line, a RADAR and COMMUNICATIONS system designed to provide tactical warning of penetrations of North American airspace by aircraft coming over the North Pole, i.e., from the Soviet Union. The DEW Line perimeter extends along the Arctic Circle from the Aleutian Islands, across Alaska and Canada, to the eastern coast of Greenland. Two separate radar systems were sited in 31 locations, with the FPS-19 for high-altitude surveillance, and the FPS-30 for low altitude coverage. Information from the radar sites is transmitted via a dedicated communications network to the North American Air Defense Command (NORAD) Combat Operations Center beneath Cheyenne Mountain, Colorado, where it is integrated with data from other surveillance systems.

Soviet aircraft flying circuitous routes could pass around the DEW Line; to detect them, a variety of early warning systems are meant to cover the East, West, and Gulf coasts of the U.S. The most recent additions are the OVER-THE-HORIZON BACKSCATTER radars, which extend surveillance out to a range of 1000 n.mi. from each coast.

The U.S. and Canada are now deploying a replacement for the DEW Line, the NORTH WARNING SYSTEM (NWS), which will consist of nine Long Range Radars (LRR) located at existing DEW sites, and several dozen remotely operated Short Range Radars (SRR). NWS will not extend into Greenland; radar sites will instead be built in Newfoundland to cover that approach. NWS radars are being developed under the U.S. Air Force "Pave Igloo" program. See also ADD; ADTAC; AIR DEFENSE.

D/F: Direction Finding, a navigational technique and form of collection for signals intelligence (SIGINT), in which a directional antenna determines the azimuth of a particular radio emitter, by determining the angle at which the signal strength is greatest.

When employed for navigation, D/F can determine the bearings of two known emission sources, so that the user's position can be fixed by triangulation. When used for target location, two or more D/F stations determine the azimuth of a single emitter, again enabling its position to be triangulated. Modern D/F systems have electronic scanners and digital signal processors, to quickly determine an accurate fix and overcome burst transmitters. Radio Direction Finding (RDF) is synonymous. See also ELECTRONIC WARFARE.

DIA: See DEFENSE INTELLIGENCE AGENCY.

DIRECTED ENERGY WEAPONS (DEW): Any weapon which operates by depositing energy on its target, in the form of a tightly focused beam of light, or atomic particles, or radio waves. DEWs fall into three basic categories: LASERS, PARTICLE BEAMS, and radio frequency (RF) weapons.

Lasers are high-energy beams of coherent, monochromatic light emitted between the infrared and X-ray frequency bands. Their primary effect is the heating of the irradiated area; if the heating is very rapid, shock waves are generated in the irradiated material, causing spalling and catastrophic failure.

Particle-beam weapons (PBWs) generate either charged-particle beams (CPBs) or neutral-particle beams (NPBs). The beams are streams of subatomic particles or light atoms accelerated electromagnetically at velocities close to the speed of light. CPBs consist of protons or electrons, which being electrically charged (ionized) are easily accelerated, but are also difficult to focus due to the effects of the earth's magnetic field. NPBs consist of light atoms, such as deuterium (H^2) or lithium, which are not ionized. To be accelerated, they must first be stripped of their electrons, thus giving them a positive charge. Before they are ejected from the accelerator, the electrons are reattached to the atoms, neutralizing their charge; hence they are not distorted by the earth's magnetic field. But NPBs do interact with atmospheric molecules, generating heat and dissipating the beam. Thus NBPs could be practical only outside the atmosphere.

Particle-beam weapons destroy targets from the inside out. When their high-velocity, high-energy atomic particles penetrate a material, the magnetic fields of the particles can shatter the bonds between the atoms in the target's molecules; with that, the tensile strength of the material deteriorates, leading to structural failure. The same interaction also generates heat, further weakening the structure. Additionally, as the atomic particles in the beam interact with the atoms in the target material, their energy is transferred to the latter, causing them to resonate and generate heat. This heating takes place very rapidly, creating shock waves in the material, which can lead to catastrophic failure. As a secondary effect, low-energy particles can split or join with atoms in the material, causing it to change its characteristics, with devastating effects upon semiconductor devices.

In RF weapons, radio beams would be phased and then focused on the target. In theory, if tuned to the proper wavelength, the beam would cause the target to resonate, generating heat. Not much is known in the West about RF weapons, which have been a subject of Soviet military research since the 1960s.

DEWs are rapidly emerging as the prospective weapons of the next millennium. Because they attack at or near the speed of light, fire control even over very long ranges and against high-speed targets is greatly simplified. The most discussed applications of DEW technology to date have been for BALLISTIC MISSILE DEFENSE, but lasers are now being tested as battlefield weapons in anti-personnel, ANTI-AIRCRAFT, and anti-missile roles. The Soviet navy is expected to deploy laser defenses against ANTI-SHIP MISSILES in the early 1990s. A Soviet ground-based laser is believed to have been developed as an ANTI-SATELLITE weapon.

DIRECT FIRE: The employment of guns or missiles in a direct visual or radar line of sight (LOS) between the firing unit and the target. Most infantry, ANTI-TANK, and ANTI-AIRCRAFT WEAPONS require an uninterrupted LOS, and hence are direct-fire weapons. Rocket and tube ARTILLERY, and even MORTARS, can also be used for direct fire, but usually operate without an LOS to the target, in an INDIRECT FIRE mode. Direct artillery fire is both more effective than indirect fire (because the gunners can themselves observe the fall of shot to quickly make corrections), and more costly (because operators are exposed to the enemy's own direct-fire weapons). In most cases, moreover, artillery cannot exploit its full range when employed for direct fire.

DIRECT SUPPORT: The support of one specific combat unit by another, with the latter being authorized to respond directly to the supported unit's requests for assistance; as opposed to "general support," provided for a large military formation as a whole, rather than for any one of its units. Thus an artillery unit in direct support of, say, a tank battalion responds to fire mission requests from that particular battalion, while an artillery unit in general support assists the operation as such, and can also respond to fire mission requests from any unit.

DISARMAMENT: A reduction in the personnel and/or equipment of armed forces, whether unilateral or international, total or partial. The usual response to a perceived ARMS RACE is much talk about disarmament. Especially if third parties can profitably be affected, measures of ARMS LIMITATION may be agreed to by the principal arms racers, to limit unwanted categories of weapons. ARMS CONTROL agreements have been impeded in the past by interested parties who have called for "general and complete disarmament" instead. Seriously conceived plans for general and complete disarmament usually provide for the retention of small forces for internal security purposes, as well as a multinational "world police" force. Historical evidence shows a low correlation between arms races and war, but a high correlation between de facto unilateral disarmament and war.

DISINFORMATION: From the Russian *dezinformatsiya*, a form of DECEPTION based on the dissemination of false or misleading information, by broadcast and printed PROPAGANDA, by the circulation of forged official documents, and by false defectors and double agents. Disinformation is one component of Soviet ACTIVE MEASURES.

DIVAD: Divisional Air Defense System, i.e., the M247 Sergeant York, a mobile battlefield AIR DEFENSE gun system, once intended as a replacement for the M163 VULCAN in U.S. Army divisional air defense battalions. DIVAD combined twin 40-mm. L70 BOFORS GUNS mounted in a fully traversable, power-driven armored turret installed on an M48A5 PATTON main battle tank chassis, with a very complex RADAR and FIRE CONTROL system based on the APG-66 radar of the F-16 FALCON. Escalating costs combined with persistent and publicized technical failures caused the cancellation of the entire program in 1986. In lieu of DIVAD, the U.S. Army immediately initiated the FAADS (Forward Area Air Defense System) program to select an off-the-shelf replacement for Sergeant York.

DIVISION: A ground-force formation of two or more BRIGADES or REGIMENTS plus supporting units, often subordinate in large armies to multi-divisional CORPS or ARMY echelons. In some armed forces, notably the U.S. Army and Marine Corps, divisions are very large formations (16,000–20,000 men) with much organic logistic support, which are normally the focus of operational decision making. In the Soviet and other WARSAW PACT armies, the division is smaller, and often functions as a component of larger multi-divisional formations, with limited capability for sustained, independent operations. Support is instead concentrated at the army and FRONT levels, which are the focal points of decision making.

In other armed forces, the effective operational unit is the independent brigade (or "brigade group" in British-style forces), with its own organic support. The division is then mainly an administrative headquarters, to which brigades are attached as needed.

Traditionally, divisions are classified by the predominant combat arm they contain within them, as infantry, motorized, mechanized, and armored (or "tank") divisions. INFANTRY divisions usually consist of foot infantry battalions, supported by towed ARTILLERY battalions, ANTI-TANK, RECONNAISSANCE, and ENGINEER battalions, and perhaps a tank unit as well. MOTORIZED divisions are similar, except that the infantry is equipped with trucks or lightly armored vehicles for transport to the battlefield, while it would still dismount for combat. MECHANIZED divisions are combined-arms formations of tank and mechanized infantry battalions, usually supported by self-propelled artillery, armored reconnaissance, anti-tank, AIR DEFENSE, and engineer battalions. Mechanized infantry ride in ARMORED PERSONNEL CARRIERS (APCs) or INFANTRY FIGHTING VEHICLES (IFVs), and may fight mounted in some cases. (Despite their name, Soviet MOTORIZED RIFLE divisions are actually mechanized divisions.) Ratios between mechanized and tank battalions vary between 1:1 and 3:2 in mechanized divisions. Armored (or "tank") divisions are similar to mechanized divisions, except that the ratio of mechanized to tank battalions is reversed.

Mountain, light, AIRBORNE, and AIR MOBILE (or AIR ASSAULT) divisions are in fact specialized infantry divisions, with lighter equipment tailored to their missions (Soviet airborne divisions, however, are actually light mechanized forces). CAVALRY divisions are normally light armored divisions often with some main battle tanks replaced by light tanks, APCs, IFVs, armored cars, and armored reconnaissance vehicles. Some armies also

have static, garrison, or fortress divisions, which are generally similar to infantry divisions, but without vehicles, and normally manned by lower-grade troops. Only the Soviet army has ARTILLERY divisions.

The organization of divisions has traditionally been either "triangular" or "rectangular"; i.e., they are composed of either 3 or 4 brigades or regiments, each in turn composed of either 3 or 4 maneuver battalions. Two-regiment divisions lack staying power, while those with more than 4 are too unwieldy. In the 1950s, the U.S. Army tried a 5-brigade "Pentomic" organization (with very small brigades), but returned to a 3-brigade organization in the 1960s ROAD (Reorganized Objective Army Division). Soviet divisions are rectangular, with 3 tank and 1 Motorized Rifle regiment in tank divisions, and the inverse ratio in motorized rifle divisions. The French army now has small rectangular divisions, of 4 very small regiments (each not much larger than a U.S. battalion); hence, French divisions amount to large independent brigades in U.S. terms.

The size of divisions varies widely (from 6000 to 26,000 men), partly in reflection of preferred operational methods. Armies that rely on attritional methods generally field large divisions with substantial organic logistic support to process the vast amounts of ammunition and equipment that such methods consume. Maneuver-oriented armies tend to prefer smaller and more flexible divisions with less of a logistic tail.

DIVISION 86: The new Tables of Organization and Equipment (TO&E) of U.S. armored and mechanized divisions, intended to make them more compatible with the army's maneuver-oriented AIR-LAND BATTLE operational method. Division 86 replaces the attrition-oriented ROAD organization of the 1960s. U.S. Army divisions began converting to Division 86 in 1982, and the process was completed in 1987.

Officially known as the "J-Series" TO&E, Division 86 is intended to reduce the command span of junior officers (while increasing it at the battalion level), to make the brigade the primary operational unit, and to integrate new equipment such as the M1 ABRAMS main battle tank (MBT), the M2/M3 BRADLEY fighting vehicle, and the MULTIPLE LAUNCH ROCKET SYSTEM (MLRS).

The number of tanks in tank PLATOONS was reduced from 5 to 4, reducing the number of tanks in a COMPANY from 17 to 14, while the number of companies in a tank battalion went from 3 to 4, thus increasing the number of tanks per BATTALION from 54 to 58. Similarly, each mechanized battalion has 4, rather than 3, rifle companies. Each battalion also received a SCOUT platoon equipped with M3 Bradley Cavalry Fighting Vehicles. To simplify battalions and make them more agile ("lean and mean"), logistic and maintenance support that was previously organic to them has been centralized at the division level, within a composite support battalion.

On the other hand, both mechanized and armored divisions have one maneuver battalion fewer: the mechanized division now has 5 mechanized and 5 tank battalions, while an armored division has 6 tank and 4 mechanized battalions. But because of the increased size of the battalions, there is no loss in firepower: mechanized divisions have a total of 290 MBTs and 381 Bradleys, while armored divisions have 348 MBTs and 327 Bradleys.

Divisional RECONNAISSANCE, aviation, and artillery were also reorganized. All helicopters are consolidated in a divisional "air cavalry combat BRIGADE," with a composite air and armored cavalry squadron, a general support squadron, and 2 attack helicopter squadrons; and the divisional artillery brigade now has 3 medium battalions of M109 self-propelled HOWITZERS (each with 24 M109s in 3 batteries) and 1 heavy battalion with 12 8-in. (203-mm.) M110 SP howitzers, and 9 MLRSs. Each brigade retains its own direct support battalion with 24 M109s.

In spite of the declared emphasis on agility, the personnel of U.S. Army divisions has actually increased under Division 86, to a total of 16,597 in mechanized divisions, and 16,295 in armored division (as compared to 12,695 men in Soviet Motorized Rifle divisions and 11,470 men in Soviet tank divisions).

DM: "Adamsite," a nonlethal CHEMICAL WARFARE agent meant for harassment and riot control, whose odorless yellow-green crystals are dispensed as an aerosol smoke. Doses of 2–3 milligrams per cubic m. induce headache, sneezing, coughing, nausea, and vomiting; these effects last for approximately 30 minutes.

DMZ: See DEMILITARIZED ZONE.

DOCTRINE: Officially enunciated principles meant to guide the employment of military forces under specified conditions. Unlike POLICY, doctrine does not necessarily demand uniform conduct, and it may invite flexibility under the broadest guidelines.

In the Soviet case, however, doctrine refers to the officially defined policies of the Communist party, which are supposed to be derived from "objective" Marxist-Leninist principles, and as such are not open to discussion or change (by the armed forces). What in the West is usually termed military doctrine is defined in Soviet texts as MILITARY SCIENCE. See also STRATEGY; TACTICS.

DoD: Abbreviation for the U.S. Department of Defense, the executive agency responsible for all U.S. military activities. DoD was created under the National Security Act of 1947, which began the consolidation of the War and Navy departments into a single organization. It is headed by the secretary of defense (SECDEF), a cabinet official who acts as the principal advisor to the president for military affairs, and as the executive officer of DoD responsible for the implementation of presidential policy directives. The department is organized into the following major elements: the Office of the Secretary of Defense (OSD); the JOINT CHIEFS OF STAFF (JCS); the Departments of the Army, Navy, and Air Force; the headquarters of each service, including the Marine Corps; and the UNIFIED and SPECIFIED commands of the armed forces.

OSD, the secretary's staff, is manned mostly by civilians; the SECDEF is assisted by a Deputy Secretary, the Under Secretary for Policy, the Under Secretary for Research and Engineering, the Comptroller, and nine other assistant secretaries: for Manpower and Reserve Affairs; Logistics; Communications, Command, Control, and Intelligence; International Security Affairs; International Security Policy; Program Analysis and Evaluation; Public Affairs; Health Affairs; Acquisition; and Special Operations.

The JCS are supposed to be the secretary's immediate military staff responsible for force planning and the coordination of the four services. Under the 1987 Defense Reorganization Act, the chairman of the JCS has greatly enhanced powers, and direct access to the president.

The Departments of the Army, Navy, and Air Force are responsible for the administration of their respective services, and the implementation of SECDEF policy directives.

The specified and unified commands (European Command, Pacific Command, Central Command, etc.) are responsible for operational control of U.S. military forces and for the planning of their operations. Any forces assigned to these commands are supposed to be under the control of the designated commander, who is directly responsible to the president, the SECDEF, and the JCS as directed.

DoD also establishes and controls "agencies" to deal with specific aspects of defense management, including: the NATIONAL SECURITY AGENCY (NSA), the Defense Nuclear Agency (DNA), the Defense Advanced Research Projects Agency (DARPA), the Defense Communication Agency (DCA), the DEFENSE INTELLIGENCE AGENCY (DIA), the Defense Logistics Agency (DLA), the Defense Audit Agency, the Defense Contract Audit Agency, the Defense Assistance Agency, the Defense Investigative Service (DIS), the Defense Mapping Agency (DMA), and the Defense Audio-Visual Agency.

DOG HOUSE: A very large Soviet PHASED ARRAY RADAR south of Moscow, associated with the GALOSH anti-ballistic missile system. In conjunction with the similar CAT HOUSE radar, Dog House provides 360° surveillance and tracking coverage for the Galosh system, and reportedly has sufficient resolution to discriminate between hostile REENTRY VEHICLES and PENAIDS. Tracking data is handed off to the TRY-ADD fire control and missile guidance radars of the different Galosh sites. It appears probable that both Dog House and Cat House will be replaced by a single, new 360° "Pill Box" phased array radar now under construction at Pushkino, north of Moscow. See also BALLISTIC MISSILE; BALLISTIC MISSILE DEFENSE.

DOLFIJN: A class of four Dutch diesel-electric SUBMARINES commissioned between 1960 and 1965; they are now being replaced by the ZEELEEUW class, and only three remain in service. Though they have a conventional cigar shape, the Dolfijns are unique in having three separate but interconnected pressure hulls in a "triple-bubble" configuration, with one large pressure hull above two smaller ones. The large upper hull contains crew quarters, the torpedo room, and most control facilities, while the two lower hulls contain the batteries, engines, and stores. This arrangement is strong and compact, but the two smaller hulls are cramped, making maintenance more difficult and increasing manufacturing costs, and the design was not repeated on subsequent Dutch submarines. A retractable snorkel allows diesel operation from shallow submergence, while two 168-cell storage batteries power the electric motors when fully submerged. Armament consists of eight 21-in. (533-mm.) tubes (four forward and four aft), for a variety of anti-ship and anti-submarine torpedoes.

Specifications Length: 260.4 ft. (79.4 m.). Beam: 25.8 ft. (7.87 m.). Displacement: 1494 tons surfaced/1826 tons submerged. **Powerplant:** single-shaft diesel-electric, 2 1550-hp. SEMT-Pielstick diesel generators, 2 2200-hp. electric motors. **Speed:** 14.5 kt. surfaced/17 kt. submerged. **Max. operating depth:** 985 ft. (300 m.). **Collapse depth:** 1640 ft. (500 m.). **Crew:** 67. **Sensors:** 1 bow-mounted medium-frequency sonar, 1 surface-search radar, 1 electronic signal monitoring array, 2 periscopes.

DOPPLER (EFFECT): The frequency shift in propagated electromagnetic energy (light, sound, or radio waves) caused by the relative motion of the source of the energy and of a reflecting object, i.e., the target. If two objects are converging, the Doppler effect will cause an increase in frequency; if they are moving apart, it will cause a decrease in frequency. This Doppler "shift" is widely exploited in military RADAR, SONAR, and navigation systems; a key application is in the Moving Target Indicator (MTI) of some radars. MTI suppresses ground clutter by filtering out reflections from stationary objects, thus showing only those with a significant Doppler shift. Doppler techniques are also used as ELECTRONIC COUNTER-COUNTERMEASURES (ECCM), to counter CHAFF or BARRAGE JAMMING. In SIDE-LOOKING AERIAL RADARS, SYNTHETIC APERTURE RADARS, and imaging radars, Doppler techniques are used to generate photolike images of objects. In sonars, the Doppler shift is used to determine if a target is closing or moving away, and can also be used to classify reflecting objects as either submarines, marine life, or sound propagation anomalies.

Doppler navigation systems use a nadir-pointing radar to measure the drift of an aircraft from its intended course on the basis of the Doppler shift between the aircraft and the ground. Drift data can be used for dead reckoning and to calibrate INERTIAL NAVIGATION systems.

DOSAAF: *Dobrovol'noye obshchestvo sodeystviya armii, aviatsii, i flotu* (The Volunteer Society for Cooperation with the Army, Air Force, and Fleet), a Soviet organization responsible for the paramilitary training of young men (and women) prior to their conscription into the armed forces. A DOSAAF unit must, by law, be established at every school, and although the organization is officially described as "voluntary," all schoolchildren are expected to join at age 14.

DOSAAF is especially valuable to the Soviet armed forces for the training of technical specialists because even the most ordinary technical skills, such as radio operation, automotive repair, or even driving, are relatively rare among Soviet youth. Because most conscripts serve for only 18 months, most of their technical training is supposed to be performed by DOSAAF prior to induction. About 33 percent of all conscripts receive some form of DOSAAF specialist training.

DOSAAF also operates flying and skydiving schools, which screen potential pilot trainees and parachute troops. It also has its own publishing house for technical texts, training manuals, and journals.

DOSAAF is thus a significant element of Soviet military power, but one frequently overlooked by Western analysts; most notably, capability estimates based on the time allocated for military training ignore its preinduction training, which in the West would be performed by the armed forces themselves.

DP: See DUAL PURPOSE (Weapon).

DRAGON: The M47 medium anti-tank weapon (MAW), a man-portable, wire-guided, ANTI-TANK GUIDED MISSILE (ATGM). Dragon was developed in the 1960s to replace the M67 90-mm. RECOILLESS RIFLE as the standard company-level ANTI-TANK weapon.

The Dragon system consists of the missile, its launch tube, a bipod mount, and a tracker unit. With three fixed wraparound tail fins for lift and stabilization only, the missile relies on a novel propulsion and steering technique: instead of a main motor, 60 very small "puff jet" rocket motors are distributed around the circumference of the missile body to provide both forward momentum and (by varying the number and order of their firing) steering. The 5.4-lb. (2.45-kg.) shaped-charge (HEAT) warhead is capable of penetrating some 24 in. (610 mm.) of rolled homogeneous steel armor. The missile is packed in a glassfiber canister which serves as both carrying case and launcher. The tracker unit consists of a telescopic sight, a trigger mechanism, an infrared tracking sensor, and a guidance electronics package. A bipod is attached to the muzzle of the launch tube to provide support and stability. The total weight of the missile, launcher, tracking unit, and bipod is 30.4 lb. (13.81 kg.).

In action, the operator attaches the tracker and bipod to the missile canister; when he identifies a target within range, he centers it in his sight and pulls the trigger. The missile is then ejected from the launcher by a gas generator in the tube, and the puff jets ignite in sequence only at a safe distance from the launcher, accelerating the missile towards its target. After the missile has flown a short distance downrange, the IR sensor of the tracking unit detects a flare in the base of the missile, and thus any deviation between the position of the missile and the operator's line of sight to the target. The guidance electronics then generate steering commands, which are transmitted through two thin copper wires which pay out from spools inside the missile. Steering changes are accomplished by variations in the ignition sequence of the puff jets, thus inducing asymmetrical thrust. So long as the operator keeps the target in the crosshairs, the missile should therefore hit the target.

In practice, Dragon is inferior to comparable missiles such as MILAN. First, the missile generates a powerful backblast at launch, precluding its use inside buildings or enclosed bunkers and often revealing the launcher's position. Second, the missile tends to drop after leaving the launcher, and unless the operator can compensate, the missile may fly into the ground. Third, the use of puffs jets for both steering and propulsion means that any radical maneuvers will significantly reduce maximum range. Fourth, the 1000-m. range of the Dragon brings it within the range of enemy machine guns—a key shortcoming. Finally, the warhead is too small to be effective against the frontal armor of latest MAIN BATTLE TANKS protected by CHOBHAM, SPACED, or REACTIVE ARMOR. Initial production of Dragon ended in 1980. Many have been upgraded to Dragon II standard, with an improved warhead. A new version (Dragon III) is presently in development.

Specifications (missile): Length: 27 in. (685.8 mm.). **Diameter:** 5 in. (127 mm.). **Weight launch:** 13.7 lb. (6.22 kg.). **Speed, max.:** 328 ft./sec. (100 m./sec.). **Range, max.:** 1200 yd. (1000 m.). **Range, min.:** 66 yd. (60 m.). **Operators:** Iran, Isr, Jor, Mor, Neth, N. Ye, S. Ar, Sp, Swi, Thai, US, Yug.

DRAGONFLY: The Cessna A-37 two-seat, twin-engine light ATTACK AIRCRAFT, developed in 1962 by USAF as a low cost COUNTERINSURGENCY aircraft. Cessna initially converted 41 T-37 trainers to an A-37A light attack configuration by installing more powerful engines and eight wing pylons. The A-37A entered service in May 1967, and was much used in Vietnam. The similar purpose-built A-37B entered service in May 1968. Of the of 577 built between 1967 and 1977, many were transferred to the South Vietnamese air force in 1972, and others were sold or given under the Military Assistance

Program. The U.S. and South Korea air forces have converted most of their Dragonflys to OA-37B FORWARD AIR CONTROL (FAC) aircraft.

Externally identical to the T-37, the A-37B has a tadpole-shaped fuselage with a GAU-2B/A 7.62-mm. "minigun" and a retractable AERIAL REFUELING probe in the nose. The pilot and observer/copilot sit side-by-side in a rather simple cockpit protected by ballistic nylon "flak curtains." Normally, attack missions are flown by a single pilot. The engines, mounted in the wing roots, are fitted with thrust reversers for short-field landings.

The unswept, low-mounted wing has two wing-tip hardpoints for fuel tanks. Eight additional hardpoints have a total capacity of 4100 lb. of fuel and ordnance. A variety of weapons can be carried, including 500-lb. (227-kg.) LDGP bombs, NAPALM, CLUSTER BOMBS, LAU-3 and LAU-32 rocket pods for 2.75-in. (70-mm.) FFARS, flares, and smoke bombs.

Specifications Length: 28.3 ft. (8.63 m.). **Span:** 35.9 ft. (10.93 m.). **Powerplant:** 2 General Electric J85-GE-17A turbojets, 2850 lb. (1295 kg.). of thrust each. **Fuel:** 2720 lb. (1236 kg.). **Weight, empty:** 6211 lb. (2823 kg.). **Weight, max. takeoff:** 14,000 lb. (6364 kg.). **Speed, max.:** 524 mph (875 kph) at 16,000 ft. (4878 m.). **Speed, Cruising:** 489 mph (816 kph) at 25,000 ft. (7622 m.). **Range, max.:** 460 mi. (768 km.) w/max. payload; 1012 mi. (1690 km.) w/4 100-gal. (398-lit.) drop tanks. **Operators:** Chi, Ecu, El Sal, Gua, ROK, Thai, US.

DRAGUNOV (SVD): The standard Soviet sniper rifle, also widely exported. The Soviet army has always emphasized the importance of SNIPERS to disrupt attacks and lower enemy morale. In World War II, a number of standard infantry rifles were fitted with telescopic sights, but the Dragunov SVD is the first Soviet weapon designed especially for snipers, with particular attention paid to ergonomics and high-quality workmanship. Introduced in the mid-1960s, the SVD has a semi-automatic (single-shot) modification of the Kalashnikov gas-operated action of the AK family of ASSAULT RIFLES. Fed from a 10-round box magazine, the SVD fires a full-power 7.62- × 54-mm. rimmed cartridge with a very flat trajectory.

The SVD has an ergonomically designed skeleton stock for balance, and a hair trigger. Its PSO-1 4x telescopic sight fitted over the barrel can be used as both a passive or semi-active INFRARED night sight. The barrel has a combination muzzle brake–flash suppressor, and can accommodate a standard Kalashnikov bayonet. A trained sniper has a 50 percent chance of hitting a man-sized target at 800 m., or a 20 percent chance at 1000 m.

The Soviet army selects snipers from recruits that demonstrate sharpshooting aptitude in DOSAAF preinduction training. Each Soviet Motorized Rifle platoon has one SVD-armed sniper whose primary targets are enemy scouts, officers, tank commanders, and ANTI-TANK GUIDED MISSILE crews.

Specifications **Length OA:** 48.23 in. (1.225 m.). **Length, Barrel:** 21.53 in. (547 mm.). **Weight, loaded:** 9.95 lb. (4.52 kg.). **Muzzle velocity:** 830 m./sec. **Effective range:** 1200 m.

DRAKEN: 1. The Saab 35 single-seat, single-engine, multi-role FIGHTER in service with Sweden, Austria, Denmark, and Finland. Designed from the early 1950s as a Mach 2 interceptor, the prototype first flew in 1955, and the initial J35A entered service in 1960. Produced in several successively more powerful versions, culminating with the J35F fighter/ground attack/reconnaissance aircraft, more than 550 Drakens were produced for Sweden alone. The J35X, generally similar to the J35F but with more internal fuel capacity and hardpoints for 9000 lb. of external stores, was produced for Denmark in three separate versions: the F35 fighter; TF35 trainer; and RF35 RECONNAISSANCE fighter. Finland assembled the Draken under license, bringing production to a grand total of 606 aircraft. Austria purchased a number of ex-Swedish Drakens in the early 1980s, rebuilt to J35O standard without provision for air-to-air missiles. Highly maneuverable and still effective despite its age, the Draken was the first European fighter to achieve Mach 2 (in 1960). Very popular with all four air forces, it is likely to remain in service until the end of the 1990s.

Because of the Swedish requirement for dispersed operations from short fields and stretches of highway, short takeoff performance was greatly stressed in the design. Draken has a unique double-delta configuration with the engine inlets incorporated into the wing roots, and the leading edges swept back at a sharp angle to a point nearly two-thirds back from the nose, at which point the wings flare out at a smaller angle to the tail. This planform improves takeoff performance over conventional deltas, and decreases landing runs by allowing aerodynamic braking.

The J35F's area-ruled ("coke bottle") fuselage has a pointed nose radome housing a multimode PSO-1 pulse-Doppler RADAR, supplemented by a Hughes S71N INFRARED tracking sensor in a chin fairing. The pilot sits immediately behind the radome under a clamshell canopy faired into the fuselage spine, which obstructs rearward vision. Cockpit AVIONICS include an optical gunsight (replaced in some models by a head-up display, or HUD), a head-down radar display, a TACAN navigational beacon receiver, a Saab Scania S7B FIRE CONTROL computer, a Saab BT9 bombing system, a RADAR WARNING RECEIVER, and a tactical DATA LINK to the Swedish STRIL-60 air defense network. An avionics compartment is located behind the cockpit, and the rest of the fuselage houses landing gear, fuel tanks, and the engine.

The double delta wing provides considerable volume for fuel and avionics. Slats are fitted to the outer wing leading edges, while the trailing edges have flaps and elevons. One ADEN 30-mm. cannon with 100 rounds of ammunition is mounted in the starboard wing root behind the engine inlet.

The Draken has three fuselage pylons and six wing pylons, with a total payload capacity of 9921 lb. (4500 kg.). In addition to the cannon, air-to-air armament consists of four Hughes AIM-26A semi-active radar-homing or AIM-4D infrared-homing FALCON (built under license as the Rb.27 and Rb.28), or four AIM-9 SIDEWINDER (license-built Rb.24) missiles, carried on two fuselage and two wing pylons. Air-to-ground ordnance includes 75-mm. and 135-mm. Bofors rocket pods, or up to 9000 lb. (4091 kg.) of bombs.

Specifications Length: 50.4 ft. (15.4 m.). **Span**: 30.83 ft. (9.4 m.). **Powerplant**: (F) 1 Svenska Flygmotor RM6C (license-built Rolls Royce Avon) afterburning turbojet, 17,635 lb. (8016 kg.) of thrust. **Weight, empty**: 18,180 lb. (8250 kg.). **Weight, max. Takeoff**: (F) 27,050 lb. (12,270 kg.); (X) 35,270 lb. (16,000 kg.). **Speed, max.**: Mach 2.0 (1200 mph/2125 kph) at 36,000 ft. (11,000 m.) clean/Mach 1.4 (924 mph/1487 kph) w/stores. **Initial climb**: 34,450 ft./min. (10,500 m./min.). **Service ceiling**: 65,000 ft. (20,000 m.). **Combat radius** (hi-lo-hi, 2 1000-lb. [454-kg.] bombs): 621 mi. (1037 km.). **Range, max.**: 2019 mi. (3250 km.).

2. A class of six Swedish diesel-electric attack SUBMARINES, commissioned in 1961–62. Two boats, *Draken* and *Gripen*, were stricken in 1981; the rest will be replaced by the VASTERGOTLAND (A-17) class in the early 1990s. See also SUBMARINES, SWEDEN.

DROP ZONE (DZ): An area designated for the parachute insertion of AIRBORNE FORCES and/or stores. A drop zone should be relatively flat, without large stones, fences, ditches, or pits, and with firm but not rocky ground. In addition, it should be close to the force's objective, but not under enemy observation—but such ideal drop zones are hard to find. For a BATTALION-size force, a DZ of 400 × 400 m. can be adequate, if the force can be dropped accurately. For that purpose, airborne forces have relied on elite pathfinders, precursor parachutists who are to locate the DZ and mark it with radar and/or visual devices for the pilots bringing in the main force.

DSAT: Defensive Satellite, any satellite equipped for self-defense (or the defense of other satellites) from ANTI-SATELLITE WEAPONS (ASATs), whatever its primary role may be. DSATs may also have a BALLISTIC MISSILE DEFENSE (BMD) capability. See also SATELLITES, MILITARY; SPACE, MILITARY USES OF; STRATEGIC DEFENSE INITIATIVE.

DSHK: Soviet heavy MACHINE GUN (HMG), in service with Soviet and other Warsaw Pact forces, Soviet-supplied forces, communist guerrillas, and, recently, anti-communist insurgents in Afghanistan, Angola, Mozambique, and Nicaragua. Known as familiarly as the Dashika, the DShK is a 12.7-mm. (.50-caliber) weapon with a Kalashnikov gas-operated action and a rotating breech. The original DShK38 was the standard Soviet HMG of World war II; many are still in service.

The current DShK46 (or DShKM), introduced in 1946, has a an especially simple belt-feed mechanism and a quick-change barrel. The practical rate of fire is limited by barrel heating and ammunition supply to only 80 rounds per minute. A variety of ammunition is produced for this weapon, including ball, tracer, and API; the latter can penetrate approximately 8 mm. of steel ARMOR at 500 m. Widely fitted as secondary armament on tanks and other armored vehicles, it can also be fired from a tripod (the DShKT tank version cannot be dismounted). Much used as an ANTI-AIRCRAFT weapon in the Vietnam War, it has been superseded in that role by the 14.5-mm. KPV machine gun.

Specifications Length OA: 62.5 in. (1.588 m.). **Length, barrel**: 38.1 in. (967 mm.). **Weight, loaded**: (w/o tripod) 74.8 lb. (34 kg.). **Muzzle velocity**: 830–850 m./sec. **Cyclic rate**: 540–600 rds./min. **Effective range**: 1500 m. vs. ground targets/1000 m. vs. aircraft.

DTST: Defense Technologies Study Team, a "blue ribbon" panel established by President Reagan in 1982 to investigate the technical requirements and feasibility of BALLISTIC MISSILE DEFENSES based on advanced technologies; also known as the Fletcher Commission, after its chairman, Dr. James Fletcher. As a result of its findings, President Reagan authorized the STRATEGIC DEFENSE INITIATIVE and the creation of the Strategic Defense Initiative Organization. See also SPACE, MILITARY USES OF.

DUAL CAPABLE (WEAPON): Any weapon that can fire or deliver both conventional and nuclear munitions. Most larger ARTILLERY pieces, many guided missiles, and some ANTI-SUBMARINE WARFARE weapons are dual capable. See also DELIVERY SYSTEM.

DUAL PURPOSE (WEAPON): A gun or missile capable of attacking both air and surface targets. Dual-purpose guns were initially developed as secondary armament for BATTLESHIPS and CRUISERS, but currently most warship guns are dual-purpose weapons ranging in caliber from 3 in. (76.2 mm.) to 5 in. (127 mm.) and more. Ground-based dual-purpose weapons are less common, but heavy ANTI-AIRCRAFT guns were also used as ANTI-TANK guns in World War II (notably, the famous German 88-mm.). Actually, the design requirements of anti-tank and anti-aircraft guns are diametrically opposed, because the former should be low-slung, easily concealed, and highly mobile, while the latter need a high mounting for high-angle fire, and a heavy base for stability. The primary characteristics of dual-purpose guns are a high muzzle velocity (for flat trajectory and armor penetration), a high rate of fire (to engage aircraft), elevation above 45°, and the ability to fire ARMOR PIERCING ammunition.

A number of sea- and land-based surface-to-air missiles also have a secondary surface-to-surface role, making them effective dual-purpose weapons. These include the NIKE HERCULES, STANDARD, the Oerlikon-Martin Marietta ADATS (Air Defense Anti-Tank System), and the British Aerospace SEA DART missiles. A dual-purpose missile must have a guidance system capable of detecting, locking onto, and tracking fast-moving aerial targets, without filtering out slow-moving or stationary ground targets, and also a warhead and fuze suitable for both types of target.

DUKE: A class of eight British FRIGATES, the first of which was commissioned in 1989; three more now under construction are scheduled for completion in the early 1990s. Ordered in 1984 as replacements for the LEANDER-class frigates, the Dukes were initially conceived as rather austere ANTI-SUBMARINE WARFARE (ASW) escorts, but the design was considerably enlarged to incorporate lessons from the 1982 Falklands War.

The Dukes are flush-decked, with a boxy superstructure forward and a separate deckhouse/hangar aft; gyro-controlled fin stabilizers are fitted to improve seakeeping. Armament consists of a 4.5-in. (114-mm.) DUAL PURPOSE gun at the bow, a 32-round VERTICAL LAUNCH SYSTEM (VLS) for SEA WOLF short-range surface-to-air missiles behind the gun, 2 quadruple launchers for HARPOON anti-ship missiles before the bridge, 2 30-mm. RARDEN anti-aircraft guns amidships, a GOALKEEPER 30-mm. radar-controlled gun for anti-missile defense aft, and 2 twin tubes for lightweight ASW homing TORPEDOES. An EH-101 ASW helicopter will operate from a landing deck and hangar on the fantail; a haul-down system is installed to allow safe landings in rough seas. See also FRIGATES, BRITAIN.

Specifications Length: 436.25 ft. (133 m.). Beam: 52.8 ft. (16.1 m.). Draft: 14.1 ft. (4.3 m.). Displacement: 3000 tons standard/3700 tons full load. Powerplant: twin-shaft CODLAG: 4 Paxman-Valenta diesel generators driving 2 2000-hp. electric motors (cruise), 2 18,000-shp. Rolls Royce SM1A Spey gas turbines (boost). Speed: 28 kt. (17 kt. on electric motors). Fuel: 800 tons. Range: 8000 n.mi. at 15 kt. Crew: 153. Sensors: 1 Type 1007 navigation/surface-search radar, 1 Type 996 long-range surveillance radar, 2 Type 911 Sea Wolf guidance radars, 1 GSA-8 Sea Archer electro-optical gun director, 1 Type 2050 bow-mounted low-frequency sonar, 1 Type 2031 passive towed array sonar. Electronic warfare equipment: 1 Decca UAF-1 "Cutlass" electronic signal monitoring array, 1 Type 675 active jamming array, 4 DLB chaff launchers.

DURANDAL: French runway-penetration bomb, built by Matra in cooperation with Thomson-Brandt and SAMP, designed to disable runways, taxiways, roads, and similar targets for extended periods by cratering and surface heaving.

Durandal can be carried at speeds up to 1200 mph (2004 kph), and released at speeds up to 625 mph (1044 kph) at altitudes as low as 200 ft. (61 m.). Immediately after release, a drogue parachute deploys from the tail of the bomb to stabilize and decelerate it; after a few seconds, a larger main parachute deploys to further decelerate the bomb, and rotate its nose downward to 90°. When this angle is reached, the parachutes are jettisoned, and a small rocket in the tail of the bomb ignites, so as to accelerate the weapon to a high impact velocity in order to penetrate more than 16 in. (406 mm.) of concrete. A time-delay fuze detonates the weapon after it has buried itself beneath the concrete, so that the explosion can leave a crater as much as 45 ft. (13.72 m.) in diameter. Durandal has been in service with the French air force since 1978 and has been widely exported; it was adopted by U.S. Air Force in 1986 as the BLU-107. See also COUNTERAIR.

Specifications   Length: 8.75 ft. (2.67 m.). **Diameter:** 9 in. (229 mm.). **Weight:** 429 lb. (195 kg.).

DZ: See DROP ZONE.

E

E-: U.S. designation for ELECTRONIC WARFARE aircraft, often used in combination with a second alphabetical prefix to indicate conversions from other types of aircraft, e.g., the EB-66 conversion of the B-66 light bomber. Models currently in service include: E-2 HAWKEYE; E-3 SENTRY; E-4 NEACP; E-6 HERMES; E-8A J-STARS; EA-3B SKYWARRIOR; and EA-6B PROWLER.

EAGLE: McDonnell-Douglas F-15 FIGHTER, in service with the U.S. Air Force (USAF) and the air forces of Israel, Japan, and Saudi Arabia. The F-15 was developed from 1965 as a pure AIR SUPERIORITY fighter optimized for combat with other fighters; as such, it incorporates old lessons learned anew in air combat over North Vietnam, including the value of internal guns, cockpit visibility, and agility—and a new lesson, the need for control at high angles of attack (AOA). McDonnell Douglas was awarded a development contract in 1969, flight tests began in 1972, and the first F-15As were delivered to USAF in 1974. Easily the best air-to-air fighter in the world when introduced, the F-15 was the first combat aircraft with a thrust-to-weight ratio exceeding 1 to 1 (for outstanding maneuverability and rate of climb), a head-up display (HUD), and a solid-state pulse-Doppler RADAR with LOOK-DOWN/SHOOT-DOWN capability. In 1975 it set no fewer than eight time-to-height records. Its major drawback was the cost, then of some $20–30 million.

USAF acquired 340 F-15As from 1974, as well as 47 F-15B two-seat models, of which 99 were transferred to the Air National Guard in the late 1980s. Israel acquired 51 F-15A/Bs from 1976; in combat against the Syrian air force (mostly in 1982), Israeli Eagles have destroyed some 50 MiG-21 FISHBEDS and MiG-23 FLOGGERS without loss.

In 1979, McDonnell-Douglas introduced the F-15C and the two-seat F-15D, both with increased internal fuel, provision for conformal FAST (Fuel and Sensor Tactical) packs, and (as with Israeli F-15s) a programmable radar signal processor with multiple TRACK-WHILE-SCAN capability. To date, USAF has acquired some 520 F-15C/Ds, and production will continue through the mid-1990s. From 1980, Japan acquired 100 F-15s under license; only the first 2 F-15Js (comparable to the F-15C) and 12 F-15DJ two-seaters were built by McDonnell-Douglas, with another 86 built Mitsubishi Heavy Industries. In 1982, Saudi Arabia purchased 62 F-15Cs and 15 F-15Ds (plus two spares), which were all delivered by 1984; in 1987 Saudi Arabian Eagles shot down two Iranian F-4 Phantoms. The F-15E Strike Eagle is a two-seat interdiction variant intended to replace the aging F-111 in the mid-1990s; the first was delivered in 1988, and USAF plans to buy some 300–400 by 1995.

Large for a fighter, the F-15 has a "pod-and-boom" configuration, with the cockpit and avionics in a central pod between twin booms housing the engines. The three structures are faired together to form a very broad, flat fuselage which adds considerably to lift, particularly at high AOA. Construction is quite conventional, though the rear fuselage is mainly of titanium alloy and the tailplanes are made of carbon composites. The F-15 was the first fighter stressed to withstand up to 9 G (i.e., 9 times the force of gravity).

The nose is taken up by a Hughes APG-63 pulse-Doppler radar with a maximum range of more than 90 mi. (150 km.). Optimized for aerial combat, it has search, boresight, look-up, and illumination modes, in addition to secondary air-to-ground and ground mapping modes. The F-15C has a digital, programmable signal processor with additional memory, allowing the radar to track and illuminate two targets simultaneously while still searching for others.

The pilot sits in an elevated cockpit under a long, bubble canopy which allows unexcelled 360° visibility (the need for an unobstructed view, especially to the rear, was a relearned lesson of the Vietnam War). Cockpit avionics are perhaps the most comprehensive in any fighter: a HUD, a head-down Vertical Situation Display (VSD) with computer-generated symbology, a Litton ASN-109 INERTIAL NAVIGATION unit, an IBM digital air data computer, a TACAN receiver, and (in late-model Eagles), a JTIDS (Joint Tactical Information Distribution System) data link terminal, which allows the display of information from E-3A SENTRY airborne warning and control (AWACS) aircraft. ELECTRONIC WARFARE equipment includes several Loral ALR-56 RADAR WARNING RECEIVERS linked to a Northrop ALQ-35 ELECTRONIC COUNTERMEASURES (ECM) suite with ACTIVE JAMMING transmitters and CHAFF/flare dispensers. The F-15 cockpit was the first to have a HOTAS (Hands on Throttle and Stick) configuration that allows the pilot to select all essential radar, display, and weapon modes without removing his hands from the flight controls.

The well behind the pilot's seat is almost empty, leaving space for a second seat without sacrificing internal fuel (though single-seat Eagles cannot easily be converted to two-seaters). Two-seat F-15B/Ds are fully combat capable, with performance almost identical to that of the single seaters. Behind the cockpit, the central pod houses two foam-filled fuel cells and is fitted with a large, distinctive dorsal speed brake. An arresting hook is mounted under the tail for short-field landings.

The two side booms house engine intakes and the engines, originally two Pratt and Whitney F100-PW-100 afterburning turbofans. The first of the new generation of high-power jet engines, the F100 offered an exceptional thrust-weight ratio and fuel efficiency, but had initial reliability prob-

lems. The F-15C/D can accept either uprated F100-PW-200 or General Electric F110-GE-100 engines. The F-15A/B has a power-to-weight ratio of 1.15 to 1, while the F-15C/D has a power-to-weight ratio of 1.12 to 1 with F100-PW-220 engines, or 1.24 to 1 with F110s. The wedge-shaped intakes are hinged at the top, and automatically droop at high AOA to provide a smooth flow of air to the engines. The port intake structure contains an aerial refueling receptacle, while the starboard intake houses an M61 VULCAN 6-barrel 20-mm. cannon, whose 950-round drum magazine is housed in the central pod. The F-15C/D can be fitted with "FAST" packs, conformal fuel tanks attached to the sides of the air intakes, which actually reduce drag, have several weapon hardpoints, can carry additional sensors and ECM gear, and are stressed to the same 9-g level as the airframe.

The high-mounted wing, swept at 45° with a straight trailing edge and distinctive notches cut out of the tips, is blended into the upper fuselage to form a single, uninterrupted lifting surface. Hence there are no high-lift devices, only plain inboard flaps and outboard ailerons. The wing's 608-square-ft. (56.48-m.²) area and consequent low wing loading of some 68.25 lb. per square foot (333.28 kg./m.²) ensure the F-15's agility. The slab tailplanes are mounted on the booms behind the engine nozzles and below the wings in order to maintain elevator effectiveness at high AOA. Similarly, the Eagle has tall twin vertical stabilizers, to maintain directional stability at high AOA.

Because of its high thrust-to-weight ratio, the F-15 can take off in only 900 feet with the standard air-to-air payload. Its combination of high thrust-to-weight and low wing loading also allows the F-15 to out-turn most other fighters (with the exception of the U.S. F-16 FALCON and FA-18 HORNET, and possibly the Soviet MiG-29 FULCRUM and Su-27 FLANKER).

The F-15 has one centerline pylon rated at 4500 lb. (2041 kg.) and two inboard wing pylons rated at 5100 lb. (2041 kg.) each. There are also four hardpoints below the air intakes for AIM-7F/M SPARROW or AIM-120 AMRAAM radar-guided air-to-air missiles, and two optional, 1000-lb. outboard wing pylons for ECM pods. A typical load for the air superiority mission is four Sparrows or AMRAAMs, four AIM-9 SIDEWINDER infrared-homing missiles under the inboard pylons, and a 600-gal. (2690-lit.) fuel tank on the centerline. While not ordinarily used for attack missions, F-15s can carry up to 16,000 lb. (7273 kg.) of free-fall bombs, CLUS-

TER BOMBS, LASER-GUIDED BOMBS, NAPALM, or rocket pods.

The F-15E Strike Eagle has the same airframe as the F-15D two-seater, but F110 engines are standard equipment. The avionics, however, have been drastically changed for all-weather ground attack capability. The APG-63 radar has been modified for high-resolution SYNTHETIC-APERTURE ground mapping, with an additional TERRAIN-FOL-LOWING mode. The cockpit has a wide-angle HUD, and multi-functional display screens in place of most instrument dials. The rear cockpit is configured for a WEAPON SYSTEM OPERATOR to manage the sensors and ECM equipment. LANTIRN night vision and target acquisition pods can be fitted to the forward Sparrow hardpoints to provide INFRA-RED imagery for the HUD. Internal fuel is slightly reduced to accommodate the new avionics, but FAST packs are standard. Payload has been increased to 24,000 lb. (10,909 kg.) and now includes GBU-15 and AGM-130 glide bombs, AGM-65 MAV-ERICK air-to-ground missiles, AGM-88 HARM anti-radiation missiles, and free-fall nuclear bombs, but without ordnance the F-15E's performance is similar to that of other Eagles.

Specifications **Length:** 63.8 feet (19.43 m.). **Span:** 42.8 ft. (13.05 m.). **Powerplant:** (A/B) 2 Pratt and Whitney F100-PW-100 afterburning turbofans, 23,930 lb. (10,855 kg.) of thrust each; (C/D) 2 F100-PW-220 afterburning turbofans, 25,-000 lb. (11,363 kg.) of thrust each, or 2 General Electric F110-GE-100 afterburning turbofans, 27,-600 lb. (12,545 kg.) of thrust each. **Fuel:** (A/B) 11,635 lb. (5227 kg.); (C/D/E) 13,455 lb. (2268 kg.) plus 10,000 lb. (4545 kg.) in FAST packs. **Weight, empty:** 28,000 lb. (12,700 kg.). **Weight, normal loaded:** (A/B) 41,500 lb. (18,824 kg.); (C/D) 44,500 lb. (20,185 kg.). **Weight, max. takeoff:** (A/B) 56,500 lb. (25,628 kg.); (C/D) 68,000 lb. (30,845 kg.); (E) 81,000 lb. (36,818 kg.). **Speed, max.:** 921 mph (1482 kph) at sea level/Mach 2.5 (1653 mph/2660 kph) at 36,000 ft. (10,973 m.). **Initial climb:** 50,-000+ ft./min. (15,239+ m./min.). **Service ceiling:** 65,000 ft. (19,817 m.). **Combat radius:** 600 mi. (966 km.) w/drop tank; 1500 mi. (2505 km.) w/FAST packs. **Range, max.:** 3450 mi. (5562 km.).

EARTH PENETRATOR (WARHEAD): A conventional or nuclear warhead with a reinforced and fusiform casing and a delay action fuze, which is intended to explode after penetrating earth or concrete to a considerable depth. Earth penetrators are designed to crater roads and runways, and to attack subterranean facilities such as command bunkers, storage tanks, submarine pens, aircraft shelters, etc. The U.S. Pershing 2 medium-range ballistic missile was armed with an earth-penetrator nuclear warhead specifically to attack Warsaw Pact underground command posts.

EBR: *Engin Blindé de Reconnaissance* (Armored Reconnaissance Vehicle), the French Panhard 8 × 8 ARMORED CAR, phased out of French service by the mid-1980s, but still serving with the armies of Mauritania, Morocco, Portugal, and Tunisia. Actually a wheeled TANK DESTROYER with a large-caliber, medium-velocity gun, the EBR was developed (from a 1930s design) in response to 1948 French army specification for a fast armored car with anti-tank capability. A total of 1,174 EBRs were built from 1950; they have been replaced in the French army by the AMX-10RC 6 × 6 armored car.

The EBR's all-welded steel hull has a maximum armor thickness of 40 mm. (heavy for an armored car) and is divided into a forward driving compartment, a center fighting compartment with turret, and a rear engine/driving compartment. Both drivers have a complete set of controls, allowing the EBR to back out of firefights without turning. Each driver has three observation periscopes and each controls a fixed 7.5-mm. machine gun mounted in the lower hull.

The EBR has a distinctive, two-part oscillating turret: the gun is mounted rigidly in the upper half, which elevates on trunnions set in the rotating lower half. Early models had the FL-10 turret of the AMX-13 light tank with a 75-mm. gun; the standard 90-mm. turret has power traverse with manual backup. The commander/gunner sits on the left side of the turret, with the loader to his right. Each has a separate hatch with five periscopes for 360° observation, while the commander also has a simple stadiametric gun sight. The EBR does not have night vision sights or NBC filters.

The main armament, a 90-mm. medium-velocity gun, can fire HEAT, HE, canister, and smoke ammunition beyond 1000 m., but the lack of adequate FIRE CONTROLS degrades long-range accuracy. Turret elevation limits are −10° and +15°. Forty-three rounds of 90-mm. ammunition are stored in the hull and turret. Secondary armament includes a 7.5-mm. coaxial machine gun (in addition to the hull machine guns). Twin smoke grenade launchers are attached to each side of the turret.

The EBR has a unique drive train with normal axles and pneumatic tires at each end, but with vertically elevated steel-rimmed wheels in the two middle positions. On roads, the vehicle rides on its tires as a 4 × 4 vehicle; cross-country, the middle wheels can be lowered for additional traction.

Specifications **Length:** 18.24 ft. (5.56 m.). **Width:** 7.94 ft. (2.42 m.). **Height:** 7.61 ft. (3.23 m.). **Weight, combat:** 13.5 tons. **Powerplant:** 200-hp. Panhard 12-cylinder diesel. **Fuel:** 86 gal. (380 lit.). **Speed, road:** 65 mph (105 kph). **Range, max.:** 403 mi. (650 km.).

ECCM: See ELECTRONIC COUNTER-COUNTER-MEASURES.

ECHELON: 1. Operationally, a sequence of formations placed one behind the other.

2. Organizationally, a level of structure and command (i.e., battalion, regiment, division, etc.).

In ground warfare, when attacking forces are organized into successive waves, the leading "first echelon" is usually intended to break through to allow the "second echelon" to exploit the breach; more echelons may be inserted to maintain the pressure of relatively fresh forces against a progressively weakened defender.

ECHO: NATO code name for a class of 34 Soviet nuclear-powered guided-missile SUBMARINES (SSGNs) completed between 1960 and 1967. The first Soviet SSGNs, the Echoes were built in two distinct subclasses. The 5 Echo is built between 1960 and 1962 were armed with SS-N-3C SHADDOCK land-attack missiles as make-do STRATEGIC weapons. With the growth in Soviet submarine-launched ballistic missile (SLBM) forces, the Echo Is were converted into (torpedo-armed) attack submarines (SSNs). Two were decommissioned in 1984–85. The 29 lengthened Echo IIs completed between 1962 and 1967 were armed with SS-N-3A/B ANTI-SHIP Shaddocks for attacks on U.S. carrier battle groups. One was converted to a research submarine (AGSSN), and 2 others were scrapped in the late 1980s; 15 have been modified since 1986 for the improved SS-N-12 SANDBOX anti-ship missile.

The hull, reactor, and machinery of the Echo class were all derived from the NOVEMBER-class SSNs, the first Soviet nuclear submarines. They have cigar-shaped double hulls with ballast tanks between the outer casing and the pressure hull. There is a large, streamlined sail amidships, and control surfaces are conventional: retracting bow planes, a single rudder, and stern planes mounted behind the propellers. As with the Novembers, the high length-to-beam ratio makes submerged maneuvers awkward.

In the Echo Is, 6 SS-N-3C Shaddock missiles were housed in 3 forward-pointing twin launchers flush with the upper deck until erected to a 30° angle for launch. Large cutouts in the sides served as blast deflectors, but caused much drag and turbulence (the missile tubes were removed and the cutouts plated over for the SSN conversion). The Echo IIs have 8 missiles in 4 launchers, 1 forward, 1 abreast, and 2 aft of the sail. Secondary armament in both subclasses consists of 10 21-in. (533-mm.) torpedo tubes (6 in the bow, 4 in the stern), with a basic load of 22 nuclear and conventional homing TORPEDOES.

Shaddock, a large, turbojet-powered airplane-configured missile with a launch weight of 6614 lb. (3000 kg.), can be launched only from the surface, thus exposing the launching submarine to radar detection. The SS-N-3C land attack missile relies on INERTIAL GUIDANCE alone, but the SS-N-3A/B anti-ship missile is controlled by a combination of inertial guidance, midcourse command guidance updates, and active radar homing for the terminal phase. Accordingly, the Echo IIs have a "Front Door/Front Piece" guidance RADAR in the front of the sail, but for long-range attack, Shaddock must rely on over-the-horizon targeting from Tu-95 Bear D reconnaissance aircraft, which can send a radar picture of the target zone to the submarine via a "Video Data Link" (VDL); the submarine then designates the target, and transmits its data to the missile through the Front Door/Front Piece radar. The SS-N-12 Sandbox improves on Shaddock in both speed and range, but it too can be launched only from the surface. The need to remain surfaced throughout the engagement (upwards of 15 minutes) while the Front Door/Front Piece is transmitting makes the Echo especially vulnerable to air attack.

As in the November class, the powerplant is notoriously unreliable; over the years, there have been numerous breakdowns, fires, and reactor accidents. Acoustic silencing is also poor, because of the turbulence generated by the blast deflectors, a lack of internal insulation, and the cavitation of the small, high-speed propellers. See also SUBMARINES, SOVIET UNION.

Specifications **Length:** (I) 360.83 ft. (110 m.); (II) 380 ft. (116 m.). **Beam:** 32.8 ft. (10 m.). **Displacement:** (I) 4500 tons surfaced/5500 tons submerged; (II) 5200 tons surfaced/6200 tons submerged. **Powerplant:** 2 pressurized-water reactors, 2 sets of geared steam turbines, 30,000 shp. **Speed:** 20 kt. surfaced/24 kt. submerged. **Max. operating depth:** 985 ft. (300 m.). **Collapse depth:**

1640 ft. (500 m.). **Crew:** 90–100. **Sensors:** 1 Front Door/Front Piece missile guidance radar, 1 *Feniks* bow-mounted medium-frequency passive sonar, 1 *Herkules* high-frequency active sonar, 1 "Snoop Slab" long-range air-search radar, 1 "Stop Light" electronic signal monitoring array, 1 "Quad Loop" radio direction-finder (D/F), 2 periscopes.

ECM: See ELECTRONIC COUNTERMEASURES.

ECONOMIC WARFARE: The manipulation of a foreign economy by boycotts, embargoes, and financial measures. More direct measures, notably blockades, are acts of war.

ECONOMY OF FORCE: A universal principle of war: no more than the minimum force should be allocated to cope with secondary objectives or threats, to leave the maximum possible force for the main effort. Maneuver warfare implies such risk taking also.

EF-111: Electronic warfare variant of the F-111 supersonic attack aircraft. See RAVEN.

EFFECTIVE RANGE: The range at which a weapon has a reasonable probability of both hitting and inflicting serious damage on a target—normally only a fraction of the weapon's maximum range. Thus typical tank guns have a maximum ranges of more than 10,000 m., but effective ranges (to penetrate enemy tanks) of only some 3000 m. For missiles, guidance constraints are the limiting factor.

EFP: See EXPLOSIVE-FORMED PENETRATOR.

ELECTROMAGNETIC LAUNCHER (EML): Also known as "rail gun", a experimental weapon which shoots projectiles at very high velocity by means of electromagnetic impulses instead of expansion of gases as in ordinary guns. A series of annular electromagnets arranged in line forms the "barrel" of the gun; projectiles of ferrous alloy are "loaded" by suspending them in the field of the initial electromagnet, and when all the magnets are electrified in sequence, forward momentum is imparted to the projectile. With no physical contact between projectile and "barrel," friction is virtually eliminated. Further, the impulse applied to the projectile as it moves down the barrel remains constant and does not decrease as in ordinary guns. Thus much greater acceleration and higher muzzle velocities are feasible. Small experimental rail guns have achieved muzzle velocities of 3000 m. per second (vs. 1600 m. per second for the latest tank guns); advanced designs are said to be capable of more than 10,000 m. per second. Possible applications include TANK, ANTI-TANK, and ANTI-AIRCRAFT weapons, as well as space-based BALLISTIC MISSILE DEFENSE.

The main technical barriers to the development of electromagnetic weapons are the low efficiency of ordinary electromagnets, and the need for a high-output pulse-power source. The advent of "high-temperature" superconductors could potentially rectify the first problem, and considerable resources have been devoted in the U.S. to the second, focusing on the development of high-speed, high-power capacitors.

ELECTROMAGNETIC PULSE (EMP): An intense pulse of radio-frequency energy generated by high-altitude nuclear explosions; GAMMA RAYS generated by the explosion interact with atmospheric oxygen and nitrogen molecules to "liberate" large numbers of electrons, which in turn react with the earth's magnetic field by emitting a pulse of electromagnetic radiation. Though the pulse would last only some 200 nanoseconds, its peak power might be in the range of 500 billion megawatts, enough to damage electronic devices at long range. Moreover, EMP occurs in the frequency bands (10 KHz to 100 MHz) of most military radios.

EMP can be especially damaging to solid-state electronics. High-power electrical fields can alter the molecular composition of metal oxide/silicon semiconductors, disrupting circuit functions. The more complex the circuit, the greater the number of sensitive components vulnerable to EMP. On the other hand, vacuum tubes are relatively immune to EMP.

Much effort has been devoted to EMP "hardening," usually accomplished by sealing vulnerable components in an electrically grounded box, or "Faraday cage." However, all cables and electrical connections that pass through the box must be shielded by surge suppressors to prevent EMP. Lately, self-protection microchips have been developed with integral cutout circuits to detect EMP and ground the chip before damage can occur. Gallium-arsenide chips, now in development, are considerably more resistant to EMP than the usual silicon-based chips.

EMP could be exploited as a weapon by deliberately detonating a nuclear weapon at high altitude over enemy territory to disrupt radar, communications, computers, and other vital electronics. But EMP effects are difficult to predict in advance, and the risk of escalation is obvious. See also NUCLEAR WARHEADS, EFFECT OF.

ELECTROMAGNETIC SPECTRUM: The frequency range of electromagnetic radiation from

TABLE E1

Radio-Frequency Spectrum

Designation	Frequency	Wavelength	Application
ELF	300 Hz–3 KHz	1000–100 km.	Submarine communication
VLF	3 KHz–30 KHz	100–10 km.	" "
LF	30–300 KHz	10–1 km.	Long-range communication
MF	300 KHz–3 MHz	1 km.–100 m.	" "
HF	3–30 MHz	100–10 m.	" "
VHF	30–300 MHz	10–1 m.	FM radio, TV
UHF	300 MHz–3 GHz	1 m.–10 cm.	Radar, TV
SHF	3–30 GHz	10–1 cm.	Radar, TV, satellite communication
EHF (MMW)	30–300 GHz	1 cm.–100 mm.	Radar, short-range communication

zero through infinity. Electromagnetic radiation is emitted, reflected, and absorbed by all matter, with the frequency of the radiation directly proportional to its energy state. For convenience, the spectrum is divided into "regions"; from the lowest to the highest frequencies, these are: radio-frequency (RF), INFRARED (IR), visible light, ultraviolet (UV), X-RAYS, and GAMMA RAYS.

The radio-frequency region, exploited for radio communications, radar, and navigation aids, is subdivided into frequency bands (see table E1). The infrared region, subdivided into short, medium, and long wavelength bands, is exploited in thermal sights, weapon guidance devices, and most military LASERS. Ultraviolet radiation has only recently been exploited in sensors, most notably in the STINGER-POST (Passive Optical Seeker Technology) missile and for space-based BALLISTIC MISSILE DEFENSE. X-rays and gamma rays generated by nuclear explosions could be used to destroy reentry vehicles as part of ballistic missile defenses, but X-RAY LASERS are also under development. See also ELECTRONIC WARFARE.

ELECTRONIC COUNTER-COUNTERMEASURES (ECCM): Techniques and equipment meant to neutralize the effects of ELECTRONIC COUNTERMEASURES (ECM) on radar and radio transmissions and weapon guidance. Inherent ECCM capabilities are key evaluation criteria for all electronic equipment. The simplest ECCM technique is "burnthrough," increasing the power of the transmission in order to overwhelm a jamming signal; but that is impractical for smaller, low-power radars, will not defeat sophisticated DECEPTION JAMMING, and is generally ineffective against CHAFF. These ECM techniques can, however, be countered by signal processing to filter out jamming, by DOPPLER moving target indicators (MTI) to suppress chaff clutter, and by "frequency

hopping"—rapid shifts across a broad frequency band to evade SPOT JAMMING. In addition, certain frequency bands and signal waveforms are relatively difficult to jam or spoof.

HOME-ON-JAM (HOJ) guidance is a form of active ECCM, whereby radar-guided missiles automatically home on the source of a jamming signal when the jamming becomes more powerful than the guidance signal from the missile's radar.

ELECTRONIC COUNTERMEASURES:
ELECTRONIC WARFARE equipment and techniques intended to degrade the performance of hostile electromagnetic emitters, notably radars and radio transmitters and weapon-guidance systems. ECM has been practical since the Russo-Japanese War (1904–5), when the Japanese navy jammed Russian radio-telegraph signals; but the introduction of RADAR and radio-navigation systems in World War II elevated ECM to a major form of warfare. In early moves, the British used ECM on German radio-beam navigation equipment to impede bombing, and later deployed "Window" (now "chaff") against German air defense radars. The German response inaugurated ELECTRONIC COUNTER-COUNTERMEASURES (ECCM), tactics and equipment to neutralize ECM, including the use of a wider range of operating frequencies to evade jamming, and filtering techniques to suppress chaff. By the end of World War II, the RAF had an entire bomber group dedicated to ECM, including complex deception techniques. The German introduction of radio-controlled air-to-surface missiles in 1943 led to the installation of ECM equipment on warships as well.

After 1945, ECM to protect heavy bombers against radar-directed guns and, later, surface-to-air missiles (SAMs) continued to be developed, but ECM for tactical aircraft was rather neglected until the Vietnam War, when the U.S. developed compact jamming and chaff-dispensing pods as well as specialized ECM aircraft with powerful jamming transmitters to suppress hostile radars at long ranges. Heavy Israeli air losses, both to SAMs and radar-directed guns, during the early phases of the 1973 Yom Kippur War reinforced the emphasis on ECM. Almost every combat aircraft now has internal ECM equipment and can carry jamming pods on weapon pylons, as well as chaff dispensers. These reduce the effectiveness of air-defense weapons, but the latter thereby impose "virtual attrition" by diverting payload from offensive weapons to self-defense.

A similar revolution at sea occurred after the Israeli destroyer *Eilat* was sunk by relatively simple Egyptian STYX anti-ship missiles in 1967. Surface-skimming anti-ship missiles are especially difficult to detect, evade, or intercept, and the primary defense against them is ECM. Many warships are lavishly equipped with RADAR WARNING RECEIVERS (RWRs) or more elaborate ELECTRONIC SIGNAL MONITORING (ESM) arrays to detect missiles by their own radar emissions; receivers can be linked directly to chaff launchers and jammers to blind or confuse missile guidance radars.

As new airborne target-acquisition radars and radar-guided, autonomously homing ground-to-ground and air-to-ground weapons are introduced, ECM also becoming critical in land warfare, through the jamming of tactical communications. Because of the increasing reliance of modern armies on electronic communications and remote sensing, ECM and ECCM together account for a rising percentage of military equipment costs. (Some experts argue that the ECM-ECCM cycle should be broken by reducing the reliance on the entire RF portion of the ELECTROMAGNETIC SPECTRUM, thus eliminating the problem at its source.)

ECM is conventionally classified as active or passive. The former is defined as the transmission of radio-frequency (RF) energy on enemy wavelengths to blind or deceive hostile systems, while the latter relies on obscurants, decoys, evasion tactics, and design criteria calculated to make radar detection more difficult ("STEALTH").

ACTIVE ECM

ACTIVE JAMMING is subdivided into NOISE JAMMING and DECEPTION JAMMING. Noise jamming, high-power CONTINUOUS WAVE (CW) transmissions on enemy operating frequencies (to drown the true signal in CW "noise"), includes BARRAGE JAMMING and SPOT JAMMING. In barrage jamming, CW transmissions over a broad frequency band affect all radar and radio transmissions (including friendly signals). Spot jamming is more selective: only specific frequencies are jammed to cope with a specific threat. Both barrage and spot jamming require powerful transmitters, and are relatively easy to counter by basic ECCM techniques such as coherent signal processing and "frequency agility," the rapid shifting of the emissions from one frequency to another to evade jamming. Moreover, the source of both barrage and noise jamming is relatively easy to detect, and can then be attacked by weapons with HOME-ON-JAM guidance.

Deception jamming is more subtle and more difficult to counter, requires less power, and is more difficult to detect. On the other hand, it requires much more complex and expensive equipment, including an RWR and a signal processor. For instance, repeat jammers (a common type) detect the hostile radar signal and retransmit it in a carefully modified form. Depending on the specific deception technique employed, the false signal can simulate a series of targets indistinguishable from the true target, or the wrong range or bearing for the right target. A BLIP ENHANCER is a subtype: it reradiates hostile signals with enhanced power, to make small decoys appear much larger, or any object much closer than it is, to divert the enemy's attention from his true target.

The standard ECCM response to deception jamming is the use of complex signal waveforms and frequency modulations which are difficult to replicate, making false echoes easier to identify.

PASSIVE ECM

The earliest and still predominant form of passive ECM is CHAFF, bundles of metal foil or metallized plastic strips cut to half the wavelength of enemy radar signals. This causes them to resonate when "painted" by enemy radar beams, generating a multitude of false echoes ("clutter") which mask true targets nearby. Chaff can be dispensed in small packets from aircraft or projected by rockets and mortars from ships, for self-defense against radar-guided missiles; it can also be dispensed in bulk from aircraft to form chaff "corridors" for other aircraft. Nowadays, chaff can be easily countered by Moving Target Indicators, or MTI, which suppress the echoes from slow-moving chaff using the DOPPLER effect, and by similar signal processing techniques.

Radar decoys, another form of passive ECM, simulate false targets to mask the true target. BALLISTIC MISSILE penetration aids (PENAIDS) often include erectable or inflatable decoys which replicate the radar "signature" (including the RADAR CROSS SECTION or RCS) of REENTRY VEHICLES to confuse BALLISTIC MISSILE DEFENSES. Similar decoys have also been developed for ships, and will undoubtedly be developed for ground forces in response to the introduction of the U.S. J-STARS and other airborne surveillance radars. Passive decoys are more difficult to design for aircraft, because they must have the aircraft's speed as well as "signature." The U.S. GAM-63 Quail, developed in the 1960s, could replicate the signature of a B-52 STRATOFORTRESS many times its size, but only with the aid of an on-board blip enhancer (which required a power source, etc.).

STEALTH

Another form of passive ECM is signature reduction, popularly called STEALTH, now a high-priority approach. The RCS of aircraft can be reduced more than one thousand times through designs that eliminate reflective corners, and by the application of RADAR-ABSORBANT MATERIALS. Stealth features have been incorporated in some aircraft since the Second World War, but only since the late 1970s have advances in aircraft control (e.g., FLY-BY-WIRE techniques) and materials allowed the design of stealth-optimized aircraft, notably the F-117A NIGHTHAWK and B-2 "Stealth Bomber." While virtually invisible to most current radars, such aircraft are extremely expensive. Moreover, counterstealth technologies (e.g., INFRARED detectors, LADAR, and multi-static radar) are already beginning to erode their advantage in detectability.

In contests between ECM and ECCM, the advantage generally resides with ground- or ship-based systems, whose power supplies are generally greater than those of aircraft. Of course, this is not true when the technological level of the two sides is very unequal. The EA-6B PROWLER or EF-111A RAVEN ECM aircraft could suffice to neutralize the radar-ECCM resources of most countries.

ELECTRONIC ORDER OF BATTLE (EOB):
A compilation of known or suspected enemy radar and radio transmitters, with data on locations, characteristics, and the conditions under which they could be used in battle. Mostly collected with electronic intelligence (ELINT) techniques, EOB data are especially important for the location and attack of enemy headquarters, and to route air strikes away from concentrated air defenses. See also ORDER OF BATTLE.

ELECTRONIC SIGNAL MONITORING (ESM): The passive interception, identification, and analysis of electromagnetic emissions (i.e., RADAR and radio signals) for the purposes of threat recognition and evasion, TARGET ACQUISITION, and electronic intelligence (ELINT) gathering.

Electromagnetic emissions can often be detected at considerable ranges; in the case of radar, the passive detection range greatly exceeds the effective range for the user, because of the attenuation of the returning echo. Thus radar can be a double-edged sword: while it allows surveillance and observation, especially at night or in bad weather, it can also act as a beacon for the enemy's benefit. With effective ESM equipment, electromagnetic signals can be detected; classified by frequency, waveform, pulse rate, and other characteristics; and located by bearing and signal strength. Because radar and radio signals are so easily detected, ESM is increasingly being used as a primary sensor, while radar and radio are used sparingly, to minimize the probability of detection (see EMCON).

All ESM systems are based on arrays of broadband radio-frequency receivers tuned to the wavelengths of enemy emitters. ESM equipment differs from simple RADAR WARNING RECEIVERS (RWRs) in the range of frequency coverage and the complexity of signal processing. While most RWRs are tuned to hostile acquisition and weapon-guidance radar frequencies, ESM also covers surveillance radars and communication frequencies. And while RWR displays indicate only azimuth and signal strength, ESM systems include powerful signal processors and computer libraries of known emitter characteristics, to allow the rapid classification and location of transmitters. ESM can also be integrated with ELECTRONIC COUNTERMEASURE (ECM) units, to form systems which use ESM data to automatically prioritize threats and initiate appropriate responses (e.g., NOISE JAMMING, DECEPTION JAMMING, and CHAFF dispensing).

In peacetime, ESM is used to intercept and record the emissions of potential adversaries. The primary source of ELINT, these signals can be analyzed in detail to determine their operational characteristics and limitations, develop new countermeasures, and update "signature" libraries. See also ELECTRONIC WARFARE.

ELECTRONIC SUPPORT MEASURES: The obfuscatory U.S. term (now widely used) for ELECTRONIC SIGNAL MONITORING. Also called Electronic Warfare Support Measures.

ELECTRONIC WARFARE (EW): Equipment, techniques, and mehods intended to ensure the use of the ELECTROMAGNETIC SPECTRUM for communications, surveillance, and weapon control, and deny those uses to the enemy. The first recorded application occurred in the Russo-Japanese War (1904–5), when the Japanese navy jammed Russian radio signals. Jamming was already common in World War I, but during World War II, the much more widespread use of radio and the introduction of RADAR systems greatly expanded the importance of electronic warfare (Churchill's "Wizard War"). Most armed forces are now even more heavily dependent on electronics for command, control, and communications (c^3), SURVEILLANCE, and FIRE CONTROL, while many modern weapons (e.g., air-to-air and surface-to-air missiles) employ radar and other electromagnetic sensors or command-guidance links. Hence EW has emerged as a distinct military speciality, which absorbs a steadily greater percentage of total military resources each year.

In the West, electronic warfare is conventionally subdivided into ELECTRONIC COUNTERMEASURES (ECM) and ELECTRONIC COUNTER-COUNTERMEASURES (ECCM), ELECTRONIC SIGNAL MONITORING (ESM), INFRARED COUNTERMEASURES (IRCM), and ELECTRO-OPTICAL COUNTERMEASURES (EOCM). The Soviet equivalent, "Radioelectronic Combat" (*Radioelekronnaya Bor'ba*, REB) adds to that a range of tactical and operational deceptions methods (see MASKIROVKA) and also signals intelligence (SIGINT).

ECM includes both ACTIVE JAMMING and passive techniques. The former includes NOISE JAMMING, to suppress hostile radars and radios, and DECEPTION JAMMING, intended to mislead enemy radars. Passive ECM includes the use of CHAFF to mask targets with multiple false echoes, as well as the reduction of radar signatures (see RADAR CROSS SECTION) through the use of RADAR-ABSORBANT MATERIALS and other STEALTH technologies.

ECCM includes all methods and equipment intended to defeat hostile ECM, including the use of signal processing to suppress jamming or chaff clutter, and broad-band "frequency agility," to evade jamming entirely.

Electronic signal monitoring (also called electronic support measures) focuses on the passive detection, identification, and analysis of hostile electromagnetic emissions for the purposes of threat evasion and targeting, as well as electronic intelligence (ELINT) gathering. ESM provides much of the information required to initiate ECM or ECCM responses.

IRCM, for the jamming or deception of INFRARED surveillance sensors and INFRARED HOMING missiles, includes the use of decoy flares and IR jamming beacons. EOCM at present is limited to CAMOUFLAGE and the use of smoke for masking. Future developments could include the use of sensor-blinding LASERS for both IRCM and EOCM.

ELECTRO-OPTICAL (SENSOR): A sensor that operates in the visible and short-wavelength infrared portions of the ELECTROMAGNETIC SPECTRUM, combining optics with electronic amplification, transmission, and display. Television, the most elementary electro-optical sensor, consists of a series of focusing lenses, a vidicon tube, a transmitter, and a cathode-ray display screen. Low-light television (LLTV) amplifies ambient light by several orders of magnitude; THERMAL IMAGING and IMAGING INFRARED (IIR) systems are similar, but amplify and image heat (infrared) energy instead of visible light.

Electro-optical sensors are now used for surveillance, as backups for radar tracking systems, and for weapon guidance. The magnification provided by their optics allows target acquisition at much greater ranges than with the naked eye, while LLTV and IIR sensors can overcome darkness, and the the latter can also penetrate fog and smoke.

Most TV-guided weapons operate on the principle of contrast-edge seeking. The weapon operator observes the target through the weapon's seeker on his video display screen; he then uses a cursor to lock the seeker onto a sharp contrast edge between target and background, or between two parts of the target; when released, the weapon homes autonomously on the contrast edge.

Imaging infrared guidance is similar, except that the seeker normally locks onto the centroid of the target's thermal emissions.

TV guidance works well only with good visibility targets that stand out from the background. Fog, smoke, haze, or even cloud shadows can prevent or break the lock on the target. IIR guidance can be adversely affected by precipitation, jammed by flares, confused by multiple hot spots in close proximity, or blocked by special anti-IR smoke.

For examples of specific electro-optical and IIR-guided weapons, see GBU-8 HOBOS, GBU-15, AGM-130 and AGM-62 WALLEYE glide bombs, and the AGM-65 MAVERICK air-to-ground missile. See also ELECTRO-OPTICAL COUNTERMEASURES; INFRARED COUNTERMEASURES.

ELECTRO-OPTICAL COUNTERMEA-SURES (EOCM): Equipment and techniques intended to counter ELECTRO-OPTICAL (TV and IIR) guided weapons and surveillance sensors. Passive EOCM consists mainly of camouflage and conceal-ment. TV-guided weapons can be defeated by blending their targets into the background and by eliminating the sharp contrast edges on which such weapons home; for this, netting and camouflage paints can suffice. IIR-guided weapons can be de-feated by IR-absorbant paints, reflective tarpau-lins, and the suppression of exhausts and other heat sources.

Active EOCM is now limited to the use of smoke screens, including "infrared" smoke opaque in the IR bands. In the future, radar-controlled, fast-re-action lasers may also be used to burn out the sensitive optics of TV- and IIR-guided weapons.

ELF: Extremely Low Frequency, radio signals in the 300 Hz to 2 KHz wave band, used mainly for communications between ground stations and sub-merged SUBMARINES. ELF transmissions tend to follow the curvature of the earth, are highly jam-resistant, and can penetrate seawater to consider-able depths. Messages are routinely received at 100 m. (328 ft.), and communications can be inter-preted down to 400 m. (1312 ft.) with advanced signal processing.

ELF transmissions use very little power (less than two watts in some cases), but require very large antennas: the U.S. ELF facilities at Clam Lake, Wisconsin, and K.I. Sawyer AFB, Michigan, have buried X-shaped arrays some 7 mi. (11.25 km.) across. To receive ELF messages, submarines must deploy trailing wire antennas some 1000 ft. (304.8 m.) long.

The main drawback of ELF is its extremely low data rate: some 15 minutes are needed to send a simple three-letter code group. By using com-pressed code groups, it is possible to transmit as many 17,500 three-letter messages, but the main purpose of ELF is to serve as a "bell ringer," to summon submarines up to a shallower depth where they can receive more detailed messages on higher frequency bands.

ELINT: Electronic Intelligence, a subdivision of signals intelligence (SIGINT) which covers the monitoring of enemy RADARS and other sources of ELECTROMAGNETIC-SPECTRUM emissions, in order to pinpoint their location, and determine their pur-pose and operating characteristics (range, fre-quency, sensitivity, etc.). Routinely employed to compile ELECTRONIC ORDER OF BATTLE (EOB) up-dates, ELINT does the analysis of enemy commu-nications patterns and transmitters, but the con-tents of message traffic are the province of "communications intelligence" (COMINT).

ELINT is essential for the development of ef-fective ELECTRONIC COUNTERMEASURES. In peace-time, ELINT is mainly conducted by unarmed ships and aircraft, and by "Ferret" satellites in orbit, as well as ground stations; in war conditions, receivers on combat platforms would be more im-portant. Sometimes "accidental" penetrations of their territory force adversaries to activate track-ing and weapon-guidance radars and command networks, while ELINT platforms are monitoring and recording their emissions. See also INTELLI-GENCE.

EML: See ELECTROMAGNETIC LAUNCHER.

EMCON: Emission Control, the selective use of electromagnetic and acoustic transmitters (e.g., RADARS, radios, and SONARS) to reduce the probabil-ity of detection by enemy sensors, and of interfer-ence with friendly transmissions. Essential in most deception plans, EMCON is of particular impor-tance in naval warfare, because emissions from fleet radars and communication can be detected by hostile receivers at much greater distances than their effective range.

EMP: See ELECTROMAGNETIC PULSE.

ENFIELD L85A1: A 5.56-mm. ASSAULT RIFLE, now replacing the L1A1 (a British version of the FN-FAL) as the standard infantry weapon of the British army. Also known as the Individual Weapon (IW), the L85A1 originated from the 1960s XL65E5 4.85-mm. assault rifle, which was rejected when the 5.56 × 45 mm. round was adopted as the NATO standard cartridge. Re-chambered for 5.56 mm., the L85A1 entered ser-vice in late 1985.

The L85A1's "bullpup" configuration (with the trigger ahead of the receiver) makes it very com-pact, and especially handy within confined areas (e.g., armored personnel carriers, helicopters, and buildings). It has a straight plastic forestock and pistol grip, and the backstock and butt are integral with the trigger housing. A 4× telescopic sight is normally fitted over the barrel, with fixed open sights for for backup. The muzzle is fitted with a flash suppressor which can accept a bayonet or a variety of bullet-trap rifle GRENADES. A conventional, gas-operated weapon fed from 30-round box magazines, the L85 can fire both M193 ball and the new, heavier SS109 steel-core round.

The L86A1 "Light Support Weapon" is a squad automatic weapon (SAW) variant, with a longer, heavier barrel and a bipod. It does not have a quick-change barrel, but because the L86A1 can only be fed from 30-round box magazines and not from belts (limiting the practical rate of fire), overheating from sustained automatic fire is not much of a problem.

Specifications **Length OA:** (L85A1) 30.9 in. (785 mm.); (L86A1) 35.43 in. (900 mm.). **Length, barrel:** (L85A1) 20.4 in. (518 mm.); (L86A1) 25.43 in. (645 mm.). **Weight, loaded:** (L85A1) 10.98 lb. (4.98 kg.); (L86A1) 14.48 lb. (6.58 kg.). **Muzzle velocity:** 940 m./sec. (3084 ft./sec.). **Cyclic rate:** 650–850 rds./min. **Effective range:** 400 m.

ENGINEERS, MILITARY: Troops trained and equipped to variously support ground forces by performing general construction, by creating and clearing obstacles, and by bridging water barriers. Construction engineers, bridging engineers, and "combat" or "assault" engineers (SAPPERS) are normally organized as distinct units.

Construction engineers, usually attached to higher echelons, are normally used only in rear areas to build roads, airfields, and other facilities. Bridging engineers maintain, restore, and supplement bridges along LINES OF COMMUNICATION, but tend to be more combat-oriented for opposed river crossings. Both construction and bridging engineers can play an important role in COUNTERINSURGENCY operations through civil-military activities, by building roads, schools, hospitals, housing, water pipelines and sewers, etc.

Assault engineers are trained as elite infantry in some armies; in all, they provide direct support for front-line units. As a rule, one company of assault engineers is either organic or attached to each regiment or brigade, while each division usually has an organic battalion. They lay and clear minefields, supervise the construction of defensive positions and field fortifications, and erect or clear roadblocks and other obstacles. Equipped with FLAMETHROWERS and DEMOLITION charges, assault engineers are especially important in URBAN WARFARE, and for attacks on fortified positions.

Engineer-specific equipment includes combat engineer vehicles (CEVs), tank chassis fitted with a dozer blade, a winch, and a short-barreled demolitions howitzer; mine-clearing tanks; bridge-laying tanks with girder or scissors bridges; portable ferries and pontoon bridges; prefabricated girder bridges and road matting; trench-digging and mine-laying vehicles; and assault boats. See also MINES, LAND.

ENHANCED RADIATION (WEAPON): A thermonuclear device (popularly "neutron bomb"), designed to yield up to 80 percent of its energy in the form of high-energy neutrons, as opposed to normal nuclear weapons, whose yield is mainly in the form of blast and heat. Developed from the 1960s, enhanced-radiation (ER) warheads are "fission-fusion" devices in which a low-yield (subkiloton) plutonium (fission) bomb triggers a fusion reaction within a deuterium-tritium jacket. This fusion reaction liberates up to six times as many neutrons as an equivalent pure fission or fission-fusion-fission device, and these have an energy level of some 14 million electron-volts. That allows neutrons to pass through a considerable thickness of metal or concrete; earth and water, on the other hand, slow and absorb neutrons. If they pass through human tissue, the heavy neutrons cause massive cell disruption, and thus incapacitation or death from radiation "poisoning" in large doses.

ER warheads are essentially anti-personnel weapons. In normal nuclear weapons, the blast radius is much greater than the radius of lethal radiation. Blast can be quite effective against exposed troops and "soft" vehicles, but its effective radius against armored vehicles and their crews extends only slightly beyond the radius of the weapon's fireball. Because the blast radius only increases as the cube root of the yield, relatively large warheads (in the 5- to 10-KT range) are deemed necessary to achieve significant results against armored forces.

Because high-energy neutrons can easily penetrate armor to incapacitate the crews of tanks and armored personnel carriers, a 1-kT ER warhead can deliver an instantaneous lethal dose of 8000 rads (the standard measurement of radiation) against these targets over a radius of 690 m., while the blast radius would be limited to only 550 m. In contrast, a normal 10-kT nuclear warhead would have the same lethal radiation radius, but with a blast radius of 1220 m. Thus ER weapons could be used in much closer proximity to friendly forces (who, presumably, would be warned to take cover) and built-up areas, a critical factor if the fighting is on friendly territory.

By the late 1970s, the U.S. had developed ER warheads for the LANCE battlefield missile and the M110 203-mm. howitzer. Both weapons were well suited to counter Soviet armored forces on the NATO Central Front, but the proposed deployment of the new warheads caused widespread protests in Europe, partly because of a misunderstanding of their purpose ("They kill people but leave property intact") which was much exploited by Soviet propaganda and other ACTIVE MEASURES. As a result, the U.S. did not deploy ER warheads to Europe; but they are stockpiled in the U.S., and could be shipped to Europe in an emergency. France has also developed an ER weapon, and the U.S.S.R. is assumed to have a similar capability. See also NUCLEAR WARHEADS, Effects of; NUCLEAR WEAPONS.

ENTERPRISE: A U.S. nuclear-powered AIRCRAFT CARRIER (CVN-65), commissioned in 1961. The second nuclear-powered surface warship (after the cruiser LONG BEACH), and the largest U.S. carrier when completed, the *Enterprise* was built to a modified KITTY HAWK design, but with virtually unlimited range, and much more aviation fuel stowage (9520 tons vs. 6650 tons, enough for 12 days of fairly intensive air operations). From 1979 through 1981, *Enterprise* was overhauled and extensively modernized with new radar, electronics, and armament. A 36-month Service Life Extension Program (SLEP) overhaul is scheduled for 1991.

The armored flight deck is 1101.25 ft. (335.75 m.) long and 257.16 ft. (76.4 m.) wide, with an angled landing deck cantilevered 10.5° to port and a boxy, rectangular island superstructure to starboard. There are four steam catapults, two on the angled deck and two at the bow, and four deck-edge elevators, two forward and one aft of the island to starboard, and one on the port quarter. The hangar deck is 860 ft. (262.2 m.) long, 107 ft. (32.62 m.) wide, and 25 ft. (7.62 m.) high.

As completed, *Enterprise* had no defensive armament, though a TERRIER long-range surface-to-air missile (SAM) system had been planned. In 1967, 2 8-round POINT DEFENSE MISSILE SYSTEMS with SEA SPARROW short-range SAMs were installed on sponsons aft. During the 1979–81 overhaul, these were replaced by 2 lightweight NATO Sea Sparrow launchers, and 3 PHALANX 20-mm. radar-controlled guns were added for anti-missile defense. A third Sea Sparrow launcher is to be added during the 1991 SLEP.

At present, the air wing consists of two fighter squadrons with a total of 20 F-14 TOMCATS, 2 fighter-attack squadrons with 20 FA-18 HORNETS, 2 medium attack squadrons with 20 A-6E and KA-6A INTRUDERS, an anti-submarine patrol squadron with 10 S-3A/B VIKINGS, and a composite squadron with 5 EA-6B PROWLER electronic warfare aircraft, 5 E-2C HAWKEYE airborne early warning aircraft, and 6 SH-3H SEA KING anti-submarine helicopters.

When first commissioned, *Enterprise* had unique SPS-32 and SPS-33 fixed-array "billboard" RADARS, also fitted on the *Long Beach*. Difficult to maintain, they were replaced during the 1979–81 overhaul by more conventional radars. For flagship duties, *Enterprise* has an NTDS data link and several satellite communications terminals. See also AIRCRAFT CARRIERS, UNITED STATES.

Specifications Length: 1040 ft. (317.2 m.). **Beam:** 133 ft. (40.5 m.). **Draft:** 39 ft. (11.9 m.). **Displacement:** 75,700 tons standard/90,970 tons full load. **Powerplant:** 8 Westinghouse A2W pressurized-water reactors, 4 sets of geared turbines, 280,000 shp. **Speed:** 33+ kt. **Crew:** 3319 + 2625 air wing = 5944. **Sensors:** 1 SPS-48C 3-dimensional and SPS-49 2-dimensional air-search radar, 1 SPS-10F surface-search radar, 1 SPS-65 threat-warning radar, 1 SPS-58 low-level air-search radar, 2 Mk.91 Sea Sparrow guidance radars. **Electronic warfare equipment:** 1 WLR-1 and WLR-11 radar warning receiver, 1 SLQ-29 active jamming unit, 4 Mk.36 SRBOC chaff launchers.

EOB: See ELECTRONIC ORDER OF BATTLE.

EOCM: See ELECTRO-OPTICAL COUNTERMEASURES.

EOD: See EXPLOSIVE ORDNANCE DISPOSAL.

ER: See ENHANCED RADIATION WEAPON.

ERAM: Extended-Range Anti-Armor Mine, an "intelligent" submunition under development for the U.S. Air Force. See SKEET.

ERC: *Engin de Reconnaissance Canon* (Cannon-Armed Reconnaissance Vehicle), a French 6 × 6 amphibious ARMORED CAR in service with the armies of Argentina, Chad, France, Gabon, the Ivory Coast, and Mexico. Developed from 1975 as a private venture by Panhard, the ERC shares many automotive components with the company's VCR armored personnel carrier. It entered production in 1979, and was adopted by the French army as a lower-cost supplement to the AMX-10RC.

The ECR's all-welded steel hull with a maximum armor thickness of 10 mm. (adequate protection against small arms) is divided into a forward crew compartment and a rear engine compartment. Most versions have a crew of three. The driver sits behind the glacis plate and has an armored vision slit, three observation periscopes, and a door on the left. Controls include automatic transmission and power steering to reduce driver fatigue.

The ERC can be fitted with a variety of turrets. The standard French army version, the ERC-90 F4 Sagaie, has a two-man GIAT TS-90 gun turret in which the commander (on the left) has six periscopes for 360° observation, while the gunner (on the right) has four periscopes and 5.9 × stadiametric gunsight. Passive night-vision sights, an NBC filter system, and air conditioning can be added.

Main armament is a GIAT long-barrel 90-mm. gun with HE, HEAT, smoke, canister, and APFSDS rounds. The latter round has a muzzle velocity of 1350 m./sec. (4430 ft./sec.) and can penetrate up to 240 mm. of homogeneous steel armor at 1000 m., making the ERC an effective TANK DESTROYER against all but the latest tanks. The turret has 360° power traverse, and elevation limits of −8° and +15°. Twenty rounds of 90-mm. ammunition are stowed in the hull and turret. Secondary armament consists of a coaxial 7.62-mm. machine gun and a second, pintle-mounted machine gun on the turret roof. A pair of smoke grenade launchers are mounted on each side of the turret.

Fully amphibious, the ERC can be propelled through the water by tire action at 2.8 mph (4.5 kph), or by a pair of optional water-jets at 5.9 mph (9.5 kph). In an unusual feature, the center pair of wheels can be disengaged and raised off the ground for improved road mileage when their additional traction is unnecessary.

Variants include the ERC-90 F1 Lynx, with the Hispano-Suiza turret and shorter 90-mm. gun of the AML armored car; the ERV-60-20, with a 60-mm. breech-loading MORTAR, a 20-mm. CANNON, and a 7.72-mm. machine gun; the ERC-60-12, with a 60-mm. mortar and a 12.7-mm. machine gun; and several anti-aircraft versions with twin 20-mm. cannons.

In 1985, Panhard introduced the enlarged ERC-90 F4 Sagaie II. Now in service with Gabon, it has the same 90-mm. gun, but its enlarged TTB-1900 turret has more armor protection, provisions for advanced FIRE CONTROLS (including a LASER rangefinder), and stowage for 32 rounds of 90-mm. ammunition.

Specifications **Length:** (Sag) 16.67 ft. (5.083 m.); (II) 19.25 ft. (5.87 m.). **Width:** (Sag) 8.16 ft. (2.49 m.); (II) 8.83 ft. (2.7 m.). **Height:** (Sag) 6.75 ft. (2.07 m.); (II) 7.6 ft. (2.3 m.). **Weight, combat:** (Sag) 7.65 tons; (II) 10 tons. **Powerplant:** (Sag)

153-hp. Peugeot V-6 diesel; (II) 2 93-hp. Peugeot XD3T 4-cylinder diesels. **Fuel:** (Sag) 55 gal. (242 lit.). **Speed, road:** (Sag) 56 mph (90 kph); (II) 62 mph (100 kph). **Range, max.:** (Sag) 500 mi. (800 km.); (II) 373 mi. (600 km.).

ERIS: Endoatmospheric Reentry Intercept System, a long-range, ground-based ballistic missile interceptor under development by the U.S. Army Strategic Defense Command as part of the STRATEGIC DEFENSE INITIATIVE (SDI). As currently planned, ERIS would be launched from underground silos to intercept incoming reentry vehicles at the end of their free-fall (midcourse) phase, when they begin the reentry phase of their trajectories. A two-stage, solid-fuel missile, ERIS would exploit miniaturization and advanced propulsion technologies to hold down size, weight, and costs. Slant range is estimated to be more than 450 mi. (751 km.), with an engagement envelope of 250,-000–450,000 ft. (76,220–137195 m.). Unlike earlier ANTI-BALLISTIC MISSILE interceptors such as the U.S. SPARTAN or Soviet GALOSH, ERIS is nonnuclear. Its warhead, a miniature kill vehicle the size of a gallon paint can, would use long wavelength INFRARED (LWIR) and ultraviolet (UV) sensors and lateral "puff-jet" thrusters to obtain a direct hit on the target, which would be destroyed by the kinetic energy of the collision.

ERIS could form part of a "layered defense," together with SPACE-BASED INTERCEPTORS to attack ICBMs in their boost and midcourse phases, and ground-based High Endoatmospheric Defense Interceptors (HEDIS) to attack surviving RVs inside the atmosphere. See also BALLISTIC MISSILE DEFENSE.

ESCALATION: A deliberate increase in the scope or intensity of armed conflict. Its purpose is to bring the fighting to a higher level wherein the escalating party believes it has superiority. What is seen as escalation may simply reflect the consequences of structural differences between the forces on each side. If, e.g., an advanced military power is engaged in a REVOLUTIONARY WAR of GUERRILLA tactics and SUBVERSION, what may be intended as a matching response could be seen as escalation (e.g., when bombing or defoliation are used in response to a campaign of assassination or sabotage). Thus the term is meaningful only if there are clearly recognized THRESHOLDS between various kinds of military action, and if those thresholds are available options for all parties to the conflict.

ESCORT JAMMING: An ELECTRONIC COUNTERMEASURES tactic whereby an ELECTRONIC WARFARE aircraft accompanies strike aircraft to the target, protecting them from radar-directed air defenses with noise and/or deception jamming. The only aircraft specialized for the purpose are the U.S. Navy's EA-6B PROWLER and the U.S. Air Force's EF-111A RAVEN. Because of the high cost of such aircraft, other air forces rely on standard fighters with jamming pods, or on transports equipped for standoff jamming, or on very low-level penetration to evade rather than suppress air defenses.

ESM: See ELECTRONIC SIGNAL MONITORING (a.k.a. electronic support measures).

ESPIONAGE: A covert method for the collection of human intelligence (HUMINT), wherein the sources are either intelligence officers or other nationals operating under cover in a foreign country, or coopted foreign nationals. Espionage provides a very small fraction of the total intelligence data collected by the major powers, but unlike "technical" intelligence collection by aircraft or satellites, it can reveal low-contrast human activities (e.g., by terrorists) and can also provide critical insights into the motives and intentions of potential enemies (in which role it competes with overt diplomatic monitoring). See also INTELLIGENCE.

ETENDARD AND SUPER ETENDARD: Carrier-based ATTACK and RECONNAISSANCE aircraft built by Dassault-Breguet for the French *Aeronavale* and for export.

Originally developed in response to a 1955 NATO requirement for a land-based, light attack aircraft, the Etendard was based on Dassault's successful Mystere design, retaining its same wings and tailplane on a revised forward fuselage. In 1956, after NATO selected the Fiat G.91 instead, Dassault redesigned the aircraft as the Etendard IV for carrier operations. The prototype flew in 1958, and the first of 69 Etendard IVM attack aircraft were delivered in 1962. In 1960, Dassault introduced the Etendard IVP for photo-reconnaissance, 21 of which were delivered through 1965. A competent if undistinguished aircraft, the Etendard IV served aboard the carriers CLEMENCEAU and *Foch* until the late 1970s. A few still serve as land-based trainers.

In the mid-1960s, the *Aeronavale* issued a requirement for an Etendard replacement. Candidates included a version of the McDonnell-Douglas A-4 SKYHAWK, the LTV A-7 CORSAIR, a naval variant of the SEPECAT JAGUAR, and even a naval-ized MIRAGE F.1. For several reasons, some political, the French navy eventually chose an upgraded and modernized "Super Etendard." Heavier and slightly faster, with more power, better fuel efficiency, and greatly improved avionics, it also retains roughly 90 percent commonality with the Etendard IV, which greatly reduced its development costs. The prototype flew in 1974, and the first production aircraft entered service in 1977. Originally, 100 Super Etendards were ordered by the *Aeronavale*, but the number was cut back to 71 by inflation. That forced the retention of one squadron of Etendard IVPs in the reconnaissance role. In 1981, the Argentine navy ordered 14 Super Etendards, 5 of which saw combat in the 1982 Falklands War. EXOCET AM.39 anti-ship missiles launched from Super Etendards were responsible for the sinking of the destroyer HMS *Sheffield* and the merchant ship *Atlantic Conveyor*. In 1983, Iraq leased 5 Super Etendards which were much used in the Iran-Iraq War for anti-shipping missions. One was lost in action, and the 4 survivors were returned to France in 1988. Despite its combat successes, the Super Etendard is a mediocre performer, even when compared to older attack aircraft such as the Skyhawk or Corsair.

The Super Etendard has a Thomson-CSF Agave multi-mode RADAR and a retractable AERIAL REFUELING probe in the nose. As in other Dassault designs of its generation, the canopy is faired into the fuselage, obstructing rear-quarter visibility. AVIONICS are quite comprehensive, and include a Thomson-CSF VE120 head-up display (HUD), a SAGEM-Kearfott ETNA navigation/attack system linked to an SK602 INERTIAL NAVIGATION unit and a Crouzet 97 navigational display, TACAN and VOR receivers, a radar altimeter, and several Thomson-CSF BF RADAR WARNING RECEIVERS.

Immediately behind the cockpit is an avionics bay for navigational equipment, below which are two 30-mm. DEFA 553 cannons with 125 rounds each. The rear fuselage houses two bag-type fuel cells and the engine, which is fed by simple cheek inlets behind the cockpit. Like all carrier aircraft, the Super Etendard has an arresting hook under the tail, but it also carries a braking parachute for landings ashore. Swept at 45°, the low-mounted wing has full-span leading-edge slats, double-slotted Fowler flaps, outboard ailerons, and spoilers, which combine to provide the low-speed handling needed for arrested carrier landings. The outboard wing sections fold upward for compact deck parking. The slab tailplane is mounted halfway up the vertical stabilizer, a Dassault design trademark.

The Super Etendard has two side-by-side fuselage pylons rated at 551 lb. (250 kg.) each, and four wing pylons rated at 822 lb. (400 kg.) each. The inboard wing pylons are plumbed for 242-gal. (1100-lit.) drop tanks, and a "buddy store" can be mounted under the fuselage, to allow one aircraft to refuel another. Typical payloads include one Exocet AM.39 and one drop tank, or one AN.52 free-fall nuclear bomb and one drop tank, or two drop tanks and two Matra R.550 MAGIC air-to-air missiles, or up to 4630 lb. of free-fall bombs, CLUSTER BOMBS, LASER-GUIDED BOMBS, NAPALM, and rocket pods. From 1988, 53 Super Etendards are being equipped to launch the ASMP nuclear stand-off missile.

Specifications (Super Etendard) Length: 46.98 ft. (14.31 m.). **Span:** 31.48 ft. (9.5 m.). **Powerplant:** 1 SNECMA Atar 8K-50 turbojet, 11,265 lb. (5110 kg.) of thrust. **Fuel:** 5085 lb. (2311 kg.). **Weight, empty:** 14,330 lb. (6500 kg.). **Weight, max. takeoff:** 26,455 lb. (12,000 kg.). **Speed, max.:** 745 mph (1200 kph) at sea level/Mach 1 (650 mph/1085 kph) at 36,000 ft. (10,975 m.). **Initial climb:** 24,600 ft./min. (7500 m./min.). **Service ceiling:** 45,000 ft. (13,715 m.). **Combat radius:** (hi-lo-hi, 1 Exocet, 1 drop tank) 930 mi. (1553 km.). **Range, max.:** 930 mi. (1553 km.).

EW: See ELECTRONIC WARFARE.

EXOCET: Ship-, air-, submarine-, and ground-launched ANTI-SHIP MISSILE developed by Aerospatiale from the late 1960s in response to a French navy requirement. Flight tests began in 1972, and the first Exocets entered French service in 1974. Since then, more than 2500 missiles have been produced for some 30 countries.

Exocet has been produced in several versions. The original MM.38, for shipboard launch, has an ogival nose housing an ADAC active RADAR seeker, a cylindrical body, fixed midmounted cruciform wings, and four smaller tail fins for steering. Armed with a 364-lb. (165-kg.) high-explosive warhead fitted with delay and PROXIMITY FUZES, and powered by a dual-impulse (booster/sustainer) solid-fuel rocket motor, it is launched from a rectangular container, and has been fitted to ships displacing only 250 tons. The AM.38 air-launched derivative, introduced in 1973, is identical to the MM.38, but a one-second delay is provided by the engine ignition circuits to allow for safe separation from the aircraft. It was superseded in 1979 by the

AM.39, a shortened, lighter version more suitable for helicopter launch. The MM.40 is an enlarged, long-range ship-launched variant with improved guidance electronics and folding wings for launch from a compact, cylindrical canister, allowing more to be carried despite its greater weight. The submarine-launched SM.39, externally similar to the AM.39, is enclosed in a watertight capsule some 19 ft. (5.8 m.) long and 21 in. (533 mm.) in diameter, so that it can be launched from standard torpedo tubes at depths down to 394 ft. (120 m.). After the capsule rises to the surface and breaks open, the missile's engine ignites to begin flight. Several ground-launched coastal defense versions are in service, with up to four launch canisters mounted on the back of a flatbed truck or trailer.

All versions are controlled by INERTIAL GUIDANCE through midcourse, with active radar homing in the terminal phase. In action, the target's range and bearing are fed from the launch platform's FIRE CONTROLS to the missile's guidance computer. After launch, the missile flies an inertially guided flight path to the target area, relying on a radar altimeter to keep it at a low altitude. When approximately 6.2 mi. (10 km.) from the last known target position, the radar seeker is activated and begins searching (the MM.40 radar has a wider scanning angle, to compensate for larger target position changes over its longer range). If the seeker acquires and locks onto the target, the missile descends to one of three preset surface-skimming heights (the lowest only 8 ft./2.5 m. above the waves), homing autonomously until impact.

Exocet has been much used in combat. In the 1982 Falklands War, Argentine MM.39s struck and sank the British destroyer HMS *Sheffield* and the large cargo ship *Atlantic Conveyor*, while the destroyer HMS *Glamorgan* was damaged by an MM.38 fired from a jury-rigged ground launcher. In the Iran-Iraq War, more than 100 Exocets were launched by Iraqi Super ETENDARD attack aircraft and SUPER FRELON helicopters, damaging many civilian tankers and the guided missile frigate USS *Stark*. In combat, Exocet has proven only moderately effective, with a hit probability of some 33 percent (mostly against rather unsophisticated targets). Because of its widespread use, the characteristics of its radar seeker are well known, and effective electronic countermeasures have been developed. Moreover, its warhead is too small to sink larger ships, and most of the damage inflicted on *Sheffield* and *Stark* was caused by the combustion of unspent rocket fuel. Despite this, Exocet is commercially successful because of its relatively low price and because it can arm almost any ship or aircraft.

Specifications **Length:** (MM.38) 17.1 ft. (5.21 m.); (AM.39) 15 ft. (4.69 m.); (MM.40) 18.98 ft. (5.78 m.). **Diameter:** 13.75 in. (350 mm.). **Span:** 39.5 in. (1004 mm.). **Weight, launch:** (MM.38) 1635 lb. (750 kg.); (AM.39) 1473 lb. (652 kg.); (MM.40) 1874 lb. (852 kg.). **Speed, max.:** Mach 0.93 (709 mph/1183 kph). **Range, max.:** (MM.38) 26 mi. (42 km.); (AM.39) 31–43.5 mi. (51.73–72.65 km.); (MM.40) 40.4 mi. (65 km.). **Operators:** Ab Dh, Arg, Bah, Bel, Bra, Bru, Cam, Chi, Col, Ecu, Egy, FRG, Gre, Indo, Iraq, Ku, Mali, Mor, Nig, Oman, Pak, Peru, Phi, Qat, ROK, Sing, Thai, Tur, UK.

EXPENDABLE JAMMER: A cheap, disposable transmitter launched from ships or aircraft to jam enemy RADAR and radar-guided missiles. Sized to fit into a dispenser pod or a CHAFF launcher, an expendable jammer usually consists of a cylindrical body housing a squib battery, a transmitter, an antenna, and a parachute to slow its rate of descent. Of limited power, they are effective only at short range, and can cover only a limited, preselected frequency band. See also ACTIVE JAMMING; ELECTRONIC COUNTERMEASURES.

EXPLOSIVE-FORMED PENETRATOR: Alternative term for SELF-FORGING FRAGMENT warheads.

EXPLOSIVE ORDNANCE DISPOSAL (EOD): The detection, identification, disarming, removal, and final disposal of unexploded explosive devices ("duds" and delayed-action weapons), and the "demilitarization" and disposal of explosive devices which have become unstable through damage or deterioration. EOD is performed by personnel trained and equipped to locate, identify, analyze, and disarm both friendly and hostile ordnance; the latter are often fitted with anti-tamper fuzes specifically to hamper EOD. See also EXPLOSIVES, MILITARY.

EXPLOSIVES, MILITARY: Compounds that undergo rapid chemical reactions when ignited, forming high-pressure gases and releasing energy in the form of heat. Military explosives are generally classified as either "low" or "high," depending upon their rate of combustion. Low explosives burn by oxidation at well below the speed of sound, producing gases at a stable and manageable rate; they are used mainly as propellants for gun cartridges and solid-fuel rockets.

High explosives burn at hypersonic speeds (on the order of 2000–8500 m. per second), so that the released gases form shock waves which shear and shatter nearby objects. Oxidation has only a minor role in high explosive reactions; most of the energy is released by the breaking of molecular bonds between the elements of the original compound, and their recombination into simpler compounds.

Most modern military explosives are nitrate-based compounds mixed with inert "moderators" (to improve their stability and reduce their sensitivity to heat and shock); most therefore require a strong shock or detonation to initiate the explosive reaction. In propellants, this is supplied by a "primer," which generates intense heat to initiate combustion, while high explosives require a "detonator" (e.g., a blasting cap) to produce the shock wave needed to start their chemical decomposition. The ignition of large propellant charges and insensitive high explosives may even require an intermediate charge, called an "ignition charge" for the former and a "booster charge" for the latter.

The purpose of propellants is to generate gases under pressure which can be used for thrust, but this pressure must be controlled to avoid bursting the gun barrel or rocket casing in which it is contained. The earliest known propellant is "black powder" (gunpowder), a mechanical mixture (not a chemical compound) of charcoal, sulfur, and potassium nitrate (saltpeter). Introduced to Europe in the late 13th century, it is now used mostly for ignition charges, having been superseded as a propellant by nitrocellulose-based compounds such as cordite and other "smokeless" powders. Nitrocellulose is composed mainly of nitrated cotton ("pyrocotton"), a high explosive which must be moderated with ether and alcohol. There are three basic types in use today: single-base, double-base and multi-base. Single-base powders are almost pure nitrocellulose, with some inert ingredients to facilitate their milling into grains (to control the rate of combustion), while double-base and multi-base powders consist of nitrocellulose mixed with nitroglycerin to increase their power. Single-base powders are relatively cool-burning and cause less barrel wear than double-base powder. Modern multi-base powders contain cool-burning explosives such as nitroguanadine, to combine the power of double bases with the low barrel wear of single bases. Most modern military guns use single- or multi-base powder, while double-base powders are used mainly in rockets.

High explosives, used in artillery shells, bombs, missile warheads, and demolition charges, must be relatively insensitive to prevent premature detonation, must have high shearing power ("brissance"), and require stability for long shelf life and resistance to heat and humidity. The following is a list of the principal explosives now in service.

TNT (trinitrotoluene), the most common military high explosive, is manufactured by treating organic toluene with nitric acid. At temperatures below 170° F, TNT is a stable, white crystalline substance with a combustion rate of some 7000 m. per second. Though normally loaded into warheads or cast into sticks while molten, TNT is granulated for use in booster charges. TNT is also the primary ingredient of other explosives, such as amatol, a mixture of TNT and ammonium nitrate widely used in aerial bombs, and tritinol, a mixture of TNT and aluminum powder for additional brissance.

RDX, also known as cyclonite, is produced by the nitration of complex organic compounds. Considerably more powerful and also more sensitive than TNT, it must be coated with wax for handling and must be mixed with other compounds to be stable enough for military use. The three forms in general use are Composition A, a mix of 81 percent RDX and 9 percent wax; Composition B, 60 percent RDX, 39 percent TNT, and 1 percent wax; and Composition C, a plastic explosive which is 90 percent RDX and 10 percent emulsifying oil.

HBX-1 (High Brissance Explosive) is a mixture of RDX, TNT, aluminum powder, and wax. Considerably more powerful than TNT, stable, and relatively insensitive to shock, it is often used in rocket and missile warheads. HBX-3 has a higher percentage of aluminum for greater brissance, and is most commonly used in torpedoes, depth charges, and naval mines.

Explosive D (ammonium picric) is a phenolic crystalline compound, stable at temperatures up to 150° F when not exposed to bare metal. Nearly as powerful as TNT and quite insensitive to shock, it is often used to fill armor-piercing shells and bombs.

Tetryl (trinitrophenylmethylnitramine) is produced by nitration of aniline-based organic compounds. Some 30 percent more powerful than TNT, it is frequently used as a booster in small-caliber shells. Mixed with TNT to form tetrytol, it can be cast in blocks and used for demolitions.

FUEL AIR EXPLOSIVES, a recent development, are combustible aerosols dispersed from bombs or rocket warheads. The aerosol mixes with the atmosphere to form an explosive cloud, which can generate a blast wave comparable to a small nuclear warhead when detonated.

EXTENDER: McDonnell-Douglas KC-10A AERIAL REFUELING tanker aircraft, in service with the U.S. Air Force STRATEGIC AIR COMMAND (SAC). Intended to supplement SAC's fleet of KC-135 STRATOTANKERS, the KC-10A was chosen in the 1978 Advanced Tanker/Cargo Aircraft competition over a modified Boeing 747. A derivative of the DC-10 wide-body airliner, the first Extender was flown on 30 July 1980, and a total of 60 were acquired through 1989.

The pilot, copilot, and flight engineer sit on the flight deck in the nose. Behind the cockpit, the fuselage has two levels, an upper cabin and a hold. The cabin can accommodate up to 60 passengers or 27 cargo pallets, loaded through large side doors behind the cockpit. The hold houses seven transfer fuel cells. 117,829 lb. (93,446 kg.). An advanced, digitally controlled flying boom under the rear fuselage can transfer fuel at a rate of 8750 lb. (3977 kg.) per minute. In addition, the Extender also has a hose-and-drogue unit to refuel U.S. navy and foreign aircraft without flying boom receptacles.

Specifications **Length:** 181.6 ft. (55.35 m.). **Span:** 165.4 ft. (50.41 m.). **Powerplant:** 3 General Electric CF6-GE-50C2 high-bypass turbofans, 52,-500 lb. (23,636 kg.) of thrust each. **Fuel:** 356,065 lb. (161,508 kg.). **Weight, empty:** 240,065 lb. (108,891 kg.). **Weight, max. takeoff:** 590,000 lb. (267,620 kg.). **Speed, max.:** 600 mph (966 kph). **Speed, cruising:** 564 mph (908 kph). **Operating Radius:** (200,000 lb. [90,700 kg.] transfer fuel) 2200 mi. (3540 km.). **Range, max.:** (w/100,000 lb. cargo) 6905 mi. (11,112 km.)/(max. fuel) 11,500 mi. (18,-500 km.).

F

F-: U.S. designation for fighter aircraft, often used in combination with a second prefix to identify multi-role aircraft (e.g., FB- for fighter-bomber and FA- for fighter/attack) or specialized conversion missions (e.g., EF for ELECTRONIC WARFARE, RF for tactical RECONNAISSANCE, TF for trainer). Current fighters include: F-4 PHANTOM; F-5 TIGER II; F-8 CRUSADER; F-14 TOMCAT; F-15 EAGLE; F-16 FALCON; F-20 TIGERSHARK; F-100 SUPER SABRE; F-104 STARFIGHTER; F-11I; F-117 NIGHTHAWK (the "Stealth Fighter"); and FA-18 HORNET.

F-111: General Dynamics twin-engine, two-seat, all-weather, long-range ATTACK aircraft, in service with the U.S. Air Force (USAF) and the Royal Australian Air Force (RAAF). Developed from 1960 as the TFX (Tactical Fighter, Experimental) meant to replace the F-105 Thunderchief supersonic nuclear strike aircraft, the design was to combine high speed and heavy payload with the ability to operate from short, unpaved fields. To reconcile these contradictory requirements, USAF proposed for the first time to use a variable geometry ("swing") wing, to provide an efficient planform over a wide variety of flight conditions, with the wings fully extended for short takeoffs and landings, partially swept for efficient cruising, and completely swept for supersonic sprints. In addition, the TFX was to be the first aircraft to have afterburning turbofan (vs. turbojet) engines for increased fuel efficiency, to make extensive use of titanium forgings to reduce weight and resist atmospheric heating, and to be equipped with fully

automatic TERRAIN-FOLLOWING RADAR (TFR) to allow high-speed, low-level precision attacks at night and in bad weather.

The program was sidetracked in 1961, when Secretary of Defense Robert S. McNamara, seeking increased commonality between air force and navy weapons, decreed that the TFX should also be developed as a carrier-based interceptor. Both Boeing and General Dynamics submitted proposals, and the latter's design was selected for development as the F-111 in 1962.

Development of the navy's F-111B interceptor was assigned to Grumman, and the first prototype flew in 1965. Changes from the original USAF design included the provision of a long-range AWG-9 search and tracking radar for AIM-54 PHOENIX missiles, lengthened wings, and reinforced landing gear. That increased weight beyond the limits of U.S. carrier decks, and a series of weight-reduction programs merely served to increase costs while reducing the commonality between the navy and USAF aircraft. The F-111B program, opposed by the navy from the first, was finally terminated in 1968, but some of the experience gained (as well as the AWG-9 and Phoenix system) were later incorporated into Grumman's F-14 TOMCAT.

Development of USAF's F-111A at first proceeded more smoothly. The first of 18 preproduction aircraft were delivered in December 1964, and the first production aircraft entered service in 1967. Powered by two Pratt and Whitney TF30 turbofans, the F-111A was equipped with semi-solid-state Phase I AVIONICS offering better blind-bombing and low-level navigation capabilities than any other aircraft of that time (except possibly the U.S. Navy's A-6A INTRUDER, a much slower aircraft).

After operational testing and evaluation, a detachment of six F-111s was sent to Southeast Asia for combat testing over North Vietnam. Two aircraft were lost without a trace in the course of 55 sorties, but wreckage from a third crash was recovered, revealing a design flaw in the horizontal stabilizer. As soon as this defect was repaired, an F-111 in the U.S. lost a wing because of cracks in the wing-pivot box; the entire force was grounded while this defect too was repaired. Because of these mechanical problems, the failure of the F-111B, rising costs, and its poor initial showing over North Vietnam, the F-111 was nearly canceled. In 1972, a wing of F-111As was redeployed to Southeast Asia for the Linebacker and Linebacker II bombing campaigns over North Vietnam. Operating at night, alone or in pairs, the F-111s completed more than 4000 sorties with a loss of only six aircraft (the best record of the Vietnam War), bombing their targets with great accuracy.

In all, 141 F-111As were built through 1969, when production switched to the F-111E, 94 aircraft with more powerful engines and enlarged engine inlets, but otherwise similar to the F-111A. They were followed in 1970 by the F-111D, with the same engines, but much more advanced, solid-state Phase II avionics. Because of rising costs, only 96 of a planned 300 were produced. The final U.S. tactical variant, the F-111F, has much more powerful TF30-P-100 engines, simplified Phase IIB avionics, longer range, and a higher maximum takeoff weight. First deployed in 1972, they are undoubtedly the best of all F-111s, but because of mid-1970s budget restrictions, production was terminated after only 106 had been completed.

Because of its high cost and initial troubles, the F-111 could not have attracted many export customers, but was in any case too powerful an aircraft to be freely exported. Australia ordered 24 F-111Cs directly off the drawing board in 1963. Fitted with the P-3 engines and Phase I avionics of the F-111A, the extended wings of the F-111B, and intended mainly for maritime strike, they were first delayed by a series of modifications and then placed in storage while Australia delayed payment; they were finally delivered in 1973.

The RAF ordered 50 F-111Ks in 1966, but canceled the order two years later. The USAF STRATEGIC AIR COMMAND (SAC) had selected the F-111 in 1965 as an "interim" replacement for the B-58 Hustler supersonic bomber, and the 18th preproduction was converted to the prototype FB-111A in 1967. The 50 ex-British aircraft were also completed as FB-111s, together with 25 new aircraft, for a total of 76, of which 67 equipped two SAC bomber wings through 1990. Powered by TF30-P-7 engines, the FB-111 has its own special avionics suite, which later served as the basis for the Phase IIB suite of the F-111F. In the 1990s they will be modified for nonnuclear attack as "F-111Gs." The latest development is the EF-111A RAVEN, an ELECTRONIC WARFARE conversion of the F-111A. Equipped with a powerful ALQ-99 ACTIVE JAMMING system, 42 F-111As were converted and completely rebuilt by Grumman between 1981 and 1985.

Only 562 F-111s were built, of which some 440 remain in service. Even now, a quarter-century after its introduction, the F-111 remains the most capable strike/interdiction aircraft in the world; only the Panavia TORNADO and Soviet Su-24 FENCER (both strongly influenced by the F-111) are in its class. With successive modernizations and overhauls, F-111s will probably remain in service through the turn of the century, though supplemented from the 1990s by the F-15E STRIKE EAGLE.

Very large for a tactical aircraft, the F-111 has a broad, area-ruled fuselage and a long, tapering nose ending in a large radome housing both a Doppler attack RADAR and a smaller TFR. In the Phase I avionics of the A, C, and E, the attack radar is an analog, liquid-cooled, electromechanical APC-113, with ground automatic and ground-velocity attack modes, and an air-to-air mode for self-defense. The Phase I TFR is a Texas Instruments APQ-110, with duplicate scanners for reliability: built-in test equipment performs a diagnostic check every 0.7 seconds, and switches to the backup scanner in the event of a failure. The radar normally operates in an automatic mode, whereby it is connected directly to the autopilot so that the latter can maintain a constant height of some 200 ft. (61 m.) off the ground; in a secondary "manual" terrain-avoidance mode, the radar provides steering cues to the pilot.

In the Phase II avionics of the F-111D, the attack radar is a digital, solid-state APQ-130 with DOPPLER-beam sharpening for higher resolution, a MOVING TARGET INDICATOR to suppress background clutter, and a continuous-wave illumination mode for SEMI-ACTIVE RADAR-HOMING air-to-air missiles (which were never used with the F-111). The TFR is a solid-state APQ-128 with greater accuracy and reliability than the earlier model. The F-111F has simplified Phase IIB avionics, based on an APQ-144 attack radar with reduced pulse width for even better target resolution, while the TFR is an upgraded APQ-146. The FB-111A has an APQ-114 attack radar, similar to the APQ-113, but with an additional "beacon" mode, a data recorder, and a north-oriented display.

The pilot and WEAPON SYSTEM OPERATOR (WSO) sit side-by-side in a unique cockpit module, with built-in parachutes and survival equipment, which can be separated from the aircraft in an emergency. Adopted to give the crew better protection than could be provided by 1960s ejection seats, the module is rather heavy and complex; it is now rendered unnecessary by modern, lightweight, "zero-zero" ejection seats.

Cockpit avionics, among the most comprehensive of any aircraft, include a pilot's lead-computing optical weapon sight, a TFR display, a WSO head-down attack radar display, a threat-warning display, a multi-functional video monitor, a radar altimeter, an INERTIAL NAVIGATION unit, a doppler navigation radar, a UHF DATA LINK, and a digital navigation/weapon delivery computer. The FB-111 also has an ASQ-119 astro-compass for celestial updates of its inertial navigation unit. Electronic warfare equipment includes an ASP-109 RADAR-HOMING AND WARNING RECEIVER, an ALR-62 RADAR WARNING RECEIVER, and an AAR-44 INFRARED warning receiver, all linked to an ALQ-137 active jammer and an ALE-37 countermeasures dispenser for chaff and flare cartridges.

The space behind the cockpit module houses a large fuel cell over an internal rotary weapons bay. Originally intended for nuclear weapons, the bay is now mostly used for additional fuel tanks, though in the F-111D it normally houses an M61 VULCAN 6-barrel 20-mm. cannon with 2000 rounds of ammunition, while in the F-111F it can accommodate a PAVE TACK forward-looking infrared (FLIR) target tracking and LASER DESIGNATOR pod (Pave Tack FLIR images are displayed on the cockpit video monitor, allowing the WSO to identify targets visually and illuminate them for LASER-GUIDED BOMBS). Additional fuel cells occupy the midfuselage area. An aerial refueling receptacle (for the USAF "flying boom" system) is located just behind the cockpit. Designed as it was for rough-field operations, the F-111 has short, reinforced landing gear with large, low-pressure tires that retract into the belly; the large main-gear

door doubles as an air brake. The engines in the rear fuselage are fed through short, quarter-cone inlets well back along the sides of the fuselage. The TF30 is prone to compressor stalls caused by turbulent air flow, and the position of the inlets makes it difficult to separate turbulent "boundary layer" near the fuselage from smooth air going to the engines. The F-111A's inlets were modified during development with a large splitter plate ("Triple Plow I"); later models have larger inlets ("Triple Plow II").

The shoulder-mounted wings, pivoted inboard from large, fixed "gloves," can be swept manually from 16° to 72.5°, to match the airspeed and angle of attack. The wing has full-span leading-edge slats and full-span Fowler flaps for short takeoff and landing performance, and large spoilers for roll control with the wings extended. The wing gloves and outboard wing sections all contain integral fuel tanks. The large slab tailplanes, mounted in line with the wings, act together as conventional elevators with the wings extended, or as differential "tailerons" to control both pitch and roll at high sweep settings. The broad vertical stabilizer has a small tip pod housing the scanner for the IRWR.

The F-111 has four pivoting wing pylons, each rated at 5000 lb. (2268 kg.). Most tactical variants also have provisions for two additional fixed 5000-lb. pylons outboard, while the FB-111A and F-111C also have provisions for a fourth pair of 5000-lb. pylons near the wingtips. The fixed pylons are rarely, if ever, used in action because they limit the wing to its 16° position. Thus, while the theoretical maximum payload is 30,000 lb. (13,636 kg.), in practice the normal maximum load is 19,800 lb. (8981 kg.), still more than any other tactical aircraft (except for the F-15E). In addition to the wing pylons, there are several light fuselage pylons for ELECTRONIC COUNTERMEASURE and weapon data link pods.

The F-111 can accommodate a wide variety of weapons. For tactical nuclear strike, it can carry up to 6 free-fall weapons, 2 in the weapons bay and 4 on the wings. The FB-111A is normally armed with 4 AGM-69A SRAM nuclear missiles and 2 600-gal. (2271-lit.) drop tanks, while Australian F-111Cs are equipped to deliver up to 4 AGM-84 Harpoon anti-ship missiles. The normal bomb load is 24 Mk.82 500-lb. (227-kg.) bombs, for delivery by low-level or "toss" attacks. For precision attack, all F-111s can carry PAVEWAY laser-guided bombs, while the F-111F has also been modified to deliver GBU-15 and AGM-130 electro-optical GLIDE BOMBS. Recently, the outboard pivoting pylons have been modified with shoulder-mounted launch rails, allowing AIM-9 SIDEWINDER infrared-homing air-to-air missiles to be carried for self-defense without displacing other weapons. Other potential payloads include CLUSTER BOMBS, BLU-108 DURANDAL runway cratering bombs, and BGM-109 TOMAHAWK cruise missiles.

Specifications Length: 73.5 ft. (22.4 m.). **Span:** (A, D, E, F) 63 ft. (19.2 m.) at 16°/31.98 ft. (9.74 m.) at 72.5°; (C, FB) 70 ft. (21.33 m.) at 16°/33.95 ft. (10.33 m.) at 72.5°. **Powerplant:** (A, C) 2 Pratt and Whitney TF30-P-3 turbofans, 18,500 lb. (8390 kg.) of thrust each; (D, E) TF30-P-9s, 19,600 lb. (8890 kg.) of thrust each; (F) TF30-P-100s, 25,100 lb. (11,385 kg.) of thrust each; (FB) TF30-P-7s, 20,350 lb. (9230 kg.) of thrust each. **Fuel:** 29,330 lb. (13,332 kg.). **Weight, empty:** (A) 46,172 lb. (20,943 kg.); (C) 43,700 lb. (21,455 kg.); (D, E) 49,000 lb. (22,226 kg.); (F) 47,175 lb. (21,398 kg.); (FB) 50,000 lb. (22,680 kg.) **Weight, max. takeoff:** (A) 91,500 lb. (41,500 kg.); (C) 114,300 lb. (51,846 kg.); (D, E) 99,000 lb. (44906 kg.); (F) 100,000 lb. (45,359 kg.); (FB) 114,300 lb. (51,846 kg.). **Speed, max.:** (clean) Mach 2.5 (1653 mph/2660 kph) at 36,000 ft. (11,000 m.)/Mach 1.2 (913 mph/1469 kph) at sea level. **Speed, cruising:** 571 mph (919 kph) w/ext. stores. **Service ceiling:** 50,000–60,000 ft. (15,250–18,300 m.). **Combat radius:** (hi-lo-hi w/ext. fuel) 918 mi. (1480 km.). **Range, max.:** 2925 mi. (4704 km.).

FA: See *Frontovaya Aviatsiya.*

FAADS: Forward Area Air-Defense System, U.S. Army program for its battlefield air defenses, initiated after the failure of the DIVAD program. FAADS has five components:

(1) Line-of-Sight, Forward, Heavy (LOS-F-H), a mobile gun/missile system for the brigade and divisional air defense units of armored and mechanized formations. The Oerlikon-Buhle/Martin-Marietta ADATS (Air-Defense Anti-Tank System) was selected for this role in late 1987 and should enter service in 1990–91. (2) Non Line-of-Sight (NLOS), a system intended to engage helicopters. FOG-M, the Boeing Fiber-Optic Guided Missile, is under development for this role and could enter service in the mid-1990s. (3) Line-of-Sight, Rear (LOS-R), a lightweight, mobile weapon intended for the defense of rear-echelon units. The Boeing Avenger, a pedestal-mounted variant of the FIM-92 STINGER shoulder-fire missile, is now undergoing evaluation. (4) Air defense upgrades of the M1 ABRAMS main battle tank and the M2/M3 BRADLEY fighting vehicle, including modifications to their fire controls and the development of new ammunition with anti-aircraft applications. (5) Improvements in battlefield air-defense command, control, communications, and intelligence (C³I) systems.

FAC: 1. See Forward Air Control.

2. See FAST ATTACK CRAFT.

FAE: See FUEL AIR EXPLOSIVE.

FAILSAFE: 1. In engineering, design features meant to ensure that a system will degrade gracefully (rather than fail catastrophically), by providing redundancy or automatic shut-down circuits, or both (as in nuclear reactors with redundant coolant circuits and automatic shut-down provisions in the event of total coolant failure).

2. A popular term for U.S. POSITIVE CONTROL procedures for nuclear weapons. Positive control is intended to prevent unauthorized release (or detonation), either by requiring the positive consent of two or more qualified operators, or by PERMISSIVE ACTION LINKS (PALs), electronic interlocks which inhibit the firing circuits of nuclear warheads. PALs can be unlocked only by key codes held exclusively by the president or his duly authorized delegates.

FALCON: 1. Hughes AIM-4 and AIM-26 series of air-to-air missiles. Developed in response to a 1947 U.S. Air Force (USAF) requirement, the AIM-4 entered service in 1954 as the world's first operational air-to-air missile. No longer in USAF service, Falcons manufactured under license by Saab as the Rb.27 and Rb.28 are still in service with the Swedish and Swiss air forces.

The original AIM-4 had a cylindrical body with four broad-chord delta wings with separate rectangular control surfaces, and four small nose strakes that housed radiometric receiver aerials. Glass-reinforced plastic was used extensively in the wings and body to reduce weight and manufacturing costs. Powered by a solid rocket engine and armed with a 29-lb. (13-kg.) high-explosive warhead fitted with both contact and radar-proximity (VT) fuzes, the AIM-4 was guided by SEMI-ACTIVE RADAR HOMING (SARH), with target illumination provided by the launch aircraft's radar. PROPORTIONAL NAVIGATION allowed collision-course engagements, a critical requirement for an anti-bomber interceptor missile. A total of 4080 AIM-4s were produced between 1954 and 1956.

In 1956, the AIM-4 was superseded by the heavier AIM-4A, with larger control surfaces for improved maneuverability and longer range; some

12,100 were delivered. Also in 1956, Hughes introduced the AIM-4B, with passive INFRARED HOMING. Slightly longer, the AIM-4B had a glass nose housing an uncooled lead-sulfide (PbS) seeker capable of attacks from the rear quarter only; speed and range were essentially those of the AIM-4A. Some 16,000 were eventually produced. The similar AIM-4C of 1957 had an improved IR seeker; some 13,500 were produced, including 1000 HM.58s for Switzerland and 3000 Rb.28s for Sweden.

In 1958, Hughes introduced the AIM-4E Super Falcon, with a longer-burning solid-rocket engine, improved SARH guidance, and a 40-lb. (18-kg.) high-explosive warhead. Only some 300 were produced in 1958–59, before they were superseded by 4000 AIM-4Fs, with dual-impulse solid rocket engines and improved guidance with ELECTRONIC COUNTER-COUNTERMEASURE features. The AIM-4F was followed by the AIM-4G, essentially an F with IR homing; 2700 were produced in 1959–60.

Though still called Falcon, the AIM-26 of 1960 was a completely new and much larger SARH missile intended to increase kill probabilities in head-on engagements. Armed with a 1.5 kT W54 NUCLEAR WARHEAD, some 1900 AIM-26As were produced exclusively for the USAF; all were retired in the early 1980s. The AIM-26B, essentially the same missile with a 60-lb. (27.27-kg.) high-explosive warhead, was the last version of Falcon in U.S. service. It was finally retired in 1987 along with the last of the F-106 Delta Dart interceptors. Some 2000 were built, including 400 HM.55s for Switzerland and 800 Rb.27s for Sweden.

The last Falcon produced was the IR-homing AIM-4D of 1963, basically an AIM-4C with the IR seeker and rocket engine of the AIM-4G. Optimized against fighters, the D was used in the Vietnam War and is credited with five kills. Some 4000 were produced, and 8000 AIM-4As and -4Cs were rebuilt to the D standard, even if it was inferior in both cost and performance, to the AIM-9 SIDEWINDER.

The ultimate Falcon development was the AIM-47 Eagle, a long-range missile intended for the abortive F-111B and YF-12A interceptors, which later evolved into the AIM-54 PHOENIX.

Specifications Length: (4, 4A) 6.5 ft. (1.98 m.); (4B) 6.6 ft. (2.02 m.); (4E) 7.16 ft. (2.18 m.); (26) 7 ft. (2.13 m.). **Diameter:** (4) 6.4 in. (163 mm.); (26) 11 in. (279 mm.). **Span:** (4) 20 in. (508 mm.); (26) 24.4 in. (620 mm.). **Weight, launch:** (4) 110 lb.

(50 kg.); (4A, B) 120 lb. (54 kg.); (4C) 134 lb. (61 kg.); (4E) 150 lb. (68 kg.); (26) 203 lb. (92 kg.). **Speed:** (4, 4A, B, C) Mach 2.5 (1625 mph/2714 kph); (4D, E, F, G) Mach 4 (2600 mph/4342 kph); (26) Mach 2 (1300 mph/2171 kph). **Range:** (4) 3 mi. (5 km.); (4A, B, C, 26) 6 mi. (9.7 km.); (4D, E, F, G) 7 mi. (11.3 km.).

2. Officially, "Fighting Falcon"; General Dynamics F-16 single-engine, multi-role FIGHTER, in service with the U.S. Air Force (USAF) and other air forces.

Concerned that the growing size, complexity, and cost of 1960s fighters threatened its ability to equip its squadrons, USAF issued a 1972 requirement for a "Lightweight Fighter" (LWF), i.e., a simple aircraft with simple AVIONICS, armed with short-range INFRARED-HOMING (IR) air-to-air missiles and a 20-mm. cannon, optimized for close-in, clear-weather dogfighting. Two competing prototypes were developed: the single-engine General Dynamics YF-16, and the twin-engine Northrop YF-17. Both featured electronic FLY-BY-WIRE controls, advanced aerodynamics, and a high thrust-to-weight ratio for agility, maneuverability, and acceleration. After flight tests throughout 1974, USAF selected the F-16 in early 1975 (the YF-17 later evolved into the U.S. Navy's FA-18 HORNET).

USAF's original intention was to acquire only a few Lightweight Fighters for evaluation, because it already had the larger, more capable F-15 EAGLE in production. But when escalating costs precluded the acquisition of enough F-15s, the F-16 was adopted as the low end of a "high-low mix" of roughly two F-16s for every F-15. The first of six YF-16A preproduction aircraft were delivered in 1976, and the first operational F-16As entered service in 1978.

Because of its high performance and relatively low cost, the F-16 was especially attractive to smaller air forces. In June 1975, Belgium, Denmark, the Netherlands, and Norway all selected the F-16 to replace their aging F-104G STARFIGHTERS, with most of the aircraft to be coproduced by a European consortium. Since then, the F-16 has been widely exported. USAF plans to acquire some 2600 F-16s through 1994. Total delivered and confirmed orders now exceed 4000 aircraft, making the F-16 by far the most numerous Western fighter now in service.

The original concept of a very simple, lightweight fighter did not survive for long, as USAF kept adding new capabilities to the design. The F-16A and the similar, two-seat F-16B have a sophisticated multi-mode RADAR, an air-to-ground weapon-delivery computer, and advanced ELECTRONIC COUNTERMEASURE (ECM). Powered by a single Pratt and Whitney F100 afterburning turbofan, it has a power-to-weight ratio better than 1 to 1 for truly remarkable agility and sustained turning performance in dogfights, while its advanced avionics make it a formidably accurate attack aircraft. Much used by the Israeli air force, the F-16 completely outclassed Syrian MiG-21 FISHBEDS and MiG-23 FLOGGERS in the 1982 Lebanon War, after executing the 1981 long-range, precision raid against the Iraqi Osirak nuclear reactor in the attack role.

In 1984, the F-16A/B was superseded by the F-16C/D, with a new radar featuring multiple TRACK-WHILE-SCAN (TWS) and look-down capabilities, provisions for medium-range, radar-guided air-to-air missiles and LANTIRN night vision/target acquisition pods, improved avionics, and a heavier payload within a higher gross weight. In addition, the C/D have a revised rear fuselage which can accommodate either an improved F100-PW-220 or a more powerful General Electric F110-GE-100 afterburning turbofan, to maintain the Falcon's high power loading and much of its original agility, despite the increased weight. The combat power-to-weight ratio varies, from 1.01 to 1 in the F-16A, to 0.95 to 1 for the F-16C with the F100-PW-220 or 1.05 to 1 with the F110-GE-100.

In 1986, USAF decided to modify some 270 F-16As for air defense, to replace the F-106A DELTA DART interceptor. Equipped with a modified radar, the F-16ADF (Air Defense Fighter) can also launch radar-guided missiles, but it has neither the range nor the true all-weather capability needed in a dedicated interceptor. USAF's decision was apparently prompted by the desire to replace "tactical" F-16As with more advanced variants of the aircraft. It has now proposed a dedicated CLOSE AIR SUPPORT variant, the A-16, as a replacement for the A-10A THUNDERBOLT, a decision similarly motivated, which has also been criticized on the grounds that the F-16 is too fast and lightly built for the role. The U.S. Navy has also acquired some 26 F-16Ns (essentially stripped-down F-16Cs) for Dissimilar Air Combat Training, to simulate late-model Soviet fighters at the TOP GUN fighter weapons school.

There have also been two experimental variants. The F-16XL has a lengthened fuselage and a much larger "cranked arrowhead" compound-delta wing for greatly improved turn radius, range,

cruise speed, and payload. Evaluated as the F-16E for the USAF "dual-role fighter," it was rejected in favor of the F-15E Strike Eagle. The F-16AFTI (Advanced Fighter Technology Integrator) is an F-16A with chin-mounted vertical canards and a flight control system modified for radical "decoupled" maneuvers, whereby the aircraft can alter its flight path without changing its attitude. Much of the technology tested in the AFTI is now being incorporated into the YF-22/23 ATF (Advanced Technology Fighter) program. Japan is also developing a new fighter (FSX) based on the AFTI.

Small by the standards of modern fighters, the F-16 uses composite materials throughout the airframe to reduce weight. The shark-nosed fuselage has a flat, drooping radome housing a compact, solid-state Westinghouse pulse-Doppler radar. The F-16A/B has an APG-66 radar with four air-combat modes: dogfight, boresight, look-up, and look-down. Maximum detection range against fighters varies from 45 mi. (75.15 km.) in look-up to 34.5 mi. (55.6 km.) in look-down, and the radar can simultaneously track one target while scanning for others. There are also seven air-to-ground modes: continuously computed impact point (CCIP), continuously computed release point (CCRP), smooth-water sea search, rough-water sea search, dive/toss, ground-mapping, and Doppler beam-sharpening for enhanced resolution.

The F-16C/D has an APG-68 radar, essentially an upgraded APG-66 with increased range, better look-down resolution, improved electronic counter-countermeasures (ECCM), and a mono-pulse illumination mode for AIM-7M SPARROW semi-active radar-homing (SARH) and AIM-120 AMRAAM active-homing missiles.

The pilot sits right behind the radar in an elevated cockpit under a clear-blown, single-piece, clamshell bubble canopy, which allows a completely unobstructed 360° view. To help the pilot resist the high g-forces generated during very tight turns, the ejection seat is reclined 30°, thus bringing the pilot's feet closer to the level of his head to help prevent "blackout." Cockpit layout is optimized for air-to-air combat. The F-16A/B has a head-up display (HUD), a head-down multi-functional video monitor, a threat warning display, a Singer-Kearfott SKN-2400 inertial navigation unit, VHF, UHF, and secure voice radios, an instrument landing system, and a TACAN beacon receiver, but still relies on analog instrument dials. The F-16C/D has a revised cockpit with two multifunctional displays instead of dials, and a wide-angle "holographic" HUD. In both models, all essential radar and weapon controls are laid out in a Hands on Throttle and Stick (HOTAS) configuration, to minimize the need to go "head-down" in the cockpit during combat. Other avionics include a Dalmo-Victor ALR-69 RADAR WARNING RECEIVER and an Advanced Self-Protection Jammer.

Uniquely, the F-16 has an isometric side-stick controller instead of the traditional central control stick. Mounted to the right of the pilot, the side stick moves only a fraction of an inch: instead of mechanical action, electronic sensors measure the pressures exerted by the pilot, and convert them into electrical impulses that reach the various actuators; the rudder pedals are also isometric. The first operational fly-by-wire fighter, the F-16 is an inherently unstable "control-configured vehicle" that relies on a Sperry digital flight-data computer to keep the aircraft on course. The pilot's control inputs are interpreted by the computer, which then generates control commands for fast-acting electronic actuators, making the F-16 very responsive. The flight control computer has preprogrammed limits of 9 g and 26° angle of attack (AOA), to keep the aircraft well within its performance envelope, allowing the pilot to concentrate on the fighting, instead of the flying.

The space behind the cockpit houses a large fuel tank, with additional fuel and the landing gear in the midfuselage, and the engine in the rear. An aerial refueling receptacle is mounted dorsally, amidships (two-seat versions have a second cockpit behind the pilot, replacing one fuselage fuel tank). The engine is fed from a simple, fixed inlet positioned below the cockpit. This limits the Falcon's maximum speed to Mach 2, but high absolute speed was not considered as important as acceleration at typical combat speeds (400–700 mph/668–1169 kph).

The midmounted wing, blended into the fuselage to reduce both drag and radar signature, has large leading-edge root extensions (LERX) that generate lift-enhancing vortices and improve handling at high AOA. The leading edge, swept at 40°, has full-span, leading edge flaps, while the straight trailing edge has large, inboard "flaperons" that act as both flaps and ailerons; both are automatically positioned by the flight data computer to provide the optimum wing profile ("camber") throughout the flight envelope. A 20-mm. M61 VULCAN 6-barrel cannon is housed in the port leading edge extension, with a 500-round drum magazine inside the fuselage behind the cockpit. The

small slab tailplane, mounted in line with the wing, has pronounced anhedral, while the vertical stabilizer is very tall, to provide good directional control at high AOA. Belgian F-16s have an extended housing at the base of the stabilizer for a Rapport III internal ECM system, while Norwegian aircraft have a similar housing for a braking parachute.

The F-16 has one centerline pylon rated at 2200 lb. (1000 kg.), a pair of inboard wing pylons rated at 4500 lb. (2041 kg.) each, two midwing pylons rated at 3500 lb. (1587 kg.) each, two outboard wing pylons rated at 700 lb. (318 kg.) each, and two wingtip pylons rated at 425 lb. (193 kg.) each. The total payload capacity, a remarkable 20,450 lb. (9276 kg.), would never be fully used in combat; normal operational payloads seldom exceed 11,950 lb. (5420 kg.). The wingtip pylons are always used for AIM-9 SIDEWINDER, Rafael SHAFRIR and PYTHON, or similar IR-homing air-to-air missiles. The outboard wing pylons are also used air-to-air missiles, which in the F-16C can be Sparrows or AMRAAMs instead of IR types. The midwing pylons can also carry air-to-air missiles, but are normally used for air-to-ground weapons, while the inboard wing pylons (which can also be used for air-to-ground weapons), are plumbed for 600-gal. (2271-lit.) drop tanks. Limited ground clearance diminishes the utility of the centerline pylon, which is normally used for either a 300-gal. (1135-lit.) drop tank, or an ALQ-131 ECM pod. The F-16C also has two light pylons near the mouth of the engine inlet for LANTIRN pods. The LANTIRN night vision pod has a forward-looking infrared (FLIR) sensor which displays images of the terrain ahead on the holographic HUD, allowing the pilot to fly at low level even at night.

Ground-attack weapons include free-fall and CLUSTER BOMBS, LASER-GUIDED BOMBS, NAPALM, chemical bombs, cannon pods, 19-round LAU-3A rocket pods for 2.75-in. (70-mm.) FFARS, AGM-65 MAVERICK air-to-ground missiles, and AGM-88 HARM anti-radiation missiles.

Specifications Length: 47.6 ft. (14.52 m.). Span: 31 ft. (9.45 m.). Powerplant: (A) 1 Pratt and Whitney F100-PW-200 afterburning turbofan, 23,-830 lb. (10,810 kg.) of thrust; (C) 1 F100-PW-220, or 1 General Electric F110-GE-100 afterburning turbofan, 27,600 lb. (12,545 kg.) of thrust. Fuel: 6972 lb. (3162 kg.). Weight, empty: (A) 15,586 lb. (7070 kg.) (C) 18,335–19,100 lb. (8316–8663 kg.).; Weight, normal loaded: (A) 23,375 lb. (10,594 kg.); (C) 18,335 lb. (8334 kg.) w/F100-PW-220; 19,100 lb. (8682 kg.) w/F110-GE-100. **Weight, max. take-off:** (A) 35,400 lb. (16,057 kg.); (C) 42,300 lb. (19,-227 kg.). **Speed, max.:** Mach 2 (1350 mph/2172 kph) at 36,000 ft. (11,000 m.)/Mach 1.2 (913 mph/1472 kph) at sea level. **Initial climb:** 50,000 ft./min. (15,239 m./min.). **Service ceiling:** 50,000 ft. (15,239 m.). **Combat radius:** (air combat) 575 mi. (925 km.); (hi-lo-hi attack) 340 mi. (547 km.). **Range, max.:** 2415 mi. (3890 km.). **Operators:** Bah, Bel, Den, Egy, Gre, Indo, Isr, Neth, Nor, Pak, ROK, Sing, Thai, Tur, US, Ven.

FANFARE: A British TORPEDO COUNTERMEAS-URE for surface ships, essentially a noisemaker sled which is towed behind a ship to decoy acoustic homing TORPEDOES. First introduced in 1944 to counter primitive German acoustic torpedoes, Fanfare has been superseded by the U.S.-designed SLQ-25 NIXIE.

FARMER: NATO code name for the Mikoyan MiG-19 twin-engine FIGHTER, the first Soviet aircraft capable of exceeding Mach 1 in level flight. Developed from the late 1940s, the prototype flew in 1952, and the first production MiG-19Fs (with Mikulin AM-5F afterburning turbojets) entered service in 1953. Underpowered, the F was soon superseded by the MiG-19S, with better Turmanskiy RD-9B engines and an all-moving slab tailplane for improved trans-sonic control. The MiG-19S entered service in 1955, and some 2500 were produced in the U.S.S.R., Czechoslovakia, and Poland through 1962. The only major Soviet variant was the MiG-19 PM "limited all-weather" INTERCEPTOR. Superseded in the Soviet air force by the MiG-21 FISHBED, the MiG-19 received a new lease on life in China, where an additional 2500 were produced as the "Shenyang F-6" from 1958 to 1978. The MiG-19 airframe is also the basis for the Chinese Nanzhang Q-5 strike fighter. Rugged, reliable, easy to maintain, and an excellent dogfighter, the MiG-19 and F-6 remain in service with many smaller air forces (Chinese versions are particularly valued for their outstanding workmanship). The Farmer's major drawbacks are its relatively short range, and the need for engine overhauls every 300 flying hours; but given Third World air force flying times, that amounts to one overhaul every two or three years, an acceptable rate.

The MiG-19S has a cylindrical fuselage with a circular nose intake divided by a vertical splitter; the MiG-19PM has a small *Izimrud* (Emerald) RADAR in the upper intake lip to provide some night interception capability. The pilot sits under a sliding bubble canopy, which provides an excellent all-around view (unlike *later* Soviet fighters, until the 1980s). AVIONICS are limited to a lead-computing optical gunsight and tactical radios. Armament consists of three powerful 30-mm. NR-30 cannons, one in the nose and two in the wing roots, each with 70 rounds per gun. Chinese versions frequently do without the nose gun, because its exhaust gases cause engine stalls. The PM variants normally do not have guns; originally they relied on now-obsolete AA-1 Alkalai radar-guided missiles. The midmounted wing, swept very sharply at 55°, has a fixed leading edge, outboard ailerons, and large Fowler flaps to reduce takeoff distance and landing speed. Large wing fences are fitted at midspan to provide adequate stall warning. The MiG-19 has a combat power-to-weight ratio of 0.89 to 1 (comparable to far more recent fighters) which, along with relatively low wing loading, provides both excellent acceleration and agility. The initial climb rate is better than either the contemporary U.S. F-104 STARFIGHTER or the later Fishbed.

The MiG-19 has four wing pylons for up to 2200 lb. (1000 kg.) of ordnance, most commonly 19-round UV-16-57 rocket pods for 57-mm. FFARS, with one pod carried on each inboard pylon; the outboard pylons reserved for fuel tanks. Pakistan has equipped its F-6s with rails for AIM-9 SIDE-WINDER air-to-air missiles, and other MiG-19s have been modified to launch the Soviet AA-2 ATOLL.

Specifications **Length:** 41.33 ft. (12.6 m.). **Span:** 29.51 ft. (9 m.). **Powerplant:** 2 Turmanskiy RD-9B afterburning turbojets, 7165 lb. (3250 kg.) of thrust each. **Weight, empty:** 11,400 lb. (5172 kg.). **Weight, normal loaded:** 16,300 lb. (7400 kg.). **Weight, max. takeoff:** 19,600 lb. (8900 kg.). **Speed, max.:** Mach 1.4 (900 mph/1450 kph) at 36,000 ft. (11,000 m.)/715 mph (1160 kph) w/ext. stores. **Initial climb:** 22,640 ft./min. (6900 m./min.). **Service ceiling:** 57,400 ft. (17,500 m.). **Combat radius:** 421 mi. (680 km.) w/2 drop tanks. **Range, max.:** 1350 mi. (2200 km.). **Operators:** Alb, Cuba, Iraq, N. Kor, Pak, PRC, Tan, Ug, Viet.

FAST ATTACK CRAFT (FAC): A small, highly maneuverable warship, with a displacement of less than 1000 tons, intended mainly for coastal operations. Steam-powered torpedo boats were introduced from the 1880s for harbor defense, but these continued to grow in size, and by the First World War had evolved into the modern destroyer (originally "torpedo-boat destroyer"). The First World War also saw the introduction of the motor torpedo boat (MTB), a much smaller vessel powered by internal-combustion engines. By the Second World War, they had evolved into the famous U.S. PT-boats, German E-boats, British MTBs, and a variety of small gunboats and submarine chasers. Devoid of armor, they relied on stealth to approach their target undetected and launch TORPEDOES from very short range, and on their speed for post-attack evasion. Naval RADAR made this mode of attack increasingly risky, and after 1945 most navies discarded FACs except for coast guard, harbor security, and riverine types.

Since the 1960s, however, there has been a resurgence of interest in FACs because of the introduction of light yet long-range ANTI-SHIP MISSILES. Like torpedoes, missiles allow FACs to strike effectively against targets many times larger than themselves, but the much greater range of missiles also allows FACs to strike from beyond the range of most target weapons (and indeed, beyond the range of FAC on-board sensors in many cases), necessitating some form of OVER-THE-HORIZON TARGETING (OTH-T).

This has made FACs very attractive to smaller powers which cannot afford larger warships, in spite of their limitations in range, seakeeping, and flexibility. Many FACs are definitely short-range vessels intended for sorties of several hours only; and even larger FACs have an endurance of only a few days, which limits their combat radius to a thousand miles or so. In general, therefore, FACs are best suited for navies with limited geographical commitments, or with access to forward basing. Moreover, while some FACs are quite seaworthy, they are all less stable than larger ships, and even if physically capable of open-ocean crossings, their effectiveness as weapon platforms is easily degraded in rough seas.

Size also limits tactical flexibility, because FACs cannot carry a wide variety of sensors, fire controls, and weapons. In theory, the low unit cost of FACs should allow their construction in large numbers, variously optimized for specific missions. Current FACs, however, are becoming increasingly complex and expensive. There are four main types: fast CORVETTES, patrol missile boats, torpedo boats, and gunboats.

Displacing around 1000 tons, with maximum speeds near 30 kt., corvettes can serve either as large missile boats or a small frigates. Their size allows them to carry more sensors and larger COMMAND AND CONTROL (C²) facilities than smaller FACs, so they are often employed as flotilla leaders for smaller missile boats. Armament usually includes 1 or 2 76- to 100-mm. DUAL PURPOSE guns, 8 or more anti-ship missiles, 1 or more 30- to 40-mm. ANTI-AIRCRAFT guns, and, in some cases, ANTI-SUBMARINE WARFARE (ASW) torpedoes and even a light helicopter (for ASW and OTH-T).

The most common type of FAC, patrol guided-missile boats (PGMS), displace between 100 and 500 tons, and have maximum speeds in the 30- to 40-knot range. Most carry between 2 and 8 canister-launched anti-ship missiles, 1 or 2 40- to 76-mm. guns, and several light cannons or MACHINE GUNS. Some also carry 2 anti-ship homing torpedoes for use against targets disabled by missiles.

Patrol torpedo boats, successors to the World War II PTs, MBTs, and E-boats, generally displace less than 100 tons, and have speeds in the 40- to 50-knot range. Most carry between two and six acoustic or wire-guided homing torpedoes, supplemented by light cannon and machine guns. While long-range homing torpedoes allow them more maneuvering room than their World War II predecessors, modern torpedo boats are still best employed in archipelagos offering ambush opportunities and refuges from counterattack.

Patrol gunboats (PGs) are essentially missile or torpedo boats without those types of main armament. Intended mainly for coastal surveillance and border control, their typical armament can include 1 or 2 dual-purpose guns, 1 or 2 anti-aircraft guns, and a variety of machine guns, MORTARS, GRENADE LAUNCHERS, and unguided ROCKETS.

Three major hullforms are currently employed: hard chine, round bilge, and HYDROFOILS. Hard-chine (V-bottom) hulls, designed for planing, are capable of very high speeds in smooth water, but pound heavily in rough seas and are not fuel-efficient at low speeds; unsuitable for vessels over 200 tons or so, their use is limited to torpedo boats and some smaller missile boats. Round-bottom (displacement) hulls have lower maximum speeds (40 kt.), but are much faster in heavy seas and more fuel-efficient over a wide range of speeds. They are found in most of the larger missile boats and corvettes. Hydrofoils, which ride on winglike foils at high speeds, are very fast (50 + kt.), but are also complex and very expensive.

Originally, most FACs were built of wood, which is cheap and light, but deteriorates rapidly without expensive maintenance. Most now have hulls of steel, aluminum, or fiberglass, with plastic or light alloy superstructures. Most World War II FACs were powered by gasoline aircraft engines. Though light and powerful, their volatile fuel presented a major explosive hazard. The predominant powerplants are now lightweight marine diesels and gas turbines, often used together in a CODAG or CODOG arrangement. See also separate FAST ATTACK CRAFT entries for FEDERAL REPUBLIC OF GERMANY; FRANCE; ISRAEL; ITALY; NORWAY; SOVIET UNION; SWEDEN; UNITED STATES.

FAST ATTACK CRAFT, FEDERAL REPUBLIC OF GERMANY:

Especially for its Baltic operations, the West German navy (*Bundeskriegsmarine*) has a large force of fast attack craft.

The Lurssen shipyard has been a world leader in the development of round-hulled, oceangoing fast missile boats (PGMs), but initially had to rely on a French shipyard to build its designs. These included the Israeli SA'AR class, and the French COMBATTANTE and Combattante II classes. Twenty of the latter were commissioned by West Germany (as the Type 148) between 1972 and 1975, with 10 built by TNC and 10 by Lurssen.

They were followed by 10 Lurssen Type 143 and 10 Type 143A missile boats commissioned between 1976 and 1984. Displacing some 390 tons, these wooden-hulled vessels have a maximum speed of 36 kt. and an armament of 4 Exocets, 2 76.2-mm. guns, and 2 21-in. wire-guided torpedoes for launch over the transom; the Type 143A also has provisions for a RIM-116A RAM surface-to-air missile launcher. All Type 143s are being modified to 143A standard, and redesignated Type 143B.

Other Lurssen designs have been intended only for export. The 410-ton FPB-57 class, four boats commissioned by Turkey in 1976–77, were the first missile boats with HARPOON anti-ship missiles. Capable of 38 kt., they resemble enlarged Combattantes and have 1 76.2-mm. gun and 2 35-mm. anti-aircraft guns, in addition to 8 Harpoons. The most widely exported design has been the TNC-45 class, sold to Ecuador (3), Argentina (2), and Singapore (6), which has also built 3 under license for Thailand. Similar to the Combattante II, these 230-ton vessels have a maximum speed of 34 kt. Armament varies, but most have a 76.2-mm. or 57-mm. gun, a 40-mm. anti-aircraft gun, and 4 Exocets (though the Singapore boats have 5 Israeli GABRIEL missiles instead).

FAST ATTACK CRAFT, FRANCE:

Though its own navy does not use fast attack craft (FACs), France is a major builder of FACs for export. This began with the construction of the COMBATTANTE, a 200-ton, wooden-hulled boat designed by the West German Lurssen shipyard. An enlarged, steel-hulled version based on the Lurssen TNC-45

class was then placed in production as the 260-ton Combattante II. Since their introduction in 1972, Combattante IIs have been sold to West Germany (10, as the S148 class), Greece (4), Libya (10), Malaysia (4), and Iran (12). The 480-ton Combattante III is an enlarged version armed with 2 76.2-mm. guns. 4 30-mm. guns, and 4 to 8 missiles. Greece has 10 (including 6 built locally), while Nigeria and Tunisia have 3 each. The 12 250-ton Israeli SA'AR missile boats, based on another Lurssen design, were also built in France. Completed in 1968–69, the last 6 were embargoed (after final payment had been made), and were finally sailed from Cherbourg to Israel on Christmas Eve 1969 without official permission.

FAST ATTACK CRAFT, ISRAEL:

In the late 1950s, the Israeli navy became the first noncommunist navy to determine that fast, missile-armed patrol boats were highly effective for a small state with geographically limited interests. While the development of the GABRIEL anti-ship missile was proceding, Israel therefore contracted the French CMN shipyard to build 12 250-ton SA'AR-class patrol boats. Completed in 1968–69, the last 6 were embargoed by the French government but were sailed from Cherbourg to Israel on Christmas Eve 1969 without permission. Built with 3 40-mm. anti-aircraft guns or 1 OTO-MELARA 76.2-mm. DUAL PURPOSE gun and 1 40-mm. gun, all were rearmed in the early 1970s with up to 8 Gabriel anti-ship missiles (Sa'ar II and III).

After operational experience with the Sa'ars, 10 much larger 450-ton Sa'ar IV, or RESHEF-class, missile boats were built by the Israeli Haifa shipyard between 1973 and 1980. Sacrificing 10 kt. for greater range, seakeeping, and habitability, the Reshefs were originally armed with 4–6 Gabriels, but many were fitted with 4 HARPOON anti-ship missiles in addition to Gabriel, and a 20-mm. PHALANX radar-controlled anti-missile gun. Having been used successfully in the 1973 Yom Kippur War, 2 Reshefs were later sold to Chile, and 3 were built in Israel for South Africa, which in turn built 9 more under license as the President class.

After experimentally fitting a Reshef with a helicopter landing deck, Israel constructed the Sa'ar 4.5, or Alia class, three 500-ton missile boats commissioned in 1980. Fitted with a hangar and landing deck, they have one light helicopter for scouting and OVER-THE-HORIZON TARGETING for groups of Sa'ars and Reshefs. The Alia hull was also adopted for the two Romat-class missile boats commissioned in 1981–82, in which the helicopter facilities are replaced by a heavier gun and missile armament, including 8 Harpoons, 6 Gabriels, 1 76.2-mm. gun, 1 Phalanx, and 2 20-mm. anti-aircraft guns. Israel is now acquiring four 1150-ton Sa'ar V guided missile corvettes, with Harpoon and Gabriel missiles, a Barak surface-to-air missile launcher, a 76.2-mm. gun, Phalanx, and remotely piloted helicopters.

In 1978, Israel ordered a patrol HYDROFOIL boat based on the Grumman Flagstaff design from a U.S. boatyard. Commissioned in 1982 as the Shimrit, it was followed by two more built in Israel under license in 1983 and 1985. Displacing 105 tons, the Shimrits have a maximum speed of more than 50 kt., and carry an armament of 4 Harpoons, 2 Gabriels, and 2 30-mm. anti-aircraft guns. They are expensive to maintain and operate, and no more will be acquired.

Israel also has some 37 Dabur-class motor gunboats based on a U.S. design. Displacing only 35 tons, and armed with a twin 20-mm. cannon and a twin 12.7-mm. machine gun, they have a maximum speed of 25 kt. and are used mainly for coastal security. The 47-ton Dvora is an enlarged Dabur built as a prototype; fitted with two Gabriels, it is the world's smallest missile boat; a number have been exported. The 54-ton Super Dvora, now entering service, can also be fitted with Gabriel or Harpoon.

FAST ATTACK CRAFT, ITALY:

In the Second World War, the Italian MAS torpedo boats were renowned for their speed and for the daring of their crews, but only a few fast attack craft were acquired by the Italian navy in the postwar period, and most have been discarded.

In the early 1970s, Italy was a participant (with West Germany) in the U.S. Patrol Hydrofoil Missile Boat (PHM) program that led eventually to the U.S. PEGASUS class. When rising costs became prohibitive, Italy withdrew from the program and developed its own HYDROFOILS, the 63-ton SPARVIERO class, seven of which were commissioned between 1974 and 1984. Armed with two OTOMAT anti-ship missiles and an OTO-MELARA 76.2-mm. DUAL PURPOSE gun, they have a maximum speed of more than 50 kt. Though their range is quite short, their speed makes them ideal for quick sorties in the Adriatic and the Straits of Tunis.

FAST ATTACK CRAFT, NORWAY:

With its long coastline deeply indented by fjords, Norway relies mainly on SUBMARINES and fast attack craft for coastal defense.

Of the latter, the oldest are the Tjeld-class motor torpedo boats, 20 of which were commissioned between 1960 and 1964, in addition to 6 sold to Greece in 1967 and 14 more sold to the U.S. (as the Nasty class) between 1962 and 1964. Displacing 82 tons, these 80-ft. (24.5-m.) wooden-hulled planing boats have a maximum speed of 45 kt. and an armament of 4 21-in. (533-mm.) TORPEDOES, 1 40-mm. BOFORS GUN, and 1 20-mm. ANTI-AIRCRAFT gun. Only 8 are still active.

Norway's oldest missile boats (PGMs) are the 125-ton Storm class, 20 of which were commissioned between 1965 and 1967. Capable of 37 kt., they carry 6 PENGUIN Mk.1 anti-ship missiles, 1 76.2-mm. DUAL PURPOSE gun, and 1 40-mm. gun. Seventeen remain in service. The Storm hull was also used for 6 Snogg-class missile boats commissioned in 1970–71. Armed with 1 40-mm. gun and only 4 Penguins, they also carry 4 21-in. wire-guided torpedoes for use against ships disabled (but not sunk) by the small Penguin warheads.

The 155-ton Hauk class are enlarged versions of the Snoggs, with a heavier armament of 6 Penguins, 2 torpedoes, and 1 40-mm. and 1 20-mm. gun. Capable of 35 kt., 14 were commissioned between 1978 and 1980.

FAST ATTACK CRAFT, SOVIET UNION:

The U.S.S.R. maintains a large force of fast attack craft (FACs), mainly for coastal defense and border patrol, including CORVETTES, patrol missile boats (PGMs), motor torpedo boats (MTBS), and patrol gunboats (PGS), in service both with the navy and the KGB Border Guards. Soviet FACs have also been widely exported.

The Soviet navy received a number of U.S. patrol torpedo boats (PTs) and British MTBs during the Second World War under the Lend-Lease program. These influenced the design of its postwar P4 and P6 class torpedo boats. The former, mass-produced between 1954 and 1958, were stepped-hydroplane boats, 62 ft. (19.1 m.) long with 2 18-in. (457-mm.) TORPEDO tubes, 2 14.5-mm. machine guns, and a maximum speed of 55 kt. All were phased out of the Soviet navy by the early 1970s, but some are still in service with Bangladesh, Cuba, and North Korea; China has at least 50 locally produced copies.

The P6 is a wooden-hulled 84.25-ft. (25.3-m.) planing-hull boat with 2 21-in. (533-mm.) torpedo tubes, 2 twin 25-mm. anti-aircraft guns, and a maximum speed of 41 kt. More than 80 were completed for the Soviet navy in the 1950s, but all were discarded in the 1970s. Many P6s were exported to Algeria, Cuba, Egypt, East Germany,

Guinea, Indonesia, Iraq, Poland, Somalia, and Vietnam. In addition, China produced more than 80 copies. Because of its wooden hull, the P6 requires much maintenance to prevent deterioration; only a few are still in service. The P6 was later developed into the P8, with surface-piercing HYDROFOILS, and the P10, with a remodeled superstructure; few, if any, remain in service.

The most important development of the P6, however, was the KOMAR-class missile boat, essentially a P6 with two SS-N-2 STYX anti-ship missiles instead of torpedoes. More than 100 were built in the U.S.S.R. from 1959, of which some 75 were exported; China still has more than 40 locally produced copies. An Egyptian Komar became the first missile boat to destroy an enemy warship when two Styx sank the Israeli destroyer *Elath* on 21 October 1967. The Komar was phased out of the Soviet navy during the 1970s, but many remain in service elsewhere.

The small and fragile Komar was soon superseded by the 127.4-ft. (38.6-m.) OSA, by far the most numerous class of missile boats in the world, with more than 290 constructed in the U.S.S.R. between 1959 and 1969, and at least 100 more built in China. Some 90 are still in the Soviet navy, while some 185 have been transferred to Warsaw Pact and other countries, including Algeria, Angola, Cuba, Egypt, Finland, India, Iraq, Syria, Vietnam, North and South Yemen, and Yugoslavia. Armed with 4 Styx and 2 twin 30-mm. ANTI-AIR-CRAFT guns, the Osas have an all-steel planing hull and a maximum speed of 36 kt.

The Osa hull was adapted in turn for the Shershen-class torpedo boats, some 80 of which were built between 1959 and 1970, with 12 more assembled in Yugoslavia. All were retired from Soviet service in the early 1980s, with many transferred to other countries, including Angola, Bulgaria, Cape Verde Islands, Egypt, East Germany, Guinea, North Korea, Vietnam, and Yugoslavia. Armed with 4 21-in. torpedoes and 2 twin 30-mm. guns, they have a maximum speed of 45 kt.

The TURYA-class torpedo boats are essentially Shershens with added surface-piercing hydrofoils, a revised superstructure, and an armament of 4 21-in. torpedoes, a twin 57-mm. anti-aircraft gun, and a twin 30-mm. gun. Fitted with dipping SONAR, they are apparently intended for inshore ANTI-SUB-MARINE WARFARE (ASW) in conjunction with shore-based helicopters. More than 40 were completed between 1974 and 1979, of which 31 are still in the Soviet navy, with 9 transferred to Cuba, and

several others transferred to Kampuchea and Vietnam.

The MATKA-class missile boats are essentially Osas fitted with surface-piercing hydrofoils added, as in the Turyas. Armament consists of only 2 Styx missiles, plus a 76.2-mm. DUAL PURPOSE gun on the bow and a 30-mm. radar-controlled gun on the fantail for anti-missile defense. Maximum speed is approximately 40 kt. Sixteen were completed between 1978 and 1981. These were the last missile boats constructed by the Soviet navy, which apparently prefers larger, corvette-type ships for both ASW and ANTI-SURFACE WARFARE (ASUW).

Also derived from the Osa hull are the Stenka-class patrol gunboats, some 100 of which are operated by the KGB Border Guards. Intended mainly for ASW, they have a dipping sonar, 2 twin 30-mm. guns, 4 16-in. (406-mm.) torpedo tubes for ASW homing torpedoes, and two DEPTH CHARGE racks.

The Soviet navy and the Border Guards also operate larger, slow patrol vessels usually classified as corvettes. In addition to slow converted minelayers of little combat value, there are also substantial numbers of fast, modern ASW and missile corvettes. The oldest are the 580-ton POTI class, some 70 of which were completed between 1961 and 1967, of which about 60 remain in the Soviet navy, with 3 transferred to Romania and 3 to Bulgaria. Classified as "Small Anti-Submarine Vessels" (*Maly Protivolodochnyy Korabl'*, MPKs), they have 1 twin 57-mm. anti-aircraft gun, 2 RBU-6000 ASW rocket launchers, and 2 or 4 16-in. torpedo tubes.

The Potis are now being superseded by the 580-ton PAUK-class MPKs, at least 14 of which have been completed since 1980, including a number operated by the KGB Border Guards. Armed with a 76.2-mm. dual-purpose gun, a 30-mm. radar-controlled gun, an SA-N-5 GRAIL surface-to-air missile (SAM) launcher, 2 RBU-1200 rocket launchers, and 4 16-in. torpedo tubes, they have a maximum speed of 32 kt.

The 770-ton NANUCHKA-class missile corvettes (Soviet designation "Small Missile Ship," *Maly Raketnyy Korabl'*, MRK) are heavily armed for coastal ASUW at longer ranges than the earlier missile boats. More than 35 have been completed to date in three distinct subclasses. The Nanuchka I has 6 SS-N-9 STARBRIGHT anti-ship missiles, an SA-N-4 GECKO SAM launcher, and a twin 57-mm. gun. The Nanuchka II, for export, has 4 Styx instead of the more advanced Starbright. Three have been sold to Algeria, 3 to India, and 4 to Libya, which lost at least one to U.S. forces in 1987. The Nanuchka III is an improved Nanuchka I with a single 76.2-mm. gun in place of the 57-mm. and a radar-controlled 30-mm. gun. The Nanuchkas have a very broad-beamed planing hull and a maximum speed of 32 kt.

The 540-ton TARANTUL class, smaller and less capable than the Nanuchkas, are more direct successors to the Osa. Based on the Pauk hull, they have a 76.2-mm. gun, 2 30-mm. radar-controlled guns, an SA-N-5 Grail launcher, and 4 anti-ship missiles. The first two ships (Tarantul I) had Styx, but later units have the much more capable SS-N-22 SUNBURN. At least 12 Tarantuls have been completed since 1979; 1 has been transferred to Poland, and 3 to East Germany.

FAST ATTACK CRAFT, SWEDEN: Following the lead of the Israeli navy, Sweden decided to discard all its large surface warships in the mid-1970s, to rely instead on submarines and fast attack craft (FACs). Sweden already had 12 T-42-class motor torpedo boats (MTBs) built between 1956 and 1959. Displacing only 45 tons, these 75.5-ft. (23-m.), 45-kt., wooden-hulled planing boats had 2 21-in. (533-mm.) TORPEDOES and 1 40-mm. BOFORS GUN; all were discarded in the early 1970s. In addition, 11 Plejad-class MTBs were bought from West Germany between 1954 and 1958. Based on the German Jaguar class, they displaced 170 tons, and had a maximum speed of 37.5 kt. and an armament of 6 21-in. torpedoes and 2 40-mm. guns. All were discarded in the 1970s, but influenced the design of subsequent Swedish FACs.

The first FACs built under the new naval policy were 6 Spica I–class MTBs commissioned in 1966–67. Displacing 230 tons, these 134.5-ft. (41-m.) steel-hulled boats are capable of 40 kt. and carry an armament of 6 21-in. wire-guided torpedoes, 1 57-mm. anti-aircraft gun, and 6 57-mm. flare rockets; 3 remain in service. They were followed between 1973 and 1976 by 12 Spica IIs, identical except for their sensors. They were rearmed with 2 RBS-15 anti-ship missiles and 2 wire-guided torpedoes between 1982 and 1985. Yugoslavia purchased 6 modified to mount 2 Soviet SS-N-2c STYX anti-ship missiles.

Officially classified as CORVETTES, the 2 320-ton Spica III (Stockholm) -class boats of 1985 are capable of 32 kt., and armed with 6 or 8 RBS-15s, 1 57-mm. and 1 40-mm. gun, and 2 wire-guided torpedoes.

The 150-ton Hugin or Jageren class of 17 missile boats, commissioned between 1972 and 1982,

were designed and built in in Norway. Smaller and handier than the Spicas, they have a maximum speed of 35 kt. and an armament of 6 PENGUIN anti-ship missiles and 1 57-mm. gun.

Sweden is currently building 6 Goteborg-class corvettes, the first of which will be completed in 1990. Displacing 425 tons, they will carry eight RBS-15s, 1 57-mm. and 1 40-mm. gun, and 4 400-mm. anti-submarine homing torpedoes.

FAST ATTACK CRAFT, UNITED STATES: The U.S. Navy has never maintained a large force of fast attack craft (FACs) in peacetime, on the argument that such vessels do not contribute much to the navy's peacetime missions, and could be mass-produced quickly in wartime.

During the 1950s, the navy did maintain a developmental squadron to evaluate FAC designs and test new tactics, but it acquired large numbers of small craft only during the Vietnam War, for coastal interdiction and riverine warfare. Some 24 80-ft. (24.4-m.) Norwegian-designed Nasty (Tjeld) class torpedo boats were operated as gunboats, and a variety of smaller boats for were built for riverine warfare, including the 50-ft. (15.24-m.) Swift and 31-ft. (9.3-m.) PBR (Patrol Boat, Riverine), all armed with machine guns, mortars, and grenade launchers. Most of these were either scrapped or transferred to South Vietnam by 1973.

From 1966 to 1970, the navy also commissioned 17 Asheville-class patrol gunboats, excellent 245-ton vessels with aluminum hulls and aluminum/fiberglass superstructures. Powered by a four-shaft CODAG plant for a maximum speed of 45 kt, they were originally armed with a 3-in. 50-caliber dual-purpose gun forward and a 40-mm. Bofors gun aft. In the mid-1970s, a number were converted to missile boats, with four RGM-78 STANDARD ARM missiles used in an anti-ship role. At one time, the navy planned to reequip these boats with eight RGM-84 HARPOON anti-ship missiles, but instead 4 were sold (1 to South Korea, 2 to Turkey, and 1 to Colombia), several were converted to research vessels, and the rest scrapped.

By the late 1960s, the U.S. Navy did recognize that anti-ship missiles such as Harpoon had at last provided FACs with a standoff weapon of great effectiveness against larger ships, but chose to develop very complex hydrofoils rather than conventional-hulled missile boats. Several prototypes were evaluated, including the Flagstaff class (later produced under license in Israel as the Shimrit), but in 1972, the U.S., West Germany, and Italy selected the Boeing PEGASUS-class missile hydrofoil

(PHM) as the NATO standard missile boat. Armed with an OTO-Melara 76.2-mm. gun and 8 Harpoon, the Pegasus is capable of more than 50 kt. while foil-borne. The U.S. was to build 30 PHMs, while Germany and Italy would build 8 each; but escalating costs caused Germany and Italy to cancel their orders, and the U.S. finally commissioned only a single squadron of 6 boats. Rather expensive to operate, the PHMs are disliked by the navy high command, which does not value the capabilities of fast attack craft (though recent experiences in the Persian Gulf and the Caribbean have been enlightening). At present, the PHMs are the only FACs in the U.S. Navy, with the exception of a few surviving PBRs, and some two dozen patrol boats used by Navy SEALs.

U.S. shipyards do, however, continue to build FACs for export. The Tacoma Shipyard, for instance, produces an improved Asheville, the 240-ton PSMM (Patrol Ship, Multi-Mission)-5 class, 8 of which have been sold to Korea (4 built under license), 4 to Indonesia, and 2 to Taiwan. Peterson Shipyards also built 8 revised Ashevilles (PGG class) for Saudi Arabia between 1980 and 1982.

Finally, at least 33 100-ft. (30.5-m.), 71-ton CPIC (Coastal Patrol Interdiction Craft), with a very seaworthy deep-V hull, a maximum speed of 43.5 kt., 2 Emerson 30-mm. twin turrets, and provisions for 2 or 4 anti-ship missiles, have been sold to South Korea under a license agreement.

FAST BURN BOOSTER: A hypothetical counter to space-based BALLISTIC MISSILE DEFENSES (BMD). While normal ballistic missile boosters have burn times of 180–360 seconds, so that they burn out well above the atmosphere (where they could easily be detected and attacked by space defenses), a fast-burn booster with a burn time of one minute or less would burn out below 180,000 ft. (54,878 m). In theory, this would deprive BMD TARGET ACQUISITION and FIRE CONTROL sensors of the easily tracked booster exhaust plume, while space-based interceptors (SBIs) or high-energy lasers would be unable to penetrate the upper atmosphere to destroy them, thus making "boost-phase" interception impossible (or at least much more difficult).

There are several impediments to the development of FBBs. First, to reach the ICBM velocity of some 15,000 mph (25,000 kph) in less than one minute, they would require a peak acceleration of some 100 G, instead of the 10–15 g of current boosters. That in turn would necessitate much stronger and heavier booster structures, reducing FBB pay-

loads as compared to current boosters. New, higher-energy fuels would also be needed (quite likely unstable and difficult to contain). Moreover, the high velocities achieved at low altitude would cause atmospheric heating that would destroy the boosters, unless an ablative heat shield were applied to their outer surfaces. That would add several tons of dead weight, further reducing the payload.

Finally, if the warheads were released within the atmosphere, aerodynamic forces would severely degrade accuracy; hence they would have to be kept inside a POST-BOOST VEHICLE (PBV) for release well outside the atmosphere. PBV thrusters could compensate for aerodynamic perturbations, but their exhaust plumes are almost as easy to detect as the larger booster plumes.

Thus FBBs would offer dubious advantages in survivability against space-based BMD, at a greatly increased cost and reduced payload efficiency. For these reasons it is no longer considered a credible BMD countermeasure within the scientific community.

FB-111A: Strategic bomber variant of the General Dynamics F-111 attack aircraft, in service with the U.S. Air Force STRATEGIC AIR COMMAND.

FBB: See FAST BURN BOOSTER.

FBM: Fleet Ballistic Missile, a U.S. Navy designation for SLBM. See also POLARIS; POSEIDON; TRIDENT.

FEBA: Forward Edge of the Battle Area, the forward limits of the area in which friendly ground forces are deployed, excluding the outer zone in which a screen or COVERING FORCE can operate. The FEBA is defined in operational orders, to coordinate fire support and the positioning of forces.

FEL: See FREE ELECTRON LASER.

FENCER: NATO code name for the Sukhoi Su-24 long-range, all-weather, variable-geometry ("swing-wing") INTERDICTION aircraft, in service with Soviet Frontal Aviation (FRONTOVAYA AVIATSIYA, FA) and Naval Aviation (AV-MF); two dozen were sold to Libya and Iraq in 1989–90. Comparable to the U.S. F-111, the Fencer is the only Soviet tactical aircraft with enough range, speed, and avionics capability for deep penetration raids throughout NATO-Europe territory at night and in bad weather; as such, it poses a serious threat to NATO's rear-area establishments, notably airfields, ports, and depots. Given its range, the Su-24 would allow Libya to threaten both Southern Europe and Israel, while Iraq could attack targets in Israel and throughout the Middle East.

Developed from the late 1960s as a replacement for the II-28 Beagle and Yak-28 BREWER light bombers, the Fencer was undoubtedly inspired by the F-111, as well as by earlier Soviet variable-geometry experiments. The Su-19 prototype was flown in 1969–70, and the first production Su-24s entered service around 1974–75. For many years the Fencer was deployed only in the Soviet interior, and it was 1979 before clear photographs were published in the West. The initial Fencer A, built in small numbers for operational evaluation, was soon followed by the Fencer B, with a higher gross weight and more powerful engines, and, from 1981, by the Fencer C, with improved avionics, more effective ELECTRONIC COUNTERMEASURES (ECM), and provisions for air-to-ground missiles. The current Fencer D, which entered service in 1983, has a retractable AERIAL REFUELING probe, a new radar, and an electro-optical sensor system. Fencer E, a maritime strike/reconnaissance variant in service with the AV-MF, has a SIDE-LOOKING AERIAL RADAR (SLAR), INFRARED line scanners, low-level cameras, and provisions for ANTI-SHIP MISSILES. Fencer F is a dedicated ELECTRONIC WARFARE and defense suppression (WILD WEASEL) variant intended to replace the Yak-28 Brewer E. Some 800 Fencers are in service, including 500 assigned to "Strategic Air Armies" for theater nuclear strike, and 65 with AV-MF squadrons in the Baltic.

Superficially similar to the F-111, the Fencer is in some ways a better aircraft, in part because Soviet designers were able to learn from the F-111's shortcomings. The broad, flat-sided fuselage has a very large nose radome housing both a multi-mode, pulse-Doppler RADAR and a smaller TERRAIN-FOLLOWING RADAR (TFR); the latter is linked, as in the F-111, to an automatic pilot for "hands-off" low-level flight. Fencer's weapon delivery computer/radar system is reported to be quite accurate, with a single-pass blind attack error radius (CEP) of some 65 m. Fencer C also has an electro-optical sensor in a chin fairing, probably a FLIR (Forward-Looking Infrared).

As in the F-111, the pilot and WEAPON SYSTEM OPERATOR (WSO) sit side-by-side under a broad canopy faired into the fuselage. In contrast to the F-111, which has a heavy and complex cockpit ejection capsule, the Fencer has conventional ejection seats. Cockpit AVIONICS include complete blind-flying instruments, a pilot's head-up display (HUD), a WSO head-down radar display, a FLIR monitor, a threat warning display, an INERTIAL NAVIGATION unit, a DOPPLER navigation radar, a radar altimeter, a TACAN-type navigational beacon receiver, and an integrated flight data/weapon delivery computer. Immediately behind the cockpit is a large avionics bay, below which are two 23-mm. or 30-mm. cannon. The midfuselage houses several bladder-type fuel tanks, with the engines at the rear. The engines are fed by vertical slot cheek inlets located just behind the cockpit, thereby avoiding the airflow problems experienced with the F-111's aft-mounted intakes. Intended for rough-field operations, the Fencer has large, twin-bogie main landing gear.

As with the F-111, the shoulder-mounted wings are pivoted inboard from large fixed "gloves" swept at 68°. The wings are not swept automatically for aerodynamic optimization; instead they have three sweep settings: 16° for takeoff and landing, 45° for cruising, and 68° for supersonic dash. Liberally fitted with high-lift devices, the wing has full-span leading-edge slats and trailing-edge "flaperons," hinged surfaces which act as both flaps and ailerons. The outboard wing sections house large integral fuel tanks. The large slab tailplane, mounted in line with the wings, acts as a differential "taileron," controlling both pitch and roll with the wings at full sweep. The broad vertical stabilizer has a data link antenna for missile control at the base of the leading edge, a RADAR WARNING RECEIVER and braking parachute in the base of the trailing edge, and an ELECTRONIC SIGNAL MONITORING array at the tip.

Fencer has two fixed wing-glove pylons, two pivoting wing pylons, and six fuselage pylons; all are rated at 2200 lb. (1000 kg.), for a maximum payload of some 22,200 lb. (10,000 kg.). On the Fencer D, the wing-glove pylons are extended over the leading edge to form a distinctive wing fence. Payloads include TN-1000 free-fall nuclear bombs, free-fall high-explosive and CLUSTER BOMBS, LASER-GUIDED BOMBS, NAPALM, chemical bombs, and AS-7 KERRY, AS-10 KAREN, AS-12 KEGLER, and AS-14 KEDGE air-to-ground missiles. Large 660-gal. (3000-lit.) drop tanks are often carried on the outboard fuselage and wing-glove pylons, while the pivoting wing pylons are often fitted with AA-8 Aphid infrared-homing air-to-air missiles for self-defense.

Specifications Length: 69.83 ft. (21.29 m.). Span: (16°) 57.4 ft. (17.5 m.); (68°) 34.45 ft. (10.5 m.). Powerplant: (A) 2 Ly'ulka AL-21F afterburning turbojets, 24,692 lb. (11,200 kg.) of thrust each; (B and later) 2 Turmanskiy R-29B afterburning turbojets, 25,196 lb. (11,453 kg.) each. Fuel: 22,928 lb. (10,400 kg.). **Weight, empty:** 41,888 lb. (19,000 kg.). **Weight, max. takeoff:** 90,390 lb. (41,000 kg.). **Speed, max.:** Mach 2.4 (1590 mph/2560 kph) at 36,000 ft. (11,000 m.)/Mach 1.14 (870 mph/1400 kph) at sea level. **Speed, cruising:** Mach 0.8 (620 mph/1000 kph). **Service ceiling:** 54,100 ft. (16,500 m.). **Combat radius:** 200 mi. (322 km.) lo-lo-lo, 17,600-lb. (8000-kg.) payload/1,115 mi. (1800 km.) hi-lo-hi, 5500-lb. (2500-kg.) payload. **Range, max.:** 4000 mi. (6440 km.).

FERRET: 1. Daimler FV701 4 × 4 ARMORED CAR, in service with the British army and widely exported. Developed in response to a 1946 British army specification, the Ferret is a direct descendant of the famous Daimler scout cars of World War II. The first prototypes were produced in 1949, and the Ferret entered service in 1950. A total of 4409 were delivered through 1971.

The all-welded steel hull has a conventional layout, with the driver up front, a fighting compartment in the middle, and the engine in the rear. Maximum armor thickness is 12 mm., sufficient to resist small arms fire and shell splinters. The driver has three windows with drop-down armored shutters and three observation periscopes. There is a spare tire attached to the left side of the vehicle, with an emergency escape hatch on the right. With the flotation screen erected, the Ferret is propelled though the water by tire action at a speed of 3 mph (5.22 kph). Light, handy, and economical, the Ferret remains useful for security operations.

The Ferret has been produced in many different versions, all with the same basic hull, engine, and drive train. The original Mk. 1 had a two-man crew, with an open-topped casemate for the commander and a pintle-mounted 7.62-mm. machine gun. The Mk.2/3 has a one-man turret with a 7.62-mm. machine gun. The field-modified Mk.2/2 had an extended collar between the hull and the turret base, to give the commander a better field of view. The Mk.2/4 was a Mk.2/3 with additional armor, while the Mk.2/5 was a 2/3 brought up to Mk.4 standard. The Mk.2/6 was a 2/3 with one Vigilant ANTI-TANK GUIDED MISSILE (ATGM) mounted on each side of the turret. The Mk.4 of 1970 introduced larger tires, a stronger suspension, and a flotation screen for amphibious use. The Mk.3 is a Mk.2/3 modified to Mk.4 standard, and the final Mk.5 variant was a Mk.4 with a modified turret and launchers for four SWINGFIRE ATGMs. Further development of the Mk.4 resulted in the larger, more heavily armed FV721 FOX.

Specifications **Length:** 12.6 ft. (3.835 m.) **Width:** 6.25 ft. (1.905 m.). **Height:** 6.16 ft. (1.879 m) **Weight, combat:** 4.4 tons. **Powerplant:** 129-hp. Rolls Royce B60 Mk.6A 6-cylinder gasoline. **Hp./weight ratio:** 29.35 hp./ton. **Fuel:** 22 gal. (97 lit.). **Speed, road:** 58 mph (93 kph). **Range, max.:** 190 mi. (306 km.). **Operators:** Bah, B Fas, Bur, Cam, CAR, Indo, Ku, Mad, N Ye, NZ, Por, Qat, S Af, Seng, Sri L, Su, UAE, UK, Zim.

2. Any ship, aircraft, or other vehicle equipped for the collection of signals intelligence (SIGINT), but the term is used most often for SIGINT-gathering satellites, designed to intercept radar, radio, telemetry, and other electromagnetic emissions.

The U.S. began launching dedicated ferret satellites in 1962 as part of the Discoverer program, but later switched to SIGINT packages "piggybacked" on KH-9 BIG BIRD and KH-11 photo-reconnaissance satellites. These packages monitor radio traffic and radar signals to determine operational characteristics and provide ELECTRONIC ORDER OF BATTLE updates; they also monitor the telemetry of Soviet missile tests, to determine the characteristics of Soviet ballistic missiles and verify compliance with the SALT I and SALT II treaties. The U.S.S.R. tends to rely on specialized satellites for this function.

The U.S. also operates specialized ferrets, code-named Rhyolite, intended specifically to monitor microwave communications. Because their beams are narrow and follow the line of sight, microwave transmissions are difficult to intercept. Rhyolite, first deployed in 1970, and reported to have a collapsible high-gain antenna some 70 ft. (21 m.) in diameter, was first placed in a geosynchronous orbit over the Horn of Africa to monitor communications in Eastern Europe and the European U.S.S.R.; a second Rhyolite, launched in 1973, was placed in orbit over Borneo to monitor the Eastern U.S.S.R. and China. The antenna is sufficiently sensitive to detect the very faint "side-lobe" signals which leak from microwave antennas. Because Soviet engineers did not believe such a large antenna could be placed in orbit, microwave communications were considered secure, and Rhyolite provided much information until it was compromised in 1977 by espionage. A follow-on system code-named Argus would have had an even larger 140-ft. (43-m.) antenna, but was canceled in 1975. A successor, code-named Aquacade, entered service in 1984, but its capabilities remain secret. See also SATELLITES, MILITARY; SPACE, MILITARY USES OF.

FFAR: Folding-Fin Aerial Rocket, a type of unguided ROCKET widely used as an air-to-ground weapon for fixed-wing aircraft and helicopters. The first FFAR, the 57-mm. R4M, was developed by the German *Luftwaffe* in 1944 as an air-to-air, anti-bomber weapon. After World War II, the U.S. Air Force developed the R4M into the 2.75-in. (70-mm.) Mighty Mouse, the main weapon of U.S. interceptors until the advent of guided missiles in the early 1960s. The 2.75-in. rocket was then adapted for air-to-ground attack, launched from 7- and 19-round pods compatible with underwing pylons. The U.S. Navy developed the 5-in. (127-mm.) ZUNI FFAR for use against bunkers and other "hard" targets.

In France, Matra and Thomson-Brandt developed the SNEB-family of 68-mm. (2.68-in.) and 100-mm. (3.9-in.) FFARs, the former launched from 19- and 21-round pods, the latter from a twin-rail launcher. Most Western air forces have adopted either 2.75-in. or 68-mm. rockets (many manufactured locally), but Switzerland uses Oerlikon 81-mm. SNORA rockets, and Sweden has 75-mm. and 135-mm. Bofors rockets.

The U.S.S.R., Warsaw Pact, China, and their customers rely mainly on the 57-mm. S-5 series of FFARs, launched from 8-, 16-, 19-, and 32-round pods. In addition, the 80-mm. S-8, 130-mm. S-13, 160-mm. S-16, 210-mm. S-21, 240-mm. S-24, 280-mm. S-28, and 325-mm. S-32 have been produced (few of the larger types remain in service).

FFARs are simple; each consists of a solid rocket motor, a warhead, and a fin assembly. Most have spring-loaded pop-out fins, normally canted to impart spin stabilization (some also have "scarfed" rocket nozzles for that purpose). Warheads are often interchangeable: the 2.75-in. U.S. Hydra-70 family, for example, includes HE, HEAT, fléchette, smoke, illumination, and minelet warheads, all weighing around 7.5 lb. (3.4 kg.).

FFARs are not accurate, but they are cheap, and when fired in salvo can cover a wide area, making them excellent suppression weapons. Performance varies with caliber and rocket motor characteristics. The Hydra-70 has a burnout velocity of 1000 m./sec., and is reasonably accurate at ranges up to 2000 m., depending on the speed and altitude at launch. The latest SNEB 68-mm. and 100-mm. rockets have a burnout velocity of 600–800 m./sec. and are less accurate, while the Soviet S-5 has a burnout velocity of 620 m./sec.

FH-70: Towed 155-mm. HOWITZER developed jointly by Britain, Italy, and West Germany in response to a 1968 requirement for a common replacement for their old U.S. M114 155-mm. howitzers and British 5.5-in. guns. The FH-70 entered service in 1978, with Britain ordering 71, West Germany 216, and Italy 164. Saudi Arabia began acquiring FH-70s in 1986, and Japan recently acquired licensed production rights.

The FH-70 has an overall length of 32.1 ft. (9.8 m.), with a 39-caliber (19.75-ft./6.02-m.) barrel and a firing weight of 20,503 lb. (9300 kg.). As in most modern towed artillery, the wheels can be jacked up so that the weapon rests on a flat turntable which can be traversed 360°. Elevation limits are −5° and +70°. The split-trail carriage has an auxiliary power unit (APU) which drives small wheels at the end of the trails; with this, the FH-70 can move across country at 10 mph (16 kph) for short distances. The APU also drives a semi-automatic loader, which allows burst fire at a rate of 3 rounds in 13 seconds and a normal rate of 6 rounds/min. (twice the manual rate of fire).

The FH-70 can fire the full range of NATO 155-mm. ammunition, including high explosive, IMPROVED CONVENTIONAL MUNITIONS (ICMs), ROCKET-ASSISTED PROJECTILES (RAP), smoke, illumination, and the M712 COPPERHEAD guided projectile, as well as nuclear and chemical rounds. Muzzle velocity is 2723 ft./sec. (830 m./sec.), for a maximum range of 24,000 m. with conventional rounds, or 30,000 m. with RAP rounds. The FH-70 has a crew of eight and can be towed by a variety of 6 × 6 trucks. See also ARTILLERY, BRITAIN; ARTILLERY, FEDERAL REPUBLIC OF GERMANY.

FIDDLER: NATO code name for the Tupolev Tu-28P, a large, long-range supersonic INTERCEPTOR of 1960s vintage. A few remain in service with the Soviet IA-PVO (Fighter Aviation of the Air Defense Troops) but are now being replaced by MiG-31 FOXHOUNDS and Su-27 FLANKERS.

FIELD ARTILLERY: Light and medium-caliber artillery intended to accompany and support mobile units. See ARTILLERY.

FIGHTER (AIRCRAFT): A combat aircraft designed primarily to engage other aircraft, including other fighters. That specific role requires agility (the ability to change direction quickly), maneuverability (generally equated with sustained turning performance), fast acceleration, a high rate of climb, and high speed, though very high absolute speed is not particularly useful except for disengagement and escape. Designs were long dictated by the needs of AIR COMBAT MANEUVERING with guns and tail-chase missiles, which require fighters to achieve a position behind their target to ensure a high probability of kill (P_K); today's all-aspect missiles have yet to influence fighter configurations. The same performance characteristics also allow modern fighters to perform a variety of other missions, including interception, ground attack, and tactical reconnaissance.

An INTERCEPTOR is a specialized anti-bomber fighter optimized for speed and a high rate of climb, often at the expense of maneuverability, while ATTACK AIRCRAFT (also called ground-attack fighters or fighter-bombers) are optimized for carrying heavy air-to-ground payloads over long distances, and may lack the speed and maneuverability needed for aerial combat. Because of the high development cost of modern aircraft, specialized interceptors and attack aircraft have tended to give way to general-purpose "multi-role" fighters, a process facilitated by recent advances in aircraft propulsion and AVIONICS. Current, high-power, fuel-efficient engines give the latest fighters thrust-to-weight ratios greater than 1 to 1, which yield rates of climb equal or superior to those of the best interceptors of the last generation. Such engines also allow fighters to carry large payloads comparable to those of attack aircraft, while advanced weapon delivery computers make them just as accurate. Moreover, multi-role fighters can immediately revert to the fighter role after completing their air-to-ground missions.

Fighters are still classified as either "day" or "all-weather," or, more meaningfully, light or heavy. The former are relatively simple aircraft with a single pilot, mainly intended to operate in good visibility conditions. Avionics may be simple, limited to tactical radios, an optical gunsight or head-up display (HUD), a few RADAR WARNING RECEIVERS (RWRs), and a range-only RADAR. Their armament likewise is normally limited to cannon and INFRARED-HOMING air-to-air missiles. Typical day fighters include the 1950s F-86 SABRE, most early MIGs, the 1970s F-5 TIGER, the original F-16 FALCON, and the Israeli KFIR. Though relatively cheap to operate, day fighters are being displaced by multi-role fighters.

The all-weather (heavy) fighters evolved from the interceptors of the 1960s. With a one- or two-man crew, they can be distinguished by their large nose radars. Originally, these were airborne-intercept (AI) sets similar to those of interceptors; currently, however, they are multi-mode pulse-Doppler radars with navigation, ground-mapping, and air-to-ground modes. Such radars allow fighters to find and engage aerial targets with radar-guided missiles (e.g., the AIM-7 SPARROW), and, in their air-to-ground modes, they allow precision low-level navigation and target acquisition at night and in bad weather. Originally, the size of such radars dictated the size of all-weather fighters (as in the F-15 EAGLE or Su-27 FLANKER), but electronic miniaturization has allowed the development of compact multi-mode radars, so that late-model all-weather fighters (e.g., the FA-18 HORNET and MiG-29 FULCRUM) are not very large. "Limited all-weather" fighters are essentially day fighters with an augmented radar, but they lack the avionics needed for blind interception or true night operations. Examples include the F-104G STARFIGHTER, the MiG-23 FLOGGER, and later versions of the F-16 FALCON.

An "AIR SUPERIORITY fighter" is a long-range all-weather fighter intended either for prolonged defensive patrols, or for operations deep into enemy airspace; the F-15 Eagle and Su-27 Flanker are paragons of this type.

Many fighter aircraft are also used for tactical reconnaissance. Some have modified noses housing cameras, infrared line scanners, and other sensors, but the current trend is to package reconnaissance equipment in pods which can equip any standard fighter.

FINAL PROTECTIVE FIRE (FPF): A pre-planned linear ARTILLERY barrage laid across likely avenues of approach to prevent enemy forces from entering a friendly defensive position. FPF has a higher priority than any other type of fire mission. See also INDIRECT FIRE.

FIRE AND FORGET: Also called launch and leave, describes weapons which require no external guidance after launch, either because they are unguided, or because they employ some form of autonomous guidance or homing. Thus an unguided anti-tank rocket, an *active*-HOMING radar-guided missile (e.g., PHOENIX or AMRAAM), or a passive, INFRARED-HOMING missile (e.g., Sidewinder) are all "fire and forget" weapons.

The main advantage of fire and forget over COMMAND-GUIDED or SEMI-ACTIVE homing weapons is that the launch platform can engage additional targets or initiate evasive action immediately after launch. On the other hand, fire-and-forget guided weapons (especially long-range, radar-guided missiles) tend to be both much more complex and expensive than command-guided or semi-active homing missiles.

FIRE AND MOVEMENT: Infantry TACTICS evolved in response to World War I trench-fighting conditions, wherein the high volumes of defensive fire made frontal assaults impossible or very costly in lives. From the initial idea that DIRECT FIRE from MACHINE GUNS and MORTARS was needed to force front-line defenders to keep their heads down, various assault schemes have emerged. All involve the division of the attacking force into two or more teams. While one (the "fire team") suppresses the defenders with machine-gun and other heavy weapon fire, the other (the "movement team") dashes forward to a position nearer to the enemy. Then it lays down suppressive fire in turn, while the original fire team advances. The process is repeated until the enemy position is overrun. Sometimes a third, smaller, team is added, to act in a scouting and command role, as in the now-discarded U.S. Able-Baker-Charlie squad system. For competitive tactics, see INFILTRATION; MARCHING FIRE; OVERWATCH.

FIREBAR: The Soviet Yakolev Yak-28P, interceptor variant of the BREWER light bomber. Few if any remain in service.

FIRE CONTROL: Equipment and procedures meant to facilitate the application of firepower. The simplest form is of course the human eye, sometimes augmented by telescopic sights or reticles to assist range estimation. This can be effective at short ranges against stationary targets, but to attack moving targets at longer ranges, more complex automated or semi-automated fire control systems are generally required, whose nature varies greatly between unguided gun or rocket weapons on the one hand, and guided missiles on the other.

Fire control systems for guns or rockets must allow for all the factors that affect both projectile and target, including interior ballistics, barrel wear, atmospheric conditions (particularly barometric pressure and wind velocity), the rate at which the projectile drops (a function of gravity and drag coefficient), the rate at which it decelerates (a function of the ballistic coefficient), the tilt or inclination of the firing platform, and relative target motion (a product of target and firing platform movement). Because projectiles decelerate and drop after firing or launch, tending to follow a parabolic trajectory, the gun or launcher must be aimed at a point above the target's position. Similarly, because of the time required by projectile to travel to a moving target ("dead time"), the fire control system must also compute the "lead," i.e.,

aim the gun or rocket at a point ahead of the target.

To accomplish these functions, gun and rocket fire control systems require a sight unit, a rangefinder, and some form of computer. The oldest types rely on optical sights and optical rangefinders (either stadiametric, stereoscopic, or coincidence), and on manual, electro-mechanical or analog computers. More modern systems have RADAR or ELECTRO-OPTICAL sighting, radar or LASER rangefinding, and digital computers. Fire control radars are normally pencil-beam types (or multifunctional radars with pencil-beam modes), to provide accurate range and angular measurements.

To determine relative target motion, a "predictor" (either a separate apparatus or part of the main computer) combines continuous target range, speed, and bearing (and sometimes apparent heading) inputs to compute the target's expected position at a given point in the future. The fire control computer then determines the correct azimuth, elevation, and firing time for the weapon. But of course the predictor must assume that the target will maintain a constant course and speed; thus even the best gun and rocket fire control systems have relatively low hit probabilities against moving targets, except for high-velocity projectiles at very short ranges, when the "dead time" between firing and hitting is minimal, as, e.g., for tank guns with muzzle velocities of some 1500-1600 m./sec., over ranges up to 1500 m. Such high-velocity projectiles, moreover, drop relatively little at first, so that 1500 m. is in effect "point-blank" range for modern tank guns: the gunner need only point the gun directly at the target. For lower-velocity weapons, point-blank range is correspondingly shorter, and the probability of hitting a moving target correspondingly less.

Aircraft fire controls have to contend with higher relative speeds, and the complication that both the target and the firing platform can be moving in three dimensions. Modern fighter gunsights with radar ranging automatically compute lead, and provide the pilot with steering and range cues, but the probability of hitting is still relatively low unless the pilot closes to point-blank range (some 500 m. or less) and minimizes relative motion by attacking from dead astern at the same altitude as the target. Modern aircraft cannon compensate for low single-shot hit probabilities by high rates of fire (up to 6000 rds./min.), but most fighters can carry only enough ammunition for a few seconds of continuous fire.

BALLISTIC MISSILE fire controls require only the coordinates of the launch point and target, and the desired time of flight. Because there are generally no provisions for postlaunch guidance corrections, ballistic missiles can only be aimed at stationary targets.

Fire controls for COMMAND-GUIDED weapons depends on their mode: with manual command to line-of-sight (MCLOS), the missile is steered manually with a joystick; with semi-automatic command to line-of-sight (SACLOS), the operator keeps his sights fixed on the target while a computer steers the missile onto the operator-target line of sight; and with automatic command to line-of-sight (ACLOS), a computer uses sensor data (usually radar) to track both missile and target, so as to generate the necessary steering commands.

BEAM-RIDING guidance also requires the fire control unit to track both missile and target, using two separate pencil-beam radars. With data from the target tracking beam, the computer derives successive predicted target positions, and then points the missile guidance beam at that point in space.

SEMI-ACTIVE RADAR HOMING (SARH) requires only that the fire control unit keep the launch platform's illumination radar pointed at the target, so that the missile may home on its return echoes. Before launch, however, the unit indicates when the missile has achieved lock-on, and when the target is within effective range. SEMI-ACTIVE LASER HOMING (SALH) systems are similar, but use laser designators instead of radar to illuminate the target.

ACTIVE HOMING missiles do not require any guidance from the launch platform, but before launch the fire control unit usually provides cues for lock-on and optimal launch ranges.

Fire control systems for unguided TORPEDOES generally resemble those for guns or rockets, while fire controls for acoustic homing torpedoes resemble those for active or passive homing missiles. WIRE-GUIDED torpedoes, on the other hand, are essentially command-guided missiles. The very low speed of the torpedoes as compared to missiles or projectiles, and the much longer "dead times" involved, greatly complicate their fire control.

FIRE DIRECTION CENTER (FDC): U.S. term for a command post through which requests for ARTILLERY support are processed. Generally found at battalion level or higher, an FDC consists of gunnery and communications personnel with the equipment (plotting boards and fire control computers) needed to convert target intelligence and requests for fire support into specific fire directions (azimuth, elevation, type and number of rounds, etc.) for artillery or mortar batteries. See also INDIRECT FIRE.

FIREPOWER: A theoretically quantitative but usually qualitative measure of the destructive power of the weapons or arrays of weapons available to a military unit (not a measure of the power of the unit itself). Firepower is rarely expressed numerically, but if so, this is done either by the weight of shell that can be fired per shooting cycle, or the total rounds of stated caliber that can be fired in a given time period.

FIRE SUPPORT: Direct or indirect artillery or naval gunfire delivered to assist or protect a ground force in combat.

FIRE SUPPORT COORDINATION LINE: A line established by a ground commander to ensure the coordination of ARTILLERY, naval gunfire, and CLOSE-AIR SUPPORT not under his operational control, but which may affect the execution of his plans. The fire support coordination line is supposed to follow well-defined terrain features, and is established in cooperation with the commanders of the appropriate forces. Targets forward of the fire support coordination line may be attacked without prior notification to the ground force commander. Targets behind the line can be attacked only after coordination with the ground force commander.

FIRE TEAM: A subunit of an infantry rifle SQUAD, usually consisting of 3–4 men under the command of a corporal. Two or three fire teams form a squad.

FIRST STRIKE (CAPABILITY, "STRATEGY"): A "first strike" is the first use of NUCLEAR WEAPONS in an armed conflict. "First strike capability" refers to the possession of DELIVERY SYSTEMS for nuclear weapons which are presumed to be unable to survive a (nuclear) attack upon them, a vulnerability which in reality depends on the specific capabilities of the enemy, but which is often perceived unilaterally in relation to the number and nature of available delivery systems (e.g., unprotected missiles requiring lengthy prelaunch preparation, or bomber forces whose short range requires their basing within range of enemy attack).

Such a "first-strike-only" capability therefore creates an incentive to first use if there is an expectation of enemy attack, because the force could not survive to deliver a post-attack retaliatory blow.

Confusingly, the phrase also has a second, completely different meaning: a strike capability sufficiently effective to destroy enemy (nuclear) forces in one blow; in other words, a disarming COUNTERFORCE capability. Thus while one meaning implies the possession of few and/or low-quality delivery systems, the other implies the exact opposite.

FISHBED: NATO code name for the Mikoyan MiG-21 FIGHTER/INTERCEPTOR, the first Soviet fighter capable of Mach 2 in level flight, and the most widely deployed of all post-1945 fighters. Developed from 1954 as a replacement for the MiG-19 FARMER, the MiG-21 was optimized for maneuverability and rate of climb. Two competitive wings were evaluated, one with a 62° sweep (later used on the Su-7 FITTER), and the other a tailed delta. After flight tests in June 1956, the latter wing was selected, and the first production aircraft entered service in 1958. The initial MiG-21F (Fishbed-C) was a day fighter with a small ranging radar, 1 or 2 30-mm. Nudelmann-Rikter NR-30 cannons, and 2 wing pylons for drop tanks or AA-2 ATOLL infrared-homing air-to-air missiles. The F was also produced in Czechoslovakia, and copied in China as the Shenyang J-7. Much used by the Vietnamese air force, it proved a formidable opponent for heavier, faster, but less maneuverable U.S. F-4 PHANTOMS.

The MiG-21 has been produced in more variants than any other jet fighter worldwide. Fishbed-C was superseded in 1962–63 by the MiG-21PF (Fishbed-D), with an R-1L "Spin Scan A" search radar offering some "all-weather" capability. The PF is distinguished from its predecessor by an enlarged centerbody radome, a revised canopy faired into an enlarged fuselage spine, and a broader vertical stabilizer. Other less noticeable changes included a more powerful engine and removal of the internal guns. Normally armed with two or four Atolls, it can also carry a 23-mm. twin-barrel GSh-23L cannon in an external ventral pack. Later models with blown flaps and provisions for rocket-assisted takeoff (RATOG) were designated Fishbed-E. The MiG-21FL, an export variant without RATOG or flap blowing, was built under license in India as the HAL "Type 77."

The MiG-21PFM (Fishbed-F) of 1965 had a better R2L "Spin Scan B" radar, an enlarged fin area, and yet more power. The MiG-21PFMA (Fishbed-J) is a multi-role development of the PFM, with an improved "Jay Bird" radar, a better ejection seat, a deeper fuselage spine, and four wing pylons with provisions for up to 2200 lb. (1000 kg.) of bombs and rocket pods. With reinforced wings, it was significantly faster than earlier models at low altitude. It was extensively used in the last years of the Vietnam War, in the Israeli-Egyptian War of Attrition (1968–70), and in the 1973 Yom Kippur War. The MiG-21M is an Indian-built variant of the PFMA with an Indian-built R-11F2S-300 engine, while the MiG-21R, a RECONNAISSANCE variant supplied to Egypt, has a ventral camera pack in place of the gun pod. The MiG-21R (Fishbed-H) is a tactical reconnaissance version of the PFMA with a centerline drop tank, a multi-sensor pod, and radar warning receivers in wingtip pods.

The MiG-21MF of 1969–70 is similar to the PFMA, but has a lighter, more powerful Turmanskiy R-13 turbojet. The MiG-21RF (Fishbed-H) is a tactical reconnaissance variant of the MF, with the same sensor package as the MiG-21R.

The MiG-21SMT (Fishbed-K) of 1971 was optimized for ground attack. It has a much longer and deeper fuselage spine which houses additional fuel, avionics, and electronic countermeasures.

By the early 1970s, however, the MiG-27 FLOGGER was selected as the principal Soviet ground attack aircraft, and in the MiG-21 bis (Fishbed-L) of 1973–74 the emphasis was shifted back to air combat. Externally similar to the SMT, Fishbed-L has much better avionics. The latest MiG-21bis (Fishbed-N) has a Turmanskiy R-25 afterburning turbojet, structural refinements, and even better avionics. Produced under license in India, it is still being built in the Soviet Union for export.

There has also been a series of two-seat MiG-21U trainers (NATO code name Mongol) corresponding to the various single-seat variants.

More than 10,000 MiG-21s have been produced since 1958, and large numbers remain in service with many countries. Still phasing out of front-line Soviet squadrons, many MiG-21s are being kept in reserve. Though never equal to the best of Western fighters, the MiG-21 could be more cost-effective for many users, being cheap, simple, easy to operate, and above all, available in large numbers.

Rather small for a modern fighter, the MiG-21 is difficult to see—a definite advantage in aerial combat, especially when jamming degrades radar observation. The cylindrical fuselage has an "area-rule" (coke-bottle) shape to reduce trans-sonic drag, and a circular nose intake with a conical centerbody housing for a modified R1L "Jay Bird" RADAR. With a narrow scan pattern and a detection range of only 18 mi. (30 km.), Jay Bird is not comparable to the large, multi-mode radars of late-model Western and Soviet fighters (e.g., it has no "look-down" capability), but it does provide illumination for SEMI-ACTIVE RADAR HOMING (SARH) versions of Atoll.

The pilot sits in a cramped cockpit under a clamshell canopy which is faired into the fuselage spine, obstructing rearward visibility. The windscreen is small and heavily framed, partially obstructing the pilot's forward view as well. Though the MiG-21bis is better than earlier variants, cockpit AVIONICS are still crude and limited by Western standards: a rudimentary head-up display (HUD), a head-down radar display with equally rudimentary signal processing, a Sirena-3 RADAR WARNING RECEIVER, a radar altimeter, a TACAN-type navigational beacon receiver, a DOPPLER navigation radar, an instrument landing system, and a UHF DATA LINK for ground-controlled interception (the latter is a necessity because of radar limitations and poor cockpit visibility).

Immediately behind the cockpit, the fuselage houses a small avionics bay. The twin-barrel GSh-23L cannon is mounted externally under the belly, with a 200-round magazine inside the fuselage. The rest of the fuselage houses several fuel cells (with additional fuel and avionics in the spine), with the engine at the rear. Though interchangeable with the earlier R-13, the R-25 is a lighter, more advanced engine with much better fuel economy and reliability (though still not up to Western standards). The Soviet Air Force (VVS) frequently operates from unpaved fields, and the MiG-21 is equipped with large, low-pressure main landing gear which retract into the lower half of the fuselage.

The midmounted wing is a 52° delta with a fixed leading edge, outboard ailerons, and inboard flaps "blown" with engine bleed air to improve low-speed handling. Two small ECM pods can be attached to the wingtips. Though the main landing-gear struts occupy the inboard wing sections, there is still sufficient space outboard for two fuel tanks. But in all variants, fully 30 percent of internal fuel cannot in fact be used, because it would make the aircraft too tail-heavy for a safe landing (!). The sharply swept slab tailplane is mounted in line with the wings; the vertical stabilizer has a large conformal aerial at the tip, and a braking parachute at the base. A large, fixed ventral fin below

the tail provides directional stability at supersonic speeds and also acts as a safety bumper during takeoff and landing. Fishbed-N has a combat thrust-to-weight ratio of more than 1 to 1, as in the latest Western and Soviet fighters. Because of the drag rise of its delta wing, the sustained turn rate is not even as good as that of the Phantom, but fast acceleration, a high rate of climb, and an excellent instantaneous turn rate still make the MiG-21 a dangerous dogfighter.

The MiG-21bis has one fuselage and two inboard wing pylons rated at 1100 lb. (500 kg.) each, and a pair of outboard wing pylons rated at 550 lb. (250 kg.) each, for a total payload of 4400 lb. (2000 kg.). But for an adequate tactical radius, the centerline pylon almost always carries a 108-gal. (490-lit.) drop tank, limiting the weapons load to only 3300 lb. (1500 kg.). For air combat, typical armament would be 2 AA-2-2 SARH Advanced ATOLL and 2 AA-8 APHID infrared-homing (IR) missiles; Indian MiG-21s carry up to 4 Matra R.550 MAGIC IR missiles, while Chinese J-7s have the PL-7, a copy of the AIM-9 SIDEWINDER. For ground attack, the 21bis can carry 550- and 1100-lb. (250- and 500-kg.) free-fall and CLUSTER BOMBS, NAPALM, chemical bombs, and 16-round UV-16-57 rocket pods for 57-mm. FFARS.

Specifications Length: 51.75 ft. (15.75 m.). **Span:** 23.5 ft. (7.16 m.). **Powerplant:** (C) 1 Turmanskiy R-11 afterburning turbojet, 11,244 lb. (5100 kg.) of thrust; (D) R-11F, 13,118 lb. (5950 kg.) of thrust; (F) R-11-300, 13,668 lb. (6200 kg.) of thrust; (MF) R-13-300, 14,550 lb. (6600 kg.) of thrust; (N) R-25, 16,536 lb. (7500 kg.) of thrust; (bis) R-25, 19,500 lb. (8863 kg.) of thrust. **Fuel:** (N) 4608 lb. (7695 kg.). **Weight, empty:** (N) 13,500 lb. (6200 kg.). **Weight, normal loaded:** 19,300 lb. (8750 kg.). **Weight, max. takeoff:** 22,000 lb. (10,000 kg.). **Speed, max.:** (N) Mach 2.1 (1385 mph/2730 kph) at 36,000 ft. (11,000 m.)/Mach 1.05 (800 mph/1290 kph) at sea level. **Initial climb:** (N) 58,000 ft./min. (17,677 m./min.). **Service ceiling:** 59,055 ft. (17,999 m.). **Combat radius:** (N) 200 mi. (320 km.). **Range, max.:** (N) 682 mi. (1100 km.). **Operators:** Afg, Alg, Ang, Bang, Bul, Cuba, Cze, DDR, Egy, Eth, Fin, Hun, Iraq, Laos, Lib, Mad, Mon, Nig, N Kor, Pol, PRC, Rom, Som, Sud, S Ye, Syr, USSR, Viet, Zam.

FIST: Fire Support Team, U.S. Army term for a squad-size artillery forward observer unit, consisting of three FORWARD OBSERVERS (FOs) and three radio operators under the command of a FIST chief, usually a second lieutenant. The FIST is usually assigned to a maneuver company by the battalion FIRE DETECTION CENTER (FDC) to provide target intelligence and formulate requests for ARTILLERY and CLOSE AIR SUPPORT, for relay back to the FDC. In tank and mechanized infantry battalions, the FIST is mounted in an M981 Fire Support Team Vehicle (FISTV). See also INDIRECT FIRE.

FISTV: U.S. M981 Fire Support Team Vehicle, a modified M113 armored personnel carrier designed to allow artillery FORWARD OBSERVERS with armored and mechanized units to acquire and designate targets while operating under armor protection. Meant to equip the TARGET ACQUISITION companies of self-propelled artillery battalions, the first FISTVs were delivered in 1984, and some 340 M113s were converted to FISTVs through 1986.

The FISTV is an M113A2/A3 with a modified version of the Emerson Electric erectable missile launcher (of the M901 IMPROVED TOW VEHICLE), with the missile launch tubes replaced by low-light television (LLTV) and THERMAL IMAGING sensors, and a LASER DESIGNATOR/rangefinder. With them, the FISTV can provide accurate target ranges in all weather and also illuminate targets for the M712 COPPERHEAD guided artillery projectile, as well as for HELLFIRE and MAVERICK laser-homing missiles and PAVEWAY laser-guided bombs. Other specialized equipment includes four tactical radios (one a 4-channel digital message unit), a north-seeking gyrocompass, and a Position Location and Reporting System (PLRS) radio-navigation unit. The FISTV has a combat weight of some 13 tons and a crew of five; it is armed with a pintle-mounted 7.62-mm. machine gun for self-defense. See also FIST; INDIRECT FIRE.

FITTER: NATO code name for the Sukhoi Su-7/-17/-20/-22 series of single-engine ATTACK AIRCRAFT. Developed from the late 1950s and continually updated, the Fitter has remained the backbone of Soviet Frontal Aviation (FRONTOVAYA AVIATSIYA, FA) attack regiments for more than 35 years. Hundreds of Fitters have also been exported to Warsaw Pact and Third World countries.

The original Su-7B Fitter A was initially designed as an air-superiority fighter, but when that role was assigned to the smaller, more agile MiG-21 FISHBED, the design was recast for ground attack. The first prototype flew in 1955, and the first production aircraft began entering service in 1959. They were soon superseded by the Su-7BM, with a more powerful engine, which was in turn replaced by the Su-7BMK with large, low-pressure tires for rough-field operations. More than 2000 were produced through 1971; only a few remain active in the Soviet air force.

Simple and exceptionally rugged, the Su-7 has a circular nose intake with a small centerbody that houses a simple SRD-5M range-only RADAR. The pilot sits just behind the nose, under a sliding bubble canopy with a bulletproof windscreen and a solid rear end. AVIONICS are rudimentary by Western standards: an ASP-5 radar-ranging optical gunsight, a Sirena-3 RADAR WARNING RECEIVER, and SRO-2M "Odd Rods" IFF, and the usual tactical radios; that effectively limits the Su-7 to day-only, clear weather operations. Its Ly'ulka AL-7F afterburning turbojet is notoriously inefficient, and the entire fuselage between the cockpit and the engine houses a single large fuel tank. Because of the tank's location, and the great weight of the engine, some of that fuel cannot be used without shifting the aircraft's center of gravity too far aft for a safe landing.

The large, midmounted wings, swept at 62°, have fixed leading edges, power-operated ailerons, and inboard slotted flaps. Prominent fences are fitted at midspan and at the tips to provide stall warning. Wing-loading is high, and the Su-7BMK is noted for its long takeoff run (7872 ft./2400 m.) and "hot" landings, for which a braking parachute is housed at the base of the vertical stabilizer. On the other hand, the high wing loading makes the Su-7 a very steady weapon platform, especially at low altitude. Two powerful 30-mm. NR-30 cannons are mounted in the wing roots, each with 70 rounds of ammunition. The main gear retract into the inboard wing sections, while the outboard sections house integral fuel tanks. The slab tailplane is power-assisted, but control forces are very high by Western standards; nonetheless, the Su-7 has excellent aerobatic characteristics.

The Su-7BMK has two side-by-side fuselage pylons rated at 1100 lb. (500 kg.) each, and four wing pylons, with the inboard pair also rated at 1100 lb. (500 kg.) each, the outboard pair at 550 lb. (250 kg.) each. While the maximum weapon load is theoretically some 5500 lb. (2500 kg.), to achieve a useful combat radius the belly or inboard wing pylons are usually fitted with 132-gal. (600-lit.) drop tanks, leaving a useful payload of only 4400 lb. (2000 kg.). Typical ordnance includes 19-round UV-16-57 rocket pods for 57-mm. FFARS, 1100-lb. (500-kg.) free-fall or CLUSTER BOMBS, NAPALM, and chemical bombs.

Short range combined with the long takeoff and

landing runs were unacceptable to the Soviet air force, which often flies from short and unpaved fields. In the late 1960s, Sukhoi equipped an Su-7 with variable geometry ("swing") wings, which could be set to 28° for takeoff and landing, or at the original 62° for high-speed cruise. That greatly reduced takeoff and landing speeds and distances, reduced control forces, and provided a smaller turn radius and much better range. With additional aerodynamic refinements, the resulting aircraft entered service as the Su-17 Fitter C in 1970–71. Since then, successive variants of the Su-17 have almost completely replaced the Su-7, becoming the most numerous attack aircraft in FA.

The initial Fitter C had the same fuselage as the Su-7, but the wingspan was extended. Powered by a new, slightly more efficient Ly'ulka AL-21F-3, the Su-17 also has a slightly greater internal fuel capacity. Eight weapon pylons are provided: two tandem pairs on the fuselage, and four on the wings, for a maximum payload of 8800 lb. (4000 kg.). In addition to the standard ordnance of the Su-7, Fitter C can also carry the AS-7 KERRY, AS-9 KYLE, and AS-10 KAREN air-to-surface missiles, greatly increasing its combat effectiveness.

Fitter C was soon followed by the Su-17M Fitter-D, with a longer nose and a deep fuselage spine, which houses additional fuel and avionics, and with the sliding bubble canopy replaced by a clamshell type faired into the spine. Avionics were greatly improved, to include a TERRAIN-FOLLOWING RADAR under the nose and a laser rangefinder in the intake centerbody. The Su-17UM Fitter E was a two-seat trainer, originally codenamed "Mongol." Fitter H has a redesigned wing and revised avionics, including a head-up display (HUD) and an internal chaff/flare dispenser. Combat radius is further extended, with a heavier payload. Fitter G is the two-seat trainer version of Fitter H. Introduced in 1984, the latest Soviet variant (Fitter K) has a redesigned vertical stabilizer with a dorsal ram-air cooling inlet. At present, there are some 650–850 Su-17s of all variants with FA, plus an additional 100 assigned to Soviet Naval Aviation (AV-MF).

The first export variant, the Su-20 (also designated Su-17MK), is similar to Fitter C, but has only two fuselage pylons and more austere avionics. The Su-22 Fitter F, supplied to Peru in 1977, is generally equivalent to the Fitter D, but has a more efficient Turmanskiy R29B afterburning turbofan and provisions for AA-2 ATOLL infrared-homing air-to-air missiles. (These were subse-

quently modernized with U.S. avionics.) The Su-22UM two-seat trainer with the Turmanskiy engine is otherwise similar to Fitter G.

The Su-22 Fitter J is an export variant of the Soviet Fitter H, supplied to Libya and Peru. Powered by the Turmanskiy engine, it has more internal fuel and provisions for Atoll missiles (two Libyan Fitter Js were shot down by U.S. F-14 Tomcats over the Gulf of Sidra in 1981). The Su-22M-4, supplied to Czechoslovakia, East Germany, and Poland, is generally similar to Fitter K. Including export models, more than 4000 variable-geometry Fitters have been produced to date.

Specifications **Length:** (A) 57 ft. (17.37 m.) (C) 61.5 ft. (18.75 m.). **Span:** (A) 29.3 ft. (8.93 m.); (C, D, H) 45.95 ft. (14 m.) at 28°/34.8 ft. (10.6 m.) at 62°. **Powerplant:** (A) 1 Ly'ulka AL-7F afterburning turbojet, 22,046 lb. (10,000 kg.) of thrust; (C) 1 Ly'ulka AL-21F-3, 24,700 lb. (11,200 kg.) of thrust; (F) 1 Turmanskiy R-29B afterburning turbofan, 25,350 lb. (11,500 kg.) of thrust. **Fuel:** (A) 5187 lb. (2353 kg.); (C) 6800 lb. (3091 kg.); (J) 11,260 lb. (5118 kg.). **Weight, empty:** (A) 19,000 lb. (8620 kg.); (C) 22,050 lb. (10,000 kg.). **Weight, normal loaded:** (A) 26,455 lb. (12,025 kg.); (C) 30,865 lb. (14,029 kg.). **Weight, max. takeoff:** (A) 29,750 lb. (13,495 kg.); (C) 39,020 lb. (17,736 kg.). **Speed, max.:** Mach 1.6 (1056 mph/1699 kph) at 36,000 ft. (11,000 m.)/Mach 1.1 (840 mph/1352 kph) at sea level. **Initial climb:** 29,900 ft./min. (9120 m./min.). **Service ceiling:** 49,700 ft. (15,150 m.). **Combat radius:** (A) 215 mi. (345 km.); (C) 391 mi. (630 km.); (H) 435 mi. (726 km.). **Range, max.:** (A) 900 mi. (1450 km.). **Operators:** (7) Afg, Alg, Cze, Egy, Hun, Ind, Iraq, N Kor, Pol, Rom, S Ye, Syr, Viet; (17/20/22) Alg, Cze, DDR, Egy, Iraq, Lib, Peru, Pol, USSR.

FLAGE: Flexible, Agile, Guided Experiment, a U.S. Army program sponsored by the STRATEGIC DEFENSE INITIATIVE to test the feasibility of designing a BALLISTIC MISSILE DEFENSE interceptor missile from "off-the-shelf" components. With active millimeter wave (MMW) guidance and lateral "puff-jet" steering rockets, a FLAGE prototype launched from a U.S. Navy F-4 Phantom on 27 June 1986 successfully intercepted a high-speed target missile launched 30 seconds earlier from the same aircraft. FLAGE thus demonstrated that sufficient accuracy for the nonnuclear interception of ballistic missile reentry vehicles was possible with existing technology. Much of the information obtained from FLAGE is being incorporated into the HEDI and ERIS ground-launched interceptor missiles now in development.

FLAGON: NATO code name for the Sukhoi Su-15/21 twin-engine, all-weather INTERCEPTOR, in service with the Soviet IA-PVO (Fighter Aviation of the Air Defense Troops), best known as the aircraft that shot down Korean Airlines Flight 007 in 1983. Developed from 1959 to meet the anticipated threat of U.S. high-altitude supersonic bombers (a threat that did not materialize), Flagon was originally developed as a further improvement of the Su-9/11 Fishpot interceptor, with twin engines and a large airborne search and tracking radar in a revised nose (early Flagons had the same delta wing, tailplane, landing gear, engine, and rear fuselage panels as the Su-11). The first prototypes flew in 1965, and the Su-15 Flagon A entered service on a limited basis in 1968. It was followed by Flagon B, an experimental V/STOL aircraft, and the Su-15U Flagon C, a two-seat trainer. Introduced in the late 1960s, Flagon D was similar to the A, but with a new compound-delta wing adopted on all subsequent models.

In service from 1973, Su-21 Flagon E was in fact the first major production version, with an increased wingspan, more powerful engines, greater internal fuel capacity, and improved avionics. Flagon F, the last production model, has a revised radome and slightly uprated engines. Flagon G is a two-seat trainer derivative of the F, with limited combat capabilities. In all, some 1400 Flagons were built through the late 1970s, forming the backbone of the IA-PVO into the early 1980s. Now being replaced by the MiG-31 FOXHOUND and the Su-27 FLANKER, some 300 remain active, though as many as 900 may be in reserve.

The area-ruled fuselage is dominated by a very long nose that terminates in a large radome. Early models were equipped with an *Uragan* 5B RADAR (NATO code name Skip Spin) in a conical radome. Flagon E introduced an improved "Twin Scan" pulse-Doppler radar which may have limited look-down/shoot-down capability, while Flagon F has a more aerodynamically refined ogival radome. In all variants, the space between the radome and the cockpit houses the bulky vacuum-tube electronics of the radar receiver, transmitter, and signal processor.

The pilot sits well back from the nose under a heavily framed sliding bubble canopy with a solid rear end. Cockpit visibility is only marginal, especially over the nose, but this is not much of a handicap given Soviet GROUND CONTROLLED INTERCEPT

(GCI) and blind-firing techniques. Cockpit AVIONICS include a reflector gunsight, a head-down radar display, a Sirena-3 RADAR WARNING RECEIVER, a radar altimeter, an HF long-range radio, and a UHF data link. The fuselage behind the cockpit is believed to house large fuel tanks, and behind them the engines. Early Flagons had two Ly'ulka AL-21Fs, while the E and F have more powerful AL-21F-3s. Fed by rectangular cheek, the AL-21 is not very fuel efficient, degrading range.

In Flagon A, the midmounted wing was a simple 60° delta, similar to that of the Su-11 and MiG-21 Fishbed. Flagon D introduced a compound delta wing of equal span, extended to 34.5 ft. in Flagon E. The inboard sections have a leading edge sweep of 53°, while the outboard sections are swept at only 37°, for better maneuverability and low-speed handling. The wing has a fixed leading edge, with outboard ailerons and simple split flaps on the trailing edge. The main landing gear retract into the inboard wing sections, but there is still sufficient volume for two integral fuel tanks. The slab tailplane is mounted in line with the wing, and the vertical stabilizer has a braking parachute housed in its base.

Flagon has two side-by-side fuselage pylons and four wing pylons. The former are used either for two GSh-23L twin-barrel 23-mm. CANNON pods, or for drop tanks. The wing pylons are used exclusively for air-to-air missiles. Early Flagons carried two or four AA-3 ANABs, large anti-bomber missiles with a range of 15 mi. (24 km.), with both SEMI-ACTIVE RADAR HOMING (SARH) and INFRARED HOMING (IR) variants. Flagon generally carry one or two of each. The E and F have been seen with Anab on the outboard wing pylons, and short-range missiles (either AA-2 ATOLLS or AA-8 APHIDS) on the inboard pylons.

Specifications **Length:** 67.25 ft. (20.5 m.). **Span:** (A, C, D) 30 ft. (9.14 m.); (E, F, G) 34.5 ft. (10.53 m.). **Powerplant:** (A, B, C) 2 Ly'ulka AL-21F afterburning turbojets, 22,000 lb. (10,000 kg.) of thrust each; (E, F, G) 2 AL-21F-3s, 22,200 lb. (10,070 kg.) of thrust each. **Fuel:** 11,023 lb. (5000 kg.). **Weight, empty:** 27,007 lb. (12,250 kg.). **Weight, max. takeoff:** 39,680 lb. (18,036 kg.). **Speed, max.:** Mach 2.5 (1650 mph/2655 kph) at 36,000 ft. (11,000 m.) clean/Mach 2.3 (1519 mph/2445 kph) with 4 missiles. **Initial climb:** 44,950 ft./min. (13,700 m./min.). **Service ceiling:** 65,615 ft. (20,000 m.). **Combat radius:** 450 mi. (725 km.). **Range, max.:** 1400 mi. (2250 km.).

FLAILS AND ROLLERS: Mine-clearing devices attached to the bow of a TANK or other armored vehicle. A flail consists of a rotating axle to which several lengths of chain are attached. Powered by a drive train, the axle rotates rapidly, causing the chains to beat the ground immediately ahead of the vehicle, to detonate mines before the vehicle runs over them. Rollers consist of weighted wheels or drums pushed ahead of the vehicle. Both flail and roller tanks can clear lanes through minefields more quickly than engineers working on foot, but both are easily spotted and quickly attract enemy fire. Moreover, with delay fuzes, mines activated by flails or rollers can be set to explode under their carrying vehicle. See also ENGINEERS; MINES, LAND.

FLAMETHROWER: A weapon that projects and ignites a flammable liquid. Developed during World War I, flamethrowers are still well suited for the reduction of bunkers and other field fortifications, as well as for URBAN WARFARE. Generally assigned to assault ENGINEERS (sappers), there are three basic types: portable, fixed, and vehicular.

Portable flamethrowers consist of a backpack with one or two fuel tanks and a pressure tank, which are connected by a flexible hose to a nozzle through which fuel can be projected over distances of some 50 m. Typically weighing about 40 lb. (20 kg.), they have enough fuel for six seconds of fire; most can also project a stream of unignited fuel (e.g., to flood the ventilators of a bunker) which can then be ignited by a subsequent burst. Replaced in U.S. service by the M202 incendiary rocket launcher, portable flamethrowers are still widely issued to Soviet and other Warsaw Pact forces, which value their devastating morale effect.

Fixed flamethrowers are employed most often as part of prepared defenses, emplaced along likely avenues of approach to be set off by tripwire or remote control. Used mainly by the Soviet army, fixed flamethrowers can be considered as a form of "off-route" mine. Field-expedient equivalents ("fougasses") are improvised from ammunition canisters filled with gasoline.

Large, longer-range flamethrowers, built into modified tanks or other armored fighting vehicles, are meant mainly for the reduction of field fortifications. Effective ranges of some 150 m. can be achieved, with enough fuel for ten seconds of fire. Many armies retain at least a few with assault-engineer units, though they are extremely vulnerable because of the large volume of fuel on board.

FLANKER: NATO code name for the Sukhoi Su-27 long-range, twin-engine air superiority FIGHTER, in service with Soviet Frontal Aviation (FRONTOVAYA AVIATSIYA, FA) and the IA-PVO (Fighter Aviation of the Air Defense Troops). Developed from the mid-1970s to replace the Tu-28P FIDDLER and Yak-28 BREWER, the Flanker resembles an enlarged MiG-29 FULCRUM and is roughly comparable in size and performance to the U.S. F-15 EAGLE. Flight tests began in 1977, and the first operational units were probably formed in 1985. Little was known about Flanker in the West until it was displayed at the 1989 Paris Air Show, where it demonstrated extraordinary maneuverability. Clearly intended mainly for long-range patrol in conjunction with the Il-76 MAINSTAY (the Soviet AWACS), the Su-27 also has the range to perform strike escort and air superiority missions deep into hostile territory. One of the aircraft being evaluated for the Soviet aircraft carrier TBILISI, a modified Su-27 has made many successful carrier takeoffs and landings in the Black Sea. At present, more than 250 Flankers are in service, with production continuing at a rate of several hundred per year. None had been exported through 1989.

As in the MiG-29, the fuselage consists of two widely separated engine nacelles faired into a central cockpit pod to form a single broad structure which generates considerable body lift, especially at high angles of attack (AOA). Titanium alloy is used extensively throughout the airframe, but advanced composites are apparently limited to control surfaces and a few body panels.

The nose is dominated by a large radome housing a 48-in. (1.22-m.) antenna for a powerful pulse-Doppler RADAR with LOOK-DOWN/SHOOT-DOWN and TRACK-WHILE-SCAN capabilities and a detection range of some 200 mi. (334 km.). Though it probably incorporates Western technology, the radar apparently can track and engage only one target at a time. In addition to the radar, the Flanker has an INFRARED Search and Track (IRST) sensor and a LASER rangefinder mounted just ahead of the windscreen. These allow long-range target acquisition, identification, and ranging without generating easily detected radar emissions, a valuable tactical advantage still lacking in most Western fighters.

The pilot sits in an elevated cockpit under a long, clamshell bubble canopy with a single-piece windscreen. While cockpit visibility is better than in earlier Soviet fighters, the cockpit roofline is still too low to allow an unobstructed view to the rear.

The Su-27UB operational trainer has an extended canopy, with a second seat behind the pilot.

Cockpit AVIONICS are comprehensive, including a head-up display (HUD) that integrates data from both the radar and the IRST, a head-down radar display, a Sirena-3 threat-warning display, full blind-flying instruments, an instrument landing system, a TACAN-type navigational beacon receiver, a tactical DATA LINK for communications with Mainstay and GROUND-CONTROLLED INTERCEPT (GCI) stations, and a flight-data/weapons-delivery computer. Moreover, the Su-27 has the world's first integrated helmet-mounted sight (in an operational fighter), which is linked to the radar, IRST, and the laser rangefinder. The seeker heads of the Flanker's air-to-air missiles can be slaved to the helmet sight, allowing the pilot to lock onto targets by simply looking at them, eliminating the need to turn the aircraft directly at them. On the other hand, the Su-27 still has analog instrument dials, rather than the multi-functional video displays of the latest Western aircraft.

Flanker is a CONTROL-CONFIGURED VEHICLE (CCV) with relaxed static stability. A quadruplex FLY-BY-WIRE system controlled by the digital flight data computer keeps the aircraft on course, and provides outstanding agility; the system has AOA and G-load limiters to keep the aircraft inside its performance envelope, but there is also a manual override (lacking in similar Western systems) which allows the pilot to exceed the normal 30–35° AOA limits in emergencies.

Behind the cockpit the fuselage contains a large avionics bay; the midfuselage houses several large fuel tanks, and the engines are mounted in the rear of their nacelles. A large dorsal airbrake fitted behind the cockpit resembles a similar installation on the F-15 Eagle. The engines are fed through underslung horizontal wedge intakes fitted with internal debris screens to prevent foreign-object ingestion during rough-field operations. The Flanker has strong main landing gear and large, low-pressure tires which retract into the wing roots.

The moderately swept, shoulder-mounted wing has large leading-edge root extensions (LERXs) blended into the fuselage. The LERXs generate lift-enhancing vortexes and also reduce drag and improve handling at high AOA. A 30-mm. cannon of unknown type is mounted in the starboard leading-edge extension, with its drum behind the cockpit. The leading and trailing edges have computer-controlled "adaptive" flaps which automatically maintain the optimal wing profile throughout the flight envelope. It is believed that the entire wing forms a large, integral fuel tank, giving Flanker a greater internal fuel capacity than any other fighter worldwide, except the Tu-28 Fiddler. Like the MiG-29, the Su-27 has slab tailplanes mounted at the ends of the engine nacelles, and twin vertical stabilizers, but these stabilizers are vertical (as in the F-15), and not canted outward as in the MiG-29. A combat thrust-to-weight ratio of 1.1 to 1, combined with its large wing area and broad fuselage, allows outstanding acceleration, sustained turn rate, and rate of climb.

Flanker has 6 wing pylons (including 2 at the tips) and 2 fuselage pylons under the engine intakes, all for air-to-air missiles. The standard load is 6 AA-10 ALAMO medium-range radar-guided missiles on the fuselage and underwing pylons, and 2 AA-8 APHID or AA-11 Archer short-range infrared-homing missiles on each wingtip. Though not meant for ground attack, the Flanker could carry up to 13,225 lb. (6011 kg.) of bombs or air-to-surface missiles.

Specifications Length: 71.9 ft. (21.92 m.). Span: 48.2 ft. (14.7 m.). Powerplant: 2 Ly'ulka AL-31F afterburning turbofans, 27,500 lb. (12,500 kg.) of thrust each. Fuel: 22,000 lb. (10,000 kg.). Weight, empty: 40,000 lb. (18,181 kg.). Weight, normal loaded: 48,400 lb. (22,000 kg.). Weight, max. takeoff: 66,000 lb. (30,000 kg.). Speed, max.: Mach 2.35 (1527 mph/2550 kph) at 36,000 ft. (11,-000 m.)/Mach 1.1 (910 mph/1520 kph) at sea level. Initial climb: 50,000 ft./min. (15,243 m./min.). Service ceiling: 60,000 ft. (18,293 m.). Combat radius: 930 mi. (1553 km.). Range, max.: 2580 mi. (4000 km.).

FLEET: A higher naval command of ships, aircraft, marine forces, and shore-based support facilities, under a single commander, who exercises operational as well as administrative control. The U.S. Navy has four fleets: Second Fleet in the Atlantic, Third Fleet in the Eastern Pacific, Seventh Fleet in the Western Pacific and Indian Ocean, and Sixth Fleet in the Mediterranean. The Soviet navy is also organized into four fleets: the Northern Fleet, headquartered at Severomorsk, the Baltic Fleet, headquartered at Kaliningrad, the Black Sea Fleet, headquartered at Sebastopol, and the Pacific Fleet, headquartered at Vladivostok. Most other navies are too small or too geographically limited for subdivision into fleets, but the French navy operates an Atlantic and a Mediterranean fleet.

FLEET AIR ARM: Naval aviation component of Britain's Royal Navy, responsible for the operation of all ship-based helicopters and fixed-wing aircraft (land-based maritime strike and patrol aircraft are operated by the Royal Air Force). Highly professional and very well trained, the FAA gave an excellent account of itself in the 1982 Falklands War. Its main shortcoming is the lack of long-range aircraft, due to the Royal Navy's lack of large carriers. See also AIRCRAFT CARRIERS, BRITAIN.

FLETCHER: A class of 181 U.S. DESTROYERS commissioned between 1942 and 1945. The primary U.S. destroyers of World War II, the last Fletchers were retired from the U.S. Navy in the early 1970s. Of the many transferred to foreign navies, 15 are still in service with Brazil (3), Greece (6), Spain (2), and Taiwan (4).

FLIR: Forward-Looking Infrared, a type of IMAGING INFRARED (IIR) sensor, mostly for fixed-wing aircraft and helicopters. Its core is a focal plane array which converts optically received thermal emissions from objects and background for display on a cockpit video screen, or for direct projection onto the pilot's head-up display (HUD). That allows aircraft to be flown at high speeds and low altitude at night or in bad weather.

FLIR can be built into larger aircraft (e.g., the TRAM turret of the A-6E INTRUDER and TVS of the B-52 STRATOFORTRESS), but fighters and attack aircraft mostly use pod-mounted FLIR units such as the U.S. LANTIRN. Entirely passive, FLIR does not generate detectable emissions and is in this regard superior to RADAR. Hence it is likely to become the primary sensor for all STEALTH aircraft. Its main drawback is a relatively short range, which is further degraded by dense fog or heavy precipitation.

FLOGGER: NATO code name for the Mikoyan MiG-23 and MiG-27 series of variable geometry ("swing-wing") FIGHTERS and ATTACK AIRCRAFT, in service with Soviet Frontal Aviation (*Frontovaya Aviatsiya*, FA), the IA-PVO (Fighter Aviation of the Air Defense Troops), and with the air forces of more than 20 countries.

Developed from the mid-1960s as a replacement for the MiG-21 FISHBED, the Flogger was initially conceived as a fighter-INTERCEPTOR emphasizing speed, range, payload, avionics, and all-weather capability at the expense of agility. As compared to earlier Soviet fighters, it is large, heavy, and relatively sophisticated. Variable geometry was adopted as the only practical way of reconciling conflicting requirements for range, payload, and short-field performance (which require a long, narrow wing), and high speed (which requires a short, broad one), despite the mechanism's considerable penalty in weight and complexity. Flight tests began in 1967, and some 50 MiG-23SM (Flogger-A) preproduction aircraft were formed into an operational evaluation unit in 1971. Powered by a Ly'ulka AL-7F afterburning turbojet rated at 22,000 lb. (10,000 kg.) of thrust, the surviving Flogger-As are still in service as testbeds for new sensors and armament.

The first production variant, the MiG-23M (Flogger-B) of 1972, was equipped with improved avionics and a Turmanskiy R-27 afterburning turbojet, and was lighter, shorter and more fuel efficient than the AL-7F. Installation of the R-27 required moving the wings forward and shortening the rear fuselage to compensate for changes in the center of gravity. The MiG-23UM (Flogger-C) is a two-seat trainer derivative with a raised second cockpit behind the pilot.

From 1975, the MiG-23M was superseded in production by the MiG-23MF (also designated Flogger-B), with a more powerful Turmanskiy R-29 engine and an improved pulse-Doppler RADAR (NATO code name High Lark), the first on a Soviet fighter. The MiG-23MS (Flogger-E) is an export variant of the MiG-23M, with the less capable R2L "Jay Bird" radar of the MiG-21, and generally simplified avionics. During the 1982 Lebanon War, Syrian Flogger-Es were badly mauled by Israeli fighters. Introduced in 1977–78, the MiG-23ML (Flogger-G) is generally similar to the MF, but has a lightweight version of High Lark and other avionic improvements. The last Soviet version, the MiG-23bis (Flogger-K) of 1981, is a further development of the ML, with dogtooth notches at the wing roots to improve stability at high angles of attack (AOA).

Some 5200 MiG-23 fighters were built through the early 1980s (when production was shifted to the MiG-29 FULCRUM), half for the U.S.S.R. and half for export. MiG-23s still equip a majority of Soviet fighter-interceptor squadrons in both the IA-PVO and FA.

Flogger-G's long, pointed nose is dominated by a large radome housing a High Lark pulse-Doppler radar. Believed to be based on the AWG-10 radar of the F-4J PHANTOM (some were recovered from crash sites in North Vietnam), High Lark has a detection range of some 53 mi. (85 km.) and a lock-on range of 34 mi. (54 km.); it also has ground mapping and TERRAIN-AVOIDANCE modes, and was the first Soviet radar with even limited LOOK-

DOWN/SHOOT-DOWN capability. The R2L Jay Bird radar of the MiG-23MS (Flogger-E) has detection and lock-on ranges of only 18 and 12 mi. (30 and 20 km.) respectively, and no look-down capability. Immediately behind the radome, a large AVIONICS bay houses the radar transmitter, receiver, and signal processor, together with a liquid cooling system for its vacuum-tube components. In addition to radar, Flogger-G has a laser-ranging and marked-target seeker (LRMTS) in a small pod under the nose; this is mainly for use with air-to-ground ordnance.

The pilot sits in a rather cramped cockpit under a clamshell canopy faired into the fuselage spine, which completely obstructs rearward visibility. The windscreen is small and heavily framed, further obstructing the pilot's forward view. Rear-view mirrors and a canopy periscope do little to alleviate this problem. Cockpit avionics were advanced by Soviet standards, including a true head-up display (HUD), the first in a Soviet fighter, a head-down radar display with rudimentary signal processing, a threat-warning display, a video display screen, an instrument landing system, and a tactical DATA LINK to ground-controlled intercept stations.

An avionics compartment is located immediately behind the cockpit, with fuel tanks and the wing-pivot mechanism in the midfuselage, and the engines in the rear. A twin-barrel 23-mm. GSh-23L cannon is mounted externally in a ventral pack, with a 200-round magazine inside the fuselage. Intended for operations from rough, unpaved fields, the MiG-23 has robust main landing gear with large, low-pressure tires that retract into the fuselage under the wing-pivot box. The R-29 engine is fed by large cheek inlets just behind the cockpit. Fitted with large, perforated splitter plates similar to those of the F-4 Phantom, the inlets have complex variable ramps which allow efficient operations at speeds beyond Mach 2.

The shoulder-mounted wing, pivoted inboard from large, fixed gloves, has sweep settings of 16° for takeoff and landing, 45° for cruising, and 72° for high-speed sprints. There is no automatic sweep mechanism activated by a flight data computer as in the F-14 TOMCAT, which limits the usefulness of the variable geometry in dogfights. The wing has a pronounced dogtooth in the outboard sections to generate lift-enhancing vortices and improve handling at high AOA; they also house SRO-69 Sirena-3 RADAR WARNING RECEIVERS. Many Floggers have an ELECTRO-OPTICAL (TV) sensor in the starboard wing for long-range visual target identification. High-lift devices include outboard leading-edge slats, full-span flaps, and spoilers for roll-control with the wings extended. The outboard wing sections also house integral fuel tanks. The slab tailplane acts differentially as a "taileron," to control both pitch and roll with the wings swept back. The vertical stabilizer, which also has a tip notch (a standard MiG design feature), is very broad, with a long dorsal extension to provide directional stability over a broad range of sweep settings. To further enhance stability, Flogger has a large ventral fin, folded to the side for takeoff and landing, extended in high-speed flight. Both the stabilizer and fin incorporate conformal UHF aerials and other antennas for communications and ELECTRONIC WARFARE. A braking parachute is also housed at the base of the vertical stabilizer. As compared with the MiG-21 (or even the F-4 Phantom), turning performance is poor, and the MiG-23 is best suited for high-speed slashing attacks, not low-speed dogfights.

The MiG-23 has one centerline pylon rated at 2200 lb. (1000 kg.), two inlet duct pylons rated at 1650 lb. (750 kg.) each, and two wing glove pylons rated at 2200 lb. (1000 kg.) each, for a total payload capacity of 7250 lb. (4500 kg.). The centerline pylon usually carries a 176-gal. (800-lit.) drop tank. For air-superiority missions, a typical load would consist of 8 AA-8 APHID short-range INFRARED-HOMING (IR) missiles on two tandem launch rails under the inlets, and 2 AA-2-2 Advanced ATOLL short-range SEMI-ACTIVE RADAR HOMING (SARH) missiles under the wing gloves, while interceptors would carry 2 AA-7 APEX IR or SARH medium-range missiles under the gloves, and 4 Aphids under the inlets. Export aircraft normally carry up to 4 AA-2 Atoll IR missiles, though Indian and Iraqi aircraft have also been equipped to launch Matra R.550 MAGIC IR missiles. Though not meant for ground attack, the MiG-23 can carry a wide variety of free-fall and guided bombs.

Development of the MiG-27 specialized ground attack variant began in 1971. While it retains the basic MiG-23 airframe, the MiG-27 has a completely new nose and cockpit, different avionics, and a simplified engine installation. Flight tests began in 1972, and the first production aircraft (Flogger-D) entered service in 1974. It is still the principal ground attack aircraft in the FA, with more than 850 in service, and has been exported to Poland, East Germany, and Cuba. Flogger-D was superseded in 1981 by the MiG-27M (Flogger-J), with a modified nose and additional sensors; it is also produced under license in India as the HAL "Bahadur." Two attack variants have been produced specifically for export: the MiG-23BN (Flogger-F), with the nose and cockpit of the MiG-27 grafted onto the rear fuselage of the MiG-23MS (Flogger-E); and the similar Flogger-H, with additional radar warning receivers. More than 5000 ground attack variants have been built to date, bringing total Flogger production to some 10,200 aircraft.

The MiG-27M's (Flogger-J) most distinctive feature is a broad, sloping "ducknose" (*utkanos*), which affords the pilot a much better forward view than in fighter models. The nose houses a small ranging radar, with a LASER DESIGNATOR/rangefinder offset to starboard. A ground mapping and TERRAIN-AVOIDANCE RADAR is mounted in the chin position, with a radar altimeter and DOPPLER navigation radar behind. The cockpit has an armored windscreen and is protected by armor plate. Cockpit avionics are considerably simplified by the removal of the MiG-23's air-search radar, reducing cockpit clutter. The MiG-27's engine installation is also changed. The inlets have simple, fixed ramps; that limits maximum speed at high altitude, not a requirement for a low-level attack aircraft. The engine has a short, fixed nozzle, also optimized for low-altitude performance. The MiG-27 has a 6-barrel 23-mm. Gatling gun mounted externally, instead of the GSh-23L of the MiG-23.

In addition to the weapon pylons of the MiG-23, the MiG-27 also has two small rear-fuselage pylons, for a total payload capacity to some 10,000 lb. (4500 kg.). Typical ordnance includes 1100-lb. (500-kg.) free-fall and CLUSTER BOMBS, LASER-GUIDED BOMBS, NAPALM and chemical bombs, TN-1200 free-fall nuclear weapons, and AS-7 KERRY, AS-9 KYLE, AS-11 KILTER, and AS-14 KEDGE air-to-surface missiles. Drop tanks can be carried on the centerline and wing glove pylons, while two Aphids can be carried under the inlets for self defense.

Specifications Length: (23) 56.55 ft. (18.15 m.); (27) 52.49 ft. (16 m.). Span: 46.75 ft. (14.25 m.) at 16°; 26.8 ft. (8.17 m.) at 72°. Powerplant: (23M) Turmanskiy R-27 afterburning turbojet, 22,500 lb. (11,200 kg.) of thrust; (23MF) Turmanskiy R-29 afterburning turbofan, 27,502 lb. (12,475 kg.) of thrust; (27) Turmanskiy R-29-300, 25,352 lb. (11,-500 kg.) of thrust. Fuel: 10,140 lb. (4600 kg.). Weight, empty: (23) 22,000 lb. (10,000 kg.); (27) 23,788 lb. (10,970 kg.). Weight, normal loaded: (23) 32,000 lb. (14,515 kg.); (27) 34,172 lb. (15,500 kg.). Weight, max. takeoff: (23) 41,667 lb. (18,900 kg.); (27) 44,313 lb. (20,100 kg.). Speed, max.: (23) Mach 2.31 (1520 mph/2446 kph) at 36,000 ft. (11,-000 m.)/Mach 1.2 (913 mph/1469 kph) at sea level; (27) Mach 1.7 (1123 mph/1807 kph) at 36,000 ft. (11,000 m.) clean/Mach 1.1 (836 mph/1345 kph) at sea level/Mach 0.95 (723 mph/1163 kph) w/ext. stores. Initial climb: 30,000 ft./min. (9145 m./min.). Service ceiling: (23) 61,000 ft. (18,595 m.); (27) 52,500 ft. (16,000 m.). Combat radius: (23) 805 mi. (1300 km.); (27) 240 mi. (390 km.) lo-lo-lo, 5000-lb. (2272-kg.) payload + 1 drop tank; 576 mi. (920 km.) w/3 drop tanks. Range, max.: (23) 1740 mi. (2800 km.); (27) 1550 mi. (2500 km.). Operators: (23) Alg, Bul, Cuba, Cze, DDR, Egy, Eth, Hun, Lib, Rom, Syr, USSR; (27) Alg, Bul, Cuba, Cze, DDR, Egy, Eth, Ind, Iraq, Lib, N Kor, Pol, Syr, USSR, Viet.

FLOT: Forward Line of Own Troops, a line that connects the foremost positions of friendly forces at any given time. FLOT information is essential for the coordination of artillery fire and close-air support, to avoid accidental attacks on friendly units.

FLUID FOUR: An air-combat formation developed by the U.S. Air Force (USAF) during the Korean War, consisting of four aircraft flying in two pairs ("elements"), with one pair several hundred meters to the side and several hundred meters above the other. Each pair in turn flies as a WELDED WING, with a designated leader and wingman. In contrast to the earlier "finger four" formation, the leader of either pair can initiate an attack, with the other pair providing cover and support.

The Fluid Four was retained into the Vietnam War, though the spacing between aircraft and between the pairs was increased dramatically because of higher speeds and larger turn radii. Not flexible enough for modern combat conditions, the Fluid Four was superseded during the 1970s by the LOOSE DEUCE or "Fluid Two" formation. See also AIR COMBAT MANEUVERING.

FLY-BY-LIGHT: An aircraft control system similar to FLY-BY-WIRE, but with the commands relayed from the flight data computer to the control actuators by fiberoptic cable, which is lighter and has higher data rates than electrical wiring.

FLY-BY-WIRE: 1. An aircraft flight control system in which the pilot's control actions are interpreted by a digital flight data computer, which converts them into electronic commands for fast-acting, electrically powered control actuators. Fly-by-wire controls are considerably lighter than traditional hydraulic actuators controlled mechanically; they provide instantaneous response, and the circuits can be widely separated and be multiply redundant for enhanced survivability. Finally, the computer can be programmed to prevent the aircraft from exceeding the limits of its flight envelope, allowing the pilot to concentrate on his mission, rather than safety margins. The advent of fly-by-wire has allowed the development of CONTROL-CONFIGURED VEHICLE (CCV) aircraft with relaxed static stability. Such aircraft (e.g., the U.S. F-16 FALCON, Soviet Su-27 FLANKER, and French MIRAGE 2000) are inherently unstable and would be unflyable with manual controls; it is the flight data computer that makes automatic control adjustments (at a rate of some 60/sec.) to keep the aircraft on course. Because they are unstable, CCV aircraft require very little control force to change attitude, and so are considerably more agile than conventional (stable) designs.

The latest development is FLY-BY-LIGHT control, in which commands from the computer are transmitted to the actuators by fiberoptic cables. Fly-by-light systems are even lighter than fly-by-wire and can transmit control data at much higher rates.

2. A Soviet form of preprogrammed INERTIAL GUIDANCE for ballistic missiles, applied in the SS-11 SEGO, SS-18 SATAN, SS-19 STILETTO, SS-N-6 SAWFLY, and SS-N-8. In fly-by-wire guidance, the missile is constrained to a precomputed pitch and velocity profile by the variable throttling of the main engines; thus it is applicable only in missiles that have liquid-fuel propulsion. Fly-by-wire allows precise TIME-ON-TARGET control, a prerequisite for avoiding FRATRICIDE in the course of COUNTERFORCE attacks against a large number of targets (notably missile silos), but it is less accurate than "navigating" inertial guidance because of the cumulative effects of random errors. In the SS-18 and SS-19, fly-by-wire is applied in the booster, with navigating inertial guidance for the POST-BOOST VEHICLE, so as to compensate for errors accumulated in the boost phase.

FLYING BOOM: AERIAL REFUELING technique developed by the U.S. Air Force (USAF) in the late 1940s, whereby a rigid boom (whose position is controlled by movable winglets) is lowered from the tail of a tanker aircraft into a receptacle on the upper fuselage of the receiving aircraft. While large and heavy, the flying boom has a higher fuel transfer rate than "probe-and-drogue"

units; moreover, it does not require the receiving aircraft to make radical connection maneuvers, but only to maintain formation with the tanker. Thus this method is much better suited for refueling less maneuverable bombers and heavy transport aircraft (though all USAF fighters are also equipped with flying boom receptacles).

FN: *Fabrique Nationale d'Armes de Guerre*, the Belgian national arms company, best known for its small arms, notably the FN-FAL (*Fusil Automatique Legere*) ASSAULT RIFLE and FN-MAG (*Mitrailleuse d'Appui General*) GENERAL-PURPOSE MACHINE GUN.

FN-FAL (FUSIL AUTOMATIQUE LEGERE):

Belgian 7.62-mm. automatic rifle, first produced in 1946 and still in service with more than 70 countries. It was produced under license by Britain, Australia, Canada, and other Commonwealth countries as the L1A1 SLR (Self-Loading Rifle).

The FAL has a conventional design with a straight wooden forestock, pistol grip, and stock (though plastic and folding-stock models were produced under license). The muzzle has a flash suppressor which can be fitted with a bayonet and which accepts a variety of rifle GRENADES. Quite sturdy and reliable, the FAL is one of the last precision-tooled infantry weapons, with machined (vs. stamped) metal parts and high tolerances.

A gas-operated weapon, the FAL fires NATO-standard 7.62- × 51-mm. ammunition from 20-round box magazines. In keeping with the British army's 1950s preference for deliberate, aimed fire, the L1A1 does not have a full-automatic setting, thus limiting its rate of fire to some 30–40 rds./min. Variants include a heavy-barreled squad automatic weapon (SAW) with a bipod, and the FN-CAL, essentially an FAL rechambered for 5.56- × 45-mm. ammunition. The FAL has been replaced in many armies by ASSAULT RIFLES, shorter, lighter, and chambered for reduced-power/smaller-caliber ammunition.

Specifications **Length OA:** 45 in. (1.143 m.). **Length, barrel:** 21.81 in. (554 mm.). **Weight, loaded:** 11 lb. (5 kg.). **Muzzle velocity:** 2750 ft./sec. (838 m./sec.) **Cyclic rate:** 650–700 rds./min. **Effective range:** 500 m.

FN-MAG (MITRAILLEUSE D'APPUI GENERAL):

Belgian GENERAL-PURPOSE MACHINE GUN (GPMG), first produced in 1950, and presently in service in more than 20 countries. Sturdy and very reliable, the MAG differs from many other GPMGs in its extensive use of precision-machined (as opposed to stamped) parts. It has a quick-change barrel to cope with overheating, and a gas regulator which allows the gunner to vary the rate of fire. Normally fired from a bipod, the MAG can also be mounted on a tripod, or on vehiclular coaxial and pintle mounts. A gas-operated weapon fed from 50-round belts, the MAG fires NATO-standard 7.62- × 51-mm. ammunition.

Specifications **Length OA:** 49.61 in. (1.26 m.). **Length, barrel:** 21.46 in. (545 mm.). **Weight, Loaded:** 22.27 lb. (10.1 kg.). **Muzzle velocity:** 32756 ft./sec. (840 m./sec.). **Cyclic rate:** 600–1000 rds./min. **Effective range:** 800 m. **Operators:** Arg, Bel, Can, Fra, Gre, Ire, Isr, Neth, NZ, S Af, UK.

FO: See FORWARD OBSERVER.

FOBS: See FRACTIONAL ORBIT BOMBARDMENT SYSTEM.

FOFA: Follow-On Forces Attack, a joint U.S.-NATO operational scheme for the attack of Soviet reinforcing second-echelon forces before they can reach the battlefield. The objective of FOFA is to disrupt the momentum of a Soviet offensive by constraining the ratio of forces engaged on the front within levels the defense can manage, and more specifically, by preventing the timely arrival of Soviet reinforcements meant to exploit a breakthrough. FOFA is thus a form of battlefield INTERDICTION.

U.S. and European conceptions of FOFA tend to differ. U.S. planners normally emphasize the direct attack and attrition of combat units with long-range artillery, aircraft, and tactical "cargo" missiles containing precision-guided "smart" submunitions; that in turn requires real-time SURVEILLANCE and TARGET ACQUISITION with systems such as the J-STARS airborne radar (FOFA-specific weapons were in fact first developed for the 1970s ASSAULT BREAKER program, and have been criticized for their high costs, complexity, and vulnerability to relatively simple countermeasures).

European conceptions of FOFA tend to emphasize the attack of fixed CHOKEPOINTS (bridges, tunnels, narrow passes) and the use of scatterable MINES to impose delay while inflicting only low-level attrition. This approach is much less technologically demanding and does not require expensive target acquisition or command and control systems.

FOFA has become an accepted element of NATO's FORWARD DEFENSE posture, but there are still considerable doubts about its practicality. Its Soviet counterpart ("reconnaissance-strike combat") is even more ambitious, but still in the study phase.

FOG-M: Fiberoptic Guided Missile, a ground-launched anti-tank/anti-aircraft missile under development by Boeing for the "Non Line-of-Sight" (NLOS) component of the U.S. Army's Forward Area Air-Defense System (FAADS) program. FOG-M originated in a research program meant to test the feasibility of transmitting guidance commands and television pictures to and from a missile via fiberoptic cable. After the cancellation of the DIVAD program, FOG-M was selected to provide air defense against low-flying helicopters. Currently in the test and evaluation phase, the missile could enter production in the early 1990s.

Six large, pop-out wings at midbody provide lift, and there are four small tail fins for steering. The nose houses a Hughes IMAGING INFRARED (IIR) sensor, with relay via two fiberoptic cables released by inertialess reels in the tail. Powered by a Sundstrand Turbomach turbojet engine, FOG-M is armed with a 13-lb. (5-kg.) shaped-charge (HEAT) warhead, equally effective against tanks and helicopters.

FOG-M combines IIR homing and COMMAND GUIDANCE, with lock-on after launch (LOAL) capability. The IIR picture of the terrain ahead is relayed back to the operator's video display via the fiberoptic cable. The operator, in turn, transmits his steering commands over the same cables. Once the operator acquires a target on his video monitor, he can lock onto it by positioning a cursor, or else steer the missile manually with a joystick. The crucial advantage of FOG-M is that it does not need a direct line-of-sight to the target at launch; the operator's eye is in effect in the nose of the missile, and he can search for targets instead of having to see them beforehand; nor does he need continuous intervisibility—a great advantage. Thus FOG-M can engage helicopters and tanks that are using terrain masking to evade direct-fire weapons such as wire-guided ANTI-TANK GUIDED MISSILES. As now configured, FOG-M would be launched vertically from an 8-cell launcher mounted on the back of a 4 × 4 High Mobility, Multi-purpose Vehicle (HMMV).

Specifications **Length:** 5.5 ft. (1.7 m.). **Diameter:** 6 in. (152 mm.). **Weight, launch:** 83 lb. (38 kg.). **Speed:** Mach 0.6 (457 mph/763 kph.). **Range:** 6 mi. (10 km.).

***FORCE D'ACTION RAPIDE* (FAR):** Rapid action force of the French army, consisting of the 9th Marine Infantry Division, 27th Alpine Division, 6th Light Armor Division, and 11th Parachute Division. Intended mainly for interventions in Africa and the Middle East, but with a secondary NATO role, the FAR is composed of elite troops admirably trained and equipped for combat against all but heavy forces in highly trafficable terrain. The capacity of French air and amphibious transport limits the number of troops deployable a long distances from France without third-party assistance. See also AIRBORNE FORCES, FRANCE.

FORCE DE DISSUASION: Formerly the *Force de Frappe* ("Strike Force"), official designation of French long-range nuclear forces. These now consist of 18 silo-based S-3 IRBMS, 112 M-4 and M-20 SLBMS aboard 6 REDOUTABLE- and one INFLEXIBLE-class nuclear-powered ballistic missile submarines, and 18 MIRAGE IV bombers with ASMP standoff missiles.

France has maintained an independent nuclear force since 1964, under De Gaulle's policy of self-reliance. This has contributed to French prestige, though the costs of the *Force de Dissuasion* are very high in relation to its capabilities. Because of the small number of its delivery systems, France has perforce adopted the COUNTERVALUE targeting of (Soviet) cities. The accuracy of Soviet ballistic missiles now holds most of the French force at risk, while increasingly effective Soviet air and BALLISTIC MISSILE DEFENSES could severely reduce the number of French warheads that reach their targets. See also IRBMS, FRANCE; MSBS; SUBMARINES, FRANCE.

FORCE MULTIPLIER: U.S. term for new tactics or equipment which are meant to increase a unit's combat effectiveness in a manner equivalent to an increase in its size; e.g., a target acquisition sensor which allows firing to commence at longer range allows a unit to engage more targets in a given time period, thus making it equivalent to a larger force.

FORGER: NATO code name for the Yak-38 vertical/short takeoff and landing (V/STOL) fighter/attack aircraft, in service with Soviet Naval Aviation (AV-MF) aboard the four KIEV-class aircraft carriers. The U.S.S.R. began V/STOL experiments in the late 1950s, and built several prototypes, including derivatives of the MiG-21 FISHBED and Su-15 FLAGON. One experimental aircraft, the tadpolelike Yak-36 Freehand, was deployed briefly in 1967 aboard the helicopter carrier MOSKVA. In the early 1970s, the Yakolev design bureau was given the task of building a jet fighter for the Kievs that would be capable of maintaining air superiority over Soviet battle groups, with secondary reconnaissance, anti-ship, and ground-attack capabilities.

The prototype Yak-38 flew in 1971, and the first production aircraft entered service in 1975. In contrast to the British HARRIER, which has a single engine with vectored nozzles, Forger has separate lift engines in addition to a vectored-thrust main engine. This arrangement is very inefficient (because the dead weight of the lift engines must be carried at all times); further, it reduces the benefits of rolling takeoffs to extend range, and does not allow "vectoring in forward flight" (VIFF) maneuvers to enhance maneuverability. In terms of range, payload, performance, and avionics, Forger is vastly inferior to the Harrier and Sea Harrier, let alone conventional carrier-based fighters like the F-14 TOMCAT. Nonetheless, it does provide the Soviet navy with some shipboard counterair and strike capability (thus denying hostile aircraft a "free ride"), and has provided valuable experience in carrier operations. In all, some 74 Forgers were produced through 1986, with possibly some limited production thereafter to replace losses. A follow-on to Forger, tentatively designated Yak-41, has been identified.

Forger A is the basic attack variant; Forger B, a two-seat trainer version, has a slightly longer nose and tail. The fuselage has a sloping nose (to provide a good forward view during landings) that houses a small ranging RADAR. The pilot sits behind the nose under a clamshell canopy faired into the fuselage, which obstructs rearward visibility. Cockpit AVIONICS include a head-up display (HUD), an instrument landing system, a TACAN-type navigational beacon receiver, and the usual tactical radios. A unique (if dubious) feature is a system that automatically triggers the ejection seat if the rate of descent during vertical flight exceeds safety limits—an oblique indication of the Forger's ease of handling.

Two Koliesov turbojet lift engines are located immediately behind the cockpit, and fed through an inlet in the upper fuselage; their exhaust nozzles are inclined forward at 13° from the vertical. The forward thrust engine, fed by two cheek inlets, has two vectored-thrust nozzles in the tail which can be rotated from horizontal to vertical for takeoff and landing. In the mid-1980s, a control system was installed to allow short rolling takeoffs by automatically rotating the thrust nozzles from horizontal to vertical to propel the aircraft off the deck after it has gathered speed. Puff-jet controls fed by engine bleed air are mounted in the nose, tail, and wingtips to provide attitude control while hovering. Given all the equipment housed in the fuselage, it has very little room for fuel.

The small, high-mounted wing has a fixed, tapered leading edge and a straight trailing edge, with outboard ailerons and plain inboard flaps. The outer wing sections fold up for deck parking. There are four wing pylons, rated at 882 lb. (440 kg.) each, for a maximum payload of some 3500 lb. (1590 kg.). Typical payloads include AA-2 ATOLL or AA-8 APHID infrared-homing air-to-air missiles, 550-lb. (250-kg.) free-fall, NAPALM, and CLUSTER BOMBS, 19-round UV-16-57 rocket pods for S-5 57-mm. FFARS, and two 133-gal. (600-lit.) fuel tanks.

Specifications **Length:** (A) 52.5 ft. (16 m.); (B) 58 ft. (17.68 m.). **Span:** 24 ft. (7.32 m.). **Powerplant:** 2 Koliesov lift engines, 7875 lb. (3580 kg.) of thrust each; 1 Ly'ulka AL-21 turbojet, 17,985 lb. (8175 kg.) of thrust. **Weight, empty:** 15,000 lb. (6804 kg.). **Weight, VTOL:** 16,500 lb. (7500 kg.). **Weight, max. takeoff:** 23,700 lb. (10,750 kg.). **Speed, max.:** Mach 1.1 (725 mph/1170 kph) at 36,-000 ft. (11,000 m.)/Mach 0.9 (700 mph/1125 kph) at sea level. **Initial climb:** 14,750 ft./min. (4500 m./min.). **Service ceiling:** 40,000 ft. (12,200 m.). **Combat radius:** (hi-lo-hi attack) 230 mi. (370 km.). **Range, max.:** 1800 mi. (2900 km.).

FORRESTAL: A class of four U.S. AIRCRAFT CARRIERS commissioned between 1955 and 1959. Funded during the Korean War buildup, the Forrestals were the first "supercarriers" designed from the outset for jet aircraft, with enclosed "hurricane" bows, angled landing decks, and automatic landing aids. Three have undergone SLEP (Service Life Extension Program) overhauls; the fourth, USS *Ranger,* is scheduled to begin its SLEP in 1993. All are to remain in service until after the year 2000.

The armored flight deck is 250.25 ft. (76.29 m.) wide in the first two (*Forrestal* and *Saratoga*), and 270 ft. (82.3 m.) wide in the last two (*Ranger* and *Independence*). The angled landing deck is cantilevered 10.5° to port, with a large island superstructure to starboard which houses ship controls, flight operation facilities, and uptakes for the engines. There are 4 steam catapults (2 on the bow and 2 on the angled deck), and 4 deck-edge elevators: 1 forward and 2 aft of the island to starboard, and 1 at the forward end of the angled deck (the latter cannot be used during landing operations, and was relocated on subsequent carriers). The hangar deck is 740 ft. (225.6 m.) long, 101 ft. (30.8

m.) wide, and 25 ft. (7.62 m.) high. The Forrestals can carry 1650 tons of aerial ordnance and 789,000 gal. (2.99 million lit.) of jet fuel.

As built, the Forrestals had a defensive armament of 8 5-in. 54-caliber DUAL-PURPOSE guns mounted on sponsons below the flight deck; in the early 1970s these were replaced by 3 8-round NATO SEA SPARROW short-range surface-to-air missile launchers, which were supplemented in the early 1980s by 3 PHALANX radar-controlled guns for anti-missile defense.

At present, the air group consists of two interceptor squadrons with a total of 20 F-14A TOMCATS, 2 fighter/attack squadrons with 20 FA-18 HORNETS, 2 medium attack squadrons with 20 A-6E INTRUDERS, an anti-submarine patrol squadron with 10 S-3A VIKINGS, and a composite support squadron with 5 E-2C HAWKEYE airborne early warning aircraft, 5 EA-6B PROWLER electronic warfare aircraft, and 6 SH-3H SEA KING anti-submarine helicopters. See also AIRCRAFT CARRIERS, UNITED STATES.

Specifications **Length:** 1063–1086 ft. (326.1–331 m.). **Beam:** 137 ft. (41.75 m.). **Draft:** 29.5 ft. (39.5 m.). **Displacement:** 59,060 tons standard/79,300 tons full load. **Powerplant:** 4-shaft steam: 8 oil-fired boilers, 4 sets of geared turbines, 280,000 shp. **Fuel:** 7800 tons. **Speed:** 33 kt. **Range:** 4000 n.mi. at 30 kt./8000 n.mi. at 20 kt. **Crew:** 2790 + 2150 air group = 4940. **Sensors:** 1 Litton LN-66 navigational radar, 1 SPS-10 surface-search radar, 1 SPS-48 3-dimensional air search radar, 1 SPS-49 2-dimensional air search radar, 1 SPS-58 low-level search radar, 3 Mk.91 Sea Sparrow guidance radars. **Electronic warfare equipment:** WLR-1, WLR-3, and WLR-11 radar warning receivers, one SLQ-29 electronic countermeasures array, three Mk.36 SRBOC chaff launchers.

FORWARD AIR CONTROLLER (FAC): A qualified pilot assigned to a ground force to direct and coordinate CLOSE AIR SUPPORT (CAS). Introduced by the U.S. and Britain in World War II, forward air controllers originally operated on the ground, frequently in radio-equipped armored cars to accompany the front-line units.

During the Korean War, the U.S. Air Force (USAF) began to fly FACs in light observation aircraft, adding artillery SPOTTING to their tasks. Targets located visually were marked with colored smoke bombs or rockets for attacking aircraft vectored to the target by the FAC, who would also assess the results.

Airborne FACs operated throughout the Vietnam War, but by the early 1970s the proliferation of air defense weapons made slow-moving light aircraft too vulnerable. FACs then began flying in the rear seat of jet fighters (USAF code name Misty FAC), originally F-100F SUPER SABRES, and later OA-37B DRAGONFLIES and OA-10A THUNDER-BOLTS. But to spot and mark enemy units in close proximity to friendly forces, Misty FACs must fly at relatively low speeds (not more than 350 mph/585 kph) and are thus as vulnerable as light aircraft once were. In the future, airborne FACs are likely to be confined to low-threat areas, with REMOTELY PILOTED VEHICLES taking over their role in high-threat settings.

FORWARD DEFENSE: Unofficial term for the official NATO strategy for the defense of Western Europe (now rendered largely out of date by German Reunification). It required the defense of West Germany as far forward as possible—almost up to the Inner German Border (IGB)—as opposed to an earlier strategy which prescribed a defense on the Rhine with West Germany left as a (nuclear) no-man's land. Discussed from 1952 as a result of German rearmament, and soon accepted as NATO policy, Forward Defense was announced only in 1962. Implementation of this strategy was understood to require the use of tactical nuclear weapons so long as the disparity in the NATO–Warsaw Pact military balance persisted. For a different (proposed) NATO strategy, see TRIPWIRE.

FORWARD OBSERVER (FO): A specialist attached to front-line units to spot and adjust ARTILLERY fire and naval GUNFIRE SUPPORT, and to pass back information on enemy troop movements. In the absence of a FORWARD AIR CONTROLLER, the FO may also be authorized to direct close-air support strikes. In the U.S. Army, FOs are usually noncommissioned officers who form a FIST (Fire Support Team) under the command of a junior officer. See also INDIRECT FIRE.

FOX: Daimler FV721 4×4 ARMORED CAR, in service with the British army and the armies of Malawi and Nigeria. Developed from 1965 from the smaller FERRET Mk.4 scout car, the Fox was accepted in 1966 as the "Combat Vehicle, Reconnaissance (Wheeled)" to supplement the SCORPION tracked vehicle. The first prototype was built in 1967, and the first production vehicles were delivered in 1973.

Fox's all-welded aluminum hull has the driver's position in front, a fighting compartment in the middle, and the engine in the rear. Maximum armor thickness is some 15-20 mm., sufficient to defeat small arms fire and shell splinters. The driver has a large window with an armored shutter and a single wide-angle observation periscope (which can be replaced by a passive night scope). The fighting compartment is topped by a manually operated two-man turret. The commander/loader, seated on the left, is provided with seven periscopes for 360° observation, and a 10x surveillance sight with passive night vision capability. On the left, the gunner has two observation periscopes, a binocular periscopic sight, and a separate passive night sight.

Main armament is a 30-mm. RARDEN cannon, firing APDS, HEI, and HE ammunition from six-round clips. The APDS round has a muzzle velocity of some 1200 m./sec., and can penetrate 40 mm. of armor at 1000 m., enough to cope with all light armored vehicles, and to attack the side and rear armor of older main battle tanks. Elevation limits of $-14°$ and $+40°$ allow the engagement of slow, low-flying aircraft. The Fox carries 99 rounds of 30-mm. ammunition. Secondary armament is a coaxial 7.62-mm. machine gun. Four smoke grenade launchers are mounted on each side of the turret. The Fox does not have a COLLECTIVE FILTRATION unit for NBC defense. As initially delivered, the Fox had an erectable canvas flotation screen for amphibious river crossings (since removed from British army vehicles).

Specifications Length: 13.65 ft. (4.16 m.). Width: 7 ft. (2.134 m.). Height: 7.2 ft. (2.2 m.). Weight, combat: 6.12 tons. Powerplant: 190-hp. Jaguar XK 6-cylinder 4.2-liter gasoline. Hp./wt. ratio: 30.04 hp./ton. Fuel: 33 gal. (145 lit.). Speed, road: 64.6 mph (104 kph). Range, max.: 270 mi. (450 km.).

FOXBAT: NATO code name for the Mikoyan MiG-25 twin-engine, all-weather, Mach 3 INTERCEPTOR/RECONNAISSANCE aircraft, presently in service with the Soviet IA-PVO (Fighter Aviation of the Air Defense Troops) and the air forces of Algeria, India, Iraq, Libya, and Syria. Developed from the late 1950s to counter the abortive U.S. XB-70 Valkyrie Mach 3 high-altitude bomber, the MiG-25 was designed as a pure interceptor, optimized for very high speed, rate of climb, and ceiling, at the expense of maneuverability and low-altitude performance. The prototype Ye-266, which flew in 1965, set several speed and time to height records which stood until the mid-1970s. Full-scale production began in 1968, but the first operational squadrons were not formed until 1970–71. By that time the XB-70 had been canceled, but the MiG-

25 remained in production, ostensibly to counter the SR-71 BLACKBIRD reconnaissance aircraft. Given its limited utility as an interceptor, the MiG-25 was itself developed into a reconnaissance aircraft, the MiG-25R Foxbat B.

The advent of the Foxbat caused consternation in Western air forces. Because of the secrecy surrounding its performance, other than its exceptional Mach 3 speed, it gained a reputation as a "superfighter"; it was partly in response to the MiG-25 that the U.S. Air Force initiated development of the F-15 EAGLE. When Lieut. Viktor Belenko defected to Japan with his MiG-25 interceptor in September 1976, Western analysts were surprised to discover that the Foxbat was a rather limited aircraft which relied on early 1960s technology and brute force to achieve very high speeds. While some analysts saw the Foxbat as an example of Soviet technological backwardness, for others it demonstrated what clever design can achieve, even within severe technology limits.

The U.S.S.R. continued to improve the Foxbat throughout the 1970s. The MiG-25U Foxbat C is a two-seat trainer, with a second cockpit stepped behind the first, while the MiG-25R Foxbat D is a strategic reconnaissance variant with a large side-looking aerial radar (SLAR). The MiG-25M Foxbat E is a modernized A, with a new LOOK-DOWN/ SHOOT-DOWN radar and modified engines for improved low-altitude performance. In all, some 500 Foxbats were produced through the early 1980s, of which some 250 A/E interceptors remain in service with the IA-PVO; these are now being replaced by the derivative MiG-31 FOXHOUND. Some 130 B/D reconnaissance variants are also in service with Frontal Aviation (FRONTOVAYA AVIATSIYA, FA). Only a handful of A and B models have been exported.

Most of the MiG-25, a large aircraft, is built of stainless steel: aluminum could not withstand the heat generated by atmospheric friction at Mach 3, while fabricating large titanium structures was apparently beyond Soviet technology of the late 1950s. Titanium was used only in a few critical areas, such as the wing leading edges. Steel can withstand the heat, but is of course heavy.

The massive, slab-sided fuselage is dominated by a very long, pointed nose. This terminates in a large radome housing a powerful search and tracking radar in the interceptor variants. The Foxbat A's I-band "Firefox" RADAR (with vacuum tubes and liquid freon cooling) has five operating modes—ground mapping, long-range search, look-

up, and continuous wave (CW) illumination—but no look-down capability against low-flying targets. Though it has an average power of 600 kW, its detection range is only some 75 mi. (120 km.), with a lock-on range of 43 mi. (70 km.), roughly half the ranges of the contemporary AWG-9 radar of the F-14 TOMCAT. It has only limited ELECTRONIC COUNTER-COUNTERMEASURES, and would rely mainly on brute-force "burn-through" methods to defeat jamming. The MiG-25M has an entirely new pulse-Doppler radar (perhaps an early development of the MiG-31's radar), with limited look-down/shoot-down and multiple TRACK-WHILE-SCAN capabilities. Foxbat E also has an infrared search and track (IRST) unit in a chin fairing for long-range passive detection and target identification.

The Foxbat B reconnaissance model has only a small "Jay Bird" navigational radar and a DOPPLER navigation radar, with the rest of the nose housing five cameras and a small SLAR. Foxbat D is similar, but has a larger SLAR and various ELECTRONIC SIGNAL MONITORING arrays for signal intelligence (SIGINT) gathering.

The pilot sits behind the radar electronics in a small cockpit under a heavily framed clamshell canopy faired into the fuselage spine. Visibility is severely limited in most directions, but this is only a minor handicap for a high-altitude bomber interceptor that must rely on GROUND-CONTROLLED INTERCEPT (GCI) vectoring to its target. Cockpit AVIONICS include a head-down radar display and an analog fire control computer with automated target vectoring via a "Markham" ground environment DATA LINK (in a system similar to the old U.S. SAGE). Other avionics include a Sirena-3 radar warning receiver, an SRO-2 "Odd Rods" IFF, a TACAN-type navigational beacon receiver, and an HF long-range radio.

The rear fuselage houses two very large Turmanskiy R-31 afterburning turbojets fed by massive horizontal-wedge intakes behind the cockpit. Though powerful, the R-31 is not fuel efficient, and fuselage between the cockpit and the engines is fully occupied by fuel tanks; moreover, the engines cannot sustain maximum speed for more than a few minutes without incurring severe damage. A braking parachute is housed in the tail between the engines, while the main landing gear retract into the sides of the air intakes.

The high-mounted wing, moderately swept at 37°, has fixed leading edges, outboard ailerons, plain inboard flaps, and a large integral fuel tank. Two wingtip pods house small CW illuminators for

SEMI-ACTIVE RADAR HOMING (SARH) missiles. The slab tailplane, mounted below the level of the wing, has a notch cut out from the trailing edge roots and another from the tips. The Foxbat has twin vertical stabilizers canted outward; each has a notch cut out of the tip, a typical Mikoyan feature. Maneuverability of the lightly built Foxbat is severely constrained by a maximum 5 G load factor (most modern fighters are stressed to 7–9 g). The MiG-25M Foxbat E has been structurally reinforced, presumably to improve low-altitude performance.

There are four wing pylons for large air-to-air missiles. Early Foxbats carried 2 or 4 AA-5 ASH (also used by the Tu-128 FIDDLER), but most now carry 2 or 4 AA-6 ACRID and AA-7 APEX long-range missiles, or even AA-8 APHID short-range INFRA-RED-HOMING (IR) missiles. Large, unwieldy anti-bomber missiles, Acrid and Apex have both SARH and IR versions; Foxbats normally carry 1 or 2 of each.

Specifications Length: 63.6 ft. (19.39 m.). Span: (interceptor) 45.75 ft. (13.94 m.); (recce) 44 ft. (13.49 m.). Powerplant: (A/B) 2 Turmanskiy R-31 afterburning turbojets, 21,007 lb. (12,850 kg.) of thrust each; (D/E) 2 R31Fs, 30,865 lb. (14,030 kg.) of thrust each. Fuel: 31,575 lb. (14,322 kg.). Weight, empty: 44,090 lb. (20,000 kg.). Weight, max. takeoff: 82,500 lb. (37,425 kg.). Speed, max.: Mach 3.2 (2112 mph/3526 kph) at 36,000 ft. (11,-000 m.) clean; Mach 2.83 (1868 mph/3006 kph) w/4 missiles; Mach 0.85 (650 mph/1050 kph) at sea level. Initial climb: 41,000 ft./min. (12,495 m./min.). Service ceiling: 80,000 ft. (24,400 m.). Combat radius: (interceptor) 702 mi. (1130 km.); (recce) 900 mi. (1503 km.).

FOXHOUND: NATO code name for the Mikoyan MiG-31 twin-engine, long-range INTER-CEPTOR, in service with the Soviet IA-PVO (Fighter Aviation of the Air Defense Troops). Derived from the MiG-25M FOXBAT, the Foxhound was the first Soviet interceptor with a true LOOK-DOWN/SHOOT-DOWN radar; it has much better low-altitude performance and maneuverability than the Foxbat, and nearly twice the combat radius. Developed from the late 1970s, the prototype flew in 1981, and the first production aircraft entered service in 1983. More than 250 Foxhounds are in service, and this number is expected to increase as the remaining MiG-25s are replaced in the 1990s.

Foxhound is externally similar to the MiG-25 (and probably has the same welded stainless-steel construction), but internally the two aircraft are quite different. The large nose radome houses a powerful pulse-Doppler multi-mode RADAR reported to use technology "transferred" from the APG-65 radar of the U.S. FA-18 HORNET. It has a search range of some 190 mi. (317.3 km.) and a lock-on range of 167 mi. (279 km.), more than twice the range of the MiG-25's Firefox radar.

Unlike the MiG-25, the Foxhound has a two-man crew in a tandem cockpit under separate, clear-blown clamshell canopies. While the rearward view is still poor, overall cockpit visibility is much better than in the Foxbat. Cockpit avionics apparently include a head-up display (HUD) for the pilot, and a head-down radar display, vertical situation display, and weapon controls for the rear-seat crewman. Other avionics include an RSO-2 "Odd Rods" IFF, a Sirena-3 RADAR WARNING RE-CEIVER, an HF long-range radio, and a UHF data link for GROUND-CONTROLLED INTERCEPTS.

As in the Foxbat, most of the midfuselage houses fuel tanks, with the engines at the rear. Probably the same those in the MiG-25M, they are fed from modified horizontal wedge inlets similar to those of the MiG-29 Fulcrum, to allow safe operations over a wide range of speeds and angles of attack (AOA). The afterburner nozzles have also been modified for greater efficiency at low altitude. The shoulder-mounted wings resemble those of the Foxbat, with the addition of small leading-edge root extensions to improve maneuverability at high AOA, while the Foxbat's wingtip illuminator pods have been deleted. Foxhound is much slower than the Foxbat at height, but Foxhound is much faster at low altitudes and is stressed to 8 or 9 g for air combat maneuverability. Though not as agile as air superiority fighters such as the F-15 EAGLE or Su-27 FLANKER, the Foxhound is not a lumbering juggernaut like its predecessor.

Foxhound has four fuselage pylons and two wing pylons. The former are normally used for the AA-9 AMOS, a long-range, radar-homing air-to-air missile with a maximum range of some 45–90 mi. (75–150 km.), while each wing pylon can carry a pair of AA-8 APHID short-range INFRARED HOMING missiles.

Specifications Length: 70.5 ft. (21.5 m.). Span: 45.95 ft. (14.01 m.). Powerplant: 2 Turmanskiy R-31F afterburning turbojets, 30,865 lb. (14,-030 kg.) of thrust each. Fuel: 31,575 lb. (14,322 kg.). Weight, empty: 48,115 lb. (21,870 kg.). Weight, max. takeoff: 90,725 lb. (41,238 kg.). Speed, max.: Mach 2.4 (1586 mph/2553 kph) at 36,000 ft. (11,000 m.). Initial climb: 41,000 ft./min.

(12,500 m./min.). **Service ceiling:** 75,000 ft. (22,865 m.). **Combat radius:** 1305 mi. (2180 km.).

FOXTROT: NATO code name for a class of 82 Soviet diesel-electric attack SUBMARINES completed between 1958 and 1984. An incremental development of the Zulu class, 62 Foxtrots were built for the Soviet navy in two groups: 45 between 1958 and 1968, and 17 more between 1971 and 1974; some 45 remain active, including several converted to oceanographic research vessels. At least two were scrapped after operational accidents. Twenty additional Foxtrots were built between 1976 and 1984 for export to India (8), Libya (6), and Cuba (6).

Like the Zulu, the Foxtrot has a cigar-shaped double hull with fuel and ballast tanks between the outer casing and pressure hull; the numerous free-flooding holes in the casing cause considerable drag and flow noises at high speeds. There is a tall, streamlined sail (conning tower) amidships, with a fin-shaped exhaust manifold at its rear. Control surfaces are conventionally arranged, with retractable bow diving planes, a single rudder, and fixed stern planes behind the propellers. A retractable snorkel allows diesel operations from shallow submergence, while large storage batteries give the Foxtrot a submerged (nonsnorkeling) endurance of 5 to 7 days at very low speeds.

Armament consists of 10 21-in. (533-m.) torpedo tubes (6 in the bow and 4 in the stern), for a basic load of 22 conventional and nuclear acoustic-homing TORPEDOES. MINES can replace torpedoes on a 2-for-1 basis. Generally considered among the most successful Soviet diesel submarines, the Foxtrots are quiet, well armed, and reliable, but are hampered by obsolescent sensors and fire controls. See also SUBMARINES, SOVIET UNION.

Specifications **Length:** 300.2 ft. (91.5 m.). **Beam:** 26.25 ft. (8 m.). **Displacement:** 1950 tons surfaced/2500 tons submerged. **Powerplant:** triple-shaft diesel-electric: 3 2000-hp. Type 37D diesel generators, 3 electric motors, 5500 hp. **Speed:** 18 kt. surfaced/16 kt. submerged. **Range:** 11,000 n.mi at 8 kt. (snorkeling)/350 n.mi at 2 kt. (batteries). **Endurance:** 70 days. **Max. operating depth:** 984 ft. (300 m.). **Collapse depth:** 1640 ft. (500 m.). **Crew:** 78. **Sensors:** 1 bow-mounted *Feniks* medium-frequency passive sonar, 1 *Herkules* high-frequency active sonar, 1 "Snoop Tray" surface-search radar, 1 "Stop Light" electronic signal monitoring array, 2 periscopes.

FPF: See FINAL PROTECTIVE FIRE.

FRACTIONAL ORBIT BOMBARDMENT SYSTEM (FOBS): A technique for the delivery of nuclear weapons whereby a warhead is placed into low-earth orbit by a modified ICBM, and then commanded to reenter the atmosphere over the target before completing a full revolution.

Under the terms of the 1966 OUTER SPACE TREATY, nuclear weapons cannot be placed in orbit, but anything less than a full revolution does not violate the treaty. The FOBS delivery method takes advantage of the orbital attack configuration while staying within the letter of the treaty. While a normal ICBM follows a high elliptical trajectory with an apogee of some 1200 km. to reach the target, and is thus easily detected by early warning RADARS, a weapon in low earth orbit (100 km.) can make a quick descent to earth, thus cutting radar warning times from 20 to roughly 3 minutes. Moreover, most early warning radars are positioned to cover the most likely (i.e., northern) approaches of conventional ICBMs, but a FOBS attack would come in "through the back door," and could thus evade detection altogether. This makes FOBS suitable as a precursor to attacking enemy command and control centers, to disrupt any defensive or retaliatory response.

The U.S.S.R. tested an operational FOBS in the mid-1960s, using a modified SS-9 Scarp ICBM as a booster. Though technically successful, the operational penalties, including degraded accuracy, reduced payload, and the difficulty of coordinating FOBS with other delivery systems, made it impractical as a COUNTERFORCE weapon, and the system was dismantled between 1967 and 1972.

FRANKLIN: Designation sometimes applied to the last 12 of the U.S. LAFAYETTE-class ballistic-missile submarines. As compared to earlier Lafayettes, the Franklins have improved electronics and acoustical silencing; all have been modified to launch the TRIDENT I (C-4) SLBM in place of the earlier POSEIDON (C-3). See also SUBMARINES, UNITED STATES.

FRAS-1: Free Rocket, Anti-Submarine, NATO identifier for a Soviet ANTI-SUBMARINE WARFARE (ASW) rocket derived from the Soviet army's FROG unguided artillery rockets, and broadly similar to the U.S. ASROC. FRAS-1 entered service in 1967 and is now deployed aboard the MOSKVA-class helicopter carriers and KIEV-class V/STOL aircraft carriers.

Stabilized in flight by four tail fins, the missile has an ogival nose which contains either a 16-in. (406-mm.) homing TORPEDO or a 15-kT nuclear

DEPTH CHARGE (the latter may give the FRAS-1 a secondary anti-surface role). Powered by a simple solid-rocket engine, the FRAS-1 is fired from a twin-arm launcher (NATO designation SUW-N-1) with a 20-round magazine immediately below it. Range is controlled by adjusting the elevation of the launch rails.

In action, target range, speed, and bearing, derived from the ship's SONAR, are fed to a FIRE CONTROL computer which calculates the target's future position, and generates the required azimuth, elevation, and launch time for the weapon. The FRAS-1 then flies a ballistic trajectory to the predicted target location. As the rocket descends, the payload separates and deploys a braking parachute to slow its water entry. The torpedo would then initiate an autonomous search pattern, while the nuclear depth charge would be detonated at a preset depth by a hydrostatic or time fuze.

Specifications **Length:** 20.3 ft. (6.2 m.). **Diameter:** 2.29 ft. (700 mm.). **Span:** 4.3 ft. (1.3 m.). **Weight, launch:** 1764 lb. (800 kg.). **Speed:** Mach 1 (762 mph/1273 kph.) **Range:** 5.8–18.6 mi. (9.25–30 km.).

FRATRICIDE: 1. Damage or casualties accidentally inflicted by friendly fire; also called "amicide."

2. The destruction or neutralization of one or more NUCLEAR WEAPONS by the detonation of other nuclear weapons launched by the same side. Fratricide is a crucial problem in counterforce attacks by ballistic missiles against hardened missile silos, because two or more warheads must be assigned to each silo to assure a (very) high probability of kill (P_K). The main causes of fratricide would be the fireball, prompt radiation (X-RAYS and GAMMA RAYS), and debris thrown up by the previous explosion. Computer modeling suggests that the cloud of debris would be the most serious problem: a REENTRY VEHICLE (RV) passing through the cloud might disintegrate, or at least be deflected off course. There appears, however, to be a "window" of some ten seconds between the subsidence of prompt radiation and the spread of the debris cloud, through which RVs could pass at relatively low risk. Therefore, accurately timing the arrival of reentry vehicles ("TIME-ON-TARGET") would be essential for the proper execution of a two-on-one counterforce attack. The guidance units of Soviet ballistic missiles (e.g., SS-18 SATAN and SS-19 STILETTO) can provide precise time-on-target control.

FREE ELECTRON LASER: A type of high-energy LASER which is generated by passing an electron beam through a series of electromagnets ("wigglers").

FRESCO: NATO code name for the Mikoyan-Gureyvich MiG-17, a single-engine FIGHTER of 1950s vintage. Widely exported, and also built in China as the Shenyang F-4, many remain in service with Third World air forces. **Operators:** Afg, Alg, Ang, Bul, Con, Cuba, Cze, DDR, Eq, Gui, Gui, Mad, Mali, Maur, Moz, N Kor, Pol, Rom, Som, Syr, Viet, Yem.

FRIGATE: A destroyerlike vessel displacing between 1000 and 4000 tons. Once there was a clear distinction between frigates (or DESTROYER ESCORTS) intended mainly for ANTI-SUBMARINE WARFARE (ASW) and larger, more versatile DESTROYERS. The difference is now minimal, and some of the latest frigates are considerably larger than many earlier destroyers. By convention, however, frigates are now relatively low-capability, single-purpose (ASW or ANTI-AIR WARFARE) warships meant for merchant-ship CONVOY escort, as opposed to high-capability, multi-purpose destroyers for battle-group escort.

During the early 1970s, the U.S. Navy had further confused matters by reclassifying its "destroyer-leaders" (large guided-missile carrier escorts) as "frigates," a classification later changed to "guided-missile CRUISERS" when the destroyer escorts were reclassified as frigates. The French navy, on the other hand, classifies some of its frigate-type vessels as CORVETTES, a term otherwise applied to warships under 1000 tons. The Soviet navy does not have a distinct frigate category; its frigatelike vessels are classified either as "Large Anti-Submarine Ships" (*Bol'shoy protivolodochnyy korabl'*, BPK), "Small Anti-Submarine Ships" (*Malyy protivolodochnyy korabl'*, MPK), or "Patrol Ships" (*Storozhevoy korabl'*, SKR). See also separate FRIGATES entries for BRITAIN; CANADA; FEDERAL REPUBLIC OF GERMANY; FRANCE; ITALY; JAPAN; NETHERLANDS; SOVIET UNION; UNITED STATES.

FRIGATES, BRITAIN: The first post-1945 British frigates, laid down from 1952, included 6 2560-ton Whitby (Type 12) class ASW frigates (plus 2 more for India), commissioned in 1955–56; 12 austere 1456-ton Blackwood (Type 14) class commissioned in 1957–59; and 4 2500-ton Leopard (Type 41) class anti-aircraft frigates (plus 3 more for India), and 4 2500-ton Salisbury (Type 61) class fighter-direction frigates, all commissioned in

1959–60. Of these early frigates, the Whitbys were the most successful: rugged and seaworthy, they had a raised forecastle to cope with heavy North Atlantic seas. Two Whitbys and two Leopards remain in service with the Indian navy; all others have been retired.

Nine 2800-ton Rothesay (Modified Type 12) class ASW frigates commissioned in 1960–61 were quite similar to the Whitbys, but they were modernized between 1966 and 1972 with SEA CAT surface-to-air missiles and a helicopter landing deck (the first on a British frigate); all have been retired. The seven Tribal (Type 81) class frigates commissioned between 1961 and 1964 were an attempt to build more economic escorts. Equipped with a complex COSAG (Combination of Steam and Gas) powerplant, they had a maximum speed of only 28 kt., and at 2700 tons were too small to accommodate modern weapons and sensors. All have been retired.

The Tribals were followed by the 2800-ton LEANDER class, a further development of the Type 12. The single largest class of warships built by the Royal Navy since 1945, a total of 26 were commissioned between 1963 and 1973, with an additional 5 for Australia (RIVER class), 2 for Chile, 6 for India, 6 for the Netherlands (van Speijk class), and 2 for New Zealand. The Leander is also the basis for the Indian Godavari-class frigates

The first British frigates designed with helicopter facilities, the Leanders were built in two batches: 16 with the standard hull, and 10 with a "broad beam" hull. Nineteen remain in service, including 5 converted for ANTI-AIR WARFARE (AAW) with SEA WOLF surface-to-air missiles; four ANTI-SURFACE WARFARE (ASUW) conversions with EXOCET anti-ship missiles and a TOWED ARRAY SONAR; 3 more austere Exocet conversions; 2 ASW escorts with IKARA anti-submarine missiles; 5 unmodified broad-beam Leanders; and 1 unmodified standard-hull ship.

The eight 3200-ton AMAZON (Type 21) class frigates commissioned between 1974 and 1978 were originally intended as Leander replacements. Built to a new Vosper-Thorneycroft design with COGOG propulsion, they lacked sufficient stability to carry their intended weapons and sensors, and have also proved to be too fragile and vulnerable. Two were lost to air attack in the 1982 Falklands War.

The subsequent BROADSWORD (Type 22) class, another attempted Leander replacement, was originally intended as a class of 26 ASW escorts with secondary AAW and ASUW capabilities. The first 4 ships (Batch 1), which displace 4400 tons, have 2 LYNX helicopters, 2 Sea Wolf launchers, and 4 Exocet. After the Falklands War, the next 6 ships (Batch 2) were completed to a modified design. Lengthened to improve seaworthiness, habitability, and endurance, and to provide space for a towed array sonar, they displace 4850 tons, but have the same armament as Batch 1. The last 4 Broadswords (Batch 3) combine the lengthened Batch 2 hull with a general-purpose armament that includes a dual-purpose gun, eight HARPOON anti-ship missiles, Sea Wolf, and a GOALKEEPER 30-mm. radar-controlled gun for anti-missile defense. Displacing 5250 tons, they are considerably larger than the SHEFFIELD-class guided-missile destroyers.

Because of their increased size and cost, the Broadswords could not be built in the numbers originally envisaged. In yet another attempt to design a Leander replacement, the Royal Navy is presently building eight DUKE (Type 23) class general-purpose frigates, the first of which entered service in 1989. Displacing 3700 tons, this has a balanced armament with 1 dual-purpose gun, 1 Sea Wolf vertical launch unit, a Goalkeeper gun, 8 Harpoon, and 1 ASW helicopter.

FRIGATES, CANADA: The first post-1945 Canadian frigates were the seven 2860-ton ST. LAURENTS commissioned between 1955 and 1957. Though derived from the British Type 12 (Whitby) class, they have a rather different hullform and internal arrangements for better seakeeping in arctic waters. Originally armed with 2 twin 3-in. 50-caliber ANTI-AIRCRAFT guns, 2 40-mm. BOFORS guns, and 2 British Mk. 10 LIMBO depth charge mortars, they were modernized in the early 1960s with a helicopter flight deck and hangar, 2 sets of Mk.32 triple tubes for lightweight ASW homing TORPEDOES, and a VARIABLE DEPTH SONAR (VDS). The St. Laurent was retired in 1974, but the others received a second major overhaul in 1981.

The St. Laurent design was developed into the 2900-ton RESTIGOUCHE class of seven ASW frigates commissioned in 1958–59. Originally armed with a British twin 3-in. 70-caliber gun on the bow, a U.S. twin 3-in. 50-caliber gun amidships, and two Limbos, four Restigouches were modified between 1968 and 1974, with a 8-round ASROC launcher and a VDS in place of the 3-in. 50-caliber gun and one Limbo. Budget constraints prevented modification of the other three ships, which were placed in reserve in 1974. The modified ships were further modernized in the early 1980s, receiving two sets of Mk.32 triple torpedo tubes and improved ELECTRONIC COUNTERMEASURES (ECM) equipment.

The Restigouche class was followed by the 2890-ton MACKENZIE class. Essentially repeats of the Restigouche, but with improved habitability and more arctic gear, these four ships have the same armament as that originally fitted to their predecessors, plus two sets of Mk.32 triple torpedo tubes added in 1980. All four received new sensors during major overhauls in the mid-1980s.

Two 3000-ton ANNAPOLIS-class frigates commissioned in 1964 were the final development of the original St. Laurent design. While retaining the basic hull and machinery of the earlier classes, they have a completely different armament of 1 twin 3-in. 50-caliber gun, 2 sets of Mk.32 triple torpedo tubes, 1 Limbo, 1 SEA KING helicopter, plus a VDS. When overhauled and modernized in the mid-1980s, they received improved sensors and improved ECM equipment.

The St. Laurent, Restigouche, Mackenzie, and Annapolis classes will all reach the end of their useful lives in the mid-1990s. As replacements, the Canadian navy is now receiving HALIFAX-class frigates, the first of which was commissioned in late 1989. Displacing 4254 tons, they have a completely new hullform and powerplant, as well as greatly improved weapons and sensors. Armament includes a twin 57-mm. anti-aircraft gun, eight HARPOON anti-ship missiles, a 16-round SEA SPARROW vertical launch system, a 20-mm. PHALANX radar-controlled anti-missile gun, and an SH-60B SEA HAWK ASW helicopter. With twin-shaft CODOG propulsion, they have a design speed of 29 kt.

FRIGATES, FEDERAL REPUBLIC OF GERMANY:
The reconstituted West German navy (*Bundeskriegsmarine*) initially relied on a variety of ex-U.S. and ex-British DESTROYERS, DESTROYER ESCORTS, and frigates, all of which were retired by the mid-1970s.

Optimized for Baltic operations, the first postwar German frigates were the Koln class, six of which were commissioned between 1961 and 1964. Displacing 2600 tons, they have a CODAG powerplant capable of 32 kt. Armament includes 2 100-mm. DUAL PURPOSE guns, 3 twin 40-mm. BOFORS GUNS, a 4-barrel Bofors 375-mm. ANTI-SUBMARINE WARFARE (ASW) rocket launcher, and 4 21-in. (533-mm.) tubes for large ASW homing TORPEDOES. Only three remain in service with West Germany; two others were sold to Turkey, and the last cannibalized for spare parts.

The Kolns have been superseded by the BREMEN-class guided-missile frigates (modified versions of the Dutch KORTENAER class), eight of which were commissioned between 1982 and 1990. Six "Improved Bremens" are planned for the mid-1990s.

FRIGATES, FRANCE:
The French navy does not employ many frigates, preferring instead a mix of large DESTROYERS and smaller, more economical CORVETTES; this reflects Mediterranean requirements, where long range is not a concern. Immediately after World War II, the French navy relied on ex-U.S. destroyer escorts and ex-British Hunt-class ANTI-SUBMARINE WARFARE (ASW) destroyers, but from 1951 and 1960 it commissioned 4 Le Corse and 14 Le Normand frigates of similar design. Displacing some 1800 tons, with steam powerplants capable of 28 kt., they were armed with 3 twin 57-mm. and 2 20-mm. ANTI-AIRCRAFT guns, 4 triple tubes for ASW homing TORPEDOES, and a quadruple DEPTH CHARGE mortar; all were retired in the 1970s.

The Normands were superseded by the nine 2200-ton COMMANDANTE RIVIERES commissioned between 1962 and 1968. Intended mainly for "colonial" operations in African and Pacific waters, they have diesel engines for long range, air conditioning for tropical cruises, and accommodations for an 80-man COMMANDO unit. The last seven in service are now scheduled for retirement.

The 1250-ton D'ESTIENNE D'ORVES class are smaller, more specialized warships intended mainly for inshore ASW. Seventeen were commissioned between 1976 and 1984, including three originally intended for South Africa and later sold to Argentina.

France is presently considering construction of three new frigates of the FL-25 class. Intended as general-purpose escorts, they would displace some 3200 tons and carry eight Exocet MM.40, a 100-mm. gun, and an ASW helicopter.

FRIGATES, ITALY:
After World War II, the Italian navy relied on the remnants of its wartime destroyer force and some ex-U.S. DESTROYERS and DESTROYER ESCORTS for ANTI-SUBMARINE WARFARE (ASW). Postwar construction initially concentrated on destroyers, and the first new Italian frigates, the four 2250-ton Centauro-class vessels, were not commissioned until 1957–58. They were followed by four 1650-ton Bergamini-class frigates commissioned in 1961–62; two remain in service, but are scheduled for retirement in the early 1990s. The latest Italian ASW frigates, the 2700-ton Alpino class, two of which were commissioned

in 1968, are capable of 27 kt. and are much better armed than their predecessors.

The 2525-ton LUPOS were Italy's first guided-missile frigates. Extremely successful, 4 were commissioned for the Italian navy between 1977 and 1980, and 6 were sold to Venezuela, 4 to Peru, and four to Iraq (completed but not delivered). The eight 3040-ton MAESTRALE-class guided-missile frigates commissioned between 1982 and 1985 are basically "stretched" Lupos.

FRIGATES, JAPAN: The Japanese Maritime Self-Defense Force (JMSDF) maintains some 17 modern frigates (designated destroyer escorts), intended for ANTI-SUBMARINE WARFARE (ASW) along Japan's SEA LINES OF COMMUNICATION (SLOCs). When first reconstituted in the mid-1950s, the JMSDF relied mainly on ex-U.S. DESTROYER ESCORTS, and the first postwar Japanese frigates were heavily influenced by U.S. designs. Subsequent classes strongly reflect purely Japanese requirements and operational preferences.

The oldest frigates still in service are the four 1800-ton Isuzus commissioned between 1961 and 1964. Optimized for ASW, they have economical diesel powerplants and a maximum speed of 25 kts. Only three remain in service as DEs; the fourth has become a training vessel.

They were superseded by 11 1800-ton CHIKUGO-class ASW frigates commissioned between 1970 and 1977. Diesel-powered and capable of 25 kt., they have considerably better sensors than do the Isuzus, and include an SQS-35 VARIABLE-DEPTH SONAR. (VDS).

The 1450-ton *Ishikari* (an experiment in size reduction), comissioned in 1981, is a general-purpose escort, more lightly armed than the Chikugos. Too cramped to carry an adequate sensor suite, the Ishikari was superseded by two 1760-ton YUBARI-class frigates, essentially "stretched" Ishikaris with the same armament, but with additional sensors, greater range, and improved habitability. The Yubaris are themselves now being superseded by six "1900-ton class" frigates, the first of which was completed in 1989. Considerably more powerful than the Yubaris, they displace some 2300 tons fully loaded, and are armed with 1 OTO-Melara 76.2-mm. gun, 1 20-mm. Phalanx radar-controlled anti-missile gun, 8 Harpoon, an ASROC launcher, and 2 sets of triple torpedo tubes. A RIM-116A RAM (Rolling Airframe Missile) launcher will be added in the 1990s. See also DESTROYERS, JAPAN.

FRIGATES, NETHERLANDS: The Royal Netherlands Navy operates 14 frigates, with 8 more under construction. The oldest belong to the 2835-ton VAN SPEIJK class, originally of 6 ships commissioned in 1967–68. Based on the British LEANDER class, only two remain in the Dutch navy; the others were sold to Indonesia in 1988.

Ordered in the 1960s to replace aging ASW destroyers, the 3786-ton KORTENAER class have proven quite successful, with ten commissioned by the Netherlands between 1978 and 1983, and two more sold to Greece. The two 3750-ton HEEMSKERCK-class guided-missile frigates commissioned in 1986 are ANTI-WAR WARFARE (AAW) derivatives of the Kortenaers, with a Mk.13 launcher for 40 RIM-66 STANDARD-1MR SAMs. They will be supplemented from 1992 by eight 3320-ton KAREL DOORMAN–class frigates.

FRIGATES, SOVIET UNION: The Soviet navy does not have a frigate category per se; rather, frigatelike warships are classified either as "Large Anti-Submarine ships" (*Bol'shoy protivolodochnyy korabl'*, BPK, a designation also assigned to some DESTROYERS), or as "Small Anti-Submarine ships" (*Malyy protivolodochnyy korabl'*, MPK, a designation also assigned to some CORVETTES), and "Patrol ships" (*Storozhevoy korabl'*, SKR, again a designation assigned to some corvettes). Most MPKs and SKRs, intended mainly for coastal defense and border patrol, can be considered frigates.

The first postwar frigates were the 8 1500-ton Kola class SKRs, completed in 1951–52, based on a German light destroyer design; all were retired in the early 1970s. They were superseded by the 1400-ton RIGA-class SKRs. Though smaller than their predecessors, they had better weapons and sensors, and a more efficient powerplant. Some 66 Rigas were completed between 1954 and 1959, including eight for export. Some 35 remain active in the Soviet navy, and 17 Rigas have been transferred to other navies; they are of marginal value except for coast guard duties.

In 1961 the Soviet navy introduced the 1150-ton PETYA-class MPKs specifically for inshore ANTI-SUBMARINE WARFARE (ASW). Some 52 Petyas were built for the Soviet navy in two distinct subclasses: 24 Petya Is completed between 1961 and 1964, and 27 Petya IIs completed between 1964 and 1969. Seventeen modified versions were built for export to India (10), Vietnam (4), Syria (2), and Ethiopia (1). The first Soviet warships with gas turbine engines, some 45 Petyas remain in Soviet service.

Eighteen 1150-ton MIRKA-class SKRs, completed between 1964 and 1966, were apparently unsuccessful competitors for the Petyas, from which they differ mainly in having an unusual form of propulsion: a twin-shaft CODAG powerplant with both propellers and also pumpjets for rapid acceleration. Built in two subclasses, the first 9 (Mirka Is) differ from the last 9 (Mirka IIs) only in details of armament; all 18 remain in service.

The 1200-ton GRISHA-class MPKs are considered successors to the Petyas. Introduced from 1968, more than 51 have been completed to date in five separate subclasses. Sixteen Grisha Is were completed between 1968 and 1974. The Grisha II is a border patrol ship (*Pogranichnyy Storozhevoy Korabl'*, PSKR) operated by the KGB Border Guards; at least 9 have been completed since 1974. The Grisha III, the single largest subclass, with more than 30 completed since 1975, is an improved I. The Grisha IV is similar to the III, but with heavier armament, while the Grisha V is a refined version of the Grisha IV. Resembling enlarged Grishas, the 1900-ton Koni-class SKRs are intended specifically for export, though one is operated by the Soviet navy for training and evaluation. At least 10 have been sold since 1975, to Algeria (2), Cuba (2), East Germany (2), Libya (2), and Yugoslavia (2).

The 3700-ton KRIVAK class are the Soviet equivalent to large Western ASW escorts such as the U.S. KNOX or British LEANDER classes, but much more heavily armed. Originally designated MPKs but later reclassified as SKRs, 32 Krivaks were completed for the Soviet navy in two subclasses between 1975 and 1979: 21 Krivak Is and 11 Krivak IIs. At least 4 more modified Krivak III PSKRs have been built for the KGB Border Guards since 1984. The world's first warships with all-gas turbine propulsion, the Krivaks have a maximum speed of 32 kt., but a range of only 3900 n.mi at 20 kt.

FRIGATES, UNITED STATES: The term "frigate" has had several different meanings in the modern U.S. Navy. Through World War II, it referred to small ANTI-SUBMARINE WARFARE (ASW) vessels displacing less than 1000 tons (a type now generally called CORVETTES); in the 1960s, it was applied to large, missile-armed escorts formerly known as "destroyer-leaders," and later redesignated "guided-missile cruisers" in the 1970s; the ships now classified as frigates (FFs) and guided-missile frigates (FFGs) are by contrast single-purpose ASW *or* ANTI-AIR WARFARE (AAW) escorts intended mainly for convoy (not carrier group) escort. As such, they are the direct descendants of

the destroyer escorts (DEs) mass-produced in World War II to counter the German U-boats.

During the 1950s, the U.S. Navy relied on its large number of wartime destroyers for ASW escort. Many FLETCHER-class destroyers (DDs) were converted to ASW escort destroyers (DDEs), a role later filled by SUMNER- and GEARING-class FRAM (Fleet Rehabilitation and Modernization) conversions. It was realized, however, that "block obsolescence" would become a problem in the 1960s, when the warships built in World War II would reach the end of their useful lives. Thus there were repeated attempts to design an "austere" ASW escort for mass production.

The first of the postwar escort designs was the 1875-ton DEALEY (DE-1006) class, 13 of which were commissioned between 1954 and 1957. In due course the navy judged them too small for effective ASW, yet too expensive for mass production. All were retired from the U.S. Navy in the early 1970s; 2 have been transferred to Uruguay and Colombia, and the remainder scrapped. Four Dealeys were also built for Portugal, while the Dealey hull served as the basis for 5 OSLO-class frigates built by Norway in 1966–67. The subsequent 1900-ton Claud Jones (DE-1033) class of 4 ships, commissioned between 1957 and 1960, were intended as an even more economical version of the Dealeys, with a diesel powerplant capable of only 21 kt. Considered useless against modern submarines, all 4 were sold to Indonesia in 1973–74.

After these experiences, the navy predictably concluded that only much larger, more complex (and more expensive) ships could be effective against nuclear submarines. The first of these "second generation" escorts were the BRONSTEIN (DE-, later FF-1037) class, two of which were commissioned in 1963. Displacing 2650 tons, they were as large as World War II destroyers. Equipped with a large, bow-mounted SQS-26 low-frequency SONAR for long-range detection and tracking, their maximum speed was only 24 kts.; but because of their size and a new hullform, they could maintain that speed in much higher seas than earlier destroyers. Both were retired in 1988–89.

They were followed by the larger and faster GARCIA (DE-, later FF-1040) class and the similar BROOKE (DEG-, later FFG-1) guided-missile escorts. Commissioned between 1964 and 1968, the 10 3500-ton Garcias were retired in 1989, but 2 were subsequently leased to Pakistan, 2 to Turkey, and 4 to Brazil. The 6 3600-ton Brookes, commissioned in 1966–67, are essentially Garcias with a

Mk.22 launcher with 16 RIM-66A STANDARD-1MR surface-to-air missiles. All 6 were retired in 1989, with 2 also leased to Pakistan and 2 to Turkey. Based on the Garcia class, the 3630-ton USS *Glover* (DE-, later FF-1098), built in 1965 as a testbed for new ASW weapons and sensors, is scheduled for retirement in the early 1990s.

By the mid-1960s, block obsolescence had become a reality with the retirement of many FRAM Sumners and Gearings. The 4260-ton KNOX (DE-, later FF-1052) class was selected for large-scale production to partially fill this gap, and 46 were commissioned by the U.S. Navy between 1969 and 1974 (five BALEARES-class guided-missile variants were also built for Spain). Larger than the destroyers of most navies, and by no means inexpensive, they were criticized at first for their low speed, single propeller, and light armament, but are now considered excellent ASW vessels. All 46 remain in service, including 8 assigned to the Naval Reserve Force (NRF).

The Knoxes only partially compensated for the retirement of the World War II destroyers, and the U.S. Navy therefore proposed two new classes for the 1970s: the high-cost, high-capability SPRUANCE (DD-963) class destroyers for carrier battle group escort, and the lower-cost, low-capability PERRY (FFG-7) class guided-missile frigates for convoy escort. Originally designated "patrol frigates" (PFs), 51 were commissioned by the U.S. Navy between 1987 and 1989 (with 6 more built for Australia and 5 for Spain). Displacing 3658 tons, they have a gas turbine powerplant and are quite maneuverable. Intended mainly for AAW, they have been criticized for their lack of a clearly defined mission: their AAW armament is insufficient to counter high-intensity threats, while their ASW capabilities are inferior to that of the Knox class.

With the completion of the last Perry in 1989, the U.S. Navy is now designing its next generation of frigates to replace the Knoxes at the turn of the century. There is still considerable disagreement within the navy over their mission and specifications. Several advanced hullforms have been proposed (including a deep-V and a catamaran), but the design has not been finalized, nor have funds been allocated for construction. See also DESTROYERS, UNITED STATES.

FROG: Free Rocket Over Ground, NATO designation for the Soviet R-75 *Luna* series of unguided artillery ROCKETS. Broadly similar to the U.S. Honest John, the FROG—initially deployed in the early 1950s—was progressively upgraded through the 1970s. Once the standard nuclear and chemical delivery system of Soviet tank and motorized rifle divisions (each has or had a battalion of 4 launchers and 16 rockets), FROGs are now being replaced by the much more accurate SS-21 SCARAB tactical ballistic missile. Numerous FROGs armed with high-explosive warheads have been supplied to the other Warsaw Pact countries, and to Cuba, Egypt, Iraq, Kuwait, Libya, North Korea, Syria, and Yugoslavia. China is reported to have produced a copy.

All FROGs are spin-stabilized solid-fuel rockets. The original FROG-1 had a cylindrical body, a conical nose, and four large, canted control fins. Armed with a large nuclear warhead, it was fielded on a single rail launcher mounted on an IS-3 heavy tank chassis. As with all unguided rockets, range was varied by altering the elevation of the launch rail. All FROG-1s were retired in the early 1970s. The FROG-2, first observed in 1957, was a smaller rocket, armed with either a nuclear or a high-explosive warhead in a bulbous nose section. It was fielded on a single rail launcher mounted on a modified PT-76 light tank chassis. The FROG-2 has also been retired.

The FROG-3, 4 and 5 were all variants of a single design. First observed in 1960, they introduced a tandem-motor configuration for extended range: both rocket motors were ignited at launch, with the forward motor's exhaust vented diagonally to protect the rear motor. Armed with a 992-lb. (450-kg.) high-explosive or 25-kT nuclear warhead in a drum-shaped 21.5-in. (550-mm.) warhead section, the FROG-3 is also launched from a modified PT-76. No longer in front-line service, a few are still in use as training rounds. The FROG-4 was generally similar, but with a slim warhead of the same diameter as the missile. Believed to be a chemical delivery system, the FROG-4 was retired in the late 1970s. The FROG-5, introduced in 1964, was a FROG-3 with a shorter, wider body; all were retired in the late 1970s. The FROG-6 was a concrete-filled, inert training round similar to the FROG-3.

Introduced in 1965, the current FROG-7 has a cylindrical body, a slim conical nose section, and four large tail fins. Because of improved solid propellants, the FROG-7 has reverted to the single-motor configuration. The basic FROG-7 is armed with a 10-, 100-, or 200-kT nuclear warhead. The FROG-7a has a 1213-lb. (550-kg.) high-explosive warhead, while the FROG-7b is a chemical delivery system with 860 lb. (390 kg.) of the persistent

NERVE AGENT VR-55. Quite inaccurate at longer ranges, conventionally armed FROGs can be used only for area bombardments. The FROG-7 is fired from a lightweight rail launcher mounted on a ZIL-135 8 × 8 truck. See also ROCKET ARTILLERY.

Specifications **Length:** (1) 32.8 ft. (10 m.); (2) 29.5 ft. (9 m.); (3) 34.5 ft. (10.5 m.); (5) 29.83 ft. (9.1 m.); (7) 29.85 ft. (9.1 m.). **Diameter:** (1) 33.5 in. (850 mm.); (2) 23.6 in. (600 mm.); (3) 15.75 in. (400 mm.); (5) 21.5 in. (550 mm.); (7) 21.7 in. (550 mm.). **Weight, launch:** (1) 6614 lb. (3000 kg.); (2) 5219 lb. (2400 kg.); (3) 4960 lb. (2250 kg.); (5) 6614 lb. (3000 kg.); (7) 5071 lb. (2300 kg.). **Range:** (1) 20 mi. (32 km.); 15.5 mi. (25 km.); (3) 25 mi. (40 km.); (5) 34 mi. (55 km.); (7) 43 mi. (70 km.). **CEP:** (7) 450–700 m.

FROGFOOT: NATO code name for the Sukhoi Su-25, a specialized CLOSE AIR SUPPORT (CAS) aircraft roughly equivalent to the U.S. A-10A THUNDERBOLT. Developed from the early 1970s, apparently to supplement attack HELICOPTERS in the anti-tank role, Frogfoot is in effect a jet-propelled version of the famous IL-2 *Shturmovik* of World War II. The prototype flew in 1977, and the first operational squadrons were formed in 1980–81. From 1982, several squadrons were deployed to Afghanistan, where 23 were shot down by ground fire over nine years. Some 250 Frogfoots have been delivered to Soviet Frontal Aviation (FRONTOVAYA AVIATSIYA, FA), and a number have been exported to Bulgaria, Czechoslovakia, Hungary, and Iraq. It is believed that production of the Su-25 is continuing at a low rate, though not many more are likely to be procured for the Soviet air force.

Frogfoot is smaller than the A-10, and bears a superficial resemblance to the Northrop XA-9, the A-10's unsuccessful competitor. The fuselage has a downward-sloping "duck nose" *(utkanos)* similar to that of the MiG-27 FLOGGER, which affords the pilot an excellent view of the ground below. The nose houses a LASER DESIGNATOR/rangefinder and a Doppler navigation radar, with a twin-barrel 30-mm. CANNON mounted on the port side. Equipped with 250 rounds of ammunition, the cannon reportedly has a rate of fire of 3000 rds./min.

As in the A-10, simplicity and survivability are stressed, with redundant, widely separated flight controls and other critical equipment also duplicated. The pilot sits under a clamshell canopy with a heavy, bullet-proof windscreen, while the cockpit has been fabricated of 24-mm. titanium armor capable of resisting 20-mm. anti-aircraft fire. Cockpit AVIONICS are limited to a head-up display (HUD), several RADAR WARNING RECEIVERS, and a weapon-delivery computer. The rear fuselage is believed to house several self-sealing fuel tanks and additional avionics, including a strike camera and two 32-round CHAFF/flare dispensers. The engines are mounted in pods blended into the sides of the fuselage and fed by plain cheek inlets behind the cockpit. The engines and fuel tanks are protected by 5 mm. of steel armor plate, and two 32-round countermeasures dispensers have been added over each engine pod, allowing the Su-25 to carry up to 256 chaff or flare cartridges for defense against radar-guided or INFRARED-HOMING missiles. The main landing gear, with large low-pressure tires for rough-field operations, retract into the sides of the engine pods.

The shoulder-mounted wing has full-span leading-edge slats, large outboard ailerons, and inboard Fowler flaps for low-speed maneuverability and short takeoff and landing (STOL) performance. The wing probably houses several self-sealing fuel tanks, and has two large wingtip ACTIVE JAMMING pods, whose rear halves are split to act as air brakes.

Frogfoot has no fewer than ten wing pylons, for a total payload of 9700 lb. (4410 kg.), including 19-round UV-57-16 rocket pods for 57-mm. FFARS, free-fall and CLUSTER BOMBS, LASER-GUIDED BOMBS, NAPALM, and AS-11 KILTER air-to-surface missiles. Two small outboard pylons can accommodate AA-2 ATOLL or AA-8 APHID infrared-homing air-to-air missiles for self-defense.

Specifications **Length:** 50.5 ft. (15.4 m.). **Span:** 46.95 ft. (14.31 m.). **Powerplant:** 2 Turmanskiy R-195 turbojets, 9340 lb. (4245 kg.) of thrust each. **Weight, empty:** 20,950 lb. (9522 kg.). **Weight, normal loaded:** 32,100 lb. (14,500 kg.). **Weight, max. takeoff:** 38,790 lb. (17,632 kg.). **Speed, max.:** 608 mph (1015 kph) at sea level. **Speed, cruising:** 428 mph (715 kph.). **Service ceiling:** 22,965 ft. (7000 m.). **Combat radius:** (hi-lo-hi) 345 mi. **Range, max.:** 800 mi. (1336 km.).

FROGMAN: Colloquial term for a combat diver. See UDT.

FRONT: Soviet term for a higher military echelon loosely equivalent to a Western ARMY GROUP. Commanded by a full general *(General Armii)* or a Marshal of the Soviet Union, a Front consists of two or more assigned ARMIES, a variety of support and service units (including artillery and anti-aircraft forces), one or more "tactical air armies" of Frontal Aviation (FRONTOVAYA AVIATSIYA), plus detachments of Transport Aviation (VTA), and naval forces. Front boundaries tend to coincide with with those of MILITARY DISTRICTS and Groups of Forces overseas; those administrative entities would be converted into operational Fronts for war. Fronts are in turn subordinate to a Theater of Military Operations (TVD).

FRONTOVAYA AVIATSIYA (FRONTAL AVIATION): The branch of the Soviet Air Force (VVS) responsible for operations in support of the army, including CLOSE AIR SUPPORT (CAS), battlefield air INTERDICTION (BAI), deep strike, tactical RECONNAISSANCE, and AIR SUPERIORITY. In the early 1980s, the Soviet army's Ground Forces (*Sukhoputnyye voyska*, SV) acquired its own aviation branch with ex-FA helicopter squadrons. This formalized the de facto organization under which many FA helicopters were placed under the direct operational control of ground formations, but it is unclear if administrative control of helicopters, air crews, and support personnel has also passed to the SV.

FA aircraft are organized into SQUADRONS of 12; 3 squadrons of the same type form an air regiment (equivalent to a U.S. Air Force WING). Three regiments in turn comprise an air division, and 2 or more air divisions form a tactical air army (equivalent to a U.S. numbered "air force"), of which there are presently 16. In peacetime, one air army is assigned to each MILITARY DISTRICT and Group of Forces abroad. In wartime, air armies would be assigned to FRONTS as required, with a strategic reserve (from interior Military Districts) kept under the direct control of the Soviet high command. See also ADD; PVO; VTA.

FUEL AIR EXPLOSIVE (FAE): A weapon designed to dispense an aerosol cloud of highly volatile fuel, which can be detonated to generate a very powerful explosion after it is mixed with air. While the blast of conventional high explosives is highly concentrated, and wanes rapidly with distance (and is thus inefficiently distributed), the fuel-air mix can be spread evenly over a wide area to achieve a uniform blast wave; the effect is many times greater than that obtainable from an equivalent weight of TNT.

For example, a 1000-lb. (500-kg.) methane-based FAE can generate peak overpressures of 12.8 psi (0.9 kg./cm.^{2}) over a radius of some 200 m., and 6 psi (0.42 kg./cm.^{2}) at 300 m. (a peak overpressure of 5.2 psi, sufficient to collapse brick and concrete buildings, is considered "severe" blast).

Thus the blast effect of FAEs is roughly equivalent to that of small nuclear weapons.

So far, most FAEs have relied on gaseous carbohydrate fuels which decompose explosively, ignite spontaneously in contact with moisture or soil, and do not require oxygen for combustion; their combustion is propagated at hypersonic speeds, thereby generating a powerful shock wave. Ethylene oxide, one of the most commonly used fuels, has 2.7 to 5 times the energy level of TNT. Propylene oxide, methyl acetelene/propadiene/propane (MAPP), acetic peroxide, propyl nitrate, and diobrane fuels are even more powerful, with ratios of 10-to-1 theoretically possible; i.e., a 1-ton FAE could deliver a blast equivalent to 10 tons of TNT.

So far, FAEs have been used mainly for attacks against bunkers or for clearing minefields. But they could also be extremely effective against tanks, armored fighting vehicles, and parked aircraft. Moreover, FAEs can be detonated underwater or in a vacuum, and could thus find applications in ANTI-SUBMARINE WARFARE, ANTI-SATELLITE (ASAT) weapons, or even BALLISTIC MISSILE DEFENSE.

FAEs can be variously delivered by artillery shells, aerial bombs, and rocket and missile warheads. Typical of FAEs now in service, the U.S. CBU-72 FAE-I 500-lb. (227-kg.) CLUSTER BOMB, consists of three separate canisters which can dispense a cloud of ethylene oxide over a radius of 50 ft. (15.25 m.), generating a peak overpressure' of 300 psi at the edge of the cloud.

FAEs have not been more widely used by U.S. forces because technical problems still impede even cloud distribution at higher delivery speeds. But if those obstacles are overcome, FAEs could replace conventional high explosives and even low-yield nuclear weapons for attacks on hardened targets, including command bunkers and missile silos. See also EXPLOSIVES, MILITARY.

FULCRUM: NATO code name for the Mikoyan MiG-29 twin-engine multi-role fighter, widely considered to be the first Soviet fighter comparable to contemporary Western fighters. Developed from the mid-1970s, the MiG-29 is roughly equivalent in size, performance, and mission to the U.S. FA-18 HORNET. Flight tests began in 1977, the first production aircraft were delivered in 1982, and the first operational squadrons were formed in 1985. Little was known in the West about the MiG-29 until a detachment of six aircraft visited Finland in 1986. The Fulcrum was demonstrated at the Farnborough Air Show in 1988, and impressed observers with its superb maneuverability and advanced avionics.

Fulcrum is presently in service with Soviet Frontal Aviation (FRONTOVAYA AVIATSIYA, FA) and the IA-PVO (Fighter Aviation of the Air Defense Troops), as well as with the air forces of East Germany, India, Iraq, North Korea, Syria, and Yugoslavia; the diffusion of such an advanced fighter among Soviet Third World clients could yet have profound effects. More than 600 Fulcrums have been delivered, and production continues at a high rate.

The MiG-29 is almost exactly the same size as the FA-18 Hornet. The fuselage consists of two widely separated engine nacelles and a central cockpit module, faired into a single broad structure which generates considerable body lift, especially at high angles of attack (AOA). The nose houses a multi-mode pulse-Doppler RADAR, apparently a copy of the APG-65 radar of the FA-18. With a detection range reported to be more than 70 mi. (117 km.), the radar has both LOOK-DOWN/SHOOT-DOWN and TRACK-WHILE-SCAN capabilities; if indeed a copy of the APG-65, then it would also have very accurate ground mapping and air-to-ground modes.

In addition, the Fulcrum also has a 7.26-micron SH-1 INFRARED Search and Track (IRST) unit mounted on the nose in front of the windscreen. The IRST allows the MiG-29 to track and identify targets from beyond visual range (BVR) without generating detectable radar emissions, a considerable tactical advantage still lacking in most Western fighters. The seekers of the Fulcrum's air-to-air missiles can be slaved to the IRST for "off-boresight" engagements.

The pilot sits in an elevated cockpit under a large clamshell bubble canopy with a single-piece windscreen. Cockpit visibility is considerably better than in earlier Soviet fighters, but the canopy roof is still too low to afford the pilot an unobstructed view to the rear. The MiG-29U Fulcrum-B, a two-seat operational trainer, has a raised canopy with a second seat behind the pilot.

Cockpit AVIONICS are quite comprehensive by Soviet standards, and include a head-up display (HUD) integrating data from both the radar and IRST, a head-down radar display, an SO-69 Sirena-3 threat warning display, an instrument landing system, and an air-data/weapon-delivery computer. In contrast to Western fighters, however, the MiG-29 still has analog instrument dials instead of multifunctional video displays, and cockpit ergonomics are poor.

The space behind the cockpit is an avionics bay, while the midfuselage probably houses several large fuel tanks. Mounted at the rear of their nacelles, Fulcrum's engines are fed by underslung horizontal-wedge intakes and unique auxiliary intakes meant to prevent damage from foreign object ingestion during takeoff from unpaved fields: the main inlets can be closed, and the engines fed instead through louvers in the upper surfaces of the wing roots. To further facilitate rough-field operations, the MiG-29 has large, low-pressure main landing gear (which retract into the wing roots), and a braking parachute housed in the tail between the engines.

The moderately swept, shoulder-mounted wing has large leading-edge root extensions (LERXs) faired into the sides of the fuselage. Generating lift-enhancing vortices, the strakes also reduce drag and improve performance at high AOA. A 30-mm. CANNON of unknown type is mounted in the port leading-edge extension, with its drum magazine housed in the fuselage behind the cockpit. The wing has full-span leading-edge flaps, plain inboard flaps, and outboard ailerons. The inboard wing sections house the main landing gear, but the outboard sections are believed to contain integral fuel tanks. The slab tailplane, mounted at the ends of the engine nacelles in line with the wings, have a slight anhedral and a distinctive notch in the tips. Fulcrum has widely spaced, outward-canted twin vertical stabilizers with long dorsal strakes extending over the wing roots, yet another feature meant to improve high AOA performance. The stabilizers have the tip notches typical of recent MiG designs, with conformal fin-tip antennas and a variety of RADAR WARNING RECEIVERS. Unlike the latest Western fighters, the MiG-29 has conventional hydraulic controls, rather than an electronic FLY-BY WIRE system; according to Soviet officials, future versions will have fly-by-wire, as well as a revised cockpit with multi-functional displays. A combat thrust-to-weight ratio of better than 1-to-1 gives the aircraft excellent acceleration and climb rate, and, in conjunction with its large wing and broad fuselage, allows an excellent sustained turn rate of some 16°/sec.

Fulcrum has one centerline and six wing pylons. The centerline, inboard, and midwing pylons are believed to be rated at 2200 lb. (1000 kg.) each, while the outboard pylons are rated at 550 lb. (225 kg.) each, for a total payload of some 12,100 lb. (6000 kg.). The centerline pylon can accommodate a large drop tank or a conformal fuel pallet similar to the FAST packs of the F-15 EAGLE. In an air-to-air role, the Fulcrum can carry up to six missiles, including the long-range, radar-guided AA-9 AMOS, the medium-range AA-10 ALAMO, and the short-range, INFRARED-HOMING AAA-8 APHID or AA-11 ARCHER (export aircraft are usually supplied with the earlier AA-7 APEX medium-range missile instead of Amos and Alamo). For ground attack, the MiG-29 can carry a variety of free-fall and CLUSTER BOMBS, LASER-GUIDED BOMBS, and AS-7 KERRY air-to-surface missiles.

Specifications Length: 56.45 ft. (17.2 m.). **Span:** 37.7 ft. (11.5 m.). **Powerplant:** 2 Isotov RD-33 afterburning turbofans, 18,298 lb. (8300 kg.) of thrust each. **Weight, empty:** 18,000 lb. (8165 kg.). **Weight, max. takeoff:** 36,000 lb. (16,330 kg.). **Speed, max.:** Mach 2.3 (1539 mph/2477 kph) at 36,000 ft. (11,000 m.)/Mach 1.2 (913 mph/1469 kph) at sea level. **Initial climb:** 50,000 ft./min. (15240 m./min.). **Service ceiling:** 56,000 ft. (17,073 m.). **Combat radius:** 715 mi. (1150 km.). **Range, max.:** 2000 mi. (3340 km.).

G

G (FORCE): A measure of acceleration, equal to the force of gravity at sea level, approximately 32.2 ft. (9.81 m.)/sec./sec. Acceleration in aircraft and missiles is generally stated as a multiple (2 g, 3 g, etc.), as is radial acceleration, i.e., the turn rate, a crucial performance variable.

The g force generated during a turn depends on the angle of bank incurred ($g = 1/\text{cosine } \theta$, where θ is the angle of bank). Because g is a multiple of gravity, an aircraft in a tight turn must generate additional lift to maintain level flight; this can be accomplished by increasing the angle of attack (AOA), but that in turn increases drag. If the engine cannot generate enough extra thrust to overcome the increased drag, the aircraft will lose energy in the form of speed, or altitude, or both. Moreover, stall speed (the minimum speed required to continue flying) increases with the square root of g (e.g., 4 g doubles the minimum flying speed); hence high-g maneuvers can only be attempted if the aircraft has a margin of unused energy (thrust, speed, or altitude).

At present, most fighter aircraft are structurally limited to +6 and −3 g. Some fighters, such as the F-16 FALCON and FA-18 HORNET, can sustain up to 9 g, but pilots can tolerate such levels only briefly before suffering "blackout" because of blood flowing from the brain. Efforts are being made to increase the g tolerance of fighter pilots.

G3 (RIFLE): A 7.62-mm. ASSAULT RIFLE produced since 1959 by Hechler und Koch for the West German army, also in service with other countries, including Denmark, Indonesia, Norway, Pakistan, Portugal, Sweden, and Turkey. The G3 is a modified version of the CETME rifle, developed by fugitive Germans in Spain after 1945; that design was in turn based on the Mauser Stg.45, the World War II prototype of the modern assault rifle. The CETME version, issued by the Spanish army, has also been widely exported. Designed for ease of production with many plastic and stamped metal parts, the G3 has a straight plastic foregrip, pistol grip, and stock, with simple fixed sights and a muzzle brake/flash suppressor which can accommodate a bayonet and various rifle grenades. The G3 has a "delayed blowback" action and fires 7.62- × 51-mm. NATO standard ammunition from a 20-round box magazine.

Specialized variants include a paratroop model (G3A4) with a telescoping stock, a short-barrel carbine, and a semi-automatic sniper rifle (G3 SG/1). Hechler und Koch have also produced two versions chambered for the 5.56- × 45-mm round (HK33 and HK53); externally similar to the G3, they are slightly smaller and much lighter. A choice of 20-, 30-, or 40-round box magazines is available. As with the G3, folding-stock, carbine, and sniper variants have been developed. The HK33 is in service in Brazil, Malaysia, and Thailand, among other countries.

Specifications **Length OA:** (G3) 40.35 in. (1.025 m.); (HK53) 36.21 in. (919 mm.). **Length, barrel:** (G3) 17.72 in. (450 mm.); (HK53) 15.03 in. (382 mm.). **Weight, loaded:** (G3) 11.08 lb. (5.03 kg.); (HK53) 7.7 lb. (3.5 kg.). **Muzzle velocity:** (G3) 800 m./min. (2625 ft./min.); (HK53) 960 m./min.

(3149 ft./min.). **Cyclic rate:** 600 rds./min. **Effective range:** (G3) 500 m. (546 yd.); (HK53) 400 m. (437 yd.).

G11 (RIFLE): An experimental 4.7-mm. AS-SAULT RIFLE under development since 1969 by Hechler und Koch. Its most revolutionary feature is the use of caseless ammunition—i.e., the propellant is cast as a solid block, eliminating the need for a brass cartridge case; this reduces weight, manufacturing costs, and weapon complexity by eliminating the cartridge extraction mechanism. All previous attempts to adopt caseless ammunition have failed due to the susceptibility of rounds to impact damage, propellant deterioration from exposure to humidity, and "cook-off" (premature ignition), and also because of the difficulty of achieving a gas-tight seal in the firing chamber (normally accomplished by expansion of the cartridge case). H&K claims to have solved these problems with a Dynamit Nobel "round" consisting of a rectangular propellant block in which the 4.7-mm. bullet is completely embedded.

The rifle's outer casing is made of a carbon-fiber casting, with the pistol grip directly below the center of gravity. An optical sight is built into a carrying handle mounted over the barrel. The magazine, located above and parallel to the barrel, can hold one or two sealed 50-round clips. After the chambered round is fired, a unique rotating cylinder breech accepts a fresh round from the magazine, then rotates 90° into line with the barrel. An emergency ejection port for clearing misfires is the only opening in the receiver. The G11 has selectable full automatic, semi-automatic, and 3-round burst fire. Field trials began in 1986; if successful, the G11 could be the prototype of all future military rifles.

Specifications **Length OA:** 29.53 in. (750 mm.). **Length, barrel:** 21.26 in. (540 mm.). **Weight, loaded:** 8.7 lb. (3.95 kg.). **Cyclic rate:** 600 rds./min. (auto)/2000 rds./min. (burst). **Effective range:** 300 m. (328 yd.).

GA: See TABUN.

GABRIEL: Israeli ANTI-SHIP MISSILE, also in service with the navies of Chile, Ecuador, Kenya, Singapore, and Thailand, and produced under license in South Africa (as the *Skorpioen*) and Taiwan (as the *Hsiung Feng*).

Developed from the mid-1960s specifically to arm the SA'AR-class missile boats, Gabriel entered service in 1970, and was credited with sinking several Egyptian and Syrian patrol boats in the 1973 Yom Kippur War. The Gabriel Mk.I was super-

seded in 1976 by the Mk.II, with twice the range and improved ELECTRONIC COUNTER-COUNTERMEASURES. The Mk.III, introduced in the 1980s, has FIRE-AND-FORGET guidance and was developed into two air-launched variants, the Mk.III A/S and the longer-range Mk.III A/S ER. A turbojet-powered Mk.IV, now in development, would reportedly have a range of some 125 mi. (200 km.).

Gabriel has a cylindrical body with an ogival nose housing a radar-seeker antenna. All ship-launched versions have four narrow rectangular wings for lift, and four smaller tail fins for steering. The air-launched variants are externally similar, except that the rectangular wings have been replaced by cropped delta wings. The Mk.III A/S has the same launch weight as its shipboard equivalent, but the Mk.III A/S ER has a slightly longer and heavier motor. All ship-launcher versions carry a 397-lb. (180-kg.) high-explosive warhead, while air-launched versions have a slightly smaller 331-lb. (150-kg.) warhead.

Powered by a dual-impulse solid-rocket motor, Gabriel has several different guidance modes. The Mk.I/II is controlled initially by an autopilot and radar altimeter, switching to SEMI-ACTIVE RADAR HOMING (SARH) in the terminal phase, with target illumination provided by the launch vessel's radars. In action, the target position would be fed from the launch vessel's fire controls to the missile's guidance unit before launch. The missile would then fly its preprogrammed course, first at a height of 328 ft. (100 m.), then at some 66 ft. (20 m.). When the missile is roughly 1 mi. (1.6–2 km.) from the estimated target position, the semi-active radar seeker is activated and locks onto the target, and the missile descends to a wave-top height of 3–10 ft. (1–3 m.) for the terminal engagement. A backup guidance mode, used mainly at shorter ranges, relies on radio COMMAND GUIDANCE, with radar or optical target-tracking from the launch vessel.

The Mk.III and the air-launched versions also have autopilot control through the midcourse phase, but rely on a frequency-agile I-band active radar seeker for terminal guidance, with secondary command-guidance and SARH modes.

Ship-launched Gabriels are launched from reusable fiberglass boxes available in either fixed single or trainable triple units (the former are light enough to be mounted on the 70-ton Super Dvora patrol boats). Air-launched variants have been fitted to the weapon pylons of Israeli A-4 SKYHAWKS,

F-4E PHANTOMS, and F-16 FALCONS. See also FAST ATTACK CRAFT, ISRAEL.

Specifications Length: (I/II) 11 ft. (3.35 m.); (III) 12.6 ft. (3.84 m.). **Diameter:** (I) 12.8 in. (325 mm.); (II/III) 13.4 in. (340 mm.). **Span:** (ship) 4.4 ft. (1.34 m.); (air) 3.6 ft. (1.1 m.). **Weight, launch:** (I) 882 lb. (400 kg.); (II) 1151 lb. (522 kg.); (III) 1230 lb. (558 kg.); (A/S ER) 1323 lb. (600 kg.). **Speed:** Mach 0.7 (533 mph/890 kph). **Range:** (I) 13.5 mi. (22 km.); (II, III, III A/S) 25.5 mi. (42.6 km.); (A/S ER) 37.3 mi. (60 km.).

GADFLY (SA-11): NATO code name for a Soviet mobile surface-to-air missile (SAM) introduced in 1983 as a replacement for the SA-6 GAINFUL; a naval version (NATO designation SA-N-7) may have preceded the SA-11 into service.

Gadfly is a single-stage, solid-fuel missile similar to the U.S. RIM-66B STANDARD-MR, with four narrow-span, broad-chord body strakes for lift, and four small cruciform tail fins for steering. The missile is believed to rely on INERTIAL GUIDANCE followed by SEMI-ACTIVE RADAR HOMING (SARH) in the terminal phase. Its 110-lb. fragmentation warhead is fitted with a radar proximity (VT) fuze.

The four-round mobile launcher, mounted on the same tracked chassis as the SA-6, has 360° traverse and a "Fire Dome" continuous-wave illumination radar mounted on its face. Four launch vehicles grouped with a separate "Tube Arm" acquisition radar form a battery; five batteries plus headquarters and service units form an air defense regiment—one of which is organic to every Soviet tank and Motorized Rifle division. Separate SA-11 regiments deployed at army and Front levels are replacing SA-3 GOA units for the medium-range defense of fixed targets.

The SA-N-7 naval version was first installed (experimentally) on the KASHIN-class destroyer *Provornyy;* since 1982 it has been the principal air defense weapon of the SOVREMENNY-class destroyers. It is fired from a compact single-arm launcher resembling the U.S. Mk.13, with a 20-round rotary magazine below decks. Target illumination is provided by "Front Dome" guidance radars, six of which are mounted on the Sovremennys.

Specifications Length: 17.55 ft. (5.35 m.). **Diameter:** 14.95 in. (380 mm.). **Span:** 3.93 ft. (1.2 m.). **Weight, launch:** 1485 lb. (675 kg.). **Speed:** Mach 3.3 (2150 mph). **Range envelope:** 4–30 km. (2.5–18.5 mi.). **Height envelope:** 50–20,000 m. (164-65,600 ft.).

GAINFUL (SA-7): NATO code name for the Soviet RK-5D *Kub* mobile surface-to-air missile,

introduced in the early 1960s to replace the S-60 57-mm. towed ANTI-AIRCRAFT gun in divisional air-defense regiments. Not displayed publicly until 1967, Gainful attracted much attention during the 1973 Yom Kippur War, when large numbers were used by Egypt and Syria. Since then, it has been widely exported.

Gainful has four cropped-delta cruciform wings at midbody for lift, and four movable tail fins for steering. The missile has a unique integral rocket/ramjet engine; the solid-fuel rocket initially burns off to boost the missile to ramjet ignition speed (Mach 1.5); then four air intakes open between the wings, and liquid fuel is injected to start the ramjet. This arrangement achieves both rocket boost and the range of the ramjet without the weight penalty of a separate rocket booster.

Gainful relies on radio COMMAND GUIDANCE through midcourse, and SEMI-ACTIVE RADAR HOMING (SARH) for terminal guidance; it is armed with a 121-lb. (55-kg.) fragmentation warhead (lethal radius 5 m.) and fitted with a radar proximity (VT) fuze. Gainful's triple launcher, mounted on a tracked, self-propelled chassis derived from the MT-S tractor, has 360° traverse and a maximum elevation of 85°. Four launch vehicles, two Zil-131 6 × 6 missile reload trucks, and a self-propelled "Straight Flush" guidance radar form a battery; five batteries plus a "Thin Skin" height-finding radar, two "Long Track" surveillance radars, and a headquarters unit form an air defense regiment, one of which is organic to each Soviet tank and Motorized Rifle division.

In action, targets are initially detected by the Long Track surveillance radars, which pass range, speed, bearing, and height data to the firing battery. The Straight Flush radar then carries out a focused search. If it acquires and locks onto a target, a missile is fired under command guidance via the Straight Flush, before it switches to SARH in the terminal phase.

A new SA-6b version introduced in 1979 (Soviet designation 9M9M) has a new launch vehicle with an on-board illumination radar. This allows engagements without the Straight Flush, reducing reaction times and enabling the battery to engage multiple targets. The new missile also has better low-altitude performance and ELECTRONIC COUNTER-COUNTERMEASURES. A firing battery now comprises one SA-6b and three standard launch vehicles. This also reduces the vulnerability of the battery, which previously could be put out of action by destroying the Straight Flush.

Gainfuls destroyed a dozen Israeli aircraft in the opening days of the Yom Kippur War, partly because the Israeli air force lacked effective jammers; some jamming equipment and ad hoc tactics greatly reduced their effectiveness within a matter of days. Throughout the three-week war, some 840 Gainfuls were fired, scoring some 20 kills (a P_k of 2.3%); their greatest achievement was that they forced Israeli aircraft to fly low to evade them, thereby coming within the effective range of highly lethal zsu-23-4 Shilka SP anti-aircraft guns. Israeli countermeasures were perfected to such a degree by 1982 that all 16 Syrian SA-6 batteries in the Bekaa Valley were destroyed without the loss of a single aircraft. The U.S. has acquired SA-6 batteries from both Israel and Egypt to study their performance, and effective jammers have been produced. Gainful is now being replaced by the SA-11 GADFLY.

Specifications **Length:** 20.33 ft. (6.2 m.). **Diameter:** 13.2 in. (335 mm.). **Span:** 5.0 ft. (1.52 m.). **Weight, launch:** 1215 lb. (552 kg.). **Speed:** Mach 2.5 (1650 mph/2755 kph). **Range envelope:** 4–27 km. **Height envelope:** 50–12,000 m. **Operators:** Alg, Ang, Bul, Cuba, Cze, DDR, Egy, Eth, Fin, Gui-Bis, Guy, Hun, Ind, Iraq, Ku, Lib, Mali, Moz, N Ye, Peru, Pol, Rom, Som, Syr, Tan, USSR, Viet, Yug, Zam.

GAL CLASS: British-built version of the West German TYPE 206 diesel-electric submarine in service with the Israeli navy.

GALAXY: The Lockheed C-5 wide-body, long-range transport aircraft, in service with the MILITARY AIRLIFT COMMAND (MAC) of the U.S. Air Force (USAF). At the time of its introduction in 1968, the C-5 was the largest aircraft in the world (a distinction it lost to the superficially similar Soviet An-124 CONDOR, and later to the even larger An-225). Development proceeded rapidly from 1965, but costs were seriously underestimated, and only 81 of a planned 115 could be built. Despite the initial cost scandal, the Galaxy has been a success, being highly reliable and especially useful for emergency supply missions (e.g., to Vietnam in 1972 and to Israel in 1973). It remains the only U.S. transport (pending delivery of the C-17) capable of carrying outsize cargo such as tanks and unloading it without special facilities. These aircraft were used so intensively that structural fatigue became a problem; 72 of 77 surviving aircraft were therefore given new wings between 1981 and 1987, to extend their service lives by some 15 years. Moreover, in 1982 it was decided to procure 50 new C-5Bs (with structural improvements and modified engines), delivered between 1986 and 1989 at a total cost of some $8 billion.

The fuselage is dominated by a huge unobstructed cargo bay. For roll-on/roll-off loading, the entire nose of the aircraft swings upward, the tail has full-width clamshell doors, and both ends have integral vehicle ramps. The cargo deck can accommodate a wide range of payloads: 2 MAIN BATTLE TANKS or M109 SP howitzers, 3 CH-47 CHINOOK heavy transport helicopters, 36 standard cargo pallets, or 270 fully equipped troops. Maximum payload is 220,967 lb. (100,439 kg.).

The pilot, copilot, engineer, navigator, and loadmaster all sit in an airline-style cockpit above the cargo bay. The comprehensive AVIONICS include a large Norden multi-mode RADAR in the nose, INERTIAL and DOPPLER navigation units, and a TACAN beacon. A cabin with a galley and accommodations for a relief crew of 15 is immediately behind the flight deck, and a second cabin in the rear fuselage above the cargo bay has an additional 75 seats. An AERIAL REFUELING receptacle is located on top of the fuselage behind the cockpit. The high-mounted wings, swept at only 25°, are fitted with a variety of high-lift devices, including full-span leading-edge flaps, large Fowler flaps, and spoilers for roll control.

Designed to operate even from unpaved fields, the C-5's landing gear has four nose wheels, and four 6-wheel main units mounted in sponsons alongside the fuselage; all have oleo-pneumatic shock absorbers which can be adjusted to different heights for loading/unloading, and for takeoff. The pressure of all 28 tires can be adjusted from the cockpit to provide greater support on soft fields, and all wheels can be turned up to 20° left or right, to facilitate crosswind landings. Despite its great size, it has quite good short-field performance, with a takeoff roll of 8400 ft. (3818 m.) at 769,000 lb. (349,545 kg.), and a landing roll of only 3675 ft. (1120 m.) at 635,850 lb. (289,022 kg.).

Specifications **Length:** 247.83 ft. (75.55 m.). **Span:** 222 ft. (67.68 m.). **Cargo hold:** 121.1 × 19 × 13.5 ft. (36.92 × 5.79 × 4.11 m.). **Powerplant:** 4 General Electric TF39-GE-1C high-bypass turbofans, 43,000 lb. (19,545 kg.) of thrust each. **Fuel:** 318, 500 lb. (144,772 kg.). **Weight, empty:** 374,000 lb. (170,000 kg.). **Weight, max. takeoff:** 837,000 lb. (380,454 kg.). **Speed, max.:** 571 mph (919 kph). **Speed, cruising:** 552–564 mph (922–942 kph). **Initial climb:** 1725 ft./min. (526 m./min.). **Service ceiling:** 35,750 ft. (10,900 m.). **Range, max.:** (max.

payload) 3749 mi. (6260 km.); (12,600-lb./51,181-kg. payload) 6529 mi. (10,903 km.); (ferry) 7890 mi. (13,176 km.).

GALIL: A series of Israeli ASSAULT RIFLES and LIGHT MACHINE GUNS (LMGs) developed after the 1967 Six-Day War by Israeli Military Industries. Based on the Soviet AK and the Finnish Valmet (itself a modified AK), the basic Galil was adopted by Israel in 1972, and has since been exported to several countries, including South Africa, and copied by others, including Sweden.

The Galil is produced in three main versions: the ARM, a light machine gun with a bipod and carrying handle; the AR assault rifle; and the SAR short-barrel "commando" rifle. All three have folding stocks. The ARM has a wire cutter built into the bipod, and all three versions have a built-in bottle opener, to dissuade soldiers from using magazine lips for that purpose.

Galils have been produced in both 7.62- × 51-mm. NATO and 5.56- × 45-mm. calibers. The 5.56-mm. ARM has a straight foregrip, a pistol grip, and a tubular stock which folds to one side (now widely copied). The simple fixed sights can be supplemented by an add-on scope, and the muzzle brake/flash suppressor can accommodate a bayonet and doubles as a grenade launcher. The cocking handle is on top of the weapon, to suit both right- and left-handed soldiers. It has a modified Kalashnikov gas action, and is fed from 35- or 50-round box magazines. The 7.62-mm. version is slightly larger, with a 20-round box magazine, while the 5.56-mm. SAR is shorter and has slightly less range.

Specifications Length **OA:** (5.56-mm. ARM) 38.54 in. (979 mm.); (5.56-mm. SAR) 32.28 in. (820 mm.); (7.62-mm) 41.34 in. (1.05 m.). **Length, barrel:** (5.56-mm. ARM) 18.1 in. (560 mm.); (5.56-mm. SAR) 12.98 in. (330 mm.); (7.62-mm.) 20.98 in. (533 mm.). **Weight, loaded:** (5.56-mm. ARM) 10.19 lb. (4.63 kg.); (5.56-mm. SAR) 7.7 lb. (3.5 kg.); (7.62-mm.) 10.3 lb. (4.68 kg.). **Muzzle velocity:** (5.56-mm. ARM) 980 m./min. (3215 ft./min.); (5.56-mm. SAR) 920. m./min. (3017 ft./min.); (7.62-mm.) 850 m./min. (2790 ft./min.). **Cyclic rate:** 650 rds./min. **Effective range:** (5.56-mm. ARM) 600 m. (656 yd.); (5.56-mm. SAR) 400 m. (437 yd.); (7.62-mm.) 400 m. (437 yd.).

GALOSH (ABM-1): NATO code name for a Soviet ANTI-BALLISTIC MISSILE, introduced in 1966 (when the abortive U.S. SAFEGUARD ABM system was being developed). A total of 64 Galosh missiles were originally deployed at four 16-missile sites around Moscow. They have since been modernized several times.

The Galosh missile has been displayed only within its cylindrical launch/transport canister, but based on the dimensions of the latter, it is believed to be a three-stage, solid-fuel missile. The first stage is believed to consist of a cluster of four rocket engines, with four folding fins for stabilization during the initial seconds of flight. The second and third stages form a conical body with an ablative, heat-resistant nose cone, shaped for hypersonic flight. Both upper stages are finless and must have some form of thrust vector control for steering. The third stage is believed to be a sort of POST-BOOST VEHICLE (PBV), which can "loiter" at altitude while waiting for enemy warheads to be "sifted" from PENAIDS by atmospheric deceleration. The missile is armed with a huge 2 to 3-MT nuclear warhead, configured for X-RAY KILL or NEUTRON KILL effects in space. Galosh is controlled by radio COMMAND GUIDANCE, and may also have some form of terminal homing.

As first completed, the Galosh system consisted of 2 CAT HOUSE and 2 DOG HOUSE phased array target acquisition radars, 32 2-launcher firing batteries, 1 TRY ADD-A tracking radar, and 2 Try Add-B guidance radars. The entire system is linked by computers to a central command center, which would receive early warning data from the 11 HEN HOUSE phased array radars sited on the perimeters of the U.S.S.R.

Targets initially detected by Hen House would be acquired by Cat House or Dog House if they approached Moscow, and would then be assigned by the command center to specific Galosh batteries, with tracking transferred to the Try Add-A, which would provide targeting data to the fire control unit. If a target were to come within range, a missile would be launched, with guidance commands transmitted by one of the two Try Add-Bs. The warhead would be command-detonated at the closest approach to the target. Apparently two missiles could be guided simultaneously against separate targets, each by a Try Add-B.

From the late 1970s, the Galosh system was extensively modernized. Its above-ground launchers were replaced by hardened silos (possibly with rapid-reload capability, a violation of the 1972 ABM TREATY). Dog House and Cat House were replaced by a single "Pill Box" phased array radar with 360° coverage, sited at Pushkino, north of Moscow. The Hen House network was supplemented by nine large phased array radars with improved range resolution and discrimination. Finally, the entire system is being converted to a two-layer configuration by the introduction of the ABM-3 GAZELLE, a short-range, hypervelocity missile for terminal defense. The emplacement of 36 Gazelles in hardened silos would bring the system up to the 100-launcher limit set by the ABM Treaty. See also BALLISTIC MISSILE DEFENSE.

Specifications Length: 65 ft. (19.82 m.). Diameter: 8.4 ft. (2.56 m.). Weight, launch: 72,000 lb. (32,727 kg.). Speed: Mach 5+. Range: 250 km. (150 mi.). Ceiling 100,000 m. (328,000 ft.).

GAMMA RAYS: Very short electromagnetic waves, part of the "prompt" radiation produced by nuclear explosions. Gamma rays have wavelengths of 0.01–0.001 angstrom, or 1/1,000,000 of a millimeter; this ensures very good penetration of all but the densest materials. The lethal effect of gamma rays on human tissue is caused by ionization of body cell atoms. The amount of gamma radiation released by a nuclear explosion varies with the intensity of the heat produced, so that fusion (hydrogen) weapons ordinarily produce much more than fission weapons; as energy yield increases, gamma ray emissions increase more than proportionately. See also NUCLEAR WEAPON.

GAMMON (SA-5): NATO code name for a Soviet long-range surface-to-air missile. Introduced in the late 1950s to counter U.S. high-altitude strategic bombers, Gammon has been deployed since 1963 at more than 100 fixed sites around high-value targets throughout the U.S.S.R. In the early 1980s, Gammon systems were deployed in East Germany, Hungary, and Czechoslovakia, and also exported to Syria and Libya (each has three sites); until 1985, these may have been manned by Soviet "advisors." There has been much confusion on the identity and configuration of the SA-5. At one time, the designation was assigned to a missile resembling a scaled-up SA-2 GUIDELINE (NATO code name Griffon), which was believed to be an ANTI-BALLISTIC MISSILE (ABM); this was deployed around Tallin and Leningrad in the early 1960s.

Intelligence sources now agree that Gammon is a large missile with four long-chord, cruciform delta wings, and four smaller cruciform tail fins. Steering is probably by aerodynamic control surfaces on the wings. The missile has a single liquid-fuel rocket main engine, and four strap-on solid-rocket boosters; the latter accelerate the missile to flight speed, then drop away. Gammon relies on radio COMMAND GUIDANCE combined with active or

SEMI-ACTIVE RADAR HOMING (SARH) in the terminal phase. There are also reports of an anti-radiation variant, developed specifically to home on the radar of U.S. E-3 Sentry AWACS aircraft. The missile has two alternative payloads: a 135-lb. high-explosive fragmentation warhead with a radar proximity (VT) fuze; and a 5 to 10-kT nuclear warhead (only the former has been exported). Gammon may have some limited ABM capability, but was not considered an ABM missile under the 1972 ABM TREATY.

Gammons are deployed on single-rail launchers, many of which have been hardened with reinforced-concrete revetments. Up to 15 launchers comprise a launch complex, along with two "Square Pair" fire control radars, a "Back Net" acquisition radar, and a "Side Net" height finding radar.

Because of its size and command guidance, the missile is relatively ineffective against small, maneuvering targets such as fighters; its main purpose is to dissuade strategic bombers from flying at high altitudes, forcing them to rely instead on more difficult and fuel-consuming low-altitude penetration. Because of its poor low-altitude capabilities, Gammon is usually deployed in conjunction with the medium range SA-3 GOA or the newer SA-11 GADFLY. Gammons were fired by Libya against U.S. aircraft during the 1986 Gulf of Sidra incident and again during the Tripoli and Bengazi air strikes. No hits were recorded, and the missiles may have been neutralized by radar jamming. Gammons have also been fired at U.S. SR-71 BLACKBIRD reconnaissance aircraft, again without effect. Gammon may be replaced in the 1990s by the SA-10 GRUMBLE and SA-12A GLADIATOR.

Specifications Length: 34.75 ft. (10.59 m.). Diameter: 1.8 ft. (548 mm.). Span: 8.5 ft. (2.59 m.). Weight, launch: 25,912 lb. (11,778 kg.). Speed: Mach 3.5 (2300 mph). Range envelope: 12–300 km. Ceiling: 30,000 m. (98,000 ft.).

GANEF (SA-4): NATO code name for the Soviet ZRK-SD *Krug* mobile surface-to-air missile, introduced in 1964 to provide high-altitude air defense for the forward edge of the battle area (FEBA). Ganef is also in service with Bulgaria, Czechoslovakia, East Germany, and Poland; a small number deployed to Egypt under Soviet control in 1970 were withdrawn in 1973.

A rather large missile, the SA-4 has four movable wings for steering, and four fixed tail fins. Powered by a ramjet engine fed by an air intake at the nose, the missile has a slender, pointed centerbody housing a SEMI-ACTIVE RADAR HOMING (SARH) receiver and a 300-lb. (136-kg.) high-explosive fragmentation warhead fitted with a radar proximity (VT) fuze. Four strap-on rocket boosters accelerate the missile to ramjet ignition speed (Mach 1.3) before being jettisoned; the ramjet then propels the missile to its maximum speed.

Ganef is controlled by radio COMMAND GUIDANCE through midcourse, then switches to SARH in the terminal phase. Because of its size and guidance mode, Ganef is only marginally effective against maneuvering targets such as fighters, especially at low altitude. An improved version, the SA-4b Ganef Mod 1 (Soviet designation 9M8M2), was introduced in 1973 specifically to improve low-altitude performance.

Mounted on a tracked, self-propelled (SP) chassis, Ganef's twin launcher has 360° traverse. Three launch vehicles, three reload vehicles, and a self-propelled "Pat Hand" fire control radar form a battery. Three batteries, plus a "Long Track" surveillance radar, a battery of ZSU-23-4 *Shilka* SP anti-aircraft guns (for close-in defense), and a headquarters unit comprise a battalion. Three battalions, a Long Track, a "Thin Skin" height-finding radar, a headquarters, and service units comprise an air defense brigade (two are assigned to each FRONT and one to each ARMY).

In action, targets would be detected initially by the Long Track, which would pass range, bearing, speed, and height data to the battery's Pat Hand radar. Once the target comes within range (and height is confirmed by the Thin Skin), a missile would be launched, to be guided by commands transmitted by the Pat Hand, until switching to SARH in the terminal phase. Ganef is more than 25 years old, and much information about it is likely to have been collected; hence effective jammers and other countermeasures are probably available. It is now being replaced by the SA-12A GLADIATOR.

Specifications Length: 28.9 ft. (8.81 m.); (4b) 27.2 ft. (8.29 m.). Diameter: 35.4 in. (900 mm.). Span: (wings) 7.5 ft. (2.28 m.); 8.5 ft. (2.59 m.). Weight, launch: 5500 lb. (2500 kg.). Speed: Mach 4.0 (2600 mph/4342 kph). Range envelope: 9.3–80 km; (4b) 7–40 km. Height envelope: 100–27,400 m. (328–89,872 ft.); (4b) 25,000 m. (82,000 ft.) max.

GARCIA: A class of ten U.S. ANTI-SUBMARINE WARFARE (ASW) FRIGATES commissioned between 1964 and 1967. They and the nearly identical BROOKE guided-missile frigates are enlarged versions of the earlier BRONSTEINS, the first modern U.S. escorts. The Garcias are flush-decked with a sharply raked bow and a small, boxy superstructure concentrated amidships. They tend to be rather wet in rough weather, though fin stabilizers are fitted to improve seakeeping.

Armament, optimized for ASW with only limited ANTI-AIR WARFARE (AAW) and ANTI-SURFACE WARFARE (ASUW) capabilities, consists of 2 single 5-in. 38-caliber DUAL PURPOSE guns (fore and aft), an 8-round ASROC pepperbox launcher amidships (with 8 reloads in the last 5 ships), and 2 sets of Mk.32 triple tubes for MK.46 lightweight homing TORPEDOES. As built, they had a landing pad and hangar for two DASHS (Drone Anti-Submarine Helicopters), but these were replaced between 1972 and 1975 by enlarged facilities for one SH-2F SEA SPRITE LAMPS I helicopter (not fitted to two ships, *Garcia* and *Brumby*).

All 10 Garcias were placed in reserve in 1988–89 as a cost-saving measure; 4 have since been leased to Brazil, 2 to Turkey, and 2 to Pakistan. See also FRIGATES, UNITED STATES.

Specifications Length: 415.5 ft. (126.67 m.). Beam: 44.2 ft. (13.47 m.). Draft: 14.5 ft. (4.42 m.). Displacement: 2620 tons standard/3560 tons full load. Powerplant: single-shaft steam: 2 oil-fired boilers, 1 set of geared steam turbines, 35,000 shp. Fuel: 600 tons. Speed: 27.2 kt. Range: 4000 n.mi. at 20 kt. Crew: 247. Sensors: 1 SPS-40 air search radar, 1 SPS-10 surface search radar, 1 SPG-35 fire control radar, 1 LN66 navigation radar, 1 SQS-26 bow-mounted low-frequency sonar; (*Garcia* and *Brumby*) 1 SQR-15 TASS passive towed array sonar. Electronic warfare equipment: 1 WLR-1 and WLR-3 radar warning receiver, 1 ULQ-6 deception jammer, 2 6-barrel Mk.35 RBOC chaff launchers.

GARIBALDI: Italian AIRCRAFT CARRIER commissioned in 1985. Authorized as the flagship of an ANTI-SUBMARINE WARFARE (ASW) group, *Garibaldi* was actually meant all along to serve as a light carrier with V/STOL fighters such as the HARRIER. But the procurement of fighters was delayed by an old law vesting control of all fixed-wing aircraft with the Italian air force. This impediment was removed by legislation in 1988, and Harriers are now being acquired. A second carrier (*Mazzini*) is planned.

A flight deck measuring 570.16 × 68.9 ft. (173.83 × 21 m.), with a 6° ski-jump bow suitable for Harrier rolling takeoff, takes up the full length of the ship. A large, boxy island superstructure offset to starboard incorporates air intakes and exhaust vents for the engines. Two elevators, fore and aft of the island, connect the flight deck with the hangar, which is 360.9 ft. (110.03 m.) × 49.25 ft. (15 m.) × 19.67 ft. (6 m.). Maintenance facilities are sufficient not only for the Garibaldi's own air group, but for the helicopters of smaller escorting vessels as well.

Defensive armament consists of 2 octuple Albatros box launchers for ASPIDE point defense missiles, 3 twin Breda 40-mm. ANTI-AIRCRAFT guns connected to Dardo FIRE CONTROLS for anti-missile defense, and 2 sets of triple tubes for light ASW homing TORPEDOES. In addition, the ship has a powerful ANTI-SURFACE WARFARE (ASUW) battery of 4 twin launchers for OTOMAT Mk.2 anti-ship missiles.

The current air group normally consists of 12 Agusta SH-3D SEA KING ASW helicopters, but up to 8 Harriers and 4 Sea Kings could be carried instead for strike missions. The flight deck is strong enough to support CH-47 CHINOOK heavy transport helicopters, and two 250-man LANDING CRAFT are normally carried for commando, SAR, and disaster-relief missions.

Specifications Length: 587.25 ft. (175.38 m.). **Beam:** 99.75 ft. (30.41 m.). **Draft:** 22 ft. (6.7 m.). **Displacement:** 10,100 tons standard/13,139 tons full load. **Powerplant:** 4 20,000-shp. Fiat/GE LM2500 gas turbines, 80,000 shp., 2 shafts. **Speed:** 30 kt. **Range:** 7000 n.mi at 20 kt. **Crew:** 560 + 250 commandos. **Sensors:** 1 RAN-3L 3-dimensional air search radar, RAT-20S and RAN-10S 2-dimensional air search radars, 1 SPS-702 surface search radar, 3 RTN-20X Dardo fire control radars, 2 RTN-30X missile guidance radars, 1 SPN-703 air traffic control radar, 1 SPN-749 navigation radar, 1 DE-1160 hull-mounted medium-frequency sonar. **Electronic warfare equipment:** 4 Elettronica electronic signal monitoring arrays, 1 active jamming system, 2 20-barrel SCLAR chaff launchers.

GASKIN (SA-9): NATO code name for the Soviet ZRK-BD *Strela-1* mobile surface-to-air missile system, introduced in 1968 to provide (very) short-range, (very) low-altitude battlefield air defense at the regimental level to supplement the ZSU-23-4 *Shilka* self-propelled anti-aircraft gun.

Gaskin has four cruciform canards for steering and four fixed tail fins. The missile relies on passive INFRARED HOMING with an uncooled lead sulfide (PbS) seeker head, reported to be capable of rear-quarter engagements up to 30° off the target's tail; it may also be capable of locking onto sunlight ("glint") reflected off the cockpit canopies of helicopters. Because it can engage only from behind, the missile must pursue and overtake its target, making aircraft flying at more than 650 mph (1085 kph) relatively safe; it is most effective against helicopters and attack aircraft flying at speeds below 300 mph (500 kph) at ranges between 3 and 6.5 km., and at heights up to 1900 m. (6232 ft.). Gaskin is armed with a 15.5-lb. (7-kg.) high-explosive fragmentation warhead fitted with both proximity and contact fuzes.

A modified version, the SA-9b (Soviet designation 9M31M), introduced in 1971, has a cryogenically cooled seeker of greater sensitivity, increasing the lock-on range and possibly allowing head-on engagements in some circumstances.

The 4-round launcher Gaskin launcher is mounted on a BRDM 4 × 4 reconnaissance vehicle chassis. The missiles are carried in sealed storage/launch containers which are fitted to racks on a 360° manually-operated turret. Gaskin is a daylight-only, clear-weather system, because it depends on visual search, target acquisition, and range estimation. When a target is detected, the operator slews the turret to its bearing and launches the missile upon receiving an audio tone indicating lock-on. After all four missiles are expended, the empty containers must be removed and the racks reloaded manually. One platoon of four launch vehicles and a command vehicle are organic to the air-defense battery of every Soviet and Warsaw Pact tank and motorized rifle regiment, along with a platoon of ZSU-23-4s.

Because of their simple IR seekers, both versions of Gaskin can be spoofed by flares, jamming beacons, and other INFRARED COUNTERMEASURES (IRCM). The missile was totally ineffective against the Israeli air force during the 1982 Lebanese War due to IRCM and and the use of terrain-masking tactics. Gaskin is now being replaced in Soviet service by the similar SA-13 GOPHER.

Specifications Length: 5.9 ft. (1.8 m.). **Diameter:** 4.72 in. (120 mm.). **Span:** 15 in. (380 mm.) **Weight, launch:** 66 lb. (30 kg.). **Speed:** Mach 2 (1400 mph/2338 kph). **Range envelope:** 0.5–4.0 mi. (0.8–6.5 km.) (b) 0.56–8 km. **Height. envelope:** 20–5000 m. (66-16,400 ft.); (b) 6100 m. (20,000 ft.) max. **Operators:** Alg, DDR, Egy, Hun, Ind, Iraq, Lib, Pol, S Ye, Syr, USSR, Viet, Yug.

GAU-8: The General Electric "Avenger" 7-barrel 30-mm. CANNON, introduced in 1975 as the primary armament of the A-10A THUNDERBOLT attack aircraft, and later included in the GE/Signaal

GOALKEEPER shipboard missile-defense weapon. The largest and most powerful aircraft gun now in service, the GAU-8 is 21 ft. (6.4 m.) long (including its drum magazine and ammunition feed) and weighs 3000 lb. (1364 kg.) loaded; i.e., it is the size of a small truck.

The gun operates on the Gatling principle, whereby an electric motor successively rotates each barrel into line with the single breech for firing. This delays overheating and permits very high rates of fire: 2100 or 4200 rds./min. in the GAU-8. The preferred armor-piercing incendiary (API) ammunition (with a depleted uranium STABALLOY penetrator core) is the size of a milk bottle and weighs 1.52 lb. (0.69 kg.); the projectiles alone weigh 12.5 ounces (0.35 kg.). A tungsten-core AP round and an HEI round have been developed for Goalkeeper (Staballoy is unacceptable to some users).

The GAU-8 has a muzzle velocity of 3500 ft./min. (1067 m./min.) firing API, and an effective range of 1500 m. (1640 yd.) against tanks. Much longer ranges are possible against softer targets.

The GAU-8's major drawback is its sheer size and weight, which precludes installation in smaller aircraft (indeed, the A-10 was designed *around* the gun). GE also has developed a much lighter, pod-mounted derivative, the GAU-13 GEPOD, with only four barrels and a rate of fire of 3000 rds./min. The GAU-13 is mounted in the GPU-5 gun pod, which is 14.1 ft. (4.3 m.) long and weighs only 1900 lb. (863 kg.) with 353 rounds of ammunition (carried in two helical layers around the gun to reduce overall length). It can be carried by most combat aircraft, and is cleared for supersonic speeds.

GAZELLE: 1. NATO code name for the Soviet ABM-3 short-range ANTI-BALLISTIC MISSILE. Introduced in the early 1980s to supplement the older, long-range ABM-1 GALOSH, Gazelle is intended for the endoatmospheric interception of ballistic reentry vehicles, thus adding a terminal defense layer to the Moscow ABM system. Although never displayed publicly, Gazelle is believed to be a single-stage, solid-fuel, conically shaped missile (with an ablative coating to resist the heat of hypersonic flight in the atmosphere), similar in design to the U.S. SPRINT and LOADS. That would imply a range of about 80 km. (48 mi.), a ceiling of 50,000 m. (164,000 ft.), and maximum acceleration of some 100 G.

Gazelle is believed to rely on radio COMMAND GUIDANCE, but may also have some form of terminal homing. Thirty-six Gazelles have been in-

stalled in hardened, reloadable silos around Moscow, together with their associated "Flat Twin" tracking and guidance RADARS, and "Pawn Shop" FIRE CONTROL units. A mobile version of Gazelle may be in development, but this would be a clear violation of the ABM TREATY, for it could make it possible to deploy a national ABM system at short notice ("breakout"). See also BALLISTIC MISSILE DEFENSE.

2. The Aerospatiale SA.340 utility/scout HELICOPTER, developed to a French Army Air Force (ALAT) requirement, and widely exported. First flown in 1967, the Gazelle was derived from the Aerospatiale Alouette but has a new fuselage and rotor assembly.

The Gazelle has a conventional "penny-farthing" configuration, but with a ducted Aerospatiale "fenestron" tail rotor. The main cabin is egg-shaped and extensively glazed, with side-by-side seating for the pilot and copilot, and a bench for three passengers in the rear. Alternatively, two stretchers can carried internally, or up to 1540 lb. (700 kg.) of cargo slung externally from a belly hook. The engine is mounted in a nacelle over the rear cabin. The main rotors have a rigid hub assembly (first developed for the Messerschmidt BO-105), for improved maneuverability and reduced maintenance. Tabular skid landing gear is fitted, as are—in attack versions—outrigger pylons for 2–4 gun/rocket pods or ANTI-TANK GUIDED MISSILES (ATGMs). AVIONICS include comprehensive night flying instruments, VHF, UHF, and HF radios, an autopilot, an INERTIAL NAVIGATION unit, and a DOPPLER navigation radar. Many Gazelles have a roof-mounted APX-397 stabilized sight for the Euromissile HOT ATGM, while Yugoslav helicopters have a sight for the Soviet AT-3 SAGGER ATGM.

Specifications Length: 31.25 ft. (4.04 m.). Span: 34.45 ft. (10.5 m.). Powerplant: 590-shp Turbomeca Astazou IIIA turboshaft. Weight, empty: 2022 lb. (919 kg.). Weight, max. takeoff: 3968 lb. (1804 kg.). Speed, max.: 164 mph (274 kph). Speed, cruising: 148 mph (247 kph). Initial climb: 1772 ft./min. (540,25 m./min.). Service ceiling: 16,404 ft. (5001 m.). Range, max.: (1100-lb. [500-kg.] payload) 224 mi. (374 km.); (max. fuel) 416 mi. (695 km.). Operators: Ab Dh, Buru, Cam, Chad, Ecu, Egy, Gui, Iran, Ire, Jor, Ken, Leb, Lib, Mor, Qat, Rwa, Sen, Syr, Trin, UK, Yug.

GB: See SARIN.

GBU-8: The HOBOS TV-guided glide bomb.

GBU-10: The 2000-lb. PAVEWAY laser-guided bomb.

GBU-12: The 500-lb. PAVEWAY laser-guided bomb.

GBU-15: U.S. ELECTRO-OPTICAL (TV) or IMAGING INFRARED (IIR) guided GLIDE BOMB in service with the U.S. and Israeli air forces. Initiated in 1972 as part of the Pave Strike program (to develop standoff weapons to attack high-value point targets), GBU-15 was meant as a longer-range successor to the GBU-8 HOBOS. Its key design feature was a DATA LINK to allow the bomb to "lock on after launch" (LOAL), thereby enabling the launching aircraft to release it in the general direction of the target, without having to fly close (and thus within range of enemy air defenses) to acquire it—a major drawback of HOBOS.

In addition, GBU-15 was to be modular, so that alternative guidance units, payloads, and lifting surfaces could be matched to specific targets and conditions. Proposed guidance modules included TV, IIR, and Distance Measuring Equipment (DME); payloads included the Mk.84 2000-lb. LDGP bomb and the CBU-75 "Pave Storm" 2000-lb. (909-kg.) CLUSTER BOMB. Two different wing configurations were proposed: a cruciform wing weapon (CWW) for low-level release from fighter-type aircraft, and a Planar Wing Weapon (PWW) for release from high-altitude bombers.

The program languished during the 1970s because of a lack of funds, and also suffered numerous technical setbacks (mainly related to the data link). By the early 1980s it became apparent that the entire project had been too ambitious, and it was scaled back considerably. The PWW and DME modules were canceled and the cluster bomb payload deferred, leaving only two production versions: TV-guided and IIR-guided CWWs with Mk.84 warheads. Flight tests began in 1980 and an initial operating capability (IOC) of the TV version was achieved in 1983. The first IIR versions were delivered in 1987. Some were supplied to Israel and to the Australian air force (for evaluation). First used operationally in the 1991 Persian Gulf conflict, the GBU-15 proved highly effective against bridges, bunkers, and other hardened point targets.

The GBU-15 kit consists of modules which are bolted to the nose and tail of a standard Mk.84 bomb. An improved version built around a BLU-109 2000-lb semi-armor piercing bomb was introduced in 1991. Each kit comprises a seeker module (TV or IIR), a nose adaptor, a guidance and control module, a data link module, and a lifting surface (wing) unit. It originally had four broad-chord delta wings and four small, fixed nose canards. Later versions have short-chord rectangular wings of the same span, and four larger canards (thus keeping wing area the same). Steering is accomplished by pneumatically actuated control surfaces on the wing trailing edges.

The seeker module is attached to the nose adaptor, which in turn is bolted to the nose of the bomb; the guidance and control unit, which is bolted to the tail of the bomb, houses an autopilot, a battery, and a flask of compressed air to drive the control actuators. The data link module, attached to the guidance and control unit, communicates with a pod-mounted transceiver on the launch aircraft. The GBU-15 has been cleared for operations on F-4E PHANTOM and F-111 attack aircraft and B-52 STRATOFORTRESS bombers.

There are three launch modes: lock-on before launch (LOBL); LOAL direct; and LOAL indirect. In the LOBL mode, the pilot or weapon system operator of the launch aircraft must acquire the target visually, using either his head-up display (HUD) or a video screen displaying the view from the GBU-15's nose camera. He then locks the TV camera onto the target by placing a cursor over a high-contrast portion of the target and releases the weapon, which homes autonomously while the aircraft can maneuver freely. This method would be used only in special circumstances, because it negates the advantage of the GBU-15 by requiring the aircraft to fly close to the target.

In the LOAL direct mode, the operator releases the weapon in the general direction of the target (without prior visual acquisition), usually pitching up the aircraft to toss the bomb in a long-range parabolic trajectory. The autopilot stabilizes the weapon and keeps it on course. As it pitches over in the descent portion of its trajectory, the view from the TV camera is transmitted back to the launch aircraft over the data link. The operator can then acquire the target and lock on, after which the weapon would home autonomously. Alternatively, he can steer the bomb manually with a small joystick, if conditions make lock-on difficult.

The LOAL indirect mode is similar, except that the bomb is guided from a second aircraft, usually orbiting at a higher altitude and at a considerable distance from the target. This allows the launch aircraft to toss the bomb at high speed and at very low altitude, before exiting the area as quickly as possible. Contact with the data link is ensured by a stabilized antenna in the transceiver pod, but range is dependent on line-of-sight; hence the greater the altitude of the controlling aircraft, the greater the standoff distance. Operation of the IIR version is essentially similar, except that the seeker locks onto the thermal emissions of the target.

The speed and range of the GBU-15 are inadequate to avoid the latest Soviet air defense weapons (hence the development of a rocket-boosted version, the AGM-130). The GBU-15 requires special assembly and test facilities, and considerable preparation time before loading. In addition, the data link has proven susceptible to jamming (a more secure data link is under development). Moreover, at $195,000 per kit, the weapon is too expensive to procure in large numbers. The AGM-130 has the same faults, and USAF is now developing a common successor to both, the Modular Standoff Weapon (MSOW).

Specifications **Length:** 12.83 ft. (3.91 m.). **Diameter:** 18 in. (457 mm.). **Span:** 4.9 ft. (1.49 m.). **Weight, launch:** 2617 lb. (1189.5 kg.). **Range:** 5 mi. (8 km.).

GBU-16: The 1000-lb. PAVEWAY laser-guided bomb.

GCD: See GENERAL AND COMPLETE DISARMAMENT.

GCI: See GROUND-CONTROLLED INTERCEPT.

GD: See SOMAN.

GEARING: A class of 98 U.S. World War II–era DESTROYERS commissioned between 1945 and 1951. Essentially enlarged versions of the SUMNER class (itself a more heavily armed FLETCHER), the Gearings are flush-decked, with relatively low freeboard, twin stacks amidships, and a rather small superstructure forward. As built, the Gearings were armed with 6 5-in. 38-caliber DUAL PURPOSE guns in 3 twin mounts (2 forward, 1 aft), 16 40-mm. BOFORS GUNS in 2 twin and 3 quadruple mounts, 10 twin 20-mm. ANTI-AIRCRAFT guns, 10 21-in. (533-mm.) TORPEDO tubes in 2 quintuple mounts; and two DEPTH CHARGE racks and 6 depth charge projectors (some also had a HEDGEHOG anti-submarine mortar).

The Gearings accounted for the bulk of the postwar U.S. destroyer force, but by the mid-1950s they were approaching obsolescence, with no possibility of replacement by new construction. The navy's response was the Fleet Rehabilitation and Modernization (FRAM) program, implemented from 1960 to 1963. Two different modernization schemes were planned: FRAM I, a complete overhaul of the hull and machinery, and an upgrade of sensors and weapons, primarily for anti-submarine

warfare (ASW); and FRAM II, a less extensive overhaul with less capable weapons and sensors. Seventy-nine Gearings received the expensive FRAM I upgrade, while 16 (mainly older ships previously converted to radar PICKETS) received the cheaper FRAM II.

Under FRAM I, extensive sound insulation was added around the engine rooms and auxiliary machinery spaces, and a new SQS-23 hull-mounted medium-frequency SONAR was installed. All 21-in. torpedo tubes, 20-mm. and 40-mm. guns, and one of the 5-in. mounts were removed in favor of a DASH (Drone Anti-Submarine Helicopter) landing deck and hangar aft, an 8-round ASROC pepperbox launcher amidships, and 2 sets of Mk.32 triple tube mounts for lightweight ASW homing torpedoes. A FRAM II refit retained the original SQS-4 high-frequency sonar and all 6 5-in. guns. Secondary armament and torpedo tubes were removed in favor of 2 trainable Hedgehog projectors, 2 sets of Mk.32 triple tubes, and an SQS-35 VARIABLE DEPTH SONAR.

Sensors and electronics were later upgraded several times; a typical suite may include one SPS-10 surface-search RADAR, one SPS-40 air search radar, one Mk. 25 FIRE CONTROL radar, as well as ELECTRONIC SIGNAL MONITORING arrays and a ULQ-6 ACTIVE JAMMING unit. No Gearings remain in U.S. service, but 51 are still active. With locally modified electronics and weapons, many of these ships will remain in service for at least a decade. See also DESTROYERS, UNITED STATES.

Specifications **Length:** 383 ft. (116.67 m.). **Beam:** 41 ft. (12.5 m.). **Draft:** 18.8 ft. (5.73 m.). **Displacement:** 2425 tons standard/3480 tons full load **Powerplant:** twin-shaft steam: 4 oil-fired boilers, 2 sets of geared turbines, 60,000 shp. **Fuel:** 720 tons. **Speed:** 32 kt. **Range:** 2400 n.mi. at 25 kt./ 4800 n.mi. at 15 kt. **Crew:** 275. **Sensors:** see text. **Electronic warfare equipment:** see text. **Operators:** Bra (2 FRAM I); Ecu (1 FRAM I); Gre (1 FRAM I/6 FRAM II); Mex (2 FRAM I); Pak (6 FRAM I); ROK (5 FRAM I/2 FRAM II); Sp (5 FRAM I); Tai (12 FRAM I/2 FRAM II); Tur (7 FRAM I/1 FRAM II).

GECKO (SA-8): NATO code name for the Soviet ZRK-SD *Romb* mobile surface-to-air missile system; a naval version (NATO designation SA-N-4) is also widely deployed as a point defense weapon. Introduced in 1974 as a replacement for the S-60 towed 57-mm. ANTI-AIRCRAFT gun, the Gecko is an alternative to the SA-6 GAINFUL and SA-11 GADFLY in the air-defense regiments of Soviet tank and Motorized Rifle divisions. While the SA-6 and SA-11 have greater range, an SA-8 battery can engage a larger number of targets simultaneously. The choice of missile assigned to a specific division may reflect local conditions, mission-related considerations, or the aim of complicating enemy countermeasures.

Compact and quite formidable, the Gecko missile is similar in configuration to the British RAPIER and Euromissile ROLAND, with four small cruciform canards near the nose for steering, and four fixed tail fins. Powered by a single-stage, dual-pulse (boost-sustainer) solid-fuel rocket engine, the SA-8 relies on radio command guidance via a jam-resistant microwave DATA LINK, and is armed with a 35.2-lb. (16-kg.) high-explosive fragmentation warhead fitted with both contact and radar proximity (VT) fuzes.

The Gecko system is self-contained on a Zil-167 6 × 6 truck chassis. The four-round launcher is mounted on a power-operated turret with a pair of launch rails on either side of an elaborate "Land Roll" FIRE CONTROL unit. Land Roll itself consists of a folding G/H-band surveillance radar, a large, flat-plate J-band tracking radar, two rapid-gathering acquisition radars (to track the missiles in flight), and two I/J-band command guidance transmitters (allowing two missiles to be directed simultaneously). A low-light television (LLTV) tracker allows optical guidance in case of jamming or other ELECTRONIC COUNTERMEASURES against radar.

In action, targets are first detected by the surveillance radar. After IFF interrogation, the commander selects the target to be engaged and can initiate either automatic or manual tracking. The turret then slews to the target's bearing, and the tracking radar is locked onto the target. Two missiles are usually ripple-fired at the same target (using different guidance frequencies), with steering commands transmitted over the data link; there is no terminal homing. "Land Roll" has considerable ELECTRONIC COUNTER-COUNTERMEASURES capability. Gecko has been used in combat by Syrian forces against Israel, reportedly downing one aircraft in 1982, as well as by Angola against South Africa and by Libya against U.S. aircraft, in both cases without success.

A modified version, designated SA-8b, was introduced in the late 1980s with a 6-round launcher and the missiles carried in sealed storage/launch containers. The missiles may now have either passive INFRARED HOMING or SEMI-ACTIVE RADAR HOMING for terminal guidance. A follow-on system des-

ignated SA-X-15 is now in development. Believed to have either improved command guidance or radar active homing, apparently it incorporates illegally acquired Western technology.

The SA-N-4 is one of the most widely deployed Soviet naval SAMs, and now equips the KIEV aircraft carriers, KIROV battle cruisers, KARA and SVERDLOV cruisers, KRIVAK frigates, GRISHA and NANUCHKA corvettes, and IVAN ROGOV amphibious assault ships. Launched from a compact, retractable twin-arm launcher fed by an 18-round rotary magazine, the SA-N-4 is guided by a "Pop Group" fire control radar, generally similar to "Land Roll." A new vertically launched point defense missile, the SA-N-9, is fitted to UDALOY-class destroyers and has replaced Gecko aboard the Kirov-class battlecruiser *Frunze*. The SA-N-9 may be a naval variant of the SA-X-15.

Specifications Length: 10.2 ft. (3.11 m.). Diameter: 8.7 in. (220 mm.). Span: 23.9 in. (607 mm.). Weight, launch: 374 lbs. (170 kg.). Speed: Mach 3 (2100 mph/3507 kph). Range envelope: 1.6–12 km. (1–7.44 mi.). Height envelope: 10–13,000 m. (32.8–42,640 ft.). Operators: Alg, Ang, Ind, Iraq, Jor, Ku, Lib, Syr, USSR and other Warsaw Pact forces.

GEMS: General Energy Management Steering, a BALLISTIC MISSILE guidance technique for solid-fuel missiles, whereby the missile makes excursions above and below the nominal (minimum energy) trajectory, thus wasting energy so that the final stage may burn to exhaustion while still achieving the velocity vector required to hit the intended target. This eliminates the need for blowout panels or other thrust-termination devices, which are generally less reliable and less accurate. GEMS is employed in all current U.S. ICBMs and SLBMs, and also on the latest Soviet solid-fuel missiles, the SS-24 SCALPEL and SS-25 SICKLE.

GENERAL AND COMPLETE DISARMAMENT (GCD): Proposals for the total elimination of all weapons and military forces, usually coupled to proposals for the renunciation of war as an instrument of policy. Plans for general and complete disarmament were, until the 1970s, often proposed to camouflage the rejection of less ambitious (but more feasible) ARMS CONTROL agreements. Since the enunciation of "agreed principles" by the U.S. and Soviet Union, frivolous GCD proposals have given way to detailed studies within the framework of the Eighteen Nation Disarmament Conference (ENDC), convened in Geneva under a UN mandate. Recent U.S. and Soviet proposals reflect extensive research into GCD intermediate stages, inspection and verification requirements, post-agreement police forces, etc. Despite such intellectual progress, GCD is as unlikely as ever, though the ENDC has been put to good use in analyzing arms control measures such as the PARTIAL TEST BAN TREATY, OUTER SPACE TREATY, and NON-PROLIFERATION TREATY. Also see, more generally, DISARMAMENT.

GENERAL-PURPOSE MACHINE GUN (GPMG): A man-portable, belt-fed, air-cooled, rifle-caliber (still usually 7.62-mm.) MACHINE GUN, generally issued with alternative bipod and tripod mounts, to function both as a light and medium machine gun. The GPMG concept was developed by the German army with the MG38 of 1938 and perfected with the MG42 of World War II. Since then, GPMGs have displaced more specialized light and medium weapons to become the dominant form of machine gun. GPMGs generally weigh between 30 and 45 lb. (excluding the mount), are served by a crew of two, and have heavy barrels with quick-change fittings for sustained automatic fire. Current models include the U.S. M60, the German MG3 (a slightly modernized MG42), the Soviet PK, and the Belgian FN-MAG.

GENERAL SUPPORT: Support provided to a force as a whole; for example, the artillery brigade of an infantry division supporting the division as a whole is said to be in general support. On the other hand, if the brigade or a part of it is assigned to support a particular subunit of the division, it is in DIRECT SUPPORT.

GENERAL WAR: A war between the superpowers (plus allies and satellites), in which all available weapons would be used, including nuclear weapons. It is the professional term for TOTAL WAR, now only in literary use. For an armed conflict between the superpowers below that threshold, CENTRAL WAR is the preferred term. See also WAR.

GENEVA CONVENTION: Common usage for various codifications of humanitarian laws of war, of which the latest dates from 12 August 1949 (with 1975 amendments). The Geneva conventions define rules for the treatment of noncombatants, prisoners of war (POWS), and populations under military occupation; they also outlaw the use of certain weapons, such as barbed arrows and dum-dum bullets, as well as chemical and biological agents. They do not, however, outlaw the use of nuclear weapons. The extent to which these rules are observed depends on the power relation-

ships that obtain. Thus Anglo-American prisoners were treated largely according to the convention by the Nazis (even though indiscriminate Allied strategic bombing violated the convention) because of the ease of retaliation against German POWs in Allied hands. Where retaliation could not be direct, as in the case of "occupied" populations, the latter were not treated in accordance with the convention. More recently, U.S. prisoners in Southeast Asia were systematically denied Geneva-convention rights even though North Vietnam was a signatory to the convention. In 1991, Iraq also systematically violated the convention by physically abusing U.S. and allied POWs, using them for propaganda broadcasts, and denying the International Red Cross access to them. See LAWS OF WAR for more details.

GEORGES LEYGUES: A class of seven French ANTI-SUBMARINE WARFARE (ASW) destroyers, the first of which was commissioned in 1979. Developed from the TOURVILLE class, the George Leygues share the hull and machinery of the CASSARD class, essentially an ANTI-AIR WARFARE (AAW) version of the same ship.

The Georges Leygues are flush-decked, with the low freeboard typical of French warships, and a large, boxy superstructure amidships. Armament consists of 1 100-mm. DUAL PURPOSE gun forward, a 6-round Sea CROTALE point defense missile launcher on the superstructure aft, 4 EXOCET (MM.38 or MM.40) anti-ship missiles in single launch canisters amidships, 2 single, manually operated 20-mm. ANTI-AIRCRAFT guns behind the bridge, and 2 catapults for lightweight ASW homing TORPEDOES. The primary ASW weapons are two Westland LYNX helicopters operated from a landing deck and hanger aft. See also DESTROYERS, FRANCE.

Specifications Length: 455.9 ft. (139 m.). **Beam:** 45.9 ft. (14 m.). **Draft:** 18.7 ft. (5.7 m.). **Displacement:** 3830 tons standard/4170 tons full load. **Powerplant:** twin-shaft CODOG: 2 26,000-shp. Rolls Royce Olympus TM3B gas turbines (sprint), 2 5100-hp. SEMT-Pielstick 16 PA 6CV diesels (cruise). **Speed:** 30 kt. **Range:** 1000 n.mi. at 30 kt./9500 n.mi. at 17 kt. (diesels). **Crew:** 216. **Sensors:** 1 DRBV 51C or DRBV 15 surface-search radar, 1 DRBC 32C fire control radar, 2 Decca 1226 navigation radars, 1 DUBV 32 hull-mounted medium-frequency sonar, 1 DUBV 43 variable depth sonar; (last three ships) 1 DSBV passive towed array sonar. Electronic warfare equipment: 1 ARBR 11B, ARBR 16, or ARBR 17 electronic signal monitoring array, 1 ARBB 32 active jamming unit, 2 Dagaie chaff launchers.

GEPARD: West German self-propelled ANTI-AIRCRAFT gun based on the chassis of the LEOPARD 1 main battle tank. Developed from 1966 as a replacement for the old U.S.-built M42 Duster, the first Gepard was delivered in 1976. By 1980, the *Bundeswehr* had taken delivery of 420 Gepards, with 55 more for the Belgian army, and 95 for the Dutch army (in a modified version designated CA.1).

The Gepard is a self-contained turret-mounted twin gun and FIRE CONTROL unit mounted on a modified Leopard 1 hull with reduced armor. The large turret increases overall height to 9.9 ft. (excluding radar antennas); it can also be mounted on most other main battle tanks. Two Oerlikon KDA 35-mm. CANNONS are mounted externally in pods attached to either side of the turret. Each gun has a cyclic rate of fire of 550 rds./min. and is supplied with 320 rounds of anti-aircraft and 20 rounds of APDS ammunition (for use against armored vehicles). The muzzle velocity of 1175 m./min. allows an effective range of up to 4000 m. The turret has 360° traverse, and the guns have elevation limits of −10° and +85°.

A Siemens search RADAR with a folding antenna is mounted on the rear of the turret roof, while a tracking radar is mounted on the turret face between the guns. German Gepards have laser rangefinders and backup optical sights. The Dutch CA.1s differ in having somewhat better Dutch-made radars with moving target indicators (MTI) and other refinements. Inside the turret, Gepard has a computer-assisted fire control unit, an inertial navigation unit, and a COLLECTIVE FILTRATION unit for NBC defense. The crew consists of a driver in the bow, with a commander and a gunner in the turret. Automotive performance is similar to the Leopard 1.

GESCHWADER: A WING in the West German *Luftwaffe*. Each *Geschwader* comprises up to three SQUADRONS with 16 to 20 aircraft each. *Geschwadern* are classified by mission. FIGHTER wings are known as *Jagdgeschwader*, fighter-bombers as *Jagdbombergeschwader,* and reconnaissance as *Aufklarungsgeschwader.*

GEV: See GROUND EFFECT VEHICLE.

GIANT (SA-12B): NATO code name for a Soviet mobile, long-range surface-to-air missile, introduced in 1988 as a further development of the SA-12A GLADIATOR. Giant is larger and heavier than Gladiator, with considerably higher perform-ance, particularly in speed, range, and ceiling. These far exceed the requirements of a purely anti-aircraft system, leading some U.S. intelligence analysts to speculate that the SA-12B is in fact an ANTI-BALLISTIC MISSILE (and as such in violation of the ABM TREATY).

Like Gladiator, Giant is a two-stage missile with a solid-fuel rocket booster and liquid-fuel second stage. Based on the size of its launch canister, it is thought that the second stages of both missiles are (at least) externally identical, but that Giant has a longer and wider booster. Thrust vector control is used for steering in both stages. Giant combines radio COMMAND GUIDANCE via a jam-resistant DATA LINK with some form of terminal homing. Performance is thought sufficient to engage at least tactical ballistic missiles and their reentry vehicles. Two warheads are available: a 5 to 10-KT nuclear warhead, or a 200-lb. high-explosive warhead with radar proximity (VT) fuze.

Giant is mounted on the same self-propelled transporter-erector-launcher (TEL) as Gladiator, but its TEL carries only two (vs. four) missiles in their sealed launch canisters. Stowed horizontally for travel, the missiles are elevated vertically off the rear of the vehicle for firing. The TEL has its own on-board missile guidance radars and fire controls. Giant and Gladiator are employed in mixed batteries of one Giant and two Gladiator TELs, with a separate self-propelled phased array surveillance and acquisition radar, a reload vehicle, and a command vehicle. Three batteries together with another surveillance and acquisition radar and 2 command vehicles comprise a battalion, while 3 battalions, 2 more surveillance and acquisition radars, and 3 command vehicles form an air defense brigade; two are assigned to each FRONT and one to each ARMY. In addition, the PVO deploys Giants and Gladiators for homeland defense.

Specifications Length: 33.75 ft. (10.29 m.). **Diameter:** 4.25 ft. (1.3 m.). **Weight, launch:** 7050 lb. (3205 kg.). **Speed:** Mach 4 (3000 mph). **Range envelope:** 12–150 km. (7.44–93 mi.). **Ceiling:** 40,-000 m. (131,200 ft.).

GIUK GAP: The narrow seas between Greenland, Iceland, and the United Kingdom, specifically the Denmark Strait, the Iceland-Faeroes Ridge, and the Norwegian Sea between the Faeroes and the Orkneys. The GIUK Gap is strategically important because it must be crossed by Soviet naval forces attempting to enter the North Atlantic. Shallow, relatively confined, and completely covered by land-based aircraft patrols, it is a crucial CHOKEPOINT behind which the Soviet navy could be bottled up; as a result, it is closely monitored by surveillance ships, reconnaissance aircraft, satellites, and the underwater SOSUS network.

GKO: *Gosudarstvennyy komitet oborony* (State Committee of Defense), the predecessor of the current Soviet DEFENSE COUNCIL (*Sovet oborony*). In time of war the latter could constitute the nucleus of a new GKO with responsibility for the overall direction of the Soviet war effort.

GLADIATOR (SA-12A): NATO code name for a Soviet mobile, long-range surface-to-air missile, introduced in 1986 as a replacement for the SA-4 GANEF in Army and Front air-defense brigades. Gladiator is an extremely powerful missile designed to intercept supersonic CRUISE MISSILES and aircraft flying anywhere from treetop height to 100,000 ft. (30,487 m.).

The first stage of the two-stage missile is believed to be a solid rocket booster, and the second stage a liquid-fuel rocket with either aerodynamic controls or, more probably, thrust vector control for steering. The missile relies on a combination of radio COMMAND GUIDANCE and some form of terminal homing, and is armed with a 200-lb. (90.9-kg.) high-explosive warhead with a radar proximity (VT) fuze.

Four sealed missile canisters are mounted on a tracked, self-propelled transporter-erector-launcher (TEL) derived from the MT-T artillery tractor. Stored horizontally for traveling, the four canisters are elevated vertically off the rear of the vehicle for launching. The TEL has a guidance radar mounted on a telescopic mast. Gladiator is believed to fly a pop-up trajectory, rising vertically to altitude before diving down against low-flying targets; this flight profile is more energy efficient than a more direct approach would be.

A modified version, the SA-12B (NATO code-name GIANT), was introduced in 1988 with an enlarged booster and increased speed, range, and ceiling; believed to have some ANTI-BALLISTIC MISSILE capability, it has the same TEL as Gladiator, albeit modified to carry only two of the larger missiles. Two Gladiator TELs and one Giant TEL form a battery together with a command vehicle, a reload vehicle, and a phased array surveillance/acquisition radar. The latter has a maximum range of 250 km. (155 mi.), and is believed to have multiple track-while-scan capability. Three batteries plus another surveillance/acquisition radar and 2 command vehicles comprise a battalion, while 3

battalions, 2 more surveillance/acquisition radars, and 3 command vehicles form an air defense brigade. In addition to Army and Front units, Gladiator is deployed by the PVO for the homeland defense of the U.S.S.R.

Specifications Length: 27.25 ft. (8.2 m.). Diameter: 2.95 ft. (0.9 m.). Weight, launch: 3450 lb. (1568 kg.). Speed: Mach 3 (2100 mph/3507 kph). Range envelope: 5–90 km. (3.1–55.8 mi.). Ceiling: 30,000 m. (98,400 ft.).

GLCM: Ground-Launched Cruise Missile, a long-range CRUISE MISSILE launched from fixed or mobile launchers, generally meant for intermediate range (1000–1500 km.) nuclear strike. The U.S. GLCM was derived from the TOMAHAWK (ship- and submarine-launched) cruise missile (SLCM), while the Soviet GLCM was developed from the SS-N-21 SAMPSON SLCM. GLCMs were banned by the 1987 INF TREATY; all existing missiles and launchers are being dismantled and destroyed.

GLIDE BOMB: An unpropelled air-launched warhead which descends to its target in wing-supported flight; speed and range depend entirely on speed and altitude at release. To overcome the inherent inaccuracy of this form of delivery, all glide bombs have some form of guidance, the most common being radio COMMAND GUIDANCE, ELECTRO-OPTICAL (TV), IMAGING INFRARED, and semi-active LASER homing. For specific examples, see GBU-15; HOBOS; PAVEWAY; WALLEYE.

GLOBAL POSITIONING SYSTEM (GPS): A navigation aid that consists of satellite-mounted transponders. With (economical) distance measuring equipment (DME) and the use of time difference of arrival (TDOA) calculators, any ship, vehicle, or aircraft equipped with a suitable receiver can determine its position with great accuracy. Two systems are currently in use: the U.S. NAVSTAR and the Soviet GLONASS. GPS was meant primarily to assist commercial shipping, but it could be exploited by third parties to provide accurate guidance for long-range missiles at very low cost.

GLONASS: Global Navigation Satellite System, a Soviet GLOBAL POSITIONING SYSTEM analogous to the U.S. NAVSTAR. GLONASS consists of 12 satellites placed in 12,000-mi. (20,000-km.) circular orbits inclined at 63°; the first satellite was launched in 1982, and the last in 1987. GLONASS is very similar in design and operation to NAVSTAR, and in fact the two use the same transmission frequencies (1.2 and 1.6 GHz), indicating a considerable degree of technology transfer. Whether GLONASS provides as much accuracy as its U.S. counterpart is not known, but it is definitely superior to earlier Soviet satellite navigation systems. See, more generally, SPACE, MILITARY USES OF.

GOA (SA-3): NATO code name for the Soviet S-125 *Pechora* surface-to-air missile (SAM), also widely employed as a shipboard weapon (NATO designation SA-N-1). Introduced in 1961 as a medium-low-altitude supplement to the high-altitude SA-2 GUIDELINE, Goa is a static area-defense SAM used by the PVO to defend major targets (such as airfields) in both the Soviet Union and Eastern Europe. Approximately 1200 Goa launchers have been deployed in the Soviet Union and Eastern Europe, and the system has been widely exported.

Goa is a two-stage, solid-fuel missile. Its first (booster) stage has 4 rectangular cruciform fins; the second stage has 4 small delta canards at the nose for steering, and 4 fixed wings at its base, 2 of which have ailerons for roll control. Goa relies on radio COMMAND GUIDANCE and is armed with a 132-lb. (60-kg.) high-explosive fragmentation warhead (lethal radius 12.5 m.) fitted with a radar proximity (VT) fuze. At launch, the booster thrusts for three seconds, accelerating the missile to Mach 2.5 (1600 mph/2672 kph) before being jettisoned; the second stage brings the missile to maximum speed. Command guidance does not begin until second-stage ignition.

The twin or quadruple rail launchers can be towed, but they are rather cumbersome and are more usually left at fixed sites. The launcher has 360° traverse and a maximum elevation of 75°. Four launchers, together with a "Low Blow" guidance radar, a command center, and four reload trucks (each with two missiles), comprise a battery. Late-model batteries also have a telescopic ELECTRO-OPTICAL (TV) director with a range of 30 km. (18.6 mi.), for use in case of radar jamming. A firing battery, a "Flat Face" surveillance radar, a "Squat Eye" low-altitude radar, a "Side Net" height-finding radar, a FIRE CONTROL unit, and a headquarters form a battalion. Three battalions, a headquarters, and service units form an air defense brigade, several of which are assigned to each FRONT to defend airfields and other high-value targets, often paired with the high-altitude long-range SA-5 GAMMON.

In action, targets are detected initially by the Flat Face radar, which has a maximum range of 250 km. (155 mi.). At 83 km. (51.5 mi.), the target is passed to the Low Blow, which continues track-ing down to 23 km. (14.25 mi.), at which point a missile may be launched by a command from the fire control unit; batteries with quadruple launchers may fire two missiles at one target. Guidance commands are passed to the missile by a UHF radio DATA LINK from second-stage ignition through intercept.

A modified version (SA-3b), introduced in 1964, has a considerably less restricted minimum range, at the cost of slightly less maximum range; later versions also have SEMI-ACTIVE RADAR HOMING (SARH) for terminal guidance, with the Low Blow providing target illumination.

The Goa was first used in combat by Egypt during the War of Attrition (1968–70); after some initial successes, it was effectively neutralized by Israeli jamming. A few were used by North Vietnam in 1972–73, again with limited success. Since then the missile has been almost completely ineffective in the Iran-Iraq War, the 1982 Lebanese War, and the 1986 U.S. air strikes against Libya. Despite this, additional launch sites were still being constructed in 1988 in both Eastern Europe and the U.S.S.R. (their supplementary radars may defeat jamming).

In its SA-N-1 version, Goa is the most widely deployed Soviet naval SAM, arming the KYNDA and KRESTA cruisers, and KASHIN, KANIN, and KOTLIN destroyers. It is fired from a twin-arm launcher which rotates vertically for loading from a 16-round magazine immediately below it (as in U.S. SAMs). Guidance is provided by a "Peel Group" fire control radar. In contrast to the SA-3, the SA-N-1 can also be fitted with a 10 KT nuclear warhead, and may have a secondary anti-ship role. The original SA-N-1 was similar to the SA-3a, but it has been replaced by the SA-N-1b with SARH terminal guidance. Now obsolescent, it is being replaced on newer ship classes by the SA-N-7 GADFLY, a naval derivative of the SA-11.

Specifications Length: 21.95 ft. (6.7 m.). Diameter: (booster) 23.62 in. (600 mm.); (sustainer) 17.72 in. (450 mm.). Span: (booster) 4.95 ft. (1.5 m.); (sustainer) 4.05 ft. (1.23 m.). Weight, launch: (sustainer) 1402 lb. (637.27 kg.). Speed: Mach 3.5 (2300 mph/3841 kph). Range envelope: 6–29 km. (3.72–18 mi.); (b) 2.4–18.3 km (1.5–11.33 mi.). Height envelope: 1500–12,200 m. (4920–40,000 ft.); (b) 300 m. (984 ft.) min. Operators: Afg, Alg, Ang, Cuba, Egy, Eth, Fin, Ind, Iraq, Lib, Moz, N Kor, Peru, Som, Syr, Ug, Viet, USSR and other Warsaw Pact forces, Yug.

GOALKEEPER: A shipboard radar-controlled gun (CLOSE-IN WEAPON SYSTEM) developed as a joint venture by General Electric and Hollandse Signaal Apparaten, now in service with the Royal Navy and the Royal Netherlands Navy. Goalkeeper is a self-contained, autonomous system consisting of a GE 7-barrel, 30-mm. GAU-8 cannon (originally developed as an anti-tank gun for the A-10 THUNDERBOLT aircraft) mounted together with an I/J-band TRACK-WHILE-SCAN (TWS) surveillance/acquisition radar, an I/K-band tracking radar, a backup ELECTRO-OPTICAL (TV) director (in case of radar jamming), and a FIRE CONTROL computer.

The GAU-8 fires tungsten-penetrator APDS ammunition with a muzzle velocity of 3350 m./min. (for an effective range of 3000 m.) at a rate of 4200 rds./min. A total of 1190 rounds can be stowed in a drum magazine beneath the gun. The mount has 360° traverse and elevation limits of −25° and +85°. The entire unit weighs 14,837 lb. (6744 kg.) loaded, and requires a reinforced deck space with adequate fields of fire.

Operation is similar to that of the U.S. PHALANX: whenever the system is activated, the search/acquisition radar scans continuously and locks onto the first target that comes within acquisition range; if multiple targets are detected, the fire control computer prioritizes them to engage the nearest/fastest first. The mount then slews to the target bearing, and the gun opens fire as soon as the target comes within firing range. The tracking radar follows both the target and the bullet stream from the gun; the fire control computer then generates aiming commands to obtain convergence between the two tracks. In tests, Goalkeeper has destroyed supersonic targets within one second; it is superior to Phalanx in the greater range and lethality of the 30-mm. GAU-8 as compared to the 20-mm. Vulcan cannon of the latter, but Phalanx is cheaper.

Goalkeeper is currently mounted on Dutch KORTENAER- and HEEMSKERCK-class frigates and Poolster-class UNDERWAY REPLENISHMENT ships. The Royal Navy has selected Goalkeeper to replace Phalanx on its four ILLUSTRIOUS-class aircraft carriers and four BROADSWORD Batch 3 frigates; it will probably also be fitted to the new DUKE-class frigates.

GOBLET (SA-N-3): NATO code name for a Soviet shipboard surface-to-air missile (SAM), introduced in 1967 as a replacement for the SA-N-1 GOA in the medium-range, low- to medium-altitude air-defense role aboard the KIEV aircraft carriers, MOSKVA helicopter carriers, and KRESTA II and KARA cruisers. Unlike other Soviet naval SAMs, it is not derived from a land-based system (although it was at first believed to derive from the SA-6 GAINFUL).

Powered by a two-stage solid-fuel rocket, the SA-N-3 has four cropped-delta cruciform wings at midbody, and four smaller rectangular tail fins; it is not known whether it is the wings or the fins that provide steering. Goblet can carry either a 331-lb. (150-kg.) high explosive or a 25-KT nuclear warhead; the proportion of missiles fitted with the latter may be as high as 25 percent. The missile could also have a secondary anti-ship role, at least in its nuclear version.

Goblet relies on radio COMMAND GUIDANCE and may also have SEMI-ACTIVE RADAR HOMING in the terminal phase. It is directed in flight by a "Headlight"-series FIRE CONTROL radar group (which includes separate target-tracking and missile-guidance radars), which also controls the SS-N-14 SILEX anti-submarine missile. Goblet's twin-arm launcher rotates vertically for reloading from a 36-round magazine immediately below it. Since 1978, Goblet has been superseded on newer ships by the SA-N-6 GRUMBLE, a derivative of the SA-10 long-range SAM.

Specifications Length: 21 ft. (6.4 m.). **Diameter:** 2.3 ft. (700 mm.). **Span:** 5.55 ft. (1.7 m.). **Weight, launch:** 1210 lb. (550 kg.). **Speed:** Mach 2.8 (1850 mph/3090 kph). **Range envelope:** 6–30 km. (3.72–18.6 mi.); (lmp) 55 km. (34.1 mi.) max. **Height envelope:** 90–24,500 m. (295–80,360 ft.).

GODAVARI: A class of three Indian FRIGATES based upon the British LEANDER frigates.

GOPHER (SA-13): NATO code name for the Soviet ZRK-BD *Strela-10* mobile surface-to-air missile (SAM), introduced in 1978 as a replacement for the similar SA-9 GASKIN. Gopher is longer and heavier than Gaskin, with greater range and a more sensitive guidance unit; mounted on a self-propelled, radar-equipped launch vehicle, it has night and limited bad-weather capabilities. Gopher is replacing the SA-9 on a one-for-one basis in Soviet and Warsaw Pact regimental air defense batteries, and has been exported to Libya and Syria.

The single-stage, solid-fuel, INFRARED-HOMING missile has the same general configuration as the SA-9, with four cruciform control canards at the nose and four fixed tail fins. The 13.2-lb. (6-kg.) high-explosive warhead is fitted with both contact and radar proximity (VT) fuzes. The cryogenically cooled infrared seeker in the nose is believed to operate in two separate frequency bands; this gives it some all-aspect capability and a greater resistance to flares and other INFRARED COUNTER-MEASURES (IRCM).

The Gopher system is mounted on a tracked vehicle derived from the MT-LB multipurpose tractor; its pedestal launcher, which replaces the vehicle's machine-gun turret, has racks for two pairs of storage/launch containers with ready-to-fire missiles on each side; the racks can also accommodate the smaller SA-9. A range-only radar is mounted on the pedestal between the missiles, but some vehicles also have a Hat Box ELECTRONIC SIGNAL MONITORING array to detect the emissions of aircraft radars. Four reload rounds are carried inside the vehicle; external racks are sometimes added for four more reloads.

The vehicles do not have their own surveillance radars, but are linked by radio to the divisional air-defense radars, for early warning and target acquisition; the system can also use target data supplied by the "Gun Dish" radars of the ZSU-23-4 anti-aircraft guns of the regimental air-defense battery. In independent action, once a target is acquired visually, the turret is slewed to the proper bearing and the ranging radar locked on. An audio tone indicates to the operator when the missile seeker has achieved a lock, whereupon two missiles can be launched in salvo.

Gophers have been used by Libyan forces in Chad, apparently without success. Several examples were captured intact, hence it is likely that effective U.S. countermeasures have now been developed. Although better than earlier Soviet IR-homing missiles, Gopher is probably inferior to Western equivalents (e.g., STINGER), and vulnerable to IRCM.

Specifications Length: 7.16 ft. (2.18 m.). Diameter: 5 in. (127 mm.). Span: 15.75 in. (400 mm.). Weight, launch: 121 lb. (55 kg.). Speed: Mach 2 (1400 mph/2338 kph). Range envelope: 0.5–8 km. (0.31–4.96 mi.). Height envelope: 9–3200 m. (29.5–10,500 ft.).

GPMG: See GENERAL-PURPOSE MACHINE GUN.

GPS: See GLOBAL POSITIONING SYSTEM.

GRAIL (SA-7): NATO code name for the Soviet 9M32 *Strela-2* man-portable, shoulder-fired, INFRARED-HOMING surface-to-air missile (SAM), also made in a shipboard version (NATO designation SA-N-5). Grail was introduced in 1966 to provide short-range, low-altitude air defense for Mo-

torized Rifle companies, each of which has a section of three launchers (usually assigned to each of the component platoons). In service throughout the Warsaw Pact, the SA-7 has also been widely exported to Third World forces and pro-Soviet guerrillas. A number have found their way to terrorists, and to anti-Soviet MUJAHIDEEN in Afghanistan. Improved copies are being built without license in China (HN-9) and Egypt (Sakr Eye). In spite of its limitations, Grail will remain in service because of its low price and the vast numbers available.

Grail has two pop-out control canards near the nose, and four pop-out stabilizer tail fins. The simple, uncooled lead-sulfide (PbS) infrared seeker head is relatively insensitive and easily decoyed by flares and other INFRARED COUNTERMEASURES (IRCM). Capable of rear-quarter engagements only, the SA-7 is very much a "revenge weapon": it can be launched only *after* an attacking aircraft has passed overhead. The 3.96-lb. (1.8-kg.) contact-fuzed high-explosive warhead is too small to reliably destroy high performance aircraft, but is generally lethal against helicopters. Grail's relatively low speed and short range makes it difficult to hit aircraft moving at more than 600 mph (1000 kph). The improved SA-7b (Soviet designation 9M32M), introduced in 1972, has longer range and somewhat better speed and ceiling; in addition, the warhead has been enlarged to 5.5 lb. (2.5 kg.), increasing launch weight to 22 lb. (10 kg.). But because the seeker has been only marginally improved, the SA-7b is still relatively ineffective.

Grail is fired off the shoulder from a tube launcher equipped with a simple optical sight and a trigger. To launch the missile, the operator must acquire the target visually and remove protective caps from both ends of the launch tube. Acquisition is assisted by a small RADAR WARNING RECEIVER carried on the operator's helmet, or (in later models) mounted on the launcher itself. Pointing the missile at the target, the operator waits for the seeker to achieve a lock (indicated by an audio tone), then pulls the trigger. A cold gas generator ejects the missile from the tube before the the rocket motor ignites at a safe distance from the operator.

The navalized SA-N-5 is employed as a POINT DEFENSE MISSILE on fast attack craft, corvettes, and amphibious landing ships of the PAUK, TARANTUL, POLNOCNY, and OSA classes. It is fired from a manually operated 4-round pedestal launcher with target acquisition data passed verbally from the ship's radar operators. The firing sequence is identical to that of the SA-7.

Grail is now being replaced by the externally similar but much improved SA-14 GREMLIN; a naval variant, the SA-N-8, is being retrofitted to ships equipped with the SA-N-5.

Specifications Length: 4.25 ft. (1.29 m.). Diameter: 2.3 in. (58 mm.). Span: 8 in. (203 mm.). Weight, launch: 20.24 lb. (6.17 kg.). Weight, system: 42.25 lb. (19.2 kg.) Speed: (7a) Mach 1.4 (1000 mph/1670 kph); (7b) Mach 1.7 (1100 mph/ 1837 kph). Range: (7a) 3700 m. (2.3 mi.); (7b) 500–5600 m. (0.3–2.67 mi.). Height envelope: (7a) 150–3000 m. (492–9840 ft.); (7b) 23–4300 m. (75–14,104 ft.).

GREMLIN (SA-14): NATO code name for the Soviet *Strela*-3 man-portable, shoulder-fired INFRARED-HOMING surface-to-air missile, introduced in 1988 as a replacement for the SA-7 GRAIL; a shipboard variant (SA-N-8) has also been deployed. In service with Soviet and Warsaw Pact forces, Gremlin has also been exported to Angola, Cuba, Finland, Nicaragua, and Syria.

Externally similar to the SA-7, the missile has the same pop-out control canards and tail fins, but an entirely new infrared seeker differentiated by a conical (vs. blunt) multifaceted window. Cryogenically cooled by a gas bottle in the launcher, the seeker is believed to have sufficient sensitivity for ALL-ASPECT engagements (a vast improvement over the rear-quarter-only SA-7). On the other hand, it seems to retain the rather small 5.5-lb. (2.5-kg.) warhead of the older missile. The engagement sequence is identical to that of the SA-7.

The SA-N-8 is replacing the SA-N-5 (the naval variant of the SA-7) aboard numerous CORVETTES, FAST ATTACK CRAFT, and amphibious warfare vessels, offering the major improvement of all-aspect capability.

Although better than its predecessor, the Gremlin is still believed inferior to the U.S. STINGER, particularly in its resistance to flares and other INFRARED COUNTERMEASURES. It has been used in Angola and Nicaragua, with mixed results. A similar missile, designated SA-16, was introduced in 1987; it appears to be identical to Gremlin, but is guided by semi-active LASER homing (as in the British JAVELIN).

Specifications Length: 4.3 ft. (1.31 m.). Diameter: 2.3 in. (58 mm.). Span: 8 in. (203 mm.). Weight, launch: 23.1 lb. (10.5 kg.). Weight, system: 41.15 lb. (18.7 kg.). Speed: Mach 1.8 (1170 mph/1954 kg). Range: 7000 m. (4.34 mi.). Ceiling: 5000 m. (16,400 ft.).

GRENADE: A small bomb thrown by hand, ejected from a rifle, or fired by a special-purpose grenade launcher. "Defensive" pineapple- or ball-shaped fragmentation grenades are the most common. Their effect is obtained from the splinters produced by the explosion of a small charge within a thick metal case, which is is corrugated internally or externally to ensure an even distribution of the fragments. Most grenades weigh between 8 and 16 ounces and have a lethal radius of 10–20 m.

Grenades that can be thrown to a distance greater than their burst radius are called "offensive," because they can be used by troops advancing without cover. "Defensive" grenades, by contrast, are most suitable for use by troops in trenches, foxholes, or other fortified positions. Offensive grenades have thinner casings with a larger explosive charge, and rely on blast effects to stun enemy personnel. A specialized variant, the British "Flash-Bang" grenade, for use in hostage situations, produces a very loud but generally harmless explosion and a blinding flash intended to incapacitate briefly without causing permanent injury.

Chemical hand grenades include incendiary, illumination, and smoke types; the first are sometimes known as "phosphorus," the second as "magnesium" grenades.

All hand grenades have pin-type chemical time fuzes, which normally give the user 4–10 seconds after the pin is pulled out in which to throw the bomb. In some armies, notably the German and Soviet, throwing sticks are added; in these "potato masher" grenades, the stick provides greater leverage and extends the throwing range.

Rifle grenades are small fragmentation, incendiary, illumination, smoke, or anti-tank bombs launched from the muzzle of a rifle. Until recently, rifle grenades required cumbersome adapters, and the use of special blank cartridges to eject the grenade. The latest types, however, can be inserted directly into the muzzle flash suppressor of modern rifles; their "bullet trap" ejection system relies on a normal bullet fired into a soft metal rod in the base of the grenade. The bullet's energy is transferred to the grenade, projecting it towards the target. Rifle grenades of this type, typified by the Belgian Energa series in 40- to 75-mm. diameter, have effective ranges of 100–500 m. Anti-tank versions have shaped-charge (HEAT) warheads which can penetrate several inches of armor, but are effective against tanks only in especially favorable circumstances, because their low muzzle velocity makes it difficult to hit a moving target beyond point-blank range. All rifle grenades have contact fuzes with a built-in safety delay.

Specialized grenade launchers are tube weapons that fire conical grenades with brass cartridges that resemble outsize pistol rounds; like rifle grenades, they usually have contact fuzes. The U.S. M79 amounts to a large break-breach shotgun. Others have short tubes attached below the barrels of standard assault rifles, as in the U.S. M203, a grenade launcher attached to an M16A1 rifle. Most grenade launchers are of 40-mm caliber, and have maximum ranges of 300 m. In the early 1960s they were undoubtedly more effective than rifle grenades, but today the modern bullet trap grenades have better range, more hitting power, and much greater flexibility.

Automatic grenade launchers are specialized weapons resembling large machine guns. See GRENADE LAUNCHERS, AUTOMATIC, for details.

GRENADE LAUNCHERS, AUTOMATIC:
Crew-served weapons resembling large machine guns, which fire small explosive or chemical bombs (GRENADES) at a rapid rate over ranges of several hundred meters. First developed by the U.S. during the Vietnam War as anti-personnel weapons for helicopters and riverine patrol vessels, and later neglected, automatic grenade launchers are now being acquired in larger numbers as suppressive weapons, especially for use for use against anti-tank weapon crews. Some lightweight automatic grenade launchers have also been developed, but are not widely used. The two most common models in service are the Soviet AGS-17 *Plamya* (Flame) and the U.S. Mk.19, both tripod- and pintle-mounted (on vehicles).

The AGS-17 30-mm. grenade launcher was introduced in the early 1970s, but not widely seen until the 1979 invasion of Afghanistan. It is primarily a suppression weapon for combat vehicles, but has also been effective against point targets such as bunkers and foxholes. The *Plamya* is 33.06 in. (840 mm.) long, with a stubby barrel and rectangular receiver. The launcher alone weighs 38.85 lb. (17.65 kg.); with its tripod, total weight is approximately 66 lb. (30 kg.). It is fed from an underslung 29-round drum magazine weighing an additional 33 lb. (15 kg.) loaded. Because of its weight the *Plamya* has limited mobility on foot, and is usually pintle-mounted on BTR-60 or BMP personnel carriers (it can be dismounted for use in defensive posi-

tions). One *Plamya* is issued to each platoon to provide additional fire support to dismounted infantry, though in Afghanistan it was also useful in suppressing fire directed against road convoys. A short-barrel version was apparently used in Afghanistan from Mil-8 HIP assault helicopters.

The *Plamya* has a blowback action with a cyclic rate of fire of 350–400 rds./min. (the practical rate, with reloading, is 50 rds./min.). Muzzle velocity is 185 m./min. (607 ft./min.), for an effective range of up to 1200 m. (122 yd.); a minimum range of 50 m. (164 ft.) is imposed by the safety fuzing of the grenades. Several types of grenade are available, including HE-fragmentation, incendiary, and HEAT. The first is the most common; it weighs 14 ounces and has a lethal radius of 7 m. (23 ft.).

The U.S. Mk.19 40-mm. grenade launcher was developed in the early 1960s for use in Vietnam aboard riverine and coastal patrol boats. After the war, it was phased out of the inventory, but was reintroduced by the U.S. Marine Corps in the early 1980s with vehicle or tripod mounts. The Mk.19 (officially designated a "heavy machine gun") resembles an enlarged M2 BROWNING with a short barrel. It is 34 in. (864 mm.) long with a 12-in. (305-mm.) barrel, and weighs 46 lb. (20.9 kg.), or 72.5 lb. (32.9 kg.) with its tripod. The Mk.19 has a blowback action, and fires belt-fed high-velocity 40-mm. rounds which resemble outsize pistol bullets, at a cyclic rate of 375 rds./min. It has a muzzle velocity of 244 m./min. (800 ft./min.), for an effective range of 1600 m. (1750 yd.). HE, fragmentation, phosphorus, HEAT, and anti-personnel (flechette) grenades are issued; all weigh roughly 1 lb. (0.4 kg.) and have a lethal radius of 10–15 m. (33–50 ft.).

GRIPEN: The Saab JAS.39 fighter/attack/reconnaissance aircraft under development for the Swedish air force. Designed from the early 1970s as a replacement for the Saab 37 VIGGEN, the Gripen incorporates state-of-the-art technology, closely tailored to specific Swedish requirements, including the need to take off from, and land on, short stretches of highway even in very poor weather conditions. The Gripen program, however, seems to have pushed the Swedish aircraft industry beyond its limits, with concomitant delays and cost overruns. Originally intended to fly in 1987, the first prototype did not in fact begin flight tests until late 1988. When it was destroyed in a landing accident in February 1989, the entire program was placed in jeopardy; if it does proceed, current plans call for a force of 140 aircraft (including 25 two-seaters) by the mid-1990s.

Much of the fuselage of the single-seat, single-engine Gripen and the entire wing are fabricated of carbon-fiber composites to reduce weight. The fuselage has an area-ruled "coke bottle" shape to reduce trans-sonic drag, and wedge-shaped lateral engine intakes. The nose houses an Ericcson-Ferranti multi-mode pulse Doppler RADAR with a lightweight carbon-fiber planar-array antenna that has SYNTHETIC APERTURE, ground-mapping, and TERRAIN-AVOIDANCE ability.

The pilot sits immediately behind the radar in a well-appointed cockpit beneath a bulged clamshell canopy with a single-piece windscreen. Visibility is generally good except in the rear hemisphere, which is obstructed by a fuselage spine. Cockpit AVIONICS are extremely advanced, and include three full-color multi-functional displays instead of multiple dials, a wide-angle "holographic" HUD, an INERTIAL NAVIGATION unit, and a digital weapon delivery/air data computer. A 27-mm. Mauser BK-27 cannon is mounted on the left below the cockpit. The fuselage behind the cockpit houses an avionics bay, fuel tanks, and the engine, a Svenska Flygmotor RM12 engine, a license-built version of the General Electric F404. Gripen has rugged "high flotation" landing gear for rough-field operations, housed in fuselage wells.

The midmounted wings have a cropped-delta planform with full span leading-edge flaps and full span elevons for outstanding short takeoff performance. The wings are unusual in having a "supercritical" cross-section for efficient transsonic cruising, and because they do not extend to the rear of the fuselage. Two large, power-operated canards with landing flaps, mounted above and ahead of the wings, give the aircraft a double-delta planform reminiscent of the Viggen, but the Gripen is a "control-configured vehicle" (ccv) with relaxed static stability (i.e., it is inherently unstable in level flight) and computer-driven, triply-redundant FLY-BY-WIRE (FBW) controls make it very agile. FBW software difficulties have delayed the program.

The Gripen has one fuselage, two wingtip, and four wing pylons. The fuselage pylon is meant for a FORWARD-LOOKING INFRARED (FLIR) pod, and the wingtip pylons for RB24 (AIM-9) SIDEWINDER infrared-homing air-to-air missiles. Total payload capacity is approximately 14,000 lb. (6363 kg.), and the four wing pylons could accommodate AIM-120 AMRAAM or British Aerospace SKYFLASH radar-guided air-to-air missiles, free-fall and LASER-GUIDED BOMBS, RBS-15 anti-ship missiles, AGM-65 MAVERICK TV- and laser-guided air-to-surface missiles, jamming pods, or multi-spectral reconnaissance pods.

Specifications Length: 45.9 ft. (14 m.). **Span:** 26.25 ft. (8 m.). **Powerplant:** Svenska Flygmotor RM12 afterburning turbofan, 18,000 lb. (8182 kg.) of thrust. **Weight, normal:** 17,637 lb. (8017 kg.). **Weight, max. takeoff:** 25,000 lb. (11,363 kg.). **Speed, max.:** Mach 1 (760 mph/1270 kph) at sea level/Mach 2 (1323 mph/2209 kph) at 36,000 ft. (11,000 m.). **Initial Climb:** na. **Service ceiling:** na. **Combat Radius:** na. **Range, max.:** na.

GRISHA: NATO code name for a class of 65 Soviet ANTI-SUBMARINE WARFARE (ASW) CORVETTES, the first of which was completed in 1968. Designated "Small Anti-Submarine Ships" (*Malyy protivolodochnyy korabl'*, MPK), the Grishas, intended for inshore ASW and border patrol, are gradually replacing the older MIRKA and PETYA classes.

There are four distinct subclasses, which differ in armament and sensors. The Grisha I ASW corvettes, 15 of which were built between 1968 and 1974, were followed from 1974 by 12 Grisha II border patrol vessels (*Pogranichnyy Storozhevoy Korabl'*, PSKR) manned by the KGB Maritime Border Guard. From 1975 to 1986, the principal production model was the Grisha III ASW corvette, of which 31 were built. These were superseded in 1985 by the Grisha V ASW corvette, of which approximately 7 have been delivered to date (some to the KGB). All subtypes share the same hull and machinery. Flush-decked, with extreme shear at the bow and a broad transom stern, they are quite seaworthy for their size. The rather boxy superstructure is concentrated forward, except for a small deckhouse aft.

The Grisha Is are armed with a twin SA-N-4 GECKO surface-to-air missile launcher on the bow, a twin 57-mm. ANTI-AIRCRAFT gun aft, 2 RBU-6000 ASW rocket launchers before the bridge, 2 twin 21-in. (533-mm.) tubes for ASW homing TORPEDOES amidships, and 2 DEPTH CHARGE racks on the fantail. The Grisha II has a second twin 57-mm. gun forward, replacing the SA-N-4 launcher. The Grisha III is similar to the Grisha I, but has a 30-mm. radar-controlled gun atop the aft deckhouse. The Grisha V is similar to the III, but has a single 76.2-mm. DUAL PURPOSE gun in place of the twin 57-mm., and only one RBU-6000. See also FRIGATES, SOVIET UNION.

Specifications Length: 236.2 ft. (72 m.).
Beam: 32.8 ft. (10 m.). **Draft:** 12.1 ft. (3.7 m.).
Displacement: 850–860 tons standard/1100–1150
tons full load. **Powerplant:** triple-shaft CODAG: 1
19,000-shp. gas turbine, 2 19,500-hp. M504 diesels. **Speed:** 31 kt. **Range:** 450 n.mi. at 30 kt./4000
n.mi. at 16 kt. **Crew:** 60. **Sensors:** 1 "Strut Curve"
air-search radar, 1 "Pop Group" SA-N-4 missile
guidance radar, 1 "Muff Cob" fire control radar, 1
hull-mounted medium frequency sonar, 1 high-
frequency dipping sonar; (II) no Pop Group; (III)
1 "Bass Tilt" fire control radar in place of Muff
Cob; (V) 1 "Strut Pair" air-search radar instead of
Strut Curve. **Electronic warfare equipment:** 2
"Watch Dog" electronic countermeasure arrays, 1
"High Pole" IFF set; (V) 2 16-barrel chaff launchers.

GROUND-CONTROLLED INTERCEPT: A
FIGHTER control technique whereby the pilot receives detailed instructions from a ground-based
controller to guide him to his target. The pilot is
left little (if any) latitude for deviation with regard
to speed, course, or altitude; the technique therefore lacks the flexibility to deal with a fluid or
confused situation (especially if the controller's
radar or communications are jammed or otherwise
disrupted). On the other hand, it is the only
method whereby aircraft without (or with very
limited) radar can accomplish interceptions at
night or in bad weather; it also allows radar-
equipped fighters to operate without emitting
radar signals until the last minute, and is generally
useful when pilot quality is low. The effectiveness
of GCI largely depends on the skill of the controller, his ability to interpret radar data, and his intui-
tive grasp of the tactical situation. GCI can be fully
or partically automated, as in the U.S. SAGE (Semi-
Automatic Ground Environment), in which a computer directed the aircraft via DATA LINKS to the
intercept point, and even indicated the correct moment to launch missiles.

GCI is not much used by U.S. or other Western
air forces (although controllers do assist pilots by
providing vectors), but it was, until recently, the
norm in Soviet and Warsaw Pact air forces, and is
still used throughout the Third World.

GROUND EFFECT VEHICLE: A generic term
for HOVERCRAFT and other machines which fly exclusively within the ground effect, or on an air air
cushion (i.e., up to several m. in height). See also
WING-IN-GROUND-EFFECT.

GROUND ZERO: Sometimes "designated
ground zero," the point on the surface of the earth

(or vertically above or below it) that is the epicen-
ter of a nuclear detonation. Nuclear effects are
usually measured as radii from ground zero at
which certain phenomena occur. See also NUCLEAR
WARHEADS, EFFECTS OF.

GROUP: 1. In U.S. Air Force terminology, a
formation of several squadrons, without the organic ground support of a WING; usually pertains to
Air National Guard or Air Force Reserve units. See
also AIR FORCE, UNITED STATES.

2. In the Royal Air Force, a major operational
command dedicated to a single mission or area of
operations.

3. In most other air forces, a multi-squadron
formation, generally equivalent to a USAF wing.

**GROUP OF SOVIET FORCES IN GERMANY
(GSFG):** Lately redesignated "Western Group of
Forces," GSFG is the largest of the so-called
"Groups of Forces Abroad," and includes all So-
viet army and air units deployed in East Germany.
GSFG is commanded by a full general or a marshal
of the Soviet Union. In war, GSFG would form the
lead echelon of the Western Theater of Military
Operations (TVD), which controls all Soviet and
Warsaw Pact forces in East Germany, Poland,
Czechoslovakia, and Poland, as well as the Baltic,
Byelorussian, and Carpathian MILITARY DISTRICTS
of the U.S.S.R.

Positioned in East Germany, GSFG was admirably situated for multiple strategic roles. First, it
represented a standing threat to NATO, and
thereby provided leverage in Soviet negotiations
with Western Europe; second, it secured Eastern
Europe for the Soviet empire more effectively than
if its forces were distributed as occupation troops,
because its concentrated military power exceeded
that of any single client state; third, GSFG was a
physically interposed between potentially rebel-
lious subjects and any aid that might come to them
from the West. The elimination of GSFG in the
wake of German reunification is therefore especially significant.

GRU: *Glavnoe Razvedyvatelnoye Upravlanie,*
or Main Intelligence Directorate, the Soviet military INTELLIGENCE organization directly controlled
by the General Staff. GRU has primary responsibility for the collection of military intelligence and
of technology with military applications. Unlike
the KGB, it has no internal security function, but the
two organizations are unfriendly competitors in
external espionage, for which the GRU reportedly
has even greater resources than the KGB.

Controlled by a chief who usually holds the rank

of full general, the GRU is divided into several directorates, geographic "directions," and independent departments. The first deputy chief controls the 1st–6th Directorates and the 1st–4th Directions, with primary responsibility for human intelligence (HUMINT), including ESPIONAGE. The First Directorate carries out HUMINT in Europe (except the UK) and consists of five "directions," each of which is responsible for a specific country or region. The Second Directorate is responsible for HUMINT in the Western hemisphere, the UK, and Australia; the Third Directorate for Asia; and the Fourth for Africa and the Middle East. The Fifth Directorate coordinates the operational intelligence activities of Soviet Military Districts, Fleets, and Groups of Forces abroad; through their Second (Intelligence) Departments, it also indirectly controls SPETSNAZ reconnaissance and diversionary activities. The Sixth Directorate controls electronic intelligence (ELINT) collection.

The four "directions" under the first deputy chief perform the same general functions. The First and Second directions control HUMINT in Moscow and Berlin; the Third runs agents in "National Liberation Movements" and terrorist organizations; the Fourth runs agent networks in the Americas from a headquarters in Cuba (and can also draw upon the resources of the Cuban intelligence service, the DGI). These agent networks have both "legal" agents with some form of diplomatic cover, who operate out of a "Residence" in Soviet embassies and consulates, and "illegal" agents, who are foreign nationals recruited by the GRU, or Soviet nationals operating under "deep cover."

The GRU chief directly controls the Directorate of Cosmic Intelligence, responsible for satellite surveillance (which also establishes requirements for the design, construction, and launch of surveillance satellites), as well as the analytical, administrative, and internal security departments of the GRU.

GRUMBLE (SA-10): NATO code name for a Soviet mobile, long-range surface-to-air missile (SAM), which also has a shipboard variant (SA-N-6). Introduced in 1980, Grumble is a very fast, highly maneuverable missile, capable of turns of up to 100 g, reportedly developed to attack low-flying aircraft, CRUISE MISSILES, and supersonic targets such as the U.S. Short-Range Attack Missile (SRAM). It is currently deployed in two versions, one static (SA-10a) for the defense of fixed targets, and the other mobile (SA-10b) for the air defense

of field forces. It is believed that the Grumble, together with the SA-12A/B GLADIATOR/GIANT, will eventually replace most older SAMs (SA-2 GUIDELINE, SA-3 GOA, and SA-5 GAMMON) for homeland defense; in forward areas it would also replace the SA-3 for airfield defense. Grumble also has some inherent anti-tactical ballistic missile (ATBM) capability, complementing the more capable SA-12B Giant. Very little is known as yet about Grumble's effectiveness or its vulnerability to ECM. U.S. analysts believe that the system incorporates "transferred" Western technology.

No photographs of Grumble have been released to date. Reportedly, it is a two-stage, solid-fuel missile with four narrow-span, broad-chord delta wings and four small tail fins. Grumble relies on TRACK-VIA-MISSILE (TVM) guidance, an unusual and advanced technique also used with the U.S. PATRIOT, whereby the target is illuminated by a ground-based radar, and the radar reflections received by an antenna in the nose of the missile are transmitted via DATA LINK to a FIRE CONTROL computer on the ground, which generates steering commands that are transmitted back to the missile via the data link. This form of guidance can be highly accurate and is extremely resistant to ELECTRONIC COUNTERMEASURES (ECM) as compared to systems where the missile's own (small) guidance unit processes the radar reflections. In addition, Grumble also has active radar homing for terminal guidance, and to provide a look-down capability against low-flying cruise missiles. There are two alternative payloads: a 200-lb. (90.9-kg.) high-explosive warhead with radar proximity (VT) fuze, or a 5- to 10-KT nuclear warhead (the latter most likely for strategic defense).

Grumble is launched from a self-propelled or towed 4-round transporter-erector-launcher (TEL). The self-propelled version is based on a MAZ-7310 8 × 8 truck chassis, with an armored cab housing power supplies and fire controls. The missiles are housed in sealed storage/launch canisters stowed horizontally for travel and elevated vertically off the rear of the TEL for launch. Three or four TELs and a "Flap Lid" PHASED ARRAY engagement radar (used to track the missile in flight) form a battery. The radar can also be either trailer-mounted or self-propelled on a MAZ-7310 chassis. Three batteries plus a "Clam Shell" 3-dimensional search and acquisition radar comprise a battalion. One battalion, a "Big Bird" long-range surveillance radar, a fire control unit, a headquarters, and service units form an air defense regiment.

The navalized SA-N-6, the primary anti-air armament of the KIROV battle cruisers and SLAVA cruisers, is fired from a VERTICAL LAUNCH SYSTEM with an 8-round rotary magazine. The Kirovs have a battery of 12 launchers, while the Slavas have 8. Missile guidance is provided by a "Top Dome" fire control group, which includes one tracking radar with multiple TRACK-WHILE-SCAN capability, and up to four separate illuminating radar/command guidance links. The naval Grumble provides an unprecedented degree of coverage against anti-ship missiles and aircraft from wavetop height to 27,500 m. (90,000 ft.).

Specifications   **Length:** 23 ft. (7 m.). **Diameter:** 1.65 ft. (503 mm.). **Weight, launch:** 4950 lb. (2250 kg.). **Speed:** Mach 6 (4000 mph/6680 kph). **Range envelope:** 2.5–100 km. (1.55–62 mi.). **Height envelope:** 300–30,400 m. (984–99,712 ft.).

GSFG: See GROUP OF SOVIET FORCES IN GERMANY.

GSG-9:   *Grenzschutzgruppe* (Border Police Group)-9, an elite West German counterterrorist and hostage-rescue unit established in 1973 after the massacre of Israeli athletes at the 1972 Munich Olympics demonstrated the inability of ordinary police forces to cope with terrorist incidents. Rated as one of the best units of its type (together with the British SAS, U.S. DELTA FORCE, and Israeli SAYARET MATKAL), GSG-9 is technically part of the Border Police (*Bundesgrenzschutze*) of the German Ministry of the Interior.

The organization and tactics of GSG-9 were established by its first commander, Col. (now Brig. Gen.) Ulrich Wegener, who had personally trained with both the U.S. FBI and the Israeli army. The unit is believed to have a total of 180–220 men, divided into a headquarters, a communication and intelligence team, engineer, weapons, training, and support team, and four strike groups (*Specialeinsatztruppe*) of 30–42 men each.

Personnel are selected from volunteers already in the Border Police. Those who pass physical, intelligence, and attitude tests undergo a five-month initial training course stressing physical fitness, firearms, and martial arts. Those who pass proceed to a final three-month training course emphasizing specialized skills, teamwork, and assault tactics.

GSG-9 has participated in several counterterrorist operations, but its most famous exploit was the storming of a highjacked Lufthansa airliner at Mogadishu, Somalia, on 17 October 1977, in which 91 hostages were rescued.

GUERRILLA (WARFARE):   Military operations conducted by irregular forces operating within territory nominally controlled by the enemy (often the country's government), whether independently or as an adjunct to regular military operations elsewhere.

Guerrilla (Spanish for "little war") tactics are based on temporary concentrations on the offensive, followed by dispersion on the defensive, either in a "clandestine" mode with concealment by natural cover (often jungle) or in a "covert" mode, i.e., in civilian disguise. Guerrilla operational methods are those of any interior force; they are characterized by the avoidance of pitched battles and the slow attrition of superior forces, by repeated attacks on exposed outposts, and by ambushes along lines of communication. The key factor in guerrilla is the source of supplies, given the lack of a logistic base. In REVOLUTIONARY WAR, the guerrilla is only an adjunct to SUBVERSION, and supplies are extracted from the local population by the covert "administration." When the guerrillas do have a logistic base provided by allies, supply lines are the problem. The solution may be either clandestine land lines (as with the Vietcong and MUJAHIDEEN) or air drops (as with many resistance forces in World War II).

Subversion is also a source of the other essential commodity needed by guerrillas intent on more than mere survival: local intelligence. Thus in the revolutionary war equation (guerrilla warfare + subversion + national political aims), subversion is the crucial factor.

GUIDED MISSILE: See MISSILE, GUIDED.

GUIDELINE (SA-2):   NATO code name for the Soviet V75 *Dvina*, an early long-range surface-to-air missile (SAM), also deployed in small numbers in a shipboard version (SA-N-2). Introduced in 1957 as a replacement for the original SA-1 Guild, the SA-2 first attracted worldwide attention on 1 May 1960 when a salvo of several missiles shot down a U.S. U-2 reconnaissance aircraft over the Soviet city of Sverdlovsk. Later the SA-2 was much used in combat, forming the backbone of the North Vietnamese and Egyptian air defense networks. Now obsolescent, Guideline is still fielded in large numbers by Soviet and Warsaw Pact forces, and has been widely exported.

Guideline is a large, two-stage missile designed primarily to engage high-altitude bombers. The first stage, a solid-fuel rocket booster, burns for about three seconds after launch, accelerating the missile to flying speed before being jettisoned. The second stage is a liquid fuel rocket with four small, fixed nose canards, four cruciform delta wings, and four small cruciform tail fins for steering. Armed with a 286-lb. (130-kg.) high-explosive warhead (lethal radius 13.5 m.) fitted with both radar proximity (VT) and contact fuzes, it is controlled by radio COMMAND GUIDANCE. Several variants were introduced between 1958 and 1968 (SA-2b–SA-2f), with slightly better performance in the later models.

All versions are fired from single-rail launchers which can be towed but are generally emplaced in fixed positions around important targets. Six launchers form a battery, inevitably deployed in a distinctive "Star of David" configuration. One battery, plus a headquarters, a FIRE CONTROL unit with a "Fan Song" missile guidance radar, and a service unit form an SA-2 battalion. Three battalions, a headquarters, service units, a "Spoon Rest" surveillance radar, and a "Side Rest" height-finding radar form an SA-2 regiment, several of which were attached to each Soviet Army or Front.

Incoming targets are initially detected by the regimental Spoon Rest radar, which passes range and bearing data to the battery Fan Song radar via land lines or radio. Each Fan Song can track up to six targets simultaneously, but can engage only one at a time. The Fan Song passes tracking data back to the battalion fire control unit; once the target is in range, a missile is launched. The Fan Song then tracks both the target and the missile, passing the data to the fire control unit, which generates guidance commands that are transmitted to the missile over an analog, narrow-band UHF DATA LINK; the missile must pick up the data link within six seconds of launch or it will "go ballistic." The guidance system is relatively inaccurate, the median error radius (CEP) being about 75 m. (roughly five times the lethal radius of the warhead).

Guideline had some initial success after its combat debut in North Vietnam in 1965, but the kill probability (P_K) declined to 1.0–1.5 percent as soon as the U.S. developed specific ELECTRONIC COUNTERMEASURES. The Fan Song radar and the UHF guidance link were especially vulnerable to jamming and could be attacked by ANTI-RADIATION MISSILES. In addition, most fighters and attack aircraft could outmaneuver a Guideline if their pilots could spot it in time. Combat experience in the Middle East and Indo-Pakistani Wars yielded similar results. But Guideline's deterrent value persists: it forces aircraft to fly at lower than optimal altitudes (and into the lethal radius of more effec-

tive weapons), or to jettison ordnance in order to perform evasive maneuvers; i.e., the missile denies the enemy a "free ride" to the target.

The SA-N-2 navalized variant, intended to complement the medium-altitude SA-N-1 Goa, was fitted to the converted Sverdlov-class cruiser *Dzerzhinsky* in 1962. Fired from a massive twin-rail launcher with a 10-round horizontal magazine behind the launcher, the system was found to be too bulky for shipboard use.

Specifications **Length:** 35.45 ft. (10.8 m.). **Diameter:** 25.98 in. (660 mm.). **Span:** (booster) 7.2 ft. (2.2 m.); (sustainer) 5.6 ft. (1.7 m.). **Weight, launch:** 5071 lb. (2305 kg.). **Speed:** Mach 3 (2000 mph/3340 kph). **Range envelope:** 9.3–35 km (5.76–21.7 mi.). **Height envelope:** 4500–28,000 m. (14,-760–91,840 ft.). **Operators:** Afg, Alg, Cuba, Eth, Ind, Iraq, Lib, Mon, N Kor, Som, Sud, Syr, Viet, USSR and other Warsaw Pact forces, Yug.

GUN: An ARTILLERY piece with a barrel length greater than 40 CALIBERS, intended to fire at relatively low angles with a high muzzle velocity and a flat trajectory. On both towed and self-propelled mounts, guns are employed for long-range indirect fire, but specialized types such as ANTI-AIRCRAFT guns, ANTI-TANK guns, and the high-velocity guns that form the main armament of TANKS are employed for DIRECT FIRE.

GUNBOAT: A small combat vessel (of less than 1000 tons displacement), meant for inshore and riverine patrol, and low-intensity warfare in general; the type is now in fact obsolete, having been replaced by CORVETTES and various kinds of FAST ATTACK CRAFT, including motor gunboats (PG) and missile boats (PGM).

GUNFIRE SUPPORT, NAVAL: Shore bombardment by naval vessels in support of ground forces, usually thought of as an adjunct to AMPHIBIOUS operations, but also employed for RECONNAISSANCE, demonstrations and feints, RAIDS, the suppression of enemy air defenses (SEAD), and INTERDICTION of coastal roads, railroads, airfields, and assembly areas.

Historically, naval guns have been of heavier caliber than most field artillery (up to 18.1 in.), and generally have higher rates of fire because of their power-operated loading gear. Except for the 16-in. (406-mm.) guns of the U.S. IOWA-class battleships and the 6-in. (152-mm.) guns of old Soviet SVERDLOV-class cruisers, most naval guns nowadays are of 4- to 5-in. (100- to 127-mm.) caliber, but these are DUAL PURPOSE guns with high muzzle velocities and very high rates of fire (40–60 rds./min.), which can fire a weight of shells equivalent to that of an entire battalion of 155-mm. HOWITZERS (18 howitzers × 3 rounds per tube/min. = 54 rds./min.). In addition, naval guns are usually radar-directed and can therefore be extremely accurate.

At short range, with a clear line of sight between ship and target (as when beach defenses are engaged), DIRECT FIRE is employed, using the ship's own sights and fire controls; such fire can be devastating because of its rapidity and accuracy. But the range of naval guns (typically 15 mi. for 5-in. guns, but as much as 24 mi. for 16-inchers) also allows ships to engage targets far inland; such INDIRECT FIRE relies on FORWARD OBSERVERS for spotting (U.S. battleships would now employ their Israeli-made REMOTELY PILOTED VEHICLES).

Firing at stationary land targets is much the same as firing at a ship dead in the water, except for the factor of terrain elevation. For indirect fire, however, the ship's position must be plotted very accurately; this has become relatively easy with the use of INERTIAL NAVIGATION, satellite GLOBAL POSITIONING SYSTEMS, and radio navigation aids.

GUN-HOWITZER: An ARTILLERY piece with a barrel length of 30–40 CALIBERS—i.e., an intermediate type between short-barrel, low-velocity HOWITZERS and long-barrel, high-velocity GUNS, with the high-angle fire of the former and much of the range of the latter. The type was pioneered by the British 25-pounder of 1938, and has since become the predominant form of field artillery (both towed and self-propelled) between 122 mm. and 203 mm. in bore.

GUNSHIP: A transport aircraft heavily armed with automatic weapons (or even light howitzers) to provide abundant and cheap fire support in permissive environments safe from fighters and antiaircraft weapons. The first gunship, the AC-47 "Puff the Magic Dragon," was a converted DAKOTA armed with six side-firing 7.62-mm. GAU-2A miniguns. Used by the U.S. in Vietnam from 1967 to 1973, a few still serve with the El Salvador air force. The AC-47 was superseded by the larger AC-119, and finally by the AC-130 Spectre, a converted HERCULES transport fitted with 20-mm. VULCAN cannon, 40-mm. BOFORS GUNS, a 105-mm. howitzer, and various night-vision sensors. These serve in U.S. Air Force Special Operations squadrons.

GUPPY: Greater Underwater Propulsion Program, a late-1940s U.S. Navy program to improve the submerged speed and endurance of World War II "Fleet" SUBMARINES; boats modified under the program were inevitably known as "Guppies." Several remain in service with smaller navies.

Postwar evaluation had revealed that the German Type XXI U-boats of 1944–45 were much superior in submerged speed and endurance, maximum depth, maneuverability, and acoustical silencing; they also incorporated valuable innovations such as retractable snorkels (for diesel operation from shallow submergence) and low-frequency passive sonars. Funds being unavailable for new construction, the GUPPY program was initiated in 1947 with the conversion of two Tench-class boats to the Guppy I configuration, which involved the removal of deck guns, streamlining of the hull and conning tower, and doubling the capacity of the storage batteries. Surface speed was slightly reduced (from 20 to 17.8 kt.), but submerged speed was doubled to 18.2 kt. Essentially experimental boats with only one periscope and no snorkel, the Guppy I was superseded by the fully operational Guppy II, to which standard 24 Tench and Balao boats (including the two Guppy Is) were converted between 1948 and 1950.

The Guppies retain their original cigar-shaped, double-hull configuration, but all surface protuberances were removed or made retractable to minimize drag. The conning tower was rebuilt as a streamlined, stepped structure ("sail") enclosing the bridge and periscopes. Control surfaces remained unchanged: folding bow planes, fixed stern planes, and a single rudder behind the propellers. Main armament remained 10 21-in. (533-mm.) torpedo tubes (6 forward, 4 aft) with 24 TORPEDOES. Sensors were extensively updated to include a circular medium-frequency SONAR array under the bow (with a separate sonar compartment in the forward torpedo room), and new FIRE CONTROLS. The powerplant remained the standard Tench unit, but four 126-cell batteries doubled underwater endurance, while a retractable snorkel allowed submerged diesel operation. All these modifications left them very cramped.

The Guppy II conversion proved too expensive ($2.5 million each in 1950 dollars), so several more-austere schemes were also implemented. The Guppy IA configuration (10 Balao/Tench boats converted in 1951–52) was similar to the Guppy II but had only two storage batteries of increased capacity, while sensor and fire control upgrades were not as extensive. The Guppy IIA configuration (16 boats converted between 1952 and 1954) removed one diesel engine to make space for additional sonar and fire control equipment. Speed and endurance were slightly reduced, but habitability and maintainability improved greatly. The Guppy IB, an even more austere conversion intended for transfer to foreign navies, was generally similar to the IA, but with simplified equipment. Four boats were modified to this standard between 1953 and 1955.

The Guppy III program was initiated in 1959 as a service life extension refit for Guppy IIs (nine were so modified between 1959 and 1962) pending the introduction of nuclear attack submarines. A 15-ft. (4.57-m.) extension was inserted in the hull to provide space for additional sonar and other electronic equipment, and to allow for additional reloads in the forward torpedo room. The stepped sail was replaced by a tall rectangular structure of glass-reinforced plastic with a lengthened conning tower compartment for additional fire controls; and all auxiliary machinery was refurbished. A BQG-4 PUFFS passive ranging sonar was also installed. Despite these modifications, the Guppy IIIs remained in service with the U.S. for only a few years, the last being retired in 1975. See also SUBMARINES, UNITED STATES.

Specifications **Length:** (II) 307 ft. (93.6 m.); (III) 322 feet (98.1 m.). **Beam:** 27.33 ft. (8.33 m.) **Displacement:** (II) 2075 tons standard/2420 tons submerged; (III) 2870 tons submerged. **Powerplant:** twin-shaft diesel-electric: 4 Fairbanks-Morse diesels, 4610 hp., 2 direct-drive electric motors, 2740 hp. **Speed:** 18 kt. surfaced/16 kt. submerged. **Fuel:** 330 tons. **Range:** 10,000 n.mi. at 10 kt. (snorkel); 95 n.mi. at 9 kt. (batteries). **Max. operating depth:** 400 ft. (122 m.). **Collapse depth:** 600 ft. (183 m.). **Crew:** 85. **Sensors:** (typical) 1 BQR-2B medium-frequency passive sonar, 1 BQS-4 active sonar, 1 SS-2A surface-search radar, 2 periscopes. **Operators:** Bra (2 Guppy I, 2 Guppy III), Gre (1 Guppy IIA, 1 Guppy III), Peru (2 Guppy IA), Tai (2 Guppy II), Tur (5 Guppy IIA, 2 Guppy III), Ven (2 Guppy II in reserve).

GURKHAS: Nepalese tribesmen who serve as volunteers in special regiments of the British and Indian armies. Renowned for their courage, loyalty, discipline, endurance, and ferocity in battle, Gurkhas were first recruited by the British East India Company in 1815, then transferred to the (British) Indian Army after the Great Sepoy Mutiny of 1857. After Indian independence in 1947, four of the ten Gurkha regiments (2nd, 6th, 7th, and 10th) were taken into the British army, while the rest were absorbed by the Indian army. Of the British Gurkhas, four battalions organized into the Gurkha Brigade form the main garrison of Hong Kong; one battalion is "on loan" under contract to the Sultan of Brunei; and one battalion is stationed in Britain as part of the 5th Brigade.

Gurkhas are among the world's finest infantry, having served with distinction in India, in various places during both world wars, in Korea, in Malaya, and, most recently, in the Falklands. After prolonged basic training (40+ weeks), Gurkhas are armed, trained, and organized on standard British infantry lines. Though pay is limited to Indian rates under the original tripartite agreement, there is no shortage of volunteers for British service. The transfer of Hong Kong to China in 1996 has once again raised questions about the future of the Gurkha Brigade, and the force has its critics; but all previous attempts to eliminate these tough little warriors from the British army have failed.

GWEN: Ground-Wave Emergency Network, a U.S. radio system meant to ensure communications between the NATIONAL COMMAND AUTHORITIES (NCA) and the strategic nuclear forces in the event of a nuclear attack upon the United States. The GWEN program, initiated in 1981, will eventually comprise some 200 individual, unmanned transmitter/receiver relay sites across the continental United States. Each site consists of a 300-ft.-tall transmitter tower and three hardened shelters housing communications equipment and electrical generators. The GWEN system employs low frequency (LF) radio ground waves, which follow the curvature of the earth and are largely immune to nuclear BLACKOUT and ELECTROMAGNETIC PULSE.

H

H&I: See HARASSMENT AND INTERDICTION FIRE.

HADES: French nuclear-armed short-range (tactical) BALLISTIC MISSILE, under development to replace the Pluton missile, with deployment scheduled for 1992. France originally planned to replace its five operational Pluton regiments with four Hades regiments grouped together into an artillery division (complete with howitzer, anti-tank, and air defense battalions) under the direct control of the armed forces chief of staff. Due to budgetary constraints, the planned number of regiments has been reduced to three, each placed under the operational control of an army CORPS.

As compared to Pluton, Hades offers longer range, improved accuracy, and greater warhead yield (60 κT). The missile is housed in sealed box canister/launchers on a semi-trailer transporter-erector-launcher (TEL) towed by a 4 × 4 truck. The missiles are stowed horizontally for traveling and raised vertically off the rear of the vehicle for launch. The trailer also carries all necessary FIRE CONTROL equipment, including a data link to the NATIONAL COMMAND AUTHORITIES. The cab of the vehicle contains a COLLECTIVE FILTRATION system for NBC protection. Flight tests of Hades began in 1988–89.

HALIFAX: A class of six Canadian FRIGATES (FFs) scheduled for completion in 1989–90. Ordered in 1977 as replacements for six ST. LAURENT-class frigates commissioned in 1956–57, the Halifax program has been delayed more than four years by political and economic difficulties.

Intended primarily as ANTI-SUBMARINE WARFARE (ASW) escorts with secondary ANTI-SURFACE WARFARE (ASUW) capability, they are armed with 1 twin 57-mm. anti-aircraft gun, a 16-round VERTICAL LAUNCH SYSTEM (VLS) for SEA SPARROW short-range surface-to-air missiles, a 20-mm. PHALANX radar-controlled gun for anti-missile defense, 2 quadruple launch canisters for HARPOON anti-ship missiles, and 2 sets of Mk.32 twin torpedo tubes for lightweight ASW TORPEDOES. The primary ASW weapon is an SH-60B SEA HAWK helicopter, operated from a quarterdeck landing deck with a hangar amidships. A Canadian "Bear Claw" haul-down device allows safe helicopter landings even in rough seas. See also FRIGATES, CANADA.

Specifications **Length:** 439.5 ft. (134 m.). **Beam:** 48.2 ft. (14.7 m.). **Draft:** 14.2 ft. (4.33 m.). **Displacement:** 3886 tons standard/4254 tons full load. **Powerplant:** 2 25,000-shp. LM-2500 gas turbines, 2 shafts. **Fuel:** 479 tons. **Speed:** 29.2 kt. (27 kt. sustained). **Range:** 4500 n.mi. at 20 kt./5700 n.mi. at 15 kt. **Crew:** 226. **Sensors:** 1 Raytheon 1629C navigational radar, 1 CMR-1820 air search radar, 1 Type 1031 surface-search radar, 1 WM-25 fire control radar, 1 SQS-505 hull-mounted medium-frequency sonar, 1 SQR-18 TACTASS towed sonar array. **Electronic warfare equipment:** 1 CANEWS radar warning receiver, 4 Mk.36 SRBOC chaff launchers.

HALO: NATO code name for the Mil Mi-26, a Soviet heavy lift transport and cargo HELICOPTER. Introduced in 1979, and still the world's largest and most powerful helicopter, the Mi-26 was designed as a replacement for the Mi-6 HOOK. Halo is currently in production for Soviet Military Transport Aviation (VTA) and Aeroflot, and several hundred may eventually be built. India is the only export customer so far.

The fuselage has an unpressurized cargo hold equipped with two 5500-lb. (2500-kg.) electric winches running on overhead rails for cargo handling. In the rear of the hold, a fold-down cargo ramp and two hydraulically powered clamshell doors permit the roll-on loading of vehicles and palletized cargo. Maximum payload capacity is 44,100 lb. (20,045 kg.), carried either internally or slung from underbelly cargo hooks; an overload of up to 55,115 lb. (25,052 kg.) can be carried for short distances. Typical payloads include BMD airborne armored fighting vehicles, ASU-85 airborne assault guns, a wide range of trucks, trailers, artillery pieces, and missile systems, or up to 104 passengers in 44 permanent tip-up seats mounted along the cabin sides and 60 temporary seats bolted to the floor.

The crew consists of a pilot, copilot, navigator, and engineer seated in an airline-type cockpit above, and a loadmaster in the hold. The cockpit windscreens have electric deicing (essential for operations in Siberian winters), and the cockpit side windows are bulged to provide visibility downward and to the rear for the observation of slung loads. Cockpit AVIONICS are quite comprehensive, and include a multi-mode mapping and weather RADAR, DOPPLER moving-map displays, an advanced, multi-channel digital FLY-BY-WIRE flight control system with autostabilization and autohover capability, and several television monitors for cameras in the nose, tail, and belly. A flight deck indicator monitors grounded gross weight through sensors in the landing gear struts, which are hydropneumatically controlled to raise or lower the height of the cabin floor for easier loading. A four-seat passenger compartment for relief crews is located behind the cockpit.

To achieve its enormous lifting capability, Halo incorporates much advanced technology, including a forged titanium alloy hub for eight blades. Fabricated from steel-tube spars and Nomex-filled glass fiber airfoil sections, the blades have titanium leading edges and electric deicing strips. The cargo floor is also fabricated from titanium alloy for great strength at reduced weight. Fuel is stored in eight integral tanks beneath the cabin floor. The engines are equipped with centrifugal dust filters and bleed air deicing.

Specifications Length: 110.65 ft. (33.75 m.). **Rotor diameter:** 104.95 ft. (32 m.). **Cargo hold:** 39.33 × 10.67 × 10.4 ft. (12 × 3.25 × 3.17 m.). **Powerplant:** 2 11,400-shp. Lotarev D-136 turboshafts. **Weight, empty:** 62,169 lb. (28,258 kg.). **Weight, normal loaded:** 109,127 lb. (49,500kg.). **Weight, max. takeoff:** 123,457 lb. (56,000kg.). **Speed, max.:** 183 mph (305 kph). **Speed, cruising:** 158 mph (264 kph). **Ceiling:** (hover) 5900 ft. (1799 m.) (service) 15,100 ft. (4604 m.). **Range:** (max. payload) 497 mi. (830 km.).

HALO: High Altitude-Low Opening, a parachute insertion technique employed by some elite forces, whereby troops jump from aircraft flying at more than 20,000 ft. (6098 m.), but delay opening their parachutes until they have descended below 1000 ft. (305 m.). Using skydiving methods, HALO-parachuted troops can glide a significant distance from their jump point to evade detection. Rectangular controllable parachutes, or "parafoils," also allow trained users to land with pinpoint accuracy. See also SAS; SEAL; SPETSNAZ.

HAMBURG: A class of four West German DESTROYERS commissioned between 1964 and 1968. The first destroyers built in Germany after World War II, the Hamburgs are general-purpose escorts optimized for operations in the Baltic.

Attractive flush-decked vessels with low freeboard and a voluminous superstructure, as completed the Hamburgs were armed with 4 100-mm. DUAL PURPOSE guns (2 forward, 2 aft), 4 twin 40-mm. BOFORS GUNS, 5 fixed 21-in. (533-mm.) tubes (3 in the bow, 2 in the stern) for anti-ship TORPEDOES, 2 4-barrel 375-mm. DEPTH CHARGE rocket launchers, and 2 depth charge racks, as well as deck rails for laying MINES over the stern. Two trainable torpedo tubes for ANTI-SUBMARINE WARFARE (ASW) torpedoes were later added amidships. Between 1974 and 1977, the third 100-mm. gun mount on each ship was replaced by 4 single launch canisters for EXOCET MM38 anti-ship missiles, the Bofors guns were replaced by newer models of the same gun, and the 21-in. anti-ship torpedo tubes were replaced by 2 more ASW torpedo tubes.

The German navy intends to continue operating the Hamburgs until the end of the century. Their air defense capability is to be enhanced by the addition of two RAM (Rolling Airframe Missile) launchers and a new computer system. See also DESTROYERS, FEDERAL REPUBLIC OF GERMANY.

Specifications Length: 439 ft. (133.85 m.). **Beam:** 44 ft. (13.41 m.). **Draft:** 17 ft. (5.18 m.). **Displacement:** 3500 tons standard/4700 tons full load. **Powerplant**: twin-shaft steam: 4 oil-fired boilers, 2 sets of geared turbines rated at 72,000 shp. **Fuel**: 810 tons. **Speed**: 35 kt. **Range**: 924 n.mi. at 34 kt./6000 n.mi. at 13 kt. **Crew**: 280. **Sensors**: 1 LW-04 long-range air surveillance radar, 1 DA-08 air/surface surveillance radar, 1 SRG-103 surface-search radar, 3 M45 gun fire-control radars, 1 Atlas ELAC hull-mounted medium-frequency sonar. **Electronic warfare equipment**: 1 WLR-6 radar warning receiver, 2 SCLAR chaff rocket launchers.

HAMMER AND ANVIL: A tactical (or operational) scheme whereby enemy forces are pinned down frontally and simultaneously attacked on the flanks (or rear) by mobile forces. Hammer and anvil methods can be applied by both attackers and defenders. An attacker would divide his forces into a pinning force (usually with much infantry), to engage the enemy and prevent his withdrawal, and a striking force (usually with tanks) to maneuver against the enemy's flanks and/or rear, and drive them back upon the pinning force.

On the defensive, the momentum of the enemy's attack would be absorbed and dissipated by a DEFENSE-IN-DEPTH "anvil" of fixed positions, obstacles, and local counterattacks. If and when the enemy is immobilized, an armored counterattack can be mounted by mobile operational reserves against the enemy's flanks to drive him upon the fixed defenses. See also OPERATIONAL METHOD; TACTICS.

HAN: NATO code name for a group of four Chinese nuclear-powered attack SUBMARINES (SSNs) completed between 1974 and 1987. Very little information is available on the design and performance of these submarines. It is believed that the first in the class was laid down as early as 1965, but was not launched until 1971 due to the disruptions of the Cultural Revolution. The Hans have a modified teardrop hull, with a tall streamlined sail (conning tower) forward. Control surfaces consist of "fairwater" diving planes on the sail, and a cruciform arrangement of rudders and stern planes mounted ahead of the propellers. Armament is estimated as six 21-in. TORPEDO tubes. Nothing has been revealed about their RADAR, SONAR, and FIRE CONTROL systems. See also SUBMARINES, CHINA.

Specifications **Length**: 295.2 ft. (90 m.). **Beam**: 26 ft. (7.92 m.). **Displacement**: 4500 tons submerged. **Powerplant**: single-shaft nuclear, 1 pressurized-water reactor, geared turbines. **Speed**: 20 kt. surfaced/30 kt. submerged.

HARASSMENT AND INTERDICTION (H & I) FIRE: INDIRECT FIRE by tube ARTILLERY, MORTARS, or ROCKET ARTILLERY against known or suspected enemy troop concentrations, supply dumps, or transportation CHOKEPOINTS, to inflict low-level attrition and damage MORALE.

HARD, HARDNESS, HARDENING: Protection (and protective techniques) to shield military equipment and installations against weapon effects, especially nuclear effects. BALLISTIC MISSILES housed in underground concrete SILOS mounted on steel springs and covered with armored lids, as well as COMMAND AND CONTROL(C²) facilities similarly protected are described as "hardened." Although any structure which can resist peak blast overpressures of 100 lb. per square in. (PSI) is described as "hard," U.S. and Soviet strategic missile silos are commonly designed to resist overpressures of 1000–2000 psi. Much more than that is precluded by tumbling effects with standard concrete silos, but "superhardening" (to resist 3000 to 10,000 psi) has been tried. However, the cost is very high, and highly accurate, terminally guided missiles could defeat superhardening anyway. Hence, the U.S. and Soviet Union are developing road- and rail-mobile ballistic missiles to achieve a degree of protection by dispersal and concealment. See also COUNTERFORCE; FIRST STRIKE.

"HARD KILL": The neutralization of a target by physical destruction, as opposed to electronic or other disruption, as in "SOFT KILL."

HARDPOINT: (1) A HARDENED point target, such as a ballistic missile SILO, (2) A reinforced position in the wings or fuselage of aircraft, to which external loads can be attached.

HARDPOINT DEFENSE: A form of BALLISTIC MISSILE DEFENSE designed to protect only ballistic missile SILOS, COMMAND AND CONTROL bunkers, and similar high-value, HARDENED point targets, rather than larger "soft" targets such as cities; short-range terminal-defense systems may thus be sufficient for the task. See also ANTI-BALLISTIC MISSILE SYSTEM.

HARM: High-Speed Anti-Radiation Missile, the Texas Instruments AGM-88A air-to-ground defense supression missile, in service with the U.S. Air Force and the West German *Luftwaffe*. Designed from 1972, HARM was intended to replace the AGM-45 SHRIKE and AGM-78 STANDARD ARM anti-radiation missiles, both improvised to meet the urgent requirements of the Vietnam War, and subject to a number of operational limitations.

Specific requirements included a much higher flight speed (hence the name), the ability to lock onto enemy emitters before they could shut down or initiate other countermeasures, relatively low manufacturing cost, improved sensitivity, an expanded launch envelope, and the provision of an autopilot to keep the missiles aimed at targets, even if these ceased to emit after lock-on. Development encountered both technical and budget difficulties, which led to much higher unit costs ($300,000), and HARM did not enter service until 1983.

Similar in size and configuration to the AIM-7 SPARROW air-to-air missile, HARM has an ogival nose cap housing a passive radar receiver antenna; a cylindrical body; four movable, tapering cruciform wings mounted at midbody to control pitch, roll and yaw; and four smaller, fixed tail fins to stabilize the missile. The 146-lb. (66.4-kg.) blast-fragmentation warhead, fitted with an advanced laser-proximity fuze, is designed to inflict maximum damage on radar antennas and unarmored missile controls. HARM is powered by a Thiokol single-grain two-pulse (boost/sustainer) solid-fuel motor which uses nonaluminized fuel to reduce smoke, and thus impede visual detection.

HARM's completely digital guidance and control system can interface directly with launching aircraft's ELECTRONIC WARFARE (EW) and FIRE CONTROL systems. The simple fixed receiver antenna, which provides broad-band frequency coverage, is linked to a digital guidance computer with three different operating modes: self-protect, target-of-opportunity, and prebriefed. In the primary self-protect mode, enemy emitters detected by the aircraft's RADAR WARNING RECEIVERS are passed to an associated on-board launch computer, which prioritizes and locates the threat, and passes the data to the missile in a few milliseconds, after which the missile may be fired. In the target-of-opportunity mode, the missile seeker itself locks onto detected radar signals. Unlike Shrike or Standard ARM, HARM can cover a wide range of frequencies, and the parameters of selected hostile radars are stored in its computer. In the prebriefed mode, the missile can therefore be fired in the general direction of suspected emitters to find them on its own. If no signals are detected by engine burnout, the missile self-destructs, but if a radar transmits even briefly, the missile can immediately lock on and attack. The autopilot will guide the missile towards the emitter's last known position even if it shuts down.

HARM can be carried by a variety of tactical aircraft, including the F-4E/F PHANTOM and its F-4G WILD WEASEL variant, the EF-111 RAVEN, the A-7 CORSAIR, A-6E INTRUDER, F-16 FALCON, FA-18 HORNET, and Panavia TORNADO. Despite their high cost, the United States has ordered 15,000 missiles to date, while West Germany has ordered 388, with an option for an additional 576. HARM was used successfully in combat against Libyan missile and radar sites by U.S. carrier-based aircraft in March 1986 and was instrumental in the suppression of Iraqi air defenses in 1991. See also ANTI-RADIATION MISSILE; SEAD.

Specifications Length: 13.7 ft. (41.8 m.). Diameter: 10 in. (254 mm.). Span: 44.5 in. (1.13 m.). Weight, launch: 807 lb. (367 kg.). Speed: Mach 2+. Range: 10 mi. (16 km.).

HARPOON: The McDonnell-Douglas AGM/RGM/UGM-84, air-, ship- and submarine-launched ANTI-SHIP MISSILE. The Harpoon program began in 1973; development and testing were relatively uncomplicated, and the missile entered service with the U.S. Navy in 1976.

Harpoon's ogival nose cap houses an active RADAR seeker antenna. The ship- and submarine-launched (RGM/UGM-84) versions are equipped with a tandem booster rocket. Four tapering cruciform wings mounted at midbody provide lift during flight, with steering accomplished by four small, movable cruciform tail fins. The wings and tail fins can be folded for tube launching from ships and submarines. The payload is a 488-lb. (221.5-kg.) high-explosive warhead fitted with contact, proximity, and delay fuzes.

The missile is powered by a Teledyne CAE J402-CA-400 turbojet rated at 661 lb. (300 kg.) of thrust, fed by a flush-mounted air intake between the lower pair of wings. In ship- and submarine-launched versions, the solid-fuel rocket booster propels the missile from the launcher; at flying speed, the booster is jettisoned and the turbojet takes over.

The missile is controlled by INERTIAL GUIDANCE from launch through the midcourse phase, with radar ACTIVE HOMING for terminal guidance. Initial targeting data can be provided before launch by over-the-horizon targeting (OTH-T) platforms such as other aircraft, helicopters, ships, and submarines. The data is fed to the strap-down inertial navigation system, which can steer the missile towards the target even if it is launched as much as 90° off the target bearing. A radar altimeter holds the missile some 50 ft. (15.25 m.) above wave height until it approaches the preprogrammed target location, when Texas Instruments PR-53/DSQ-58 active radar is activated to search for and acquire the target; a few seconds before impact, the missile automatically executes a steep pullup, followed by a diving attack—a maneuver intended to defeat terminal defenses focused on wave heights, and inflict maximum damage to the target. The "B" version omits this "pop-up" attack profile.

The submarine-launched (UGM) version is launched in a watertight capsule that fits standard 21-in. (533-mm.) torpedo tubes. Once ejected from the tube, the capsule rises to the surface and breaks apart; the booster motor then ignites, thrusting the missile into the air. Air-launched (AGM) Harpoons are mated to standard missile pylons. Of the various ship launchers, the most common is the Mk.141, consisting of four sealed storage/launch canisters mounted in a fixed steel frame which can be bolted to the decks of most types of warship, down to FAST ATTACK CRAFT. Harpoon can also be fired from Mk.13 and Mk.26 STANDARD missile launchers, from modified 8-round ASROC pepperbox launchers, and from Mk.41 VERTICAL LAUNCH SYSTEMS.

More than 5000 Harpoons have been produced to date. Ship- and submarine-launched versions are in service with the U.S. and 20 other navies, including those of Britain, Canada, West Germany, Israel, Italy, and the Netherlands. The air-launched variants are presently in service only with U.S. forces; it can be carried by the A-6E INTRUDER, A-7 CORSAIR, FA-18 HORNET, P-3 ORION, S-3 VIKING, F-111, and B-52G STRATOFORTRESS. Harpoons were used successfully by the U.S. against Libyan vessels in the Gulf of Sidra in 1986, and against Iranian vessels in the Persian Gulf in April 1988.

A land attack version under development as the AGM-84E SLAM (Standoff Land Attack Missile) has an IMAGING INFRARED seeker from the AGM-65D MAVERICK air-to-surface missile, and the DATA LINK of the AGM-62 WALLEYE. Intended for precision attack against heavily defended land targets at ranges out to 60 mi. (100 km.), SLAM was rushed into production in 1991 and was used successfully against various Iraqi targets.

Specifications Length: (AGM) 12.5 ft. (3.84 m.); (RGM/UGM) 15.16 ft. (4.62 m.). Diameter 13.5 in. (324 mm.). Span: 36 in. (914 mm.). Weight, launch: (AGM) 1168 lb. (530.9 kg.); (RGM) 1470 lb. (668 kg.); (UGM) 1530 lb. (695.45 kg.). Speed:

570 mph (960 kph). Range: (A) 60 mi. (100 km.); (B/C) 70 mi. (117 km.); (D) 80 mi. (134 km.).

HARRIER AND HARRIER II: A vertical/short takeoff and landing (V/STOL) FIGHTER/ATTACK AIRCRAFT originally developed by British Aerospace, various models of which are in service with the Royal Air Force (RAF), the FLEET AIR ARMY (FAA), the U.S. MARINE CORPS (USMC), and the Spanish and Indian navies.

The Harrier is derived from the Hawker-Siddeley P.1127 experimental aircraft, whose most revolutionary feature was the use of a single jet engine with four movable exhaust nozzles which could be pivoted to vector thrust for both horizontal and vertical flight. Previous V/STOL aircraft had separate lift and flight engines, an inherently inefficient arrangement (still found on the Soviet Yak-36 FORGER). The P.1127 began flight tests in 1960, and by 1964 Hawker produced a slightly larger version (the Kestrel), which could carry a useful payload. In 1965, the RAF ordered a larger, more powerful version of the Kestrel, which became the Harrier, and the first operational squadron was formed in 1969. The initial Harrier GR.1 was a simple light attack aircraft, with only an optical gunsight and rudimentary AVIONICS. The RAF ordered 132, plus 19 T.2 two-seat trainers. Its original Rolls Royce Pegasus 101 turbofan, rated at 19,000 lb. (8636 kg.) of thrust, was soon replaced by the 20,000-lb. (9091-kg.) Pegasus 102. The GR.1 was adopted by the USMC as the AV-8A, of which 110 were purchased (plus 8 two-seat trainers) between 1971 and 1976. Many were destroyed in flight accidents; in the early 1980s, 36 of the surviving Marine Harriers were upgraded to AV-8C standard with improved avionics, but all were retired by 1987. The Spanish navy purchased 11 GR.1s and 2 trainers from the U.S. (designated AV-8S and TAV-8S Matadors) for service aboard the light AIRCRAFT CARRIER *Dedalo*. The RAF subsequently upgraded all its Harriers to GR.3 standard, with a LASER DESIGNATOR system; a RADAR HOMING AND WARNING system, and a more powerful Pegasus 103 engine.

Of conventional aluminum monocoque construction, the GR.3 is stressed to +7.8 and −4.2G. The nose has an extended thimble housing for the Marconi laser ranger and marked target seeker. The cockpit, well up in the nose, has a flat sliding canopy faired into the fuselage spine, and thus obstructing rearward visibility; the pilot is provided with a Martin-Baker "zero-zero" ejection seat. Cockpit avionics include a Smith Industries

6-50 head-up display (HUD), a Ferranti FE541 IN-ERTIAL NAVIGATION system with a rolling-map display, a TACAN navigational beacon receiver, and a threat-warning receiver.

The engine, located immediately behind the cockpit, is fed by two large cheek inlets. The exhaust is routed to four movable nozzles which can be swiveled from 0° (for forward flight) to 110° (for backing and lift), by using a lever in the cockpit. When taking off, hovering, or flying at very low speeds, the aircraft is controlled by small jets in the nose, tail, and wingtips fed by engine bleed air, and actuated by the rudder pedals and control stick. To take off vertically, the nozzles are positioned to 90° and the throttle advanced to full power; at a safe height the nozzles are slowly rotated aft, and the aircraft transitions smoothly to forward flight. Whenever possible, takeoff with a short forward roll is preferred, because it is much more fuel-efficient.

The area behind the engines contains a fuel cell and an avionics bay, with Marconi ARI 18223 RADAR WARNING RECEIVER (RWR) in a tail stinger. The Harrier has bicycle landing gear with the nose wheel retracting into a bay behind the cockpit, and the main gear housed behind the engine. Two small "pogo" wheels under the wing tips provide stability on the ground, and retract backwards into the trailing edge while in flight.

The small, shoulder-mounted, moderately swept wing provides a smooth ride at high speed and low altitude; it has pronounced anhedral intended to catch and hold the air cushion ("ground effect") generated by the nozzles during vertical takeoff; the slab-tailplane also has considerable anhedral. Extended wing tips can be bolted on for added lift on long-range ferry flights. The relatively small vertical stabilizer has yet another RWR mounted on its leading edge near the tip. Normally, the small wing area would limit maneuverability, especially in tight turns, but the USMC has developed the VIFF (Vectoring In Forward Flight) technique, subsequently adopted by the RAF, whereby the nozzles are rotated in flight to increase the aircraft's rate of turn, decelerate rapidly, and displace vertically without pitching over; with VIFF, the Harrier is surprisingly effective in aerial combat.

The GR.3 has 1 centerline pylon rated at 1000 lb. (454 kg.), 2 inboard wing pylons rated at 2000 lb. (909 kg.) each, and 2 outboard pylons rated at 1000 lb. each, but total payload capacity is limited to some 5000 lb. (2268 kg.), and most missions are flown with 3000 lb. (1364 kg.). A variety of ordnance can be carried, including Matra rocket pods for SNEB 65-mm. FFARS, 1000-lb. (424-kg.) free-fall and LASER GUIDED BOMBS, CLUSTER BOMBS, and NAPALM canisters. Since the 1982 Falklands War, the outboard pylons have also been wired for AIM-9 SIDEWINDER air-to-air missiles. Two additional belly hardpoints on either side of the centerline are used for 30-mm. ADEN cannon pods, each with 125 rounds of ammunition.

The potential of V/STOL aircraft for shipboard operations is obvious, because they can operate without the long landing decks, catapults, and arresting gear of conventional aircraft carriers. In 1975, the Royal Navy commissioned the development of a navalized Sea Harrier for its INVINCIBLE-class carriers. Essentially a GR.3 with a Ferranti Blue Fox multi-mode RADAR linked to the HUD and an elevated cockpit and bubble canopy for better visibility, the Sea Harrier is slightly longer; its Pegasus 104 engine, otherwise identical to the 103, has some new parts made from corrosion-resistant materials. Finally, the outboard pylons were wired for Sidewinder right from the start, because air defense is the Sea Harrier's primary mission. At present, a new radar is being installed, to provide illumination for SKYFLASH and AMRAAM radar-guided missiles.

Invincible-class carriers normally operate 5 Sea Harriers (which can increase to 8 in wartime), primarily for fleet air defense, but also for anti-ship and land attack and for reconnaissance. In addition to the GR.3's basic ordnance, the Sea Harrier can also carry SEA EAGLE anti-ship missiles. The Indian navy has ordered 12 Sea Harriers for operations from its two carriers, VIKRANT and VIRAAT.

Although the Harrier is not dependent on long (and vulnerable) runways, and can be dispersed to any short road segment, or even clearings the size of tennis courts, its logistic support must follow, at considerable cost, and the aircraft's autonomy is particularly limited. For CLOSE AIR SUPPORT (CAS), Harriers can "loiter" on the ground near the battle area until needed, thus conserving fuel, and thanks to VIFF, the Harrier is also quite potent in air combat. Both the GR.3 and Sea Harrier were much used in the Falklands, and Sea Harriers recorded 25 Argentine aircraft kills with no air combat losses of their own; even so, the Harrier's lack of range and payload is severe compared to conventional aircraft.

British Aerospace and McDonnell Douglas accordingly collaborated to design an improved Harrier II (AV-8B for the USMC and GR.5 for the RAF). Harrier II has a new and larger wing, revised avionics, and a lighter General Electric F402-GE-406 engine (Pegasus 11-21E). The aircraft has a completely revised nose section housing an angle-rate bombing system (ARBS), the raised cockpit of the Sea Harrier, an improved HUD, a multi-mode display screen and map display, and an ALE-40 chaff/flare launcher in the aft fuselage.

The new wing is fabricated in one piece from graphite epoxy composites, has a supercritical profile to minimized trans-sonic drag, and coincidentally, can hold more fuel. The wing, which gives greater lift at all speeds, has large, slotted flaps to trap the ground effect, increasing vertical lift, and leading edge root extensions (LERX) to improve maneuverability. There are now six wing pylons, including an outboard pair specifically for Sidewinders, freeing the inboard pylons for air-to-ground ordnance. U.S. aircraft have a 25-mm. GAU-12A cannon pod, with a separate ammunition pod, while British versions retain the two Aden gun pods. Air dams (Lift Enhancement Devices, or "LIDs") between the pods trap the ground effect cushion, resulting in a VTOL lift gain of 6700 lb. (3045 kg.). The typical mission payload has risen to nearly 7000 lb. (3181 kg.). Maximum speed is slightly reduced by the larger wing, but turn rates and lift are greatly improved.

The Marine Corps is acquiring 336 AV-8Bs to replace all of its AV-8A/Cs and A-4M SKYHAWKS. The RAF has ordered 60 GR.5s and Spain has ordered 12 AV-8Bs for the new V/STOL aircraft carrier PRINCIPE DE AUSTURIAS. A two-seat version, the TAV-8B, is under development. See also AIRCRAFT CARRIERS, BRITIAN; AIRCRAFT CARRIERS, INDIAN.

Specifications Length: (GR.3) 46.83 ft. (14.28 m.); (Sea) 47.6 ft. (14.51 m.); (II) 46.33 ft. (14.125 m.). Span: (GR.3, Sea) 25.25 ft. (7.7 m.)/29.67 ft. (9.05 m.) ferry; (II) 30.33 ft. (9.25 m.). Powerplant: (GR.3) Pegasus 103 turbofan, 21,500 lb. (9772 kg.) of thrust (II) General Electric F402-GE-406 turbofan (Pegasus 11-21E), 22,000 lb. (10,000 kg.) of thrust. Fuel: (GR.3, Sea) 5060 lb. (2295 kg.); (II) 7759 lb. (3519 kg.). Weight, empty: (GR.3) 12,200 lb. (5545 kg.); (Sea) 13,000 lb. (5909 kg.); (II) 12,750 lb. (5795 kg.). Weight, VTOL: (GR.3, Sea) 20,000 lb. (9091 kg.); (II) 20,000 lb. (9091 kg.). Weight, max. takeoff: (GR.3, Sea) 26,000 lb. (11,818 kg.); (II) 29,750 lb. (13,523 kg.). Speed, max.: (GR.3, Sea) 730 mph (1175 kph) at sea level/Mach 1.3 (900 mph/1503 kph) in shallow dive; (II) 673 mph (1124 kph). Initial climb: 40,000 ft./min. (12,195 m./min.). Service ceiling: 50,000 ft. (15,243 m.). Combat radius: (GR.3, Sea) 260 mi. (434 km.) hi-lo-hi; (II) 748 mi. (1250 km.). hi-lo-hi Range, max.: (GR.3, Sea) 2070 mi. (3457 km.); (II) 3310 mi. (5528 km.).

HARUNA: A class of two Japanese helicopter-carrying DESTROYERS (DDHs) commissioned in 1973–74. Designed in the late 1960s as flagships for ANTI-SUBMARINE WARFARE (ASW) escort groups, the Harunas have a traditional destroyer superstructure and weapons forward, while the entire aft end is dedicated entirely to helicopter facilities. For the Japanese Maritime Self-Defense Force, the Harunas are substitutes for ASW AIRCRAFT CARRIERS (deemed prohibited by the Japanese constitution).

The large flight deck extends from amidships to the fantail. A hangar built into the aft superstructure facing the flight deck can accommodate three large Mitsubishi-built SEA KING ASW HELICOPTERS, the ships' primary weapons. For close-range ASW, the Harunas have an 8-round ASROC launcher mounted in front of the bridge, and two Mk.32 triple tubes for lightweight ASW TORPEDOES amidships. As built, their only other armament consisted of two Mk.42 5-in. 54-caliber DUAL PURPOSE guns forward of the ASROC launcher, with the second gun mounted on a raised platform to superfire the first gun. Between 1983 and 1988, their ANTI-AIR WARFARE (AAW) and ANTI-SURFACE WARFARE (ASUW) capabilities were upgraded by the addition of an 8-round NATO SEA SPARROW surface-to-air missile system, two PHALANX radar-controlled guns for anti-missile defense, and two quadruple launchers for HARPOON anti-ship missiles.

The ships form the nucleus of two ASW escort groups, each with two destroyer "divisions" of two or three TAKATSUKI- or YAMAGUMO-class destroyers. Popular and effective vessels, the Harunas were the basis for the enlarged and improved SHIRANE DDHs. See also DESTROYERS, JAPAN.

Specifications Length: 502 ft. (153.04 m.). Beam: 57.5 ft. (17.53 m.). Draft: 16.75 ft. (5.1 m.). Displacement: 4700 tons standard/6300 tons full load. Powerplant: twin-shaft steam: 4 oil-fired boilers, 2 sets of geared turbines, 70,000 shp. Speed: 32 kt. Crew: 340. Sensors: 1 OPS-11 air-search radar, 1 OPS-17 surface-search radar, 2 GFCS-1 gun fire control radars, 1 MFCS-1 missile guidance radar, 1 OQS-3 hull-mounted medium-frequency active/passive sonar, 1 SQS-35 variable depth sonar. **Electronic warfare equipment:** 1 NOLQ-1 active jamming array, 1 OLR-9B radar warning receiver, 2 Mk.36 SRBOC chaff launchers.

HATAKAZE: A class of two new Japanese guided-missile DESTROYERS (DDGs) commissioned in 1986 and 1988. An incremental development of the TACHIKAZE class, with gas turbine propulsion instead of the steam powerplant of the earlier ships, the Hatakazes are are flush-decked, with a sharp "clipper" bow, a blocky superstructure amidships, and a single large stack.

Though optimized for ANTI-AIR WARFARE (AAW), the Hatakazes retain considerable ANTI-SUBMARINE WARFARE (ASW) and ANTI-SURFACE WARFARE (ASUW) capabilities. The primary AAW weapon is a single-arm Mk.13 missile launcher for 40 STANDARD MR surface-to-air missiles (SAMs) on the bow. Other AAW weapons include two PHALANX 20-mm. radar-controlled guns for anti-missile defense, and a Mk.42 5-in. 54-caliber DUAL PURPOSE gun at each end of the ship. For ASUW, the guns supplement two quadruple launchers for HARPOON anti-ship missiles amidships. ASW weapons consist of an 8-round ASROC pepperbox launcher in front of the bridge, and two sets of Mk.68 triple tubes for lightweight ASW homing TORPEDOES. There is a HELICOPTER pad on the fantail, but no hangar or maintenance facilities. See also DESTROYERS, JAPAN.

Specifications Length: 492 ft. (150 m.). **Beam:** 54 ft. (16.46 m.). **Draft:** 15.75 ft. (4.8 m.). **Displacement:** 4450 tons standard/5400 tons full load. **Powerplant:** twin-shaft COGAG: 2 Rolls Royce Spey SM1A and 2 Rolls Royce Olympus TM3D gas turbines, 72,000 shp. **Speed:** 32 kt. **Crew:** 260. **Sensors:** 1 OPS-12 surveillance radar, 1 OPS-11 air-search radar, 1 OPS-28 surface-search radar, 2 SPG-51C missile guidance radars, 1 GFCS-2 gun fire control radar, 1 OQS-4 bow-mounted low-frequency sonar. **Electronic warfare equipment:** 1 NOLR-1 active jamming array, 1 OLR-9B radar warning receiver, 2 Mk.36 SRBOC chaff launchers.

HATSUYUKI: A class of 12 Japanese DESTROYERS commissioned between 1983 and 1987. A further 8 ships, built to an "Improved Hatsuyuki" design, are presently under construction; the first ship (ASAGIRI) was completed in 1988.

Specialized ANTI-SUBMARINE WARFARE (ASW) vessels of a completely new design, the Hatsuyukis derive little if anything from their immediate predecessors of the TAKATSUKI and YAMAGUMO classes. They have a sharp "clipper" bow, a long, boxy superstructure amidships, and a single large stack. The hull is fitted with gyro-controlled fin stabilizers to improve seakeeping.

The weapons suite is optimized for ASW with secondary ANTI-AIR WARFARE (AAW) and ANTI-SURFACE WARFARE (ASUW) capabilities. Primary ASW armament is a Mitubishi-built SEA KING helicopter, operated from a large landing deck and hangar amidships (the Hatsuyukis are the first general-purpose Japanese destroyers to operate manned HELICOPTERS). On-board ASW weapons include an 8-round ASROC pepperbox launcher ahead of the bridge, and two sets of Mk.32 triple tubes for lightweight homing TORPEDOES amidships. AAW weapons are meant purely for self-defense, with an 8-round NATO SEA SPARROW short-range surface-to-air missile launcher on the fantail, and two PHALANX radar-controlled guns for anti-missile defense mounted on the superstructure amidships. The primary ASUW systems are two quadruple launchers for HARPOON anti-ship missiles, also amidships. An OTO-Melara 76.2-mm. DUAL PURPOSE gun ahead of the ASROC launcher can supplement the AAW and ASUW weapons.

The Improved Hatsuyukis are 21.3 ft. (6.5 m.) longer than the original design, and their beam is greater by 3.4 ft. (1.03 m.). The major difference from the original design is the replacement of COGOG propulsion with a COGAG arrangement. The new powerplant also required the rearrangement of the exhaust stacks and engine intakes. Other changes include a new surface search radar and a more comprehensive ECM suite. See also DESTROYERS, JAPAN.

Specifications Length: 426.4 ft. (130 m.); **Beam:** 44.6 ft. (13.6 m); **Draft:** 13.4 ft. (4.08 m.) **Displacement:** 3050 tons standard/3800 tons full load; **Powerplant:** twin-shaft COGOG: 2 Rolls Royce Tyne RM1C gas turbines, 10,680 shp. (cruise), 2 Rolls Royce Olympus TM3B turbines, 56,780 shp. (sprint). **Speed:** 30 kt. **Crew:** 190; **Sensors:** 1 OPS-14B air-search radar, 1 OPS-18 surface-search radar (OPS-28 in Asagiri), 1 GFCS-2-21/21A gun fire control radar, 1 MFCS-2-12A missile guidance radar, 1 OQS-4(II) hull-mounted medium-frequency sonar, 1 SQR-19 TACTASS towed sonar array. **Electronic warfare equipment:** 1 NOLR-6C radar warning receiver, 1 OLT-3 active jamming system, 2 Mk.36 SRBOC chaff launchers.

HAVOC: NATO code name for the Soviet Mil Mi-28 twin-engine attack HELICOPTER now in service with Soviet Army Aviation units. Havoc is apparently intended to be the successor to the Mi-24/25 HIND, whose excessive size and lack of maneuverability reflect its derivation from the Mi-8 HIP transport helicopter. The Havoc is an entirely new design, rectifying the Hind's deficiencies and incorporating a number of state-of-the-art features.

Very little was known about the Mi-28 until its appearance at the 1989 Paris Air Show. Superficially similar to the U.S. Ah-64 Apache, the Havoc has a narrow fuselage with a stepped tandem cockpit. Following traditional attack helicopter practice, the WEAPON SYSTEM OPERATOR (WSO) sits in the front seat, with the pilot behind and above him. The cockpit sides are armored to resisist 12.7-mm. projectiles, and the canopy is fabricated of flat, nonglint, armored glass panels. A small, thimble-shaped radome located immediately ahead of the cockpit houses a ranging RADAR antenna. A small window in the chin position below the radar houses twin forward-looking infrared (FLIR) sensors and a LASER DESIGNATOR. A flexible gun turret on the belly below the WSO's seat is armed with a 30-mm. cannon identical to that of the BMP-2 infantry fighting vehicle. The turret, with 300 rounds of ammunition, has traverse limits of 55° left and right, and elevation limits of −45 and +13. Two stub wings behind the pilot's seat each have 4 hardpoints for up to 8 AT-6 SPIRAL anti-tank guided missiles (ATGMs); or 8 unguided rocket pods, free-fall high explosive, and chemical bombs; or four modified SA-14 GREMLIN surface-to-air missiles for self-defense and use against enemy helicopters (the wings also contribute up to 15 percent of total lift at high speeds). The Havoc has a crash-resistant, fixed landing gear.

Though its five-bladed main rotor is the same size as that of the Hind, the rotor hub, transmission, and dynamic parts are all new. The rotor blades are made entirely of composites, and the hub is manufactured of titanium to withstand combat damage. Though much more agile than the Hind, the Havoc is still not as maneuverable as the latest Western helicopters. Its rotor system is still of the fully articulated "flapping-blade" type, which limits its negative-g maneuverability, a critical constraint when contour-flying at nap-of-the-earth heights. The engines, mounted in pods on either side of the fuselage just above the stub wings, are fitted with dust filters and infrared suppressors. The helicopter also carries an INFRARED COUNTERMEASURES beacon and several chaff/flare dispensers as protection against infrared and radar-guided missiles.

Specifications Length: 52.5 ft. (16 m.). **Rotor diameter:** 55.75 ft. (17 m.). **Powerplant:** 2 2200-shp. Isotov TV3 turboshafts. **Weight, empty:** 15,430 lb. (7013 kg.). **Weight, max. takeoff:** 22984 lb. (10,447 kg.), inc. 8800-lb. (400-kg.) payload. **Speed, max.:** 186 mph (300 kph). **Combat radius:** 150 mi. (250 km.).

HAWK: Homing All the Way Killer, the Raytheon MIM-23 land-based SURFACE-TO-AIR MISSILE (SAM). Development of HAWK was initiated in 1954, and the missile became operational with the U.S. Army in 1959; with many modifications and improvements, it is still in production today.

The original requirement called for a medium-range SAM capable of engaging targets even at low altitude, and with sufficient mobility for field army use. As compared to modern battlefield SAM systems such as ROLAND, RAPIER, or CROTALE, HAWK is rather large, cumbersome, and expensive, but also much more capable; it has been purchased by 21 countries. More than 38,000 HAWK missiles were produced by 1987.

The Improved HAWK (I-HAWK) system, introduced in 1972, has an improved missile (MIM-23B), with longer range and generally higher performance, better RADAR, and enhanced FIRE CONTROL systems. I-HAWK has supplanted the original system with most users.

The original MIM-23A missile has a cylindrical body with an ogival nose radome housing a SEMI-ACTIVE RADAR HOMING (SARH) antenna. Four cruciform cropped delta wings extend from midbody to the tail, with moving control surfaces on their trailing edges for steering and roll control. Its dual-pulse solid rocket motor provides an initial boost which rapidly accelerates the missile, and a second, sustainer pulse which maintains speed over the remainder of the trajectory. HAWK is armed with a 120-lb. (54.5-kg.) high-explosive CONTINUOUS ROD WARHEAD equipped with contact and radar proximity fuzes.

Slightly longer but otherwise almost unchanged externally, the I-HAWK (MIM-23B) has an improved motor for greater range, and a gap between the wing trailing edges and the movable control surfaces to improve maneuverability. In addition, the MIM-23B is handled as a "certified round" in a depot-sealed storage canister, eliminating the need for field maintenance and testing prior to launch.

The basic HAWK firing "section" consists of three triple launchers, an Information Control Center (ICC), a Pulse Acquisition Radar (PAR), a Continuous Wave Acquisition Radar (CWAR), a Range-Only Radar (ROR), and a High Power Illumination Radar (HiPIR). BATTERIES of 2 or 3 sections, commanded from a Battery Control Center (BCC), are grouped into BATTALIONS: standard battalions of 4 batteries, each with 2 sections, or "Triad" battalion of 3 batteries, each with 3 sections.

In action, area surveillance is provided by the PAR for long-range high altitude coverage, and the CWAR for the coverage of low altitudes at shorter ranges. Both radars are synchronized to present target data on a composite tactical display. The ROR, a simple ranging radar that operates on a different frequency band from the two acquisition radars, is used for fast reaction, and when ELECTRONIC COUNTERMEASURES (ECM) have neutralized the PAR and CWAR. The BCC assigns targets to fire control operators who slave a launcher and HiPIR to the target bearing via the ICC. When lock-on is achieved, the missile is launched to home on the radiation reflected off the target by the illuminator.

HAWK has been much used in combat by Israel and Iran, with a reported probability of kill (P_K) of over 40 percent, remarkable for any surface-to-air missile. One operational shortcoming is the difficulty of engaging short-range, low-altitude "pop-up" targets, due to the need to achieve radar lock-on before launch. To remedy this, the Israelis have developed an ELECTRO-OPTICAL (EO) sensor which is mounted directly over the HiPIR, thereby circumventing the need to first acquire the target with the PAR or CWAR; if the target can be observed with the EO sensor, the illuminator is automatically locked on. This system has been adopted by the U.S. Army and Marine Corps. A second shortcoming is the large size and limited mobility of the sections. Although all its elements can be towed and airlifted by helicopter, it takes approximately one hour to emplace a single firing battery. To remedy this, the Marines have developed a streamlined "Assault Firing Unit" for AMPHIBIOUS ASSAULTS, with a PLATOON Command Post (PCP), essentially a reduced BCC, a CWAR (but no PAR), a HiPIR, and a section of three triple launchers. A U.S. Army attempt to develop a complete self-propelled unit based on the M113 armored personnel carrier chassis was not successful.

In the early 1980s, Hughes developed a solid-state, PHASED ARRAY radar to replace the PAR, CWAR, and ROR, greatly simplifying battery organization and enhancing mobility. So far, this system has been adopted only by the Belgian army. See also AIR DEFENSE; ANTI-AIRCRAFT WEAPONS.

Specifications Length: (A) 16.5 ft. (5.03 m.); (B) 16.8 ft. (5.12 m.). Diameter: 14 in. (356 mm.). Span: 47.5 in. (1.2 m.). Weight, launch: 1295 lb. (588 kg.). Speed: Mach 2.5. Range: (A) 22 mi. (36.75 km.); (B) 25 mi. (41.75 km.). Height envelope: 100–36,000 ft. (30.5–11,000 m.). Operators: Bel, Bra, Den, Fra, FRG, Gre, Iran, Isr, It, Jap, Jor, Ku, Neth, Phi, ROK, S Ar, Sp, Swe, Tai, Thai, UK, US.

HAWK: A single-engine, two-seat trainer and light ATTACK AIRCRAFT built by British Aerospace, in service with the RAF, the U.S. Navy, Abu Dhabi, Dubai, Finland, Indonesia, Kenya, Kuwait, and Zimbabwe. The Hawk was selected by the Royal Air Force in 1971 to replace its fleet of Gnat and HUNTER advanced trainers. An order for 176 aircraft, designated Hawk T.1, was placed in 1972. The Hawk has proven to be maneuverable, economical, reliable, and generally forgiving in the hands of student pilots. It can be fitted with a variety of gun pods and aerial ordnance for weapon training and light attack. The airframe is cleared for up to 6000 hours of low-altitude flight, and is stressed to withstand $+8$ and -4 g. Simplicity of design keeps costs down and availability high.

The area-ruled ("coke bottle") fuselage has small, cheek-mounted engine inlets. The student and instructor sit in a tandem cockpit in the nose, under a single-piece clamshell canopy. The instructor's rear seat is raised over the front seat to provide adequate forward vision during landings. Both positions are provided with Martin-Baker zero-zero ejection seats, and lead-computing optical gunsights. The low-mounted wing has very moderate sweep and a supercritical profile to reduce trans-sonic drag. Two small fences on the outboard leading edge provide stall warning and improve low-speed handling. The trailing edge has outboard ailerons, inboard flaps, and spoilers. The tailplane is more sharply swept, and the all-moving horizontal stabilizer has a pronounced degree of anhedral.

The Hawk has one centerline and four wing hardpoints for external fuel tanks and ordnance.

The centerline pylon usually carries a 30-mm. ADEN gun pod with 125 rounds of ammunition, and the inboard wing pylons are plumbed for drop tanks. A maximum of 6500 lb. (2955 kg.) of ordnance can be carried, including 1000-lb. (454-kg.) LDGP bombs, FFAR pods, NAPALM, and CLUSTER BOMBS. To remedy the RAF's shortage of interceptors for home air defense, the British have modified 72 Hawks to T.1A standard with AIM-9 SIDEWINDER air-to-air missiles on the two outboard pylons. Most export versions also have this capability.

The U.S. Navy is acquiring a carrier-capable trainer (T-45 Goshawk), which will be built under license by McDonnell-Douglas. The T-45 has reinforced landing gear, a catapult tow bar, a revised cockpit, and a new rear fuselage incorporating an arrester hook and small twin speed brakes mounted on the sides of the fuselage. Many of the fuselage panels are manufactured of carbon-fiber composites to reduce weight.

A specialized single-seat light attack variant (Hawk 200), with uprated engines, four wing pylons rated at 2000 lb. (909 kg.) each, two internal guns, enhanced AVIONICS, and a Ferranti Blue Fox multi-mode RADAR mounted in a large nose radome, has been developed for export.

Specifications Length: 38.95 ft. (11.87 m.). Span: 30.7 ft. (9.36 m.). **Powerplant:** 1 Rolls Royce-Turbomeca Adour turbofan: (T.1) 5340 lb. (2427 kg.) of thrust; (export) 6250 lb. (2840 kg.) of thrust. **Weight, empty:** (T.1) 7100 lb. (3227 kg.); (export) 8500 lb. (3863 kg.). **Weight, max. takeoff:** (T.1) 11,100 lb. (5055 kg.); (export) 18,890 lb. (8586 kg.). **Speed, max.:** 625 mph (1043 kph) at sea level/ Mach 1.02 (750 mph/1252 kph) in shallow dive. **Time to 30,000 ft. (9150 m.):** 6 min. **Service ceiling:** 50,000 ft. (15,245 m.). **Range, max.:** (internal fuel) 1520 mi. (2538 km.); (two drop tanks) 2530 mi. (2530 km.).

HAWKEYE: The Grumman E-2 twin-turboprop, carrier-based airborne warning and control (AWACS) aircraft, in service with the U.S. Navy and the air forces of Israel, Egypt, and Japan. Developed from the late 1950s to replace the piston-engined Grumman E-1 Tracer, the prototype E-2A began flight tests in January 1960, and the first production variant (E-2B) entered service in 1964. It was superseded by the present E-2C version in 1971.

The primary mission of the Hawkeye is the long-range detection of hostile aircraft and CRUISE MISSILES, and the direction of friendly FIGHTER aircraft

against them. The most important element of the E-2 system is a large air search RADAR housed in a streamlined, disc-shaped rotodome mounted above the fuselage. In the E-2B, the radar was a General Electric APS-96. The E-2C has the more advanced APS-125 and its Advanced Radar Processing System (ARPS), which provide sufficient resolution and discrimination capabilities to detect surface skimming ANTI-SHIP MISSILES and (very) low-flying aircraft. The rotodome rotates at 6 revolutions/min., and can provide surveillance over a radius of 300 mi. (500 km.). The ARPS signal processor can track up to 250 individual targets at one time, and can detect objects as small as cruise missiles out to a range of 115 mi. (192 km.), as well as surface ships and vehicles on land. In addition to the radar, the E-2C also has an ALR-59 ELECTRONIC SIGNAL MONITORING array for the passive detection and classification of radio-frequency emissions.

The pilot and copilot sit side-by-side on an airline-style flight deck, with a cabin behind it for the Airborne Tactical Data System (ATDS), manned by a COMBAT INFORMATION CENTER (CIC) officer and a radar officer. Numerous display screens and communication links enable them to track targets and control up to 30 friendly aircraft. A powerful air-conditioning unit between the ATDS compartment and the flight deck cools the radar and the other electronic systems on board. The rotodome, mounted over the rear fuselage behind the wings and supported by streamlined tripod struts, can be lowered slightly to fit inside carrier hangar decks. The dome itself is an effective airfoil, which generates enough lift to support its own weight; but it does adversely affects the aircraft's lateral stability. To counteract this, the Hawkeye has four vertical stabilizers along a large horizontal tailplane, with very pronounced dihedral.

The shoulder-mounted wings have a high aspect ratio for economical cruising performance. The trailing edges are fitted with Fowler flaps and drooping ailerons, to improve low-speed handling during carrier landings. The engines are mounted in underwing nacelles, the wings contain intergral fuel tanks, and the outer wing sections fold back for carrier deck parking. The aircraft's nose landing gear has a tow bar for catapult launches, and an A-frame arrester hook housed under the tail. Since 1986, U.S. Hawkeyes have been retrofitted with more powerful engines to cope with the added weight of new electronic equipment.

The U.S. Navy's 90 Hawkeyes have all been

brought up to E-2C standard. (In 1986–87, two aircraft were transferred to the U.S. Customs service for anti-drug operations.) Each U.S. aircraft carrier has a detachment of four Hawkeyes, at least one of which is always on station. Israel operates four Hawkeyes, which played an important role in the June 1982 Lebanon War, when their support facilitated the suppression of Syrian air defenses and the interception of Syrian fighters. Japanese Air Self-Defense Force Hawkeyes are integrated into its Base Air Defense Ground Environment (BADGE). Egypt received four Hawkeyes in late 1985.

The Hawkeye's wings, engines, and tailplane are used in the C-2 (Greyhound) Carrier On-board Delivery (COD) transport, which has a boxy fuselage with a fold-down ramp door under the tail. It can carry up to 10,000 lb. (4540 kg.) of cargo or 39 passengers. The U.S. Navy operates a total of 56 Greyhounds. See also AIRBORNE EARLY WARNING; ANTI-AIR WARFARE.

Specifications Length: 57.6 ft. (17.56 m.). Span: 80.6 ft. (24.57 m.). Powerplant: 2 4910-shp. Allison T56-A-425 turboprops. Fuel: 12,400 lb. (5636 kg.). Weight, empty: 37,678 lb. (17,126 kg.). Weight, max. takeoff: 51,569 lb. (23,440 kg.). Speed, max.: 375 mph (626 kph). service ceiling: 30,800 ft. (9390 m.). Mission radius: (3–hr. loiter) 200 mi. (334 km.). Endurance: 6 hr. Range, max.: 1600 mi. (2672 km.).

HAZE: NATO code name for the Mil Mi-14 shore-based ANTI-SUBMARINE WARFARE (ASW) HELICOPTER in service with Soviet Naval Aviation (AV-MF). Haze is a derivative of the ubiquitous Mi-8/17 HIP transport helicopter, with a watertight, boat-shaped hull for amphibious landings, and more powerful engines; the main rotors, rotor hub, and transmission are also identical to the Mi-17. More than 100 Haze A/B are in service with the AV-MF; low-rate production is continuing.

The fuselage has a central radome under the nose for a surface-search RADAR, and a ramp-door in the tail. The four-poster landing gear retract into into sponsons extending from the fuselage sides, which also serve as stabilizer floats; a smaller float mounted under the tail rotor supports the tail boom when the helicopter is waterborne. Haze has fairly comprehensive ASW equipment, including an active/passive dipping SONAR, SONOBOUYS, a MAGNETIC ANOMALY DETECTOR (MAD), DEPTH CHARGES, homing TORPEDOES, and MINES. Weapons are carried in two ventral bays in the hull bottom with watertight doors. A MINE COUNTER-MEASURES (MCM) version (Haze B) has a small pod in place of the MAD (whose exact purpose is unknown), as well as provisions for towing a minesweeping sled meant to detonate both magnetic and acoustical mines.

Specifications Length: 60.5 ft. (18.45 m.). Rotor diameter: 69.85 ft. (21.29 m.). Powerplant: 2 2200-shp. Isotov TV3-117 turboshafts. Weight, empty: 19,400 lb. (8818 kg.). Weight, max. takeoff: 30,864 lb. (14,029 kg.). Speed, max.: 143 mph (230 kph). Speed, cruising: 124 mph (208 kph). Range, max.: 500 mi. (835 km.).

HBX: A high-explosive compound, consisting primarily of TNT, with additives to improve both its explosive effect and stability. See also EXPLOSIVES, MILITARY.

HE: High Explosive (round, shell), a form of ammunition in which a hollow steel projectile is filled with EXPLOSIVES, detonated by a nose or base fuze, and meant to inflict damage through blast and fragmentation effects. HE rounds are most effective against exposed troops and other soft targets, unless fitted with time-delay fuzes, to allow them to penetrate hardened structures. APHE (Armor Piercing High Explosive) ammunition has a thick, armor-piercing case and a reduced explosive charge. HEI (High Explosive Incendiary) ammunition contains flammable substances such as phosophorous or thermite, mixed in with the explosive filler, in order to start fires as well as inflict blast and fragmentation damage. HEAT (High Explosive Anti-Tank) and HESH (High Explosive Squash Head) ammunition are specialized ARMOR-PIERCING rounds.

HEADQUARTERS (HQ): The decision-making center of an organization, usually a military organization. General Headquarters (GHQ) defines a decision-making center which exercises COMMAND AND CONTROL over other headquarters. Many ground formations maintain two separate headquarters, a small, mobile, "tactical" HQ, and a larger, fixed, "main" HQ responsible for administrative and support functions.

HEAT: High Explosive Anti-Tank, a type of ARMOR-PIERCING ammunition which exploits the "hollow" or SHAPED CHARGE principle: when an inverted copper cone backed by high explosives is detonated, a phenomenon known as the "Munroe Effect" focuses the force of the explosion and transforms the copper into a thin, hypervelocity (8250 m./sec.) jet capable of penetrating a thickness of steel ARMOR equivalent to roughly five times the cone's original diameter, leaving a hole eight to ten times the diameter of the jet. Shock and heat, as well as the penetration, often cause secondary explosions inside armor vehicles.

Unlike KINETIC ENERGY projectiles fired from high-velocity guns, which rely on impact for effect, HEAT rounds are insensitive both to range and velocity. Thus they are the key to the armor-piercing capability of light guns, RECOILLESS RIFLES, anti-tank rocket launchers (BAZOOKAS), and ANTI-TANK GUIDED MISSILES (ATGMs). HEAT warheads arm folding fin aerial rockets (FFARS), anti-tank CLUSTER BOMBS, and anti-vehicle MINES.

The performance of HEAT rounds, however, is degraded by the spin imparted by rifled guns, because the centrifugal force dissipates their jet. Thus HEAT rounds are most effective when fired from smoothbore guns. The French AMX-30 main battle tank, which has a 105-mm. rifled gun, is issued with a complex round (OBUS-A), in which the HEAT charge is mounted on bearings inside the shell's outer casing, thereby reducing its spin to only 20–30 revolutions/min.

HEAT charges require a standoff distance to allow the jet to form for maximum effect. Many HEAT rounds therefore have a long probe to initiate detonation before impact. SPACED ARMOR reduces the effectiveness of small HEAT charges by detonating them at a greater than optimal standoff distance from the main armor, but is not particularly useful against larger charges because the optimal standoff distance is usually much greater than the spacing between armor layers.

Two new forms of protection have greatly reduced the effectiveness of HEAT charges. CHOBHAM ARMOR, consisting of laminated layers of steel, ceramics, exotic alloys, and KEVLAR mesh, absorbs and dissipates the energy of the jet. With REACTIVE ARMOR, boxes of low-sensitivity explosives bolted over a vehicle's main armor explode on contact with the jet, deflecting and dissipating it before it can penetrate the main armor.

In response, "top attack" warheads are now being developed to attack the thin roof and upper deck of armored vehicles. The Bofors BILL, for instance, has a proximity fuze to detonate a HEAT warhead angled obliquely downward.

HEAVY MACHINE GUN (HMG): A fully automatic weapon firing a round larger than infantry small arms (up to 7.62 mm.), but smaller than CANNON (20 mm. or more), often mounted on ARMORED FIGHTING VEHICLES for use against light armor, field works, and aircraft, but also serving as tripod-mounted INFANTRY weapons in fixed positions. In the past, they were widely used as aircraft weapons, but have generally been replaced by cannon of 20 mm. or greater caliber, except on some helicopters. HMGs weigh between 60 and 200 lb. (27 and 100 kg.), hence they cannot be carried by infantry on the attack. The most common types in service include the Browning M2 .50-caliber (12.7-mm.), the DSHK 12.7-mm., and the KPV 14.5-mm. See also GENERAL PURPOSE MACHINE GUN; LIGHT MACHINE GUN.

HEDGEHOG: 1. An independent defensive perimeter protected by ANTI-TANK and ANTI-AIRCRAFT weapons, minefields, and field fortifications, usually containing ARTILLERY. Forces inside the hedgehog are organized and supplied to resist on their own, even if temporarily cut off and isolated from other friendly forces. Hedgehog positions can be organized into a DEFENSE-IN-DEPTH intended to resist BLITZKRIEG-style offensive; by using their artillery to interdict the passage of "soft" transport and INFANTRY units between them, hedgehogs can separate enemy TANK and MECHANIZED forces from their supporting infantry and supply, so that they can be more easily defeated by armored counterattacks, especially after they are weakened by LOGISTIC starvation. See also ARMORED WARFARE.

2. An ANTI-SUBMARINE WARFARE (ASW) weapon developed by the British navy in 1943, a multiple-spigot MORTAR capable of throwing 24 contact-fuzed DEPTH CHARGES in a circular pattern 285 yards (260 m.) ahead of the firing vessel. Each charge weighed approximately 40 lb. (18 kg.), and the impact pattern was sized to give a high statistical probability of at least one hit on a submarine-sized target. The standard World War II Mk.10 version was fixed to fire straight ahead, and stabilized to compensate for the ship's roll. The postwar Mk.15 version can be trained to either side and was widely adopted by the U.S. and allied navies. The Mk.15 mount weighs 17,425 lb. (7920 kg.), has a crew of eight, and requires about five minutes to load.

At the time of its introduction, Hedgehog was a radical improvement over the traditional method of dropping hydrostatically fuzed depth charges over the stern. By firing ahead, sonar contact with the target could be maintained throughout the engagement, and the use of small charges with contact fuzing avoided the turbulence created by much larger hydrostatic depth charges. A 1944 U.S. Navy study indicated that Hedgehog had a 28 percent probability of kill (P_K) per attack compared to 6 percent for traditional depth charges. Against modern submarines, Hedgehog is only marginally effective due to its short range, and the relatively long "dead time" until the charges reach current submarine operating depths. Hedgehog remains in service with some navies, especially for use in shallow and confined waters.

HEDI: High Endoatmospheric Defensive Interceptor, a ground-launched ANTI-BALLISTIC MISSILE (ABM) being developed under the STRATEGIC DEFENSE INITIATIVE (SDI). HEDI is designed to intercept incoming enemy REENTRY VEHICLES (RVs) during their final seconds of flight, after they have entered the atmosphere. The relatively small missile has very powerful engines providing acceleration on the order of 100 to 500 G, with a range of approximately 60–120 mi. (100–200 km.) and intercept envelope of 50,000–250,000 ft. (15,250–76,220 m.). Unlike earlier high-acceleration interceptor missiles, such as SPRINT and LOADS, whose radar command guidance required nuclear warheads to offset poor accuracy, HEDI would have long wavelength infrared (LWIR), ultraviolet (UV), or millimeter wave (MMW) homing to provide extreme accuracy, thus relying entirely on KINETIC ENERGY as a kill mechanism without any warhead at all. Steering would be controlled lateral thrusters, to enable the missile to make radical course adjustments at very high velocity. HEDI could be one part of a two-tiered terminal defense system together with the longer-ranged ERIS (Exoatmospheric Reentry Vehicle Interceptor System), to intercept RVs outside the atmosphere. See also BALLISTIC MISSILE DEFENSE.

HEEMSKERCK: A class of two Dutch guided-missile FRIGATES (FFGs) commissioned in 1985–86. Modified versions of the KORTENAER class, the Heemskercks were ordered to replace two of those ships sold to Greece. The hull, machinery, and basic sensors are unchanged, but the Heemskercks have a Mk.13 launcher for STANDARD MR surface-to-air missiles aft in place of the Kortenaers' helicopter landing pad, and the latter's bow-mounted 76.2-mm. DUAL PURPOSE gun was removed to compensate for the weight of the Mk.13 system. A GOALKEEPER 30-mm. radar-controlled gun for anti-missile defense is also mounted on the fantail. Intended as flagships, the Heemskercks are equipped with an NTDS Link 11 DATA LINK. Other systems include Nixie torpedo countermeasures, and an SQR-18A TACTASS towed SONAR array. See also FRIGATES, NETHERLANDS.

HELICOPTER: An aircraft lifted and propelled entirely by power-driven rotary wings (called rotors). Each rotor blade is actually a very narrow airfoil, i.e., a wing. As the blades rotate, they generate lift, which enables the helicopter to rise vertically or hover at lower power settings; forward, backwards, and sideways movement is achieved by altering the blade pitch.

The first practical helicopters were developed in the late 1930s by Igor Sikorsky; small numbers of primitive helicopters were actually used by U.S. forces in 1944–45. The main technical difficulties that had to be overcome included the development of strong, reliable rotor blades, transmission linkages from engine to rotor, the neutralization of rotor torque forces (i.e., the tendency of the helicopter to rotate in the opposite direction of the rotor), and the arrangement of control linkages to the rotor.

Neutralization of rotor torque is accomplished by a number of alternative rotor configurations. In the 1930s, Sikorsky developed what remains the standard remedy—the tail rotor ("penny-farthing") configuration, whereby a large main rotor mounted over the helicopter's center of gravity is counterbalanced by a smaller rotor mounted at the extreme tail, set perpendicular to the main rotor and parallel to the helicopter's longitudinal axis. Linked to the main engine by a drive shaft, the tail rotor produces lateral thrust to offset main rotor torque.

Another popular remedy is a twin tandem rotor configuration with two main rotors fore and aft, which rotate in opposite directions to neutralize torque. This method is used mainly in Boeing Vertol designs such as the CH-46 SEA KNIGHT and CH-47 CHINOOK.

The Soviet Kamov design bureau favors a coaxial solution, whereby two main rotors counter-rotate on a common shaft. That reduces overall length, a most valuable characteristic for shipboard helicopters, such as the KA-25 HORMONE and Ka-27 HELIX.

Side-by-side and side-by-side intermeshing configurations, in which counter-rotating main rotors are mounted perpendicular to the helicopter's longitudinal axis, have also been tried. McDonnel Douglas has recently developed a NOTAR (No Tail Rotor) configuration, in which the tail rotor is replaced by compressed air jets.

The ability of helicopters to take off and land vertically, to hover and fly at very low speeds, and to make radical turns as well as lateral maneuvers is obtained at the cost of severe disadvantages as compared to fixed-wing aircraft: mechanical complexity (with costs two or three times greater per pound), and—because they rely entirely on power lift—relatively inefficiency in terms of load carrying capability for the fuel used, as well as relatively low speed (maximum speed is less than 300 mph due to the limitations of rotor aerodynamics). Finally, helicopters are aerodynamically unstable, and place heavy demands on their pilots (though automatic systems are now alleviating some of the workload).

Helicopters nevertheless are ideally suited to a number of specialized military applications; the U.S. employed them in the Korean War for point-to-point transport, liaison, SCOUTING and RECONNAISSANCE, and casualty evacuation (CASEVAC). These roles remain predominant, but post-Korea roles include ANTI-SUBMARINE WARFARE (ASW), search and rescue (SAR), naval minesweeping, and ELECTRONIC WARFARE (EW), as well as fire support for AIR MOBILE forces, much used in the Vietnam War. Most recently, the helicopter emerged as an important ANTI-TANK weapon. To carry out these diverse roles, helicopter designs have become more specialized.

Transport configurations include light or "utility" helicopters (e.g., the UH-1 HUEY and UH-60 BLACKHAWK), medium lift helicopters (e.g., the CH-47 Chinook or Aerospatiale PUMA), and heavy lift types (e.g., the CH-53 SEA STALLION, Mi-12 HOOK, and Mi-26 HALO). Utility helicopters have also been outfitted for ASW, ambulance, light attack, and special operations (with silenced engines and rotors, night vision devices, and aerial refueling equipment).

For reconnaissance, liaison, and spotting, specialized "scout" helicopters have been developed (e.g., the Bell OH-58 KIOWA and McDonnell Douglas MD-500 DEFENDER); generally smaller and more maneuverable than utility helicopters, they often have light armament and TARGET ACQUISITION equipment.

Attack helicopters evolved from utility helicopters armed in improvised fashion with rocket pods and MACHINE GUNS to clear landing zones and provide fire support for transport helicopters. The first purpose-built design was the AH-1 COBRA of the mid-1960s, whose higher speed, greater maneuverability, reduced frontal area, armor protection, and integrated weapon/sensor systems still defines the type, except for the later Soviet "flying tank," the Mi-24 HIND—a much larger machine with heavy armament, high speed, and much armor, at the expense of stealth and maneuverability. Most

attack helicopters are now armed with ANTI-TANK GUIDED MISSILES, to serve as mobile anti-tank weapons. The effectiveness of the anti-tank helicopter has not yet been proven in combat, but a number of studies and exercises seem to indicate that under certain circumstances, helicopters may be very deadly indeed.

A new role now emerging is anti-helicopter warfare. The very fast and maneuverable Soviet helicopter code-named HOKUM may have been designed specifically to engage other helicopters.

HELIX: NATO code name for the Soviet Kamov Ka-27 shipboard HELICOPTER, meant for ANTI-SUBMARINE WARFARE (ASW), OVER-THE-HORIZON TARGETING (OTH-T), and search and rescue (SAR). Helix is essentially an enlarged and improved Ka-25 HORMONE with more powerful engines and a larger fuselage. The basic layout is the same, with the pilot and copilot seated side-by-side in the nose, and the main cabin in the rear. A large chin radome houses a surface-search RADAR, and two smaller radomes mounted on either side of the nose house ELECTRONIC SIGNAL MONITORING arrays for passive target detection and classification. Much of the fuselage and the tail boom are constructed of composite materials, and the main fuselage members are made of titanium for strength at reduced weight. The fuselage is sealed for buoyancy.

The ASW variant (Helix A), has the same weapons load as Hormone A, in an internal weapons bay built into the fuselage belly, but can carry 10 or 12 SONOBUOYS, compared to 3 for the Hormone. The OTH-T variant (Helix B) has a radar and DATA LINK system compatible with the SS-N-3 SHADDOCK, SS-N-12 SANDBOX, and SS-N-19 SHIPWRECK antiship missiles. The SAR version (Helix C) has a crew of three including a loadmaster; a power winch is fitted to the cabin's left door.

As compared to Hormone, Helix offers 225 percent more power and 167 percent greater range, without greatly increased dimensions. It is gradually replacing Hormone as the standard Soviet shipboard helicopter, and 18 have been acquired by the Indian navy for its two aircraft carriers, VIKRANT and VIRAAT, and its KASHIN-class DESTROYERS.

Specifications Length: 37.1 ft. (11.31 m.). **Rotor diameter:** 52.16 ft. (15.9 m.). **Cargo hold:** 14.83 × 4.25 × 4.45 ft. (4.52 × 1.3 × 1.36 m.). **Powerplant:** 2 2225-ship. Isotov TV3-117V turboshafts. **Weight, empty:** 14,220 lb. (6463 kg.). **Weight, normal loaded:** 24,250 lb. (11,022 kg.).

Weight, max. takeoff: 27,775 lb. (12,625 kg.). **Speed, max.:** 155 mph (259 kph). **Speed, cruising:** 143 mph (190 kph). **Range, max.:** 497 mi. (830 km.). **Endurance:** 4.5 hr.

HELLFIRE: The Rockwell AGM-114 ANTI-TANK GUIDED MISSILE (ATGM). Developed from 1971 in association with the AH-64 APACHE attack HELICOPTER, which can carry up to 16 on four quadruple launch racks, it can also be launched from the AH-1T Sea COBRA and AH-1W Super Cobra, the UH-60 BLACKHAWK, and the Westland LYNX utility helicopters. In full-scale production since the early 1980s, Hellfire has a cylindrical body with an ogival nose housing a laser seeker head; four small cruciform canards mounted behind the seeker head are used for steering, and four broad-chord cruciform wings near the tail provide lift and stability. Powered by a low-smoke Thiokol TX657 solid-fuel rocket motor, Hellfire is armed with a 7-in. (178-mm.), 20-lb. (9.09-kg.) HEAT warhead capable of penetrating all known MAIN BATTLE TANK armor.

Hellfire differs from most ATGMs in having SEMI-ACTIVE LASER HOMING rather than WIRE GUIDANCE: the seeker detects and homes on LASER radiation reflected off the target from a LASER DESIGNATOR. There are two guidance modes: Lock-On Before Launch (LOBL), when the missile has a direct line of sight (LOS) from the launch platform to the target; and Lock-On After Launch (LOAL), when the missile locks onto the target only when it is midway through its trajectory. In the LOAL mode, an autopilot stabilizes the missile until the target is acquired. LOAL allows the launch platform to fire from behind cover; a wire-guided missile, by contrast, must have an uninterrupted LOS. Hellfire targets may be illuminated by a laser designator aimed from the launch platform itself, other aircraft, or from the ground. Hence Hellfire can be ripple-fired in LOAL mode, to engage several targets almost simultaneously.

The U.S. Army is presently implementing a phased upgrade of the basic laser-homing variant, the "Hellfire Optimized Missile System" (HOMS), with a tandem-HEAT warhead capable of penetrating Chobham-type composite armor, an improved seeker hardened against laser countermeasures, and more reliable electronics. There is also a new "fire-and-forget" variant (Longbow), with active millimeter wave (MMW) homing, intended to provide the AH-64 with a better "all-weather" weapon. Brimstone is a similar variant being developed by Marconi for Royal Air Force Harrier attack aircraft.

In 1987, Sweden adopted a ground-launched, laser-homing derivative of Hellfire (RBS-17) for coastal defense. The U.S. Army is now developing a similar missile (Crossbow) for launch from a 4 × 4 HMMV ("Hummer") light truck. A shipboard variant for light patrol boats has also been proposed. Future developments could include a variety of new seekers, including a "fire-and-forget" IMAGING INFRARED (IIR) seeker, and a dual-mode seeker with both MMW and IIR.

Specifications **Length:** 5.33 ft. (1.625 m.). **Diameter:** 7 in. (178 mm.). **Span:** 13 in. (330 mm.). **Weight, launch:** 98.86 lb. (44.84 kg.). **Speed:** Mach 1.17 (890 mph/1486 kph). **Range:** 6000 m.

HEN HOUSE: NATO code name for a family of Soviet PHASED ARRAY tracking RADARS associated with the GALOSH anti-ballistic missile system. Hen House radars are large, and bistatic; i.e., the transmitting and receiving antennas are separately located, with each element housed in a building approximately 300 m. long and 20 m. wide. Hen House is believed to operate on a frequency of 150 megahertz, with a peak power of more than 10 megawatts. It is designed to detect incoming ballistic missile REENTRY VEHICLES out to a range of 3680 mi. (6145 km.) as they approach Soviet territory; this information is then passed to the DOG HOUSE and CAT HOUSE tracking radars of the Galosh system around Moscow. The 11 Hen House radars are now deployed at 6 sites along the periphery of the Soviet Union are being supplanted by a family of new, very large phased array radars deployed at 7 locations, including the one at Krasnoyarsk in central Siberia, which was cited by the U.S. government as a violation of the 1972 ABM TREATY. See also BALLISTIC MISSILE DEFENSE; BATTLE MANAGEMENT.

HEP: High Explosive, Plastic; U.S. Army term for HESH (High Explosive Squash Head) ANTI-TANK ammunition.

HERCULES: The Lockheed C-130 four-engine, turboprop-powered medium transport aircraft. Hercules is the most widely used transport aircraft produced in the West since World War II, with more than 1800 delivered to 57 countries, including 1115 to the U.S. forces, 582 to foreign military services, and 107 to commercial users.

Developed in response to a 1951 U.S. Air Force (USAF) requirement for a turboprop transport for use by the TACTICAL AIR COMMAND (TAC), the prototype YC-130A flew in 1954, and the C-130A production version entered service in 1956 (the first USAF Hercules is still in service with the New York Air National Guard). A total of 231 were built, followed by 320 C-130Bs (with increased fuel capacity, uprated engines, and strengthened landing gear) delivered between 1959 and 1961. Of the next major variant, the C-130E, with increased fuel capacity, a total of 503 were built, including the similar C-130F used by the U.S. Marine Corps (USMC). The current C-130H version has uprated engines. More than 470 have been delivered to date, with a further 230 on order.

All basic Hercules variants are externally similar, with a weather RADAR housed in the extreme nose, and the pilot, copilot, and navigator accommodated in a pressurized airline-type cabin equipped with a galley and fold-down bunks for long flights. The cargo hold behind the flight deck (also pressurized) has two pairs of side doors, and a large, fold-down ramp door in the rear. The tail section is swept upward to provide clearance for loading. Payload varies by model; 27,000 lb. (12,272 kg.) is a typical load. As a troop transport the Hercules can carry 64 paratroops, 92 passengers, or 70 stretcher cases. Jeeps, 2½-ton trucks, and light armored vehicles (including the M551 SHERIDAN) can be loaded though the ramp door, which can be opened during flight; cargo can be delivered by parachute or dropped from a few feet above the ground using the Low Altitude Parachute Extraction System. The tricycle landing gear is designed for rough field operations; the main wheels retract into characteristic sponsons on the lower sides of the fuselage. The wings, mounted above the hold to provide an unobstructed cargo floor, have a high aspect ratio for economical cruising, outboard ailerons, and large Fowler flaps which give good short takeoff and landing (STOL) performance. Wing pylons between the inboard and outboard engines can carry 1360-gal. (5984-lit.) fuel tanks or other podded payload. The agility of the Hercules allows its use in tactically demanding situations as an "assault transport."

One shortcoming of the Hercules is the limited volume of the cargo hold: operators often run out of space long before reaching maximum weight limits. To remedy this problem, the RAF version (C-130K) has a fuselage stretched 15 ft. (4.57 m.), allowing them to carry 7 standard cargo pallets (vs. 3); 92 paratroops (vs. 64); 128 passengers (vs. 92); or 97 stretchers (vs. 70). Lockheed now offers this stretched version as the L-100 series. A number of RAF C-130Ks have been equipped with aerial refueling probes and long-range navigation equipment to support British forces in the Falkland Islands.

Specialized variants include the EC-130H Compass Call (ELECTRONIC WARFARE), the EC-130Q TACAMO (submarine communications), WC-130 (meteorological), HC-130N (search and rescue, SAR), HC-130P (helicopter refueling tanker), HC-130H (long-range SAR), MC-130E/H COMBAT TALON (special operations), KC-130R/T (USMC tanker), LC-130R (ski-equipped arctic transport), KC-130H (tanker), DC-130H (drone carrier), HC-130N (space capsule recovery), and C-130H-MP (maritime patrol). Lockheed is also developing an AIRBORNE EARLY WARNING variant as a private venture.

The excellence of the Hercules design is demonstrated by the continuing inability of USAF and other users to develop a successor; production will probably continue indefinitely.

Specifications Length: 97.75 ft. (29.8 m.). Span: 132.6 ft. (40.42 m.). Cargo hold: 41.45 × 10.35 × 9.2 ft. (12.63 × 3.16 × 2.8 m.). Powerplant: (A) 4 Allison 3750-shp. T56-A-1A turboprops; (H) 4 4508-shp. T56-A-15s. Fuel: 46,970 lb. (21,350 kg.). Weight, empty: 76,469 lb. (34,758 kg.). Weight, max. takeoff: 175,000 lb. (19,545 kg.). Speed, cruising: 386 mph (645 kph). Initial climb: 1900 ft./min. (580 m./min.). Service ceiling: 33,000 ft. (10,060 m.). Range: (max. payload) 2356 mi. (3945 km.); (max. fuel) 4894 mi. (8173 km.).

HESH: High Explosive Squash Head, a type of ANTI-TANK ammunition consisting of a thin-walled shell filled with plastic explosives, with a base-detonating contact fuze. Upon hitting the target, the thin casing ruptures, spreading the plastic explosive on the ARMOR in "pancake" fashion, before the base fuze detonates. The shock wave generated by the explosion is transmitted to the armor, knocking large scabs off the inside of the plate ("spalling") which ricochet at high velocity inside the target, damaging equipment and personnel. HESH has been particularly favored by the British army; it is versatile, being effective against personnel and concrete bunkers, as well as armor, but its anti-armor effects have always been inferior to HEAT rounds, and can now be dissipated by SPACED ARMOR or CHOBHAM-type composite armor. HESH is designated HEP (High Explosive, Plastic) by the U.S. Army. See also ARMOR PIERCING.

HF: High Frequency, a band of the electromagnetic spectrum between 3 and 30 megahertz, used mainly for long-range COMMUNICATIONS. HF signals can travel for thousands of miles by repeated bouncing off the ionosphere and the earth's surface. Still used widely, especially by naval forces, HF is slowly being supplanted by satellite communications because of its severe drawbacks: it requires a large and elaborate antenna array, the signals are omnidirectional and thus relatively easy to intercept, and signal quality is greatly affected by atmospheric and solar phenomena (notably sunspots).

HIND: NATO code name for the Soviet Mil-24/25, twin-engine attack HELICOPTER, in service with Soviet Army Aviation and widely exported. First observed in 1973, the Hind is a specialized attack helicopter derived from the Mi-8 HIP transport helicopter, with the same engines, transmission, and other components. The initial Hind A version has an extensively glazed "greenhouse" nose equipped with a flexibly mounted, manually operated 12.7-mm. MACHINE GUN and stub wings for a variety of rocket and gun pods, and ANTI-TANK GUIDED MISSILES (ATGMs). Unlike other attack helicopters, Hind A has a large cabin behind the cockpit which can accommodate up to eight soldiers. The next two versions (Hind B and C) were essentially similar, with only minor improvements to the avionics and dynamic systems.

Introduced in 1978, the Hind D has a completely redesigned forward fuselage with a stepped tandem cockpit (providing better visibility for the rear seat pilot), a power-operated ventral gun turret, and various TARGET ACQUISITION and weapon aiming systems. The cockpit, in which the WEAPON SYSTEM OPERATOR (WSO) sits ahead of the pilot, and the fuselage belly are both very heavily armored for a helicopter (they have been known to resist penetration by 12.7-mm. machine-gun fire). The eight-man cabin remains, but is occasionally used for additional fuel or ammunition. Two large stub wings on the fuselage sides have pronounced anhedral and a high angle of incidence; at cruising speeds they can generate up to 25 percent of total lift, unloading the rotor system and improving both speed and maneuverability. Each wing has three hardpoints for rocket pods, 500-lb. high-explosive and chemical bombs, up to four AT-2 SWATTER or AT-3 SAGGER ATGMs, or a variety of submunition dispensers. A 4-barrel 12.7-mm. Gatling gun mounted in a flexible ventral turret under the nose is controlled by the WSO's sight, and linked to a prominent low-speed sensor which compenates for crosswinds.

Hind D has a comprehensive AVIONICS suite which includes electronic flight control and engine management systems, air-ground communications, and all-weather navigation equipment. ELECTRO-OPTICAL and forward-looking infrared (FLIR) sensors are mounted to the right of the gun turret, with a ranging RADAR on the left. A LASER DESIGNATOR is often fitted to the left wing tip. RADAR WARNING RECEIVERS and ELECTRONIC COUNTERMEASURE systems, including CHAFF/flare dispensers and an INFRARED COUNTERMEASURES (IRCM) beacon are standard equipment.

As in the Mi-8 Hip, the engines are mounted side-by-side in a nacelle over the cabin; the intakes can be fitted with dust and debris filters. The engine exhausts are now shrouded with infrared suppressors, reflecting combat experience in Afghanistan. The engines drive a fully articulated, five-bladed rotor, with severely limited high-g and negative-g maneuver envelopes. Each rotor blade incorporates a leading-edge anti-erosion strip and electrical deicing. The latest Hind E version is essentially identical to Hind D, but has a fixed 23-mm. cannon in place of the gun turret, and is armed with AT-6 SPIRAL ATGMs in place of Swatter or Sagger. Export versions (Mi-25) are generally equipped with simplified avionics and weapon.

Hind is the Soviet Union's most important attack helicopter; more than 1000 are in service, with high-volume production continuing for both Hind D and E. Its design reflects the Soviet Union's view of the helicopter as a "flying tank," which stresses speed, firepower, and armor in place of the stealth and agility emphasized in Western designs. The cabin, unique to Hind among attack helicopters, allows it to carry scout and anti-tank teams, or reload missiles, but adds considerably to the size and weight of the helicopter.

Hind has been used in combat in Afghanistan, Angola, and Nicaragua. As a result, several operational deficiencies have been revealed, most attributable to the Hind's derivation from the Mi-8: a lack of agility, a low power-to-weight ratio, poor high-altitude and hot-weather performance, and a large silhouette that makes it a relatively easy target. In spite of its IRCM suite, Hind has proved to be quite vulnerable to STINGER heat-seeking missiles. In response, the Soviet Union is developing a new attack helicopter, the Mi-28 HAVOC, which is smaller, faster, and more agile than Hind.

Specifications Length: 60.67 ft. (18.5 m.). **Rotor diameter:** 55.67 ft. (17 m.). **Powerplant:** 2 2200-shp. Isotov TV3-117 turboshafts. **Weight,**

empty: 16,534 lb. (7517 kg.). **Weight, max. takeoff:** 24,250 lb. (11,022 kg.). **Speed, max.:** (clean) 220 mph (305 kph); (w/stores) 199 mph (333 kph). **Speed, cruising:** 183 mph (305 kph). **Max. climb:** 2953 ft./min. (900 m./min.). **Combat radius:** 100 mi. (167 km.). **Operators:** Afg, Alg, Cuba, Eth, Ind, Iraq, Laos, Lib, Moz, Nic, Peru, S Ye, USSR and other Warsaw Pact forces.

HIP: NATO code name for the Mil Mi-8 and Mi-17 medium transport HELICOPTERS, in service with Soviet other Warsaw Pact forces, and widely exported; it is also produced in China as the Z-6. Hip is the world's most widely used helicopter, except for the ubiquitous Bell HUEY; total production has exceeded 11,000, exclusive of Chinese-built models.

Designed as a replacement for the Mi-4 Hound in the role of utility helicopter and assault transport, the first Hip A version flew in 1961, but technical difficulties delayed its service introduction until 1966. Since then, Hip has been produced in a number of essentially similar variants, including the Hip C (utility), Hip E (assault), and Hip J/K (ELECTRONIC WARFARE). The Mi-8 model was replaced in production in the early 1980s by the Mi-17, externally similar but equipped with more powerful engines.

The pilot and copilot sit side-by-side in the nose, in front of the flight engineer. The cabin is located immediately behind the cockpit. Passenger variants can carry up to 32 troops or 12 stretcher cases. Cargo variants, with large sliding side doors and a full-width clamshell door and ramp at the rear of the cabin, can carry up to 8800 lb. (4000 kg.) internally or 6600 lb. (3000 kg.) slung externally. The left cargo door can be equipped with a 331-lb. (150-kg.) electric hoist for search-and-rescue (SAR) missions. The main fuel tank, located beneath the cabin floor, can be supplemented by two external fuel tanks and two ferry tanks carried in the cabin; maximum fuel capacity is more than 5500 lb. (2500 kg.).

All military versions carry some form of armament. The Hip E assault variant (the most common) is often described as the most heavily armed helicopter in the world; it has a manually operated 12.7-mm. MACHINE GUN in the nose, and two weapon booms extending from the fuselage, each of which can carry three rocket pods with 32 57-mm. FFARS, or 3 550-lb. (250-kg.) high-explosive, NAPALM, or chemical bombs, plus 2 AT-2 SWATTER or AT-3 SAGGER anti-tank guided missiles.

The Mi-8 is powered by engines mounted side-by-side in a nacelle over the cabin, which drive a 5-bladed main rotor. The engine intakes can be fitted with dust and debris filters. Military Hips are equipped with an INFRARED COUNTERMEASURES beacon and CHAFF/flare dispensers. The newer Mi-17 Hip H has more powerful engines for better "hot-and-high" performance.

Hip has been used in combat in the Middle East, Afghanistan, Ethiopia, and Central America. It has proven to be rugged, reliable, and easy to maintain, but lacking in range, payload, and maneuverability. Except where high-altitude, hot-weather performance it needed, its operators seem quite satisfied with the Mi-8/17; no replacement is under development. The Mi-8/17 has served as the design basis for the Mi-14 HAZE, an ANTI-SUBMARINE WARFARE helicopter, and the Mi-24/25 HIND attack helicopter.

Specifications **Length:** 60.5 ft. (18.45 m.). **Rotor diameter:** 69.85 ft. (21.3 m.). **Cargo hold:** 17.55 × 7.67 × 5.85 ft. (5.34 × 2.34 × 1.8 m.). **Powerplant:** (8) 2 1700-shp. Isotov TV2-117A turboshafts; (17) 2 1900-shp. (2200-shp. emergency) Isotov TV3-117MTs. **Weight, empty:** (8) 16,005 lb. (7275 kg.); (17) 15,653 lb. (7106 kg.). **Weight, max. takeoff:** (8) 26,455 lb. (12,025 kg.); (17) 28,660 lb. (13,027 kg.). **Speed, max.:** (8) 143 mph (239 kph); (17) 155 mph (259 kph). **Speed, cruising:** (8) 112 mph (187 kph); (17) 149 mph (249 kph). **Range, max.:** (8) 276 mi. (460 km.); (17) 289 mi. (483 km.). **Operators:** Alg, Ang, Bang, Cuba, Egy, Eth, Fin, Gui-Bis, Iraq, Laos, Lib, Mal, Moz, N Ye, Nic, Peru, PRC, Som, Su, USSR and other Warsaw Pact forces, Viet, Zam.

HIRAM (DECOY): Hycor Infrared Anti-Missile, an expendable, shipboard INFRARED COUNTERMEASURE dispensed from the U.S. Mk.36 Super-Rapid Blooming Off-Board Chaff (SRBOC) launcher. HIRAM consists of a parachute flare attached to an inflatable float which simulates the INFRARED signature of warships, to "seduce" INFRARED HOMING anti-ship missiles. See also CHAFF; ELECTRONIC COUNTERMEASURES; ELECTRONIC WARFARE.

HIRAM (INFANTRY): Israeli army term for all elite LIGHT INFANTRY units, i.e., Israel's five parachute brigades, the 1st (Golani), and the Givati infantry brigades. Hiram troops are specially trained in mountain and URBAN WARFARE, as well as for less-demanding SPECIAL OPERATIONS and armor escort in close terrain. See also AIRBORNE FORCES, ISRAEL.

HMG: See HEAVY MACHINE GUN.

HOBOS: Homing Optical Bomb System, the Rockwell International GBU-8 ELECTRO-OPTICAL (EO) GLIDE BOMB, now in service with the U.S. Air Force (USAF) and the Israeli air force. Developed from 1966 in response to a requirement for a precision-guided standoff weapon for use against heavily defended point targets in North Vietnam, limited numbers were tested in Southeast Asia in 1968–69, and the weapon was used extensively in the 1972–73 "Linebacker" air campaigns.

HOBOS consists of a modular conversion kit (KMU-355A/B) for standard Mk.84 2000-lb. (909-kg.) or M118 3000-lb. (1364-kg.) free-fall bombs, with an EO seeker head bolted to the nose, a guidance unit bolted to the tail, four narrow body strakes, and four rectangular, cruciform tail fins with movable control surfaces. In 1972, Rockwell introduced an IMAGING INFRARED (IIR) seeker for use at night and in poor visibility. In action, the pilot or weapon operator on the attacking aircraft must first acquire the target visually, align it in his gunsight or HUD, and lock on the HOBOS seeker head by positioning the target in the crosshairs of a cockpit video monitor which displays the seeker field of view. When the pilot confirms the lock-on, he can release the weapon, which then homes autonomously onto a contrast edge in the target.

HOBOS is highly reliable and very accurate, with a CEP of under 10 ft., but the weapon has three serious drawbacks: first, it will only home on a high-contrast target, and does not perform well in hazy conditions or against camouflaged targets (the optional IIR seeker reduces but does not eliminate this problem); second, it has a glide range well within the range of enemy AIR DEFENSE weapons; finally, it must be locked on before launch, for which the aircraft must fly a straight approach at some altitude, during which it is vulnerable. To remedy these problems, USAF developed the GBU-15, which has a longer range and a DATA LINK which enables the attacking aircraft to release the weapon without visual target acquisition, for lock-on after launch.

Specifications Length: (Mk.84) 12.4 ft. (3.78 m.); (M118) 12.1 ft. (3.69 m.). Diameter: (Mk.84) 18 in. (457 mm.); (M118) 24 in. (610 mm.). Span: (Mk.84) 44 in. (1.12 m.); (M118) 52 in. (1.32 m.). Weight, launch: (Mk.84) 2240 lb. (1018 kg.); (M118) 3404 lb. (1547 kg.). Range: 2–3 mi. (3–5 km.).

HOE: See HOMING OVERLAY EXPERIMENT.

HOJ: See HOME-ON-JAM.

HOKUM: NATO code name for a Soviet attack HELICOPTER now under development by the Kamov design bureau, initially believed to be a fighter-helicopter designed specifically to attack other helicopters. At present, however, two theories predominate: (i) that Hokum was developed by Kamov as an unsuccessful competitor to the Mil-28 HAVOC, and (ii) that it is a specialized skip-board attack helicopter intended to support Soviet NAVAL INFANTRY. According to U.S. sources, the Hokum weighs approximately 12,000 lb. (5454 kg.), and has a slender fuselage with a two-seat tandem cockpit and twin turboshaft engines. Artists' renderings depict a helicopter similar to the Sikorsky XH-59 Advancing Blade Concept (ABC) experimental aircraft, with two coaxial main rotors (a favorite Kamov configuration), large stub wings, and prominent tail surfaces. Armament includes a nose-mounted gun turret, and wing-mounted air-to-air missiles, possibly a development of the SA-14 GREMLIN. U.S. sources credit Hokum with a maximum speed of 217 mph (363 kph), and a COMBAT RADIUS of about 150 mi. (250 km.), but the first photographs released in the West showed a conventional Kamov coaxial rotor, which makes such high performance unlikely. The helicopter may now be in full-scale production.

HOLY WAR: A war regarded by at least one party as having been inspired by, or fought for the benefit of, a deity, religion, or religious group. A JIHAD is the Islamic equivalent. The Second World War is described in the Soviet Union as the Great Patriotic War, fought on behalf of the (holy) entity of "Mother Russia" (Rodina), not the Union of Soviet Socialist Republics. Because of their ostensible motivation, holy wars are generally characterized by unusual fanaticism (leading to heavy casualties, frequent atrocities, and the possible obliteration of one or both parties, unless preceded by mutual exhaustion). The Iran-Iraq War was a typical example of a holy war, as was the MUJAHIDEEN effort against the Soviets in Afghanistan.

HOME-ON-JAM (HOJ): A missile guidance technique whereby data from a passive RADAR receiver is processed to direct the missile towards the emissions of an ACTIVE JAMMING system. Home-on-jam frequently serves as a backup mode for active or SEMI-ACTIVE RADAR HOMING missiles, selected when ELECTRONIC COUNTERMEASURES render the primary guidance inoperable. When operating in the HOJ mode, the missile essentially acts as an ANTI-RADIATION MISSILE (ARM). See also ELECTRONIC WARFARE.

HOMING OVERLAY EXPERIMENT: An important demonstration conducted by the U.S. STRATEGIC DEFENSE INITIATIVE Organization (SDIO) on 10 June 1984 to prove that BALLISTIC MISSILE REENTRY VEHICLES (RVs) can be intercepted by nonnuclear means. The HOE interceptor consisted of a MINUTEMAN II ICBM booster, an attitude control system, and an INFRARED HOMING system. Instead of a warhead, the interceptor had a set of 15-ft. (4.57-m.) folding radial spokes (much like the ribs of an umbrella), to extend its frontal area in order to increase the probability of hitting the target, which was to be destroyed by the force of impact only.

Launched from Kwajalein Atoll in the Pacific, the interceptor achieved a direct hit at a height of 100 mi. (167 km.) against an inert Minuteman III RV launched from Vandenberg Air Force Base in California. The target had deliberately been aimed 20 mi. (33.4 km.) off course, to test the ability of the interceptor to track and home out-of-plane. Relative velocity at impact was more than 10,000 mph (16,700 kph). Technology developed for HOE was incorporated into HEDI and ERIS, two proposed SDI terminal defense weapons. See also BALLISTIC MISSILE DEFENSE.

HONEST JOHN: The MGR-1, U.S.-built unguided, spin-stabilized battlefield bombardment rocket of 1950s vintage. No longer used by the U.S. Army, it is still in limited service in South Korea.

HOOK: NATO code name for the Soviet Mil Mi-6 heavy transport HELICOPTER. At the time of its introduction in 1957, Hook was the largest helicopter in the world and also set a number of speed records which stood for many years. Designed to a 1954 civil/military requirement for the transport of heavy equipment such as bulldozers, light armored vehicles, and missile launchers, in an internal cargo hold, Hook proved to be very successful, with more than 800 in service with Aeroflot and Soviet Military Transport Aviation (VTA), and the air forces of Algeria, Egypt, Iraq, Peru, and Vietnam. It is only now being replaced by the even larger, more capable Mi-26 HALO.

The fuselage has a glazed nose position for the navigator/observer, who also mans manually operated 12.7-mm. DSHK machine gun in military versions. The pilot, copilot, radio operator, and flight engineer are accomodated on an airline-style flight deck above and behind the navigator. The cargo hold extends from the flight deck to a pair of clamshell doors at the rear, which incorporate a hydraulically actuated vehicle ramp. A 1764-lb. (802-

kg.) cargo winch is standard equipment in the hold. Hook can carry a maximum payload of 26,-455 lb. (12,025 kg.) internally, or 19,841 lb. (9019 kg.) slung externally. Tip-up seats along the cabin walls can accommodate up to 70 fully equipped troops; an additional 20 seats can be bolted to the cabin floor. Alternatively, up to 40 stretcher cases and two attendants can be carried in the CASEVAC role.

Of unique size in a helicopter, stub wings mounted near the top of the fuselage generate nearly 25 percent of total lift at cruise speed, thereby unloading the rotor and greatly enhancing both speed and range (at the expense of maneuverability). Hook is powered by a pair of engines mounted side-by-side in nacelles above the fuselage, driving a fully articulated five-bladed rotor. It was the first helicopter in the world with twin turboshaft propulsion, and the first to have its engines above the cabin, in order to leave it unobstructed. The Mi-6 also served as the design basis for the Mi-10 Harke flying crane.

Specifications **Length:** 108.85 ft. (33.2 m.). **Rotor diameter:** 114.83 ft. (35 m.). **Span:** 50.2 ft. (15.3 m.). **Cargo hold:** 39.33 × 8.7 × 8.5 ft. (12 × 2.65 × 2.6 m.). **Powerplant:** 2 5500-shp. Soleviev D-25V turboshafts. **Fuel:** 25,800 lb. (11,727 kg.). **Weight, empty:** 60,053 lb. (27,297 kg.). **Weight, max. takeoff:** 93,695 lb. (42,588 kg.). **Speed, max.:** 186 mph (310 kph.) **Speed, cruising:** 155 mph (259 kph). **Range, max.:** (2/payload) 373 mi. (623 km.).

HORMONE: NATO code name for the Kamov Ka-25, the standard Soviet shipboard HELICOPTER, in service with Soviet Naval Aviation (AV-MF) and the navies of India, Syria, and Yugoslavia. First observed in 1961 (and initially assigned the code name Harp), the Ka-25 entered service in 1965 and was redesignated Hormone. Its distinctive coaxial rotor configuration minimizes overall length, facilitating its operation from the relatively small flight decks of Soviet FRIGATES and DESTROYERS.

Hormone has a squat fuselage and twin vertical stabilizers on a short tail boom. The pilot and copilot sit side-by-side in the nose. Immediately behind them, the main cabin has a large sliding door on the left side. A large chin radome housing a surface-search RADAR is mounted below the cockpit. The helicopter has fixed, four-poster landing gear to which inflatable floats can be attached. The fuel tanks and a weapons bay are beneath the cabin floor. The engines are mounted side-by-side in nacelles above the cabin, driving two 3-bladed, counter-rotating main rotors that fold hydraulically for hangar storage.

Hormone has been built in three variants, for over-the-horizon targeting (OTH-T), ANTI-SUBMARINE WARFARE (ASW), and search and rescue (SAR). The ASW version (Hormone A) is armed with two 17.7-in. (450-mm.) homing TORPEDOES or 1–2 conventional or nuclear DEPTH CHARGES, plus three SONOBUOYS, dye markers, and smoke pots. AVIONICS include a "Big Bulge" I/J-band radar in the chin radome, a data link system, a "Tee Rod" ELECTRO-OPTICAL sensor, a radar altimeter, DOPPLER radar, and an autohover system. A MAGNETIC ANOMALY DETECTOR and dipping SONAR are standard equipment. Three ASW operators and their various displays are in the main cabin.

The OTH-T version (Hormone B) can remotely provide TARGET ACQUISITION for ships or submarines armed with SS-N-3 SHADDOCK, SS-N-12 SANDBOX, or SS-N-19 SHIPWRECK anti-ship missiles. Its chin radome houses a "Short Horn" surface-search radar; a second radar may be housed in a smaller radome beneath the main cabin. The SAR version (Hormone C) has the same radar as the ASW version, and is equipped with an electric rescue hoist mounted by the cabin door.

More than 400 Hormones of all types were produced between 1965 and 1975. Since 1975, Hormone has been replaced in production by the vastly improved Ka-27 HELIX. As Helix becomes available in larger numbers, Hormones are being retired or transferred to training or utility duties.

Specifications Length: 32 ft. (9.75 m.). **Rotor diameter:** 51.64 ft. (15.75 m.). **Cargo hold:** 12.95 × 4.95 × 4.10 ft. (3.95 × 1.51 × 1.25 m.). **Powerplant:** 2 990-shp. Glushinkov GTD-3BM turboshafts. **Weight, empty:** 10,500 lb. (4773 kg.). **Weight, max. takeoff:** 16,535 lb. (7516 kg.). **Speed, max.:** 137 mph (229 kph). **Speed, cruising:** 120 mph (200 kph). **Range, max.:** 400 mi. (668 km.).

HORNET: The McDonnell-Douglas FA-18 twin engine, single-seat FIGHTER/ATTACK AIRCRAFT, in service with the U.S. Navy and Marine Corps, and the air forces of Australia, Canada, and Spain.

The Hornet was derived from the Northrop YF-17, a lightweight, clear-weather-only AIR SUPERIORITY fighter designed in the early 1970s for the U.S. Air Force lightweight fighter competition (eventually won by the F-16 FALCON). The U.S. Navy originally specified a less costly supplement to the F-14 TOMCAT fighter, and commissioned a carrier-based development of the YF-17 to be built by McDonnell-Douglas, with Northrop as subcontractor. Along the way, the navy expanded its requirement to include a comprehensive ground-attack capability, to enable the aircraft (now designated FA-18) to replace the A-7E CORSAIR and F-4N/S PHANTOM in navy and marine corps attack squadrons. The navy concurrently demanded all-weather avionics and the radar electronics required for radar-guided AIM-7 SPARROW (and later AIM-120 AMRAAM) air-to-air missiles. Thus the final Hornet design was much heavier and far more complex than the YF-17. The prototype YF-18A flew in late 1978, the Hornet entered service in early 1983, and it has since replaced the A-7E in all active navy light attack squadrons (now redesignated fighter/attack squadrons), and is replacing the F-4 in marine fighter/attack squadrons. The navy plans to equip 6 fighter and 24 attack squadrons with the Hornet, while the Marines have a requirement for 12 fighter and 20 attack squadrons; or a total of 1366 aircraft including spares (of which more than 400 have been delivered to date).

The Hornet at first attracted much criticism due both to escalating costs and its inferiority to the A-7E in terms of range and payload. With greater service experience, however, the Hornet was better appreciated as reliable, highly maneuverable, and extremely accurate as a weapon platform.

The design of the FA-18 reflects the state of the art in aerodynamics as of the late 1970s. The fuselage center section, containing the cockpit and AVIONICS, merges towards the tail into twin engine pods. The nose houses a Hughes APG-65 multimode RADAR with DOPPLER beam sharpening to improve resolution in the ground mapping mode. Other modes include air search, boresight, tracking, and ranging. Capable of tracking ten targets simultaneously, and displaying eight, the radar is compatible with both Sparrow and AMRAAM, and is linked to the Hornet's digital navigation and weapon delivery system.

The nose section behind the radar contains an avionics compartment, an M61 VULCAN 20-mm. cannon, and a retractable AERIAL REFUELING probe mounted on the right. The pilot sits in an elaborately equipped cockpit under a two-piece bubble canopy with all-round visibility. Cockpit avionics include a large Head-Up Display (HUD) and three multi-functional video displays which variously show radar and other sensor data, and monitoring data for weapons and systems. Essential switches are configured for "Hands on Throttle and Stick"

(HOTAS) operation, i.e., the pilot need never look down during combat.

The section behind the cockpit contains additional avionics and fuel cells. A large, paddle-type speed brake is located on top of the fuselage, and an arrester hook is mounted under it. The Hornet's tricycle landing gear is stressed for carrier landings, and the nose gear incorporates a catapult tow bar. The engines, mounted side-by-side in pods blended into the main fuselage, are fed by simple fixed inlets under the wing roots.

The unswept, tapered wings have exceptionally large leading-edge root extension (LERX), which generate additional lift, reduce drag, and improve control at high angles of attack (AOA). The wings, which can fold up near the tips to facilitate deck parking, have full-span leading edge flaps and double-sloted trailing edge flaps (which automatically match wing camber to AOA), as well as outboard ailerons for roll control. The Hornet has a unique swept tailplane, with twin vertical stabilizers at the wing trailing-edge roots, canted outboard. The all-moving horizontal stabilizer is at the extreme tail, behind the vertical stabilizers; this unusual configuration reduces interference drag and improves control at high AOA.

The Hornet can carry up to 17,000 lb. (7272 kg.) of ordnance and fuel on 6 wing and 3 fuselage pylons. Two wingtip pylons are exclusively for AIM-9 SIDEWINDER air-to-air missiles. In the fighter role, standard armament is 2 to 4 infrared-homing Sidewinders, and 2 Sparrow or AMRAAM carried on hardpoints under the engine intakes. In the attack role, the aircraft can carry a variety of free-fall and retarded bombs, LASER-GUIDED BOMBS, NAPALM, CLUSTER BOMBS, AGM-65 MAVERICK and AGM-12 BULLPUP air-to-ground missiles, AGM-65 WALLEYE glide bombs, AGM-45 SHRIKE and AGM-88 HARM anti-radiation missiles, and AGM-84 HARPOON anti-ship missiles. The centerline and inboard wing pylons are also plumbed for 300-gal. (1320-lit.) drop tanks. Typical mission payloads include 4 Maverick or Harpoon missiles, or up to 9000 lb. (4070 kg.) of bombs. Two additional light hardpoints under the LERX can be fitted with a Forward Looking Infrared (FLIR) and an ASQ-173 LASER Spot Tracker (LST) for low-level attacks at night and in other conditions of poor visibility. The ELECTRONIC COUNTERMEASURES include internal RADAR HOMING AND WARNING RECEIVERS, ACTIVE JAMMERS, and CHAFF/flare dispensers.

A two-seat operational trainer version (FA-18B) has the same physical dimensions and perform-

ance as the single-seater, but carries 6 percent less fuel. It retains full combat capabilities and may be used as a pathfinder for night attacks.

Canadian, Australian, and Spanish Hornets are more austere, without carrier capability and equipped with simplified avionics. The U.S. Navy has initiated development of the FA-18C, an improved single-seater, while the marines are acquiring the FA-18D, a two-seater configured for forward air control (FAC) and RECONNAISSANCE. See also AIRCRAFT CARRIERS, UNITED STATES.

Specifications Length: 56 ft. (17.07 m.). Span: 37.5 ft. (11.43 m.). Powerplant: 2 General Electric F404-GE-400 afterburning turbofans, 16,-000 lb. (7272 kg.) of thrust each. Fuel: 10,860 lb. (4936 kg.). Weight, empty: 20,583 lb. (9356 kg.). Weight, normal loaded: 35,000 lb. (15,910 kg.). Weight, max. takeoff: 50,064 lb. (22,756 kg.). Speed, max.: Mach 1.83 (1190 mph/1987 kph) at 36,000 ft. (11,000 m.)/800 mph (1336 kph) at sea level. Initial climb: 50,000 ft./min. (15,244 m./min.). Service ceiling: 49,000 ft. (14,939 m.). Combat radius: (fighter) 460 mi. (768 km.); (attack) 550 mi. (918 km.). Range, max.: 2300 mi. (3841 km.).

HOT: *Haute-subsonique Teleguide Optiquement Tire d'un Tube* (literally, High Subsonic, Optically Guided, Tube Fired), a wire guided ANTITANK GUIDED MISSILE (ATGM) produced by the Franco-German Euromissile consortium. The missile originated in a 1964 joint Franco-German requirement, and entered mass production in 1977. HOT is not man-portable, but is mounted on a variety of vehicles and helicopters, including the BO-105 P/C, SA.342M GAZELLE, SA.361H Dauphin, and Westland LYNX helicopters, and the French VAB and West German JAGBPANZER *Rakete* anti-tank vehicles.

Four tail-mounted cruciform wings provide lift and stabilization, while steering is accomplished by thrust vector control (TVC) of engine exhaust. Powered by a dual-impulse (booster/sustainer) solid rocket motor, it was originally armed with a 5.3-in. (135-mm.) shaped-charge (HEAT) warhead (HOT 1) that could penetrate 31.5 in. (800 mm.) of steel ARMOR; the current HOT 2 version has has a 6.5-in. (165-mm.) warhead capable of penetrating more than 50 in. (1.27 m.) of steel armor.

HOT is prepackaged in a sealed container/launch tube, which is attached to a rail mounted on the firing platform. Its semi-automatic command to line-of-sight (SACLOS) guidance only requires the operator to keep the target centered in the optical sight; by tracking the missile's infrared signal, the guidance system automatically directs the missile to the point in the crosshairs, i.e., the target, transmitting its steering commands through two wires paid out from an inertialess spool in the base of the missile.

When the operator fires the missile, the booster ignites inside the tube, burning for 0.9 sec. and accelerating the missile to cruising speed. The sustainer pulse then ignites and burns for 17.4 sec. The missile takes 8.7 sec. to reach 2000 m., 12.5 sec. to reach 3000 m., and 16.3 sec. to reach 4000 m. Warhead fuzing and the need to gather the infrared signal impose a minimum range of 400 m. An electronic fuze detonates the warhead as soon as the streamlined nose cap is deformed by an impact.

Specifications Length: 4.18 ft. (1.27 m.). Diameter: 6.5 in. (165 mm.). Span: 12.3 in. (312 mm.). Weight, launch: 55 lb. (16.76 kg.). Speed: 560 mph (260 m./sec.). Range: 4000 m. Operators: Egy, Fr, FRG, Iraq, Ku, Lib, S Ar, Sp, Syr.

HOVERCRAFT: Also called air cushion vehicle (ACV) and ground effect vehicle (GEVs); a craft that rides on a cushion of air generated by ducted fans in a space beneath the hull, the plenum chamber. The latter is normally formed either by solid walls that extend below the sides of the vehicle, or by inflatable rubber skirts. While the ducted fans direct air downward into the plenum chamber to maintain the air cushion which raises the hovercraft a few inches off the surface, aircraft-type propellers induce motion. Thus hovercraft are the only ground vehicles that can transit over water, land, and soft terrain indifferently. Steering is by aircraft-type rudders, or by pivoting the propellers, which also stop (or reverse) the vehicle by pitch variation. Most hovercraft are powered by gas turbine engines.

Although in widespread commercial use, hovercraft have not been acquired in large numbers by military forces, except for the Soviet Union. The U.S. did experiment with hovercraft during the Vietnam War (they proved ideal for riverine warfare and traversing swamps and rice paddies), but the U.S. Navy's rejection of all small units precluded further development. The Soviet Union, by contrast, operates more than 60 hovercraft, including the 27-ton Gus class, the 86-ton Lebed class, and the 250-ton AIST class, which can carry two MAIN BATTLE TANKS (MBTs) or up to 120 troops at a speed of more than 80 kt. over a range of 200–300

n.mi, and which are armed with two twin 30-mm. radar-controlled anti-missile cannons.

In spite of their obvious suitability for AMPHIBIOUS WARFARE, the U.S. Marine Corps only began procurement of 90 Landing Craft, Air Cushion (LCACS) in 1984. They weigh roughly 88 tons and can carry up to 60 tons of cargo (including one MBT), at a speed of 50 kt. over a range of 120 n.mi.

Hovercraft have the potential to revolutionize amphibious warfare by virtue of their speed and their ability to deliver without the need for an unobstructed beach. Unlike conventional landing craft, which must be loaded and sent off fairly near the shore, hovercraft can be loaded well over the radar horizon and still arrive very quickly. Other possible military applications include ANTI-SUBMARINE WARFARE, MINE COUNTERMEASURES, search and rescue (SAR), and casualty evacuation (CASEVAC). Hovercraft are far more expensive than ships on a per-ton basis, but are very much cheaper than helicopters, and their military uses will undoubtedly expand.

HOWITZER: The most common type of ARTILLERY, designed to fire high-explosive (HE) projectiles in a lofted trajectory. Howitzers have relatively short barrels compared to GUNS (ranging in length between 23 and 39 CALIBERS), which impart low muzzle velocities to their shells. Bores range between 75 mm. and 203 mm., with 105 mm. and 155 mm. being the most common in Western armies, and 122 mm. and 152 mm. the most common in Soviet-style forces. Most modern howitzers are self-propelled.

True howitzers have a relatively short range, and are being supplanted in most armies by gun-howitzers, hybrids with barrels longer than the classic howitzer limit.

HQ: See HEADQUARTERS.

HUD: Head-Up Display, a cockpit device for aircraft which projects key flight data, system status, RADAR symbology, and weapon aiming cues on a transparent glass plate directly in the pilot's line of sight, thereby enabling him to fly the aircraft, monitor systems, and aim weapons without having to look down to examine cockpit instruments at critical moments. HUDs have become priority equipment for all tactical aircraft (FIGHTERS, INTERCEPTORS, and ATTACK AIRCRAFT), and they are now being installed in other types as well.

In the now-standard technique that dates from the early 1960s, a glass or acrylic plate, mounted at an oblique angle on the top of the instrument panel facing the pilot, acts as a see-through display. Symbology and alphanumeric data generated by an on-board computer are projected onto the inclined plate, where they appear as bright lines and letters focused at infinity. The information displayed usually includes airspeed, altitude, height, attitude, rate of climb or descent, heading, fuel state, weapon status, radar mode, and, above all, weapon aiming cues.

The latest HUDs have wide-angle screens and employ high-resolution "raster" scanning which can also project video images from forward looking infrared (FLIR) or low light television (LLTV) sensors, facilitating flight at low altitude at night or in poor visibility. Such "holographic" HUDs are becoming standard equipment on latest-generation combat aircraft (including the F-16C FALCON, FA-18 HORNET, and Saab JAS.39 GRIPEN).

HUEY: Unofficial but universally recognized nickname for the Bell 204/205/212/214/412 family of utility HELICOPTERS, in service with military and civil operators worldwide. The nickname is derived from the original U.S. Army designation, HU-helicopter, utility-1, changed to UH-1 in 1962. The most widely used helicopter in the world, more than 9000 Hueys of all models have been built, including large numbers produced under license by Agusta in Italy, Dornier in West Germany, Fuji in Japan, and AIDC in Taiwan.

The Huey was originally developed in response to a U.S. Army requirement for a CASEVAC and utility helicopter. The resulting 1956 prototypes (Model 204) were the first "jet" (i.e., turboshaft-powered) helicopters ordered by a U.S. service. After minor modifications, it was placed in production as the HU-1A Iroquois (later changed to UH-1A), which could carry a crew of 2, plus 6 troops or 2 stretcher cases. It was quickly superseded by the UH-1B, with a larger main rotor, an enlarged cabin, and room for 7 passengers or 3 stretcher cases. In 1965, Bell introduced the UH-1C, with revised rotor blades for improved speed and maneuverability. Other versions of the basic Model 204 included the UH-1E for the U.S. Marine Corps (with a rescue hoist and extra avionics), the UH-1L, with further engine upgrades, and the UH-1M, with night vision equipment. The Model 204 remains in production in Italy, Japan, and Taiwan.

In 1960, Bell developed the Model 205, with a longer cabin, relocated fuel cells, and room for a pilot and up to 14 troops or 6 stretcher cases. First flown in 1961, the Modle 205 entered service in 1963 as the UH-1D. An improved version, the UH-1H, produced since 1967, has a more powerful engine and became the single most numerous version of the Huey, with 3573 built for the army alone, plus 1357 for export. An additional 118 were built in Taiwan. The Model 205 also remains in production in Italy and Japan.

The representative UH-1H has a pod-shaped fuselage and oval-section tail boom. The pilot and copilot/observer sit side-by-side in the nose, each with an automobile-type side door. The main cabin, contiguous to the cockpit, has an internal volume of 220 cubic ft. (6.23 m³). Bench seating can accommodate up to 14 troops or 6 stretchers. With seats removed, up to 3880 lb. (1763 kg.) of cargo can be carried internally. Large sliding side doors allow rapid entry and exit, and easy cargo loading. All versions of the Huey have tubular landing skids, which can be fitted with ground-handling wheels or flotation bags.

The engine, mounted in a nacelle above the cabin, drives a 2-bladed rotor using the well-proven Bell "teeter-totter" rotor system, in which the blades are counterbalanced by a pair of "stabilizer bars" mounted rigidly on the hub at right angles to the blades. The engine also powers the tail rotor through a drive shaft running along the top of the tail boom. All versions of the Huey have a horizontal stabilizer mounted on the tail boom, which is hydraulically actuated by an automatic flight control system to keep the fuselage level during forward flight.

AVIONICS suites vary, but usually include at least UHF and VHF radio and blind flying instruments. (The U.S. Army plans to equip 2700 UH-1Hs with a completely new avionics system including HF communications, a RADAR altimeter, DOPPLER radar, ELECTRONIC COUNTERMEASURE equipment, an INFRARED COUNTERMEASURES beacon, a CHAF/fl/flare dispenser, and an INFRARED exhaust suppressor.)

Most Hueys are unarmed transports, but the army began arming some Hueys early in the Vietnam War to provide fire support for AIR MOBILE operations. Known as "Hogs," they carried pintle-mounted door machine guns, gun and rocket pods bolted to the skids, and, by 1972, TOW anti-tank guided missiles. Other Huey operators have added their own choice of light weapons. Although not as fast, maneuverable, or well protected as specialized attack helicopters, armed Hueys provide an economical substitute for many users.

Agusta Bell developed shipboard ANTI-SUBMARINE WARFARE (ASW) variants equipped with dipping SONAR and homing TORPEDOES, used by the

Italian and other navies, mainly because their size suits smaller FRIGATES and CORVETTES. Other specialized variants include SAR, electronic intelligence (ELINT) gatherers, and communications-relay platforms.

In 1968, Bell developed the much more powerful Model 212, which combined the 205 airframe with a more powerful engine, for improved lift capability and better "hot and high" performance. The U.S. armed forces ordered 300 of these as UH-1Ns between 1970 and 1979. The Model 212 production continues in Italy, where Agusta Bell has produced an ASW variant equipped with a roof-mounted search radar and dipping sonar, with provisions for SONOBUOYS, depth charges, homing torpedoes, and anti-ship missiles.

In 1970, Bell introduced the more powerful Model 214 "Huey Plus," with twin engines and a stretched cabin with room for 19 troops. Iran was the single largest purchaser, with the shah ordering 272 in 1972, of which many are still in service. Others went to China, Thailand, Uganda, and Venezuela.

The Model 412, introduced in 1978, has the 212 fuselage, but with an advanced, 4-bladed rotor. The 412 is in service with Bahrain, Indonesia, Nigeria, and Venezuela. Agusta is producing a modified AB.412 Griffon for the Italian army, with crash-resistant landing skis, self-sealing fuel tanks, armored, crash-resistant seats, and hardpoints for a variety of ordnance. Turkey has ordered a naval version armed with British SEA SKUA anti-ship missiles. Other users include Spain, Lesotho, Singapore, and Zimbabwe.

The Huey has never been an outstanding performer; its main virtues have been ruggedness, reliability, ease of maintenance, and relatively low cost. Huey variants will probably remain in production until the next century. The rotors, engine, and transmission of the Huey served as the basis for the AH-1 COBRA attack helicopter, originally known as the Huey Cobra.

Specifications Length: (H) 41.9 ft. (12.77 m.); (212/412) 42.4 ft. (12.92 m.); (214) 50 ft. (15.24 m.). **Rotor diameter:** (H) 48 ft. (14.63 m.); (212) 48 ft. (14.63 m.); (214) 52 ft. (15.85 m.); (412) 46 ft. (14 m.). **Powerplant:** (H) 1 1400-shp. Lycoming T53-L-11 turboshaft; (212) 1 1800-shp. Pratt and Whitney Canada PT6T-3 twin turbine; (214) 2 1625-shp. General Electric CT7-GE-2A turboshafts; (412) 1 1308-shp. Pratt and Whitney Canada PT6T-3B-1 twin turboshaft. **Fuel:** (H) 1302 lb. (592 kg.); (212) 3065 lb. (1384 kg.). **Weight, empty:**

(H) 5210 lb. (2368 kg.); (212) 6144 lb. (2792 kg.). **Weight, max. takeoff:** (H) 9500 lb. (4318 kg.); (212) 11,200 lb. (5091 kg.); (214) 15,000 lb. (6818 kg.). **Speed, max.:** (H) 127 mph (212 kph); (212) 142 mph (238 kph); (412) 142 mph (238 kph). **Speed, cruising:** 127 mph (204 kph). **Range, max.:** (H) 318 mi. (531 km.); (212) 261 mi. (436 km.); (412) 230 mi. (384 km.). **Operators** (all versions): Arg, Aus, Bah, Bang, Bol, Bra, Bur, Can, Chi, Col, Cos R, Dom Rep, Dub, Ecu, El Sal, FRG, Gha, Gre, Gua, Guy, Hon, Indo, Iran, Isr, Jam, Jap, Lib, Mex, Nor, NZ, Pak, Pan, Peru, Phi, ROK, Sing, Sp, Sri L, Tai, Tan, Thai, Tur, Ug, Ur, US, Ven, Viet.

HUMINT: Human Intelligence, U.S. official jargon for information collected from human sources, including travelers, agents-in-place, spies, defectors, refugees, and prisoners of war. HUMINT represents only a small portion of the total spectrum of INTELLIGENCE sources, but is the only source capable of providing insight into the insubstantial—e.g., enemy plans, intentions, and attitudes. See also ELINT, SIGINT, TECHINT.

HUNTER: British single-seat, single-engine subsonic FIGHTER/ATTACK AIRCRAFT of 1950s vintage; widely exported, a few remain in service in smaller air forces. **Operators:** Chi, Ind, Iraq, Leb, Oman, Qat, Sing, Som, Swi, Zim (Britain retains a few as trainers and calibration aircraft).

HVAP: Hyper-Velocity, Armor Piercing, a form of ARMOR-PIERCING ammunition also called "rigid composite shot," which consists of a dense tungsten core or penetrator surrounded by a light alloy jacket or "carrier" to make up the full bore. Because the overall weight of HVAP is thus less than that of homogenous (monobloc) shot, muzzle velocity is considerably higher for any given gun. HVAP has excellent armor-penetrating capability at short to medium ranges (500–1000 m.) but performance falls off rapidly at longer ranges due to its low ballistic coefficient (the ratio of mass to frontal area). HVAP has therefore been superseded by "discarding sabot" ammunition (APDS and APFSDS), in which the carrier or "sabot" separates upon leaving the muzzle of the gun, raising the ballistic coefficient to the penetrator's level.

HVM: Hyper-Velocity Missile, an ANTI-TANK missile under development by LTV for the U.S. Army and Air Force. Unlike most anti-tank missiles, the HVM has no warhead and relies entirely on the kinetic energy of a long-rod penetrator of tungsten or depleted uranium (STABALLOY), which—unlike normal missile shaped-charge warheads—is unaffected by spaced or composite

ARMOR. The HVM would be launched from multitube pods, mounted on ground vehicles or aircraft, which would be laid on target by an electronically scanning laser radar (LADAR). The extremely high speed of the missile, more than 3400 mph (1520 m./sec.), comparable to the muzzle velocity of tank guns, would keep the time of flight to under one second in most cases, eliminating any need for post-launch guidance.

LTV is currently developing a 40-round launcher aimed by a CO_2 LASER, which would also provide 3-dimensional range and DOPPLER data. Tests have been conducted at ranges up to 3.7 mi. (6.2 km.). HVM is meant as a simple, reliable, low-cost weapon (the 1989 claimed unit cost was under $5000).

HYDROFOIL: A surface vessel configuration employed in some FAST ATTACK CRAFT (FACs), in which winglike foils lift the hull out of the water at speed, thereby overcoming surface drag and the bow-wave effect to achieve very high speeds with considerable maneuverability. There are two basic types: those with completely submerged foils mounted on struts protruding from the hull, and those with semisubmerged "surface-piercing" foils mounted directly on the hull. The latter is the preferred type in Soviet and Chinese designs.

In both types, the foils act on the same principles as airplane wings, generating lift from the flow of water over their surfaces, thereby lifting the entire vessel. Surface-piercing foils, usually in a V or W shape, may or may not be fitted with movable control surfaces, or "flaps." In the latter case, lift depends entirely on speed, and the vessel acts as a hydroplane. Surface-piercing hydrofoils thus do not require complex stabilization systems, but like hydroplanes, they tend to pound heavily in rough seas, and are most suitable for calmer coastal waters.

Fully submerged foils can either be rotated horizontally or are fitted with movable flaps to control lift. Most are fitted with automated control systems similar to aircraft autopilots, to compensate for pitch and roll due to wave action, and to keep the vessel at a constant height above the wave crests, thereby providing a smooth ride. If the seas are particularly rough, such craft tend to follow wave contours, forcing a reduction in speed to avoid overstressing the hull and crew. Their control systems, with stabilized inertial platforms and/or radar altimeters as well as computer-directed hydraulic actuators, multiply the cost per ton of hydrofoils as compared to conventional displace-

ment vessels, but such hydrofoils are much more stable in higher sea states.

Alternative foil configurations include the "airplane" type, with a long foil forward and a shorter foil aft, which is less effective when the vessel is hull-borne; the tandem, in which two foils of equal length are spaced equidistant from the vessel's center of gravity, and compatible with a normal boat hull; and the canard, with a short foil forward with a longer foil aft, which is most suitable for seakeeping when hull-borne.

Submerged foils protrude deep into the water, and some form of retraction is therefore necessary for berthing and shallow-water operations. By contrast, surface-piercing foils (which are fixed rigidly) adversely affect the motion of the vessel when hull-borne, and expose the foils to damage from debris and wave motion.

Surface-piercing hydrofoils are usually powered by conventional propellers mounted on long shafts. Fully submerged hydrofoils, which ride too far out of the water for that, are usually powered by water jets mounted on the foil struts.

Military hydrofoils have a number of operational advantages over conventional hulls. Because their speed can exceed 50 kt., they are ideal as fast attack craft for hit-and-run operations; they are also less vulnerable to TORPEDOES and MINES while foil-borne, and can use their maneuverability to evade ANTI-SHIP MISSILES. Finally, they may have considerable potential for anti-submarine warfare, with dipping sonar or towed arrays, drifting quietly until a target is detected, then sprinting swiftly to attack.

In the West, rigid service distinctions have blocked the application of this technology. The Soviet Union, by contrast, is the largest user of military hydrofoils, with more than 100 in service, including the MATKA, TURYA, Sarancha, and Pchela classes, all with surface-piercing foils. China also has some 100 Hu Chwan surface-piercing hydrofoils, several of which have been sold abroad. The United States operates only six highly sophisticated and very costly missile-armed hydrofoils of the PEGASUS class, while Italy has seven smaller SPARVIERO-class boats—both with canard-configured submerged foils. Israel has three hydrofoils of the U.S.-designed Shimrit class, with airplane-configured submerged foils.

HYDROGEN PEROXIDE: H_2O_2, a colorless, odorless, strongly oxidizing liquid, extremely unstable in concentrated form, and capable of reacting explosively with combustible materials, which is used as fuel for spacecraft and missile rocket engines, and for naval TORPEDO propulsion.

In spacecraft and missiles, H_2O_2 is used both as a monopropellant or bipropellant with a catalyst to decompose it into hydrogen and oxygen, which are then burned in a combustion chamber; and as a bipropellant, in combination with another compound (e.g., nitrogen tetroxide, hydrazine, or red-fuming nitric acid), with which it reacts hypergolically.

In peroxide-powered or "oxygen" torpedoes, a catalyst decomposes the compound into oxygen and water, which are mixed with alcohol in a combustion chamber, generating steam which drives a turbine that turns the propellers. Oxygen torpedoes have much greater power and range than steam or electric torpedoes of equivalent size; they were much used by the Japanese navy in World War II, but the potential hazard of hydrogen peroxide in enclosed submarine hulls has dissuaded other users.

The same obstacle has impeded the development of peroxide-powered SUBMARINES. The Walther-cycle engine, invented in Germany during World War II, decomposes hydrogen peroxide to burn constituent hydrogen and oxygen for steam to drive a turbine. A few small Walther submarines were built in Germany before 1945, but proved to be unreliable. After World War II, the U.S., Britain, and the U.S.S.R. all experimented with Walther-cycle designs, but the explosive hazard posed by the fuel and the advent of nuclear power forestalled further development.

I

IA-PVO: *Istrebitel'naya aviatsiya-Protivo-vozdushnoy oborony,* Fighter Aviation of Air Defense, the component of the Air Defense Troops (PVO) which operates manned interceptors in defense of Soviet airspace. IA-PVO aircraft are integrated into the ground control units of the PVO's AIR DEFENSE network. Since the early 1980s, however, OPERATIONAL control of interceptor forces has been decentralized, and now comes under the individual MILITARY DISTRICTS. See also VVS.

ICBM: Intercontinental Ballistic Missile; by negotiated U.S.-Soviet convention, any land-based BALLISTIC MISSILE with a range greater than 5000 mi. ICBMs are still the principal long-range nuclear DELIVERY SYSTEMS of the United States and Soviet Union, which have 1050 and some 1400, respectively. China also has a small number of operational ICBMs. The Soviet Union long favored liquid fuel for its ICBMs, but has recently developed two solid-fuel ICBMs, the SS-24 SCALPEL and SS-25 SICKLE. The U.S. also built liquid-fuel ICBMs at first, but soon switched to solid fuel, beginning with the Minuteman I ICBM of 1963. Liquid-fuel ICBMs generally have two stages, while most solid-fuel ICBMs have three.

ICBMs are armed with nuclear warheads housed inside aerodynamic REENTRY VEHICLES (RVs). Modern ICBMs usually have three or more RVs mounted on a maneuvering POST-BOOST VEHICLE (PBV)—a final substage which sequentially directs RVs towards their impact points, which may be widely separated; such RVs are called MULTI-PLE INDEPENDENTLY TARGETED REENTRY VEHICLES, or MIRVs.

ICBMs are usually based in SILOS, underground vertical launchers with heavy lids, which are reinforced (HARDENED) to resist the blast effects of NUCLEAR WEAPONS. Since the late 1970s, the accuracy of ICBMs and other long-range missiles (SLBMs, cruise) has improved to the point where even the hardest silos are vulnerable, giving attackers a COUNTERFORCE capability. In response, the U.S. and U.S.S.R. are both beginning to deploy road- and rail-mobile ICBMs which can be dispersed and concealed for protection. See also ICBMS, CHINA; ICBMS, SOVIET UNION; ICBMS, UNITED STATES.

ICBMS, CHINA: Through all its upheavals, the People's Republic of China has kept up a relatively costly ICBM development and production effort ever since the 1960s. Very little is known about the Chinese program, but it is believed that much of its early technology was derived from pre-1962 Soviet ballistic missiles. The first Chinese ballistic missile was a single-stage, liquid-fueled MRBM (U.S. designation CSS-1), of which some 100 were deployed along the Sino-Soviet border in the early 1970s. This was followed by a single-stage, liquid-fueled IRBM (U.S. designation CSS-2), of which about two dozen were deployed between 1971 and 1979. In 1988, China sold a number of CSS-2s to Saudi Arabia, without nuclear warheads. The CSS-2 also served as the basis for a "limited" ICBM (U.S. designation CSS-3) which consists of a CSS-2 with an added second stage to achieve a range of 3500–4000 mi. (vs. the 5000+ mi. of

ICBMs as defined in U.S.-Soviet agreements). Only a few CSS-3s were deployed in the mid-1970s; they are believed to carry a single large 3-MT warhead.

In the late 1970s, China began flight testing a large, 2-stage, liquid-fuel missile (U.S. designation CSS-4), which is said to be similar in size and configuration to the Soviet SS-9 Scarp of mid-1960s vintage. If so, the CSS-4 would be some 118 ft. (36 m.) long and 10.16 ft. (3.1 m.) in diameter, with a launch weight of 400,000 lb. (181,181 kg.). Maximum range is estimated at 4000–6200 mi. (6400–10,000 km.); this ICBM may carry a 4 to 5-MT warhead with moderate accuracy, for a median error radius (CEP) of at least 1000 m. A few CSS-4s were deployed in hardened silos in Sinjiang Province by 1983. The Chinese ICBM force is not expanding rapidly, because of economic and technological constraints, but Chinese ICBM technology is now being used for the "Long March" series of space boosters. See also IRBMS, CHINA.

ICBM, SOVIET UNION: In spite of the rising importance of other strategic weapons, notably submarine-launched ballistic missiles (SLBMs), Soviet leaders still regard ICBMs as the decisive weapon in any nuclear war with the U.S., and therefore as a critical measure of military power.

Soviet military planners recognized the potential of BALLISTIC MISSILES for nuclear delivery long before their bomber-fixated U.S. counterparts; they started an intense development program immediately after World War II with captured German equipment and personnel. Despite serious technical obstacles, the U.S.S.R. actually managed to deploy an ICBM in 1957, some two years before the U.S. This missile (U.S. code name SS-6 Sapwood) was scarcely a practical weapon, being too large, too costly, and very inaccurate; it also required 72 hours of preparation before launch. Only a dozen Sapwoods were actually deployed operationally.

Pending development of more effective ICBMs, the Soviet leadership decided to magnify this token deployment by launching a campaign of strategic deception (MASKIROVKA), under which false claims were carefully orchestrated to give the impression that the U.S.S.R. already had a large and effective ICBM force. The Soviet Union duly gained much prestige, but the resulting myth of a MISSILE GAP induced the U.S. to undertake multiple ICBM programs, so that by mid-1962 it had more than four times as many ICBMs as the Soviet Union. After the failure of strategic deception (and of the rede-

ployment gambit with shorter-range missiles which resulted in the 1962 Cuban missile crisis), Soviet ICBM deployments became more systematic. By then, the "Strategic Rocket Forces" (*Raketnyy Voiska Stratecheskogo Naznacheniya*, RVSN) had been established as a separate service in charge of long-range missiles. However, as table I1 shows, the U.S.S.R. did not attempt to match the United States in overall numbers until the U.S. unilaterally stabilized the size of its force at 1050 missiles. So long as the contest remained open-ended, the Soviet leadership apparently saw no practical means of matching U.S. production. But once the U.S. froze the size of its force in accordance with the doctrine of ASSURED DESTRUCTION, the Soviet leadership apparently saw the opportunity of achieving ICBM superiority.

Soviet ICBMs have been designed for different operational purposes than their U.S. counterparts, because Soviet planners have not recognized the existence of DETERRENCE as a phenomenon independent of war-fighting capabilities: the enemy's "weapons of mass destruction" are to be attacked to destroy his will to fight, and thus secure the Soviet homeland against attack. Soviet ICBM development efforts have therefore emphasized COUNTERFORCE capabilities. Moreover, while U.S. efforts have emphasized constant readiness with high availability rates, Soviet planners have been willing to compromise continuous readiness on the assumption that the ICBM force could be "surged" for a FIRST STRIKE. The post-1987 changes in Soviet military doctrine have yet to affect ICBM development priorities.

The first fully operational Soviet ICBM (U.S. code name SS-7 Saddler) was a large missile fueled

TABLE I1

Soviet ICBM Deployments 1960–1979

Year	Total Deployed	Comments
1960	30+	
1965	200+	
1966	250+	U.S. ICBM force stabilized at 1050
1967	570+	
1968	700+	
1969	1200+	
1970	1300+	
1972		SALT I limits—US: 1050, USSR: 1408
1979	1400	SALT II limits
1989	1460+	

by storable liquid propellants; some 200 were deployed between 1962 and 1965. Most Soviet ICBMs until the early 1980s retained the basic SS-7 features: storable liquid fuel, preprogrammed INERTIAL GUIDANCE, and a large "throw weight" (payload) compared to U.S. ICBMs. Relatively inaccurate, Saddler compensated for this with a large 20-MT warhead. By 1965, the second generation of Soviet ICBMs entered service, codenamed SS-9 Scarp, SS-11 SEGO, and SS-13 SAVAGE.

The SS-9, an enlarged and much improved development of the SS-7, was built in four separate variants ("Mods"). Mod 1 had a 20-MT warhead, Mod 2 had a 25-MT warhead and improved accuracy, Mod 3 was used to test a round-the-world attack profile (FRACTIONAL ORBIT BOMBARDMENT SYSTEM, FOBS), and Mod 4 was armed with three MULTIPLE REENTRY VEHICLES (MRVs). Accuracy improved with each version, from a median error radius (CEP) of 1000 m. in Mod 1 to only 650 m. in Mod 4. The SS-11, a much lighter missile using the same basic technology, was built in three Mods: Mods 1 and 2 with large unitary warheads, and Mod 3 with three MRVs.

The SS-9 and SS-11 formed the backbone of the RVSN through the mid-1970s, with over 300 Scarps and 1016 Segos deployed. The SS-13, by contrast, was the first Soviet attempt to develop a solid-fuel ICBM and was not successful because of the incompatibility of preprogrammed guidance with solid propulsion (yet the SS-13 remained in service until the 1980s).

By the early 1970s, the Soviet third-generation ICBMs were ready for deployment: the light SS-16 SINNER, the medium SS-17 SPANKER, the ultra-heavy SS-18 SATAN, and intermediate SS-19 STILETTO. These were the first Soviet ICBMs with MULTIPLE INDEPENDENTLY TARGETED REENTRY VEHICLES (MIRVs), and their CEPs were much smaller than those of their predecessors. The SS-18, with ten MIRVs, and the SS-19 with six, are both assessed as having HARD TARGET kill capability, the prerequisite for a counterforce strike. All are storable liquid-fuel missiles, except for the SS-16, which was the solid-fuel successor of the SS-13. Deployment of the SS-16 was never confirmed (the missile violated per se the terms of the SALT II Treaty), but its first two stages were developed into the SS-20 SABER IRBM.

In the mid-1980s, the deployment of two fourth-generation ICBMs, the SS-24 SCALPEL and SS-25 SICKLE, revealed a radical shift in configuration. Both are propelled by solid-fuel rockets, and both

are suitable for mobile deployment. The SS-24, equivalent in size to the U.S. PEACEKEEPER (MX), has ten MIRVs, while the SS-25, a light missile equivalent to MINUTEMAN, has only one RV. But Soviet planners have not yet opted for an all-mobile force: in late 1987, flight tests revealed a successor to the ultra-heavy SS-18, the even larger SS-X-26. See also ICBMS, UNITED STATES.

ICBMS, UNITED STATES: U.S. planners have long viewed ICBMS as part of a deterrent TRIAD, along with submarine launched ballistic missiles (SLBMS) and manned BOMBERS; the purpose of this diversity is to ensure the survival of a retaliatory force sufficient to inflict "unacceptable" damage on any attacker, even after a would-be disarming counterforce attack. That is the essence of the policy of ASSURED DESTRUCTION, officially repudiated, yet still in effect.

Because the U.S. Air Force preferred bombers, the U.S. did not begin working in earnest to develop an ICBM until 1956, when rumors of a Soviet ICBM program leaked out. U.S. efforts were much accelerated by the Soviet launch of the first satellite *Sputnik* in 1957, and even more after Soviet propaganda successfully magnified the reality of a few crude ICBMs to create the impression of a large and formidable missile force. The resulting fear of a MISSILE GAP resulted in the funding of three concurrent ICBM crash programs. By 1960, the U.S. had deployed its first operational ICBM, the Atlas; this was followed in 1962 by the TITAN I, the first ICBM housed in a protective SILO, and in 1963 by the Titan II, the first ICBM with storable liquid propellents. By late 1963, the U.S. had more than four times as many ICBMs as the Soviet Union, with many more in production. The development of a solid-fuel missile, the MINUTEMAN, had begun in 1958 and the Minuteman I also entered service in 1963. As many as 3000 were to be built, but by 1965 it was concluded (wrongly) that the U.S.S.R. had given up any numerical competition, and Assured Destruction had become official U.S. policy: its assumptions rationalized a drastic reduction in the Minuteman program. (It was further believed in official quarters that if the U.S. voluntarily restrained itself, the U.S.S.R. would stabilize its own force, in obedience to the universal logic of Assured Destruction.)

Accordingly, the size of the U.S. force was limited to 1000 Minutemen and 50 Titan IIs—a level which was maintained until the early 1980s when the Titans were retired. On the other hand, the Minuteman I was replaced by the improved Minuteman II in 1965, and the Minuteman III, armed with three MULTIPLE INDEPENDENTLY TARGETED REENTRY VEHICLES (MIRVs), was deployed from 1970. With the gradual retirement of the Titan II between 1982 and 1985, the ICBM inventory was stabilized at 1000—450 Minuteman II and 550 Minuteman IIIs, until 50 of the latter were replaced by MX Peacekeepers in 1986.

As it turned out, U.S. restraint did not evoke Soviet emulation, but rather inspired a major effort to achieve strategic-nuclear superiority. During the 1970s, the Soviet Union reached and then passed the U.S. 1000-ICBM level, but the most alarming aspect of the buildup was qualitative: Soviet ICBMs were acquiring combinations of throw-weight and accuracy theoretically sufficient for COUNTERFORCE targeting.

In response, the U.S. initiated the development of a new missile, the MX (later named PEACE-KEEPER) heavy ICBM with ten MIRVs and high accuracy. There was, however, much domestic opposition to the MX, both from prodisarmament voices opposed to all new strategic weapons, and from strong-defense elements that focused on the theoretical vulnerability of any silo-based ICBM to Soviet attack. From the mid-1970s, several alternative basing schemes were proposed for MX, including multiple protected shelters for a "shell-game" deployment, underground railways ("racetracks"), DENSE PACK, and deep underground silos. None was satisfactory, and by 1980 it seemed that the U.S. might retreat to a "biad" of SLBMs and bombers, with ICBMs deemed too vulnerable to keep. In 1981, however, the Reagan administration resurrected silo basing as an interim solution for the MX, and the deployment of the first 50 (in Minuteman III silos at Warren Air Force Base) began in 1986 while a long-term remedy was sought.

A radical alternative to Peacekeeper was proposed by the bipartisan Scowcroft Commission in 1987: the Small ICBM (SICBM), better known as "Midgetman"—a 37,000-lb., single-warhead missile that could be mounted on a wheeled transporter-erector-launcher (TEL) to achieve protection by dispersal and concealment. Disputes about Midgetman's final size, configuration, and (rising) costs continued, while the Reagan administration proposed a rail-mobile Peacekeeper as a cheaper alternative on a per-warhead basis. By 1989, the Gorbachev revolution in Soviet politics had further eroded support for a Midgetman force.

The ICBMs of the United States are controlled by the Air Force STRATEGIC AIR COMMAND. See also ICBMS, SOVIET UNION.

ICM: See IMPROVED CONVENTIONAL MUNITIONS.

IFF: Identification, Friend or Foe, a RADAR transponder installed in COMBAT AIRCRAFT and warships, which authentically responds with a coded message identifying the aircraft or ship as friendly, when "interrogated" by the proper radar signal. Interrogator devices are fitted to search, surveillance, and acquisition radars, and also in the guidance systems of certain anti-aircraft missiles.

In theory, all aircraft and ships fitted with IFF could freely be engaged if they fail to respond correctly when detected by radar; but IFF transponders are less than 100 percent reliable, especially under combat conditions. The danger of IFF errors has restricted the utility of beyond-visual-range (BVR) weapons; U.S. RULES OF ENGAGEMENT (ROEs) have notably mandated the visual confirmation of target identity (despite IFF indications) in recent combat situations (including the Vietnam War).

IFV: See INFANTRY FIGHTING VEHICLE.

IIR: See IMAGING INFRARED.

IKARA: Widely deployed, Australian-designed shipboard ANTI-SUBMARINE WARFARE (ASW) guided missile, an all-weather torpedo-delivery system analogous to the French MALAFON and Soviet SS-N-14 SILEX. In service with the Australian, Royal, and several other navies, Ikara is also produced under license in Brazil as the Branik. A compact, airplane-configured missile with stub delta wings and a vertical stabilizer, Ikara has a rectangular fuselage that can house either the Mk.44 or MK.46 lightweight acoustical homing torpedo (which weigh 425 and 568 pounds, respectively), and is powered by a dual thrust solid-rocket motor.

Stored in a below-decks magazine, Ikara is launched from a trainable, single-arm launcher after the wings and stabilizer have been manually attached before loading. Usually launched on the target bearing at an elevation of 45° or more, Ikara is stabilized in flight by an autopilot and a radar altimeter. Midcourse corrections are transmitted by radio COMMAND GUIDANCE from the ship's FIRE CONTROLS (or from those of a second ship or helicopter). Midcourse updates eliminate the "dead time" between launch and impact, which is a major shortcoming of unguided rocket-boosted torpedoes such as the U.S. ASROC or Soviet FRAS-1.

In action, the guidance system steers the missile towards the target zone on the basis of SONAR or SONOBUOY data; once the missile is in position, a command releases the torpedo, which descends while suspended by a braking parachute. Upon entering the water, the parachute is jettisoned, the torpedo activates its on-board sonar and initiates a search for the target.

Specifications **Length:** 11.25 ft. (3.43 m.). **Cross section:** 14 × 21 in. (533 × 356 mm.). **Span:** 5 ft. (1.53 m.). **Weight, launch:** varies w/payload. **Speed, max.:** Mach 0.8 (600 mph/960 kph). **Range, max.:** 15 mi. (24 km.).

IKV.91: *Infanteriekanonenvagn,* Infantry Support Gun ("ASSAULT GUN"), a Swedish armored fighting vehicle, also variously described as a LIGHT TANK or TANK DESTROYER, which is intended mainly to provide armor support for infantry in terrain too soft for full-size tanks. The Ikv.91 was developed from 1968 as a multipurpose vehicle to replace the Strv.74 light tank, the Ikv.102 and Ikv.103 infantry support guns, and the *Panservarnskanone* m/63 self-propelled gun. Prototypes were produced in 1969, and the vehicle entered service in 1975.

Armor protection on the Ikv.91, a tanklike vehicle with a fully tracked hull and a rotating turret, was compromised to retain amphibious capability without flotation devices. The all-welded steel hull is divided into a driver's compartment up front, a fighting compartment including the turret in the center, and the engine compartment in the rear. The commander sits in a cupola on the right side of the turret, above and behind the gunner, with the loader on the left. The commander has five observation periscopes, including an infrared night scope, while the gunner has an optical gunsight and a laser rangefinder.

Primary armament is a Bofors L/54 medium-velocity 90-mm. gun which can fire high-explosive (HE) and HEAT but not kinetic energy rounds (ineffective without high velocity); a total of 59 rounds are carried. The turret has power elevation and traverse with manual backup; elevation limits are +15° and −10°. Secondary armament consists of a coaxial 7.62-mm. MACHINE GUN and a second, pintle-mounted machine gun by the loader's hatch. Six smoke GRENADE launchers are mounted on the sides of the turret. The Ikv.91 is equipped with a COLLECTIVE FILTRATION system for NBC defense.

The Ikv.91's broad tracks reduce ground pressure to 6.96 psi (0.49 kg./cm.²), for excellent mobility in soft terrain, and can be fitted with 2-in. (51-mm.) grips to improve traction on ice. The vehicle is fully amphibious with minimal preparation, and can swim (by track actions only) at a maximum speed of 4.4 mph (7.1 kph). The Ikv.91's ability to cope with snow and ice in winter, and also with summer swamps, is essential for operations in Lapland, the critical sector for Sweden's defense on land.

Specifications **Length:** 20.1 ft. (6.13 m.). **Width:** 9.83 ft. (3 m.). **Height:** 7.75 ft. (2.36 m.). **Weight, combat:** 36 tons. **Powerplant:** 350-hp. Volvo Penta TD120A 6-cylinder turbocharged diesel. **Speed, road:** 43 mph (69 kph). **Range, max.:** 342 mi. (548 km.).

Il-: Designation of the Soviet aircraft design bureau founded by S. V. Ilyushin, noted mainly for its light bombers, transports, and maritime patrol aircraft. Models in service include the Il-38 MAY and Il-76 CANDID.

IMAGE INTENSIFIER: A night-vision device that operates by amplifying low levels of visible light up to 100,000 times to produce a daylike image, even on the darkest of nights. The dim light reflected by objects is collected by a lens and focused on a photo-cathode, which releases electrons when it absorbs photons of light. The electrons are then accelerated by an electrical field (thus raising their energy levels) and projected onto a phosphor screen to generate a bright image. The earliest types, first used in the Vietnam War, have only a single stage, allowing amplification factors of 20,000–40,000 times the ambient light. To achieve higher gain, several image intensifier tubes can be coupled in series (a "cascade" array), but these have relatively poor resolution, because of the cumulative distortion of the original image as it passes through each stage of amplification. In addition, their phosphor screens can be "saturated" by exposure to bright lights.

Channel electron amplification, an alternative method of achieving higher light gain, relies on tubes lined with semiconductor glass, formed into fiberoptic mosaics and inserted between a photocathode and a phosphor screen. Electrons generated by the photo-cathode collide with the semiconductor tubes, releasing additional electrons, which are accelerated by an electrical field and projected onto the phosphor screen. As compared to cascade arrays, channel tubes are lighter and smaller, have higher resolution, and are less prone to saturation.

Because of their light weight and low power requirements, image intensifiers are often used as night sights for rifles and crew-served weapons, as battlefield surveillance devices, and as personal night-vision goggles. However, because they operate only in the visible spectrum, their effectiveness is degraded by smoke, fog, dense foliage, and heavy precipitation.

IMAGING INFRARED (IIR): A type of INFRARED (IR) night-vision sensor identical to THERMAL IMAGING. Infrared radiation (heat emissions) from objects and background are focused by optics onto a focal plane array of IR detectors. By electronic signal processing, these emissions are then converted into televisionlike video images. IIR can function in the total absence of visible light, and can also penetrate smoke and haze better than light-amplification devices such as Low Light Television (LLTV). IIR sensors are used on some armored fighting vehicles, many combat aircraft, and warships for SURVEILLANCE and TARGET ACQUISITION. "Forward Looking Infrared" (FLIR) systems for combat aircraft are a specific application of IIR. IIR is also used as a guidance technique in missiles such as the AGM-65D MAVERICK and AGM-130(V)2. After the weapon operator designates a target, the weapon homes on the target's image generated by the IIR sensor in its nose, in the same way as ELECTRO-OPTICAL weapons home on TV images, but IIR remains effective even in low visibility. See also INFRARED COUNTERMEASURES.

IMPROVED CONVENTIONAL MUNITIONS (ICMS): Artillery ammunition, missile warheads, and aerial bombs whose lethality is enhanced by the distribution of the explosive payload into submunitions (bomblets or minelets). While unitary warheads waste much of their energy in "overkill" at the point of impact, submunitions can be sized to match target characteristics (hard, soft, or mixed), and are scattered over a much wider area, thereby increasing the probability of a hit. Examples of ICMs include small artillery-scattered MINES such as ADAM and RAAMS; CLUSTER BOMBS and submunition dispensers; TERMINALLY GUIDED SUBMISSILES (TGSMs); sensor-fuzed, SELF-FORGING FRAGMENT submunitions such as SADARM, ERAM, and SKEET; and, above all, bomblet rounds for artillery in the standard calibers. In each case, the ICBM consists of a "cargo" round warhead, a dispensing mechanism, and the submunitions.

The impact of ICMs on the military equation remains to be assessed, but standard 155-mm. howitzers firing ICMs can be several times as effective as with standard HE rounds—so that artillery (with ICMs) has suddenly outpaced the capability increases achieved by armor and infantry since 1945.

IMPROVED TOW VEHICLE (ITV): U.S. Army M901 TANK DESTROYER variant of the M113 armored personnel carrier. Developed from 1975 to allow the launching and guidance of TOW anti-tank guided missiles from under armor protection, the first ITVs entered service in 1979, and now some 2900 equip the anti-tank units of U.S. armored and mechanized divisions.

The M901 is essentially an M113A2 with an Emerson Electric twin TOW launcher attached to a modified commander's cupola. The launcher, which also contains television and THERMAL IMAGING sights, folds flat for traveling, and is erected for launch, allowing the ITV to fire from protected "hull-down" positions with only the launcher exposed. Twelve additional missiles are carried inside the vehicle. To reload, the launcher is folded partially backwards, and new missiles are inserted manually through a roof hatch.

The ITV crew consists of a driver, a commander, a gunner, and a loader. Performance is generally similar to that of the M113. The M901 is also the basis for the M981 Fire Support Team Vehicle (FISTV).

INDIRECT APPROACH: A concept of war, whose essence is the avoidance of frontal attacks. *Tactically*, the Indirect Approach prescribes surprise moves to disrupt the enemy's defenses, as opposed to their destruction by sheer attrition. The concept was developed and promoted by Sir Basil H. Liddell Hart, the noted military theorist, whose slogan was "Natural hazards, however formidable, are less dangerous than fighting hazards."

Operationally, the Indirect Approach prescribes the penetration in depth of the enemy's front, in order to cut his lines of supply, disrupt command and communication links, and "create data"—i.e., the mass of sighting reports, which along with communication failures, can confuse the enemy's command, prevent his identification of the actual axes of advance, and thus inhibit the enemy's attempts at interception. By such means, in theory, an absolutely weaker attacker can win by achieving brief but decisive local superiorities at successive points of his advance. One precondition is the ability of the attacking columns to maintain momentum; a second precondition is air superiority.

Strategically, the Indirect Approach embraces all strategies whereby the enemy's strengths are circumvented, e.g., by blockading at sea a superior land power; or by developing new types of weapons to neutralize enemy superiority in the old (as the U.S.S.R. did in the 1950s by pioneering ICBMs, instead of competing with U.S. bomber superiority); or by using nonmilitary means to defeat superior military power, as in REVOLUTIONARY WAR.

At each level, there is a price to be paid for doing the unexpected, because the expected is what is more economical and prudent. Risks increase with indirectness, and moreover, circumvention can easily degenerate into avoidance, as in the case of most naval "crisis interventions," whose results are usually insignificant. See also BLITZKRIEG; MANEUVER; STRATEGY; TACTICS.

INDIRECT FIRE: Fire brought against a target without an uninterrupted line of sight (LOS) from weapon to target. Indirect fire is the usual mode of operation for tube ARTILLERY, MORTARS, and longer-range ROCKET ARTILLERY. When "blind" fire is directed at fixed geographic coordinates, its effectiveness is reduced by geodetic inaccuracies and ballistic and meteorological variables, as well as by any target designation error (including target movement). Indirect fire is therefore usually conducted with the aid of ground or airborne FORWARD OBSERVERS (FOs). The FO transmits engagement instructions ("fire missions") by radio or field telephone to the artillery command post (a FIRE DIRECTION CENTER or FDC in the U.S. Army), and then fire corrections.

The fire mission consists of a warning notice, the target's location and composition, the FO's location and his bearing to the target, and the type of ammunition and number of rounds required. Target location is given by map grid, polar reference, or shift from a known point.

Target locations given relative to the FO's position must be converted into coordinates relative to the firing battery's location. The conversion can be made on a manual plotting board with two superimposed transparent grids, or by computer. Target locations are then translated into gun elevation and traverse settings, either manually with a ballistic table or by computer. Because of errors in estimating the target position, and the effects of wind, air density, propellant irregularities, and internal ballistics on projectile trajectories, it is customary to fire one or two trial rounds ("registration" shots). The FO calls in corrections ("adjusts fire") until the shells are landing on target, at which point "fire for effect" can begin.

To preserve secrecy, silent registration techniques may be used; e.g., a battery may fire against a hypothetical target on a reciprocal bearing at the same range as the real target. Another method employs specialized radars to track the flight path of a trial shell time-fuzed to explode after a short portion of its trajectory. A ballistic computer can then extrapolate the remainder of the trajectory to generate the required corrections, without revealing impacts in the target area itself.

BARRAGE fire is a form of preplotted indirect fire intended to saturate or isolate an extensive zone. In "neutralization fire," the impact area, or "sheaf," is sized so as to ensure that enough shells hit the target to suppress it. In "destructive fire," pioneered by the Israeli army, computer-directed pinpoint concentrations are intended to shatter targets with a minimum number of rounds. HARASSMENT AND INTERDICTION (H&I) fire is directed without observation or correction against suspected enemy targets or geographic chokepoints to inflict low-level attrition and damage enemy morale. FINAL PROTECTIVE FIRE (FPF) is a prearranged and preregistered linear barrage, intended to cover the last line of defense of a position under attack. Nuclear and chemical missions are variations on these standard procedures. See also DIRECT FIRE.

INDOCTRINATION: The process whereby ideas, methods, or information are transferred to a controlled group. The word is used, with negative connotations, to describe activities which, if they were our own, might be described as education or training.

INERTIAL GUIDANCE: A form of missile guidance which relies on the measurement and cumulative calculations of accelerations accomplished by devices within the missile, without need of external intervention. Inertial guidance is employed both in BALLISTIC MISSILES and CRUISE MISSILES, as well as in many kinds of tactical missiles. At its simplest, an inertial guidance system consists of an inertial platform and a computer. The inertial platform has three gimbal-mounted gyroscopes aligned with the missile's roll (x), pitch (y), and yaw (z) axes; and either 3, 6, or 9 accelerometers likewise oriented. The inertial platform can thus register attitude and acceleration changes; on that basis, the computer can compare the intended and actual courses, to generate corrective steering commands to the missile's controls.

The inertial guidance systems of ballistic missiles employ one of two basic techniques: preprogrammed (explicit) inertial guidance, or navigating (implicit) inertial guidance. In a preprogrammed system, the desired ("nominal") trajectory of the missile from a given launch point to a given impact point is precomputed on the ground, and loaded into the guidance computer before launch, in the form of a pitch, heading, and velocity vs. time profile. During the missile's boost phase, the computer compares the actual flight path and velocities with the nominal parameters for the same time frames, and generates guidance commands to resolve the differences.

Thus, preprogrammed guidance can be used only with static missiles aimed at known, fixed targets for which nominal trajectories can be computed. In addition, this system requires some means of controlling velocity without deviations from the nominal pitch and heading profile. That constraint limits the application of preprogrammed inertial guidance to liquid-fueled ballistic missiles, in which velocity can be altered by controlling the fuel flow. Most Soviet liquid-fuel ICBMs have "fly-by-wire" systems, in which thrust magnitude control (engine throttling) keeps the missile on the nominal trajectory. Fly-by-wire and similar techniques cannot be accurate with solid-fuel missiles, which have no reliable method of thrust control.

In "navigating" inertial guidance systems, the computer is programmed with three input variables: the launch coordinates, the target coordinates, and the time of flight, and also with a set of quadratic equations ("guidance laws"). During the flight, the computer continuously generates position and velocity readings with data from the inertial platform. Applying the guidance equations, the computer calculates the instantaneous ("correlated") velocity vector required to hit the target, and generates steering commands accordingly; this "Velocity Gained" (V_g) technique, employed in most solid-fuel ballistic missiles, is inherently more flexible than preprogrammed methods, but requires far more capable guidance computers.

Because even small guidance errors can result in large miss distances over intercontinental ranges, in some newer ballistic missiles celestial navigation is used to correct inertial platform data. An ELECTRO-OPTICAL star sensor is programmed to lock onto selected stars or constellations once the missile has left the atmosphere; on that basis the computer can compare the predicted and actual positions of the missile to detect and correct for gyro drift and other inertial navigation errors. Such "stellar-inertial" guidance is especially useful in submarine-launched ballistic missiles (SLBMs),

because their launch positions are determined by the ship's inertial navigation system (SINS), which is itself prone to gyro drift errors.

Cruise missiles employ a form of navigating inertial guidance in which the missile is continuously directed towards the target without regard to its velocity (relevant only in ballistic missiles). Inertially guided, long-range cruise missiles often have DOPPLER radar or radio navigation (e.g., LORAN) devices to update inertial navigation data during their long flights.

Many short-range tactical missiles rely on simpler "strap-down" inertial guidance units, in which the gyros and accelerometers are mounted rigidly in the airframe, instead of being suspended on a stable platform. See also INERTIAL NAVIGATION.

INERTIAL NAVIGATION: Dead reckoning performed automatically by a device which continuously integrates all successive accelerations since departure from a point of known position. An inertial navigation system consists of an inertial platform and a computer. The inertial platform has two or three gyroscopes to measure attitude, and three accelerometers to measure acceleration longitudinally, laterally, and vertically. From these attitude and acceleration data, the computer derives the course and distance traveled, and calculates the current position in latitude and longitude, for display either numerically or graphically (on a map).

Inertial navigation has the advantage of being completely passive and self-contained; hence its users are not vulnerable to detection by ELECTRONIC SIGNAL MONITORING, or disruption by ELECTRONIC COUNTERMEASURES. But gyroscopes tend to lose their alignment and "drift" over time, because of mechanical imperfections. Inertial platforms thus become progressively less accurate over time, a major problem on long-endurance flights, and even more for ballistic-missile submarines on patrol. Hence many inertial navigation systems have provisions for external position updates by RADAR, celestial observation, or with radio navigation aids such as LORAN or the GLOBAL POSITIONING SYSTEM (GPS).

Inertial navigation systems are used on most modern COMBAT VESSELS, especially SUBMARINES, on long-range COMBAT AIRCRAFT, and, increasingly, on combat vehicles such as MAIN BATTLE TANKS, self-propelled ARTILLERY, and mobile AIR-DEFENSE weapons. See also INERTIAL GUIDANCE.

INF: Intermediate Nuclear Forces, a term applied to nuclear delivery systems with ranges between 300 and 3000 mi. (500 and 5000 km.). See also LONG-RANGE INTERMEDIATE NUCLEAR FORCES.

INFANTRY: Ground troops equipped and trained to fight primarily on foot. Traditional or "straight leg" infantry forces both fight and march on foot, having little or no organic armor or even transport in their units, and only towed ARTILLERY in support. Aside from China, no major power now retains this type of infantry, except for some low-readiness reserves and home-defense units.

MOTORIZED INFANTRY, otherwise organized and trained on the same lines as foot infantry, is equipped with trucks and motorized support forces. The U.S., British, French, and several other armies retain some motorized infantry for use in "close" terrain (mountains, forests, or urban areas), and as air-portable INTERVENTION forces.

MECHANIZED INFANTRY is mounted in ARMORED PERSONNEL CARRIERS or INFANTRY FIGHTING VEHICLES, and organized into COMBINED ARMS formations with tanks and self-propelled artillery. Trained to fight both mounted and dismounted, mechanized infantry is optimized for combat in open terrain, and generally lacks the rifle strength required for operations in forests, mountains, or cities. Despite their name, Soviet MOTORIZED RIFLE divisions are actually mechanized formations with much armor.

LIGHT INFANTRY, as the name implies, has a lighter establishment, particularly in regard to artillery. One type of light infantry is simply normal "straight leg" infantry deprived of heavy weapons for the sake of economy or mobility; as a result, its combat value is generally low. But light infantry can also be a quasi-elite arm of troops trained in stealth-and-stalking tactics for operations in close terrain; such agile forces are very versatile, being equally suitable for low-intensity warfare, and for combat against armored and mechanized forces hampered by close terrain.

ALPINE TROOPS are light infantry of the second kind, specifically trained and equipped for mountain operations. AIRBORNE FORCES are likewise light infantry, but trained for insertion behind enemy lines by parachute, fixed-wing aircraft, or helicopters (Soviet airborne divisions, however, are now organized as light mechanized forces). RANGERS, COMMANDOS, and SPECIAL FORCES are all elite light infantry trained for SPECIAL OPERATIONS or UNCONVENTIONAL WARFARE. MARINES or NAVAL INFANTRY trained and equipped for AMPHIBIOUS WARFARE

may be organized as light, motorized, or mechanized infantry.

Despite the prevalence of mechanization since 1945, infantry remains dominant in close terrain. Only infantry forces can actually hold territory, and must normally be introduced to consolidate the results of armored attack, as well as to anchor any defense. In combined-arms operations, infantry is needed to protect tanks, especially against enemy infantry armed with BAZOOKA-type weapons. In close terrain, tanks are normally ineffective without infantry support, and agile light infantry can easily prevail over more conventionally trained forces (even if much more heavily armed). See also URBAN WARFARE.

INFANTRY FIGHTING VEHICLE (IFV): A tracked ARMORED FIGHTING VEHICLE for MECHANIZED INFANTRY forces. While ARMORED PERSONNEL CARRIERS (APCs) are only "battle taxis" intended to transport infantry to the battlefield for dismounted combat, IFVs are meant to operate across the battlefield in the face of enemy fire. Thus they tend to be more heavily armored than APCs (but not nearly as heavily armored as MAIN BATTLE TANKS), and their cross-country mobility should be matched to that of the tanks with which they would normally operate. Moreover, while APCs are lightly armed, often with only machine guns for self-defense, IFVs have turret-mounted 20-mm. to 30-mm. automatic CANNON to fight opposing light armor (including other IFVs), and some also carry ANTI-TANK GUIDED MISSILES (ATGMs). In addition, most IFVs also have firing ports that allow infantrymen aboard to fire their weapons from behind armor.

In some armies, it was believed that with IFVs, the mechanized infantry would rarely have to fight dismounted. Partly because of Egyptian and Syrian experience with Soviet BMPs in the 1973 Yom Kippur War, it is now universally recognized that because IFVs cannot withstand heavy anti-tank weapons, their on-board infantry must dismount to fight on foot, albeit with IFVs nearby to provide fire support. Mounted combat remains feasible only to pursue disorganized and/or lightly armed opposition.

In addition to the BMP, current IFVs include upgrades of the U.S. M113 APC, the British MICV-80 WARRIOR, the West German MARDER, and the U.S. M2/3 BRADLEY. See also ARMORED FORCES.

INFILTRATION: The COVERT (disguised) or CLANDESTINE (hidden by its nature) penetration of enemy-held zones by land, sea, or air, in peace or in war. Tactically, infiltration defines the use of stealth and/or concealment to pass around, or between, enemy forces in order to achieve deep penetrations into rear areas—often to disrupt enemy COMMAND AND CONTROL, lines of supply, and COMMUNICATIONS, or to attack high-value facilities.

Infiltration can occur at all levels of conflict, from long-range RAIDS and GUERRILLA-type operations to the penetration of main enemy defenses by large, combined-arms forces in ARMORED WARFARE. See also BLITZKRIEG; INDIRECT APPROACH; OPERATIONAL MANEUVER GROUP.

INFLEXIBLE: French nuclear-powered ballistic-missile submarine commissioned in 1985, essentially a modified version of the REDOUTABLE class.

INFRARED: The portion of the electromagnetic spectrum delimited by wavelengths of 1.5 to 14 microns. Infrared radiation is emitted by all objects warmer than absolute zero ($-273°$ C); the hotter the object, the greater the magnitude of IR energy emitted. The IR spectrum is divided into three broad bands: near or short-wavelength infrared (SWIR, 1.5 to 4.5 microns); midwavelength infrared (MWIR, 4.5 to 8 microns); and far or long-wavelength infrared (LWIR, 8 to 14 microns). Objects radiate over a number of frequency bands; generally speaking, the hotter the object, the shorter the wavelength of its IR emissions.

Water vapor, carbon dioxide, and other constituents of the atmosphere absorb IR radiation at several specific frequencies, limiting the militarily useful IR spectrum to two "windows" between 2–3 microns and 8–14 microns. The SWIR window is exploited by ground and airborne surveillance systems designed to detect objects with temperatures greater than 80° F, such as vehicles, aircraft, ships, and troops. LWIR devices, which can detect objects cooler than 80° F, is used for the space-based detection and tracking of reentry vehicles, and could form part of a ballistic missile defense system.

Infrared sensors function by optically focusing IR radiation on a focal plane of photo-electric detectors "tuned" to specific wavelengths. The simplest detectors of the lead sulfide (PbS) type are relatively insensitive but easy to manufacture. More advanced types require cryogenic cooling by a gas bottle or cryomotor, in order to suppress the "noise" generated by the heat of their own optics and electronics.

Military applications of IR devices include SUR-

VEILLANCE, TARGET ACQUISITION and tracking, and weapon guidance.

Surveillance devices fall into two categories: semi-active and passive. The former require an infrared searchlight to illuminate objects in their field of view. Passive IMAGING INFRARED (IIR) or THERMAL IMAGING devices require no illumination, and convert received infrared emissions into video images. Space-based systems, such as the U.S. DEFENSE SUPPORT PROGRAM (DSP) satellites, have nonimaging passive sensors to detect missile launches and nuclear detonations. High-resolution airborne and space-based IR surveillance systems rely on infrared line scanners (IRLS).

The most common weapon-guidance application is in passive INFRARED HOMING seekers, employed in most short-range air-to-air and surface-to-air missiles. Air-to-ground missiles, on the other hand, can be guided by IIR homing, in which the weapon is locked onto the target's video image, as relayed by an IIR sensor in the nose of the missile.

The performance of infrared systems is degraded by clouds, fog, and precipitation, all of which diffuse and attenuate IR emissions. In addition, IR devices are vulnerable to INFRARED COUNTERMEASURES (IRCM).

INFRARED   COUNTERMEASURES (IRCMS): Devices and techniques designed to jam, suppress, deceive, or otherwise neutralize INFRARED (IR) surveillance, tracking, or weapon guidance. The most common IRCMs are flares used to decoy INFRARED HOMING missiles; many aircraft and warships are equipped with flare dispensers, and in the future they will probably be added to armored fighting vehicles as well. Flares emit more intense IR radiation than the platforms they are to protect, to cause missiles to home on them rather than on the intended target. But the latest IR homing missiles have seekers tuned to two or more frequency bands, specifically to discriminate between flares and valid targets. In response, the latest flares are designed to emit IR radiation that more closely approximates actual target signatures.

Infrared sensors are also vulnerable to deception countermeasures, mainly rotating beacons and "hot brick" generators. Rotating beacons are intense IR strobe lights; they generate scintillating infrared emissions which missile guidance units tend to interpret as indicating that the missile is not properly aimed at the target. When the missile turns towards the perceived (false) target location, the true target passes out of the seeker's field of view, thereby breaking lock. The U.S. ALQ-128, a pod-mounted device for attack aircraft, the ALQ-157 for helicopters, and the AAQ-4, built into larger aircraft, are of this type. Hot brick generators burn flammable gas to heat refractory tiles in a pod, and have shutters to generate scintillating IR emissions.

A third form of IRCM is infrared smoke, i.e., smoke also opaque in the IR spectra of THERMAL IMAGING and IMAGING INFRARED sights. IR smoke can be released by aerial bombs, rockets, artillery shells, and mortar bombs.

INFRARED HOMING: A passive form of missile guidance, used mainly in air-to-air and surface-to-air missiles, whereby an INFRARED (IR) seeker detects and locks onto a heat source, such as the tailpipe of a jet aircraft. Signals from the seeker are passed to a guidance unit, which generates steering commands for the missile's flight controls. The earliest IR homing missiles could only detect aircraft directly ahead of them, within an arc of 20–30° from the axis of the jet exhaust; the latest missiles, with more sensitive, cryogenically cooled seekers, can detect the contrast even between relatively cool areas of aircraft and their background, and are therefore capable of ALL-ASPECT attack.

IR homing missiles are much cheaper than, and also have a number of tactical advantages over, ACTIVE or SEMI-ACTIVE RADAR HOMING (SARH) missiles: they do not require illumination of the target, so that launching aircraft can make radical maneuvers immediately after release, while surface-to-air missile launchers can engage multiple targets without waiting for a free illumination radar; they do not generate any emissions, hence they do not reveal their attack to RADAR WARNING RECEIVERS on board the target aircraft; and they are easier to use in short-range dogfights. On the other hand, their performance is degraded by clouds and precipitation, they cannot be used beyond visual range (BVR), and they can be vulnerable to relatively simple INFRARED COUNTERMEASURES (IRCM), such as decoy flares—though the latest IR homing missiles incorporate counter-countermeasures, such as two-frequency seekers which can discriminate between flares and aircraft. Infrared homing missiles in service include the U.S. AIM-9 SIDEWINDER, the Israeli SHAFRIR and PYTHON, and the French MAGIC; Soviet AAMs are generally built in both radar- and IR-homing versions. Most short-range surface-to-air missiles (SAMs) also rely on IR homing, as e.g., the U.S.

REDEYE and STINGER and the Soviet SA-7 GRAIL and SA-14 GREMLIN; some larger Soviet SAMs, such as the SA-6 GAINFUL, are believed to have infrared homing as a backup for terminal guidance.

INF TREATY: An ARMS CONTROL treaty between the United States and Soviet Union that prohibits land-based missiles (both ballistic and cruise) with ranges in excess of 500 km. (300 mi.) but under 5500 km. (3300 mi.). Under the terms of the treaty, all such missiles must be dismantled and destroyed within three years of ratification (the so-called ZERO OPTION). Compliance with the terms of the treaty is to be ensured by on-site verification; detailed rules regulate the activities of the inspection teams of both signatories. The prohibited weapons are the U.S. PERSHING 2 and 1a medium-range ballistic missiles and TOMAHAWK ground-launched cruise missiles; and the Soviet SS-4 SANDAL medium-range ballistic missiles, SS-20 SABER intermediate-range ballistic missiles, and the SS-12 SCALEBOARD and SS-23 SPIDER short-range ballistic missiles. The treaty was signed on 8 December 1987 by President Ronald Reagan and Soviet General Secretary Mikhail Gorbachev, and was ratified by the Senate of the United States in 1988, after some debate over verification. See also ARMS LIMITATION; DISARMAMENT; NATO; SALT; START.

INITIAL OPERATING CAPABILITY: The stage in a weapon's acquisition cycle when enough of them become available to equip one combat unit, and when logistic support suffices for combat operations. The term more precisely defines the date of service introduction.

INSURGENCY: A localized INTERNAL WAR between a constituted government and rival elements originating in the same national territory, which may be GUERRILLA, civilian-insurrectional, or terrorist in nature. REVOLUTIONARY WAR may begin as an insurgency, but one need not develop into the other. The term correctly applies to localized conflicts, often caused by ethnic or regional demands for autonomy or secession.

INTELLIGENCE: The collection, collation, analysis, and dissemination of information on the capabilities and intentions of actual or potential adversaries. Intelligence activities are classified as either strategic or tactical depending on the level of decision making at which the information is to be used.

Strategic intelligence, the ostensible basis of national policy formation and military planning, nowadays embraces economic, social, political, demographic, scientific, and technical data, as well as data on military strength, policies, and plans.

Tactical intelligence defines information on enemy forces in a given area, and on that terrain or area as such. Combat intelligence is almost synonymous, but applies to intelligence activities conducted at the level of a single formation, e.g., a division or brigade, rather than at higher levels of command.

Among the means of collection, human intelligence (HUMINT) exploits diplomatic, commercial, journalistic, and tourist sources; it also includes the monitoring of foreign press and electronic media, and the interrogation of defectors and refugees, as well as classic espionage, which generally accounts for only a small part of HUMINT activities. SCOUTING and POW interrogations are military HUMINT sources useful mostly for tactical intelligence in wartime. Technical intelligence (TECHINT) is information collected by machines, including photographic reconnaissance by aircraft or surveillance satellites, and signals intelligence (SIGINT), obtained both by monitoring analyzing enemy communications (COMINT), and the interception and analysis of enemy radar or sonar characteristics (ELINT). TECHINT is extremely valuable for both strategic and tactical purposes, but can provide only a partial picture, often more revealing of capabilities than of intent.

The analysis of raw data begins with interpretation processes such as photographic interpretation, the cryptanalysis of codes and cyphers, the statistical analysis of communications, and electronic impulse discrimination, as well as simple linguistic translation. Next comes collation to concentrate the data from all sources into subject categories, and finally the analysis itself, to fit fragmented information into patterns deemed meaningful (it is at this stage that self-deception can intervene). Finally, evaluation presents the analyzed information to decision makers in terms relevant to the problems at hand, or to introduce new problems.

It is dissemination that makes intelligence productive, and the entire activity has no purpose without it (except to the intelligence bureaucracy itself); but dissemination also implies exposure to criticism, and the possible disclosure of sources and methods; hence it is always restricted. Analysis at the strategic level is generally the responsibility of national intelligence organizations, such as the CENTRAL INTELLIGENCE AGENCY, DEFENSE INTELLIGENCE AGENCY, NATIONAL SECURITY AGENCY, KGB,

and GRU, as well as policy officials with or without relevant experience.

Tactical intelligence evaluation and dissemination are conducted at many levels, with Intelligence (G-2) staffs at the general headquarters (GHQ) level; at army, navy, air force, and other service headquarters; at theater level and with subtheatre commands (e.g., Soviet TVDS); in higher formations such as fleets, numbered air forces, ARMY GROUPS, and numbered armies; and so on down to air wings, divisions, naval task groups, and lesser formations. In contrast to the loose scope of strategic intelligence, tactical intelligence tends to be focused on the ORDER OF BATTLE (ORBAT), deployment posture, logistic situation, and morale of enemy forces.

INTERCEPTOR: A type of combat aircraft designed to locate, intercept, identify, and destroy hostile aircraft, especially bombers penetrating friendly airspace. Interceptors can be interchangeable with FIGHTER aircraft, but nowadays they require long-range RADARS and missiles, as well as high speed and a high rate of climb, even at the expense of maneuverability. Aircraft designed specifically as interceptors include the Soviet Su-15 FLAGON, MiG-25 FOXBAT, and MiG-31 FOXHOUND, and the U.S. F-106 Delta Dart. Only the Soviet Union still produces specialized interceptors; other air forces use suitable standard fighters, such as the F-4 PHANTOM, F-15 EAGLE, and MIRAGE F.1 (or even unsuitable light fighters such as the F-16 FALCON). See also ADTAC; AIR DEFENSE; IA-PVO.

INTERDICTION: The attack of LINES OF COMMUNICATION to disrupt supply flows, and, if possible, isolate specific zones or enemy forces therein.

Battlefield interdiction (BI), intended to cut off enemy front-line formations by preventing the arrival of reinforcements and supplies, would usually be carried out within 60 mi. or so of the front lines by long-range ARTILLERY, aircraft, or tactical missiles. Deep interdiction, normally intended to disrupt the transport of war materiel, can be executed throughout the depth of enemy territory by longer-range ATTACK AIRCRAFT such as the TORNADO, F-111, or Su-24 FENCER. Blockade is the naval form of interdiction, carried out operationally by surface ships, submarines, and aircraft; or strategically by a system of administrative controls. In both world wars, Germany and its allies were cut off from raw materials by "distant blockade," enforced by the inspection of all oceanic traffic not authorized by the Allies at the source.

The success of an interdiction campaign depends on enemy LOGISTIC requirements, the density of transportation networks, and the availability of substitutes for critical materiel. Historically, the results obtained have varied between mild and serious inconvenience, with more decisive results most unusual. See also ASSAULT BREAKER; DEEP STRIKE.

INTERNAL WAR: Organized armed conflict between parties that mainly originate from, and are based in, the same territory. If two or more parties acknowledge a common nationality while they openly wage war, the conflict is a civil war. If two or more parties acknowledge a common nationality, but one party relies mainly on GUERRILLA WARFARE and SUBVERSION, the conflict is a REVOLUTIONARY WAR. If two or more parties are fighting for control of less than the totality of the national territory, the conflict is an INSURGENCY (normally ethnic or regional).

The terms "low-intensity conflict," "sublimited warfare," "special warfare," "UNCONVENTIONAL WARFARE," "stability operations," etc. normally refer to counterguerrilla operations in the framework of revolutionary war, unless they are euphemisms for civil war. Insurrection defines mass civilian action against the established power, a mode of conflict that can succeed only if the armed authorities refrain from using the force at their disposal because of political inhibitions.

INTERVENTION: The insertion of external forces, often into zones of ongoing or potential conflict, to prevent the outbreak or spread of hostilities; or to resolve the situation to the advantage of a favored party; or more broadly to take advantage of the situation in order to acquire territory or influence.

INTRUDER: The Grumman A-6 twin-engine, two-seat, subsonic, carrier-based ATTACK AIRCRAFT, in service only with the U.S. Navy and Marine Corps. The Intruder originated in a 1956 Marine Corps requirement for a heavy attack aircraft capable of precision bombing at night and in poor weather. Prototypes first flew in April 1960, and the first production version (A-6A) entered service in 1963.

When first introduced, the Intruder had a number of unique capabilities, notably its radar-based DIANE (Digital Integrated Navigation and Attack Equipment) system, which incorporated one of the first airborne digital computers. DIANE allowed the Intruder to bomb with precision at night and in bad weather (a capability shared even today only with the F-111, the Panavia TORNADO, and the Su-24 FENCER).

On the other hand, the A-6 exceeded the state of the art, and suffered accordingly from severe reliability problems as well as high costs. Nonetheless, the A-6A served throughout the Vietnam War, flying more than 35,000 sorties, often against heavily defended targets; some 65 were lost to enemy action, and only 2 were shot down by fighters.

A total of 503 A-6As were built between 1963 and 1969, of which 21 were converted to EA-6A ELECTRONIC WARFARE aircraft, 19 to A-6Bs with provisions for the launch of the AGM-78 STANDARD ARM anti-radiation missile, 12 to A-6Cs with forward looking infrared (FLIR) and low-light television (LLTV) sensors, and 62 into KA-6D AERIAL REFUELING TANKERS.

The A-6E, introduced from 1970, incorporated a number of improvements, including updated AVIONICS, more powerful engines, and slightly refined aerodynamics. More than 350 A-6Es now equip navy and marine corps medium-attack squadrons, of which 120 are new aircraft, and the rest are conversions.

An ungainly tadpole-shaped aircraft, the A-6E is dominated by a large, bulbous radome housing a Norden APQ-148 multi-mode radar (which replaced the separate search and tracking radars of the A-6A). The APQ-148 has at least four different modes, including TRACK-WHILE-SCAN (TWS), TERRAIN AVOIDANCE, ground mapping, and airborne moving target indicator (AMTI). Solid-state technology has improved availability from 35 percent to 85 percent. Mounted ventrally behind the radome is a retractable TRAM (Target Recognition and Attack Multisensor), a turret which contains a FLIR and a LASER DESIGNATOR.

The pilot and WEAPON SYSTEM OPERATOR (WSO) sit side-by-side under a single-piece clamshell canopy. The cockpit is equipped with a Kaiser AVA-1 Head-Down Display, a video monitor on which flight data, navigation information, and weapon aiming cues can all be displayed. The aircraft's navigation/attack system can generate synthetic terrain/sea and sky images, by combining radar, FLIR images, and other sensor data. The WSO has a separate set of displays, including a video monitor linked to the TRAM, which can be used to designate targets for LASER GUIDED and ELECTRO-OPTICAL bombs and missiles. Other avionics include a Litton INERTIAL NAVIGATION unit, an ASQ-133 navigation/attack computer, an APN-153

DOPPLER navigation radar, and an ALQ-41 or ALQ-100 DECEPTION JAMMING system.

An aerial refueling probe is located at the base of the windscreen, while the fuselage behind the cockpit is filled with avionics compartments, generators, air conditioning units for the radar, and integral fuel tanks. The Intruder is powered by a pair of Pratt and Whitney J52 turbojets mounted in the underside of the fuselage and fed by two large air intakes located just behind the radome.

With a high aspect ratio and only a modest amount of sweep, the midmounted wing is optimized for high subsonic cruising and low-speed maneuverability. The outer portions can fold upward to save space on carrier decks. With full-span leading edge slats, full-span flaps, and spoilers for lateral control, landing speed is only 132 mph, a key consideration for arrested carrier landings. The A-6A had two large paddle-type airbrakes mounted on the fuselage; in the A-6E, these have been replaced by two wingtip-mounted split-type brakes. The wing also houses several large integral fuel tanks.

The Intruder has one centerline and four wing pylons, each with a capacity of 3600 lb. (1637 kg.), for a total payload of 18,000 lb. (8182 kg.); all five pylons are plumbed for external fuel tanks. Since 1986, many Intruders have also been fitted with light pylons on the outer wing sections for AIM-9 SIDEWINDER air-to-air missiles, carried for self-defense. Offensive ordnance can include free-fall bombs, CLUSTER BOMBS, PAVEWAY laser-guided bombs, AGM-62 WALLEYE glide bombs, AGM-65 MAVERICK air-to-surface missiles, AGM-84 HARPOON anti-ship missiles, BGM-109 TOMAHAWK cruise missiles, and a variety of naval mines. The aircraft is also wired for nuclear delivery, and can carry one or two B63 free-fall nuclear bombs.

By the late 1980s, the Intruder still had unmatched all-weather attack capabilities, but its ability to survive against air defenses was becoming marginal. Wedded to the subsonic/heavy-payload formula—which no other air force deems viable—the navy's response was an incrementally improved A-6F, with more powerful and fuel-efficient General Electric F404-GE-400D turbofans rated at 10,700 lb. (4864 kg.) each, a new cockpit with five full-color multifunctional displays and a HUD, a new radar with inverse SYNTHETIC APERTURE and air-to-air modes compatible with the new AIM-120 AMRAAM, and a new wing fabricated from composite materials to reduce weight and increase fatigue life. The prototype A-6F flew in late 1987, and was scheduled for squadron service by 1990, but budgetary pressures caused its cancelation in favor of the Advanced Tactical Aircraft (ATA), which ironically was cancelled in 1991. In the interim, A-6Es may be modernized with elements of the A-6F program.

The Intruder airframe also served as the basis of the EA-6B PROWLER, a specialized 4-seat electronic warfare aircraft which went into production in 1971 as a replacement for the earlier EA-6A. See also AIRCRAFT CARRIERS, UNITED STATES.

Specifications Length: 54.6 ft. (16.65 m.). Span: 53 ft. (16.15 m.). Powerplant: (A/B/C/KA-6D) 2 J52-P-6 turbojets, 8300 lb. (3773 kg.) of thrust each; (E) 2 J52-P-8As, 9300 lb. (4228 kg.) of thrust each. Fuel: 15,935 lb. (7230 kg.) plus 8020 lb. (3638 kg.) in 5 drop tanks. Weight, empty: 26,-746 lb. (12,132 kg.). Weight, max. takeoff: 60,400 lb. (27397 kg.). Speed, max.: 644 mph (1037 kph) at sea level/625 mph (1000 kph) at altitude. Speed, cruising: 450 mph (720 kph). Range: (max weapons) 1077 mi. (1724 km.); (max fuel) 3100 mi. (4960 km.).

INVINCIBLE: A class of three British V/STOL AIRCRAFT CARRIERS commissioned between 1980 and 1984. The refusal of British governments to fund a new class of conventional aircraft carriers to replace the old HMS *Eagle* and *Ark Royal* prompted the Royal Navy in the late 1970s to design a large, air-capable "cruiser" instead, nominally for ANTI-SUBMARINE WARFARE (ASW) and ostensibly similar to the Soviet MOSKVA and Italian VITTORIO VENETO hermaphrodite cruisers. Originally these ships were presented as helicopter carriers, but provisions to operate SEA HARRIER V/STOL fighters were predictably added, yielding a design with a carrier-type flight deck that left little room for cruiser armament. A final design change in 1977 added "commando carrier" facilities for AMPHIBIOUS ASSAULT operations. The final design was obviously an aircraft carrier, albeit a small one; but having circumvented its masters, the Royal Navy politely classified the ships as "through-deck cruisers"—a pretense which was maintained until 1980, when they were reclassified as ASW aircraft carriers (CVS). Prior to 1982 it had been planned to sell HMS *Invincible* to the Australian navy, but the ship's performance during the Falklands War was impressive, the need to defend the islands remained, and the sale was therefore canceled.

The flight deck, 550 ft. (167.68 m.) long and 44 ft. (13.41 m.) wide, extends from the extreme stern but ends well short of the bow, leaving the forecastle clear for defensive weapons. The dominant feature of the flight deck is a "ski-jump" ramp at the extreme forward end on the port side. Used by Sea Harriers for rolling start takeoffs, it radically improves their COMBAT RADIUS and payload as compared to both vertical and flat rolling takeoffs. On the first two ships, HMS *Invincible* and *Illustrious*, the ski-jump has a 7° slope; on the last ship, HMS *Ark Royal*, it has a 12° slope which is to be retrofitted into the other ships as well. The hangar deck, equal in size to the flight deck, is reached by two middeck elevators, one amidships and one aft. A large island superstructure offset to starboard incorporates the usual ship-handling and flight-operation facilities, two large engine intakes and exhaust stacks, and a large, tower mast for sensors. To serve as flagships for ASW escort groups and other task forces, the Invincibles have extensive COMMAND AND CONTROL facilities, an ADAWS-5 DATA LINK, and two SCOT satellite communications terminals.

The Invincibles have some autonomous ANTI-AIR WARFARE (AAW) capability with a twin-arm SEA DART surface-to-air missile launcher with 22 missiles (which also have a secondary ANTI-SURFACE WARFARE capability). The launcher is mounted on the ships' centerline at the forward end of the flight deck, limiting the aircraft handling area near the bow. After Sea Dart proved ineffective against low-flying aircraft during the 1982 Falklands War, U.S.-made PHALANX 20-mm. radar-controlled guns were installed for anti-missile defense, two aboard *Invincible* and *Illustrious* and three aboard Ark Royal; the latter also has two twin 30-mm. BMARC cannons amidships. The older ships will be brought up to this standard.

The air group usually consists of 7 Westland SEA KING ASW helicopters, at least 2 of which are equipped with a Searchwater radar for AIRBORNE EARLY WARNING (a capability absent in the Falklands War), and 8 Sea Harriers. Up to 12 Sea Harriers can be carried. See also AIRCRAFT CARRIERS, BRITAIN.

Specifications **Length:** 677 ft. (206.4 m.). **Beam:** 90 ft. (24.43 m.). **Draft:** 26 ft. (7.93 m.). **Displacement:** 16,000 tons standard/19,500 tons full load. **Powerplant:** 4-shaft COGAG: 4 28,000-shp. Rolls Royce Olympic TM3B gas turbines. **Speed:** 28 kt. **Range:** 5000 n.mi. at 18 kt. **Crew:** 670 + 284 air group. **Sensors:** 1 Type 1022 air surveillance radar, 1 Type 992 search radar, 1 Type 1006 navigational radar, 2 Type 909 Sea Dart guidance radars, 1 Type 2016 hull-mounted medium-frequency active/passive sonar. **Electronic warfare equipment:** 1 UAA-1 Abbey Hill electronic signal monitoring array, 2 8-barrel Corvus chaff launchers, 4 Mk.36 SRBOC chaff launchers.

IOC: See INITIAL OPERATING CAPABILITY.

IOWA: A class of four U.S. BATTLESHIPS (USS *Iowa, New Jersey, Wisconsin,* and *Missouri*) commissioned in 1943–44, subsequently decommissioned, and again recommissioned (and partially modernized) between 1982 and 1987. Ordered in 1940, the Iowas were designed as fast battleships meant to operate with carrier task forces. Extremely graceful, flush-decked ships, the Iowas are the largest warships in the world, aside from aircraft carriers. Very heavily armored, they have 12.1-in. (308-mm.) side belts, 17-in. (432-mm.) turret faces, 6-in. (152-mm.) armored decks, and elaborate anti-torpedo protection. As built, the Iowas were armed with 9 16-in. 50-caliber guns in 3 triple turrets (2 forward, 1 aft) and 20 5-in. 38-caliber DUAL PURPOSE guns in 10 twin mounts; during World War II, as many as 76 40-mm. BOFORS GUNS and 52 20-mm. Oerlikon guns were added for anti-aircraft defense.

After World War II, the *Iowa, New Jersey,* and *Wisconsin* were laid up in reserve, while the *Missouri* continued to serve as a flagship. The other three ships were reactivated in 1950 for the Korean War, but by 1956 all four had again been mothballed. In 1967, the *New Jersey* was reactivated to provide GUNFIRE SUPPORT off the coast of Vietnam, but was again decommissioned in 1969. By the late 1970s, however, the retirement of the U.S. Navy's remaining World War II cruisers and destroyers left few ships capable of providing fire support for amphibious operations, and in 1980 the Reagan administration decided to reactivate and modernize all four Iowas, both for the fire support role and also to serve as the nucleus of missile-armed SURFACE ACTION GROUPS (SAGs). In 1990, it was decided to decommission Iowa and Wisconsin as a cost-cutting measure.

The post-1981 modernization included a thorough overhaul of the ships' machinery, and the replacement of obsolete electronics with modern, fleet-compatible systems. All 16-in. guns were retained, but 4 5-in. mounts were replaced by 8 quadruple armored box launchers for a total of 32 TOMAHAWK cruise missiles, 4 quadruple launchers for HARPOON anti-ship missiles, and four PHALANX 20-mm. radar-controlled guns for anti-missile defense. The 16-in. guns can fire armor piercing (AP) and high-explosive (HE) ammunition out to a range of 23 mi. (36.8 km.); discarding-sabot rounds and rocket assisted projectiles (RAP) have more than twice that range.

Up to four SH-60B SEAHAWK helicopters can be accommodated on a fantail landing pad, though there is no hangar; they would be needed to provide OVER-THE-HORIZON TARGETING (OTH-T) for Tomahawk and Harpoon missiles, and perform aerial SPOTTING for the 16-in. battery. Since 1984, the Iowas have also been provided with Israeli-made Mastiff REMOTELY PILOTED VEHICLES for OTH-T and spotting over heavily defended targets. The ships have been equipped with two WSC-3 satellite communications terminals and an NTDS Link 11 receive-only DATA LINK.

The reactivation of the Iowas was hotly debated at the time (with more debate in 1989, after a catastrophic fire in the Iowa's no. 2 16-in. turret). It is widely believed, however, that these ships represent an effective means of providing gunfire support, and can also supplement aircraft carriers with their heavy missile armament. Their armor protection should suffice to render them virtually invulnerable to most existing anti-ship missiles, giving them more combat staying power than most modern warships. Nevertheless, plans for a more extensive modernization, with the removal of the after turret and installation of a large flight deck for V/STOL fighters, have been shelved. Despite their excellent condition and usefulness in Third World contingencies, the Iowas will probably be deactivated as a cost-saving measure. See also ANTI-SURFACE WARFARE.

Specifications Length: 887.2 ft. (270.5 m.). **Beam**: 108.2 ft. (33 m.). **Draft**: 38 ft. (11.6 m.). **Displacement**: 45,000 tons standard/58,000 tons full load. **Powerplant**: 4-shaft steam: 8 oil-fired boilers, 4 sets of geared turbines, 212,000 shp. **Speed**: 33 kt. **Range**: 5000 n.mi. at 30 kt./15,000 n.mi. at 17 kt. **Crew**: 1500. **Sensors**: 1 SPS-67 surface-search radar, 1 SPS-49 air-search radar, 1 LN-66 navigational radar, 1 Mk.25 and 1 Mk.13 gun fire control radar. **Electronic warfare equipment**: 1 SLQ-32 (V)3 electronic signal monitoring and active jamming array, 2 Mk.36 SRBOC chaff launchers.

IRAN: Inspect and Repair as Necessary, a logistic support procedure for major weapon systems, whereby thorough, depot-level inspections are performed at regular intervals, and not merely in response to breakdowns. IRANs are convenient scheduling dates for repainting, refurbishing, and

the installation of any new components under ongoing modernization programs.

IRBM: Intermediate-Range Ballistic Missile; any land-based BALLISTIC MISSILE with a range between 1500 mi. (2400 km.) and 3300 mi. (5500 km.). IRBMs have armed the LONG-RANGE INTERMEDIATE NUCLEAR FORCES of the United States and Soviet Union, but they are the chief strategic-nuclear weapons of France and China. See also INF TREATY; IRBMS, CHINA; IRBMS, FRANCE; IRBMS, SOVIET UNION; IRBMS, UNITED STATES.

IRBMS, CHINA: The People's Republic of China has made continuous, if low-key, efforts to develop its land-based BALLISTIC MISSILE force within the limits of an inadequate technological base. Some observers claim that Chinese IRBM designs are still derivative of 1950s Soviet missiles, notably the SS-4 SANDAL.

The first operational Chinese ballistic missile (U.S. designation CSS-1), a single-stage, liquid-fuel medium range ballistic missile (MRBM) with a maximum range of about 1000 mi. (1600 km.), may have been a direct copy of the Sandal. The CSS-1 became operational in the early 1970s, and some 100 are deployed near the Soviet border.

The CSS-1 was followed by an intermediate-range single-stage liquid-fuel missile (U.S. designation CSS-2), with a maximum range of approximately 1700 mi. (2720 km.). A force of 15–20 CSS-2s was deployed in the mid-1970s, and CSS-2s were also sold to Saudi Arabia in 1989, without nuclear warheads. With the addition of a second stage, the CSS-2 may also have served as the basis of the first Chinese ICBM (CSS-3), which had a range of about 3200 mi. (5120 km.).

Little is known of the payloads of these missiles. Warhead yields between 20 KT and several megatons (MT) have been cited. Nothing has been released on the accuracy of the CSS-1 and CSS-2, but if Soviet experience is any guide, CEPs could be on the order of 1000 to 2000 m. See also ICBMS, CHINA.

IRBMS, FRANCE: France initiated the development of a land-based BALLISTIC MISSILE in 1959, under President de Gaulle's policy of national self-sufficiency in nuclear weapons and delivery systems. By the mid-1960s, France was flight testing the S-2, a 2-stage solid-fuel IRBM, and by 1971 two *Escadres de Missiles Strategiques* (EMS), or "Strategic Missile Squadrons," each of nine missiles, were deployed in silos located on the Plateau d'Albion in southeast France. The S-2 was 48.6 ft. (14.82 m.) long and 4.95 ft. (1.5 m.) in diameter, with a launch weight of 70,574 lb. (32,-

079 kg.); its casing was fabricated of rolled steel, and the solid propellant was cast in it as a single block. The first stage had four gimballed nozzles with a combined thrust of 121,252 lb. (55,115 kg.); the second stage also had four nozzles, with a combined thrust of 99,206 lb. (45,094 kg.). The payload consisted of a single ablative REENTRY VEHICLE (RV) housing a 150-KT nuclear warhead. Maximum range was some 1709 mi. (2735 km.), with a median error radius (CEP) of 1000 m.

In 1973 the French began development of the S-3 second-generation IRBM, with a new second stage consisting of the lightweight would-glassfiber casing of the M-2 SLBM (see MSBS), and a new 1.2-MT reentry vehicle with a high ballistic coefficient for improved accuracy. This RV is hardened against nuclear radiation and contains PENAID devices to counter the Soviet GALOSH anti-ballistic missile system. The S-3 is 44.9 ft. (13.7 m.) long and has the same body diameter as the S-2, but the launch weight has been reduced to 56,879 lb. (25,-854 kg.). Range is extended to 2175 mi. (3480 km.) through a combination of the reduced structural weight and more efficient engines. Accuracy figures have not been released, but CEP is estimated at 500 m. The S-3 began flight testing in 1973, and replaced the S-2 in 1980–81.

IRBMS, SOVIET UNION: The Soviet Union developed IRBMS during the mid-1950s in a logical progression from post-1945 tests of captured V-2s, to the deployment of functional ICBMS. The first Soviet IRBM, the R-14 (U.S. designation SS-5 Skean), essentially an enlarged version of the earlier R-12 (SS-4 Sandal) MRBM, was a single-stage, liquid-fuel missile which burned a storable combination of undimensional dimethyl hydrazine (UDMH) and nitrogen tetroxide (NTO). Thought to carry a 1-MT nuclear warhead, the SS-5 had an effective range of 2175 mi. (3480 km.) and very poor accuracy, with a presumed CEP of some 2000 m. The missile entered service in 1962, and by 1966 several hundred were deployed in hardened silos in the western U.S.S.R. The Skean remained the principal Soviet IRBM until the late 1970s, and (with the addition of a liquid-fueled upper stage) became a workhorse of the Soviet space program as the B-series booster.

The shortcomings of Skean included its poor accuracy (which limited its usefulness to attacks against cities) and its need for elaborate launch facilities. Development of a successor began in the mid-1960s, with the specific requirement of mobility to provide security against a PREEMPTIVE

STRIKE—increasingly a possibility from a technical standpoint, as U.S. ICBMs acquired some COUNTERFORCE capability. Improved accuracy to allow attacks against military targets was also a requirement, but definitely secondary to mobility.

The emphasis on mobility precluded the standard Soviet preference for incremental improvements. Liquid-fuel missiles are too delicate to withstand the rigors of road or rail transport (any fuel leak can lead to a catastrophic explosion), and Soviet designers were thus forced to rely on solid propellant, with which they had little experience. Moreover, solid-fuel missiles did not lend themselves to the standard Soviet "Fly-by-Wire" technique of preprogrammed INERTIAL GUIDANCE. The only Soviet solid-fuel missile in service at that time, the SS-13 SAVAGE ICBM, had poor accuracy even by Soviet standards and a relatively small payload. Nonetheless, an attempt was made to develop an IRBM by using the two lower stages of the SS-13, and the resulting missile, designated SS-14 Scapegoat, weighed 26,500 lb. (12,045 kg.), carried a 1-MT warhead, and had a presumed CEP of some 1500 m. It was carried on a tracked transporter-erector-launcher (TEL) based on a heavy tank chassis. Apparently, the system was not a success, and only a few were deployed in the late 1960s.

By the early 1970s, the U.S.S.R. produced its first successful solid-fuel ICBM, the SS-16 SINNER, with more efficient propellants, lighter casings, and a "navigating" inertial guidance system better suited to solid fuel missiles than the Fly-by-Wire technique. Following the precedent established by the SS-13/14, the U.S.S.R. concurrently developed the RSD-10 *Pioneer* (U.S. designation SS-20 SABER) from the first two stages of the SS-16. Within the diameter of its upper stage, the SS-20 accommodated a POST-BOOST VEHICLE with three MULTIPLE INDEPENDENTLY TARGETED REENTRY VEHICLES (MIRVs), each with a yield of some 500 kT. The SS-20 is carried in a sealed launch canister on a wheeled TEL, and uses the COLD LAUNCH technique. It has a range of 3542 mi. (5668 km.), with a CEP that varies between 340 and 440 m., depending upon whether it is launched from a fixed (presurveyed) or expedient site. The SS-20 went into series production in 1979, and by 1988, more than 900 had been built, completely replacing the SS-5. Because of its mobility, the U.S. could not accurately estimate SS-20 numbers with surveillance satellites (before the INF Treaty negotiations, the CIA estimated the number of SS-20s at approximately 400; during the negotiations, the U.S.S.R. declared some 650).

Because of its combination of mobility, accuracy, and throw weight, the SS-20 gave the U.S.S.R. the capability of destroying military installations throughout Western Europe from well inside the U.S.S.R. That raised the specter of "decoupling" the U.S. nuclear deterrent from NATO. In response, the U.S. and NATO decided to deploy long-range intermediate nuclear forces (LRINF), consisting of 108 PERSHING II MRBMs and 464 TOMAHAWK ground-launched cruise missiles (GLCMs)—a response which, in the context of the Gorbachev revolution in Soviet policy, eventually led to the the INF TREATY of December 1987, under which both sides agreed to dismantle all LRINF systems including the SS-20. (During the mid-1980s, however, the Soviet Union had initiated flight tests of a new IRBM based on the lower stages of the solid-fuel ICBM code-named SS-25 SICKLE.) See also ICBMS, SOVIET UNION.

IRBMS, UNITED STATES: The United States developed a number of IRBMS during the mid-1950s, and actually deployed the liquid-fueled, single-stage Jupiter and Thor missiles. All were seen as interim weapons, pending the introduction of U.S. ICBMs. After that, the U.S. chose not to develop IRBMs, leaving the theater-nuclear role to manned aircraft, until the advent of the PERSHING II medium-range ballistic missile (MRBM) and TOMAHAWK ground-launched cruise missile (GLCM) programs. The INF TREATY of December 1987 now precludes the development of any new American IRBMs.

IRCM: See INFRARED COUNTERMEASURES.

IROQUOIS: 1. A class of four Canadian DE-STROYERS (DDs) commissioned in 1974–75. Distinguished by a sharply raked bow, a blocky superstructure, and twin stacks in a V configuration, the Iroquois were the first Western destroyers with gas turbine propulsion. Like most Canadian warships, they are very seaworthy and unusually robust, to withstand the rigors of arctic operations.

The Iroquois are optimized for ANTI-SUBMARINE WARFARE (ASW), with scant ANTI-AIR WARFARE (AAW) and ANTI-SURFACE WARFARE (ASUW) capabilities. Their primary weapons are two large SH-3D SEA KING ASW helicopters operated from a flight deck and hangar amidships. The deck is equipped with a "Bear Claw" haul-down landing system, which allows helicopter operations even in very rough seas. For short-range ASW, the ships are armed with two sets of Mk.32 triple tubes for

MK.46 lightweight torpedoes amidships, and a 3-barrel LIMBO depth charge mortar on the fantail. The only other weapons are an OTO-MELARA 5-in. DUAL PURPOSE gun on the foredeck, and two 4-round SEA SPARROW short-range surface-to-air missile (SAM) launchers in front of the bridge.

In 1986 the ships began major overhauls under the Tribal-classes Update and Modernization Program (TRUMP), which is to be completed in 1991. Under TRUMP, the ships are to receive more efficient engines to improve range, and will be fitted with vastly improved weapons and sensors, to convert them into guided-missile destroyers. The 5-in. gun is to be replaced by a Mk.41 VERTICAL LAUNCH SYSTEM (VLS) for STANDARD SM-2MR SAMs and HARPOON anti-ship missiles, the Sea Sparrow launcher is to be replaced by an OTO-MELARA 76.2-mm. dual purpose gun, and a PHALANX 20-mm. radar-controlled gun is to be added on top of the hangar as defense against anti-ship missiles. A completely new, Dutch-made radar suite will replace the current radars. New tactical displays and DATA LINKS will also be added to enable the ships to serve as flagships. ASW weapons and sensors are not changed. See also FRIGATES, CANADA.

Specifications **Length:** 423 ft. (129 m.). **Beam:** 50 ft. (15.2 m.). **Draft:** 14.5 ft. (4.4 m.). **Displacement:** 3551 tons standard/4200 tons full load. **Powerplant:** twin-shaft COGOG: 2 3700-shp. FT4A2 gas turbines (cruise), 2 25,000-shp. Pratt and Whitney FT12H sprint turbines (sprint). **Speed:** 29 kt. **Range:** 4500 n.mi. at 20 kt. **Crew:** 285. **Sensors:** 1 SQS-505 hull-mounted medium-frequency active/passive sonar, 1 medium-frequency variable depth sonar, 1 SPS-501 air surveillance radar, 1 SPQ-2D surface-search radar, 2 HSA M.22 fire control radars; (post-TRUMP) 1 LW-08 air-search radar, 1 DA-08 surface-search radar, 1 STIR 1.8 fire control and missile-guidance radar. **Electronic warfare equipment:** 1 WLR-1 radar warning receiver, 1 ULQ-6 active jamming unit, 1 6-rail chaff launcher, 1 Knebworth-Corvus chaff launcher.

2. Official U.S. Army designation for the UH-1 utility helicopter. See HUEY.

IVAN ROGOV: A class of two Soviet AM-PHIBIOUS WARFARE ships completed in 1978 and 1983. Classified as "Large Landing Ships" (Bol'-shoy Desantnyy Korabl, BDK) by the Soviet navy, they are combination dock landing ships (LPDS) and tank landing ships (LSTS) in U.S. Navy terms. The hull has a raised forecastle and a massive superstructure amidships; one helicopter landing deck is in the waist ahead of the superstructure, with another on the fantail. A large hangar built into the superstructure can accommodate up to 4 Ka-25 HORMONE helicopters, and inclined ramps at each end of the hangar allow them to be moved from one landing pad to the other. The docking well under the aft helicopter pad is capable of holding 3 Lebed-class AIR CUSHION VEHICLE landing craft, which are launched and retrieved by flooding the well and opening a large sea door in the stern. The vehicle deck in the forward half of the ship can accommodate 10 MAIN BATTLE TANKS or 30 ARMORED PERSONNEL CARRIERS, and there are accommodations for 550 troops, i.e., for an entire MOTORIZED RIFLE battalion. The Ivan Rogovs are flat-bottomed, and with their clamshell bow doors and articulated vehicle ramp, they can ground themselves to disembark vehicles into shallow water or directly onto a beach.

Armament includes a twin 76.2-mm. DUAL PURPOSE gun on the forecastle, a 40-tube 122-mm. multiple rocket launcher on the forward superstructure, one SA-N-4 GECKO short range surface-to-air missile system, and four 30-mm. radar-controlled guns amidships.

Specifications **Length:** 521.4 ft. (159 m.). **Beam:** 79.16 ft. (24.13 m.). **Draft:** 27 ft. (8.23 m.). **Displacement:** 11,000 tons standard/13,000 tons full load. **Powerplant:** 2 25,000-shp. gas turbine engines, 2 shafts. **Speed:** 23 kt. **Range:** 8000 n.mi. at 20 kt./12,500 n.mi. at 14 kt. **Crew:** 400. **Sensors:** 1 "Head Net" air search radar, 2 "Don Kay" navigational and surface search radars, 1 "Owl Screech" 76.2-mm. fire control radar, 1 "Pop Group" SA-N-4 guidance radar, 2 "Bass Tilt" 30-mm. fire control radars. **Electronic warfare equipment:** 2 "Bell"-type electronic signal monitoring arrays, 2 chaff launchers.

J

JAGDPANZER: Literally, "Hunting Tank," a family of West German TANK DESTROYERS in service with the West German and Belgian armies. Developed on the chassis also incorporated in the MARDER infantry fighting vehicle, the first prototypes were produced in 1960. These vehicles entered service with the *Bundeswehr* in 1965 to replace the first (French-built) postwar tank destroyers. There are two basic versions with a common hull and drive train: the *Jagdpanzer Kanone,* armed with a 90-mm. ANTI-TANK gun, and the *Jagdpanzer Rakete,* armed with ANTI-TANK GUIDED MISSILES (ATGMs).

The *Jagdpanzer Kanone* (Jpz.4-5)—a direct descendant of World War II German tank destroyers—is designed to stalk tanks, relying on speed and a low silhouette to achieve advantageous (static) firing positions and evade return fire. The vehicle has an all-welded steel hull, but armor protection is limited to a maximum of 50 mm. (concentrated mainly on the well-sloped frontal glacis plate), in order to reduce the combat weight. A high power-to-weight ratio, combined with a low ground pressure of 10.67 lb. per square ft. (0.75 kg./cm.²), gives the *Jagdpanzer* good cross-country mobility; the vehicle is especially designed to move quickly in reverse in order to disengage rapidly from firing positions. *Jagdpanzera* are fitted with a COLLECTIVE FILTRATION for NBC protection, and with a wading kit they can ford streams as deep as 6.9 ft. (2.1 m.).

The *Jagdpanzer Kanone* has a combined driving/fighting compartment at the front, with the engine compartment in the rear. Of the four-man crew, the driver, seated on the left front of the hull, is provided with three periscopes; the gunner, seated slightly behind and to the right of the driver, has a simple stadiametric gunsight; the commander, in a raised cupola behind the gunner, has several periscopes for all-around observation; and the loader sits behind the driver, at the rear of the fighting compartment. All periscopes are fitted with semi-active INFRARED night vision devices, used in conjunction with an infrared searchlight.

The 90-mm. M36 main gun in the bow of the vehicle has a ball-socket mount with traverse limited to 15° to either side, and elevation limits of −8° and +15°. With HESH and HEAT rounds, the gun is effective against all but the latest MAIN BATTLE TANKS out to a maximum range of 2000 m. (2200 yd.). The weapon is also very effective against soft targets and field fortifications in an infantry-support (ASSAULT GUN) role. A high rate of fire (as much as 12 rounds per minute) can be achieved by a well-trained crew; a total of 51 rounds are stowed in the fighting compartment. Secondary armament consists of a coaxial 7.62-mm. MACHINE GUN, and a pintle-mounted 7.62-mm. machine gun by the commander's hatch. Eight smoke GRENADE launchers are fitted on the rear deck behind the fighting compartment.

The West German *Bundesheer* acquired 750 *Jagdpanzer Kanone* between 1965 and 1967. The Belgian army purchased 80 more in 1975. These differ from the West German versions in the suspension and transmission (derived from the

Marder), and their far more elaborate ancillaries: they have a Belgian SABCA computerized FIRE CONTROL system with a LASER rangefinder, gun stabilization, and passive night vision sights.

The *Jagdpanzer Rakete* was designed to complement the *Jagdpanzer Kanone* by providing long-range ATGM fire. It has the same hull, chassis, and powerplant, but the main gun is replaced by missile launchers and sights mounted on top of the hull. Originally armed with two launch rails for a total of 14 French SS-11 missiles, with manual command to line-of-sight (MCLOS) guidance (whereby the operator steers the missile with a joystick), the *Rakete* has been rearmed with 20 Euromissile HOT ATGMs, with semi-automatic command to line-of-sight (SACLOS) guidance, a larger warhead, and a maximum range of 4000 m. (4300 yd.). (*Jagdpanzers* with HOT are designated Jaguar 1.) Secondary armament consists of a 7.62-mm. machine gun in the bow, and a pintle-mounted 7.62-mm machine gun at the commander's position. The *Bundesheer* acquired some 370 *Jagdpanzer Rakete* in 1967–68. Most *Jagdpanzer Kanone* were to be converted to missile carriers with TOW ATGMs and a thermal imaging sight (Jaguar 2), but the introduction of REACTIVE ARMOR in the Soviet tank force has greatly reduced the effectiveness of ATGMs, and the program's future is now in doubt.

Specifications **Length:** 20.5 ft. (6.238 m.). **Width:** 9.75 ft. (2.98 m.). **Height:** 6.83 ft. (2.085 m.). **Weight, combat:** 27.5 tons. **Powerplant:** 500-hp. Daimler-Benz MB837Aa 8-cylinder, water-cooled diesel. **H./wt. ratio:** 19.4 hp./ton. **Fuel:** 106 gal. (470 lit.). **Speed, road:** 43.5 mph (70 kph). **Range, max.:** 250 mi. (400 km.).

JAGER: Literally, "hunter," German term for a type of LIGHT INFANTRY that relies on stealth and surprise more than massed firepower (originally, *Jagers* were recruited from huntsmen, to serve as scouts and skirmishers in close terrain). In today's West German *Bundesheer*, the term is more honorific than functional: it designates the motorized infantry units of the TERRITORIAL ARMY (TA), a reserve organization of twelve *Heimatschutzebrigaden* (HSBs), or Home Defense Brigades, each of one or two *Panzer* (tank) battalions, and one or two *Jager* battalions. The latter have 950 men, organized into a headquarters COMPANY, a heavy company, and three *Jager* companies of 170 men each. The troops are mounted either in light trucks or M113 armored personnel carriers, but these serve exclusively "battle taxis" for transportation: *Jagers* always fight on foot.

To counter the threat of Soviet AIRBORNE, AIR ASSAULT, and SPETSNAZ units in rear areas, the Territorial Army has been augmented by fifteen *Heimatschutzeregimentsen* (Home Defense Regiments), each with three truck-mounted *Jager* battalions of older reservists (not to be compared with true *Jagers*). The West Germany *Bundesheer* also has one brigade of *Gebirgsjager* (ALPINE infantry) and three brigades of *Fallschirmjager* (airborne infantry). Unlike the Territorials, these elite forces employ classic *Jager* tactics. The *Jager* tradition also persists in the alpine infantry units of the Austrian army, and in the better Finnish and Swedish infantry units. See also AIRBORNE FORCES, FEDERAL REPUBLIC OF GERMANY.

JAGUAR: Single-seat, twin-engine ATTACK AIRCRAFT built by the Anglo-French consortium SEPECAT (British Aerospace and Breguet). The Jaguar originated in a 1965 Anglo-French joint requirement for a two-seat trainer with supersonic speed and some attack capability, roughly on the lines of the Northrop T-38 Talon. The first prototype flew in 1968, and the first operational version, the Jaguar E (*École*) 2-seat trainer, entered service with the French air force in 1972. But the Jaguar proved to be too expensive to serve as a dedicated trainer, and all 2-seat versions have been given full combat capability. (The need for cheaper advanced jet trainers remained: France combined with West Germany to produce the ALPHA JET, while Britain produced the HAWK.)

The first single-seat attack variant, the Jaguar A (*Appui*), serves in the French air force for CLOSE AIR SUPPORT (CAS), battlefield INTERDICTION (BI), and nuclear strike. Without RADAR, it has a navigation/attack system based on a DOPPLER-updated, two-gyro INERTIAL NAVIGATION unit. Armed with two DEFA 30-mm. cannons in the fuselage, the Jaguar A can carry up to 10,000 lb. (4545 kg.) of external stores on one centerline and four wing pylons. Compatible ordnance includes SNEB 65-mm. folding fin aerial rocket (FFAR) pods, free-fall bombs, Beluga CLUSTER BOMBS, DURANDAL runway-cratering bombs, MARTEL anti-radiation missiles, Matra MAGIC infrared homing air-to-air missiles for self-defense, and AN.52 tactical nuclear bombs. In 1978, French Jaguars were modified to carry Thomson-CSF Atlis electro-optical TARGET ACQUISITION and LASER DESIGNATOR pods to augment their built-in avionics; photo-reconnaissance and Super Cyclops INFRARED line scanner (IRLS) pods are also available. France acquired a total of 190 Jaguars between 1972 and 1978.

The Royal Air Force GR.1 attack variant, intended for all-weather interdiction and tactical RECONNAISSANCE, was built with a much more comprehensive sensor and AVIONICS suite than the Jaguar A, with a digital inertial navigation and weapon-aiming system (INAVWAS), a laser rangefinder/designator in a chisel-shaped nose window, a head-up display (HUD), a 3-gyro inertial platform, and a Ferranti RADAR WARNING RECEIVER. The INAVWAS employs the continuously computed impact point (CCIP) weapon-aiming technique, to provide a CEP of some 50 ft. (15.25 m.). British aircraft have 30-mm. ADEN cannon in place of the similar DEFA. External stores include 1000-lb. (455-kg.) free fall bombs, PAVEWAY GBU-10/13 laser-guided bombs, BL.755 cluster bombs, and a British Aerospace multi-spectral reconnaissance pod (with cameras and IRLS) linked directly to the INAVWAS. All British aircraft also received more powerful engines in 1978. The British T.2 trainer lacks the laser designator and has only one cannon, but is otherwise combat capable. The RAF acquired 165 GR.1s and 37 T.2s between 1975 and 1978, but is already phasing out the Jaguar in favor of the Panavia TORNADO and British Aerospace HARRIER.

The Jaguar was not a success and not many were exported. In 1977, Ecuador purchased ten attack aircraft and two trainer Jaguars, both similar to the British variants, but most buyers opted for the special export version (Jaguar International), which has more powerful engines; overwing pylons for Sidewinder or Magic air-to-air missiles (leaving all underwing and fuselage pylons for offensive weapons and fuel); either the British-style nose or one with a French Agave multi-mode air-to-ground radar and a laser designator in a chin fairing; a low-light television (LLTV) night acquisition system; and additional weapon options, including KORMORAN and HARPOON anti-ship missiles. India is the single largest customer, having ordered 136 in 1979, of which 45 were assembled by SEPECAT, 60 assembled in India, and 31 built under license in India. Oman purchased 23 Jaguars in 1978, and Nigeria 18 in 1983.

The slab-sided fuselage is area-ruled to reduce trans-sonic drag. The nose is either chisel-shaped for aircraft with the laser rangefinder/designator, or has a conical radome for the Agave radar, with a small fairing below it for the Ferranti laser designator. Avionics occupy much of the space between the nose sensor and the cockpit. A retractable AERIAL REFUELING probe is mounted on the right side of the nose. Cockpit layouts vary: the British and International variants have a HUD, a video display screen, and a zero-zero ejection seat. Rearward visibility is restricted in all variants by the prominent dorsal spine. Cockpit noise levels are rather high, leading to pilot fatigue on long missions.

The two ventral 30-mm. cannons, immediately behind the nose gear well, have 125 rounds per gun. The main landing gear retract into wells located immediately behind the guns, and have large twin bogies with strong struts for rough-field operations. The rear fuselage houses the engines and a large integral fuel tank. A braking parachute (stowed in the extreme tail) and an arrester hook are standard, to reduce landing roll.

The Jaguar has a shoulder-mounted swept wing with an area of only 260 square ft. (24.18 m.²). Its small wing limits range and maneuverability, but gives the Jaguar a smooth high-speed ride at low altitudes, a prerequisite for a modern attack aircraft. Full-span double-slotted flaps and outboard leading-edge slats improve low-speed control and short takeoff and landing (STOL) performance; spoilers provide roll control. The wing has four hardpoints, with the inboard pair plumbed for external fuel tanks (the International version has two additional overwing pylons located at midspan for air-to-air missiles). Each wing houses a large integral fuel tank. The hydraulically actuated slab tailplane has very pronounced anhedral (downward angle), while the vertical stabilizer has a fin-tip ELECTRONIC COUNTERMEASURES fairing and a conformal UHF/VHF antenna. Two small ventral fins under the rear fuselage improve stability. The Jaguar is powered by two Rolls Royce-Turbomeca Adour afterburning turbofans mounted in the rear of the fuselage and fed by slab-sided cheek inlets. Even with the more powerful engines of later versions, the Jaguar is somewhat underpowered.

Specifications Length: 55.2 ft. (16.83 m.). Span: 28.5 ft. (8.69 m.). Powerplant: (early A/E) 2 Adour Mk. 102s, 7305 lb. (3314 kg.) of thrust each; (GR.1) 2 Adour Mk. 104s, 8040 lb. (3647 kg.) of thrust each; (Int'l) 2 Adour Mk.811s, 9270 lb. (4215 kg.) of thrust each. Fuel: 7213 lb. (3279 kg.) plus up to 6182 lb. (2810 kg.) in 3 drop tanks. Weight, empty: 15,432 lb. (7000 kg.). Weight, max. takeoff: 34,612 lb. (15,700 kg.). Speed, max.: (A) 990 mph (593 kph) at 36,000 ft. (11,000 m.); (Int'l) 1056 mph (1699 kph) at 36,000 ft. (11,000 m.)/(all)

840 mph (1352 kph) at sea level. **Combat radius:** (lo-lo-lo) 335 mi. (537 km.); (hi-lo-hi) 875 mi. (1460 km.). **Range, max.:** 2190 mi. (3500 km.). **Operators:** Ecu, Fra, Ind, Nig, Oman, UK.

JAVELIN: British man-portable, shoulder-fired surface-to-air missile, in service with the British army, the Royal marines, and several other armies. Developed from the early 1970s, Javelin is an improved version of the Shorts BLOWPIPE, with more advanced guidance, a more powerful rocket motor, a larger warhead, and improved fuzing. Externally similar to Blowpipe, the missile has four delta canards for steering and four folding tail fins. Powered by a dual-thrust solid-fuel rocket, and armed with a proximity-fuzed 6.05-lb. (2.75-kg.) high-explosive fragmentation warhead, Javelin has a unique feature that allows the operator to disable the warhead in flight, to avoid accidental attacks on friendly aircraft.

The missile is housed in a sealed storage/launch canister, which is attached to a sight/guidance unit. In contrast to the manually controlled Blowpipe, Javelin relies on semi-automatic command to line-of-sight (SACLOS) guidance. The operator lines up the target in his sight and launches the missile when it enters effective range. An INFRARED sensor in the guidance unit tracks the missile and measures the angle between the missile and the operator-target line; the guidance electronics then automatically generate steering commands, which are transmitted to the missile via a radio DATA LINK. All the operator needs to do is keep the target centered in his sight until impact.

Specifications (missile) **Length:** 4.6 ft. (1.39 m.). **Diameter:** 3 in. (76.2 mm.). **Span:** 10.6 in. (270 mm.). **Weight, launch:** 20.67 lb. (9.5 kg.). **Speed, max.:** Mach 1.8 (1368 mph/2285 kph). **Range envelope:** 300–4500 m. (0.2–2.8 mi.). **Height envelope:** 10–3000 m. (32.8–9840 ft.).

JCS: See JOINT CHIEFS OF STAFF.

JEANNE D'ARC: French helicopter carrier (CVH) and cadet training vessel, commissioned (as *La Resolue*) in 1964. Shortly thereafter, the ship acquired its present name, replacing the CRUISER *Jeanne d'Arc* as the cadet training ship of the French navy. The *Jeanne d'Arc* has a conventional, cruiserlike bow, with a 203.4- × 68.25-ft. (62- × 21-m.) flight deck from amidships to the stern. A 12-ton elevator at the aft end of the flight deck provides access to and from the hangar deck. When configured as a training ship, the hangar deck is partially occupied by accommodations and

classrooms, which can be removed for full flight operations.

Defensive armament consists of 4 100-mm. DUAL PURPOSE guns, 2 on the fantail and 2 at the forward corners of the flight deck. In the late 1970s, the ship was also armed with 6 EXOCET MM.38 anti-ship missiles in individual launch canisters mounted at the forward edge of the flight deck, to provide a considerable ANTI-SURFACE WARFARE (ASUW) capability. The air group normally consists of 4 SUPER FRELON helicopters, but up to 8 helicopters can be accommodated by removing the cadet quarters. Combat missions for the *Jeanne d'Arc* include ANTI-SUBMARINE WARFARE (ASW) support, AMPHIBIOUS ASSAULT, and troop transport. Two LCVP landing craft are carried on davits amidships for COMMANDO raiding parties.

Specifications **Length:** 597.1 ft. (182 m.). **Beam:** 78.7 ft. (24 m.). **Draft:** 24 ft. (7.3 m.). **Displacement:** 10,000 tons standard/12,365 full load. **Powerplant:** twin-shaft steam: 4 oil-fired boilers, 2 sets of geared turbines, 40,000 shp. **Fuel:** 1360 tons. **Speed:** 32 kt. (26.5 kt. sust.). **Range:** 3000 n.mi. at 26.5 kt./3750 n.mi. at 25 kt./5500 n.mi. at 20 kt./6800 n.mi. at 16 kt. **Crew:** 627. **Sensors:** 1 DRBV 22D air-search radar, 1 DRBV 50 surface-search radar, 1 DRBN 32 navigational radar, 3 DRBC 32A gun fire control radars, 1 hull-mounted SQS-503 medium frequency active/passive sonar. **Electronic warfare equipment:** 2 8-barrel Silex chaff rocket launchers.

JINKING: An evasive aerial maneuver which can be employed by ATTACK AIRCRAFT to evade AIR DEFENSE fire after a bombing run. Jinking generally involves several rapid changes of heading combined with concurrent altitude changes, and rolling maneuvers performed at maximum speed. The aim is to deny a predictable course to enemy gunners or missiles.

JOAO COUTINHO: A class of six Portuguese FRIGATES commissioned in 1970–71 as general-purpose patrol and ANTI-SUBMARINE WARFARE (ASW) vessels. Three were built by the Spanish Bazan shipyard, and three by Blohm und Voss in West Germany.

Weapons and sensors are austere. Armament consists of 1 twin 3-in. 50-caliber DUAL PURPOSE gun at the bow, 2 40-mm. BOFORS GUNS, 1 Mk.10 HEDGEHOG ASW mortar, 2 DEPTH CHARGE projectors, and 2 depth charge racks on the stern. A landing deck on the aft superstructure can accommodate only a small ASW helicopter such as the Agusta Bell 212 HUEY. A marine raiding detachment of 34 men can also be accommodated, along with several inflatable landing boats.

Economical, albeit rather limited vessels, the Joao Coutinhos have served as the basis for two improved designs. Blohm und Voss built four Baptiste de Andrade–class frigates for Portugal between 1974 and 1976, which are essentially identical to the Coutinhos, but have a French 100-mm. dual-purpose gun in place of the 3-in. guns, and improved electronics. The Coutinhos were also the design basis of the larger, faster, and more capable DESCUBIERTA-class frigates.

Specifications **Length:** 279.9 ft. (84.59 m.). **Beam:** 33.5 ft. (10.3 m.). **Draft:** 10.7 ft. (3.3 m.). **Displacement:** 252 tons standard/1401 tons full load. **Powerplant:** twin-shaft diesel: 2 OEW-Pielstick 12PC2V280 engines, 10,586 hp. **Speed:** 24.4 kt. **Range:** 5900 n.mi at 18 kt. **Crew:** 93. **Sensors:** 1 TM626 surface-search radar, 1 MLA-1B air-search radar, 1 SPG-34 fire control radar, 1 QCU-2 hull-mounted medium-frequency sonar.

JOINT CHIEFS OF STAFF (JCS): The primary U.S. military planning body, responsible for the coordination of the four armed services. Established ad hoc in World War II, and placed on a permanent footing by the 1947 National Security Act, the JCS consists of a chairman, the chiefs of staff of the army, navy and air force, and the commandant of the marine corps. JCS is not a "general staff" in the true sense of the term; it does not exercise operational control over military forces (a function reserved for UNIFIED and SPECIFIED COMMANDS responsible directly to the president), but rather acts as the immediate military staff of the secretary of defense, reconciling administrative and budget directives.

Under the 1986 Goldwater-Nichols Defense Reorganization Act, the chairman was given much broader powers. Not only is he now the primary military advisor to the president, but he is responsible for furnishing strategic direction to the armed services, strategic and contingency planning, establishing budget priorities, and developing joint doctrine for all four services. Of course, the chairman must still rely on the advice and cooperation of the service chiefs, which often results in compromise solutions based on service interests.

The chairman and the Joint Chiefs are supported by a "Joint Staff" controlled by a director and subdivided into eight directorates: Manpower and Personnel (J-1), JCS Support (DIA), Operations (J-3), Logistics (J-4), Strategic Plans and Policy (J-5), Command, Control, and Communications (J-6), Operational Plans and Interoperability (J-7), and Force Structure, Resources, and Assessment (J-8), as well as a secretary; an inspector general; an advisor for mapping, Charting, and Geodesy; and a director for Information and Resource Management. See also DOD.

JOINT SURVEILLANCE SYSTEM (JSS): An AIR DEFENSE control network operated by the U.S.-Canadian NORAD command to monitor the air sovereignty of the U.S. and Canada. Successor to both the SAGE and BUIC ground-controlled intercept systems, JSS was developed from the late 1970s and became operational in 1983. In contrast to its predecessors, which relied solely on military radars, JSS is designed to incorporate both military and civilian (i.e., air-traffic control) radars to reduce costs and increase coverage.

JSS consists of 46 radar sites in the continental United States (CONUS) connected to three Reporting and Operations Control Centers (ROCCs); 14 radars connected to one ROCC in Alaska; and 24 radars connected to two ROCCs in Canada. The ROCCs automatically process data from their associated radars to detect, identify, and track air targets. In peacetime, the ROCCs can assign and direct fighters to intercept unknown aircraft; in war, this function would be performed by six E-3 SENTRY AWACS aircraft assigned to NORAD, while the ROCCs would provide a backup capability.

ROCC track data are passed back to the NORAD Combat Operations Center under Cheyenne Mountain, Colorado, where they are correlated with data from other sensors and displayed to provide an overall picture of the North American air defense situation.

J-STARS: Joint Surveillance and Target Attack Radar System, a side-looking aerial radar (SLAR) and data processing system under development by the U.S. Army and Air Force (for which it was originally known as "Pave Mover"). Initiated in the late 1970s as part of the DEEP STRIKE and ASSAULT BREAKER technology programs, J-STARS is intended to provide "real-time" target-finding capabilities against enemy forces operating well beyond the Forward Edge of the Battle Area (FEBA).

Developed by a team of Grumman, Norden, Boeing, and Cubic, the radar is a fixed, side-looking PHASED ARRAY capable of scanning 60° to either side of boresight. It uses advanced SYNTHETIC APERTURE RADAR (SAR) and DOPPLER Moving Target Indicator (MTI) techniques to overcome ground clutter so as to generate detailed high-resolution ground maps, and detect, track, and classify vehicles and other moving targets (armor, helicopters, etc.), at ranges in excess of 60 mi. (100 km.). A programmable on-board signal processor extracts track data on likely targets for relay to ground stations through a secure, wide-band DATA LINK. The radar can also guide ground- and air-launched missiles against selected targets, transmitting steering commands through encrypted radar beams. J-STARS is a key element in the proposed NATO Follow-On Forces Attack (FOFA) program for attacking Soviet second echelon forces, and is being incorporated in the U.S. Army's Tactical Missile System (TACMS). To reduce electronic vulnerability, J-STARS uses advanced signal processing techniques against the effects of ACTIVE JAMMING.

Two aircraft were originally suggested as potential J-STARS platforms: the Lockheed TR-1 (a variant of the U-2 reconnaissance aircraft) and the Boeing 707-300 airliner. The latter, designated E-8A, was selected because of its greater payload, and ample accommodations for on-board technicians and a tactical coordination team. Operationally, E-8As would fly a racetrack pattern at a safe distance behind friendly lines, while scanning up to 60 mi. (100 km.) into enemy territory. The prototype J-STARS aircraft was deployed to Saudi Arabia in 1991 to track Iraqi ground forces in Kuwait.

JTIDS: Joint Tactical Information Distribution System, a high-volume DATA LINK under development by the U.S. armed forces. JTIDS originated in a 1960s air force tactical control project; it was merged with a similar navy project in 1974, and the army and marine corps were brought into the program in the early 1980s.

JTIDS is intended to provide secure (jam-resistant) data and voice COMMUNICATIONS between the tactical COMMAND AND CONTROL (C²) points and individual combat units of all four services. It uses a Time Division Multiple Access (TDMA) technique, whereby each JTIDS user is assigned a discrete period of time ("time slot") in which to transmit information. During each user's time slot, all other users are placed in a receive-only mode. The TDMA provides a maximum of 128 time slots per second per net, for as many as 128 nets, i.e., it provides grand total of 16,384 slots per second. Each slot contains 225 bits of data, split into 5-bit pulses which are spread over a frequency band of 960–1215 megahertz. With such a broad spectrum and ample scope for frequency-hopping techniques, JTIDS is very resistant to DECEPTION JAMMING. Simpler NOISE JAMMING on specific frequencies (SPOT JAMMING) is also relatively ineffective against JTIDS, because each 5-bit pulse is transmitted twice on different frequencies. BARRAGE JAMMING of the entire JTIDS bandwidth remains possible, but it would also render many enemy communications inoperable. Each JTIDS transceiver can also serve as a message relay point, to extend the system's range and avoid areas of intensive jamming or nuclear BLACKOUT. JTIDS messages have been successfully transmitted over ranges greater than 300 mi. (480 km.).

The first JTIDS transceivers to enter service, the Class 1 terminals (URQ-33[V]), intended for large ship or air platforms, are currently being installed on Boeing E-3A SENTRY airborne warning and control (AWACS) aircraft, and the navy's TICONDEROGA-class Aegis cruisers. Smaller Class 2 terminals, now in development, are intended for individual tanks, tactical aircraft, etc. The first Class 2 terminals will be installed in the F-15E Strike Eagle variant of the F-15 EAGLE fighter.

If JTIDS works as advertised, a number of intriguing operational schemes will be made possible by the integration of all available information sources into a single command, control, communications, and intelligence (C³I) network. Aircraft on attack missions, for example, could display the locations of all known enemy air defense systems on their cockpit terminals in "real time," exploiting all available signal intelligence (SIGINT), aerial RECONNAISSANCE, and satellite surveillance data. Fighter aircraft operating in conjunction with AWACS could determine the relative positions of all friendly and hostile aircraft, relying on AWACS radar data alone, while keeping their on-board radar in a standby mode, to avoid being detected by enemy RADAR WARNING RECEIVERS. The enemy would then know the location of the AWACS, but not that of the associated fighters. Higher commanders, on the other hand, could access and integrate data from all available sources to watch the battle unfold in real time, so as to allocate resources more effectively.

There would be a real danger, however, in relying too heavily on JTIDS. Although the system is *currently* believed to be secure and jam resistant, this may not be so in the future, especially if JTIDS is perceived as the critical node in the U.S. command and control system, thus becoming the key enemy target. U.S. forces could then be paralyzed, while enemy forces, with less efficient but more decentralized communications, would have a relative advantage.

JULIETT: NATO code name for a class of 16 Soviet diesel-electric guided-missile SUBMARINES (SSGs) completed between 1962 and 1969. The Julietts are similar to the ECHO-class nuclear-powered guided-missile submarines (SSGNs): both are armed with SS-N-3 SHADDOCK anti-ship missiles to counter U.S. CARRIER BATTLE GROUPS (CVBGs). But while the Echo has eight missiles, the smaller Juliett has only four.

In the early 1960s, Soviet production capacity for nuclear submarines was limited by low reactor production rates; by supplementing the Echos with the Julietts, a larger number of missile platforms could be deployed by the Soviet navy, given its ample shipyard capacity for conventionally powered submarines. As many as 72 Julietts were originally planned, but the program was terminated after 16 were built, probably because of the introduction of more advanced anti-ship missiles.

Derived from the FOXTROT-class attack submarine (SS), which is in turn an advanced version of the Zulu class (itself influenced by the German Type XXI U-boat of 1944), the Juliett has a streamlined, cigar-shaped hull in a double hull configuration, with ballast and fuel tanks installed between the pressure hull and outer casing. A rather long, streamlined sail is positioned amidships. Control surfaces consist of retractable bow planes, a single rudder, and stern planes mounted behind the propellers. A snorkel allows diesel operation from shallow submergence; its exhaust vent is faired into the aft end of the sail and forms a distinctive notch in its profile. A number of Julietts have been spotted with bulged fairings on the sail, similar to those of Echo-class submarines refitted with the SS-N-12 SANDBOX, an advanced derivative of Shaddock.

As noted, the main armament consists of four SS-N-3 Shaddock missiles housed in twin launchers, which are flush with the upper deck casing until elevated for launch. The submarine must surface to launch and must remain surfaced until im-

pact, to provide guidance to the missile during its flight. A large, supersonic missile, Shaddock carries either a 2200-lb. (1000-kg.) high-explosive or a 300-kT nuclear warhead. Terminal guidance is by active radar homing, but to fully exploit its maximum range of some 250 mi. (400 km.), some form of OVER-THE-HORIZON TARGETING (OTH-T) is required. Currently, a BEAR D maritime SURVEILLANCE aircraft would transmits a radar picture of the target area to the submarine via a video DATA LINK; the submarine in turn would transmit course corrections to the missile via a "Front Door/Front Piece" guidance radar, mounted in the forward edge of the sail (it deploys outward by rotating 180°).

Other armament consists of 6 bow-mounted 21-in. (533-mm.) tubes for a total of 12 acoustical homing TORPEDOES, and 4 16-in. (400-mm.) stern tubes for countermeasures and lightweight anti-submarine torpedoes.

Originally, the Julietts were assigned to the Northern Fleet, to defend Soviet coastal waters. Since the early 1980s, however, a number have been deployed into the Mediterranean, to shadow U.S. carrier battle groups. The Shaddock's major drawback, the need for the submarine to remained surfaced and exposed throughout its flight (as much as 25 minutes), would be nullified if the missiles were used preemptively at the outset of hostilities; hence the Echos and Julietts remain in service for the shadowing role. See also SUBMARINES, SOVIET UNION.

Specifications Length: 284.4 ft. (86.7 m.). **Beam:** 33.1 ft. (10.1 m.). **Displacement:** 3000 tons surfaced/3750 tons submerged. **Powerplant:** twin-shaft diesel-electric: 2 2000-hp. diesel generators, 2 1700-hp. electric motors. **Speed:** 6 kt. surfaced/ 12 kt. submerged. **Range:** 19,000 n.mi. at 7 kt. **Max. operating depth:** 984 ft. (300 m.). **Collapse depth:** 1640 ft. (500 m.). **Crew:** 80. **Sensors:** 1 "Front Door/Front Piece" missile guidance radar, 1 bow-mounted *Feniks* medium-frequency passive sonar array, 1 high-frequency *Herkules* active sonar, 1 "Snoop Slab" surveillance radar, 1 "Stop Light" electronic signal monitoring array, 2 periscopes.

K

K-: U.S. designation for AERIAL REFUELING tanker aircraft. Tankers are invariably conversions of other types of aircraft, so the designation is normally used in conjunction with others; as, e.g., KA-6 for the tanker variant of the INTRUDER attack aircraft. The most important tanker aircraft in U.S. service are the KC-10 EXTENDER and KC-135 STRATOTANKER. The primary Soviet tankers are the Il-76 MIDAS and the Mya-4 Bison.

KA-: Prefix of a Soviet HELICOPTER design bureau founded by N. I. Kamov. Kamov designs favor the twin-turbine, coaxial-rotor solution to the torque problem, obtaining compact helicopters well suited for shipboard operations. Kamov supplies the Soviet navy with most of its ANTI-SUB-MARINE WARFARE and TARGET ACQUISITION helicopters, notably the Ka-25 HORMONE and the more advanced Ka-27 HELIX. The Kamov bureau may also be responsible for the fast, powerful attack helicopter code-named HOKUM by NATO (the actual Soviet designation is not known).

KALASHNIKOV: Family of Soviet automatic ASSAULT RIFLES and LIGHT MACHINE GUNS, most notably the AK-47. See AK-.

KANGAROO (AS-3): NATO code name for a Soviet long-range air-to-surface CRUISE MISSILE. First spotted in 1961, the Kangaroo is a turbojet-powered missile similar in both size and configuration to a small fighter aircraft. The missile body is cylindrical, with an engine intake in the nose which contains a small centerbody radome; the wings and tailplane are very sharply swept. The missile is powered by a Turmansky R-11 or R-15

afterburning turbojet rated at 15,000 lb. of thrust and armed either with an 800-kT nuclear warhead or 5000 lb. (2272 kg.) of high explosive to attack large area targets. The Kangaroo is believed to rely on BEAM RIDING guidance during the initial phase of flight, followed by preprogrammed autopilot (or INERTIAL GUIDANCE); it is not particularly accurate.

The Kangaroo was the primary payload of the Tu-95 BEAR B bomber. Due to its very large size, only one missile can be carried, semi-recessed in the bomber's belly. The Kangaroo has been superseded by the AS-4 KITCHEN on Tu-95s upgraded to the Bear G configuration.

Specifications Length: 49.1 ft. (14.96 m.). **Diameter:** 2.9 ft. (884 mm.). **Span:** 29.5 ft. (9 m.). **Weight, launch:** 17,600 lb. (8000 kg.). **Speed, max.:** Mach 1.8 (1170 mph/1875 kph). **Range, max.:** 400 mi. (650 km.).

KANIN: NATO code name for a class of eight Soviet guided-missile DESTROYERS (DDGs), actually ANTI-AIR WARFARE/ANTI-SUBMARINE WARFARE (AAW/ASW) conversions of Krupnyy-class ANTI-SURFACE WARFARE (ASUW) missile destroyers, whose hulls and machinery were themselves derived from the earlier gun-and-torpedo KOTLIN class. The Krupnyys were originally completed between 1960 and 1962 with an armament of 2 SS-N-1 Scrubber anti-ship missiles, 6 21-in. TORPEDO tubes, and 4 quadruple 57-mm. ANTI-AIRCRAFT guns. They were radically converted to the Kanin configuration between 1968 and 1977.

Graceful, flush-decked vessels, the Kanins carry

a balanced ASW/AAW armament, but are classified as Large Anti-Submarine Ships (*Bol'shoy Protivolodochnyy Korabl*, BPK) by the Soviet navy. One twin SA-N-1 GOA surface-to-air missile launcher is mounted aft, with 2 quadruple 57-mm. anti-aircraft guns forward, 4 twin 30-mm. radar-controlled anti-missile guns amidships, 3 RBU-600 ASW rocket launchers, and 2 quintuple 21-in. torpedo tubes. An elevated landing deck over the fantail can accommodate a Ka-25 HORMONE ASW helicopter, but there is no hangar. See also DESTROYERS, SOVIET UNION.

Specifications **Length:** 458.75 ft. (140 m.). **Beam:** 49.4 ft. (15 m.). **Draft:** 16.5 ft. (5.03 m.). **Displacement:** 3700 tons standard/4750 tons full load. **Powerplant:** twin-shaft steam: 4 oil-fired boilers, 2 sets of geared turbines, 80,000 shp. **Speed:** 35 kt. **Range:** 1000 n.mi. at 32 kt./4500 n.mi. at 15 kt. Crew: 300–350. **Sensors:** 2 "Don Kay" navigational radars, 1 "Head Net" 3-dimensional air search radar, 2 "Drum Tilt" 30-mm. fire control radars, 1 "Hawk Screech" 57-mm. fire control radar, 1 "Peel Group" SA-N-1 missile guidance radar, 1 bow-mounted medium-frequency active/passive sonar. **Electronic warfare equipment:** several "Bell"-series electronic signal monitoring arrays, 1 "Top Hat" active jamming system.

KARA: NATO code name for a class of seven Soviet guided-missile CRUISERS completed between 1973 and 1980. Enlarged and refined versions of the earlier KRESTA class, the Karas are optimized for ANTI-SUBMARINE WARFARE (ASW) and classified by the Soviet navy as Large Anti-Submarine Ships (*Bol'shoy Protivolodochny Korabl'*, BPK). The Karas have extensive command and control facilities, and often serve as flagships. At the time of their completion, the Karas were the largest gas-turbine-powered ships in the world.

Primary armament consists of 2 quadruple launchers for SS-N-14 SILEX ASW guided missiles. Mounted on either side of the bridge facing forward, the launchers are fixed in elevation and azimuth. Other ASW weapons include 2 RBU-6000 and 2 RBU-1000 rocket launchers, and 2 sets of quintuple 21-in. TORPEDO tubes. A landing deck and hangar incorporated into the fantail can accommodate one Ka-25 HORMONE ASW HELICOPTER. The Karas have a more powerful ANTI-AIR WARFARE (AAW) armament than the Krestas, with a twin SA-N-3 GOBLET surface-to-air missile (SAM) launcher at each end of the ship, supplemented by 2 retractable launchers for SA-N-4 GECKO short-range SAMs, 2 twin 76.2-mm. DUAL PURPOSE guns

amidships, and 4 30-mm. radar-controlled anti-missile guns. Goblet is believed to have some secondary ANTI-SHIP capability. One ship, the *Azov*, has been fitted with a VERTICAL LAUNCH SYSTEM for the SA-N-6 GRUMBLE SAM in place of one of the Goblet launchers. See also CRUISERS, SOVIET UNION.

Specifications **Length:** 571 ft. (174 m.). **Beam:** 51.33 ft. (15.7 m.). **Draft:** 22.16 ft. (6.75 m.). **Displacement:** 8200 tons standard/9700 tons full load. **Powerplant:** twin-shaft COGAG: 4 30,000-shp. gas turbines, 120,000 shp. **Speed:** 34 kt. **Range:** 3000 n.mi. at 32 kt./8800 n.mi. at 15 kt. **Crew:** 525. **Sensors:** 2 "Don Kay" navigational radars, 1 "Palm Frond" surface-search radar, 1 "Top Sail" and 1 "Head Net" 3-dimensional air-search radar, 2 "Owl Screech" 76.2-mm. fire control radars, 2 "Bass Tilt" 30-mm. fire control radars, two "Pop Group" SA-N-4 missile guidance radars, 2 "Head Light" SA-N-3/SS-N-14 missile guidance radars, 1 large bow-mounted low-frequency active/passive sonar, 1 medium-frequency variable depth sonar. **Electronic warfare equipment:** several "Bell"-series electronic signal monitoring arrays, 2 "Side Globe" active jamming units.

KAREL DOORMAN: A class of eight Dutch guided-missile FRIGATES (FFGs), the first of which will be commissioned in 1992. Intended as general-purpose patrol and escort vessels, the Doormans are flush-decked vessels with a blocky superstructure amidships. These ships will have a computer-controlled rudder stabilization system (instead of conventional fin stabilizers) to improve seaworthiness in North Sea conditions.

Armament consists of 1 OTO-MELARA 76.2-mm. DUAL PURPOSE gun, one GOALKEEPER 30-mm. radar-controlled gun for anti-missile defense, 8 HARPOON anti-ship missiles in 2 quadruple launchers, a 16-round vertical launcher for SEA SPARROW short-range surface-to-air missiles, and 2 sets of Mk.32 twin tubes for lightweight anti-submarine TORPEDOES. A landing deck and hangar aft can accommodate one LYNX anti-submarine helicopter. See also FRIGATES, NETHERLANDS.

Specifications **Length:** 401 ft. (122.25 m.). **Beam:** 47.23 ft. (141.4 m.). **Draft:** 19.85 ft. (6.05 m.). **Displacement:** 2800 tons standard/3320 tons full load. **Powerplant:** twin-shaft CODOG: 2 4225-hp. SEMT-Pielstick diesels (cruise), 1 28,560-shp. Rolls Royce SM-1A Spey gas turbine (sprint); (later ships) 48,927-shp. SM-1C gas turbine. **Speed:** 29 kt. (turbine); 21 kt. (diesels). **Range:** 5000 n.mi. at 18 kt. **Crew:** 100. **Sensors:** 1 DA-08 medium-range air-search radar, 1 WM-25 fire control radar, 1 PHS-36 hull-mounted medium-frequency active/passive sonar; (later ships) 1 SQR-19A TACTASS passive towed array. **Electronic warfare equipment:** 1 Argo APECS-2 active/passive unit, 2 Mk.36 SRBOC chaff rocket launchers.

KAREN (AS-10): NATO code name for a Soviet short-range air-to-surface missile, generally equivalent to the U.S. AGM-65 MAVERICK. Developed in the early 1970s from the AS-7 KERRY, Karen has a cylindrical body with an ogival nose, four small control canards, and four cruciform tail fins. As with Maverick, both SEMI-ACTIVE LASER HOMING and ELECTRO-OPTICAL (TV)–guided versions have been developed. An IMAGING INFRARED (IIR)–guided version may also be in service. All versions are powered by a solid-fuel rocket and armed with a 132-lb. (60-kg.) high-explosive warhead. Karen is reported to be carried by MiG-27 FLOGGER, Su-17 FITTER, and Su-24 FENCER attack aircraft.

Specifications **Length:** 11.5 ft. (3.5 m.). **Diameter:** 11.81 in. (300 mm.). **Span:** 35.4 in. (900 mm.). **Weight, launch:** 660 lb. (300 kg.). **Speed:** Mach 0.8 (560 mph/896 kph). **Range:** 6.2 mi. (10 km.).

KASHIN: NATO code name for a class of 23 Soviet guided-missile DESTROYERS (DDGs), completed between 1963 and 1982. Though they have a well-balanced, general-purpose armament, the Kashins are classified as Large Anti-Submarine Vessels (*Bol'shoy Protivolodochnyy Korabl'*, PBK) by the Soviet navy. Graceful, flush-decked vessels, the Kashins were the world's first gas-turbine powered warships and are exceptionally fast. But early Soviet gas turbines were rather inefficient, and their range is limited.

Armament consists of 1 twin SA-N-1 GOA surface-to-air missile (SAM) launcher and 1 twin 76.2-mm. DUAL PURPOSE gun at each end of the ship, 2 RBU-6000 and 2 RBU-1000 ANTI-SUBMARINE WARFARE (ASW) rocket launchers, and two sets of quintuple 21-in. (533-mm.) TORPEDO tubes. A small landing deck on the fantail can accommodate 1 Ka-25 HORMONE ASW HELICOPTER, but there is no hangar. Deck rails can be fitted for up to 20 MINES.

From 1973, six Kashins were extensively rebuilt to improve their ANTI-SURFACE WARFARE (ASUW) capabilities; these "Modified Kashins" have 4 rearward-firing SS-N-2c Improved STYX anti-ship missiles in individual launch canisters aft. Other changes include the addition of 4 30-mm. radar-controlled guns for anti-missile defense and their associated "Bass Tilt" fire control radars, the provision of an enlarged and elevated landing pad on the fantail, and the removal of the RBU-1000 rocket launchers. The sonar system is entirely new, consisting of a bow-mounted medium-frequency array and a passive VARIABLE DEPTH SONAR suspended under the landing pad. The Modified Kashins are thus much more capable than the original ships, without changing their general dimensions, machinery, or performance. In 1974, one unmodified Kashin, the *Otvazhnyy*, sank in the Black Sea with heavy loss of life after an internal explosion and catastrophic fires. Another ship, the *Provornyy*, was modified to serve as the trials ship for the SA-N-7 GADFLY SAM system.

In 1978, India ordered three Kashin II variants. Delivered between 1980 and 1982, these have an elevated landing deck extending from the fantail to the after SA-N-1 launcher (in place of the aft 76.2-mm. mount). In addition, they carry two fixed SS-N-2 Styx launch canisters abreast of the bridge. Displacement has risen by 450 tons, but overall dimensions, machinery, and performance appear to be identical to those of the Soviet Kashins. Three additional Kashin IIs ordered in 1982 were delivered between 1986 and 1989. See also DESTROYERS, SOVIET UNION.

Specifications **Length:** 475.16 ft. (144.86 m.). **Beam:** 52.16 ft. (15.9 m.). **Draft:** 15.83 ft. (4.83 m.). **Displacement:** 3750 tons standard/4500 tons full load; (II) 4950 tons full load. **Powerplant:** twin-shaft COGAG: 4 24,00-shp. gas turbines. **Speed:** 37–38 kt. **Range:** 1500 n.mi. at 36 kt./4000 n.mi. at 20 kt. **Crew:** 220–330. **Sensors:** 1 or 2 "Big Net" or "Head Net" 3-dimensional air-search radars, 2 or 3 "Don-2" navigational and surface-search radars, 2 "Owl Screech" 76.2-mm. fire control radars, 2 "Peel Group" SA-N-1 missile guidance radars, 1 hull-mounted high-frequency active sonar. **Electronic warfare equipment:** 1 "Watch Dog" radar warning receiver.

KASPUTIN YAR MISSILE TEST CENTER: Soviet missile test center on the Volga River at 48°30′ N 46° E. Associated mainly with the development of IRBMs and tactical BALLISTIC MISSILES, Kasputin Yar is also the site of small military satellite and sounding rocket launches. See also SPACE, MILITARY USES OF.

KC-10: U.S. AERIAL REFUELING tanker aircraft, based on the DC-10 airliner. See EXTENDER.

KC-135: U.S. AERIAL REFUELING tanker aircraft. See STRATOTANKER.

KEDGE (AS-14): NATO code name for a Soviet short-range air-to-surface missile introduced in the late 1970s. Believed to be a follow-on to the AS-10 KAREN, Kedge has a cylindrical body, an ogival nose, four small control canards, and four cropped-delta tail fins. Like Karen, Kedge has been produced in SEMI-ACTIVE LASER HOMING, ELECTRO-OPTICAL (TV), and IMAGING INFRARED (IIR) versions. Some may employ a DATA LINK for midcourse guidance, giving the missile a "launch and leave" capability. All are armed with a 132-lb. (60-kg.) high-explosive warhead. Powered by a solid-fuel rocket engine, Kedge has been seen on MiG-27 FLOGGER attack aircraft, and is probably cleared for use by Su-24 FENCERS and Su-25 FROG-FOOTS.

Specifications   Length: 12.46 ft. (3.8 m.). Diameter: 11.8 in. (300 mm.). Span: 35.5 m. (900 mm.). Weight, launch: 770 lb. (350 kg.). Speed: Mach 0.8 (560 mph/896 kph). Range: 18 mi. (30 km.).

KEGLER (AS-12): NATO code name for a Soviet ANTI-RADIATION MISSILE (ARM), introduced in 1978 as successor to the AS-9 KYLE missile. Possibly derived from the AS-14 KEDGE, Kegler has a cylindrical body, an ogival nose, four small control canards, and four cruciform tail fins. The nose houses a passive radar receiver, which is probably linked to a broad-band signal scanner and a guidance unit tuned to a variety of Western radars. Like most modern ARMs, it probably has an inertial guidance unit which allows it to keep flying towards the target even if the hostile radar stops transmitting. Powered by a solid-rocket engine, Kegler is reportedly armed with a 132-lb. (60-kg.) high-explosive fragmentation warhead fitted with contact and PROXIMITY FUZES. Kegler has been seen on Su-24 FENCER and Su-25 FROGFOOT attack aircraft, as well as on Tu-22M BACKFIRE bombers; with the latter, the missile is probably intended for attacks on warships.

Specifications   Length: 12.63 ft. (3.85 m.). Diameter: 11.81 in. (300 mm.). Span: 35.4 in. (900 mm.). Weight, launch: 770 lb. (350 kg.). Speed: Mach 0.8 (560 mph/896 kph). Range: 21 mi. (35 km.).

KELT (AS-5): NATO code name for a Soviet long-range air-to-surface ANTI-SHIP and land attack missile. First observed in 1967, the Kelt has an airplane configuration with a short, cigar-shaped body, sharply swept wings and tailplane, and a large vertical stabilizer. Kelt bears a superficial resemblance to the earlier AS-1 Kennel, and to the SS-N-2 STYX surface-to-surface missile. But while those missiles are turbojet-powered, Kelt has a liquid-fuel rocket motor, eliminating the need for air intakes and allowing a larger radar to be mounted in the nose. Kelt relies on preprogrammed autopilot guidance combined with active radar terminal homing for anti-ship missions. An ANTI-RADIATION variant with passive radar homing is also in service. Both versions are armed with a 2200-lb. (1000-kg.) high-explosive warhead, but an optional nuclear warhead has also been reported.

The Kelt is a standard weapon of the Tu-16 BADGER G anti-shipping bombers of Soviet Naval Aviation (AV-MF), which have also been supplied to a number of other countries, including Libya and Egypt. One missile can be carried under each wing. Some 25 Kelts were launched into Israel by Egyptian Badgers during the 1973 Yom Kippur War; only 5 managed to penetrate Israeli air defenses, but 1 destroyed a radar station. See also CRUISE MISSILE.

Specifications   Length: 28.16 ft. (6.45 m.). Diameter: 2.95 ft. (900 mm.). Span: 15.75 ft. (4.8 m.). Weight, launch: 6600 lb. (3000 kg.). Speed: Mach 1.2 (780 mph/12250 kph) at 36,000 ft. (11,000 m.)/Mach 0.9 (675 mph/1080 kph) at sea level. Range: 100 mi. (160 km.) at sea level/200 mi. (320 km.) at high altitude.

KENT (AS-15): NATO code name for a Soviet long-range, air-launched CRUISE MISSILE. Generally similar to the U.S. BGM-109 TOMAHAWK, the Kent became operational in 1984, probably for both land-attack and anti-ship roles. Derived from on the SS-N-21 SAMPSON submarine-launched cruise missile, Kent has a cylindrical body with straight, pop-out wings and small cruciform tail fins. The payload is believed to be a 272-lb. (123-kg.), 200-KT nuclear warhead. Kent relies on DOPPLER-aided INERTIAL GUIDANCE, and must also have some form of terminal homing, because the median error radius (CEP) in the land attack role has been estimated at 150 m. Powered by a small turbojet or turbofan engine, after launch Kent cruises at an altitude of 10,000 ft. (3050 m.), diving to a height of 160 ft. (50 m.) for the final run to the target. Kent is currently deployed with the Tu-95 BEAR H bomber, but is expected to become the primary armament of the Tu-160 BLACKJACK supersonic bomber. It provides the Soviet air force with significantly increased capabilities for intercontinental attack and anti-ship strike. See also ADD; ALCM; AV-MF.

Specifications Length: 20 ft. (6.1 m.). **Diameter**: 21 in. (533 mm.). **Span**: 10.67 ft. (3.25 m.). **Weight, launch**: 3740 lb. (1700 kg.). **Speed**: Mach 0.8 (600 mph/960 kph). **Range**: 1800 mi. (3000 km.).

KERRY (AS-7): NATO code name for a Soviet short-range, air-to-surface tactical missile, in service with the Soviet Air Force (vvs) and Naval Aviation (av-mf) since the early 1970s. A single-stage, solid-fuel missile, Kerry has a cylindrical body with four small control canards and cruciform tail fins. Kerry relies on radio COMMAND GUIDANCE, whereby the pilot or WEAPON SYSTEM OPERATOR of the launch aircraft steers the missile to the target with a small joystick; it is thus similar to the US AGM-12 BULLPUP of 1950s vintage. Kerry can be carried by a variety of Soviet tactical aircraft, including the MiG-27 FLOGGER, Su-17/22 FITTER, and Su-25 FROGFOOT, but has been superseded in Soviet service by the AS-10 KAREN.

Specifications Length: 11.5 ft. (3.5 m.). **Diameter**: 11.8 in. (300 mm.). **Span**: 35.43 in. (900 mm.). **Weight, launch**: 649 lb. (295 kg.). **Speed**: Mach 1. **Range**: 6.8 mi. (11 km.).

KEVLAR: A synthetic (aramid) fiber of very high tensile strength (greater than steel), which can be woven into a tough cloth that provides ARMOR protection in bulletproof vests. With an epoxy resin, Kevlar can be molded into solid sheets of lightweight armor, for aircraft and other applications. The U.S. Army now issues a helmet molded from Kevlar. The material was originally developed for the U.S. space program. See also ARMOR, BODY.

KEW: Kinetic Energy Weapon, any weapon that attacks the target by transferring energy through impact, via a projectile. The term is now used with reference to BALLISTIC MISSILE DEFENSES. See also ELECTROMAGNETIC LAUNCHER; KKV.

KFIR: An Israeli-built FIGHTER/ATTACK AIRCRAFT derived from the French Dassault MIRAGE III/5. Israel purchased 73 Mirages from France in the early 1960s; in 1967 they were the only Israeli fighters capable of meeting Arab MiG-21 FISHBEDS on equal terms. When France embargoed further sales in 1968, and would not deliver 50 Mirage 5s already paid for, the Israelis began the indigenous production of components. By the early 1970s, they were producing Mirage copies designated Nesher, with French-supplied engines and parts. Some 50 Neshers were completed by 1973, and fought in the Yom Kippur War alongside the remaining Mirages. All surviving Neshers were later sold to Argentina (as "Daggers"); several were shot down in the 1982 Falklands War.

The Mirage/Nesher was essentially a mid-1950s design, with a low thrust-to-weight ratio, inadequate range, and poor handling at high angles of attack (AOA). The first Israeli remedy was to mate the Mirage airframe with the General Electric J79-GE-17 afterburning turbojet of the F-4E PHANTOM, which has a maximum thrust rating of 17,900 lb., as compared to 15,873 for the Atar 9C of the basic Mirage; in addition, it is both lighter and more fuel efficient. The resulting Kfir required a complete redesign of the rear fuselage, but its combat thrust-to-weight ratio of 0.87 to 1 (compared to the Mirage's 0.72 to 1) gives better acceleration and sustained turn rates.

The first C.1 model, produced from 1975, retained the Mirage III/5 configuration, but the fuselage is shorter, and the nose section more slender (as in the Mirage 5) to house a small Elta EL/M-200-1B range-only RADAR. Completely reequipped with modern AVIONICS, the C.1 has a central navigation and weapon-aiming system based on an Elbit Systems-80 digital computer, and a U.S. Singer-Kearfott KT-70 inertial platform. Both flight and weapon aiming data are fed to a head-up display (HUD), and ELECTRONIC COUNTERMEASURES are built into the airframe. Because the J79 requires more cooling than the Atar, several auxiliary intakes have been added, the largest at the base of the dorsal fin. The wing retains the span and 60° delta planform of the Mirage, but has been modified by the addition of an outboard leading edge dogtooth (with increased camber) to improve handling at low speeds and high AOA. Internal armament is unchanged: two 30-mm. DEFA 552 cannon with 280 rounds of ammunition mounted under the engine intakes, which have been redesigned to prevent ingestion of muzzle gases throughout the flight envelope. The Kfir has one centerline pylon, four wing pylons, and two pairs of tandem fuselage pylons with a total capacity of 8500 lb. (3864 kg.). External stores include 374-gal. (1645-lit.) fuel tanks, two PYTHON, SHAFRIR or AIM-9 SIDEWINDER infrared homing air-to-air missiles, MAVERICK air-to-surface missiles, AGM-45 SHRIKE anti-radiation missiles, electro-optical GLIDE BOMBS, LASER-GUIDED BOMBS, folding fin aerial rocket (FFAR) pods, free-fall high-explosive bombs, CLUSTER BOMBS, or NAPALM.

In 1976, the improved Kfir C.2 added two fixed delta canards just behind the cockpit and two small horizontal strakes on the nose; these greatly improve the turn radius and performance at high AOA. The C.1 and C.2 are otherwise identical, and all earlier aircraft have been brought up to C.2 standard. In the early 1980s, the C.7 was introduced with a more powerful engine and new avionics optimized for ground attack. A two-seat combat-capable trainer (TC.2) is also used for ELECTRONIC WARFARE missions.

The Kfir has seen extensive combat service over Syria and Lebanon, with solid results. By 1987, more than 200 Kfirs had been produced. Although it is cheap and effective, limitations imposed on the U.S.-made engine restricted Kfir exports while its technology was still current. In 1986, however, the U.S. Navy leased 24 Kfir C.2s to serve as "aggressor" trainers at the TOP GUN fighter weapons school. Ecuador has purchased 11 Kfirs, and Colombia 8, having also contracted to have all of its Mirages converted to Kfirs. See also AIR COMBAT MANEUVERING.

Specifications **Length:** 50.33 ft. (16.87 m.). **Span:** 27 ft. (8.23 m.). **Powerplant:** 1 General Electric J79-GE-17 afterburning turbojet, 17,900 lb. (8136 kg.) of thrust. **Fuel:** 5964 lb. (2711 kg.). **Weight, empty:** 16,060 lb. (7300 kg.). **Weight, normal loaded:** 20,470 lb. (9305 kg.). **Weight, max. takeoff:** 32,340 lb. (14,700 kg.). Speed, max.: 863 mph (1380 kph) at sea level/Mach 2.3 (1516 mph/2425 kph) at 36,000 ft. **Initial climb:** 45,950 ft./min. (14,009 m./min.). **Service ceiling:** 58,000 ft. (17683 m.). **Combat radius:** (intercept) 330 mi. (528 km.)/(hi-lo-hi) 477 mi. (764 km.).

KGB: *Komitet gosudarstvennoy bezopasnosti*, or Committee for State Security, the organ of the Soviet government responsible for internal security, foreign intelligence (including espionage and counterespionage), PROPAGANDA, SUBVERSION, and other ACTIVE MEASURES. It shares with the GRU (Soviet Military Intelligence) the responsibility for the collection, analysis, and distribution of military, political, and technological INTELLIGENCE.

The KGB is overseen by a chairman, who is normally a member of the Politburo of the Central Committee of the Communist Party of the Soviet Union (CPSU), and of the DEFENSE COUNCIL, and who reports directly to the general secretary of the CPSU (and now the president of the Supreme Soviet).

The KGB is organized into four Chief Directorates, seven independent Directorates, and six independent Departments, each of which has subordinate departments, directorates, services, and geographic "directions."

The branches of greatest military significance are the First Chief Directorate, the Border Guards Chief Directorate, and the Armed Forces Independent Directorate.

The First Chief Directorate is responsible for all foreign *operations*, including espionage, counterespionage, subversion, propaganda, and active measures. It focuses on the acquisition of military, political, and technological intelligence, both by technical means and by the espionage of Soviet operatives under diplomatic or commercial cover, subverted foreign nationals, and "illegal" Soviet agents under foreign cover. It also uses the resources of other Warsaw Pact and other client-state intelligence services, notably the Cuban DGI. Active measures are principally political, but include the masking of Soviet intentions and plans (see *Maskirovka*).

The Border Guards Chief Directorate is responsible for patrolling Soviet frontiers on land and at sea, to a unique standard of control. It controls large military forces, including more than 200,000 troops with tanks, armored fighting vehicles, artillery, combat vessels, and aircraft. KGB "Kremlin Guard" forces also insure the regime against the regular military forces, and serve as part of the strategic reserve.

The Armed Forces Independent Directorate oversees the political reliability of the Ministry of Defense, the GRU, the regular military forces, and especially the Strategic Rocket Forces (it physically controls all nuclear weapons), as well as the Border Guards, the Militia (police), civil aviation, and the Moscow Military District. Its primary function is to protect the CPSU against military conspiracies; it serves to saturate the armed forces with spies and informants who report on the political reliability of officers and men (including the Political Deputies, or *Zampoliti*). See also CENTRAL INTELLIGENCE AGENCY; DEFENSE INTELLIGENCE AGENCY; NATIONAL SECURITY AGENCY.

KH-9: U.S. high-resolution RECONNAISSANCE satellite. See BIG BIRD.

KH-11: U.S. high-resolution RECONNAISSANCE satellite, successor to the KH-9 BIG BIRD. Most technical details are classified, and much of what is revealed publicly is misleading. The first KH-11 was launched by a Titan IIID booster in December 1976, and placed in a 310 × 155 mi. (496 × 248 km.) polar orbit; subsequent satellites have had similar orbital parameters. The KH-11 differs from the Big Bird in having multi-spectral ELECTRO-OPTICAL (EO) sensors rather than film cameras, thus eliminating the need to return film capsules to earth for development. The KH-11 is reportedly capable of transmitting high-resolution sensor data to ground stations in near "real time," though the resolution of EO sensors is not as good as that of film cameras: the KH-11 is believed to have a spatial resolution on the order of 1 m., whereas KH-9 film is said to have a resolution of some 15 cm. Hence, KH-11s were often employed as "quick look" satellites, scanning for objects of interest, for subsequent KH-9 photography. In addition to its optical sensors, the KH-11 reportedly has a number of electronic intelligence (ELINT) modules to detect radar and radio emissions.

The KH-11 was designed to have much greater endurance than the KH-9; the latter had an average life expectancy of six months, whereas a KH-11 lasts an average of more than two years. But the actual endurance of a satellite depends on the extent of its maneuvering to shift orbits. It was therefore the usual practice to keep at least two operational satellites on station at all times (plus at least one on orbital standby, a "dark spare"). But by the mid-1980s, the KH-11 was scheduled for replacement by the more advanced KH-12, so that only one satellite was kept in orbit. The KH-12 could only be launched by the Space Shuttle; when the shuttle fleet was grounded in 1986, it became necessary to launch an additional KH-11 in late 1987, with a Titan 34D booster. See also SATELLITES, MILITARY; SPACE, MILITARY USES OF.

KH-12: U.S. high-resolution RECONNAISSANCE satellite, successor to the KH-11. Reportedly, the KH-12 has ELECTRO-OPTICAL sensors combining the "real time" data transmission capability of the KH-11 with the very high spatial resolution of the KH-9 BIG BIRD. In addition, the KH-12 carries several electronic intelligence (ELINT) modules as "piggyback" payload. As originally designed, the KH-12 could be carried into orbit only by the Space Shuttle. When the shuttle fleet was grounded in 1986, the KH-12 was redesigned for launch by the Titan IV rocket. See also SATELLITES, MILITARY; SPACE, MILITARY USES OF.

KIDD: A class of four U.S. guided-missile DESTROYERS (DDG-992 to DDG-995), essentially ANTI-AIR WARFARE (AAW) versions of the SPRUANCE class (DD-963). Due to escalating costs, the U.S. Navy was unable to procure any ships of this type under the original Spruance program, but four were ordered by the Shah of Iran. After his overthrow in 1979, Iran refused to accept delivery and the U.S. government purchased the nearly completed ships at bargain prices—they are unofficially known in the navy as the "Ayatollah class." The Kidds were commissioned in 1979–80.

They differ from the Spruances in their weapons and sensors, as well as in design details such as improved air conditioning and the provision of dust filters for the engine air intakes (essential for Persian Gulf operations). In place of the Spruance's 8-round ASROC launcher and 8-round NATO SEA SPARROW short-range surface-to-air missile (SAM) system, the Kidds have one twin-arm Mk.26 launcher at each end of the ship, which can launch both ASROC and STANDARD SM-2 MR SAMs. The forward Mk.26 Mod 0 launcher, with a magazine capacity of 22 missiles, is used primarily for ASROC, while the aft Mk.26 Mod 1 launcher, with a capacity of 44 missiles, is used mainly for Standard. Other armament is identical to the Spruance class. The main sensor changes are the replacement of the Spruance's SPS-40B 2-dimensional air-search RADAR with an SPS-48A 3-dimensional air-search set, and the addition of two SPG-55D Standard missile guidance radars, along with two Mk.74 missile FIRE CONTROL units. The hull, superstructure, and machinery of the Spruance class are retained intact, but displacement has risen to 8140 tons fully loaded from the Spruance's 7810 tons; performance is essentially unchanged.

These ships are now the U.S. Navy's most powerful destroyers, with weapons and sensors equivalent to those of the VIRGINIA-class nuclear-powered CRUISERS. The Mk.26 launchers will eventually be replaced by two Mk.41 VERTICAL LAUNCH SYSTEMS (VLS), increasing both the volume and flexibility of the ships' firepower. See also DESTROYERS, UNITED STATES.

KIEV: A class of four Soviet V/STOL AIRCRAFT CARRIERS (CVV), classified by the Soviet navy as *Takiticeskaya Avianostny Kreysera*, or "Tactical Air-Capable Cruisers." The Kievs represent the middle step in the logical progression from the earlier MOSKVA-class helicopter carriers towards the first Soviet full-scale carrier for conventional takeoff aircraft, the TBILISI. As compared to the Moskvas, the Kievs are larger, more heavily armed, and can operate Yak-38 FORGER V/STOL fighters, not just ANTI-SUBMARINE WARFARE (ASW) helicopters.

The forward half of the ship has a CRUISER configuration with heavy ASW, ANTI-SURFACE WARFARE (ASUW), and ANTI-AIR WARFARE (AAW) armament, while the aft half has a carrier configuration, with a large island superstructure offset to starboard, and an angled flight deck 601.25 ft. (183.3 m.) long by 65.6 ft. (20 m.) wide, canted 4.5° to port. Portions of the flight deck are covered with refractory tiles to protect it from jet blast. The long hangar deck is connected to the flight deck by two large middeck elevators.

Armament is both heavy and balanced. The primary AAW weapons are 2 twin SA-N-3 GOBLET surface-to-air missile (SAM) launchers, one on the bow, the other at the aft end of the island. They are supplemented by 2 SA-N-4 GECKO short-range SAM launchers and 8 ADG6-30 30-mm. radar-controlled guns for anti-missile defense. The primary ASW weapon is a twin-arm launcher for the FRAS-1 (Free Rocket Anti-Submarine, a rocket-boosted TORPEDO analogous to ASROC). Other ASW weapons include 2 RBU-6000 ASW rocket launchers, and 2 sets of quintuple 21-in. (533-mm.) torpedo tubes. The main ASUW battery consists of 4 twin launchers for SS-N-12 SANDBOX anti-ship missiles, with 12 reload rounds (Goblet also has a secondary anti-ship capability). Finally, 2 twin 76.2-mm. DUAL PURPOSE guns on the bow can support both the AAW and ASUW batteries. (By contrast, U.S. carriers are armed only for close-in defense against aircraft and missiles.)

The Kievs' air group consists of 13 Yak-38 Forger V/STOL fighters and 16 Ka-27 HELIX helicopters, the latter used for both ASW and OVER-THE-HORIZON TARGETING (OTH-T) for the Sandbox missiles. The Forger, a short-range aircraft with limited performance and payload, is much less capable than the catapult-launched aircraft of U.S. carriers, and also inferior to the SEA HARRIER V/STOL fighters of British carriers; nonetheless, it can provide a modicum of air defense which the Soviet navy never had before. The Kievs are fitted with extensive COMMAND AND CONTROL equipment to serve as flagships, including a "Vee Bars" high-frequency antenna and two "Punch Bowl" satellite communications terminals.

Though hardly comparable to U.S. attack carriers, the Kievs still have much offensive capability in their anti-ship missile batteries, while their air group can provide both limited air defense and considerable ASW coverage. The Soviet experience with shipboard air operations will no doubt be reflected in the design of the 60,000-ton TBILISI carriers. See also AIRCRAFT CARRIERS, SOVIET UNION.

Specifications Length: 895.7 ft. (273.1 m.). Beam: 107.3 ft. (32.72 m.). Draft: 32.8 ft. (10 m.). Displacement: 30,000 tons standard/38,000 tons full load. Powerplant: 4-shaft steam: 8 oil-fired boilers, 4 sets of geared turbines, 200,000 shp. Speed: 32 kt. Range: 4000 n.mi. at 30 kt./13,500 n.mi. at 18 kt. Crew: 1300 + 400 air group. Sensors: 1 "Top Sail" 3-dimensional air-search radar, 2 "Don Kay" navigational sets, 1 "Top Knot" air traffic control radar, 1 "Top Steer" surveillance radar, 1 "Trap Door" SS-N-12 missile guidance radar, 2 "Pop Group" SA-N-4 missile guidance radars, 2 "Head Light" SA-N-3 guidance radars, 2 "Owl Screech" 76.2-mm. fire control radars, 4 "Bass Tilt" 30-mm. fire control radars, 2 "Tee Plinth" electro-optical gun directors, 1 bow-mounted low-frequency active/passive sonar, 1 medium-frequency variable depth sonar. **Electronic warfare equipment:** 1 "Rum Tub" electronic signal monitoring array, 2 "Side Globe" electronic countermeasures units, 4 chaff launchers.

KILDIN: NATO code name for a class of four Soviet guided-missile DESTROYERS (DDGs) completed between 1958 and 1960. The Kildins are modifications of the KOTLIN-class gun-and-torpedo destroyers, optimized for ANTI-SURFACE WARFARE (ASUW) with ANTI-SHIP MISSILES; they are classified as Large Missile Ships (*Bol'shoy Raketnyy Korabl'*, BRK) by the Soviet navy. Built on Kotlin hulls redesigned while still under construction, they have the same machinery as the earlier destroyer. As built, the primary ASUW armament consisted of 2 SS-N-1 Scrubber anti-ship missiles in a trainable launcher mounted on the stern. ANTI-AIR WARFARE (AAW) armament consisted of 4 quadruple 57-mm. anti-aircraft guns, 2 forward and 2 amidships, while ANTI-SUBMARINE WARFARE (ASW) armament consisted of 2 sets of 21-in. (533-mm.) triple TORPEDO tubes and 2 RBU-2000 ASW rocket launchers.

Between 1973 and 1975, three Kildins were extensively rebuilt as "Modified Kildins," with the SS-N-1 launcher replaced by 2 twin 76.2-mm. DUAL PURPOSE guns, and 4 SS-N-2c STYX anti-ship missiles installed in fixed launch canisters amidships, firing aft. The AAW and ASW batteries are unchanged, except for one ship, the *Bedovyy*, which has 4 quad 45-mm. guns in place of the 57-mm. battery. See also DESTROYERS, SOVIET UNION.

Specifications Length: 417.5 ft. (127.28 m.). Beam: 43 ft. (13.11 m.). Draft: 15.16 ft. (4.62 m.). Displacement: 2800 tons standard/3500 tons full load. Powerplant: twin-shaft steam: 4 oil-fired boilers, 2 sets of geared turbines, 72,000 shp. Speed: 38 kt. Range: 1000 n.mi. at 32 kt./3600 n.mi. at 18 kt. **Crew:** 300. **Sensors:** 1 "Don-2" navigational/surface-search radar, 1 "Head Net" 3-dimensional air-search radar, 1 "Owl Screech" 76.2-mm. fire control radar (in Modified ships), 2 "Hawk Screech" 57-mm. fire control radars, 1 hull-mounted *Herkules* or *Pegas* medium-frequency active sonar. **Electronic warfare equipment:** 1 "Watch Dog" radar warning receiver.

KILO: NATO code name for a class of at least 27 Soviet medium-range diesel-electric attack SUBMARINES (SSs), the first of which was completed in 1982. The Soviet navy operates some 14 Kilos; new production is intended for export. The Kilos are in a sense successors to the WHISKEY class, built in large numbers during the 1950s, but the Kilo is an entirely new design with a completely different hullform. While the Whiskeys had a cigar-shaped hull (derived from the German Type XXI U-boat), the Kilos have a modified teardrop-shaped hull optimized for underwater performance. Following standard Soviet practice, the Kilos have a double hull configuration with fuel and ballast tanks installed between the pressure hull and outer casing. They have a flat upper deck and a rather long streamlined sail amidships. Control surfaces consist of retractable bow diving planes and a cruciform rudder and stern planes; uniquely, the Kilos do not have the upper rudder otherwise typical of submarines with teardrop hulls. A retractable snorkel allows diesel operation from shallow submergence, while a large battery gives the Kilos long underwater endurance. Armament consists of six 21-in. (533-mm.) torpedo tubes in the bow, for a total of some 12 free-running and acoustical homing TORPEDOES. MINES may be substituted for torpedoes on a 2-for-1 basis. See also SUBMARINES, SOVIET UNION.

Specifications **Length:** 239.45 ft. (73 m.). **Beam:** 32 ft. (9.9 m.). **Displacement:** 2500 tons surfaced/3000 tons submerged. **Powerplant:** single-shaft diesel-electric: 2 2000-hp. diesel generators, 1 5000-hp. electric motor. **Speed:** 12 kt. surfaced/20 kt. submerged. **Max. operating depth:** 1476 ft. (1450 m.). **Collapse depth:** 2133 ft. (650 m.). **Crew:** 60. **Sensors:** 1 bow-mounted low-frequency active/passive sonar, 1 conformal array sonar, 1 "Snoop Tray" surveillance radar, 1 "Brick Group" electronic signal monitoring array, 1 "Quad Loop" D/F antenna, 2 periscopes. **Operators:** Ind (5 + 3 on order), Alg (2), Rom (2), Pol (4), USSR (14+).

KILTER (AS-11): NATO code name for a Soviet short-range, air-to-surface missile. Devel-

oped in the late 1970s, Kilter is believed to be an advanced version of the AS-10 KAREN, with greater range and improved lethality against tanks and other hard targets. Externally, the two missiles are quite similar. Both have a solid-fuel rocket motor, a cylindrical body, an ogival nose, four small control canards, and four cruciform tail fins, but Kilter is some 110 lb. (50 kg.) heavier. The payload, believed to be larger than the 132-lb. (60-kg.) high-explosive warhead of the AS-10, may be in the 220-lb. (100-kg.) class. Kilter is believed to have IMAGING INFRARED guidance (IIR), possibly the same guidance unit developed for the AS-10. Reportedly, the missile is carried by MiG-27 FLOGGER, SU-24 FENCER, and Su-25 FROGFOOT attack aircraft.

Specifications **Length:** 11.5 ft. (3.5 m.). **Diameter:** 11.81 in. (300 mm.). **Span:** 35.43 in. (900 mm.). **Weight, launch:** 770 lb. (350 kg.). **Speed:** Mach 0.8 (560 mph/896 kph). **Range:** 9 mi. (15 km.).

KINGFISH (AS-6): NATO code name for a very large Soviet long-range air-launched land attack and ANTI-SHIP MISSILE. First spotted in the late 1960s, Kingfish has an airplane configuration with a cylindrical body, pointed nose, stubby delta wings, and a small, conventional tailplane. Superficially similar to the earlier AS-4 KITCHEN, but in fact entirely different, Kingfish is powered by a liquid-fuel rocket engine. The missile flies a high-altitude, supersonic approach, followed by a steep diving attack on the target.

Kingfish relies on INERTIAL GUIDANCE through midcourse, with active radar homing in the terminal phase. This guidance system, combined with the diving attack profile, assures excellent accuracy against ships. In service with Soviet Long Range Aviation (ADD) for land attack, and with Naval Aviation (AV-MF) for anti-ship attack, Kingfish can be armed either with a 200-kT nuclear or 2200-lb. (1000-kg.) high-explosive warhead. The Tu-16 BADGER can carry 1 missile under each wing, while the Tu-26 BACKFIRE can carry up to 3 missiles, 1 semi-recessed in the belly, and 1 under each wing. See also CRUISE MISSILE.

Specifications **Length:** 34.5 ft. (10.51 m.). **Diameter:** 2.95 ft. (900 mm.). **Span:** 8.2 ft. (2.5 m.). **Weight, launch:** 11,000 lb. (5000 kg.). **Speed:** Mach 3 (1950 mph/3120 kph) at 36,000 ft. (11,000 m.). **Range:** 135 mi. (216 km.).

KIOWA: U.S. Army OH-58 light observation HELICOPTER, a militarized version of the Bell 206 Jet Ranger, first flown in 1961. The Kiowa was chosen as the army's standard scout/observation helicopter after a 1967 competition against the Hughes OH-6 Cayuse (now MD-500 DEFENDER). Some 2200 OH-58As were ordered in 1968, followed by the similar OH-58B and the definitive OH-58C, with a flat, nonglint windscreen, a more powerful engine, and improved avionics.

The Kiowa has an aluminum monocoque fuselage and slender tail boom. In addition to the pilot and copilot/observer, the cabin can accommodate up to three passengers or two stretcher cases. Tubular skid landing gear are mounted under the cabin. A belly-mounted cargo hook with a capacity of 1500 lb. (682 kg.) is usually fitted. The Kiowa is powered by a single Allison turbo-shaft engine mounted in a nacelle over the cabin. The main rotor has two aluminum alloy blades in the classic Bell "teeter-totter" configuration, whereby the rotors are counterbalanced by a stabilizer bar set at right angles to the blades. The small, two-bladed counter-torque tail rotor is mounted at the end of the tail boom.

The OH-58 is old, but the U.S. Army had no available replacement (the follow-on LHX is years behind schedule). Some 375 Kiowas are now being upgraded to OH-58D standard under the Army Helicopter Improvement Program (AHIP), with a more powerful engine driving a 4-bladed, composite construction rotor, and a slightly longer fuselage. The most notable addition is a stabilized, mast-mounted sight system with both ELECTRO-OPTICAL (TV) and IMAGING INFRARED sensors. The OH-58D also has a hydraulic stability augmentation system to facilitate precision hovering and NAP-OF-THE-EARTH (NOE) flying. Increased weight negates the performance increase of the more powerful engine.

Variants for the export market include the Model 406CS Combat Scout, with most of the AHIP improvements and a 730-shp. Allison 250-C34 engine. Most Kiowas are unarmed, but can be fitted with a variety of (light) gun and rocket pods, or ANTI-TANK GUIDED MISSILES. Different operators use the Kiowa in varied armed roles, including ANTI-SUBMARINE WARFARE with lightweight homing TORPEDOES. Despite its mediocre performance, the Kiowa has been a great commercial success because of its reliability and economy. It is also produced under license by Agusta in Italy.

Specifications Length: (A) 32.6 ft. (9.98 m.); (D) 42.16 ft. (12.85 m.). **Rotor diameter:** (A) 35.33 ft. (10.77 m.); (D) 35 ft. (10.67 m.). **Powerplant:** (A) 1 317-shp. T63-A-700 turboshaft; (C) 1 420-

shp. T63-A-720; (D) 650-shp. Allison Model 250-C30R. **Weight, empty:** (A) 1818 lb. (826.4 kg.); (D) 2825 lb. (1285 kg.). **Weight, max. takeoff:** (A) 3200 lb. (1455 kg.); (D) 4500 lb. (2045 kg.). **Speed, max.:** 150 mph (250 kph). **Speed, cruising:** 138 mph (230 kph). **Range, max.:** 305 mi. (488 km.). **Endurance:** 3.5 hr. **Operators:** Arg, Aus, Aut, Bra, Bru, Can, Chi, Col, Dub, Fin, Guy, Iran, Isr, It, Jam, Jap, Ku, Liber, Mall, Malt, Mex, Mor, Peru, S Ar, Sp, Sri L, Swe, Tan, Thai, Tur, UAE, Ug, US, Ven.

KIPPER (AS-2): NATO code name for a large Soviet long-range, air-launched ANTI-SHIP MISSILE first spotted in 1961. An airplane-configured missile with a cylindrical body, large pointed nose radome, and sharply swept wings and tail surfaces, Kipper is powered by a 12,000-lb.-thrust Lyulka AL-5 afterburning turbojet engine in a nacelle beneath the missile body. It relies on a combination of inertial or autopilot guidance with radio COMMAND GUIDANCE updates, and active radar homing in the terminal phase. Armed with a 2200-lb. (1000-kg.) shaped-charge (HEAT) warhead designed to penetrate the armored decks of AIRCRAFT CARRIERS, Kipper is still the standard weapon of the TU-16 BADGER C anti-shipping bomber of Soviet Naval Aviation (AV-MF), with a single missile carried semi-recessed in the aircraft's belly. See also CRUISE MISSILE.

Specifications Length: 32.83 ft. (10 m.). **Diameter:** 2.95 ft. (900 mm.). **Span:** 16 ft. (4.88 m.). **Weight, launch:** 9260 lb. (4209 kg.). **Speed:** 840 mph (1344 kph). **Range:** 132 mi. (212 km.).

KIROV: A class of four very large, nuclear-powered Soviet warships. Classified as Nuclear Missile Cruisers (*Atomnaya Raketnyy Kreysera*, ARKR) by the Soviet navy, the Kirovs are considered BATTLE CRUISERS in Western terms because of their size and firepower. They represent a major step in the evolution of Soviet naval construction towards larger ships with greater endurance and survivability—warships for sea control rather than sea denial. The Kirovs can serve both as powerful ANTI-AIR WARFARE (AAW) and ANTI-SUBMARINE WARFARE (ASW) escorts in CARRIER BATTLE GROUPS, and as ANTI-SURFACE WARFARE (ASUW) flagships for independent surface action groups.

The largest combat vessels (other than aircraft carriers) built anywhere since World War II, the Kirovs have a sharply raked bow with a raised forecastle and pronounced shear; the stern has a wide transom with a sunken quarterdeck. Because of their size and broad beam, the Kirovs are excellent sea boats and very stable weapon platforms.

The Kirovs have a unique CONAS (Combination of Nuclear and Steam) powerplant: two nuclear reactors feed two sets of geared turbines, providing base power for economical long-range cruising; for higher speeds, steam from the reactors is passed through two oil-fired superheaters to boost its temperature and pressure for higher turbine output. This arrangement is much more economical than the reactor capacity needed for equivalent all-nuclear propulsion. The ships are also fitted with two "Big Ball" satellite communications terminals.

The lead ship *Kirov* has a powerful and balanced weapon suite. Primary ASUW armament consists of 20 SS-N-19 SHIPWRECK long-range, supersonic ANTI-SHIP MISSILES, housed in a VERTICAL LAUNCH SYSTEM (VLS) in the bow, which probably has some form of armor protection. For long-range AAW, the *Kirov* has 12 vertical launchers for the SA-N-6 GRUMBLE, the naval variant of the advanced SA-10 long-range surface-to-air missile (SAM); each launcher is fed from an 8-round rotary magazine. For short-range air defense, the Kirov has 2 SA-N-4 GECKO SAM launchers and 8 30-mm. ADG6-30 radar-controlled anti-missile guns. Two 100-mm. DUAL PURPOSE guns supplement both AAW and ASUW armament. ASW armament includes a twin SS-N-14 SILEX ASW missile launcher at the bow, 1 RBU-6000 ASW rocket launcher on the forecastle, 2 RBU-1000s, and 2 sets of quintuple 21-in. (533-mm.) TORPEDO tubes. A landing pad on the quarterdeck has a hangar below it, reached by a large elevator. The hangar can accommodate 3 Ka-27 HELIX helicopters for both ASW and over-the-horizon targeting (OTH-T) for the Shipwreck missiles.

In later ships of the class, the 2 100-mm. guns are replaced by a more powerful twin 130-mm. dual-purpose gun, identical to those on the SLAVA and SOVREMENNY classes; the 2 Gecko launchers are replaced by 8 vertical launchers for the advanced SA-N-9, and instead of the Silex launcher, space has been reserved for a new weapon, possibly a vertically launched ASW missile. See also CRUISERS, SOVIET UNION.

Specifications **Length:** 814 ft. (248.2 m.). **Beam:** 92 ft. (28.05 m.). **Draft:** 29 ft. (8.85 m.). **Displacement:** 24,000 tons standard/28,000 tons full load. **Powerplant:** twin-shaft CONAS, 150,000 shp. **Speed:** 20 kt. (nuclear)/33 kt. (superheat). **Range:** 3000 n.mi. at 30 kt. **Crew:** 900. **Sensors:** 2 "Palm Frond" navigational and surface-search radars, 1 "Top Pair" 3-dimensional air-search radar, 1 "Top Steer" surveillance radar, 2 "Top Dome" SA-N-6 missile guidance radars, 2 "Pop Group" SA-N-4 missile guidance radars (*Kirov* only), 1 "Eye Bowl" SS-N-14 missile guidance radar (*Kirov* only), 4 "Bass Tilt" 30-mm. fire control, 1 "Kite Screech" fire control radar, 2 "Foot Ball" radomes (purpose unknown), 2 "Tin Man" electro-optical sensors, 1 bow-mounted low-frequency active/passive sonar, 1 low-frequency variable depth sonar. **Electronic warfare equipment:** 4 new "Bell"-type electronic signal monitoring arrays, several radar warning receivers, a variety of active jamming units, two chaff launchers.

KITCHEN (AS-4): NATO code name for a Soviet long-range, air-launched, land attack and ANTI-SHIP MISSILE first seen in 1962. An airplane-configured missile with a cylindrical body, ogival nose, short delta wings, and a conventional tailplane (similar in planform to the MiG-21 FISHBED fighter), Kitchen is powered by a liquid-fuel rocket, and has been built in several different versions. Kitchen A, intended for land attack, is controlled by INERTIAL GUIDANCE and is armed with a 350-KT nuclear warhead; Kitchen B, an anti-ship missile controlled by a combination of inertial guidance plus active radar homing in the terminal phase, has with a 2200-lb. (1000-kg.) high-explosive (HE) warhead. A third version is an ANTI-RADIATION MISSILE armed with a 2200-lb. HE warhead. Still an important Soviet standoff weapon, Kitchen is in service with both Long Range Aviation (ADD) and Naval Aviation (AV-MF). Among the aircraft that carry it are the Tu-95 BEAR G (one missile under each wing), the Tu-22 BLINDER (one missile semi-recessed in the fuselage), and the Tu-26 BACKFIRE, which can carry either one missile under the fuselage or one under each wing. See also CRUISE MISSILE.

Specifications **Length:** 37 ft. (11.28 m.). **Diameter:** 2.95 ft. (900 mm.). **Span:** 9.83 ft. (3 m.). **Weight, launch:** 3,225 lb. (6011 kg.). **Speed:** Mach 3.5 (2275 mph/3640 kph). **Range:** (lo) 185 mi. (296 km.); (hi) 285 mi. (456 km.).

KITTY HAWK: A class of four U.S. AIRCRAFT CARRIERS (CV-63, -64, -66, and -67), commissioned between 1961 and 1968. The first three (*Kitty Hawk, Constellation*, and *America*), were upgraded versions of the FORRESTAL class, with an improved flight deck layout, better aircraft handling facilities, and enlarged fuel bunker and magazine capacities. After the nuclear-powered carrier ENTERPRISE (CVN-65) was commissioned in 1961, the U.S. Navy announced its intention to build only nuclear carriers thereafter. But in 1964 funds for a nuclear carrier were refused, and the navy opted to build the conventionally powered *John F. Kennedy* (CV-67) to a modified Kitty Hawk design.

The four Kitty Hawks differ slightly in their external dimensions. The flight deck varies between 1047 and 1062 ft. (319.2 and 323.8 m.) in length, and 250 and 267 ft. (76.21 and 81.4 m.) in width. The angled deck section, cantilevered some 10° to port, is 744.5 ft. (227 m.) long on all four ships. There are 4 steam catapults, 2 at the bow and 2 on the angled deck; elevator layout is greatly improved over that of the Forrestals, with all four elevators on the deck edge, away from the takeoff and landing areas. Two portside elevators forward of the island service the bow catapults; one elevator behind the island and one on the starboard quarter serve the angled deck catapults. This successful arrangement has been repeated on all subsequent U.S. carriers. The hangar deck on the first three ships is 740 ft. (225.6 m.) long, 101 ft. (30.8 m.) wide, and 25 ft. (7.62 m.) high; in the *Kennedy*, the hangar is only 688 ft. (210 m.) long, but 106 ft. (32.31 m.) wide.

The rather compact island superstructure incorporates the usual ship-handling and flight-control facilities, the engine exhaust stacks, and various masts for RADAR and COMMUNICATIONS antennas. A lattice mast on the flight deck behind the island supports additional radars. The *John F. Kennedy* has a slightly different stack, designed to divert turbulent exhaust gases away from the flight deck. At one time it was planned to equip the Kitty Hawks with a bow-mounted SQS-23 SONAR, on the (mistaken) assumption that they could conduct autonomous ASW operations. Only the *America* was actually equipped with the SQS-23, which was removed in 1981.

The first three Kitty Hawks were originally armed with 3 Mk. 10 launchers for TERRIER medium-range surface-to-air missiles (SAMs) on sponsons, but they consumed a great deal of internal space, and were in any case ineffective against the major threat of surface-skimming ANTI-SHIP MISSILES. The *Kennedy* was fitted instead with 3 NATO SEA SPARROW short-range SAM launchers, and the other ships were modified likewise. In the early 1980s, all 4 received three PHALANX 20-mm. radar-controlled guns for anti-missile defense to supplement the Sea Sparrows. Long-range air defense is left to aircraft and the escort vessels of the CARRIER BATTLE GROUP (CVBG).

The air group now consists of two fighter squadrons with a total of 24 F-14 Tomcat INTERCEPTORS, a medium attack squadron with 10 A-6E INTRUDER all-weather ATTACK AIRCRAFT, two fighter/attack squadrons with a total of 24 FA-18 HORNETS, an ASW patrol squadron with 10 S-3A/B VIKINGS, an ASW HELICOPTER squadron with 6 SH-3 SEA KINGS, and a composite support squadron with 4 EA-6B PROWLER electronic warfare aircraft, 4 KA-6 AERIAL REFUELING tankers, 4 E-2C HAWKEYE airborne early warning aircraft, and 1 C-2A Greyhound Carrier On-board Delivery (COD) aircraft. A few more aircraft could be added in wartime. The Kitty Hawks carry 5919 tons of aviation fuel and 1250 tons of aerial ordnance—considerably more than the Forrestals—allowing them to conduct air operations for more extended periods without UNDERWAY REPLENISHMENT. All four carriers are scheduled for the Service Life Extension Program (SLEP)—a 28-month overhaul meant to keep them on active service through the year 2010. See also AIRCRAFT CARRIERS, UNITED STATES.

Specifications **Length:** 1047.5–1062.5 ft. (319.36–324 m.). **Beam:** 129.5–130 ft. (39.48–39.63 m.). **Draft:** 36 ft. (11 m.). **Displacement:** 60,100–61,000 tons standard/80,800–82,000 tons full load. **Powerplant:** 4-shaft steam: 8 oil-fired boilers, 4 sets of geared turbines, 280,000 shp. **Fuel:** 7800 tons. **Speed:** 32 kt. **Range:** 4000 n.mi. at 30 kt./ 8000 n.mi. at 20 kt. **Crew:** 2000 + 2150 air group = 4150. **Sensors:** 1 SPS-49 2-dimensional air-search radar, 1 SPS-48 3-dimensional air-search radar, 1 SPS-10 surface-search radar, 1 SPN-35 air traffic control radar, 3 Mk.91 Sea Sparrow guidance radars. **Electronic warfare equipment:** 1 SLQ-29 electronic signal monitoring array w/radar homing and warning receivers, active jamming units, 2 Mk.36 SRBOC chaff launchers.

KKV: Kinetic Kill Vehicle, a small, rocket-powered, terminally guided interceptor for use in BALLISTIC MISSILE DEFENSES, lately redesignated SPACE-BASED INTERCEPTORS.

KNOX: A class of 46 U.S. FRIGATES (FF-1052 through -1097), commissioned between 1969 and 1974—the largest single post-1945 class of warships built in the U.S. until the advent of the PERRY class (FFG-7) from 1971. The Knox class was an attempt to design an economical ANTI-SUBMARINE WARFARE (ASW) escort to replace the large number of World War II GEARING- and SUMNER- class destroyers reaching block obsolescence in the late 1960s. But the design quickly grew to what was then considered destroyer size, becoming too expensive for mass production. When first introduced, these ships were much criticized for their low speed and seemingly weak armament. Since then, however, they have been reappraised as excellent ASW platforms.

Flush-decked, with a sharply raked bow, a blocky superstructure, and a distinctive conical "mack" (combination of mast and stack), the Knox hull is an enlarged version of the GARCIA class, which in turn was a development of the BRONSTEIN class, progenitor of all postwar U.S. escorts. Though quite seaworthy and equipped with gyro-controlled fin stabilizers, the Knoxes have always had a tendency to plow into waves (because of the lack of flare in the bow), a problem only partly solved by the addition of raised bulwarks and spray strakes at the bow. The Knoxes have a single-shaft steam powerplant, also much criticized as lacking speed and survivability. However, navy studies had indicated that frigate-sized ships rarely survived a torpedo hit in any case, and moreover, single-shaft propulsion was meant to make the Knoxes suitable as a "mobilization design" for eventual wartime mass production. While speed does compare poorly with the 32 kt. of the World War II Sumners and Gearings, the Knoxes can maintain much higher speeds in rough seas, a major consideration for North Atlantic operations.

Optimized for ASW, the Knoxes have only modest ANTI-AIR WARFARE (AAW) and ANTI-SURFACE WARFARE (ASUW) capabilities. In fact, at first glance, their armament seems absurdly light (especially when compared to Soviet ships of equivalent tonnage such as the KRIVAK frigates): 1 5-in. 54-caliber DUAL PURPOSE gun at the bow, with an 8-round ASROC pepperbox launcher behind it, 4 Mk.32 tubes for lightweight ASW TORPEDOES hidden in the superstructure, and an 8-round NATO SEA SPARROW short-range surface-to-air missile (SAM) launcher on the fantail. Appearances, however, are deceptive: the 5-in. gun is a rapid-fire Mk.42, equivalent to twin 5-in. mounts of the previous generation; the ASROC launcher has 16 reloads in an automatic loader, modified to fire HARPOON anti-ship missiles as well, providing an unseen ASUW capability (most Knoxes carry at least four Harpoons). From 1984, the Sea Sparrow launcher has been progressively replaced by a PHALANX 20-mm. radar-controlled gun for anti-missile defense. But the most important weapon on the ship is an SH-2F SEA SPRITE (LAMPS I) helicopter, flown from a landing deck amidships and kept in a large, telescoping hangar. Armed with SONOBUOYS, DEPTH CHARGES, and MK.46

torpedoes, the Sea Sprite can engage submarines out to the "second sonar convergence zone" some 70–100 mi. (112–160 km.) from the ship. Moreover, it has a surface-search radar, and can also be used for the OVER-THE-HORIZON TARGETING (OTH-T) of Harpoon missiles.

The U.S. Navy has transferred several Knoxes to the Naval Reserve Force (NRF), to serve as training vessels and as a rapid mobilization base. Spain commissioned five modified Knoxes between 1973 and 1976 as the BALEARES class, with a single-arm Mk.22 missile launcher instead of a landing deck and hanger. See also FRIGATES, UNITED STATES.

Specifications Length: 438 ft. (133.5 m.). Beam: 46.75 ft. (14.3 m.). Draft: 24.75 ft. (7.55 m.) Displacement: 3011 tons standard/4100 tons full load. Powerplant: single-shaft steam: 2 oil-fired boilers, 2 geared turbines, 35,000 shp. Fuel: 750 tons. Speed: 28 kt. Range: 4300 n.mi. at 20 kt. Crew: (active) 169; (NRF) 136. Sensors: 1 SQS-26CX bow-mounted low-frequency active/passive sonar, one SQR-18 TACTASS towed array sonar, 1 SPS-10 surface-search radar, one SPS-40 air-search radar, 1 SPG-9 fire control radar. Fire controls: 1 Mk.1 target designation system, 1 Mk.114 ASW fire control system, 1 Mk.68 gun fire control system, (Sea Sparrow) 1 Mk.115 missile fire control system. Electronic warfare equipment: 1 SLQ-32(V)½ electronic signal monitoring array, 2 Mk.36 SRBOC chaff launchers.

KOMAR: NATO code name for a large class of Soviet missile-armed FAST ATTACK CRAFT (FACs), introduced from 1963. Derived from the P6 class motor torpedo boats (in fact, the first were P6 conversions), the Komars have a hard-chine, V-bottom planing hull; constructed mainly of wood, the hull is subject to rapid deterioration if not properly maintained. The Komars are armed with two SS-N-2 STYX anti-ship missiles housed in separate launchers at the stern, aimed forward. A twin 25-mm. ANTI-AIRCRAFT gun is mounted on the bow.

Because of their very small size, the Komars are suitable only for operations in relatively smooth waters. Superseded in the Soviet navy by the larger, more capable OSA class, more than 100 remain in service in other navies. China alone has 75 boats, most produced indigenously as the steel-hulled Haku class. The Komar was the first missile boat ever used in combat: an Egyptian Komar sank the Israeli destroyer *Elat* with two Styx on 21 October 1967. See also FAST ATTACK CRAFT, SOVIET UNION.

Specifications **Length:** 83.65 ft. (25.5 m.). **Beam:** 20 ft. (6.1 m.). **Draft:** 4.9 ft. (1.5 m.). **Displacement:** 80 tons. **Powerplant:** 4 1200-hp. diesels, 4 shafts. **Speed:** 40 kt. **Range:** 400 n.mi. at 30 kt. **Crew:** 16. **Sensors:** 1 small surface-search radar.

KORMORAN: West German air-launched ANTI-SHIP MISSILE developed by Messerschmidt-Bolkow-Blohm, in service with the West German *Marineflieger* (Naval Aviation) and the Italian air force. The design is based on the French Nord AS.34 air-to-surface missile, but with more advanced guidance. Flight tests began in 1970, but the Kormoran did not enter service until 1977.

Kormoran has a cylindrical body with a pointed nose radome, four small tail fins, and four cropped delta wings at midbody. The 352-lb. (160-kg.) high-explosive warhead consists of 16 individual subcharges with delayed-action fuzes, to allow the missile to penetrate deep into the target before exploding. The first (booster) stage of the two-stage, solid-fuel missile has two solid-rocket motors which burn for only one second to provide initial acceleration, while the second-stage sustainer motor provides continuous thrust throughout the flight. Kormoran is controlled by INERTIAL GUIDANCE through midcourse, with active radar homing in the terminal phase. After launch, the missile flies at a constant height of 98 ft. (29.87 m.) above the sea (with the aid of a radar altimeter), descending to wave-top height as it nears the target when it switches to radar terminal guidance. Supposedly very reliable, Kormoran arms Italian and *Marineflieger* TORNADO strike aircraft.

Specifications **Length:** 10.83 ft. (3.3 m.). **Diameter:** 13.39 in. (340 mm.). **Span:** 39.37 in. (1.0 m.). **Weight, launch:** 1323 lb. (601 kg.). **Speed:** 700 mph (1120 kph) at sea level. **Range:** 23 mi. (37 km.).

KORTENAER: A class of 20 Dutch-designed FRIGATES (FFs) commissioned between 1975 and 1983, and now in service with the navies of the Netherlands, Greece, and West Germany. The Kortenaers were designed as successors to the Dutch navy's 6 Modified LEANDER-class ANTI-SUBMARINE WARFARE (ASW) frigates; the Greek navy ordered 2 in 1980–81, and the West German Navy 6 more (to replace its FLETCHER-class destroyers and Koln-class frigates) between 1982 and 1984.

Armament consists of 1 OTO-MELARA 76.2-mm. DUAL PURPOSE gun at the bow, an 8-round NATO SEA SPARROW short-range surface-to-air missile (SAM) launcher on the forward superstructure, 8 HARPOON anti-ship missiles in 2 quadruple launch canisters on the superstructure amidships, 2 sets of Mk.32 triple tubes for lightweight ASW TORPEDOES on the deck amidships, and a GOALKEEPER 30-mm. radar-controlled gun for anti-missile defense mounted on the superstructure aft. A landing deck and hangar on the fantail can accommodate a LYNX ASW helicopter.

The Greek (Eli-class) ships differ in having a second 76.2-mm. gun in place of the Goalkeeper, and their hangar has been extended by 6.6 ft. (2 m.) to accommodate an Agusta-Bell AB.212 HUEY ASW helicopter. The German (BREMEN-class) ships, built in West Germany, include various modifications, including a CODAG powerplant and German-made sensors optimized for Baltic conditions. The Dutch are now building two modified Kortenaers, as the HEEMSKERCK class, to replace the two ships sold to Greece. Optimized for ANTI-AIR WARFARE, they have a Mk.13 missile launcher with 40 STANDARD MR SAMs on the superstructure aft in place of the hangar, while the 76.2-mm. gun is deleted as weight compensation. These ships are also equipped with two SCOT satellite communications terminals. See also FRIGATES, FEDERAL REPUBLIC OF GERMANY; FRIGATES, NETHERLANDS.

Specifications **Length:** 428 ft. (130.5 m.). **Beam:** 47.9 ft. (14.6 m.). **Draft:** 20.3 ft. (6.2 m.). **Displacement:** 3050 tons standard/3630 tons full load. **Powerplant:** twin-shaft COGOG: 2 4900-shp. Rolls Royce Tyne RM-1C gas turbines (cruise), 2 25,800-shp. Rolls Royce Olympus TM-3B turbines (sprint). **Speed:** 30 kt. **Range:** 4700 n.mi. at 16 kt. **Crew:** 176. **Sensors:** 1 ZW-06 navigational/surface-search radar, 1 WM-25 track-while-scan fire control radar, one STIR fire control radar, 1 LW-08 air-search radar, 1 SQS-505 hull-mounted medium-frequency active/passive sonar; (Heemskerck) 1 modified STIR fire control radar, 1 DA-05 3-dimensional air-search radar, 1 PHS-36 bow-mounted sonar. **Electronic warfare equipment:** 1 Sphinx electronic support measures array or 1 Ramses active jamming array; two Mk.36 SRBOC chaff launchers.

KOTLIN: NATO code name for a class of 27 Soviet DESTROYERS completed between 1954 and 1958. Successors to the Skoryy class, the Kotlins were the last destroyers designed on classic World War II lines, with heavy gun and anti-ship TORPEDO batteries. Flush-decked, with a graceful shear and two stacks, the Kotlins were originally armed with 1 twin 130-mm. DUAL PURPOSE gun at each end of the ship, 4 quadruple 45-mm. and 2 twin 25-mm. ANTI-AIRCRAFT guns, 2 sets of quintuple 21-in. (533-mm.) tubes for anti-ship torpedoes, 2 DEPTH CHARGE racks on the fantail, and deck rails for up to 55 MINES.

In 11 Kotlins modified between 1960 and 1962, 1 set of torpedo tubes was replaced by a deck house, 2 RBU-600 ANTI-SUBMARINE WARFARE (ASW) rocket launchers replaced the forward 45-mm. mount, and 2 more 25-mm. guns were added. One modified Kotlin, the *Moskovskiy Komsomolets*, received more powerful RBU-6000s instead of RBU-600s, and was also fitted with a VARIABLE DEPTH SONAR. Another ship, the *Svetlyy*, was fitted with a HELICOPTER pad on the fantail.

Between 1966 and 1972, nine unmodified Kotlins were converted to guided-missile destroyers with surface-to-air missiles (SAMs). In these "Kotlin SAM" vessels, the aft 130-mm. guns were replaced by a twin launcher for the SA-N-1 GOA SAM with 22 missiles, and its "Peel Group" guidance radar. ASW capabilities were also upgraded by the installation of RBU-6000 or RBU-2500 rocket launchers in place of the depth charge rack. Four of these ships have also been fitted with 30-mm. radar-controlled guns for anti-missile defense, and their associated "Drum Tilt" fire control radars. Hull, machinery, and general performance are unchanged.

The Soviet navy originally planned to build a total of 36 Kotlins, but 4 were converted to KILDIN-class guided-missile destroyers while still under construction. An additional 8 Kotlin hulls were completed as Krupnyy-class destroyers armed with ANTI-SHIP MISSILES, later converted to KANIN-class guided-missile destroyers. See also DESTROYERS, SOVIET UNION.

Specifications **Length:** 417.5 ft. (127.3 m.). **Beam:** 43 ft. (13.1 m.). **Draft:** 15.16 ft. (4.62 m.). **Displacement:** 2600 tons standard/3500 tons full load. **Powerplant:** twin-shaft steam: 4 oil-fired boilers, 2 sets of geared turbines, 72,000 shp. **Speed:** 38 kt. **Range:** 1000 n.mi. at 34 kt./3600 n.mi. at 18 kt. **Crew:** 300–336. **Sensors:** 2 "Don-2" navigational and surface-search radars, 2 "Egg Cup" 130-mm. fire control radars, 2 "Hawk Screech" 45-mm. fire control radars, 1 "Sun Visor" 130-mm. fire control radar, 1 hull-mounted *Herkules* high-frequency sonar. **Electronic warfare equipment:** 1 Watch Dog radar warning receiver.

KRESTA: NATO code name for a class of 14 Soviet guided-missile CRUISERS completed between 1967 and 1978 in two distinct subtypes. The first four Kresta Is, essentially an interim design to improve the hull and machinery, are considerably more capable than their immediate predecessors of the KYNDA class. Like the latter, the Kresta Is are designed for ANTI-SURFACE WARFARE (ASUW) against U.S. CARRIER BATTLE GROUPS (CVBGs), and are classified as Missile Cruisers (*Raketnyy Kreysera*, RKR) by the Soviet navy. Radar and EW antennas are mounted on a large, pyramidal tower structure and a short "mack" (a combination mast and exhaust stack).

Primary armament consists of 2 twin SS-N-3 SHADDOCK anti-ship missile launchers mounted on each side of the bridge, which can be elevated for firing. The Krestas also have a powerful ANTI-AIR WARFARE (AAW) armament of 1 twin SA-N-1 GOA surface-to-air missile (SAM) launcher with 44 missiles at each end of the ship, supplemented by 2 twin 57-mm. ANTI-AIRCRAFT guns. Four 30-mm. radar-controlled guns were later added to defend against anti-ship missiles. ANTI-SUBMARINE WARFARE (ASW) armament is limited to 2 RBU-6000 and 2 RBU-1000 ASW rocket launchers, and 2 sets of quintuple 21-in. (533-mm.) torpedo tubes. A landing deck and hangar on the fantail can accommodate 1 Ka-25 HORMONE helicopter for OVER-THE-HORIZON TARGETING (OTH-T).

The ten definitive Kresta IIs, completed between 1970 and 1978 as specialized "Large Anti-Submarine Ships" (*Bol'shoy Protivolodochny Korabl'*, BPK), are slightly larger than their predecessors, but the hull and machinery are otherwise unchanged. The weapons and sensors, however, are completely different. The Shaddocks have been replaced by 2 quadruple launchers for SS-N-14 SILEX ASW missiles, and 4 30-mm. radar-controlled guns are standard. The SA-N-1 Goa has been replaced by the newer SA-N-3 GOBLET SAM, which has a secondary anti-ship capability, and the helicopter is the ASW version of Hormone. See also CRUISERS, SOVIET UNION.

Specifications **Length:** (I) 513.16 ft. (156.45 m.); (II) 524.5 ft. (160 m.). **Beam:** 56 ft. (17.1 m.). **Draft:** 19.75 ft. (6.02 m.). **Displacement:** (I) 6000 tons standard/7600 tons full load; (II) 7700 tons full load. **Powerplant:** twin-shaft steam: 4 oil-fired boilers, 2 sets of geared turbines, 100,000 shp. **Speed:** 35 kt. **Range:** 2400 n.mi. at 32 kt./10,500 n.mi. at 14 kt. **Crew:** 380. **Sensors:** (I) 1 "Big Net" air-search radar, 2 "Don-2" navigational radars, 2 "Plinth Net" surface-search radars, 2 "Peel Group" SA-N-1 missile guidance radars, 1 "Scoop Pair" SS-N-3 missile guidance radar, 2 "Muff Cob" 57-mm. fire control radars, 2 "Bass Tilt"

30-mm. fire control radars, 1 *Herkules* hull-mounted medium-frequency active/passive sonar; (II) 1 "Top Sail" 3-dimensional air-search radar, 2 "Head Light" SA-N-3/SS-N-14 guidance radars, 1 bow-mounted medium-frequency sonar. **Electronic warfare equipment:** several "Bell"-type electronic signal monitoring arrays, 2 "Side Globe" electronic countermeasure systems.

KRIVAK: NATO code name for a class of 36 Soviet guided-missile FRIGATES completed between 1968 and 1976. Specialized ANTI-SUBMARINE WARFARE (ASW) vessels, the Krivaks were originally classified as Large Anti-Submarine Ships (*Bol'shoy Protivolodochny Korabl'*, BPK) by the Soviet navy, but redesignated in 1978 as Patrol Ships (*Storozhevoy Korabl'*, SKR), no doubt due to their limited range. The first large warships fitted with gas turbine engines, the Krivaks have been built in three distinct subclasses.

Krivak I has a very wide transom stern to improve stability and provide ample space for sensors including a variable depth sonar (VDS) and weapons. The main armament is a trainable, quadruple launcher in the bow for SS-N-14 SILEX ASW missiles, roughly analogous to the Western IKARA system. Other ASW weapons include 2 RBU-6000 rocket launchers, and 2 sets of quadruple 21-in. (533-mm.) TORPEDO tubes. Air defense is provided by a twin SA-N-4 GECKO short-range surface-to-air missile (SAM) launcher at each end of the ship, and 2 twin 76.2-mm. DUAL PURPOSE guns aft. There are also rails for up to 20 mines. Twenty-two Krivak Is were completed between 1968 and 1976.

The Krivak II has 2 100-mm. dual-purpose guns in place of the 76.2-mm. guns, improved "Owl Screech" fire control radars, and an improved "Spin Trough" navigational/surface-search radar. Eleven were built between 1973 and 1982.

A major deficiency of the Krivak design was its lack of facilities for an ASW HELICOPTER, but in the Krivak III, the SS-N-14 launcher has been replaced by a 100-mm. gun, 1 SA-N-4 launcher has been replaced by 2 30-mm. radar-controlled guns for anti-missile defense, and the aft gun mounts have been replaced by a landing deck and hangar for 1 Ka-25 HORMONE ASW helicopter. The VDS is retained under the flight deck. Hull, powerplant, and performance are otherwise unchanged. The Krivak III is being built for KGB naval border guards stationed on the Soviet Union's Pacific coast. At least three were built between 1981 and 1986. See also FRIGATES, SOVIET UNION.

Specifications **Length:** 405.2 ft. (123.6 m.). **Beam:** 45.9 ft. (14 m.). **Draft:** 16.4 ft. (5 m.). **Displacement:** 3100 tons standard/3400 tons full load. **Powerplant:** twin-shaft COGAG: 2 12,100-shp. gas turbines (cruise), 2 24,300-shp. gas turbines (boost). **Speed:** 32 kt. **Range:** 700 n.mi. at 30 kt./ 4500 n.mi. at 16 kt. **Crew:** 220. **Sensors:** 1 medium-frequency hull-mounted sonar, 1 medium-frequency variable depth sonar, 1 "Don-2" navigational and surface-search radar, 1 "Head Net C" air search radar, 2 "Eye Bowl" SS-N-14 guidance radars, 2 "Pop Group" SA-N-4 guidance radars, 1 "Owl Screech" 76.2-mm. fire control radar. **Electronic warfare equipment:** 2 "Bell Shroud" electronic signal monitoring arrays, two chaff rocket launchers.

KT: Abbreviation for kiloton, the blast equivalent of 1000 tons of TNT, and the standard measure of the energy yield of NUCLEAR WEAPONS. One thousand kilotons are equal to one megaton, or MT.

KYLE (AS-9): NATO code name for a Soviet air-launched ANTI-RADIATION MISSILE (ARM) developed in the late 1960s. Roughly equivalent to the U.S. AGM-78 STANDARD ARM, Kyle has a cylindrical body, an ogival nose, four small control canards, and four delta tail fins. Powered by a solid rocket engine, Kyle is armed with a 440-lb. (200-kg.) high-explosive fragmentation warhead fitted with contact and PROXIMITY FUZES.

Kyle probably has a simple crystal monopulse receiver which must be pretuned to the specific frequency of target radars. Like other early ARMs, it probably does not have INERTIAL GUIDANCE, hence it cannot continue to home on the target radar if it stops transmitting. Kyle is intended for high-altitude release, and has a diving attack profile. It has been seen on the MiG-25 FOXBAT, MiG-27 FLOGGER, Su-17 FITTER, Su-24 FENCER, and Tu-22M BACKFIRE. Although Kyle is now superseded in production by the AS-12 KEGLER, large numbers remain in service.

Specifications **Length:** 19.68 ft. (6 m.). **Diameter:** 19.68 in. (500 mm.). **Span:** 4.92 ft. (1.5 m.). **Weight, launch:** 1650 lb. (750 kg.). **Speed:** Mach 2 (1300 mph/2080 kph). **Range:** 45 mi. (75 km.).

KYNDA: NATO code name for a class of four Soviet guided-missile CRUISERS, completed between 1962 and 1965. Designated "Missile Cruisers" (*Raketnyy Kreysera*, RKR) by the Soviet navy, the Kyndas were the first large combat vessels designed primarily for ANTI-SURFACE WARFARE (ASUW) with ANTI-SHIP MISSILES. Because the primary mission of the Kyndas was sea denial, i.e., to attack convoys and U.S. CARRIER BATTLE GROUPS in short operations, they have many guns and missiles in a relatively small hull, at the expense of endurance and survivability.

Primary armament consists of 1 trainable quadruple launcher for SS-N-3 SHADDOCK anti-ship missiles at each end of the ship. The primary ANTI-AIR WARFARE (AAW) weapon is a twin SA-N-3 GOA surface-to-air missile (SAM) launcher with 24 missiles at the bow (ahead of the SS-N-3 launcher), which is supplemented by 2 twin 76.2-mm. DUAL PURPOSE guns aft. One ship, the *Varyag*, has been fitted with 4 30-mm. radar-controlled guns for defense against anti-ship missiles. ANTI-SUBMARINE WARFARE (ASW) armament is limited to 2 RBU-6000 ASW rocket launchers and 2 sets of triple 21-in. (533-mm.) TORPEDO tubes. A landing deck on the fantail can accommodate a Ka-25 HORMONE helicopter for OVER-THE-HORIZON TARGETING (OTH-T) of the Shaddock missiles, but there is no hangar. The multitude of radar, ELECTRONIC WARFARE, and communication antennas are mounted on two tall, pyramidal towers, a design feature that became typical of all subsequent Soviet cruisers. See also CRUISERS, SOVIET UNION.

Specifications Length: 476.6 ft. (145.3 m.). Beam: 52.2 ft. (16 m.). Draft: 17.5 ft. (5.33 m.). Displacement: 4400 tons standard/5500 tons full load. Powerplant: twin-shaft steam: 4 oil-fired boilers, 2 sets of geared turbines, 100,000 shp. Speed: 36 kt. Range: 2000 n.mi. at 34 kt./7000 n.mi. at 14.5 kt. Crew: 375. Sensors: 2 "Don-2" navigational radars, 2 "Head Net" air-search radars, 2 "Plinth Net" surface-search radars, 1 "Peel Group" SA-N-1 guidance radar, 2 "Scoop Pair" SS-N-3 guidance radars, 1 "Owl Screech" 76.2-mm. fire control radar, one *Herkules* hull-mounted high-frequency active sonar; (*Varyag* only) 2 "Bass Tilt" 30-mm. fire control radars. **Electronic warfare equipment:** several "Bell"-series electronic signal monitoring arrays, 1 "Top Hat" electronic countermeasure system.

L

LABS: See LOW-ALTITUDE BOMBING SYSTEM.

LADAR: Acronym for Laser Radar, an experimental technique analogous to RADAR, which relies on an electronically or mechanically scanned LASER beam, rather than radio or microwave radiation, for the detection and ranging of targets. The laser beam is reflected back from target objects and received by an array of photo-electric detectors. As in radar, the time delay between transmission and reception reveals target range, the beam direction reveals azimuth, while the phase difference between transmitted and received signals (the DOPPLER shift) indicates the target's relative velocity. Because lasers operate at much higher frequencies than radar signals (micron wavelengths, as opposed to millimeters or more), much higher resolution and data rates are theoretically possible; for the same reason, however, atmospheric attenuation now limits the effective range of LADAR to short distances—but ADAPTIVE OPTICS may resolve that problem in the future. LADAR could be superior to radar in its resistance to ELECTRONIC COUNTERMEASURES, and could not be detected with current RADAR WARNING RECEIVERS. Because atmospheric attentuation is not a problem in space, LADAR is under active consideration as a space-based sensor for BALLISTIC MISSILE DEFENSE. LIDAR, or Laser Imaging Radar, would perform complex signal processing of Doppler data to obtain photolike images of target objects.

LADDER-DOWN: A hypothetical tactic for blinding RADAR-directed ground-based BALLISTIC MISSILE DEFENSES (BMDs), with successive high-altitude nuclear explosions. The detonation of one salvo would temporarily blind local BMD radars by NUCLEAR BLACKOUT, allowing subsequent salvos to penetrate more closely before being detonated; by repeating the process, the target area is eventually reached and destroyed. The tactic has, of course, never been tested, and is likely to be precluded by FRATRICIDE and other problems of implementation. See also ANTI-BALLISTIC MISSILE; NUCLEAR WEAPONS.

LAFAYETTE: A class of 31 U.S. nuclear-powered ballistic-missile submarines (SSBNs) commissioned between 1963 and 1967. The Lafayettes, improved versions of the Ethan Allen class, were the standard SSBNs of the U.S. Navy throughout the 1960s and '70s. The 16-tube layout of these submarines was a major influence on the design of the British RESOLUTION, the French REDOUTABLE, and the Soviet YANKEE classes. The Lafayettes are longer than the Ethan Allens, and have improved acoustical silencing, electronics, sensors, and weapons, while retaining the same general configuration.

The Lafayettes have a modified "teardrop" hullform, with an ogival bow, cylindrical center section, and tapering stern; the single hull is fabricated of HY-80 steel. A tall, streamlined sail (conning tower) is located well forward, and a flat-decked, "whaleback" casing behind it is lined with the covers of the missile tubes. The control surfaces are arranged in the standard U.S. configuration, with the "fairwater" diving planes mounted on the sail, and cruciform rudders and stern planes mounted just ahead of the propeller. All machinery is mounted on sound-isolation rafts to reduce radiated noise, and the propeller is shaped with extreme precision to reduce cavitation noise.

The Lafayettes are armed with 16 ballistic missiles mounted in two rows of 8 tubes behind the sail, in a compartment known to their crews as "Sherwood Forest." The first 8 boats in the class were originally armed with the POLARIS A-2 missile, which carried a single nuclear warhead and had a range of 1725 mi. (2880 km.). The remaining 23 boats were armed from the beginning with the later Polaris A-3, which carried three multiple REENTRY VEHICLES (MRVs) over a range of 2300 mi. (3841 km.), and the A-3 was eventually retrofitted into the first 8 submarines. Beginning in 1970, all 31 boats were gradually rearmed with the POSEI-DON C-3, a much larger missile with ten MULTIPLE, INDEPENDENTLY TARGETED REENTRY VEHICLES (MIRVs) and a range of 2875 mi. (4800 km.). Since 1979, the last 12 boats built (unofficially labeled the "BENJAMIN FRANKLIN" class), have been rearmed once again, with the TRIDENT C-4 missile, which carries 8 MIRVs and has a range of 4600 mi. (7682 km.). The Lafayettes' missile tubes are too small to accommodate the even larger Trident II D-5 missile carried aboard the newest OHIO-class SSBNs.

Secondary armament consists of four bow-mounted 21-in. (533-mm.) tubes which can launch MK.48 wire-guided acoustical homing torpedoes, plus a variety of TORPEDO COUNTERMEASURES such as the Mobile Submarine Simulator System (MOSS).

The Lafayettes, like other SSBNs, are designed to hide, not seek; accordingly, their sonars are optimized for the long-range passive detection of surface ships or submarines. With accurate navigation essential to their mission, the Lafayettes rely on a Ships INERTIAL NAVIGATION System (SINS) to determine missile-launch coordinates. Submerged COMMUNICATIONS with shore bases depend on VLF and ELF radios whose long antennas are towed behind the submarines; because the transmission rate of such systems is low, and because even the briefest transmission to acknowledge orders can reveal the SSBN's position, the dilemma between controllability and security cannot be resolved, and remains the weak point of ballistic missile submarines.

In accordance with standard U.S. Navy policy for SSBNs, each boat has two crews ("Blue" and "Gold"), which man alternating 60 to 72-day patrols, with a two-week overhaul period in between, to maximize time on station without degrading crew effectiveness (boredom is the great affliction of SSBN crews).

As new submarines of the Ohio class are commissioned, the Lafayettes are being taken out of service to conform to SALT II treaty limitations. To date, two submarines, USS *Nathan Hale* and *Nathaniel Greene*, have been placed in reserve; one, USS *Sam Rayburn*, has had its missile system deactivated and is being used as a dockside trainer. Several alternative roles for retired SSBNs have been proposed, including conversion to attack submarines, CRUISE MISSILE carriers, or COMMANDO transports. See also SUBMARINES, UNITED STATES.

Specifications Length: 425 ft. (129.6 m.). **Beam:** 33 ft. (10.06 m.). **Displacement:** 7350 tons surfaced/8250 tons submerged. **Powerplant:** single-shaft nuclear: 1 S5W pressurized-water reactor, 1 set of geared turbines, 15,000 shp. **Speed:** 20 kt. surfaced/25 kt. submerged. **Max. operating depth:** 1150 ft. (350 m.). **Collapse depth:** 1525 ft. (465 m.). **Crew:** 140. **Sensors:** 1 BQR-21 bow-mounted low-frequency passive sonar, 1 BQR-7 conformal hydrophone array, 1 BQS-4 medium-frequency active/passive sonar, 1 BQR-19 navigational sonar, 1 BQR-15 passive towed array, 1 BPS-11/11A or BPS-15 surface-search radar, 1 electronic signal monitoring array, 2 periscopes.

LAMPS: Light Airborne Multipurpose Platform System, U.S. Navy term for HELICOPTERS operated from the flight decks of FRIGATES, DESTROYERS, and CRUISERS. LAMPS helicopters were intended originally as replacements for the unsuccessful DASH (Drone Anti-Submarine Helicopter) for long-range ANTI-SUBMARINE WARFARE (ASW); in addition, they now provide OVER-THE-HORIZON TARGETING (OTH-T) for ANTI-SHIP MISSILES, as well as search and rescue (SAR) and standoff ELECTRONIC COUNTERMEASURES (ECM) capabilities. In the near future, LAMPS helicopters will also be equipped with PENGUIN anti-ship missiles for ANTI-SURFACE WARFARE (ASUW).

The LAMPS program began in 1970 with the conversion of 105 UH-2 SEA SPRITE utility helicopters to SH-2D and SH-2F configuration, under the designation LAMPS I; another 48 new production SH-2Fs were procured in the late 1980s.

LAMPS II, a proposal for an improved Sea Sprite, was superseded in 1974 by the LAMPS III requirement calling for a larger and more capable aircraft. In 1977 the SH-60B SEAHAWK, a navalized version of the UH-60A BLACKHAWK, was selected as the LAMPS III helicopter, and began entering service in 1983. The navy plans to acquire a total of 200 SH-60Bs for TICONDEROGA-class cruisers, KIDD- and SPRUANCE-class destroyers, and PERRY-class frigates. Because some older navy escorts are too small to carry the Seahawk, the SH-2F will be retained in service, with periodic upgrades of its sensors and weapons.

LANCE: The Vought MGM-52 short-range BALLISTIC MISSILE, primarily a battlefield nuclear delivery system, although conventional warheads have also been produced. Developed from 1962 to replace both the unguided, spin-stabilized HONEST JOHN rocket and the inertially guided Sergeant missile, Lance was first test-fired in 1965, entered production in 1971, and has since been the standard battlefield nuclear missile for NATO armies. The U.S. Army has 8 battalions of 8 launchers each, of which 6 battalions are deployed with V and VII Army Corps in West Germany. The British army has 1 regiment of 18 launchers; West Germany has 4 battalions, each with 6 launchers; and Belgium, the Netherlands, and Italy each have 1 battalion of 9 launchers. British, Belgian, German, and Italian missiles are armed with nuclear warheads under dual-key arrangements; the Dutch have only conventional warheads. Israel has also been supplied with 18 launchers, now organized into a Lance brigade of 3 6-launcher battalions; only conventional warheads were supplied to Israel.

The Lance missile consists of an ogival nose housing the warhead and guidance system, a cylindrical center section, and a tapering tail section ending in the rocket nozzle. The missile is powered by a Rocketdyne storable liquid bipropellant rocket engine, which has two concentric nozzles. The outer nozzle provides 50,000 lb. of thrust during the first few seconds of flight to boost the missile to cruising velocity, after which the inner (sustainer) nozzle takes over; its thrust can be throttled from 5000 lb. (22727 kg.) down to zero for precise range control. The missile employs a simple form of INERTIAL GUIDANCE combined with spin stabilization, provided at launch by venting propellant gas through four canted nozzles at the top of the motor casing, and sustained in flight by four canted triangular tail fins.

Lance can be equipped with a variety of warheads, though the 467-lb. (212-kg.) W70 Mod ½ nuclear device (with selectable yields of 10, 50, or 100 kT) is standard. The W70 Mod 3 ENHANCED RADIATION (ER) warhead (or "neutron bomb") with a selectable yield of 0.8 or 1.6 kT is also in U.S. service, but is not currently deployed in Europe because of political considerations. The

1000-lb. (454-kg.) nonnuclear M251 warhead contains 836 BLU-63 anti-personnel/anti-materiel (APAM) submunitions which weigh 0.95 lb. (0.43 kg.) each; when airburst by a proximity fuze, it can saturate an area of 900 yards. This warhead has been used by the Israeli army to suppress air defenses. Israel may also have developed nuclear warheads for its Lance missiles, although a longer-range missile would be a more likely delivery system.

Lance is usually launched from the M572 tracked self-propelled transporter-erector-launcher, derived from the M113 armored personnel carrier. Each M572 can carry one ready-to-fire missile, and is usually accompanied by the similar M688 loader-transporter vehicle, with two reload missiles and a hoist. Both vehicles are amphibious. U.S. AIRBORNE and AIR-MOBILE forces have a lighter trailer-launcher which can be delivered by helicopter or parachuted.

Throughout the 1970s, Vought proposed a number of Improved Lance variants, the most notable of which was the T22 missile (under the ASSAULT BREAKER concept), which was to have midcourse guidance update, increased range, and an IMPROVED CONVENTIONAL MUNITIONS payload of SKEET sensor-fuzed weapons or TERMINALLY GUIDED SUBMISSILES (TGSMs). Development of a Lance replacement now proceeds under the Follow-on to Lance (FOTL) program; because of the 1987 INF TREATY and the collapse of the WARSAW PACT, the renewal of NATO Lance units has become controversial.

Specifications **Length:** 20.25 ft. (6.17 m.). **Diameter:** 22 in. (558 mm.). **Span: weight, launch:** (nuc.) 3373 lb. (1533 kg.); (ICM) 3920 lb. (1782 kg.) **Range, max:** (nuc.) 75 mi. (125 km.); (ICM) 43 mi. (72 km.). **Range, min.:** 3 mi. (5 km.). CEP: 400 m.

LANDING CRAFT: A flat-bottomed boat for AMPHIBIOUS WARFARE. Most have a ramp door in the bow, to discharge troops, vehicles, or cargo directly onto the beach. The largest landing craft displace several hundred tons and are capable of short, open-ocean voyages, but most lack range and seaworthiness, and must be carried to the scene of operations by larger ships. Specific types include the LCT (Landing Craft, Tank), also called LCU (Landing Craft, Utility); the LCM (Landing Craft, Mechanized); LCVP (Landing Craft, Vehicles and Personnel); and LCPL (Landing Craft, Personnel, Light). Recently, the United States and Soviet Union have both begun using HOVERCRAFT to supplement more conventional landing craft. See also LANDING SHIP.

LANDING SHIP: A large AMPHIBIOUS WARFARE vessel with a flat bottom and shallow draft, designed to beach itself to discharge troops, vehicles, and cargo directly ashore, usually through large bow doors or articulated vehicle ramps. Specific types of landing ship include the LST (Landing Ship, Tank), LSM (Landing Ship, Medium), and LSD (Landing Ship, Dock). The LSD is actually a different type of vessel with a deep draft and incapable of beaching. Instead, LSDs have a large, floodable well aft, from which smaller LANDING CRAFT and amphibious assault vehicles can float out to the beach.

LANDING ZONE (LZ): A cleared area used for the takeoff and landing of HELICOPTERS. In AIR ASSAULT operations, the LZ is often in close proximity to enemy positions. An ideal landing zone is fairly level, firm, close to the objective, large enough to receive the entire force in a short time, and free of enemy fire or observation; some of these requirements are of course mutually exclusive. During the Vietnam War, LZs were often blasted out of the jungle with large high-explosive bombs or FUEL AIR EXPLOSIVES (FAEs).

LANTIRN: Low Altitude Navigating and Targeting Infrared for Night (operations), a U.S.-developed night vision and TARGET ACQUISITION system for FIGHTERS and ATTACK AIRCRAFT. The highly ambitious LANTIRN was developed during the 1970s and '80s for the U.S. Air Force (USAF) by Martin Marietta Corporation. Due to many technical difficulties, the system did not enter squadron service until 1987, at a cost that greatly exceeded early estimates.

LANTIRN consists of two separate pods for suitably wired aircraft pylons; each pod can also function as a separate unit. The "navigation pod" contains a wide field-of-view (FOV) Forward Looking Infrared (FLIR) sensor and a TERRAIN-FOLLOWING RADAR (TFR), while the "targeting pod" contains a narrow FOV FLIR and a LASER DESIGNATOR. The navigation pod is 78.2 in. (1.986 m.) long and 12 in. (305 mm.) in diameter; the targeting pod is 98.5 in. (2.5 m.) long and 15 in. (380 mm.) in diameter. The weight and drag of the two pods do not significantly degrade aircraft performance, but of course they diminish payload in proportion.

Data from the navigation pod's FLIR and TFR are displayed as images on the aircraft's Head Up Display (HUD), to give the pilot a visual presentation of the terrain ahead together with appropriate steering cues, thus enabling him to fly low-level

missions at night or in bad weather even at very high speeds. The FLIR and TFR both slew to lead the aircraft into turns, and are stabilized to compensate for aircraft altitude.

The targeting pod's narrow FOV has two different magnification settings, for target acquisition and identification. This pod also has an automatic target recognition device which scans the scene autonomously for preset target types (e.g., tanks). When an appropriate target signature is detected, the device automatically locks on to it and passes the information to IMAGING INFRARED (IIR) guided air-to-surface missiles (at present, AGM-65D MAVERICKS). This information is also displayed on the HUD, so that missiles may be launched either manually or automatically (with pilot override). Because the device can (in theory) identify and lock on to targets much more quickly than a human operator, multiple missile launches may be possible on a single pass. The targeting pod can also operate in a cued mode, using preset target coordinates from the aircraft's INERTIAL NAVIGATION system. In either case, the pod's laser designator can provide target ranging for free-fall weapon delivery and to illuminate targets for LASER-GUIDED BOMBS (LGBs) or AGM-65E laser-guided Mavericks. The pilot identifies the target using the pod's FLIR, and then places a cursor over the target image on his HUD; this automatically slaves the laser onto the target, which is then tracked automatically from weapon release to impact.

It was originally planned to use LANTIRN with most tactical aircraft; due to the cost overrun, however, the U.S. Air Force now plans to wire only the F-15E STRIKE EAGLE and F-16 FALCON for LANTIRN. A-10A THUNDERBOLT close air support aircraft therefore remain without an effective night and all-weather capability. Employed by F-15E squadrons in Saudi Arabia in 1991, LANTIRN was apparently quite effective against Iraqi targets.

LARS: Light Artillery Rocket System, a West German multiple rocket launcher (MRL). Developed during the early 1960s, LARS entered service in 1969, and a total of 209 are currently in inventory. LARS consists of a 36-tube launcher for 110-mm. unguided, fin-stabilized rockets mounted on a modified 6×6 truck chassis. The launcher has traverse limits of 105° right and left, and elevation limits of 0° to 55°. The rockets can be fired individually or in ripples at a rate of two rockets per second. Reloading is manual, requiring approximately 15 minutes.

The solid-fuel rockets, which weigh 55 lb. (25 kg.) apiece, have a minimum range of 6000 m. and a maximum range of 15,000 m., determined by launcher elevation. A variety of warheads are available, including high-explosive (HE), incendiary, smoke, and IMPROVED CONVENTIONAL MUNITIONS (ICMs) of two different types, one armed with eight AT-1 Pandora anti-vehicle MINES, and the other with five AT-2 Medusa anti-tank mines. The latter is now considered the primary round for LARS.

LARS has been built in two different versions: LARS-1, mounted on a Magirus Jupiter 6 × 6 chassis with an armored cab; and LARS-2 on a MAN 6 × 6 chassis with better cross-country mobility. In addition, LARS-2 has an improved FIRE CONTROL system and a wider variety of rocket warheads. Both versions weigh approximately 17.5 tons in firing order.

LARS equips West German divisional rocket launcher battalions, each consisting of a headquarters company, a fuze company, and 2 firing batteries. A battery in turn consists of 8 launchers, a reload vehicle with 144 rockets, and 2 truck-mounted "Fieldguard" fire control radars, which track outgoing rockets to correct the aim of subsequent launches.

Like all MRLs, LARS is relatively inaccurate, and best employed for area saturation. As the U.S. MULTIPLE LAUNCH ROCKET SYSTEM (MLRS) enters service with the West German army, LARS will probably be relegated to reserve units. See also ARTILLERY, FEDERAL REPUBLIC OF GERMANY; ROCKET ARTILLERY.

LASER: Light Amplification by Stimulated Emission of Radiation, a technique used to generate greatly intensified beams of light—probably the most important military innovation of recent decades. Lasers have two principal characteristics: the light they emit is both coherent (i.e., phase synchronized) and monochromatic (of a uniform frequency, or "color"). Thus laser energy does not dissipate as ordinary "incoherent" light does, and can be focused into very narrow beams to send large amounts of energy over long distances.

TECHNOLOGY

Laser beams are generated by *stimulated emission*: the atoms within a suitable substance (the *lasing medium*) are excited, or *pumped* by an external energy source, causing their electrons to assume higher energy states; once in that higher energy state, atoms tend to return to their normal ("base") state by emitting *photons*—the source of the laser beam. (The *brightness*, or energy deposited per photon, is directly proportional to the frequency; i.e., the higher the frequency, the more energy transferred by the photon.) Within the lasing medium, atoms are bombarded by photons whose frequency corresponds to the energy level of the atoms themselves, which therefore do not absorb the photons, but instead emit other photons, identical in frequency and phasing to the first. This process is repeated with each collision between photons and atoms, in a geometrically progressing chain reaction.

The first working laser, built in 1958, was a ruby crystal device. The ruby acts as the lasing medium, being formed into a rod with reflective silver ends, one of which is fully mirrored, while the other is only partially mirrored. The ruby is pumped by light from a xenon flash lamp, which excites the chromium atoms within the crystal, causing them to emit photons which collide with other chromium atoms, a percentage of which engage in stimulated emission. The emitted photons are reflected back and forth between the reflective ends of the rod, gaining energy with each traverse and colliding with other atoms, increasing the photon flow. When the energy level is sufficiently high, the photons escape through the partially mirrored end of the rod, forming the laser beam.

Many other types of crystal can be used for lasing, the frequency of the beam varying with the composition of the crystal. Commonly used types include Neodymium-doped Yttrium-Aluminum-Garnet (Nd-YAG); Neodymium-doped Glass (Nd-G); and undoped Yttrium-Aluminum-Garnet (YAG). All are temperature-limited to relatively low power levels, making them unsuitable as weapons, and being dependent on a high-intensity flash lamp for pumping, they can be used only in a pulsed mode. (Laser diodes are very small semiconductor devices which emit low-power laser radiation when excited by a small electrical current.)

To achieve higher, weapon-level outputs, measured in kilowatts (kW) or megawatts (MW), alternative lasing media must be used. In the earliest type of high-energy laser (HEL), the gas laser, carbon dioxide (CO_2), carbon monoxide (CO), argon, krypton, or deuterium fluoride is stimulated by a powerful electric current inside a sealed glass tube, or *optical chamber*. The inner walls of the chamber are silvered or semi-silvered in the same manner as the ends of a lasing crystal. The stimulated photon emissions are reflected back and forth in the optical chamber, finally emerging through a semi-silvered aperture. Gas lasers can produce high-power beams in frequencies ranging from infrared through the visible portion of the spectrum, and because the electric current used for pumping can be maintained indefinitely, gas lasers can operate in a continuous wave (CW) mode; i.e., for extended periods. The gas dynamic laser is similar, but relies on a continuous flow of gas rather than a sealed tube, with the lasing medium pumped through the optical chamber at supersonic velocities. The major advantage of gas dynamic lasers is that the rapidly flowing gas dissipates the heat generated by the lasing process. On the other hand, gas dynamic lasers require much power and use very large volumes of lasing gas.

Chemical lasers are a third form of HEL, in which two compounds (commonly CO and nitrous oxide, or hydrogen and fluorine) are burned in a combustion chamber to generate CO_2 with electrons already excited to a very high energy state. This state is maintained by passing the CO_2 though supersonic expansion nozzles into the optical chamber, where the stimulated emission of photons occurs. Mirrors at either end of the chamber reflect the photons, which are finally emitted through a semi-mirrored aperture. The hot gases are then vented into the environment through rocketlike nozzles. Chemical lasers can generate very high power levels: the U.S. Navy's MIRACL (Mid-Infrared Chemical Laser) is rated at 2.5 MW, while the Alpha laser now under development has a projected output of 5 MW. But the endurance of chemical lasers is limited by the supply of the lasing medium; in addition, the disposal of exhaust gasses can be a significant problem.

Eximer (Excited Dimer) lasers are similar to gas lasers, but use combination of an inert (noble) gas (such as argon, krypton, or xenon) and halogens (such as chlorine or fluorine), which, when excited by a high-power pulsed electron beam, form a highly unstable two-atom molecule, or *dimer*. When the molecule returns to its base state, it rapidly disassociates into its constituent atoms, releasing high-energy photons. The beam produced by an eximer laser is emitted in rapid pulses, corresponding to the impulses of the electron beam which pumps the lasing medium. Eximers have much higher frequencies than do chemical lasers, ranging from blue-green to near-ultraviolet,

and therefore have a much higher energy density, other things being equal.

FREE ELECTRON LASERS (FELs) are a type of HEL in which the lasing medium is an electron beam passed through a series of alternating electromagnets or *wigglers*, which distort the beam; no supply of gas is needed at all. As the beam passes through the magnetic field of the wigglers, it gives off photons in the infrared frequencies; these are captured by mirrors in an optical chamber, and as the photons pass between the mirrors, they gain energy through stimulated emission to form a laser beam. A unique feature of FELs is that their operating frequency can be varied (over a narrow bandwidth) by "tuning" the fields of the electromagnetic wigglers, and this may help to overcome atmospheric attenuation. FELs have the great advantage of using only electricity, not gas; but they require very high-voltage power.

X-RAY LASERS (XRLs), a very powerful type of HEL now being studied on paper, would amount to a sort of directional nuclear explosive. A number of thin metal rods, each a lasing medium, would be attached to a small nuclear warhead. When the latter is detonated, the high-energy X-rays generated by the nuclear explosion would pump the atoms in the rods, which would then emit laser beams in X-ray frequencies along their longitudinal axis (which would be pointed at the target). All this would happen in less than 1/1000 of a second; i.e., before the nuclear fireball vaporizes the rods. XRL research is now conducted at the Los Alamos National Laboratory as Project Excalibur.

APPLICATIONS

Lasers are now most widely used for rangefinding and target designation. Laser rangefinders are handy and very accurate ($\pm$ 3 m. at 9999 m.); they measure distance by the time required for a laser pulse to travel to the target and back. LASER DESIGNATORS illuminate targets for bombs, missiles, or shells with semi-active laser homing guidance, e.g., LASER-GUIDED BOMBS such as PAVEWAY and missiles like HELLFIRE and MAVERICK. Laser diodes are used in advanced types of proximity fuze, as notably in the active optical fuze used of AIM-9L/M versions of the SIDEWINDER air-to-air missiles. Lasers diodes are also used in fiberoptic communications, which are now replacing conventional copper wire. Fiberoptic links have much higher data rates than electrical or microwave communications, and are largely immune to ELECTROMAGNETIC PULSE (EMP)—a nuclear effect which can disrupt radio and wire links. Fiberoptic FLY-BY-LIGHT systems are expected to replace electronic FLY-BY-WIRE systems in future aircraft and weapons.

Laser radar, or LADAR, is an incremental development of the rangefinder, which relies on a scanning laser (and digital signal processing) to determine the azimuth and radial velocity of objects, in addition to their range. LADAR is still immune to most forms of ELECTRONIC COUNTERMEASURES and is virtually undetectable with existing RADAR WARNING RECEIVERS. In addition, LADAR could, in theory, have much better spatial and range resolution than existing RADARS. LADAR has been suggested as an active sensor for STEALTH aircraft and for space-based ballistic missile defenses.

Lasers can also be used for line-of-sight wireless communications with amplification-modulated beams from point-to-point. Blue-green lasers are being developed by the U.S. Navy as a means of communicating between satellites and submerged SUBMARINES, seawater being transparent to lasers in that frequency band. Submarine laser communications would be much more reliable and faster than current VLF and ELF radios; detection by the same means is now being explored as a revolutionary (and destabilizing) ant-submarine warfare application.

The most controversial military application of the laser technique so far is in DIRECTED ENERGY WEAPONS (DEWs). When the beam from an HEL hits a target, it rapidly vaporizes the target's surface material, reciprocally generating a destructive shock wave which damages its internal structure; against thinner materials, HELs may burn through completely. Thermal damage is a secondary effect. Even lasers of relatively low power can damage sensitive optics and cause temporary blindness in humans. Finally, because laser beams travel at the speed of light, FIRE CONTROL is simplified to a matter of pointing and firing; there are no ballistics to consider.

The most frequently discussed strategic applications of HELs are for BALLISTIC MISSILE DEFENSE (BMD) and ANTI-SATELLITE (ASAT) weapons. Lasers seem ideally suited to BMD because they would have little difficulty in hitting and destroying BALLISTIC MISSILE boosters and POST-BOOST VEHICLES (PBVs), provided they have adequate power. REENTRY VEHICLES may be harder to destroy since they are protected by heat-resistant

coatings. In one proposed scheme, space-based HELs would be mounted on orbital satellites equipped with beam pointing and fire control systems. In another, the beam-generating equipment would remain on the ground, and would be pointed at a series of relay and "fighting" mirrors in space, which would aim and focus the beam against the targets. It has been estimated that a practical BMD laser must have a power output of 20–25 MW, or ten times greater than the largest existing U.S. systems. ASAT laser weapons could function in much the same manner as BMD lasers, though much less power would be needed to disable the optics of reconnaissance satellites, or destroy sensitive solar cells. There is some evidence that the Soviet Union is developing a ground-based laser with that capability.

Laser weapons for tactical combat are still in their infancy because of the power requirements of existing HELs. Intercepts of aircraft and missiles with aircraft- or vehicle-mounted lasers have been tried, but the most likely tactical use of lasers at present is for shipboard anti-missile defense. Unlike aircraft and ground vehicles, ships are large enough to carry adequate supplies of lasing media (for gas or chemical lasers) and have the electrical power needed to pump large lasers. The Soviet navy has already fitted experimental laser devices on warships. While there is no evidence that these lasers are powerful enough to destroy aircraft or missiles, crewmen aboard U.S. patrol aircraft have been temporarily blinded in a number of cases.

The major technical problem that inhibits the development of all ground-based laser weapons is atmospheric attenuation. The beam emitted by the laser is coherent and monochromatic, but atmospheric molecules and particulate matter tend to diffuse the beam and throw it out of phase, thereby reducing its power drastically as range increases. The problem may be at least partially solved by the use of "tunable" FELs and by ADAPTIVE OPTICS, or "rubber mirrors," whose articulated surfaces can alter the beam's focus to neutralize attenuation. Due to the size and cost of these mirrors, they could be used only (at least initially) for large, ground-based BMD or ASAT lasers.

LASER DESIGNATOR: A device used to illuminate targets with a LASER beam so that they can be detected by the seekers of LASER-GUIDED BOMBS, missiles, or other laser-sensing devices. Laser designators can be hand-held, as well as mounted on ships, aircraft, and ground vehicles. They frequently incorporate a range finder, which calculates the distance from designator to target by the delay between the transmission of the beam and the reception of laser radiation reflected back from the target.

LASER-GUIDED BOMB (LGB): A glide bomb equipped with a SEMI-ACTIVE LASER HOMING (SALH) device, which guides the bomb towards laser radiation reflected off a target when it is illuminated by a LASER DESIGNATOR. Such bombs were first developed by the U.S. during the Vietnam War under the PAVEWAY program. Since then, France, Israel, and the Soviet Union have introduced their own laser-guided bombs, all essentially similar to Paveway.

A typical LGB is a conversion of a standard free-fall bomb. A modular kit containing a laser seeker head, guidance electronics, and canard control surfaces is attached to the nose, while enlarged fins (sometimes of the pop-out variety for compact stowage), are attached to the tail, to provide lift and extend glide range.

The seeker head contains a silicon detector array divided into quadrants; laser radiation received by the detectors is converted into electrical impulses from which the guidance computer generates steering commands for the canard control surfaces. The computer directs the bomb in such a way as to equalize the signal received by all four detector quadrants, which implies that the weapon is headed straight towards the target.

In a typical LGB attack, the target is illuminated by a laser designator on the attacking aircraft, on a second aircraft, on a REMOTELY PILOTED VEHICLE (RPV), or on the ground. Once the target is illuminated, the weapon is released in its general direction. LGB ranges vary with release speed and height; for the latest types (with large, pop-out wings), a range of 2–5 mi. for a low-level attack is typical. The target must be illuminated throughout the bomb's flight; for this reason, illumination from a platform other than the releasing aircraft is preferred, so that the bombing aircraft can take evasive action immediately after release.

LATERAL SHIFT: A method of concentrating defending forces to meet enemy attacks by moving units from the flanks to the threatened sector, rather than by moving up a mobile RESERVE in the rear. Sometimes known as the "Leavenworth Concept," lateral shift was a major element of the U.S. Army's "Active Defense" DOCTRINE between 1976 and 1982. A key assumption underlying the concept was the belief that the Soviet army would employ massive, preplanned, and rigid ("steamroller") attacks which could be detected well in advance because of the massing of formations, and which could not maneuver once launched. In reality, Soviet doctrine had become much more flexible by then, calling for initial broad-front attacks to be followed up by large reserves ready to be sent forward to reinforce units which had achieved the deepest advances. Thus, U.S. forces shifting laterally to stop a Soviet attack elsewhere would be opening holes in the frontage for other Soviet units, and the result would be the envelopment and eventual destruction of the defending force. Lateral shift was replaced in 1982 by the AIR-LAND BATTLE doctrine, which relies more on maneuver, counterattack, and the retention of operational reserves. See also DEFENSE-IN-DEPTH.

LAV: Light Armored Vehicle, a family of 8×8 wheeled ARMORED FIGHTING VEHICLES developed for the U.S. Marine Corps. The LAV program was initiated in the late 1970s to meet an army and marine corps requirement for a vehicle that could be transported by air, and which would provide both fire support and light armor protection. Prototypes were submitted by the British Alvis Company, Cadillac Gage, and General Motors of Canada. The winning entry was the GM-Canada vehicle, an 8×8 version of the Swiss MOWAG PIRANHA, built under license in Canada. The marines were originally to buy 289 vehicles and the army 680. The army withdrew from the program in 1985, and the marines then ordered a total of 758 vehicles, all of which were delivered by 1988.

The marines operate six different LAV variants: the LAV-25 ARMORED PERSONNEL CARRIER (APC); a transport vehicle; a mortar carrier (with an 81-mm. MORTAR); an anti-tank vehicle with an erectable TOW launcher; an armored recovery and maintenance vehicle; and an armored command vehicle (ACV). All have the same basic chassis and drive train.

The LAV-25's all-welded steel hull is divided into a driver/engine compartment up front, a fighting compartment in the middle, and a troop compartment in the rear. Maximum ARMOR thickness is 10 mm., which provides protection against shell splinters and small arms. The basic vehicle crew consists of a driver, a commander, and a gunner. The driver sits on the left side with the engine mounted to his right, while the commander and gunner sit in a powered turret armed with a Hughes M249 Bushmaster 25-mm. CHAIN GUN and a 7.62-mm. coaxial MACHINE GUN. The commander's cupola has six vision blocks, and both the commander and the gunner have THERMAL IMAGING gun sights for night combat. The Bushmaster gun, fully stabilized to permit firing on the move, is capable of penetrating armored personnel carriers and the flank and rear armor of most main battle tanks. The vehicle carries 210 rounds of 25-mm. and 420 rounds of 7.62-mm. ammunition. A 7.62-mm. or 12.7-mm. machine gun can also be pintle-mounted by the commander's hatch for ANTI-AIRCRAFT defense. Eight smoke grenade launchers are attached to the sides of the turret. The troop compartment can carry up to six fully equipped soldiers, seated back-to-back. Access is provided through two swing-out doors in the rear and two roof hatches. Two firing ports are provided on each side, plus one in each rear door. Surprisingly, there is no NBC protection. Amphibious with quick preparation, the LAV is propelled in water by two rear-mounted screws at 6 mph (10 kph).

The LAV transport variant has a two-man crew, no turret, a higher roof, and a cargo crane; the mortar carrier has the same hull as the LAV-25 minus the turret, and the mortar fires through an enlarged roof hatch; the anti-tank version has the two-round Emerson erectable launcher of the M901 IMPROVED TWO VEHICLE (ITV) in place of the turret; the maintenance/recovery variant has a five-man crew and an A-frame derrick; and the armored command vehicle (ACV), similar to the transport variant, carries a variety of COMMUNICATIONS gear. AIR DEFENSE and ELECTRONIC WARFARE (EW) variants are also under development. The U.S. Air Force plans to procure a large number of LAVs for base defense and EXPLOSIVE ORDNANCE DISPOSAL. In June 1988, the marines also began developing an "assault gun" variant armed with a SOFT RECOIL 105-mm. gun; this could enter production in 1990.

The LAV provides U.S. INTERVENTION forces with much-needed mobility and fire support; it has almost the same capability as the M2/3 BRADLEY Fighting Vehicle at a lower cost, but none of its variants have tanklike armament as mounted in other wheeled armored cars.

Specifications **Length:** 20.88 ft. (6.36 m) **Width:** 8.2 ft. (2.5 m) **Height:** 8.83 ft. (2.69 m.). **Weight, combat:** 14 tons. **Powerplant:** 275-hp. GM-Detroit 6V-53T 6-cylinder diesel. **Hp./wt. ratio:** 21.3 hp./ton. **Fuel:** 75 gal. (336 lit.). **Speed, road:** 60 mph (98 kph). **Range, max.:** 400 mi. (668 km.).

LAW: Light Anti-tank Weapon (sometimes called Light Assault Weapon), a term applied generally to shoulder-fired BAZOOKA-type (unguided) rocket launchers, and specifically to the U.S. M72 66-mm. rocket fired from a disposable carrier/launcher tube. To use the weapon, the operator removes protective caps from each end of the tube, pulls a safety pin, and extends a telescoping section to bring the tube to its full length. The weapon is aimed with adjustable peep sights at each end of the tube, and the rocket is fired by depressing a lever. The M72's small shaped-charge (HEAT) warhead is supposed to penetrate 260 mm. of armor, but that is an optimistic figure.

The Soviet RPG-18, a reverse-engineered copy, now supplements the much older, clumsier, but more powerful RPG-7 rocket launcher. The SPACED ARMOR, REACTIVE ARMOR, and CHOBHAM ARMOR of the latest tanks seriously degrades the capabilities of small shaped-charge weapons such as the M72, and recent attempts to produce man-portable weapons with a high probability of kill (P_K) against modern MAIN BATTLE TANKS have not been very successful. So far, all U.S. Army attempts to develop a successor to the M72 have failed, and the Swedish AT-4 84 mm rocket launcher is being purchased as an interim weapon. The M72 remains useful against field fortifications and light armored vehicles.

Specifications (M72A1) **Length:** (launcher) 2.14 ft. (731 mm.) folded/2.9 ft. (884 mm.) ready to fire. **Diameter:** 66 mm. **Weight (rocket)** 2.2 lb. (1 kg.); (system) 4.69 lb. (2.13 kg.). **Velocity, max.:** 144 m./sec. **Range, max.:** 200 m. (100 m. eff.).

LAWS OF WAR: International treaties, conventions, traditions, and tacit agreements which are supposed to govern the military treatment of civilians and other noncombatants, private property, neutrals, the sick and wounded, and PRISONERS OF WAR. In addition, the Laws of War would limit or prohibit outright the use of weapons regarded as excessively cruel or barbaric, notably that 19th-century horror "dum-dum" ammunition, as well as chemical weapons and incendiary devices.

Restraints on the ferocity of war date back in Western tradition at least to the Middle Ages, but they became more detailed and authoritative in the 18th century in reaction to the excesses of the religious wars of the previous century. An international code of conduct did in fact develop, and was generally respected by most European armies; it included safeguards for noncombatants and the wounded, cartels for the exchange of prisoners, and customs such as the "honors of war" granted to surrendering forces.

The emergence of more destructive weapons during the 19th century, and economic industrialization of warfare, led to calls for more rules. The Geneva Convention of 1864 established the Red Cross movement, and gave Red Cross personnel the status of protected neutrals with the right to provide care for sick and wounded combatants and civilians. With broader scope, if not effect, the 1899 and 1906 conferences at The Hague established the modern canon of the Laws of War. These were supplemented by the Geneva Conventions of 1925, 1949, and 1977, which covered the treatment of civilians, neutrals, and prisoners of war. The first Geneva Convention of 1925 also outlawed CHEMICAL WARFARE, a prohibition reiterated by the Chemical and Toxin Weapons Convention of 1977.

The obvious defect of the Laws of War is that they can be enforced only by the belligerent parties, the only other restraint being domestic public opinion—hardly manifest in totalitarian states, and distorted by emotions and propaganda even in democracies. The fear of retaliation during a conflict, and the fear of retribution in the event of defeat afterwards, do appear to have a moderating effect upon international behavior, and they perhaps explain the avoidance of chemical warfare during World War II. But the systematic atrocities committed against civilians and prisoners during and after that conflict by signatories to various international agreements (e.g., Nazi Germany, the Soviet Union, and North Vietnam) indicate the general ineffectiveness of legal restraints. When victory is at stake, leaders are generally willing to do whatever they think will ensure victory; as Gen. William Tecumseh Sherman put it, "War is cruelty, you cannot refine it." The characteristic of World War II, however, was the prevalence of atrocities that were irrelevant to victory, or even counterproductive; that experience undermines hope in rational legal constructs, derived from considerations of ultimate self-interest. See also ARMS CONTROL; ARMS LIMITATION.

LCAC: Landing Craft, Air Cushion, U.S. Navy HOVERCRAFT developed for amphibious assault operations. The LCAC, built by Bell Aerospace Textron, was derived from an experimental hovercraft (JEFF-B) delivered to the navy in 1977. The first production LCACs were delivered in 1984 and the navy currently plans to acquire at least 90 by 1993, with Lockheed Corporation as a second-source producer.

With an all-around, inflatable rubber skirt designed to contain the air cushion, the LCAC is fully amphibious and can clear vertical obstacles up to 4 ft. (1.22 m.) high. Ramp doors at both ends allow roll-on/roll-off loading and unloading. Lift and propulsion are provided by four gas turbines. Two drive four fans to generate the air cushion, while the remaining two drive ducted, variable-pitch propellers in the stern. Rudders behind the propeller ducts provide directional control. LCACs are designed to carry a normal load of 60 tons: i.e., 1 M60 PATTON or M1 ABRAMS main battle tank, or 5 Light Armored Vehicles (LAVs), or 2 M198 155-mm. towed HOWITZERS. The cargo deck has an area of 1809 square ft., and can carry an emergency overload of up to 75 tons. The LCAC is a very costly vehicle on a per-ton-delivered basis. On the other hand, it is much faster than normal landing craft. It can be carried by all of the larger U.S. amphibious assault ships, including the WASP (LHD-1), 3 each; TARAWA (LHA-1), 1 each; LSD-41, 2 or 3 each; LSD-36, 4 each; LSD-28, 3 each; and LPD, 2 each. A total of 83 LCACs are required to land a division-sized MARINE EXPEDITIONARY FORCE (MEF), while 35 to 42 can land a MARINE EXPEDITIONARY BRIGADE (MEB).

Hovercraft such as LCAC could revolutionize amphibious assault by allowing the rapid delivery of troops and equipment to the beach from beyond the radar horizon, thus increasing the probability of achieving surprise. Moreover, the LCAC's air cushion renders it virtually immune to most types of naval and land MINES, while enabling it to cross surf to discharge troops and cargo on dry land. Because hovercraft are not restricted by tides, beach gradients, or surf conditions, the LCAC can exploit four times as many landing beaches worldwide as conventional LANDING CRAFT.

Specifications **Length:** 87.9 ft. (26.8 m.). **Beam:** 47 ft. (14.33 m.). **Weight:** 88 tons empty/ 200 tons full load. **Powerplant:** 4 Avco TF40B gas turbines, 12,444 shp. **Fuel:** 6.2 tons. **Speed:** 50 kt. empty/40 kt. w/max. payload. **Range:** 200 n.mi. at 40 kt. **Crew:** 5. **Sensors:** 1 navigational radar.

LCM: Landing Craft, Mechanized, a beachable, flat-bottomed LANDING CRAFT capable of carrying troops, trucks, and armored vehicles. Most are 60–80 ft. (18.3–24.4 m.) long, have a draft of 4–6 ft. (1.22–1.83 m.), and displace between 60 and 120 tons. They are generally open boats, with square ramp-door bows. Top speed is typically 10–12 kt., with ranges under 200 mi. They cannot undertake long ocean voyages, and hence are usually carried to the operational area inside the floodable docks of larger amphibious assault ships (LHAS, LHDS, LSDS, etc.).

LCPL: Landing Craft, Personnel, Light, a U.S.-made, small, flat-bottomed open boat featuring a square bow with a ramp door for the discharge of troops onto a beach. They are some 36 ft. (10.97 m.) long, with a beam of 13 ft. (3.96 m.) and draft of 3.5 ft. (1.07 m.), weighing 13 tons at full load. They have a capacity of 17 men and are generally used as command transports, for the control of other LANDING CRAFT, inshore patrol, or the delivery of SPECIAL OPERATIONS forces. They are usually carried as deck cargo, or hanging from the davits of larger amphibious-assault ships.

LCT: Landing Craft, Tank, a flat-bottomed, beachable LANDING CRAFT capable of carrying one or two MAIN BATTLE TANKS, or an equivalent cargo (60–120 tons). LCTs are typically open boats measuring 130–150 ft. (39.63–45.73 m.) in length, with a beam of about 30 ft. (9.15 m.), draft of 6 ft. (1.83 m.), and a displacement of 200–400 tons. They have square ramp-door bows and a small superstructure (often offset to one side). Most have top speeds of 10–11 kt., and are capable of short open-ocean voyages. For longer voyages, they are usually carried inside the floodable docks of larger amphibious assault ships (LHAS, LHDS, or LSDS). The U.S. Navy has 37 such vessels, designated as LCUs (Landing Craft, Utility). Similar craft are in service with many navies, especially those bordering on narrow seas where larger vessels are not required.

LCU: Landing Craft, Utility, a term used by the U.S. and British navies to describe LCTS (Landing Craft, Tank).

LCVP: Landing Craft, Vehicles and Personnel, a ubiquitous type of light, flat-bottomed LANDING CRAFT capable of carrying jeeps, light trucks, or up to 40 fully equipped troops. LCVPs were the principal type of landing craft used in World War II and have changed very little since then. They are open boats some 36 ft. (10.97 m.) long, with a beam of 10–11 ft. (3.05–3.35 m.) and draft of 3–4 ft. (0.91–1.22 m.), weighing up to 14 tons fully loaded. They have relatively short ranges and are usually carried on davits aboard larger amphibious assault ships.

LD-50: Lethal Dose–50 Percent, the concentration of a CHEMICAL WARFARE agent, measured in milligrams per minute per cubic m. (mg./min./m.^{3}) required to kill half of all unprotected personnel exposed. This is a general index of the relative lethality of chemical agents: the more potent the agent, the lower the LD-50. The most lethal known chemical weapon, the nerve agent VX, has an LD-50 of 5, as compared, e.g., to the blood agent hydrogen cyanide (AC) which has an LD-50 of 2000.

LDGP: Low-Drag, General-Purpose (bombs), the most common type of air-to-ground munition in service today. A general-purpose (GP) bomb is a medium-case weapon in which the explosive filler amounts to some 40–50 percent of total weight, as opposed to 60–75 percent in light-case "demolition" bombs or as little as 20 percent in heavy-case armor-piercing bombs. Relying on a combination of blast and fragmentation, GP bombs are effective against most soft and "semi-hardened" targets.

As opposed to the short, blunt-nosed GP bombs of World War II, modern LDGPs have long, slender profiles to reduce drag at high speeds. Developed by the U.S. Navy in the 1950s, the much-used (and copied) Mk.80-series includes the 250-lb. (113-kg.) Mk.81, the 500-lb. (227-kg.) Mk.82, the 1000-lb. (454-kg.) Mk.83, and the 2000-lb. (909-kg.) Mk.84. All have small cruciform tail fins, but can be fitted with SNAKEYE or "ballute" air brakes for low-level delivery. The Mk.82 is 90 in. (2.28 mm.) long and 10.6 in. (269 mm.) in diameter, with an actual weight of 531 lb. (241 kg.), of which 192 lb. (87.3 kg.) is Minol, Tritinol, or H6 explosives. The Mk.84 is an enlarged version, 154 in. (3.91 m.) long and 18 in. (457 mm.) in diameter, with an actual weight of 1972 lb. (896 kg.), including 945 lb. (430 kg.) of explosives.

Other types of LDGP bombs in service include the French SAMP-series (50–1000 kg.), the Soviet FAB-series (100, 250, 500, 750, and 1000 kg.), and the British Mk.13 1000-lb. (454-kg.) bomb.

LEAHY: A class of nine U.S. guided-missile CRUISERS (CGs) commissioned between 1962 and 1964. Designed as ANTI-AIR WARFARE (AAW) escorts for CARRIER BATTLE GROUPS, the Leahys have some ANTI-SUBMARINE WARFARE (ASW) and ANTI-SURFACE WARFARE (ASUW) capabilities as well. The Leahys are the smallest vessels in the U.S. Navy designated as cruisers, smaller in fact, than SPRUANCE-class destroyers.

As designed, the Leahys had an all-missile armament, but a secondary gun battery was added while they were still under construction, because of well-founded concerns about the general reliability of missiles and their particular inability to cope with close-in attacks. Their main armament consists of twin Mk.10 missile launchers fore and aft, each with 40 missiles. Originally armed with TERRIER, the ships have been modified to launch RIM-67B STANDARD ER-2 surface-to-air missiles (SAMs). For ASW they have an 8-round ASROC pepperbox launcher forward and 2 sets of Mk.32 triple tubes for MK.46 lightweight ASW homing torpedoes. As completed, the Leahys had 2 twin 3-in. 50-caliber ANTI-AIRCRAFT guns amidships, now replaced replaced by two quadruple launch canisters for HARPOON anti-ship missiles. In the early 1980s, two PHALANX 20-mm. radar-controlled guns were fitted as defense against surface-skimming missiles. The Leahys have no helicopter facilities except for a VERTREP landing pad. The nuclear-powered cruiser BAINBRIDGE has a Leahy weapon and sensor suite in an enlarged and slightly modified hull. See also CRUISERS, UNITED STATES.

Specifications Length: 533 ft. (162.5 m.). **Beam:** 54.8 ft. (16.7 m.). **Draft:** 25 ft. (7.6 m.). **Displacement:** 6670 tons standard/8203 tons full load. **Powerplant:** twin-shaft steam: 4 oil-fired boilers, 2 sets of geared turbines, 85,000 shp. **Fuel:** 1800 tons. **Speed:** 32 kt. **Range:** 2500 n.mi. at 30 kt./8000 n.mi. at 14 kt. **Crew:** 423. **Sensors:** 1 SPS-10F or SPS-67 surface-search radar, 1 SPS-48A 3-dimensional air-search radar, 1 SPS-49(V)3 air-search radar, 4 SPG-55B missile guidance radars, 1 bow-mounted SQS-23 low-frequency active/passive array; *(Yarnell)* 1 SQQ-23 PAIR sonar. **Fire controls:** 1 Mk.14 weapon-direction system, 4 Mk.76 missile fire control systems, 1 Mk.114 ASW fire control system. **Electronic warfare equipment:** 1 SLQ-32(V)3 electronic countermeasures array, 2 Mk.36 SRBOC chaff launchers.

LEANDER: A large and commercially successful class of British-designed FRIGATES commissioned between 1963 and 1973, now in service with the Royal Navy and the navies of Australia, Chile, the Netherlands, and New Zealand. Improved versions of the Rothesay and Type 12 (Whitby) frigates of the 1950s, the Leanders were built in two basic types: standard hull and "Broad Beam," the latter introduced from 1968 to improve seakeeping and add volume; both have been subjected to modification programs since completion.

Both types have a raised forecastle (essential in

very rough seas) and a short, boxy superstructure concentrated amidships. With gyro-controlled fin stabilizers and twin rudders, which give them excellent maneuverability, the Leanders were optimized for North Atlantic operations and have proven seaworthy, reliable, and economical. Like other British warships, their greatest weaknesses are the scanty electronic warfare provisions for anti-missile defense, and relatively poor DAMAGE CONTROL arrangements.

The original weapon suite, on both standard and Broad Beam ships, consisted of a twin 4.5-in. (114-mm.) DUAL PURPOSE gun mount on the bow, a twin 40-mm. BOFORS GUN, 2 20-mm. ANTI-AIRCRAFT guns, and a 3-barrel LIMBO depth charge mortar in a well aft of the superstructure. There is also landing deck for a light ANTI-SUBMARINE WARFARE (ASW) helicopter on the fantail; the Leanders were the first British frigates designed from the start with a helicopter deck. Beginning with the Broad Beam types, the Bofors guns were replaced by 2 4-round SEA CAT short-range surface-to-air missile (SAM) launchers.

From 1971, major armament modifications have included:

1. "Ikara Leander" (Batch 1): A specialized ASW version with an Australian IKARA anti-submarine missile launcher in place of the 4.5-in. guns. These ships have both Sea Cat and a twin Bofors gun, but only one 20-mm. gun.

2. "Exocet Leander" (Batch 2A): An ANTI-SURFACE WARFARE (ASUW) enhancement, with 4 launch canisters for EXOCET MM38 anti-ship missiles in place of the 4.5-in. guns. These ships also have 2 sets triple tubes for lightweight ASW homing torpedoes in place of Limbo, and can operate larger LYNX ASW helicopters. Batch 2B ships are similar but have a twin Bofors mount, 1 20-mm. gun, and 3 Sea Cat launchers.

3. Broad Beam Leander (Batch 3A): In these ships, the 4.5-in. mount has been replaced by a 6-round SEA WOLF SAM launcher and 4 Exocet canisters. Other armament is limited to 2 triple torpedo tubes, 2 20-mm. guns, and a Lynx helicopter.

4. Broad Beam Leander (Batch 3B): These ships retain the 4.5-in. guns and are otherwise unchanged, except for the addition of 3 20-mm. guns.

Sensors vary. Unmodified standard and Broad Beam Leanders have 1 Type 978 navigational RADAR, 1 Type 993 surface-search radar, 1 Type 965 air-search radar, 1 Type 903 gun fire control radar, and (in Broad Beams only) 1 Type 904 Sea Cat missile-guidance radar. The SONARS include a Type 184M hull-mounted medium-frequency active/passive array, a Type 170B high-frequency active attack sonar, and a Type 162 high-frequency side-scanning classification array. Ikara Leanders have an Ikara control radar in place of the Type 903 set, and their sonar suite includes a Type 199 VARIABLE DEPTH SONAR (VDS). Exocet Batch 2A ships have a Type 2031 passive TOWED ARRAY SONAR in place of the Type 170 attack sonar, and Batch 2B ships have a second Type 904 Sea Cat guidance radar, with only the Type 184M and Type 162A sonars. Broad Beam Batch 3A ships with Sea Wolf have an entirely new radar suite with a Type 1006 navigational set, a Type 967/968 pulse doppler air search and tracking set, and one Type 910 Sea Wolf guidance set. The sonar suite has also been revised, and now consists of a Type 2016 hull-mounted medium-frequency array and a Type 2008 active sonar. Batch 3B ships retain the basic sensor suite.

The Royal Navy built 26 Leanders, including 10 Broad Beam types, of which 19 remain in service, including 2 Ikara (Batch 1); 4 Exocet (Batch 2A); 4 Exocet (Batch 2B); 4 Broad Beam Sea Wolf (Batch 3A); and 5 Broad Beam (Type 3B). These will probably remain in service until the year 2000.

Australia built 4 standard Leanders (as the RIVER class) between 1961 and 1971. These have modified sensors, retain the 4.5-in. guns forward, and have an Ikara launcher aft in place of the helicopter pad. Chile operates 2 modified Broad Beam types built in Britain in 1973–74, which have both Exocet and the 4.5-in. guns, while retaining the helicopter pad aft. India built 6 Broad Beam types locally between 1972 and 1981, with Dutch sensors and a slightly altered weapon suite. In addition, India is building 6 frigates derived from the Leander design (as the Godavari class). These ships, with a displacement of 3600 tons at full load, have Soviet SS-N-2c STYX anti-ship missiles, an SA-N-4 GECKO SAM system, and Soviet 57-mm. and 30-mm. guns. The Netherlands built 6 ships to a modified standard Leander design (as the Van Speijk class), armed with an OTO-MELARA 76.2-mm. dual-purpose gun forward, 2 HARPOON anti-ship missiles, 2 Sea Cat systems, 2 triple torpedo tube mounts, and a Lynx helicopter. The sensors are entirely Dutch-made, and include the widely used LW-03 3-dimensional air-search radar. New Zealand has 2 Broad beam types essentially identical to Batch 3B, 1 Standard Leander, and 1 Ikara Leander. See also FRIGATES, BRITAIN.

Specifications **Length:** 372 ft. (113.4 m.). **Beam:** (std.) 41 ft. (12.5 m.); (Brd. Beam) 43 ft. (13.1 m.). **Draft:** 4.8 ft. (1.5 m.). **Displacement:** (std.) 2450 tons standard/2860 tons full load; (Brd. Beam) 2500 tons standard/2962 tons full load. **Powerplant:** twin-shaft steam: 2 oil-fired boilers, 2 sets of geared turbines, 30,000 shp. **Fuel:** (std.) 460 tons; (Brd. Beam) 500 tons. **Speed:** (std.) 30 kt; (Brd. Beam) 28 kt. **Range:** 4500 n.mi. at 12 kt. **Crew:** 260–270. **Sensors:** see text. **Electronic warfare equipment:** 1 UA-8/9 radar warning receiver, 2 Knebworth Corvus chaff launchers.

LEOPARD 1: West German MAIN BATTLE TANK (MBT) in service with the armies of West Germany, Australia, Canada, Denmark, Greece, Italy, the Netherlands, Norway, and Turkey. Upon the reestablishment of the West German army (*Bundeswehr*), its newly raised PANZER and PANZERGRENADIER brigades were equipped with U.S.-built M48 PATTON tanks, which the Germans judged as too heavy. In 1956, the *Bundeswehr* issued a requirement for an indigenously designed MBT with a combat weight of 30 tons, a height of not more than 7.2 ft. (2.2 m.), a road speed of 40 mph (67 kph), a multi-fuel engine, and a 105-mm. gun.

In 1957, France and West Germany signed an agreement to develop a joint-venture tank with the intention of producing a standard design that would be offered to all NATO armies. Four prototypes were delivered in 1961, but after competitive trials against the French AMX-30 version, West Germany decided unilaterally to place its own design in production as the Leopard, while the French proceeded with the somewhat lighter and less heavily armored AMX-30. Krauss Maffei of Munich, the prime contractor, delivered the first production vehicle in September 1965, and since then, the Leopard 1 has been purchased by nine other armies, and produced under license by OTO-Melara of Italy. The German production run of 4281 vehicles was completed in 1979, but the assembly line was reopened in 1981 to produce Leopards for Greece.

The Leopard has undergone numerous modifications, with many early production vehicles updated in turn. In the original Leopard 1, the all-welded steel hull is divided into a crew compartment (including the turret) forward and an engine compartment in the rear. Armor thickness varies between 70 mm. (2.75 in.) on the frontal glacis to 25–35 mm. (1.0–1.37 in.) on the sides and rear. German armored warfare doctrine at the time stressed mobility, and such relatively limited protection was deemed adequate. The glacis plate, however, is sloped 60° from the vertical, effectively doubling armor protection. Turret and side armor are also well sloped.

The Leopard 1 has a conventional MBT layout with a crew of four. The driver has 3 periscopes; the center scope can be replaced with an INFRARED (IR) or IMAGE INTENSIFICATION night vision device. The other crewmen sit in the all-cast turret, with the gunner on the right, the commander above and behind him, and the loader on the left. The commander has a single-piece hatch surrounded by 8 periscopes; the center scope can be replaced by an IR or image intensification scope. The gunner has 1 observation periscope, while the loader has 2. The commander is also provided with a TRP 2A zoom periscope, with magnification settings between 6x and 20x, which can be trained in azimuth and elevation, and also slaved to the main gun. This too can be replaced by an IR device. The gunner is equipped with a TEM 2A optical rangefinder which can be used in both coincidence and stereoscopic modes; until the advent of laser devices, this was the best rangefinder in service on any tank. It has magnification settings of 2x and 8x, is mechanically linked to the main gun, and has separate settings for different types of ammunition. A backup TZF 1A 8X stadiametric sight is mounted coaxially with the main gun.

Main armament is the British-designed L7A3 105-mm. rifled gun, also used on the M48 and M60 PATTON, late-model CENTURIONS, and the M1 ABRAMS. Highly accurate out to more than 2000 m., it fires APDS, APFSDS, HEAT, HESH, and smoke rounds. The Leopard 1 carries 55 rounds of 105-mm. ammunition, 42 in the hull and 13 in the turret. The original Leopard 1 did not have a gun stabilization system, hence it had to stop to fire with any accuracy. Secondary armament consists of a coaxially mounted Rheinmetall MG3 7.62-mm. machine gun and a pintle-mounted G3 by the commander's hatch. Four smoke grenade launchers are mounted on either side of the turret.

The Leopard 1 was one of the first tanks with a positive overpressure COLLECTIVE FILTRATION system for NBC protection mounted in the front of the hull. An XSW-30-U white light/IR searchlight is mounted over the main gun; used in conjunction

with the driver's and commander's IR periscopes, the IR mode has an effective range of 1200 m. The Leopard 1 has the best power-to-weight ratio of any tank of its generation. The engine and transmission are a single unit, which can be field-changed in 20 minutes. The vehicle can ford streams up to 13 ft. (3.96 m.) deep using a snorkel kit to provide air for the engine and crew.

From 1971, the West German army initiated a series of modification programs. In the first, all Leopard 1s were retrofitted with thermal sleeves for the main gun barrel to reduce barrel warp, thus improving accuracy; a gun stabilization system to permit firing on the move, rectifying the Leopard's greatest deficiency, aside from inadequate protection; improved tracks; and armored side skirts (for more protection against light ANTI-TANK weapons). The resulting vehicles were redesignated Leopard 1A1 (most have since been fitted with add-on armor for the turret and gun mantlet, and redesignated once again as Leopard 1A1A1).

A late production batch, designated Leopard 1A2, featured most of the improvements of the Leopard 1A1, plus a stronger turret, a new NBC filter, and passive night vision sights for the commander and driver. Another 110 tanks for the the the Bundeswehr, designated Leopard 1A3, have a new welded turret with spaced armor to resist ANTI-TANK GUIDED MISSILES (ATGMs). This is also the version ordered by the Greek and Turkish armies. The final production variant, the Leopard 1A4, of which 250 were built, is generally similar to the 1A3, but has an integrated FIRE CONTROL system with a ballistic computer.

Beginning in 1986, some 1300 Leopard 1s were retrofitted with a Krupp-Atlas EMES-18 fire control system and THERMAL IMAGING sights. Many Leopards 1s are also being fitted with additional applique armor and fast-acting automatic fire-suppression systems. The Leopard 1 seems to have accommodated all these changes easily: road speed, cross-country mobility, and range were only marginally reduced.

In West German Panzer divisions, the Leopard 1 is now being replaced by the all-new LEOPARD 2, with late-model Leopard 1s being given to Panzergrenadier divisions (previously equipped with the M48 Patton), while early Leopard 1s are being passed down to reserve and TERRITORIAL ARMY units. The Leopard 1 chassis has also served as the basis for a variety of specialized vehicles, including an armored recovery variant, an armored engineer variant, the *Biber* bridgelaying tank, a driver training tank, and the GEPARD anti-aircraft tank.

Specifications Length: 23.25 ft. (7.09 m.). Width: 10.66 ft. (3.25 m.). Height: 8.75 ft. (2.67 m.). Weight, combat: (1) 40 tons; (1A4) 42 tons. Powerplant: 830-hp. MTU MB838 M500 10-cylinder multi-fuel diesel. Hp./wt. ratio: 20.75 hp./ton. Fuel: 240 gal. (1076 lit.) Speed, road: 39 mph (65 kph). Range, max.: (roads) 400 mi. (668 km.); (cross-country) 270 mi.

LEOPARD 2: An advanced, German-designed MAIN BATTLE TANK (MBT) still in production, and now in service with the armies of West Germany, the Netherlands, and Switzerland. The Leopard 2 program started in 1970 following the collapse of the joint U.S.-German MBT-70 project for a successor to both the German LEOPARD 1 and the U.S. M60 Patton. Utilizing Leopard 1 and MBT-70 components, a new vehicle was developed reflecting German tank design priorities, and this became the Leopard 2. Sixteen prototypes were completed between 1972 and 1974 to test a variety of layouts. As a result, and also because of the influence of Israeli advice, by 1974 the German army had abandoned its predilection for mobility at the expense of armor, coming much closer to U.S. design preferences. In that year, the U.S. and Germany duly signed a Memorandum of Understanding (MOU) announcing the intention to standardize tank production on a single design, after competitive tests of the Leopard 2 and the M1 ABRAMS. For political and doctrinal reasons, however, this effort also failed, and each side produced its own vehicle independently (although the U.S. did eventually adopt the Leopard 2's main gun for the improved M1A1 version of the Abrams). In 1977, West Germany selected Kraus Maffei as prime contractor for the Leopard 2, placing an initial order for 2050 vehicles. The first was delivered in 1979, and production for the West German army was completed in 1987. All Leopard 2s are assigned to the tank battalions of active *Panzer* divisions.

In March 1979, the Dutch army placed an order for 445 Leopard 2s, which were delivered between 1982 and 1986 (with some 60 percent of all their components produced in the Netherlands). In August 1983, the Swiss army placed an order for 385 vehicles, of which 35 were built by Krauss Maffei and the rest being built under license in Switzerland. Like the M1 Abrams, the Leopard 2 represents a quantum improvement in vehicular performance, weapon effectiveness, and protection over previous Western MBTs. Both tanks are believed to be superior to the latest generation of Soviet tanks, the T-64, T-72, and T-80.

The slab-sided hull and turret are constructed of an advanced armor, believed to consist of a combination of SPACED ARMOR and CHOBHAM composite armor, giving protection equivalent to more than 600 mm. (24 in.) of homogeneous steel plate, and thus obviating the need for steeply sloped armor (a characteristic of all other modern tanks). The Leopard 2 retains the conventional MBT layout and a four-man crew. The hull is divided into three compartments: a driver's compartment forward, a fighting compartment (including the turret) in the middle, and an engine compartment in the rear. The driver sits in the front of the hull on the right, and is provided with three periscopes; the center scope can be replaced by a passive night vision device. Twenty-six rounds of main gun ammunition are stowed to the driver's left. The remaining crewmen sit inside the turret, with the gunner on the right, the commander above and behind him, and the loader on the left. The commander's position is surrounded by periscopes providing 360° observation. In addition, he is provided with a stabilized PERI-R17 panoramic periscope, which can be traversed through 360°, has magnification settings of 2x and 8x, and can be used by the commander to aim the main gun.

The gunner has a dual-magnification, stabilized EMES 15 gun sight with an integral LASER rangefinder (accurate out to 10,000 m.) and THERMAL IMAGING sight linked to a FIRE CONTROL computer. As a backup, the gunner also has an 8x stadiametric sighting telescope. The fire control computer combines range, ammunition type, vehicle tilt, relative target motion, and lateral wind data to arrive at a ballistic solution, giving a high first-round hit probability.

Main armament is a Rheinmetall 120-mm. high-velocity smoothbore gun, which now also arms the M1A1 Abrams. Two types of ammunition have been developed so far: Armor-Piercing Fin-Stabilized Discarding Sabot (APFSDS), and High Explosive Anti-Tank (HEAT). The APFSDS or "arrow" round has a total weight of 41.8 lb. (19 kg.), a penetrator weight of 16 lb. (7.27 kg.), and a muzzle velocity of more than 1600 m. (1750 yd.) per second. This round is believed capable of penetrating all known Soviet tank armor at ranges greater than 2000 m. (2185 yd.). The HEAT round weighs 37.4

lb. (17 kg.), with a 12.33-lb. (5.6-kg.) projectile. Effective out to 3000 m., it is believed capable of defeating most types of spaced and reactive armor. Both rounds employ partially combustible cartridge cases (with metal base stubs) to reduce clutter in the turret. The main gun is fully stabilized to permit firing on the move; elevation limits are +20° and −9°. A total of 42 rounds are stowed: 26 in the driver's compartment and 16 in the left side of the turret bustle, which has blow-out panels to vent ammunition explosions. Secondary armament consists of a coaxial MG3 7.62-mm machine gun, a pintle-mounted G3 by the loader's hatch, and eight smoke grenade launchers mounted on either side of the turret.

A positive overpressure COLLECTIVE FILTRATION system is provided for NBC protection. As in the Leopard 1, the engine and transmission are designed as a single unit which can be replaced in the field in under 30 minutes. The Leopard 2 can ford streams up to 13 ft. deep using a snorkel kit to provide air to the engine and crew. An armored recovery vehicle based on the Leopard 2 chassis is now in production (recovery vehicles based on earlier tanks cannot cope with its 60-ton weight).

Specifications **Length:** 25.33 ft. (7.72 m.). **Width:** 12.16 ft. (3.7 m.). **Height:** 9.25 ft. (2.82 m.). **Weight, combat:** 60 tons. **Powerplant:** 1500-hp. MTU MB873 KA501 12-cylinder multi-fuel diesel. **Hp./wt. ratio:** 27.72 hp./ton. **Fuel:** 325 gal. (1457 lit.). **Speed, road:** 45 mph (75 kph). **Range, max.:** 330 mi. (551 km.).

LF: Low Frequency, radio signals in the 30–300 KHz frequency band, used mainly for reliable long-range (but low-quality) communications and radio-navigation aids such as LORAN. LF waves tend to follow the curvature of the earth, and can be received by submarines at shallow submergence.

LGB: See LASER-GUIDED BOMB.

LHA: Assault Landing Ship, Helicopter, a very large AMPHIBIOUS ASSAULT ship, intended to combine the capabilities of several earlier types of amphibious ship in a single hull. LHAs have an AIRCRAFT CARRIER-type flight deck for transport and attack HELICOPTERS; a large, floodable well deck for LANDING CRAFT, amphibious assault vehicles, and HOVERCRAFT; vehicle storage decks; troop accommodations; and extensive COMMAND AND CONTROL facilities. The U.S. Navy's TARAWA class (LHA-1) is a maximal example. Because of their

size and capabilities, LHAs are natural flagships for amphibious assault groups. The can also perform as helicopter or V/STOL aircraft carriers for CONVOY protection and ANTI-SUBMARINE WARFARE (ASW).

LHD: Helicopter Landing Ship, Dock, a type of AMPHIBIOUS ASSAULT vessel primarily designed to operate HELICOPTERS, but also incorporating a floodable well deck for LANDING CRAFT and HOVERCRAFT. The U.S. Navy's WASP class (LHD-1), the first ships of this type, differ from the earlier TARAWA class (LHA-1) in having larger aviation and docking facilities at the expense of vehicle storage space. Like LHAs, they can also perform as V/STOL aircraft carriers for CONVOY protection and ANTI-SUBMARINE WARFARE (ASW).

LIGHT INFANTRY: A form of INFANTRY that is supposed to rely on better training to offset a lighter establishment of weapons and vehicles, so as to have more strategic air MOBILITY and also the tactical versatility to fight in close terrain such as mountains, forests, jungles, or urban areas. Traditionally, light infantry has also been used for SCOUTING, and SCREENING in close terrain, roles now often performed less well by ARMORED CAVALRY.

Modern light infantry falls into two broad categories. The first, exemplified by most airborne forces, is trained and organized for the same linear defense and FIRE-AND-MOVEMENT tactics as normal infantry, but lacks its heavy ARTILLERY and other fire support. Because fire-and-movement tactics depend on supporting arms, and are not generally suited for combat in close terrain, this type of light infantry has little operational utility against competent and more heavily armed opposition; its only advantage is air portability. Such forces, however, have been used successfully by Western armies for interventions against poorly trained Third World forces.

The second form of light infantry is trained and organized to employ nonlinear tactics, stressing the use of terrain for concealment and cover from enemy fire on the defense, and infiltration tactics on the offense. The virtues of this form of quasi-elite light infantry are being rediscovered in many quarters. New formations are being raised in several countries (e.g., the U.S. "Light" divisions) for operations in all terrains and situations unsuitable for armored forces.

AIRBORNE, AIR-MOBILE, and air-landed forces are generally equipped as light infantry of one type or the other, with the exception of the Soviet Airborne Troops, which are organized as light MECHANIZED forces.

LIGHT MACHINE GUN (LMG): A fully automatic, rifle-caliber (5.56-mm. to 7.62-mm.) infantry weapon, usually weighing under 30 lb. (13.63 kg.) loaded, which is used as a support weapon at the SQUAD and PLATOON level. LMGs were first developed during the First World War, to provide portable automatic fire for advancing infantry units. To keep their weight at a manageable level, LMGs have lightweight, air-cooled barrels which must be changed frequently to allow sustained, high-rate fire. Classic LMGs, such as the British BREN, Soviet RPK, or American BAR (Browning Automatic Rifle), are fed from box or drum magazines holding between 20 and 75 rounds. The LMG was largely superseded after World War II by GENERAL PURPOSE MACHINE GUNS (GPMGs), first developed by the Germans in the 1930s and exemplified by the MG42 (still in service today as the MG3); and by ASSAULT RIFLES, which gave automatic (but not sustained) fire to each soldier. GPMGs have slightly heavier, quick-change barrels and belt fed ammunition, but the LMG is now reemerging as the SAW, or Squad Automatic Weapon, a heavy-barreled assault rifle optimized for more sustained automatic fire. GPMGs are being retained at the platoon level and above as full-power (7.62-mm.) weapons, while SAWs fire reduced-caliber (4.85 to 5.56-mm.) ammunition. See also AK.

LIGHT TANK: A tracked, turreted vehicle with the same general configuration and appearance as a MAIN BATTLE TANK (MBT), but with much less armor protection and a combat weight of around 20 tons (as compared to 40–60 tons for an MBT). Light tanks evolved from the CAVALRY tanks of the 1920s, which were intended as direct replacements for the horse cavalry in their RECONNAISSANCE, SCOUTING, and PURSUIT duties. Lightly armed and thinly armored, light tanks relied on speed for protection—with diminishing success as gun fire controls improved to allow the engagement of moving targets. During World War II, successive light tanks increased in size and capability, until, by 1945, the British Comet and U.S. M41 Walker Bulldog were merely (somewhat) scaled-down battle tanks. In the German and Soviet armies, by contrast, more intense armored warfare led to the early demise of the type.

Light tanks were used as ersatz MBTs by poorer armies after 1945, but for first-class armies the emergence of the main battle tank as a type combining the mobility of light tanks with the armor and firepower of the heavy tank reduced the useful scope of the light tank to reconnaissance, AMPHIBIOUS ASSAULT, and AIRBORNE/AIR-MOBILE roles. Typical examples of postwar (1950s) light tanks are the Soviet PT-76 amphibious tank, the troubled U.S. M551 SHERIDAN airborne assault vehicle, the French AMX-13 (used mainly as a heavily armed but lightly armored TANK DESTROYER), and the British SCORPION (officially described as a "tracked armored reconnaissance vehicle"). Until recently, light tanks were armed either with high-velocity guns up to 76.2 mm. in caliber, or with medium-velocity guns up to 90 mm. firing shaped-charge HEAT rounds—neither of which could cope with MBT armor in most cases. The constraint was recoil force: high-velocity guns larger than 100 mm. required a chassis of more than 30 tons. But this constraint has now been lifted by the development of SOFT RECOIL guns, which rely on a combination of hydraulic recoil mechanisms and muzzle brakes. Thus it is now possible to mount tanklike 105-mm. high-velocity guns on chassis weighing as little as 20 tons, thereby giving light tanks (and heavy ARMORED CARS) the firepower to engage and destroy MBTs, at least from behind cover to offset the protection imbalance. Examples of this new breed of light tank are FMC's Close Combat Vehicle (Light), weighing 18 tons, and Cadillac Gage's Stingray, weighing 17 tons, both armed with a high-velocity 105-mm. gun; they have been offered to the U.S. Army. But in its former reconnaissance, fire support, and tank destroyer roles, the light tank must now compete with its wheeled equivalent, the gun-armed heavy armored car with the same firepower and lower costs.

LIMBO: British-designed, shipboard DEPTH CHARGE mortar, in service with the Royal Navy and the navies of Australia, Canada, Indonesia, India, Iran, Malaysia, and Thailand. Developed by the Admiralty Underwater Weapons Establishment in the early 1950s as a successor to the World War II–vintage Squid, Limbo consists of three 12-in. (304-mm.) mortars mounted on a pitch-and-roll stabilized platform. The three barrels can be trained individually to fire a 3-dimensional pattern of depth charges with the size and shape of the pattern varied according to the depth, size, and speed of the target. Each depth charge weighs 375 lb. (170.5 kg.), and can be fitted with hydrostatic (depth), contact, delay, or magnetic-proximity fuzes.

Limbo has a maximum range of 900 m. (984 yd.) and a maximum effective depth of 375 m. (1230 ft.). In operation, the ship's SONAR would provide target range, depth, azimuth, and course. These data are fed to an electrical predictor which computes the barrel elevation and lateral deflection needed to drop a pattern around the target. An automatic loader and a reload magazine are located alongside the weapon. Although considered obsolescent, Limbo remains useful in shallow waters, in which more advanced weapons such as acoustical homing TORPEDOES are relatively ineffective. See also ANTI-SUBMARINE WARFARE.

LIMITED STRATEGIC WAR: A conceptualized war in which nuclear strikes would be aimed at the enemy's homeland, but with deliberate restraint as to the number and power of the warheads, and, above all, in regard to the nature of the targets. Such limits are envisaged as part of a bargaining process, under the assumption that communications could continue amidst the detonations if military targets and not cities are attacked. See also ESCALATION.

LIMITED WAR: A military conflict in which at least one party is exercising restraint in the weapons used, the targets attacked, or the areas involved. The objectives of limited war are not necessarily limited, since the conflict may remain at a particular THRESHOLD only because one party cannot, while the other need not, escalate. In particular, what is "limited" for a great power may be "total" for a small power. In the common usage of the 1950s and 1960s, a limited war was one in which a party with nuclear capability (i.e., the U.S.) refrained from using it; e.g., the Korean and Vietnam wars. Clausewitz held that all wars are limited in some way, with their conduct constrained by self-interest, customary rules, habits, and professional preferences.

LINES OF COMMUNICATION (LOCS): Physical routes that connect forces in the field (or at sea) with their supply depots, and, ultimately, with the initial sources of supplies and reinforcements in their home country. Because modern ARMORED and MECHANIZED divisions require up to 6000 tons of supplies daily (mainly fuel and ammunition), the severing or disruption of their lines of communication can result in their rapid neutralization by LOGISTIC starvation (unless local stockpiles are available). Thus, in ARMORED WARFARE the aim of operations is often to penetrate into the enemy's rear, in order to cut, or at least threaten, lines of communication so as to induce withdrawal. Ships

and aircraft are likewise dependent on LOCs and bases, not only for fuel and ordnance, but also for critical electronic and mechanical replacement parts.

The purpose of INTERDICTION campaigns by air or naval forces is to disrupt lines of communication, by bombardment or blockade. Forces such as GUERRILLAS, LIGHT INFANTRY, and SPECIAL OPERATIONS units, which do not engage in continuous attritional combat, have correspondingly reduced logistical requirements; hence they can be independent of lines of communication for sustained periods, and cannot be defeated by interdiction in most cases.

LLTV: Low-Light Television (sometimes called Low-Light-Level Television, or LLLTV), a passive ELECTRO-OPTICAL night-vision device which operates in the visible portion of the electromagnetic spectrum. LLTV consists of a standard video camera linked to a signal processor which amplifies ambient light by several orders of magnitude (as much as 40,000 times in some cases), thereby providing a visual image of its field of view without any form of artificial illumination; that image can be displayed on a cathode ray tube (CRT) or a normal video monitor. Because it operates in the visual spectrum, LLTV cannot penetrate foliage, camouflage, smoke, fog, or other barriers to normal vision. LLTV equip ships, aircraft, vehicles, or crew-served weapons with a power supply. Smaller, man-portable IMAGE INTENSIFIERS operate on different principles. See also FLIR.

LMG: See LIGHT MACHINE GUN.

LOADS: Low-Altitude Defense System, a short-range, high-velocity, high-acceleration ANTI-BALLISTIC MISSILE projected by the U.S. Army Ballistic Missile Defense Command (now Strategic Defense Command) as a successor to the SPRINT interceptor missile developed in the 1960s (together with the long-range SPARTAN missile) for the SAFEGUARD anti-ballistic missile system.

Like Sprint, LOADS was intended to destroy incoming enemy warheads after they had reentered the atmosphere. The proposed design had the same configuration as Sprint: a sharply conical missile with a high ballistic coefficient, covered with an ablative heat shield to dissipate the intense heat generated by very high-speed flight through the atmosphere. Again like Sprint, LOADS was controlled by radar COMMAND GUIDANCE, and was armed with a low-yield, ENHANCED RADIATION nuclear warhead. The primary difference between the two missiles was the levels of technology applied. As compared to Sprint, LOADS was smaller, lighter, cheaper to produce, and more accurate. Moreover, it relied on small, mobile tracking and guidance RADARS to reduce the system's vulnerability to nuclear BLACKOUT and LADDER-DOWN tactics. Since the establishment of the STRATEGIC DEFENSE INITIATIVE in 1983, work on LOADS has been terminated, with much of its technology transferred to the nonnuclear HEDI system. See also BALLISTIC MISSILE DEFENSE.

LOCAL WAR: A Soviet concept which defines a war limited only geographically; i.e., a war confined to a given theater, but within which all weapons, including NUCLEAR WEAPONS, might be used. Soviet doctrine still anticipates the use of nuclear weapons on the battlefield at some point in a NATO–WARSAW PACT war, but during the late 1950s and early 1960s, Soviet doctrine actually prescribed the use of nuclear weapons in the earliest stages of combat. The later Soviet expectation of a prolonged, conventional (or at least nonnuclear) war amounted to a repudiation of the Local War concept and parallels NATO policy, which assumes a sustained, nonnuclear "holding" phase. In any case, by 1970 or so, Soviet theory was out of phase with Soviet artillery forces, organized at great cost organized for sustained nonnuclear bombardment, implying the abandonment of Local War. See also FORWARD DEFENSE.

LOGISTICS: All activities that pertain to the supply and upkeep of armed forces, including acquisition, storage, transport, distribution, and repair. In modern armed forces, a wide range of equipment and supplies are employed in widely varied mixes, and logistics thus involves a great deal of planning and ongoing calculation, as well as physical activity. The aim is to provide each echelon and unit with the optimum quantity of each item, in order to minimize both overstocking (which restricts mobility and causes waste) and shortages. In modern military establishments, the operations-research calculations required to achieve this goal are usually assisted by automated data processing systems, which are now employed down to the battalion, ship, and aircraft-squadron level. It is customary to divide logistics into strategic and tactical: the former covers the acquisition, stocking, and transport of supplies to a theater of military operations; the latter covers the distribution of supplies within a theater, and local repair activities. Because modern military forces consume materiel in vast amounts (an armored or mechanized division in combat can consume up to

6000 tons of supplies per day), the availability of supplies is a salient constraint on military operations. Both sides will strive to safeguard their LINES OF COMMUNICATIONS while attempting to cut or disrupt the enemy's. See also INTERDICTION.

LONG BEACH: A U.S. nuclear-powered guided-missile CRUISER (CGN-9) commissioned in 1961. The *Long Beach* was the world's first nuclear-powered surface warship, and the first cruiser of any kind built by the United States since World War II. Intended as an ANTI-AIR WARFARE (AAW) escort for the nuclear-powered aircraft carrier ENTERPRISE (CVN-65), the ship was initially planned as a 7800-ton DESTROYER, before being redesigned with additional missile launchers as a cruiser twice as large (a decision driven by the available reactor design). Unlike later guided-missile cruisers, which are essentially large destroyers, the *Long Beach* has a true cruiser hull, although the weight of the reactor precluded the provision of an armored belt.

As designed, the *Long Beach* had an all-missile armament consisting of 2 forward-mounted Mk.10 twin launchers for TERRIER surface-to-air missiles (SAMs), each with a 60-round magazine, and a Mk.12 twin launcher for TALOS long-range SAMs aft, with a 52-round magazine. An 8-round ASROC launcher with 12 reloads, mounted amidships, and 2 sets of Mk.32 triple tubes for lightweight homing TORPEDOES provide a secondary ANTI-SUBMARINE WARFARE (ASW) capability. In addition, space was reserved in the design for either Regulus II land attack CRUISE MISSILES or POLARIS ballistic missile tubes, but neither were ever installed.

After completion, the ship was fitted with 2 5-in. 38-caliber DUAL PURPOSE guns for close-in defense against aircraft and fast attack craft (FACs). In 1979, the obsolete Talos launcher was removed, along with its associated SPG-49B and SPW-2B guidance radars. During an extensive 1980 overhaul, 2 quadruple launch canisters for HARPOON anti-ship missiles were mounted over the Talos magazine, while 2 PHALANX 20-mm. radar-controlled guns were installed on the pedestals of the Talos guidance radars. Finally, in 1985, the Harpoon canisters were moved amidships, and 2 armored box launchers for 8 TOMAHAWK long-range cruise missiles were mounted in their place. Concurrently, the Terrier launchers were also modified for RIM-67 STANDARD ER SAMs.

The most distinctive characteristic of the *Long Beach* is the high, square forward superstructure, originally designed to hold fixed-array SPS-32 and

SPS-33 "billboard" RADARS (also fitted on the *Enterprise*); but these radars proved unreliable in service, and were replaced with more conventional air-search radars. To replace the weight removed, 44 mm. of aluminum armor was applied to the superstructure, and the radar foundations and waveguides were also armored. The *Long Beach* is fitted out as a flagship with several satellite communication terminals and an NTDS data link.

In the late 1970s, the conversion of the *Long Beach* to an AEGIS-equipped "strike cruiser" was proposed. This would have involved the installation of SPY-1 PHASED ARRAY radars and Mk.26 missile launchers. But the plan was canceled due to the fear that it would divert funding from new construction. The *Long Beach* will probably remain in service into the next century. See also CRUISERS, UNITED STATES.

Specifications Length: 721.25 ft. (219.9 m.). **Beam:** 73.25 ft. (22.33 m.). **Draft:** 29 ft. (8.84 m.). **Displacement:** 4,200 tons standard/17,100 tons full load. **Powerplant:** twin-shaft nuclear: 2 C1W pressurized-water reactors, 4 sets of geared turbines, 80,000 shp. **Speed:** 30+ kt. **Range:** 90,000 n.mi. at 30 kt./360,000 n.mi. at 20 kt. **Crew:** 1160. **Sensors:** 1 SPS-48C 3-dimensional air-search radar, 1 SPS-49 2-dimensional air-search radar, 1 SPS-67 surface-search radar, 4 SPG-55D missile guidance radars, 2 Mk.35 gun fire control radars, 2 SCS-23 PAIR hull-mounted low-frequency active/passive sonar. **Electronic warfare equipment:** 1 SLQ-32(V)3 electronic countermeasures array, 4 Mk.36 SRBOC chaff launchers.

LONG-RANGE INTERMEDIATE NUCLEAR FORCES: Sometimes called Intermediate Nuclear Forces, or INF; a term applied to nuclear DELIVERY SYSTEMS with ranges between 500 and 1500 mi. (835 and 2505 km.), including BALLISTIC MISSILES, CRUISE MISSILES, and tactical aircraft. These fill the gap between "tactical" or "battlefield" nuclear weapons on the one hand, and intercontinental or "strategic" nuclear forces on the other. Under the terms of the INF TREATY signed by the United States and Soviet Union in 1987, all ballistic missiles and ground-launched cruise missiles within the INF range limits are to be totally eliminated from the inventories of both powers. See also ZERO OPTION.

LONG-RANGE MARITIME PATROL (AIRCRAFT): Large, land-based aircraft used for ANTI-SUBMARINE WARFARE and for long-range reconnaissance over water. For specific types, see ATLANTIC; MAI; MAY; NIMROD; ORION.

LOOK-DOWN/SHOOT-DOWN: The ability of a FIGHTER aircraft's RADAR and missiles to detect, track, and engage low-flying targets from above—a difficult feat because of ground clutter, spurious radar reflections from the ground which masks targets. Clutter can be suppressed by pulse-DOPPLER techniques and complex signal processing, capabilities present only in the newest and most sophisticated (hence most expensive) aircraft. Among Western fighters, only the F-14 TOMCAT, F-15 EAGLE, F-16 FALCON, FA-18 HORNET, TORNADO F.2, and some late-model F-4 PHANTOMS have look-down/shoot-down capability. Likewise, only the latest Soviet fighters and INTERCEPTORS— the MiG-29 FULCRUM, MiG-31 FOXHOUND, and Su-27 FLANKER—are believed to have true look-down/shoot-down capability.

LOOSE DEUCE: U.S. Navy term for an AIR COMBAT MANEUVERING (ACM) formation of two aircraft flying in rough line abreast about 1 mi. apart, and staggered only slightly in altitude. Unlike earlier pairings, such as those within the Finger Four, or WELDED WING, the roles of leader and covering "wingman" are not specified. Both pilots are free to initiate attacks, either simultaneously or sequentially, based upon the tactical situation and aircraft positions. In a sequential attack, the aircraft in the more favorable position becomes the leader, with the other aircraft flying cover. In a simultaneous attack, each aircraft selects a separate target, and mutual support is provided as required, either by presence or by active intervention. If attacked, the Loose Deuce will maneuver to achieve an offensive position, often by having one aircraft "drag" (i.e., act as bait) the enemy into the sights of the other.

Loose Deuce is extremely flexible and has been used successfully in hundreds of dogfights by the Israeli air force as well as the U.S. Navy (historically, both have lacked the numbers for larger formations). It has also been adopted by many NATO air forces, but it requires truly well-trained pilots imbued with initiative. Third World and Warsaw Pact air forces have therefore relied on larger and more rigid formations (though this may be changing in the Soviet air force).

LOS ANGELES (SSN-688): A class of 69 U.S. nuclear-powered attack SUBMARINES (SSN) already commissioned, under construction, or planned. Their design was in part a response to the appearance of the Soviet VICTOR-class SSNs, especially their considerable speed advantage over their U.S. contemporaries of the PERMIT and STURGEON classes. U.S. attack submarines had increased significantly in size from the 3500-ton SKIPJACK class to the 4311-tons Permits and 4777-ton Sturgeons, to accommodate additional fire controls and sensors, as well as improved acoustical silencing. But they all had the same 15,000-shp. S5W pressurized-water reactor of the Skipjacks, with a concomitant loss in speed from 33+ kt. to barely 26 kt. (whereas the Victors are capable of at least 30 kt.). The Los Angeles SSNs introduced the new S6G reactor, rated at 30,000 shp., which restored the lost speed margin while allowing a considerable increase in displacement over previous classes. Adapted from the D2G reactor of U.S. CRUISERS, the S6G employs convection (i.e., natural circulation) cooling at low power settings, eliminating the need for noisy pumps at low speeds. The single, low-speed 7-bladed propeller is designed to reduce cavitation noise, and all machinery is individually soundproofed and mounted on sound-isolation "rafts" within the hull, making the Los Angeles class the quietest of attack submarines.

The Los Angeles class have a modified teardrop hull with an ogival bow, cylindrical midsection, and tapered stern, and a single-hull configuration with ballast tanks concentrated fore and aft of the pressure hull, which is fabricated of HY-80 steel. The bow houses a large spherical sonar array, displacing the torpedo tubes amidships. They have a rather short, streamlined "sail" (conning tower) located well forward, and the control surfaces are arranged according to standard U.S. practice: "fairwater" diving planes on the sail, and cruciform rudders and stern planes just ahead of the propeller. The stern planes have large, rectangular endplates which house CONFORMAL ARRAY hydrophones and also improve maneuverability response. Because of the small size of the sail, the fairwater planes cannot be rotated completely vertically, preventing these submarines from surfacing through polar ice.

The Los Angeles class are armed with 4 21-in. (533-mm.) torpedo tubes amidships, angled outward from the hull, for MK.48 wire-guided acoustical homing TORPEDOES, UGM-84 HARPOON, TOMAHAWK nuclear and conventional cruise missiles, and a variety of MINES. Some 24 weapons of all types can be carried, so that the proliferation of weapon types has forced a reduction in the basic torpedo load. To rectify this situation, the last 35 boats (beginning with USS *Providence*, SSN-719)

are being equipped with 15 vertical launch tubes for Tomahawk, mounted in the forward ballast tank area. Older SSN-688s will be retrofitted with 12 vertical tubes during their midlife overhauls. In the future, they will also carry SEA LANCE antisubmarine missiles.

The Los Angeles SSNs have a very effective sensor suite optimized for long-range passive detection and classification (active sonar reveals the emitter's position). The main sonar has Digital Multi-Beam Steering (DIMUS), with electronic rather than electro-mechanical scanning, for multi-target TRACK-WHILE-SCAN capability. All sonars are integrated into a BSY-1 digital FIRE CONTROL system.

For all its virtues, these SSNs have been criticized for their cost ($579 million each in 1988), small weapons load, large size (which makes them easier to detect, albeit with active sonar), and inability to surface through ice. Certainly the design was not as innovative as the Soviet ALFA, AKULA, and MIKE classes. Much of the blame is attributable to the leaden hand of Adm. Hyman G. Rickover, who in fact controlled the design of U.S. submarines through his domination of nuclear engineering until his retirement in 1982. Their successors, the SSN-21 or SEA WOLF class, will not enter service until the mid-1990s, and thus the SSN-688s will remain in service well into the 21st century.

Specifications **Length:** 360 ft. (109.75 m.). **Beam:** 33 ft. (10.06 m.). **Displacement:** 6080 tons surfaced/6927 tons submerged. **Powerplant:** see text. **Speed:** 30+ kt. **Max. operating depth:** 1475 ft. (450 m.). **Collapse depth:** 2460 ft. (750 m.). **Crew:** 133. **Sensors:** 1 BQQ-5A bow-mounted low-frequency active/passive spherical sonar, 1 BQR-15 passive towed array, 1 flank-mounted conformal hydrophone array, 1 BQS-15 underice sonar, 1 WLR-9 sonar intercept array, one BPS-15 surface-search radar, 1 WLR-8 radar warning receiver, 1 WLR-12 electronic signal monitoring array, 2 periscopes.

LOW-ALTITUDE BOMBING SYSTEM (LABS): An automated weapon-delivery device, developed in the U.S. during the 1950s to enable low-flying tactical aircraft to release nuclear weapons safely (and accurately). There are two modes of operation: TOSS BOMBING and "over-the-shoulder" bombing. In both, the aircraft rotates away from the target as the weapon is released, in order to avoid its blast effects. Once programmed with the target coordinates, aircraft position and velocity, barometric pressure, and wind velocity, LABS automatically initiates the required maneuvers at the proper point and time to hit the target, and provides weapon release cues to the pilot. See also BOMBING TECHNIQUES.

LPD: Landing Platform, Dock, a large AMPHIBIOUS ASSAULT ship that contain a large, floodable well deck in which LANDING CRAFT, HOVERCRAFT, and other small vessels can load or discharge troops and cargo, and undergo maintenance. Access to the well is provided by a large folding door in the transom of the stern. A platform deck over the well serves as a helicopter landing pad. The forward part of the ship contains crew and troop accommodations, cargo hold, and vehicle parking decks. LPDs are generally similar to the earlier LSD (Landing Ship, Dock), but have increased troop and vehicle capacity at the expense of a reduced well deck.

LPH: Landing Platform, Helicopter, a type of AMPHIBIOUS ASSAULT ship designed primarily for the operation of HELICOPTERS. LPHs resemble small AIRCRAFT CARRIERS, but lack the equipment to operate fixed-wing aircraft, except for V/STOL types such as the HARRIER. LPHs also have accommodations for a battalion-size landing force and its vehicles and heavy equipment. Some carry several small landing craft to supplement the helicopters.

LRINF: See LONG-RANGE INTERMEDIATE NUCLEAR FORCES.

LRMP: See LONG-RANGE MARITIME PATROL.

LSD: Landing Ship, Dock, a large AMPHIBIOUS ASSAULT vessel with a floodable well deck in which LANDING CRAFT, HOVERCRAFT, and other small vessels can load and unload troops and cargo, and also undergo maintenance. The well deck is accessed through a large folding door in the transom of the stern. The area above the well is usually decked over and serves as a landing pad for HELICOPTERS. The forward end of the ship contains accommodations for the crew and landing force, cargo holds, and storage decks for armored vehicles, trucks, and other heavy equipment. Unlike other LANDING SHIPS, LSDs do not have flat bottoms and cannot be beached.

LST: Landing Ship, Tank, a large, flat-bottomed AMPHIBIOUS ASSAULT vessel which can be beached to discharge vehicles and cargo directly ashore. Developed by the U.S. and Great Britain during World War II, the LST became the essential instruments of amphibious warfare at that time. LSTs have also served in a variety of support-

ing roles; e.g., as hospitals and as tenders for PT boats and other small craft.

The standard World War II LST was an austere, mass-produced design, 328 ft. (100 m.) long and displacing 4080 tons at full load. With a small forecastle and aft superstructure (including the pilot house), LSTs were and are characterized by a large, open tank deck. The bow has large clamshell doors with a fold-down ramp over which tanks and other vehicles can roll on or off. Each LST could carry some 17 tanks or up to 400 troops; powered by twin-shaft diesel engines, they had a top speed of 11.6 kt., and carried 569 tons of fuel for a range of 15,000 n.mi. at 9 kt. Though uncomfortable, LSTs were quite seaworthy, and many made ocean crossings under their own power. More than 1050 were built between 1942 and 1945, and many are still in military or civilian use. Several countries have produced indigenous designs based on the wartime LST, the Soviet ALLIGATOR class being an excellent example.

Since the 1960s, the U.S. Navy has acquired 20 LSTs of the Newport class, a completely new design that reflects the navy's preference for larger ships which are more efficient in peacetime. The Newports are are in fact very large, at 522.16 ft. (159.2 m.) in length, 69.5 ft. (21.2 m.) in beam, and 17.5 ft (5.33 m.) in draft, for a displacement of 4793 tons empty and 8450 tons fully loaded. Their twin-shaft diesel powerplant allows a maximum speed of 22 kt., twice as fast as wartime LSTs. They have a large superstructure amidships with a full-length vehicle deck beneath.

Because beaching is impractical for ships so large, the Newports do not have bow doors; instead, a pair of derricks extending over the bow deploy articulated ramps running from the tank deck, over the bow, and down to the beach. There is a large ramp door in the stern, to allow roll-on/roll-off loading and unloading from docks, and these can also be opened at sea to launch or retrieve amphibious assault vehicles (e.g., LVTP-7). The Newports have 17,300 square ft. of vehicle parking space, for a capacity of up to 21 MBTs and 17 2.5-ton trucks; or 25 LVTP-7s and 17 trucks; or 42 LAVs (Light Armored Vehicles); or 5000 tons of general cargo. In addition four LCVP and one LCPL landing craft are carried on davits amidships. Though the Soviet IVAN ROGOV class is often described as an LST, it is actually a hybrid combining LST features with the well-deck and helicopter pad of LSDS or LPDS.

LUCHS: A West German 8 × 8 ARMORED CAR, officially designated *Spahpanzer* (SPz) 2, which equips the armored RECONNAISSANCE units of the *Bundeswehr*. Developed in response to a mid-1960s requirement for an armored amphibious reconnaissance vehicle and designed by Daimler Benz, it was built under a December 1973 contract by Rheinmetall *Wehrtechnik* (now Thyssen Henschel); the total production run of 408 vehicles was delivered between 1975 and 1978.

A direct descendant of the large, powerful armored cars of the World War II German army, with several unique automotive features, the Luchs has greater mobility and less armament than any other heavy armored car. The all-welded steel hull has well-sloped front and rear glacis plates and upper sides. The forward glacis and the turret are armored against rounds up to 20 mm. (but not APDS); the rest of the vehicle is armored against small arms fire and shell splinters. The hull is divided into a forward crew compartment and a rear engine compartment. In the four-man crew, the driver sits on the left side of the vehicle and has three periscopes; the commander and gunner sit in a power-operated Rheinmetall TS-7 turret, with the commander on the left and the gunner on the right. The commander has three periscopes while the gunner has a stadiametric optical gunsight. Driver, commander, and gunner are also provided with passive INFRARED night-vision scopes. A second, rear-facing driver in the engine compartment has a full set of vehicle controls, to enable the vehicle to back out of trouble without delay; the second driver also acts as a radio operator. The Luchs has a positive overpressure COLLECTIVE FILTRATION system for NBC defense.

The primary armament, disproportionately light, is a 20-mm. Rheinmetall Rh 202 CANNON with 375 rounds of ammunition and a dual feed system allowing the selection of ARMOR PIERCING or high explosive (HE) ammunition as required. The AP rounds have a range of 1000 m. and can penetrate light armored vehicles. With elevation limits of −15° and +60°, the gun can also be used against (low-flying) aircraft. Other armament includes a coaxial 7.62-mm. MG3 machine gun, and a second G3 on the turret roof. Four smoke grenade launchers are mounted on each side of the turret.

The engine is mounted on the right side of the engine compartment with the second driver alongside. Maximum road speed is possible in both forward and reverse gears. All 8 wheels are powered, and the driver can select steering with either the front 4 or all 8 wheels; in the latter mode, the vehicle has a turning radius of only 37.7 ft. (the rear driver steers using only the rear 4 wheels). Power steering is provided at both ends, making the Luchs easy to handle despite its size. Fully amphibious, the Luchs has two steerable propellers mounted under the rear hull, which allow a top speed of 5.6 mph (9.3 kph).

Specifications **Length:** 25.4 ft. (7.75 m.). **Width:** 9.8 ft. (2.98 m.). **Height:** 9.5 ft. (2.9 m.). **Weight, combat:** 19.5 tons. **Powerplant:** 390-hp. Daimler Benz OM403A 10-cylinder multi-fuel diesel. **Hp./wt. ratio:** 20 hp./ton. **Fuel:** 150 gal. (672 lit.). **Speed, road:** 56 mph (94 kph). **Range, max.:** 480 mi. (802 km.).

LUPO: A class of 18 Italian-designed FRIGATES in service with the navies of Italy, Iraq, Peru, and Venezuela. The Lupos are relatively small frigates designed for CONVOY escort in the Mediterranean. The Italian navy commissioned 4 Lupos between 1977 and 1980; Iraq ordered 4 in 1982, whose delivery was delayed by the Iran-Iraq War (due to the Iraqi invasion of Kuwait and the subsequent international arms embargo, it is unlikely that these ships will ever be delivered); Peru commissioned 2 in 1979, and an additional 2 in 1984–85; and finally, Venezuela commissioned 6 between 1980 and 1982.

The Lupos are equipped with gyro-controlled fin stabilizers to enhance seakeeping. Armament consists of 1 OTO-Melara 5-in. 54-caliber DUAL PURPOSE gun forward, 2 twin Breda 40-mm. ANTIAIRCRAFT guns aft, 8 Teseo launchers for OTOMAT Mk.2 anti-ship missiles amidships, an 8-round box launcher for ASPIDE surface-to-air missiles atop the aft superstructure, and 2 sets of Mk.32 triple tubes for lightweight anti-submarine homing TORPEDOES. A landing deck and hangar for an ASW HELICOPTER are located on the fantail (most operators use the Agusta-Bell AB212 ASW variant of the HUEY utility helicopter). The 40-mm. guns can be controlled by a Dardo FIRE CONTROL system to provide some anti-missile capability. Iraqi Lupos are equipped with EXOCET instead of OTOmat. See also FRIGATES, ITALY.

Specifications **Length:** 371.3 ft. (113.2 m.). **Beam:** 37.1 ft. (11.31 m.). **Draft:** 12.1 ft. (3.69 m.). **Displacement:** 2208 tons standard/2525 tons full load. **Powerplant:** twin-shaft CODOG: 2 25,000-shp. GE/Fiat LM-2500 gas turbines (sprint), 2 GMT A230-20M diesels, 7900 hp. (cruise). **Speed:** 35 kt. **Range:** 900 n.mi. at 35 kt./3450 n.mi. at 20 kt. **Crew:** 186. **Sensors:** 1 RAN-10S air- and surface-search radar, 1 SPQ-2F surface-search radar, 1 RAN-11/LX surveillance radar, 1 Orion RTN-10X 5-inch gun fire control radar, 1 RTN-20X 40-mm. fire control radar, 1 SPN-703 navigational radar, 1 Mk.91 missile-guidance radar, 1 Raytheon DE1160B hull-mounted medium-frequency sonar. **Electronic warfare equipment:** active jamming units, electronic signal monitoring arrays, 2 SCLAR 20-round chaff launchers.

LUTJENS: A class of four U.S.-built West German guided-missile DESTROYERS (DDGs), which are modified versions of the U.S. ADAMS class. Commissioned in 1969–70, they differ from the Adams class most visibly because their radar and communications antennas are mounted on "macks" (combined mast and exhaust stack structures), rather than on freestanding lattice masts. The Lutjens have also been modified to launch HARPOON anti-ship missiles from their Mk.13 STANDARD missile launchers, and were given an extensive ANTI-AIR WARFARE (AAW) upgrade in 1986–87, applied to only a few of the U.S. ships. Dimensions, performance, and weapons are otherwise similar to the original. See also DESTROYERS, FEDERAL REPUBLIC OF GERMANY.

LVTP-7: Landing Vehicle, Tracked, Personnel (recently redesignated AAVP7A1, for Amphibious Assault Vehicle, Personnel), a large, U.S.-built, fully tracked, amphibious ARMORED PERSONNEL CARRIER (APC) designed to deliver troops through heavy surf and over the beach for AMPHIBIOUS ASSAULTS. Built by FMC, the LVTP-7 is a direct descendant of the LVT and LVTA amphibious tractors (or AMTRACs) of World War II, by way of the LVTP-5 and LVTH-6 vehicles of the 1950s. The USMC acquired a total of 965 LVTP-7s between 1970 and 1974, not including a number of specialized command and recovery variants.

The LVTP-7's rather boxy, watertight all-welded aluminum hull has a boat-shaped bow for seaworthiness. The maximum armor thickness of 45 mm. is sufficient to defeat small arms, shell fragments, and machine guns of up to 14.5-mm. caliber. The hull is divided into a driver/engine compartment up front, and a fighting/troop compartment behind. The driver sits on the left side of the bow next to the transversely mounted engine, under a cupola equipped with six observation periscopes, one of which can be replaced by an INFRARED (IR) night-vision scope. The commander sits in a separate cupola behind the driver. Opposite the commander's cupola, on the right side of the hull, is a fully traversable powered turret mounting a

12.7-mm. (.50-caliber) MACHINE GUN; with its elevation limits of − 15° and + 60°, it can also be used against aircraft. The large troop compartment can accommodate a maximum of 21 fully equipped marines. Access is through a large, fold-down ramp door in the rear, and two large roof hatches. A Mk.19 40-mm. automatic GRENADE LAUNCHER can be pintle-mounted by the roof hatches. The LVTP-7 is propelled in water by two steerable pumpjets.

The command variant, the LVTC-7, is similar but is equipped with command, control, and communications (c^3) equipment in the passenger compartment; approximately 100 were built. The repair and recovery LVTR-7 variant has a 6000-lb. (2727-kg.) capacity telescoping crane, a 30,000-lb. (13,636-kg.) capacity cable winch, and maintenance equipment; some 60 are in service with the USMC.

Under a Service Life Extension Program (SLEP) initiated in 1983, a total of 853 LVTPs, 77 LVTCs, and 54 LVTRs were upgraded to AAVP7A1, AAV.C, and AAVR standard, respectively, with new engines, smoke generating equipment, passive night vision devices, and improved vehicle electronics (vetronics). After the SLEP, these vehicles will remain in service until the turn of the century.

Specifications Length: 26 ft. (7.92 m.). **Width:** 10.75 ft. (3.29 m.). **Height:** 10.25 ft. (3.12 m.). **Weight, combat:** 25 tons. **Powerplant:** 400-hp. Cummings VT400 turbocharged diesel. **Fuel:** 145 gal. (650 lit.). **Speed:** (land) 40 mph (67 kph); (water) 8.4 mph (14 kph). **Range, max.:** (land) 300 mi. (500 km.); (water) 55 mi. (92 km.). **Operators:** Arg, It, Phi, ROK, Sp, Thai, Ven.

LYNX COMMAND AND RECONNAISSANCE VEHICLE: A fully tracked armored reconnaissance vehicle based on the ubiquitous, U.S.-built M113 armored personnel carrier (APC), which is now in service with the armies of Canada and the Netherlands. Developed by FMC as a private venture replacement for the unsuccessful M114 reconnaissance vehicle, Lynx has the basic chassis and many automotive components of the M113 to reduce costs. The first prototypes were built in 1963, and in 1968 the Canadian Forces purchased 174 vehicles and the Netherlands a total of 250.

Lynx is shorter and lighter and has a lower silhouette than the M113. The all-welded aluminum hull retains the boxy shape of the M113, but has been cut down by 15.1 in. (384 mm.) and drastically rearranged, with the engine moved from the front to the rear, occupying what had been the passenger compartment on the M113. The vehicle has a 3-man crew of driver, commander/gunner, and radio operator, all seated in a single compartment at the front of the hull. The driver and radio operator have their own hatches and observation periscopes, and the driver's cupola is also equipped with an INFRARED night-vision scope. Canadian vehicles are armed with a remotely controlled 12.7-mm. MACHINE GUN fired by the commander/gunner from inside his larger cupola, which is encircled by observation periscopes. Secondary armament consists of a pintle-mounted 7.62-mm. machine gun mounted by the radio operator's hatch. In the Dutch vehicles, the commander/gunner sits in a power operated Oerlikon-Buhle GBD-AOA turret armed with a 25-mm. KBA-B CANNON, with 200 rounds of ammunition. The cannon has two rates of fire, 175 and 570 rounds per minute, and can fire a mix of ARMOR PIERCING (80 rounds) and high explosive (120 rounds) ammunition against both ground and aerial targets. In both Canadian and Dutch vehicles, three smoke grenade launchers are mounted on each side of the hull.

Amphibious without extensive preparation, Lynx is propelled through water by track action at a top speed of 3.5 mph (5.85 kph), but its lack of freeboard limits it to relatively calm waters. Neither Canadian nor Dutch vehicles are provided with an NBC protective system.

Specifications **Length:** 15.1 ft. (4.6 m.). **Width:** 7.9 ft. (2.4 m.). **Height:** 7.1 ft. (2.16 m.). **Weight, combat:** 8.77 tons. **Powerplant:** 215-hp. GMC-Detroit 6V53 6-cylinder diesel. **Fuel:** 75 gal. (336 lit.). **Speed, road:** 44 mph (73.5 kph). **Range, max:** 325 mi. (543 km.).

LYNX, WESTLAND: A British light army utility and shipboard ANTI-SUBMARINE WARFARE (ASW) helicopter. Designed by the British Westland company and produced jointly by Westland and the French Aerospatiale group, Lynx was designed in response to the 1967 WG.13 requirement for a joint Anglo-French general-purpose naval and civilian helicopter. The program was later expanded to include land-based assault transport, SCOUT, and ANTI-TANK missions. The shipboard HAS.2 variant entered service in 1976, while the army AH Mk.1 variant was placed in production in 1977.

Both aircraft have the same airframe and power train, differing mainly in their avionics, weapons, and ancillary, mission-specific equipment. Pilot and copilot sit side-by-side in the nose. The main cabin is used for troops and cargo in the AH Mk.1, and houses fire controls and tactical displays in the HAS.2. A tubular tail boom extends from the upper rear of the cabin to support the tail rotor and vertical stabilizer. The engines, mounted side-by-side in a nacelle over the main cabin, drive a four-bladed main rotor; the rotor blades can fold for storage in ASW versions. Both versions are very maneuverable and can be handled violently when lightly loaded.

The AH.Mk.1 utility variant can carry a maximum of 12 fully equipped soldiers or 2000 lb. (909 kg.) of internal cargo, or a 3000-lb. (1364-kg.) slung load. It has very comprehensive AVIONICS (as compared with other helicopters of this class), including a GEC autopilot/autostabilization system, a gyrocompass and radio compass, and a wide range of radio navigation aids and ELECTRONIC COUNTERMEASURES. For the attack role, Lynx can carry gun and rocket pods, pintle-mounted door guns, and mine/submunition dispensers. Its most common armament, however, consists of eight TOW anti-tank guided missiles (ATGMs), carried on outriggers and directed by a stabilized, roof-mounted sight controlled by the copilot.

The ASW variant, which also has secondary ANTI-SURFACE WARFARE (ASUW) and search-and-rescue (SAR) roles, has wheeled, rather than tubular-strut, landing gear. In addition, the HAS.2 has a horizontal stabilizer mounted on top of the vertical fin opposite the tail rotor. The entire tail boom can fold forward for more compact deck stowage.

ASW variants have the same basic avionics as the utility version, plus a TACAN beacon, an automatic direction-finding (D/F) system, and an I-band ship transponder. Mission avionics include a Ferranti Sea Spray surface-search RADAR housed in a nose radome, an IFF set, and an ELECTRONIC SIGNAL MONITORING array. ASW equipment includes a Texas Instruments or Crouzet MAGNETIC ANOMALY DETECTOR (MAD), a Bendix or Alcatel dipping SONAR, and a tactical computer and displays located in the main cabin. Armament options include two DEPTH CHARGES or lightweight homing TORPEDOES, SONOBUOYS, or rocket and gun pods. For attacks on surface vessels, Lynx can carry up to four SEA SKUA anti-ship missiles or a similar number of ATGMs.

Westland is currently developing an enlarged Lynx 3 version, with more powerful engines and enhanced avionics, including night-vision sensors, a mast-mounted sight, and an INFRARED COUNTER-MEASURES system.

Specifications **Length:** 43.2 ft. (13.17 m.). **Rotor diameter:** 42 ft. (12.8 m.). **Cargo hold:** 6.75 × 5.83 × 4.67 ft. (2.05 × 1.77 × 1.42 m.). **Powerplant:** 2 Rolls Royce Gem turboshafts rated at 900 shp. or 1120 shp. **Fuel:** 1616 lb. (734.5 kg.). **Weight, empty:** (HAS) 6040 lb. (2745 kg.); (AH) 5683 lb. (2583 kg.). **Weight, max. takeoff:** (HAS) 10,000–10,500 lb. (4545–4773 kg.); (AH) 10,000 lb. (4545 kg.). **Speed, max.:** 190 mph (317 kph). **Speed, cruising:** (HAS) 144 mph (240 kph); (AH) 160 mph (267 kph). **Climb, max.:** (HAS) 2170 ft./min. (662 m./min.); (AH) 2480 ft./min. (1127 m./min.). **Combat radius:** (HAS) 111 mi. (185 km.). **Range, max.:** (HAS) 368 mi. (645 km.); (AH) 336 mi. (561 km.). **Mission endurance:** (HAS) 2.5 hr.

LZ: See LANDING ZONE.

M

M1 MAIN BATTLE TANK: See ABRAMS.

M2 HB BROWNING HEAVY MACHINE GUN: A .50-caliber (12.7-mm.) air-cooled, recoil-operated MACHINE GUN in service with U.S. and many other Western forces. First introduced in 1921, the M2 (popularly known as "Ma Deuce" or "50-cal"), is the standard U.S. HEAVY MACHINE GUN, fitted on tanks and other ARMORED FIGHTING VEHICLES (AFVs), on aircraft and helicopters, and in power-driven or manual shipboard mounts, and also used by infantry on a tripod mount.

At higher rates of fire the barrel can easily overheat and must be changed frequently. The M2 is fed by 100-round disintegrating-link metallic belts. Ammunition types include ball, tracer, HEI, and API. Armor penetration with the latter is 12 mm. at 500 m.—enough to penetrate the side armor of light AFVs, and devastating against trucks and aircraft. See also DSHK.

Specifications Length OA: 64.7 in. (1.64 m.). **Length, barrel:** 45 in. (1.14 m.). **Weight:** (w/o tripod) 84 lb. (38.18 kg.); (tripod) 20–30 lb. (9–14 kg.). **Muzzle velocity:** 890 m./sec. **Cyclic rate:** (normal) 400–500 rds./min.; (anti-aircraft) 700 rds./min. **Effective range:** 1400 m.

M2 INFANTRY FIGHTING VEHICLE: See BRADLEY.

M3 CAVALRY FIGHTING VEHICLE: See BRADLEY.

M4 MEDIUM TANK: See SHERMAN.

M4 SLBM: See MSBS.

M16 ASSAULT RIFLE: A U.S. 5.56-mm., gas-operated ASSAULT RIFLE used by the armed forces of the United States and at least 12 other countries. The first of the modern reduced-caliber light rifles, privately designed in the late 1950s (as the Armalite AR-15), adopted by U.S. Air Force security police in 1961, subsequently used by the British army, and supplied to the South Vietnamese army, the M16 was adopted by the U.S. Army in 1966 as its standard infantry weapon, replacing the 7.62-mm. M14. U.S. production of the M16 has been conducted by Colt Firearms under license from Armalite; it is also produced in South Korea, the Philippines, and Singapore. The M16 has been copied as the Chinese CQ automatic rifle, the Taiwanese Type 65, and the Singaporean SAR80.

As compared to earlier M14, the M16 is almost 4 lb. lighter at 8.02 lb. (3.65 kg.) loaded, has a higher rate of fire, and is considerably cheaper to manufacture. The 5.56- × 45-mm. M198 ball cartridge fired by the M16 weighs only half as much as the 7.62-mm. round, yet can cause larger and more incapacitating wounds. On the other hand, the M16 is a precision-designed weapon, whose early models were prone to jamming and misfires under combat conditions, especially in mud. In Southeast Asia, the weapon was not popular with U.S. troops (who showed a disconcerting preference for the Soviet AK-47). In 1966, Colt introduced the M16A1, with a bolt-closure mechanism to ensure the seating of rounds in the chamber, and this remained in production until 1985, when an improved version, the M16A2, was adopted by the U.S. Marine Corps, and later by the U.S. Army.

The M16A1 is long, with a straight stock, a pistol grip, and a straight, oval foregrip. Fabricated from stamped metal and injection-molded plastic to simplify production and reduce costs, the M16 is actually cheaper to replace than to have repaired by an armorer. A carrying handle over the receiver incorporates an adjustable rear peep sight; the foresight is of the fixed-post type. A flash suppressor on the muzzle can accommodate a bayonet and also launch a variety of rifle grenades. The foregrip is made of plastic, with a triangular cross-section and large ventilation holes at the top. The stock, also made of plastic, contains a shock-absorbing recoil cylinder. The weapon is fed by 20- or 30-round box magazines which are inserted ahead of the trigger guard.

The M16A2 has a new, round foregrip and a more effective flash suppressor. The barrel is heavier, and has rifling matched to the higher-velocity Belgian SS109 cartridge (U.S. designation M855). The M16A2 has selectable single-shot and 3-round burst, but no fully automatic setting.

In addition to the A1 and A2 versions, Colt has also produced a short-barrel "carbine" variant (Commando), which was used in Vietnam without much success. The only other variants in service are the M203 Dual Purpose Weapon (an M16A1 with a 40-mm. grenade launcher attached below the barrel); and the M231 Firing Port Weapon, a fully automatic rifle without front sight or fore grip, used exclusively on the M2/M3 BRADLEY Fighting Vehicle. The M16 is nearly 30 years old and has reached the end of its development. It is considerably longer, more awkward, and somewhat heavier than newer "bullpup" designs, notably the British ENFIELD L85A1 and Austrian Styr AUG.

Specifications Length OA: (A1) 39.98 in. (1.01 m.); (A2) 39.37 in. (1.0 m.). **Length, barrel:** 20 in. (508 mm.). **Weight, loaded:** (A1) 8.02 lb. (3.65 kg.); (A2) 8.5 lb. (3.8 kg.). **Muzzle velocity:** (A1) 1000 m./sec. **Cyclic rate:** (A1) 700–950 rds./min. **Effective range:** (A1, M198 round) 400 m.; (A2, M855 round) 800 m.

M20 SLBM: French submarine-launched ballistic missile. See MSBS.

M47 MEDIUM ANTI-TANK WEAPON: U.S. anti-tank guided missile. See DRAGON.

M47 MAIN BATTLE TANK: A U.S. main battle tank of 1950s vintage, the second member of the PATTON family; still in service with Greece, Iran, Italy, Pakistan, Somalia, Spain, Sudan, Tai-wan, Turkey, Yugoslavia, and South Korea, but slowly being phased out.

M48 MAIN BATTLE TANK: Third member of the PATTON series, the M48 was a radical development of the original M46 meant to rectify the manifest shortcomings of its immediate predecessor, the M47, with a new hull and turret, revised armor, and numerous automotive improvements, while retaining (initially) the M47's main armament and FIRE CONTROL system. Developed from 1950, the first production M48s entered service in 1953, and a total of 11,703 of these tanks were produced between 1952 and 1959, serving initially with the U.S. Army and Marine Corps, and later exported to numerous U.S. allies. The M48 has proven well suited to improvements and modifications, and is still used in large numbers.

The original M48, with an 810-hp. Continental AV-1790-5B gasoline engine, was rapidly superseded by the M48A1, with an upgraded AV-1790-7C engine, and a fully enclosed cupola/machine-gun mount resembling a small, secondary turret. The M48A2, which entered service in 1954, had an 825-hp. fuel-injected AV-1790-8 engine, as well as larger fuel tanks, improved fire controls, and a T-shaped muzzle brake on the main gun. The M48A3 was a rebuild of earlier A1s and A2s, with a Continental AVDS-1790-2A diesel engine, a simplified coincidence rangefinder, and an improved cupola. The M48A4 was the original designation of the later M60 tank. The Israeli army, which received 200 M48A2s from West Germany in 1965 and an additional 450 from other sources, began a modernization program in 1968 which included the installation of the AVDS-1790 diesel and an automatic transmission, replacement of the 90-mm. gun with a British 105-mm. L7A1, new fire controls, and a new, low-profile cupola. Most of these improvements were subsequently adopted by the U.S. Army as the M48A5; 22064 surviving M48s were rebuilt to that standard, and many other users have made the same changes.

The hull is cast as a single piece, with additional armor plates welded in place. The glacis has a rounded front, with a thickness of 120 mm sloped at 60° (effectively doubling its ballistic protection). As in all Pattons, the hull is divided into a driving compartment up front, a fighting compartment in the middle, and an engine compartment in the rear. The M48 was the first U.S. tank designed from the outset with a 4-man crew: driver, commander, gunner, and loader. The driver is provided with three M27 observation periscopes and an M24 INFRARED (IR) night scope. The remaining crewmen sit in the turret, a one-piece egg-shaped casting with a maximum armor thickness of 110 mm. The gunner sits at the front on the right, with the commander above and behind him, and the loader on the left. The commander has a rotating cupola, which in earlier versions incorporated an M2 BROWNING heavy machine gun, five observation periscopes, and an M28C stadiametric magnifying periscope. In the M48A5, this has been replaced by an Israeli-designed low-profile cupola with an externally mounted 7.62-mm. M60 MACHINE GUN. The commander also has a coincidence rangefinder with $10 \times$ magnification, accurate to a range of 2000 m. The rangefinder is mechanically connected to an electro-mechanical fire control computer with manual inputs for ammunition type and target motion. The gunner has an $8 \times$ roof-mounted periscope and an $8 \times$ coaxial telescope, both used to lay the gun once the correct elevation angle has been generated by the computer. Both commander and gunner have IR night sights.

The original main armament was a 90-mm. M41 rifled gun, virtually identical to the M36 gun of the M47, with a total of 62 rounds stowed in the hull and turret. By the mid-1960s, the 90-mm. gun was only marginally effective against the latest Soviet tanks; in the M48A5 and similar upgrades, it has been replaced by the British-designed 105-mm. L7A1 rifled gun (U.S. designation M68), which also arms the M60, M1 ABRAMS, Israeli MERKAVA, German LEOPARD 1, and late-model British CENTURIONS. The L7A1/M68 fires APDS at a muzzle velocity of 1458 m./sec., APFSDS at roughly 1600 m./sec., and HEAT at 1170 m./sec.; it has an effective range in excess of 2000 m., and can penetrate the armor of all Soviet MBTs with the APFSDS round. A total of 54 rounds are carried. The gun has elevation limits of $+19°$ and $-9°$, but does not have any form of stabilization, making accurate firing on the move impossible.

Secondary armament consists of a 7.62-mm. coaxial machine gun, the 7.62-mm. M60 machine gun on a ring mount around the commander's hatch, and a pintle-mounted 7.62-mm. for the loader. Most users have also added smoke grenade launchers on each side of the turret. A white light/infrared searchlight mounted over the main gun is slaved to it in elevation. The M48A5 has a COLLECTIVE FILTRATION unit to which each crewman can connect his personal gas mask.

Because it is reliable, relatively cheap, and available in large numbers, many armies have cho-

sen to upgrade their M48s. Israel has also fitted its 650 M48s with blocks of BLAZER reactive armor to increase protection against HEAT rounds, and is reported to have installed digital fire controls. Other users can be expected to follow suit. The M48 chassis has also served as the basis for several specialized vehicles, including the M48 AVLB bridgelayer, the M88 armored recovery vehicle, the M67 flamethrower tank, and the ill-fated Sergeant York DIVAD anti-aircraft tank.

Specifications Length: 21.05 ft. (6.42 m.). Width: 11.9 ft. (3.63 m.). Height: 10.15 ft. (3.09 m.). Weight, combat: (A1/2) 45 tons; (A3/5) 49 tons. Powerplant: (A1/2) 810- or 825-hp. AV-1790 gasoline engine; (A3/5) 750-hp. AVDS-1790-2A diesel. Hp./wt. ratio: (A1/2) 18 hp./ton; (A3/5) 15.3 hp./ton. Fuel: 320 gal. (1408 lit.). Speed: (A1/2) 25.2 mph (42.08 kph) road/7.5 mph (12.5 kph) cross-country; (A3/5) 30 mph (50 kph) road. Range, max: (A2) 154 mi. (257 km.); (A3/5) 296 mi. (494 km.). Operators: FRG, Gre, Iran, Isr, Jor, Leb, Mor, Nor, Pak, Por, ROK, Som, Sp, Tai, Tun, Tur, US (Reserves and National Guard), Viet.

M60 (MACHINE GUN): The standard U.S. GENERAL-PURPOSE MACHINE GUN (GPMG), fired from bipod, tripod, vehicle, and helicopter mountings. The M60 design is based (loosely) on the German FG42 assault rifle and MG42 machine gun (now known as the MG3), but has a lower rate of fire, a more complex barrel changing mechanism, and a disturbing propensity to jam.

A gas-operated weapon, the M60 fires NATO standard 7.62- × 51-mm. (.308-caliber) ammunition, including ball, tracer, and AP. The M60 is fed by 50-round disintegrating link belts, which can be carried in a box magazine attached to the left side of the gun. As originally designed, the barrel did not have a handle, requiring the use of an asbestos glove while changing the barrel (a glove which often disappeared when needed); in addition, the gas cylinder was attached to the barrel, a needless complexity. These faults were rectified in later versions, but the M60 has remained a mediocre weapon. A lightweight version built under license by Singapore Arms is somewhat better, but has not been adopted by U.S. forces. See, more generally, MACHINE GUN.

Specifications Length OA: 47.7 in. (1.21 m.). Length, barrel: 22.6 in. (574 mm.). Weight, loaded: 22.9 lb. (10.4 kg.). Muzzle velocity: 850 m./sec. Cyclic rate: 850 rds./min. Effective range: 900 m.

M60 MAIN BATTLE TANK: The fourth and final member of the PATTON series (which started in 1948 with the M46, and which evolved through the M47 and M48). The M60 originated in 1956 as the M48A4, an up-armored, up-gunned, and die-sel-engined variant of the M48 developed in direct response to the Soviet T-54/55. The first production vehicles entered service in 1960, and the M60 be-came the standard U.S. MBT throughout the 1960s and 70s. By the time production ended in 1988, more than 15,000 M60s had been delivered, in-cluding several hundred built under license abroad.

The original M60 was superseded in 1962 by the M60A1, whose "needle nose" turret has improved ballistic protection. The M60A2, also introduced in 1962, had a totally redesigned turret mounting a revolutionary, short-barreled 152-mm. gun which could fire both HEAT rounds and SHILLELEIGH anti-tank guided missiles. The M60A2 was intended as a very long-range tank killer, but the Shilleleigh and its associated FIRE CONTROLS were extremely complex (crews called it the "starship"), and notoriously unreliable. Only 540 were produced, and these were withdrawn from service in the early 1980s to be placed in storage. The M60A3 is a greatly modernized A1 with digital, integrated fire controls, night-vision equipment, and automo-tive refinements; most U.S. M60A1s have been brought up to A3 standard, supplementing new production.

The hull is fabricated from cast sections and forged floor plates welded together. The glacis plate is more than 100 mm thick, sloped at 60° to double ballistic protection; side and rear armor vary between 76.2 and 25 mm. As in all Pattons, the hull is divided into a driving compartment in the front, a fighting compartment in the middle, and an engine compartment in the rear. The driver has three observation periscopes and an INFRARED (IR) night vision scope; the remaining crewmen sit in the turret, with the gunner up front on the right, the command above and behind him, and the loader on the left.

In the original M60, the turret was an egg-shaped, single-piece casting similar to that of the M48. The needle nose turret of the A1 and A3, which is also cast, has reduced frontal area, and a maximum thickness of more than 127 mm. The commander sits under a hand-operated rotating cupola, which incorporates a 12.7-mm. (.50-cali-ber) M2 BROWNING heavy machine gun; the cupola also has eight vision blocks for all-around observa-

tion. In the A1, the commander has an M28C opti-cal gunsight, which can be replaced by an M36 IR night sight, while the gunner has a roof-mounted M31 8 × periscope and a coaxial M36E1 8 × tele-scope which can be replaced by an M32 IR night sight. The gunner also has an M17A1 or M17C coincidence rangefinder, accurate between 550 and 1500 m. The M60A3, on the other hand, has a Hughes integrated LASER-rangefinder/sight and a THERMAL IMAGING night sight for the commander, and a Hughes VVG-2 laser rangefinder and VGS-2 thermal imaging sight for the gunner. The laser rangefinder, accurate to 5000 m., is linked to an M21 digital fire control computer which incorpo-rates range, vehicle and target motion, vehicle tilt, ammunition type, gun wear, and meteorological data into the ballistic solution, giving a very high first-round hit probability. Both the commander and gunner can traverse the turret, control the ranging laser, and fire the gun.

Main armament is the 105-mm. M68 rifled gun, a license-built version of the L7A3 gun, which also arms the M48A5, M1 ABRAMS, Israeli MERKAVA, German LEOPARD 1, the Swedish S-TANK, and late-model British CENTURIONS. The M68 can fire APDS, APFSDS, HEAT, anti-personnel, and smoke rounds, with an effective range of more than 2500 m. Muz-zle velocities are 1600 m./sec. for APFSDS, 1450 m./sec. for APDS, 1170 m./sec. for HEAT, and 821 m./sec. for anti-personnel rounds.

The M68, which has seen much combat, can defeat the armor of all known Soviet MBTs with APFSDS rounds. In the M60, the gun has eleva-tion limits of + 20° and − 10°, and a maximum rate of fire of 6 to 8 rounds per minute. The turret has electro-hydraulic power traverse, enabling the gun to rotate 36° in 15 seconds. Its high-pressure hydraulic lines, however, were found to be a fire hazard in the 1973 Yom Kippur War, and the sys-tem was modified by the Israelis and later retrofit-ted on U.S. M60s. The main gun was not stabilized in the M60 and M60A1, but stabilization is stan-dard on the A3 and has been retrofitted into most older M60s, allowing them to fire fairly accurately on the move. In addition, the M60A3 has an in-sulated thermal sleeve around the gun to prevent barrel warp, and a muzzle reference system for use with the laser rangefinder. The M60A3 carries a total of 60 rounds of main gun ammunition: 26 in the driving compartment, 13 in the turret, 21 in the turret bustle, and 3 under the gun.

Secondary armament includes the commander's M2, a coaxial 7.62-mm. machine gun, and, in some cases, a 7.62-mm. machine gun by the loader's hatch. A six-round smoke grenade launcher is fitted on each side of the turret. A white light/infrared searchlight was mounted over the main gun on the M60A1, but this has been removed from many A3s. All M60s have a COLLECTIVE FILTRATION unit to which the crew can connect their personal gas masks. The M60A3 also has a rapid-acting, automatic HALON fire extinguisher (another Israeli-developed feature).

U.S. intelligence assessments have rated the M60A1 superior to the Soviet T-62 MBT, and marginally inferior to the later T-64 and T-72; and the M60A3 as equal to the T-64 and T-72, and marginally inferior to the T-80. To maintain the effectiveness of the M60 against rapidly improving Soviet tanks, several users have initiated modernization programs to improve mobility, protection, and fire controls. The Israeli army has produced a radically modified variant (Megach 7) which has modular CHOBHAM armor (as used on the Merkava III) added to the turret and frontal glacis, Chobham armor side skirts, a more powerful engine, and Elbit "Matador" computerized fire controls. Teledyne Continental offers a "High Performance M60," with a 1200-hp. AVCR-1790-1B engine, a reinforced suspension, spaced applique armor, and armored side skirts. This vehicle is reported to be protected against the heaviest Soviet tank guns at normal battle ranges (800–1500 m.), to have a maximum speed of 45 mph (75 kph) on roads and 28 mph (46.75 kph) cross-country, and to have a power-to-weight ratio of 23.1 hp./ton. Given the high cost of the latest MBTs (e.g., the LEOPARD 2 and M1 Abrams), it appears probable that some M60 users will opt for a comprehensive upgrade program in the 1990s.

The M60 chassis has been used for a number of specialized vehicles, including the M60 AVLB bridgelayer and the M728 COMBAT ENGINEER VEHICLE (CEV).

Specifications Length: 22.8 ft. (6.95 m) Width: 11.95 ft. (3.64 m) Height: 10.7 ft. (3.26 m) Weight, combat: 51 tons. Powerplant: (A1/3) 750-hp. Continental AVDS-1790-2A diesel; (Megach 7) 900-hp. AVCR-1790-6A diesel. Hp./wt. ratio: (A1/3) 14.7 hp./ton; (Megach 7) 17.6 hp./ton. Fuel: 370 gal. (1628 lit.). Speed: 30 mph (50 kph) roads/8.5 mph (14.2 kph) cross-country. Range, max.: 300 mi. (500 km.). Operators: Aus, Egy, Iran, Isr, It, Jor, Leb, N Ye, Oman, ROK, S Ar, Sud, Tun, US.

M72A1 LIGHT ANTI-TANK WEAPON:
See LAW.

M107: U.S. 175-mm. self-propelled (SP) GUN, whose chassis is shared by the M110 203-mm. SP HOWITZER. Designed in 1952 as a replacement for existing heavy artillery, the M107 and M110 share the same chassis, hull, and drive train, and have interchangeable gun mounts. The first production vehicles were delivered in 1961, and were soon used in Vietnam, where their extremely long range was offset by inaccuracy, a much lighter shell than the 203-mm. howitzer, and rapid barrel wear. When the range of the M110 was extended, the U.S. Army decided to phase out its M107s by replacing the 175-mm. guns with lengthened 203-mm. howitzers, converting the vehicles to M110s. But the M107 was also widely exported, and still serves with Great Britain, Greece, Iran, Israel, Italy, South Korea, the Netherlands, Spain, Turkey, and West Germany.

The M107 consists of a 175-mm. M113 gun in an open mount atop a light, fully tracked chassis. The hull is of all-welded steel construction, with a driving compartment up front on the left, and the engine compartment on the right. The back of the vehicle houses the gun elevation and traversing mechanisms, a recoil mechanism, and a platform for the gun crew. A large spade that extends across the rear of the vehicle can lowered hydraulically before firing, to transfer some of the recoil load to the ground. The gun is served by a crew of 13, 5 of whom (the commander, driver, and 3 gunners) travel on the gun mount, while the remainder accompany the gun in an M548 tracked cargo vehicle. The gun can traverse 65° to the left and right, and has elevation limits of −2° and +65°.

The M107 is equipped with a hydraulic rammer and loader mechanism to reduce the crew's workload. The gun can fire its 147-lb. (66.8 kg.) shell to a range of 40,000 m. at a rate of one or two rounds per minute. Only two ready rounds are carried on the vehicle, with more in the M548. A major disadvantage of the M107 (and M110) is the lack of overhead cover, which leaves the gun crew totally exposed to COUNTERBATTERY fire and NBC effects. A KEVLAR splinter shield can be erected over the gun, but is no substitute for an armored turret. The vehicle is not amphibious, but can ford streams to a depth of 3.5 ft. (1.06 m.). See also ARTILLERY, UNITED STATES.

Specifications Length: 20.33 ft. (6.22 m.). Width: 8.85 ft. (7.2 m.). Height: 6.83 ft. (2.085 m.). Weight, combat: 17.4 tons. Powerplant: 405-hp.

Detroit 8V-71T diesel. **Hp./wt. ratio:** 14 hp./ton. **Fuel:** 100 gal. (44 lit.). **Speed, Road:** 36 mph (60 kph). **Range, Max.:** 435 mi. (726 km.).

M108: U.S. 105-mm. self-propelled (SP) HO-WITZER, no longer in U.S. service, but still in service with the armies of Belgium, Brazil, and Spain. See also ARTILLERY, UNITED STATES.

M109: A U.S. 155-mm. self-propelled (SP) HOWITZER which has the same chassis, hull, drive train, and turret as the M108 105-mm. SP howitzer. The first M109s entered service in 1961, and were much used in Vietnam. The M109 has been by far the most successful of all SP howitzers, and has been widely exported. Still in production, more than 4000 M109s have been delivered to date.

The physical dimensions and automotive characteristics of the M109 are virtually identical to those of the M108, but combat weight is three tons greater because of the much heavier gun barrel, and the addition of two hydraulically powered spades at the rear, which can be lowered before firing to transfer recoil forces to the ground.

The initial M109 had a short M126 howitzer tube, which could fire a 94-lb. (47.72-kg.) projectile to a maximum range of 14,600 m. Seriously outranged by Soviet guns, the M126 was replaced in 1973 by the longer M185 tube (with the resulting vehicle designated M109A1). The M185 is actually a GUN-HOWITZER, distinguished by its longer (39-caliber) barrel with a prominent fume extractor and large, multi-baffle muzzle brake. It has a maximum range of 18,000 m., which can be extended to 24,000 m. with ROCKET-ASSISTED PROJECTILES (RAPs). It can fire a wide variety of ammunition, including high explosive (HE), bomblet-filled IMPROVED CONVENTIONAL MUNITIONS (ICMs), unitary and BINARY chemical, smoke, illumination, and tactical nuclear rounds. Fourteen rounds are stowed in the vehicle. The turret has 360° traverse and elevation limits of −3° and +75°. The maximum rate of fire is two rounds per minute.

The M109A2, introduced in 1979, has numerous detailed improvements, including an enlarged turret bustle with stowage for an additional 22 rounds of ammunition (for a total of 36); the current production version (M109A3) has additional minor refinements. The M109A5, now in development, is a radically modified vehicle with new FIRE CONTROLS, and an automatic loader (increasing the rate of fire to six rounds per minute in short bursts). See also ARTILLERY, UNITED STATES.

Specifications Length: 20.3 ft. (6.19 m.). Width: 10.16 ft. (3.1 m.). Height: 10.75 ft. (3.28 m.). Weight, combat: 24 tons. Powerplant: 405-hp. Detroit 8V-71T turbocharged diesel. Hp./wt. ratio: 16.23 hp./ton. Fuel: 116 gal. (511 lit.). Speed, road: 35 mph (56.3 kph). Range, max.: 216 mi. (349 km.). Operators: Aut, Bel, Can, Den, Egy, Eth, FRG, Gre, Iran, Isr, It, Ku, Lib, Mor, Neth, Nor, Pak, Peru, Por, ROK, S Ar, Sp, Swi, Tai, Tun, Tur, UK, US.

M110: U.S. self-propelled (SP) 203-mm. GUN-HOWITZER, which shares the chassis of the M107 175-mm. SP GUN. Designed in 1952, the M110 entered service in 1961 and was much used in Vietnam. The M110 has been very successful and has become the standard corps-level artillery piece in many Western armies. Still in production, more than 750 have been delivered to date.

The hull, chassis, drive train, and gun mount of the M110 are identical to those of the M107; only the gun tubes differ. As originally delivered, the M110 had an M2A1 203-mm. HOWITZER tube of pre–World War II vintage, which could fire a 203.5-lb. (92.5-kg.) projectile to a range of 16,800 m. at a rate of one round per minute. Available ammunition included high-explosive (HE), chemical, and nuclear rounds. The weapon proved satisfactory in every respect except range. To rectify that deficiency, the M2A1 was replaced with the M201 GUN-HOWITZER tube, distinguished by its much longer barrel. New vehicles, and older models retrofitted with the new tube, were redesignated M110A1 (without muzzle brake) or M110A2 (with muzzle brake).

The M110A2 uses separate-loading (bag charge) ammunition, and can fire its 198-lb. (90-kg.) projectile to a range of more than 30,000 m. Because that range was competitive with that of the problem-ridden 175-mm. gun of the M107, all U.S. M107s were converted to M110A1/A2s beginning in 1976. In addition to HE and nuclear rounds, the M110A1/A2 can also fire bomblet-filled IMPROVED CONVENTIONAL MUNITIONS and BINARY chemical rounds. Only three ready rounds can be carried on the vehicle, with more inside an accompanying M548 tracked cargo vehicle. The M110 has a crew of 13, 5 of whom (driver, commander, and 3 gunners) ride on the vehicle, with the remainder in the M548.

The principal drawback of the M110 (and the M107) is the lack of overhead cover for the gun crew, which leaves them exposed to counterbattery fire and NBC effects. A KEVLAR splinter shield can be erected over the gun, but is no substitute for an enclosed armored turret. See also ARTILLERY, UNITED STATES.

Specifications **Length:** 20.33 ft. (6.22 m.). **Width:** 8.85 ft. (7.2 m.). **Height:** 6.83 ft. (2.085 m.). **Weight, combat:** 17.4 tons. **Powerplant:** 405-hp. Detroit 8V-71T diesel. **Hp./wt. ratio:** 14 hp./ton. **Fuel:** 100 gal. (44 lit.). **Speed, road:** 36 mph (60 kph). **Range, max.:** 435 mi. (726 km.). **Operators:** Bel, Gre, Iran, Isr, It, Jap, Jor, Neth, Pak, ROK, S Ar, Sp, Tai, UK, US.

M113: U.S.-developed tracked ARMORED PERSONNEL CARRIER (APC), now in service worldwide. Developed in the late 1950s by FMC as a replacement for the M59 APC, the M113 was to be smaller, cheaper, lighter, faster, and amphibious. The original M113, equipped with Chrysler gasoline engines, entered service in 1960; it was followed in 1963 by the M113A1, with a General Motors Detroit diesel engine, which increased range and reduced the risk of fire. The M113A2, which entered service in 1978, introduced a more robust suspension and various small improvements, but was otherwise similar. The current production version (M113A3), which entered service in 1987, has spaced applique armor for extra protection against the shaped-charge (HEAT) warheads of light anti-tank weapons. To date, more than 73,000 M113s of all models have been delivered, (including 4500 produced under license by OTO-Melara in Italy)—more than any other APC. In fact, the mechanization of the infantry in all Western armies has largely been accomplished with the M113.

The hull is a box of aluminum armor, with a glacis plate sloped at 60°, vertical sides, and a square rear end which frames a large powered ramp-door. Armor thickness varies from 12 to 38 mm., sufficient for protection against shell splinters and small arms fire (up to 12.7 mm. frontally). The hull is divided into an engine compartment up front and a crew/troop compartment in the rear. In addition to its driver and a commander, the M113 can carry a rifle squad of 11 men. The driver, up front on the left with the engine to his right, is provided with four observation periscopes, one of which can be replaced by a passive or INFRARED (IR) night scope. The vehicle commander sits in the center of the vehicle just behind the driver, under a cupola equipped with five observation periscopes. The troops sit behind the commander on inward-facing bench seats along the sides of the compartment. In addition to the ramp-door, the

vehicle has a large rectangular roof hatch. Although the M113 does not have gun ports, infantrymen can fire their weapons over the side by standing in the opened roof hatch. The drive trains of all three versions are similar, but the A2 and A3 have a more rugged suspension. The M113 is amphibious, being propelled through the water by track action at 3.5 mph (5.85 kph); a fold-down trim vane mounted on the glacis is erected before entering the water.

The M113 is essentially a "battle taxi" intended to transport troops to the battlefield for dismounted action, and not an INFANTRY FIGHTING VEHICLE (IFV) designed for some degree of mounted combat, like the M2 BRADLEY or Soviet BMP. Standard armament is limited to a single M2 HB BROWNING .50-caliber (12.7-mm.) machine gun on a ring mount around the commander's cupola. Most users also add pintle mounts for two or three 7.62-mm. machine guns by the roof hatch. Some Australian M113s used for fire support have a SALADIN armored car turret with a 76.2-mm. gun in place of the commander's cupola, while some Swiss M113s have power-operated 20-mm. cannon. Smoke grenade launchers are also becoming standard equipment, while a variety of COLLECTIVE FILTRATION units for NBC defense have been fitted by different users. Israeli M113s have been radically modified to reduce vulnerability: fuel tanks have been removed from under the floor to external panniers on either side of the ramp door (to cope with mines); an armored box on the floor is installed to contain ordnance, and a complex form of spaced armor has been added to the sides for protection against HEAT warheads.

The M113 chassis is the basis of many specialized vehicles, including the M125 (81-mm.) MORTAR carrier; the M106 (107-mm.) mortar carrier; the M577 armored command vehicle or ACV (with enlarged cabin and extra radio equipment); the M548 tracked cargo carrier; the M163 VULCAN air defense vehicle (with 6-barrel 20-mm. cannon); a light armored recovery vehicle; the LYNX armored reconnaissance vehicle, M901 IMPROVED TOW VEHICLE (ITV); the M981 FISTV artillery spotting vehicle; the CHAPARRAL surface-to-air missile (SAM) launcher; the British tracked RAPIER SAM launcher; and the Canadian version of ADATS.

Another derivative of the M113 is the Armored Infantry Fighting Vehicle (AIFV), developed as a private venture by FMC in competition with the more complex and expensive Bradley. Essentially an M113A2 with additional applique armor, a cut-down troop compartment, and sloping sides, the AIFV has a power-operated turret with a 25-mm. Oerlikon cannon and five firing ports. It carries a crew of three (driver, commander, and gunner) and a squad of seven men. The Dutch army ordered 850 vehicles in 1975, and it now uses the AIFV chassis for an entire range of specialized vehicles, including a TOW missile launcher (with ITV turret), a command vehicle, a radar vehicle, a cargo carrier, a 120-mm. mortar tractor, and an ambulance, in addition to the basic APC. In 1981, Belgium ordered 514 vehicles, with many built under license. The Philippines also have about 30 AIFVs.

Specifications Length: (A1/2/3/4) 16 ft. (4.88 m.); (AIFV) 17.25 ft. (5.26 m.). **Width:** 8.81 ft. (2.686 m.); (AIFV) 9.25 ft. (2.82 m.). **Height:** (A1/2/3/4) 8.16 ft. (2.49 m.); (AIFV) 9.16 ft. (2.79 m.). **Weight, combat:** (A1/2/3/4) 11.5 tons; (AIFV) 13.6 tons. **Powerplant:** 215-hp. GM Detroit 6V-53 diesel. **Hp./wt. ratio:** (A1/2/3/4) 18.51 hp./ton; (AIFV) 15.8 hp./ton. **Fuel:** 81.85 gal. (360 lit.). **Speed, road:** (A1/2/3/4) 42 mph (70 kph); (AIFV) 38 mph (63.5 kph). **Range, max.:** (A1/2/3/4) 370 mi. (618 km.); (AIFV) 305 mi. (570 km.). **Operators:** Aus, Bel, Bol, Bra, Can, Chi, Cos R, Den, Ecu, Egy, Eth, FRG, Gre, Gua, Hai, Iran, Isr, It, Jor, Ku, Laos, Leb, Lib, Mor, Neth, NZ, N Ye, Nor, Pak, Peru, Phi, Por, ROK, S Ar, Sing, Som, Sp, Su, Swi, Tai, Thai, Tun, Tur, Ur, US, Zai.

M551 ARMORED RECONNAISSANCE VEHICLE: See SHERIDAN.

MAC: See MILITARY AIRLIFT COMMAND.

MACHINE CANNON: An older term used for very heavy machine guns (in excess of 15 mm.) and automatic CANNON (from 20 mm).

MACHINE GUN: Weapons in calibers up to 15 mm., in which the entire cycle of loading, chambering, firing, extraction, and reloading is fully automatic. Since modern shoulder arms are now ASSAULT RIFLES also capable of automatic fire, machine guns are distinguished by their capacity for *sustained* fire, achieved by heavier (or water-cooled) barrels, which in turn require the use of bipods, tripods, or other forms of support. Machine guns fire cartridges mostly between 7.62 mm (.30 caliber) and 14.5 mm. (.57 calibre); Squad Automatic Weapons (SAWs), which fire smaller rounds, are actually heavy-barreled assault rifles, while weapons larger than 14.5 mm are classified automatic CANNON.

LIGHT MACHINE GUNS (LMGs) are rifle-caliber weapons designed to be carried by one man, capable of firing from the hip or shoulder for brief periods, and usually assigned for direct support within infantry squads. They are usually fed from box or drum magazines holding from 20 to 75 rounds, and have relatively light barrels limiting them to burst fire unless the barrel is changed frequently; quick-change barrel-fittings are usually provided. The LMG has now generally been supplanted by the SAW, except in the Soviet Union, which still uses its RPK and RPD models.

GENERAL-PURPOSE MACHINE GUNS (GPMGs) are heavier rifle-caliber weapons served by two men, almost always fired from a bipod or tripod, fed from ammunition belts and capable of sustained fire with their quick-change barrels. A German concept of World War II, the GPMG is suitable for both the light and medium roles, and has become the predominant type on the modern battlefield. The most important models in service are the U.S. M60, the West German MG3, the Begian FN-MAG, and the Soviet PK.

HEAVY MACHINE GUNS (HMGs) fire larger-caliber cartridges (mostly 12.7-mm. or 14.5-mm.), and have heavy barrels for sustained fire (older HMGs were water-cooled). Weighing from 80 to 120 lb. (24.4 to 54.5 kg.), they are often manned by a crew of three, must be fired from tripod, vehicle, or powered mounts, and are normally used to arm vehicles or aircraft, and as ANTI-AIRCRAFT guns. The most important models in service are the U.S. M2 HB BROWNING and the Soviet DSHK and KPV.

A number of different loading and firing mechanisms ("actions") are employed in current machine-gun design. Blowback actions rely on the gas pressure generated by firing to drive the bolt backwards, open the breech and eject the spent casing. The bolt is then driven forward again by a spring, pushing a round from the belt or magazine into the chamber as it advances; only the inertia of the bolt and the recoil spring hold the bolt closed. Pure blowback actions are suitable only for SUBMACHINE GUNS, but machine-gun variants include retarded or delayed blowback, which allows the full gas pressure to propel the bullet out of the barrel before the breech is opened. In recoil actions, the rearward travel of the barrel opens the breech, and the barrel is returned to its firing position by a spring. In gas actions, the pressure of combustion gases drives back a piston which mechanically opens the bolt; a port at the end of the barrel shunts gas to a cylinder (over or under the barrel) which houses the piston. Rotary-barrel or Gatling guns have three or more barrels fed by a single breech. The barrels rotate around a common axis, coming successively into line with the chamber; very high rates of fire can thus be achieved without overheating. Gatlings are the oldest practical type of machine gun, first introduced in 1858, with a hand crank to rotate the barrels; modern versions use electric motors. Gatling actions are used mainly for cannon, but the U.S.-made rifle-caliber 7.62-mm. GAU-2A minigun is often mounted on helicopters. The CHAIN GUN principle has so far been limited to cannon, although several rifle-caliber and 12.7-mm. models have been proposed.

MACHINE PISTOL: Another term for SUB-MACHINE GUN (derived from the German *Machinepistole*).

MACKENZIE: A class of four Canadian FRIGATES commissioned in 1962–63. Modified versions of the earlier RESTIGOUCHE class, the MacKenzies have improved habitability and better cold-weather performance. The MacKenzies are flush decked, but have a rounded "turtleback" foredeck designed to prevent ice buildup in arctic conditions. Like all Canadian warships, they are extremely rugged, with very good seakeeping for Arctic operations. Under current plans, the MacKenzies will be retained by Canada until 1993–94.

Armament is rather light for the tonnage: 2 twin 3-in. 50-caliber DUAL PURPOSE guns at the bow and stern, 2 Mk.10 LIMBO depth charge mortars aft, and 2 sets of triple tubes for lightweight anti-submarine homing TORPEDOES. See also FRIGATES, CANADA.

Specifications Length: 371 ft. (168.6 m.). Beam: 42 ft. (12.8 m.). Draft: 13.5 ft. (4.11 m.). Displacement: 2380 tons standard/2890 tons full load. Powerplant: twin-shaft steam: 2 oil-fired boilers, 2 sets of geared turbines, 30,000 shp. Speed: 28 kt. Range: 4750 n.mi. at 14 kt. Crew: 210. Sensors: 1 SPS-12 air-search radar, 1 SPS-10 surface-search radar, SPG-34 and SPG-48 fire control radars, 1 Sperry Mk.2 navigation radar, 1 SLQ-505 hull-mounted medium-frequency sonar. Electronic warfare equipment: 1 WLR-1 radar warning receiver, 1 SLQ-25 Nixie torpedo countermeasures unit.

MAD: 1. Mutual Assured Destruction, another term for the U.S. strategic policy known officially as ASSURED DESTRUCTION.

2. See MAGNETIC ANOMALY DETECTOR.

MAESTRALE: A class of eight Italian FRIGATES commissioned between 1982 and 1985. Enlarged versions of the earlier LUPO class, the Maestrales are specialized for ANTI-SUBMARINE WARFARE (ASW). The added size has been used to provide a fixed hangar for two Agusta-Bell AB.212ASW (HUEY) helicopters and a platform for a VARIABLE DEPTH SONAR (VDS) in the stern, as well as better seaworthiness, habitability, and range. The ships also have a PRAIRIE MASKER noise suppression device.

Armament is heavy for the tonnage: a single OTO-Melara 5-in. 54-caliber DUAL PURPOSE gun on the bow, an 8-round launcher for ASPIDE air-defense missiles (with 16 reloads) before the bridge, 2 twin Breda 40-mm. ANTI-AIRCRAFT guns amidships (linked to the Dardo anti-missile fire control system), 4 single Teseo launchers for OTOMAT Mk.2 anti-ship missiles on top of the aft superstructure, 2 21-in. (533-mm.) torpedo tubes, and 2 sets of ILAS-3 triple tubes for lightweight ASW homing TORPEDOES. See also FRIGATES, ITALY.

Specifications Length: 402.6 ft. (122.75 m.). Beam: 42.3 ft. (12.9 m.). Draft: 27.6 ft. (8.41 m.). Displacement: 3040 tons standard/3200 tons full load. Powerplant: twin-shaft CODOG: 2 GMT B230 diesels, 10,146 hp. (cruise), 2 General Electric/FIAT LM2500 gas turbines, 51,600 shp. (sprint). Speed: 32 kt. Range: 1500 n.mi. at 32 kt./6000 n.mi. at 16 kt. Crew: 225. Sensors: 1 RAN10S air-and-surface search radar, 1 SPQ-2F surface-search radar, 1 Albatros missile guidance radar, two RTN-30X Dardo fire control radars, 1 SPN-703 navigation radar, 1 Galileo OG-30 electro-optical target tracker, 1 DE1160B hull-mounted medium-frequency sonar, 1 DE1164 VDS. Electronic warfare equipment: 1 Electronica active/passive electronic countermeasures suite, 2 SCLAR 20-round chaff launchers, 1 SLQ-25 Nixie torpedo countermeasure sled.

MAG: See MARINE AIR GROUP.

MAGIC: The French Matra R.550 short-range, INFRARED-HOMING air-to-air missile (AAM), developed from 1968 to compete with the U.S. AIM-9 SIDEWINDER. Flight tests began in 1973, and the first production missiles were accepted by the French *Armee de l'Air* in 1973; since then, the R.550 has been widely exported. A key factor in this success is Magic's complete compatibility with Sidewinder launchers and fire controls, without the latter's subjection to U.S. export controls. More than 8000 Magics have been delivered to date at an average price of $15,000.

Magic is broadly similar to Sidewinder, with an INFRARED (IR) seeker in the nose, a warhead section at midbody, a rocket motor in the rear, fixed tail fins, and canard controls. It differs, however, in having a double set of cruciform canards: a fixed set of winglets is mounted ahead of a second set of movable control surfaces. Four fixed tail fins provide lift and stabilization. The IR seeker is based on a lead sulfide (PbS) detector cooled by a liquid nitrogen bottle mounted within the launch rail. The seeker is effective within a 140° cone centered on the target's jet pipe (in this regard the missile is comparable to earlier Sidewinders; i.e., the AIM-9/G/H/J, rather than later, all-aspect models). The high-explosive warhead weighs 27.6 lb. (12.55 kg.) and is fitted with both contact and proximity fuzes. Magic flies a boost-glide profile: its SNPE Romeo solid rocket motor burns for 1.9 seconds, and the missile glides thereafter. Performance is roughly half that of the latest Sidewinders. Maximum launch speed is 808 mph (1350 kph), and the load limit is 6 g at launch.

In 1984, Matra introduced the Magic II, with a more powerful Richard butylene motor, a more sensitive all-aspect IR seeker, and an improved proximity fuze. The Magic II seeker can be slaved to the radar of the MIRAGE 2000 fighter for long-range lock-on; several users have ordered the new missile as the primary armament for their Mirage 2000s. Performance is otherwise similar to the original R.550.

Specifications   Length: 9.08 ft. (2.77 m.). Diameter: 6.2 in. (158 mm.). **Span:** 26.3 in. (665 mm.). **Weight, launch:** 198 lb. (90 kg.). **Speed:** Mach 3 (2100 mph/3507 kph). **Range envelope:** 0.2–6.2 mi. (0.22–10.35 km.). **Height envelope:** 300–59,000 ft. (91–17,988 m.). **Operators:** Ab Dh, Arg, Ecu, Egy, Fra, Gre, Ind, Iraq, Ku, Lib, Oman, Pak, S Af, S Ar, Syr, UAE.

MAGNETIC   ANOMALY   DETECTOR (MAD): An airborne ANTI-SUBMARINE WARFARE (ASW) sensor which detects disturbances in the earth's magnetic field caused by the steel mass of submarine hulls. MAD systems consist of very sensitive magnetometers whose data is shown on tactical displays inside the aircraft. As the aircraft approaches a submarine, the strength of the magnetic flux increases, peaking directly overhead. To avoid disruption of the magnetometers by the airframe, MAD systems are magnetically isolated by placing them on a boom ("stinger") extended from the tail of an aircraft, or by trailing them at the end of a long cable ("bird") from a helicopter. MAD systems can be extremely accurate, but have very short range; hence they are of use mainly to fix the precise position of submarines for weapon release, after they have been found by other means, including air-dropped SONOBUOYS or dipping SONAR.

MAGNETO-HYDRODYNAMIC   (MHD) POWER: An advanced form of propulsion, not yet built, but proposed for nuclear submarines. An MHD plant would consist of an open-ended, flexible tube surrounded by a sealed sleeve filled with a ferrous metal in a liquid solution (a "ferro-liquid"). Seawater would be drawn into the forward end of the tube, while a powerful electromagnetic field would induce sympathetic vibrations in the ferro-liquid, to create an undulating "traveling wave" within the tube which would accelerate and expel the seawater to the rear, thereby generating forward thrust. Because of the absence of moving parts, MHD propulsion would be extremely quiet, but it would require a huge amount of electrical energy, and could propel the submarine only at low speed. Nevertheless, MHD could be ideal for nuclear-powered ballistic-missile submarines (SSBNs), which spend most of their time cruising silently at very low speeds (a conventional propeller would be retained for high-speed evasion). Because MHD was first described in a Soviet technical journal, there have been unconfirmed reports that such a plant has been installed on TYPHOON-class SSBNs.

MAIL: NATO code name for the Beriev Be-12 *Tchaika* (Seagull), a twin-turboprop amphibious flying boat in service with Soviet Naval Aviation (AV-MF) for ANTI-SUBMARINE WARFARE (ASW), search and rescue (SAR), and maritime patrol.

MAIN BATTLE AREA: The battlefield zone selected as the primary defensive position for a unit of brigade size or larger. The main battle area is located behind the COVERING FORCE zone, which contains light forces and observation posts, to provide warning of enemy approaches, inflict casualties, and channel the enemy advance into the defense-favored sectors of the main battle area. The main battle area can have considerable depth. Depending on the OPERATIONAL METHOD of the defense, it may contain fortified positional defenses covered by obstacles and minefields; or purely mobile forces, whose mission is to attack the flanks and rear of enemy formations without attempting to hold terrain. Most plans also call for the retention of a large, mobile reserve behind the main battle area, to plug gaps in the line, contain enemy penetrations, and finally to mount large-scale counterattacks aimed at pushing the enemy back behind the covering force zone. See also AREA DEFENSE; DEFENSE-IN-DEPTH; MOBILE DEFENSE.

MAIN BATTLE TANK (MBT): During World War II, light, medium, and heavy TANKS were employed conjointly. LIGHT TANKS, often fast but lightly armed and thinly armored, were used mainly for RECONNAISSANCE and PURSUIT. Heavy tanks, slow and ponderous, strongly armed and heavily armored, were used as long-range tank killers and to spearhead breakthroughs. Medium tanks were the jacks of all trades, often with good mobility, adequate ARMOR, and medium-caliber guns. By 1943, however, a new class of tank had emerged; weighing 36–45 tons, they combined the firepower of heavy tanks with the mobility of the mediums. The first of these tanks was the pre-1941 Soviet T-34, the ancestor of all modern tanks, which was followed by the German PzKpfw.V PANTHER, the upgraded Soviet T-34/85, and the later British CENTURION, and the U.S. M26 Pershing. After 1945, it was recognized that this class of tank was capable of undertaking almost all combat missions; described as main battle tanks (MBTs), they became the predominant type, marginalizing both light and heavy tanks.

Any MBT design entails a compromise among firepower, protection, and MOBILITY; the relative emphasis placed on each differs from army to army, in accordance with varying concepts of armored warfare. During the 1960s the West German army, for example, maintained continuity with the German World War II emphasis on rapid counterattacks against enemy flanks, and thus stressed mobility, combined with reasonable armor and standard 105-mm. firepower in its LEOPARD 1. Under Israeli Influence, however, the priority of protection greatly increased in the later LEOPARD 2. The French army placed even greater emphasis on mobility at the expense of armor in its AMX-30. The British army, on the other hand, bases its armored doctrine on long-range engagements from prepared positions. Accordingly, its CHIEFTAIN and CHALLENGER MBTs are very heavily armored, and mount high-velocity 120-mm. guns, but are relatively underpowered and have only mediocre mobility. The Israeli army has similar design priorities, even though its armored doctrine emphasizes mobility, because Middle Eastern conditions expose tanks to long-range fire; its MERKAVA is one of the heaviest MBTs in service, and it too was underpowered in early versions. The U.S. PATTON series (M46, M47, M48, and M60), all originally derived from the Pershing, reflect an almost equal balance of the three elements which is maintained in the newer M1 ABRAMS. Soviet post-war tanks, including the T-44, T-54/55, and T-62 (all developed from the wartime T-34/85), reflect an armored doctrine based on high-tempo offensive operations by mass formations: they are highly mobile, heavily armed, but with relatively light armor (fractionally offset by a very low silhouette).

Until the mid-1970s, any increase in armor protection inevitably meant a reduction in mobility, and this limited most MBTs to 50 tons or less. But the latest MBTs have 1200 to 1500-hp. diesel or gas-turbine engines which allow combat weights to rise into the 60-ton range, with a concurrent *increase* in speed and mobility over older designs. The latest generation of MBTs, including the Challenger, the Abrams, the Leopard 2, the Soviet T-64, T-72, and T-80, and the Israeli Merkava III, also incorporate advanced composite armor (e.g., CHOBHAM ARMOR), providing much greater protection per pound than conventional steel.

The main disadvantage of the latest MBTs is their cost (typically more than $1 million each), which precludes acquisition in large numbers. Designers worldwide are actively pursuing radical solutions to reduce size, weight, and cost—including turretless designs such as the Swedish S-TANK, and tanks with externally mounted guns. Poorer armies, on the other hand, are upgrading existing tanks with added armor, more powerful engines, and new guns and fire controls, instead of purchasing current MBTs.

MAINSTAY: NATO code name for the AWACS (Airborne Warning and Control System) variant of the Soviet Il-76 CANDID four-engine transport aircraft. Developed in the late 1970s as a replacement for the Tu-126 MOSS, Mainstay entered service in 1985–86. The aircraft has a large search RADAR in a rotodome mounted over the fuselage, and this radar is believed to have look-down capability over land, a feature lacking in the Moss. The cabin contains a tactical center with several fighter control consoles. Physical dimensions and performance are essentially identical to the Candid's.

Mainstay has two basic missions: to provide radar coverage across the Soviet Union's northern frontiers, and to provide airborne early warning and control for FRONTAL AVIATION (FA) aircraft defending and supporting the ground forces. It is believed that Mainstay was designed specifically to work with the latest generation of Soviet interceptors: the MiG-29 FULCRUM, MiG-31 FOXHOUND, and Su-27 FLANKER. See also AIRBORNE EARLY WARNING; AIR DEFENSE; PVO.

MALAFON: French shipboard anti-submarine missile. Developed from 1956, the missile entered service in 1964 and currently equips French destroyers and frigates of the Aconit, Galisonniere, SUFFREN, and TOURVILLE classes, each of which has a single launcher with 13 missiles.

Malafon has an airplane configuration with twin rudders mounted on the ends of the tailplane. The nose section houses a 21-in. (533-mm.) L4 acoustic homing TORPEDO. When launched, two solid rocket boosters attached to the missile body burn for several seconds, accelerating the missile to its flight speed before being jettisoned; the missile completes its trajectory as a glider. Malafon is stabilized in flight by an autopilot and radar altimeter. Course corrections are affected by a radio COMMAND GUIDANCE link. In-flight optical tracking from the ship is assisted by flares mounted in the wingtips.

Targeting data is provided by the launching ship's SONAR and FIRE CONTROLS. When the missile reaches a point approximately 800 m. from the target, a braking parachute is deployed from its tail, ejecting the torpedo from the nose section. Malafon can be used in a secondary anti-ship role with fire control data provided by the ship's surface-search radar. See also ASROC; FRAS-1; IKARA; SILEX.

Specifications Length: 20.2 ft. (6.15 m.). Diameter: 25.6 in. (650 mm.). Span: 10.8 ft. (3.29 m.). Weight, launch: 3300 lb. (1500 kg.). Speed: 516 mph (862 kph). Range: 9.3 mi. (16 km.).

MAMBA: A German ANTI-TANK GUIDED MISSILE, an enlarged and improved version of the Messerschmidt-Bolkow-Blohm COBRA.

MANEUVER: A term often synonymous (in loose language) with mere movement; more meaningfully, purposeful military actions related to the location or character of specific enemy forces. The term "maneuver warfare" refers to TACTICS and OPERATIONAL METHODS which attempt to exploit presumed weaknesses in the enemy's force structure and/or tactics, so as to cut LINES OF COMMUNICATION, disrupt their COMMAND AND CONTROL, and undermine the enemy's will to resist. While it evokes visions of agile forces using hit-and-run tactics, maneuver warfare may also take the form of positional defenses or straightforward frontal assaults—for it is the unexpected that is called for, and such things are not expected of some forces. Tactics and methods based on de-

tailed analyses of the enemy's own methods to uncover their weaknesses, and on specifically tailored responses, have been described as "relational-maneuver" warfare by Edward Luttwak. Such methods can enable a small force to defeat a much larger one, but when they fail, they can fail catastrophically (such high risk/high payoff methods are therefore appropriate only for the weaker side in a war, for which low-risk/low-payoff methods could only guarantee defeat by numerical or materiel inferiority).

MANEUVERING REENTRY VEHICLE (MARV): A ballistic missile REENTRY VEHICLE (RV) equipped with its own attitude-control thrusters and/or aerodynamic control surfaces, as well as on-board guidance, to alter the ballistic trajectory—either to correct for aiming errors or to evade BALLISTIC MISSILE DEFENSES (BMDs). MaRVs have been tested by both the United States (on a TRIDENT I SLBM) and the Soviet Union (on an SS-18 Mod 5 SATAN ICBM), but neither country has deployed them to date. (The RADAR AREA CORRELATION GUIDANCE [RADAG] of the now-prohibited PERSHING 2 MRBM could impart small course corrections during the final seconds before impact, but this was not considered a true MaRV capability.)

MaRVs would be generally heavier, much more complex, and more costly than ballistic RVs, but they may become more attractive if advanced BMD systems are deployed.

MARCHING FIRE: An infantry tactic competitive with the more common FIRE-AND-MOVEMENT tactics in which troops are divided into two teams which alternately advance and provide covering fire. In marching fire, all troops advance in a single skirmish line, firing at any position which could conceal enemy forces. This tactic, pioneered in World War II by U.S. Gen. George S. Patton, uses more ammunition and implicitly accepts the casualties from an exposed advance and also a reduction in the accuracy of fire, to obtain shock effects and a controllable concentration of forces. The tactic has now generally been superseded in Western armies by either INFILTRATION or OVERWATCH tactics; a very similar method, however, is now prescribed for Soviet MOTORIZED RIFLE troops at the squad, platoon, and company levels.

MARDER: West German INFANTRY FIGHTING VEHICLE (IFV) developed in the late 1960s to replace the unsatisfactory Swiss Spz. 12-3. The first IFV worldwide, Marder remains to this day an excellent vehicle, matched (if at all) only by the much more recent U.S. M2/M3 BRADLEY and British MICV-80 WARRIOR. Marder was developed as one of a family of vehicles which includes the JAGD-PANZER (JPz) *Kanone* and JPz *Rakete* ASSAULT GUN/TANK DESTROYERS, which all share common automotive components. The prototype Marder (*Schutzenpanzer Neu-1966*) was delivered by Thyssen-Henschel in 1966, the first production vehicles were delivered in 1970, and 3000 were built by the close of production in 1975. Issued to the *Panzergrenadier* battalions of active *Panzer* and *Panzergrenadier* divisions, Marder reflects the German style in ARMORED WARFARE, with excellent speed and cross-country mobility, good armor protection, and heavy firepower, including firing ports for mounted combat.

The all-welded steel hull, with its well-sloped frontal glacis and side plates, is believed to be proof against armor-piercing ammunition up to 14.5 mm in caliber. Divided into a driving compartment, an engine compartment, and a fighting/troop compartment, Marder has a crew of three (driver, commander, and gunner) and can carry a small rifle squad of six men. The driver has three observation periscopes, one of which can be replaced by an INFRARED (IR) night scope.

The vehicle commander and the gunner sit in a two-man turret offset to the right, immediately behind the engine. Both the commander and the gunner have three observation periscopes and an optical gunsight, which can be replaced by an (IR) night sight. The turret, armed with a 20-mm. Rheinmetall Rh.202 automatic cannon and a coaxial 7.62-mm. machine gun, has 360° traverse and elevation limits of $-17°$ and $+65°$, making it effective against (slow) aircraft as well as ground targets. Firing ARMOR-PIERCING ammunition, the Rh.202 can defeat ARMORED PERSONNEL CARRIERS and reconnaissance vehicles, and may be able to penetrate the side and rear armor of some tanks. A MILAN anti-tank guided missile can be mounted on the turret to add anti-tank capability. Six smoke grenade launchers and a white light/infrared searchlight are also mounted on the turret.

The troop compartment behind the turret is accessed through a rear ramp door and two roof hatches. Two circular firing ports on each side with associated roof periscopes allow fire from under armor. As delivered, early Marders also had a remotely controlled 7.62-mm. machine gun mounted on the roof at the rear of the vehicle, but this proved impractical and was later removed. Marder has a positive overpressure COLLECTIVE FILTRATION unit, allowing its occupants to work in a shirt-sleeve environment under NBC conditions. The vehicle's only major drawback is its lack of amphibious capability. No plans to replace the Marder are known, but published upgrades include replacement of the Rh.202 with a 25-mm. cannon and installation of passive night vision equipment. Derivatives include the *Radarpanzer TUR*, with a surveillance radar mounted on an articulated mast; and the Argentine TAM medium tank.

Specifications **Length:** 22.25 ft. (6.78 m.). **Width:** 10.67 ft. (3.25 m.). **Height:** 9.83 ft. (3 m.). **Weight, combat:** 29 tons. **Powerplant:** 600-hp. MTU MB833 Ea-500 6-cylinder diesel. **Hp./wt. ratio:** 20.54 hp./ton. **Fuel:** 170 gal. (750 lit.). **Speed, road:** 46.6 mph (77.8 kph). **Range, max.:** 323 mi. (540 km.).

MARINE AIR GROUP: Tactical aviation unit of the U.S. MARINE CORPS, subordinate to a MARINE AIR WING (MAW). Each MAG consists of 3 to 5 squadrons, for a total of 30–90 fixed-wing aircraft or helicopters. Usually each MAG operates only one class of aircraft (i.e., fighter, transport, attack aircraft, etc.); MARINE EXPEDITIONARY BRIGADES, however, have an attached composite MAG with a mix of fighter, attack, transport, and helicopter squadrons. The Marine Corps has 12 active MAGs, and 4 in the Marine Air Reserve.

MARINE AIR WING (MAW): A major aviation command of the U.S. MARINE CORPS. Each MAW consists of 3 or more MARINE AIR GROUPS (MAGs), each with one class of aircraft (fighter, attack, transport, etc.). Currently there are three active MAWs, the 1st, 2nd, and 3rd, each associated with the respective Marine Division. First MAW, based in Japan, consists of MAG-12, MAG-15, and MAG-36, with a total of some 150 aircraft. Second MAW, based at Cherry Point, North Carolina, consists of MAG-14, MAG-26, MAG-29, MAG-31, and MAG-32, with a total of about 300 aircraft. Third MAW, based at El Toro, California, consists of MAG-11, MAG-13, and MAG-16, with about 200 aircraft. The Marine Air Reserve forms the 4th MAW, with MAG-41, MAG-42, MAG-46, and MAG-49.

In all, the marines operate about 1300 aircraft of all types. The primary mission of Marine Aviation is to provide CLOSE AIR SUPPORT (CAS) and fighter cover for marine ground forces; accordingly, air-ground coordination is greatly emphasized. The corps has 15 attack squadrons with A-4M SKY-HAWKS, AV-8B HARRIERS, and A-6E INTRUDERS; 13

fighter/attack squadrons with FA-18 HORNETS; an ELECTRONIC WARFARE squadron with EA-6B PROWLERS; a RECONNAISSANCE squadron with RF-4B PHANTOMS; 4 AERIAL REFUELING/transport squadrons with KC-130F/R HERCULES; 2 observation squadrons with OV-10A/D BRONCOS; 11 heavy helicopter squadrons with CH-53A/D SEA STALLIONS and CH-53E SUPER STALLIONS; 7 utility helicopter squadrons with UH-1N HUEYS and AH-1T Sea COBRAS; and 15 medium helicopter squadrons with CH-46E SEA KNIGHTS.

MARINE CORPS, UNITED STATES: A separate service under the Department of the Navy, with primary responsibility for AMPHIBIOUS WARFARE. Formed in 1775, and currently numbering about 200,000 men, the USMC is by far the largest naval landing force in the world, with its own organic armor, artillery, helicopter, and air support in addition to a core of infantry forces.

The commandant of the Marine Corps, a full general and member of the Joint Chiefs of Staff, is responsible for all administration and training. Operational control, on the other hand, resides with two Fleet Marine Forces, Atlantic and Pacific, under the respective navy commanders-in-chief of the Atlantic and Pacific fleets.

Administratively, the corps is organized into 3 Marine Infantry Divisions plus a reserve division, three MARINE AIR WINGS (MAWs) plus a reserve MAW, and 3 Service Support Groups, in addition to the central services and headquarters. Each Marine Division has approximately 20,000 men (including some attached navy personnel), in 3 Marine Infantry Regiments, an artillery regiment, a tank battalion, an amphibious assault vehicle battalion, a reconnaissance battalion, a light armored vehicle (LAV) battalion, a combat engineer battalion, and various support units. Each infantry regiment in turn has 3061 men in 3 infantry battalions, each with 847 men in 3 rifle companies and a heavy weapons company.

Operationally, on the other hand, the corps forms combined-arms Marine Air-Ground Task Forces (MAGTFs), around its battalions, regiments, and divisions. The basic MAGTF, the Marine Expeditionary Unit (MEU), consists of a battalion landing team, a composite air squadron (18 aircraft), and a MEU service-support group (for a total of 2506 men), carried in four to six amphibious assault ships. One MEU is usually afloat in the Mediterranean and another in the Pacific, to serve as rapid intervention forces.

The next larger MAGTF is the MARINE EXPEDI-TIONARY BRIGADE, consisting of a regimental landing team, a composite MARINE AIR GROUP (MAG), and a brigade service-support group, for a total of 15,670 men. A MEB requires from 21 to 26 ships for its transport.

The largest MAGTF is the MARINE EXPEDITION-ARY FORCE, consisting of a Marine Infantry Division, a MAW of up to 300 aircraft, and a force service-support group (a total of 50,600 men) for which some 50 ships are needed. Because of support element and shipping limitations, the corps can deploy only 2 MAFs, up to 6 MEBs, or 12 MEUs at any one time. Originally conceived as ad hoc groupings for specific operations, MAGTFs have become quasi-permanent organizations.

Marines, renowned for their toughness and esprit de corps, are trained and equipped primarily as heavy assault infantry for opposed landings in the classic manner of Tarawa and Iwo Jima. This specialization came under increased scrutiny during the 1980s, when many analysts argued that the marines should be reorganized as a light force for rapid intervention in the Third World, leaving major amphibious landings on the European mainland to the the U.S. Army (which in fact carried out all the European-theater landings of World War II).

MARINE EXPEDITIONARY BRIGADE: A standard formation of the U.S. MARINE CORPS, which consists of a regimental landing team, a composite MARINE AIR GROUP, and a brigade service-support unit, for a total of 15,670 men—more than in many divisions.

MARINE EXPEDITIONARY FORCE: A major operational command of the U.S. MARINE CORPS, consisting of a Marine Infantry Division, a MARINE AIR WING, and a Force Service-Support Unit, for a total of 50,600 men.

MARINE EXPEDITIONARY UNIT: A tactical formation of the U.S. MARINE CORPS, consisting of a battalion landing team, a composite air squadron, and a service-support unit, for a total of 2506 men.

MARINEFLIEGER: The air arm of the West German navy.

MARINES: Naval infantry forces primarily trained, organized, and equipped for AMPHIBIOUS WARFARE. The Royal Marines, the U.S. Marine Corps, and Soviet Naval Infantry belong to their respective navies; in other countries, such as France, most of the amphibious infantry is under army control. See also MARINE CORPS, UNITED STATES; MARINES, ROYAL; NAVAL INFANTRY, SOVIET.

MARINES, ROYAL: The naval infantry branch of the British Royal Navy, consisting of some 8000 officers and men trained and equipped for AMPHIBIOUS WARFARE. In contrast to the U.S. MARINE CORPS, which is organized as heavy assault infantry, the Royal Marines (RM) are LIGHT INFANTRY, trained and equipped for raiding, small-scale interventions, and covering-force operations on NATO's mountainous northern flank.

The main RM formation is 3 Commando (CDO) Brigade, which consists of 3 infantry battalions (called COMMANDOS to commemorate their World War II predecessors), an artillery regiment (29 CDO Regiment), 2 commando engineering companies, a helicopter squadron, and a small logistics regiment. Each of the three Commandos (Nos. 40, 42, and 45) has a total of 800 men organized into three rifle companies, a headquarters company, and a support company. No.29 CDO Regiment is actually a British army ARTILLERY unit under RM operational control. It has 544 men and 18 105-mm. towed guns in 3 6-gun batteries, plus a headquarters company.

In addition to the forces assigned to 3 CDO Brigade, the Royal Marines include the elite SPECIAL BOAT SQUADRON (SBS), a separate "Commachio Company" (for the defense of North Sea oil platforms), a training establishment, and small security detachments aboard British warships. Landing craft and helicopters of the Royal Navy are often manned by RM crews.

The RM's major operational task—in cooperation with the Royal Netherlands Marines—is to form an amphibious assault group assigned to NATO's ACLANT command for the reinforcement of Norway, Denmark, Schleswig-Holstein, or the Atlantic islands. Concurrently, the RM, along with the Army's Parachute Regiment, form the British rapid intervention force; in that capacity the RM spearheaded the retaking of the Falkland Islands in 1982. The RM also rotate units to Northern Ireland for four-month security tours, and supply some personnel to United Nations peacekeeping forces.

RM recruits are volunteers subjected to very rigorous training (*basic* training lasts 32 weeks). Their proficiency and esprit de corps make them one of the finest light infantry forces in the world for small-scale operations in all climates and in most tactical conditions. Their weakness, apart from their small numbers, is the inability of the Royal Navy to provide adequate sealift, amphibious landing craft, and gunfire support.

MARTEL: Air-to-surface ANTI-SHIP and ANTI-RADIATION MISSILE (ARM) produced by the Anglo-French consortium of Matra and British Aerospace. Development began in 1963 to produce a missile with both ELECTRO-OPTICAL (TV-guided) and anti-radar versions—hence the name, Missile, Anti-Radar and Television. With an airframe based on the earlier French AS.30, Martel has two alternative guidance heads: the Royal Air Force (RAF) acquired the TV-guided version (mainly for an anti-shipping role), under the designation AJ.168; the French acquired the ARM version, designated AS.37. Small numbers of AS.37s were also acquired by the RAF, but they are being phased out in favor of the BAe ALARM. Martel is carried by the BUCCANEER, JAGUAR, and MIRAGE III fighter/attack aircraft, and by the Breguet ATLANTIC maritime patrol plane. It is believed that several thousand were delivered before production terminated in 1978–79.

Martel has four sharply swept cruciform wings at midbody for lift, with four smaller tail fins for steering. Both versions are powered by a dual-pulse solid fuel motor: an initial booster burns for 2.4 seconds, followed by a 22.2-second sustainer.

The AS.37 has an ogival nose housing an EMD AD.37 passive radar seeker, linked to a programmable scanner which can search through a preselected frequency band. When a radar or radio transmitter is detected, the seeker locks on and the missile can be launched. Even if the emitter changes frequencies, the scanner can reacquire the target so long as the transmission remains within the selected frequency band (alternatively, the missile can be preset beforehand to the frequency of a specific enemy radar). Although the AS.37 is claimed to be highly reliable and very resistant to electronic countermeasures, the RAF was so dissatisfied with its performance that it preferred to use the much smaller and theoretically less effective U.S. AGM-45 SHRIKE during the 1982 Falklands War.

The AJ.168 has a shorter, rounded transparent nose that houses a TV camera. Prior to launch, the camera's field of view is displayed on a television screen inside the cockpit. The operator must locate the target on the screen, and lock onto it by positioning a cursor over it prior to launch. After launch, the missile descends to wave-top height, with its course and altitude maintained by an autopilot and radar altimeter. The scene from the TV is continuously relayed to the launch aircraft via a two-way DATA LINK. The missile can home autonomously, but it may also be steered manually by the operator with a small joystick. Both Martels have the same 331-lb. (150-kg.) high-explosive warhead, but the AJ.168 has a contact fuze, while the AS.37 has a proximity fuze.

Specifications Length: (AS.37) 13.51 ft. (4.12 m.); (AJ.168) 12.7 ft. (3.87 m.). **Diameter:** 15.75 in. (400 mm.). **Span:** 47.25 in. (1.2 m.). **Weight, launch:** (AS.37) 1213 lb. (551 kg.); (AJ.168) 1168 lb. (356 kg.). **Speed:** Mach 0.9 (650 mph/1085 kph). **Range:** 37.3 mi. (62.3 km.) high/ 18.6 mi. (31 km.) at sea level.

MARV: See MANEUVERING REENTRY VEHICLE.

MASCURA: A French naval surface-to-air missile (SAM) deployed aboard the two SUFFREN-class destroyers and the cruiser COLBERT. Developed from the early 1950s by Matra, Mascura is a two-stage, solid-fuel missile similar in configuration to the U.S. TERRIER and STANDARD-ER SAMs, but is considerably larger and heavier, albeit inferior in range and accuracy.

The missile has four broad-chord body strakes for lift, and four movable tail fins; the booster has four fixed tail fins. Mascura's ogival nose contains a RADAR receiver antenna, behind which is a 264-lb. (120-kg.) high-explosive fragmentation warhead fitted with both impact and proximity fuzes. Like Terrier and Standard-ER, Mascura is launched from a twin-arm launcher with a 48-round magazine behind it. At launch, the booster burns for 4.5 seconds to accelerate the missile to flying speed, then falls away; the main engine then burns for 30 seconds, accelerating the missile to its maximum speed. Mascura relies on SEMI-ACTIVE RADAR HOMING, with target acquisition provided by the launch ship's DRB123 surveillance radar, and target illumination by a DRBR51 guidance radar (an earlier BEAM-RIDING variant is no longer in service).

Specifications Length: (booster) 10.85 ft. (3.32 m.); (sustainer) 17.6 ft. (5.38 m.); (overall) 28.45 ft. (8.7 m.). **Diameter:** (booster) 22.4 in. (570 mm.); (sustainer) 16 in. (406 mm.). **Span:** (booster) 4.98 ft. (1.5 m.); (sustainer) 30.3 in. (770 mm.). **Weight:** (booster) 2531 lb. (1140 kg.); (sustainer) 2094 lb. (950 kg.); (total) 4625 lb. (1998 kg.). **Speed:** Mach 3 (2000 mph/3400 kph). **Range:** 31 mi. (50 km.). **Height envelope:** 100–75,500 ft. (30–23,000 m.).

MASKIROVKA: (Russian, lit. "masking"), a Soviet term frequently mistranslated as "camouflage," but which actually encompasses all activities connected with the preservation of secrecy, and the deception of the enemy as to the plans, capabilities, and intentions of Soviet forces. Soviet texts describe *maskirovka* as an art requiring imagination and resourcefulness, whose techniques must be tailored to each situation individually.

Maskirovka is classified as TACTICAL, OPERATIONAL, or STRATEGIC, according to the level of the activity it is designed to mask. Tactical *maskirovka* is similar to normal Western CAMOUFLAGE, CONCEALMENT, AND DECEPTION (CC&D) techniques, but is much more intensive and has a higher priority with commanders in the field. It includes all traditional forms of camouflage, including screening from aerial RECONNAISSANCE, the use of terrain masking, and extensive use of smoke to obscure movement on the battlefield. Operational *maskirovka* includes, in addition the use of SPETSNAZ or normal units for diversionary missions, the deployment of dummy and decoy equipment and the employment of ELECTRONIC WARFARE (*Radioelektron naya bor'ba*, "Radio-Electronic Combat"), including radar and communication jamming, spoofing, and false communications traffic meant to disrupt enemy command and control. Strategic *maskirovka* is meant to disguise the true nature of Soviet objectives and foreign policies in peacetime as well as in war, and to impede a timely and effective adversary response. In pursuit of strategic *maskirovka*, the full resources of the Soviet state are brought to bear, including diplomacy, propaganda, and ACTIVE MEASURES ranging from the subversion of foreign governments, political groups, and individuals to sabotage and the support of terrorism. Strategic *maskirovka* is directed by the Politburo of the Central Committee of the CPSU; in its scope and degree of integration, it far exceeds any comparable effort in the West, and greatly enhances the effectiveness of Soviet strategy. Deep-rooted cultural traits favor *maskirovka* at all levels; Soviet camouflage was unsurpassed in World War II, and the *maskirovka* of intentions was a key factor in the postwar subjection of Eastern Europe.

MASSIVE RETALIATION: The prevalent interpretation of the U.S. strategic policy enunciated by Secretary of State John Foster Dulles in January 1954 (". . . the way to deter aggression is for free communities to be willing and able to respond vigorously at places and with means of our own choosing . . ."). The essence of this interpretation is that any aggression, even if minor, or carried out by proxy, could be met by a nuclear attack on the Soviet Union. That policy had the attraction of economy, since all U.S. interests anywhere in the world could be protected vis-à-vis the Soviet Union by strategic nuclear weapons only, thus obviating the need to maintain large (and expensive) conventional forces.

As interpreted, the policy was criticized on the grounds that it would be ineffectual, since the threat was not credible; i.e., that the Soviet leadership would not be deterred from, say, Third World interventions, because it would (correctly, it turned out) refuse to believe that the U.S. response would actually take the form of a nuclear "first strike" against the U.S.S.R., thus initiating a general nuclear war. Critics therefore argued that U.S. reliance on Massive Retaliation would lead to unacceptable choices between capitulation or obliteration in local crises (see CREDIBILITY).

Massive retaliation was never fully applied in U.S. military strategy; it was in fact background noise for the diplomatic technique of BRINKSMANSHIP. As an intellectual concept, it was an attempt to equate the balance of power with the BALANCE OF TERROR in order to economize on nonnuclear forces. FLEXIBLE RESPONSE and ASSURED DESTRUCTION policies superseded Massive Retaliation in the 1960s, and still form the basis of U.S. national strategy.

MATKA: NATO code name for a class of 17 Soviet HYDROFOIL missile boats, the first of which was completed in 1978. Designed as a replacement for the OSA-class missile boats, the Matkas combine the Osa hull and powerplant with surface-piercing hydrofoils. Generally similar to the TURYA-class hydrofoil torpedo boats, the Matkas' superstructure is considerably larger and taller than the Osas', giving the boats a top-heavy appearance. The Matkas appear to be seriously overloaded, and have been superseded by the much larger TARANTUL-class missile corvettes.

Armament consists of two SS-N-2c STYX antiship missiles in single fixed launch canisters aft, a 76.2-mm. DUAL PURPOSE gun on the bow, and a 30-mm. ADG-630 radar-controlled Gatling gun for anti-missile defense aft between the missile launchers. See also FAST ATTACK CRAFT, SOVIET UNION.

Specifications Length: 131.2 ft. (40 m.). **Beam:** 25.3 ft. (7.71 m.). **Draft:** 6.2 ft. (1.89 m.) excluding foils. **Displacement:** 225 tons standard/ 260 tons full load. **Powerplant:** 3 5000-hp. M504 diesels, 3 shafts. **Speed:** 36 kt. **Range:** 400 n.mi. at 36 kt./650 n.mi. at 25 kt. **Crew:** 30. **Sensors:** 1 "Cheese Cake" surface-search radar, 1 "Plank Stave" air-search radar, 1 "Bass Tilt" fire control radar. **Electronic warfare equipment:** 1 "High Pole" IFF set, 2 "Square Head" IFF interrogators.

MATRA R.530 AND SUPER 530: The Matra R.530 is an obsolescent French medium-range air-to-air missile (AAM) developed from 1957. Available with alternative INFRARED HOMING (IR) or SEMI-ACTIVE RADAR HOMING (SARH) guidance, the R.530 entered service in the early 1960s as the primary armament of the MIRAGE III fighter. Although a mediocre performer against fast, maneuvering targets, the R.530 was sold to at least 14 countries, most of which were Mirage users. At least 44,000 were delivered at an average cost of $44,000 before production terminated in 1980.

The R.530 has an ogival nose and a cylindrical centerbody; four long-chord cruciform delta wings at midbody generate lift, and four small, movable tail fins faciltate steering. The alternative nose sections (SARH or IR) are interchangeable in the field. Matra claims that the IR seeker has all-aspect capability, while the SARH seeker is an EMD AD.26, tuned to the Cyrano 1 bis/ll radar of the Mirage III, the Cyrano IV of the MIRAGE F.1, or the APQ-94 of French F-8E(FN) CRUSADERS. Most users carry R.530s in pairs, one of each type (a practice also followed with Soviet AAMs).

Matra initiated development of the Super 530 in 1971 to redress the performance deficiencies of the R.530. Although superficially similar to the older missile, the Super 530 is actually an entirely new design of vastly superior performance, which incorporates many advanced features. Flight testing began in 1973, with initial production missiles entering French service in 1979. More than 2000 missiles have been delivered to date to more than ten nations, including most operators of the Mirage F.1 and MIRAGE 2000.

The Super 530 has an ogival nose housing an SARH seeker; a cylindrical centerbody housing the guidance electronics and a 65-lb. (29.5-kg.) Thomson-Brandt high-explosive fragmentation warhead with a proximity fuze; and a cylindrical tail section housing a solid rocket motor. The Super 530 has four broad-chord body strakes for lift, and is steered by four cruciform tail fins. The seeker head is produced in two versions: the Super 530F has an AD.26 seeker tuned to the Cyrano IV radar of the Mirage F.1; the Super 530D, introduced in 1985, has a Doppler monopulse seeker matched to the RDI radar of the Mirage 2000, and is claimed to have true LOOK-DOWN/SHOOT-DOWN capability. Both versions are powered by an SNPE Angile butylene dual-impulse motor, with a 2-second booster pulse and 4-second sustainer.

Specifications (R.530): **Length:** (SARH) 10.78 ft. (3.29 m.); (IR) 10.5 ft. (3.2 m.). **Diameter:** 10.35 in. (263 mm.). **Span:** 43.43 in. (1.1 m.). **Weight, launch:** (SARH) 423.3 lb. (192.4 kg.); (IR) 426.6 lb. (194 kg.). **Speed:** Mach 2.7. **Range:** 11.2 mi. (18 km.). **Operators:** Arg, Aus, Bra, Col, Fra, Iraq, Isr, Leb, Pak, S Af, Sp, Ven.

Specifications (Super 530): **Length:** 11.62 ft. (3.54 m.). **Diameter:** 10.35 in. (263 mm.). **Span:** (strakes) 19.7 in. (500 mm.); (fins) 35.4 in. (900 mm.). **Weight, launch:** 551–584 lb. (250–265.4 kg.). **Speed:** Mach 4.6 (3200 mph/5344 kph). **Range:** 22–25 mi. (36.75–41.75 km.). **Ceiling:** 82,000 ft. (25,000 m.).

MATRA R.550: See MAGIC.

MAVERICK: The Hughes AGM-65 series of air-to-surface tactical missiles. Maverick was initiated in 1965 as a homing successor to the command-guided AGM-12 BULLPUP, specifically for use against hardened point targets such as tanks and bunkers. After a 1968 competition with Rockwell, Hughes was awarded a contract for 17,000 missiles by the U.S. Air Force (USAF). Flight tests began in 1969, with the initial production version, the ELECTRO-OPTICAL (TV-guided) AGM-65A, entering service in 1972. The initial production run was completed in 1975, with low-volume production continuing thereafter. In that year, Hughes introduced the AGM-65B, with a scene-magnification TV camera to improve target recognition. Total production of the A and B models reached 31,022 by 1984, at an average price of $43,000 (later models average $265,000). The AGM-65C, introduced in 1977, had semi-active LASER homing (SALH); it was superseded in 1982 by the AGM-65E, which has the "tri-service laser seeker" and a heavier warhead. The AGM-65D, also introduced in 1977, has IMAGING INFRARED (IIR) homing, which is more suitable for night and bad-weather attacks; more than 50,000 of this model have been ordered. The U.S. Navy purchased the AGM-65F, similar to the D, but with the heavier warhead of the E. The AGM-65G, used by USAF, is similar to the F. All versions of Maverick, whether TV-, laser-, or IIR-guided, have a common airframe, with four broad-chord cruciform delta wings for lift, and four rectangular tail fins of equal span for steering. The missile body consists of a rounded nose section containing the guidance package, a midbody warhead section, and a rear motor section. Maverick is compatible with many different aircraft, including the F-4 PHANTOM, F-15 EAGLE, F-16 FALCON, FA-18 HORNET, A-7 CORSAIR, A-10 THUNDERBOLT, F-5 TIGER, and AJ.37 VIGGEN.

To employ the TV-guided AGM-65A, the operator must first acquire the target visually (using either the gunsight or a video display screen), and then point the missile's camera in the appropriate direction with a small joystick. The camera can then be locked onto the target, by positioning a cursor over a prominent contrast line between the target and its background, or two visibly different parts of the target. Once locked on, the missile can be launched; it then homes autonomously to the target while the aircraft can maneuver freely. The AGM-65A was used with considerable success by Israel in the 1973 Yom Kippur War, but it is extremely dependent on good visibility and will work only in high-contrast conditions. The AGM-65B is similar, but has a scene-magnification camera with a zoom lens, allowing the operator to acquire the target at longer ranges by searching with the seeker head.

The AGM-65C, meant mainly for CLOSE AIR SUPPORT (CAS), homes in on laser radiation reflected off the target by a LASER DESIGNATOR on the ground, in the launch aircraft, or in another aircraft. The designator may be pulse or frequency coded, allowing an aircraft to fire several missiles simultaneously at different targets. The AGM-65E works on the same principle, but uses the tri-service laser seeker, compatible with a wider range of designators.

The AGM-65D has an imaging infrared seeker which converts thermal emissions to video images. It operates in much the same manner as the TV-guided models, but it can be used in low light against low-contrast targets. The AGM-65D is now the standard weapon used in conjunction with the LANTIRN night vision/target designation system. The AGM-65F and G use the same seeker but have modified guidance electronics optimized for anti-shipping attacks.

The AGM-65A/B has a 130-lb. (59-kg.) shaped-charge (HEAT) warhead that can devastate a tank. The U.S. Navy and Marine Corps prefer the 250-lb. (113.6-kg.) Mk.19 blast/fragmentation warhead, which can also be fitted to the C and D models. The AGM-65F and G have a 300-lb. (136.3-kg.) penetration warhead designed to explode deep within a ship for maximum effect.

Maverick was originally powered by a Thiokol TX-481 dual-impulse solid rocket motor; this was replaced in 1981 by a TXH-633 motor with a greatly reduced smoke signature. Effective range is constrained by target acquisition. The need to designate the target before launch is the key limitation of all Mavericks; it forces the launch aircraft to operate within the air defense envelopes of many targets, which is why USAF is trying to develop lock-on after launch (LOAL) missiles, notably SLAM, a land attack derivative of the AGM-84 HARPOON.

Specifications   Length: 8.16 ft. (2.48 m.). **Diameter**: 12 in. (305 mm.). **Span**: 28.3 in. (719 mm.). **Weight, launch**: (A/B/D) 463 lb. (210.5 kg.); (E/F) 677 lb. (307.7 kg.). **Speed**: Mach 2 (1400 mph/2338 kph). **Range**: 6–10 mi. (10–16 km.) at sea level/25 mi. (41.75 km.) at high altitude. **Operators**: Egy, Gre, Iran, Isr, It, Mor, ROK, S Ar, Swe, Tur, UK, US.

MAY: NATO code name for the Ilyushin Il-38 LONG-RANGE MARITIME PATROL (LRMP) aircraft derived from the Il-18 Coot turboprop transport. First test-flown in 1967, the May entered service in 1970, and now equips ANTI-SUBMARINE WARFARE (ASW) squadrons of Soviet Naval Aviation (AV-MF) and the Indian navy.

The Il-38 has the same wing, tail assembly, and engines as the Il-18. The chief modifications are a fuselage lengthened by 12 ft. (3.66 m.) by means of a plug-in extension behind the wing, the installation of a weapons bay, and various items of ASW equipment. A ventral radome beneath the cockpit houses a "Wet Eye" J-Band surface-search RADAR with 360° scanning. The weapons bay, located ahead of the wing, can carry an estimated 10,000 lb. (3048 kg.) of ordnance, including nuclear and conventional DEPTH CHARGES, homing TORPEDOES, and SONOBUOYS. No wing stations have been observed, and the aircraft does not seem able to carry external stores. In contrast to Western LRMP aircraft, the Il-38 does not have a separate sonobuoy dispenser; thus sonobuoys must displace offensive weapons in the main bay. The concentration of so much weight ahead of the wings upset the stability of the basic Il-18 airframe, necessitating the addition of the 12-ft. fuselage plug. The main cabin now houses a tactical compartment with eight ASW system operators, data processing equipment, and display consols. A MAGNETIC ANOMALY DETECTOR (MAD) boom is housed in a tail "stinger" fairing. Various antennas and radomes on the wings and fuselage indicate the installation of an ELECTRONIC SIGNAL MONITORING array.

The effectiveness of the Il-38 as an ASW plat-form is difficult to assess in the absence of combat data. But Soviet sonobuoys are believed to be copies of older U.S. and allied models, while Soviet computers have neither the speed nor memory of their Western counterparts. Hence most naval analysts believe the Il-38 to be greatly inferior to its Western equivalents, notably the P-3C ORION, to which it bears a superficial resemblance.

Specifications   Length: 129.83 ft. (39.58 m.). **Span**: 122.67 ft. (37.4 m.). **Powerplant**: 4 4250-shp. IvchenckoAl-20M turboprops. **Weight, empty**: 88,700 lb. (40,318 kg.). **Weight, max. takeoff**: 50,-000 lb. (68,181 kg.). **Speed, max.**: 390 mph (651 kph). **Speed, cruising**: 200 mph (334 kph). **Range, max.**: 4000 mi. (6680 km.). **Endurance**: 13–16 hr.

MBFR: Mutual and Balanced Force Reductions, a continuing series of negotiations for the reduction of conventional force levels by NATO and the WARSAW PACT, later renamed the Conventional Stability Talks (CST) and again renamed (in 1988) the Conventional Armaments and Forces in Europe (CAFE) talks. MBFR talks continued for more than 15 years, without tangible results. The CAFE talks, on the other hand, trailed behind unilateral force reductions. A tentative CFE (Conventional Forces in Europe) Treaty between NATO and the Warsaw Pact was signed in December 1990.

MBT: See MAIN BATTLE TANK.

MEB: See MARINE EXPEDITIONARY BRIGADE.

MECHANIZED   INFANTRY: Sometimes called "armored infantry"; now the prevalent form of INFANTRY, trained and equipped to operate in close cooperation with tanks in combined-arms formations. Mechanized infantry is usually mounted in INFANTRY FIGHTING VEHICLES, or more commonly ARMORED PERSONNEL CARRIERS; in most armies, however, it is trained to fight dismounted against any serious opposition, using the vehicles (if at all) only for fire support. See also ARMORED FORCES; MOTORIZED RIFLE TROOPS.

MEETING ENGAGEMENT: Mainly a Soviet usage: a battle in which two opposing forces advance into contact—which is likely to happen only if the ratio of forces to space is relatively low (preventing the establishment of strong continuous lines), or if a breakthrough has already occurred.

In a meeting engagement, both sides would normally attempt to discover the size and strength of the enemy, and to turn each other's open flanks. Soviet theory holds that a meeting engagement would develop into a swirling melee, in which success would go to the side whose commanders can think more quickly, clearly, and flexibly, and who can retain the last uncommitted reserves for decisive use.

MEF: See MARINE EXPEDITIONARY FORCE.

MEKO: An innovative modular family of FRIGATES and CORVETTES designed by the West German Blohm und Voss shipyard, mainly for export. The MEKO system employs several standard hulls ranging in size from 1000 to 3600 tons, equipped with weapons, sensors, and electronics mounted on interchangeable pallets ("functional units"). This allows users to tailor the various hulls to their specific requirements without major structural modifications. In theory, the MEKO system would also allow users to change weapon and sensor suites at short notice to perform quite different missions. To date, MEKO-class ships have been acquired by Argentina, Australia, New Zealand, Nigeria, and Turkey.

The MEKO weapon units are standardized containers, 15.41 ft. (4.7 m.) long, 13.15 ft. (4 m.) wide, and 8.5 ft. (2.6 m.) high, on or in which are mounted a variety of guns and missile launchers. Each container incorporates magazines, servo motors, power connections, and data interfaces. Weapon modules currently offered include 127-mm., 120-mm., 100-mm., and 76.2-mm. DUAL PURPOSE (DP) guns; 57-mm., 40-mm., 35-mm., and 30-mm. twin ANTI-AIRCRAFT guns; and GOALKEEPER, PHALANX, and SEA ZENITH radar-controlled guns for anti-missile defense. Surface-to-air missile modules include ROLAND, ASPIDE, SEA SPARROW, Sea CROTALE, and RAM. Modules for HARPOON, EXOCET, and OTOMAT anti-ship missiles and Bofors 375-mm. anti-submarine rocket launchers are also available.

Electronic modules, 9.85 to 14.75 ft. (3.0 to 4.5 m.) long, 8 ft. (2.44 m.) wide, and 7 ft. (2.13 m.) high, each contain major electronic components, power connections, and data interfaces for a variety of RADAR, SONAR, FIRE CONTROL, communication, ELECTRONIC WARFARE, and navigation units. Aerials and antennas are mounted on standardized pole and tripod masts. Each MEKO hull has a number of openings in the deck, into which various "functional units" can be lowered and bolted down. Some weapon modules can be installed by a dockyard within one hour.

Integration of these disparate elements is accomplished by a Blohm und Voss Data Information Link (DAIL)—a dynamic information network based on standardized data buses and a multiple-interface computer, which can be programmed to accommodate all weapon and electronic "functional units." DAIL is essential for the plug-in/plug-out MEKO system.

To date, four MEKO hulls have been offered for sale. The first and largest, the MEKO 360, is a general-purpose frigate. Flush-decked, with a long, boxy superstructure and a helicopter landing pad and hangar near the stern, it has fin stabilizers to improve stability and seakeeping. Argentina has ordered four MEKO 360s as the "Almirante Brown" class, commissioned in 1983–84. They are armed with an OTO-MELARA 5-in. (127-mm.) 54-caliber DP gun on the bow, 4 twin Breda 40-mm. anti-aircraft guns, an 8-round Aspide launcher on the after superstructure, 2 quadruple launch canisters for Exocet MM.40 missiles, 2 sets of triple tubes for lightweight ASW homing TORPEDOES, and 2 Westland LYNX ASW helicopters (still unavailable due to the post-Falklands embargo). The single Nigerian MEKO 360 ("Arada"), commissioned in 1982, is nearly identical to the Argentine ships. Armament consists of an OTO-Melara 5-in. gun, 4 twin 40-mm. Bredas, an Aspide launcher, 2 sets of triple torpedo tubes, 8 OTOmat anti-ship missiles, and one Lynx helicopter.

The next largest member of the family is the MEKO 200, 4 of which were commissioned by Turkey between 1987 and 1989; 3 additional ships were completed in 1988–89. These "Yavuz class" frigates resemble scaled-down MEKO 360s. Armament consists of a U.S. Mk.45 5-in. 54-caliber DP gun on the bow, 3 Sea Zenith radar-controlled guns, 1 before the bridge and 2 on either side of the helicopter hangar, an 8-round Sea Sparrow launcher aft, 2 quadruple Harpoon launchers amidships, 2 sets of triple torpedo tubes, and an AB.212ASW HUEY helicopter. In 1989, Australia and New Zealand ordered 8 and 2 MEKO 200s, respectively, the first of which are to be delivered in 1995–96; the weapon and sensor suites have not yet been finalized.

The smallest of the family built to date is the MEKO 140 "light frigate" (actually a corvette), 3 of which were commissioned by Argentina between 1985 and 1987 (3 more left incomplete due to lack of funds are now for sale). These "Espara class" ships still resemble the MEKO 360, but the helicopter pad has been moved from the fantail to the superstructure amidships, and there is no hangar. Armament consists of an OTO-Melara 76.2-mm. DP gun on the bow, a Breda twin 40-mm. in front of the bridge and a second on the fantail, 2 pintle-mounted 12.7-mm. heavy machine guns, 2 quadruple launchers for Exocet MM.40 aft, 2 sets of Mk.32 triple torpedo tubes, and an ALOUETTE or Lynx helicopter.

A smaller corvette, the MEKO 100, displacing about 1000 tons and armed with a 76.2-mm. DP gun, a radar-controlled gun, 1 light helicopter, and 4 to 8 anti-ship missiles, has not been ordered so far. See also FRIGATES, FEDERAL REPUBLIC OF GERMANY.

Specifications Length: (360) 412.1 ft. (125.64 m.); (200) 362.5 ft. (110.52 m.); (140) 299.2 ft. (91.22 m.). **Beam:** (360) 43.1 ft. (13.14 m.); (200) 46.6 ft. (14.2 m.); (140) 40 ft. (12.2 m.). **Draft:** (360) 14.1 ft. (4.3 m.); (200) 13.1 ft. (4 m.); (140) 10.8 ft. (3.29 m.). **Displacement:** (360) 2900 tons standard/ 3600 tons full load; (200) 2700 tons standard/3000 tons full load; (140) 1470 tons standard/1700 tons full load. **Powerplant:** (360) twin-shaft CODOG: 2 MTU 20V956 TB92 diesels, 11,070 hp./2 Rolls Royce Olympus TM-3B gas turbines, 50,000 shop.; (200) 4 MTU 20V1163 diesels, 22,536 hp., or 2 LM2500 gas turbines, 56,000 shp., 2 shafts; (140) 2 11,300-hp. SEMT-Pielstick 16PC2-5V400 diesels, 2 shafts. **Fuel:** (360) 440 tons; (200) 380 tons; (140) 230 tons. **Speed:** (360) 30.5 kt.; (200) 27 kt. (diesel) or 30 kt. (turbines); (140) 27 kt. **Range:** (360) 4500 n.mi. at 18 kt.; (200) 4000 n.mi. at 20 kt.; (140) 4000 n.mi. at 18 kt. **Crew:** (360) 200–220; (200) 180–200; (140) 93. **Sensors:** (360) 1 DA-08 air-and-surface search radar, 1 ZW-08 surface-search radar, 1 Decca 1226 navigation radar, 1 WM-25 gun and missile fire control radar, 2 HSA LIROD radar/optronic 40-mm. fire control systems, 1 Atlas 80 hull-mounted medium-frequency sonar; (Arada) 1 Decca 1226 navigation radar, 1 Plessy AWS-5D air-search radar, 1 WM-25 fire control radar, 1 Search, Tracking and Illumination Radar (STIR), 1 HSA PHS-32 hull-mounted medium-frequency sonar; (200) 1 navigation radar, 1 Plessy Dolphin surface-search radar, 1 DA-08 air-search radar, 1 WM-25 fire control radar, 1 STIR, 2 Albis Sea Zenith fire control radars, 1 Siemens optronic gun director, 1 SQS-56 hull-mounted medium-frequency sonar; (140) 1 DA-05/2 air-search radar, 1 WM-28 fire control radar, 1 Decca TM1226 navigation radar, 2 LIROD optronic gun directors, 1 KAE ADS-4 hull-mounted sonar. **Electronic warfare equipment:** (360) 1 Telefunken electronic countermeasures (ECM) suite, 2 20-barrel SCLAR chaff and flare launchers; (Arada) 1 Decca RDL-2 electronic signal monitoring (ESM) array, 1 RCM-2 active jammer, 2 SCLAR chaff/flare launchers; (200) 1 passive ESM array, 2 Mk.36 SRBOC chaff launchers; (140) 1 RDC-2ABC ESM array, 1 RCM-2 ECM array, 2 Dagaie chaff launchers.

MERKAVA ("CHARIOT"): Israeli MAIN BATTLE TANK (MBT), developed from 1968 in successive versions. One man, Gen. Israel Tal, a very successful tank commander, has been in charge of the program from its inception. In production since 1977, the Merkava has a radically original configuration which reflects the intensive combat experience that shapes Israeli armor doctrine. On the premise that only crews assured of maximum protection will boldly exploit armament and mobility, the Merkava's design places primary emphasis on crew survivability, with firepower next, and mobility last. Actually, Israeli armor doctrine stresses *tactical* mobility (the ability to advance against enemy fire) rather than mechanical mobility (unopposed speed), on the further premise that a relatively slow but heavily protected tank will actually advance more quickly than mechanically faster but more lightly armored tank (e.g., the French AMX-30).

The Merkava I entered service in 1979, and by 1982, some 200 had been delivered, many of which participated with great success in the Lebanon War (only one was penetrated). In 1982, the Merkava II was introduced, with better sensors and FIRE CONTROLS, an improved transmission, and a more powerful engine (all Merkava Is have been upgraded to II standard). The Merkava III, now in production, has an Israeli version of composite CHOBHAM ARMOR (in a unique modular form that allows easy replacement), a much more powerful engine, and a new, locally produced 120-mm. gun. More than 700 Merkavas have been delivered to date, with production running at a rate of 80–100 per year.

The hull, fabricated from castings and welded sections, has a unique configuration: the entire front end is occupied by the engine compartment and fuel tanks to form a spaced-armor barrier. In the Mk.I/II, the glacis plate, sloped at some 75°, consists of an outer layer of cast steel armor, behind which are voids filled with diesel fuel (which acts as an energy trap), backed by a second, thicker layer of armor. The Israelis report that this system gives excellent protection against HEAT rounds and ANTI-TANK GUIDED MISSILES (ATGMs), as well as kinetic projectiles. Actual armor thickness is classified, but has been estimated at more than 200 mm. In the Mk.III, modules of Chobham-type composite armor bolted over a steel shell provide ballistic protection equivalent to more than 600 mm. of homogeneous steel armor. In contrast to welded armor, the modules can be replaced easily in the field to repair battle damage or upgrade protection even further. The Mk.III also has special "performated" armor added to protect the rear of the vehicle, and spring-mounted anti-bazooka side-skirts.

The engine is mounted on the right side of the hull behind the fuel tank, and the mass of the engine block provides additional protection for the crew. The driver sits to the left of the engine, and is provided with three observation periscopes, one of which can be replaced by a passive INFRARED (IR) night scope. The gunner, commander, and loader occupy the turret and a large fighting "room" in the rear of the vehicle, under the turret. In contrast to most other tanks, there is no turret basket; instead, the floor moves as the turret rotates. Another unique feature of the Merkava is its rear access hatch, which allows ammunition to be loaded straight through, much faster than with traditional hand-loading through the turret hatches. For special purposes, a ten-man rifle squad or four stretcher cases can be carried instead of main gun ammunition; more commonly, tank commanders can carry an extra man for the loss of some rounds. The turret, a welded shell with cast parts in the Mk.I/II, and with modular Chobham armor in the Mk.III, has a distinctive wedge shape which reduces frontal area while presenting only highly oblique surface to enemy shot; it was designed from the start to accept either 105-mm. or 120-mm. guns. The gunner sits on the right side of the turret, with the commander above and behind him, and the loader standing on the left. The commander does not have a cupola at all, but rather a low-profile hatch surrounded by five observation periscopes and an optical sight which can be traversed through 360°, with zoom magnification settings from 4x to 14x for long-range target acquisition. The gunner has a periscopic sight (stabilized in the Mk.III) with zoom settings from 1x to 12x, and an integral LASER rangefinder. Both the commander's and gunner's sights are linked to an Elbit digital fire control computer which incorporates ammunition type, target and vehicle motion, barrel tilt, barrel wear, and unusually complete meteorological data into the ballistic equation, providing a very high first-round hit probability. In the Merkava II and III, both commander and gun-

ner have THERMAL IMAGING sights for use at night or through smoke.

Main armament in the Mk.I/II is the British-designed 105-mm. L7A3 rifled gun, which also arms the M48 and M60 Patton, late model CENTURIONS, the M1 ABRAMS, and the LEOPARD 1. The gun has an effective range in excess 2500 m. and can fire APDS, APFSDS, HESH, HEAT, anti-personnel (APERS), and smoke ammunition. Muzzle velocities are 1458 m./sec. for APDS, 1600 m./sec. for APFSDS, 1170 m./sec. for HEAT, and 875 m./sec. for HESH, APERS, and smoke. The widely exported Israeli 105-mm. APFSDS round, with a long-rod "arrow" penetrator, can pierce more than 150 mm. of homogeneous steel armor at 2000 m., and is believed capable of destroying all known Soviet MBTs (in 1982, T-72s were penetrated by this round at 3000+ m.). The Merkava III has a 120-mm. smoothbore gun which differs from the Rheinmetall gun of the Leopard 2 and the M1A1 version of the Abrams in having a concentric recoil system. It can fire APFSDS at a muzzle velocity of 1650 m./sec., as well as fin-stabilized HEAT; armor penetration exceeds that of the L7A3 to an extent so far unknown.

The gun has elevation limits of −8.5° and +20°, and is fully stabilized to permit accurate firing on the move (although Israeli doctrine stresses firing from the halt in hull-down positions). During the 1973 Yom Kippur War, many Israeli tanks ran low on ammunition at critical moments. In response to this, the Merkava I and II were designed with an unusually large load of 62 main gun rounds. The Merkava III can carry only 50 rounds of 120-mm. ammunition, still more than most MBTs.

Secondary armament includes a coaxial 7.62-mm. machine gun, and two pintle-mounted 7.62-mm. machine guns by the commander's and loader's hatches. In addition, a 60-mm. MORTAR (unique on an MBT) snaps up from the turret roof, mainly to fire smoke and illumination bombs. Ordinary smoke grenade launchers are also mounted on the turret. A white light/IR searchlight housed *vertically* inside the turret bustle is reflected horizontally with a steel mirror. There is a positive overpressure COLLECTIVE FILTRATION unit for NBC defense. Yet another unique feature is a broadband radar-warning receiver (the first on an MBT) to alert the crew to hostile surveillance and missile-guidance radars. All ammunition is sowed in protective containers, and automatic Halon fire extinguishers with instantaneous electronic response greatly reduce the risk of explosions. In the Merkava III, the turret has all-electric control mechanisms that eliminate flammable hydraulic fluids.

In the Mk.I/II, the power-to-weight ratio is low, but cross-country mobility is still relatively good due to the careful design of its suspension and transmission. The Merkava III has 30 percent more power, with a concomitant increase in speed and cross-country mobility. The Mk.III has its 12 road wheels on independent trailing arm suspensions, allowing the increased power to be exploited in rough terrain (trailing arms are also easier to repair than torsion bar suspension). The Merkava III is now the best MBT in the world because of its superior layout, abundance of unique protective features (including all-electric turret drive, perforated applique armor in the rear, spring-mounted side-skirts, protected ammunition storage, and radar-warning receiver), powerful main gun, large ammunition load, and advanced fire controls.

Specifications **Length:** (I/II) 24.4 ft. (7.44 m.); (III) 24.93 ft. (7.6 m.). **Width:** 12.16 ft. (3.7 m.). **Height:** 9.0 ft. (2.74 m.). **Weight, combat:** (I/II) 60 tons; (III) 61 tons. **Powerplant:** (I/II) 900-hp. Teledyne Continental AVDS-1790-6A turbocharged multi-fuel diesel; (III) 1200-hp. AVDS-1790-9AR diesel. **Hp./Wt. ratio:** (I/II) 15 hp./ton; (III) 20 hp./ton. **Fuel:** 225 gal. (990 lit.). **Speed, road:** 28.5 mph (47.6 kph). **Range, max.:** (I) 240 mi. (400 km.); (II/III) 300 mi. (500 km.).

MEROKA: A short-range, shipboard CLOSE-IN WEAPON SYSTEM for anti-aircraft and anti-missile defense, produced by CETME for the Spanish navy. Meroka consists of a turret armed with six (!) pairs of 20-mm. Oerlikon CANNONS, controlled by a RAN-12L search and target designation RADAR, an on-mount PVS-2 DOPPLER tracking radar, a THERMAL IMAGING sight, and a PDS-10 FIRE CONTROL computer. Each 20-mm. gun can fire up to 750 rounds per minute, giving the mount a combined rate of fire of 9000 rds./min. A total of 720 rounds are stowed inside the turret, with an additional 720 rounds stowed in three external panniers. The Oerlikon cannons have a muzzle velocity of 1200 m. per second, and a maximum effective range of 2000 m. CETME estimates that a typical engagement could require up to ten 12-round bursts, allowing the unit to engage up to six missiles before reloading. Targets are to be automatically acquired and designated by the RAN-12L; the target is then tracked (and the turret laid on target) by PVS-2 data. In the event of heavy radar JAMMING, the thermal imaging sight can be used in a manual backup mode. The entire mount weighs 9921 lb. (4510 kg.). Meroka has been fitted to the Spanish aircraft carrier PRINCIPE DE ASTURIAS (four mounts), and to the DESCUBIERTA-PERRY- and BALEARES-class frigates (one mount each). No export sales have been reported.

MEU: See MARINE EXPEDITIONARY UNIT.

MF: Medium Frequency, radio signals in the 300 KHz–3 MHz frequency band, used mainly for commercial broadcasting, though some LORAN and automatic direction-finding equipment also operates in this band.

MG3: German GENERAL-PURPOSE MACHINE GUN (GPMG), actually the famous MG42 of World War II, rechambered for NATO standard 7.62 × 51-mm. (.308-caliber) ammunition. The original MG42, the first and easily the best of GPMGs, influenced the design of all subsequent machine guns. Made with a maximum use of steel stampings to reduce cost and complexity, it introduced a unique bolt-locking mechanism with a compound gas-assisted recoil action for an exceptionally high cyclic rate of fire of 1200 rds./min. (German doctrine stresses suppressive rather than attiritional fire.) Nearly one million of these 7.92-mm. weapons were produced between 1942 and 1945. From 1950, the MG42 was placed back in production as the 7.62-mm. MG1, and many surviving MG42s were also rechambered for the 7.62-mm. round as the MG2. The final development was the MG3, with a modified belt-feed mechanism, an enlarged ejection port to reduce (already rare) jamming, and a chrome-plated bore to reduce (common) barrel fouling. It can be fired from a bipod as an assault weapon, or from a tripod for sustained fire, or from vehicle pintle mounts. The MG3 is air-cooled, and the barrel has a quick-change fitting to cope with overheating—essential given the very high cyclic rate. When used with a bipod, the MG3 is fed by 50-round belts stored in drum magazines, or from 100-round belts in box magazines. When used on a tripod or pintle mount, the gun is fed by 50-round, hand-held belts. Ammunition types are the NATO-standard ball, tracer, and AP. See, more generally, M60 MACHINE GUN; MACHINE GUN.

Specifications **Length OA:** 48.25 in. (1.225 m.). **Length, barrel:** 22.3 in. (566 mm.). **Weight:** 25.5 lb. (11.59 kg.). **Muzzle velocity:** 820 m./sec. **Cyclic rate:** 1100–1200 rds./min. **Effective range:** 800 m. **Operators:** Aut, Chi, Den, FRG, Iran, It, Nor, Pak, Por, Sp, Su, Tur.

MHD: See MAGNETO-HYDRODYNAMIC POWER.

MICV: Mechanized Infantry Combat Vehicle. See INFANTRY FIGHTING VEHICLE.

MICV-80: British infantry fighting vehicle. See WARRIOR.

MIDAS: Ilyushin Il-78 tanker version of the Soviet Il-76 CANDID four-engine, long-range transport aircraft. Development of Midas began in the late 1970s as a replacement for Mya-4 BISON bombers converted to tanker duties. Midas has two (possibly three) AERIAL REFUELING hose-and-drogue units mounted in the fuselage and wing tips. At least 60,000 lb. (27,272 kg.) of fuel can be carried. Physical and performance characteristics are identical to those of the Candid.

MIDWAY: A class of three U.S. AIRCRAFT CARRIERS commissioned between 1945 and 1947. Only two (USS *Midway* and *Coral Sea*) remain; USS *Franklin D. Roosevelt* was retired in 1972. The largest U.S. aircraft carriers built in World War II, they were the first U.S. carriers with armored flight decks, and the only ones able to operate the first naval jet aircraft without modifications.

As built, the Midways had open bows and straight ("axial") flight decks 136 ft. (41.46 m.) wide, with one deck-edge and two centerline elevators. Defensive armament was particularly heavy, reflecting wartime experience: 18 5-in. 54-caliber DUAL PURPOSE guns, 21 quadruple 40-mm. BOFORS guns, and 34 twin 20-mm. cannons, all mounted on deck-edge sponsons. During the 1950s, all three Midways were completely modernized, receiving enclosed "hurricane" bows, angled landing decks cantilevered to port, more powerful steam catapults, and automated mirror landing systems, which allowed them to operate the larger, heavier, higher-performance jets of the 1960s. Since then, *Midway* and *Coral Sea* have been modernized several times, and now diverge in several respects.

Midway's flight deck has been widened to 258.5 ft. (78.8 m.), and there are now 2 catapults and 3 deck-edge elevators (2 to starboard at either end of the island, and 1 on the port quarter). The island itself has been extended to provide support for additional radars and electronics. Large "bulges" fitted to the hull in 1986 to enhance stability in fact had the opposite effect, to the detriment of flight operations in heavy seas.

Defensive armament now consists of 2 8-round NATO SEA SPARROW short-range surface-to-air missile launchers and 2 PHALANX 20-mm. radar-controlled guns for anti-missile defense. The air wing presently consists of 2 fighter/attack squadrons with a total of 36 FA-18 HORNETS, a medium attack squadron with 14 A-6E INTRUDERS and KA-6D tankers, and a support squadron with 4 EA-6B PROWLERS and 6 SH-3H SEA KING anti-submarine helicopters.

Coral Sea's flight deck, only 236 ft. (71.9 m.) wide, also has 2 catapults and 3 deck-edge elevators. Defensive armament is limited to 3 Phalanx. Sensors and electronic warfare equipment are similar to those of the *Midway*. Its air group consists of 40 FA-18 Hornets, 12 A-6E Intruders, 5 KA-6D tankers, 7 SH-3H Sea Kings, and 4 E-2C HAWKEYE airborne early warning aircraft.

The oldest active carriers in the U.S. Navy, the Midways are too small to operate the S-3 VIKING anti-submarine patrol aircraft or F-14 TOMCAT interceptor. Thus in the 1990s *Midway* will be retired, and *Coral Sea* converted into a training carrier. See also AIRCRAFT CARRIERS, UNITED STATES.

Specifications **Length:** (as built) 968 ft. (295.12 m.); (*Midway*) 1006.25 ft. (306.9 m.); (*Coral Sea*) 1003.67 ft. (306 m.). **Beam:** (as built) 113 ft. (34.45 m.); (*Midway*) 140.71 ft. (42.9 m.); (*Coral Sea*) 121 ft. (36.9 m.). **Draft:** (as built) 34.5 ft. (10.52 m.); (*Midway*) 36 ft. (11 m.); (*Coral Sea*) 35 ft. (10.7 m.). **Displacement:** (as built) 45,000 tons standard/53,358 tons full load; (*Midway*) 53,-400 tons standard/67,000 tons full load; (*Coral Sea*) 48,000 tons standard/65,200 tons full load. **Powerplant:** 4-shaft steam: 12 oil-fired boilers, 4 sets of geared turbines, 212,000 shp. **Speed:** 32 kt. **Range:** 15,000 n.mi. at 15 kt. **Crew:** (*Midway*) 2615 + 1800 air group = 4415; (*Coral Sea*) 2500 + 2220 air group = 4720. **Sensors:** 1 LN-66 navigation radar, 1 SPS-48C 3-dimensional air-search radar, 1 SPS-49 2-dimensional air-search radar, 1 SPS-65 surveillance radar, 2 Mk.91 Sea Sparrow guidance radars. **Electronic warfare equipment:** 1 SLQ-29 electronic countermeasures array, 2 Mk.36 SRBOC chaff launchers.

MiG: Designation of (fighter) aircraft developed by the Soviet design bureau founded by Artem Mikoyan and Mikhail Guryevich in 1939. The MiG bureau designed a series of generally mediocre single-engine fighters during World War II, then moved on to the design of better jet fighters in the late 1940s. Its first jet, the MiG-9, was not technically successful because of the primitive technology of Soviet jet engines. But by combining German wartime swept-wing research with a copy of the British Rolls Royce Nene centrifugal-flow turbojet (an early example of technology transfer), the bureau developed the revolutionary MiG-15 Faggot, which saw extensive combat in the Korean War and the 1956 Sinai War. The MiG bureau thereafter became the primary supplier of jet fighters to the Soviet air force, so that in popular terminology, MiG is synonymous with any Soviet or Warsaw Pact fighter. Models currently in service include: MiG-17 FRESCO; MiG-19 FARMER; MiG-21 FISHBED; MiG-23/27 FLOGGER; MiG-25 FOXBAT; MiG-29 FLANKER; and MiG-31 FOX-HOUND.

MIKE: NATO code name for the experimental Soviet nuclear-powered attack SUBMARINE (SSN) *Komsomolets* completed in 1982, apparently an enlarged development of the high-speed, deep-diving ALFA class, with a titanium pressure hull and (possibly) a liquid metal reactor system. It was lost at sea in 1989, following a reactor fire.

The Mike had a modified teardrop hull, and like most Soviet submarines, a double-hull configuration, with the outer hull sheathed in CLUSTER GUARD anechoic tiles to reduce radiated noise and impede active sonar. Titanium pressure hulls provide greater strength at lighter weight than an equivalent thickness of steel, and gave the submarine a maximum depth some 2000 ft. deeper than U.S. submarines. A long, low sail (conning tower) was located amidships, but in contrast to the Alfa, it was not blended into the hull. The control surfaces followed the standard Soviet pattern: retractable bow planes, with cruciform rudders and stern planes mounted ahead of the propellers. The greater size of the Mike, compared to the Alfa, allowed better acoustical silencing arrangements; the Mike was reported to be considerably more quiet than the Alpha, although not nearly as quiet as the latest U.S. SSNs.

Armament consisted of 6 21-in. (533-mm.) and 2 25.6-in. (650-mm.) tubes in the bow. The 21-in. tubes could launch a variety of anti-surface and anti-submarine TORPEDOES, and the SS-N-15 STARFISH anti-submarine missile. The 25.6-in. tubes could accommodate very heavy torpedoes, such as the Type 65 wake homer, anti-ship variants of the SS-N-21 SAMPSON cruise missile, and SS-N-16 STALLION anti-submarine missiles. A total of 24 weapons were carried. The lack of a passive TOWED ARRAY SONAR suggested that the Mike was intended mainly for anti-ship, rather than anti-submarine, operations. See also SUBMARINES, SOVIET UNION.

Specifications Length: 360.9 ft. (110 m.). Beam: 40.7 ft. (12.4 m.). Displacement: 5280 tons surfaced/6290 tons submerged. Powerplant: single-shaft nuclear: 2 compact lead/bismuth or liquid sodium reactors, 2 sets of geared turbines, 60,000 shp. Speed: 20 kt. surfaced/36–40 kt. submerged. Max. operating depth: 2625 ft. (800 m.). Collapse depth: 3973 ft. (1211 m.). Crew: 100. Sensors: 1 bow-mounted low-frequency active/passive sonar, 1 "Snoop Head" surface-search radar, 1 Bold Head electronic signal monitoring array, 2 periscopes.

MIL: 1. An angular measurement widely used in sighting and FIRE CONTROL. One mil is equal to 0.05625°; hence 6400 mils complete a (360°) circle, and 17.77 mils equal one degree. The mil is a useful measure for estimating the distance or size of an object, because 1 mil of angular separation equals one unit of length or height at 1000 units of distance; e.g., if an object known to be 20 m. long fills an angle of 2 mils, the object is 1000 m. away. Conversely, if an object is known to be a particular distance away, its size can be determined by its angular expanse. This principle forms the basis for most optical sighting (e.g., stadiametric) systems, and is also used for the adjustment of INDIRECT FIRE.

2. Designation of helicopters designed by the bureau founded by Mikhail Mil, the principal source of military helicopters for the Soviet air force and Soviet Army Aviation. Models in service or under development include: Mil-2 Hoplite; Mil-6 HOOK; Mil-8 HIP; Mil-17 HAZE; Mil-24 HIND; Mil-26 HALO; and Mil-28 HAVOC.

MILAN: Medium-range ANTI-TANK GUIDED MISSILE (ATGM) produced by the Franco-German Euromissile consortium. The Milan program was initiated in 1962 as a design study by Nord Aviation and Messerschmidt-Bolkow-Blohm. The first production missiles entered service in the mid-1970s with the French, West German, and British armies. Since then, more than 170,000 Milans have been delivered to more than 30 countries. Milan is produced under license in Britain, India, and Italy, and has been used in combat by Britain in the Falklands, by Iraq against Iran, and by Chad against Libya.

The Milan system consists of two major components: a launcher and guidance unit; and the missile itself, supplied in a sealed container/launch tube. The launcher/guidance unit includes an optical sight, a tracking unit, and a trigger. The optical sight is periscopic, allowing the operator to fire from a prone position, reducing his exposure (a major tactical advantage over most other ATGMs). A 15.4-lb. (4.7-kg.) MIRA THERMAL IMAGING sight can be attached for the acquisition and tracking of targets out to 3000 m. at night or through smoke.

The missile consists of a warhead section housing a 6.57-lb. shaped charge (capable of penetrating 25.6 in. of steel armor); a center section containing a dual-impulse solid rocket motor; and a tail section with a thrust vector control (TVC) steering unit, four pop-out fins, and an inertialess reel for two guidance wires. The missile in its combined storage/launch tube is treated as a certified ("wooden") round requiring no preflight checks. Milan is man-portable with a two-man crew, but is often launched from vehicle mounts.

To use Milan, the operator removes caps from both ends of the launch tube and attaches it to the launcher/guidance unit. He then lines up the target in the crosshairs of the sight and pulls the trigger, activating a gas generator which ejects the missile from the tube, while simultaneously ejecting the empty tube from the launcher. After a safety delay (to clear the operator), the motor ignites; its initial booster pulse rapidly accelerates the missile to 174 m./sec. (388 mph), and the sustainer then takes over, burning throughout the entire trajectory and accelerating the missile to its maximum speed. With Milan's semi-automatic command to line-of-sight (SACLOS) guidance, the operator must keep the target centered in his sight; an infrared sensor in the launch unit detects heat from the missile tracking flares and feeds missile-position data to the guidance electronics, which calculate the angle between the missile and the operator-target line-of-sight. The guidance unit generates steering commands to close the angle, which are transmitted to the missile via the wires that pay out from its tail. Steering commands are implemented by deflection of the rocket exhaust, and that is why the motor must burn for the entire time of flight. Average time of flight to maximum range is 14 sec., and the operator must keep the target in his sight throughout.

The effectiveness of light shaped-charge (HEAT) weapons such as Milan has been made questionable by the introduction of advanced composite, spaced, and reactive armors. In 1984, Euromissile introduced Milan 2, with a heavier warhead capable of penetrating 41.7 in. (1060 mm.) of steel armor.

Specifications **Length:** 2.51 ft. (765 mm.). **Diameter:** 3.54 in. (900 mm.). **Span:** 10.43 in. (265 mm.). **Weight:** (missile) 14.66 lb. (6.66 kg.); (launcher) 33 lb. (15 kg.); (system) 60.9 lb. (27.68 kg.). **Speed:** (launch) 76 m./sec. (170 mph); (max.) 200 m./sec. (450 mph). **Range:** 25–2000 m. **Operators:** Alg, Bel, Chad, Chi, Egy, Fra, FRG, Gre, Ind, Iraq, Ire, Isr, Ken, Leb, Lib, Mor, Sen, Som, Sp, Syr, Tun, UK.

MILITARY AIRLIFT COMMAND (MAC):

A major operational command of the U.S. Air Force (USAF), responsible for all strategic and tactical transport aircraft not assigned to other operational or theater commands. MAC is organized into three numbered air forces (21st, 22nd, and 23rd), and control units, notably the Air Weather Service, the Special Missions Operational Test and Evaluation Service, and the USAF Airlift Center. Active MAC units can be supplemented by Air Force Reserve and Air National Guard units, plus the commercial airliners of the Civil Reserve Air Fleet (CRAF).

MILITARY DISTRICT, SOVIET:

One of 16 administrative regions called "military districts" (Baltic, Byelorussian, Carpathian, Central Asian, Far Eastern, Kiev, Leningrad, Moscow, North Caucasus, Odessa, Siberian, Trans-Baikal, Trans-Caucasus, Turkestan, Ural, and Volga), to which Soviet armed forces, with the exception of "Groups of Forces" abroad and KGB and MVD troops, are assigned in peacetime. Military districts are responsible for the induction, training, and maintenance of personnel and equipment for all units within their geographic jurisdiction. In the event of war, the peripheral military districts would be converted into FRONTS, with the district commander normally becoming the front commander. The interior districts, notably the Moscow, Volga, and Ural Military Districts, from the nucleus of the Soviet strategic reserve, under the direct control of the Supreme High Command (GKO).

MILITARY SCIENCE:

A Soviet term for the body of military theory, commonly described as DOCTRINE by Western armies. According to the Soviet Officer's Handbook (Spravochnik ofitsera), military science ". . . is a unified system of knowledge for the preparation of armed conflict in the interests of the defense of the Soviet Union and other Socialist countries against Imperialist aggression." Within those limits, intellectual and practical debate is encouraged, to inspire the development of new operational methods. In contrast, Soviet military *doctrine* is defined as ". . . a unified system of views and aims, free from private views and estimates." In other words, it is the current formulation of official Communist party military policy, hence no dissent is allowed.

Soviet "military science" is devided into seven separate disciplines: the Theory of Military Art (i.e., STRATEGY, OPERATIONAL METHOD, and TACTICS); Force Posture (organization, materiel, personnel, and mobilization); Military Pedagogy (training and education); Party-Political Work; Military History; and Military-Technical Science. Taken as a whole, these different approaches to military problems give Soviet officers an integrated intellectual framework within which to work. In contrast to its disrepute in the West, military science is a major field of study in the Soviet Union: several thousand officers hold "candidate" degrees in the field (somewhere between a master's degree and a Ph.D.), and several hundred hold full doctorates (a degree awarded to "candidates" who have completed a rigorous course of study and defended a dissertation to become acknowledged authorities in their fields).

MINAS GERAIS: Brazilian AIRCRAFT CARRIER, originally the British light carrier HMS *Vengeance* of the Colossus class, completed in 1945, lent to the Royal Australian Navy in 1953, returned to Britain in 1955, and purchased by Brazil in 1956. The ship was comprehensively refitted and modernized in Rotterdam between 1957 and 1960, when a steam catapult, an 8.5° angled deck, a mirror landing system, new electronics, and two centerline elevators were added.

The flight deck runs the full length of the hull and has a maximum width of 121 ft. (36.9 m.); the hangar deck below it is 445 ft. (135.67 m.) × 52 ft. (13.85 m.) × 17.5 ft. (5.33 m.). The ship has a large, characteristically British island superstructure to starboard, which incorporates the engine uptakes. Defensive armament consist of 2 quadruple and 1 twin 40-mm. BOFORS GUNS, all located on sponsons below the flight deck. The air group currently consists of 6 to 8 Grumman S-2E TRACKER anti-submarine patrol aircraft, and 4 to 6 SH-3 SEA KINGS, 2 SAH-11, and 3 UH-1 (HUEY) helicopters. In 1984, Brazil announced plans to acquire 12 A-4 SKYHAWK attack aircraft for the air group, but canceled the order in the following year. Current plans call for the *Minas Gerais* to remain in service until the year 2000; a V/STOL carrier is projected for that date.

Specifications Length: 695 ft. (211.9 m.). **Beam:** 80 ft. (24.4 m.). **Draft:** 24.5 ft. (7.47 m.). **Displacement:** 15,890 tons standard/19,890 tons full load. **Powerplant:** twin-shaft steam: 4 oil-fired boilers, 2 sets of geared turbines, 42,000 shp. **Fuel:** 3200 tons. **Speed:** 24 kt. **Range:** 6200 n.mi. at 23 kt./12,000 n.mi. at 14 kt. **Crew:** 1000 + 300 air group = 1300. **Sensors:** 1 SPS-40B air-search radar, 1 SPS-4 surface-search radar, 1 Raytheon 1402 navigation radar, 2 SPG-37 fire control radars. **Electronic warfare equipment:** 1 SLR-2 radar warning receiver.

MINE COUNTERMEASURES: Equipment and techniques used to locate and destroy naval MINES. The earliest mine countermeasure technique was developed during World War I, against moored contact mines. Special vessels ("minesweepers") are equipped with "paravanes"—torpedo-shaped floats towed by cables secured to the bow. Underway, the paravanes fan out on either side of the ship, forming an inverted V with the ship at the apex. The mooring cables of the mines are snagged by the cables and dragged back to the paravanes, which are fitted with cable cutters. Once the mooring cable is cut, the mine floats to the surface, and it can then be seen and destroyed by gunfire.

Paravanes are not effective against bottom-laid "influence" mines, which have no cables; they can be cleared only by triggering their fuzes. Magnetic mines, introduced in 1939, can be detonated by towing two buoyant, electrified cables behind the minesweeper, to form a magnetic field which detonates the mine. For that reason, minesweepers are now constructed of wood, plastic, or nonferrous metals to minimize their magnetic signature. Acoustical mines can be detonated by towed noise-maker sleds. Because modern mines often combine two or more fuze mechanisms, minesweepers now usually tow both anti-magnetic and anti-acoustic devices; alternatively, these can be mounted on hydrofoil sleds towed by low-flying helicopters.

More advanced mines include sophisticated anti-sweeping devices, or rely on hydrodynamic (i.e., water-pressure) fuzes which are difficult to spoof. In response, a new type of vessel has been developed, the "minehunter"—a small ship of plastic or wood, equipped with high-resolution bottom-mapping and side-scanning SONARS to locate bottom mines. Once a mine is detected, divers or remote-controlled submersibles are dispatched either to recover it, or plant explosives to destroy it in place.

MINEGUMO: A class of three small Japanese hunter-killer DESTROYERS (DDKs) commissioned between 1968 and 1970. The Minegumos (and very similar YAMAGUMOS) are specialized ANTI-SUBMARINE WARFARE (ASW) vessels with very little ANTI-AIR WARFARE or ANTI-SURFACE WARFARE capability (indeed, they could more accurately be described as DESTROYER ESCORTS or FRIGATES).

The Minegumos are flush-decked, with sharply raked bows and rather boxy superstructures. They were originally armed with a twin 3-in. 50-caliber DUAL PURPOSE gun at each end of the ship, a 4-barrel Bofors 375-mm. ASW rocket launcher before the bridge, 2 sets of Mk.32 triple tubes for Mk.46 ASW homing TORPEDOES, and a flight deck and hangar for 2 DASH (Drone Anti-Submarine Helicopter) remotely piloted vehicles. In 1976, the DASH platform was removed and an 8-round ASROC launcher was put in its place. In addition, the 3-in. guns were replaced by OTO-MELARA 76.2-mm. Compact guns (these modifications make the Minegumos practically identical to the Yamagumos, which were equipped with ASROC from the start). See also DESTROYERS, JAPAN.

Specifications Length: 377 ft. (114.94 m.). Beam: 38.75 ft. (11.82 m.). Draft: 13 ft. (3.96 m.). Displacement: 2100 tons standard/2700 tons full load. Powerplant: 6 Mitsubishi 12 UEV 30/40 diesels, 26,000 hp., 2 shafts. Speed: 27 kt. Range: 7000 n.mi. at 20 kt. Crew: 215. Sensors: 1 OPS-11 air-search radar, 1 OPS-17 surface-search radar, 1 SPG-34 or GFCS-2 fire control radar, 1 OQS-3 hull-mounted medium-frequency sonar, 1 SQS-34J variable depth sonar. Electronic warfare equipment: 1 NOLR-5 electronic signal monitoring array.

MINES, LAND: Explosive devices placed on or under the ground, so as to be triggered by personnel or vehicles. The simplest mines are containers filled with explosives, fitted with a tripwire or pressure fuze. Mines are classified both by function and the means of deployment: anti-personnel mines are relatively small devices, intended to kill or maim people; ANTI-TANK or anti-vehicle mines are considerably larger, in order to wreck tracks and tires, or blow holes in the underside of the vehicle. Most mines are laid below the ground, by hand or by specialized minelaying plows. Scatterable mines are laid on the ground by hand, or by vehicle dispensers, or dropped by special artillery

shells, missile and rocket "cargo" warheads, fixed-wing aircraft, or helicopters.

The typical anti-personnel mine is a small fragmentation bomb weighing between 1 and 3 lb. (0.5–1.5 kg.), which is placed in a shallow hole and covered. Such mines are usually fitted with a pressure-sensitive fuze set to detonate under the weight of a man; alternatively, they may be have a pin-activated friction fuze, attached to one or more tripwires. Anti-personnel mines are often deliberately kept too small to kill outright, so as to induce the enemy to use more troops to retrieve the wounded. In addition, maiming wounds can be even more demoralizing than "clean" deaths to those nearby. An exception was the German S-mine ("Bouncing Betty") of World War II, with two separate charges: the first propelled the second (a fragmentation grenade) out of the ground to explode at waist height, disemboweling its victims within a 5 to 10-m. radius.

Anti-tank mines are generally similar, but weigh between 10 and 30 lbs., and are fitted with less-sensitive pressure fuzes meant to be triggered only by vehicles. The anti-tank equivalent of the S-mine is the "bottom attack" mine, a SHAPED CHARGE planted in the ground, so as to be directed upward at the less protected belly of armored vehicles. Bottom attack mines are mostly triggered by whisker-type electrical contact fuzes, or by magnetic proximity fuzes.

Scatterable mines are broadcast on the ground. Because they lie on and not in the ground, the larger anti-vehicle types are relatively easy to detect; hence their main role is to impose the need to clear them, thus delaying enemy movements (perhaps severely, if proper equipment is not at hand). Smaller anti-personnel types, on the other hand, can be camouflaged. A typical example is the air-scattered "butterfly" mine used by the Soviet Union in Afghanistan, whose low-contrast shape is compounded by earth colors. Weighing about 6 ounces (0.3 kg.), it contains a powerful liquid explosive and is triggered by a sensitive contact fuze activated when the mine first hits the ground. The U.S. "gravel" mine used in Southeast Asia was even smaller, resembling, as the name suggests, coarse gravel. Neither type is powerful enough to kill, but both can inflict disabling (and to others demoralizing) wounds.

Off-route mines are GRENADE LAUNCHERS, MORTARS, or rocket launchers aimed at a road or trail and controlled by contact, proximity, or photoelectric trigger.

A Claymore mine is a special "directional" anti-personnel device consisting of ball bearings embedded in a convex charge of plastic explosives. When detonated, the ball bearings are projected in a fan-shaped pattern at high velocity. Claymores can be remotely command-detonated by an operator, or triggered by a tripwire; depending on size, the lethal radius can be as much as 50 m.

Mines are employed both offensively and defensively. On the defensive, minefields planted in front of, between, or behind positions can keep the attacker away, or channel his movements into prepared ambushes or kill zones. In-ground mines are more suitable for those purposes than scatterable mines, but they too can be cleared quickly, unless the minefield is covered by fire to oppose enemy ENGINEERS. In fluid combat, scatterable mines can be broadcast to create expedient minefields directly ahead of the enemy's advance, or in his rear in order to interdict supply convoys.

Offensively, scatterable mines can be used to protect the flanks of an attack, to delay enemy counterattacks, or to impede the enemy's retreat. In GUERRILLA operations, mines can be implanted singly or in small numbers to harass, demoralize, and inflict low-level attrition. Tripwire mines, Claymores, and off-route mines are often used by guerrillas.

In the simplest mine-clearing technique, engineers or other instructed troops advance on hands and knees, probing for mines beneath the ground with bayonets. When a mine is detected, it is either marked with a small flag, excavated, and stacked, or disarmed in place (the last is the most dangerous option, because mines often have anti-tampering devices). Magnetic detectors, introduced in 1942, allow the relatively rapid and reliable sweeping of larger areas. This method is still widely used today, but has been rendered less reliable by the development of mines with nonmagnetic wood, fiberglass, or plastic casings. Specialized mine-clearing tanks, also introduced in 1942, have heavy rollers or chain flails ("scorpions") rotating on a boom that extends from the bow in order to detonate mines at some distance from the vehicle. Mine plows that can be attached for the task (without requiring a specialized tank) uproot mines, casting them on either side of a broad furrow to clear a route, leaving the mines to be destroyed or disarmed at leisure. Mines can also be cleared by intensive artillery bombardments, but they also tend to crater the terrain and reduce trafficability. Rocket-propelled line charges that sweep forward (an old technique) and FUEL AIR EXPLOSIVES (a new one) also use explosive force to detonate mines, but are much more selective than artillery barrages.

MINES, NAVAL: The oldest and simplest form of naval mine is the moored contact mine, a buoyant sphere filled with explosives, and anchored to the ocean floor at some depth below the surface (with the depth determined by the type of ship to be attacked). The casing has many contact detonators ("horns") on it, to trigger the mine. Bow waves deflect moored mines outward, but they then revert, colliding with the hull sides. Although easy to sweep, contact mines are cheap enough to be used in large numbers, and may be emplaced in water up to 400 ft. (122 m.) deep (longer cables are impractical).

Bottom-laid influence mines, first developed in Germany in 1939, are activated by some form of proximity fuze. The earliest types had magnetic influence fuzes, triggered by variations in the earth's magnetic field generated by the steel hull of a ship. This type of mine was easily countered by degaussing ships with electrically charged cables. Acoustic proximity fuzes, first used in 1942, are triggered by the sound of ship engines and propellers; they can be spoofed and predetonated by noisemakers. The latest type of fuze, however, detects the hydrodynamic pressure wave generated by the passage of ships above it, and is difficult to spoof. Modern influence mines often combine two or more fuze types, and may have on-board microprocessors to defeat MINE COUNTERMEASURES, as well as "ship counters," which keep the mine dormant until a set number of ships (presumably minesweepers) have passed by. The most sophisticated mines have computers with libraries of ship signatures, to detect the difference between minesweepers and other ships. Almost all modern mines also have "sanitizer" circuits, to disarm them after a preset period of time. Bottom-laid influence mines of all types can be extremely effective, because they explode beneath a ship's keel and can break its back. But they cannot generally be used in water deeper than 150 ft. (45 m.).

DESTRUCTORS (a U.S. term) are aerial bombs converted into mines by fitting them with magnetic, acoustic, or hydrodynamic fuzes. While not as effective as purpose-built mines, they are considerably cheaper and available in large numbers.

The Mk.60 CAPTOR (en*cap*sulated *tor*pedo) mine developed by the United States for ANTI-SUBMARINE WARFARE (ASW) consists of a lightweight acoustic-homing TORPEDO housed in a buoyant capsule moored to the ocean floor. When a submarine is detected within range, the mine launches the torpedo. "Quick Rise" mines for use against surface ships are similar: bottom-laid in deep waters, they release an explosive charge which is propelled towards the surface, to explode beneath the ship. Quick Rise mines could overcome the depth limitation of influence mines.

Naval mines are used both offensively and defensively. Defensive minefields are laid off friendly harbors and coastal waters to protect them against enemy ships and submarines. Offensively, mines are laid at the entrances to enemy ports, and in coastal waters, narrow seas, canals, and navigable rivers to interdict traffic, inflict casualties, and divert resources into mine-clearing operations.

MINISUB: A small, manned submersible vessel, literally a miniature SUBMARINE, usually designed for clandestine reconnaissance, or for raids inside enemy harbors and in inshore waters. Most have very short range and a very limited underwater endurance, and must therefore be towed or carried by a full-size submarine or a surface vessel. Soviet naval SPETSNAZ are believed to have used at least two different types of minisubs to probe the harbors and inshore defenses of Sweden and northern Japan; one type is known to have tanklike tracks, to crawl along the bottom. Minisubs used to deliver "frogmen" (combat swimmers) have lock-out compartments through which they can exit and enter. See also SWIMMER DELIVERY VEHICLE.

MINUTEMAN: A family of U.S. land-based intercontinental ballistic missiles (ICBMS) which form the bulk of the U.S. Air Force (USAF) STRATEGIC AIR COMMAND's missile force. Development began in 1956 on a proposed solid-fuel intermediate-range ballistic missile (IRBM) under Project Q. Impressed by the advantages of solid propulsion, including safety, a very quick reaction time, and a long storage life, USAF upgraded Project Q to an ICBM, awarding the design and production contract to Boeing in October 1958. Development was amazingly rapid by today's standards, and the missile reached its INITIAL OPERATING CAPABILITY (IOC) in December 1962. In the context of the MISSILE GAP anxieties, a force of some 2000 missiles was projected, with some 600 missiles on rail-mobile launchers, and the rest in hardened underground silos. The rail-mobile plan was soon abandoned, and the total force was reduced to 1000 in 1963, when it was concluded that the Soviet ICBM force was very small after all (rail-mobile basing has now been resurrected for the LGM-118 PEACEKEEPER, or MX, and is actually used with the Soviet SS-24 SCALPEL).

In the first production version, the LGM-30A Minuteman I, the first stage was powered by a Thiokol M55 engine developing 200,000 lb. (90,-910 kg.) of thrust; the second stage had an Aerojet General engine rated at 60,000 lb. (27,272 kg.); while the third stage had a Hercules engine rated at 35,000 lb. (15,910 kg.). The first and second stages were fabricated from rolled steel, while the third stage had a wound-glassfiber casing—revolutionary for the time. All stages had four gimballed nozzles for attitude control and steering. Minuteman I was armed with a single 635-lb. (288.6-kg.) Mk.5 REENTRY VEHICLE (RV) housing a 1.3-MT W59 nuclear warhead. Controlled by an Autonetics solid-state INERTIAL GUIDANCE unit, the missile had an estimated P_K of .15 against a target hardened to 1000 psi.

Minuteman I was initially deployed by the 341st Strategic Missile Wing (SMW) in 150 underground SILOS at Malmstrom Air Force Base (AFB) in Montana. The silos were grouped into "flights" of ten each, under the control of an underground Launch Control Center (LCC). Hardened to resist blast, the silos were positioned at least 4.5 mi. (7.5 km.) apart, with none closer than 3.3 mi. (5.5 km.) to its LCC, to minimize vulnerability to a counterforce strike. Five flights formed a squadron, with three or four squadrons comprising a wing-still the standard organization of SAC missile forces. With Minuteman I the "hot launch" technique was used, whereby the main engines ignited in the silo, and exhaust gases were vented upward between the missile and the walls of the silo; thus the silo had to be completely reconditioned after launch. Despite the successful development of COLD LAUNCH techniques (as used with Peacekeeper and many Soviet ICBMs), hot launch was used with all subsequent Minuteman variants, because reloading was deemed unimportant under a deterrant strategy.

In 1963, production switched to the LGM-30B (still designated Minuteman I), identical to the first version except for a lighter titanium second stage (that allowed a slightly longer range), and an improved guidance unit which contained integrated microcircuitry. This "B" version equipped the 44th SMW at Ellesworth AFB, South Dakota (150 missiles), the 91st SMW at Minot AFB, North Dakota (150 missiles), the 351st SMW at Whiteman AFB, Missouri (150 missiles), and Warren AFB, Wyoming (200 missiles), for a total of 800 missiles deployed by 1965.

In 1964, the air force began flight testing the LGM-130F Minuteman II. The missile reached IOC in 1966 with the 321st SMW at Grand Forks AFB, North Dakota (150 missiles), and by the late 1960s it had completely replaced the Minuteman I. The missile introduced a new second stage with a more powerful Aerojet SR19 engine rated at 60,-000 lb. (27,272 kg.) of thrust, whose single fixed and recessed nozzle uses liquid freon injection for thrust vector control (TVC) steering (instead of gimballed nozzles). The new stage was also filled with a more powerful fuel compound, for extended range. A new Autonetics NS-11C guidance unit reduced CEP, and had an electronic memory to store data on alternative targets. Armed with an Avco Mk.11 RV housing a 1.2-MT W56 warhead, the missile also carries a Tracor Mk.1 or Mk.1A penetration aids (PENAIDS) package to counter ANTI-BALLISTIC MISSILE SYSTEMS.

In 1970, Boeing introduced the LGM-30G Minuteman III, with a new Aerojet/Thiokol wound glassfiber third stage of expanded diameter. Its Aerojet SR73 engine has 34,400 lb. (15,636 kg.) of thrust and freon-injection TVC. Minuteman III has the same overall length and maximum diameter as Minuteman II, but is much heavier, and range is extended considerably.

The Minuteman III was the first operational ICBM with MULTIPLE, INDEPENDENTLY TARGETED REENTRY VEHICLES (MIRVs). It originally carried 3 Mk.12 RVs, each with a 200-kT W62 warhead, but the missile was later upgraded to 3 Mk.12As, each with a 350-kT W78 warhead, and a PENAID package. The RVs are mounted on a Bell Aerospace POST-BOOST VEHICLE (PBV)—essentially a substage mounted above the third stage. With a throttleable liquid-fuel engine rated at 300 lb. (136.63 kg.) of thrust, 6 pitch/yaw jets rated at 22 lb. (10 kg.) each, and 4 flush-mounted roll jets rated at 18 lb. (98.18 kg.) each, the PBV separates from the third stage after it burns out, and then imparts minor velocity adjustments to direct the MIRVs towards their separate targets. Improved guidance has reduced the presumed CEP, giving a P_k of roughly .24 against a 1000-psi target. In accordance with the policy of Assured Destruction, no attempt was made to maximize COUNTERFORCE capabilities of the Minuteman.

By 1972 (at the time of the SALT I Accords), Minuteman III had replaced 550 Minuteman IIs,

TABLE M1

1972 Minuteman Deployment

Wing	Location	Missiles
341st SMW	Malmstrom AFB	150 MMII, 50 MMIII
44th SMW	Ellesworth AFB	150 MMII
91st SMW	Minot AFB	150 MMIII
351st SMW	Whiteman AFB	150 MMII
90th SMW	Warren AFB	200 MMIII
321st SMW	Grand Forks AFB	150 MMIII

so that deployments stood as shown in table M1. The total stood at 450 Minuteman IIs and 550 Minuteman IIIs. From 1986, 50 Minuteman IIIs at Warren AFB were replaced by the LGM-118 Peacekeeper (IIIs rather than IIs were replaced because of arms control limits on MIRVed missiles). Current plans call for the long-term retention of 450 Minuteman IIs and most of the 500 Minuteman IIIs, with periodic upgrades of the guidance systems, PENAIDS, etc. The rocket engines, however, have limited lives, and the entire Minuteman force will have to be thoroughly overhauled (or replaced) around the year 2000. See also BALLISTIC MISSILES; ICBMS, UNITED STATES.

Specifications **Length:** (I) 53.95 ft. (16.45 m.); (II/III) 59.7 ft. (27.14 m.). **Diameter:** 6.33 ft. (1.92 m.). **Weight, launch:** (I) 64,815 lb. (29,461 kg.); (II) 70,000 lb. (31,818 kg.); (III) 76,058 lb. (34,572 kg.). **Range:** (I) 6214 mi. (10,000 km.); (II) 6990 mi. (11,250 km.); (III) 8078 mi. (13,000 km.). **CEP:** (I) 1000 m.; (II) 600 m.; (III) 366 m. (Mk.12)/200 m. (Mk.12A).

MIRAGE III/5: A series of French FIGHTER/ATTACK AIRCRAFT designed by Avions Marcel Dassault (now Dassault-Breguet). The original Mirage I, designed to a 1952 French air force specification for a lightweight INTERCEPTOR, was a very small, tailless delta-wing aircraft powered by two Rolls Royce Viper turbojets. First flown in 1955, the Mirage I was a good performer in its role, but with only limited range and scant growth potential. Dassault planned a slightly larger model, the Mirage II, but soon scrapped it in favor of the scaled-up Mirage III. Development proceeded rapidly: the first prototypes flew in November 1957, and by 1958 the Mirage III became the first European-built aircraft to exceed Mach 2 (1200 mph/2004 kph) in level flight. In 1960, the French air force ordered 95 Mirage IIIC interceptors, which entered service in 1961. Since then, a combination of aggressive marketing and a solid combat record with the Israeli air force (which purchased 72 Mirage IIICJs in 1961) have resulted in sales to Abu Dhabi, Argentina, Australia, Belgium, Brazil, Colombia, Egypt, Gabon, Lebanon, Libya, Pakistan, Peru, South Africa, Spain, Switzerland, Venezuela, and Zaire.

The Mirage IIIC also formed the basis of the IIIB 2-seat trainer, the IIIE attack aircraft (introduced in 1962), and the IIIR RECONNAISSANCE fighter (introduced in 1963). These subtypes in turn formed the basis of export models equipped to customer specifications: the IIICJ (Israel); IIIEA (Argentina); IIIEP and IIIRP (Pakistan); IIICZ, IIIBZ, IIIRZ, and IIIDZ (South Africa); IIIOA (Australia); IIIS and IIIRS (Switzerland); IIIEE and IIIDE (Spain); IIIEBR (Brazil); IIIEL (Lebanon); and IIIEV (Venezuela).

In 1966, Israel asked Dassault to produce Mirages optimized for visual ground attack, with more payload and simplified AVIONICS. First flown in 1967, this aircraft became the Mirage 5, essentially a IIIE with a slim nose housing only a small ranging radar. Weight saved on avionics was put back into additional fuel and payload; in addition, the simplified avionics made the aircraft more reliable and easier to maintain. Although already paid for in full, the 72 Israeli aircraft were embargoed after the Six-Day War, and were eventually resold to the French air force and Libya. Soon a number of variants were produced for export customers, including the 5R reconnaissance aircraft, the 5D trainer, and the 5E2 for Egypt. The Mirage 50, very similar but with a more powerful engine, was introduced in 1977, and soon sold to Chile, Libya, and Zaire, among other countries. In all, more than 2000 Mirage III/5s have been sold to date, not including about 50 built (without license) in Israel as the "Nesher."

The nose of the IIIC/E houses a large radome for a Thomson-CSF Cyrano I(bis) or Cyrano II RADAR; the IIIE also has a small ventral radome under the nose for a Thomson-CSF DOPPLER navigation radar. In all these aircraft, the pilot sits well forward in a cramped and austere cockpit. Visibility is quite good forward, downward, and sideways, but is obscured to the rear by a dorsal fairing, a typical aerodynamic feature for the 1960s, now unacceptable in a fighter. Cockpit avionics vary slightly from model to model, but generally include a head-down radar display, a CSF.97 reflector gunsight (actually a rudimentary HUD), a radar warning receiver, and a VHF radio. The Mirage 5 has a slender nose housing a lightweight

Agave or Aida range-only radar, sometimes backed up by a LASER rangefinder/target designator in a chin fairing. The Mirage 50 has a true HUD linked to the weapon delivery computer; some also have an IVM3 Doppler radar under the nose.

Behind the nose, the Mirage III and 5 are virtually identical. The fuselage has the "coke bottle" shape of the trans-sonic area rule for reduced drag. An avionics bay is located immediately behind the cockpit, and behind that is a large fuel tank. The rear fuselage is occupied entirely by the engine, which is fed by two cheek inlets. The original IIIC was armed exclusively with air-to-air missiles, usually the primitive (but expensive) COMMAND-GUIDED Matra R.510. This arrangement, however, was unacceptable to the Israel air force, which insisted on the installation of two 30-mm. DEFA cannons; with 250 rounds each, they are accommodated in a ventral pack under the air intakes. This configuration has since become standard.

The Mirage III/5 is a true 60° tailless delta; the low-mounted wing is filleted into the bottom of the air intakes. The wing has no high-lift devices at all on the leading edge, but a slot is cut into it at midspan to provide stall warning. The trailing edge has inboard and outboard elevons, which act together to control pitch, act differentially to control roll, and also double as flaps. The aircraft has tricycle landing gear, the nose wheel retracting under the cockpit, the main gear retracting under the inboard wing sections. The vertical stabilizer is also swept at 60°; many aircraft have an HF antenna built into a dorsal extension at the base of the fin.

The Mirage III/5 has one fuselage and four wing pylons. The centerline and inboard wing pylons are rated at 1764 lb. (800 kg.) each; the outboard pylons are rated at only 500 lb. (227 kg.) and are normally used for lightweight AAMs. Total ordnance load is about 4000 lb. (1818 kg.) for the Mirage III and 9260 lb. (4209 kg.) for the Mirage 5, including 150- and 150-gal. (660- and 1320-lit.) drop tanks, free-fall and LASER-GUIDED BOMBS, NAPALM, CLUSTER BOMBS, rocket pods, Matra R.530 radar-guided AAMs, AS.37 MARTEL anti-radiation missiles, and AIM-9 SIDEWINDER, Matra R.550 MAGIC, and IAI SHAFRIR and PYTHON infrared-homing AAMs. French IIIEs can also carry an AN.52 tactical nuclear bomb on the centerline pylon. The Mirage IIIC was designed to accept a detachable rocket booster in a ventral pack, whose fuel tank displaces the 30-mm. cannon. While this (costly) option gives an exceptional rate of climb, it has seldom been used.

The Mirage III/5's commercial success, and its notable combat record with the Israeli air force, have obscured the reality of a rather mediocre aircraft with serious deficiencies: it has the lowest power-to-weight ratio of any Mach 2 aircraft still in service, which results in low acceleration and rate of climb; and a heavy, fuel-guzzling engine, which diminishes the aircraft's payload, leaving it with a short combat radius (a very serious handicap, as the Argentine air force discovered in the Falklands War). The tailless delta design, while providing good supersonic performance, is also responsible for very long takeoff rolls (typically 6000 ft./1830 m.) and very high landing speeds. In addition, delta designs have high drag in maneuvers, causing them to decelerate rapidly in tight turns. While this can be a useful attribute in some tactical situations, the low thrust-to-weight ratio makes it difficult to regain the speed dissipated in turning. Israeli successes with Mirage IIIs against Arab MiG-21 FISHBEDS were not due to any world-beating attribute to the former, but rather to the tactical acumen of Israeli pilots, and their skill in wringing the last ounce of performance from their aircraft.

Because of the Mirage III's manifest shortcomings, Israel initiated an upgrade program, eventually developing the greatly superior KFIR. Dassault has reciprocated, by developing the Mirage IIING (*Nouveau Generation*), with canards, leading-edge extensions, and new avionics, patterned on those of the Kfir. Dassault is still offering the IIING both as new production and as an upgrade kit.

Specifications Length: (IIIC) 48.4 ft. (14.75 m.); (IIIE) 49.3 ft. (15.03 m.); (5/50) 51 ft. (15.54 m.). **Span:** 27 ft. (8.23 m.). **Powerplant:** (IIIC) SNECMA Atar 9B afterburning turbojet, 13,325 lb. (6023 kg.) of thrust; (IIIE) Atar 9C, 13,669 lb. (6213 kg.) of thrust; (5/50) Atar 9K-50, 15,873 lb. (7215 kg.) of thrust. **Fuel:** 5100 lb. (2318 kg.). **Weight, empty:** (IIIC) 13,570 lb. (6168 kg.); (IIIE) 15,540 lb. (7064 kg.); (5/50) 15,763 lb. (7165 kg.). **Weight, max. takeoff:** (IIIC) 19,700 lb. (8955 kg.); (IIIE) 29,760 lb. (13,527 kg.); (5/50) 30,203 lb. (13,-729 kg.). **Speed, max.:** Mach 1.14 (863 mph/1441 kph) at sea level/Mach 2.2 (1460 mph/2438 kph) at 36,000 ft. (11,000 m.). **Initial climb:** (IIIC) 16,400 ft./min. (5000 m./min.); (50) 36,400 ft./min. (11,-097 m./min.). **Service ceiling:** 56,000 ft. (17,000 m.). **Combat radius:** (typical) 745 mi. (1245 km.)

hi-lo-hi/391 mi. (656 km.) lo-lo-lo. **Range, max.:** 2485 mi. (4150 km.).

MIRAGE IV: A French two-seat supersonic strategic BOMBER/ reconnaissance aircraft built by Avions Marcel Dassault (now Dassault-Breguet). Essentially a scaled-up version of the MIRAGE III fighter, the Mirage IV was developed from 1957, when Dassault was awarded a contract to provide bombers for the French nuclear force. In a remarkably efficient program, the first prototype began flight tests in 1959, production commenced in 1960, the first Mirage IVAs entered service in 1964, and all 72 aircraft were delivered by 1966. The Mirage IVA initially equipped nine squadrons, reduced to six in 1971: *Escadron de Bombardement* (EB) 1/91 at Mont de Maison; EB 2/91 at Cozaux; EB 3/91 at Orange; EB 1/94 at Avard; EB 2/94 at Dizier; and EB 3/94 at Luxeil (EB 3/91 was disbanded in 1983).

The fuselage has the same configuration as the Mirage III, but is some 50 percent larger, and modified to accommodate twin engines and a two-man crew. The nose houses an air-to-ground targeting and mapping RADAR. The pilot and navigator sit in a tandem cockpit, the navigator's position being entirely enclosed save for two small windows. The area behind the cockpit contains an avionics bay, fuel tanks, an AERIAL REFUELING receptacle, and the engines, which are fed by cheek inlets behind the cockpit. The wing is a true 60° delta, with no leading edge devices. The trailing edge has inboard and outboard elevons, which control pitch and roll, and also function as flaps. The use of the delta planform gives good high-speed performance, but requires long runways for takeoff and landing.

The Mirage IVA is armed with a single 65-kT AN.22 free-fall nuclear bomb, carried in a recessed well under the fuselage. In 1979, 18 aircraft were converted to Mirage IVPs with an ASMP short-range attack missile on a revised centerline pylon, an Antelope 5 ground-mapping radar, and a new navigation/attack computer. The first of the modified aircraft entered service in 1986. For reconnaissance missions, the Mirage IVA carries a 2205-lb. (1000-kg.) CT-52 multi-sensor reconnaissance pod (with either cameras or an INFRARED line scanner) in the fuselage well. All Mirage IVAs are now being phased out; the Mirage IVP will be retained until 1996, when it will be replaced by ASMP-equipped MIRAGE 2000Ns.

Specifications Length: 77.1 ft. (23.5 m.). **Span:** 38.85 ft. (11.85 m.). **Powerplant:** 2

SNECMA Atar 9K afterburning turbojets, 14,771 lb. (6714 kg.) of thrust each. **Weight, empty:** 31,-967 lb. (14,530 kg.). **Weight, max. takeoff:** 69,666 lb. (31,666 kg.). **Speed, max.:** (clean) Mach 2.2 (1460 mph/2438 kph) at 36,000 ft. (11,000 m.); (w/payload) Mach 1.8 (1190 mph/1987 kph). **Range, max.:** 2485 mi. (4150 km.).

MIRAGE F.1: A French single-seat FIGHTER/ INTERCEPTOR produced by Dassault-Breguet. Recognizing the many deficiencies of the MIRAGE III fighter/attack aircraft, Dassault began development of the Mirage F.1 as a private venture in 1964. While retaining the Mirage name, the F.1 was an entirely new design, which discarded the traditional Dassault tailless delta configuration for a conventional swept wing and tailplane. First test-flown in 1966, the F.1 proved greatly superior to the Mirage III, with a shorter takeoff and landing rolls, improved maneuverability, slightly better acceleration and rate of climb, and rather better handling characteristics. An order for three pre-production aircraft was placed by France in 1967, followed by a production contract in 1969. The first version, the Mirage F.1C interceptor, entered service in 1973, followed shortly thereafter by the F.1A attack aircraft, which is also the predominant export version. Ironically, the F.1 never achieved the sales of the inferior delta-winged Mirage III, and has been superseded in production by yet another delta, the Mirage 2000.

The fuselage bears some resemblance to the Mirage III, with a pronounced area rule ("coke bottle") shape and cheek-mounted engine inlets behind the cockpit. The nose has a large radome housing a Thomson-CSF Cyrano IVM multi-mode RADAR, reported to have some LOOK-DOWN/SHOOT-DOWN capability. The cockpit configuration is also similar to that of the Mirage III, being faired into a dorsal spine which obstructs rearward visibility (in this regard the F.1 is markedly inferior to its near-contemporaries, the F-14 TOMCAT, F-15 EAGLE, and F-16 FALCON, all of which have bubble canopies).

Cockpit AVIONICS include a HUD, a multi-mode radar display screen, HF, VHF, and UHF radios, a TACAN receiver, an IFF set, a DOPPLER navigation set, a SAGEM INERTIAL NAVIGATION unit, a LASER DESIGNATOR, a RADAR WARNING RECEIVER, and a digital navigation/attack computer. The area behind the cockpit houses an avionics bay and a large fuel tank. Two 30-mm. DEFA cannons, each with 135 rounds, are mounted in a ventral pack under the engine intakes. The rear fuselage houses the en-

gine. The F.1 has rugged tricycle landing gear, optimized for rough field operations. Internal fuel capacity is 40 percent greater than the Mirage III's.

The shoulder-mounted wing, with a leading-edge sweep of 47.5°, is lavishly equipped with high-lift devices, including leading-edge flaps, large trailing-edge flaps, outboard ailerons, and spoilers. Takeoff runs at typical mission weights are only 2500 ft. (762 m.), as compared with more than 6,000 ft. (1830 m.) for a Mirage III; the landing roll is an excellent 1650 ft. (503 m.). The tail unit has an all-moving slab tailplane and a large vertical stabilizer, enhancing agility generally, but especially at high angles of attack (AOA).

The Mirage F.1 has one centerline and six wing pylons. Two wing-tip pylons, rated at only 280 lb. (127 kg.), are exclusively for lightweight air-to-air missiles (AAMs), usually the Matra R.550 MAGIC. The inboard pair of pylons are rated at 2000 lb. (909 kg.) each, the outboard pair at 1100 lb. (500 kg.), and the centerline pylon at 4500 lb. (2045 kg.). Payloads include drop tanks of up to 374 gal. (1645 lit.), free-fall and LASER-GUIDED BOMBS, CLUSTER BOMBS, NAPALM, DURANDAL anti-runway bombs, rocket pods, AS.30 air-to-surface missiles, AS.37 MARTEL anti-radiation missiles, and MATRA R.530 AND SUPER 530 radar-guided AAMs. Maximum payload is 8820 lb. (40009 kg.).

Specifications **Length:** 50.01 ft. (15.25 m) **Span:** 27.7 ft. (8.45 m) **Powerplant:** SNECMA Atar 9K-50 afterburning turbojet, 15,873 lb. (7215 kg.) of thrust. **Fuel:** 7140 lb. (3245 kg.). **Weight, empty:** 6,314 lb. (4415 kg.). **Weight, normal loaded:** 25,353 lb. (11,525 kg.). **Weight, max. take-off:** 33,510 lb. (15,232 kg.). **Speed, max.:** Mach 1.2 (913 mph/1525 kph) at sea level/Mach 2.4 (1442 mph/2410 kph) at 36,000 ft. (11,000 m.). **Initial climb:** 17,000 ft./min. (5183 m./min.). **Service ceiling:** 67,000 ft. (20,000 m.). **Combat radius:** 398 mi. (665 km.). **Range:** (max. wpns.) 560 mi. (935 km.); (max. fuel) 2050 mi. (3424 km.). **Operators:** Ecu, Fra, Gre, Iraq, Jor, Ku, Lib, Mor, Qat, S. Af, Sp.

MIRAGE 2000: A single-seat FIGHTER/ATTACK AIRCRAFT produced by Dassault-Breguet and in service with France, Egypt, Greece, India, Peru, and the United Arab Emirates. The Mirage 2000, which succeeded the MIRAGE F.1, represents a return to the tailless delta-wing configuration made famous by Dassault's MIRAGE III/5 series. Although bearing a superficial resemblance to the older aircraft, the Mirage 2000 was a totally new and vastly superior design incorporating 1970s state-of-the-art technology. Yet development was rapid: the prototype flew in March 1978, and the first production Mirage 2000C was delivered to the French air force in 1982 (the Mirage 2000B, a two-seat trainer, was introduced in 1983). Despite an estimated price of $50 million each, the Mirage 2000 has been a commercial success, though not in Europe, where it lost out to the F-16 Falcon in the great fighter competition. Egypt, however, has ordered 20 aircraft, India 40, and Peru 26. The Mirage 2000N, now being delivered to the French air force, is a nuclear attack variant specifically reinforced for low-level flight. Armed with the ASMP short-range attack missile, it will fully replace the MIRAGE IV bomber by the late 1990s.

The Mirage 2000 has the same basic configuration as earlier Mirages, with an area-ruled ("coke bottle") fuselage to reduce trans-sonic drag, and prominent cheek intakes for the engine. The nose has a large radome housing an advanced Thomson-CSF RDI pulse-Doppler multi-mode RADAR with LOOK-DOWN/SHOOT-DOWN capability. The cockpit, located well forward, has a clamshell canopy with limited rear-hemisphere visibility (a holdover from earlier Mirages and a major tactical shortcoming). Cockpit AVIONICS are comprehensive, including a HUD and head-down multifunctional display, an INERTIAL NAVIGATION system, TACAN, VHF and UHF radios, a RADAR WARNING RECEIVER, a radar altimeter, and a digital navigation/attack/flight control computer. Two 30-mm. DEFA cannons, each with 125 rounds, are mounted in a ventral pack under the engine intakes, as in all Dassault fighters. The area behind the cockpit houses an avionics bay and fuel tank.

The low-mounted wing, a true 60° delta partially blended into the fuselage by bulged wingroot fairings, has full-span, automatically controlled leading-edge slats which enhance controllability at low speeds and high angles of attack (AOA). The trailing edge has inboard and outboard elevons which control pitch and roll, and also function as flaps. The vertical stabilizer is very broad and tall for longitudinal stability at high AOA; two small fixed canards over the wing leading-edge roots further enhance high-AOA performance and reduce drag.

The major disadvantages of the tailless delta configuration are high landing speeds, long takeoff rolls, and a tendency to lose speed rapidly in tight turns due to control surface drag (i.e., the elevons must push the nose up by pushing the tail down, creating drag and killing lift). The Mirage 2000

overcomes the first two problems with its high-lift devices, while the last problem is solved by the use of control-configured vehicle (CCV) techniques and digital FLY-BY-WIRE controls. The aircraft's center of gravity is located near the tail, making it dynamically unstable; the flight control computer keeps the aircraft on course, but during turns it must force the tail up, increasing lift and reducing drag.

The Mirage 2000 has 4 wing and 5 fuselage pylons. The latter include 1 centerline pylon rated at 2200 lb. (1000 kg.), and two pairs of tandem pylons under the air intakes, each rated at 1100 lb. (500 kg.). The inboard wing pylons are rated at 2200 lb. (1000 kg.) each; the outboard pair, rated at only 300 lb. (136 kg.), are exclusively for light air-to-air missiles (AAMs) such as the Matra R.550 MAGIC. Total payload capacity is 11,020 lb. (5010 kg.), including external fuel tanks, free-fall and LASER-GUIDED BOMBS, CLUSTER BOMBS, NAPALM, DURANDAL runway-cratering bombs, gun and rocket pods, AS.30 air-to-surface missiles, AS.37 MARTEL anti-radiation missiles, and MATRA SUPER 530 radar-guided AAMs. As previously noted, the Mirage 2000N will carry the ASMP nuclear missile.

The Mirage 2000 is something of an enigma. Its performance as an air-combat fighter is adequate, but not comparable to the F-16 FALCON or other fighters of its generation. As an attack aircraft, its low wing loading makes for a bumpy ride and difficult weapon-aiming at low altitude. In effect, Dassault's ingenuity only partially overcomes the limitations of inefficient French turbojets and its own choice of tailless delta designs.

Specifications Length: 47.1 ft. (14.36 m.). **Span:** 29.5 ft. (8.99 m.). **Powerplant:** 1 SNECMA M53-5 afterburning turbojet, 19,240 lb. (8745 kg.) of thrust. **Weight, empty:** 16,535 lb. (7516 kg.). **Weight, normal loaded:** 20,944 lb. (9543 kg.). **Weight, max. takeoff:** 36,375 lb. (16,535 kg.). **Speed, max.:** Mach 1.2 (915 mph/1528 kph) at sea level/Mach 2.2 (1450 mph/2422 kph) at 36,000 ft. (11,000 m.). **Initial climb:** 49,000 ft./min. ft. (14,-940 m./min.). **Service ceiling:** 59,000 ft. (17,988 m.). **Combat radius:** 435 mi. (726 km.). **Range, max.:** 1118 mi. (1867 km.).

MIRKA: NATO code name for a class of 18 Soviet ANTI-SUBMARINE WARFARE (ASW) CORVETTES (Soviet designation *Storozhevoy Korabl'*, SKR, or Patrol Ship) completed between 1964 and 1966. Originally designated "Small Anti-Submarine Ships" (*Malyy protivolodochnyy korabl'*, MPKs), the Mirkas are modified versions of the earlier PETYA class, designed for inshore ASW operations in the Baltic and Black seas. The Mirkas have a very pronounced sheer, a raised poop deck, and a broad, flat transom stern.

The Mirkas were built in two separate subclasses, each of nine ships, differing only in their weapon and sensor suites. The Mirka I has 2 twin 76.2-mm. DUAL PURPOSE guns, one forward and the other amidships; 4 12-barrel RBU-6000 ASW rocket launchers; 1 DEPTH CHARGE rack aft; and 1 set of quintuple 21-in. (533-mm.) TORPEDO tubes. The Mirka II has only 2 RBU-6000s but carries 2 sets of quintuple tubes.

The Mirkas have a unique powerplant: gas turbines drive compressed air pumpjets, while diesels drive conventional propellers mounted inside the pumpjet tunnels. The turbine exhausts are vented out from the transom, contributing their residual thrust to the ship's forward motion. Some Mirkas may be in reserve; the remainder should serve well into the 1990s.

Specifications Length: 270.3 ft. (82.4 m.). **Beam:** 29.9 ft. (9.12 m.). **Draft:** 9.8 ft. (3 m.). **Displacement:** 950 tons standard/1150 tons full load. **Powerplant:** twin-shaft CODAG: 2 15,000-shp. gas turbines, 2 6000-hp. diesels. **Speed:** 35 kt. (turbines)/15 kt. (diesels). **Range:** 500 n.mi. at 30 kt./4800 n.mi. at 10 kt. **Crew:** 92. **Sensors:** 1 "Slim Net" (Mirka II: "Strut Curve") air-search radar, 1 "Hawk-Screech" fire control radar, 1 "Don 2" surface-search/navigation radar, 2 "High Pole" IFF sets and 2 "Square Head" IFF interrogators, 1 hull-mounted medium-frequency sonar, one high-frequency dipping sonar. **Electronic warfare equipment:** 2 "Watch Dog" electronic countermeasure arrays.

MIRV: See MULTIPLE INDEPENDENTLY TARGETED REENTRY VEHICLE.

MISSILE GAP: The presumed inferiority of the United States in strategic BALLISTIC MISSILE deployments as compared to the Soviet Union, which played a major role in U.S. domestic politics in 1957–61, becoming a campaign slogan in the 1960 presidential election. By 1961, U.S. intelligence estimates of the number of Soviet intercontinental ballistic missiles (ICBMS) were revised downward by more than 97 percent, but by then the U.S. had its own ballistic missile crash programs well under way.

There is little doubt now that the Missile Gap overestimate was the (counterproductive) result of a deliberate Soviet disinformation campaign, rather than of purposeful overestimation by the U.S. Air Force, or mere error. Although the Soviet Union had only a handful of operational ICBMs by 1960, its deception campaign created the perception that many more had been built, presumably in order to exert disproportionate diplomatic leverage. It was only large-scale photographic reconnaissance information (supplied first by U-2 aircraft, later by surveillance satellites) that finally exposed the Soviet deception and allowed the U.S. to determine the true number of Soviet ballistic missiles. See RVSN for current Soviet ballistic missile deployments; and ICBMS, SOVIET UNION.

MISSILE, GUIDED: Literally, a missile is a projectile of any sort (e.g., a thrown stone); in current military terminology, however, it describes an unmanned, expendable, self-propelled flying vehicle equipped with some form of guidance, which allows it to be steered towards, rather than aimed at, the target. Missiles are categorized by mission, flight profile, and launch modes. In the broadest sense, missiles can be classified either as STRATEGIC or TACTICAL. Strategic missiles are long-range DELIVERY SYSTEMS for NUCLEAR WEAPONS aimed at the enemy homeland; tactical missiles have shorter ranges, may be armed with high-explosive (HE), chemical, or NUCLEAR WARHEADS, and are used by or in support (direct or indirect) of combat forces (including naval and air forces).

BALLISTIC MISSILES, both strategic and tactical, are rocket-powered vehicles equipped with INERTIAL GUIDANCE, which are powered during their ascent ("boost phase") to a point in space from which they will coast in an elliptical, free-fall ("ballistic") trajectory to the target. Ballistic missiles are classified by range: ICBMS and IRBMS are always strategic weapons; MRBMS as well as SRBMS may be tactical. Submarine-launched ballistic missiles (SLBMS) are ICBMs or IRBMs which can be launched from (submerged) submarines.

CRUISE MISSILES are essentially small winged aircraft, generally jet propelled, which fly at subsonic or trans-sonic speeds, mostly at low altitudes. Some are meant for strategic attack, and are equipped with inertial or TERCOM guidance; others, for tactical use, have conventional warheads, and usually some form of active or passive TERMINAL GUIDANCE.

Surface-to-air missiles (SAMs) are land-based or shipboard ANTI-AIRCRAFT weapons. Large, fixed-site SAMs for homeland defense are often referred to "strategic," and may be armed with small nuclear warheads. Lightweight mobile systems are often designated "battlefield" SAMs. ANTI-BALLIS-

TIC MISSILES (ABMs) are similar to SAMs (indeed, some SAMs can be used in an ABM role) but generally have longer ranges and higher speeds, to intercept incoming REENTRY VEHICLES. Most now in service are armed with ENHANCED RADIATION nuclear warheads for X-RAY or NEUTRON KILL. The earliest SAMs (and most ABMs) are controlled by some form of radar COMMAND GUIDANCE or BEAM RIDING guidance—both ineffective against high-speed targets maneuvering at low altitude. Current SAMs rely on active radar homing (ARH) or SEMI-ACTIVE RADAR HOMING (SARH), often combined with a different form of terminal guidance. The smallest man-portable SAMs are controlled by either passive INFRARED HOMING or semi-active LASER homing (SALH). These short-range weapons are meant for the self-defense of small ground units.

Air-to-air missiles (AAMs) are now the principal armament of modern FIGHTERS and INTERCEPTORS. Long-range AAMs (70–120 mi.) are usually controlled by radar ACTIVE HOMING; medium range missiles (15–45 mi.) by SARH; and short-range ("dogfight") missiles by passive IR homing.

Air-to-surface missiles (ASMs) include strategic air-launched cruise missiles (ALCMS), but are more often short-range tactical weapons meant for attacks against point targets such as tanks, trucks, bunkers, or buildings. Many different forms of guidance are in use, including command, ELECTRO-OPTICAL, IMAGING INFRARED, and SALH. ANTI-RADIATION MISSILES (ARMs) are a specific subtype equipped with receivers to home on enemy RADAR or radio emissions.

ANTI-SHIP MISSILES may be launched from ships, aircraft (as an ASM subtype), shore emplacements, or submarines, nowadays with underwater launch; they may be armed with high explosive or tactical nuclear warheads, and, depending upon their range and sophistication, may be controlled by command guidance, or a combination of inertial guidance with active radar and/or passive infrared. Anti-submarine missiles, usually ship- or submarine-launched, deliver either a homing TORPEDO or nuclear DEPTH CHARGE to to attack their target. Their principal advantage is their speed margin over torpedoes (in-water missiles) or depth charges dropped by airplane or ships, which minimizes the "dead time" between detection and attack, during which the target can move away. Most anti-submarine missiles are radio command-guided.

ANTI-TANK GUIDED MISSILES (ATGMs) are

ground- or air-launched weapons designed specifically to attack armored vehicles with shaped-charge (HEAT) warheads. Most have ranges between 2000 and 6000 m., and are controlled by WIRE GUIDANCE, SALH, or radio command guidance.

MISSION-TYPE ORDERS: U.S. Army term for orders in which a commander is given an objective and a time limit, with few other constraints, leaving the method of implementation entirely to his discretion. This procedure is part of the theoretical basis of U.S. AIR-LAND BATTLE doctrine, but in practice U.S. Army senior officers still tend to "micro-manage" their subordinates. For a more complete exposition, see AUFSTRAGSTAKTIK.

MISTRAL: A French shoulder-launched surface-to-air missile, also available in helicopter-launched (air-to-air) and shipboard versions. Developed by Matra from the early 1980s in response to a French army requirement for a lightweight anti-helicopter missile, Mistral completed testing and evaluation in 1988, and is now entering service.

Mistral's conical nose houses an all-aspect INFRARED-HOMING seeker. Powered by a dual-pulse solid-fuel rocket motor, Mistral is armed with a 6.6-lb. (3-kg.) high-explosive warhead containing 1800 tungsten ball bearings. There are four pop-out control canards and four fixed tail fins.

The basic shoulder-launched version consists of the missile in its Kevlar launch/storage canister, and a gripstock/trigger/sight assembly. In action, the operator would first acquire the target visually, and then wait for the missile seeker to lock onto it (as indicated by an audible tone). Once lock-on is conformed, the missile can be launched to home autonomously. In the helicopter-launched version, two missiles in their canisters are attached to a twin launch rack. There are presently two naval versions: SADRAL, a 4- or 6-round pedestal launcher with a remote IMAGING INFRARED sight, and LAMA, a 4-round pedestal launcher with an on-mount sight for local control.

Specifications Length: 70.9 in. (1.88 m.). **Diameter:** 3.54 in. (90 mm.). **Weight, launch:** 38 lb. (17 kg.). **Speed:** Mach 2.6 (1820 mph/3040 kph). **Range envelope:** 0.2–3.7 mi. (300–6000 m.). **Ceiling:** 6500 ft. (2000 m.).

MK.46 (TORPEDO): A U.S. lightweight anti-submarine TORPEDO designed in the early 1960s as a replacement for the Mk.44, with superior range, speed, and operating depth. The first Mk.46 Mod O version of 1963 had a solid monopropellant (OTTO) piston engine which proved difficult to maintain. It was superseded in 1967 by the Mod 1, with more reliable liquid-fuel OTTO propulsion. The Mod 1 was designed for both ship and air launch, and could also be delivered as the payload of the RUR-5A ASROC anti-submarine missile. The Mod 2, which entered service in 1972, was designed specifically for helicopter delivery. The Mod 3 was never placed in production. The Mod 4 was designed specifically for the CAPTOR mine system.

All versions are similar in size and general configuration. The blunt nose houses an active/passive SONAR, behind which is a 95-lb. (43.2-kg.) high-explosive warhead. The cylindrical centerbody houses an acoustical-homing guidance computer and fuel tanks. The tapered tail section houses the OTTO engine, which drives a counterrotating propeller, and actuators for the tail control surfaces. Air- and ASROC-launched versions also have a parachute package attached to the tail section, which breaks away when the torpedo enters the water. On entering the water, the Mk.46 either makes a straight run to the target, or starts a preprogrammed spiral search pattern. The target is initially acquired by the sonar in the passive mode; active sonar is switched on for the final run-in.

Although by far the most widely used of Western torpedoes, the Mk.46 has serious shortcomings, including short range, relatively low speed, and a small warhead. The latest Soviet submarines may be able to outrun or outdive it; even if it achieves a hit, the 95-lb. warhead may be incapable of ensuring the destruction of large, Soviet double-hull submarines. As an interim measure, the U.S. Navy developed the Mk.46 Mod 5 NEARTIP (Near-Term Improvement Program), with an improved sonar unit, a new guidance computer, and a dual speed engine for longer range. The first Mod 5s entered service in 1980; some are of new production, but most are earlier mods updated with kits. The Mk.46 is to be replaced in the 1990s by the Mk.50 BARRACUDA ALWT (Advanced Lightweight Torpedo). The Marconi STINGRAY is replacing it in service with the Royal Navy.

Specifications Length: 8.5 ft. (2.59 m.). **Diameter:** 12.75 in. (324 mm.). **Weight, launch:** 508 lb. (231 kg.). **Speed:** 40–45 kt. (depending on depth). **Range:** 6.8 mi. (11.35 km.). **Max. operating depth:** 1500 ft. (457 m.). **Operators:** Aus, Bra, Can, Ecu, Fra, FRG, Gre, Indo, Iran, Isr, It, Jap, Mor, Neth, NZ, Pak, S Ar, Sp, Tur, Tai, UK, US.

MK.48 TORPEDO: A "heavy" submarine-launched, WIRE-GUIDED acoustic homing TORPEDO built by Gould for the U.S. Navy and the navies of Australia, Canada, and the Netherlands, among others. Development of the Mk.48 began in 1957 as Project RETORC (Research Torpedo Configuration). Originally intended to arm both surface ships and submarines, it was completed in the mid-1960s exclusively as a submarine-launched weapon, with both anti-ship and anti-submarine capabilities. Two prototypes were built, one by Westinghouse (Mk.48 Mod 0), the other by Gould (Mk.48 Mod 1). The latter was refined into the initial production version, the Mk.48 Mod 2, which entered service in 1972. The Mod 3, introduced in 1973, added two-way communication over the wire guidance link, to transmit back data from the torpedo's on-board sonar to the submarine's fire control system. The Mod 4, introduced in 1980, has a higher maximum speed and a greater operating depth, plus an autonomous homing (fire-and-forget) mode. In 1978 the navy initiated the Mk.48 ADCAP (Advanced Capability) program to counter the rapid improvement in Soviet submarine performance. Designated Mk.48 Mod 5, the ADCAP entered service in 1986, with a new acoustic homing unit that has an acquisition range of 12,000 ft. (3658 m.), an electronically scanned sonar, increased range, and underice capability. Further development is being funded under an Upgraded ADCAP program. The Mk.48 is expected to remain the standard U.S. submarine torpedo in the foreseeable future.

The round nose section houses an active/passive SONAR, behind which is a 650-lb. (295.5-kg.) high-explosive warhead. The cylindrical centerbody contains the on-board guidance computer and fuel tanks. The liquid monopropellant (OTTO) engine is in the tapered tail section, which also houses actuators for the control surfaces, and an inertialess reel for the guidance wire. The tail terminates in a shrouded pumpjet, a low-speed, multi-bladed turbine that rotates between two sets of stator vanes inside an annular duct. This method of propulsion is believed to be significantly quieter than conventional propellers, especially at high speed.

In the wire-guided mode, the torpedo is steered initially by commands from the submarine's fire control system, on the basis of data from the submarine's sonar. When the torpedo's on-board sonar (operating in the passive mode) detects the target, this information is relayed through the guidance wire to the submarine for processing.

During the terminal phase of the run the wire is severed, and the torpedo homes autonomously using its own sonar (in both passive and active modes) under the control of its own guidance computer. In the fire-and-forget mode, the torpedo homes autonomously from launch, allowing the launching submarine to make radical evasive maneuvers, but leaving the torpedo more susceptible to spoofing countermeasures.

Like its contemporary, the British Mk.24 TIGER-FISH, the Mk.48 long suffered from poor reliability, which (it is believed) has finally been rectified in the Mod 5. Its remaining shortcomings include a relatively small warhead, only marginally effective against large surface ships and submarines; and a high unit cost (more than $4.5 million for ADCAP), which precludes the acquisition of large numbers.

Specifications Length: 19.17 ft. (5.84 m.). **Diameter:** 21 in. (533 mm.). **Weight, launch:** 3480 lb. (1582 kg.). **Speed:** (1–3) 48 kt.; (4) 55 kt.; (5) 60 kt. **Range:** (1–3) 20 mi. (33.4 km.); (4) 17.5 mi. (29.22 km.); (5) 23.75 mi. (39.66 km.). **Max. operating depth:** (4–5) 3000 ft. (915 m.).

MK.50: U.S. Advanced, Lightweight Torpedo. See BARRACUDA.

MLMS: Multi-purpose, Lightweight Missile System: air-to-air derivative of the STINGER shoulder-fired surface-to-air missile, now in service with U.S. Army, Navy, and Marine Corps helicopters, primarily as an anti-helicopter weapon.

MLRS: See MULTIPLE LAUNCH ROCKET SYSTEM.

MMW: Millimeter Waves, electromagnetic radiation in the 30–300 gigahertz (GHz) frequency band (100–1 mm. wavelength), sometimes known as "Extremely High Frequency" (EHF). Because of their higher frequencies, millimeter waves allow the achievement of much higher resolutions with very small antennas; moreover, MMW beams are very narrow and do not generate spurious secondary beams ("sidelobes"), making them more difficult to detect than lower-frequency microwaves.

Their drawback is that millimeter waves in general are rapidly attenuated by atmospheric absorption; there are, however, four frequency "windows" in which attenuation is minimized (35, 94, 140, and 220 GHz). Most military applications requiring long ranges have so far exploited only the 35- and 94-GHz windows, because of the difficulty of generating higher-frequency signals.

Millimeter waves are currently employed mainly for high-volume, secure communications (especially for satellite communications), but they are now also being developed for target acquisition and weapon guidance. High-resolution MMW RADARS can be made small enough to fit into autonomous-homing "brilliant" submunitions (see SAD-ARM), allowing them to locate and identify targets (e.g., tanks) even against background clutter. So-called "passive" MMW radar, which detects the reflections of background "sky radiation," are also in development. Because MMW attenuation is less than in the INFRARED frequency band, MMW sensors offer better performance in fog, smoke, and precipitation, but their resolutions are somewhat less because of their lower frequencies. See also ELECTROMAGNETIC SPECTRUM.

MOBILE DEFENSE: A defensive operational method in which no serious attempt is made to hold territory, to leave mobile forces free to attack the flanks of the enemy advance (which become longer and potentially more vulnerable as the enemy pushes deeper into friendly territory). Eventually, logistic exhaustion may cause the enemy to halt his advance, at which time his (overextended) forces can be counterattacked most advantageously. A mobile defense thus trades space for time, and conserves the defender's forces by avoiding head-on resistance (and exposure to preparatory artillery bombardments—thus, this method is often favored by the weaker side). Some defenders, however, lack the geographic depth to implement this scheme, or are politically inhibited from abandoning territory. The most important recent example of a mobile defense is the Soviet Union's 1942 campaign. See also DEFENSE-IN-DEPTH.

MOBILITY: The ability of troops and equipment to move (or be moved) from one place to another. Mobility can be classified as STRATEGIC, OPERATIONAL, or TACTICAL. The requirements of the different types of mobility are incompatible.

Strategic mobility is the ability to transport military forces from the homeland to an overseas theater of operations, or from one theater to another, in timely fashion. This usually requires air portability, which places a premium on light weight to the exclusion of most other attributes. Thus AIRBORNE FORCES have high strategic mobility, but they tend to lack armor protection (and thus tactical mobility), as well as firepower and sustainability.

Operational mobility is the ability to move rapidly within a theater, and requires speed, endurance, and often cross-country capacity. ARMORED

CARS, trucks, and helicopters are usually superior to armored forces for this, but lack protection.

Tactical mobility is the ability to move on the battlefield, in the face of enemy fire. This requires primarily armor protection and heavy firepower—the attributes of MAIN BATTLE TANKS. Mechanical speed, by contrast, is almost irrelevant (faster and lighter forces may be completely unable to advance under fire). On the other hand, armored forces with good tactical mobility are usually too heavy to have useful strategic mobility (being dependent on ships for intertheater transport), and may lack the range and speed for adequate operational mobility. Force structures must therefore compromise among the different types of mobility, selectively sacrificing one for another according to strategic considerations.

MOPP: Mission-Oriented Protective Posture, a U.S. Army term for a progression of measures (MOPP-1 through MOPP-4) intended to protect personnel from chemical weapons with the least degradation of their combat effectiveness. MOPP status reflects the likelihood of chemical attack:

MOPP-1: Enemy forces possess chemical weapons, but the probability of their employment is estimated as low. Protective overalls are worn; rubber boots, the gas mask, protective hood, and rubber gloves are kept handy but not worn. MOPP-2: Chemical attack is somewhat more likely, but still improbable. Both overalls and boots are worn; mask, hood, and gloves are carried. MOPP-3: Chemical attack is probable. Overalls, boots, mask, and hood are worn; gloves are carried. The hood may be worn open or closed, depending on the weather. MOPP-4: Chemical attack either imminent or in progress. Full protective gear is worn and closed. See, more generally, CHEMICAL WARFARE.

MORALE: The combination of psychological and social factors that induce soldiers (or civilians) to endure hardships and risks to perform their duties. Morale is impossible to quantify reliably, yet it is often the most important determinant of actual military power. Napoleon's dictum "The moral is to the physical as three is to one" has not been invalidated by modern technology. Among many and varied imputed morale factors, perhaps the most important is small-group COHESION, the comradeship that binds men in squads or platoons, ship crews, etc. In larger groups, a degree of solidarity (or collective identity) can be fostered by seemingly inconsequential ceremonies, in-group rituals, distinctive uniforms, units songs and march

tunes, etc. If a special formation has a distinguished combat history, its members frequently identify with the past victories of their predecessors, and may fight especially hard to avoid sullying their (common) reputation. The British regimental system, so obviously inefficient in managerial terms, has nevertheless been effective precisely because of the beneficial effect of regimental traditions on morale.

It is usually essential for morale that leaders be perceived as concerned with the welfare of their subordinates—a perception that can be fostered by such rules as equal (at least) risk-sharing, subordinate precedence for meals, timely mail service, provisions for the airing of grievances, a ready ear for personal problems, and all the tricks of the officer's trade which make up "leadership," as well as the reality of tactical and managerial proficiency. Troops will only fight for an officer who demonstrates his ability on the battlefield but only if they believe that he is trying to achieve results with the minimum of losses; even competent "glory hounds" cannot normally evoke maximal efforts.

While high morale is a key ingredient on the offensive, its true state can be seen only in defeat. A demoralized army may simply disintegrate when beaten, its men passive or fleeing in panic ("inside every army is a mob waiting to get out"). A force with high morale can still be defeated, but will rally to fight again, if it can. An army that can do so *repeatedly* is a formidable opponent indeed; e.g., the German army of World War II, which fought vigorously to the last despite successive defeats over a period of two years.

MORTAR: A simple weapon designed to loft a shell (always called a "bomb") in a high trajectory, at relatively low velocities. Mortars are the oldest form of ARTILLERY, but until this century they were ponderous, large-caliber weapons used mainly in sieges. The modern mortar (originally "trench mortar") was developed in World War I to fire high-explosive and chemical bombs against opposing trench lines, while being light enough to accompany infantry in attacks. Mortars are made in a variety of calibers, ranging from 50 to 240 mm; the most common are 60, 81, 82, 107 (4.2 in.), 120, and 160 mm. Mortars smaller than 120 mm. are generally considered infantry weapons and are organic to infantry units; larger mortars are considered artillery, and manned by specialists. Mortars smaller than 60 mm. are used mainly to fire illumination or smoke rounds.

The 60-mm. mortar is a lightweight weapon frequently used by AIRBORNE or LIGHT INFANTRY forces, while 81- or 82-mm. mortars are normally assigned to infantry battalions. While mortars up to 82 mm. can be carried by two or three men, they are usually loaded on vehicles or mounted on self-propelled carriages derived from armored personnel carriers. Mortars larger than 82 mm. are too heavy to be carried, and are towed on wheeled carriages if not self-propelled.

A mortar has three basic components: a barrel (or tube), a baseplate, and a bipod. The barrel is usually a simple cylinder, open at one end (the muzzle), with a firing pin at the bottom (breech). Because of the weapon's low muzzle velocity, the barrel can be manufactured from mild steel, greatly reducing costs. The tube is usually smoothbore, and the bombs are fin-stabilized in flight; some larger mortars, however, (notably the U.S. 4.2-in. [107-mm.] M30, and some 120-mm. weapons) are rifled and fire spin-stabilized bombs for greater accuracy at long range.

The tube is usually connected to the baseplate by means of a ball-socket joint, movable to adjust elevation and traverse. Because mortars are usually fired at angles in excess of 45°, most of the recoil force is directed downward into the earth, eliminating the need for a complex recoil mechanism (a major cause of complexity in guns and howitzers). The baseplate serves merely to stabilize the barrel and spread the recoil load over a greater area to prevent the tube from digging into the ground. The bipod is used to elevate the tube by a ratchet or screw mechanism.

Mortars up to 120 mm. are usually muzzle-loaded; i.e., the bomb is dropped tail-first down the barrel to collide with the firing pin, triggering a propellant charge attached to the tail of the bomb. In smaller mortars, the pin is fixed, so the weapon can be fired (for short periods) as quickly as bombs can be dropped down the tube (the main constraint on rate of fire is the rate of heat dissipation). Most mortars of 120 mm. or greater calibres have actuated firing pins, activated by lanyard triggers. Mortars of 160 mm. or larger dimensions are usually breech-loaders, due to the difficulty of lifting heavy shells up to the muzzle. The barrel breaks near the breech (much like a shotgun); the bomb is inserted, and the barrel closed. The weapon is then elevated and fired by pulling the lanyard. Because of the complexity of this procedure, the rate of fire is relatively low.

Many different types of mortar bombs are available; the most common are high-explosive (HE), fragmentation, illumination, smoke, and CHEMICAL rounds. Because of their relatively low muzzle velocity, mortar bomb casings can be much thinner than in artillery shells; hence the explosive charge or chemical payload can be much larger for a given caliber. IMPROVED CONVENTIONAL MUNITIONS (ICMs) for mortars are currently under development, including guided anti-armor mortar projectiles, minelet bombs, and cluster bombs.

As compared to guns or howitzers, mortars are cheap and thus can be acquired in large numbers; their simplicity also makes them rugged, reliable, and easy to operate. Moreover, mortars can be carried or transported into areas inaccessible to other artillery. Finally, their high-angle trajectory makes mortars ideal for mountain or jungle warfare. But mortars lack the range of other artillery and are inherently less accurate. See also separate MORTAR entries for BRITIAN; FRANCE; ISRAEL; SOVIET UNION; UNITED STATES.

MORTARS, BRITAIN: The British army currently relies on only two mortars: the 51-mm. and the 81-mm. L16. The 51-mm. mortar, assigned to infantry platoons, is primarily meant for smoke and illumination, but can also fire small high-explosive bombs (roughly equivalent to a hand grenade). The weapon, extremely light and simple, consists of a steel tube attached to a fixed rectangular baseplate. The tube is 29.5 in. (750 mm.) long and weighs only 5.75 lb. (2.61 kg.); the baseplate assembly weighs 6.7 lb. (3.05 kg.), for an all-up weight of only 12.45 lb. (5.66 kg.). It does not have a bipod; the weapon is elevated manually by holding the tube in one hand. A simple optical sight is provided, and the weapon can be surprisingly accurate to a maximum range of 800 m.; a very low minimum range of only 50 m. is achieved by an ingenious barrel insert with an extended firing pin, which prevents the bomb from reaching the bottom of the tube. This increases the chamber volume, thereby reducing both muzzle velocity and range.

The 81-mm. L16 is a more conventional mortar issued to companies and battalions. The barrel is 50.4 in. (1.28 m.) long and weighs 27.95 lb. (12.7 kg.). Forged of high-grade steel, it is corrugated on the outside of its lower half to increase heat dissipation. The barrel is attached by a ball-socket to a circular aluminum baseplate (of Canadian design) with a diameter of 21.5 in. (546 mm.) and weight of 27.5 lb. (12.5 kg.). The bipod, which has an unusual "K" configuration caused by the incorporation of the elevating screw into one of the legs (a weight-saving measure), weighs 27 lb. (12.27 kg.), can traverse 100 miles (5.625°) left and right, and has elevation limits of +45° to +90°. The complete mount weighs 80.45 lb. (36.6 kg.) ready to fire, and can loft a 9.26-lb. (4.2-kg.) bomb at a muzzle velocity of 297 m. per second to a maximum range of 5650 m., at a rate of up to 15 bombs per minute, which can be maintained indefinitely. The L16 can fire a full range of HE, smoke, and illumination bombs. The British army also has a self-propelled versions mounted on a modified FV432 TROJAN armored personnel carrier. Considered one of the finest mortars in service, the L16 was adopted in 1984 by the U.S. Army as the M252. It is in service with Australia, Bahrain, Canada, Guyana, India, Kenya, Malawi, Malaysia, New Zealand, North Yemen, Norway, Oman, Qatar, and the United Arab Emirates.

Obsolete British mortars still used by Third World armies include the 2-in. mortar (similar to the 51-mm.) and the 3-in. mortar. The 2-in. mortar, introduced in the 1930s, weighs 9.1 lb. (4.13 kg.), has a fixed rectangular baseplate, and is hand-held for firing. The 3-in. mortar, introduced in 1936, has a 42.25-in. (1.07-m.) barrel weighing 44 lb. (20 kg.), a rectangular baseplate weighing 42.9 lb. (19.5 kg.), and a conventional bipod weighing 45.1 lb. (20.5 kg.) for a firing weight of 132 lb. (60 kg.). It has a range of 450 to 2560 m., and a maximum rate of fire of 15 rounds per minute. It fires a variety of bombs weighing approximately 9.72 lb. (4.42 kg.).

MORTARS, FRANCE: France produces a wide variety of mortars, designed by Hotchkiss and by Thomson-Brandt, which are in service with the French and many other armies. Small mortars are attached to French infantry units down to the company level, while 120-mm. mortars are a standard artillery piece at the regimental and divisional levels (frequently substituting for guns and howitzers in reserve formations).

The smallest mortar in service is the Hotchkiss-Brandt 60-mm. MO-60-63 light, introduced in 1963. It has a conventional configuration with a 28.5-in. (724-mm.) nickel-steel barrel weighing 8.36 lb. (3.8 kg.); a triangular baseplate weighing 13.2 lb. (6 kg.); and an 11-lb. (5-kg.) bipod, for a firing weight of 32.56 lb. (14.8 kg.). The maximum range is 2356 m. and the minimum range only 100 m. Its high-explosive (HE), smoke, and illumination bombs weigh 4–5 lb. (1.8–2.2 kg.). Thomson-Brandt is currently developing a long-range 60-

mm. mortar with a 53.1-in. (1.35-m.) barrel and a firing weight of 50.6 lb. (23 kg.), which has a range of 5000 m., comparable to most 81-mm. mortars.

The Thomson-Brandt 81-mm. medium mortar was introduced in 1961, in two versions: the short-barrel MO-81-61C, and the long-barrel MO-81-61L. The 61L has a 57.1-in. (1.45-m.) barrel weighing 31.9 lb. (14.5 kg.), while the barrel of the 61C is only 42.27 in. (1.07 m.) long and weighs 27.28 lb. (12.4 kg.). Both versions use the same 32.56-lb. (14.8-kg.) triangular baseplate and conventional bipods weighing 26.85 lb. (12.2 kg.), for firing weights of 91.3 lb. (41.5 kg.) in the 61L and 86.7 lb. (39.4 kg.) in the 61C. The 61C fires a 7.26-lb. (3.3-kg.) bomb to a maximum range of 4100 m., while the 61L fires a more powerful 9.46-lb. (4.3-kg.) bomb to a range of 5000 m. A complete range of ammunition is available for both versions. The maximum rate of fire for both versions is 12–15 bombs per minute. A long-range 81-mm. mortar is now in entering production, with a 66.9-in. (1.7-m.) barrel, a firing weight of 189.2 lb. (86 kg.), and maximum range of 7600 m.

The Thomson-Brandt 120-mm. MO-120-60 lightweight mortar, introduced in 1960, has a 64.25-in. (1.63-m.) barrel weighing 74.8 lb. (34 kg.); a triangular steel baseplate weighing 79.2 lb. (36 kg.); and a 52.8-lb. (24-kg.) bipod for a firing weight of 206.8 lb. (94 kg.). The barrel has a two-position firing pin which can be fixed for drop-firing or set for lanyard firing. Its HE, smoke, and illuminating bombs weigh between 28.6 and 29.5 lb. (13 and 13.4 kg.), with a maximum range of 7000 m. and minimum range of 500 m. A ROCKET-ASSISTED PROJECTILE (RAP) bomb is also available, with a maximum range of 9000 m. The maximum rate of fire is 15 rounds per minute, limited to 1 minute by barrel heating; the normal sustained rate is 8 rounds per minute.

The MO-120-M65 strengthened lightweight mortar is similar, but has a reinforced barrel weighing about 96.8 lb. (44 kg.), and is mounted on a wheeled trolley that can be towed by a jeep or other light vehicle. It has been replaced in French service by the MO-120-LT mortar, mounted on a heavy wheeled carriage. That weapon weighs 543.4 lb. (247 kg.) in traveling order and 367 lb. (166.8 kg.) ready to fire. Performance is generally similar to the MO-120-60.

The Thomson-Brandt 120-mm. MO-120-RT rifled mortar is a relatively heavy, long-range weapon which can be used as a divisional artillery piece. Its 81.9-in. (2.08-m.) barrel weighs 250.8 lb.

(114 kg.), and is attached to a massive triangular baseplate weighing 418 lb. (190 kg.). Complete with a two-wheel carriage, it weighs 1283 lb. in firing position, and requires about two minutes to emplace. The MO-120-RT fires special 41.15-lb. (18.7-kg.) rifled bombs with preengraved drive bands which must be engaged when the weapon is loaded. Maximum range is 8350 m., although a RAP round is available, extending range to 13,000 m. In an emergency, the mortar can fire standard 120-mm. smoothbore ammunition (except for types with spring-loaded fins).

France is unique in having developed a 60-mm. breech-loaded mortar specifically for ARMORED CARS and other light armored vehicles. The Hotchkiss-Brandt LR gun-mortar has a 70.86-in. (1.8-m.) barrel, and can be either muzzle-loaded for high-angle fire or breech-loaded for flat-trajectory fire within elevation limits of −11° to +75°. The gun-mortar can fire most 60-mm. mortar ammunition, at a range of up 400 m. in a flat trajectory, or up to 3000 m. as a mortar.

MORTARS, ISRAEL: The Israeli army uses large numbers of mortars; they are well suited to the Israeli style of warfare, being highly mobile for their volume of fire. To suppress ANTI-TANK GUIDED MISSILES and for laying smoke, a 60-mm. mortar is even mounted on the turret of the MERKAVA main battle tank. The Israeli gunmaker Soltam manufactures 60-, 81-, 120-, and 160-mm. mortars based on the Finnish Tampella design; Israeli Military Industries (IMI) manufactures a 52-mm. "commando" mortar.

The IMI 52-mm. mortar is a very simple weapon which consists of a 19.3-in. (490-mm.) barrel attached to a small, fixed baseplate. It does not have a bipod, and the barrel is elevated by hand. The complete weapon weighs only 17.38 lb. (7.9 kg.), is aimed visually, and has a range limit of 130 to 420 m. The maximum rate of fire is 20–35 rounds per minute; the high-explosive (HE), smoke, and illumination bombs weigh between 1.76 and 2.2 lb. (0.8–1.0 kg.). This mortar is used by dismounted infantry, paratroops, and commandos.

Soltam makes three different 60-mm. mortars, designated "standard," "long-range," and "commando." With a bipod the standard is the basic company support weapon for dismounted infantry; without the bipod it serves as a hand-held close assault weapon. The standard barrel is 29 in. (737 mm.) long and weighs 14 lb. (6.36 kg.); its elliptical baseplate weighs 10.34 lb. (4.7 kg.), and the bipod weighs 9.9 lb. (4.5 kg.), for a total firing

weight of 34.33 lb. (15.6 kg.). Without the bipod, the weapon weighs 27.5 lb. (12.5 kg.), and can be supported by a leather carrying handle. The maximum range is 2550 m. and the minimum range 150 m.

The commando mortar has a lighter 21-in. (533-mm.) barrel and no bipod; it weighs only 13.2 lb. (6 kg.). It is aimed visually, and has a minimum range of 100 m. and maximum range of 800 m.

The long-range version has a 37-in. (940-mm.) barrel weighing 15.4 lb. (7 kg.), a 10-lb. (4.55-kg.) bipod, and a 12.1-lb. (5.5-kg.) baseplate, for a firing weight of 37.6 lb. (17.1 kg.). It has a maximum range of 4000 m. All versions fire HE, smoke, and illumination bombs weighing between 3.3 and 4.4 lb. (1.5–2.0 kg.). Maximum rates of fire for all versions are up to 30 rounds per minute.

Soltam makes no fewer than four different 81-mm. mortars: long-range; long-barrel; short-barrel; and split-barrel. These are the basic company support weapons for the mechanized infantry, and are used in both dismounted and self-propelled (SP) versions (based on the M113 armored personnel carrier or the old M3 halftrack). The different barrels provide some flexibility in a variety of tactical situations. The long-range mortar has a heavy, reinforced barrel and is usually an SP weapon. The long-barrel version is issued to mechanized units, while the short-barrel version is issued to paratroops, dismounted infantry, and commandos. The split-barrel version is an attempt to combine the range of the long barrel with the portability of the short barrel.

The long-range 81-mm. mortar has a 61.4-in. (1.56-m.) barrel weighing 46.2 lb. (21 kg.); the long-barrel version has a 57.28-in. (1.45-m.) tube weighing 37.4 lb. (17 kg.); while the short-barrel version has a 45.5-in. (1.15-m.) tube weighing 25.3 lb. (11.5 kg.). The split-barrel version has the same overall length as the long-barrel, but can be broken down into two separate sections weighing 39.6 lb. (18 kg.). All four versions use the same 27.5-lb. (12.5-kg.) circular baseplate and 30.8-lb. (14-kg.) bipod with attached sights. Firing weights are 107.8 lb. (49 kg.) for the long-range mortar, 92.4 lb. (42 kg.) for the long-barrel mortar, 83.6 lb. (38 kg.) for the short-barrel mortar, and 96.8 lb. (44 kg.) for the split-barrel mortar. Maximum ranges are 6500 m. for the long-range, 4900 m. for the long-barrel, and 4100 m. for the short-barrel. The split-barrel mortar has a maximum range of 4660 m. assembled and 4100 m. broken down. All four versions fire standard HE, smoke, and illumination bombs weighing between 6.85 and 9.5 lb. (3.1–4.3 kg.).

The Soltam 120-mm. "lightweight" mortar is the standard direct-support weapon for mechanized infantry battalions, in both towed and SP versions. The barrel is 68.1 in. (1.73 m.) long and weighs 94.6 lb. (43 kg.). Its firing pin can be set for drop-firing, or spring-loaded for lanyard control. The elliptical baseplate weighs 136.4 lb. (62 kg.), and the bipod 69.3 lb. (31.5 kg.), for a total firing weight of 300.3 lb. (136.5 kg.). The towed version is carried on a light, two-wheeled carriage weighing 299 lb. (136 kg.); it can be pulled by a jeep and is both air-portable and parachutable. Firing bombs weighing some 28.6 lb. (13 kg.), it has a maximum range of 6250 m., and a minimum range of 250 m. A new 69.2-in. (1.76-m.) barrel weighs 100.2 lb. (45.5 kg.), and extends maximum range to 7200 m.

Soltam also produces a 120-mm. heavy mortar, with a firing weight of 534 lb. (242.7 kg.) and maximum range of 8500 m.; and a "standard" 120-mm. mortar with a firing weight of 496.3 lb. (225 kg.) and range of 6500 m. Both serve as a form of brigade artillery in SP versions.

The largest Israeli (and non-Soviet) mortar is the 160-mm. M-66, usually assigned to divisional artillery. The barrel, 112.2 in. (2.85 m.) long, rests on a 550-lb. (250-kg.) baseplate, and is elevated hydraulically. The tube is lowered to the horizontal plane for loading, then reelevated for firing. It fires an 88-lb. (40-kg.) bomb out to 9600 m. and thus provides very heavy firepower within modest ranges. The M-66 can be mounted on a towed two-wheel carriage, but is normally in service as a self-propelled weapon based on a SHERMAN tank chassis.

MORTARS, SOVIET UNION: The Soviet army makes much use of mortars at all levels from company to FRONT. The mortar became a favorite Soviet weapon in 1941–42, when the loss of most other artillery and, above all, of trained gunners forced the substitution of cheap, simple mortars for guns and howitzers. Ever since then, the Soviet army has made a virtue of necessity, attaching mortars to almost every unit. This practice is well suited to Soviet OPERATIONAL METHODS, which concentrate heavy artillery at Army and Front levels. Soviet mortars are also widely issued by other Warsaw Pact armies and have been much exported to the Third World; unlicensed copies are produced in China.

The 82-mm. mortar is the standard support weapon for MOTORIZED RIFLE companies and airborne battalions. The M41, first introduced in 1941, has a 48-in. (1.22-m.) barrel weighing some 42 lb. (19 kg.), a circular baseplate, and a bipod with a unique inverted Y configuration. It has a total weight of 114.5 lb. (52 kg.), and fires a 6.71-lb. (3.05-kg.) bomb to a maximum range of 2550 m. at a maximum rate of 15–20 rounds per minute. Also still in service is the 82-mm. M37, introduced in 1937, which has a similar barrel and baseplate to the M41, with a more conventional bipod. It has a firing weight of 123.2 lb. (56 kg.), and a maximum range of 3000 m. The even older M36 mortar, still used in the Third World, has a 50.7-in. (1.29-m.) barrel and a square baseplate, with a firing weight of 126 lb. (57.3 kg.) and maximum range of 3100 m.

The 120-mm. M43, the standard battalion support weapon, has a 73-in. (1.85-m.) barrel, a circular baseplate, and a firing weight of 604.5 lb. (275 kg.). On a two-wheeled, 405-lb. (184-kg.) towed carriage, it has a traveling weight of 1009.5 lb. (458.6 kg.). It can fire a 33.9-lb. (15.4-kg.) bomb to a range of 5700 m. at a rate of 12–15 rounds per minute. In addition to high-explosive (HE), smoke, and incendiary rounds, the M43 probably has chemical munitions as well. The older 120-mm. M38, still used in the Third World, is practically identical, except for a different shock absorber and elevating mechanism.

The 160-mm. M160 is generally issued to mountain-oriented Motorized Rifle divisions in place of the D30 122-mm. gun-howitzer. This breech-loaded weapon has a 179-in. (4.5-m.) barrel, a large, circular baseplate, and a two-wheeled carriage which doubles as a firing platform. It weighs 2860 lb. (1300 kg.) ready to fire, and 3234 lb. (1470 kg.) in traveling order. The barrel is elevated hydraulically. The barrel splits at the breech for loading, rotating forward to a horizontal position. A bomb is inserted, and the barrel is reattached to the breech. The M160 fires 91.3-lb. (41.5-kg.) bombs to a maximum range of 8040 m. at a rate of 2–3 rounds per minute. It can fire HE, smoke, incendiary, and chemical munitions.

The largest Soviet mortar (and largest in the world) is the 240-mm. 2S4. Introduced in the early 1980s, this self-propelled weapon is attached to Army and Front artillery units; it has a maximum range of 9700 m. and is believed capable of firing a nuclear warhead in addition to HE and chemical bombs. It is meant primarily for use against cities and fortified areas. The older M240 mortar, still in limited service, is a large, ponderous weapon with a 210.2-in. (5.33-m.) barrel, huge circular base-plate, and two-wheeled carriage which forms part of the mount; it has a firing weight of 9149 lb. (4158 kg.) and requires at least 25 minutes to emplace. The weapon is breech-loaded by rotating the barrel forward to a horizontal position, inserting a bomb, and reelevating. It can fire 286.6-lb. (130.27-kg.) bombs over a range of 9700 m., at a rate of 1 round per minute. HE, smoke, incendiary, and chemical bombs are believed to be available.

The most unusual Soviet mortar is the 82-mm. *Vasilek* AM (*Avtomaticheskiy minomet*) automatic mortar, introduced in 1973 and used extensively in Afghanistan, but not seen in the West until 1983. The *Vasilek* is mounted on a two-wheeled artillery carriage and appears to be breech-loaded; it is fed by a four-round clips for burst fire at 40 to 60 rounds per minute. Capable of low-angle fire as well, it can be used in the anti-tank role with HEAT rounds. In many respects it is similar to the French Thomson-Brandt 60-mm. LR gun-mortar, with the addition of automatic loading (see MORTARS, FRANCE). The *Vasilek* weighs some 1750 lb. (795.5 kg.), and has a maximum range of 1000 m. for low-angle and 5000 m. for high-angle fire with standard 82-mm. bombs. The *Vasilek* appears to be replacing the 120-mm. M43 in some Motorized Rifle battalions; field-expedient SP versions based on the BTR-60 armored personnel carriers were used in Afghanistan.

The 2S9 120-mm. gun-mortar, introduced in the early 1980s, is also similar to the Thomson-Brandt LR. Mounted on a modified BMD airborne personnel carrier, it has elevation limits of $-5°$ and $+80°$ and a maximum range of 8500 m. It fires HE and smoke, has some anti-tank capability with HEAT, and may be replacing the ASU-85 assault gun in Soviet airborne divisions. See also ARTILLERY, SOVIET UNION.

MORTARS, UNITED STATES: The U.S. Army and Marine Corps have relatively few mortars, reflecting a preference for longer-range artillery. Only 60-mm., 81-mm., and 4.2-in. (107-mm.) mortars are in service, pending the much-delayed purchase of the Soultam Tampella "lightweight" 120-mm. mortar (see MORTARS, ISRAEL).

The 60-mm. M224 Lightweight Company Mortar is the standard DIRECT SUPPORT weapon for infantry and airborne companies. Developed during the Vietnam War as a lighter, more portable replacement for the 81-mm. models then in service, its 42-in. (1.06-m.) barrel weighs 14 lb. (6.36 kg.), and the lower half has radial fins to improve heat

dissipation. The barrel is attached by a ball-socket joint to a 17.5-lb. (7.95-kg.) circular aluminum baseplate; the bipod (with sight) weighs 13.5 lb. (6.13 kg.), for a total firing weight of 45 lb. (20.45 kg.). The mortar can also be fitted to a small rectangular baseplate for use in the assault role; this configuration has no bipod (the mortar being supported manually) and weighs only 17.16 lb. (7.8 kg.). The M224 fires a new series of high-explosive (HE), smoke, and illumination bombs which reportedly have the same effect as standard 81-mm. bombs. Maximum range is 3500 m. with bipod, or 1000 m. in the assault configuration; rate of fire is 30 bombs per minute maximum, or 15 sustained. The older M19 60-mm. mortar remains in service with U.S. reserve and National Guard units, and with numerous U.S. allies.

The 81-mm. M29 and M29A1 mortars (essentially identical) are standard support weapons in mechanized infantry companies. The barrel is 51 in. (1.3 m.) long, weighs 27.9 lb. (12.68 kg.), and has a corrugated exterior to improve heat dissipation. The circular baseplate and bipod each weighs 39.86 lb. (18.11 kg.), for a total weight of 92.62 lb. (42.1 kg.). It fires a variety of HE, smoke, and illuminating bombs weighing from 9.25 to 11.5 lb. (4.2–5.2 kg.), out to a maximum range of 4737 m. The maximum rate of fire is 20–30 rounds per minute. The M29A1, which equips the M125A1 mortar carrier (a modified M113 armored personnel carrier) is now being replaced by the 81-mm. M252, a license-built version of the British L16 mortar (see MORTARS, BRITAIN, for specifications).

The largest U.S.-made mortar is the 4.2-in. (107-mm.) M30 rifled mortar, fielded as a battalion support weapon. Originally developed in 1919 to deliver chemical bombs, the M30 has a 60-in. (1.52-m.) barrel weighing 155 lb. (70.45 kg.), attached to a circular baseplate weighing 217 lb. (98.64 kg.). Instead of a bipod, the M30 uses a bridge and monopod elevation unit weighing 220 lb. (100 kg.), for a firing weight of approximately 650 lb. (295.45 kg.). It fires a series of spin-stabilized HE, smoke, and illumination bombs weighing between 24 and 27 lb. (10.0–12.28 kg.). Maximum range is 5650 m. The maximum rate of fire is 18 rounds per minute for 1 minute; the sustained rate is 9 per minute. The M30 is also carried on an SP mount, the M106A1 mortar carrier (also based on the M113). Soltam 120-mm. mortars are to replace the heavy and short-ranged M30.

MOSKVA: A class of two Soviet "hermaphrodite" helicopter carrier/missile CRUISERS completed in 1967–68. Classified by the Soviet navy as "Anti-Submarine Cruisers" (*Protivolodochnyy Kreyser*, PKRs), the *Moskva* and its sister ship *Leningrad* were designed in the early 1960s to counter U.S. Polaris ballistic-missile submarines (SSBNs) operating in waters adjacent to Soviet territory. They combine the superstructure and armament of a guided missile cruiser forward, with a large helicopter flight deck aft, a configuration apparently inspired by the Italian ANDREA DORIA class. The Moskvas were seemingly found to be too small, and were superseded by the much larger KIEV-class V/STOL aircraft carriers. In any case, the Moskvas were an essential first step in the development of Soviet shipboard aviation, and remain useful, if limited, vessels. The Moskvas have usually been deployed in the Mediterranean as apart of the Soviet 5th Squadron, but have occasionally been seen in the Baltic, North Atlantic, and Indian Ocean.

The Moskvas have a most unusual hullform, with the maximum beam well aft of amidships to form a very broad flight deck, but this also causes a very disturbing motion which fin stabilizers cannot ameliorate; they are reported to be very poor sea boats. The flight deck is 282 ft. (86 m.) long and 115 ft. (35.06 m.) wide, extending from amidships to the fantail. Two elevators connect the flight deck to the hangar below, which is 213.25 ft. (65 m.) long and 78.75 ft. (24 m.) wide, and can accommodate up to 18 Kamov Ka-25 HORMONE or Ka-27 HELIX ASW helicopters. The tall, pyramidal superstructure amidships incorporates an immense "mack" (combination of mast and smokestack) at its rear end. The bottom of the mack houses a small hangar for two helicopters.

The forward end of the ship accommodates powerful ASW and ANTI-AIR WARFARE (AAW) weapons. At the extreme bow are 2 12-barrel RBU-6000 ASW rocket launchers, behind which is a twin-arm launcher for nuclear-armed FRAS-1 ASW rockets. Behind the FRAS-1 launcher are 2 twin-arm launchers for SA-N-3 GOBLET surface-to-air missiles (SAMs), each with 24 missiles. Two twin 57-mm. ANTI-AIRCRAFT guns are located amidships. As completed, the ships also had 2 sets of quintuple 21-in. (533-mm.) TORPEDO tubes amidships, but these have been removed and their ports plated over (probably to improve seaworthiness). The air group, the primary ASW armament, normally consists of 14 Hormones or Helixes, plus 2 Mil Mi-14 HAZE minesweeping helicopters (which are too large for the hangar deck). See also AIR-

CRAFT CARRIERS, SOVIET UNION; CRUISERS, SOVIET UNION.

Specifications Length: 620 ft. (189 m.). **Beam:** 85.3 ft. (26 m.). **Draft:** 25.25 ft. (7.7 m.). **Displacement:** 14,500 tons standard/17,500 tons full load. **Powerplant:** twin-shaft steam: 4 turbo-pressurized, oil-fired boilers, 2 sets of geared turbines, 100,000 shp. **Speed:** 30 kt. **Range:** 4500 n.mi. at 29 kt./14,000 n.mi. at 12 kt. **Crew:** 850. **Sensors:** 1 "Top Sail" and 1 "Head Net-C" 3-dimensional air-search radar, 3 "Don-2" surface-search/navigation radars, 2 "Muff Cob" fire control radars, 2 "Head Light" missile guidance radars, 1 hull-mounted low-frequency sonar, 1 medium-frequency variable depth sonar. **Electronic warfare equipment:** 8 "Side Globe" electronic signal monitoring arrays; 2 "Bell Clout," 2 "Bell Slam," 2 "Bell Tap," and 2 "Top Hat" electronic countermeasures arrays; 2 chaff rocket launchers.

MOSS: Mobile Submarine Simulator system, the U.S. Mk.30 submarine-launched self-propelled decoy. A torpedo-shaped vehicle with a length of 20 ft. (6.1 m.), diameter of 21 in. (533 mm.), and launch weight of 2000 lb. (909 kg.), MOSS is equipped with a variety of noisemakers and deception jammers designed to simulate the acoustical and sonar signature of a full-size submarine. Launched from standard torpedo tubes, it can be programmed to run a divergent course after launch. MOSS has a maximum speed of 30 kt. and an operational depth of 2000 ft. (610 m.); range has not been revealed.

The Mk.38 Mini Mobile Target is a scaled-down version of MOSS, and more of them can be carried aboard submarines without displacing weapons. See also TORPEDO COUNTERMEASURES.

MOSS: NATO code name for the Soviet Tupolev Tu-126 AIRBORNE EARLY WARNING (AEW) aircraft. First observed in 1970–71, the Tu-126 is based on the Tu-114 commercial airliner, itself a derivative of the Tu-95 BEAR strategic bomber. The nose of the aircraft has a glazed-in navigator's position, above which is an AERIAL REFUELING probe. The flight crew sits in an airline-style cockpit above and behind the navigator. Numerous aerials and radomes on the fuselage indicate the presence of ELECTRONIC SIGNAL MONITORING arrays and fighter control DATA LINKS. The main cabin has been converted to a tactical control center where several fighter direction officers can track targets and direct interceptors using data from a "Flap Jack" air-search RADAR housed in a large rotodome over the rear fuselage. The wings and tail are identical to those of the Tu-95 (with a sweep angle of about 54°).

The Flap Jack radar of the Tu-126 is not comparable to the more advanced radars of the U.S. E-2C HAWKEYE or E-3A SENTRY (AWACS). Reportedly, it cannot pick out low-flying targets against ground clutter. It does not use advanced pulse-DOPPLER techniques, but only a simpler form of moving target indicator (MTI). Moreover, the large steel propellers must generate strong echoes which interfere with the radar. Only about ten Tu-126s remain in service; they are being replaced by the more advanced Il-76 MAINSTAY.

Specifications **Length:** 188.1 ft. (57.35 m) **Span:** 167.67 ft. (51.12 m) **Powerplant:** 4 15,000-shp. Kuznetsov NK-12MV Tuboprops. **Weight, empty:** 210,000 lb. (95,455 kg.). **Weight, max. take-off:** 415,000 lb. (188,864 kg.). **Speed, max.:** 540 mph (902 kph) at 30,000 ft. (9146 m.). **Speed, cruising:** 465 mph (777 kph). **Range, max.:** 6000 mi. (10,000 km.). **Endurance:** 20 hr.

MOTORIZED INFANTRY: INFANTRY provided with trucks for transport to the battle area. Motorized infantry troops always fight dismounted, and are usually organized and trained in the same manner as "straight let" infantry.

MOTORIZED RIFLE (TROOPS): *Motostrelkovyye Voiska*, the Soviet term for the primary INFANTRY branch of the Soviet armed forces. Despite their name, they are not MOTORIZED INFANTRY, but rather MECHANIZED INFANTRY, equipped with ARMORED PERSONNEL CARRIERS (APCs) and integrated into combined-arms divisions with numerous tanks. Formed in 1963, they now comprise more than 1.3 million men organized into more than 100 divisions.

Each Motorized Rifle DIVISION now has roughly 13,000 men organized into a divisional headquarters, 3 Motorized Rifle regiments, one tank regiment, an artillery regiment, a multiple rocket launcher battalion, a FROG rocket or SS-21 SPIDER missile battalion, an anti-tank battalion, a reconnaissance battalion, an engineer battalion, and various service units. Each Motorized Rifle REGIMENT in turn has about 2500 men organized into 3 Motorized Rifle battalions, one tank battalion (with 40 MAIN BATTLE TANKS), an artillery battalion, an anti-tank company, an anti-aircraft company, a reconnaissance company, an engineer company, and some small service units. Each Motorized Rifle BATTALION in turn has 430 to 444 men, in a small headquarters unit, 3 Motorized Rifle companies, a heavy mortar battery, and a small

service echelon. Each Motorized Rifle company has 102 men in 3 platoons with 3 10-man squads.

Motorized Rifle troops are mounted either in BMP infantry fighting vehicles or BTR-60/70/80 wheeled APCs. With their own tanks, regiments would normally operate as organic formations; the Soviet army would rarely, if every, cross-attach units to form ad hoc COMBAT TEAMS.

Each Soviet tank division also includes one Motorized Rifle regiment, while each tank regiment in a tank division has one Motorized Rifle battalion. Motorized Rifle formations of other Warsaw Pact armies are organized along similar lines.

MOUT: Military Operations in Urban Terrain, the U.S. Army term for URBAN WARFARE. The general principles of MOUT have much in common with the urban warfare doctrine of other modern armies. On the offensive, cities and built-up areas should be bypassed (and blockaded) whenever possible, rather than attacked. When an urban area simply cannot be avoided, attempts should be made to isolate it by controlling all approaches to prevent reinforcement and resupply. This would be followed by a deliberate assault, if necessary. Artillery and tanks would generally operate in *direct* support to assist dismounted infantry in reducing enemy strongpoints. After enemy resistance is broken, the area must be cleared block by block, a task often left to follow-on units.

If the enemy is taken by surprise, the urban area may be overrun by a hasty attack which may develop in the midst of a deliberate assault should forward units detect the opportunity for a rapid advance. But given the intricacy of urban terrain, hasty attack is extremely risky, especially if the enemy's strength is unknown (high-rise buildings complicate everything: it may require an entire battalion to clear an office block). When attacking shallow "strip developments," on the other hand, overwhelming force may be concentrated for a breakthrough on a narrow sector, which follow-on units can exploit by rolling up the enemy flanks.

On the defensive, the U.S. Army would prefer to use villages or suburbs along the approaches to a city as fortified outposts to slow, attrit, and channel the enemy into kill zones covered by long-range fire. If the enemy penetrates into the urban area itself, the city would be divided into battalion and company strongpoints sited in the sturdier brick and concrete buildings, connected when possible by tunnels or sewers. Tanks and artillery would be dug in to provide direct fire support. Extensive use of mines and demolitions would be made to channel the enemy into kill zones between strong-points. Finally, a strong operational reserve would be be maintained to plug holes, relieve exhausted units, and mount local counterattacks.

MOUT doctrine is sound, but the U.S. Army seldom trains for it above the tactical level, even though more than 60 percent of West Germany territory is built up in some way. The reason is partly political (to defend a city is to destroy it), and partly reflects a preference for more mobile armored warfare in more open terrain.

MRBM: Medium-Range Ballistic Missile, any BALLISTIC MISSILE with a maximum range between 1000 and 3000 km. (600 to 1800 mi.). Because MRBMs generally lack both the targeting flexibility of intermediate range ballistic missile (IRBMS) and the mobility of short-range ballistic missiles (SRBMS), they have generally passed out of active inventories. The MRBMs still in service include the U.S. PERSHING II and Soviet SS-4 SANDAL (both scheduled for destruction under the terms of the 1987 INF TREATY); and the Chinese CSS-1.

MRV: See MULTIPLE REENTRY VEHICLES.

MSBS: *Mer-Sol Ballistic Stratégique*, designation of the French submarine-launched ballistic missiles (SLBMS), which, together with S-3 land-based IRBMS and bomber-launched ASMP cruise missiles, form the FORCE DE DISSUASION, the French nuclear force. The MSBS program was initiated in the early 1960s after French President de Gaulle rejected the U.S. offer of POLARIS SLBMs, choosing instead to develop a completely indigenous missile, despite formidable technical difficulties and high costs.

The first MSBS, the M1, began flight tests in 1967 and entered service aboard the REDOUTABLE-class nuclear-powered ballistic missile submarines (SSBNs) between 1971 and 1973. The M1 was a two-stage, solid-fuel missile. The first stage had a welded steel casing and a Type 904/P10 engine rated at 100,000 lb. (45,455 kg.) of thrust, with four gimballed nozzles for steering and attitude control. The second stage had a wound glassfiber casing and a 39,682-lb. (18,037-kg.) Rita I/P4 engine with a single fixed nozzle and steering by freon injection thrust vector control (TVC). The missile was controlled by INERTIAL GUIDANCE based on the EMD Sagittaire computer, and carried a single Aerospatiale REENTRY VEHICLE (RV) housing a 500-KT warhead.

The M1 was quickly followed by the M2, externally identical but equipped with an improved Rita II/P6 second stage rated at 70,547 lb. (32,067 kg.) of thrust, increasing the launch weight while extending maximum range. The M2 entered service in 1974, when it was retrofitted into the first three Redoutables. The M20 missile, introduced in the late 1970s, is similar to the M2, but has an improved guidance unit which reduces the presumed CEP, and a new MR-60 RV with a 1-MT warhead hardened against nuclear effects. A PENAID package was also fitted to counter the Soviet ABM-1 GALOSH anti-ballistic missile system (a lighter MR-61 RV, also with a 1-MT warhead, was substituted for the MR-60 in the early 1980s).

The M20 replaced the M2 on all six Redoubtables, but is in turn being replaced on five of the submarines by the M4, an entirely new and much larger missile (the M20 will remain in service aboard the *Redoutable* itself until its retirement in 1997). In addition to five Redoutables, the M4 will also equip the new INFLEXIBLE-class SSBN.

The M4 is a three-stage solid-fuel missile. The first stage has a steel casing with a Type 401 engine rated at 154,320 lb. (70,145 kg.) of thrust. The second stage has a wound glassfiber casing and a Type 402 engine rated at 66,138 lb. (30,063 kg.) of thrust. The third stage has a lightweight wound KEVLAR casing and a Type 403 engine rated at 15,432 lb. (7015 kg.) of thrust. All three stages have single submerged nozzles to reduce overall length; the first and second have gimballed nozzles, while the third stage has freon injection TVC. In contrast to earlier missiles, the M4 has six TN-70 150-kT MULTIPLE INDEPENDENTLY TARGETED REENTRY VEHICLES (MIRVs) and an advanced PENAID package mounted on a post-boost vehicle (PBV).

A still newer missile, designated M5, is currently in development for use on a new class of French SSBNs to be completed in the late 1990s. It is reported to carry 8–10 MIRVs and to have a range of 6600 mi. (11,000 km.).

Specifications **Length:** (M1/2/20) 34.1 ft. (10.36 m.); (M4) 36.25 ft. (11.05 m.). **Diameter:** (M1/2/20) 4.95 ft. (1.5 m.); (M4) 6.3 ft. (1.92 m.). **Weight, launch:** (M1) 39,638 lb. (18,017 kg.); (M2/20) 44,091 lb. (20,041 kg.); (M4) 77,323 lb. (35,147 kg.). **Range:** (M1) 1491 mi. (2400 km.); (M2/20) 1926 mi. (3100 km.); (M4) 2485 mi. (4000 km.). **CEP:** (M1/2) 1200 m.; (M20) 930 m.; (M4) 460 m.

MT: Abbreviation for megaton, a measure of the energy yield of nuclear warheads, equivalent to the blast effect of 1 million tons of TNT.

MTB: Motor Torpedo Boat, a FAST ATTACK CRAFT normally displacing some 100 tons, armed with TORPEDOES, MACHINE GUNS, and light automatic CANNON. Patrol Torpedo (PT) boats were the U.S. Navy's equivalent.

MTI: Moving Target Indicator, a feature of many search and tracking RADARS, used to suppress background clutter. The MTI technique relies on the DOPPLER shift generated by moving objects to filter out stationary or slowly moving objects.

MT-LB: Soviet multi-purpose tracked vehicle used most often as an artillery tractor, but also as cargo vehicle and ARMORED PERSONNEL CARRIER (APC). Because of its superior mobility in soft ground or deep snow, it replaces BTR-60/-70/-80 wheeled APCs in tank and MOTORIZED RIFLE divisions assigned to arctic or swampy areas. Developed in the 1960s, the hull is of all-welded steel construction, with a boat-shaped bow to enhance mobility in mud, snow, and water. Armor thickness is on the order of 6–12 mm., sufficient to defeat small arms or shell splinters. The hull is divided into a crew compartment in front, an engine compartment in the middle, and a troop/cargo compartment in the rear. The crew consists of the commander and the driver. The driver sits on the left, with a small armored windshield and a drop-down vision slit. The commander sits on the right in a small, manually operated turret armed with a 7.62-mm. machine gun. The engine is immediately behind the driver. The troop/cargo compartment can accommodate up to 11 infantrymen on seats which fold up when cargo is carried. Access is provided by two large rear doors and two roof hatches. A positive overpressure COLLECTIVE FILTRATION unit for NBC defense is standard. Fully amphibious, the vehicle is propelled through the water by track action at a speeds up to 3–4 mph (5–6.7 kph). Extra-wide tracks can be fitted for use in snow or swamps.

The MT-LB has been developed into several specialized vehicles, including the MT-LBU command vehicle; the MT-LB SON with a surveillance and counterbattery radar; and the MTL-LB armored repair vehicle. In addition, the SA-13 GOPHER surface-to-air missile system and the SAU-122 self-propelled howitzer are both mounted on the MT-LB chassis.

Specifications **Length:** 21.16 ft. (6.45 m.) **Width:** 6.16 ft. (1.87 m.) **Height:** 9.33 ft. (2.85 m.). **Weight, combat:** 12 tons. **Powerplant:** 240-hp. V-8 diesel. **Hp./wt. ratio:** 20.16 hp./ton. **Fuel:** 102 gal. (450 lit.). **Speed, road:** 38 mph (64 kph). **Range, max.:** 310 mi. (518 km.).

MUJAHIDEEN: Literally, "Holy Warriors," a term favored by Islamic GUERRILLA forces opposing non-Islamic governments or foreign invaders. According to Islamic doctrine, warriors waging Jihad, or HOLY WAR, who die fighting the infidel are immediately transported to a paradise well provided with comforts of every description; this belief should make them fearless fighters, able to endure all hardships and risks without complaint. The term has been used most frequently used by various guerrilla factions fighting Soviet and government forces in Afghanistan, and by various armed groups in Iran, Egypt, Syria, and Lebanon.

MULTIPLE INDEPENDENTLY TARGETED REENTRY VEHICLES (MIRVS): A BALLISTIC MISSILE payload of two or more REENTRY VEHICLES (RVs) mounted on a POST-BOOST VEHICLE (PBV) or "bus" which can modify its velocity vector so as to direct the RVs against widely separated targets. Because a missile with MIRV capability can attack several different targets, it can confer a proportionate advantage to the attacker in a COUNTERFORCE strike (if one missile can destroy two or more missile silos, an equal enemy force can be neutralized with a fraction of the attacking force). Thus MIRVs are inherently destablizing in the absence of effective defenses or alternative DELIVERY SYSTEMS (e.g., SLBMS, BOMBERS, and CRUISE MISSILES). Under the terms of the SALT II TREATY, the United States and Soviet Union can deploy no more than 10 MIRVs per missile, a limit which appears to have been observed, although some Soviet missiles (e.g., the SS-18 Mod 4 SATAN) have been assessed as capable of carrying up to 14 RVs.

MULTIPLE LAUNCH ROCKET SYSTEM: A U.S. self-propelled multiple rocket launcher (MRL), also in service with several NATO allies. Development began in 1976 to meet a requirement for a long-range saturation weapon for use against enemy armor, artillery, and troop concentrations, and for the Suppression of Enemy Air Defenses (SEAD). Two competitive systems were designed by Boeing and LTV; after trials held in 1979–80, the latter's was selected for production. The first of 333 launchers were delivered to the U.S. Army in 1982, with production scheduled for completion in the early 1990s. MLRs has also been ordered by Britian (67 launchers), West Germany (202 launchers), France (56 launchers), Italy (20 launchers), and the Netherlands (30 launchers); most are to be built in Europe by a multinational consortium.

MLRs consists of a 12-round launcher mounted on the tracked M987 Self-Propelled Launcher/Loader (SPLL), derived from the chassis of the

M2/M3 BRADLEY fighting vehicle. With a lightly armored cab forward and a traversable Launcher/ Loader Module (LLM) in the rear, the SPLL weighs more than 25 tons. The LLM is a large armored box with a built-in twin-boom crane to load and unload two Launch Pod Containers (LPCs), each of which contains six ready-to-fire rockets. Each SPLL has its own fire control equipment, and can fire rockets singly, in pairs, or in ripples up to 12 rounds. The LLM can be reloaded within ten minutes from a tracked supply vehicle carrying additional LPCs.

MLRs 227-mm. unguided, fin-stabilized rockets, with a length of 12.95 ft. (3.95 m.) and a launch weight of 550–675 lb. (250–307 kg.), have four alternative warheads, designated Phase 1 through Phase IV. Phase 1 is a 338.8-lb. (154-kg.) submunition package containing 644 M7 anti-armor/anti-personnel bomblets. The entire round weighs 675 lb. (307 kg.), has a maximum range of 32,000 m., and is intended mainly for COUNTERBATTERY and SEAD missions. Phase II is a 235.4-lb. (107-kg.) dispenser with 28 West German AT-2 anti-tank MINES. The round weighs 566.6 lb.(257.55 kg.), has a maximum range of 40,000 m., and would be used to delay advancing armor (a single SPLL can lay 366 mines in a 1000- × 5000-m. area in less than one minute). Phase III, still in development, is to have six TERMINALLY GUIDED SUBMISSILES (TGSMs) to attack individual armored targets (e.g., tanks) with pinpoint accuracy. The round will weigh 550 lb. (250 kg.) and have a range of 45,000 m. Phase IV, also in development, is a BINARY chemical warhead with 92 lb. (41.82 kg) of NERVE GAS. Other warhead packages are also in development, notably the MLRS-SADARM, similar to the Phase II warhead, but with six SADARM (Search and Destroy Armor) intelligent submunitions.

The MLRS has also been selected as the launch platform for the U.S. Army Tactical Missile System (ATACMS), a planned NATO-wide, conventionally armed tactical ballistic missile. ATACMS is some 13 ft. (3.9 m.) long with a diameter of 2 ft. (600 mm.) and launch weight of 2000 lb. (909 kg.). The initial version (Block I) will have a 1000-lb. (454-kg.) ICM payload with M77 APAM submunitions for counterbattery fire; the later Block II will have up to 36 TGSMs for long-range anti-armor attacks. SADARM and unitary (high-explosive) warheads are also in development. ATACMS has an advanced solid-fuel rocket engine and a maximum range of about 100 mi. (167 km.). Two ATACMS could be carried by each SPLL with

only some modifications to the fire control software. The first ATACMS entered service in 1990–91 and were used in the Persian Gulf conflict to attack mobile Scud missile launchers and troop concentrations.

Other MLRS studies are investigating the feasibility of lightweight towed launchers for airborne and light infantry units, and navalized launchers for shore bombardment.

Although powerful and effective, the MLRS is overly complex and costly due to the U.S. preoccupation with precision targeting, which ignores the fact that MRLs are inherently less accurate than tube artillery, and are best employed as saturation weapons against area targets. This complexity makes the MLRS expensive ($1.2 million per fire unit, including rockets), and precludes their acquisition in large numbers, in contrast to Soviet MRLs, which are cheap, expendable, and ubiquitous. See also ARTILLERY, SOVIET UNION; ARTILLERY, UNITED STATES; ROCKET ARTILLERY.

MULTIPLE PROTECTIVE SITES: A proposed basing scheme for the MX PEACEKEEPER ICBM, under which several hundred SILOS would be constructed for each missile in service. The missiles would be moved from silo to silo on a random basis (a high-stakes shell game), thereby preventing Soviet surveillance satellites from detecting which silos are active at any given time; hence all potential MX silos would have to be attacked to destroy the force. The plan was rejected on several points, including the cost of building the additional silos; arms control verification problems; and the fact that the Soviet Union could easily destroy all the silos anyway given the small number of Peacekeepers (and silos) envisaged. See also DENSE PACK.

MULTIPLE REENTRY VEHICLES: A BALLISTIC MISSILE payload of two or more REENTRY VEHICLES (RVs) released shotgun-fashion, so as to fall in a fixed pattern around a common centroid. In contrast to multiple independently targeted reentry vehicles (MIRVs), MRVs do not require a POST-BOOST VEHICLE (PBV) and cannot be directed against widely separated targets. Nevertheless, MRVs can increase the efficiency of a ballistic missile by replacing large, unitary warheads, much of whose yield is dissipated upward; several smaller warheads with a much lower combined yield can destroy a much larger area. MRVs arm the U.S.-built POLARIS A-3 SLBM, and the Soviet SS-11 Mod 3 SEGO ICBM.

MULTIPLE ROCKET LAUNCHER: See ROCKET ARTILLERY.

MUSTARD "GAS": 2-chloroethyl sulfide (U.S. designation HD), a powerful vesicant, or BLISTER AGENT, for use in CHEMICAL WARFARE; so called because of the mustard taste it leaves on the tongue. Mustard is actually an oily, colorless, and nearly odorless liquid which can be dispensed as an aerosol mist from artillery shells, mortar bombs, rocket and missile warheads, aerial bombs, and spray tanks. It acts by skin contact, inhalation, or ingestion, and causes painful chemical burns followed by large, liquid-filled blisters. Mustard is a slow-acting agent; its effects are not usually manifest until several hours after exposure, depending on dosage. It has a very high boiling point (260°C), and is therefore extremely persistent (depending on temperature and wind conditions, it can take as long as a month to dissipate). Because mustard is much denser than air, it tends to concentrate in shell holes, trenches, ravines, and other low-lying terrain, precisely where troops would find shelter from direct fire weapons. Once there, it saturates the soil and contaminates food, clothing, and any equipment with which it comes in contact.

Mustard can easily penetrate ordinary clothing; troops operating in contaminated areas must therefore wear nonpermeable overalls, hoods, and gloves, as well as respirators. Once mustard comes in contact with the skin, burns and blisters become evident within 4–6 hours. Exposure to the eyes can cause temporary or permanent blindness. Inhalation causes chemical edema leading to destruction of the lungs and death from pneumonia. Ingestion of contaminated water or food can cause irritation and burning of the digestive tract, while massive doses (of all sorts) can cause systemic poisoning. Mustard has an LD-50 of 1300 mg./min./m.^{3}; but a dose of only 200–1000 mg./min./m.^{3} is sufficient to cause blindness, blisters, and disabling burns.

Introduced by the German army in April 1917, mustard was quickly adopted by all other combatants in World War I, and became known as the "King of the War Gases" because of its persistence, horrific effects, and the difficulty of decontamination. Because passage of mustard-contaminated areas required full protective clothing, troops in such areas operated at much lower levels of efficiency and tired quickly. Mustard formed the bulk of chemical weapons stockpiled (but not used) by all sides in World War II. But it has been used in several Third World conflicts: the Italo-Abyssinian and Sino-Japanese Wars of the 1930s and, more recently, in the Iran-Iraq War.

Closely related to the original mustard are lewisite (2-chlorovinyl dichloroarsine), whose U.S. designation is L; nitrogen mustard ((2-chloroethyl) amine), U.S. designation HD; and a mustard-lewisite mix known as HL. Lewisite is considerably more volatile than mustard, hence is much faster acting; effects generally appear within 30 minutes, while a massive dose can cause death within 10 minutes. Thus HL combines both rapid-acting and slow-acting agents for maximum effect. HD is similar to mustard, but is even more persistent; in addition, eye damage is almost immediate.

MX: U.S. ICBM. See PEACEKEEPER.

MYCOTOXINS: Biological toxins used in chemical warfare. See "YELLOW RAIN."

N

NACKEN (TYPE A-14): A class of three Swedish diesel-electric attack SUBMARINES, commissioned in 1980–81. Ordered in 1972 as successors to the SJOORMAN class, the Nackens are relatively small, short-range vessels optimized for operations in the confined waters of the Baltic and its approaches. The Nackens' modified teardrop hullform has a blunt, bulbous bow, cylindrical midsection, and short, tapered stern in a single-hull configuration, with a large, flat, deck casing, and a tall streamlined sail (conning tower) located slightly forward of amidships. The bow planes are on the sail, while the stern planes form an X-configuration controlled by computers; these X-planes act as both diving planes and rudders, providing redundancy and enhancing maneuverability. Pioneered on the Sjoormans, X-planes have become a distinctive feature of Swedish submarine design.

The Nackens are armed with 8 bow-mounted torpedo tubes: 6 are long 21-inch (533-mm.) tubes and 2 are short 15.7-inch (400-mm.) tubes. The long tubes are for heavy Type 61 anti-ship homing TORPEDOES, while the short tubes are for lightweight Type 42 wire-guided anti-submarine homing torpedoes. The basic load consists of 8 Type 61s and 4 Type 42s. Alternatively, the Type 61s can be replaced by 16 influence MINES. The 21-inch tubes are equipped for positive impulse discharge, allowing the Type 61s to be launched from any depth, while the Type 42 torpedoes can be launched only by the swim-out method. All sensors and weapons are integrated in an Ericsson IDPS (Integrated Data Processing System) with two Censor 932 digital computers, which provides tactical data and monitor the boat's mechanical systems; its PEAB FIRE CONTROLS have multiple-target TRACK-WHILE-SCAN capability.

A retractable snorkel allows diesel operation from shallow submergence; power for fully submerged propulsion is provided by a 168-cell Tudor battery. All machinery is mounted on resilient sound-isolation "rafts." Both the engines and the batteries are concentrated in the aft section of the hull, leaving the forward hull exclusively for the crew, weapons, and sensors.

In 1986, one Nacken was modified to serve as a testbed for an innovative Sterling closed-cycle diesel engine, which burns liquid oxygen, and thus requires no external oxygen source; it is also less noisy than conventional diesels. The Stirling can drive the boat at speeds up to 5 knots on direct drive and also maintain the battery charge, thereby increasing the submerged (nonsnorkeling) radius by up to 500 percent over unmodified Nackens. See, more generally, SUBMARINES, SWEDEN.

Specifications Length: 162.4 ft. (49.5 m.). Beam: 18.7 ft. (5.7 m.). Displacement: 1030 tons surfaced/1125 tons submerged. Powerplant: single-shaft diesel-electric: 1 1800-hp. MTU 16V652 diesel-generator, 1 1500-hp. Jeumont-Schneider electric motor. Speed: 20 kt. Max. operating depth: 984 ft. (300 m.). Collapse depth: 1640 ft. (500 m.). Crew: 23. Sensors: 1 Krupp-Atlas CSU-3 low-frequency active/passive circular sonar array, 1 surface-search radar, 1 electronic signal monitoring array, 1 search/attack periscope.

NADGE: NATO Air Defense Ground Environment, an integrated system of RADARS, computers, and COMMUNICATION links intended to coordinate all AIR DEFENSE weapons and interceptors for the defense of NATO airspace. NADGE was planned in the early 1960s to replace manual control procedures which had become almost completely ineffective against high-speed jets. It covers the entire NATO perimeter from North Cape in Norway to Mount Ararat on the Soviet-Turkish border. In NATO, only Great Britain maintains a separate automated system (UKADGE); this, however, is interoperable with NADGE. Constructed and managed by NADGECo, a multinational consortium led by Hughes (builders of the similar SAGE system for NORAD), NADGE was completed in 1975, and has been updated and extended since then. Any aircraft penetrating NATO airspace can be detected and tracked continuously, with the data transmitted automatically to COMMAND AND CONTROL centers from which interceptors and missiles are directed. The missing element in the system is a reliable IFF, to discriminate between errant or unreported friendlies and hostile aircraft. At present, air defense missiles would have to stand down in designated sectors to avoid shooting at friendly aircraft.

NANUCHKA: NATO code name for a large class of Soviet-built missiles CORVETTES in service with the Soviet navy and the navies of India, Algeria, and Libya. Designated "Small Missile Ships" (*Malyy raketnyy korabl'*, MRK) by the Soviet navy, the Nanuchkas entered production in 1969, apparently as successors to the OSA and KOMAR missile boats, but with vastly increased range and firepower. More than 30 have been built to date, and production continues at a rate of one unit per year. The Nanuchkas have been built in three distinct subclasses (Nanuchka-I, -II, and -III), varying mainly in their weapons and sensors. Seventeen Nanuchka-Is were commissioned between 1969 and 1974, followed by 11 Nanuchka-IIs, built for export with simplified systems. India acquired 3 Nanuchka-IIs from 1977, and has ordered at least 5 more; Algeria received 4 between 1980 and 1982; and Libya bought 4 from 1981, 2 of which were sunk by U.S. Navy aircraft in the Gulf of Sidra incident of March 1986. Replacements are probably on order. The Nanuchka-III, which entered production in 1977, is intended for the Soviet navy and features improved weapons and fire controls; at least 6 are now in service.

The Nanuchkas have a beamy hull with broad transom stern and hard chine bilges, features usually found in much smaller FAST ATTACK CRAFT; as a result, their seakeeping is reported to be poor. The Nanuchkas are flush-decked, with a large boxy superstructure running most of the length of the ship; a small bridge and pilothouse are located on top of the superstructure. Perhaps the most heavily armed craft of their size in the world, the Nanuchkas are armed with 6 SS-N-9 STARBRIGHT anti-ship missiles in fixed triple launch canisters amidships. A twin SA-N-4 GECKO surface-to-air missile launcher (with 20 missiles) is housed in the forecastle, and a twin 57-mm. DUAL PURPOSE gun is mounted on the fantail to provide (limited) air defense. The Nanuchka-II has 4 much less advanced SS-N-2c STYX instead of the SS-N-9s. The Nanuchka-III has a single 76.2-mm. dual purpose gun in place of the twin 57-mm., and also a 6-barrel 30-mm. radar-controlled gun for anti-missile defense on top of the aft superstructure.

Though generally considered coastal vessels, the Nanuchkas have been deployed in the North Sea and the Mediterranean by the Soviet navy, apparently for the "tattletale" mission of following Western battle groups. The Nanuchkas could be very potent warships in competent hands.

Specifications **Length:** 194.6 ft. (59.33 m.). **Beam:** 41.3 ft. (12.6 m.). **Draft:** 7.9 ft. (2.4 m.). **Displacement:** 780 tons standard/900 tons full load. **Powerplant:** 3 10,000-hp. M504 diesels, 3 shafts. **Speed:** 32 kt. **Range:** 900 n.mi. at 30 kt./ 2500 n.mi. at 12 kt. (on 1 engine). **Crew:** 70. **Sensors:** 1 "Band Stand" SS-N-9 missile guidance radar, 1 "Pop Group" SA-N-4 missile guidance radar, 1 "Muff Cob" fire control radar (for the dual-purpose battery), 1 "Bass Tilt" 30-mm. fire control radar (Nanuchka-III only), 1 "Peel Pair" air-search radar, 1 "Stump Spar" surface-search and navigational radar. **Electronic warfare Equipment:** 2 IFF arrays, 4 passive electronic signal monitoring arrays, 2 16-barrel chaff launchers.

NAPALM: Acronym for Napthenic Acid and Palmitate, a jellied incendiary used as a filler for bombs, and as fuel for FLAMETHROWERS. Developed by the U.S. towards the end of World War II, napalm is now widely produced. Napthenic acid is a petroleum product, while palmitate is an extract of palm oil; both are cheap and readily available. When mixed with gasoline, the two substances form a thick, viscous liquid which burns at very high temperatures. Aluminum powder or benzene is sometimes added to increase the volatility of the mixture. The jellied composition, high burning temperature, and low cost of napalm make it an ideal weapon against soft targets such as troops, buildings, and unarmored vehicles. It can also be quite effective against armored vehicles, not because armor acts as a conductor of heat, but because napalm burns at such a high rate that it consumes all the oxygen in the immediate area, asphyxiating the crew. Because napalm can flow as a liquid, it can also be used to attack concrete bunkers and similar structures, though most modern fortifications have overhanging lips over their gun slits, and traps in their air ventilators, specifically to prevent this. Napalm is not suitable against HARD TARGETS such as bridges. Because of its low density, napalm bombs are bulky in relation to their weight, reducing the performance of aircraft proportionately. On the other hand, napalm splatters over a wide area on impact, reducing the need for accuracy. Moreover, like all flame weapons, napalm can have a shattering effect on the morale of even the best troops.

NAP-OF-THE-EARTH (NOE):   A flight profile flown by HELICOPTERS for maximum terrain cover (only a few feet off the ground, usually *below* the treetops). NOE flight requires careful route planning to avoid hills and ridge lines; for safety, speeds must be kept low (60–80 mph), hence NOE cannot be employed by most fixed-wing aircraft. See also CONTOUR FLYING.

NATIONAL COMMAND AUTHORITIES (NCA):   A U.S. bureaucratic term for the top national security officials; i.e., the president, vice-president, and their authorized alternates or successors. The Soviet equivalent is the DEFENSE COUNCIL *(Sovyet Oborony)* of the Politburo, and its subordinate, the Main Military Council. In the U.S., only the NCA can authorize the release of NUCLEAR WEAPONS; it is believed that the *Sovyet Oborony* has a similar monopoly.

NATIONAL EMERGENCY AIRBORNE COMMAND POST: NEACP ("Kneecap"), a.k.a. the Post-Attack COMMAND AND CONTROL System (PACCS), a.k.a. the "Doomsday Plane." Officially designated E-4, it is a specially modified Boeing 747 equipped with a variety of command, control, and COMMUNICATIONS (C³) systems to serve as a nuclear-war command post. Four 747s were converted to E-4s between 1973 and 1979; the first three aircraft were interim conversions (E-4A), while the last was an E-4B, with more powerful engines and new C³ systems. The earlier aircraft were subsequently upgraded to E-4B standard. Modifications include provisions for AERIAL REFUELING, extensive shielding of electronics against ELECTROMAGNETIC PULSE (EMP), a 5-mi.-long VLF trailing wire antenna for communication with submerged submarines, and an SHF link to defense communications satellites. It may also be possible for the NEACP to receive early warning data directly from DEFENSE SUPPORT PROGRAM satellites. An Airborne Launch Control System is also be fitted, to allow the targeting and retargeting of MINUTEMAN and PEACEKEEPER ICBMs from the aircraft. The main cabin area includes workspaces for a 60-man battle staff, a briefing room, accommodations for the battle staff, and VIP quarters for the president or his NCA surrogates. Auxiliary power units and generators in the cargo hold provide electricity to the various data processing, display, and communications systems.

Fourteen ground stations can receive communications from the NEACP and link it to other surviving networks. All four NEACPs are normally based at Offut Air Force Base, Kansas, close to STRATEGIC AIR COMMAND (SAC) Headquarters at Omaha, for immediate use by the SAC commander. In a crisis, at least one would be detached to Washington, D.C. or wherever the president is located.

NATIONAL LIBERATION MOVEMENT: A Soviet team for Soviet-inspired, if not Soviet-controlled, terrorist, subversive, or guerrilla activities aimed at overthrowing anti-Soviet governments. See, more generally, REVOLUTIONARY WAR.

NATIONAL SECURITY AGENCY: A U.S. government agency, established by an executive order in 1952, to manage the the collection and analysis of ELECTRONIC INTELLIGENCE and to carry out protective cryptography, cryptanalysis, communications security, and data processing security activities. With headquarters located at Fort Meade, Maryland, the NSA operates electronic listening posts around the world; much of its raw material, however, is collected by military-operated receiving stations, and by the listening posts of allied and cooperating powers. The NSA reportedly has the greatest concentration of data processing capability in the world at Fort Meade; it is generally believed, however, that only wartime tactical needs could validate much of its product, because the Soviet use of costly "one-time" cipher techniques greatly limits the value of the NSA's peacetime output. Unlike other U.S. intelligence agencies, it operates in almost complete secrecy; indeed, until the 1970s its very existence was not officially admitted (wits had it that NSA stood for

"No Such Agency"). See also CENTRAL INTELLI-
GENCE AGENCY; DEFENSE INTELLIGENCE AGENCY.

NATIONAL SECURITY COUNCIL: The
principal advisory body of the U.S. president for
military, foreign policy, economics, and other national security issues. The NSC was created under
the National Security Act of 1947 to advise the
president and to coordinate the activities of the
State Department, Department of Defense (DOD),
and the various intelligence agencies. The council
itself is composed of the president, vice president,
the secretary of state, and the secretary of defense.
The president is usually represented at meetings
by his national security advisor (officially, the Advisor to the President for NSC Affairs), who is the
de facto head of the NSC staff. The deputy and
undersecretaries of other executive departments,
and of the army, navy, and air force, may participate at times. The council staff has grown to several hundred military and civilian personnel,
under the control of a (civilian) executive secretary. The director of Central Intelligence is responsible to the president through the NSC, but
usually reports directly to him.

The NSC, the primary forum for the discussion
of national security issues, usually operates
through a number of specialized groups including
the Senior Interagency Group (SIG) for Foreign
Policy and the Senior Interagency Group for Defense Policy (SIG membership includes deputy
secretaries and the national security advisor); and
the Interdepartmental Group (IG) for Foreign Policy and the Interdepartmental Group for Defense
Policy, which formulate and recommend policy options to their respective SIGs (IG members include
the undersecretaries of State, Defense, and other
departments). Ad hoc groups handle specific problems as required.

**NATIONAL TECHNICAL MEANS (OF IN-
TELLIGENCE):** An ARMS CONTROL term for the
complete array of INTELLIGENCE-gathering methods, other than human intelligence (HUMINT), including: photographic RECONNAISSANCE and electronic intelligence (ELINT) satellites; ground- and
ship-based ELINT; and aerial photography and
airborne ELINT. In most arms control treaties
signed by the United States and the Soviet Union,
"national technical means" are the only recognized method of verifying compliance; as such,
both parties are prohibited from interfering with
their operation.

NATIONAL TESTBED: A large-scale computer simulation facility being developed by the
STRATEGIC DEFENSE INITIATIVE Organization
(SDIO) at Falcon Air Force Base, Colorado, where
elements of BALLISTIC MISSILE DEFENSE systems,
such as sensors, weapons, and BATTLE MANAGE-
MENT architectures, can be stimulated and
analyzed to determine optimal configurations
without the need for prototypes—an approach
made necessary by the high cost of space-based
defensive systems and the need to avoid violating
restrictive interpretations of the 1972 ANTI-BALLIS-
TIC MISSILE TREATY.

NATO: North Atlantic Treaty Organization.
Formed between 1949 and 1951 to implement the
North Atlantic Treaty, NATO has not undergone
significant structural change since then. In addition to the U.S. and Canada, the members are
Belgium, Denmark, Great Britain, the Federal Republic of Germany, Greece, Iceland, Italy, Luxembourg, the Netherlands, Norway, Portugal, and
Turkey; Spain became a full member in 1987.
France is a signatory of the North Atlantic Treaty,
but withdrew from the alliance's military command structure, and thus NATO, in 1966 (France
does maintain liaison officers at the various NATO
headquarters, and is generally expected to participate in the defense of Western Europe against a
Soviet attack). Iceland, Luxembourg, Norway, and
Denmark have placed constraints on their participation in NATO activities, notably prohibiting the
deployment of nuclear weapons on their soil.

The chief governing body of the alliance is the
North Atlantic Council, formed by the permanent
ambassadors of all 16 members. Headquartered in
Brussels, the council is chaired by a secretary general (invariably a European) with a permanent,
multinational staff. A Defense Planning Committee formulates strategic policy. A Military Committee of permanent military representatives from
all members (except Iceland) supervises the various NATO military commands; the chairman is
invariably a European officer. NATO military
forces are organized into three nominally equal
territorial commands: SHAPE (Supreme Headquarters, Allied Powers in Europe) for the continent;
ACLANT (Allied Command, Atlantic) for the North
Atlantic; and (the much smaller) ACCHAN (Allied
Command, Channel), responsible for the English
Channel and the nearby North Sea. SACEUR (Supreme Allied Commander, Europe), the commander of SHAPE, and SACLANT, the commander of ACLANT, (always a U.S. general and
admiral, respectively) participate in the U.S. Joint
Strategic Planning System for nuclear targeting,

and also have planning authority over British POLARIS and U.S. POSEIDON and TRIDENT ballistic-missile submarines. Defense ministers convene to form the NUCLEAR DEFENSE AFFAIRS COMMITTEE (NDAC), which is supposed to establish general policy for the employment of nuclear weapons; detailed contingency planning for the use of "tactical" nuclear weapons is carried out by a lower-level NUCLEAR PLANNING GROUP (NPG). Total peacetime NATO forces now still include approximately 90 DIVISION equivalents with 1,100,000 combat and support troops; 20,000 MAIN BATTLE TANKS; 14,200 ARTILLERY pieces; 3,250 COMBAT AIRCRAFT; and 650 attack HELICOPTERS. In wartime, these forces could be reinforced by French forces and by U.S. formations brought in from CONUS under the REFORGER reinforcement program. The CAFE (ex-CST) negotiations are to reduce NATO as well as WARSAW PACT forces.

NAVAL INFANTRY, SOVIET: *Morskoy Pekhota*, the Soviet counterpart to the U.S. MARINE CORPS (USMC). As compared to the USMC, the Soviet Naval Infantry is much smaller, and its roles are limited to the seizure of territory adjacent to straits and other important waters, the conduct of flanking operations in support of the army, and the execution of coastal raids.

It seems that the Naval Infantry is directed by a major general attached to the staff of the commander-in-chief of the navy; it is significant that his rank is low by Naval Headquarters standards. Currently estimated to number 12,000 officers and men, the Naval Infantry is organized into five REGIMENTS; one is assigned to each of the Northern, Baltic, and Black Sea fleets, while two regiments formed into a division are attached to the Pacific fleet. Each regiment has some 2038 men, organized into three naval infantry BATTALIONS, a tank battalion, a RECONNAISSANCE company, an anti-tank BATTERY, a rocket launcher battery, an AIR DEFENSE battery, an ENGINEER company, a CHEMICAL WARFARE company, and various small maintenance and support units.

The Soviet navy has enough amphibious assault ships, landing ships, landing craft and hovercraft of the IVAN ROGOV, ALLIGATOR, POLNOCNY, AIST, and Lebed classes to deliver the entire force in one lift. Soviet doctrine calls for the landing of assault-engineer teams in advance of the main force to clear obstacles; the first assault echelon would then be put ashore from landing ships aboard their BTR-60/70/80s and PT-76s. Heavier equipment would be brought ashore in landing ships, landing craft,

and hovercraft. The introduction of the Ivan Rogovs now allows the Naval Infantry to use transport HELICOPTERS to go ashore. Naval GUNFIRE and CLOSE AIR SUPPORT are provided by the Soviet's SVERDLOV-class light cruisers (with up to 12 6-inch guns each) and by land-based aircraft, respectively. The introduction of the TBILISI-class aircraft carrier may add carrier-based air support, although it would be risky to expose such a high-value target close to shore for that purpose.

Each Soviet fleet also has a naval SPETSNAZ brigade to carry out CLANDESTINE and COVERT reconnaissance, raids, and SPECIAL OPERATIONS. Each includes a parachute battalion, 2–3 assault swimmer battalions, a MINISUB battalion, and an anti-VIP company specialized for assassination.

NAVSTAR: A.k.a. the Global Positioning System (GPS), a U.S. satellite navigation system consisting of three integrated elements: a space segment of 18 satellites placed into three polar orbits at an altitude of 11,000 mi. (18370 km.); a user segment which processes data from the satellites; and a control segment, which tracks the satellites, corrects their positions on a daily basis, and synchronizes the system using rubidium atomic clocks.

The satellites transmit very accurate position coordinates and timing data to NAVSTAR users equipped with an omnidirectional receiver, a signal processor, and a display unit. The user equipment automatically receives signals from the space segment, selects four satellites most favorably placed, locks on to their signals, and computers the approximate range (using signal delay) to each. It then solves four simultaneous equations with as many unknown variables: the three coordinates of the user's position (U_x, U_y, U_z), and the clock bias factor. A digital computer solves the equation, providing the user's position, velocity, and time to within less than 10 m. (32.8 ft.).

Because their segment is completely passive, an infinite number of users can exploit the system without saturating it; and, of course, users do not generate any electromagnetic emissions which may be detected or jammed. On the other hand, the signals from the satellites could be jammed (or spoofed); or the satellites themselves could be destroyed with ASAT weapons.

NAVSTAR receivers are being placed on ships, aircraft, and armored fighting vehicles; portable man-pack receivers have also been developed. In addition, plans are now being developed to place NAVSTAR receivers on long-range CRUISE MIS-

SILES, providing a completely passive form of midcourse update for their INERTIAL GUIDANCE systems.

The first NAVSTAR satellite was launched in 1978. Original plans called for a 24-satellite network, but the system was cut back to 18 satellites due to escalating costs. The space segment was to have been completed in 1984, but this has been delayed due to a shortage of suitable space boosters. See also SATELLITES, MILITARY; SPACE, MILITARY USES OF.

NBC: Abbreviation for Nuclear, Biological, and Chemical weapons or warfare, including defensive measures against the same. Equivalent terms include ABC (Atomic, Biological, and Chemical) and CBR (Chemical, Biological, and Radiological). See also BIOLOGICAL WARFARE; CHEMICAL WARFARE; NUCLEAR WEAPONS.

NCA: See NATIONAL COMMAND AUTHORITIES.

NCO: See NONCOMMISSIONED OFFICER.

NDAC: See NUCLEAR DEFENSE AFFAIRS COMMITTEE.

NEACP: See NATIONAL EMERGENCY AIRBORNE COMMAND POST.

NEPTUNE: Lockheed P-2 LONG-RANGE MARITIME PATROL (LRMP) and ANTI-SUBMARINE WARFARE (ASW) aircraft, once much used by the U.S. Navy and various U.S. allies, and still serving in small numbers with the Argentine navy; a modified variant (P-2J) built by Kawasaki is also in service with the Japan Maritime Self-Defense Force (JMSDF).

Developed from the early 1940s as a patrol bomber, the prototype P-2 flew in May 1945, and the initial version entered service with the U.S. Navy in 1947. Successively upgraded, it remained the backbone of U.S. Navy patrol squadrons until 1962, and was widely exported. The final U.S. version, the P-2H, entered service in 1954. During the Vietnam War, many were converted into specialized jamming or electronic intelligence (ELINT) platforms. Kawasaki produced 48 P-2Hs before switching to the P-2J, with new engines and much better sensors.

Of conventional construction, the P-2H has a glazed nose for a navigator/observer, and an airliner-type cockpit for the pilot, copilot, and flight engineer. A large surface-search RADAR is mounted ventrally behind the cockpit, and there is a large weapon bay amidships for 2 homing TORPEDOES or 4 DEPTH CHARGES. The rear fuselage houses a tactical compartment for 8 system operators, with various display screens and SONOBUOY chutes, and the tail houses a MAGNETIC ANOMALY DETECTOR (MAD) boom in a tail "stinger" fairing. The midmounted wing has fixed wing-tip tanks, with a powerful searchlight housed in the starboard one.

First flown in 1966, the P-2J is slightly longer and powered by two General Electric T64 tuboprops and two license-built J34s. A total of 82 were delivered through 1969, with a new sensor suite that includes an SPS-80 surface-search radar, a MAD boom, an ELECTRONIC SIGNAL MONITORING array, a sonobuoy data display, a digital signal processor, and an integrated tactical display.

Specifications **Length:** (H) 91.33 ft. (27.84 m.); (J) 95.85 ft. (29.23 m.). **Span:** 103.83 ft. (31.65 m.). **Powerplant:** (H) 2 3500-hp. Wright R-3350 radial piston engines and 2 Westinghouse J34 turbojets, 3400 lb. (1512 kg.) of thrust each; (J) 2 3060-shp. General Electric T64 tuboprops, 2 J34 turbojets. **Weight, empty:** 49,935 lb. (22,650 kg.). **Weight, max. takeoff:** (H) 79,989 lb. (36,240 kg.); (J) 79,989 lb. (34,000 kg.). **Speed, max.:** (H) 403 mph (649 kph) at 10,000 ft. (3050 m.); (J) 400 mph (668 kph). **Speed, cruising:** (H) 207 mph (333 kph); (J) 249 mph (400 kph). **Range, max.:** (H) 3685 mi. (5930 km.); (J) 2765 mi. (4450 km.).

NERVE AGENTS (OR "GASES"): Highly lethal chemicals, specifically anticholinesterases, which interfere with the transmission of nerve impulses by disrupting enzyme reactions in the nervous system. Symptoms of nerve agent intoxication include spasmodic muscular contractions (twitching, jerking), convulsions, asphyxia, and, finally, death by respiratory collapse. All nerve agents are, in fact, liquids which can be spread as aerosols from bombs, artillery shells, spray dispensers, and rocket or missile warheads.

The first nerve agent, TABUN (U.S. code GA), developed in Germany in 1937, was quickly followed by the even more lethal SARIN (GB) and SOMAN (GD). All three are odorless, colorless, NONPERSISTENT, and can be absorbed directly through the skin (percutaneously), as well as by inhalation. Lethal concentrations (LD-50) for Tabun are approximately 70–400 milligrams per minute per cubic meter inhaled; or 1000 milligrams per minute percutaneously. Nerve agents act very rapidly: incapacitation occurs within 1–10 minutes, and death within 10–15 minutes. The Germans had stockpiled 13,000 tons of Tabun by 1945, and had a Sarin production facility on standby. Both the United States and Soviet Union have carried out further development of these agents.

In the 1950s, British scientists developed a new series, the "V-agents." VE and vx are also odorless and colorless liquids, with much greater lethalities than the older G-agents. VX, for instance, has an LD-50 of only 36 milligrams per minute per cubic meter inhaled, or 15 milligrams percutaneously. Incapacitation occurs almost immediately, with death occurring within 15 minutes to one hour. Moreover, VX is highly persistent, and may not dissipate to safe levels for weeks (under certain conditions).

Troops can be protected against nerve agents by prophylactic treatment with a series of drugs which *reversibly* bind the enzyme acetylcholinesterase (AChE). Nerve agents act by binding *permanently* with AChE, thereby disrupting the functions of the nervous system; a reversibly bound pool of AChE within the body thus acts as a reserve. After exposure, prepared troops can be treated with drugs which unbind the protected pool of enzymes, thereby neutralizing the nerve agent. The only other first aid treatment for nerve agent intoxication is injection with atropine or other such stimulants within a few seconds of exposure. See, more generally, CHEMICAL WARFARE.

NET ASSESSMENT: U.S. term for the systematic comparison of the military capabilities of two competing states or coalitions to determine which could better achieve its own objectives in a given context. See also the Soviet concept CORRELATION OF FORCES.

NEUTRALITY: Claimed noninvolvement in an armed conflict, or third-party conflicts in general. Neutrality is usually respected by belligerents if (a) the neutral state refrains from breaches of its status (e.g., by supplying arms, intelligence, or other assistance to a belligerent), and (b) no belligerent regards the benefits of an attack on the neutral power as greater than the costs of such an attack. Heavily armed neutrality on the Swiss or Swedish model has usually succeeded, while unarmed or weakly armed neutrality (on the 1940 Belgian, Dutch, Norwegian, and Danish models) has usually failed.

NEUTRON BOMB: Popular name for ENHANCED RADIATION (ER) nuclear weapons, which produce greater prompt radiation effects and less blast for a given energy yield, as compared to "normal" nuclear weapons. See also NUCLEAR WARHEADS, EFFECTS OF.

NEUTRON (KILL): A method of destroying objects, notably enemy NUCLEAR WARHEADS, by means of high-energy neutrons emitted by a nuclear detonation, as, for example, with the SPARTAN and SPRINT interceptor missiles of the SAFEGUARD Anti-Ballistic Missile System. The high-energy neutrons generated by a (small) 1 to 2-kT explosion could initiate nuclear fission within enemy warheads (preinitiation), to prevent an actual detonation. Neutrons may also interact with the structural materials and electronics of the warhead, leading to their mechanical failure. Attempts to shield warheads from neutron effects have been only partially successful so far, and entail significant weight penalties.

NIGHTHAWK: Unofficial name of the Lockheed F-117A "stealth fighter," in service with the 37th Tactical Fighter Wing (formerly the 4550th Tactical Group) of the U.S. Air Force (USAF).

Developed from 1973, first under the "Have Blue" then under the "XST" (Experimental Stealth Tactical) and finally under the "Senior Trend" programs, the F-117A was the first aircraft designed specifically to exploit "low observables" (STEALTH) technology. Developed in extreme secrecy, its existence was finally admitted by USAF in 1989, because increasing flight operations made continued secrecy impossible. The first test flights occurred in June 1981, and the F-117A became operational with the 4450th Tactical Group in October 1983. It is believed that a total of 59 aircraft were delivered between 1981 and 1990, of which at least three have been lost in accidents.

Intended mainly for clandestine tactical reconnaissance, defense suppression, and "surgical" strikes, the F-117 would rely on its stealth characteristics for the penetration of hostile airspace at night and at low altitudes to attack high-priority point targets with advanced sensors and precision-guided weapons. Though clandestine deployments to Europe have been reported, the first documented use of the F-117A occurred during the opening phases of the U.S. intervention in Panama in December 1989. It was much used in the 1991 Persian Gulf conflict, achieving spectacular successes in night attacks against highly defended point targets.

Of a highly unusual configuration, the F-117A has an arrowhead planform with sharply swept leading edges (intended to reflect radar signals away from their source), a vestigial fuselage, and a small "butterfly" tail. its outer surfaces, fabricated of composites and RADAR ABSORBANT MATERIALS (RAM), are all highly angular. This stealth technique ("faceting"), intended to dissipate radar signals to the sides, thus reducing the aircraft's frontal RADAR CROSS SECTION (RCS), was adopted because 1970s-vintage RAM was too difficult to fabricate in complex curves (as in the later B-2 "Stealth Bomber"). While effective, faceting does exact significant aerodynamic penalties.

The pilot sits right up in the nose under a small, heavily framed canopy affording only limited visibility to the sides (and none to the rear). A small window below the windscreen apparently houses a forward-looking infrared (FLIR) sensor and LASER DESIGNATOR/rangefinder. Other avionics reportedly include an INERTIAL NAVIGATION unit, a TERRAIN-FOLLOWING RADAR, an ELECTRONIC SIGNAL MONITORING array, and comprehensive ACTIVE JAMMING transmitters. Inherently unstable, the F-117A relies on computerized FLY-BY-WIRE controls to maintain level flight. Even so, the aircraft is difficult to handle at low speeds, and lacks the maneuverability of other fighter aircraft.

The engines are mounted in upper-wing nacelles to each side of the cockpit. The engine intakes are fitted with radar-absorbant baffles to prevent the reflection of radar signals off of the engine turbine blades (a major source of RCS), and the exhausts are fitted with an INFRARED suppression system.

The F-117A is armed with LASER or ELECTRO-OPTICAL glide bombs and missiles housed in two small internal weapon bays behind the cockpit; total payload capacity is roughly 4000 lb. (1800 kg.).

Specifications Length: 50 ft. (15.5 m.). **Span:** 40 ft. (12 m.). **Powerplant:** 2 General Electric F404 turbofans, 12,500 lb. (5760 kg.) of thrust each. **Weight, empty:** 20,000 lb. (9100 kg.). **Weight, max. takeoff:** 45,800 lb. (20,400 kg.). **Speed, max.:** Mach 0.85 (595 mph/995 kph) at sea level. **Combat radius:** (hi-lo-hi) 1364 mi. (2200 km.).

NIKE HERCULES: MIM-24A/B, an obsolescent U.S.-built long-range strategic surface-to-air missile (SAM) still fielded by Belgium, Denmark, Greece, West Germany, Italy, Norway, South Korea, and Taiwan. Designed to intercept heavy bombers flying at high altitude, the first Nike Hercules batteries were declared operational in 1958; they were soon deployed around cities and military bases across the continental U.S. (CONUS). The eventual total of 73 battalions (each with 9 launchers), were all integrated into the SAGE automated weapon-control system. By 1963, the U.S. Army had a total of 134 three-launcher Hercules BATTERIES (in Europe and Asia, as well as CONUS), and

more had been sold to U.S. allies; more than 25,-500 missiles were produced between 1957 and 1978. The U.S. phased out its CONUS Hercules batteries by 1974, and transferred control of overseas batteries to the host nations. Most other operators are also retiring the Hercules (in some cases in favor of PATRIOT), with the exceptions of South Korea, Greece, and Turkey.

The Nike Hercules is a two-stage, solid-fuel missile. The first (booster) stage consists of four clustered solid-fuel rockets, each rated at 59,000 lb. (26,818 kg.) of thrust. Cruciform tail fins stabilize the missile during the initial seconds of flight. The second (sustainer) stage has a cylindrical body with a pointed nose section, four long-chord delta wings with control surfaces mounted on the trailing edges, and small cruciform fins mounted ahead of the wings (which are actually receiving aerials for the guidance system); it is powered by a solid-fuel rocket motor. At liftoff, the booster burns for 2.5 seconds before the sustainer takes over, accelerating the missile to maximum speed. Both high-explosive fragmentation and 1-KT W31 nuclear warheads are available. W31s were supplied to Belgium, Greece, Italy, the Netherlands, and West Germany under NATO dual-key arrangements.

The missile's radio COMMAND GUIDANCE requires rather cumbersome ground equipment. Each battery has three single-rail launchers; low- and high-power acquisition radars; a large tracking radar; and data processing equipment. The high-power acquisition radar can be mounted on three trucks to give it some mobility, but is usually kept at fixed sites.

When a target is detected by one of the acquisition radars, it is interrogated by an Identification, Friend or Foe (IFF) system. If it is classified as hostile, the data are transferred to the tracking radar, and the missile is launched by remote control at a fixed elevation of 85°. It flies ballistically until booster separation, when the guidance system is activated. The missile is programmed to roll onto the target azimuth and to pitch over to the correct elevation angle. Steering commands are passed over a radio DATA LINK to bring the missile to the intercept point, at which time a warhead detonation command is transmitted (command guidance is essential for POSITIVE CONTROL under U.S. nuclear weapon doctrine).

Only marginally effective against low-altitude targets, but otherwise quite accurate, Hercules even has some limited ANTI-BALLISTIC MISSILE capability, having successfully intercepted a Corporal BALLISTIC MISSILE and another Hercules in tests. South Korea has converted some of its Nike Hercules into surface-to-surface missiles, without evoking U.S. sanctions.

Specifications Length: 41–41.5 ft. (12.5–12.65 m.). **Diameter:** 2.63 ft. (801 mm.). **Span:** 6.16 ft. (1.88 m.). **Weight, launch:** 10,405–10,710 lb. (4729.5–4868 kg.). **Speed, max.:** Mach 3.6 (2400 mph/4008 kph). **Range, max.:** 87 mi. (145 km.). **Height envelope:** 3000–150,000 ft. (914–45,-732 m.).

NIMITZ: A class of 6 U.S. nuclear-powered AIRCRAFT CARRIERS; 5 were commissioned between 1975 and 1990, 1 is presently under construction, and 2 more are planned. Intended initially to replace the MIDWAY class, the Nimitzes are the largest and most powerful warships in the world, as well as the most expensive.

The Nimitzes are based on the conventionally powered KITTY HAWK; the last three (*Theodore Roosevelt, Abraham Lincoln,* and *George Washington*), of slightly modified design, are known as the "Improved Nimitz" class. The flight deck is 1092 ft. (332.8 m.) long and 252 ft. (76.8 m.) wide, with a 780-ft. (257.7-m.) angled deck cantilevered 8° to port. A compact island superstructure to starboard houses the usual ship and flight control facilities. There are 4 steam catapults (2 on the bow and 2 on the angled deck) and 4 deck-edge elevators (2 forward and 1 aft of the island, and 1 on the port quarter) connecting the flight deck and the hangar. These carriers have magazines for up to 2570 tons of aerial ordnance, and bunkers for some 2.8 million gal. (10.6 million lit.) of aviation fuel, enough for 16 days of fairly intense operations. In the last three ships, the hangar and magazine are protected by additional KEVLAR armor. Defensive armament is limited to 3 8-round SEA SPARROW pepperbox launchers, and 4 PHALANX 20-mm. radar-controlled guns for anti-missile defense.

The air group now consists of 2 fighter/interceptor squadrons with a total of 20 F-14 TOMCATS, 2 fighter/attack squadrons with 20 FA-18 HORNETS, 2 medium attack squadrons with 20 A-6E INTRUDERS, an anti-submarine patrol squadron with 10 S-3 VIKINGS, a composite support squadron with 5 EA-8B PROWLERS and five E-2C HAWKEYES, and a helicopter squadron with six SH-3 SEA KINGS. See also AIRCRAFT CARRIERS, UNITED STATES.

Specifications Length: 1092 ft. (332.8 m.). **Beam:** 134 ft. (40.8 m.). **Draft:** 37 ft. (11.3 m.).

Displacement: (1st 3) 81,600 tons standard/91,487 tons full load; (last 3) 96,836 tons full load. **Powerplant:** 4-shaft nuclear: 2 A4W pressurized-water reactors, 4 sets of geared steam turbines, 280,000 shp. **Speed:** 35 kt. **Crew:** 3660 + 2626 air group = 6286. **Sensors:** 1 SPS-10 or SPS-64 surface-search radar, 1 SPS-48 3-dimensional air-search radar, 1 SPS-49 2-dimensional air-search radar, 1 SPS-65 surveillance radar, 3 Mk.92 Sea Sparrow guidance radars. **Electronic warfare equipment:** 1 WLR-8 electronic signal monitoring array, 1 WLR-1 radar warning receiver, 1 SLQ-17 active jamming array, 4 Mk.36 SRBOC chaff launchers.

NIMROD: A LONG-RANGE MARITIME PATROL (LRMP) and ANTI-SUBMARINE WARFARE (ASW) aircraft built by British Aerospace (formerly Hawker-Siddeley) for the Royal Air Force. The world's first jet-propelled patrol aircraft (for lack of a more suitable British-made turboprop), the Nimrod was developed to replace the ancient, piston-engined Avro Shackleton. Based on the old Hawker-Siddeley Comet airliner, Nimrod (then designated HS.801) was developed from 1964 and was approved the following year. The first prototype flew in 1967 and the aircraft entered service in 1969 as the Nimrod MR.1; it now equips five squadrons of No. 18 Group of RAF, flying patrols over the eastern Atlantic and North Sea. A total of 46 MR.1s were delivered between 1968 and 1977.

Though the Comet served as its design basis, the Nimrod is very different, notably because of the addition of a second (unpressurized) compartment running the full length of the fuselage beneath the original pressurized cabin; this contains equipment and a very large weapons bay. Other changes include new engines, a large, nose-mounted Marconi "Searchwater" RADAR, an Emerson ASQ-10 MAGNETIC ANOMALY DETECTOR (MAD) boom, and an ELECTRONIC SIGNAL MONITORING (ESM) pod on the tail, and a 70-million-candlepower searchlight mounted in a wing pod. Between 1977 and 1979, 35 MR.1s were upgraded to MR.2 standard, with new AVIONICS and new digital (versus analog) data processing equipment. It was only during the 1982 Falklands War that all MR.2s were fitted with AERIAL REFUELING probes to enable them to fly patrols off the Falklands from bases on Ascension Island.

The main cabin, behind the cockpit, houses an operations center with tactical displays and data processing, fire control, and communications equipment. The center is manned by 10 men, giving the MR.2 a total crew of 13 including the pilot, copilot, and flight engineer. The aft portion of the cabin includes a galley, lavatory, and bunks, all essential for the normal 12 to 18-hour patrols.

The unpressurized lower fuselage contains electrical generators and air conditioning units for onboard electronics, as well as a cavernous weapons bay running from just behind the cockpit to just behind the wings, though the aircraft can carry just 13,500 lb. (6136 kg.) of ordnance (as compared to 20,000 lb./4091 kg. in the U.S. P-3C ORION), including SQQ-81 Barra SONOBUOYS, MK.46 or STINGRAY lightweight ASW homing TORPEDOES, and conventional or nuclear DEPTH CHARGES. During the Falklands War, the aircraft were hastily modified to carry HARPOON anti-ship missiles as well.

The wings have swept leading edges, straight trailing edges, and engine nacelles faired into the wing roots. The outer wings form integral fuel tanks. Since the Falklands War, two underwing hardpoints have been added to carry four AIM-9L SIDEWINDER air-to-air missiles for self-defense. As noted, the Nimrod's vertical stabilizer is capped with a bubble fairing housing an ESM array. Some aircraft have also been fitted with Loral Rapport ESM systems in wingtip pods.

The Nimrod's high maximum speed allows it to reach its patrol area (or to dash to investigate a contact) much more quickly than its turboprop counterparts. In addition, the low noise and vibration levels of jet engines significantly reduce crew fatigue on long flights. It is standard practice to shut down one, two, or even three engines to conserve fuel.

In 1971, the RAF took delivery of three Nimrods, designated R.1, configured for electronic intelligence (ELINT) gathering. These aircraft carry many aerials, and have spiral receivers in thimble fairings in place of the wing pods and MAD boom. In 1982, they were also fitted with Rapport ESM systems in wing-tip pods. The R.1s were fitted with aerial refueling probes during the Falklands War, and carried out many long-range ELINT missions; their performance of is generally similar to that of the MR.1/2.

In 1973, studies were initiated for the conversion of the Nimrod to an airborne warning and control (AWACS) aircraft to replace the decrepit Shackleton AEW.2s. In March 1977, the British government decided to proceed with the conversion of 11 MR.1s despite some pressure from the United States to purchase the Boeing E-3A SENTRY. The British decision was actually based on the desire to give work to its hard-pressed electronics

industry, though its declared rationale was that the E-3 was optimized for overland operations, while the British desired a "maritime" AWACS capability (a slight difference, if that).

The AEW.3 had two large, bulbous radomes at the nose and tail housing large Marconi multimode radars, each scanning a 180° sector. The interior was completely revised to house a fighter direction center with data processing equipment, tactical display screens, DATA LINKS, and several voice communication channels.

The first AEW.3 prototype began flight tests in 1977, and the conversion of some MR.1s started; but the Marconi radar system suffered from severe technical problems, running far behind schedule and over budget. As usual, RAF specifications were extreme, and the project management overconfident, if not inept: the outcome was the greatest British technical debacle of recent decades. Finally, in 1986, Marconi was given one last chance to complete the project; this failed, and the British government canceled the program and purchased the E-3A instead. It is not known if the 11 partially converted Nimrods will be restored to MR.2 standard.

Specifications Length: 127.6 ft. (38.9 m.). Span: 114.83 ft. (35 m.). Powerplant: 4 Rolls Royce Spey turbofans, 12,140 lb. (5518 kg.) of thrust each. Weapons bay: 48.5 ft. (14.78 m.). Weight, empty: 86,000 lb. (39,090 kg.). Weight, max. takeoff: 192,000 lb. (58,536 kg.). Speed, max.: 575 mph (960 kph). Speed, cruising: 230 mph (384 kph). Range, max.: 5755 mi. (9160 km.).

NITEROL: A class of six FRIGATES built by the British Vosper-Thorneycroft shipyard for the Brazilian navy between 1972 and 1980. Vosper has designed a broad family of CORVETTES and frigates, mainly for the export market, all scaled versions of a basic design with many common weapons, electronics, and mechanical systems. The first was the 500-ton Mk.1 corvette of 1964–65; this was followed by a 660-ton Mk.2; a 1540-ton Mk.5 frigate; and a 1780-ton Mk.7 frigate. Eight 3250-ton AMAZON-class (Type 21) frigates built for the Royal Navy are still recognizably members of the family. A Mk.9 corvette, displacing only 780 tons, was built for Nigeria in 1980–81.

The Niterois (Vosper Mk.10), similar to the Amazons though somewhat larger, were ordered in 1970 in two distinct subclasses: four ships optimized for ANTI-SUBMARINE WARFARE (ASW) and two general-purpose escorts. Both have the same hull, machinery, and basic electronics. Like all

Vosper frigates, they are sleek, elegant vessels, flush-decked (except for a small sunken quarter-deck), with sharply raked bows and a streamlined superstructure which gives them a yachtlike appearance.

The ASW ships are armed with a 4.5-in. (114-mm.) DUAL PURPOSE gun on the foredeck; a twin-barrel Bofors 375-mm. ASW rocket launcher on the superstructure; 2 40-mm. BOFORS GUNS, also on the superstructure; 2 sets of triple tubes for light-weight ASW TORPEDOES amidships; 2 triple launchers for SEA CAT short-range surface-to-air missiles on the aft superstructure; a "Branik" launcher for IKARA ASW missiles on the quarter-deck; and a 5-round DEPTH CHARGE rack on the fantail. A landing pad and hangar just forward of the quarterdeck can accommodate a Westland LYNX ASW helicopter.

The general-purpose ships differ in having a second 4.5-inch gun on the quarterdeck in place of Ikara, and 4 launch canisters for EXOCET MM.38 anti-ship missiles on the superstructure amidships. See also FRIGATES, BRITAIN.

Specifications **Length:** 425.1 ft. (129.6 m.). **Beam:** 44.2 ft. (13.47 m.). **Draft:** 18.2 ft. (5.55 m.). **Displacement:** 3200 tons standard/3800 tons full load. **Powerplant:** twin-shaft CODOG: 4 MTU 16V956 TB91 diesels, 15,670 hp. (cruise)/2 Rolls Royce Olympus TM3B gas turbines, 56,000 shp. (sprint). **Speed:** 32 kt. **Range:** 1300 n.mi. at 29 kt./5300 n.mi. at 17 kt. **Crew:** 209. **Sensors:** 1 Plessey AWS-2 air surveillance and warning radar, 1 Dutch Signal ZW-06 surface-search radar, 2 Italian Selena RTN-10X fire control radars, 1 U.S. EDO 610E hull-mounted medium-frequency sonar, 1 EDO 700 variable depth sonar, (ASW) 1 Ikara missile guidance radar. **Electronic warfare equipment:** 2 Decca RDL-2/3 radar warning receivers.

NIXIE: The U.S. SLQ-25 TORPEDO COUNTERMEASURE system—a noisemaker sled towed behind surface ships to decoy acoustical homing TORPEDOES. It equips most U.S. warships and those of several U.S. allies.

NOE: See NAP-OF-THE-EARTH.

NOISE JAMMING: A crude but reliable ACTIVE JAMMING technique whereby powerful radio signals ("noise") are transmitted on the frequencies used by enemy RADARS or radios. This earliest and simplest form of active jamming (it was first used in the Russo-Japanese War, 1904–6), is a brute force method: to suppress the enemy signal in a mass of noise, proportionate power levels are needed—levels difficult to provide in small air-

craft, for example. Noise jamming also precludes the use of the affected frequency to friend as well as foe.

When enemy operating frequencies are known with some certainty, a more discriminating noise jamming technique, SPOT JAMMING, can be used. A spot jammer normally transmits on a single frequency, and can thus be countered by frequency-agile system, though flexible spot jammers can be tuned to match a range of frequencies. When the enemy operates on a number of different frequencies, another noise jamming technique, BARRAGE JAMMING, can be used. A barrage jammer transmits on a broad frequency band, of course denying the use of that band to all. Swept-spot jamming is a third noise jamming technique which combines the capabilities of spot and barrage jamming: by scanning through a number of frequencies and jamming each in turn, enemy systems may be disrupted, though they can function adequately between scans, if suitably designed.

All forms of noise jamming require much more power than DECEPTION JAMMING, and can also be defeated by sophisticated digital signal processing to suppress unwanted noise. See, more generally, ELECTRONIC COUNTER-COUNTERMEASURES; ELECTRONIC COUNTERMEASURES.

NONCOMMISSIONED OFFICER (NCO): Military personnel promoted from the ranks to command small subunits and/or execute administrative and technical tasks. The role of NCOs varies; in most Western armed forces, NCOs are long-serving professionals who act as a buffer between academy-trained officers and other ranks, and who represent a source of continuity for short-term conscript forces. In Soviet and Soviet-style forces, NCOs are almost entirely conscripts, selected upon intake and given a slightly more rigorous training program; as a result, Soviet-style forces tend to be officer-led. The Israeli Defense Forces are unique in that their officers are not academy-trained, being selected instead from the best NCOs, who in turn are selected from the best recruits. This drains talent from the NCO pool; but the Israelis rely on their young, short-course officers for small-unit leadership. The Western NCO concept actually reflects a particular social structure—of sturdy peasants (soldiers), reliable yeoman/bailiffs (NCOs), and aristocrats (officers)—which hardly persists anywhere. Much effort has been needlessly expended in Third World military assistance programs to evoke into existence an NCO corps where the requisite social structure is lacking. It is actually much easier to train educated young men to serve as combined junior officer/NCOs.

NONPERSISTENT AGENT: A CHEMICAL WARFARE agent that dissipates in the atmosphere within minutes or hours at most, thus allowing the contaminated terrain to be traversed safely. Effective both to inflict casualties and spread panic, nonpersistent agents can be used as precursors: by the time the attacking forces arrive on the scene, the agent will have dissipated. See also PERSISTENT AGENT.

NON-PROLIFERATION TREATY: A U.S./ Soviet-inspired multilateral agreement meant to prevent the spread of nuclear weapons by prohibiting the manufacture or transfer of weapons-grade fissile materials, production facilities for the same, and critical ancillary equipment, as well as actual nuclear weapons. The text was first agreed to by the United States and the Soviet Union and then tabled at the Eighteen Nation Disarmament Committee (ENDC) at the beginning of 1968; it was endorsed by the ENDC with very minor changes, and presented as a motion at the General Assembly of the United Nations, which passed by a vote of 95 for, 4 against, and 21 abstentions. The treaty was then signed in Washington, London, and Moscow on 1 July 1968. Since then, another 80 states have signed the treaty, but a number of nuclear-capable powers have not, including India and the People's Republic of China, both of which have tested nuclear weapons. Some other states believed to have nuclear capabilities have also refused to sign, including Israel and South Africa.

NORAD: North American Air Defense Command, a joint U.S.-Canadian command which supervises the air defense of the continental United States, Canada, and Alaska. Its main operating headquarters are inside Cheyenne Mountain, near Colorado Springs, Colorado. NORAD controls:

1. RADAR and COMMUNICATION networks for the detection, classification, and tracking of ballistic missiles, bombers, and cruise missiles; these include the BALLISTIC MISSILE EARLY WARNING SYSTEM (BMEWS); the Distant Early Warning System (DEW LINE); PAVE PAWS phased array radars (for the detection of submarine-launched ballistic missiles); and OVER-THE-HORIZON BACKSCATTER (OTH-B) radars. The Dew Line is being replaced by the similar, but more capable, NORTH WARNING SYSTEM.

2. Sixteen FIGHTER/INTERCEPTOR squadrons of the First U.S. Air Force (ex-ADTAC, ex-ADC), in-

cluding 4 squadrons of F-15 EAGLES, 7 squadrons of F-4C/D PHANTOMS, and 4 squadrons of F-16 FALCONS.

3. Two fighter/interceptor squadrons of CF-18 HORNETS of the Canadian Fighter Group.

The previous SAGE and BUIC automated ground controlled intercept systems have now been replaced by the JOINT SURVEILLANCE SYSTEM (JSS).

Throughout the 1960s and 1970s, the United States continued to reduce NORAD forces in response to what was perceived as a diminishing Soviet bomber threat, and in accordance with the doctrine of ASSURED DESTRUCTION (which holds that only absolute vulnerability can provide security). Specifically, U.S. interceptors, which once numbered in the thousands, were reduced to barely 400 aging Delta Darts; and all air defense artillery and NIKE HERCULES surface-to-air missile batteries were dismantled. By the late 1970s, however, the increasing strength of Soviet Long-Range Aviation (ADD), with the introduction of the supersonic Tu-26 BACKFIRE bomber, the development of the longer-range Tu-160 BLACKJACK bomber, and the advent of the highly accurate AS-15 KENT air-launched cruise missile, caused a policy reversal manifest in the provision of F-15s and—more dubiously—in the replacement of aging Phantoms by the F-16 (an excellent dogfighter but hardly a suitable interceptor). Still, the air defense of North America remains very thin in comparison to the Soviet air defense force (PVO). See also NADGE and UKADGE.

NORTHAG: Northern Army Group, a NATO ground command. Subordinate to AFCENT, the Central Front command, NORTAG is responsible for all Allied ground forces deployed in West Germany north of a line centered on Kassel, the area loosely described as the North German Plain. Forces currently assigned to NORTAG include I Netherlands CORPS, I German Corps, I British Corps, and I Belgian Corps, plus the 3rd Brigade and 2nd U.S. Armored Division, for a total of 9 2/3 DIVISIONS on mobilization. These troops can be reinforced in an emergency by U.S. REFORGER units from CONUS, and (probably) by the French III Corps. Other NATO Army Groups include CENTAG and SOUTHAG.

NORTH WARNING SYSTEM: A chain of U.S.-operated early warning RADAR stations under the joint U.S.-Canadian air defense command NORAD, which will eventually replace the existing Distant Early Warning (DEW LINE) chain, to provide improved detection and tracking of low-flying

Soviet BOMBERS and CRUISE MISSILES. The North Warning System of 6–8 minimally attended Long Range Radars (LRRs) located at existing DEW Line sites, and some two dozen unmanned Short Range Radars (SRRs), will cover the airspace of the North Slope of Alaska, northern Canada, and the coast of Labrador; the radars are situated to provide overlapping coverage (even if one site fails), and will transmit their information via microwave DATA LINKS and communication satellites to NORAD headquarters at Cheyenne Mountain, Colorado. Implementation is seriously behind schedule, in part due to disagreements between the United States and Canada. See also AIR DEFENSE.

NOVEMBER: NATO code name for a class of 15 Soviet nuclear-powered attack SUBMARINES (SSNs) completed between 1958 and 1963. The Novembers, the first nuclear submarines built by the Soviet Union, served as the basis for the ECHO-class guided-missile submarines (SSGNs) and Hotel-class ballistic-missile submarines (SSBNs). They followed the first U.S. nuclear submarine, the *Nautilus*, by four years, and were rather crude in comparison to the contemporary U.S. SKIPJACK class.

The Novembers have a long, cigar-shaped, double hull; their high length-to-beam ratio makes them quite unwieldy and also unstable at high submerged speeds. As in most Soviet submarines, the outer hull has numerous free-flooding holes which cause considerable turbulence and flow noise, facilitating passive sonar detection. A small, streamlined sail (conning tower) is located forward. The control surfaces are conventional, with small, retractable bow planes, a single rudder, and stern planes mounted ahead of the propellers; a large, fixed vertical stabilizer is mounted over the stern.

Designed primarily for ANTI-SURFACE WARFARE, the Novembers are well armed for the purpose, with 8 21-inch (533-mm.) torpedo tubes in the bow and 2 16-inch (400-mm.) stern tubes. The bow tubes fire a variety of free-running and acoustical homing TORPEDOES, with a basic load of 18 conventional (high-explosive) and 6 15-kT nuclear torpedoes. The stern tubes can launch two lightweight ANTI-SUBMARINE WARFARE torpedoes and also a variety of TORPEDO COUNTERMEASURES.

The November's reactor is notoriously unreliable, and its shielding is inadequate. One November was lost in the Atlantic in 1970 following a fire, and numerous breakdowns at sea have been observed. Two Novembers were placed in reserve in the early 1980s; the remaining 12 will probably be retired in the early 1990s. See, more generally, SUBMARINES, SOVIET UNION.

Specifications Length: 359.9 ft. (109.7 m.). Beam: 29.83 ft. (9.09 m.). Displacement: 4200 tons surfaced/5000 tons submerged. Powerplant: 1 pressurized-water reactor, 2 sets of geared steam turbines, 30,000 shp. Speed: 30 kt. (22 kt. practical). Max. operating depth: 985 ft. (9300 m.). Collapse depth: 1640 ft. (500 m.). Crew: 80. Sensors: 1 *Herkules* bow-mounted medium-frequency passive sonar, 1 *Feniks* medium-frequency active sonar, 1 "Snoop Tray" surface-search radar, 1 "Stop Light" electronic signal monitoring array, 2 periscopes.

NPG: See NUCLEAR PLANNING GROUP.

NPT: See NON-PROLIFERATION TREATY.

NSA: See NATIONAL SECURITY AGENCY.

NSC: See NATIONAL SECURITY COUNCIL.

NTB: See NATIONAL TESTBED.

NTDS: Naval Tactical Data System, a key U.S. Navy combat information network of ship-to-ship DATA LINKS and compatible computer-assisted displays, intended to integrate the sensors and weapons of single ships with all other NTDS-equipped ships in a BATTLE GROUP. The concept originated in the early 1950s as a sort of "electronic bookkeeper," to enable officers manning on-board COMBAT INFORMATION CENTERS (CIC) to keep track of rapidly evolving air battles. The first prototype, the CDS (Comprehensive Display System), was a joint Anglo-American program of 1950–51. In 1953, the U.S. Navy introduced the Electronic Data System (EDS), which combined automated CIC functions with the ability to share data with other ships over a (low-rate) analog data link. In 1956, the navy added a high-capacity digital data link to EDS, forming the first working version of NTDS. Further upgrades followed.

In effect, NTDS can compensate for the range and accuracy limits of the sensors aboard any one ship; with it the battle group command can access and correlate sensor data from all ships in the force. Conversely, NTDS allows the best use of available weapons in lieu of exclusive reliance on whatever weapons are aboard any one ship; i.e., the battle group can optimize the allocation of weapons to targets force-wide (weapons can actually be controlled and fired automatically from the CIC of the battle group flagship). In addition, the complementary ATDS (Air Tactical Data System) can integrate information from AIRBORNE EARLY

WARNING AIRCRAFT, notably the E-2C HAWKEYE, into the NTDS, thereby extending the battle group's radar horizon by several hundred miles.

But NTDS must be used sparingly in combat, because its data links generate many signals easily detected by enemy ELECTRONIC SIGNAL MONITORING, thereby revealing the battle group's position. See also BATTLE MANAGEMENT.

NUCLEAR ARTILLERY: Any tube ARTILLERY capable of firing a nuclear shell. The first nuclear artillery pieces, developed during the 1950s, were very large and heavy, to accommodate the large warheads then available (e.g., the U.S. 280-mm. "Atomic Cannon," which was mounted on a railway carriage). Early nuclear artillery pieces were not, therefore, of much practical value: they fired relatively high-yield weapons over excessively short ranges, and required much time to move, set up, and dismantle, being vulnerable throughout to attack by enemy conventional artillery and air power. Contemporary rockets or ballistic missiles were much superior, but were not sufficiently "army" for the U.S. and Soviet military bureaucracies. By the early 1960s, miniaturization and the development of very low-yield warheads (down to 0.1 kT) made it possible to fire nuclear shells from standard field artillery pieces (in fact, the U.S. even developed a nuclear warhead for the jeep-mounted Davy Crockett recoilless rifle). As a result, nuclear artillery is the most abundant form of tactical nuclear delivery. The U.S. M109A2/3 155-mm. self propelled (SP) gun-howitzer, M110A2 203-mm. SP gun-howitzer, and M198 towed gun howitzer are all nuclear-capable, as are the Soviet 2S3 152-mm. SP HOWITZER, 2S5 152-mm. SP gun, M1976 152-mm. towed gun, 2S7 203-mm. SP gun, and 2S4 240-mm. SP MORTAR. See also ARTILLERY, SOVIET UNION; ARTILLERY, UNITED STATES; and, more generally, NUCLEAR WEAPONS.

NUCLEAR BLACKOUT. See BLACKOUT, NUCLEAR.

NUCLEAR DEFENSE AFFAIRS COMMITTEE: A high-level NATO policy body established in 1966. Membership in NDAC is open to all member states, but France, Iceland, and Luxembourg do not participate. NDAC is actually a meeting of defense ministers, with the NATO secretary general acting as the ex officio chairman. Of necessity, the ministers limit themselves to broad discussions, with detailed nuclear contingency planning left to the NUCLEAR PLANNING GROUP (NPG).

NUCLEAR FALLOUT: Pulverized earth and other particulate debris ejected into the atmosphere by the explosion of a NUCLEAR WEAPON and irradiated by that explosion, or otherwise contaminated by the condensation of radioactive fission products on the particles. Heavier debris forms the stem of the familiar mushroom cloud, while lighter debris is sucked up into its head and may reach very high altitudes, until it cools, disperses, and forms a cloud of fine particles of radioactive dust (and bomb material) which drifts downwind from the site of the explosion, falling back to earth to contaminate the objects on which it settles. Heavier particles fall more rapidly to land in the immediate vicinity of the explosion, while the lighter particles, falling more slowly and from higher altitudes, can drift hundreds of miles before landing. Most fallout reaches the ground within 24 hours; the rest disseminates over the earth's surface during a period of weeks or months. Precipitation during or immediately after a nuclear explosion can wash the fallout from the atmosphere much more quickly ("rainout"), limiting its spread. Most fallout material would result from ground or near-ground bursts; airbursts (optimal for attacks on cities and other soft targets) would generate little or no fallout.

The radiation in fallout emanates from highly unstable isotopes created during nuclear fission, or by the bombardment of (nonradioactive) debris by "slow" neutrons emitted by the explosion. Because the isotopes are unstable they decay quite rapidly, losing most of their radioactivity within a few days. The "Seven-Ten Rule" is a rough estimate of the rate of radioactive decay in fallout: the level of radiation will decrease to 10 percent of the current level when the fallout is 7 times older than its age at measurement (or to 1 percent of the current level when it becomes 49 times older). Much of the radiation in fallout is generated by alpha or beta particles, against which a few inches of dirt or concrete provide adequate protection. The greatest danger of radiation poisoning from fallout is due to the ingestion or inhalation of radioactive particles; thus air filtration and a supply of uncontaminated food and water are critical to survival. Fallout shelters can significantly decrease the risk, but survivors would probably have to remain in them for several days; after that, it may be possible to go out in relative safety (if thoroughly clothed) for increasingly longer periods as the fallout decays. See, more generally, CIVIL DEFENSE; NUCLEAR WARHEADS, EFFECTS OF.

NUCLEAR FREE ZONE: An area within which no NUCLEAR WEAPONS may be stored or deployed; a specialized form of ARMS CONTROL, which can be very useful to the party which is superior in nonnuclear forces. The Soviet Union has long proposed the creation of a European nuclear free zone. In recent years the concept has gained support in both the South Pacific and within the United States (where some local jurisdictions have declared themselves to be "nuclear free zones" as a symbolic gesture).

NUCLEAR FREEZE: An ARMS CONTROL scheme of the 1980s, whereby all participating parties were to agree not to add to their nuclear arsenals. The rationale of the "freeze" proposal was the belief that competition to achieve a "meaningless" nuclear superiority was the self-sustaining cause of the ARMS RACE, which, in turn, was believed to increase the probability of nuclear war. By freezing nuclear inventories, advocates hoped to eliminate all need to compete, thereby stopping the process at its source. Actually, without modernization, nuclear forces would eventually shrink as weapons reach the end of their usable shelf lives (15–20 years in most cases). Freeze advocates did not, however, explicitly advocate the merits of a reversion to a world of nonnuclear military balances.

The more obvious shortcoming of the nuclear freeze concept was that it would freeze any existing asymmetries, thus making accidents of chronology consequential. See also DISARMAMENT.

NUCLEAR PLANNING GROUP: A NATO policy body at the junior-minister level responsible for the formulation of detailed rules for the use of tactical NUCLEAR WEAPONS. It is current NATO policy that a full-scale attack that cannot otherwise be defeated would have to evoke the early use of tactical nuclear weapons, even without prior Soviet nuclear strikes against NATO territory (which would also evoke a tactical nuclear response). The NPG, established in 1966, determines the when and how of such nuclear use. All NATO members except France, Iceland, Luxembourg and Spain participate in nuclear planning, but only seven members are represented on the NPG at any one time, with the small-country members taking their place by rotation. The NPG is supposed to be supervised by the (ministerial) NUCLEAR DEFENSE AFFAIRS COMMITTEE (NDAC), but NDAC meetings are too infrequent, short, and otherwise burdened for much supervision.

NUCLEAR PROLIFERATION: A.k.a the nth nation problem, or, more cynically, the $n + 1$ nation problem. Six states are known to have devel-oped NUCLEAR WEAPONS: the United States, the Soviet Union, Great Britain, France, China, and India. All have both plutonium-239 (Pu_{239}) and uranium-235 (U_{235}) technology for weapon production; all except India have both fission and fusion weapons. Because the acquisition of nuclear weapons is in any case difficult, and because it is still generally believed to confer significant political advantage, the nuclear powers agree on the virtues of nondissemination. At the same time, they compete to export "peaceful" nuclear reactors, most of which can be used to produce Pu_{239}. Nondissemination policies include:

1. The refusal to sell, lend, or give nuclear weapons, their components, weapon-grade fissionable material, or assistance in the design and manufacture of nuclear weapons (but China has supplied design data).

2. Controls on the publication of technical data.

3. Contractual arrangements for the return of any plutonium produced in reactors supplied or fueled by members of the "club."

Israel, Pakistan, and South Africa have reportedly produced nuclear weapons, or are on the verge of doing so; another dozen or so states, including Argentina and Brazil, produce plutonium in sufficient quantity for weapon use. Many more have "research" reactors useful for training potential weapon engineers, including Taiwan, Yugoslavia, Romania, Poland, Indonesia, Iran, Iraq, South Korea, Turkey, and Venezuela. The proliferation forecasts of the 1960s have proved to be grossly exaggerated (e.g., 40 nuclear powers by 1980). If nuclear weapons have not spread, the reason is likely to have been the diminished perceived utility of nuclear weapons, rather than the effect of nonproliferation policies. In reality, all members of the nuclear "club" have competed to sell equipment and information to all (paying) comers: competitive pressures have subverted officially restrictive policies. What are in fact intermediate stages of weapon production (including plutonium-producing reactors) are treated as though their technology could be insulated from future weapons programs. Inspection arrangements intended to ensure the return of weapons-grade plutonium to the supplier of (uranium) reactor fuel are of course chimerical: a state determined to produce nuclear weapons will do so, even if contracts must be violated (Canada did not declare war on India after the latter used the plutonium of a Canadian-supplied reactor to make a bomb).

Nuclear reactors are rated in megawatts thermal (MW_{th}), in the case of research reactors; and megawatts electrical (MW_e), in the case of power-generating commercial reactors. For each MW_{th} of capacity, a reactor fueled with enriched uranium will produce approximately 0.25 kg. of plutonium *per year;* each MW_e of capacity similarly yields about 1 kg. of plutonium per year. Ratings of 500–1000 MWe are common in reactors exported to nonnuclear powers, while less than 7 kg. of plutonium suffices for each bomb. Even quite small countries have reactors rated at 25 MW_{th}, which could produce one bomb per year. Thus nondissemination policies have failed entirely for plutonium. Only the financial and technological burden of building the facilities needed to manufacture and machine plutonium into precision components persists as an obstacle to proliferation. The production of U_{235} (essential for of fusion weapons), by contrast, is much more difficult, though gas centrifuge techniques may change that.

Of course, the n + 1 nation problem cannot be isolated from broader security and political considerations. Canada, for example, has all the resources needed to produce fission bombs almost immediately, and to acquire fusion bombs quite quickly. But Canada is an unlikely nuclear power, because it faces no acute threats, nuclear or otherwise, and it is protected by formal and implicit security arrangements against most conceivable threats. Compare Israel, whose leaders are indifferent to prestige considerations (such as those which motivated the French nuclear program), but who face the ultimate threat of "politicide" without the guarantee of any military alliance. Thus Israel, with only a fraction of Canada's nuclear resources, has developed nuclear weapons. Nuclear proliferation, and its control, are thus not separable from the problem of international order in an age which sanctifies small-country sovereignty. See also NON-PROLIFERATION TREATY.

NUCLEAR PROPULSION: A naval power-plant based on one or more nuclear reactors. Nuclear propulsion was first developed concurrently by the U.S. and Soviet Union in the early 1950s, primarily to power naval vessels, and especially SUBMARINES. The U.S. launched the first nuclear-propelled warship, the submarine *Nautilus,* in 1954.

Nuclear propulsion revolutionized undersea warfare by making it possible to build the first true submarines, capable of sustained underwater operations; all their nonnuclear predecessors were actually submersibles, essentially surface ships capable of brief submergence. Nuclear propulsion has also been used by the United States and the Soviet Union to power some larger surface warships (AIRCRAFT CARRIERS and CRUISERS). In their case, high sustained speeds and unlimited endurance are relative, not absolute, advantages because crew fatigue and the range of nonnuclear escorts impose endurance limits anyway.

In the most common form of nuclear propulsion, the pressurized-water reactor (PWR), water pumped through the reactor core acts both as a coolant and as the heat-transfer medium to generate steam. Because the water flowing through the reactor (the "primary coolant loop") is under pressure, it can be heated to several hundred degrees Centigrade without boiling. It is then passed through a heat exchanger, around which unpressurized water is pumped. Heat is transferred from the primary loop to the unpressurized water (in the secondary loop), which is thus transformed into high-pressure steam; the water in the primary loop is thereby cooled and returns to the reactor core to repeat the process.

The conversion of the steam in the secondary loop into mechanical energy is most often accomplished with reduction-geared turbines. Steam passes over the blades of the turbine, which then rotates at very high speed. Several stages of reduction gearing are needed to slow the rotation rate before the propeller shaft is reached. The main alternative is turbo-electric drive, in which steam drives an electric dynamo, which in turn powers an electric motor linked to the propeller shaft. In theory, turbo-electric drive offers important advantages, including reduced noise levels and greater layout flexibility. In practice, however, dynamos and electric motors are much heavier and less reliable than reduction-geared turbines; hence the technique is not much used.

The more advanced alternative to the PWR is the liquid-metal reactor, essentially identical but employing a molten metal, generally sodium, lead, or bismuth, in the primary loop. Being denser, liquid metals can carry much more energy per unit of volume than water, hence liquid metal reactors can be much lighter and more compact. But liquid metals, and especially sodium, are highly volatile and corrosive; in addition, the metal must be kept molten at all times, otherwise it will solidify, clogging tubes and pumps irremediably. The U.S. Navy experimented with liquid sodium reactors in the 1950s, but abandoned them as dangerous and unreliable. The Soviet navy, however, relies on liquid metal reactors for its fastest attack submarines, notably the (45-knot) ALFA class; the (40-knot) MIKE, lost at sea in 1989, also had a liquid-metal reactor.

The Soviet navy has also introduced a new type of hybrid nuclear propulsion on its KIROV-class battle cruisers. In this CONAS (Combination Of Nuclear And Steam) propulsion, a typical PWR provides power for cruising at low and moderate speed; when maximum speed is needed, steam in the secondary loop is passed through an oil-fired superheater, increasing the temperature (and hence the energy) of the steam before it reaches the turbines. This enables the Kirovs to have a smaller reactor than would otherwise be needed, saving much interior volume and weight. CONAS cannot, of course, be used for submarines because of the superheater's air intake.

A major drawback of nuclear propulsion for submarines is the noise level of coolant pumps. To reduce acoustical signatures, machinery can be mounted on resilient sound-isolation rafts, or several different cooling pumps can be used in stages, so that at low speed one small pump is sufficient; at moderate speed a second pump is added, and so on until at high speeds all pumps are on line. The U.S. LOS ANGELES- and OHIO-class submarines have convection ("natural circulation") reactors, which exploit the fact that hot water rises while cold water sinks. At low speed, convection currents are sufficient to ensure adequate circulation through the primary loop. But at higher speeds (above 10 kt.), pumps must still be used.

The additional weight and cost of nuclear propulsion are severe disincentives to its use. Lead radiation shielding around reactors adds significantly to displacement as compared to conventionally powered vessels. In addition, nuclear-powered ships are much more expensive to build, because of the special materials and very close tolerances required in their manufacture, but actual life-cycle cost differentials vary with the price of fossil fuels.

NUCLEAR WARHEADS, EFFECTS OF: Although the power, or "energy yield," of a NUCLEAR WEAPON is conventionally measured in kilotons and megatons (KT and MT, the blast equivalents of 1000 and 1 million tons of TNT, respectively), the effects of nuclear weapons are not proportional to their energy yield, and there are several effects which do not occur (at all, or to a significant degree) in nonnuclear explosions. In order of military importance, the principal effects of a nuclear blast are:

1. Blast and shock, most usefully measured in pounds per square inch (psi) of pressure above the ambient atmosphere ("overpressure"); by the velocity of the winds generated; and by the size of the crater produced by a ground burst.

2. Heat, most significantly measured by the radius from the point of detonation (ground zero) within which specific materials (or environments) are ignited under given conditions.

3. Immediate or "prompt" radiation, mainly gamma rays and neutron emissions, measured in roentgen units, or, more usefully, as the distance from ground zero at which certain effects are experienced by unsheltered humans or electronic equipment.

4. ELECTROMAGNETIC PULSE (EMP), a powerful "spike" of electromagnetic energy, transmitted over a significant distance, which can disrupt radio and radar operations (nuclear BLACKOUT) and disable the semiconductors of electronic devices.

5. Residual radiation, or NUCLEAR FALLOUT, consisting of dirt, debris, and weapon materials thrown up by the explosion into the atmosphere, from which it eventually descends to the ground. Local fallout takes the form of a cloud, whose main axis corresponds to the prevailing wind direction.

To attack any given type of target, only one or two effects are useful in most cases, while the others are irrelevant or even undesirable. Relevant effects for three types of target are shown in table N1.

TABLE N1

Nuclear Weapon Damage Effects

Target	Relevant Damage Effect	Optimum Burst
Missile silos, other small hardened targets	Blast (psi overpressure)	Ground
Urban areas, industrial complexes, and other soft targets	Heat (temperature at radius from ground zero); blast (psi overpressure)	Air
Missile warheads and other electronic systems	Prompt radiation (neutrons, X-rays); electromagnetic pulse	Air

NUCLEAR WARHEADS VS. MISSILE SILOS:

Ballistic missile SILOS, command bunkers, and other HARDENED targets are the most difficult to destroy with nuclear weapons, being both small and specifically designed to withstand the primary effect, blast overpressure. The two critical variables which determine the effectiveness of nuclear weapons against hard targets are accuracy (measured in circular error probable, or CEP, the radial distance from the aim point within which half of all warheads impact), and yield. In general, the probability of kill (P_K) for a weapon of given accuracy will increase as a function of the cube root of the increase in warhead yield (because much of the energy released by a nuclear detonation is directed upward, away from the target). Thus increased yield provides marginal improvements in P_k as compared to improvements in accuracy. The relationship between CEPs and yields against a target hardened to withstand 300 psi of peak overpressure is shown in table N2 below.

Until the mid-1970s, CEPs of less than 2000 ft. (610 m.) were unobtainable by long-range ballistic missiles. Today, ballistic missiles such as the U.S. PEACEKEEPER or Soviet SS-18 SATAN regularly achieve CEPs of 750 ft. (240 m.) or less; as a result, the hardening of missile silos has been increased from 300 psi to 1000 psi, with efforts under way to develop "superhardened" silos capable of withstanding more than 2000 psi.

NUCLEAR WARHEADS VS. CITIES

Large, soft area targets, such as cities, airfields, or industrial complexes, are more efficiently destroyed by airbursts than by groundbursts, with the optimum height of burst (HOB) determined by

TABLE N2

Single-Shot Kill Probability of Selected Warheads against a 300-psi Target

Accuracy (CEP in F/feet)	5 MT	1 MT	500 kT	50 kT
10,000	0%	3%	2%	0%
5,000	34%	12%	8%	2%
2,000	93%	55%	40%	11%
1,000	99%	96%	87%	35%
500	>99%	>99%	>99%	82%

TABLE N3

Nuclear Weapon Effects on Soft Area Targets

	Groundburst		Airburst (with optimal HOB)	
	1 MT	10 MT	1 MT	10 MT
Crater depth (ft./m.)	230/70	500/152	na	na
Crater diameter (ft./m.)	950/290	2600/792	na	na
Max. radius of complete destruction: brick structures (mi./km.)	2.7/4.5	6/10	3.5/5.9	8/13.4
Max. radius of light damage (mi./km.)	7.2/12	5.5/25.9	13/21.7	26/44
Radius of lethal winds (mi./km.)	4/6.7	9/15	6.5/10.9	14/23
Radius of 2nd-degree burns, exposed skin (mi./km.)	9.4/15.7	23.5/39	11/18.4	26/44
Radius for ignition of fabrics (mi./km.)	5.6/9.4	14.5/24	6/10	17/29

weapon yield. Heat is the primary effect because combustion radii exceed those of blast and prompt radiation. Heat is compounded by high-velocity winds generated by the shock wave, which can blow down lightly built structures, rip off roofs, and spread flammable debris. Under certain conditions, secondary fires caused by ruptured gas lines and the ignition of debris spread by the wind can result in a self-sustaining "firestorm" (observed, and so named, following the July 1943 incendiary bombardment of Hamburg); the likelihood and extent of such ignition depends, *inter alia*, on the materials involved, the spacing between them, and the weather. Immediate effects relevant to soft, area targets appear in table N3.

NUCLEAR WARHEADS VS. BALLISTIC MISSILE REENTRY VEHICLES (RVS)

The nuclear warheads of anti-ballistic missiles, including the Soviet GALOSH and U.S. SAFEGUARD systems, are designed to destroy or neutralize incoming warheads mainly by prompt radiation, specifically high-energy (hard) X-rays and high-energy (fast) neutrons. The effects of prompt radiation against electronic systems, whether those of the RVs or of ABM RADARS, cannot be easily tabulated. Hard X-RAY KILL is generally thought effective for exoatmospheric engagements (i.e., above the perceivable atmosphere), while NEUTRON KILL is used for endoatmospheric engagements by short-range weapons, as in SPRINT or LOADS. ENHANCED RADIATION (ER) weapons, which generate large numbers of fast neutrons for a given yield, are ideally suited for the purpose. Nuclear blackout, caused by heat and beta radiation, has been proposed as a means of blinding ABM radars.

RADIATION EFFECTS ON HUMAN BEINGS

The radiation generated by nuclear warheads is normally divided into three categories:

1. Prompt radiation, produced during the first seconds of the explosion;

2. Local fallout, consisting of heavy particulate debris irradiated by the detonation and thrown up into the atmosphere, which falls downwind within several hundred miles of ground zero; and

3. Global fallout, consisting of lighter radioactive dust thrown into the upper atmosphere, where it disperses to eventually descend around the world.

Because the effects of ionizing radiation (the type generated by nuclear explosions) on biological processes can be cumulative, and because nuclear explosions produce a host of unstable isotopes which decay at widely different rates, no precise measure of radiation effects on humans is possible. It is known, for example, that two separate doses of radiation within a 24-hour period are cumulative without discount; e.g., a dose of 100 roentgens one day and 200 the next is equivalent to a single 300-roentgen dose. However, the same dosage spaced over longer periods (a week or more must be discounted; a 100-roentgen dose one week and 200 the next amounts to less than 300 roentgens altogether. For a summary of radiation effects on humans, see table N4.

With small "tactical" warheads, the prompt radiation doses that can be expected at various distances from ground zero are shown in table N5.

Exposure to local fallout tends to be cumulative, but can be minimized by sheltering, because the effects of fallout dissipate quite rapidly: within 7

TABLE N4

Effect of Prompt Radiation on Human Beings

Radiation Dosage (in roentgens)	Effects on Average Unsheltered Humans
0–50 r	No effects
50–200 r	Radiation sickness, level I: Nausea and vomiting within 24 hours; less than 5% fatalities within 60 days
200–450 r	Radiation sickness, level II: Vomiting and nausea within 4 hours; less than 50% fatalities within 30–60 days
450–600 r	Radiation sickness, Level III: Severe nausea and vomiting within hours; internal bleeding; more than 50% fatalities within 30 days
More than 600 r	Severe nausea and vomiting immediately; internal bleeding; neurological disorders; 100% fatalities within 7 days

TABLE N5

Prompt Radiation Doses at Distance from Ground Zero

	Weapon Yield, in kT			
	5	20	30	40
Radius for 200 r (in m.)	1200	1300	1400	1600
Radius for 1000 r (in m.)	900	1000	1100	1250

days, radiation levels will, on average, have fallen to 10 percent of their original level; within 49 days they will fall to 1 percent. The effect of low-level global fallout is also difficult to predict. Most responsible authorities believe that a large-scale nuclear exchange would result in a 5 to 20 percent increase in the rates of several different cancers (notably leukemia) over a period of 20 years. Other long-term effects may include reduced fertility and increases in stillbirths.

NUCLEAR WINTER

Several scientists (notably Dr. Carl Sagan) have postulated that large amounts of dust ejected into the upper atmosphere during a nuclear war would raise the atmosphere's albedo (reflectivity), thus reducing sunlight, which would in turn lead to a catastrophic drop in temperatures, causing widespread crop failures, starvation, and potentially the "end of all human life on earth." The hypothesis received widespread publicity when first presented, but has since been largely refuted. Sagan's study results were skewed by inflated assumptions about the employment of nuclear weapons (notably improper estimation of the prevalance of ground bursts), and did not include mitigating factors, such as the "greenhouse effect," the role of the oceans as heat sinks, and the rate at which particulate matter settles from the atmosphere. Nuclear winter theorists were unable to explain how the human race survived several volcanic eruptions which injected far more dust into the atmosphere than their estimated threshold for the nuclear winter scenario. It is currently believed that a large-scale nuclear exchange would result in an average global temperature loss of 1–3 degrees Centigrade, depending on the precise modalities of weapon use and the season.

NUCLEAR WEAPON: An explosive devices that exploits one or both of two energy-liberating nuclear phenomena: fission (as in the atomic bomb) and fusion (as in the hydrogen, a.k.a "thermonuclear," bomb).

Chemical (nonnuclear) explosives liberate energy by very rapid combustion; different compounds have different rates of combustion. Nuclear explosions, on the other hand, are produced by altering the structure of the atoms themselves; the energy thus released is many times greater per unit weight than in the case of chemical explosives. Apart from their blast power, or energy yield, nuclear explosions generate certain effects, such as ionizing radiation, which are not produced at all by chemical explosions.

The two basic methods of altering the structure of atoms are to split heavy ones (fission) or combine light ones (fusion). The former generates much less energy per unit weight, but even so-called "nominal" bombs have yields equivalent to 20,000 tons of TNT (20 KT).

Fission weapons exploit the principle of the nuclear chain reaction. When a suitable material is bombarded by neutrons, some of its atoms will split, releasing large amounts of energy, but also emitting two or more neutrons. If these neutrons in turn hit other atoms, which also split, releasing more energy and and more neutrons, the chain

reaction becomes self-sustaining. If the rate at which fission occurs is not moderated by barriers of neutron-absorbing material (as in nuclear reactors), so much energy is released that an explosion ensues.

Two materials have been used in fission weapons: uranium 235 (U_{235}), a rare isotope of the rather common element uranium; and plutonium 239 (Pu_{239}), a man-made element produced by bombarding atoms of (common) uranium-238 (U_{238}) with neutrons. A certain amount of U_{235} or Pu_{239} (the "critical mass") will spontaneously initiate a chain reaction, leading to a (fission) explosion. The critical mass has been estimated (unofficially) as 16–20 kg. of U_{235} or 7 kg. of Pu_{239} (lower weights have also been suggested, assuming a neutron-concentrating shape). A weapon therefore consists of at least two precision-milled blocks of fissionable material (each of which is "subcritical"), which are brought together at high speed by a chemical explosion, thereby exceeding the critical mass and initiating an explosive chain reaction. Of course, practical military weapons also require shielding, packaging, fuzing, and safety controls. Since 1945, U.S. nuclear weapons have sustained fragmentation, chemical explosions, fires, high-altitude drops, and electrical discharges, without any accidental detonation.

Various methods can be used to rapidly concentrate a critical mass in fission weapons. In the simplest "shotgun bomb," a large but subcritical block of U_{235} (the "target") is placed at the muzzle end of a tube which acts as a gun barrel. The target has a cavity cut into it to receive a "bullet" also machined from U_{235}. A high-explosive charge drives the bullet into the target at high velocity, initiating the chain reaction. This was the design of the Hiroshima bomb; it is easy to manufacture and practically foolproof, but requires more U_{235} than more complex designs.

Only 0.7 percent of all uranium atoms are U_{235}; the rest are nonfissionable U_{238} atoms. Because U_{235} and U_{238} are chemically identical, they cannot be separated by chemical means; complex physical techniques are needed. Three processes have been tried to date: gaseous diffusion, the only proven method so far, which requires vast quantities of electricity in a costly and technologically demanding plant; electromagnetic separation, tried during the Manhattan Project, but which did not produce usable quantities of U_{235}; and gas centrifuging, also tried during the Manhattan Project, when it failed because required centrifuge speeds could not be attained. Recent technological advances may have made gas centrifuge separation practical. Because this method is far cheaper than gaseous diffusion, the technical barrier to NUCLEAR PROLIFERATION could be lowered, making U_{235} available more broadly.

Plutonium is obtained by bombarding U_{238} with neutrons inside a nuclear reactor—a much easier process than gaseous diffusion or gas centrifuging. The production of plutonium is therefore a normal result of operating a uranium reactor, but a special separation plant is required to extract pure (weapon-grade) plutonium from spent fuel rods. Such separation plants have valid civilian uses (to extract fuel to be recycled into other reactors), but in general their acquisition by a nonnuclear state is an indicator of a probable weapons program. A plutonium separation plant, though complex, costs very much less than a gaseous diffusion plant (a 1 to 50 cost ratio is estimated). Plutonium is thus much cheaper than U_{235}; moreover, it is much better suited for lightweight, low-yield "tactical" weapons. But the shotgun method will not work with plutonium, due to the relatively low speed at which the two subcritical masses are brought together. To achieve the high speeds needed for a plutonium chain reaction, an "implosion" is required: a subcritical hollow sphere of plutonium is surrounded by an outer sphere of high-explosive "shaped charges," which, when properly detonated, act as "lenses," compressing the plutonium sphere almost instantaneously to create a critical mass. This was the design of the Nagasaki bomb; it remains typical of most fission weapons built since then. Manufacture of the plutonium sphere and the explosive "lenses" requires absolute precision; thus implosion weapons are considerably more difficult to design and manufacture than shotgun bombs.

As a nuclear explosion begins, the atoms in the fissionable mass are pushed apart, quickly dropping below the critical density. Thus only a small proportion of the fissionable material is actually exploited. The yield of a given fissionable mass can be increased by jacketing it with U_{238}, which acts as a tamper (holding the mass together longer), and, more importantly, as a neutron reflector. Only a small percentage of the neutrons released in a chain reaction actually strike and split other atoms; when "free" neutrons hit the U_{238} jacket, they are reflected back into the fissionable mass, increasing the number of atoms split. Conversely, U_{238} reflectors can lower the critical mass substan-

tially, making it possible to build smaller bombs with lower yields.

Fusion weapons, on the other hand, exploit the fact that certain isotopes of hydrogen (deuterium and tritium) can be fused together at high temperature to produce helium (in the "D-T reaction"), thereby releasing much more energy than was needed to initiate the process. In theory, the initial heat required could be produced by other means (e.g., high-energy lasers), but so far all fusion devices have been triggered by a prior fission explosion (the "trigger"), which compresses and heats the hydrogen fuel, initiating fusion. Usually the fuel is in the form of lithium-6 deuteride, a solid compound at normal temperatures. It can be incorporated directly into the fissile core, or it can be packaged separately. In the latter case, radiation from the fission explosion is contained and its energy used to compress and ignite a physically separate component containing the fusion material. The fissile core is generally referred to as the "primary," the fusion material being called the "secondary." Weapons of this type are said to have two "stages," and are described as "fission-fusion" weapons. If higher yields are desired, radiation from the secondary fusion stage can also be contained and used in turn to initiate a "tertiary" stage of U_{238}. While that isotope is not normally fissile, its fission can be induced by fast neutrons generated by the fission-fusion reaction of the primary and secondary stages. The U_{238} "jacket" also functions as a tamper to hold the bomb mass together longer, thereby increasing the efficiency of the weapon. Approximately half the energy released by a three-stage weapon is generated by the fusion (secondary) stage; the rest comes from the primary and tertiary fission stages. Such "fission-fusion-fission" devices are relatively cheap to produce, and have very efficient weight-to-yield ratios, but are very "dirty," i.e., they generate much nuclear fallout because of the dispersal of fission products and the irradiated remains of the U_{238} jacket.

ENHANCED RADIATION (ER) weapons, popularly called "neutron bombs," are two-stage (fission-fusion) thermonuclear devices designed to maximize the output of high-energy (fast) neutrons generated by the fusion of deuterium and tritium, while minimizing the blast effect of the explosion. Prompt radiation is enhanced by reducing the fission yield relative to the fusion yield (i.e., by eliminating U_{238} in the tamper material). Enhanced radiation weapons use gaseous deuterium and tritium, rather than lithium-based fusion material, in order to maximize the release of fast neutrons. For a given yield, an ER weapon can generate lethal neutron radiation over a much greater radius than a "conventional" nuclear weapon of equivalent yield, allowing smaller weapons to be used to achieve the same "kill radius."

Full-fledged nuclear powers deploy a range of nuclear weapons:

1. Pu_{239} fission devices, especially suitable for lightweight, low-yield (0.1 to 20-kT) NUCLEAR ARTILLERY shells and short-range "tactical" missile warheads;

2. U_{235} fission devices with yields between 20 and 200 kT, suitable for tactical aircraft bombs, depth charges, etc.;

3. Fission-fusion devices with yields of 40 kT–20 megatons (MT) for "strategic" weapons, notably ballistic missiles (these are relatively "clean" weapons, with the amount of fallout depending on the size of the fission trigger and the type of burst—airbursts can generate very little fallout, while groundbursts maximize it;

4. Fission-fusion-fission weapons with yields greater than 1 MT (the largest nuclear weapon ever exploded was a 57-MT Soviet device detonated in 1961)—while highly efficient, three-stage weapons are much too "dirty" (i.e., cause too much fallout) for purposeful use.

The trend over the last two decades has been to reduce the yield of nuclear weapons, while increasing the accuracy of their delivery. Weapons are considerably "cleaner" than in the past, as alternative means have been found to boost the power of secondary and tertiary stages without increasing the amount of fissionable material in the bomb. Many modern nuclear weapons have selectable yields, probably obtained by altering either the material levels in the fission primary stage or by changing the speed of the reaction.

NUDETS: U.S. nuclear-detonation detection network consisting of linked seismic detectors spread throughout U.S. territory. Intended to estimate the point of detonation and the yield of nuclear explosions, NUDETS is also used in conjunction with DEFENSE SUPPORT PROGRAM satellites to ensure compliance with the TEST BAN TREATY.

O

OBERON: A class of 27 British-built diesel-electric attack SUBMARINES, commissioned between 1960 and 1977, in service with the Royal Navy (10) and the navies of Australia (6), Brazil (3), Canada (3), and Chile (2). The Oberons are improved versions of the earlier Porpoise class, which was in turn heavily influenced by the German Type XXI U-boat of 1944–45, and by the U.S. TANG class; 13 were originally built for the Royal Navy between 1960 and 1967. The Royal Navy began phasing out its Oberons in the late 1980s, but because of delays of the new Upholder class, 9 Oberons have been upgraded with new sensors. Other users will probably keep their boats in service until the end of the century.

The Oberons have a conventional, cigar-shaped hull in a double-hull configuration, with fuel and ballast tanks installed between the outer casing and the pressure hull; the latter is fabricated of high-yield UKE steel, while the outer hull and superstructure are constructed of light alloys and glass-reinforced plastic to reduce weight. The Oberons have a large, flat, deck casing with a streamlined sail (conning tower) amidships. The control surfaces are conventional, with retractable bow diving planes, stern planes behind the propellers, and a single, low-mounted rudder. A retractable snorkel allows diesel operation from shallow submergence. The electric motors are powered by two 224-cell lead-acid batteries identical to those of U.S. GUPPY-class submarines. Designed primarily for ANTI-SUBMARINE WARFARE (ASW) in narrow waters, the Oberons are exceptionally quiet, the key attribute for ASW.

The Oberons are armed with 6 full-length, 21-in. (533-mm.) bow torpedo tubes, and 2 shorter, 21-in. stern tubes. The bow tubes can launch a variety of free-running and acoustical homing TORPEDOES, including the Mk.8 steam torpedo and the Mk.24 TIGERFISH. Twelve reloads are stowed in the forward torpedo room. The stern tubes were originally meant for short ASW homing torpedoes, which were in fact superseded by the dual-purpose Tigerfish. The aft tubes can now be used only for decoys and TORPEDO COUNTERMEASURES. All sensors are linked to DCH tactical computers, which permit two Tigerfish to be launched simultaneously. Australian and Canadian Oberons have been updated with new U.S. sensors compatible with U.S. MK.48 homing torpedoes and UGM-84 HARPOON anti-ship missiles. Brazilian and Chilean boats retain the original sensors. See also SUBMARINES, BRITAIN.

Specifications Length: 290 ft. (88.41 m.). Beam: 26.5 ft. (8.08 m.). Displacement: 2080 tons surfaced/2450 tons submerged. Powerplant: twin-shaft diesel-electric: 2 1840-hp. ASR 16VVS-AS21 diesel generators, 2 3000-hp. English Electric electric motors. Speed: 12 kt. surfaced/17 kt. submerged Fuel: 300 tons. Range: 9000 n.mi. at 9 kt. Max. Operating Depth: 656 ft. (200 m.). Collapse depth: 984 ft. (300 m.). Crew: 64. Sensors: (early) 1 Type 2007 flank-mounted passive conformal array, 1 Type 197 sonar intercept array, 1 Type 2046 clip-on passive towed array; 1 Type 1006 surface search radar, one MEL electronic signal monitoring array, 2 periscopes; (modified RN Oberons) 1 Type 2051 Triton integrated sonar.

OBSERVATION POST (OP): A military outpost overlooking enemy-controlled territory or possible enemy avenues of approach. OPs are usually sited on dominating terrain in front of the main line of resistance, to which they are linked by field telephone or radio. Manned by small detachments, usually no more than a SQUAD or FIRE TEAM, OPs are meant to detect enemy attacks, direct and adjust ARTILLERY fire, and study the terrain for attacks by friendly forces. See also FIST: INDIRECT FIRE.

OBSOLESCENT, OBSOLETE: A weapon is deemed obsolescent when it is only marginally effective *against first-rate opposition*, and obsolete when useless against such oppositions. Context is decisive: a World War II–vintage Sherman tank is obsolete on the NATO Central Front, where it could neither survive nor inflict significant damage; in the Middle East (e.g., in Iraq or Iran), Sherman tanks could still be marginally useful, hence are only obsolescent; in Third World settings where there are no modern tanks and few anti-tank weapons, the Sherman could still be a first-line weapon.

The rate at which weapons become obsolete or obsolescent depends on the capabilities inherent in the system, as well as the rate with which actual or potential enemies develop countermeasures. A weapon with only marginal advantages when first introduced will, in general, become obsolete much more quickly; but so will a weapon of truly superior performance, which therefore evokes an extraordinary countermeasures effort. Further, weapons of broad capabilities (i.e., versatile enough to accommodate modifications and improvements in response to evolving countermeasures) will remain effective longer than narrow, single-function weapons. Thus, e.g., the F-4 Phantom has remained a first-line fighter for almost 30 years, while the MiG-23 Flogger was obsolescent the day it was introduced.

Armed forces differ in their attitudes towards obsolete and obsolescent equipment. In the West, and especially the United States, it is usually scrapped when taken out of service. In the Soviet Union, on the other hand, obsolescent equipment removed from first-line formations is used to replace obsolete equipment in second-line formations, and the latter in turn is handed down to the lowest-readiness mobilization units. The Soviet approach reflects the calculation that while obsolescent or obsolete equipment may not be useful against enemy forces equipped with the latest weapons, by the time the mass of second-line formations reaches the battlefield, there may be few modern weapons remaining intact in enemy hands.

OCEANOGRAPHY: The study of the phenomenology of the world's oceans, including currents, temperature and salinity, bottom topography, magnetic and gravitational anomalies, and marine life forms. Oceanography plays an increasingly important role in ANTI-SUBMARINE WARFARE, because of the effects of ocean conditions on SONAR performance. In addition, data on bottom topography, magnetic anomalies, and gravitational perturbations can be exploited by ballistic-missile submarines for precision navigation and thus for the accurate targeting of submarine-launched ballistic missiles (SLBMs). Both the U.S. and Soviet navies maintain large oceanographic research establishments. See also THERMOCLINE.

OH-58: U.S. light observation helicopter. See KIOWA.

OHIO: The latest class of U.S. nuclear-powered ballistic-missile SUBMARINES (SSBNs), the first of which was commissioned in 1981. Since then, an additional 8 submarines have been commissioned with 6 more under construction and 5 more on order. The Ohios were designed specifically for the TRIDENT D-5 SLBM, which is too large for the missile tubes of the earlier LAFAYETTE and FRANKLIN SSBNs. Initially, the Ohios were to be slightly enlarged Lafayettes, with the same S5W pressurized-water reactor. But the increased displacement of the new submarine would have reduced maximum speed to less than 20 kt., which was deemed unacceptable by Adm. Rickover—though SSBNs normally cruise at only a few knots. In addition, the U.S. Navy wanted to apply new silencing techniques and introduce a new natural-circulation reactor of greatly increased power. Both requirements led to a complete redesign on a much larger scale. The final result was a submarine displacing more than twice as much as its predecessors, and carrying 24 missiles rather than the 16 which had been the U.S. standard. In the navy's perspective, the new reactor and silencing techniques required additional displacement; since the reactor had enough power to drive a submarine even larger than that, it was deemed wise to enlarge the scale further to accommodate more missile tubes, thereby improving "cost effectiveness" on a per-missile basis.

The first Ohio was plagued by technical delays and cost overruns, which attracted unfavorable attention; the program was also criticized for its high cost ($1.5 billion), for the larger number of missiles carried in a single hull (which increases the vulnerability of the SSBN force as a whole under the limits of the SALT I agreement), and for its highly accurate Trident D-5 missile, seen by some as a destabilizing COUNTERFORCE weapon.

Except for the Soviet TYPHOON class, the Ohios are the largest submarines in the world. Like the earlier Lafayettes, they have a modified teardrop hullform (a bulbous bow, cylindrical center section, and tapered stern) in a single-hull configuration, with ballast tanks concentrated at both ends outside the pressure hull, which is fabricated of HY-80 steel. The Ohios have a tall, streamlined sail near the bow, behind which a flat-decked whaleback casing covers the tops of the missile tubes, arranged in two rows of 12. The control surfaces are arranged in standard U.S. fashion, with "fairwater" diving planes mounted on the sail, and cruciform rudders and stern planes mounted ahead of the propeller. The stern planes have large, rectangular endplates, which improve rudder performance and serve as mountings for conformal-array hydrophones.

The Ohios have a turbo-electric nuclear powerplant: steam from the reactor feeds electrical turbo-generators, which in turn power an electric motor that turns the propeller shaft. This system is significantly more quiet than the usual steam turbine drive. Moreover, its S8G reactor relies on natural circulation cooling; i.e., at low speeds, natural convection is sufficient to ensure an adequate flow of coolant; noisy pumps are needed only at higher speeds. In addition, all machinery is mounted on resilient sound-isolation rafts, and much insulation is fitted throughout. The Ohios are the quietest submarines in service anywhere; according to unofficial U.S. Navy sources, no Soviet submarine has ever been able to track an Ohio on patrol. Like all U.S. SSBNs, they actually have two crews (Blue and Gold), which man the submarines during alternate 70-day patrols, with a 25-day refitting period in between.

The primary armament of the Ohio class is the Trident D-5 missile. The first Ohios were armed with the interim Trident C-4 until the D-5 became operational in 1990. The latter has a range of 6000 n.mi., a payload of 8 100-kT MIRV warheads, and

a presumed median error radius (CEP) of less than 125 m. The missile is 45.5 ft. (13.87 m.) long and 6.9 ft. (2.1 m.) in diameter, i.e., both longer and wider than the missile tubes of earlier SSBNs. It is launched pneumatically with high-pressure steam; an automatic hover system maintains depth control during firing by compensating for the weight of ejected missiles. Secondary armament consists of four 21-in. (533-mm.) midships-mounted tubes for MK.48 wire-guided acoustical homing TORPEDOES, decoys, and TORPEDO COUNTERMEASURES.

Like all SSBNs, the Ohios rely on stealth as their primary defense. Thus their sensors are optimized for long-range passive detection and classification: their BQQ-6 sonar system is similar to the BQQ-5 of the LOS ANGELES–class attack submarines, but lacks the active sonar component of the latter. Trailing wire antennas for VLF and ELF radios allow submerged communications with shore stations. The Ohios have two Mk.2 SINS (Ship's Inertial Navigation Systems) for precision underwater navigation; they also provide launch position data to the Trident missiles. The Mk.2 can update its inertial platform with data from the NAVSTAR global positioning system and other navigational satellites. The Trident missiles are targeted and fired with a Mk.98 digital FIRE CONTROL system, which has a rapid retarget capability. Torpedo fire control is provided by a Mk.118 system. See also SUBMARINES, UNITED STATES.

Specifications   Length: 560 ft. (170.73 m.). **Beam:** 42 ft. (12.8 m.). **Displacement:** 16,764 tons surfaced/18,750 tons submerged. **Powerplant:** single-shaft nuclear: 1 S8G natural-circulation reactor with turbo-electric drive, 30,000 shp. (rated)/60,000 shp. (max.). **Speed:** 20 kt. surfaced/30+ kt. submerged. **Max. operating depth:** 985 ft. (300 m.). **Collapse depth:** 1640 ft. (500 m.). **Crew:** 160. **sensors:** 1 BQQ-6 sonar system (1 BQS-13 passive low-frequency spherical array, 1 BQR-15 passive conformal array, 1 BQR-23 passive towed array), 1 BQS-15 underice sonar, 1 BQR-19 bottom-mapping sonar, one BPS-15 surface-search radar, 1 WLR-8(V)5 electronic signal monitoring array, 2 periscopes.

OMG: See OPERATIONAL MANEUVER GROUP.

OP: See OBSERVATION POST.

OPERATIONAL   MANEUVER   GROUP (OMG):   A Soviet term for a large, self-contained formation meant to penetrate through gaps in the enemy's defensive lines in order to execute major attacks throughout the depth of enemy territory, to disrupt the enemy's defensive plan as a whole. The OMG is a return to an operational concept pioneered by Marshal Tukhachevskiy in the early 1930s, but which was abandoned until 1944–45, when large combined-arms "mobile groups" carried out deep penetrations of German rear areas. Ignored after World War II under Stalin's influence, the concept was again revived from the late 1960s in Soviet doctrinal discussions.

As presently understood, an OMG would be built around selected and reinforced DIVISIONS; it would be held in reserve until a passage is discovered or opened by other forces through enemy lines, when it would pass through the first Soviet echelon to begin operations in depth.

Considering the shallow NATO front in Central Europe, and the paucity of operational RESERVES, OMGs, in conjunction with SPETSNAZ operations and airborne DESANTS, could pose a significant threat to the cohesion of NATO's defensive array, especially if force reductions thin out the front. See also ARMORED WARFARE; BLITZKRIEG.

OPERATIONAL METHOD: A specific application of operational art (or, in Soviet terminology, "operational science"), the area of military science or art between STRATEGY and TACTICS. For armies, operational method covers the employment of CORPS and DIVISIONS, while tactics apply to lower-echelon units (BRIGADES, BATTALIONS, COMPANIES and PLATOONS), and strategy to ARMY GROUPS and national commands. Different armies favor different operational methods (which reflect national characteristics, available resources, and specific terrain conditions) in a continuum from ATTRITION to MANEUVER, though all combine both elements to some degree. Attritional operational methods, aimed at destroying the enemy's *ability* to fight by the cumulative destruction of personnel and materiel, are preferred by armies that enjoy a preponderance of firepower. Slow and expensive (although not necessarily in human lives), such methods are also highly predictable and can be fully planned in advance. Moreover, when attritional operational methods do fail, they tend to fail gracefully—in other words, they offer small payoffs with small risks.

Maneuver operational methods, on the other hand, are aimed at destroying the enemy's *will* to resist, by disrupting command structures and causing confusion to the point of collapse. They are generally preferred by armies that are outgunned or cannot otherwise endure a long war (e.g., Israel). Maneuver methods require careful analysis

of the enemy and of *his* operational methods, in order to uncover weaknesses that can be exploited; highly trained troops to implement demanding plans; and talented commanders to adapt those plans quickly to changing circumstances. Such methods can gain spectacular success (as in the Six-Day War), but when they fail, they tend to fail catastrophically (e.g., Stalingrad). In other words, they offer high payoffs with high risks.

For specific operational methods, see ACTIVE DEFENSE; AIR-LAND BATTLE; ARMORED WARFARE; BLITZKRIEG; DEFENSE-IN-DEPTH.

OPERATIONS/OPERATIONAL: The level of warfare between the strategic and the tactical. For ground forces, the operational level deals with the conduct of specific campaigns within a THEATER OF OPERATIONS; hence, it is concerned with army CORPS and DIVISIONS, as opposed to strategic-level national commands or ARMY GROUPS, and tactical-level units below division, such as BRIGADES, battalions, companies, and platoons. See also OPERATIONAL METHOD; STRATEGY; TACTICS.

ORBAT: See ORDER OF BATTLE.

ORBITAL BOMBARDMENT SYSTEM: A hypothetical weapon consisting of a NUCLEAR WEAPON installed in a satellite held in low earth orbit, kept ready to be deorbited on command to hit a selected target target on earth. Orbital bombardment systems were seriously considered in the early 1960s as a means of circumventing single-azimuth BALLISTIC MISSILE DEFENSES. But they would be extremely inaccurate compared to BALLISTIC MISSILES, and would present considerable problems of coordination with other nuclear DELIVERY SYSTEMS. Orbital bombardment systems are now specifically banned by the OUTER SPACE TREATY. See also FRACTIONAL ORBIT BOMBARDMENT SYSTEM.

ORDER OF BATTLE (ORBAT): Information that defines the organization, numerical strength, disposition, and equipment of a military force. All armed forces define their own orders of battle in TABLES OF ORGANIZATION AND EQUIPMENT (TO&Es), and monitor their strength through situation and readiness reports. Order of battle INTELLIGENCE is compiled on the basis of open sources, espionage, defector reports, aerial and satellite imagery, and signals intelligence. In wartime, scouts and RECONNAISSANCE actions, refugee reports, and prisoner-of-war interrogations supplement other sources. Such data are collated into TO&Es to allow commanders to analyze the strength and composition of the forces facing them. Information pertaining to the identity, personality traits, and experience of enemy commanders, and the MORALE of enemy rank and file, is often appended to ORBAT data to assess the combat effectiveness of enemy forces.

An ELECTRONIC ORDER OF BATTLE (EOB) is a taxonomy of enemy RADAR and radio emitters, normally used to uncover enemy AIR DEFENSE systems and locate HEADQUARTERS units.

ORDNANCE: Explosives, chemical agents, pyrotechnics, and complete munitions, as well as guns, rockets, missiles, flares, mines, torpedoes, etc.

ORION: The Lockheed P-3 four-engine turboprop-powered LONG-RANGE MARITIME PATROL (LRMP) and ANTI-SUBMARINE WARFARE (ASW) aircraft. The Orion was developed to meet a 1957 requirement for a replacement to the P-2 NEPTUNE patrol bomber. Based on the Lockheed Model L-188 Electra commercial airliner, the Orion was selected by the U.S. Navy in 1958; the first production aircraft was delivered for testing and evaluation in 1961, and squadron service began in 1962. The initial P-3A version was superseded in 1965 (after the 157th aircraft) by the P-3B, with more powerful engines. The current P-3C version entered service in 1969, and featured better avionics, sensors, and fire control systems, which have been further improved since then by Update I, II, and III modification programs. All surviving P-3A/Bs in U.S. service have been modified to P-3C standards.

The U.S. Navy operates nearly 500 P-3s of all types, including a number modified for special roles: the RP-3A for RECONNAISSANCE; WP-3A for weather reconnaissance; EP-3B and EP-3E for electronic intelligence (ELINT) gathering; RP-3D for global magnetic survey; and WP-3D for atmospheric research. Japan has 42 P-3Cs, including 38 built under license by Kawasaki from 1979. During the 1970s, Iran received six P-3Fs configured for long-range ocean SURVEILLANCE; it is not known how many remain operational. Canada acquired a variant designated CP-140 Aurora, which combines the airframe of the P-3C with the avionics and weapon systems of the Lockheed S-3A VIKING carrier-based ASW aircraft.

The P-3C retains the wings, tail, and basic structure of the Electra, but the fuselage has been lengthened by 7.4 ft. (2.25 m.) and incorporates a large internal weapons bay, while the nose houses a large APS-115 surface-search and weather RADAR. The 7-man flight crew sits on an airline-style flight deck. AVIONICS include INERTIAL NAVIGATION; DOPPLER navigation; LORAN and TACAN; UHF, VHF, and HF radios; and an autopilot. The cabin behind the cockpit contains a tactical information center, with a tactical computer, several tactical displays which integrate data from all the aircraft's sensors, sonobuoy receivers, and a FIRE CONTROL system. The center is manned by a tactical coordinator, 3 weapon operators, and a technician, with provision for 2 additional observers, giving the Orion a crew capacity of 12 men. A compartment behind the tactical center contains a SONOBUOY launcher, with 48 externally loaded ventral dispenser tubes and 4 internal (reloadable) sonobuoy/flare chutes. The rear fuselage houses a galley, lavatory, and several bunks, all necessities on long patrols. An ASQ-81 MAGNETIC ANOMALY DETECTOR (MAD) is housed in a stinger fairing extending from the tail. The weapons bay beneath the cabin floor has a maximum capacity of 20,000 lb. (9091 kg.). Typical payloads include either 8 MK.46 ASW homing TORPEDOES, 2 2000-lb. MINES, 4 1000-lb. mines, 3 Mk.57 DEPTH CHARGES, or 8 Mk.54 depth charges.

The low-mounted wing contains large fuel tanks in its outboard sections. Four underwing hardpoints can carry either 4 Mk.46 torpedoes, 4 depth charges, 16 ZUNI 5-in. unguided rockets, or 4 HARPOON anti-ship missiles. A variety of RADAR WARNING RECEIVERS and ELECTRONIC SIGNAL MONITORING arrays are mounted outside the fuselage and on the wingtips. It is standard practice to shut down one or two engines on patrol to save fuel.

The successive upgrading of avionics and weapons makes the current P-3C Update III completely different from the P-3A/B, or even the original P-3C; only the Lockheed S-3 VIKING and British Aerospace NIMROD MR.2 are in its class. An Update IV package is now in development, but an all-new P-3C follow-on, designated P-7, is scheduled to enter service in 1990–91, with the ability to carry Harpoons internally. An AIRBORNE EARLY WARNING version with an APS-138 surveillance radar mounted in a dorsal rotodome has been purchased for the U.S. Customs Service.

Specifications **Length:** 116.83 ft. (35.62 m.). **Span:** 99.67 ft. (30.38 m.). **Powerplant:** (A) 4 4500-hp. Allison T56-A-10W turboprops; (B/C) 4 4910-shp. T56-A-14 turboprops. **Weight, empty:** (B/C) 61,491 lb. (27892 kg.). **Weight, normal loaded:** (B/C) 135,000 lb. **Weight, max. takeoff:** (B/C) 142,-000 lb. **Speed, max.:** 473 mph (790 kph). **Speed, cruising:** 237 mph (395 kph). **Combat radius:** 1550 mi. (2588 km.) w/3-hr. loiter; 2384 mi. (3960 km.) w/o loiter. **Operators:** (A/B) Aus, Neth, Nor, NZ, Por, Sp; (C) Jap, US; (F) Iran.

OSA: NATO code name for a large class of Soviet-designed missile boats, the first of which was completed in 1961. Since then, more than 275 Osas have been built in the Soviet Union, and an additional 125 have been built in China (as the Huang Feng class) from 1978. Some 100 Soviet-built Osas are in service worldwide. Designed as a larger, more capable replacement for the KOMAR-class missile boats, the Osas are somewhat more seaworthy and make better gun and missile platforms; they carry twice the missile load, and have superior FIRE CONTROL and more range and endurance. Two separate versions, Osa I and Osa II, differ in their weapons and sensors, but are otherwise identical.

The Osas' hard chine bilges and V-shaped bottoms enable them to plane at high speeds, but also make them poor sea boats in heavy weather. The hull is constructed of steel (the wooden-hulled Komars tended to deteriorate rapidly), while a blocky "citadel" superstructure amidships is equipped with a COLLECTIVE FILTRATION system for NBC defense.

Primary armament consists of four SS-N-2a/b STYX anti-ship missiles, in individual launch canisters mounted two per side abaft the bridge. The Osa I has rectangular box canisters, while Osa II has lighter, cylindrical launchers. The aft launchers have a fixed elevation of 15°, enabling them to fire over the forward launchers, which have a fixed elevation of only 12°. Secondary armament on the Osa I consists of two twin 30-mm. CANNONS mounted in small turrets at the bow and stern. The Osa II supplements these with a quadruple pedestal launcher for SA-N-5 GRAIL short range surface-to-air missiles (SAMs), a navalized version of the SA-7 shoulder-fired SAM.

Osas have been used in combat by Egypt and Syria in the 1973 Yom Kippur War, by India in the 1971 Indo-Pakistani War, and by Iraq in the Iran-Iraq War. In general, they have proven inferior to (much newer) Western missile boats such as the COMBATTANTE, RESHEF, or SA'AR classes, especially in seakeeping, range, fire control, and weapons. Nonetheless, they are cheap and available in large numbers. The basic Osa hull is also used in the TURYA- and MATKA-class HYDROFOILS; the Mol-class torpedo boat; and the Stenka-class motor gunboat. See also FAST ATTACK CRAFT, SOVIET UNION.

Specifications **Length:** 128 ft. (39 m.). **Beam:** 25.3 ft. (7.71 m.). **Draft:** 5.9 ft. (1.8 m.). **Displacement:** (I) 210 tons full load; (II) 245 tons full load. **Powerplant:** 3 12,000-hp. M503A diesels, 3 shafts. **Speed:** 38 kt. **Range:** 500 n.mi. at 34 kt./ 750 n.mi. at 25 kt. **Crew:** 30. **Sensors:** 1 "Square Ties" surface-search radar, 1 "Drum Tilt" fire control radar, 1 "High Pole-B" IFF system, 2 "Square Head" IFF interrogators. **Operators:** Alg, Bul, Cuba, DDR, Egy, Eth, Fin, Ind, Iraq, Lib, Mor, Pol, PRC, Rom, Som, Sud, S Yem, Syr, Tun, USSR, Yug.

OSCAR: NATO code name for a class of five huge Soviet nuclear-powered guided-missile SUB-MARINES (SSGNs), the first of which was completed in 1980. The class is currently in series production at rate of approximately one unit every two years. The Oscars, successors to the CHARLIE-class SSGNs, are designed to attack U.S. carrier battle groups (CVBGs) with long-range SS-N-19 SHIP-WRECK supersonic anti-ship missiles. Apparently based on both the Charlies and the experimental PAPA SSGN of 1970, the Oscars are much larger than either. The Oscars have a flattened teardrop hullform, with a broad, bulbous bow, oval center section, and tapered stern. Like most Soviet submarines, they have a double hull, with ballast tanks and missile tubes housed in the space between the outer casing and pressure hull. The need to accommodate the missile tubes outside the pressure hull accounts for the Oscar's extremely broad beam. A long, streamlined sail is located amidships, with a short, flat deck casing extending from its after end. The sail houses several radar and ELECTRONIC SIGNAL MONITORING arrays used for TARGET ACQUISITION. Control surfaces are arranged in standard Soviet fashion, with retractable bow diving planes and cruciform rudders and stern planes mounted ahead of the propellers. The entire outer surface is covered with CLUSTER GUARD anechoic tiles, which reduce both radiated noise and the effectiveness of hostile active sonar. Although not as quiet as Western attack submarines (in part due to their use of two small, high-speed propellers in place of a single large, low-speed propeller), the Oscars are much quieter than their predecessors, indicating the increased sophistication of Soviet submarine design.

Primary armament consists of 24 SS-N-19 Shipwreck missiles, with a range of some 240 mi. (400 km.). These missiles can be fired underwater from fixed tubes mounted in two rows of 12 on either side of the sail at a fixed elevation of approximately

45°. The tops of the tubes are covered by large sliding doors. Secondary armament consists of 8 bow-mounted torpedo tubes, 4 of 21-in. (533-mm.) and 4 of 26-in. (650-mm.) diameter, for a variety of free-running and homing TORPEDOES (including the Type 165 long-range wake homer), plus SS-N-15 STARFISH and SS-N-16 STALLION anti-submarine missiles. The normal load is believed to consist of 24 weapons of all types. Communications equipment includes a "Pert Spring" satellite communications terminal, a "Shot Gun" VHF system, and a trailing wire antenna for VLF or ELF radio. From 1987, production switched to a slightly longer Oscar II variant. See also SUBMARINES, SOVIET UNION.

Specifications **Length:** (I) 470 ft. (143.3 m.); (II) 505.12 ft. (154 m). **Beam:** 60 ft. (18.3 m.). **Displacement:** (I) 11,000 tons surfaced/13,000 tons submerged. (II) 15,000 tons surfaced/18,000 tons submerged. **Powerplant:** 2 pressurized-water reactors, 2 sets of geared turbines, 90,000 shp. **Speed:** 35 kt. **Max. operating depth:** 1640 ft. (500 m.). **Collapse depth:** 2725 ft. (830 m.). **Crew:** 130. **Sensors:** 1 large bow-mounted low-frequency active/passive sonar, 1 medium-frequency active ranging sonar, 1 flank-mounted conformal sonar, 1 "Snoop Pair" surface search radar, 1 "Rim Hat" electronic signal monitoring array, 2 periscopes.

OSLO: A class of five Norwegian FRIGATES commissioned in 1966–67. The largest surface warships of the Norwegian navy, the Oslos are based on the American DEALEY class, with a slightly modified hull and European-made weapons and sensors. They were ordered under a 1960 naval modernization program, with half the cost paid by the U.S.

The hull and machinery are essentially identical to the Dealeys', with slightly more freeboard forward to cope with severe North Sea conditions. As completed, the Oslos were armed with a twin 3-in. 50-caliber DUAL PURPOSE gun and a Norwegian Terne III 6-barrel ANTI-SUBMARINE WARFARE (ASW) rocket launcher forward, with a second 3-in. twin mount and a DEPTH CHARGE rack aft. The Terne III fires 265-lb. (120.5-kg.) depth charges to a range of 400–900 m. Each charge has a lethal radius of 20 m., and can be fitted with hydrostatic and acoustical proximity fuzes. Once fired, Terne can be reloaded automatically within 40 seconds.

In the late 1970s, the ships were also armed with an 8-round NATO SEA SPARROW short-range surface-to-air missile launcher, 6 single launchers for PENGUIN Mk.2 anti-ship missiles, and 2 sets of Mk.32 triple tubes for lightweight ASW TORPEDOES. Other armament remained unchanged.

Specifications **Length:** 316.9 ft. (96.6 m.). **Beam:** 36.7 ft. (11.2 m.). **Draft:** 14.4 ft. (4.4 m.). **Displacement:** 1450 tons standard/1850 tons full load. **Powerplant:** single-shaft steam: 2 oil-fired boilers, 1 set geared turbines, 20,000 shp. **Speed:** 25 kt. **Range:** 4500 n.mi. at 15 kt. **Crew:** 150. **Sensors:** 1 DRBV22 combined air- and surface-search radar, 1 WM-22 fire control radar, 1 Decca 1226 navigation radar, 1 Mk.91 Sea Sparrow guidance radar, 1 hull-mounted SQS-36 active/passive medium-frequency sonar, 1 Terne III active ranging sonar.

OSPREY: The Bell/Boeing-Vertol V-22 tilt-rotor V/STOL aircraft, developed from the Bell XV-15 research aircraft commissioned by NASA in 1973. A tilt-rotor is an otherwise conventional-looking aircraft powered by two large propellers mounted in wingtip nacelles. The nacelles are hinged to allow the propellers to pivot through 90° from vertical to horizontal and vice versa. When horizontal, the propellers act as rotary wings, generating lift like helicopter rotors, enabling the aircraft to take off and land vertically, to hover motionless, and also to maneuver at very low speeds. As the rotors transition to a vertical orientation, they impart forward motion, while lift is generated by the wing as in a conventional airplane. Thus the tilt-rotor combines the V/STOL and hovering capability of the HELICOPTER with the speed and lifting efficiency of fixed-wing aircraft.

The success of the XV-15 attracted the attention of all four U.S. military services, and a requirement was issued in 1981 for a Joint Advanced Vertical Lift Aircraft (JVX). Bell teamed with Boeing-Vertol to produce the design that became the Osprey. A total of 932 aircraft were ordered by the four services, with the U.S. Marine Corps (USMC) the single largest customer with a requirement for 552 MV-22A assault transports, capable of carrying 24 fully equipped troops or 5760 lb. of cargo. The air force ordered 80 CV-22A transports for logistic support, while the navy wanted at least 50 HV-22As for long-range SEARCH AND RESCUE (SAR), and an SV-22 ANTI-SUBMARINE WARFARE variant. The U.S. Army had a requirement for 231 utility transports similar to the MV-22A. The first production Osprey rolled off the assembly line in mid-1988, but by 1989 costs had risen and budgets had fallen, placing the program in jeopardy. Only the

USMC views the Osprey as essential, and may be the only customer, at least initially.

The Osprey has a boxy fuselage to maximize cargo and passenger space. Behind the side-by-side cockpit for the two-man flight crew, tip-up seats mounted on the sides of the fuselage provide an unobstructed cargo deck. The rear end of the cabin sweeps upward for a full-width folding ramp-door for the roll-on/roll-off loading of light vehicles and palletized cargo. The aircraft has tricycle landing gear, with the main gear retracting into sponsons mounted on the lower sides of the fuselage; in the projected navy and air force versions, the sponsons are stretched to house long-range fuel tanks. The Osprey has a shoulder-mounted stub wing fitted with ailerons and trailing-edge flaps to control horizontal flight, as well as twin vertical stabilizers mounted as endplates on the tailplane. More than 60 percent of the basic airframe is fabricated from carbon-graphite composites to save weight. The aircraft is powered by two Allison turboprops, which drive two 3-bladed rotors. The engines are mounted in self-contained wingtip nacelles with a single-point rotating hinge. The rotor tilt mechanism is controlled hydraulically, with digital FLY-BY-WIRE flight controls to facilitate the transition between vertical and horizontal flight, and provide differential control in the event of an engine failure. To facilitate parking and handling, the Osprey's wings can be rotated 90° to lie parallel to the top of the fuselage, while the rotor blades can be folded inward for storage. The entire process is performed hydraulically in under 90 seconds.

If the Osprey program lives up to its promise, the aircraft will be a most revolutionary innovation. In addition, if costs can be contained the Osprey will probably attract numerous foreign military buyers and commercial customers. There are already plans to produce an AIRBORNE EARLY WARNING (AEW) variant to provide AWACS capability for small V/STOL AIRCRAFT CARRIERS such as those of the British INVINCIBLE class, a capability now restricted to large-deck conventional carriers.

Specifications **Length:** 57.9 ft. (17.65 m.). **Span:** 46.6 ft. (14.2 m.). **Rotor diameter:** 38 ft. (11.6 m.). **Powerplant:** 2 6000-shp. Allison 501-M89C turboprops. **Weight, max. takeoff:** (VTOL) 43,000 lb. (19,545 kg.); (STOL) 59,000 lb. (26,818 kg.). **Speed, max.:** 375 mph (626 kph). **Speed, cruising:** 275 mph (460 kph). **Range, max.:** (MV) 450 mi. (751 km.); (CV/HV) 1600 mi. (2672 km.).

OTH-B: See OVER-THE-HORIZON BACKSCATTER RADAR.

OTH-R: See OVER-THE-HORIZON RADAR.

OTH-T: See OVER-THE-HORIZON TARGETING.

OTOMAT: A surface-launched ANTI-SHIP MISSILE built by the Franco-Italian consortium of OTO-MELARA and Matra. OTOmat was designed in response to a 1969 Italian navy requirement for a ship- and air-launched anti-ship missile light enough to arm very small combat vessels. Flight tests began in 1971, and OTOmat Mk.I entered service with the Italian navy in 1975. An improved OTOmat Mk.II (also known as Teseo) began flight tests in 1974 and entered service in 1978; since then, more than 850 have been sold.

OTOmat has a cylindrical body, with a blunt nose housing a radar, and a tapered tail section. Four fixed, cropped-delta wings mounted at mid-body provide lift, while the missile is steered by four small cruciform tail fins. Two solid-fuel strap-on booster rockets burn for four seconds after launch, accelerating the missile to cruising speed, at which point the missile's main engine starts and they are jettisoned. The main engine, a Turbomeca TR281 Arbizon turbojet rated at 836 lb. (380 kg.) of thrust, is fed through small air inlets in the wing roots. Both versions are armed with a 463-lb. (210-kg.) high-explosive warhead.

The missile is controlled by INERTIAL GUIDANCE, with active-radar terminal homing. Targeting data from the ship's sensors are fed to the missile's guidance computer (the autopilot permits launches up to 200° off the target azimuth); after launch, the missile cruises under inertial guidance at a height of approximately 820 ft. (250 m.) until, at a predetermined distance from the target, it descends to 65 ft. (20 m.) above the water, maintaining height with a radar altimeter. At 7.5–9 mi. (12.5–15 km.) from the target, the ACTIVE HOMING radar is turned on. It automatically searches for, acquires, and locks onto a target, after scanning 20° left and right of the missile's boresight. A short distance from the target, OTOmat I climbs to 575 ft. (175 m.) and executes a diving attack; in contrast, OTOmat Mk.II flies a surface-skimming profile all the way to the target. In addition, the Mk.II has a Marconi Italiana TG-2 DATA LINK for OVER-THE-HORIZON TARGETING which can exploit midcourse updates from helicopters, aircraft, or other surface vessels.

OTOmat's sealed container-launcher is considerably larger than those used by equivalent missiles (e.g., HARPOON, EXOCET, or GABRIEL) because its wings are fixed (instead of popping out after launch), limiting the number that can be carried on smaller vessels. As a result, both France and Italy are developing folding-wing, tube-launched versions, designated OTOmat Compact and OTOmat Mk.2, respectively. A coastal defense version, consisting of two launch canisters mounted on a trailer, has been sold to the Egyptian army. The air-launched variant was never developed. Italy is reported to be developing a supersonic replacement for OTOmat, tentatively identified as Otomach 2.

Specifications **Length:** 15.8 ft. (4.82 m.). **Diameter:** 18 in. (457 mm.). **Span:** 3.9 ft. (1.18 m.). **Weight, launch:** 1698 lb. (772 kg.). **Speed, max.:** Mach 0.9 (685 mph/1144 kph). **Range, max.:** (I) 37.5 mi. (62 km.); (II) 62 mi. (100 km.). **Operators:** Egy, Iraq, It, Ken, Ku, Lib, Nig, Peru, S Ar, Ven.

OTO-MELARA: Italian armaments manufacturer, best known for its naval guns and the OTOMAT anti-ship missiles. The 76.2-mm. OTO-Melara Compact Gun (known in the U.S. Navy as the Mk.75 76.2-mm./62-caliber gun), which entered service in 1969, has been sold to the navies of more than 40 countries, and is produced under license in Japan, Spain, and the U.S.

A DUAL PURPOSE weapon intended for installation on ships of any size from 60-ton FAST ATTACK CRAFT upward, the Compact Gun is fully automatic, and consists of two major assemblies: the shank (barbette) and the turret, or gunhouse. The shank (installed below deck) consists of a rotating platform holding an 80-round ready-use magazine, and an ammunition hoist to the turret. The latter houses the 76.2-mm. 62-caliber gun and an automatic loader, which in the standard version has a rate of fire of 85 rds./min. at a muzzle velocity of 925 m./sec. (3035 ft./sec.). It can fire a variety of HIGH EXPLOSIVE and ARMOR PIERCING ammunition equipped with point-detonating or radar proximity (VT) fuzes. The barrel is fitted with a muzzle brake to reduce recoil, and a fume extractor to prevent smoke from being sucked into the turret. Expended shell cases are ejected from the turret through a chute mounted under the base of the gun barrel. The gun has elevation limits of $-15°$ and $+85°$, and has an effective range of 8000 m. (4.8 mi.) against surface targets and up to 5000 mi. (3 mi.) against aircraft. The gun and its loader are both covered by a fiberglass gun house sealed against NBC contamination. The entire system weighs only 7.35 tons.

Italy, Denmark, and Singapore have acquired the Super Rapid Compact Gun, with a cyclic rate of 120 rds./min., for increased effectiveness against aircraft and missiles. The rate of fire is selected from the fire control station in the ship's COMBAT INFORMATION CENTER (CIC). A local fire control system with a stabilized line of sight is also available.

Operators (76.2-mm. Compact Gun): Den, FRG, It, Jap, Iran, Isr, Nor, Sing, Sp, US, 20 others.

OUTER SPACE TREATY: The Treaty on the Exploration and Use of Outer Space, signed by the United States, Great Britain, and the Soviet Union in January 1967. Most other countries, except France and China, acceded to the treaty shortly thereafter. The terms of the treaty prohibit the deployment of weapons on both natural and artificial celestial objects. Any satellite containing a weapon would therefore violate the treaty, as would any missile warheads that make a full orbit before returning to earth. Thus ORBITAL BOMBARDMENT SYSTEMS and coorbital anti-satellite (ASAT) weapons (such as the Soviet Union's current ASAT system) are illegal, but a FRACTIONAL ORBIT BOMBARDMENT SYSTEM (FOBS), which would not fly a complete orbit, is allowed. The treaty does not limit other military satellites, including those used for communications, surveillance, reconnaissance, or navigation. See also SATELLITES, MILITARY; SPACE, MILITARY USES OF.

OVER-THE-HORIZON BACKSCATTER RADAR: A type of OVER-THE-HORIZON RADAR whose transmissions follow the curvature of the earth at low altitude (the ground wave effect) for ranges of up to 1800 mi. (3000 km.). Signals that hit airborne objects are reflected back along the same path to a receiver located near the transmitter, thus revealing the position of the detected object. The United States is building a chain of OTH-B radars along its coasts to provide long-range early warning against low-flying bombers and cruise missiles. See also NORAD.

OVER-THE-HORIZON RADAR: A specialized type of bistatic surveillance RADAR, whose signals follow the curvature of the earth for distances well beyond line of sight (exploiting the ground wave effect), and then bounce off the ionosphere in a series of sawtooth waves until they reach a receiver located on the far side of the earth. Any object hit by the signals would interrupt the transmission, thereby revealing its presence. OTH radars are useful only for very general attack warnings, which would require confirmation by other means. See also OVER-THE-HORIZON BACKSCATTER RADAR.

OVER-THE-HORIZON TARGETING (OTH-T): TARGET ACQUISITION beyond the RADAR horizon of the firing platform, achieved by using information from other sensors, mostly mounted on other platforms. OTH-T is a critical requirement in naval ANTI-SURFACE WARFARE (ASUW). Many ANTI-SHIP MISSILES have ranges of several hundred miles but cannot simply be launched in the general direction of moving targets, because of the "dead time" problem: by the time the missile reaches the last known target position, the target may have moved beyond the range of the missile's own terminal homing system. Many missiles can execute preprogrammed search patterns, but this method reduces range and is very much a shot in the dark. Since the number of missiles aboard a ship is quite limited, commanders are loathe to fire unless they have accurate real-time knowledge of the enemy's position. Thus most navies rely on HELICOPTERS, fixed-wing aircraft, REMOTELY PILOTED VEHICLES (RPVs), SUBMARINES, or surface ships to maintain direct contact with enemy forces, relay firing data to launch platforms, and pass midcourse guidance updates to the missile during its flight, via DATA LINKS. Airborne platforms are normally preferred, because the RADAR HORIZON increases with height (which also increases the standoff distance from the enemy).

OVERWATCH: A variant of FIRE-AND-MOVEMENT tactics for armored/mechanized forces, developed by the U.S. Army in the early 1970s: two or more forces alternate to move forward and take up positions to observe ("overwatch") the movement, providing suppressive fire against enemy forces if they are spotted or revealed by their fire on the advancing element.

"Traveling overwatch" is prescribed when enemy contact is possible but maximum speed is desired. The force is divided into a lead and a trail element; the lead element advances continuously along the most concealed and covered route available, while the trail element follows, maintaining visual contact with the lead element, close enough to provide suppressive fire, but beyond the range of most enemy fires directed at the lead element.

"Bounding overwatch" is prescribed when enemy contact is expected. The force is divided into two elements which move forward alternately by bounds, with the second or trailing element always in position to overwatch the lead element. This is the slowest, but also the most secure method of movement, because suppressive fire can be provided immediately. The first element advances to a point from which it can provide fire support to the advance of the second element. On command, the second element advances until abreast of the first element, at which point it halts and deploys into overwatch positions to cover the resumed advance of the first element. The process is repeated until the objective is reached or contact is made with the enemy.

In a variation of bounding overwatch, the first element advances until it reaches a point from which it can overwatch the advance of the second element, which, on command, moves beyond the first element until it reaches a position from which it can overwatch the advance of the first element. The two elements alternately advance and overwatch until the objective is reached or contact is made with the enemy.

P

P-2: U.S. maritime patrol aircraft. See NEP-TUNE.

P-3: U.S. maritime patrol aircraft. See ORION.

PACAF: Pacific Air Forces, a major overseas command of the U.S. Air Force. Headquartered in Hawaii, PACAF is divided into three numbered air forces: 5th (Japan), 7th (South Korea), and 13th (the Phillippines).

PACIFISM: See WAR.

PACIFIC SECURITY TREATY: See ANZUS.

PAH: *Panzerabwehr Hubschrauber*, or anti-tank helicopter; West German army designation for the MBB BO-105P (PAH-1). PAHs are organized into three anti-tank REGIMENTS *(Panzerabwehr Regiment)*, each with two 27-helicopter SQUADRONS. One regiment is attached to each German army CORPS, as part of its ANTI-TANK reserve. A separate PAH squadron is attached to Army Aviation Regiment *(Heeresflieger Regiment)*-6, a fast-reaction unit assigned to the NATO reinforcement of Norway. The PAH-1 will be replaced in the 1990s by the PAH-2, a joint Franco-German development by MBB and Aerospatiale.

PAL: See PERMISSIVE ACTION LINK.

PAPA: NATO code name for a Soviet nuclear-powered (anti-ship) guided-missile SUB-MARINE (SSGN; Soviet designation *Podvodnaya lodka atomnaya raketnaya krylataya*, PLARK) completed in 1970. The *Papa* is viewed by Western intelligence as an interim successor to the CHARLIE class, intended to test design features and technology later incorporated into the larger, even more formidable OSCAR-class SSGNs. As compared to the Charlies, the *Papa* is displaces 2000 tons more, and is 13 kt. faster and considerably quieter. The *Papa* has apparently suffered from numerous technical problems which have required two lengthy overhauls, and has seen little operational service. The *Papa* has a modified teardrop hull-form in a double hull. The pressure hull may be fabricated of titanium alloy for exceptional depth capability. The outer casing is covered with CLUS-TER GUARD anechoic tiles to reduce radiated noise and impede active sonar. A rather short, squat sail (conning tower) is located amidships. Control surfaces are of standard Soviet pattern, with retractable bow diving planes, cruciform rudders, and stern planes ahead of the propeller.

The primary armament consists of 10 SS-N-9 STARBRIGHT anti-ship missiles carried in tubes mounted in pairs between the pressure hull and outer casing, forward of the sail. With a fixed elevation of 40°, these tubes are covered by sliding hatches in the outer hull, and may be fired while submerged. Starbright has a range of roughly 70 mi. (117 km.), and can carry either 1102 lb. (500 kg.) of high explosive, or a 200-KT (estimated) nuclear warhead. Secondary armament consists of 6 bow-mounted 21-in. (533-mm.) tubes, for a variety of conventional or nuclear acoustical homing TORPEDOES or the SS-N-15 STARFISH anti-submarine missile. The basic load is believed to be 12 weapons, which can be replaced by mines on a 2-for-1 basis. See also SUBMARINES, SOVIET UNION.

Specifications Length: 375.6 ft. (114.5 m.). Beam: 37.75 ft. (11.5 m.). **Displacement:** 6100 tons surfaced/7000 tons submerged. **Powerplant:** twin-shaft nuclear: 1 pressurized-water reactor, 2 sets of geared turbines, 60–75,000 shp. **Speed:** 20 kt. surfaced/39 kt. submerged. **Max. operating depth:** 1315 ft. (400 m.). **Collapse depth:** 2000 ft. (600 m.). **Crew:** 110. **Sensors:** 1 bow-mounted low-frequency active/passive sonar, one medium-frequency fire control sonar, 1 retractable "Snoop Tray" surface-search radar, "Brick Spit" and "Brick Pulp" electronic signal monitoring arrays, 1 "Park Loop" radio direction finder (D/F), 2 periscopes.

PAR: See PHASED ARRAY RADAR.

PARACHUTE REGIMENT, ROYAL: Administrative formation of the British army. See AIRBORNE FORCES, BRITAIN, for details.

PARTIAL TEST BAN TREATY: See TEST BAN TREATY.

PASSIVE HOMING: A form of weapon guidance which exploits emissions generated by the target itself (e.g., heat, sound, radio frequency, etc.). The most common are INFRARED HOMING (IR) technique, used mainly in air-to-air and surface-to-air missiles; acoustical homing, used in TORPEDOES; ELECTRO-OPTICAL (EO) or television guidance, used in air-to-surface missiles; IMAGING INFRARED (IIR), similar to EO, but which detects IR emissions rather than visible light; and ANTI-RADIATION MISSILES guidance, which homes on radar and radio emissions.

PATRIOT: The Raytheon MIM-104 land-based surface-to-air missile (SAM). Patriot originated in 1965 as the SAM-D, replacement for both the high-altitude NIKE HERCULES and the medium/low-altitude HAWK. Flight tests began in 1970, but for a variety of reasons, including technical difficulties and cost escalation, production was not authorized until 1980, with the first operational battalion fielded by the U.S. Army only in 1983. By late 1986, 4 battalions had been organized, out of an intended total of 11 battalions with 5360 missiles. In addition, 2 battalions with 840 missiles will be leased to West Germany for the defense of NATO airbases, while West Germany is purchasing 2 more battalions with 779 missiles. Japan and the Netherlands are also acquiring Patriots to replace their Nike Hercules units. Under the terms of a 1979 NATO agreement, Belgium, Denmark, France, and Greece may also purchase Patriot, although its high cost may prove prohibitive. Israel acquired several Patriot batteries from the U.S. in 1990 to provide an anti-tactical ballistic missile defense. Additional batteries were acquired from Germany in 1991 for defense against Iraqi Scud missile attacks.

The Patriot system is based on fire units, each with an MPQ-53 PHASED ARRAY radar, an MSQ-104 engagement control station, and up to eight M901 quadruple launchers. Three or four fire units are grouped together to form a battalion. The MPQ-53 is a multi-mode radar which performs all surveillance, acquisition, IFF, tracking, and illumination functions required by the missile (the older HAWK requires four separate radars); it is mounted with its power supply and signal processor on a single four-wheel trailer. The MSQ-104 engagement control station, mounted on a 6×6 flatbed truck, houses two weapon system operators, a FIRE CONTROL computer, and two tactical display consols. The computer controls all operations, including radar scheduling, weapon allocation, and target prioritization in several different operating modes, including fully automatic, semiautomatic, and computer-assisted manual.

The M901 launchers are mounted on semi-trailers together with their electrical generators. Their trainable, erectable frame holds four sealed containers, each holding a fully certified, flight-ready missile (i.e., no preflight preparation is needed). Communications by secure microwave DATA LINKS allow the launchers, radar, and control station to be widely dispersed.

The missile has an ogival nose, cylindrical body, and four cruciform tail fins. The nose houses a receiver antenna for a SEMI-ACTIVE RADAR HOMING (SARH) unit, behind which is a 200-lb. (90.9-kg.) high-explosive fragmentation warhead with a radar proximity fuze. Powered by a single-stage Thiokol TX-486-1 solid-fuel rocket, the original SAM-D design had inherent anti-tactical ballistic missile (ATBM) capability, a feature deleted to comply with the ABM TREATY. Ironically, in response to the threat to the NATO rear area by Soviet SS-21 SCARAB and SS-23 SPIDER ballistic missiles, a program was later initiated to modify Patriot for the ATBM role. These modified Patriots were used to intercept Iraqi Scud missiles launched against Israel and Saudi Arabia in the 1991 Persian Gulf conflict.

Patriot relies on an unusual TRACK-VIA-MISSILE (TVM) guidance technique (also employed by the Soviet SA-10 GRUMBLE), in which the target is illuminated by the MPQ-53, and target reflections received by the missile's SARH unit are relayed via data link to the engagement control station on the ground. The data is processed by its fire control computer, which transmits guidance commands back to the missile via the data link. This guidance technique is reported to be very resistant to active jamming and other ELECTRONIC COUNTERMEASURES, as compared to systems that rely on processing by (much smaller) on-board computers.

Specifications Length: 17.5 ft. (5.33 m.). Diameter: 16 in. (406 mm.). Span: 36 in. (914 mm.). Weight, launch: 2200 lb. (1000 kg.). Speed, max.: Mach 3 (2100 mph/3507 kph). Range, max.: 37.3 mi. (62.3 km.). Height envelope: 500–78,750 ft. (152–24,000 m.). Operators: FRG, Jap, Neth, US Army.

PATTON: A series of U.S. MAIN BATTLE TANKS (MBTs), originally designed in the late 1940s and still fielded in large numbers. The original Patton, the M46 of 1948, was a derivative of the World War II–vintage M26 Pershing "heavy" tank, with a 90-mm. gun, a conventional layout, and a five-man crew. It was superseded in 1951 by the M47, hastily developed for the Korean War by the placement of an improved turret on an M46 chassis.

The M48, introduced in 1953, still with a 90-mm. gun, had a new, cast-steel hull and an egg-shaped turret, together with simplified fire controls. The M48 proved to be quite adaptable: diesel engines, 105-mm. guns, and advanced fire controls were provided on later models. The last Patton, the M60, originated as a diesel-powered M48 with a 105-mm. gun and a redesigned, up-armored hull. The basic M60 had a turret similar to the M48's, but the M60A1 of 1962 introduced a new "needle nose" turret with better armor protection. The M60A2, introduced in 1964, had a totally different turret with a short 152-mm. gun capable of firing both high-explosive shells and the SHILLELEIGH anti-tank guided missile. The A2 was not successful, and only 562 were built; all were retired in the early 1980s. The M60A3 is an upgraded A1 with a laser rangefinder, ballistic computer, and passive night sights (many A1s have been modified to A3 standard). The M60A1/A3 remained the U.S. Army's standard MBT until the introduction of the M1 ABRAMS in the late 1970s. Because they were produced in such large numbers, many Pattons will remain in service well into the next century.

PAUK: NATO code name for a class of 30 Soviet anti-submarine warfare (ASW) CORVETTES (Soviet designation *Malyy protivolodochny korabl'*, MPK, or Small Anti-Submarine Ship), the first of which was completed in 1979. Intended as replacements for the Pot corvettes, 24 Pauks now serve in the Soviet navy; they are assigned to the Soviet Baltic and Pacific fleets for inshore ASW, operating in conjunction with shore-based Mi-14 Haze helicopters under the control of a shore-based command center. Several Pacific-fleet Pauks may be operated by the KGB for border patrol. India, the only export customer to date, has acquired 6.

The hull is similar to that of the TARANTUL-class missile corvettes, but has a 6.6-foot (2-m.) extension at the stern housing a dipping sonar. These ships are flush-decked, and have a large, boxy superstructure amidships. Armament consists of a 76.2-mm. DUAL PURPOSE gun on the bow, a 6-barrel 30-mm. ADG6-30 radar-controlled gun for anti-missile defense, a quadruple launcher for SA-N-5 GRAIL short-range surface-to-air missiles, 4 single 21-in. (533-mm.) tubes for ASW homing torpedoes, 2 5-barrel RBU-1000 ASW rocket launchers, and 2 DEPTH CHARGE racks with 12 charges. See also FAST ATTACK CRAFT, SOVIET UNION.

Specifications Length: 190.3 ft. (58 m) Beam: 34.4 ft. (10.5 m) Draft: 8.2 ft. (2.5 m.). Displacement: 480 tons standard/580 tons full load. Powerplant: twin-shaft diesel: 4 5000-hp M517 engines. Fuel: 50 tons. Speed: 34 kt. Range: 2000 mi. at 20 kt. Crew: 40. Sensors: 1 "Plank Shave" air-search radar, 1 "Spin Trough" surface-search/navigation radar, 1 "Bass Tilt" 30-mm. fire control radar, 1 hull-mounted medium-frequency sonar, 2 dipping sonars (1 high frequency, 1 medium frequency). Electronic warfare equipment: several passive warning arrays, 1 "High Pole B" IFF, 2 "Square Head IFF" interrogators, 2 16-barrel chaff launchers.

PAVE LOW: U.S. Air Force code name for a series of helicopters modified for special operations. Pave Low I, II, and III were variants of the Sikorsky S-65 SEA STALLION (HH-53H), equipped with night vision sensors, blind navigation instruments, ELECTRONIC COUNTERMEASURES (ECM), an AERIAL REFUELING probe, and door-mounted 7.62-mm. GAU-2A/B miniguns. Their primary mission was the insertion and extraction of clandestine forces at night, deep inside enemy territory. Pave Low III is no longer in service, having been replaced by the HH-60A Night Hawk variant of the UH-60 BLACKHAWK utility helicopter. The Night Hawk has avionics similar to those of the Pave Low III, plus a DOPPLER navigation radar, a forward-looking infrared (FLIR) sensor, and a more comprehensive ECM suite. The U.S. Army is also developing a special-operations variant of the CH-47 CHINOOK medium transport helicopter.

PAVE PAWS: A network of four large PHASED ARRAY radars sited on the coasts of the United States to provide early warning of submarine-launched ballistic missile (SLBM) attack. Pave Paws radars are located at Otis AFB, Massachusetts; Beale AFB, California; Eldorado AFB, Texas; and Robbins AFB, Georgia. Data from the radars are relayed to NORAD air defense headquarters beneath Cheyenne Mountain, Colorado. See also BALLISTIC MISSILE EARLY WARNING SYSTEM; DEW Line; OVER-THE-HORIZON BACKSCATTER RADAR.

PAVE PENNY: U.S. Air Force code name for the AAS-35 LASER target detection unit, mounted on A-7D CORSAIR and A-10A THUNDERBOLT attack aircraft. The AAS-35 is a small pod which detects laser energy reflected from targets illuminated by ground or airborne LASER DESIGNATORS. The signals are converted by the aircraft's navigation/attack computer into target location coordinates, which are displayed symbolically on the pilot's head-up display (HUD). The pod is mounted below the engine intake on the A-7D, and on a small pylon on the right side of the A-10A.

PAVE TACK: U.S. Air Force code name for the AVQ-26 ELECTRO-OPTICAL target designator. Pave Tack is a streamlined pod with a gimballed turret housing a forward-looking infrared (FLIR) sensor and a LASER DESIGNATOR/rangefinder. Controlled by the WEAPON SYSTEM OPERATOR on two-seat aircraft, the FLIR serves to identify targets at night or in poor visibility, which can then be illuminated with the laser designator for LASER-GUIDED BOMBS and guided missiles. Alternatively, the laser can be used for precision rangefinding in the delivery of free-fall bombs. The turret is fully stabilized, and has an autotrack system to keep it pointed at the target regardless of aircraft or target motion. The Pave Tack pod is carried externally by the F-4E and RF-4C PHANTOM, and in the weapons bay of the F-111F attack aircraft.

PAVEWAY: U.S. Air Force code name for a family of LASER-GUIDED BOMBS (LGBs) designed by Texas Instruments. Project Paveway was initiated during the Vietnam War to develop precision-guided munitions for attacks on hardened point targets such as bridges, tunnels, and command bunkers. The first prototypes were flight-tested in 1965; by 1971 an entire series of weapons was introduced, all based on the Mk.80-series of low-drag, general-purpose (LDGP) bombs. Under the Paveway concept, field modification kits are bolted to the nose and tail of standard LDGPs; each kit includes a LASER seeker, a guidance unit and control canards (to be attached to the nose of the bomb), and a set of enlarged tail fins to enhance lift and thus extend the bomb's glide range. The modifications add roughly 30 lb. (13.63 kg.) to the nominal weight of the bomb.

The initial Paveway I series included the GBU-10, a 2000-lb. (909-kg.) Mk.84 bomb plus a KMU-351 kit; the GBU-16, a 1000-lb. (454-kg.) Mk.83 plus a KMU-421 kit; the GBU-12, a 500-lb. (227-kg.) Mk.82 plus a KMU-388 kit; and the GBU-11, a 3000-lb. (1364-kg.) M118 demolition bomb plus a KMU-370 kit. Paveway Is were used extensively in Vietnam from 1971 to 1973, successfully destroying targets which had survived bombardment by thousands of unguided bombs (e.g., the Than Hoa Bridge). A special kit was also developed for the British 1000-lb. (454-kg.) LDGP bomb, the resulting conversion being designated Mk.13/18.

The Paveway II series, introduced in 1980, featured a simplified, less expensive seeker and guidance unit, plus folding, extended-span tail fins for longer range. Some Paveway II Mk.13/18 were used by the RAF during the 1982 Falklands War. The latest Paveway III series began entering service in 1987 after much delay and cost escalation. They have a number of refinements to improve low altitude performance, including an on-board microprocessor, proportional guidance, and a more sensitive laser seeker.

By 1987, more than 150,000 Paveway kits had been delivered, at an average cost of $2500 for early models and $15,000 for Paveway IIIs. The Paveway series has been widely exported and copied by France, Israel, and the USSR. **Operators:** Aus, Fra, Greece, Isr, Neth, PRC, ROK, S Ar, Tai, Tur, UK, US, USSR.

PBV: See POST-BOOST VEHICLE.

PBX: A high-explosive compound, consisting primarily of TNT with additives which increase its explosive effect and improve stability. See also EXPLOSIVES, MILITARY.

PDMS: See POINT DEFENSE MISSILE SYSTEM.

PEACEKEEPER: Official name of the U.S. MGM-118 land-based intercontinental ballistic missile (ICBM), more commonly known as the MX. Development began in 1974 as a replacement for the MGM-30G MINUTEMAN III. The MX was specifically intended to more accurate, to provide a degree of hard-target COUNTERFORCE capability (in order to compete with the Soviet SS-18 SATAN and

SS-19 STILETTO ICBMs). Research and development were fairly straightforward, but greatly prolonged by nontechnical considerations, including arms control controversies and basing disputes. As a result, flight testing began only in 1983, and the first 50 operational missiles were not deployed until 1986.

A three-stage, solid-fuel missile with more than twice the weight of the Minuteman III, Peacekeeper is roughly comparable in size to the SS-19. The first stage, powered by a four-nozzle Thiokol engine, burns out at 80,000 ft. (24,390 m.); the second stage has an Aerojet engine which carries the missile to 280,000 ft. (85,365 m.); while the final stage has a Hercules engine which inserts the missile into its ballistic trajectory. Peacekeeper carries a payload of ten Mk.21 MULTIPLE INDEPENDENTLY TARGETED REENTRY VEHICLES (MIRVs), each armed with a 300-KT W87 nuclear warhead. The MIRVs are mounted on a Rockwell International RS-34 POST-BOOST VEHICLE (PBV), which has its own storable liquid bipropellant rocket engine, an attitude control system, a PENAID package, and an advanced Northrop INERTIAL GUIDANCE unit with ring laser gyros, which is reported to be extremely accurate. Peacekeeper is the first U.S. ICBM to use the COLD LAUNCH technique (employed by most Soviet ICBMs), whereby a solid propellant cold gas generator ejects the missile from the cannister to a height of 80 to 100 ft. (25–30 m.) before the first-stage rocket engine ignites, thus allowing the launcher to be reloaded within hours of firing.

The Peacekeeper program has been controversial for several reasons, aside from the standard objections of anti-nuclear groups to any modernization of U.S. strategic forces. Some arms control advocates have claimed that its accuracy, heavy payload, and high-yield warheads amount to a destabilizing counterforce capability, even though the number of missiles proposed remains too small to threaten a significant portion of the Soviet missile force. Others have objected to the Peacekeeper's basing, arguing that all missiles in fixed silos are nowadays too vulnerable to attack by their Soviet counterparts to provide a stable deterrent.

At one point (c. 1979) it was planned to keep the missiles moving on a subterranean railroad (the "Racetrack" scheme) with one complete track and 23 alternative firing positions for each missile. This was dropped due to opposition by the affected states. The aim of the "Racetrack" and similar schemes was to combine verifiability (for arms control) with safety from attack, but the attempt to have both visibility and invisibility proved too costly and too intrusive environmentally. The radically different DENSE PACK scheme called for 100 missiles to be deployed in closely spaced hardened silos, so that FRATRICIDE effects would minimize losses in an attack; though ingenious, this solution was deemed unreliable. Eventually, it was decided to deploy 50 missiles in existing Minuteman III silos at Warren AFB, Wyoming, (a process which began in 1986) as an interim measure until an alternative basing mode could be developed. The latest plan, "Rail Garrison," calls for the missiles to be placed on rail cars, which would be kept on military reservations in peacetime, but which could be dispersed throughout the country in emergencies (a similar method is used by the Soviet SS-24 SCALPEL). Development funding for Rail Garrison was included in the FY 89 defense budget. See also ICBMS, UNITED STATES.

Specifications Length: 70.87 ft. (21.6 m.). Diameter: 7.67 ft. (2.33 m.). Weight, launch: 193,000 lb. (87,727 kg.). Range max.: 8700 mi. (14,000 km.). CEP: 100 m.

PEGASUS: A class of six U.S. patrol hydrofoil missile boats (PHMs) commissioned between 1977 and 1983, developed from 1972 as a proposed NATO-standard hydrofoil. The U.S. originally planned to acquire 30 of these vessels, while the West German and Italian navies ordered 8 each. Rapidly escalating costs forced Germany and Italy to withdraw from the program, and the U.S. Navy considered canceling after the launch of the *Pegasus* in 1974, but at the insistence of the U.S. Congress, the navy bought 5 more between 1981 and 1983. All 6 are now based at Key West, Florida, where they conduct tactical evaluations and drug-interdiction patrols. Though very capable, they are expensive to operate, and are disliked by senior navy commanders. who do not understand or appreciate FAST ATTACK CRAFT. Derived from the experimental Boeing Tucumcari design, they have a deep-V hull and retractable, canard-configured fully submerged foils with computer-controlled flaps to maintain stability in high seas.

Armament consists of an OTO-MELARA 76.2-mm. Compact dual-purpose gun on the bow, and two quadruple launch canisters for RGM-84 HARPOON anti-ship missiles on the fantail. There are also provisions for two manually operated 20-mm. cannons on the superstructure. They also have an OE-82 satellite communications terminal, and an SSQ-

87 automatic anti-collision system. See also FAST ATTACK CRAFT, UNITED STATES.

Specifications Length: 145.3 ft. (44.3 m.). Beam: 28.2 ft. (8.6 m.). Draft: 7.5 ft. (2.3 m.). Displacement: 198 tons standard/265 tons full load. Powerplant: CODOG w/waterjets: 2 1630-hp. MTU diesels (hull-borne), one 16,767-shp. General Electric LM2500 gas turbine (foils). Fuel: 50 tons. Speed: 12 kt. hull-borne/50+ kt. (foils). Range: 600 mi. at 40 kt. Crew: 24. Sensors: 1 SPS-63 surface-search radar, 1 Dutch-designed Mk.92 fire control radar (Mk.94 in USS *Pegasus*). Electronic warfare equipment: 1 SLR-20 electronic signal monitoring array, 2 Mk.34 SRBOC chaff launchers.

PENAID: Penetration Aid, a generic term for devices and tactics intended to assist aircraft and missiles to penetrate defense systems. Six main PENAID methods are known:

1. Masking, by means of CHAFF, aerosols (to mask INFRARED emissions), ACTIVE JAMMING, or nuclear BLACKOUT, to hide the delivery vehicle or warhead from defense radars, or at least to delay TARGET ACQUISITION, thereby reducing the reaction time for interception.

2. Saturation, by massed attack (e.g., by MIRVs) and by numerous decoys, warhead-imitating balloons, and other target simulators.

3. Shielding, by physical protection against weapon effects. (See NEUTRON KILL and X-RAY KILL.)

4. Evasion, by the use of MANEUVERING REENTRY VEHICLES in ballistic missiles (or by standoff missiles for bombers).

5. Salvage fuzing, whereby ICBM warheads are detonated when interception appears probable, or, more practically, when the first effects of an interceptor weapon are detected, thereby achieving a partial kill, or at least initiating nuclear blackout.

One envisaged form of evasion against ballistic missile defenses is to launch missiles in DEPRESSED TRAJECTORIES (to go under the defenses); another, the FRACTIONAL ORBIT BOMBARDMENT SYSTEM (FOBS), would send missiles by way of the South Pole, out of view of north-facing radars. FOBS is implicitly banned under the OUTER SPACE TREATY, and would, in any event, present significant operational and technical problems.

All PENAIDS exact a price in lost payload which could otherwise be used for weapons. Further, their performance against a BALLISTIC MISSILE DEFENSE (BMD) system must remain unknown and unreliable, especially if the defenses employ several different types of sensors and have overlapping interceptors. In particular, lightweight decoys and balloons decelerate rapidly during reentry, and are "sifted" out from reentry vehicles, rendering them useless against terminal defenses. Nuclear blackout may be very effective, but only against radars operating at low frequencies. At this time there are no PENAIDS which are both cheap (in terms of payload and development costs) and effective. See also ANTI-BALLISTIC MISSILE; CHEVALINE; STRATEGIC DEFENSE INITIATIVE.

PENGUIN: Norwegian ANTI-SHIP MISSILE developed from 1961 by A/S Kongsberg Vaapenfabrik to meet a Norwegian navy requirement for a light missile compatible with small warships operating in coastal waters. Flight tests began in 1970, with the Norwegian navy taking delivery of the first production version (Penguin Mk.1) in 1971 for deployment on Storm- and Snogg-class FAST ATTACK CRAFT. Development of the Mk.2 began in 1974, with the first deliveries in 1979 to the Norwegian and Swedish navies. The Mk.2 is also produced in coastal defense and air-launched versions. Penguin Mk.3, developed in 1980 specifically for launch from Norwegian Air Force F-16 FALCONS, entered service in 1988. The Mk.2 Mod 7 (U.S. designation AGM-119) is an air-launched version developed specifically for U.S. Navy SH-60B SEAHAWK (LAMPS III) helicopters; it combines the guidance unit and warhead of the Mk.3 with the engine of the Mk.2. An initial order for 272 missiles was placed in 1986.

All versions have four cruciform canards at the nose and swept cruciform wings at midbody; air-launched versions have folding wings for compact carriage. The missile is divided into a nose section housing an INFRARED (IR) seeker, a guidance electronics section, a midmounted 264-lb. (80.5-kg.) warhead section, and a rear-mounted rocket motor (dual-impulse in the Mks.1/2, single-impulse in the Mk.3).

Ship-launched Penguins are carried in simple container/launchers which weigh 1433 lb. (651 kg.) ready to fire. All Penguins employ a combination of INERTIAL GUIDANCE and terminal INFRARED HOMING: the missile is aimed towards the target on the basis of fire control data supplied by the launch platform and is kept on course by inertial guidance; when the missile reaches the estimated target position, it switches from inertial to IR guidance, homing passively until impact. Surface-skimming height is maintained by a radar altimeter in the Mks. 1/2; the Mk.3 has a pulsed

LASER altimeter, thereby eliminating the missile's most detectable electromagnetic emission.

Specifications **Length:** (I/II) 9.7 ft. (2.95 m.); (III) 10.4 ft. (3.17 m.). **Diameter:** 11 in. (280 mm.). **Span:** (I/II) 4.6 ft. (1.4 m.); (III) 3.28 ft. (1.0 m.). **Weight, launch:** (I/II) 750 lb. (341 kg.); (III) 794 lb. (361 kg.). **Speed:** (I/II) Mach 0.8 (575 mph/ 960 kph); (III) Mach 0.9 (650 mph/1085 kph). **Range, max.:** (I) 12.4 mi. (20.7 km.); (II) 16.9 mi. (28.2 km.); (III) 37.3 mi. (62.3 km.). **Operators:** Gre, Swe, Tur, US.

PERMISSIVE ACTION LINK (PAL): U.S. electronic safety device intended to prevent the accidental or unauthorized detonation of NUCLEAR WEAPONS. PALs are installed on all U.S. tactical nuclear weapons, aircraft bombs, and ICBMs (but not on SLBMS or naval CRUISE MISSILES), and have been incorporated into many British and French nuclear weapons as well. It is not known whether equivalent devices control Soviet and Chinese weapons.

A PAL is a code-controlled switch in the arming circuit of the weapon, which can be opened only by entering a 4- to 12-digit code. Most newer PALs will permanently disable the arming circuit if the wrong code is entered repeatedly, or if any attempt is made to tamper with the device. Knowledge of the arming codes is limited to the National Command Authorities (the president and his delegates), and is transmitted to the launch points only when firing permission is given. The U.S. president is accompanied at all times by an officer who carries the PAL codes in a special briefcase called the "Football." See also POSITIVE CONTROL.

PERMIT: A class of 14 U.S. nuclear-powered attack SUBMARINES (SSNs) commissioned between 1962 and 1968; originally the Thresher class, it was redesignated after the lead boat was lost in a diving accident in 1963. Numerous safety modifications were incorporated into the subsequent 13 SSNs under Project Subsafe. Designed as successors to the revolutionary SKIPJACK class, the Permits have better sensors, fire controls, and acoustical silencing, and also incorporated many features pioneered in the experimental hunter-killer submarine Tullibee, including a large sonar which takes up the entire bow, relegating the torpedo tubes to amidships. The Permits were built in three separate subclasses: 9 of the basic design, USS *Jack* (SSN-605), and the last 4 boats, *Flasher*, *Greenling*, and *Gato* (SSNs 613–615); the differences among them are due to the installation of slightly different sonar arrays, and the incorpora-

tion of further Subsafe modifications in the last 3 boats, including larger ballast tanks and high-pressure air flasks.

All Permits have a slightly modified version of the teardrop hullform used in the Skipjacks, with a round bow, cylindrical midsection, and tapered stern; they have a single-hull configuration with ballast tanks concentrated fore and aft, outside the pressure hull. The Permits were the first U.S. submarines fabricated from high-yield HY-80 steel with a tensile strength 80,000 psi. A rather small, rectangular sail (conning tower) is located well forward; *Flasher*, *Greenling*, and *Gato* have taller sails, testing a design for the follow-on STURGEON class. Control surfaces are arranged in accordance with standard U.S. practice: "fairwater" diving planes on the sail, with the cruciform rudders and stern planes just ahead of the propeller.

Because the Permits have the same powerplant as the Skipjack class but displace some 1000–1100 tons more, submerged speed has been reduced by 4–5 kt. *Jack* (SSN-605) has an experimental powerplant, with two counter-rotating turbines mounted on a single shaft. Intended to reduce machinery noise, this arrangement in fact offered no appreciable benefits over the standard powerplant, and was not repeated. In all Permits, machinery is mounted on resilient sound-isolation rafts, for a considerable reduction in radiated noise over earlier classes.

The Permits are armed with 4 21-in. (533-mm.) tubes amidships for MK.48 wire-guided homing torpedoes, SUBROC nuclear-armed anti-submarine missile (to be replaced in the near future by the SEA LANCE ASW-SOW), UGM-84 HARPOON anti-ship missiles, and TOMAHAWK long-range cruise missiles. The basic load usually comprises 11 Mk.48s, 4 Harpoons, and 8 Tomahawks. MINES can replace other weapons on a 2-for-1 basis. Sensors are optimized for long-range passive detection and tracking, and linked to a Mk.117 digital FIRE CONTROL system, compatible with torpedoes, Harpoon, and Tomahawk. Communications include a WSC-3 satellite terminal and a VLF trailing wire antenna. See also SUBMARINES, UNITED STATES.

Specifications **Length:** 278.5 ft. (84.9 m.); *(Jack)* 297.4 ft. (90.67 m.); (613–615) 292.25 ft. (89.1 m.). **Beam:** 31.67 ft. (9.65 m.). **Displacement:** 3750 tons surfaced/4311 tons submerged; *(Jack)* 3800 tons surfaced/4470 tons submerged; (613– 615) 3800 tons surfaced/4642 tons submerged. **Powerplant:** single-shaft nuclear: 1 S5W pressurized-water reactor, 2 sets of geared turbines, 15,-

000 shp. **Speed:** 18 kt. surfaced/26–27 kt. submerged. **Max. operating depth:** 1315 ft. (400 m.). **Collapse depth:** 1970 ft. (600 m.). **Crew:** 134–141. **Sensors:** (early) BQQ-2 sonar suite (1 bow-mounted BQS-6 passive low-frequency spherical array, 1 BQS-7 low-frequency active array, 1 BQR-7 flank-mounted conformal array/(later) BQQ-5 sonar system (1 BQS-11, -12, or -13 passive spherical array, 1 BQS-7, 1 BQR-15 clip-on passive towed array), 1 BQS-14 underice sonar, 1 WLR-1 radar warning receiver, 1 BPS-15 surface-search radar, 2 periscopes.

PERRY: A class of 58 U.S. guided-missile FRIGATES (FFGs) commissioned between 1977 and 1988, of which 51 are in service with the U.S. Navy, 4 with the Australian navy, and three with the Spanish navy. In addition, Spain has ordered a fourth ship and plans to build a fifth, while Australia plans to build a fifth ship in the 1990s. Thus total production may eventually reach 61 ships, making the Perrys the most numerous class of warships built in the West since World War II.

The Perrys originated in the early 1970s as the Patrol Frigate (PF), brainchild of then–Chief of Naval Operations (CNO) Adm. Elmo Zumwalt. Concerned about the "block obsolescence" of the many World War II destroyers then reaching the end of their operational lives, Zumwalt advocated a "high-low mix" of large, multi-mission destroyers (the SPRUANCE class), for carrier escort, with a larger number of cheaper, less capable PFs for convoy escort. The initial PF design, which emphasized ease of production, low operating costs, and low manning requirements, was criticized for its lack of firepower; as a result the design was enlarged in 1975 to carry more armament, and designation as an FFG.

Flush-decked, with a graceful shear line and raised bow bulwarks, the Perrys have a large, boxy superstructure amidships and fin stabilizers to improve seakeeping. Aluminum was used extensively in the superstructure to reduce topside weight, but the Perrys do have some splinter protection (in the form of aluminum and Kelvar armor) for critical control and machinery spaces and radar waveguides. All sensors are linked to a Mk.13 weapons-direction system, and a Link 11 NTDS data link allows these ships to use fire control data from other NTDS-equipped units in a battle group. In the event of main engine failure, the Perrys have two retractable 325-hp. "get home" motors which can propel the ships at up to 10 kt. They are also equipped with a PRAIRIE MASKER bubble generator

to reduce radiated noise. Beginning with the 27th ship, USS *Underwood* (FFG-37), an 8-foot (2.44-m.) extension was added at the stern to accommodate the SH-60B SEAHAWK (LAMPS III) helicopter. Eighteen Perrys are assigned to the U.S. Naval Reserve Force (NRF) with reduced crews.

The Perrys are intended primarily as ANTI-AIR WARFARE (AAW) escorts, with limited ANTI-SUBMARINE WARFARE (ASW) capabilities. Their main armament is a single-arm Mk.13 Mod 4 missile launcher on the bow, for STANDARD MR surface-to-air missiles (SAMs) and HARPOON anti-ship missiles. The magazine has a total capacity of 40 missiles, the normal load being 36 Standard and 4 Harpoon. Secondary AAW armament consists of an OTO-MELARA 76.2-mm. DUAL PURPOSE gun (U.S. designation Mk.75) amidships, and a PHALANX radar-controlled gun for anti-missile defense on the aft superstructure. Both the 76.2-mm. gun and the Phalanx have limited fields of fire because of interference by the superstructure; the ship must be turned stern-on to engage masked targets. Unlike other U.S. escorts, the Perrys do not carry ASROC anti-submarine missiles, close-in ASW armament being limited to 2 sets of Mk.32 triple tubes amidships for MK.46 homing torpedoes. Their primary ASW weapons, however, are 2 LAMPS helicopters operated from a fantail landing pad and hangar. The first 26 ships have 2 SH-2F SEA SPRITE (LAMPS I) helicopters; later vessels carry the larger SH-60B. The landing pad is equipped with an hydraulic haul-down system (RAST, Recovery Assistance, Securing and Traversing), which allows helicopter operations in very rough seas. Standard equipment from FFG-36 onward, RAST is being retrofitted in earlier vessels.

The Perrys have long been controversial in the U.S. Navy, and there is still considerable debate over their proper role, because they are too large and expensive to be expendable, yet too lightly armed to cope with serious attacks. Further, they do not have sufficient space and weight margins to accommodate much more equipment. During U.S. convoy escort missions in the Persian Gulf in 1987–88, the Perry-class USS *Stark* was struck by an Iraqi air-launched Exocet missile, which it could neither detect nor engage. A second Perry, the *Samuel B. Roberts*, struck an Iranian mine. Both ships however, survived heavy damage (proving more robust than the British SHEFFIELD-class destroyers in the 1982 Falklands War). Regardless of their deficiencies, the Perrys will continue to dominate the U.S. escort force during the foreseeable future. See also FRIGATES, UNITED STATES.

Specifications **Length:** (FFG-7–36) 445 ft. (135.67 m.); (FFG-37–61) 453 ft. (138.8 m.). **Beam:** 45 ft. (13.72 m.). **Draft:** 24.5 ft. (7.47 m.). **Displacement:** (FFG-7–36) 2769 tons standard/ 3605 tons full load; (FFG-37–61) 3010 tons standard/3900 tons full load. **Powerplant:** single-shaft COGAG: 2 20,000-shp. General Electric LM2500 gas turbines. **Fuel:** 587 tons. **Speed:** 29 kt. **Range:** 4500 mi. at 20 kt./5400 mi. at 16 kt. **Crew:** (active) 206; (NRF) 114 plus 76 reservists. **Sensors:** 1 SPS-49 air-search radar, 1 SPS-55 surface-search radar, 1 Mk.92 fire control radar, 1 STIR (Separate Target Illumination Radar), 1 SQS-56 hull-mounted medium-frequency sonar, 1 SQR-19 (late) or SQR-18(V)2 (early) TACTASS passive towed array. **Electronic warfare equipment:** 1 SLQ-32(V)2 electronic countermeasures array, 1 SLQ-25 Nixie torpedo countermeasures system, 2 Mk.36 SRBOC chaff launchers.

PERSHING: U.S. short- and medium-range ballistic missiles (MRBMS). Development of the original MGM-31 Pershing 1 began in 1958 in response to a Army requirement for a mobile, rapid-reaction "battlefield support" missile armed with a tactical nuclear warhead, to replace the large, unwieldy Redstone. Flight tests began in 1960, and the first operational unit was formed in 1962. The first Pershing battalion (with 4 firing batteries, each with 9 launchers) was deployed to Europe in 1964 as part of NATO's Quick Reaction Alert Force; ultimately 3 battalions with a total of 12 batteries and 108 launchers were deployed to West Germany. Pershing 1 was also fielded by the West German air force, with 72 launchers and 137 missiles organized into two GESCHWADERN. As with all NATO-issue nuclear weapons, the nuclear warheads for the German missiles remained under a dual-key arrangement.

In November 1966 the U.S. Army began development of the Pershing 1a, with better launchers, various improvements to reduce reaction time, and improved launch control facilities. The first operational Pershing 1a battery was formed in 1969, and the conversion of both U.S. and German units was completed by 1971.

The Pershing 1a is a two-stage, solid-fuel missile. The first stage is powered by a Thiokol TX-174 rocket engine delivering 26,290 lb. (11,950 kg.) of thrust for 30.8 seconds; the second stage has a TX-175 engine rated at 19,220 lb. (8736 kg.) of thrust for 39 seconds. An INERTIAL GUIDANCE unit

controls the ascent trajectory with three movable delta tail fins on the first stage, and three wedge-shaped fins on the second. The payload is a 650-lb. (295.5-kg.) ablative reentry vehicle housing a W50 nuclear warhead with selectable yields of 60,200, or 400 kT.

The missile is carried on an M575 Transporter/ Erector/Launcher (TEL), a four-wheeled semi-trailer towed by a 5-ton 8 × 8 truck. All launchers in a firing battery can be linked to the battery control center by a "sequential launch adapter"; up to three missiles can be launched without shifting power cables and pneumatic hoses. An automatic reference unit allows the Pershing 1a to be fired from unsurveyed launch sites, but many potential launch sites have been presurveyed throughout Germany.

Alarmed by the introduction of the mobile, highly accurate Soviet SS-20 SABER IRBM, in December 1979 NATO endorsed the deployment of a new missile, the Pershing 2, to replace U.S. Pershing 1as in West Germany. The Pershing 2 was a larger, longer-ranged missile that incorporated much advanced technology intended to improve range and accuracy; it was sufficiently accurate to allow the targeting of Warsaw Pact command bunkers in Poland and the western U.S.S.R. Development proceeded quickly, with flight tests beginning in 1982 and the first operational missiles delivered in 1983. Deployment of 108 missiles to Germany began in December of that year, and was completed by the end of 1986, in spite of much controversy. In late 1987 however, the U.S. agreed to withdraw and destroy all Pershing 2s under the terms of the INF TREATY; in return, the Soviet Union agreed to destroy all of its MRBMs and IRBMs, including all SS-20s.

The first two stages of the Pershing 2 are similar to the 1a, but a new propellant provides greater thrust. The missile retains the same inertial guidance unit and three-fin steering as its predecessor, but has a completely new PRECISION-GUIDED REENTRY VEHICLE (PGRV) armed with a W85 earth-penetrator warhead, with selectable yields of 5 to 50 kT. The PGRV relies on RADAR AREA CORRELATION GUIDANCE (RADAG) to compensate for errors in the ballistic trajectory: a terrain-mapping radar in the nose scans the area below as the missile descends, comparing the resulting radar map with data stored in its guidance computer; small triangular fins at the base of the PGRV alter the trajectory until the radar map and the computer map coincide. This method is extremely accurate. The

Pershing 2 is also mounted on a towed TEL based on the M575, but a new fire control unit in the battery control center has rapid retarget capability; new target data can be loaded onto the missile guidance computer in near-real time. See, more generally, BALLISTIC MISSILE.

Specifications **Length:** (1a) 34.55 ft. (10.5 m.); (2) 34.45 ft. (10.5 m.). **Diameter:** 3.3 ft. (1.0 m.). **Weight, launch:** (1a) 10,925 lb. (4966 kg.); (2) 16,400 lb. (7455 kg.). **Speed:** (1a) Mach 8 (5000 mph/8350 kph); **Range:** (1a) 112.5–450 mi. (188–752 km.); (2) 808 mi. (1300 km.). **CEP:** (1a) 450 m.; (2) 25–45 m.

PERSISTENT AGENT: A chemical weapon of low volatility, which may not dissipate to safe levels for several weeks or even months. Persistent agents, though frequently lethal, are inefficient for simply inflicting casualties: their more useful ability is to render an area impassable to unprotected personnel. Troops forced to wear protective clothing cannot use their weapons freely, move more slowly, and tire much more quickly; persistent agents can thus act as barriers, and can seriously degrade troop effectiveness. Persistent agents include MUSTARD "GAS" and its derivatives, and the nerve agent VX. See, more generally, CHEMICAL WARFARE.

PERTH: A class of three Australian guided-missile DESTROYERS (DDGs), modified versions of the U.S. ADAMS class, commissioned between 1965 and 1967. They differ from the U.S. ships mainly in details of armament and electronics. All three were extensively modified between 1985 and 1987, when they were fitted with U.S.-built SLQ-32(V)2 electronic countermeasure arrays and two Mk.36 SRBOC chaff launchers. See also DESTROYERS, UNITED STATES.

PETYA: NATO code name for a class of 45 Soviet ANTI-SUBMARINE WARFARE (ASW) CORVETTES completed between 1961 and 1969. Designated "Patrol Vessels" (*Storozhevoy korabl'*, SKR) by the Soviet navy, the Petyas were completed in three distinct subclasses, all used for inshore ASW and border patrol. Eighteen Petya Is were completed between 1961 and 1964, followed by 27 Petya IIs completed between 1964 and 1969 with heavier armament. The final subtype, Petya III, comprises 16 ships built specifically for export to India (12), Syria (2), and Vietnam (2). By the late 1980s the Petyas were being phased out of the Soviet navy, and at least 5 Petya IIs have been transferred to Vietnam (3) and Ethiopia (2).

The Petya IIs are 2.3 ft. (700 mm.) longer than the Petya Is; the Petya III is similar to the Petya II. All three subtypes have flush-decked hulls with pronounced shear and a relatively broad transom stern. Armament in the Petya I consists of 2 twin 76.2-mm. DUAL PURPOSE gun mounts (one on the bow, the other amidships); 4 16-barrel RBU-2500 ASW rocket launchers (2 forward, 2 aft); 1 set of quintuple 21-in. (533-mm.) tubes for ASW homing TORPEDOES; 2 24-round DEPTH CHARGE racks on the stern; and rails for up to 30 MINES. The Petya II has 2 RBU-6000 rocket launchers in place of the four RBU-2500s, 2 quintuple torpedo tube mounts, and no depth charge racks. The Petya III has triple torpedo tubes and 4 RBU-2500s.

Specifications **Length:** (I) 268.4 ft. (81.83 m.); (II/III) 270.7 ft. (82.53 m.). **Beam:** 29.9 ft. (9.12 m.). **Draft:** 9.5 ft. (2.9 m.). **Displacement:** (I) 950 tons standard/1150 tons full load; (II/III) 1160 tons full load. **Powerplant:** triple-shaft CODAG: 1 6000-hp. Type 61V3 diesel, 2 15,000-shp. gas turbines. **Speed:** 30 kt. **Range:** 450 mi. at 29 kt./1800 mi. at 16 kt. (diesel only). **Crew:** 92. **Sensors:** 1 "Slim Net" or "Strut Curve" air-search radar, 1 "Don-2" surface-search/navigation radar, 1 "Hawk Screech" fire control radar, 1 hull-mounted high-frequency sonar, 1 high-frequency dipping sonar (I/II only). **Electronic warfare equipment:** 1 "High Pole B" or "Square Net" IFF, 2 "Watch Dog" electronic countermeasure arrays.

PG: Patrol Gunboat, a type of FAST ATTACK CRAFT, usually displacing between 100 and 400 tons, armed with automatic CANNON and DUAL PURPOSE guns up to 76.2 mm. (3 in.) in caliber. PGs are frequently employed for coastal defense and inshore patrol, sometimes by police or coast-guard organizations.

PGM: 1. See *Precision-Guided Munitions*.

2. Patrol Guided-Missile Boat, a type of FAST ATTACK CRAFT displacing between 100 and 500 tons, armed with ANTI-SHIP MISSILES, light ANTI-AIRCRAFT guns, and DUAL PURPOSE guns up to 76.2 mm (3 in.) in caliber. For specific examples, see COMBATTANTE; KOMAR; OSA; RESHEF; SA'AR.

PGRV: See PRECISION-GUIDED REENTRY VEHICLE.

PHALANX: The General Dynamics Mk.15/16 CLOSE-IN WEAPON SYSTEM (CIWS), a shipboard, radar-controlled 20-mm. multi-barreled cannon for short-range (terminal) defense against low-flying aircraft and ANTI-SHIP MISSILES, in service with the U.S. Navy and the navies of Australia, Canada, China, Great Britain, Greece, Israel, Japan, Saudi Arabia, and Thailand.

Development of Phalanx began in the early 1970s, in response to a requirement for an autonomous weapon to defend ships against surface-skimming missiles. Prototype testing began in 1977, and the first operational weapons were installed on the nuclear-powered aircraft carrier EN-TERPRISE in 1980. Under current U.S. plans, 400 Phalanx are being installed on 250 warships. Aircraft carriers and battleships have 4 each, cruisers and large destroyers 2, and frigates 1. Most foreign navies have adopted similar allocations. Some, such as Israel, have installed Phalanx on vessels as small as 450-ton missile boats.

Phalanx is a completely self-contained system consisting of a 6-barrel, 20-mm. M61 VULCAN "Gatling gun" on a power-driven mount, with 360° traverse and elevation limits of $-25°$ to $+80°$, two radars, and autonomous fire controls. The mount can traverse at a rate of 100° per second at elevate at 86° per second, enabling it to track fast-moving targets. With a maximum rate of fire of 3000 rounds per minute, the Vulcan fires depleted uranium (STABALLOY) APDS ammunition with a muzzle velocity of 1097 m./sec. and an effective range of up to 1500 m. The original Mk.15 has 1000 rounds in a drum magazine under the cannon; the Mk.16 is believed to have a 1500-round magazine.

A radome mounted over the gun houses separate search and tracking radars (because of its shape, the Phalanx is popularly known in the fleet as "RtwoDtwo"). An electronics module attached to the back of the mount contains a ballistic computer and radar signal processors. The entire mount weights 13,500 lb. (6136 kg.) ready to fire, and can be bolted to any clear deck space with an adequate field of fire.

Phalanx normally operates in a fully automatic mode with manual override. The search radar scans the weapon's field of fire for incoming targets. If a target is acquired, the electronics module trains the gun onto it and locks on the tracking radar. If multiple targets appear, the electronics unit automatically prioritizes them to engage the most threatening first. Automated spotting corrects the aim of the gun: the tracking radar follows both the target and the path of the 20-mm. projectiles, and the ballistic computer directs aiming corrections until the two tracks converge. It is claimed that an average of only 50 rounds is required to engage and destroy surface-skimming missiles such as HARPOON or EXOCET; this equates to a total engagement time of approximately 1–3 sec. After destroying a target, Phalanx either locks on to the next target prioritized by the ballistic computer, or resumes searching for a new target. See also GOAL-KEEPER; MEROKA.

PHANTOM: The McDonnell-Douglas F-4 multi-role combat aircraft, perhaps the most successful Western fighter design since World War II. A total of 5195 Phantoms were delivered between 1961 and 1981, of which more than 3000 remain in service.

Development began in 1953, in response to a U.S. Navy (USN) requirement for a missile-armed, carrier-based INTERCEPTOR. The first prototype flew in 1958 and proved immediately that with its two powerful engines the aircraft's performance was exceptional. Until then, the inferiority of carrier-based aircraft to their land-based counterparts was generally accepted as a fact of life, but the Phantom outclassed all contemporary fighters, setting no less than ten speed, altitude, and climb records.

The first production variant (F-4B) entered service with the USN in June 1961. It had a large, multi-mode search-and-tracking RADAR, two General Electric J79-GE-8 afterburning turbojets rated at 15,000 lb. (6818 kg.) of thrust each, and carried an armament of four radar-guided AIM-7 SPAR-ROW and four INFRARED-HOMING AIM-9 SIDE-WINDER air-to-air missiles; 651 were delivered between 1961 and 1965.

In 1961–62, the U.S. Air Force (USAF) conducted ground attack trials with the Phantom, and in September 1963 (under civilian pressure) it adopted a minimum change derivative, the F-4C, of which 583 were built, including 40 later transferred to Spain. In 1965 USAF introduced the F-4D, more extensively modified for ground attack. A total of 825 F-4Ds were produced between 1965 and 1968, including 32 sold to Iran and 36 transferred to South Korea.

All three early-model Phantoms saw combat in Vietnam, where they proved to be rugged and reliable jacks-of-all-trades, flying attack, escort, and AIR SUPERIORITY missions with equal facility. Nonetheless, a number of operational shortcomings were revealed, of which the most important were the lack of an internally mounted gun and a propensity to stall and spin without warning during high angle-of-attack (AOA) maneuvers. The gun had been omitted deliberately by the navy, under the assumption that dogfighting would be impossible at high speed, and that all future air combat would consist of engagements with beyond visual range (BVR) missiles. The air war over North

Vietnam exposed this fallacy: the inability of radar to distinguish friend from foe forced U.S. pilots to close with the enemy; moreover, the North Vietnamese flew highly maneuverable, gun-armed MiG-17 FRESCOES and MiG-21 FISHBEDS, and their short-range "knife fighting" tactics made missile shots difficult. To remedy this deficiency, USAF developed the F-4E, with an extended nose housing a 20-mm. M61 VULCAN cannon. The F-4E became the most numerous version of the Phantom; a total of 1402 were delivered between 1967 and 1981, including 127 F-4EJs built under license by Mitsubishi in Japan, 35 transferred to Egypt (later resold to Turkey), 56 sold to Greece, 177 sold to Iran, 204 sold or transferred to Israel, 37 sold to South Korea, and 87 sold to Turkey (not including ex-Egyptian aircraft). West Germany adopted a simplified variant, the F-4F (without provisions for Sparrow missiles), of which 175 were built by a consortium of McDonnell-Douglas and various German companies.

The stall/spin problem proved more intractable, being inherent in the Phantom's aerodynamics. The USN and USAF adopted different solutions, the former relying on pilot training in stall-avoidance techniques, the latter installing leading-edge slats on the F-4E in the early 1970s.

After completing production of the F-4B, the navy introduced the F-4J, with a new radar and uprated engines. A total of 522 were delivered between 1965 and 1971 (of which 15 were sold to Britain in 1986). During the 1970s, the Royal Air Force (RAF) and Fleet Air Arm had acquired modified Phantoms (F-4K and F-4M), powered by Rolls Royce Spey afterburning turbofans but otherwise similar to the F-4J. The survivors of 52 F-4Ks and 118 F-4Ms are now operated by the RAF.

During the early 1970s, the USN initiated the modernization of its Phantoms, under the Conversion in Lieu of Procurement (CILOP) program. A total of 243 F-4Bs with new avionics and various improvements were redesignated F-4Ns; while 302 F-4Js with more extensive modifications, including maneuvering slats and new avionics, were redesignated F-4S. The navy began replacing Phantoms in active squadrons with the F-14 Tomcat from 1973. The last F-4N was retired in the early 1980s; only a few F-4s remain in reserve squadrons. In 1975 USAF began the conversion of 116 F-4Es to F-4G WILD WEASEL defense suppression aircraft, and is otherwise replacing its Phantoms with F-15 EAGLES and F-16 FALCONS.

In addition to the fighter models, there have been three RECONNAISSANCE variants. The RF-4B, developed for the U.S. Marine Corps, is an F-4B with an extended nose, housing forward and oblique cameras; the RF-4C was a similar conversion for USAF (also bought by Spain), while the RF-4E, developed specifically for export, was sold to Germany, Iran, Israel, Japan, and Turkey.

The F-4C/D/S/M/K have large nose radomes housing liquid-cooled Westinghouse radars with 32-in. (813-mm.) antennas (APQ-100 in the F-4C, the partially solid-state APQ-109 in the F-4D, and APR-59 pulse-Doppler radar in the F-4S, K, and M). Some Cs and Ds also have INFRARED tracking sensors in a chin fairing under the radome. The F-4E/F have smaller, solid-state APQ-120 pulse-Doppler radars whose performance is equivalent to that of the APQ-59. The F-4G Wild Weasel has a PQW-120 radar optimized for ground attack. All RF versions have only a small APQ-99 TERRAIN-AVOIDANCE RADAR.

In the E and F, an M61 Vulcan 20-mm. cannon is mounted in a fairing below the nose with its 639-round ammunition drum mounted behind the radar. The gun has been removed from the F-4G and replaced by an APR-38 RADAR HOMING AND WARNING RECEIVER (RHWR) antenna.

The Phantom's two-man crew sits in a tandem cockpit under separate clamshell canopies. In comparison to later U.S. fighters, cockpit visibility is poor, especially to the rear. The pilot is provided with a lead-computing optical gunsight, a (head-down) radar display, and full flight controls. The rear-seater, called a WEAPON SYSTEM OPERATOR (WSO) by USAF and a radar intercept officer (RIO) by the USN, operates the radar and defensive electronics (as well as providing a second set of eyes for visual search). USAF Phantoms have full flight controls in the rear cockpit; navy Phantoms do not. In addition to radar and radar warning receivers, AVIONICS include VHF and UHF radio, a TACAN receiver, an IFF, a DOPPLER navigation system, an INERTIAL NAVIGATION unit, and a navigation/attack computer.

Six or seven fuselage fuel cells behind the cockpit hold a total of 7760–8330 lb. (3527–3786 kg.); integral wing tanks hold an additional 4285 lb. (1947 kg.). USAF Phantoms have an AERIAL REFUELING receptacle on the fuselage spine behind the cockpit, while navy and foreign aircraft have retractable refueling probes on the starboard engine intake. The engines, fed by two massive intakes on either side of the cockpit, are mounted in the rear of the fuselage. The installation of the larger and heavier Spey required extensive revision of the engine intakes and aft fuselage of British Phantoms (in a slow and costly program). In all models the extreme tail houses an arrester hook and braking parachute.

The low-mounted wing has a sharply swept leading edge and straight trailing edge. The outboard sections, which have considerable dihedral and a large dogtooth, fold up for stowage. In the E, G, and S, leading-edge maneuvering slats improve the turn rate, reduce turn radius, and enhance stability at high AOA. Other models have blown leading edge flaps to reduce landing speed. Many F-4Es also have a TISEO (Target Identification System, ELECTRO-OPTICAL) sensor on the port leading edge for visual target identification at long range. The trailing edges have inboard ailerons and outboard blown flaps. The vertical stabilizer, which is broad and rather low, is responsible (in part) for poor handling at high AOA. The all-moving slab tailplane has considerable anhedral; in the F, G, and S, it also has fixed slats to improve effectiveness at high AOA.

Most Phantoms have four recessed wells under the fuselage for radar-guided air-to-air missiles, either the AIM-7 Sparrow or British Skyflash. With special adaptors, the wells can accommodate instead a variety of active jamming pods, strike cameras, or other pod-mounted equipment. The German F-4F does not carry radar-guided missiles, while the RF versions are completely unarmed. All Phantoms have one fuselage and four wing pylons. The fuselage and inboard pylons are rated at 3500 lb. (1590 kg.) each, while the outboard pylons are rated at 2240 lb. (1018 kg.) each. Inboard wing pylons have launch rails for four AIM-9 Sidewinder or other infrared homing missiles. Phantoms have been used to carry almost every known type of ordnance, including free-fall and LASER-GUIDED BOMBS, CLUSTER BOMBS, NAPALM, gun and rocket pods, HOBOS and GBU-15 glide bombs, AGM-12 BULLPUP and AGM-65 MAVERICK air-to-surface missiles, AGM-45 SHRIKE, AGM-78 STANDARD ARM, and AGM-88 HARM anti-radiation missiles, DURANDAL anti-runway bombs, AGM-84 HARPOON and Israeli GABRIEL anti-ship missiles, ALQ-119 jamming pods, and ALE-37 chaff/flare dispensers. The theoretical limit is 16,000 lb. (7272 kg.); the usual payload is roughly half that.

Although it cannot compete with the latest generation of air superiority fighters, the venerable Phantom is still a potent weapon, especially for ground attack. Several users are refitting their Phantoms to maintain their effectiveness into the next century. West German F-4Fs, for example, have been fitted with advanced pulse-Doppler radars, while Japanese F-4EJs now have the APG-66 radar of the F-16 Falcon, a head-up display (HUD), and numerous avionics improvements. The Israelis have the most ambitious rebuild program, which includes new PW-1128 turbofan engines, a new radar, a HUD, and all-new avionics.

Specifications Length: (C/D/S) 58.25 ft. (17.76 m.); (K/M) 57.6 ft. (17.56 m.); (E/F/G/RF) 63 ft. (19.2 m.). Span: 38.4 ft. (11.7 m.). Powerplant: (C/D/RF-4C) 2 J79-GE-15 afterburning turbojets, 15,000 lb. (6818 kg.) of thrust each; (E/F/G/RF-4E) 2 J79-GE-17s, 17,900 lb. (8136 kg.) of thrust each; (K/M) 2 Rolls Royce Spey 202/203 afterburning turbofans, 20,515 lb. (6255 kg.) of thrust each. Fuel: 12,045–12,615 lb. (5475–5734 kg.). Weight, empty: (C/D/S) 28,000 lb. (12,727 kg.); (E/F/RF) 29,000 lb. (13,181 kg.); (G/K/M) 31,000 lb. (14,090 kg.). Weight, max. takeoff: (C/D/K/M/S) 58,000 lb. (26,363 kg.); (E/F/G) 60,-630 lb. (27,560 kg.). Speed, max.: (J79) Mach 1.19 (910 mph/1520 kph) at sea level/Mach 2.27 (1500 mph/2505 kph) at 36,000 ft. (11,000 m.); (Spey) 920 mph (1536 kph) at sea level/1386 mph (2315 kph) at 36,000 ft. (11,000 m.). Initial climb: (J79) 28,000 ft./min. (8536 m./min.); (Spey) 31,000 ft./min. (9451 m./min.). Service ceiling: 60,000 ft. (18,292 m.). Combat radius: (hi-lo-hi) 520 mi. (868 km.). Range, max.: 1750 mi. (2922 km.) w/internal fuel/2600–2800 mi. (4342–4676 km.) w/max. fuel. Operators: FRG, Gre, Isr, Jap, Iran, ROK, Turkey, UK, US.

PHASED ARRAY (RADAR): An advanced type of RADAR mostly used for missile tracking and air defense. In a conventional radar, the electromagnetic radiation which detects, locates, and tracks objects is fed into a single-beam antenna, which must be slewed physically to project the beam in a given direction. A phased array radar has a stationary antenna (the array), which is actually composed of many (several thousand) miniature transmitter/receiver antennas (elements). A computer controls the time phasing of transmission for each element, thereby forming beams which can be scanned electronically, without moving the antenna. Clusters of elements can be phased in and out separately, thereby creating multiple beams. In the more sophisticated types, the elements can employ several different waveforms to perform different functions (e.g., search, tracking, discrimination, etc.), allowing the radar to operate in several different modes simultaneously (see TRACK-WHILE-SCAN). The AEGIS naval air defense system and the PATRIOT land-based surface-to-air missile both have phased array radars.

PHM: Patrol Hydrofoil, Missile (boat), a type of FAST ATTACK CRAFT similar in size and armament to a patrol missile boat (PGM). For specific examples, see MATKA; PEGASUS; SPARVIERO; see, more generally, HYDROFOIL.

PHOENIX: The Hughes AIM-54 long-range air-to-air missile, which arms the the U.S. Navy's F-14 TOMCAT fighter/interceptor. Development began in 1960 to provide the primary armament of the F-111B (the ill-fated naval interceptor version of the Air Force F-111 attack aircraft), and was closely linked to that aircraft's AWG-9 radar fire control system. Flight testing began in 1965, and continued successfully through 1973. When the F-111B was canceled in 1969, the AIM-54 and AWG-9 were both transplanted into the F-14, entering squadron service in 1973. The initial production version, the AIM-54A, was the most powerful air-to-air missile in the world at the time of its introduction; it was also the most expensive, at roughly $1 million each. A total of 2566 AIM-54As were delivered between 1973 and 1980, including 484 sold to Iran (few of these were ever operational). The next version, the AIM-54B, simplified to reduce costs, was not placed in production. The current version, the AIM-54C, has all new digital electronics, a solid-state RADAR, a strap-down INERTIAL GUIDANCE unit, and enhanced ELECTRONIC COUNTER-COUNTERMEASURES (ECCM). Some 2000 have been delivered since 1982, at an average cost of $1.245 million each.

The AIM-54C has a configuration is similar to that of earlier Hughes missiles (e.g., the AIM-4/AIM-26 FALCON), with an ogival nose, a cylindrical body, four long-chord cruciform delta wings, and four rectangular tail fins of equal span. The nose houses an active radar homing unit, behind which are the guidance electronics, which include an inertial platform. A 132-lb. (60-kg.) high-explosive fragmentation warhead with radar proximity fuzes is mounted at midbody. The engine, an Aerojet Mk.60 or Rocketdyne Mk.47 long-endurance solid rocket, is mounted in the tail, and provides performanced unmatched by other air-to-air missiles.

Phoenix has two guidance modes. For (relatively) short-range engagements (out to 25 mi./41.75 km.), the missile is guided by SEMI-ACTIVE RADAR HOMING (SARH) with illumination provided by the AWG-9. At longer ranges, the missile initially employs SARH, then switches to inertial guidance, finally switching to ACTIVE HOMING in the terminal phase. Although it has never been used in combat (except possibly by Iran), Phoenix has repeatedly demonstrated long-range, multiple-target kill capabilities in tests against aircraft and cruise missile–sized targets at all altitudes down to sea level, often in a jamming environment. The missile's main drawbacks (in addition to its cost) are its size and reliance on the AWG-9, which limits its use to the F-14.

Specifications **Length:** 13.15 ft. (4 m.). **Diameter:** 15 in. (380 mm.). **Span:** 36.4 in. (925 mm.). **Weight, launch:** 985 lb. (448 kg.). **Speed, max.:** Mach 4.3 (3000 mph/5010 kph). **Range:** 125+ mi. (208 km.). **Height envelope:** 100–80,000 ft. (30–24,390 m.).

PHOSGENE: Carbonyl chloride (U.S. code name CG), a CHEMICAL WARFARE lung irritant or CHOKING AGENT. Developed in 1916 as a more effective substitute for chlorine gas, phosgene was responsible for more than 80 percent of all gas deaths in World War I. A highly volatile, colorless liquid which has the odor of musty hay or green corn, when inhaled it causes chemical edema, or severe irritation of the lung tissues, causing them to fill with fluid. Symptoms include coughing, retching, frothing at the mouth, and asphyxia, leading eventually to death from respiratory collapse. The effects of phosgene intoxication can be delayed for up to three hours from initial exposure. Phosgene is highly lethal, with an LD-50 of 3200 mg./min/m³ (as compared with 19,000 mg./min/m³ for chlorine). Because of its volatility, it is a nonpersistent agent which dissipates within ten minutes to three hours, depending upon weather and terrain. Diphosgene (trichloromethyl chloroformate, U.S. code DP) is a semi-persistent variant (dissipation time 3–12 hours) whose effects and lethality are similar to phosgene.

Phosgene and diphosgene were extensively stockpiled by all sides in World War II, but both agents have been phased out by the major powers in favor of BLOOD AGENTS (hydrogen cyanide or cyanogen chloride) and NERVE AGENTS (Sarin, Soman, Tabun, and VX). Phosgene and diphosgene are quite easy to produce and store; Iraq used them against Iranian troops and Kurdish villages, and other small powers could use them in future conflicts.

PICKET SCREEN: 1. In ground combat, a line of outposts positioned ahead of the main force, to provide warning and prevent enemy scouts or RECONNAISSANCE forces from determining the location and composition of the main body. See also COVERING FORCE.

2. In naval warfare, a line of escort vessels positioned between a battle group and the expected vectors of enemy forces, to provide early warning of air attack. In carrier battle groups, picket ships can usually control one or more fighter aircraft on COMBAT AIR PATROL to intercept approaching raids. The pickets themselves, however, are exposed and very vulnerable (HMS *Sheffield* was on picket duty when attacked and hit by an EXOCET missile in the 1982 Falklands War); hence they are replaced by AIRBORNE EARLY WARNING aircraft whenever possible.

PIRANHA: A family of wheeled ARMORED FIGHTING VEHICLES developed in the 1960s by the Swiss MOWAG company. These 4×4, 6×6, and 8×8 vehicles have similar hulls and share common automotive components. The first prototypes were completed in 1972, and the first production vehicles were delivered to the Swiss army in 1976. In 1977, Canada became the first export customer, ordering the 6×6 version as the basis for three different vehicles: the Cougar fire support vehicle (with a 76-mm. low-velocity gun); the Grizzly ARMORED PERSONNEL CARRIER (APC); and the Husky armored recovery vehicle. A total of 491 Piranhas were built under license in Canada between 1979 and 1982.

All three versions have similar all-welded steel hulls, differing only in length and weight. They have a well-sloped frontal glacis and side plates, with a square rear end incorporating a folding ramp-door. Maximum armor thickness is only 10 mm. (sufficient only against shell splinters). The hull is divided into a driving/engine compartment up front and a crew/troop compartment in the rear. In most cases, the driver is provided with infrared or passive night-vision driving periscopes, as well as three day periscopes. The troop compartment can accommodate up to 9 men in the 4×4, 13 in the 6×6, and 15 in the 8×8. All vehicles can be armed a variety of weapons. The 4×4 is usually armed with one or two pintle-mounted 7.62-mm. MACHINE GUNS, or a remotely operated 7.62-mm. in a rotating turret, but it can carry weapons as heavy as a 20-mm. CANNON in a one-man turret. The 6×6 generally has a turret-mounted cannon, or guns of sizes up to the 90-mm. Cockerill. The 8×8 can be fitted with the Emerson TOW missile launcher, the Euromissile Mefisto HOT

missile launcher, or low-pressure 105-mm. guns. An NBC filter, air conditioning, and night-vision equipment can also be fitted. All versions are fully amphibious, propelled through water by two rear-mounted screws at 5.7 to 6.5 mph (9.5 to 10.8 kph).

Specifications Length: (4×4) 17.45 ft. (5.32 m.); (6×6) 19.6 ft. (5.97 m.); (8×8) 20.9 ft. (6.37 m.). **Width:** 8.3 ft. (2.53 m.). **Height:** 6.1 ft. (1.86 m.). **Weight, combat:** (4×4) 7.8 tons; (6×6) 10.5 tons; (8×8) 12.3 tons. **Powerplant:** (4×4) 216-hp. General Motors Detroit 6V-53 diesel; $(6 \times 6/8 \times 8)$ 300-hp. 6V-53T diesel. **Speed, road:** 62 mph (100 kph). **Range, max.:** (4×4) 420 mi. (700 km.); (6×6) 360 mi. (600 km.); (8×8) 468 mi. (781 km.). **Operators:** (4×4) Gha, Lib; (6×6) Can, Chi, Gha, Nig, Tai; (8×8) Gha, US Marine Corps (see LAV).

PK: *Pulemet Kalashnikova*, a Soviet 7.62-mm. GENERAL-PURPOSE MACHINE GUN (GPMG), introduced in 1946 and now in service with the Soviet and other Warsaw Pact forces, and with many Third World armies; China produces a copy as the Type 80. Fired from a bipod, a tripod, and vehicular mounts, the PK is a simple, rugged, gas-operated weapon with few moving parts; it fires powerful 7.62- x 54-mm. rimmed cartridges from 100-round belts and has a quick-change barrel to cope with overheating. Variants include the PKS tripod-mounted anti-aircraft gun, the PKT for armored vehicles, the bipod-mounted PKM, and the tripod-mounted PKMS.

Specifications Length OA: 45.67 in. (1.16 m.). **Length, barrel:** 25.91 in. (658 mm.). **Weight, loaded:** 21.91 lb. (9.98 kg.). **Muzzle velocity:** 825 m./sec. (2707 ft./sec.). **Cyclic rate:** 690–720 rds./min. **Effective range:** 800 m. (875 yd).

P_K: Probability of Kill, the statistical probability that a hit from a given weapon can destroy or at least neutralize a given target. The components of P_k include the probability of firing or launching (P_f); the probability of hitting the target (P_h); and the probability of detonating or penetrating the target to inflict lethal damage in the event of a hit (P_d). Thus:

$$P_k = P_f \times P_h \times P_d$$

Many variables affect each component, making P_k difficult to predict in the absence of *extensive* combat data. Most often, P_k levels as stated are only theoretical approximations based upon firing range tests; combat P_ks are in general substantially lower.

PKO: *Protivokosmicheskaya oborona*, or Space Defense Force, a branch of the Soviet PVO

(Air Defense Force) responsible for the destruction or neutralization of enemy military space systems (e.g., surveillance and communication satellites, space-based BALLISTIC MISSILE DEFENSES, etc.). Formed sometime in the mid-1960s, the PKO probably controls Soviet ANTI-SATELLITE (ASAT) weapons and most manned military space vehicles. See also SPACE, MILITARY USES OF.

PLATOON: An army unit subordinate to the company or battalion, comprising a number of squads or sections. Normally the smallest tactical unit with an organizational identity, it is usually commanded by a lieutenant, and varies in size from 24 to 50 men for infantry platoons, 3 to 5 tanks, and up to 4 mortars.

PLESETSK MISSILE TEST CENTER: The principal Soviet military spaceport, also known as the Northern Cosmodrome, located at 62°43′N, 40°18′ E. Generally analogous to the U.S. Air Force launch facility at Vandenburg AFB, California, it is used for the majority of Soviet photo RECONNAISSANCE, meteorological, early warning, and ELINT satellite launches, and as the principal base for the Soviet coorbital ANTI-SATELLITE (ASAT) weapon. See also SPACE, MILITARY USES OF.

PLOTTING: The transformation of sensor data and intelligence reports into a graphic representation of the relative positions of friendly and enemy units for tracking and targeting purposes. When multiple sources are involved, information must be correlated to eliminate (or at least minimize) ambiguities caused by the inherent inaccuracy of the sensors, and by uncertainties regarding their actual (vs. assumed) positions. Originally accomplished manually (on tracing paper or Plexiglas plotting boards), plotting is now performed on computer-driven displays, which can also perform automatic correlation.

POINT DEFENSE MISSILE: A short-range, shipboard surface-to-air missile intended for the self-defense of the launch platform and its immediate surroundings, often used as a supplement or alternative to gun-based CLOSE-IN WEAPON SYSTEMS. For specific examples, see ASPIDE; SA-N-4 GECKO; SA-N-5 GRAIL; SA-N-14 GREMLIN; RAM; SEA CAT; SEA SPARROW; SEA WOLF.

POINT DEFENSE MISSILE SYSTEM: A U.S. Navy short-range, surface-to-air missile system based on the RIM-7 Sea Sparrow, a navalized version of the AIM-7 SPARROW radar-guided air-to-air missile. The earliest version, known as the Basic Point Defense Missile System (BPDMS), was designed hastily in the early 1960s to counter the threat of surface-skimming ANTI-SHIP MISSILES; it was tested in 1965 and placed aboard the nuclear-powered aircraft carrier ENTERPRISE in 1966. By the early 1970s, it had been added to many U.S. warships, including most aircraft carriers, the KNOX-class frigates, and early SPRUANCE-class destroyers. The BPDMS combined a modified 8-cell ASROC pepperbox launcher with a 3-in. 50-caliber gun mount. Its RIM-7E missile was directed by a hand-operated radar illuminator, with targeting information relayed verbally from the COMBAT INFORMATION CENTER (CIC). The missile had a range of 8–12 mi., and an altitude envelope of 50 to 50,000 ft. The system proved less than effective against surface-skimming missiles (its raison d'être), and was also rather vulnerable to ELECTRONIC COUNTERMEASURES (ECM).

In 1968, the United States signed a Memorandum of Agreement with Denmark, Italy, and Norway to develop an Improved PDMS (IPDMS). The Netherlands joined the consortium in 1970, followed by West Germany in 1977. Now known as NATO Sea Sparrow, the system is fully automatic from target acquisition to engagement; it has RIM-7H missiles with folding fins, and a lightweight, compact Mk.29 8-cell launcher. Missile effectiveness is greatly enhanced (especially against low-flying targets) by the Hughes Mk.23 TARGET ACQUISITION and FIRE CONTROL system, which incorporates a pulse-DOPPLER radar and IFF; it can track and prioritize up to 54 targets simultaneously. NATO Sea Sparrow entered production in 1973 and has since replaced the PBDMS throughout the U.S. Navy; it is also been acquired in large numbers by NATO and other U.S. allies. See also ANTI-AIR WARFARE.

POL: Petrol, Oil, and Lubricants. POL comprises between 40 and 60 percent (by weight) of modern army supply requirements. See also INTERDICTION; LOGISTICS.

POLARIS: The UGM-27, the first U.S. submarine-launched ballistic missile (SLBM), no longer in U.S. Navy service, but still aboard the Royal Navy's RESOLUTION-class nuclear-powered ballistic missile submarines (SSBNs). The development of the Polaris system was a triumph of technical ingenuity and innovative management. Despite the need to pioneer several new technologies, including solid-fuel rocket propulsion, lightweight ablative REENTRY VEHICLES (RVs), and miniaturized INERTIAL GUIDANCE units, in combination with lightweight NUCLEAR WARHEADS, high-speed submarine design, and accurate underwater navigation, development proceeded very rapidly from 1957, with flight tests beginning in 1958 and the first operational missiles entering service aboard the submarine *George Washington* (SSBN-598) in November 1960. This called for the concurrent development of all key technologies, even though failure in any one of them could have halted the entire program. A System Program Office (SPO) under Rear Adm. William F. Raborn had absolute control and responsibility (as opposed to the normal multiplicity of management authorities).

Although the Soviet Union was the first power to deploy SLBMs (aboard its makeshift *Zulu, Golf,* and *Hotel* submarines) its missiles were short-ranged, inaccurate, and quite unreliable, and could be launched only from the surface. By 1967 the U.S. had commissioned a force of 31 SSBNs of the George Washington and Ethan Allen classes, each armed with 16 Polaris missiles, whose technical capabilities the Soviet navy could not begin to match until the introduction of its YANKEE class in 1967. So successful was the Polaris program that all subsequent SLBM developed by the Soviet Union, Great Britain, France, and China are essentially refinements of its original concept.

The initial Polaris A-1 (UGM-27A) was considered an interim weapon pending development of a longer-ranged model. The A-1 was a two-stage solid-fuel missile controlled by an MIT/GE/Hughes inertial platform linked to a GE Mk.80 missile fire control system. Both stages had welded steel cases with four-nozzle solid rocket engines fitted with deflector rings ("jetvator") for thrust vector control (TVC) steering. The second stage was fitted with blow-out panels for thrust termination. The A-1 was armed with a 500-kT W47 nuclear warhead.

The launch procedure established for the A-1 has become standard for all subsequent SLBMs. Target coordinates are fed to the missile's guidance unit from the submarine's fire control sys-

tem, while the launch position is determined by the Ship's Inertial Navigation System (SINS), updated, if possible, by a celestial or satellite fix. The submarine then hovers or moves at very low speed at a depth of 60–100 ft. (20–30 m.). Outer doors covering the tops of missile launch tubes are opened. High-pressure steam from the submarine's reactor then ejects the missiles with sufficient force to break the surface, the first stage engine ignites, and the missile then follows a ballistic trajectory to its target.

The A-1 was deployed on the five George Washington–class submarines, but by 1959, Lockheed had begun development of Polaris A-2 (UGM-27B), intended as the definitive missile. Flight tests began in 1962, and the first operational missiles were deployed aboard the *Ethan Allen* in the same year. Eventually, the A-2 equipped all five Ethan Allens, and the first eight of the subsequent LAFAYETTE-class submarines; by 1965 it had also replaced the A-1 aboard the George Washingtons. In 1962, the *Ethan Allen* successfully fired an A-2 with a nuclear warhead into the Pacific near Christmas Island, the one and only live-fire test of a U.S. ballistic missile. The longer and heavier A-2 filled the entire length of the launch tube. To reduce dead weight, the second stage had a wound glassfiber casing and freon-injection TVC, while the first stage had four gimballed nozzles for steering and attitude control. Combined with a high-energy propellant, these changes extended maximum range, allowing SSBNs to stand off further from Soviet coasts.

In 1960, Lockheed began development of the final Polaris A-3 (UGM-27C), which had a full diameter ogival RV (to use all available tube volume), and a new first stage with a glassfiber casing. A new Mk.2 guidance unit weighing less than half as much as the original unit further reduced dead weight and improved accuracy. Originally armed with a single Mk.2 RV, the A-3 was modified in the early 1970s to carry three MULTIPLE REENTRY VEHICLES (MRVs), each armed with a 200-kT W58 warhead, to make more efficient use of throw weight and defeat anticipated Soviet BALLISTIC MISSILE DEFENSES. The A-3 was first deployed aboard the last 23 Lafayettes, and began replacing the A-2 aboard earlier submarines in 1964 (the last A-2 was retired in 1967). The A-3 remained in the U.S. inventory until the retirement of the last Ethan

Allen from the SSBN role in 1981 (their missile tubes were too small to accommodate the later Poseidon missile). Production of the A-3 terminated in 1968.

In 1965, 102 A-3s were sold to Britain to equip the four RESOLUTION-class SSBNs; an additional 31 were acquired later to make up for attrition. Originally armed with three British-designed 200-kT MRVs, these missiles were rearmed during the 1970s with the CHEVALINE countermeasures system (to penetrate putative Soviet ballistic missile defenses). Chevaline combines three 60-kT MRVs on a maneuverable post-boost vehicle, together with a complex PENAID package. In the early 1980s, all British A-3s were refurbished to extend their service lives into the mid-1990s, when they are to be replaced by the TRIDENT I missile on a new class of SSBNs. See also SLBMS, UNITED STATES; SUBMARINES, BRITAIN; SUBMARINES, UNITED STATES.

Specifications Length: (A-1) 28 ft. (8.53 m.); (A-2) 30.75 ft. (9.37 m.); (A-3) 32.3 ft. (9.85 m.). Diameter: 4.5 ft. (1.27 m.). Weight, launch: (A-1) 28,000 lb. (8536 kg.); (A-2) 30,000 lb. (13,6376 kg.); (A-3) 35,000 lb. (15,909 kg.). Range, max.: (A-1) 1380 mi. (2305 km.); (A-2) 1727 mi. (2884 km.); (A-3) 2880 mi. (4810 km.). CEP: (A-1/2) 2000 m.; (A-3) 900–1200 m.

POLICY: Broad, generalized guidelines formulated by political leaders among others, covering international relations and the conduct of military operations. Policy is supposed to be the basis of military STRATEGY.

POLITICAL WARFARE: The manipulation of political forces within the enemy camp by SUBVERSION and other COVERT operations as well as PROPAGANDA. Political warfare may be associated with an ongoing armed conflict, but some forms of political warfare are a normal adjunct of international relations; e.g., a speaking tour by A's spokesman in B-land may improve A's bargaining position in A-B negotiations if B's demands are convincingly presented as extreme, thus stimulating internal opposition.

POLNOCNY: NATO code name for a class of some 64 Soviet medium landing ships (LSMs; Soviet designation Sredniy desantnyy korabl', SDK) completed between 1962 and 1973. An additional 23 were built for the Polish navy between 1964 and 1971, and 20 ships were built for export, for a total production run of 107. Of the 64 Soviet ships, only 37 are still in service; of the remainder, 1 has been scrapped and 23 transferred. One Iraqi vessel was lost in combat in 1980, and one Libyan Polnocny was lost by fire in 1978.

The Polnocnys were built in three distinct subclasses, designated by NATO Types A, B and C, with slight differences in size and layout; Type C is considerably larger than the others. All have flat bottoms and split bow doors that facilitate beaching to unload troops and vehicles. All have a raised forecastle and a large, rectangular superstructure aft; beneath it there is a vehicle deck measuring (A) 120 × 17 ft. (36.38 × 5.18 m.); (B) 150 × 17 ft. (45.73 × 5.18 m.); and (C) 175 × 22 ft. (53.35 × 6.10 m.). Typical loads include (A) 4–5 ARMORED PERSONNEL CARRIERS, LIGHT TANKS, or MAIN BATTLE TANKS; (B) 6–7; and (C) 8–9, in addition to a NAVAL INFANTRY detachment of 100–180 men.

Armament varies. Type A has a twin 14.5- or 30-mm. ANTI-AIRCRAFT gun, 2 SA-N-5 GRAIL short-range surface-to-air missile (SAM) launchers, and, in some cases, 2 18-tube 140-mm. multiple rocket launchers (MRLs). Types B and C have 1 or 2 14.5- or 30-mm. twin mounts, 2 140-mm. MRLs, and 4 Grail or SA-N-14 GREMLIN SAM launchers. Export variants, all based on the Type C, also have helicopter landing pads on the forecastle; many have 122-mm., rather than 140-mm., MRLs. See also AMPHIBIOUS WARFARE; LANDING SHIP.

Specifications Length: (A) 239.4 ft. (73 m.); (B) 242.75 ft. (74 m.); (C) 266.7 ft. (81.3 m.). Beam: (A/B) 28.2 ft. (8.6 m.); (C) 33.1 ft. (10.09 m.). Draft: (A) 6.35 ft. (1.94 m.); (B) 6.6 ft. (2.01 m.); (C) 6.9 ft. (2.1 m.). Displacement: (A) 770 tons full load; (B) 800 tons full load; (C) 700 tons standard/1150 tons full load. Powerplant: twin-shaft diesel, 2 5000-hp. engines. Speed: 18–19 kt. Range: (A/B) 1500 mi. at 14 kt.; (C) 3000 mi. at 14 kt. Crew: 35–40. Sensors: 1 "Spin Trough" surface-search/navigation radar, 1 "Drum Tilt" fire control radar. Operators: Alg, Ang, Cuba, Egy, Eth, Ind, Indo, Iraq, Lib, Pol, Som, S Yem, Syr, USSR, Viet.

POMORNIK: NATO code name for at least two large Soviet HOVERCRAFT landing vessels, the first of which was completed in 1985–86. The Pomorniks are the largest military hovercraft in the world, supplanting the Soviet AIST class. They have low, enclosed, boat-shaped hulls surrounded by an inflatable rubber air cushion skirt. Like the Aists, they have bow and stern doors for roll-on/roll-off loading of vehicles and cargo. Typical payloads include 2 main battle tanks or 4 PT-76 light tanks, plus a Naval Infantry detachment of 220 men, for a total of more than 100 tons. The Pomorniks are armed with 2 ADG6-30 30-mm. "Gatling" radar-controlled guns for anti-missile defense, and 2 quadruple launchers for SA-N-5 GRAIL or SA-N-14 GREMLIN short-range surface-to-air missiles. The Pomorniks have 5 engines; 2 drive 4 lift fans; the other 3, mounted on pylons at the stern, drive ducted 4-blade variable-pitch propellers. Thrust diverters provide steering. See also LCAC, the only U.S. counterpart.

Specifications Length: 183.7 ft. (56 m.). **Beam:** 72.2 ft. (22 m.). **Displacement:** 360 tons. **Powerplant:** 5 12,100-shp. NK-12 gas turbines. **Speed:** 55–60 kt. **Range:** (est.) 500 n.mi. **Crew:** 40. **Sensors:** 1 air-/surface-search radar, 1 "Bass Tilt" fire control radar, 1 modified "Squeeze Box" electro-optical gun sight.

POSEIDON (C-3): Lockheed UGM-73A submarine-launched ballistic missile (SLBM), successor to POLARIS. Development of Poseidon began in 1964, in response to a U.S. Navy requirement for a missile with greater range, payload, and accuracy than Polaris, but still able to fit into the launch tubes of the LAFAYETTE-class nuclear-powered ballistic-missile submarines (SSBNs). Development was quite rapid, with flight tests beginning in 1968 and initial production versions entering service in 1970.

A two-stage, solid-fuel missile, Poseidon is 18 in. (457 mm.) longer, 20 in. (508 mm.) wider, and fully 30,000 lb. (13,636 kg.) heavier than the Polaris A-3. The increase in diameter was made possible by the removal of a fiberglass tube liner needed with Polaris; Poseidon fills the Lafayette missile tubes completely. Both stages have wound glassfiber casings with recessed, single-nozzle engines to reduce overall length; the nozzles are gimballed for steering and attitude control.

Poseidon was the first U.S. ballistic missile equipped with MULTIPLE INDEPENDENTLY TARGETED REENTRY VEHICLES (MIRVs), in the form of 10 Mk.3 MIRVs, each with a 40-kT W68 nuclear warhead, mounted on a POST-BOOST VEHICLE (PBV), euphemistically called the "equipment section" in the navy. The PBV has its own storable liquid bipropellant rocket engine, an attitude control system, and a guidance unit to successively direct the REENTRY VEHICLES (RVs) even against widely separated targets. The PBV also carries a PENAID package to decoy putative Soviet ballistic missile defenses. The missile relies on a navigating INERTIAL GUIDANCE unit which employs General Energy Management Steering (GEMS) techniques; this allows the second stage to burn to exhaustion, eliminating the need for thrust termi-

nation ports as on Polaris. The guidance package is linked to the submarine's Mk.68 missile FIRE CONTROL system and Ships Inertial Navigation System (SINS), which continuously determine the submarine's position; launch parameters for all 16 missiles carried on board are updated continuously. As a result of guidance and RV refinements, the presumed CEP is only half that of the Polaris A-3.

Operational tests in 1973 revealed serious deficiencies, mainly in the guidance system and PBV (it was widely rumored that up to 60 percent of all Poseidons had major defects); the resulting Poseidon Modification Program apparently rectified all problems. Beginning in 1979, Poseidon was replaced by the TRIDENT I on the Franklin subgroup of the Lafayette-class submarines, but remains in service aboard the survivors of the original Lafayettes; however, these are being retired or converted to other duties as new OHIO-class SSBNs are commissioned (in order to comply with SALT II limitations). See also SLBMS, UNITED STATES; SUBMARINES, UNITED STATES.

Specifications Length: 34 ft. (10.36 m.). **Diameter:** 6.16 ft. (1.88 m.). **Weight, launch:** 65,000 lb. (29,545 kg.). **Range:** 3230 mi. (5394 km.). **CEP:** 550 m.

POSITIONAL DEFENSE: A defensive scheme based on fortified lines, or chains of strongpoints, manned by static forces or troops with limited mobility. Such defenses are usually sited on critical terrain or lines of communication, or terrain which dominates them. Positional defenses are, in general, meant to be preclusive: their purpose is to prevent enemy penetrations rather than to weaken or channel the enemy forces. Although several lines may be built one behind the other (as in the trench systems of World War I), positional defenses are still shallow in strategic terms. Because their troops lack mobility, once they're penetrated the line(s) can be rolled up. The vulnerability of positional defenses was demonstrated by the German offensives of 1918, and fully exploited by the BLITZKRIEG operational method (developed specifically to defeat World War I–style positional defenses). Since World War II, positional defenses have been out of favor, except in two specific situations: first, as the outer edge of a DEFENSE-IN-DEPTH (supplemented by mobile operational reserves); and second, when geographic depth for a more fluid defense is lacking.

POSITIVE CONTROL: The U.S. policy which governs the release of NUCLEAR WEAPONS, intended to prevent accidental or unauthorized detonations. The primary tenet of Positive Control is the two-man rule: the active cooperation of at least two authorized individuals is required to launch a nuclear weapon. When the two-man rule is inapplicable (e.g., on one-man attack aircraft), electronic interlocks called PERMISSIVE ACTION LINKS (PALs) are employed instead. In the case of guided tactical nuclear weapons (e.g., the TALOS, TERRIER, and NIKE HERCULES surface-to-air missiles and the Mk.45 ASTOR torpedo), Positive Control requires the use of command guidance with remote detonation to destroy any weapon flying off course which could explode outside the intended target zone.

POST-BOOST VEHICLE: Also called "bus," a substage mounted over the final stage of a BALLISTIC MISSILE armed with MULTIPLE INDEPENDENTLY TARGETED REENTRY VEHICLES (MIRVs). The PBV is equipped with a small rocket engine, an attitude control system, and a guidance unit, to adjust the correlated velocity vector of each REENTRY VEHICLE (RV) before release, so as to direct them against their separate impact points. After the last booster stage of the missile shuts down or burns out, the PBV separates, rotates to the desired attitude, and thrusts towards the release point for the first RV. Once there, PBVs with stop-start engines can shut down their propulsion, rotate to align themselves for 0° angle of attack (AOA) at reentry, and then perform a "soft release" maneuver, whereby clamps or explosive bolts fastening the RV to the PBV are released, and the PBV backs away at a low relative velocity, while the RV continues on its ballistic trajectory. This is the best way of minimizing perturbations which can contribute to inaccuracy at the impact point, and is used in U.S. and other Western missiles with MIRV capability. Soviet PBVs do not as yet have start-stop engines; instead, they have continuous-burning rocket motors, sometimes with dual (high-low) thrust modes. Thus Soviet PBVs must continue thrusting throughout the release sequence. To ensure accuracy, the PBV must be aligned (before RV release) along the Range Insensitive Direction (RID), i.e., an orientation in which minor variations in velocity do not change the point of impact, only the time of flight. While thrusting up the RID, Soviet PBVs soft-release their payload by opening hold-down clamps and accelerating away from the RV.

After releasing the first RV, both types of PBV rotate to the new orientation and thrust towards the release point of the second RV; this process is repeated until all RVs have been released. PBVs can also carry chaff, decoys, and other penetration aids (PENAIDS) for use before, during, and after RV release, to confuse BALLISTIC MISSILE defenses. See also ICBM; IRBM; SLBM.

POW: Prisoner of War, an individual status defined by legal concepts, customs, and specific conventions. POW status is usually claimed for armed force personnel in captivity; it is also claimed by (and sometimes extended to) parties waging REVOLUTIONARY WAR. It is usually denied to guerrillas until they establish safe areas where live enemy soldiers can be held for bargaining purposes. See also GENEVA CONVENTION; LAWS OF WAR.

PRAIRIE MASKER: U.S. noise-reduction device for surface ships, consisting of a perforated sleeve attached to the ship's bottom below the engine room and machinery spaces. The sleeve is fed by air pumps to emit a stream of small bubbles, which form an insulating cushion between the ship and the sea, reducing radiated noise. This serves two purposes: first, it reduces or disguises the ship's acoustical signature, which could be detected by the enemy; second, it reduces the interference of self-generated noise with the ship's own SONAR. See also ANTI-SUBMARINE WARFARE.

PPI: Plan Position Indicator, a form of RADAR display which shows a horizontal-plane depiction of all radar-reflective objects (including ground features) around the radar.

PRECISION-GUIDED MUNITION (PGM): A missile or projectile equipped with some form of guidance. By convention, the term is applied to all weapons with a P_k greater than 50 percent. PGMs include LASER-GUIDED BOMBS, ELECTRO-OPTICAL (EO) or IMAGING INFRARED (IIR) glide bombs, cannon-launched guided projectiles (e.g., COPPERHEAD), and "smart" submunitions (e.g., SADARM, SKEET, etc.) as well as all guided missiles. See also IMPROVED CONVENTIONAL MUNITIONS.

PRECISION-GUIDED REENTRY VEHICLE (PGRV): A ballistic missile REENTRY VEHICLE (RV) equipped with a terminal homing sensor and guidance, to compensate for the accumulated errors of the ballistic trajectory. PGRVs differ from MANEUVERING REENTRY VEHICLES (MaRVs) insofar as the latter would have considerably greater maneuver capabilities to confuse or evade enemy BALLISTIC MISSILE DEFENSES; the former, by contrast, would deviate from their ballistic trajectories only during the final seconds of flight, and only to the extent necessary to hit their target. To date,

the only PGRV to enter service armed the PERSH-
ING II MRBM, now being withdrawn under the 1987
INF TREATY. The Pershing II had a CEP (median
accuracy) of only 45 m., as compared to hundreds
of meters for purely ballistic RVs. See also BALLIS-
TIC MISSILES.

PREEMPTIVE ATTACK: An attack launched
in the belief that an enemy attack has already
entered the executive phase (i.e., that the decision
has already been made), and which actually
reduces or eliminates the effect of the enemy's
imminent attack. Not a single historical example of
this concept can be adduced, as opposed to
pseudo-preemptive and would-be preemptive attacks.

PREEMPTIVE STRIKE: A theoretical concept: a nuclear COUNTERFORCE attack launched in
the belief that an enemy nuclear attack is imminent.

PREEMPTIVE WAR: A war initiated in antic-
ipation of an attack which need not be imminent;
its purpose may be to interrupt a planned or ongoing military buildup or mobilization.

PRINCIPE DE ASTURIAS: Spanish V/STOL
AIRCRAFT CARRIER, ordered in June 1977 to replace
the *Dedalo,* a World War II–vintage U.S. light
carrier (it now appears, however, that the *Dedaldo*
will be retained until the late 1990s, when it is to
be replaced by a second Principe). Based on the
U.S. Navy's abortive Sea Control Ship, the *Principe* has two pairs of gyro-controlled fin stabilizers
to improve seakeeping. The flight deck is 574 ft.
(115 m.) long and 98.5 ft. (30 m.) wide, with an
enclosed 12° "ski-jump" bow. A large island
superstructure on the starboard side aft incorporates intakes and exhausts for the engines. Two
elevators, one dead aft, the other in front of the
island, connect the flight deck to the hangar. A
PRAIRIE MASKER bubble generator is fitted to reduce radiated noise. Defensive armament is limited to four MEROKA radar-controlled, multi-barrel
20-mm cannons for anti-missile defense. The ship
is equipped with an NTDS data link and a SATCOM
satellite communications terminal, allowing it to
serve as the flagship of an ANTI-SUBMARINE WAR-
FARE (ASW) escort group.

The air group includes approximately 20 aircraft: 6–8 AV-8B HARRIER II V/STOL fighters and
12–14 helicopters (a mix of SH-60B SEAHAWK and
SH-3D SEA KINGS for ASW, and AB.212 HUEYS).
One or two Sea Kings may be outfitted for AIR-
BORNE EARLY WARNING with the British Searchwater radar. See also AIRCRAFT CARRIERS, SPAIN.

Specifications Length: 640 ft. (195.12 m.).
Beam: 80 ft. (24.39 m.). **Draft:** 29.83 ft. (9.09 m.).
Displacement: 16,200 tons full load. **Powerplant:** 2
23,300-shp. General Electric LM2500 gas turbines,
1 shaft. **Speed:** 26 kt. **Range:** 6500 mi. at 20 kt.
Crew: 779 plus air group. **Sensors:** 1 SPS-55 surface-search radar, 1 SPS-52 3-dimensional air-
search radar, 1 SPN-35 air-control radar, 4 Meroka
PVS-1 fire control radars. **Electronic warfare
equipment:** 1 Nettunel electronic signal monitoring array, 4 Mk.36 SRBOC chaff launchers, 1 SLQ-
25 Nixie torpedo countermeasures system.

PRO: *Protivoraketnyy Oborona,* or Anti-Missile Defense Force, a branch of the Soviet PVO
formed in the 1960s, which is responsible for the
operation of ANTI-BALLISTIC MISSILE systems and
other BALLISTIC MISSILE DEFENSES.

PROBE AND DROGUE: See AERIAL RE-
FUELING.

PROJECT 124: Iraqi short-range ballistic
missile program. See AL HUSAYN.

PROJECT 395: Iraqi short-range ballistic
missile program. See AL HUSAYN.

PROPAGANDA: The deliberate manipulation of information for military or political purposes. The contents of propaganda range from selective truth to complete fabrications. Three
distinct types are recognized: "white," "gray,"
and "black." In white propaganda, no attempt is
made to conceal the source, as, e.g., in declared
radio broadcasts, acknowledged publications, and
official communications. In gray propaganda
sources are either unacknowledged or kept ambiguous. Examples include statements issued by the
Soviet Novosti press agency, which claims independence but is actually controlled by the KGB
(even after *glasnost*). While the sources are usually
well known in intelligence and political circles,
they may not be known to the target audience. In
black propaganda, the source is hidden or actively
misrepresented, to conceal the actual originator.
In World War II, British propaganda was also
broadcast by radio stations purportedly controlled
by dissident Germans. Since then, the KGB has
routinely forged purported U.S. government documents containing information calculated to embar-
rass the United States, enrage public opinion in
third countries, or confuse U.S. allies. Such documents are usually published in newspapers covertly influenced or actually controlled by Soviet
agents. For its part, during the 1950s the CENTRAL
INTELLIGENCE AGENCY planted stories in foreign
newspapers. Today much Soviet black propaganda

is disseminated through Soviet-controlled front organizations such as the World Peace Council. Propaganda is only one element in a broad range of techniques which the Soviet Union classifies as ACTIVE MEASURES. See also PSYCHOLOGICAL WARFARE.

PROPORTIONAL GUIDANCE: Also proportional navigation, a form of weapon guidance, usually applied in air-to-air and surface-to-air missiles, whereby the missile steers so as to achieve a constant line-of-sight (LOS) angle between itself and a moving target. By maintaining a constant LOS, the missile can, in theory, fly a straight collision course to intercept a nonmaneuvering target. Proportional guidance assumes that the target is flying in a straight line at any given moment; if it changes direction, the missile's course must be adjusted to maintain a constant LOS. Additional adjustments may be needed if the target or the missile changes speed.

PROWLER: The Grumman EA-6B, a specialized ELECTRONIC WARFARE (EW) variant of the A-6 INTRUDER carrier-based attack aircraft. The need for a dedicated EW aircraft was recognized by the U.S. Navy and Marine Corps in the early 1960s. As an interim measure, a number of A-6A Intruders were converted to EA-6As while development proceeded on the definitive EA-6B. The prototype Prowler began flight tests in 1968, and the initial production versions entered service in 1971; it provides the navy with a very powerful jamming platform, matched only by the much later U.S. Air Force EF-111A RAVEN. Current requirements call for the Prowler to remain in production through 1990 to provide a total of 132 aircraft for 12 navy and 18 marine corps squadrons. Because of its complex electronics and small production run, the Prowler is one of the most expensive military aircraft in service. A detachment of three or four Prowlers is assigned to each carrier air wing.

The Prowler has the same wings, rear fuselage, and tail as the Intruder, but the wings have been extended by 5 ft. (1.52 m.) at the tips to accommodate the higher gross weight. The forward fuselage is radically modified to accommodate a four-seat (2×2) cockpit for the pilot and three EW system operators. A large pod added to the tip of the vertical stabilizer houses an ALR-23 INFRARED warning system as well as RADAR HOMING AND WARNING RECEIVERS for the ALQ-99 electronic warfare system. Additional threat warning receivers are mounted on the sides of the stabilizer and on the fuselage. The Prowler has one fuselage and four wing pylons. The former usually carries an external fuel tank, while the latter can carry up to four Cutler-Hammer ALQ-99 ACTIVE JAMMING pods. The ALQ-99 (also used by the EF-111) is the world's first "smart" noise jammer: its digital signal processors are controlled by computers to automatically match jammer emissions to specific threat signals. Each ALQ-99 pod has its own windmill-driven electrical generator, and can operate in two separate wavebands. All known threat emissions are preprogrammed into the ALQ-99; the EW operators monitor system status and allocate jammers to counter the most dangerous threats.

To keep Prowlers in the active inventory until 2010, the aircraft are being modernized under an Improvement Capability (ICAP)-2 program, which provides upgraded receivers, displays, and software to cover the full range of known Soviet surveillance radars, as well as those associated with air defense weapons.

Specifications **Length:** 59.83 ft. (18.24 m.). **Span:** 58 ft. (11.68 m.). **Powerplant:** 2 Pratt and Whitney J52-PW-408 turbojets, 11,200 lb. (5091 kg.) of thrust each. **Weight, empty:** 32,161 lb. (14,618 kg.). **Weight, normal loaded:** 54,461 lb. (24,755 kg.). **Weight, max. takeoff:** 65,000 lb. (29545 kg.). **Speed, max.:** 623 mph (1040 kph). **Speed, cruising:** 481 mph (803 kph). **Range, max.:** 2082 mi. (3477 km.).

PROXIMITY FUZE: A device that detonates a warhead at a preset distance from presumed targets. The most common types are the radar proximity (VT) fuze and the INFRARED proximity fuze. LASER proximity fuzes, now fitted on some anti-aircraft missiles, have laser diodes instead of radar to determine the distance from a reflecting object. Other types include magnetic proximity fuzes (in naval MINES and TORPEDOES), which are activated by fluctuations in the earth's magnetic field caused by the passing of a ship's steel hull; and acoustical proximity fuzes (also in naval mines), activated by the sound of ship machinery.

PSI: Pounds per square inch, a measure of overpressure (pressure over the atmospheric norm), often used as a measure of the blast effect of explosions, including those of NUCLEAR WARHEADS.

PSYCHOLOGICAL WARFARE: All measures designed to influence enemy personnel (including political leaders) to serve the manipulator's purposes. The tools of psychological warfare include the presentation or distortion of images (see PROPAGANDA); the coordination of military and/or diplomatic action in order to create certain images; and the exploitation of existing tensions within the enemy camp in order to affect morale, discipline, or the decision-making context. "Brainwashing" is a journalistic dramatization of ordinary manipulative techniques applied to captive subjects; e.g., POWS or a television audience (no special techniques were employed by the Chinese captors of U.S. POWs in the Korean War, contrary to legend). Psychological warfare is always important in conflict situations, and all military and diplomatic activities have an inherent psychological dimension which can be modulated to maximize benefits. Such modulation is often more effective than propaganda as such.

PSYWAR: See PSYCHOLOGICAL WARFARE.

PT: Patrol Torpedo Boat, a U.S. Navy designation for a motor torpedo boat (MTB). See also FAST ATTACK CRAFT, UNITED STATES.

PT-76: A Soviet amphibious LIGHT TANK, widely used as a RECONNAISSANCE vehicle by Soviet, other Warsaw Pact, and many other forces; China produces an unlicensed copy. Designed in the early 1950s as a replacement for dwindling numbers of obsolete World War II–vintage light tanks, the PT-76 was first observed in 1952.

The all-welded steel hull has a boatlike bow, a sharply sloped frontal glacis, and vertical sides. Maximum armor thickness is only 14 mm, sufficient to defeat small arms and shell splinters. The hull is divided into a crew compartment forward and engine compartment in the rear. The driver, who sits up front on the centerline, is provided with three observation periscopes, one of which can be replaced by an INFRARED (IR) night scope. The remaining crewmen, the commander and gunner, sit in a small turret mounted immediately behind the driver, with the commander on the right of the main gun and the gunner on the left. Neither has a cupola; instead, the entire top of the turret is covered by a large, forward-hinged hatch. The commander's position is equipped with three observation periscopes; the gunner has a simple stadiametric gunsight. Neither has any form of (built-in) night-vision equipment (some users have added optional night-vision devices). Unlike most other Soviet armored vehicles, the PT-76 does not have a built-in NBC filter system. Fully amphibious, the vehicle is propelled in the water by two rear-mounted pump-jets at up to 6 mph (10 kph). Although the PT-76 is used by the Soviet NAVAL INFANTRY and Polish Marines, it is not truly seaworthy and cannot cross heavy surf.

The main armament, a 76.2-mm. D56T (or D56TM) medium-velocity rifled gun, can fire AP, API, HE, HEAT, or HVAP ammunition, with a typical muzzle velocity of 680 m. per sec. The HVAP round can penetrate 58 mm. of steel armor at a range of 1000 m.; the HEAT round can penetrate up to 120 mm. of armor out to the gun's maximum range of about 2500 m., though the lack of a rangefinder makes accurate fire beyond 800 m. almost impossible. The gun has elevation limits of −4.5° to +31°. A total of 40 rounds are stowed in the hull. The Chinese variant (Type 63) has an 85-mm. gun in a dome-shaped turret. The main gun is ineffective against modern tanks. Secondary armament consists of a coaxial SGMT 7.62-mm. machine gun; many vehicles also have a pintle-mounted DSHK 12.7-mm. machine gun on the turret roof.

PT-76s performed well in Vietnam and poorly in the Arab-Israeli wars; the fact the Soviet army has not introduced a better replacement reflects their quasi-sacrificial role as fire-drawing precursors for main battle tanks.

Specifications Length: 22.67 ft. (6.91 m.). Width: 10.26 ft. (3.13 m.). Height: 7.23 ft. (2.2 m.). Weight, combat: 14 tons. Powerplant: 240-hp. V-6 diesel. Fuel: 56 gal. (246 lit.). Speed, road: 27 mph (45 mph). Speed, cross-country: 18 mph (30 kph). Range, max.: 155 mi. (259 km.) on roads/10 mi. (184 km.) cross-country/40 mi. (67 km.) in water.

PUFFS: Passive Fire-Control Feasibility Study, U.S. BQG-4 passive ranging SONAR for submarines. Now obsolete, PUFFS was developed from 1953 and entered service in 1960 on the GUPPY, TANG, and DARTER class submarines.

PUFFS consisted of three 6-ft. (1.83-m.) vertical hydrophone arrays (in distinctive finlike domes) mounted on the upper deck at the bow, stern, and amidships. The bearing of target noises could be measured from each array, and the target position computed by triangulation. PUFFS has been superseded by a combination of flank-mounted CONFORMAL ARRAYS and passive TOWED ARRAY SONARS.

PURSUIT: That final phase of a battle or campaign during which the victor presses forward against a broken or retreating opponent, to prevent enemy forces from reconstituting a coherent defense, demoralize defeated troops, and cut off and encircle them. Pursuit is the culmination of battlefield success, without which victory may yield only transient results. The techniques of pursuit have remained essentially unchanged since antiquity: the enemy is pressed frontally to prevent rallying, while more mobile light forces attempt to turn his flanks, to cut off the line of retreat. Although the necessity of a vigorous pursuit is universally endorsed, military history records many failures, caused by several factors: the battle may have exhausted the victor materially and/or morally (due to the normal postbattle relaxation of effort); even the defeated retain the initiative in choosing their line of retreat (as, e.g., Rommel in North Africa in 1942 and 1943); or they may have a mobility advantage (as, e.g., the U.S. Marines in their 1950 Korean retreat), in which case the victor can follow, but not press his pursuit.

PVO: Abbreviation of *(Voyska) Protivovozdushnoy oborony*, or Air Defense Forces, the separate service of the Soviet armed forces responsible for the operation of all surface-to-air missiles (SAMs), ANTI-AIRCRAFT artillery (AAA), INTERCEPTOR aircraft, and air defense RADARS. The PVO was established in 1948 as the *PVO-Strany*, or National Air Defense Forces, in response to the threat of the manned bombers of the U.S. STRATEGIC AIR COMMAND. It was responsible only for the air defense of the Soviet Union itself; mobile air defense units with the ground forces belonged to a branch of the army, the *PVO-Sukhoputnykh voysk*, or Air Defense of the Ground Forces. The *PVO-Strany* was originally organized into several air defense districts independent of MILITARY DISTRICT commands. In addition, two special air defense regions were established around Leningrad and Baku.

In 1981 the *PVO-Strany* was reorganized, absorbing all army air defense units, and receiving its present designation. All air defense districts except Moscow's were disbanded, and their functions transferred to the military districts, no doubt to reduce the headquarters' overhead and enhance the integration of military forces. The PVO presently controls a force of some 500,000 men, organized into several branches: (1) Fighter Aviation of Air Defense (*Istrebitel'haya aviatsiya-PVO*, IA-PVO); (2) Anti-Aircraft Missile Troops (*Zenitnyye raketnyye voyska*, ZRV; lit. "Zenith Rocket Troops"); (3) Anti-Aircraft Artillery Troops (*Zenitnyye artilleriysklye voyska*, ZAV); (4) Radio-Technical Troops (*Radiotekhnicheskiye voyska*,

RTV); (5) Space Defense Troops (*Protivokosmicheskaya Oborona*, PKO), responsible for ANTI-SATELLITE (ASAT) weapons; (6) Anti-Missile Troops (*Protivoraketnaya oborona*, PRO), responsible for BALLISTIC MISSILE DEFENSES, notably the GALOSH anti-ballistic missiles; and (7) Troop Air Defenses (*Voyskavaya PVO*), which include all SAMs and AAA units assigned to the ground forces (administered by the PVO, but operationally controlled by their assigned ground formations).

PYTHON: An Israeli short-range, INFRARED HOMING air-to-air missile, developed by Rafael in the late 1970s as an incremental upgrade of the SHAFRIR. Like the earlier missile, Python is similar to the U.S. AIM-9 SIDEWINDER, with canard steering and fixed tail fins. Optimized for short-range dogfighting, it is reportedly particularly effective for off-boresight "snapshots." It was used with considerable success against Syrian aircraft during the 1982 Lebanon War.

Considerably heavier than Sidewinder, Python consists of an INFRARED (IR) seeker unit in the nose, behind which are the guidance electronics; a midmounted 24-lb. (10.9-kg.) CONTINUOUS ROD warhead with proximity and contact fuzes; and a Rafael solid rocket motor in the rear. The missile is steered by its four cruciform delta canards. Four sharply swept tail fins ("wings") contribute to the missile's outstanding maneuverability. Four slipstream-driven "rollerons" mounted on the tips of the wings act as stabilizing gyros and roll control devices (a feature copied from Sidewinder).

The IR seeker is reported to have wide-angle optics and exceptional sensitivity, for an ALL-ASPECT engagement capability. It can be operated in boresight, uncaged (autonomous scanning), and radar-slaved modes. Completely compatible with Sidewinder and Shafrir launch equipment, Python has been fitted to Israeli F-15 EAGLES, F-16 FALCONS, F-4 PHANTOMS, and KFIR C.2s, and has also been exported.

Specifications Length: 9.83 ft. (3 m.). **Diameter:** 6.25 in. (159 mm.). **Span:** 33.9 in. (861 mm.). **Weight, launch:** 264 lb. (120 kg.). **Speed:** Mach 2.5 (1625 mph/2714 kph). **Range:** 0.3–9.3 mi. (0.5–15.5 km.).

QUICKSTRIKE: A family of U.S. air-delivered naval MINES. Introduced in the 1980s as replacements for the DESTRUCTORS of the Vietnam War, most Quickstrike mines are Mk.80-series LDGP bombs fitted with TDD-57 acoustic, magnetic, and hydrodynamic proximity fuzes that allow them to be used as bottom-laid mines. Models in service include the Mk.62, based on the Mk.82 500-lb. (227-kg.) bomb; the Mk.63, based on the 1000-lb. (454-kg.) Mk.83 bomb; and the Mk.64, based on the 2000-lb. (909-kg.) Mk.84 bomb. While not as effective as purpose-built mines, they are inexpensive and available quickly from bomb inventories.

There is also a Quickstrike Mk.65, a purpose-built, 2000-lb. mine with a thinner, high-capacity case loaded with some 1500 lb. (682 kg.) of PBX explosives.

R

RAAMS: Remote Anti-Armor Mine System, two U.S. 155-mm. submunitions-dispensing ARTILLERY projectiles. Both versions of RAAMS, the M741 (short active life) and M718 (long active life), contain nine magnetically fuzed M75 anti-tank MINES which can immobilize any MAIN BATTLE TANK now in service. Unless detonated, the mines self-destruct after a predetermined time to allow counterattacks through mine-covered terrain. RAAMS could be used in conjunction with ADAM (an anti-personnel mine-dispensing round) to create hasty barriers, or interdict lines of communication. Typically, 24 RAAMS shells are required to cover an area of 400 × 400 m. A standard U.S. M109A2 howitzer can fire RAAMS shells out to an effective range of 17,400 m. See also ARTILLERY, UNITED STATES; IMPROVED CONVENTIONAL MUNITIONS.

RADAG: See RADAR AREA CORRELATION GUIDANCE.

RADAR: Radio Detection and Ranging. Radars operate by transmitting electromagnetic energy in the radio frequency (RF) portion of the spectrum and by processing that portion of it (the "echo") which is reflected back in order to determine the relative position and range of the reflecting objects. Developed more or less independently in Britain, Germany, and the U.S. during the late 1930s, radar was much used in World War II, to detect and locate aircraft, ships, and lesser targets; to direct weapons; as a navigational aid; in a "secondary" form, IFF, to identify transponder-equipped aircraft; and in simple forms, as proximity (VT) fuzes. Since then, many other applications have been developed. Radar is often the most important single element of complex weapon systems, and accounts for a substantial portion of military budgets.

BASIC OPERATING PRINCIPLES

Radar exploits three basic properties of electromagnetic radiation: (1) that it travels in "waves" at the constant speed of light (186,000 mi. per sec.), so that the time between the transmission of a signal and the return of its echo divided by two and multiplied by the speed of light determines the range from radar to object; (2) that the waves can be transmitted in "beams" of varied shape by appropriate antennas—narrow "pencil beams" can accurately locate and follow ("track") moving objects; and (3) that the waves change in frequency when reflected from objects which are moving relative to the antenna (the "Doppler effect"), so that the speed of those objects can be calculated from the frequency shift.

In most elementary form, a radar consists of an oscillator which generates RF energy of a specific frequency; a waveguide that carries the energy to an antenna, which in turn transmits the waves and collects the return echoes; a receiver which selects and amplifies echoes; and a display that presents echo data in useful form.

Radars have been operated in frequencies ranging from 25 MHz to 70 GHz, but most operate in the microwave region, i.e., in the UHF, SHF, and EHF bands between 300 MHz and 35 GHz. The size of the required antenna for a given beam width increases as the frequency decreases. Thus small antennas (such as those on aircraft or missiles) can achieve narrow-beam precision only with high frequencies. But the higher the frequency, the greater the rate of atmospheric absorption (attenuation) of the signal, which reduces the radar's effective range accordingly. Hence long-range radars (e.g., for air-defense early warning) must have large antennas and low frequencies (typically 600 MHz or less), in order to combine long range with directional precision.

The earliest (and still most common) form of radar is the "pulse" type, in which the oscillator is switched on and off at a very high rate (the pulse repetition frequency, or PRF), to generate short bursts of energy which can be timed from transmission to return in order to determine target range. The antenna is usually rotated mechanically to scan over a given sector, or in a 360° arc; and the target's azimuth is determined by the antenna's bearing at the time of reception.

Continuous wave (CW) or DOPPLER radars, by contrast, can determine azimuth and relative velocity, but not range. The oscillator generates a steady signal at a constant frequency, but echoes from moving targets will be altered in frequency by the Doppler effect; if the echo is at a higher frequency than the signal, the target is approaching; if it returns at a lower frequency, the target is receding. Relative speed is determined by the magnitude of the shift, while azimuth is again determined by antenna position. To determine the range as well, the CW technique is often used in conjunction with frequency modulation (FM). The transmitter sends out a continuous ascending sawtooth wave (i.e., one which starts at a lower frequency, rises at a constant rate to a peak, then falls back immediately to the initial frequency). Range is determined by comparing the frequency of the return echo with that of the signal being transmitted at the moment of reception. Since the sawtooth wave rises at a constant rate, this gives a time base from which range can be derived.

Pulse-Doppler radars are a relatively recent type which combines both pulse and CW techniques. The signal is transmitted in relatively long bursts of CW radiation, so that the timing of the return echoes can provide range, while the Doppler shift within each burst reveals the relative velocity. Pulse-Doppler radars are particularly suitable for aircraft and guided missiles, because Doppler shift measurement can be used to suppress unwanted echoes ("clutter") from the (stationary) ground or water background.

In most radars, the transmitter and receiver are combined in a single antenna, or side-by-side antennas are combined into a ("monostatic") array, but some long-range surveillance radars have widely separated ("bistatic") antennas, to achieve better range resolution by triangulation. It is theoretically possible to use a single transmitter antenna in conjunction with several ("multistatic") receivers, in order to derive very precise range and position data.

PHASED-ARRAY RADARS do not have conventional, mechanically scanned antennas, but rather "arrays" consisting of many (up to several thousand) individual transmitter/receiver elements, which are controlled ("phased") by computer to generate electronically scanned beams. Phased-array radars can usually generate several beams simultaneously, allowing them to follow several different targets while searching for still others (giving a TRACK-WHILE-SCAN capability).

Radar information for automated systems (e.g., missiles) can be processed directly by computers, but for manned systems the data must be transformed into intelligible formats, normally by using cathode ray display screens. (The earliest type of display was the A-scope, which shows range and signal intensity, from which the size or number of targets can be derived; the B-scope displays range and azimuth in Cartesian coordinates; the C-, D-, and E-scopes display range and elevation; the F- and G-scopes of gun-direction radars show azimuth and elevation, again in Cartesian coordinates; H-scopes display azimuth and range as two dots on a single screen, while I-scopes provides the same information in a complex conical display; J-scopes are similar to A-scopes, but show radial deflections rather than Cartesian coordinates; K- and L-scopes are similar, but are used with multiple-antenna systems; M-scopes have a range step to assist interpretation, while N-scopes combine the K and M displays.) The familiar, circular Plan Position Indicator (PPI) displays the range and position of all objects relative to the position of the radar, and is still the most commonly used display.

All early displays presented "raw" data, which only skilled operators could interpret, especially in clutter or against jamming. The most modern radars, by contrast, have digital signal processors to automatically filter incoming signals and display them as easily understood synthetic-video symbols ("icons").

TYPES

Radars are generally classified by function. Surveillance radars with long range and relatively poor resolution are used for the early detection of aircraft and missiles; they also equip AIRBORNE EARLY WARNING (AEW) and AWACS aircraft, whose coverage is not limited by the radar horizon. The similar two-dimensional air-search radars used aboard naval vessels provide target range and bearing, while three-dimensional sets provide height as well. Naval surface-search radars are used for the detection of ships (and low-flying aircraft), and also for navigation.

Acquisition and tracking radars generate narrow beams to lock onto and follow their targets. Height-finding radars are a specialized subtype which scan vertically; they are normally used in conjunction with two-dimensional radars. FIRE CONTROL radars also generate narrow beams, to provide precise range and bearing data for gunlaying.

Several different types of radar are used for missile guidance. The most common are illumination radars, which are directed at targets to obtain echoes which can be tracked by SEMI-ACTIVE RADAR HOMING (SARH) missiles (which have a radar receiver, but not their own transmitter). BEAM RIDING missiles follow the narrow "pencil-beams" which special guidance radars aim at their targets.

Battlefield surveillance radars, used to detect the movement of troops and vehicles at night or in poor visibility, rely on Doppler effects to provide generalized warning, rather than precise target location. COUNTERBATTERY and countermortar radars track incoming artillery and mortar projectiles, and their associated computers extrapolate the trajectories back to the firing battery.

TERRAIN-FOLLOWING and TERRAIN-AVOIDANCE radars, which equip bombers and fighter-bombers optimized for very low-level flight, both detect obstacles in their path; the former are connected to autompilots which allow the aircraft to maintain a steady hight, while the latter provide cues for manual evasion. Airborne ground-mapping radars provide a pictorial representation of the terrain, for target acquisition and navigation. SIDE-LOOKING AERIAL RADAR (SLAR) is a specialized type of high-resolution ground-mapping radar which is used for RECONNAISSANCE and surveillance. Many SLARs employ SYNTHETIC APERTURE RADAR techniques, whereby sequential radar "snapshots" are processed and integrated to create the effect of an antenna several hundred feet across, with proportionally high resolution. The "multi-mode" radars of advanced combat aircraft perform several different functions by varying their frequency and/or waveform.

AIR DEFENSE

Radar-assisted AIR DEFENSE systems generally consist of two parts. An interlocking chain of long-range surveillance radars, usually sited along the defended perimeter to provide early warning, produce reports that are filtered and integrated (manually or automatically) to produce target tracks. These are passed to TARGET-ACQUISITION and tracking radars, which are often directly linked to anti-aircraft guns or surface-to-air missile batteries.

Tracking-radar data is processed by fire control computers, to predict target paths and generate instructions for anti-aircraft weapons. In gun systems, tracking radars can direct servo-controls to point and fire. In surface-to-air missile systems, procedures vary: with both command-guided and beam-riding missiles, acquisition radars pass target data to specialized guidance radars, which then track both missile and target; in command-guided systems, the resulting convergence-vector calculations yield steering commands; in beam-riding missiles, they are used to point the guidance beam. In SARH missile systems, on the other hands, target data are passed to an illumination radar, which locks onto the target and "paints" it with radar beams.

In the radar control of so-called "all-weather" fighter/interceptors, an aircraft may be directed verbally towards its target by a ground controller who obtains his data from surveillance radars (GROUND-CONTROLLED INTERCEPT); or by an airborne controller aboard an AWACS aircraft. In more automated systems, surveillance-radar data may be transmitted by DATA LINKS directly to the aircraft's fire control system, which produces steering cues for the pilot. When the aircraft comes within range, the target is acquired by its on-board radar, which locks onto the target and feeds its information to the fire control computer, which in turn provides steering and firing cues to the pilot, nowadays on a head-up display (HUD); in some cases, firing is initiated automatically by the computer.

NAVIGATION

Most shipboard navigation radars have PPI displays centered on the ship, which show returns from other ships, land features (especially promontories and built-up areas), and some weather conditions (e.g., thunderstorms). Modern displays can have synthetic symbology, often overlaid with chart data (soundings, channels, traffic lanes, etc.) for easy interpretation.

For airborne navigation, radar altimeters provide very accurate height information, while Doppler navigation radars can provide information on wind drift by measuring the frequency shift of a beam directed downward. Ground-mapping radar can identify significant landmarks, especially coastlines, rivers, lakes, and built-up areas. The technique was already used for bombing in World War II as an all-weather substitute for optical bombsights. Bombers can also be tracked by friendly radars and directed to fixed targets by coded radio instructions. When the bomber is over the target, a second bomb-release signal can be transmitted, thereby achieving accuracy even at night or through cloud cover. Current techniques are considerably more complex and often rely on INERTIAL NAVIGATION.

LIMITATIONS OF RADAR

Radar cannot produce very high-resolution images, nor discriminate between colors, and most do not function well against targets very close to the ground or to agitated waters. In addition, most surface-based radars are limited in range to the radar horizon, which is little greater than the visual horizon. OVER-THE-HORIZON RADARS are the exception; their low-frequency beams follow the curvature of the earth, but their broad beams can provide only warning, not tracking. While ordinary radar cannot distinguish between friendly and enemy targets, IFF (Identification Friend or Foe) "secondary" radar transponders send out recognition signals when "interrogated" by friendly radars. IFF, however, can be unreliable in combat, resulting in "own goals."

The greatest shortcoming of radar is its vulnerability to ELECTRONIC COUNTERMEASURES (ECM). Because all radars rely on RF signals, their emissions can be detected by simple receivers. Most warships and aircraft are now equipped with one or more RADAR WARNING RECEIVERS which indicate when they are being "painted" by a hostile radar. RADAR HOMING AND WARNING RECEIVERS, more sophisticated devices, can also pinpoint the location of hostile radar emitters, for attack by ANTI-RADIATION MISSILES that home on the radar signal. Even if not attacked directly, radars can be neutralized by various ECM techniques. These include CHAFF—strips of aluminum foil that cause spurious echoes which can blank out displays—and ACTIVE JAMMING. The latter can take many forms, including BARRAGE JAMMING (the broadcasting of powerful signals to saturate an entire wave band); SPOT JAMMING (broadcasts which saturate a specific frequency); and DECEPTION JAMMING, in which radar signals are subtly altered and rebroadcast to give a false target position.

In response, many radars now incorporate ELECTRONIC COUNTER-COUNTERMEASURES, such as the ability to shift frequency rapidly to evade jamming ("frequency agility") or to filter out chaff and jamming by complex signal processing techniques. But the inherent detectability of radar remains an insurmountable problem; hence more emphasis is now placed on passive (nonemitting) infrared means of detection and tracking. See also ELECTRONIC WARFARE.

RADAR ABSORBANT MATERIAL: Substances which irreversibly convert radio frequency (RF) energy (notably from RADAR transmitters) into heat or other forms of energy, thereby minimizing the strength of any reflected signal (echo). When applied on, or incorporated into, the structure of aircraft, ships, etc., such materials can substantially reduce their RADAR CROSS SECTION (RCS), and hence the maximum range at which they can be detected by a radar of a given frequency and power. Radar absorbant materials are an essential element of STEALTH technology.

RADAR AREA CORRELATION GUIDANCE: A form of missile guidance whereby the target area as "seen" by an on-board ground-mapping radar is compared to digital map data stored in the missile's guidance computer. The computer then generates steering commands until the radar map and the computer map coincide at the target point. This extremely accurate technique is most practical for terminal guidance; it was applied in the PERSHING II medium-range ballistic missile, the most accurate ballistic missile ever built.

RADAR CROSS SECTION (RCS): A standard measurement of the radio-frequency reflectivity of objects, i.e., their radar "visibility." Usually measured in decibels per square m. (dBSM), or alternatively, square m. (m.²), RCS is especially significant in aircraft design. It is determined by measuring or calculating the amount of radar energy reflected from an object in a given aspect, with the result stated as the area of a sphere that would reflect as much. The smaller the RCS of an object, the shorter the maximum range at which it can be detected by a RADAR of given characteristics.

To reduce RCS by several orders of magnitude is the principal objective of STEALTH technology. But the RCS of complex shapes varies considerably with the aspect angle between radar and target; it also varies with the radar's frequency. Thus the broadside RCS of an aircraft can be many times larger than its frontal RCS—especially if the aircraft is boxy, because the best radar reflectors are flat plates perpendicular to incoming signals. Stealth designs reduce RCS through the use of RADAR ABSORBANT MATERIALS, and by minimizing the number of flat-plate surfaces. While a 1970s fighter—such as the F-15 EAGLE—has a frontal RCS of approximately 100 m.², a Stealth fighter (e.g., the Lockheed F-117 NIGHTHAWK) has a frontal RCS of only 0.01 m.² (roughly the same as a large bird); the B-52 STRATOFORTRESS bomber likewise has a frontal RCS of 1000 m.², while the more recent B-1B has an RCS 1 m.², and the B-2—designed for stealth—is estimated to have a frontal RCS of only 0.1 m.²

RADAR HOMING AND WARNING RECEIVER: A more sophisticated form of RADAR WARNING RECEIVER which can indicate the position as well as the range of enemy RADAR transmitters; such receivers are installed on WILD WEASEL defense-suppression aircraft, for target acquisition. See also ELECTRONIC COUNTERMEASURES; ELECTRONIC WARFARE.

RADAR HORIZON: The range beyond which a target cannot be detected by RADAR due to the curvature of the earth. The exact radar horizon of specific radars depends on several factors, including the height of the radar, the height of the target, and the presence of intervening terrain. In general, the radar horizon is defined as: $R_{ho} = 2h_2 + 2h_2$, where R_{ho} is the radar horizon in statute miles, h_1 is the height of the radar antenna in

feet, and h_2 is the height of the target in feet. Thus the higher the radar (or the the target), the longer the radar horizon: hence the desirability of airborne radar surveillance platforms such as AWACS. Usually the radar horizon is slightly longer than the visual horizon, because of the atmospheric refraction of long radar waves; certain atmospheric phenomena, however, create "surface ducts" which trap the signal close to the surface of the earth, extending detection ranges by up to several hundred miles. In addition, OVER-THE-HORIZON RADARS, which operate in the Very Low Frequency (VLF) band, produce signals which follow the curvature of the earth, thus offering very long ranges, albeit with poor resolution.

RADAR WARNING RECEIVER: An electronic device that detects the signals of hostile RADARS, and provides an audio or visual indication of their type and proximity. First developed in World War II for heavy bombers, RWRs are now standard equipment on almost all combat aircraft and warships. The simplest types are tuned to just one frequency band and provide only a rough indication of signal strength (and hence range). More advanced types have broad-band receivers which can detect signals over a wide range of frequencies, and cathode ray tube (CRT) displays to show the type and bearing of the signal, in addition to its range. The very latest types employ digital signal processors, and have built-in electronic libraries of known hostile radars to classify and prioritize many threats simultaneously for evasion, attack, or a jamming response. See also ELECTRONIC COUNTERMEASURES; RADAR HOMING AND WARNING RECEIVER.

RADIO COMMAND GUIDANCE: See COMMAND GUIDANCE.

RADIO-FREQUENCY WEAPON: A hypothetical weapon which would use concentrated high-power radio waves to destroy or disable targets by surface heating (as in a microwave oven), or by disrupting electronic circuits. Because of atmospheric attenuation, only very powerful transmitters could generate signals of sufficient strength over useful ranges. If the weapon were based in space, attenuation would not occur, but a very large power supply would still be needed. The Soviet Union has apparently been conducting research in this area for many years, to develop both short-range (tactical) weapons and longer-range devices for ANTI-SATELLITE (ASAT) and BALLISTIC MISSILE DEFENSE.

RAID: 1. In air warfare, any attack against a ground or naval target.

2. In ground combat, a surprise entry into enemy territory, not intended to hold terrain. Raids are usually conducted by commandos or other elite forces (e.g., RANGERS, SAS, SPECIAL FORCES, SPETSNAZ) to create diversions, disrupt enemy command and control, destroy critical military or industrial installations, or gather intelligence. See also SPECIAL OPERATIONS.

RAIL GUN: See ELECTROMAGNETIC LAUNCHER.

RAM: 1. See ROLLING AIRFRAME MISSILE.

2. See RADAR ABSORBANT MATERIALS.

RANGERS: Elite battalions of the U.S. Army, trained, organized, and equipped for long-range RECONNAISSANCE, RAIDS, and other SPECIAL OPERATIONS. Created in World War II, the Rangers eventually reached a strength of six battalions and served with distinction in Sicily, Italy, and Normandy, but mostly as shock troops in costly assaults (e.g., storming Point du Hoc in Normandy), rather than in commando roles. In 1953, the Rangers were disbanded and most of their missions transferred to the SPECIAL FORCES (Green Berets). After the Vietnam War, the Special Forces were drastically reduced, and in 1975 two Ranger battalions (the 1st and 2nd Battalions, 75th Infantry Regiment) were reestablished. A third battalion (3/75) was raised in 1984, and a regimental administrative headquarters was established in 1987. The Rangers were scheduled to participate in the abortive 1980 hostage rescue mission in Iran (Operation Eagle Claw), and led the attack on Point Salinas Airport during the 1982 invasion of Grenada (another frontal assault, as it turned out). The 1st, 2nd, and 3rd Rangers, based at Fort Stewart, Oregon, Fort Lewis, Washington, and Fort Benning, Georgia, respectively, are currently assigned to CENTRAL COMMAND, the U.S. rapid-intervention force.

A Ranger battalion has a total of 606 men organized into a headquarters and three rifle companies. Each battalion also has a number of attached specialists, including three U.S. Air Force FORWARD AIR CONTROL (FAC) teams, to coordinate close air support.

Rangers are volunteers selected from airborne (i.e., parachute-qualified) troops; the rigorous training course at Fort Benning concentrates on weapon handling, physical conditioning, and light infantry skills. More advanced training is conducted within the battalions, and includes patrol-ling, mountaineering, land navigation, and martial arts. Many non-Ranger personnel (particularly officers) also go through Ranger School.

RAP: See ROCKET-ASSISTED PROJECTILE.

RAPID DEPLOYMENT FORCE: Generally, any military force intended for rapid intervention at short notice (e.g., the French *Force d'Action Rapide*). More specifically, the U.S. Rapid Deployment Joint Task Force, established in 1979–80 in response to the seizure of U.S. diplomats in Iran and the Soviet invasion of Afghanistan, and now superseded by U.S. CENTRAL COMMAND.

RAPIER: British low-altitude, battlefield surface-to-air missile developed in the 1960s as a replacement for the 40-mm. BOFORS GUN. A low-cost weapon of limited capability, Rapier entered service in 1971 with the British army and the RAF Regiment; it has since been widely exported. By 1987, more than 23,000 missiles had been produced, of which roughly 10,000 have been expended in tests as well as in combat in the Falklands and in the Iran-Iraq War.

Rapier has four cropped-delta wings at midbody for lift, and four smaller tail fins for steering. Powered by a dual-pulse (booster/sustainer) solid-rocket engine, Rapier is a "hittile" that relies on a direct impact for effect. It has semi-automatic command to line-of-sight (SACLOS) guidance, and a very small 3.1-lb. (1.4-kg.) contact-fuzed warhead.

Mounted on a lightweight two-wheel trailer, the standard four-round Rapier launcher incorporates a FIRE CONTROL computer, a surveillance RADAR, and a microwave COMMAND GUIDANCE transmitter; Tracked Rapier, an 8-round launcher mounted on a modified M113 armored personnel carrier, incorporates not only the surveillance radar and guidance transmitter, but also an optical tracking unit.

The basic Rapier is a fair-weather-only system, consisting of the towed launcher ("fire unit"), an optical tracker, and a power unit. When the surveillance radar detects an aircraft, its built-in IFF interrogates it to determine if it is hostile; if so, the target range and bearing are passed to the operator, who must then acquire the target in his optical sight and track it with a joystick controller. Tracking signals are fed automatically to the fire control computer, which slews the launcher onto the target bearing. On receiving a signal from the computer, the operator launches the missile, which is guided towards the target by microwave signals as long as the operator continues to track it (i.e., the system requires both intervisibility and manual skill).

To provide a degree of all-weather capability, the Marconi DN181 Blindfire tracking radar was developed in the 1970s. Mounted on a separate two-wheel trailer, Blindfire can either supplement or replace the optical tracker. Some Tracked Rapiers, however, now have a Darkfire THERMAL IMAGING tracker in place the original optical sight. In the British army, Rapier equips three Light Air Defense Regiments, each with 4 batteries, 2 with 12 towed units, and 2 with 12 tracked units. Every towed fire unit is accompanied by 2 resupply vehicles with a total of 13 reloads; Tracked Rapier is followed by an M548 tracked cargo vehicle (another M113 derivative) with an additional 20 missiles.

Improvements planned for the 1990s include an 8-round towed fire unit, a steerable IMAGING INFRARED tracker, a three-dimensional surveillance radar, and an improved Blindfire tracking radar. Two new missiles are also in development: the Rapier Mk.2A with an enlarged, proximity-fuzed fragmentation warhead for use against REMOTELY PILOTED VEHICLES and CRUISE MISSILES; and the Mk.2B with a contact-fuzed shaped-charge (HEAT) warhead for use against helicopters and aircraft.

Specifications (missile) Length: 7.4 ft. (2.25 m.). **Diameter:** 5.2 in. (132 mm.). **Span:** 15 in. (381 mm.). **Weight, launch:** 93.7 lb. (42.6 kg.). **Speed:** Mach 2 (1400 mph/2250 kph). **Range envelope:** 200–6800 m. (220–7435 yd.). **Height envelope:** 5–3000 m. (20–9850 ft.). **Operators:** Aus, Bru, Indo, Iran, Oman, Qat, Sing, Swi, Tur, UAE, UK, Zam.

RARDEN: British 30-mm. L21 automatic CANNON, main armament of the FOX armored car and SARACEN tracked reconnaissance vehicle. Developed in the 1960s specifically for light armored vehicles, the Rarden is essentially a miniature tank gun; it is recoil-operated, with selectable semi-automatic and 6-round burst fire. The gun is 9.18 ft. (2.8 m.) long (of which only 11.02 in. is inside the turret) and weighs only 220 lb. (100 kg.). It has a cyclic rate of fire of 90 rounds per minute (but ammunition clips hold only 6 rounds) and an effective range of up to 4000 m., and can fire APDS, AP, and HE ammunition. The APDS round has a muzzle velocity of 1200 m. per second and can penetrate the side armor of older main battle tanks.

RAVEN: The Grumman/General Dynamics EF-111A, an ELECTRONIC WARFARE (EW) variant of the U.S. Air Force F-111 attack aircraft. Developed from 1975 as a comprehensive airborne jamming platform with fighterlike performance, the Raven (popularly known as the "Spark 'Vark") began flight tests in 1977 and entered service in 1981. Of the 40 delivered between 1981 and 1985, 6 are used for training, 24 are now assigned to the 366th Tactical Fighter Wing (TFW) at Mountain Home AFB, Idaho, and an additional 12 are attached to the 20th TFW at Upper Heyford, UK. The Raven has three different missions: standoff jamming (while orbiting at altitude over friendly territory), escort jamming (accompanying strike aircraft on deep penetration raids), and close support (flying over the front lines at low altitude).

The EF-111A is essentially an F-111A stripped of all attack AVIONICS, thoroughly overhauled, and fitted by Grumman with an elaborate electronic warfare suite with numerous antennas. A large fairing atop the vertical stabilizer houses the receiver array for the ALQ-99E ACTIVE JAMMING unit, an advanced version of the system first used on the Grumman EA-6B PROWLER. The same pod also houses an ALR-23 INFRARED warning sensor. The weapon bay has been replaced by a large ventral "canoe" radome for additional receivers and several jamming transmitters. More receiver and transmitter aerials are mounted on the fuselage, wings, and on the sides of the vertical stabilizer. In the completely revised cockpit, the copilot's position has been converted into an EW operator's station.

The heart of the Raven is the ALQ-99E, a completely automated ELECTRONIC COUNTERMEASURES system designed to detect, classify, and threat-prioritize enemy radar emissions, and then automatically engage them with jammers of the appropriate frequencies. The degree of automation is much higher than in the EA-6B, which needs three operators; the EW officer of the Raven usually does little more than monitor the system.

Their complete airframe overhaul is meant to allow these aircraft (originally built as F-111As in 1968–69) to remain in service for 20 years. The conversion adds roughly 6000 lb. (2728 kg.) to the basic weight of the F-111 (55,271 lb./25125 kg.), but it is aerodynamically "clean," so that performance is only slightly reduced. Its long range, terrain-following radar and supersonic low-altitude dash speed allow the Raven to operate in high-threat environments (unlike the navy's subsonic EA-6B).

RBOC: Rapid Blooming Offboard Chaff, the U.S. Mk.33 and 34 naval countermeasure dispensers, now in service with many Western navies.

Intended for destroyers, the Mk.33 consists of four six-tube mortars, arranged in pairs at elevations of 55°, 65° and 75°; the Mk.34, for frigates and smaller vessels, has only two six-tube mortars. Activated automatically by the ship's radar-warning receiver, or semi-automatically on command from the COMBAT INFORMATION CENTER (CIC), the RBOC fires Mk.171 CHAFF cartridges, which release a cloud of foil strips intended to mask the ship from missile guidance radars. Other RBOC cartridges include the HIRAM (Hycor Infrared Anti-Missile) Decoy, which dispenses parachute flares against INFRARED-HOMING missiles; and Gemini, which dispenses both flares and chaff. In the U.S. Navy, the Mk.33 and 34 have been replaced by Mk.36 Super Rapid Blooming Offboard Chaff (SRBOC) units. See also ELECTRONIC COUNTER-MEASURES.

RBU-: *Reaktivnaya bombometnaya ustanovka* (Rocket Bombardment Mount), Soviet rocket-propelled DEPTH CHARGE launchers. RBUs are found on most Soviet-designed warships, from aircraft carriers to corvettes. Each of the five models now in service is designated by its maximum effective range in meters.

The RBU-6000, which entered service in 1982, is a fully automatic 12-barrel launcher, whose 250-mm. rockets weigh 70 kg. (154 lb.) and are armed with a 21-kg. (46.3-lb.) high-explosive depth charge warhead, fitted with contact, hydrostatic, and proximity fuzes. The horseshoe-shaped launcher is reloaded automatically from a magazine immediately below it.

The RBU-2500, introduced in 1957, is a 16-barrel automatic launcher, also for 250-mm. rockets, but with an effective range of only 2500 m.

The RBU-1200, introduced in 1958, is a 5-barrel, manually loaded launcher with automatic elevation and manual traverse. Its 250-mm. rockets are of an earlier type with a 34-kg. (75-lb.) depth charge warhead.

The RBU-1000 is a modern, fully automatic 6-barrel launcher, often used in conjunction with the RBU-6000. It fires 300-mm. rockets weighing 120.5 kg. (265 lb.), including a 55-kg. (121-lb.) depth charge warhead.

Introduced in 1960, the RBU-600, the smallest in the series, is a manually loaded, manually operated 6-barrel launcher firing 300-mm. rockets.

These unguided rockets are fired in wide patterns intended to bracket submerged submarines. While only marginally effective against modern submarines in open waters, they remain useful for inshore ASW and in narrow seas. Equivalent Western systems include the Swedish Bofors 375-mm. ASW rocket launchers, the British Mk.10 LIMBO depth charge mortar, and the U.S. HEDGEHOG spigot mortar.

RCS: See RADAR CROSS SECTION.

RDF: See RAPID DEPLOYMENT FORCE.

RDX: A high-explosive compound, also known as Cyclonite and Hexogen, considerably more powerful but also more sensitive than TNT. RDX is a crystallized substance which must be mixed with wax or other materials to make it stable enough for military use. The most common RDX-based explosives are Composition A, a mixture of 91 percent RDX and 9 percent wax, often used as a projectile filler; Composition B, a mixture of 60 percent RDX, 40 percent TNT, and less than 1 percent wax, also used as a projectile and bomb filler; and Composition C, a plastic explosive of 90 percent RDX and 10 percent emulsifying oil, used mainly for demolitions. See also EXPLOSIVES, MILITARY.

REACTIVE ARMOR: Boxes containing low-sensitivity explosive sandwiched between thin steel plates, bolted onto armored vehicles to reduce their vulnerability to shaped-charge (HEAT) warheads. Detonated when hit by shaped charges, they deform and deflect the charges' explosive jets, thereby protecting the main armor. Invented by a West German, reactive armor was first developed by Israel (See BLAZER). After the Israeli army used the device in the 1982 Lebanon War, it was promptly copied by the Soviet army; reactive armor boxes have been added to its latest tanks (T-64, T-72, and T-80), and also retrofitted on many older tanks (T-54/55 and T-62). NATO armies rely heavily on HEAT-armed ANTI-TANK GUIDED MISSILES (ATGMs); yet only the largest, most sophisticated, and most expensive ATGMs can overcome reactive armor. See also ANTI-TANK.

RECOILLESS (WEAPON): A projectile weapon in which the recoil is reduced and counteracted by vents in the breech, which release up to 80 percent of the propellant gases to the rear. Recoilless weapons can therefore fire artillery-sized ammunition with only minimal recoil, eliminating the need for heavy mountings and complex hydraulic recoil mechanisms, albeit at the cost of greatly reduced range and muzzle velocity. Internal barrel stresses are also greatly reduced, thus allowing the use of thinner and lighter tubes than in conventional artillery of the same caliber. A typical 106-mm. recoilless rifle will have an effective range of only 1000–2000 m. whereas a conventional gun of the same caliber could have a range of up to 25,000 m. But the weight differential is on the order of 20 to 1.

Recoilless weapons were developed by Germany, Britain, and the U.S. in World War II, principally as infantry ANTI-TANK weapons firing shaped-charge (HEAT) ammunition. The U.S. then fielded 57-mm. and 75-mm. recoilless rifles (RCLs), which were replaced in the 1950s by the 90-mm. M67 and 106-mm. M40 (which were in turn replaced in the 1970s by DRAGON and TOW anti-tank guided missiles). The British developed two huge 120-mm. RCLs (Mobat and Wombat), now also phased out. The most widely fielded RCL in the West is now the Swedish 84-mm. CARL GUSTAV man-portable anti-tank/anti-bunker weapon. The U.S.S.R., its clients, and China still employ large numbers of recoilless anti-tank weapons, the most common being the 73-mm. SPG-9, the 82-mm. B-10, and the 107-mm. B-11.

The Finnish Tampella company has recently introduced a new series of smoothbore recoilless weapons which operate on the "countermass" principle: rather than venting combustion gases out the breech (with a conspicuous backblast), these weapons offset the recoil by ejecting a somewhat smaller mass of bagged sand to the rear. The weapons are made in various sizes from 50 mm. to 120 mm.; the latter can fire high-velocity armor-piercing fin-stabilized discarding sabot (APFSDS) rounds.

RECONNAISSANCE: 1. As intelligence gathering: The collection of visual, photographic, INFRARED, or electronic information about enemy forces or terrain (as opposed to SURVEILLANCE, the observation of a given area for activity of any sort). STRATEGIC reconnaissance seeks to uncover the major dispositions and plans of hostile states, while TACTICAL reconnaissance is for the immediate support of combat forces. The latter is particularly important for the conduct of mobile operations, when conditions are fluid, and accurate information is a prime resource. Very select, elite units (e.g., U.S. RANGERS, SEALS, and SPECIAL FORCES; British SAS and SBS; Israeli SAYARET MATKAL; and Soviet SPETSNAZ) can perform both types of reconnaissance, while the scout units of armored CAVALRY and light mechanized forces in general are trained only for tactical reconnaissance. Most ground formations contain (tactical) reconnaissance units, often manned by select personnel, with light or highly mobile equipment.

Aerial reconnaissance is also classified as strategic or tactical. The former may be undertaken by specially designed long-range aircraft, such as the SR-71 BLACKBIRD and U-2, or by converted bombers (see, e.g., the Soviet Tu-95 BEAR); the latter is usually conducted by modified fighters flying fast at very low altitudes. Naval reconnaissance is performed by SUBMARINES, MINISUBS, or small surface vessels. Because of the time delays involved, satellites are of limited value for tactical reconnaissance, but are ideal for strategic reconnaissance. For special-purpose or adapted equipment, see ARMORED CAR; ARMORED RECONNAISSANCE VEHICLE; SATELLITES, MILITARY.

2. As a form of combat: The employment of probing forces to induce the enemy to reveal his dispositions and/or weapon locations. Reconnaissance thus involves combat, unlike scouting.

RED FLAG: Code name for air combat training exercises conducted by the U.S. Air Force Fighter Weapons Center at Nellis AFB, Nevada. Established in 1975 in reaction to the dismal performance of Air Force fighters during the Vietnam War, and patterned after the U.S. Navy's TOP GUN program, Red Flag is conducted on a much larger scale over an instrumented range, which records and displays all aircraft positions and maneuvers for subsequent analysis. Pilots in training fly against an "Aggressor Squadron" whose pilots are trained in Soviet air combat tactics, and whose F-5 TIGERS and modified F-16 FALCONS are supposed to match the performance of the MiG-21 FISHBED and MiG-29 FULCRUM. See also AIR COMBAT MANEUVERING.

REDEYE: The General Dynamics FIM-43A man-portable (shoulder-fired) INFRARED-HOMING surface-to-air missile, developed from 1958 to provide anti-aircraft self-defense for infantry and mechanized platoons. Redeye entered service in 1968 with the U.S. Army and Marine Corps, and was widely exported; more than 85,000 were produced between 1967 and 1970. Redeye has been superseded in U.S. service by the more advanced STINGER, but may still be found in some reserve units.

The Redeye missile has two pop-out nose canards for steering and four pop-out tail fins. A gas-cooled infrared seeker in the nose can only home on hot exhausts, making Redeye a "retribution weapon" that can normally be fired only at aircraft that have already attacked the site. Powered by a dual-pulse (booster-sustainer) solid-fuel rocket engine, the missile comes in a sealed launcher/storage canister, which attaches to the rear of a trigger/sight unit. In action, the operator acquires the target visually, and points the launcher at the target. If the IR seeker acquires and locks onto a target, a buzzer sounds in the sight unit; the operator then pulls the trigger, initiating the booster charge which expels the missile from the launch tube. The canards and rear fins then extend, and the sustainer charge ignites at a safe distance.

Redeye is only marginally effective against high-performance aircraft; its range and speed make it difficult for it to overtake targets moving at more than 500 mph (800 kph). In addition, its rather simple IR seeker can easily be decoyed by flares and other INFRARED COUNTERMEASURES, while its small contact-fuzed warhead may often fail to inflict lethal damage.

Specifications (missile) Length: 4 ft. (1.22 m.). Diameter: 2.75 in. (70 mm.). Span: 5.5 in. (140 mm.). Weight, launch: 18 lb. (8.2 kg.). Speed: Mach 1.6 (1056 mph/1690 kph). Range: 2.1 mi. (3.36 km.). Ceiling: 5000 ft. (1525 m.). Operators: Aus, Den, FRG, Gre, Isr, Jor, Swe, US (reserves).

REDOUTABLE: A class of six French nuclear-powered ballistic-missile SUBMARINES (SSBNs) commissioned between 1971 and 1985. Developed as the core of the French nuclear deterrent force (FORCE DE DISSUASION), quite independently of U.S. ballistic missile submarine technology, these submarines represent a considerable technical achievement; their cost, however, could have been much lower if U.S. assistance had been accepted. Five of the submarines were completed between 1971 and 1980; the sixth, L'Inflexible, was completed in 1985 to a modified design.

The Redoutable's design imitates the U.S. LA-FAYETTE class, with a modified teardrop hullform in a single-hull configuration, a streamlined sail located well forward, and 16 missile tubes amidships, arranged in two rows of 8 under a flat-decked "whaleback" casing. The control surfaces are also arranged in standard U.S. fashion, with the "fairwater" diving planes on the sail, and cruciform rudders and stern planes mounted ahead of the propeller. The pressure hull is fabricated of high tensile strength steel; L'Inflexible has a stronger hull for greater depth capability.

The first two submarines, Le Redoutable and Le Terrible, were initially armed with M1 two-stage, solid-fuel ballistic missiles, with a single 500-KT warhead and a range of 2500 km. The third vessel, Le Foudroyant, was armed with the improved M2

missile, with a range of 3000 km. The last two submarines, L'Indomptable and Le Tonnant, were at first armed with a mix of M2 and M20 missiles, the latter identical to the M2 but armed with a 1-MT warhead. From 1977 all five Redoutables were gradually rearmed with the M20 missile. The sixth and last boat, L'Inflexible, is armed with the M4 missile, with six 150-kT MIRV warheads and a range of 4000 km.; the M4 will eventually be retrofitted into the older submarines, except for Le Redoutable. (For details of French SLBMs, see MSBS.)

Secondary armament in the first five Redoutables comprises four 21.7-in. (550-mm.) torpedo tubes in the bow for L5 Mod 3 anti-submarine and F17 dual-purpose homing TORPEDOES. L'Inflexible has 21-in. (533-mm.) tubes and can also launch EXOCET MM.39 anti-ship missiles. Eighteen weapons of all types can be carried. All sensors are linked to a DLT D3 torpedo FIRE CONTROL unit.

The Redoutables have nuclear turbo-electric drive: steam from the reactor feeds two turbo-generators, which in turn power an electric motor. This system is considerably quieter than the geared steam turbine drive of most nuclear submarines. As in the British and U.S. navies, each French ballistic missile submarine has two crews (Blue and Amber) which serve on alternating 70- to 90-day patrols, to maximize time on station.

The first five Redoutables are scheduled for retirement between 1997 and 2006; L'Inflexible is to remain in service until 2012. All six are due to be replaced by a new class of SSBNs displacing between 14,000 and 15,000 tons. See, more generally, SUBMARINES, FRANCE.

Specifications Length: 422.2 ft. (128.7 m.). Beam: 34.8 ft. (10.6 m.). Displacement: 8045 tons surfaced/8940 tons submerged. Powerplant: single-shaft nuclear: 1 pressurized-water reactor, turbo-electric drive, 16,000 shp. Speed: 18 kt. surfaced/25 kt. submerged. Max. operating depth: 820 ft. (250 m.); (L'Inflexible) 1085 ft. (330 m.). Collapse depth: 1085 ft. (330 m.); (L'Inflexible) 1525 ft. (465 m.). Crew: 135. Sensors: 1 DUUV 23 bow-mounted panoramic passive sonar, 1 DUUX 2 active ranging sonar; (L'Inflexible) 1 DSUX 21 multi-functional array, 1 DUUX 5 digital ranging sonar, 1 DRUA 33 surface-search radar, 1 electronic signal monitoring array, 2 periscopes.

REENTRY VEHICLE (RV): An aerodynamic vehicle which contains a warhead (generally nuclear) and constitutes the payload of a BALLISTIC MISSILE. The RVs can also serve to protect the warhead from radiation and other weapon effects during the free-fall (midcourse) phase of the trajectory, but their main purpose is to shield the warhead from the heat of reentry, and facilitate a stable and predictable trajectory through the atmosphere to the point of impact.

The earliest RVs were heavy and cumbersome "heat sinks" of copper or tungsten, which absorbed the heat of reentry within their own mass. By the early 1960s, lightweight "ablative" RVs were developed; these dissipate heat by the progressive evaporation of their protective coating.

RV design must compromise between the need for a stable trajectory and the aim of passing through the atmosphere as quickly as possible, to minimize aerodynamically induced impact errors. Both characteristics are a function of the hypersonic ballistic coefficient ("beta"), very roughly the ratio between mass and frontal area. In general, a body with a high beta is long and narrow, whereas a low-beta body is broad and blunt. Low-beta bodies are dynamically stable, but decelerate rapidly, hence are exposed to aerodynamic effects for a longer period, with concomitant inaccuracy. High-beta bodies are inherently unstable, but decelerate relatively little, hence are much more accurate. The designer's goal is to maximize the ballistic coefficient without exceeding the point where the RV is so unstable that it tumbles and disintegrates from aerodynamic stresses.

Most U.S. RVs, such as the Mk.12 and Mk.21 of the MINUTEMAN and PEACEKEEPER ICBMs, have simple "sphere-cone" configurations; they resemble inverted ice cream cones with slightly rounded tips. The nose cones are fabricated from heat-resistant carbon-carbon composites, while the rest of the RV is covered by an ablative carbon-phenolic heat shield over an aluminum shell (conformal patch antennas are mounted around the base for radar-proximity [VT] fuzes). These RVs have very high beta and require very precise alignment for a 0° angle of attack (AOA) at reentry. That in turn requires a start-stop engine on the POST-BOOST VEHICLE (PBV) from which the RVs are released. U.S. RVs are spin-stabilized during free-fall by small rocket motors in their base, which ignite shortly after release.

The Soviet Union has not yet developed a PBV with a start-stop engine and cannot align its RVs for a 0° AOA at reentry. Hence it must use relatively low-beta bodies with rather blunt noses and complex biconic or triconic configurations, which are more stable but less accurate than high-beta

sphere-cones. Although the latest Soviet ICBMs (the SS-24 SCALPEL and SS-25 SICKLE) appear to have sphere-cone RVs, these are still rather blunt and have relatively low ballistic coefficients. Soviet designers, it seems, have different priorities than their U.S. counterparts. The latter seek to design the most accurate RV possible; the former only to make the RV as accurate as necessary to destroy a given target with a warhead of specified yield. Soviet RVs also tend to be heavier and more robust, making them inherently more resistant to ANTI-BALLISTIC MISSILE warheads and FRATRICIDE. See also MANEUVERING REENTRY VEHICLE; MULTIPLE INDEPENDENTLY TARGETED REENTRY VEHICLES; MULTIPLE REENTRY VEHICLES.

REFORGER: Return of Forces to Germany, annual exercises (suspended in 1989) intended to test the ability of U.S. forces to redeploy at short notice from the continental U.S. to Western Europe, in order to reinforce the NATO CENTRAL FRONT. Current NATO plans require the U.S. to transfer up to six divisions of troops to Europe in the first week of a war.

REGIMENT: An army formation usually subordinate to a DIVISION, normally equivalent to a BRIGADE, and comprising three or four BATTALIONS. The regiment is the standard subdivisional formation of Soviet, other Warsaw Pact, and Chinese armies, usually with 2000–2500 men (fewer in tank regiments). In the British army, the regiment is the traditional focus of loyalties and social activities, but is essentially an administrative echelon for recruitment and career management, with operational control running from brigade to individual battalions. In the U.S. Army, the regiment is merely symbolic, except for ARMORED CAVALRY regiments, brigade-size mechanized formations usually attached to a CORPS for SCREENING and RECONNAISSANCE.

REM: Roentgen Equivalent in Man, a measure of the effects of ionizing radiation (other than X-RAYS or GAMMA RAYS) on humans, derived by multiplying the energy yield of the radiation by a ratio which expresses that radiation's effect on the human body. See, more generally, NUCLEAR WARHEADS, EFFECTS OF.

REMOTELY PILOTED VEHICLE (RPV): An unmanned aircraft usually meant to be recoverable, and controlled in flight by a (remote) operator through a DATA LINK. The term is often used interchangeably with "drone," though the latter properly defines any land, sea, or air vehicle that is *either* remotely *or* automatically controlled; in addition, drones are usually meant to be expendable.

RPVs come in many different sizes, from relatively large supersonic vehicles for strategic RECONNAISSANCE to small, slow units that resemble enlarged radio-controlled model airplanes (e.g., the Israeli IAI Scout and Tadiran Mastiff), used for battlefield SURVEILLANCE and artillery SPOTTING. RPVs are rapidly becoming an important element in air warfare as well. Because of their small size, they are difficult to detect and hit; being also much cheaper than manned aircraft, they can be used for tasks too dangerous for the latter.

During the 1982 Lebanon War, Israeli forces used some RPVs to decoy Syrian air defenses, and others to spot targets for artillery and for tactical reconnaissance in general. Both the U.S. and Israeli air forces are now developing attack RPVs to replace manned aircraft in strikes against particularly well-defended targets. The main drawbacks of RPVs are the limited view and low situational awareness of operators, and the vulnerability of data links to jamming.

REPLICA DECOY: A ballistic missile penetration aid (PENAID), intended to deceive BALLISTIC MISSILE DEFENSE radars and infrared sensors by emulating the radar cross section, thermal emissions, and dynamic movement of REENTRY VEHICLES (RV). A replica decoy that could accurately simulates all the signature characteristics of an RV would be as costly in payload as an RV; hence PENAID decoys normally replicate only one or two RV characteristics, to foil a specific sensor. Thus a ballistic missile defense system with different types of sensors should (in theory) be able to discriminate between RVs and decoys. "Active" or "responsive" replicas with small thruster rockets, heating elements, and active jammers have been proposed, but the cost and weight of such sophisticated decoys would make their value dubious; large numbers of simple inflatable decoys intended to swamp data processing capabilities are likely to be both cheaper and more effective.

RESERVES: 1. Forces deliberately kept out of battle, to subsequently relieve depleted units, meet sudden or especially strong attacks, or intervene offensively. Such reserves are termed TACTICAL, OPERATIONAL, or STRATEGIC, depending upon the level of control. Tactical reserves, often of BRIGADE size, are retained for the immediate support of front-line units under divisional command. Operational reserves range in size from a single DIVISION upwards, for use under ARMY or ARMY GROUP command. Strategic reserves may be of any size, but are controlled by the national command authorities, as the ultimate contingency force.

It is an axiom of warfare that the side which retains the last uncommitted reserve also retains the final initiative, and therefore the opportunity to achieve decisive results.

2. Trained but unmustered forces of any type, manned by recently discharged personnel or part-time volunteers. In general, reserve forces are armed with older weapons passed down from active units. In ground warfare, this need not be a severe handicap, because the active forces on both sides may be worn down by the time reserves are mobilized and reach the front. As a result, the numerous reserve formations of the Soviet army have long been the chief variable in the worldwide, all-service military balance.

In the case of smaller armies that rely upon mobilized reserves for their front-line strength (e.g., the Israeli army), reserve formations require training and equipment roughly comparable to that of the active forces. In the case of the United States, each service maintains its own reserve contingents (army, navy, air force, and marine corps), and each state maintains army and air force "National Guard" units which are normally controlled by their state governors (for disaster relief and civil action) but which can be "federalized" by the president in an emergency. Britain and West Germany both maintain "Territorial Armies" which augment their regular forces in wartime. In the former, the troops are all volunteers, while in the latter they are time-expired conscripts. See also TERRITORIAL ARMY, BRITAIN; TERRITORIAL ARMY, FEDERAL REPUBLIC OF GERMANY.

RESHEF: A class of 13 Israeli fast missile patrol boats (PGMS) commissioned between 1973 and 1980. Officially known as the Sa'ar IV class, the Reshefs are enlarged versions of the original French-built SA'AR-class boats, with numerous improvements that reflect Israeli operational experience. For instance, endurance and seakeeping have been improved, at some cost in maximum speed, while the decks are composed of two thin steel sheets with neoprene rubber in between to dampen vibration and reduce crew fatigue. Like the Sa'ars, they have steel hulls with round bilges, and a small light alloy deckhouse forward.

The first 2 boats, *Reshef* and *Keshet*, were completed in time for the 1973 Yom Kippur War, during which they destroyed one Syrian OSA and four KOMAR missile boats. Two boats were transferred to Chile in 1979–80, and 3 additional boats were built for South Africa in 1977–78 (with 9 more built under license as the Minister class). The Israeli navy currently operates 8 Reshefs.

Armament is modular. The most common configuration includes 2 OTO-MELARA 76.2-mm. automatic DUAL PURPOSE guns (fore and aft), 2 manually operated 20-mm. CANNONS, 3 12.7-mm. (.50-caliber) MACHINE GUNS, and 2 or 4 HARPOON and 4 or 6 GABRIEL anti-ship missiles in launcher/storage containers. Variants include a 40-mm. cannon in place of one or both 76.2-mm. guns. Since 1985, most Reshefs have carried a U.S. 20-mm. PHALANX radar-controlled gun for anti-missile defense, in place of the forward 76.2-mm. mount. The ELECTRONIC WARFARE suite is one of the most comprehensive fitted to any warship, let alone one of this size.

In 1977, one boat, the *Tarshish*, was experimentally fitted with a helicopter landing deck on the fantail, the helicopter being used mainly for OVER-THE-HORIZON TARGETING (OTH-T). This led to the construction in 1980 of two Sa'ar 4.5 (Aliyah-class) boats, essentially enlarged Reshefs with a hangar and landing deck aft of the deckhouse, in place of the gun and missiles. The two Aliyehs are intended as flotilla leaders, to provide OTH-T and COMMAND AND CONTROL for groups of Sa'ars and Reshefs. Armament, concentrated on the bow and atop the hangar, consists of a Phalanx, 2 20-mm. cannons, 4 12.7-mm. machine guns, 4 Harpoons, and 4 Gabriels. The Aliyah hull was adopted for the two Romat-class missile boats commissioned in 1981–82, with the helicopter facilities replaced by a heavy gun and missile armament, including 8 Harpoons, 6 Gabriels, 1 76.2-mm. gun, 1 Phalanx, and 2 20-mm. anti-aircraft guns. See also FAST ATTACK CRAFT, ISRAEL.

Specifications Length: (Reshef) 190.6 ft. (58.1 m.); (Aliyah) 202.4 ft. (61.7 m.). **Beam:** 24.9 ft. (7.6 m.). **Draft:** 8 ft. (2.4 m.). **Displacement:** (Reshef) 415 tons standard/450 tons full load; (Aliyah) 500 tons full load. **Powerplant:** 4 3500-hp. MTU 16V956 TB1 diesels, 4 shafts. **Speed:** 32 kt. **Range:** 1500 n.mi. at 30 kt./4000 n.mi. at 17.5 kt. **Crew:** (Reshef) 45; (Aliyah) 53. **Sensors:** 1 Thomson-CSF TH-D1040 surface-search radar, 1 Selenia Orion RTN-10X fire control radar, 1 EDO sonar (optional). **Electronic warfare equipment:** 1 Israeli-made Elta electronic countermeasures array controls, 4 24-tube and 4 single-tube chaff launchers.

RESOLUTION: A class of four British nuclear-powered ballistic-missile SUBMARINES (SSBNs) commissioned between 1967 and 1969. Built to replace the V-Bombers of the British nuclear force, the Resolution class was designed with the help of the U.S. Navy to employ POLARIS missiles, and is similar to the U.S. LAFAYETTE class. Like the Lafayettes, they have a modified teardrop hullform in a single-hull configuration. A tall, streamlined sail is located well forward, with 16 missile tubes mounted behind it in two rows of 8, under a flat-decked "whaleback" casing. Control surfaces are arranged in the standard British fashion, with folding bow planes, and cruciform rudders and stern planes ahead of the propeller.

Main armament consists of 16 UGM-27C Polaris A3 missiles, with a range of 4600 km. Originally fitted with a single large warhead of British design, they were modified under the slow, costly, and controversial CHEVALINE program to carry six 150-kT MULTIPLE REENTRY VEHICLES (MRVs) and a complex PENAID package to defeat Soviet ballistic missile defenses. Secondary armament comprises six 21-in. (533-mm.) torpedo tubes in the bow for 18 Mk.24 TIGERFISH wire-guided homing torpedoes. Mobile decoys and other TORPEDO COUNTERMEASURES are also carried.

The original SONAR suite was identical to that of the VALIANT-class nuclear attack submarines, but in the late 1970s the Resolutions were fitted with improved sensors. A U.S.-supplied Ships Inertial Navigation System (SINS) is fitted to provide the precision navigation needed for missile targeting; VLF and ELF trailing wire antennas allow submerged communication with shore facilities. As with U.S. and French ballistic missile submarines, there are actually two crews (Port and Starboard) which man the boats on alternating 70- to 90-day patrols, with a two-week overhaul period in between. The Resolutions are due to be replaced in the late 1990s by the Vanguard-class TRIDENT missile submarines. See, more generally, SUBMARINES, BRITAIN.

Specifications Length: 425 ft. (129.5 m.). **Beam:** 33 ft. (10.1 m.). **Displacement:** 7500 tons surfaced/8400 tons submerged. **Powerplant:** single-shaft nuclear: 1 PWR.1 pressurized-water reactor, two sets of geared turbines, 15,000 shp. **Speed:** 20 kt. surfaced/25 kt. submerged **Max. operating depth:** 1150 ft. (350 m.). **Collapse depth:** 1525 ft. (465 m.). **Crew:** 143. **Sensors:** 1 chin-mounted Type 2001 low-frequency active/passive sonar, 1 Type 2007 flank-mounted conformal array sonar, 1 Type 2019 passive sonar intercept array, 1 Type 2024 passive towed array sonar, 1 Type 1007 surface-search radar, 1 electronic signal monitoring array, 2 periscopes.

RESTIGOUCHE: A class of seven Canadian ANTI-SUBMARINE WARFARE (ASW) FRIGATES commissioned in 1958–59. Four have been extensively modified since then and will remain in service until 1996, while the other three were placed in reserve in 1974.

The Restigouches are derived from the earlier ST. LAURENT class but lack their helicopter facilities. Flush-decked, they have a rounded "turtleback" foredeck to prevent the accumulation of ice. Like all Canadian warships, they are extremely rugged and seaworthy. As completed, armament consisted of a British twin 3-in. 70-caliber DUAL PURPOSE gun forward, a U.S. twin 3-in. 50-caliber gun amidships, and 2 Mk.10 LIMBO depth charge mortars aft. From 1968, the 4 modernized vessels received an 8-round ASROC launcher with 8 reloads and a 4-round Canadian SEA SPARROW surface-to-air missile launcher, in place of the aft gun and 1 Limbo.

From 1980, the four active ships began major overhauls under the DELEX (Destroyer Life Extension) Program, receiving new sensors, two sets of Mk.32 triple tubes for lightweight ASW TORPEDOES, and two WSC-3 SATCOM satellite communications terminals. See also FRIGATES, CANADA.

Specifications **Length:** 371.1 ft. (113.1 m.). **Beam:** 42 ft. (12.8 m.). **Draft:** 14 ft. (4.3 m.). **Displacement:** 2390 tons standard/2900 tons full load. **Powerplant:** twin-shaft steam: 2 oil-fired boilers, 2 sets of geared turbines, 30,000 shp. **Speed:** 28 kt. **Range:** 4750 n.mi. at 14 kt. **Crew:** 214. **Sensors:** 1 SPS-503 air-search radar, 1 navigation radar, 1 Mk.60 fire control radar, 1 SQS-501 bottom-mapping sonar, 1 SQS-503 hull-mounted medium-frequency sonar, 1 SQS-505 variable depth sonar. **Electronic warfare equipment:** 1 CANEWS electronic signal monitoring array, 1 ULQ-6 active jamming unit, 6 4-barrel Mk.36 SRBOC chaff launchers, 1 Mk.25 Nixie torpedo countermeasures unit.

RETARDED BOMB: An aerial bomb equipped with some form of braking device in its tail to cause it to decelerate quickly, thereby allowing its release closer to the aim point. In low-altitude attacks, retarded bombs enable aircraft to accelerate away from the bombs' blast and fragmentation envelope. The earliest retarded bombs

of World War II had parachutes in their tails. During the Vietnam War the U.S. introduced SNAKEYE, an ordinary low-drag general-purpose (LDGP) bomb fitted with spring-driven fins, which opened after release to form a cruciform air brake. Snakeye has been widely copied. The latest retarded bombs have "ballute" brakes, a combination of balloon and parachute contained in the tail, which inflates after release. Lighter and cheaper, ballute bombs are now displacing Snakeyes.

REVOLUTIONARY WAR: Armed conflict between a government and opposing forces, wherein the latter rely mainly on GUERRILLA warfare and SUBVERSION rather than formal warfare. The revolutionary side operates by establishing a rival state structure which embodies a political ideology, and which is intended to replace the existing order. This competing administration is itself the chief instrument of warfare. In revolutionary war, the winning side out-administers, rather than out-fights, the loser; the covert "administration" collects taxes, conscripts, and information—all of which can be extracted from the population even if the government is in apparent military control of the area in question. These resources are supplied to the guerrilla arm, which strives to erode the government's control and undermine its prestige. That in turn facilitates subversion (propaganda + terror) to extend the reach of the covert administration, which sustains the guerilla, and so on. The main threat to the government is subversion, not the guerrilla, and concentration on fighting the latter is normally an error. Conventional warfare against guerrilla forces is in any case uneconomical (and may be counterproductive), because it requires a very high degree of superiority in order to succeed. For antidotes to revolutionary war, see COUNTERINSURGENCY. For a classification of various forms of internal conflict, see INTERNAL WAR.

RGM-84: U.S. anti-ship missile. See HARPOON.

RHAW: See RADAR HOMING AND WARNING RECEIVER.

RIFLE: The basic personal firearm of soldiers, a shoulder-fired weapon which shoots a small-caliber bullet through a relatively long rifled barrel by means of the explosion of smokeless powder. Most rifles used in both world wars were manually operated bolt-action weapons with 5- to 10-round magazines. Slightly shorter versions ("carbines") were issued to the cavalry, artillery, and other support forces. These weapons fired 7.62- to 7.92-mm. (0.30- to 0.32-caliber) bullets weighing more than

150 grams, and generating muzzle energy levels in excess of 2500 ft.-lb. In addition to producing a powerful kick, these rounds had very long ranges (up to 2000 m.) and were quite heavy, limiting the number that were normally carried by foot soldiers to 100 or less.

The first semi-automatic rifles in general issue, the U.S. M1 Garand and the Soviet Tokarev (both introduced before 1940), fired the same high-power ammunition but at a much higher rate of fire (up to 30 rounds per minute). "Automatic rifles," such as the famous Browning BAR, were actually LIGHT MACHINE GUNS (LMGS).

ASSAULT RIFLES, first introduced by the German army during World War II, fire an intermediate-power cartridge, to combine the lighter weight of rifles with the automatic fire of LMGs. Their effective range is under 800 m., but most infantry firefights occur at less than 400 m. After World War II the German assault rifle concept became the standard infantry firearm worldwide. The latest models fire higher-velocity lightweight 4.85- to 5.56-mm. rounds, normally effective only out to 350 m. or so, but capable of inflicting massive wounds, because their bullets "tumble" after penetrating instead of going right through, as with the old full-power rounds.

Bolt-action rifles survive in use as specialized SNIPER weapons equipped with telescopic sights.

RIGA: NATO code name for a class of 65 Soviet CORVETTES completed between 1952 and 1958. The Soviet navy still retains some 30 Rigas in active service, and 17 Rigas were transferred to other navies; 2 to Bulgaria, 5 to East Germany (1 since destroyed by fire), 2 to Finland, and 8 to Indonesia; 4 Rigas were built in China in 1958–59, with components supplied by the U.S.S.R.

Designated "Patrol Vessels" (*Storozhevoy korabl'*, SKR) by the Soviet navy, the Rigas are flush-decked, with graceful shear line, a broad transom stern, a single stack, a small superstructure forward, and a detached deckhouse aft. Armament consists of 3 single 100-mm. DUAL PURPOSE guns (2 forward, 1 aft), 2 twin 57-mm. ANTI-AIRCRAFT guns, 2 16-barrel RBU-2500 anti-submarine rocket launchers, 1 twin or triple 21-in. (533-mm.) torpedo tube mount, 2 DEPTH CHARGE racks, and rails for up to 80 MINES. A few ships also have 2 twin 25-mm. anti-aircraft guns.

Specifications Length: 300.2 ft. (91.5 m.). Beam: 33.1 ft. (10.1 m.). Draft: 10.5 ft. (3.2 m.). Displacement: 1260 tons standard/1510 tons full load. Powerplant: twin-shaft steam: 2 oil-fired boilers, 2 sets of geared turbines, 20,000 shp. Fuel: 230 tons. Speed: 28 kt. Range: 550 n.mi. at 28 kt./2000 n.mi. at 13 kt. Crew: 175. Sensors: 1 "Slim Net" air-search radar, 1 "Don-2" or "Neptune" navigational/surface-search radar, 1 hull-mounted high-frequency sonar. Electronic warfare equipment: 1 "High Pole-B" IFF, 2 "Square Head" IFF receivers, 2 "Watch Dog" electronic countermeasure arrays.

RIO TREATY: Formally, the Inter-American Treaty of Reciprocal Assistance, signed in 1947 by almost all Latin American countries and the United States. Ecuador and Nicaragua did not sign, but participate in the treaty's executive arm, the Organization of American States (OAS). Under the terms of the treaty, all signatories are pledged to intervene on behalf of any member attacked by an outside power. No form of intervention is specified, and any intervention would be subject to a decision by a conference of foreign ministers convened within the framework of the treaty. By contrast, detailed provisions for the peaceful settlement of intra-American disputes are embodied in the OAS Charter. Cuba was expelled from the OAS (and from treaty membership) in 1962. The current authority of the Rio Treaty is problematic (as demonstrated by the situation in Central America).

RIO: Radar Intercept Officer, U.S. Navy term for the radar operator/navigator in two-seat fighter aircraft. Unlike WEAPON SYSTEM OFFICERS (WSOs) in the U.S. Air Force, RIOs are not rated pilots but specialists trained solely for the "backseat" functions.

RIVER: A class of six Australian FRIGATES, modified versions of the British LEANDER class, commissioned between 1961 and 1971. One ship, HMAS *Yarra*, was stricken in 1985 and cannibalized for spare parts.

The Rivers have a raised forecastle, a recessed quarterdeck, and a low, blocky superstructure amidships. Armament consists of one twin 4.5-in. DUAL PURPOSE gun behind the forecastle, a SEA CAT short-range, surface-to-air missile launcher with 24 missiles on the aft superstructure, and an IKARA anti-submarine missile launcher aft; a 3-barrel LIMBO depth charge mortar has been replaced by two sets of Mk.32 triple tubes for lightweight anti-submarine homing TORPEDOES. See also FRIGATES, BRITAIN.

Specifications: Length: 369 ft. (112.75 m.). Beam: 41 ft. (12.5 m.). Draft: 12.8 ft. (3.9 m.). Displacement: 2100 tons standard/2750 tons full load. **Powerplant:** twin-shaft steam: 2 oil-fired boilers, 2 sets of geared steam turbines, 34,000 shp. **Fuel:** 400 tons. **Speed:** 30 kt. **Range:** 4500 n.mi. at 12 kt. **Crew:** 251. **Sensors:** 1 Type 978 navigation radar, 1 LW-02 air-search radar, 1 WM-22 fire control radar, 1 Ikara guidance radar, 1 Mulloka hull-mounted medium-frequency sonar. **Electronic warfare equipment:** several electronic signal monitoring arrays.

ROCKET: Any vehicle or projectile propelled by a reaction motor whose fuel incorporates its own oxidizer, thus rendering it independent of atmospheric oxygen. In military terms, a rocket is an unguided weapon, usually fin or spin-stabilized, and powered by a rocket engine (as opposed to missiles, which have some form of guidance). The most common types now in service include barrage rockets and "free" rockets (e.g., FROG and HONEST JOHN), both forms of ROCKET ARTILLERY; and aerial rockets carried by aircraft, often in multi-round pods (e.g., FFAR and ZUNI). Barrage rockets can also be used by naval forces for shore bombardment. Light anti-tank weapons (LAWS), such as the classic BAZOOKA, launch short-range, tube-fired rockets. All rockets are inherently less accurate than tube artillery.

ROCKET ARTILLERY: Ground-to-ground unguided rockets with payloads. The two basic types are barrage rockets, normally launched by multiple rocket launchers (MRLs), and "free" rockets.

MRLs were first developed by the Soviet Union and much used during World War II as a cheap and simple substitute for conventional artillery, as well as for their shock value. With the standard wartime model, the BM-13 (the famous *Katyusha* or "Stalin's Organ"), 16 very crude, fin-stabilized 132-mm. rockets were fired off the back of a flatbed truck from an angled rail launcher. The BM-13 had a range of only 9000 m., and was extremely inaccurate, but a battalion of 18 launchers could fire 200 rockets in 30 seconds, blanketing a large area. German troops found its weight of fire and infernal howling seriously demoralizing. The German response from 1943 was the *Nebelwerfer* series of rocket weapons, which formed an ever-larger share of the German artillery during the last year of the war. The U.S. Army introduced the 4.5-in. (114-mm.) "Screaming Mimi" rocket launcher in 1944.

Seen as wartime expedients, MRLs were abandoned by Western armies, but not by the Soviet Union, which valued their suppressive effect, and

their suitability for delivering CHEMICAL WARFARE agents. Current Soviet models include the 40-round, 122-mm. BM-21; the 12-round, 240-mm. BM-24; and the 16-round, 220-mm. BM-27. In the 1960s the West German army developed the 110mm LARS (Light Artillery Rocket System), and in the 1980s the U.S., Britain, Italy, and West Germany all adopted the Vought MULTIPLE LAUNCH ROCKET SYSTEM (MLRS), an extremely sophisticated and very expensive MRL with a variety of high-explosive, submunition, and chemical warheads.

"Free" rockets are simply scaled-up, relatively long-range rockets fired from single-rail launchers, with heavy explosive, chemical, or nuclear warheads. Though outwardly similar to tactical BALLISTIC MISSILES, they do not have any form of guidance but are merely fin- or spin-stabilized. Range is controlled by altering the elevation of the launcher, as with guns. Free rockets are susceptible to crosswinds, as well as variations in engine burn time and thrust, and most have been supplanted by ballistic missiles. See the U.S. HONEST JOHN and the Soviet FROG (Free Rocket Over Ground) series.

ROCKET-ASSISTED PROJECTILE (RAP): An artillery shell with a rocket engine in its base. While they can extend range by up to 33 percent, RAP rounds are inherently less accurate, because of variations in the thrust of the rocket, and changes in the round's center of mass as the propellant burns off; in addition, RAP rounds contain a smaller explosive charge, because of the weight of the rocket motor. Nonetheless, RAP rounds are valued for long-range COUNTERBATTERY fire, the suppression of enemy air defenses (SEAD), and harassment attacks against rear areas. See, more generally, ARTILLERY.

ROCKEYE: A series of anti-tank CLUSTER BOMBS developed by the U.S. Naval Weapons Development Center at China Lake. The 500-lb. (227-kg.) Mk.20 Rockeye II, which entered service in 1968, is 95 in. (2.41 m.) long and 13 in. (330 mm.) in diameter, with a hemispheric nose, a cylindrical body, and a tapered tail with four pop-out stabilizer fins. The original Mk.20 Mod 0 was filled with 247 Mk.118 shaped-charge (HEAT) bomblets, each weighing 1.34 lb. (0.60 kg.) and capable of penetrating the thin top armor of main battle tanks. In 1974 the navy introduced the 750-lb. (340-kg.) Mk.20 Mod 1 (CBU-59), with 717 BLU-77 APAM (anti-personnel/anti-materiel) submunitions weighing 1.02 lb. (0.46 kg.) each, which act as shaped charges against armor, and as fragmentation bomblets against trucks, troops, and other soft targets. A barometric fuze in the nose of the bomb detonates a burster charge at a preset altitude, to scatter the submunitions evenly over a an area of 328 × 157 ft. (100 × 50 m.). Rockeye can be released at a minimum altitude of 100 ft. (30.5 m.) at speeds up to Mach 1 (760 mph/1216 kph).

The standard U.S.anti-tank cluster bomb, Rockeye has been sold to Israel and several other countries.

ROE: See RULES OF ENGAGEMENT.

ROLAND: A low-altitude, battlefield surface-to-air missile, developed by the Franco-German Euromissile consortium to a joint 1964 Franco-German requirement in two versions: the daylight-only Roland 1 and the all-weather Roland 2. The former was bought only by the French army, and has since been displaced by the all-weather version. Roland entered service with French and German forces in 1977; it has since been widely exported. More than 600 launchers and 24,300 missiles have been produced to date.

A maneuverable and compact missile, Roland has four cropped delta wings at midbody and four small nose canards; the wings fold flat against the body, allowing the missile to be housed in a cylindrical launch/storage canister. Roland relies on a form of COMMAND GUIDANCE, and is steered by thrust-vector control. Powered by a dual-pulse (booster/sustainer) solid-fuel rocket motor, it is armed with a small, radar-proximity (VT)-fuzed high explosive warhead.

Roland's twin-arm launcher is mounted on a power-operated turret which can be installed in a variety of vehicles: France, Nigeria, and Spain use the chassis of the AMX-30 main battle tank; Germany and Brazil that of the MARDER infantry fighting vehicle. Rolands mounted on 8 × 8 MAN cross-country trucks are used to defend air bases in West Germany; this cheaper "shelter" version has also been supplied to Iraq, Venezuela, and Argentina. The latter used them against British aircraft in the Falklands, destroying at least two aircraft. Iraqi Rolands were used against Iranian aircraft, with mixed results.

The turret is completely self-contained, with onboard surveillance and tracking RADARS, a backup optical tracker, a FIRE CONTROL computer, a microwave command guidance link, and an automatic loader fed from two 4-round rotary magazines. To load, the launcher is aligned over two vehicle ports through which launch canisters are raised onto the launcher arms.

The pulse-Doppler surveillance radar mounted atop the turret can detect aircraft and helicopters out to ranges of 15–18 km. (9–10.8 mi.). Upon detection, an on-board IFF interrogates the target; if it is confirmed as hostile, tracking data is fed to the fire control computer, which aligns the turret on the target's bearing so that the target can be acquired by the tracking radar mounted on the turret face. When the target comes within range, a missile is launched. The tracking radar has two separate antennas: one follows the target, while the other tracks the missile (an INFRARED detector on the missile tracking antenna captures the missile at a range of 500–700 m.). The fire control computer compares the angle between the two antennas and generates steering commands which are transmitted to the missile over the DATA LINK. The Roland 1's optical tracker is retained in Roland 2 for use when jamming or other ELECTRONIC COUNTERMEASURES make radar tracking ineffective. (Actually, optical target tracking is preferred because it is more accurate than radar tracking and minimizes detectable radar emissions.) Roland is still the most effective Western weapon in its class.

Specifications Length: 7.85 ft. (2.4 m.). Diameter: 6.3 in. (160 mm.). Span: 19.7 in. (500 mm.). Weight, launch: 139 lb. (63 kg.). Speed: Mach 1.5 (1050 mph/1680 kph). Range envelope: 500–6300 m. Height envelope: 20–3000 m. Operators: Arg, Bra, Fra, FRG, Iraq, Nig, Sp, US (abortive program), Ven.

ROLLBACK (Attack): The progressive destruction or neutralization of opposing forces, starting at the periphery and working inward. The term is generally applied to the suppression of enemy air defenses (SEAD), but has recently been applied to the attack of space-based BALLISTIC MISSILE DEFENSES. During the early 1950s, the term was used in a strategic sense, to imply the liberation of Eastern Europe from Soviet rule.

ROLLING AIRFRAME MISSILE (RAM): A shipboard POINT DEFENSE MISSILE developed under a joint U.S.–West German–Danish program (U.S. designation RIM-116A) to supplement both the NATO SEA SPARROW missile and the 20-mm. PHALANX radar-controlled gun. RAM entered flight testing in 1977, and the first production missile was delivered in 1987. Initial operating capability (IOC) for the U.S. Navy was scheduled for 1989, but rising costs and technical problems have delayed deployment.

A simple, lightweight missile, RAM has four folding delta nose canards for steering, and four folding tail fins. The missile is spin-stabilized at launch and continues to roll during flight (hence the name). This permits the use of a simplified single-channel control system, because the same command will at one point in the roll induce yaw, and in another generate pitch. To reduce development costs, RAM uses the rocket engine, the 22.5-lb. (10.2-kg.) CONTINUOUS-ROD fragmentation warhead, and the laser PROXIMITY FUZE of the AIM-9L SIDEWINDER, and the INFRARED (IR) seeker of the STINGER shoulder-fired surface-to-air missile. The missile has a unique dual-mode guidance unit. Immediately after launch, it is controlled by semi-active (radiometric interferometry) guidance via two nose-mounted probe antennas. Its radar seeker has a broad frequency range and can home on signals reflected off the target by any number of surveillance, tracking, or fire control radars, or on the guidance radar signals of hostile anti-ship missiles. The IR seeker is slaved to the radar seeker; when it acquires the "glint" of the target, the missile switches to INFRARED HOMING until impact.

RAM can be accommodated by several different launchers. West German, Danish, and some smaller U.S. ships will be fitted with the EX-41, a lightweight, 24-round pepperbox launcher fitted to a modified Phalanx gun mount. In other ships, an 8-round NATO Sea Sparrow launcher can be modified, with two of the launcher cells for that much larger missile replaced by 5-round RAM launcher modules. Thus each modified launcher would carry six Sea Sparrow and ten RAM.

Specifications Length: 9.16 ft. (2.794 m.). **Diameter:** 5 in. (127 mm.). **Weight, launch:** 159 lb. (72.1 kg.). **Speed:** Mach 2 (1400 mph/2240 kph.). **Range:** 5.85 mi. (9.4 km.). **Height envelope:** 50–15,000 ft. (12–4500 m.).

ROMEO: NATO code name for a class of 20 Soviet diesel-electric attack SUBMARINES completed between 1958 and 1962. An additional 97 were built by China between 1960 and 1982, including 6 for Egypt and 7 for North Korea. A further 10 have been built in North Korea, for a total production run of at least 127 boats. Of the Soviet-built units, 12 have been transferred: 6 to Egypt between 1966 and 1969, 2 to Bulgaria in 1971–72, 2 to Algeria in 1982–83, and 2 to Syria in 1986; the remaining 8 are no longer in service.

The Romeos were developed out of the prolific WHISKEY class, which was based in turn on the German Type XXI U-boat of World War II. The Romeos retain the cigar-shaped hullform and double-hull configuration of the Whiskeys (including the numerous free-flooding holes in the outer hull, which are a major source of drag and flow noise at high speeds). A streamlined sail (conning tower) is located amidships. Control surfaces are the same as in the Whiskeys: retractable bow planes, a single rudder, and stern planes mounted behind the ducted propellers. A retractable snorkel permits diesel operation from shallow submergence.

Armament consists of 8 21-in. torpedo tubes, 6 in the bow and 2 in the stern. The basic load is 18 free-running or acoustical homing TORPEDOES, which can be replaced by bottom-laid MINES on a 2-for-1 basis. The Romeos have outdated sensors and fire controls, but they are maneuverable and relatively quiet, and hence still useful for inshore anti-submarine warfare, and in narrow seas. Egyptian Romeos are being refitted with Singer Librascope FIRE CONTROLS, and a similar upgrade may be applied to some Chinese boats as well. See also SUBMARINES, SOVIET UNION.

Specifications Length: 252.6 ft. (77 m.). **Beam:** 22 ft. (6.7 m.). **Displacement:** 1330 tons surfaced/1700 tons submerged. **Powerplant:** twin-shaft diesel-electric: 2 2000-hp. diesels, 2 1500-hp. electric motors. **Speed:** 15.5 kt. surfaced/13 kt. submerged. **Range:** 7000 n.mi at 5 kt. **Max. op. depth:** 984 ft. (300 m.). **Collapse depth:** 1640 ft. (500 m.). **Crew:** 55. **Sensors:** 1 *Feniks* bow-mounted medium-frequency passive sonar, 1 *Herkules* high-frequency active sonar, 1 "Snoop Plate" surface-search radar, 1 "Quad Loop" radio direction finder. (D/F), 1 "Stop Light" electronic signal monitoring array, 2 periscopes.

RPD: *Ruchnoy pulemet Degtyareva*, a Soviet LIGHT MACHINE GUN (LMG) introduced in the late 1950s. Once the standard squad-level weapon in the Soviet and other Warsaw Pact armies, it has been superseded by the RPK, but remains in service with reserves and many Third World armies; China produces a copy designated "Type 56."

Of conventional design, the RPD has a straight wooden stock and pistol grip, a small foregrip, and a metal bipod. Gas-operated, the RPD fires the same 7.62- × 39-mm. round as the AK-series assault rifles. Fed from 100-round belts housed in drum magazines, its practical rate of fire is limited by overheating (the RPD does not have a quick-change barrel). Other shortcomings include a tendency to jam when the ammunition or gas-feed mechanism get dirty.

Specifications Length, OA: 40.78 in. (1.036 m.). **Length, barrel:** 20.5 in. (521 mm.). **Weight, empty:** 15.65 lb. (7.1 kg.). **Muzzle Velocity:** 740 m./sec. (2427 ft./sec.). **Cyclic rate:** 700 rds./min. **Effective range:** 800 m. (875 yd.).

RPG-7: *Reaktivnyy protivotankovyy grana-tomet*, or Rocket-Propelled Anti-Tank Grenade Launcher, a Soviet man-portable anti-tank weapon introduced in 1962 as a replacement for the RPG-2, which was in turn a close copy of the German *Panzerfaust* of World War II. The RPG-7, now the most common anti-tank weapon world-wide, is still issued to many Motorized Rifle and airborne squads in the Soviet and other Warsaw Pact armies; it is also in service with numerous Third World forces, and is much used by guerrillas and terrorists. It has been produced in China (as the Type 69), Romania, and elsewhere. Captured and purchased RPG-7s also equip Israeli infantry squads.

There are two standard Soviet versions, the RPG-7V for Motorized Rifle troops and the RPG-7D (which can be broken down into two parts) for airborne troops. In both cases, the weapon consists of an open-ended launch tube (with two pistol grips, a trigger, and an optical sight), and an un-guided rocket-propelled grenade. The pistol grip/trigger assembly is mounted near the muzzle, while the 2.5x stadiametric sight is attached to the top of the tube. The rear end of the tube has a flared venturi to dissipate the backblast of the rocket grenade.

The basic rocket grenade is the PG-7, a fin-stabilized shaped-charge (HEAT) round capable of penetrating up to 11.8 in. (300 mm.) of homogeneous steel armor. It is being replaced in Soviet service by the PG-7M, with an improved HEAT warhead and a more powerful rocket motor; its armor penetration is estimated at 14.75 in. (375 mm.). For attacks on soft targets, the weapon can fire the OG-7 HE-fragmentation grenade. All three rounds consist of a larger warhead section, a 40-mm. dual-pulse solid-fuel rocket motor, and a folding-fin unit.

The RPG-7 is loaded by inserting the rocket motor/fin assembly into the muzzle of the launch tube until only the warhead protrudes. The operator places the launcher on his shoulder, lines up the target in the sight, and estimates the range using the stadiametric recticle. When the target is in range, he squeezes the trigger, activating the booster charge, which ejects the grenade safely; stabilizer fins then pop open and the sustainer

charge ignites, accelerating the projectile to its maximum velocity. If it does not hit, the warhead self-destructs after five seconds of flight.

Only the most skilled operators can hit a moving target at anything beyond point-blank range (the U.S. Army estimates an 18 percent chance of a first-round hit at 300 m.). In particular, the stadiametric sight, which derives range from target size, is quite difficult to use in combat, and the grenade is very sensitive to crosswinds. In addition, the performance of its warhead—as with all unsophisticated HEAT warheads—can be seriously degraded by CHOBHAM, spaced, or REACTIVE ARMOR (wire mesh was hung from armored vehicles in Vietnam, to short-circuit its piezoelectric nose fuze). Nonetheless, the RPG-7 can be very effective in short-range ambush situations, and when many can be fired to compensate for inaccuracy. It is also highly effective against bunkers, and in house-to-house fighting (it is the favorite weapon in Beirut). The RPG-7 is being superseded in Soviet service by the handier RPG-16.

Specifications Length: (tube) 38.9 in. (990 mm.); **Diameter:** (tube) 1.57 in. (40 mm.); (rocket) 85 mm. (PG-7)/70 mm. (PG-7M)/50 mm. (OG-7). **Weight:** (tube) 17.4 lb. (7.9 kg.); (rocket) 5.5 lb. (2.5 kg.); (system) 20.2 lb. (10.4 kg.). **Velocity, muzzle:** 117 m./sec. (303.75 ft./sec. **Velocity, max:** 294 m./sec. (964.33 ft./sec.) **Range:** (max.) 300 m.; (eff.) 100 m.

RPG-16: *Reaktivnyy protivotankovyy grana-tomet*, Rocket-Propelled Anti-Tank Grenade Launcher, Saret weapon introduced in the mid-1970s to replace the RPG-7. The RPG-16 has a longer launch tube than the older weapon, and fires a smaller grenade with improved range and armor penetration. As in the RPG-7, the weapon consists of an open-ended launch tube with a pistol grip/trigger assembly, and a muzzle-loaded rocket grenade. The RPG-16 has only one pistol grip, and the complex stadiametric optical sight of the RPG-7 has been replaced by simple fixed sights. In the airborne version (RPG-16D) the tube can be broken down into two parts. The RPG-16 fires a shaped-charge (HEAT) grenade that consists of a warhead section—believed to be an advanced double shaped charge capable of penetrating more than 14.75 in. (375 mm.) of homogeneous steel armor—plus a dual-pulse solid-fuel rocket motor and a folding-fin assembly.

The weapon is loaded by inserting the grenade base-first into the muzzle of the launcher. Unlike the RPG-7, the warhead does not protrude from the end of the tube. The operator places the loaded launcher on his shoulder, lines up the target in his sights, and pulls the trigger. A booster charge ejects the grenade, the folding fins then pop out, and the sustainer charge accelerates it to maximum velocity. A higher maximum velocity, combined with improved fin design, makes the RPG-16 much easier to aim than its predecessor (although only the most skilled operators can hit a moving target at more than 100 m.).

Specifications **Length:** (tube) 3.6 ft. (1.1 m.). **Diameter:** (tube) 2.29 in. (58.3 mm.); (rocket) (58.3 mm.). **Weight:** (tube) 22.6 lb. (10.2 kg.); (rocket) 6.6 lb. (3 kg.); (system) 29.2 lbs. (13.2 kg.). **Velocity, muzzle:** 130 m./sec. (426.5 ft./sec.). **Velocity, max:** 350 m./sec. (1140 ft./sec.). **Range:** (max.) 500 m.; (eff.) 100 m.

RPG-18: *Reaktivnyy protivotankovyy granatomet*, Rocket-Propelled Anti-Tank Grenade Launcher, a Soviet tube-launched, disposable 66-mm. anti-tank rocket, basically a copy of the U.S. M72A1 LAW, introduced in 1977 to complement the more powerful RPG-7 and RPG-16. Like the U.S. weapon, it consists of a rocket housed in a telescoping fiberglass storage/launch canister. To fire, the operator simply extends the tube to its full length, removes protective caps from each end, points, and pulls the trigger. The empty launcher is then discarded.

The RPG-18 rocket is somewhat heavier than the M72A1. It has been reported that the Soviet rocket has an improved shaped-charge (HEAT) warhead with better armor penetration than the U.S. original. Otherwise, its performance is essentially identical to the M72A1. Like the M72A1, it is only marginally effective against modern tanks, but is still useful against lighter vehicles and bunkers. The RPG-18 is believed to be an interim weapon, due to be superseded by the RPG-22, with an 80-mm. rocket.

Specifications **Diameter:** (rocket) 66 mm. **Weight:** (rocket) 3.17 lb. (1.44 kg.). **Muzzle velocity:** 114 m./sec. (373.92 ft./sec.). **Effective range:** 135 m.

RPK: *Ruchnoy pulemet Kalashnikova*, a light machine gun variant of the AK-47 assault rifle, and the standard squad-level support weapon in the Soviet and other Warsaw Pact armies. See AK for details.

RPV: See REMOTELY PILOTED VEHICLE.

RUBIS: A class of four French nuclear-powered attack SUBMARINES (SSNs) commissioned between 1983 and 1988. The smallest operational nuclear submarines anywhere, the Rubis are highly automated and have a very compact powerplant; they displace no more than some diesel-electric submarines, and less than one-third as much as the U.S. LOS ANGELES—class SSNs. The single hull has a modified teardrop form, and is fabricated of HLES-80 steel (equivalent to the HY-80 of the latest U.S. submarines). The Rubis has a flat deck casing, and a tall, streamlined sail amidships. The control surfaces comprise forward diving planes mounted on the sail, and cruciform rudders and stern planes mounted ahead of the propeller.

The Rubis have a very compact form of NUCLEAR PROPULSION: their ingeniously combined pressurized-water reactor/heat exchanger requires less internal volume and less radiation shielding than other reactors. At low speeds, the reactor relies on natural circulation cooling, thus minimizing the use of noisy pumps (which are still needed at higher speeds). Noise is further reduced by turbo-electric drive, in lieu of geared steam turbines: steam from the reactor feeds two large and one small turbo-generator. The latter supplies electricity to the submarine's auxiliary systems, while the former power the electric motor rated that turns the propeller.

Armament consists of 4 550-mm. (21.7-in.) torpedo tubes in the bow for F17 wire-guided anti-ship TORPEDOES, L5 Mod 3 anti-submarine homing torpedoes, and EXOCET SM.39 anti-ship missiles. The basic load is 10 torpedoes and 4 Exocets; MINES can replace other weapons on a 2-for-1 basis. All sensors are linked to a DLT D3 digital torpedo FIRE CONTROL unit and a SADE combat data computer.

A follow-on design (Amethyste) is generally similar, but slightly longer with a hydrodynamically improved hull and better sensors. See also SUBMARINES, FRANCE.

Specifications **Length:** 236 ft. (72.1 m.). **Beam:** 24.9 ft. (7.6 m.). **Displacement:** 2350 tons surfaced/2630 tons submerged. **Powerplant:** single-shaft nuclear, turbo-electric drive, 9500 hp. **Speed:** 18 kt. surfaced/25 kt. submerged. **Max. operating depth:** 985 ft. (300 m.). **Collapse depth:** 1640 ft. (500 m.). **Crew:** 66. **Sensors:** 1 DSUV 22 bow-mounted circular passive sonar, 1 DUUA 2B medium-frequency attack sonar, 1 DUUX 5 Fenelon passive ranging sonar, 1 DUUG/AUUD sonar intercept array, 1 DRUA 33 surface-search radar, ARUR and ARUD electronic signal monitoring arrays, 2 periscopes.

RULES OF ENGAGEMENT: Detailed instructions issued by military authorities to specify binding limits on combat operations. In modern times, ROEs have become an inevitable aspect of all military operations. During the Vietnam War, very stringent ROEs governed the U.S. air war over North Vietnam, and seriously hampered the conduct of operations. Overly strict ROEs were likewise blamed for the nonreaction of the USS *Stark*, when attacked by Iraqi aircraft in the Persian Gulf in 1988. On the other hand, ROEs relaxed because of the *Stark* incident later contributed to the destruction of an Iranian airliner by the USS *Vincennes*. Given the potential political consequences of any military action, and the prevalence of ambiguous situations in the nuclear era, the tension between political control and military expediency will likely remain unresolved.

RUR-5A: U.S. anti-submarine missile. See ASROC.

RV: See REENTRY VEHICLE.

RVSN: *Raketnye Voiska Strategicheskogo Naznacheniya* (literally, Rocket Troops of Strategic Designation), or Strategic Rocket Troops, the branch of the Soviet armed forces responsible for all land-based ballistic missiles. Established as a separate service in 1959, the RVSN is ranked first in precedence among all Soviet services, with the first choice in selecting conscripts and the highest priority for materiel. See also ICBMS, SOVIET UNION; IRBMS, SOVIET UNION.

RWR: See RADAR WARNING RECEIVER.

S

S-2 IRBM: See IRBMS, FRANCE.

S-3 IRBM: See IRBMS, FRANCE.

SA-: NATO designation for Soviet land-based surface-to-air missiles (SAMs), further identified by code names beginning with the letter G. The original SA-1 Guild, now phased out, was a large, single-stage missile. Soviet land-based SAMs in service include: SA-2 GUIDELINE; SA-3 GOA; SA-4 GANEF; SA-5 GAMMON; SA-6 GAINFUL; SA-7 GRAIL; SA-8 GECKO; SA-9 GASKIN; SA-10 GRUMBLE; SA-11 GADFLY; SA-12A GLADIATOR; SA-12B GIANT; SA-13 GOPHER; and SA-14 GREMLIN.

SA'AD 16: Iraqi short-range ballistic missile program. See AL HUSAYN.

SA'AR: A class of 12 Israeli fast missile patrol boats (PGMS) commissioned in 1970. Built in France to a modified German Lurssen design, and then embargoed shortly after their completion in 1969, the Sa'ars were seized overtly by Israeli receiving crews, and sailed from Cherbourg to Haifa on Christmas Eve 1969. They served with distinction in the 1973 Yom Kippur War, destroying several Syrian and Egyptian OSA and KOMAR missile boats, and have since been active in numerous anti-guerrilla operations and the 1982 Lebanon War.

Like all Lurssen designs, the Sa'ars have flush-decked steel hulls with rounded bilges and a small, light-alloy superstructure forward. They are quite seaworthy, and have excellent endurance for their size. Subclasses are differentiated by armament. All 12 were originally completed in France with an all-gun armament of 3 single 40-mm. ANTI-AIR-CRAFT guns. Once in Israel, all were upgraded to Sa'ar II standard, with one OTO-MELARA 76.2-mm. DUAL PURPOSE gun on the foredeck, a 40-mm. gun aft, 2 manually operated 12.7-mm. (.50-caliber) MACHINE GUNS, and either 5 or 6 GABRIEL anti-ship missiles in twin and triple launchers. The last 6 boats were later upgraded once again to Sa'ar III standard, with 3 Gabriels replaced by a pair of U.S. RGM-84 HARPOON missiles. Four Sa'ar IIs have been further modified by the addition of 2 324-mm. tubes for lightweight homing TORPEDOES and other anti-submarine warfare equipment.

In 1973, Israel introduced an enlarged version designated Sa'ar IV, but better known as the RE-SHEF class. The Sa'ar V is now being built as a 1150-ton CORVETTE of radically new design. See also FAST ATTACK CRAFT, ISRAEL.

Specifications **Length:** 147.6 ft. (44.9 m.). **Beam:** 23 ft. (7 m.). **Draft:** 8.2 ft. (2.5 m.). **Displacement:** 220 tons standard/250 tons full load. **Powerplant:** 4 3500-hp. hp. MTU MD871 diesels, 4 shafts. **Fuel:** 30 tons. **Speed:** 40 kt. **Range:** 1000 n.mi. at 30 kt./1600 n.mi. at 20 kt./2500 n.mi. at 15 kt. **Crew:** 35–40. **Sensors:** 1 Thomson-CSF Neptune TH-D 1040 surface-search radar, 1 Selenia Orion RTN-10X fire control radar, 1 EDO 780 variable depth sonar (optional), 1 VHF direction finder (D/F). **Electronic warfare equipment:** 1 Elta NM-53 or NS-9000 electronic signal monitoring array, 6 24-tube chaff launchers, 4 larger chaff rocket launchers.

SABER (SS-20): NATO code name for the Soviet RSD-10 *Pioner* mobile, intermediate-range ballistic missile (IRBM). Developed in the late 1960s as a replacement for both the SS-4 SANDAL and SS-5 Skean, the SS-20 caused much alarm within NATO because its payload of three MULTIPLE INDEPENDENTLY TARGETED REENTRY VEHICLES (MIRVs), greatly improved accuracy, and mobility meant that the SS-20 could be used to attack military targets in a discriminating fashion (as opposed to SS-4s and SS-5s, usable only to attack large cities).

In response to the SS-20, NATO agreed to deploy 562 LONG-RANGE INTERMEDIATE NUCLEAR FORCE (LRINF) missiles. In the context of Gorbachev's new policy, however, in December 1987, the U.S. and U.S.S.R. signed the INF TREATY, which prescribed the elimination of all missiles with ranges between 500 and 5500 kilometers, including the SS-20 and the LRINF missiles.

The SS-20 was a two-stage, solid-fuel missile derived from the first two stages of the SS-16 SINNER intercontinental ballistic missile (ICBM). Both stages are believed to be fabricated from rolled steel. The first stage has two gimballed nozzles for steering and attitude control; the second stage has a single gimballed nozzle. Introduced in 1977, the initial SS-20 Mod 1 was armed with a single 650-kT REENTRY VEHICLE (RV). The Mod 2, which also entered service in 1977, had three 150-kT MIRVs mounted on a solid-fuel POST-BOOST VEHICLE (PBV). Introduced in 1985, the Mod 3 was considerably more accurate than earlier versions, and carried three 50-kT MIRVs. The missile's accuracy depends on whether or not it is launched from a presurveyed site and on the time available to align the guidance system. All three versions are controlled by INERTIAL GUIDANCE, but the Mod 3's system is significantly better than its predecessors'.

The SS-20 was fielded in a sealed launch/storage canister mounted on an 8x8 wheeled transporter/erector/launcher (TEL). The canister was stowed horizontally for traveling, and raised vertically off the rear of the vehicle for launch with the COLD LAUNCH technique, whereby the missile is ejected from the canister before the main engine ignites.

SS-20s were deployed in brigades of 9 TELs, 9 reload vehicles, and 9 launch control vehicles. All were normally based in permanent facilities but were to be dispersed to concealed launch sites as needed. According to Soviet figures released under the 1987 treaty, some 800 missiles had been deployed by 1988—nearly twice as many as estimated by U.S. intelligence (the huge error reveals the difficulty of monitoring mobile missiles with surveillance satellites). See also IRBMS, SOVIET UNION.

Specifications Length: 54.08 ft. (16.5 m.). Diameter: (1st stage) 5.87 ft. (1.8 m.); (2nd stage) 4.82 ft. (1.46 m.). Weight, launch: 40,000 lb. (18,181 kg.). Range: (½) 3105 mi. (5000 km.); (3) 4350 mi. (6700 km.). CEP: (3) 340 m. presurveyed/425 m. unsurveyed.

SABRE: North American F-86, 1950s-vintage FIGHTER, once the backbone of the U.S. fighter force, and still in limited service with Argentina, Bolivia, the Philippines, South Korea, Taiwan, Tunisia, Uruguay, and Yugoslavia.

SAC: See STRATEGIC AIR COMMAND.

SACEUR: Supreme Allied Commander, Europe. See SHAPE.

SACLOS: Semi-Automatic Command to Line-of-Sight, a form of COMMAND GUIDANCE most often employed in wire-guided ANTI-TANK GUIDED MISSILES (ATGMs). With the SACLOS technique, the operator tracks the target through an optical or ELECTRO-OPTICAL sight. An INFRARED sensor built into the sight unit tracks the missile after launch; on the basis of the angle between the missile and the line of sight from operator to target, a guidance computer generates steering commands to bring the missile onto the line of sight. There is no problematic "gathering" process as in manual systems (see MCLOS): all the operator must do is keep the sight centered on the target until the missile hits. That is, however, the main shortcoming of SACLOS: the need for a *continuous* line of sight throughout the engagement. This means that a target can evade a missile by moving into masking terrain or because smoke interrupts the line of sight. And of course, the operator must remain exposed to enemy counterfire during the engagement, unless the launcher is mounted on an armored vehicle or housed in a fortification. For specific SACLOS systems, see DRAGON; HOT; MILAN; SPANDREL; SPIRAL; SPIGOT; TOW.

SADARM: Sense and Destroy Armor, an artillery-delivered "smart" submunition under development for the U.S. Army. Actually intended more for COUNTERBATTERY than ANTI-TANK missions, SADARM was initiated in the early 1970s as an 8-in. (203-mm.) howitzer round, but is now being developed for the 155-mm. M109 howitzer and the 227-mm. MULTIPLE LAUNCH ROCKET SYSTEM (MLRS) rocket launcher.

Both versions are 10 in. (328 mm.) long and weigh some 30 lb. (13.65 kg.). The 155-mm. version, however, is 5.8 in. (147.33 mm.) in diameter, while the MLRS version is 6.9 in. (175.26 mm.) in diameter. The main difference between the two versions is the warhead's diameter; sensors, fuzes, and electronics are virtually identical. Two SADARM submunitions are carried in a 155-mm. cargo round, while an MLRS rocket can hold six.

SADARM's SELF-FORGING FRAGMENT (SFF) warhead consists of a concave tungsten disk backed by plastic explosives. Upon detonation, the force of the explosion converts the disk into a penetrator bolt which is projected at 6000–8000 ft. (1830–2440 m.) per second along the axis of the disk. Two different sensors are under development. One employs active millimeter wave (MMW) radar and two-color INFRARED (IR), while the other uses a combination of active and passive MMW and one-color IR. The use of two different sensor techniques allows the submunition to identify targets against a confused background, and is meant to resist countermeasures.

In the engagement sequence, a 155-mm. cargo round or MLRS rocket is launched into the general target area. At a preset height, a PROXIMITY FUZE ejects the submunitions (in the 155-mm. round, SADARM is base-ejected; in MLRS, it is side-ejected). In each case, a drogue parachute deploys to decelerate and despin the submunition. Once the SADARM is slowed, the drogue is jettisoned and a parasail is deployed in its place. This further slows the descent while causing the submunition to spin, thereby allowing the MMW and IR sensors to scan a 30° cone on the ground below. If the sensors detect a tank, self-propelled gun, or other vehicle, the warhead is activated at a height of about 75 ft. (22 m.), driving the SFF through the target's thin top armor. If no target is detected, the submunition settles on the ground to act as a time-delay anti-personnel MINE.

A ground-emplaced version of SADARM is under test as part of the Wide Area Mine program. The U.S. Air Force has a similar SFF submunition, called SKEET.

SADEYE: Mk.15 700-lb. CLUSTER BOMB developed by the U.S. Navy in the 1960s. Sadeye is 92 in. (2.33 m.) long and 16 in. (406 mm.) in diameter; it holds a total of 2100 anti-personnel bomblets.

SAFEGUARD: A U.S. BALLISTIC MISSILE DEFENSE system projected in the late 1960s to protect MINUTEMAN ICBM silos, bomber bases, and nuclear command centers, and to provide a ''thin'' area defense of U.S. population centers against a small and/or unsophisticated nuclear strike (Chinese or accidental). A single site completed at Grand Forks, North Dakota, was deactivated the day after completion in 1972, when the program was abandoned in the wake of the 1972 ABM TREATY.

The system consisted of Perimeter Acquisition radars (PARs), Missile Site radars (MSRs), and nuclear-armed SPARTAN long-range and SPRING short-range interceptor missiles. A picket line of PARs was to detect ICBMs approaching from over the poles, or SLBMs coming in from the oceans, and perform initial target designation; tracking would then be passed to the MSRs, which could also discriminate between REENTRY VEHICLES (RVs) and PENAIDS. Upon target acquisition, the MSRs were to launch Spartans for exoatmospheric interception, followed as necessary by endoatmospheric Sprints for terminal interception.

The technical reason for the cancellation of Safeguard was the development of MULTIPLE INDEPENDENTLY TARGETED REENTRY VEHICLES (MIRVs), which make it cheaper for the attacker to add more warheads than for the defender to add more interceptors. The political reason was the successful negotiation of the ABM Treaty, which limits the U.S. and U.S.S.R. to two (later one) 100-missile sites. Whereas the Soviet Union chose to deploy and maintain its GALOSH ABM system around Moscow, the U.S. dismantled the one Safeguard site to save costs.

SAGE: Semi-Automated Ground Environment, a GROUND-CONTROLLED INTERCEPT (GCI) system formerly part of the U.S.-Canadian NORAD air defenses. SAGE consisteds of 13 overlapping radar control centers which could track intruding aircraft and direct Bomarc surface-to-air missiles or manned interceptors against them. Automated data links provided interceptor pilots with both steering and weapon-release cues, giving a high theoretical kill probability against nonmaneuvering targets (i.e., bombers). SAGE was supplemented by BUIC, the Back-Up Interceptor Control system. Both have since been replaced by the JOINT SURVEILLANCE SYSTEM (JSS).

SAGGER (AT-3): NATO code name for the Soviet PTUR-64 *Malyutka* ANTI-TANK GUIDED MISSILE (ATGM). First introduced in the mid-1950s and perfected in the 1960s, Sagger was the standard Soviet ATGM during the 1970s, and was widely exported. China produces an unlicensed copy, and Israel has captured large stocks, which it supplies to Lebanese Christian militias. Sagger has been replaced in first-line Soviet units by the AT-4 SPIGOT and AT-5 SPANDREL.

Sagger consists of a detachable 6.6-lb. (3-kg.) shaped-charge (HEAT) warhead capable of penetrating 15.75 in. (400 mm.) of homogeneous steel armor, a guidance unit, a dual-pulse solid rocket engine, wire spools that pay out to maintain the connection to the launcher, and four folding, cruciform control fins. A simple weapon, Sagger is controlled by WIRE GUIDANCE with manual command to line-of-sight (MCLOS); the operator steers the missile with a joystick. Sagger has a maximum flight time of 27 sec.; an improved version (Sagger B), introduced in 1973, is slightly faster, reducing the maximum time of flight to 21 sec.

To engage a target, the operator must first line it up in an 8x periscopic optical sight. When the target is within range, the missile is launched, and flies straight ahead to its minimum range of 300 m. The operator must then "gather" the missile (bring it onto the line of sight to the target) by using his joystick. Once the missile is aligned with the target, the operator must continue to track both until impact, making course corrections as needed. The entire procedure requires considerable skill and training: in the Soviet army, an operator must fire 23,000 simulated missiles before being considered qualified. Even so, hitting a moving target can be quite difficult.

Sagger can be fired from a number of man-portable, vehicle, and helicopter launchers. The basic infantry version, the "suitcase Sagger," consists of two missiles, a simple rail launcher, and a sight unit, all housed in a light alloy backpack. The sight unit includes cabling which allows the operator to be offset up to 49 ft. (15 m.) from the launcher; this can increase minimum range to 500–800 m. because of the increased difficulty of gathering the missile. Vehicle mounts include a sextuple rail launcher for the BRDM-1 (no reloads) and BRDM-2 (with 8 reloads); in the BMP and BMD infantry fighting vehicles, a single rail is mounted over the 73-mm. gun. Helicopter mounts are available for the Mi-2 Hoplite, Mi-8 HIP, and Mi-24 HIND. In the late 1970s, an improved version (Sagger C) with semi-automatic command to line-of-sight (SACLOS) guidance was introduced for helicopters and vehicles.

Specifications **Length:** 2.9 ft. (885 mm.). **Diameter:** 4.69 in. (119 mm.). **Weight, launch:** 24.9 lb. (11.3 kg.). **Speed:** (A) 268 mph (120 m./sec.); (B) 335 mph (150 m./sec.). **Range:** 300–3000 m. **Operators:** Afg, Alg, Ang, Egy, Eth, Ind, Iran, Iraq, Lib, Moz, Nic, N Kor, Syr, Ug, USSR and other Warsaw Pact forces, Viet

ST. LAURENT: A class of 7 Canadian ANTI-SUBMARINE WARFARE (ASW) FRIGATES commissioned in 1956–57. The St. Laurent herself was stricken in 1974, and 3 others were placed in reserve in 1987, while the remaining 3 have been extensively modified and remain in active service. Developed in response to a 1949 NATO requirement, the St. Laurents were derived from the British Type 12 (Whitby) class, but differ externally because of specifically Canadian operational requirements. The St. Laurents are flush-decked, with a rounded "turtleback" foredeck to reduce ice buildup; they have a squat, blocky superstructure forward and side-by-side stacks amidships. Like all Canadian warships, the St. Laurents are very rugged and excellent sea boats, well suited to operations in the Arctic and North Atlantic.

Armament as completed consisted of 2 U.S. 3-in. 50-caliber twin DUAL PURPOSE guns mounted forward and amidships, 2 single 40-mm. BOFORS GUNS, and 2 British Mk.10 LIMBO depth charge mortars. These weapons, already obsolete at the time of their introduction, were replaced between 1961 and 1963. One Limbo and the aft 3-in. gun were replaced by a large hangar and landing deck for an SH-3D SEA KING ASW helicopter, and two sets of Mk.32 triple tubes for lightweight ASW homing TORPEDOES were added for close-in defense. The three active ships are now scheduled for a major overhaul during which the VARIABLE-DEPTH SONAR and Limbo will be replaced by a passive TOWED ARRAY SONAR. See also FRIGATES, CANADA.

Specifications **Length:** 371 ft. (113.1 m.). **Beam:** 42 ft. (12.8 m.). **Draft:** 13.25 ft. (4.2 m.). **Displacement:** 2260 tons standard/2860 tons full load. **Powerplant:** twin-shaft steam: 2 oil-fired boilers, 2 sets of geared turbines, 30,000 shp. **Speed:** 28 kt. **Range:** 4750 n.mi. at 14 kt. **Crew:** 228. **Sensors:** 1 SPS-10 surface-search radar, 1 SPS-12 air-search radar, 1 SPG-48 fire control radar, 1 SQS-50B hull-mounted medium-frequency sonar, 1 SQS-501 bottom-mapping sonar, 1 SQS-504 variable depth sonar. **Electronic warfare equipment:** 1 WLR-1 radar warning receiver.

SALADIN: The British Alvis 6 × 6 ARMORED CAR, in limited service with the British army and widely exported. Developed as a successor to the World war II Daimler armored car, the Saladin was designed in the late 1940s as part of a family of vehicles which also includes the SARACEN armored personnel carrier, but did not enter service until 1958. By 1972, a total of 1177 Saladins had been delivered; most have been phased out of the

British army in favor of the SCORPION tracked reconnaissance vehicle.

Saladin's all-welded steel hull is divided into a crew compartment up front and an engine compartment in the rear. Armor thickness varies from 8 to 16 mm., only enough to resist small arms and shell splinters. The driver is provided with a vision slit and two periscopes. The remaining crewmen sit in a manually traversed, all-welded steel turret, which has a maximum armor thickness of 32 mm. The commander, who must also act as loader, sits on the right, with the gunner on the left. The gunner is provided with a simple optical (stadiametric) sight.

The main armament is a low-velocity 76.2-mm. L5 gun with 42 rounds of ammunition. The gun can fire HE, HESH, and smoke ammunition, with a muzzle velocity of only 533 m./sec. and maximum effective range of some 1000 m. Secondary armament consists of a 7.62-mm. coaxial MACHINE GUN, often supplemented by a second, pintle-mounted machine gun on the turret roof. Two six-round smoke grenade launchers are mounted on either side of the hull.

Specifications **Length:** 16.16 ft. (4.92 m.). **Width:** 8.33 ft. (2.54 m.). **Height:** 9.6 ft. (2.93 m.). **Weight, combat:** 11.6 tons. **Powerplant:** 170-hp. Rolls Royce B80 8-cylinder gasoline. **Hp./wt. ratio:** 14.65 hp./ton. **Fuel:** 100 gal. (440 lit.). **Speed, road:** 45 mph (72 kph). **Range, max.:** 250 mi. (400 km.). **Operators:** Bah, Gha, Hon, Indo, Ken, Ku, Leb, Lib, N Ye, S Leo, Sri L, Sud, S Ye, Tun, UAE, UK.

SALT I (ACCORDS): Strategic Arms Limitation Talks accords, a series of arms control agreements between the U.S. and Soviet Union which set limits on the types and numbers of offensive and defensive "strategic" weapons. Signed in Moscow on 26 May 1972 and ratified by the U.S. Senate in October 1972, the SALT I accords actually include four separate agreements, including the ABM TREATY, the "Interim Agreement on Certain Measures with Respect to the Limitation of Offensive Arms," a "Protocol" to the Interim Agreement, a list of "Agreed Interpretations and Unilateral Statements," there was also a separate statement of "Basic Principles of Relations" between the U.S. and U.S.S.R.

In common usage, however, SALT I usually refers only to the Interim Agreement and its attached Protocol. Under the terms of the Interim Agreement, both parties agreed not to begin construction of new ICBM silos after 12 July 1972, and not to convert existing "light" ICBM silos for use by "heavy" missiles. The latter clause effectively limited the U.S.S.R. to 308 heavy ICBMs, and the U.S. to only 54. The main accomplishment of the Interim Agreement, however, was to limit the number of SLBM launchers and ballistic missile submarines (SSBNs). Under the terms of the Protocol, the U.S. was limited to 44 SSBNs and 710 launch tubes, while the U.S.S.R. was limited to 62 submarines and 950 launchers. The modernization of existing forces was allowed, provided that the maximum ceilings were not exceeded.

For their verification, the accords relied exclusively on "national technical means," i.e., surveillance satellites and electronic intelligence (ELINT) gathering. Each party agreed not to interfere with the other's national technical means, or to encrypt missile test telemetry data. A Standing Consultative Commission was established to resolves differences of interpretation and to adjudicate claims of violations. The Interim Agreement was to last for five years, pending a definitive arms control agreement.

SALT I was modified by the "Joint Statement on Strategic Offensive Arms," better known as the Vladivostok Accord. Signed on 24 November 1974, it extended the Interim Agreement through 1977, and established the goal of reaching a definitive agreement by then. It also stated that the prospective definitive agreement (SALT II) would place limits on the aggregate numbers of all long-range nuclear delivery systems (ICBMS, SLBMs, and bombers), and specific sublimits on the number of MULTIPLE INDEPENDENTLY TARGETED REENTRY VEHICLES (MIRVs).

From a U.S. perspective, SALT I's major flaw was that it allowed the Soviet Union a COUNTERFORCE potential with its heavier missiles. Moreover, the agreement was overly optimistic on the efficacy of national technical means, as shown by the ensuing disputes over the deployment of the Soviet SS-16 SINNER ICBM and other possible violations.

SALT II (ACCORDS): Specifically, the "Treaty Between the United States and USSR on the Limitation of Strategic Offensive Arms," signed at Vienna on 18 June 1979. Negotiated according to the principles established by the 1974 Vladivostok Accord, SALT II was to supersede the SALT I "Interim Agreement" of 1972. The accords consist of the treaty proper, a "Protocol" to the Treaty, a "Memorandum of Agreement" establishing current force levels, a Soviet "Unilateral Statement" on the capabilities of the Tu-22M

BACKFIRE bomber, and a "Joint Statement of Principles" for further arms control negotiations.

Under the terms of the treaty, both parties agreed: (1) to a ceiling of 2400 long-range nuclear delivery systems, including ICBMS, SLBMS, and "strategic" BOMBERS; (2) not to begin construction of new ICBM silos or to convert existing "light" silos for use by heavy ICBMs; and (3) not to provide ICBM silos with rapid reload capability. Further, both parties agreed to a sublimit of 1320 for ICBMs and SLBMs with MULTIPLE INDEPENDENTLY TARGETED REENTRY VEHICLES (MIRVs), including not more than 820 MIRVed ICBMs, with no more than ten MIRVs per missile. In addition, both parties agreed not to deploy more than one new type of ICBM within the span of the agreement, not to deploy CRUISE MISSILES with ranges of more than 600 km. (360 mi.), and not to deploy nuclear missiles in space or on the seabed. Ample warning was also to be given before all ballistic missile test flights, and each side agreed not to interfere with the other's "national technical means" of verification (i.e., surveillance satellites). Finally, a Standing Consultative Commission was established to resolve ambiguities and adjudicate disputes.

Ratification of SALT II was strongly opposed in the U.S. Senate for various reasons, including the method of counting heavy bombers and the treaty's exclusive reliance on national technical means for verification. All U.S. bombers, including unflyable B-52 Stratofortresses in open-air "boneyards," were to be counted against the total of U.S. delivery systems, while the new Soviet BACKFIRE was not not deemed "strategic," and not counted against Soviet limits; the unilateral Soviet statement that Backfire was not a "strategic" bomber did not persuade many senators. The verification issue was more intractable, because the U.S.S.R. had been encrypting missile test telemetry for some time (in violation of SALT I). More fundamentally, critics of the agreement believed that it perpetuated Soviet strategic superiority, and left the U.S. vulnerable to a disarming COUNTERFORCE attack. Ratification by the U.S. Senate was doubtful, and after the Soviet invasion of Afghanistan in December 1979, the treaty was withdrawn from consideration. Nonetheless, the U.S. announced that it would abide by the ceilings of SALT II, so long as the U.S.S.R. did likewise. This situation persisted until the mid-1980s, when new Soviet missile and bomber deployments exceeded the limits for MIRVs and total delivery systems. See also START.

SALVO: Two or more shots fired simultaneously (or nearly so) from the same battery at the same target.

SALYUT: A series of Soviet manned space stations launched in the 1970s and early 1980s. Two of these, *Salyut 3* and *Salyut 5*, were military spacecraft placed in low, 217-mi. (350-km.) orbits and employing military telemetry channels. Each was equipped with a 32.8-ft. (10-m.) focal-length telescope (ostensibly for solar observation), with a ground resolution of roughly 1 ft. (30 cm.); film and magnetic tapes were regularly returned to earth from both in recovery capsules. See also SATELLITES, MILITARY; SOYUZ; SPACE, MILITARY USES OF.

SAM: Surface-to-air missile. See ANTI-AIRCRAFT; MISSILE, GUIDED.

SAMPSON (SS-N-21): NATO code name for a Soviet submarine-launched CRUISE MISSILE (SLCM), believed to be a variant of the AS-15 KENT air-launched cruise missile (ALCM) and the SS-C-4 ground-launched cruise missile (GLCM).

Introduced in the early 1980s, and broadly similar to the U.S. BGM-109 TOMAHAWK, Sampson has a cylindrical body with pop-out wings and cruciform tail fins. Payload is believed to be a 270-lb. (123-kg.), 500-KT nuclear warhead. It relies on DOPPLER-aided INERTIAL GUIDANCE, and may also have some form of terminal homing. Sampson is powered by a 6-ft. (1.83-m.) solid rocket booster and a small turbojet or turbofan sustainer engine. After launch, the missile climbs to a cruising altitude of 10,000 ft. (3050 m.), diving to 160 ft. (50 m.) on the final run to the target.

Although it can be launched from standard 21-in. (533 mm.) torpedo tubes, the SS-N-21 has been associated only with the latest Soviet nuclear attack submarines (AKULA and SIERRA). In addition, several YANKEE-class ballistic-missile submarines (SSBNs) have been converted to cruise missile carriers ("Yankee Notch") with up to 40 Sampsons. The ground-launched SS-C-4, mounted on a 4-round, wheeled launcher, was to be scrapped under the 1987 INF TREATY.

Specifications **Length:** 26 ft. (7.92 m.). **Diameter:** 21 in. (533 mm.). **Weight, launch:** 3740 lb. (1700 kg.). **Speed:** Mach 0.8 (600 mph/960 kph). **Range:** 1800 mi. (2880 km.). **CEP:** 150 m.

SA-N-: NATO prefix designator for Soviet shipboard surface-to-air missiles (SAMs), many of which are navalized versions of land-based systems. Models currently in service include: SA-N-1

GOA; SA-N-2 GUIDELINE; SA-N-3 GOBLET; SA-N-4 GECKO; SA-N-5 GRAIL; SA-N-6 GRUMBLE; SA-N-7 GADFLY; SA-N-8 GREMLIN; and SA-N-9.

SA-N-9: NATO designation for a modern Soviet short-range, vertically launched naval surface-to-air missile (SAM), currently deployed on UDALOY-class destroyers, KIROV-class battle cruisers, and the KIEV-class aircraft carrier *Novorossiysk*.

Believed to be the successor to the ubiquitous SA-N-4 GECKO, the SA-N-9 is thought to have a maximum range of some 9 mi. (14.5 km.) and a ceiling of 60,000 ft. (18,250 m.). It apparently relies on SEMI-ACTIVE RADAR HOMING (SARH), combined with active radar or passive INFRARED terminal homing. Guidance is provided by a "Cross Swords" radar system, which incorporates both an acquisition and tracking radar and two separate illumination radars, and by a secondary ELECTRO-OPTICAL tracking unit.

SANDAL (SS-4): NATO code name for the Soviet R-12 medium-range ballistic missile (MRBM), developed in the early 1950s; it achieved notoriety during the 1962 Cuban Missile Crisis. Only a few remain in service; all are to be eliminated under the terms of the 1987 INF TREATY.

SANDBOX (SS-N-12): NATO code name for the Soviet P-35 ship- and submarine-launched ANTI-SHIP MISSILE introduced in the early 1970s, essentially an incremental development of the SS-N-3 SHADDOCK, with a higher speed, longer range, and improved accuracy.

A large, airplane-configured missile, Sandbox has sharply swept, midmounted wings, a small, swept tailplane, and a small vertical rudder. Powered by a turbojet engine fed by an annular air inlet ahead of the wings, Sandbox can carry a 2200-lb. (1000-kg.) high-explosive and a 350-KT nuclear warhead. Sandbox relies on radio COMMAND GUIDANCE through the midcourse phase, combined with active RADAR or ANTI-RADIATION homing for the terminal phase. The missile usually cruises at high altitude, and then executes a high-speed diving attack.

Sandbox is launched from large individual canisters aboard the KIEV aircraft carriers and SLAVA guided-missile cruisers, or from erectable, watertight twin launchers on modified ECHO and JULIETT guided-missile submarines. Because the missile can be launched only on the surface, these submarine would be extremely vulnerable through the midcourse phase.

The missile is launched in the general direction of the target, using data supplied by the ship's sensors or from OVER-THE-HORIZON TARGETING (OTH-T) aircraft. Its in-flight guidance requires the transmission of a radar picture of the target area from the spotter aircraft to the launch platform, via a video DATA LINK (VDL). The launch platform then designates the target and relays the radar picture up to the missile via a "Trap Door" guidance radar in the Kievs, or a "Front Door/Front Piece" radar in the Julietts, Echos, and Slavas. As with the earlier Shaddock, the missile's weaknesses include its over-reliance on OTH-T, its high-altitude approach (which makes it easier to detect and intercept), and the need for the launch platform to continue passing in-flight guidance updates until the terminal phase of the engagement. Launching submarines would have to remain surfaced for as long as 15 minutes while transmitting easily detected radar signals. On the other hand, Sandbox is a very powerful missile, which could destroy or disable all but the largest warships with a single hit.

Specifications **Length:** 38.4 ft. (11.7 m.). **Diameter:** 2.8 ft. (853 mm.). **Span:** 8.2 ft. (2.5 m.). **Weight, launch:** 11,025 lb. (5011 kg.). **Speed:** Mach 2.5 (1620 mph/2600 kph). **Range:** 330 mi. (550 km.).

SAP: Semi-Armor-Piercing; artillery shells or aerial bombs with a reinforced nose and casing, designed to penetrate concrete and (nonarmor) steel plate before exploding. SAP munitions are an intermediate type with lighter casings and larger explosive charges than true ARMOR-PIERCING rounds, but smaller charges than "high capacity" rounds and demolition bombs. The term is now obsolete, having been superseded by "common shell" and "general purpose" bombs.

SAPPER: Another term for assault engineers, i.e., troops trained and equipped to support combat forces by laying or clearing mines, emplacing demolitions, erecting field fortifications, creating obstacles, and operating specialized equipment such as FLAMETHROWERS, fougasses (fixed flame projectors), line charges, and armored engineer vehicles (AEVs). See also ENGINEERS.

SAR: 1. Search and Rescue; operations undertaken to locate and recover survivors, especially aircrew. SAR is often carried out by specially trained personnel with commando and paramedic skills, equipped with specialized helicopters fitted with rescue hoists and precision navigation equipment. The use of helicopters for SAR was pioneered by the U.S. during the Korean War and perfected during the war in Southeast Asia. Effective SAR is now considered vital to aircrew morale.

2. See SYNTHETIC APERTURE RADAR.

SARACEN: Alvis 6x6 wheeled ARMORED PERSONNEL CARRIER (APC), based on a chassis shared with the SALADIN armored car. Developed in the late 1940s and produced from 1952, some 1838 Saracens were delivered through 1972.

Resembling an armored truck, the Saracen has the engine compartment up front, with a driver/passenger compartment in the rear. The all-welded steel hull has a maximum thickness of 16 mm., only enough to resist small arms fire and shell fragments. The driver, seated up front in the center, is provided with three armored vision slits. The commander sits immediately to his rear in a manually rotated turret armed with a 7.62-mm. machine gun. The passenger compartment can accommodate up to ten troops on bench seats. Access is provided by twin doors in the rear of the vehicle. There are three firing ports on each side, and a roof hatch at the rear with a 7.62-mm. machine gun on a ring mount. Like the Saladin, the vehicle is not amphibious, nor is it equipped with an NBC filter unit or night-vision equipment. Variants include a command vehicle and an ambulance.

Specifications Length: 17.16 ft. (5.37 m.). Width: 8.33 ft. (2.54 m.). Height: 8.1 ft. (2.47 m.). Weight, combat: 8.6 tons. Powerplant: 170-hp. Rolls Royce B80 Mk.6A 8-cylinder gasoline. Hp./wt. ratio: 19.76 hp./ton. Fuel: 55 gal. (242 lit.). Speed, road: 45 mph (72 kph). Range, max.: 248 mi. (397 km.). Operators: Indo, Jor, Ku, Leb, Lib, Nig, Qat, S Af, Sud, Thai, UAE, UK, Hong Kong Police.

SARH: See SEMI-ACTIVE RADAR HOMING.

SARIN: Lethal NERVE AGENT (U.S. designation GB) developed in Germany during the Second World War. A colorless, odorless, percutaneous (skin-permeable) liquid, Sarin, like all nerve agents, acts by binding permanently with aceto-chlinesterase (AChE), an enzyme essential for the central nervous system. Symptoms of Sarin intoxication include involuntary contractions of the large muscles, convulsions, frothing at the mouth, and death from respiratory collapse within 15 minutes.

Sarin is extremely potent, with an LD-50 (the dose required to kill 50 percent of all exposed personnel) of 100 milligrams per min. per cubic m. $(mg./min/m.^3)$ if inhaled or 10,000 mg./min./m.^{3} percutaneously. Because Sarin is skin-permeable,

complete protective suits are needed for protection, in addition to respirators.

Troops can be "hardened" against exposure by a prophylactic regime of synergistic drugs. One set of drugs, administered before exposure, *temporarily* binds with some AChE, forming a protected pool of the enzyme. A second series of drugs, administered after exposure, releases the enzyme. The only other form of first aid consists of massive doses of adrenaline or atropine, which must be administered within several minutes of exposure.

Relatively nonpersistent, Sarin has a low boiling point, and will dissipate within 15 minutes to 4 hours, depending on the weather. It can be delivered by artillery or mortar shells, rocket and missile warheads, or aerial bombs, or from aerosol dispensers. Sarin forms a substantial portion of the chemical arsenals of both the U.S. and U.S.S.R. See also CHEMICAL WARFARE.

SARY SHAGAN: Soviet missile test center in Central Asia, now used mainly for BALLISTIC MISSILE DEFENSE research and development activities. In 1979, U.S. surveillance satellites detected a very large facility (codenamed "Tora") at Sary Shagan, which was later assessed to be an explosive-driven MAGNETO-HYDRODYNAMIC powerplant for a powerful iodine LASER. This device was used in some 30 tests against incoming reentry vehicles through the 1980s.

SAS: Special Air Service, an elite British SPECIAL OPERATIONS force first established during the Second World War, and later emulated by the armies of Australia, Canada, and New Zealand (and, formerly, Rhodesia). The British SAS is regarded, man for man, as the finest force of its type anywhere. The U.S. DELTA FORCE began as a direct copy.

Raised in 1942 during the North Africa campaign by Cap. David Sterling to conduct raids against German airfields, the SAS later took on roles including long-range scouting, sabotage, counterinsurgency, and, since the late 1960s, counterterrorism and hostage rescue. Since World War II, the SAS has seen action in all British wars, including the Malayan insurgency, the Borneo campaign, and the civil war in Aden. In recent years, SAS troops have been active in operations against the IRA in Northern Ireland, and have participated in several hostage rescues, most notably the storming of the Iranian embassy in London on 5 May 1980. During the 1982 Falklands War, the SAS reverted to more traditional roles, including a raid on the Pebble Island airfield and long-range scouting.

Once a brigade of five regiments, the British SAS now consists of three small "regiments" with 600 men or less. One, No. 22 SAS, belongs to the Regular Army (RA), while the other two (21 SAS and 23 SAS) are part of the part-time TERRITORIAL ARMY (TA). There are also two signal "squadrons" (i.e., companies), one attached to 22 SAS, and the other in the TA. Each regiment consists of a headquarters squadron with an intelligence element, an administrative "wing" and workshop, a training wing, and four combat squadrons. Each squadron consists of 6 officers and 66 men, organized into an amphibious "troop" (platoon), an air troop, a mobility and survival troop, and a mountain troop, each with 1 officer and 15 men. In 22 SAS, one squadron is assigned to Counter-Revolutionary Warfare (CRW), with responsibility for anti-terrorist and hostage-rescue actions. It too consists of four troops, but each troop is further subdivided into 4-man "operating units" of two 2-man teams. Each SAS man has a specialty (communications, first aid, demolitions, etc.), but is cross-trained in other skills.

The SAS is armed and equipped as LIGHT INFANTRY, with U.S. M16 rifles and standard British weapons, but the CRW squadron also uses specialized equipment such as frame charges (for blowing out doors) and "flash-bang" stun GRENADES for hostage rescue. Long-range patrols rely on portable satellite communication terminals, while raiders are equipped with DEMOLITION charges and silenced Heckler and Koch SUBMACHINE GUNS.

SAS personnel are volunteers selected from other British regiments. Applicants are first passed through a screening process, and then subjected to rigorous physical tests in the Brecon Beacons of Wales, including ten days of fitness training and map reading, followed by ten days of solitary cross-country orienteering, culminating in a 40-mi. (64-km.) hike in 20 hours with a rifle and a 55-lb. (25-kg.) pack. Those who pass then begin 14 weeks of training which includes parachute jumping and survival techniques. Only some 20 percent of all applicants eventually become full members of the SAS. Many serve for only three years before rotating back to their original regiments. The SAS training cycle includes individual and group training in such specialties as unarmed combat, demolitions, mountaineering, small boat handling, etc.

The British SAS works closely with the Special Boat Squadron (SBS) of the Royal Marines, an amphibious reconnaissance and raiding force similar to the U.S. Navy's SEALS.

SATAN: (SS-18): NATO code name for the Soviet RS-20 intercontinental ballistic missile (ICBM). Introduced in 1974 as the successor to the SS-9 Scarp, the SS-18 is the largest ICBM ever placed in service (considerably larger than the U.S. TITAN II), and caused much anxiety through the 1970s because its accuracy and payload threatened U.S. ICBM silos. At least 308 SS-18s have been deployed in hardened silos at six missile bases in Soviet Central Asia, at Aleysk, Dombarovskiy, Imeni Gastello, Kartaly, Uzhur, and Zhangiz Tobe. Over the years, the missile has been progressively upgraded. The original SS-18 Mod 1 had a single REENTRY VEHICLE (RV) with a huge 27-MT thermonuclear warhead. It was followed in 1976 by the Mod 2, with 10 650-kT MULTIPLE INDEPENDENTLY TARGETED REENTRY VEHICLES (MIRVs), and by the Mod 3 of 1977, with improved guidance and a single 20-MT RV. Currently, all SS-18s in service are believed to be Mod 4s, introduced in 1982. They have greatly improved accuracy and a payload of 10 to 14 450-kT MIRVs. There have also been several experimental variants, including one that conducted tests of a MANEUVERING REENTRY VEHICLE (MaRV) in 1985–86. In 1989 the Strategic Rocket Forces (RVSN) began flight tests of the Mod 5, with 10 900-kT MIRVs. During the late 1990s, the SS-18 may be superseded by an incremental development, currently designated SS-X-26.

Stored in a canister which is loaded directly into the silo, the SS-18 employs the cold-launch technique, whereby the missile is ejected from the silo by a cold gas generator; the first stage only ignites at a height of some 328 ft. (100 m.) above the ground. Because the silo is not damaged by the rocket exhaust, it can be reloaded within hours.

A two-stage liquid-fueled missile, the SS-18 burns the standard Soviet storable bipropellant mix of undimensional dimethyl hydrazine (UDMH) and nitrogen tetroxide (NTO). The first stage has four gimballed main engines generating more than 1,007,144 lb. (4480 kN) of thrust, while the second stage has a single fixed engine rated at 203,452 lbs. (905 kN) of thrust, plus two gimballed vernier engines for steering and attitude control. All engines have thrust magnitude control for precise velocity adjustment. The Mod 4 also has a liquid-fuel (UDMH/NTO) POST-BOOST VEHICLE (PBV) with dual thrust modes. Its payload, estimated at some 17,820 lb. (8100 kg.) is limited by the SALT II Treaty to 10 MIRVs, but it is believed that there is enough volume for as many as 14. No decoys or other PENAIDS have been observed so far, but apparently space is also reserved for them.

The SS-18 Mod 4 has a hybrid INERTIAL GUIDANCE system: the first two stages employ the Soviet "fly-by-wire" technique (whereby the missile is constrained to a pitch and velocity profile), while the PBV employs "navigating" inertial guidance. The combination of both methods enhances accuracy and provides better time-on-target control. Because its engine cannot be shut down, the PBV must be aligned along the "range-insensitive direction" (RID) for RV release. Because excursions up the RID increase the time of flight without changing the impact point, it is not possible for the SS-18 to direct two RVs against the same target so that they arrive simultaneously (as they must, to avoid warhead FRATRICIDE). Thus "Two-on-One Cross-Targeting" must be used instead, whereby two different missiles each release an RV against the same target to increase the probability of its destruction. Given a 450-kT warhead, each SS-18 RV has a kill probability (P_K) of destroying a Minuteman silos, while with two RVs the P_k rises to .85. In theory, therefore, most of the U.S. land-based missile force could be destroyed by only 200 SS-18s, a fact which prompted the U.S. to develop mobile missiles to evade attack. See also COUNTERFORCE; ICBMS, SOVIET UNION.

Specifications Length: 114.83 ft. (35 m.). **Diameter:** 9.83 ft. (3 m.). **Weight, launch:** 470,800 lb. (214,000 kg.). **Range:** 6200 mi. (10,000 km.). **CEP:** 250 m.

SATELLITES, MILITARY: At present, military satellites perform six primary functions: early warning and "attack assessment": SURVEILLANCE and RECONNAISSANCE; COMMUNICATIONS; navigation; meteorology; and geodesy.

EARLY WARNING AND ATTACK ASSESSMENT

The short flight times of ICBMS (30 minutes) and SLBMS (as little as ten minutes) require that missile launches be detected and assessed as quickly as possible; ground-based ballistic missile early warning radars (e.g., BALLISTIC MISSILE EARLY WARNING SYSTEM or HEN HOUSE) can detect missiles only after they pass over the RADAR HORIZON, while satellites with INFRARED (IR) sensors can detect missiles by their hot exhaust plumes within seconds of liftoff. Since the early 1960s, both the U.S. and the U.S.S.R. have employed constellations of early warning satellites, which can also monitor missile tests and space launches.

The U.S. maintains 6 early warning satellites (3 active and 3 in reserve) in geosynchronous orbits (GEO) under the DEFENSE SUPPORT PROGRAM (DSP). DSP is to be replaced in the early 1990s by the Boost Surveillance and Tracking System (BSTS), with higher resolution and even faster response time.

The U.S.S.R. maintains a nine-satellite constellation for early warning. Because GEO is difficult to achieve from Soviet launch sites, these satellites are in highly elliptical, semi-synchronous orbits inclined at 62°. While not as efficient as the DSP constellation, this arrangement allows each satellite to observe U.S. ICBM sites for 6 hours in each 12-hour orbit, and the spacing of the satellites ensures uninterrupted coverage.

Both the U.S. and U.S.S.R. have also deployed satellites to detect nuclear detonations on the surface or in space, in order to verify compliance with the TEST BAN TREATY in peacetime, and to provide rapid damage assessment in war.

SURVEILLANCE AND RECONNAISSANCE

"Spy satellites" are used to obtain information on a wide range of military and economic activities. Surveillance satellites maintain regular or continual observation, while reconnaissance implies a deliberate search for specific and often time-critical information. These functions were formerly performed by separate, specialized satellites, but the recent U.S. practice is to combine both functions in a single, large platform.

Photographic (more properly, imaging) surveillance and reconnaissance satellites have optical, infrared, ultraviolet, and imaging radar sensors to produce high-resolution pictures. "Quick-look" satellites have video sensors for real-time observation of time-critical targets and general surveillance, with the images relayed continuously to the ground. When a target of interest is detected, or if greater detail is required, a "close-look" photographic satellite can be directed to fly over the area within several hours; the film is returned to earth in small recovery capsules for processing, which can cause a delay of several days between collection and analysis of their data. Furthermore, the amount of film and the number of recovery capsules is limited. This dual-platform approach is still used by the Soviet Union, but the U.S. has exploited advances in video image processing to combine the quick- and close-look functions in a single satellite.

The first U.S. photo-reconnaissance satellites were the Discoverer series of the early 1960s, which provided valuable information on Soviet nuclear forces. These were followed by the "Keyhole" (KH) series, of which the best known are the KH-9 BIG BIRD and the KH-11. Big Bird, a high-resolution close-look satellite, was supplanted in the 1970s by the KH-11, a combined quick-look/close-look platform with a variety of multi-spectral sensors. The latest U.S. imaging satellite is the KH-12, whose photogrammetric sensors are said to combine very high resolution with real-time data transfer.

The standard Soviet photo-reconnaissance satellites are the Cosmos 758 series, apparently a development of the *Vostok* manned spacecraft. Introduced in 1978, they have a mission life of some 60 days (as compared to two or three years for a KH-11). In addition to Cosmos, the U.S.S.R. has used its SALYUT space stations as manned observation platforms.

Because they must operate in very low orbits (75–100 mi.), imaging reconnaissance satellites must use maneuvering engines to overcome residual atmospheric drag, as well as to change position so as to fly over different targets. The large U.S. satellites have much more fuel than their Soviet counterparts, and it is this that has allowed the U.S. to limit satellite launches to one or two per year, whereas the U.S.S.R. must launch as many as 35 of its short-lived satellites per year.

The exact resolution of surveillance and reconnaissance satellites is a closely guarded secret, and public statements on the subject can be deliberately misleading. Early U.S. satellites had resolutions of 1–3 m.; i.e., objects of that size could be discerned against the background, while precise size, shape, and other details could only be determined for objects 2–3 times as large. Current satellites are reported to have resolutions of roughly 0.3 m. (1 ft.), so that they can provide detailed information on objects 2–3 m. (6.6–9.85 ft.) across; sig-

nal processing and image enhancement techniques can significantly improve clarity.

Current civilian imaging satellites (e.g., the French SPOT) have resolutions of some 5 m. (16.5 ft.), and their successors will have resolutions of 1 m. or less. Such satellites are already useful for general surveillance, and may soon be able to provide images equivalent to those of military satellites to any paying customer.

Less well known, but equally important, are signals intelligence (SIGINT) satellites ("Ferrets"), equipped with large, sensitive receiver antennas which can detect even faint emissions from a variety of electronic emitters. They are used to intercept communications, determine the signal characteristics and performance of radars, and to locate emitters for eventual attack. To obtain better lines of sight, Ferrets are usually placed in 300- to 900-mi. (480- to 1440-km.) orbits, but the very large U.S. "Rhyolite" series are placed in GEO in order to monitor telemetry from Soviet missile tests and intercept microwave communications.

Electronic Intelligence (ELINT) Ocean Reconnaissance Satellites (EORSATs) are Ferrets specifically intended to monitor naval radars and naval communications, while Radar Ocean Reconnaissance Satellites (RORSATs) track surface ships and provide OVER-THE-HORIZON TARGETING data to Soviet naval units.

COMMUNICATIONS

The concept of communications relay satellites was anticipated by Arthur C. Clarke in the late 1940s: three satellites placed in GEO at 120° intervals have a direct line of sight with each other, and with any point on the earth. The concept was proven in the early 1960s, and today more than 70 percent of all long-distance U.S. message traffic is transmitted via satellite.

U.S. military communications satellite networks include the Defense Satellite Communications System (DSCS), the Air Force Satellite Communications System (AFSATCOM), the Fleet Communications Satellite System (FLTSATCOM), and the Satellite Data System (SDS). The Pentagon's primary high-volume communication system, DSCS consists of four active and two spare satellites with 1300 voice channels each and a 100-megabit/sec. data rate. The current DSCS-II satellites are being replaced by the jam-resistant, ELECTROMAGNETIC-PULSE-hardened DSCS-III.

FLTSATCOM, the U.S. Navy's primary network, also has four satellites, with 900 data links each, to connect surface ships, aircraft, submarines, and shore bases. AFSATCOM consists of communications packages "piggy-backed" on DSCS, FLTSATCOM, and other satellites. It links E-4A NEACP, TACAMO, and other airborne command posts to various ground stations. SDS, a 3-satellite network in highly elliptical orbits, fills gaps in the polar regions, and may also be able to relay data from KH-11s.

All these systems are to be replaced in the late 1990s by MILSTAR, the Military Strategic/Tactical and Relay System, employing high-capacity millimeter-wave links. It will comprise 8 satellites: 4 in GEO, 3 in elliptical orbits for polar coverage, and 1 on-station spare. Additional "dark spares" will be parked in super-synchronous, 110,000-mi. (176,000-km.) orbits as a wartime reserve.

The basic Soviet system is Molniya, six satellites in 62.8° elliptical orbits which provide 24-hour, point-to-point coverage of the U.S.S.R. The U.S.S.R. also has a tactical network of eight satellites in 930-mi. (1490-km.) orbits inclined at 74°, for real-time command and control in Eastern Europe.

NAVIGATION

Transit, the first U.S. satellite navigation system, has been operational since 1964. Developed to allow ballistic missile submarines to update their inertial navigation systems (SINS), Transit consists of a constellation of satellites in 680-mi. (1090-km.) circular orbits. Each transmits an oscillating signal, on the basis of which an appropriate receiver can determine its position to within 150 m. by measuring the Doppler shift. The U.S.S.R. maintains a similar system in 620-mi. (995-km.) orbits.

The new GLOBAL POSITIONING SYSTEMS (GPS) employ some 18 satellites in 12,425-mi. (19,880-km.) orbits inclined at 63°. By using very precise clocks and "time difference of arrival" (TDOA) techniques, GPS can be used to establish receiver positions to within 15 m., and can also provide a precise velocity reading. GPS receivers are light and compact, and thus ideal for a wide range of applications, including on-board missile guidance. The U.S. system, NAVSTAR, has been operational since 1988. An almost identical Soviet system, GLONASS, has been in operation since the early 1980s.

METEOROLOGY

Accurate weather prediction is critical to many military activities, particularly air and naval operations. Before the advent of weather satellites, armed forces were dependent on ground stations, aircraft, and weather balloons, none of which could provide wide coverage. The U.S. uses civilian weather satellites, but also has a high-resolution Defense Meteorological Satellite Program (DMSP), which consists of two satellites in 12-hour orbits, equipped with optical and IR sensors of 0.37-mi. (0.6-km.) resolution, as well as sensors to determine water temperature, ozone levels, and air temperature. The also have a gamma ray detector to locate atmospheric nuclear explosions. The equivalent Soviet system, Meteor II, consists of three satellites in slightly lower orbits.

GEODESY

Accurate mapping is essential for long-range weapon targeting, and satellites can provide mapping data on remote or hostile regions in only a fraction of the time required by aircraft, and without the risk of interception. The U.S. Defense Mapping Agency initiated the Geodetic Satellite Program (GSP) in the mid-1960s, and the U.S.S.R. has a similar program. In addition to traditional map data, these satellites collect precise measurements of the earth's gravitic and magnetic fields, and also determine the oblateness (variations from the spherical) of the earth's shape, all critical parameters for ballistic missile guidance.

See also SPACE, MILITARY USES OF.

SATKA: Surveillance, Acquisition, Tracking, and Kill Assessment, a program under the U.S. STRATEGIC DEFENSE INITIATIVE (SDI), meant to develop the sensor technology for BALLISTIC MISSILE DEFENSE (BMD). Specific SATKA programs include the Boost Surveillance and Tracking System (BSTS) and the Space Surveillance and Tracking System (SSTS), both satellite-based infrared sensors; the Ground Surveillance and Tracking system (GSTS), an IR sensor mounted on a suborbital sounding rocket; the Airborne Optical Adjunct (AOA); and the TERMINAL IMAGING RADAR (TIR).

The major challenge of SATKA is to develop long-range sensors of sufficient resolution to discriminate between REENTRY VEHICLES and decoys (PENAIDS), and to point DIRECTED ENERGY WEAPONS or guide SPACE-BASED INTERCEPTORS, and also assess the effects of their attacks.

SAURO: A class of 6 Italian diesel-electric attack submarines commissioned between 1980 and 1988. Successors to the TOTI-class coastal submarines, the larger Sauros are intended for longer-range operations. Though they were initially ordered in 1967, budget shortfalls delayed the start of construction until 1974, and completion of the first 2 was further delayed by a lack of suitable batteries (the last 2, completed to a slightly modified design, are sometimes called the Pelosi class). Small, quiet, and highly maneuverable, the Sauros are ideally suited for operations in the Mediterranean.

All six submarines have a modified teardrop hullform in a single-hull configuration. The pressure hull is fabricated of HY-80 steel, and flat deck casing extends from the bow to just behind the tall, streamlined sail (conning tower) amidships. The forward diving planes are mounted on the sail, while the cruciform rudders and stern planes are just ahead of the propeller. A retractable snorkel allows diesel operation from shallow submergence, and two 148-cell batteries provide an underwater endurance of some 65 hours at 4 kt. All machinery is resiliently mounted, and the engine rooms are lined with acoustic insulation to reduce radiated noise.

Armament consists of six 21-in. (533-mm.) bow-mounted torpedo tubes, equipped with a positive discharge system to allow safe torpedo launches at any depth. The basic load of 12 Whitehead A184 wire-guided homing TORPEDOES can be replaced by bottom-laid influence MINES on a 2-for-1 basis. The last two Sauros have longer torpedo tubes which can also accommodate the UGM-84 HARPOON anti-ship missile. All sensors are linked to a SEPA CCRG torpedo FIRE CONTROL unit, which can track four targets simultaneously. See also SUBMARINES, ITALY.

Specifications Length: (Sauro) 209.5 ft. (63.9 m.); (Pelosi) 211.2 ft. (64.4 m.). **Beam:** 22.4 ft. (6.8 m.). **Displacement:** (Sauro) 1460 tons surfaced/1650 tons submerged; (Pelosi) 1476 tons surfaced/1662 tons submerged. **Powerplant:** single-shaft diesel-electric: 3 1070-hp. GMT A210 16 diesel-generators, 1 3210-hp. Marelli electric motor. **Speed:** 14 kt. surfaced/20 kt. submerged. **Fuel:** 144 tons. **Range:** 6500 n.mi. at 11 kt. surfaced/2500 n.mi. at 12 kt. snorkeling. **Endurance:** 45 days. **Max. operating depth:** 820 ft. (250 m.). **Collapse Depth:** 1345 ft. (410 m.). **Crew:** 49. Sen-

sors: 1 ELSAG/USEA IPD-70 bow-mounted low-frequency active/passive sonar, 1 sonar intercept array, 1 ELSAG/USEA passive flank-mounted conformal array sonar (Pelosi), 1 IPD-70S integrated sonar, 1 SMA SPS-70 surface-search radar, 1 ELT-724S electronic signal monitoring array, 2 periscopes.

SAVAGE (SS-13): NATO code name for a Soviet intercontinental ballistic missile (ICBM). Introduced in 1969, as the first Soviet solid-fuel ICBM, Savage was never deployed in quantity, because its range, payload, and accuracy were all inferior to its liquid-fuel contemporary, the SS-11 SEGO. Only 60 SS-13s were deployed in fixed silos at Yoshkar Ola in the central U.S.S.R. The two upper stages of the SS-13 were developed into the SS-14 Scapegoat intermediate-range ballistic missile (IRBM), but it too was unsuccessful and was not deployed in quantity. It is believed that the SS-13 was replaced by the more advanced solid-fuel SS-16 SINNER in the mid-1970s, and that the Sinner was superseded in turn by the mobile, solid-fuel SS-25 SICKLE in the mid-1980s. See also ICBMS, SOVIET UNION.

SAW: Squad Automatic Weapon.

1. In general, a heavy-barrel version of an ASSAULT RIFLE intended for more sustained automatic fire, to function as a LIGHT MACHINE GUN at squad level, but chambered for reduced-caliber (5.45- to 5.56-mm.) ammunition.

2. The U.S. M249, a license-built variant of the Belgian FN Minimi, introduced in 1974, adopted by the U.S. Army and Marine Corps in 1979. The Minimi is also in service with the armies of Australia, Canada, and Indonesia. In addition to the standard SAW, FN offers two specialized variants: an airborne version with a short barrel and retractable stock, and a pintle-mounted version for vehicles, without a stock.

Experience in Vietnam had shown that the M16 assault rifle lacked range and penetrating power, and that infantry squads needed a specialized automatic weapon lighter than the standard M60 machine gun, to provide a base of fire for "fire-and-movement" tactics. Two M249s are now organic to each army and marine infantry squad.

A gas-operated weapon similar to the 7.62-mm. FN-MAG, the M249 has a straight, tubular stock, a carrying handle over the receiver, and a bipod. A PSV-4 IMAGE INTENSIFIER night-vision sight can be fitted, and the trigger guard can be removed to facilitate operation with winter or chemical-protection gloves. The quick-change barrel has a flash suppressor, but no provision for a bayonet or rifle grenades. The rifling has a much greater twist than in the M16, in order to stabilize the heavy SS109 (U.S. designation M855) high-performance 5.56-× 45-mm. round. The SS109 has a slightly lower muzzle velocity than the standard M198 round of the M16, but its effective range is extended because of its greater weight. The M249 is fed from a 200-round belt stowed in a plastic box magazine attached to the left side of the receiver. In an emergency, the M249 can also fire M198 ammunition from standard M16 30-round box magazines. The weapon has two rates of fire, normal and "adverse." Continual operation at the higher adverse rate causes considerable barrel wear, but studies of field operations have revealed that it is often preferred by troops.

Specifications Length OA: 41.34 in. (1.05 m.). **Length, barrel:** 18.31 in. (465 mm.). **Weight, empty:** 14.33 lb. (6.5 kg.). **Weight, loaded:** 21.38 lb. (9.7 kg.). **Muzzle velocity:** (SS109) 915 m./sec. **Cyclic rate:** (normal) 750 rds./min. (adverse) 1000 rds./min. **Effective range:** (SS109) 600 m.; (M198) 350 m.

SAWFLY (SS-N-6): NATO code name for a Soviet submarine-launched ballistic missile (SLBM). Developed from the early 1960s, the SS-N-6 was deployed from 1968 aboard the YANKEE-class nuclear-powered ballistic-missile submarines (SSBNs). The original SS-N-6 Mod 1 carried a single 700-kT REENTRY VEHICLE (RV). The much lighter Mod 2, with longer range and a 600-kT reentry vehicle, replaced the Mod 1 from 1972. The final Mod 3 version, which entered service in 1974, has the same range as the Mod 2 but a payload of two 350-kT MULTIPLE REENTRY VEHICLES (MRVs). Though obsolescent, Mod 3s will remain in service as long as Yankees are retained in the SSBN role.

A two-stage liquid-fuel missile, Sawfly uses a storable fuel mix of undimensional dimethyl hydrazine (UDMH) and nitrogen tetroxide (NTO), which is highly corrosive and extremely unstable, and gives off toxic fumes. The risks of carrying liquid-fuel missiles inside submarines are obvious, and in fact a Yankee was lost in October 1986 when a Sawfly exploded inside its launch tube. But Soviet guidance technology of the early 1960s could not function with solid-fuel rockets with any accuracy. Sawfly relies on the Soviet "fly-by-wire" technique of preprogrammed INERTIAL GUIDANCE, whereby the missile is constrained to a pitch and velocity profile, using thrust magnitude control for precision velocity adjustments. Accuracy is relatively poor, thus the Sawfly could only be used against soft area targets (e.g., cities). The SS-N-6 is ejected from its launch tube by steam or compressed air with sufficient force to break the surface; only then does the main engine ignite. See also SLBMS, SOVIET UNION.

Specifications Length: 32.83 ft. (10 m.). **Diameter:** 5.9 ft. (1.8 m.). **Weight, launch:** 14,580 lbs. (6627 kg.). **Range:** (1) 1490 mi. (2400 km.); ($^2/_3$) 1800 mi. (2880 km.). **CEP:** 800–1000 m.

SAXHORN (AT-7): NATO code name for the Soviet *Metis* ANTI-TANK GUIDED MISSILE (ATGM). Introduced in the early 1980s, this missile is believed to be a medium-range man-portable weapon similar in concept and performance to the U.S. M47 DRAGON. It may be intended as a replacement for the SPG-9 and B-10 recoilless rifles.

Saxhorn is believed to weigh some 15 lb. (6.8 kg.), with a 100-mm. shaped-charge (HEAT) warhead capable of penetrating up to 400 mm. (15.75 in.) of homogeneous steel armor. Powered by a dual-impulse solid rocket engine, it has a maximum velocity of 200 m./sec. (447 mph) and a maximum range of 1000 m., for a maximum flight time of only 5 seconds. A minimum range of 50 m. is imposed by safety fuzing.

Saxhorn relies on wire guidance with semi-automatic command to line-of-sight (SACLOS). After the operator lines up the target in his sight, he can launch the missile. As the missile flies downrange, a sensor in the launcher detects infrared emissions from a flare in its base. A guidance computer determines the angle between the missile's course and the operator's line of sight to the target, and generates steering commands to bring the missile onto that line. These commands are transmitted through two thin wires which pay out from an inertialess reel in the base of the missile. All the operator needs to do is keep the target centered in his sight until the missile impacts.

Saxhorn probably shares Dragon's shortcomings, including a substantial backblast which can reveal the operator's position the need for the operator to remain exposed throughout the engagement, and a relatively small warhead of only marginal effectiveness against REACTIVE ARMOR or modern composite (CHOBHAM) armor.

SAXON: British AT-105 4x4 wheeled ARMORED PERSONNEL CARRIER (APC), initially developed in the 1970s by GKN as a private venture, and later adopted by the British army (in 1976) to equip second-line motorized infantry battalions based in the U.K. but assigned to reinforce the British Army of the Rhine (BAOR). The British army ordered more than 1000 vehicles, which were delivered from 1984. Saxon is also in service with the armies of Bahrain, Kuwait, Malaysia, Nigeria, and Oman. Intended exclusively as a "battle taxi," Saxon is meant to deliver troops to the battlefield, where they would dismount to fight.

Derived from the Bedford MK 4x4 4-ton truck, Saxon's rather boxy, all-welded steel hull has some 12–15 mm. of armor, only enough for protection against small arms and shell fragments. The driver sits on the right behind an armored glass windshield with drop-down steel shutter; the commander sits behind the driver under a cupola with four periscopes and a pintle-mounted 7.62-mm. machine gun. The troop compartment in the rear seats eight; no firing ports are provided. Access is through twin doors in the rear and a single door on each side. Variants include the AT-105 ambulance, the AT-105Q command vehicle, an armored recovery vehicle, and an internal security vehicle with a riot-control water cannon.

Specifications **Length:** 16.95 ft. (5.16 m.). **Width:** 8.16 ft. (2.5 m.). **Height:** 9.4 ft. (2.85 m.). **Weight, combat:** 10.6 tons. **Powerplant:** 164-hp. Bedford 500 6-cylinder diesel. **Hp./wt. ratio:** 15.47 hp./ton. **Fuel:** 40 gal. (176 lit.). **Speed, road:** 60 mph (100 kph). **Range, max.:** 317 mi. (507 km.).

SAYARET MATKAL: "General Staff Reconnaissance," also known as "Unit 269," the elite Israeli special operations, counterterrorism, and hostage-rescue unit at the top of the hierarchy of Israeli commando units. Formed after the 1967 Six-Day War, and under the direct control of the chief of Military Intelligence, the unit conducts covert and clandestine SCOUTING and RAIDS as well as special operations. During the 1973 Yom Kippur War, it conducted raids behind Syrian lines to ambush tank columns. Otherwise, its primary tasks are hostage rescue and preemptive counterterrorist strikes (including the abduction and assassination of terrorist leaders). In 1972, *Sayaret Matkal* conducted the first successful assault on a highjacked airliner (at Lod Airport), with its men approaching the aircraft disguised as mechanics to rescue 90 passengers. Its most famous exploit among dozens of operations remains the 1976 Entebbe rescue, in which *Sayaret Matkal* was supported by Paratroops and Golani commandos. Since 1975, the unit has also been complemented by an elite 150-man Border Police hostage rescue

unit for internal counterterrorist missions. *Sayaret Matkal* personnel also serve as armed sky marshals on El Al commercial flights.

Sayaret Matkal personnel are volunteers selected from the annual conscript intake, supplemented by transfers from the Paratroop, Golani, and naval commandos. Though the Israeli air force receives the pick of the conscripts, *Sayaret Matkal* still gets very high-quality manpower, including rejected pilot candidates. Officers go on to regular line commands and often reach the highest ranks. Individual training emphasizes long-range infiltration and desert survival, close combat with pistols and knives, unarmed combat, parachuting (including HALO techniques), demolitions, rappelling, small-boat handling, combat swimming, communications, and foreign languages. Much group training is conducted "on-the-job," in less demanding raids. The unit plans its own operations. *Sayaret Matkal* uses standard Israeli infantry weapons, including the 9-mm. UZI submachine gun and the 5.56-mm. GALIL assault rifle, as well as U.S. and Soviet small arms. See also AIRBORNE FORCES, ISRAELI; HIRAM INFANTRY.

SBI: See SPACE-BASED INTERCEPTOR.

SBS: Special Boat Squadron, the SPECIAL OPERATIONS unit of the British Royal Marines (RM), trained and equipped for raiding and clandestine SCOUTING on enemy-held shores, roughly comparable to the U.S. Navy's SEALS. Formed during the Second World War, the SBS became famous for its sabotage missions conducted with collapsible kayaks ("cockleshells"). Since 1945, the SBS has seen action in Borneo, Oman, and, most recently, the 1982 Falklands War. A related RM unit, the Comacchio Company, is responsible for the security of North Sea oil platforms.

The SBS is a company-size unit of some 120 men, but usually operates by independent 4-man teams. Personnel are selected from Royal Marine volunteers, who must first pass a 3-week screening test, followed by a 15-week training program which includes scuba diving, small-boat handling, navigation, demolitions, unarmed combat, and advanced weapon training. After that, the recruits undergo 4 weeks of parachute training, including HALO techniques. In contrast to the SAS (with whom the SBS frequently cooperates), SBS personnel are not compelled to return to their parent units after a given term; many do leave, however, in order to obtain further promotion.

SBS weapons include the M16 assault rifle and a silenced version of the STERLING 9-mm. submachine gun. SBS members are also issued a variety of DEMOLITION charges and limpet MINES. For clandestine penetrations of enemy waters, the SBS uses Royal Navy submarines, Klepper Mk.13 collapsible kayaks, Gemini inflatable boats with muffled outboard motors, and militarized dories with 140-hp. outboards (manned by the Royal Marine Rigid Raider squadron). See also MARINES, ROYAL.

SCADS: Shipboard, Containerized Air-Defense System, a British proposal for converting merchant container ships into V/STOL AIRCRAFT CARRIERS, similar in concept to the U.S. Arapaho program.

SCADS would consists of standard 8- x 8- x 40-ft. (2.43- x 2.43- x 12.2-m.) cargo containers, modified to house the systems and facilities required to support helicopters and HARRIER V/STOL fighters. A typical SCADS ship could carry as many as 6 Harriers and 2–4 SEA KING helicopters. In war, container ships could be quickly converted to SCADS to provide CONVOYS with a modest degree of fighter and anti-submarine support, thereby freeing more capable carriers for other duties. Though a simple and inexpensive solution to the convoy escort problem, SCADS has not been adopted to date.

SCALEBOARD (SS-12): NATO code name for the Soviet OTR-22 short-range (tactical) BALLISTIC MISSILE. Introduced in 1969, 1 Scaleboard brigade with 9 launchers and 18 missiles was attached to each Soviet FRONT. A modified version (SS-12M) introduced in 1979 was initially thought to be a new missile designated SS-22. As compared to the original version, the SS-12M had greater range, improved accuracy, and reduced reaction time. By 1987, more than 80 Scaleboard launchers had been modified to SS-12M standard, but all Scaleboards were to be dismantled under the terms of the 1987 INF TREATY.

SCALPEL (SS-24): NATO code name for a Soviet mobile intercontinental ballistic missile (ICBM). Developed in the late 1970s, the solid-fuel SS-24 is a medium complement to the lightweight SS-25 SICKLE, and is roughly comparable in size to the U.S. PEACEKEEPER (the SS-24 and SS-25 are the first Soviet solid-fuel ICBMs to be placed in serial production since the SS-13 SAVAGE). Flight tests began in the early 1980s and Scalpel was declared operational in 1985. The initial SS-24 Mod 1 was deployed in 100 hardened silos previously occupied by the old SS-11 SEGO, while the SS-24 Mod 2 is a rail-mobile missile deployed on special 85-ft. (26-m.) rail cars with a 264,000-lb. (120,000-kg.)

capacity. These are normally based at centralized control and maintenance complexes, but can be dispersed to presurveyed remote sites for launch, thereby enhancing survivability (a similar scheme has been proposed for the U.S. Peacekeeper).

All three stages of the missile are believed to be of wound glassfiber construction for reduced weight. The first stage has four gimballed nozzles for steering and attitude control; the second and third stages each has a single gimballed nozzle, possibly recessed to reduce overall length. The payload consists of ten MULTIPLE INDEPENDENTLY TARGETED REENTRY VEHICLES (MIRVs), each with a 300-kT warhead, and all mounted on a liquid-fuel POST-BOOST VEHICLE (PBV). The PBV engine has one large longitudinal thruster and four smaller gimballed nozzles. The former is used for large velocity increments, the latter for small adjustments and attitude control. As with previous Soviet PBVs, the engine is of the continual-thrust type, hence the PBV must be aligned along the range-insensitive direction (RID) during REENTRY VEHICLE (RV) deployment. The RVs themselves are believed to be simple sphere-cones with relatively high ballistic coefficients ("beta") for increased accuracy. High-beta RVs are dynamically unstable, and require precise alignment at release to avoid uncontrolled tumbling; release errors have apparently been responsible for several failures during test flights.

Scalpel's inertial guidance unit employs General Energy Management Steering (GEMS), whereby the missile makes excursions above and below its nominal trajectory, to achieve the required velocity vector while expending all available fuel (this eliminates the need for blow-out ports and other inherently inaccurate thrust-termination devices). Accuracy depends on whether the missile is launched from a silo or a remote site, but would be only marginally sufficient against missile silos.

For both silo and rail launch, Scalpel employs the cold-launch technique. The missile is stored in a sealed launch canister, which is lowered into a silo or stowed on a railway car. For launch, the missile is ejected from the canister by a cold gas generator, rising to a height of 328 ft. (100 m.) before the main engines ignite. This protects the silo and rail car from exhaust damage, and permits rapid reloading. See also ICBMS, SOVIET UNION; RVSN.

Specifications Length: 69.72 ft. (21.25 m.). Diameter: 6.56 ft. (2 m.). Weight, launch: 220,000 lb. (100,000 kg.). **Range:** 6200 mi. (10,000 km.). **CEP:** (silo) 250 m.; (rail) 340 m.

SCARAB (SS-21): NATO code name for the Soviet *Tochka* tactical BALLISTIC MISSILE, introduced in 1976 as a replacement for the FROG series of unguided rockets. Intended mainly as a division-level nuclear delivery vehicle, but also capable of carrying high-explosive, cluster, or chemical warheads, Scarab is roughly comparable to the U.S. LANCE battlefield missile. The SS-21 is now replacing the FROG-7 on a 1-for-1 basis within the Soviet army; by the late 1980s, more than 250 launchers were in service with Soviet tank and motorized rifle divisions. Each SS-21 battalion has 4 launchers and 16 missiles. The SS-21 has also been supplied to the other Warsaw Pact armies, and exported to Cuba, Egypt, Iraq, North Korea, Libya, Syria, and Yugoslavia, without nuclear warheads.

A single-stage solid-fuel missile, Scarab has four tail fins for aerodynamic control. Payload options include 10- and 100-kT nuclear and 1540-lb. (700-kg.) high-explosive, cluster, and chemical warheads. Scarab's INERTIAL GUIDANCE system is quite accurate. The SS-21 is mounted on a fully amphibious, self-propelled transport-erector-launcher (TEL) derived from the ZIL-167 6x6 truck. The missile is stowed horizontally within the TEL and is erected vertically off the rear of the vehicle for launch.

Specifications Length: 30.26 ft. (9.22 m.). **Diameter:** 1.5 ft. (457 mm.). **Weight, launch:** 5490 lb. (2495.5 kg.). **Range envelope:** 8.7–74.6 mi. (14–120 km.). **CEP:** 50–100 m.

SNORKEL: 1. In SUBMARINES, a retractable tubular mast designed to supply air to diesel engines, allowing submerged diesel operation for propulsion or battery charging down to a depth of 40–60 ft. (12.2–18.3 m.). A standard feature on all modern diesel-electric submarines, snorkels usually incorporate an exhaust vent, and a stop-valve to prevent water ingestion should the mast be momentarily submerged.

2. In TANKS, an erectable tube meant to provide air to the engine during deep fording (as deep as 21 ft./6.4 m. in some cases). Many tanks also have a second tube to provide air for the crew compartment. Standard equipment on Soviet tanks, snorkels are not popular with crews because if they collapse or leak, catastrophic consequences ensue.

SCHWERPUNKT: German military concept, frequently (but inadequately) translated as the "point of main effort" or "center of gravity"; a more meaningful definition is "thrust vector." First used by Clausewitz in his *On War*, *Schwerpunkt* always refers to the principal effort, but contrary to a popular misunderstanding, the *Schwerpunkt* does not unfold on a fixed axis of advance, but rather shifts in response to changing circumstances in seeking the line of least enemy resistance. In mobile warfare, the maintenance of the *Schwerpunkt* thus requires leadership capable of exploiting fleeting opportunities (see AUFSTRAGSTAKTIK).

In the classic German doctrine, the *Schwerpunkt* principle applies simultaneously at all levels (strategic, operational, and tactical); though the *Schwerpunkt* at each level may differ at any one time, all are coordinated towards a common objective. Thus the enemy force is kept off-balance by the rapid shifting in direction of the main thrust.

The *Schwerpunkt* principle applies on the defensive as well, especially the German-style elastic defense and DEFENSE-IN-DEPTH formats based on a combination of delaying actions and escalating counterattacks. In such cases, the *Schwerpunkt* would be aimed at the vulnerable flanks of enemy penetrations, to disrupt the offensive and force a retreat.

The *Schwerpunkt* principle is the core of both the Israeli and Soviet operational styles, though the details of application vary. The U.S. Army's AIR-LAND BATTLE doctrine also acknowledges the *Schwerpunkt* principle, but its actual operational and tactical practices tend to ignore it. See also ARMORED WARFARE; BLITZKRIEG.

SCIMITAR CVR(T): British FV107 light armored vehicle, a variant of the SCORPION CVR(T), armed with a 30-mm. RARDEN cannon.

SCORPION CVR(T): The British FV101 tracked reconnaissance vehicle, lead member of a family of light armored vehicles with common automotive components. Developed by Alvis from the early 1960s in response to a British army requirement for an air-portable "Tracked Combat Vehicle, Reconnaissance" (CVR[T]) to replace the SALADIN armored car, Scorpion was required to have both ANTI-TANK and RECONNAISSANCE capabilities. A prototype was produced in 1969, and the first production vehicles were delivered in 1970. To date more than 3500 Scorpions have been built for the British army, the RAF Regiment, and export.

With a ground pressure of only 5.1 lb. per square in. (0.36 kg./cm.^{2}), actually less than a man, Scorpion can traverse bogs, swamps, and soft sand.

Armor protection is minimal: the all-welded aluminum hull has a maximum thickness of only 0.5 in. (12.7 mm.), barely enough to resist some small arms and (some) splinters. The hull is divided into a driving/engine compartment up front, and a turreted fighting compartment in the rear. The driver sits on the left, with the commander and gunner in the all-welded aluminum turret. The driver's two observation periscopes can be replaced by night vision scopes. The commander, on the left side of the turret, has an optical gunsight and five periscopes for 360° observation. On the commander's right, the gunner has a similar sight and two periscopes. Fire control options include a LASER rangefinder, passive and active night vision sights, and a FIRE CONTROL computer. Mounted transversely to the driver's right, the Jaguar V-6 engine is almost the only gasoline engine in a 1970s combat vehicle. Export customers, however, have generally opted for a diesel. The vehicle can cross water with the aid of a collapsible flotation screen, being propelled by track action at a maximum speed of some 3.7 mph (6 kph).

The basic main armament is an antiquated L23A1 low-pressure 76.2-mm. gun (a lightened version of the Saladin's gun), which can fire HESH, HE, canister, and smoke ammunition. The muzzle velocity is only 533 m./sec., making the claimed maximum effective range of some 2000 m. rather doubtful. The gun can penetrate only light armored vehicles, and the side armor of older MAIN BATTLE TANKS (MBTs). Another antiquated feature is the manually traversed turret, and the manual elevation of the gun within limits of −10° and +35°. Forty rounds of ammunition are stowed in the hull and turret. Secondary armament comprises a coaxial 7.62-mm. machine gun and 4-round smoke grenade launchers on each side of the turret. Scorpion is fitted with a COLLECTIVE FILTRATION unit for NBC defense. Export customers have generally opted for the FV101 (90) Scorpion-90, with a Belgian 90-mm. Cockerill Mk.II medium-velocity gun and power traverse.

The second member of the CVR(T) family, the FV107 Scimitar, entered service in 1978. Intended mainly for fire support, it has the same hull and drive train as Scorpion, but with a revised two-man turret armed with a 30-mm. L21 RARDEN cannon (also fitted to the Fox armored car). RARDEN can fire APDS, HE, and HEI rounds at a cyclic rate of 90 rds./min. The APDS round has a muzzle velocity of 1200 m./sec. and can therefore penetrate all light armored vehicles and the side armor of most MBTs; it also has a marginal anti-aircraft capability. Secondary armament is the same as for Scorpion.

The FV102 Striker is a tank destroyer armed with SWINGFIRE anti-tank guided missiles (ATGMs), introduced in 1978 in both the British and Belgian armies. In place of a turret, it has a rectangular 5-round Swingfire launcher, which is stowed horizontally while traveling and erected to a 35° angle for firing. Five reloads are carried inside the vehicle. Swingfire has a maximum range of 4000 m. and can penetrate up to 31.5 in. (800 mm.) of homogeneous steel armor. A pintle-mounted 7.62-mm. machinegun is also included.

The FV103 Spartan, also introduced in 1978, is a small ARMORED PERSONNEL CARRIER which can transport a 5-man FIRE TEAM in addition to a crew of three; it is operated by Britain, Belgium, and Oman. The vehicle commander sits in a cupola behind the driver with a pintle-mounted 7.62-mm. machine gun, while the infantry section leader has a roof hatch in the center. The troop compartment is accessed through a large rear door. Some British Spartans have a 2-round MILAN ATGM launcher in place of the section leader's hatch.

The FV104 Samaritan ambulance variant is slightly longer and taller than Spartan, to accommodate a 3-man crew and up to five stretcher cases. The FV105 Sultan is an armored command vehicle (ACV) similar to Spartan, but with a raised roof and an erectable tent "penthouse" which extends to the rear. Fitted with various radios, display screens, and map tables, it can accommodate a command group of five men. The FV106 Samson is an armored recovery vehicle with a 12-ton winch, a light crane, and a small workshop section. The latest CVR(T) variants are the FV4233 Stormer, a stretched version of Spartan for a full 10-man infantry section; and the Streaker high-mobility load carrier and artillery tractor.

Specifications Length: (Scorpion/Scimitar) 15.7 ft. (4.79 m.); (Striker) 15.83 ft. (4.83 m.); (Spartan) 16.16 ft. (4.93 m.); (Samaritan) 16.6 ft. (5.07 m.). Width: 7.33 ft. (2.24 m.). Height: (Scorpion/Scimitar) 6.85 ft. (2.1 m.); (Striker) 5.67 ft. (1.73 m.); (Spartan) 7.4 ft. (2.26 m.); (Sultan) 8.33 ft. (2.56 m.); (Samaritan) 7.95 ft. (2.41 m.). Weight, combat: (Scorpion/Scimitar) 8.5 tons; (Scorpion 90) 9.2 tons; (Spartan) 8.1 tons. Powerplant: 190-hp. Jaguar J60 6-cylinder, 4.2-lit. gasoline; or 155-hp. Perkins 5.8-lit. T6-3544 diesel. Hp./wt. ratio: (gas) 22.35 hp./ton; (diesel) 18.23 hp./ton. Fuel: 150 gal. (660 lit.). Speed, road: 50 mph (90.5 kph);

(Scorpion-90) 45 mph (72.5 kph). **Range, max.:** 400 mi. (844 km.). **Operators:** Bel, Bru, Hon, Iran, Ire, Ku, Mali, NZ, Nig, Phil, Oman, Tan, UAE, UK.

SCORPION: Soviet anti-ship missile. See ss-n-24.

SCOUTING: The use of light forces, aircraft, or small vessels to observe terrain and determine the location and composition of enemy forces. Unlike RECONNAISSANCE, which often requires some fighting to force the enemy to reveal his forces and weapons, scouting implies no fighting other than in self-defense.

SCREENING: 1. The use of light forces to prevent enemy scouts and RECONNAISSANCE elements from determining the location and composition of the main body of forces.

2. In naval warfare, the use of DESTROYERS, FRIGATES, or other escorts to defend a convoy or battle group from air, surface, or submarine attack.

SCUD (SS-1): NATO code name for the Soviet R-17 short-range (tactical) BALLISTIC MISSILE, intended as a nuclear, chemical, or "conventional" delivery system for Army and FRONT echelons. First deployed in 1957, Scud has been progressively upgraded to improve its range, accuracy, and reliability. The initial SS-1 a Scud A had a range of 81 mi. (130 km.); armed with a 40-kT nuclear warhead, it was mounted on a modified IS-III (Stalin) heavy tank chassis. It was followed in 1965 by the SS-1b Scud B, mounted on a more agile wheeled transporter. Introduced in 1970, the current SS-1c Scud C (R-17E) has longer range with some loss of accuracy. There are currently some 620 Scud launchers in Soviet service, organized into brigades of 12 or 18 launchers, with an equal number of reload missiles. One brigade is attached to each Army, while Fronts can have several brigades. Scud A and B have been exported to other Warsaw Pact countries, as well as Egypt, Syria, Libya, Iraq, and South Yemen. All export versions have conventional high-explosive warheads, but several countries have been developing their own chemical warheads. Scuds were launched by Egypt against Israeli forces in the Sinai during the 1973 Yom Kippur war, and by both sides in the Iran-Iraq War (the Iranian missiles were acquired from Libya). Scud achieved international notoriety when numerous missiles were launched by Iraq against Israel and Saudi Arabia during the 1991 Persian Gulf conflict. They caused relatively little damage, but much political and personal anxiety. Because of their poor accuracy, conventional Scuds can be used with any

effect only against large area targets, notably cities.

A single-stage liquid-fuel missile, Scud has four tail fins to provide stabilization and aerodynamic control, supplementing control vanes at the rocket nozzle. Payload options for Soviet Scud Cs include 40- to 100-kT nuclear, 2200-lb. (1000-kg.) high-explosive, and chemical warheads. Scud is mounted on a self-propelled transporter-erector-launcher (TEL) based on the MAZ 543 8x8 truck. The missile is stowed horizontally for travel and erected vertically off the rear of the vehicle for launch. Scud is being superseded by the more accurate SS-23 SPIDER. Iraq has produced a modified version as the AL HUSAYN.

Specifications **Length:** 37.4 ft. (11.4 m.). **Diameter:** 2.75 fet. (840 mm.). **Weight, launch:** (A) 9700 lb. (4409 kg.); (B) 14,050 lb. (6370 kg.); (C) 14,043 lb. (6370 kg.). **Range envelope:** 50–112 mi. (80–180 km.) nuclear; 174 mi. (280 km.) HE or chemical. **CEP:** 930 m.

SDI: See STRATEGIC DEFENSE INITIATIVE.

SDV: See SWIMMER DELIVERY VEHICLE.

SEA CAT: British shipboard, short-range surface-to-air missile (SAM), developed by Short Brothers during the late 1950s as a replacement for the 40-mm. BOFORS GUN, and first deployed by the Royal Navy in 1962. Relatively inexpensive, it is also almost completely ineffective against high-performance aircraft and all anti-ship missiles; its residual effect is psychological. A land-based derivative, TIGER CAT, is also of nuisance value only.

The Sea Cat missile has a cylindrical body with a bulbous warhead, four fixed rectangular tail fins, and four sharply swept, hydraulically actuated wings for steering. Powered by a dual-pulse solid-fuel rocket, the basic Sea Cat relies on manual radio COMMAND GUIDANCE with purely optical tracking facilitated by flares at the tips of the tail fins; but the Royal Navy also employs a variety of tracking radars for night and low-visibility engagements. The missile is armed with a 33-lb. (15-kg.) CONTINUOUS ROD fragmentation warhead fitted with both radar-proximity (VT) and delay-contact fuzes. Sea Cat is most commonly fired from a manually loaded quadruple launcher; a lighter triple launcher is also available.

Because of its low speed and limited range, Sea Cat cannot chase high-speed targets or follow their maneuvers, except under the most advantageous conditions. Moreover, its unsecured command guidance link is highly vulnerable to jamming. A number of Sea Cats were launched during the

1982 Falklands War, without scoring a single confirmed kill. The missile is now being superseded by the far more formidable SEA WOLF.

Specifications Length: 4.85 ft. (1.48 m.). **Diameter:** 7.5 in. (190 mm.). **Span:** 2.2 ft. (650 mm.). **Weight, launch:** 149.9 lb. (68 kg.). **Speed:** Mach 1 (762 mph/1220 kph). **Range:** 3.4 mi. (5.5 km.). **Height envelope:** 100–3000 ft. (30–915 m.). **Operators:** Aus, Bra, Chi, Ind, Indo, Iran, Neth, NZ, Nig, Pak, UK.

SEA COBRA: U.S. Marine Corps version of the AH-1 COBRA attack helicopter.

SEAD: Suppression of Enemy Air Defenses, a U.S. term for weapons, tactics, and operations whose aim is to destroy or otherwise neutralize ANTI-AIRCRAFT guns and surface-to-air missiles, in order to allow attack aircraft to operate more freely. SEAD includes the use of standoff jamming and escort jamming aircraft, chaff curtains, and other ELECTRONIC COUNTERMEASURES; direct attacks on air defense weapons by ordinary attack aircraft or specialized WILD WEASELS armed with ANTI-RADIATION MISSILES (ARMs) and CLUSTER BOMBS; and the use of long-range artillery or tactical BALLISTIC MISSILES. In U.S. Air Force doctrine, SEAD operations should normally precede other missions in priority, if not in time.

SEA DART: British shipboard surface-to-air missile system, developed by British Aerospace from the mid-1960s as the area defense weapon for the BRISTOL-class destroyers; when these were canceled after only one ship, a lighter version was developed for the smaller SHEFFIELD class. The missile was later also installed on the INVINCIBLE-class V/STOL aircraft carriers. Intended to intercept medium to high-altitude targets, See Dart was used in the 1982 Falklands War and credited with 8 kills for 31 missiles launched; however, it was only marginally effective against maneuverable targets at low altitudes.

Sea Dart has four broad-chord wings for lift, and four movable tail fins for steering. A two-stage missile, Sea Dart is powered by a tandem solid-fuel rocket booster and a liquid-fuel ramjet sustainer; the latter is fed through a nose air intake. After accelerating to ramjet-ignition speed (Mach 1.5), the booster drops away and the ramjet accelerates the missile to maximum speed. Sea Dart is guided by SEMI-ACTIVE RADAR HOMING (SARH); the missile homes on radar reflections from targets illuminated by the associated Type 209 tracking and guidance radar. Armed with a large high-explosive fragmentation warhead fitted with contact and radar proximity (VT) fuzes, it can also be used against surface targets at ranges of up to 18.5 mi. (30 km.). A Sea Dart Mk.2 with improved low-altitude performance was proposed in the late 1970s to counter Soviet ANTI-SHIP MISSILES, but the program was canceled in 1981.

Sea Dart is launched from a twin-arm launcher with a vertical magazine below. On the Bristol and Sheffield Batch 3 destroyers, the magazine holds 40 rounds, while on the Sheffield Batches 1 and 2 and the Invincibles, the magazine has only 20 missiles. A lightweight launcher for ships as small as 300 tons has also been developed, with deck-mounted launch canisters and simplified fire controls; none has been sold to date.

Specifications Length: 14.3 ft. (4.36 m.). **Diameter:** 1.4 ft. (420 mm.). **Span:** 2.9 ft. (0.91 m.). **Weight, launch:** 1213 lb. (550 kg.). **Speed:** Mach 3 (2100 mph/3360 kph). **Range:** 40.5 mi. (65 km.). **Height envelope:** 100–60,000 ft. (30–18,290 m.).

SEA EAGLE: British air-launched ANTI-SHIP MISSILE. Developed in the early 1970s as an all-weather replacement for the TV-guided MARTEL, Sea Eagle began flight testing in 1980 and entered service in 1985 with RAF BUCCANEER maritime strike squadrons. The missile has also been cleared for launch from the Panavia TORNADO and BAe SEA HARRIER. Saudi Arabia became the first export customer in 1987, ordering the missile along with Tornado strike aircraft. A helicopter-launched version with two strap-on boosters, was recently sold to the Indian navy for use from modified SEA KING anti-submarine helicopters. A ship-launched version was canceled when the Royal Navy ordered the U.S. RGM-84 HARPOON instead; a ground-launched coastal-defense variant has also been developed.

Derived from Martel, the airframe has four swept, cruciform wings for lift and and four smaller tail fins for steering. The most noticeable change is in propulsion: Martel is a solid-fuel rocket, while Sea Eagle is powered by a Microturbo TRI-60 turbojet, fed from a ventral air intake. Armed with a 330-lb. (150-kg.) high-explosive warhead, Sea Eagle relies on a combination of INERTIAL GUIDANCE with radar ACTIVE HOMING in the terminal phase. After release, the missile flies towards the estimated target position on inertial guidance; its advanced autopilot can be programmed to fly a variety of flight profiles (including dogleg patterns), and a radar altimeter can hold the missile at wave-top height. The Marconi radar seeker provides data for the guidance computer in the terminal phase; the computer said to be able to discriminate between target radar signatures, so that it can be programmed to pass over the nearest target to attack a larger target beyond.

Specifications **Length:** 13.6 ft. (4.14 m.). **Diameter:** 1.3 ft. (400 mm.). **Span:** 3.95 ft. (1.2 m.). **Weight, launch:** 1213 lb. (550 kg.). **Speed:** Mach 0.9 (620 mph/1000 kph). **Range:** 62 mi. (100 km.).

SEA HARRIER: Naval version of the British Aerospace HARRIER V/STOL attack aircraft.

SEAHAWK: U.S. shipboard anti-submarine helicopter. See BLACKHAWK.

SEA KING: Sikorsky S-61/SH-3 amphibious transport and ANTI-SUBMARINE WARFARE (ASW) helicopter. The most successful helicopter of its type, the U.S.-built Sea King is in service with U.S. and allied forces; Sea Kings are also built under license in Italy by Agusta and in Japan by Mitsubishi. In Britain, Westland produces both an ASW version and the "Commando" assault helicopter under license.

Built to a 1957 U.S. Navy requirement for a large, carrier-based ASW helicopter, the initial SH-3A first flew in 1959 and entered service in 1961. A number were also assembled by Canadair as the CH-124, while Mitsubishi built 118 for the Japan Maritime Self-Defense Force. The SH-3A was soon superseded by the more powerful SH-3D, which remains in production, though many U.S. Navy Sea Kings have been upgraded to an improved SH-3H standard. Other models include the VH-3A VIP transport, the SH-3G utility helicopter, the HH-3A search-and-rescue (SAR) helicopter, the CH-3C assault helicopter, the HH-3E SAR helicopter (the "Jolly Green Giant" of the Vietnam War), and the closely related HH-3F "Pelican" of the U.S. Coast Guard. More than 1200 Sea Kings are presently in worldwide, including some 750 built by Sikorsky and 400 built overseas.

The SH-3D has a boat-shaped, watertight fuselage, and a short, oval boom to support the stabilizing tail rotor. Its retractable "tail-dragger" landing gear is housed in watertight sponsons, which also serve as stabilizing floats. The pilot and copilot sit side-by-side in the nose, and in ASW versions, the main cabin houses a two-man tactical team to operate weapons and sensors. The Sea King is powered by two General Electric T58 turboshaft engines mounted in a nacelle over the main cabin, which drive a five-bladed rotor that folds back for deck stowage.

AVIONICS include complete blind-flying instruments, an autopilot and auto-hover system, a TACAN receiver, a DOPPLER navigation radar, a radar altimeter, and various RADAR WARNING RECEIVERS. Most Agusta-built models also have a chin-mounted surface-search RADAR. For ASW, the Sea King has an AWS-13 or -18 active/passive dipping SONAR, a MAGNETIC ANOMALY DETECTOR (MAD) "bird" extended from the starboard side of the fuselage, and SONOBUOY dispensers in the landing gear sponsons. Armament normally includes up to four MK.46 or Mk.50 BARRACUDA lightweight homing torpedoes, or four DEPTH CHARGES. For anti-shipping missions, Sea Kings carry HARPOON, EXOCET, SEA SKUA, or OTOMAT missiles, as well as MINES, rocket pods, flares, and smoke pots. Total payload capacity is some 8000 lb. (3629 kg.).

Agusta and Mitsubishi versions are virtually identical to the SH-3D, while the Westland Sea King has British engines and ASW sensors. Many have a dorsal MEL AW.391 or ARI.5991 "Sea Searcher" radar behind the engines, and a few have been converted to Mk.2 AIRBORNE EARLY WARNING platforms with a Thorn-EMI "Searchwater radar" in an inflatable dome attached to the starboard side. The Westland Commando version has simple fixed landing gear without sponsons; its main cabin is configured for up to 28 troops, and a variety of door guns and rocket pods can be fitted.

The CH- and HH-3E "Jolly Green Giant" versions retain the engines, main rotors and transmission of the Sea King, but have a completely different fuselage with a large rear ramp door, a narrow, rectangular tail boom, retractable tricycle landing gear, and a cabin configured for up to 25 troops, 15 stretcher cases, or 5000 lb. (2268 kg.) of cargo. The HH-3E has a number of specialized SAR features, including a rescue hoist, extensive armor protection, self-sealing fuel tanks, provision for external fuel tanks, an AERIAL REFUELING probe, and three door-mounted 7.62-mm. GAU-2 "Miniguns." The HH-3F Pelican is similar, but has no armament, armor, or in-flight refueling capability.

Specifications **Length:** (3D) 54.75 ft. (16.69 m); (3E) 57.25 ft. (17.45 m.). **Rotor diameter:** 62 ft. (18.9 m.). **Cargo hold:** (3E) 25.85 x 6.5 x 5.9 ft. (7.89 x 1.98 x 1.91 m.). **Powerplant:** (3A) 2 1250-shp. T58-GE-8B tuboshafts; (3D) 2 1400-shp. T58-GE-10s; (Westland) 2 1660-shp. Rolls Royce Gnome H.1400-1 turboshafts; (3E) 2 1500-shp. T58-GE-5 turboshafts. **Fuel:** 6867 lb. (3124 kg.). **Weight, empty:** (3D) 11,865 lb. (5382 kg.); (3E) 13,255 lb. (6012 kg.). **Weight, max. takeoff:** (3D) 20,500 lb. (9299 kg.); (3E) 22,050 lb. (10,002 kg.). **Speed,**

max.: 166 mph (267 kph). **Speed, cruising:** 136 mph (219 kph). **Initial climb:** 2200 ft./min. (671 m./min.) **Range, max.:** (3D) 625 mi. (1006 km.); (3E) 465 mi. (748 km.). **Operators:** (Sikorsky) Arg, Aus, Bel, Can, Den, Egy, FRG, Ind, Indo, Jap, Malay, Nor, Pak, Sp, US; (Augusta) Arg, Bra, Iran, Iraq, Italy, Lib, Mor, Per, S Ar, Syr; (Mitsubishi) Jap; (Westland) UK.

SEA KNIGHT: The Boeing-Vertol Model 107/CH-46 medium transport and utility HELICOPTER, in service with the U.S. Navy and Marine Corps, the Canadian Forces, the Japanese Ground and Air Self-Defense Forces, the Swedish Navy, and Saudi Arabia. Developed in the mid-1950s to a U.S. Army requirement, the Boeing-Vertol 107 first flew in 1958. Three prototypes were evaluated by the army, but that service chose the larger Ch-47 CHINOOK instead. In 1961, however, the 107 was selected as an assault helicopter by the marine corps, which ordered some 160 CH-46As, as well as 14 UH-46As (for the U.S. Navy) from 1962. These were soon superseded by 266 of the more powerful CH-46Ds and 10 UH-46Ds, which were in turn replaced by 146 CH-46Fs with improved avionics. Some 270 CH-46D/Fs were rebuilt from 1977 as CH-46Es, with up-rated engines, improved crashworthiness, and fiberglass rotor blades. Since 1965, Kawasaki has had worldwide rights to the Boeing-Vertol 107, and has built some 85 for Japan, 8 for Sweden, and 6 for Saudi Arabia. All are essentially similar to the CH-46D, except for the Swedish HKP.4s, which have Rolls Royce Gnome engines.

The CH-46E has the typical Vertol tandem-rotor configuration, with one 3-bladed rotor at the front and another at the rear. The pilot and copilot sit side-by-side, ahead of the main cabin; there is a powered ramp door in the rear. Most versions can carry up to 25 fully equipped troops, or 7000 lb. (3175 kg.) of cargo. The fuselage is watertight, and the main landing gear retract into sealed sponsons which also house tanks, yet can still act as stabilizer floats. AVIONICS include an autopilot and a stability augmentation system, a radar altimeter, a DOPPLER navigation radar, a TACAN receiver, and a RADAR WARNING RECEIVER. Most Sea Knights are unarmed, but Swedish HKP.4s used for ANTI-SUBMARINE WARFARE are equipped with dipping SONAR, a MAGNETIC ANOMALY DETECTOR (MAD) "bird," SONOBUOYS, lightweight homing TORPEDOES, and DEPTH CHARGES.

Specifications Length: 44.83 ft. (13.66 m.). Rotor diameter: 50 ft. (15.24 m.). Cargo hold:

24.16 x 6 x 6 ft. (7.37 x 1.83 x 1.83 m.). **Powerplant:** 2 1870-shp. General Electric T58-GE-16 turboshafts. **Fuel:** 2037 lb. (926 kg.). **Weight, empty:** 11,585 lb. (5255 kg.). **Weight, max. takeoff:** 21,400 lb. (9707 kg.). **Speed, max.:** 158 mph (254 kph). **Speed, cruising:** 150 mph (240 kph). **Max. climb:** 2050 ft./min. (625 m./min.). **Range:** (w/payload) 633 mi. (1019 km.).

SEAL: Sea, Air and Land (teams), the U.S. Navy's special operations force, trained and equipped for the clandestine SCOUTING of enemy coasts, harbors, and inshore waters, and also for RAIDS. SEALs may be inserted by parachute or small boat, but more typically they are transported into action by submarine, and then either swim or ride to the shore in submersible SWIMMER DELIVERY VEHICLES (SDVs).

SEAL personnel are volunteers selected from navy Underwater Demolition Team (UDTS) "frogmen." Thus they have already completed a demanding 24-week training program which includes physical conditioning, survival and escape procedures, demolitions, and combat swimming. To this, the SEAL program adds foreign languages, unconventional warfare, and parachute techniques, including HALO.

SEALs are organized into "Teams" of 27 officers and 156 enlisted men. Each team is divided into 5 platoons, each of which in turn is capable of self-contained operations. SEALs generally use standard U.S. equipment as well as German submachine guns, but there are also some specialized weapons, such as a waterproofed, silenced 9-mm. automatic pistol known as the "Hushpuppy."

The six SEAL Teams are controlled by two Naval Special Warfare Groups (NAVSPECWAR-GRUs, one) for the Pacific (based at San Diego, California) and one for the Atlantic (at Norfolk, Virginia). One SEAL Team (No. 6) is specialized for hostage rescue and more land-oriented operations in general.

SEA LANCE: The Anti-Submarine Standoff Weapon (ASW-SOW), a U.S. Navy program to develop a common successor to both the surface-launched RUR-5A ASROC and the submarine-launched UUM-44 SUBROC. Initiated in the 1970s—because both ASROC and SUBROC were aging, incompatible with new digital FIRE CONTROL systems, and in any event only marginally effective against the latest Soviet submarines—the ASW-SOW was intended to overcome the deficiencies of both by achieving supersonic speed and by having a homing TORPEDO instead of a DEPTH CHARGE warhead. The surface-launched version, designed to fit the Mk.41 vertical launch system (VLS) of the SPRUANCE, BURKE, and TICONDEROGA classes, was to incorporate a secure DATA LINK for midcourse guidance corrections, so as to eliminate the "dead time" which undermined the effectiveness of ASROC. A variety of different missile designs were proposed, including variants of the TOMAHAWK cruise missile and HARPOON anti-ship missile, as well as a totally new design powered by an integral rocket/ramjet. Technical problems with the surface version forced the Navy to develop a vertically launched ASROC (VLA) instead, with ASW-SOW, now named Sea Lance, as the SUBROC replacement only. In 1990, the navy canceled the entire program to cut costs.

Powered by a dual-impulse solid-fuel rocket with thrust-vector control (TVC) steering, Sea Lance has a torpedo-shaped body with wraparound tail fins; the payload, a MK.46 Mod 5 or Mk.50 BARRACUDA anti-submarine homing torpedo, is housed in a streamlined nose fairing. In action, target data from the submarine's SONAR would be passed to a Mk.117 digital torpedo FIRE CONTROL unit, which would compute the required impact position and time of flight for the missile's INERTIAL GUIDANCE unit. After ejection from a standard torpedo tube, Sea Lance rises to the surface; the rocket engine then ignites, the tail fins deploy, and the missile flies a ballistic trajectory towards the predicted target area. After burnout, the booster is jettisoned, and the torpedo coasts to the impact point. A braking parachute slows the weapon for water entry, after which the torpedo executes an autonomous search for the target. See also ANTI-SUBMARINE WARFARE; SUBMARINES, UNITED STATES.

Specifications Length: 20.5 ft. (6.25 m.). Diameter: 21 in. (533 mm.). Weight, Launch: 3093 lb. (1406 kg.). Speed: Mach 1.5 (1125 mph/1800 kph). Range: 63–103.5 mi. (101–165.5 km.).

SEA LINES OF COMMUNICATION (SLOCS): Maritime routes from a naval force to its bases and home country, and from sources of raw materials and/or reinforcements to the home country.

SEARCH AND RESCUE.: See SAR.

SEA SKUA: British helicopter-launched ANTI-SHIP MISSILE, developed by British Aerospace in the 1970s as a replacement for the AS.12 wireguided missile. Intended specifically for the Westland LYNX anti-submarine helicopter, and closely associated with the Ferranti Sea Spray search and

acquisition RADAR, Sea Skua began flight tests in 1979 and was used in the 1982 Falklands War before it had actually been declared operational. In four attacks on small Argentine patrol ships, Sea Skuas scored eight hits with eight missiles. Since then, they have been exported to Brazil, West Germany, and Turkey.

Sea Skua has a rounded nose and a cylindrical body, which tapers to a reduced-diameter tail section. Four movable delta wings mounted near the nose provide lift and steering, while four swept tail fins provide stabilization. Powered by a BAJ-Vickers dual-pulse solid-fuel rocket engine, the missile is armed with a 77-lb. (35-kg.) semi-armor-piercing (SAP) warhead designed to penetrate deep within the target before exploding.

Sea Skua relies on a combination of INERTIAL GUIDANCE and radar ACTIVE HOMING in the terminal phase. An autopilot and radar altimeter can be set to four different surface-skimming heights, and the missile can be programmed to perform a "pop-up" maneuver for target acquisition before the terminal phase (alternatively, performed on command from the helicopter); it then performs a diving attack.

Specifications Length: 9.33 ft. (2.85 m.). Diameter: 8.75 in. (222 mm.). Span: 24 in. (620 mm.). Weight, launch: 325 lb. (147 kg.). Speed: Mach 0.9 (620 mph/1000 kph). Range: 12.5 mi. (20 km.).

SEA SPARROW: Shipboard version of the AIM-7 SPARROW missile.

SEA SPRITE: Kaman SH-2, a U.S. Navy shipboard ANTI-SUBMARINE WARFARE (ASW) helicopter. Originally developed as a utility helicopter, the Sea Sprite first flew in 1959, and Kaman delivered some 190 single-engine UH-2A/Bs from 1962; starting in 1967, all were converted to twin-engine UH-2Cs. Also in 1967, Kaman produced a few HH-2C search-and-rescue versions, with a 7.62-mm. chin turret, 7.62-mm. door guns, a rescue hoist, and some armor protection; the later HH-2D was similar but unarmed. In the late 1960s, the U.S. Navy began to consider manned helicopters to replace the unsuccessful DASH (Drone Anti-Submarine Helicopter) aboard cruisers, destroyers, and frigates. In 1971, two HH-2Ds were converted for ASW, and in the following year the Sea Sprite was selected as the Lightweight, Airborne Multi-Purpose System (LAMPS)-I helicopter. A number of HH-2Ds were converted to SH-3D LAMPS, with chin-mounted search radar, ASW sensors, and lightweight homing torpedoes. In 1973, Kaman converted 185 UH-2Ds to SH-2Fs;

all surviving SH/HH-2Ds were later brought up to that standard. Some 190 UH-2s were built through 1965, of which about 90 SH-2F conversions are still in service. The navy intended to replace the Sea Sprite with the more advanced SH-60B Seahawk (LAMPS-III), but because of rising costs, it was decided instead to build 48 new SH-2Fs from 1980. With additional sensor and engine upgrades, (SH–2G) the Sea Sprite is to remain in service through the year 2000.

A compact yet powerful helicopter, the Sea Sprite's fuselage and tail boom are neatly blended, and sealed to provide some flotation. Mounted in a pod over the main cabin, twin engines drive a four-bladed composite rotor, whose blades can fold back for deck stowage. A Litton LN-66 surface-search radar is mounted in a chin radome. The pilot and copilot (who doubles as "tactical coordinator") sit side-by-side in the nose, with excellent visibility both forward and to the sides. AVIONICS include an ANS-123 DOPPLER navigation unit, complete blind-flying instrumentation, a TACAN receiver, and an ALR-66 RADAR WARNING RECEIVER. The main cabin houses a systems operator and several tactical display consols. AN ASQ-81 MAGNETIC ANOMALY DETECTOR (MAD) "bird" extends from the right side of the fuselage, and there is a dispenser for 15 SONOBUOYS on the left. The Sea Sprite is not equipped to analyze sonobuoy data on-board, but is fitted with a secure two-way DATA LINK allowing the data to be processed aboard the mother ship and relayed back to the helicopter. Two MK.46 or Mk.50 BARRACUDA homing torpedoes can be carried on two fuselage hardpoints. For anti-ship missions, the Sea Sprite can carry two AGM-119 PENGUIN missiles.

Specifications Length: 38.33 ft. (11.69 m.). Rotor diameter: 44 ft. (13.41 m.). Powerplant: (F) 2 1350-shp. General Eelctric T58-GE-8F turboshafts; (G) two 1625–Shp. T700-GE-401 turboshafts. Fuel: 2772 lb. (1260 kg.). Weight, empty: 7040 lb. (3193 kg.). Weight, max. takeoff: 13,500 lb. (6124 kg.). Speed, max.: 150 mph (241 kph). Speed, cruising: 138 mph (222 kph). Max. climb: 2440 ft./min. (744 m./min.). Range, max.: 411 mi. (61 km.).

SEA STALLION AND SUPER STALLION:

The Sikorsky H-53 series of heavy assault transport HELICOPTERS, in service with the U.S. Air Force, Navy, and Marine Corps and the armed forces of Austria, West Germany, Iran, and Israel. Developed to a 1960 Marine Corps specification for a heavy-lift helicopter, the H-53 combined the engines, transmission, and rotor of the CH-54 "Flying Crane" with a new, enclosed and watertight fuselage for amphibious use. The prototype first flew on 16 October 1964, and the first CH-53As were delivered to the marine corps in 1966; they were much used in Vietnam from 1967. In 1966, the U.S. Air Force ordered the search-and-rescue (SAR) HH-53B variant as a replacement for the HH-3E "Jolly Green Giant" (see SEA KING). These were followed by the improved HH-53C "Super Jolly" and the CH-53C mobile command post. (In the 1980s, nine surviving HH-53Cs were converted to MH-53H/J PAVE LOW IIIs for special operations.) From 1969, the CH-53A was replaced by the CH-53D, with more powerful engines, and in 1973, the navy took delivery of 30 RH-53Ds for airborne minesweeping. Some 464 of all models were produced for U.S. forces. West Germany purchased 112 CH-53Gs in 1969, some 90 of which were assembled by Fokker VFW in Germany, while Israel operates some 30 CH-53Ds received in the mid-1970s.

The large, rectangular fuselage ends in a short tail boom which supports a four-bladed tail rotor; deep sponsons on the sides house the main landing gear and fuel tanks, yet also act as stabilizer floats during water operations. Pilot and copilot sit side-by-side in the nose. The main cabin, equipped with a power-operated ramp door in the rear, can accommodate up to 37 fully equipped troops, 24 stretcher cases, or 8000 lb. (3629 kg.) of cargo internally. Alternatively, a 20,000-lb. (9072-kg.) external laod can be slung from a belly hook. For long-range flights, two 3150-lb. (1432-kg.) drop tanks can be attached to the sponsons, for a total fuel load of 13,650 lb. (6205 kg.); most Sea Stallions also have an AERIAL REFUELING probe on the right side of the nose. The CH-53D is powered by twin engines mounted on either side of a tall "hump" over the main cabin; the MH-53D, used to tow minesweeping sleds, has uprated engines. The engines drive six titanium rotor blades, which can fold back for compact stowage.

AVIONICS include comprehensive blind-flying instruments, an autopilot and auto-hover system, a TACAN receiver, a DOPPLER navigation unit, and a variety of RADAR WARNING RECEIVERS. CH-53s in U.S. service are normally unarmed, but HH-53s have three door-mounted 7.62-mm. GAU-2 "Miniguns" for defense suppression, and foreign operators add various weapons to their machines.

From 1971, Sikorsky began the development of a more powerful Sea Stallion to meet USMC heavy lift requirements. First flown in 1974, the CH-53E Super Stallion is the largest Western helicopter (though much smaller than the Soviet Mi-26 HALO). Development was protracted by budgetary shortfalls, and the first Super Stallions did not enter service until 1981. The marines currently operate some 90 CH-53Es, while the navy has 35 RH-53E "Sea Dragon" minesweepers.

Though externally similar to the CH-53D, the Super Stallion has nearly three times the horsepower and four times the payload of its predecessor. Three engines drive a seven-bladed, titanium-nomex composite rotor. Enlarged sponsons house additional fuel, and two 4550-lb. (2068-kg.). drop tanks can be added for a total fuel capacity of 16,219 lb. (7372 kg.) The RH-53E has even larger sponson tanks, which allow a mission endurance of some 20 hours. Both versions have aerial refueling probes. The CH-53E can carry up to 55 fully equipped troops, 7 cargo pallets, or a 36,000-lb. slung load. Avionics include all-digital flight controls, night-vision equipment, radar warning receivers, and chaff/flare dispensers.

Specifications **Length:** (D) 67.16 ft. (20.48 m.); (E) 74.33 ft. (22.33 m.). **Rotor diameter:** (D) 72.25 ft. (22.02 m.); (E) 79 ft. (24.08 m.). **Cargo hold:** 30 x 9.5 x 6.5 ft. (9.14 x 2.29 x 1.98 m.). **Powerplant:** (D) 2 3925-shp. General Electric T64-GE-7 turboshafts; (MH-53D) 2 4380-shp. T64-GE-415s; (E) 3 4380-shp. General Electric T64-GE-416 turboshafts. **Fuel:** (D) 3675 lb. (1670 kg.); (E) 7119 lb. (3236 kg.); (RH-53E) 22,400 lb. (10,181 kg.). **Weight, empty:** (D) 23,485 lb. (10,653 kg.); (E) 32,226 lb. (15,071 kg.). **Weight, normal loaded:** (D) 36,400 lb. (16,511 kg.); **Weight, ma. takeoff:** (D) 42,000 lb. (19,051 kg.); (E) 69,750 lb. (33,340 kg.). **Speed, max.:** 196 mph (315 kph): **Speed, cruising:** 173 mph (278 kph). **Max. climb:** (D) 2180 ft./min. (664 m./min.); (E) 2500 ft./min. (762 m./min.). **Combat radius:** (E) 60 mi. (96 km.) w/max. payload/500 mi. (800 km) wi/20,000-lb. (9091-kg.) payload. **Range, max.:** (D) 257 mi. (412 km.) int. fuel/600 mi. (960 km) w/ext. tanks; (E) 1290 mi. (2076 km.) w/max. fuel.

SEA WOLF: 1. British shipboard, short-range surface-to-air missile (SAM) system. Developed by British Aerospace from 1962 (as the successor to SEA CAT), to engage high-performance aircraft and ANTI-SHIP MISSILES, Sea Wolf entered service in 1979 aboard the BROADSWORD-class frigates. It was later also fitted to converted LEANDER-class frigates, and is being installed on the new DUKE-class frigates. Complex and rather expensive, Sea Wolf is also highly capable, judged on the basis of the 1982 Falklands War, when it shot down five Argentine aircraft. In tests, it has also successfully intercepted a 4.5-in. (114-mm.) shell and EXOCET surface-skimming anti-ship missiles.

The Sea Wolf missile has a conical nose and cylindrical body, with four broad-chord delta wings for lift, and four smaller tail fins for steering. Powered by a dual-pulse solid-rocket engine, Sea Wolf is armed with a small 29.5-lb. (13.4-kg.) high-explosive warhead fitted with a contact fuze, and relies on radio COMMAND GUIDANCE with ELECTRO-OPTICAL or radar tracking. With its semi-automatic command to line-of-sight (SACLOS) technique, the operator need only keep the target centered in his TV or radar sight. The system then senses the angular difference between the missile's position and the operator-target line of sight, and the computer generates steering commands until intercept.

The basic launcher is a twin-arm, 6-cell unit, as installed on the Broadswords and Leanders. The missiles are housed in closed canisters for protection from the weather, and a 30-round reload magazine is located immediately below the launcher. A vertical launch system, first tested in 1968, is being installed on the Duke class. As a private venture, British Aerospace has also developed a conversion kit to allow Sea Wolf to be fired from 4-round Sea Cat launchers. An evolutionary development, the GWS Mk.27, is to have an ACTIVE HOMING seeker and longer range.

Specifications **Length:** 6.25 ft. (1.9 m.). **Diameter:** 11.8 in. (300 mm.). **Span:** 1.5 ft. (4350 mm.). **Weight, launch:** 180.4 lb. (82 kg.). **Speed:** Mach 2 (1400 mph/2240 kph). **Range:** 4.04 mi. (6.5 km.). **Height envelope:** 15–10,000 ft. (4.7–3050 m.).

2. (SSN-21). A proposed class of U.S. nuclear attack SUBMARINES, the first of which is to be laid down in 1990. Current navy plans call for as many as 28 of these submarines to be completed from 1995 to 2010. Successors to the LOS ANGELES (SSN-688) class, the Sea Wolfs are intended to combine higher speed, greater operating depth, and better acoustical silencing with enhanced sensors and armament to counter projected Soviet submarine developments well into the next century. Despite these objectives, their design is technically conservative; they have been criticized for both their large size and high cost ($1 *billion* each). In particular, the limited use of automation and the retention of a pressurized-water reactor have been criti-

cized. Recent Soviet submarines, by contrast, are highly automated and have compact liquid-metal reactors to reduce overall size (and cost). On the other hand, recent Soviet submarine accidents have confirmed the navy's prejudice against such innovations.

The length-to-beam ratio has been reduced to only 8 to 1, vs. 10 to 1 in the Los Angeles, in order to improve maneuverability. The Sea Wolfs are to retain the modified teardrop hulform (bulbous bow, cylindrical centerbody, and tapered stern) and single-hull configuration of their predecessors. It had been planned to use super-strong HY-100 steel in the pressure hull, but cost and technical problems have caused a reversion to standard HY-80 steel instead, but of greater thickness. The outer hull surfaces are to be covered with ANE-CHOIC tiles to reduce radiated noise and the effective range of hostile active sonar, a technique not applied before in U.S. submarines, but which as been standard on Soviet and British submarines for many years. A tall, streamlined sail is to be located well forward and reinforced for surfacing through polar ice. A small wedge in the sail's forward base is intended to improve hydrodynamic flow and reduce drag. Control surfaces depart from previous U.S. practice by having retractable bow planes which allow surfacing through polar ice (instead of sail-mounted "fairwater" planes), and six stern fins (instead of four) to improve low-speed maneuverability.

The Sea Wolf will have twice the power of the Los Angeles class. Instead of a propeller, the Sea Wolf will have a pumpjet unit, a low-speed, multi-blade turbine inside an annular duct. Pioneered on the British TRAFALGAR-class submarines, pump jets are significantly quieter than propellers, particularly at high speeds. All machinery will be individually soundproofed and the entire powerplant plant will be mounted on a resilient sound-isolation raft, allowing the SSN-21 to travel quite silently at speeds up to 20 kt. Maximum speed is 5 kt. faster than the Los Angeles, but still slower than the fastest Soviet submarines.

Armament will consist of eight large-diameter 26-in. (650-mm.) tubes amidships, which will be able to accommodate MK.48 ADCAP wire-guided homing torpedoes, UGM-84 HARPOON anti-ship missiles, UGM-109 TOMAHAWK cruise missiles, and SEA LANCE ASW-SOW anti-submarine missiles, as well as future weapons of increased diameter (Soviet submarines have had 26-in. tubes for many years). The basic load is to be 36–40 weapons of all

types, as compared to 24 in the Los Angeles. All sensors are to be linked to a new digital Submarine Advanced Combat System (SUBACS), a fully integrated FIRE CONTROL unit. See also SUBMARINES, UNITED STATES.

Specifications Length: 326.1 ft. (99.42 m.). **Beam:** 4 ft. (12.2 m.). **Displacement:** 7770 tons surfaced/9150 tons submerged. **Powerplant:** single-shaft nuclear: 1 S6W pressurized-water reactor, 2 sets of geared turbines, 60,000 shp. **Speed:** 20 kt. surfaced/35 kt. submerged. **Max. operating depth:** 1805 ft. (550 m.). **Collapse depth:** 2789 ft. (850 m.). **Crew:** 149. **Sensors:** 1 BSY-2 sonar array, 1 passive conformal array sonar, 1 passive towed array sonar, 1 surface-search radar, 1 electronic signal monitoring array, 2 periscopes.

SECOND STRIKE (CAPABILITY, STRATEGY): A "second strike" is a strategic nuclear strike that follows an enemy nuclear attack. The less ambiguous term "strike-back" is often recommended but seldom used.

To launch a nuclear strike after having already suffered the effects of one—i.e., to have a "second strike capability"—bombers must be able to evade attack by dispersal, airborne alert, or quick-reaction ground alert, while missiles must be HARDENED or otherwise protected by mobility, and associated control facilities must also survive. Of course, vulnerability depends as much on the enemy's capabilities, and cannot be reliably deduced unless the latter are fully anticipated; a force well protected against "officially recognized" threats may still be vulnerable (e.g., to commando raids). If a party believes that it has a second strike capability, it may decide to use it only in reprisal for a prior nuclear attack; that amounts to a "second strike strategy." But possession of a second strike capability does not necessarily imply a "strike-back only" strategy. The U.S. ASSURED DESTRUCTION concept is contextual, and therefore more useful.

SEGO (SS-11): NATO code name for the standard Soviet lightweight ICBM of the later 1960s and early 1970s. Introduced in 1966, Sego was the "light" counterpart of the heavyweight SS-9 Scarp, and armed the bulk of the Strategic Rocket Forces (RVSN) through the mid-1970s. At one time, some 1020 Segos were deployed in fixed hardened silos at 12 sites across the U.S.S.R., but since the 1970s the missile has been progressively supplanted by the SS-17 SPANKER and SS-19 STILETTO (and more recently by the SS-24 SCALPEL). At present, less than 400 remain in service at six sites: Drovyannaya, Gladkaya, Olovyannaya, Prem, Svobodnyy, and Teykovo.

A two-stage, liquid-fuel missile, Sego burns a storable, hypergolic fuel mix of undimensional dimethyl hydrazine (UDMH) and nitrogen tetroxide (NTO), used in most other Soviet ICBMs and SLBMs. The first stage has four fixed main engines and four gimballed vernier engines for steering and attitude control; the second stage has a single fixed main engine and two gimballed verniers. Sego relies on INERTIAL GUIDANCE, employing the traditional Soviet "fly-by-wire" technique, whereby the missile is constrained to a preprogrammed pitch and velocity profile by thrust magnitude control. This method provides reasonable accuracy without the need for high-speed computers.

Sego employs the "hot launch" technique, whereby the main engines ignite within the silo, and the missile flies out under its own power, with the exhaust gases vented upwards between the missile and the silo walls.

Sego has been built in four distinct subtypes. The Mod 1 was armed with a single 950-kT REEN-TRY VEHICLE (RV), while the Mod 2, introduced in the late 1960s, was similar, but carried an improved RV. The Mod 3, first deployed in 1973, has a longer range, better accuracy, and a payload of three 250-kT MULTIPLE REENTRY VEHICLES (MRVs). An experimental Mod 4, spotted in the mid-1980s, had three RVs of complex shape and low terminal velocity; this is believed to be a chemical weapon delivery system. All remaining SS-11s are believed to be Mod 3s. See also ICBMS, SOVIET UNION.

Specifications Length: 65.6 ft. (20 m.). Diameter: 8.2 ft. (2.5 m.). Weight, launch: 99,205 lb. (45,095 kg.). Range: (½) 6200 mi. (10,000 km.); (3) 6600 mi. (10,560 km.). CEP: (½) 1400 m.; (3) 850 m.

SELF-FORGING FRAGMENT (WARHEAD):

Also known as Explosive-Formed Penetrators (EFPs), these innovative armor-piercing munitions consist of a concave disc of tungsten or some other very dense metal, backed by plastic explosives. In a novel application of the SHAPED CHARGE principle, detonation of the explosives converts the disc into a long, narrow bolt, which is projected along the original axis of the disc at several thousand m./sec. (much faster than gun-fired penetrators). Even a relatively small warhead is thus capable of penetrating several inches of steel armor. In the U.S., Self-Forging Fragment warheads are being developed in several forms for top attack anti-armor

weapons, including the air-delivered SKEET and the artillery-delivered SADARM.

SEMI-ACTIVE LASER HOMING: A technique used in LASER-GUIDED BOMBS and missiles (e.g., PAVEWAY, MAVERICK), whereby the bomb or missile homes on radiation from a laser designator reflected off the target; the designator can be mounted on the launch aircraft or aimed by another operator or aircraft.

SEMI-ACTIVE RADAR HOMING: A form of guidance usually applied in surface-to-air or air-to-air missiles, whereby the missile homes on radiation reflected off the target from a RADAR not built into the missile itself. The nose of the missile houses a radar receiving antenna, but there is no transmitter; the target must therefore be "illuminated" by a special pencil-beam radar, normally located on the launch platform. The missile's receiver detects the target echo, and the missile's guidance computer then steers a collision course (see PROPORTIONAL GUIDANCE). Often, there is a "reference antenna" in the missile's tail to sample the signal frequency from the illumination radar. When compared to the signal received by the nose antenna, the revealed "DOPPLER shift" can determine the target's velocity relative to the missile.

The advantages of SARH include all-weather and ALL-ASPECT capabilities. Moreover, by taking advantage of powerful radar transmitters on the launch platform, SARH missiles can be effective at much longer ranges than INFRARED-HOMING missiles, without the added expense and complexity of ACTIVE-HOMING missiles (which have both a radar transmitter and receiver on-board). On the other hand, the target must be illuminated throughout the engagement, making the illuminating platform vulnerable to anti-radiation missiles. In addition, the illumination beam can be detected by hostile RADAR WARNING RECEIVERS, and this may allow the target to initiate evasive maneuvers, jamming, or other ELECTRONIC COUNTERMEASURES. For fighter aircraft, there is an additional disadvantage: since the illumination radar in the fighter's nose has only a limited field of view, the aircraft must continue flying towards the target, exposing itself to counterattack. In surface-to-air missile systems, the number of targets which may be engaged concurrently is limited by the number of illuminating radars available; that in turn limits the ability to deal with mass attacks. The U.S. Navy's AEGIS system overcomes this problem by using computer-controlled time-sharing techniques to illuminate each target in turn only a few seconds before missile impact.

SEMI-AUTOMATIC WEAPON: Any self-loading weapon (but usually a small arm) which fires a single shot each time the trigger is pulled.

SENTINEL: A U.S. BALLISTIC MISSILE DEFENSE program proposed in late 1967 and later superseded by SAFEGUARD. Both were based on the same components, but whereas Sentinel was primarily oriented against a notional "Chinese" (light, ICBM-only) attack against 25 major U.S. cities, Safeguard was intended to provide all-around protection, so as to defend MINUTEMAN silos and bomber bases, as well as Washington, D.C., and other command centers, and also to provide a marginal area defense capability against a small or unsophisticated ICBM attacks. See also ANTI-BALLISTIC MISSILE.

SENTRY: Boeing E-3 Airborne Warning and Control System (AWACS) aircraft, in service with the U.S. Air Force (USAF), the Royal Air Force, Royal Saudi Air Force, and a NATO multi-national unit. Developed to a 1970 USAF specification, the Sentry is a modified Boeing 707-320B airliner equipped with a large surveillance RADAR and COMMAND AND CONTROL (C^2) facilities both to detect hostile aircraft and to direct fighters against them. The prototype flew in 1972, and the first production E-3As entered service with USAF's TACTICAL AIR COMMAND (TAC) in 1977. Since 1979 the E-3 has also served with NORAD, to provide airborne early warning for the continental U.S. and Canada.

USAF acquired 24 E-3As through 1981. These were followed by the "US/NATO Standard E-3A," with improved data processing and a maritime surveillance capability. Nine (plus one converted prototype) were produced for USAF, while 18 were acquired by NATO under a special agreement whereby the aircraft are registered in Luxembourg and operated by multi-national crews. From 1984, USAF converted its 24 E-3As to a new E-3B standard with upgraded computers, jam-resistant communications, additional tactical displays, and an "austere" maritime surveillance capability. In 1984, USAF also modified its 10 US/NATO Standard aircraft to E-3Cs, with additional C^2 facilities. In 1985, Saudi Arabia purchased 5 modified "E-3A/Saudis" with high-bypass turbofan engines and extended range. The Royal Air force E-3s purchased to replace the abortive NIMROD AEW have E-3B-type avionics, but with more maritime surveillance capability.

The Sentry is undoubtedly one of the most capable airborne early warning aircraft extant; it can serve as a valuable FORCE MULTIPLIER by detecting hostile aircraft at much greater ranges than ground-based surveillance radars, and then vectoring fighters into advantageous intercept positions. Its main drawbacks are the cost (some $60 million each) and the inherent vulnerability of a transport which emits powerful radar signals and is also so valuable a target. Even if protected by friendly fighters at all times, the E-3 remains vulnerable to long-range ANTI-RADIATION MISSILES.

The E-3's wings, fuselage, and tail assembly are virtually identical to those of the commercial Boeing 707-320B, except that an AERIAL REFUELING receptacle is added over the cockpit. The main cabin is configured as an airborne command post, with a communications console, radar maintenance console, and nine tactical display consoles. A galley and three bunks for relief crews are provided in the rear fuselage; the Sentry can remain on station for up to 22 hours with aerial refueling. The cargo hold below the cabin houses communications equipment, the radar transmitter, and air conditioning for the avionics.

The heart of AWACS is a Westinghouse surveillance RADAR whose antenna is contained in a characteristic, disc-shaped rotodome mounted on twin pylons over the rear fuselage. Rotating at a rate of six revolutions/min., the dome has an airfoil cross-section, and thus generates sufficient lift at cruising speed to support its own weight. The original APY-1 radar had four operating modes: low pulse rate for long-range surveillance; high-resolution pulse-Doppler; pulse-Doppler height finding; and passive (receive-only)—the latter being meant for the detection of hostile aircraft by their own radar emissions. The US/NATO Standard E-3A, the E-3B, and the E-3C all have the improved APY-2 radar with automatic track initiation, and an additional maritime surveillance mode capable of detecting ships and slow-flying helicopters against surface "clutter." Modes can be switched from scan to scan, or even within scans. When the E-3 cruises at a height of 29,000 ft. (8850 m.), the radar has a range of some 230 mi. (368 km) against low-flying, fighter-size targets.

Radar signals are processed by an IBM 4Pi computer, a Model CC-1 in the basic E-3A or a CC-2 (with twice the memory) in later aircraft. Other AVIONICS include a weather radar in the nose, TACAN, LORAN, and INERTIAL NAVIGATION units, various UHF and VHF fighter control radios, and (in USAF and NATO aircraft) a JTIDS (Joint Tactical

Information Distribution System) data link. USAF Sentrys are also being fitted with ELECTRONIC COUNTERMEASURES equipment, including ACTIVE JAMMERS and CHAFF/flare dispensers. See also AIR-BORNE EARLY WARNING.

Specifications Length: 231.33 ft. (70.51 m.). Span: 195.67 ft. (59.64 m.). Powerplant: (U.S. and NATO) 4 Pratt and Whitney TF33-PW-100 (commercial JT3D) low-bypass turbofans, 21,000 lb. (9525 kg.) of thrust each; (Saudi) 4 SNECMA CFM56-2-A2 high-bypass turbofans, 22,000 lb. (9979 kg.) of thrust each. Weight, empty: 162,000 lb. (73,482 kg.). Weight, max. takeoff: 325,000 lb. (147,410 kg.). Speed, max.: 530 mph (853 kph). Speed, cruising: 468 mph (781 kph). Mission radius: (6 hr. loiter) 1000 mi. (1609 km.).

SH-2: U.S. anti-submarine helicopter. See SEA SPRITE.

SH-3: U.S. anti-submarine helicopter. See SEA KING.

SH-60: The U.S. Seahawk anti-submarine helicopter. See BLACKHAWK.

SHADDOCK (SS-N-3): NATO code name for a Soviet ship- and submarine-launched ANTI-SHIP MISSILE introduced in the early 1960s; a ground-launched derivative, the SS-C-1 Sepal, is also in service. The initial SS-N-3c of 1960, developed for the nuclear attack of land targets, was first deployed aboard "Twin Cylinder" and "Long Bin" modifications of the WHISKEY-class diesel-electric attack submarines, and later armed the ECHO I nuclear-powered submarines (SSGNs). It was followed in 1962 by two anti-ship variants, the SS-N-3a, for the JULIETT and Echo II–class submarines, and the SS-N-3b for the KYNDA and KRESTA II class cruisers.

A large, airplane-configured missile with mid-mounted cropped-delta wings and two ventral stabilizers, Shaddock is powered by two strap-on solid rocket boosters and a turbojet. The missile cruises at high altitude, but then executes a high-speed diving attack. Payload options are either a 2200-lb. (1000-kg.) high-explosive or a 300-KT nuclear warhead; the SS-N-3c carries a 5071-lb. (2305-kg.), 800-kT nuclear warhead.

The original SS-N-3c relied solely on INERTIAL NAVIGATION, and was not very accurate; even with its large warhead, it could be used only against very large targets (i.e., cities). The SS-N-3a/b combines inertial guidance with midcourse radio COMMAND GUIDANCE updates and radar ACTIVE HOMING in the terminal phase. For long-range attacks, these missiles require target data from OVER-THE-

HORIZON TARGETING (OTH-T) platforms (either aircraft, ships, or submarines) in direct radar contact with the target. The OTH-T platform relays a radar picture of the target area to the launch platform via a video DATA LINK (VDL). The launch platform then designates the target and relays this data to the missile using either a "Snoop Pair" (surface ship) or "Front Door/Front Piece" (submarine) guidance radar.

Shaddock is launched from trainable quadruple launchers on surface ships, or from erectable twin launchers on submarines. The missile must be fired from the surface; hence, submarines are extremely vulnerable throughout the engagement sequence, but especially after launch, when emissions from their guidance radar can be easily detected. Additional weaknesses include an over-reliance on OTH-T and the missile's high-altitude approach (which makes it easy to detect and intercept). On the other hand, Shaddock is far more powerful than most anti-ship missiles, and a direct hit, even from a conventional version, could sink or disable all but the largest vessels. Shaddock has been superseded in newer ships by the SS-N-12 SANDBOX (now retrofitted to some Julietts and Echo IIs) and the SS-N-19 SHIPWRECK.

Specifications Length: (a/b) 33.5 ft. (10.21 m.); (c) 38.7 ft. (11.8 m.). Diameter: 2.8 ft. (854 mm.). Span: 16.4 ft. (5 m.). Weight, launch: (a/b) 11,905 lb. (5411 kg.); (c) 12.786 lb. (5811 kg.). Speed: Mach 1.4 (810 mph/1296 kph). Range: (a/b) 286 mi. (458 km.); (c) 460 mi. (736 km.).

SHAFRIR: Israeli INFRARED HOMING (IR) air-to-air missile developed by Rafael in the mid-1960s. Similar in configuration to the AIM-9 SIDE-WINDER, Shafrir is slightly shorter, but has a greater diameter and weight. While generally interchangeable with Sidewinder, the Israeli missile is optimized for short-range dogfights and "snapshot" engagements. The prototype Mk.1 did not enter production, but the Mk.2, which entered service in 1969, has been credited with destroying more than 200 aircraft. This performance, combined with a unit cost of some $20,000, has resulted in sales to Chile, South Africa, Taiwan, and several other countries. Shafrir Mk.2 has now been superseded by the all-aspect Rafael PYTHON, originally designated Shafrir Mk.3.

Shafrir has a blunt nose housing an IR seeker, and a cylindrical body which contains the guidance unit, a 24.3-lb. (11-kg.) high-explosive fragmentation warhead fitted with contact and PROXIMITY FUZES, and a rocket motor. As with Sidewinder, the missile has four cruciform canards for steering, and four fixed tail fins for lift. Slipstream-driven "rollerons" (another Sidewinder feature) fitted to the tail fins act as control gyroscopes. The cryogenically cooled IR seeker, with wide-angle Cassegrain optics, is capable of rear-hemisphere and beam attacks, and may have some limited ALL-ASPECT capability. In combat it has proven quite accurate and reliable, with a claimed kill probability (P_K) of 60 percent.

The engagement sequence is similar to that of sidewinder. The pilot selects a missile and acquires the target in his gunsight. When the seeker locks onto the infrared emissions of the target, it provides the pilot with visual and aural launch cues. After launch, the missile homes autonomously, using proportional navigation to achieve a collision course.

Specifications Length: 8.1 ft. (2.47 m.). Diameter: 6.3 in. (160 mm.). Span: 33.9 in. (861 mm.). Weight, launch: 205 lb. (93.2 kg.). Speed: Mach 2.5 (1625 mph/2600 kph). Range: 3.1 mi. (5 km.).

SHAPE: Supreme Headquarters, Allied Powers Europe, one of two major commands within NATO, the other being ACLANT (Allied Command, Atlantic). From headquarters in Brussels, its commander, SACEUR (Supreme Allied Commander, Europe)—who has always been a U.S. Army general—controls ACE (Allied Command, Europe), in charge of land, naval and air forces. SHAPE is also involved in strategic-nuclear planning because of the commitment of some U.S. and all British nuclear forces to SACEUR. SHAPE is responsible for the defense of all NATO European territory except Britain and Portuguese coastal waters.

ACE controls directly a small mobile force of seven augmented battalions, a light armored unit, and several fighter/attack squadrons, and also supervises the NATO air defense system (NADGE). But ACE's main combat forces are those of NATO-Europe's three area commands: AFNORTH, AF-CENT, and AFSOUTH.

AFCENT (Allied Forces, Central Europe), is responsible for all NATO forces on the "Central (European) Front." Its subordinate commands are the Northern (NORTHAG) and Central (CENTAG) Army Groups, plus Allied Air Forces Central Europe (AAFCE). AFNORTH (Allied Forces, Northern Europe) is responsible for all NATO forces in Northern Europe, i.e., Norway, the North Sea, and the Arctic Ocean. AFSOUTH (Allied Forces, Southern Europe) is responsible for all NATO forces in Southern Europe, including Portugal, Spain, Italy, Greece, Turkey, and the Mediterranean.

SHAPED CHARGE: Also called "hollow charge," a directional explosive charge which exploits the "Munroe Effect" to penetrate armor and concrete. Shaped charges consist of cylindrical blocks of explosives with a conical cavity at one end, lined with copper or an alloy. When the charge is detonated, the force of the explosion is focused along the axis of the cone towards its base, forming a narrow high-velocity "jet" of very hot gases which can burn through steel plate and concrete that would resist the blast of conventional explosives. Shaped charges are incorporated into High-Explosive Anti-Tank (HEAT) munitions, most ANTI-TANK GUIDED MISSILE warheads, some mines, and many demolition devices.

SHEFFIELD (TYPE 42): A class of 14 British guided-missile DESTROYERS commissioned between 1972 and 1985. Two additional ships were built for the Argentine navy between 1971 and 1981; both were put up for sale due to a lack of spare parts after the 1982 Falklands War and the resulting British embargo. The Sheffields were designed as anti-air warfare (AAW) escorts after the cancellation of the larger and more costly BRISTOL-class escorts. It combines a cheaper version of the Bristol's weapon suite with a smaller, more economical hull. From the beginning, they were criticized for their scant armament, sensors, and ELECTRONIC WARFARE (EW) equipment, as well as for their poor DAMAGE CONTROL provisions. All these concerns were validated by the experience of the Falklands War, when 2 Type 42s, *Sheffield* and *Coventry*, were sunk due to such design deficiencies (*Sheffield* succumbed to a single hit from an Exocet missile, while the U.S. PERRY-class frigate *Stark* survived two Exocet hits in roughly the same area). The Falklands experience led to a major redesign of the last 4 ships, designated Type 42 Batch 3 (sometimes called the "Manchester" class).

The Type 42s were built in three subclasses, or batches. Batches 1 and 2 are generally identical except for slight differences in armament and internal arrangements. The Batch 3 ships are similar in appearance, but have been lengthened to to improve seakeeping, habitability, damage control, and access to equipment, with only minimal changes in armament and sensors. Flush-decked, the Sheffields have a boxy superstructure forward, a large funnel amidships, and a detached helicopter hangar aft. All three batches have two pairs of gyro-controlled fin stabilizers.

Armament in Batches 1 and 2 consists of a twin-arm SEA DART surface-to-air missile launcher with 20 missiles at the bow; a 4.5-in. (114-mm.) DUAL PURPOSE gun ahead of the Sea Dart launcher; 2 manually operated 20-mm. ANTI-AIRCRAFT guns behind the bridge; and 2 sets of triple tubes for lightweight anti-submarine TORPEDOES aft. Batch 3 armament is identical, except that the Sea Dart launcher has 40 missiles. After the Falklands War, the Royal Navy added 2 power-operated twin 30-mm. anti-aircraft guns amidships and 2 twin 20-mm. guns aft on all ships. All subclasses operate a LYNX anti-submarine helicopter from a landing deck and hangar on the fantail. All ships also have a SCOT satellite communications terminal. See also DESTROYERS, BRITAIN.

Specifications Length: (½) 412 ft. (125.5 m.); (3) 462.75 ft. (141.1 m.). **Beam:** (½) 47 ft. (14.5 m.); (3) 49 ft. (14.9 m.). **Draft:** 19 ft. (5.8 m.). **Displacement:** (½) 3150 tons standard/4100 tons full load; (3) 4100 tons standard/4775 tons full load. **Powerplant:** twin-shaft COGOG: 2 Rolls Royce Tyne RM1C gas turbines, 9700 shp. (cruise), 2 25,000-shp. Rolls Royce Olympus TM3B turbines (sprint). **Speed:** (½) 29 kt.; (3) 30 kt. **Range:** 650 n.mi. at 29 kt./4500 n.mi. at 18 kt. (4750 n.mi. for Batch 3). **Crew:** (½) 253; (3) 301. **Sensors:** (½) 1 Type 965M long-range air-search radar or (3) 1 Type 1002 air-search radar, 1 Type 1006 navigation radar, 1 Type 992Q surface/air-search radar, 2 Type 909 fire control and Sea Dart guidance radars, (½) 1 Type 184 or (3) Type 2016 hull-mounted medium-frequency sonar, 1 Type 162M side-looking classification sonar (Batch 3 may eventually be refitted with a bow-mounted Type 2050 low-frequency sonar). **Electronic warfare equipment:** 1 UAA-1 Abbey Hill electronic signal monitoring array, 2 8-barrel Knebworth-Corvus chaff rocket launchers; (post-Falklands) 2 Type 670 active jamming units, 2 Mk.36 SRBOC 12-barrel chaff launchers.

SHERIDAN: M551 armored reconnaissance vehicle. Developed in response to a 1961 U.S. Army requirement for a lighter, air-portable replacement for the M41 Walker Bulldog light tank, the M551 was an unsuccessful attempt to use an excessively advanced missile armament in order to combine the speed and mobility of a LIGHT TANK with the firepower of a MAIN BATTLE TANK (MBT). Rushed into production in 1966 and immediately dispatched to Vietnam, it suffered from chronic

engine overheating, while its ammunition and fire controls deteriorated rapidly in the heat and humidity. Moreover, its thin armor offered scant protection against the MINES and RPG-7 rocket launchers of the Communist forces. Even when later deployed in more temperate European environments, the Sheridan remained mechanically unreliable and operationally almost useless. It was finally withdrawn from service in 1979 (except with the 82nd Airborne Division), after 1729 had been produced. In the early 1980s, some 300 were externally modified to resemble Soviet tanks, ant-aircraft vehicles, and armored personnel carriers for training purposes.

The hull, fabricated of cast aluminum armor sections, has a maximum thickness of only 40–50 mm., enough to defeat armor-piercing rounds up to 14.5 mm. The glacis plate is sloped at 40° to improve ballistic protection, but the thinner side and rear plates are flat. Internally, the hull is divided into a driving compartment forward, a fighting compartment in the middle, and an engine compartment in the rear.

The driver is provided with a vision slit and three periscopes, one of which can be replaced by an INFRARED (IR) night scope. The remaining crew sit in the welded steel turret, with the commander on the right, above and behind the gunner, and the loader on the left. The commander has a cupola surrounded by periscopes for 360° observation; the center periscope can be replaced by an IR night scope. The gunner has a day/night optical sight with an integral LASER rangefinder (one of the first in service) linked to a complex analog FIRE CONTROL computer.

Main armament is the unique M81 152-mm. gun/missile launcher, designed specifically for the MGM-51 SHILLELEIGH anti-tank guided missile (ATGM), as well as low-velocity shells. Also meant to arm the abortive MBT-70 and the soon-with-drawn M60A2 PATTON, Shilleleigh is launched with a low-velocity cartridge from the short-barreled M81 gun. With a maximum range of 4000 m. and a 15-lb. (6.8-kg.) shaped-charge (HEAT) warhead, it can penetrate up to 430 mm. (17 in.) of homogeneous steel armor. The missile relies on infrared COMMAND GUIDANCE from an IR transmitter incorporated into the gun sight. To engage armored targets below Shilleleigh's 1000-m. minimum range, the gun can also fire a low-velocity HEAT shell with a combustible cartridge case. Unfortunately, this ammunition proved to be quite fragile, and could be a significant explosive hazard.

Moreover, the gun's recoil frequently disabled the complex fire control computer needed to use the missile, while the low muzzle velocity of the HEAT round made accurate shooting beyond 800 m. almost impossible. Finally, the size of Shilleleigh meant that only 8 missiles (together with 20 HEAT rounds) could be stowed in the hull.

Secondary armament consists of a coaxial 7.62-mm. machine gun, and an M2 BROWNING heavy machine gun on a ring mount over the commander's cupola. Four smoke grenade launchers are mounted on each side of the turret. For NBC defense, each crewman can connect his individual respirator to a COLLECTIVE FILTRATION unit. A flotation screen can be erected for amphibious river crossings, using track action to propel the vehicle at 3.5 mph (6 kph).

Specifications **Length:** 20.36 ft. (6.2 m.). **Width:** 7.6 ft. (3.32 m.). **Height:** 7.45 ft. (2.27 m.). **Weight, combat:** 17 tons. **Powerplant:** 300-hp. Detroit 6V-53T 6-cylinder turbocharged diesel. **Hp./ wt. ratio:** 17.65 hp./ton. **Fuel:** 150 gal. (660 lit.). **Speed, road:** 42 mph (68 kph). **Range, max.:** 360 mi. (576 km.).

SHERMAN: U.S. M4 medium TANK of World War II, still serving in modified form with Third World armies and used by the Israeli army until quite recently.

SHF: Super-High Frequency, electromagnetic radiation with wavelengths between 0.1 m. and 1 cm. (5–50 GHz frequency band); SHF is commonly used in TARGET ACQUISITION radars and satellite communication links.

SHILLELEIGH: U.S. MGM-51A gun-launched ANTI-TANK GUIDED MISSILE (ATGM). Developed in the early 1960s as part of the abortive MBT-70 program, and later used to arm both the M60A2 PATTON main battle tank and the M551 SHERIDAN airborne reconnaissance vehicle, Shilleleigh was an attempt to overcome the inherent limits of contemporary fire controls in order to provide long-range precision fire, Flight tests began in 1964, and the missile entered service in 1966; some 88,000 were produced through 1970. The critical element of the Shilleleigh system is the M81 152-mm. low-velocity gun (actually a HOWITZER), which serves as a missile launcher for long-range attack, and can fire a shaped-charge (HEAT) round for attacks out to a range of 800 m.

Shilleleigh has a round nose, a cylindrical body, and four pop-out tail fins. A small explosive cartridge attached to its base of the ejects the missile with a muzzle velocity of 689 m./sec. (1540 mph);

its solid rocket engine then ignites to burn for 1.18 seconds, accelerating the missile to its maximum speed. Shilleleigh is armed with a 15-lb. (6.8-kg.) HEAT warhead which can penetrate up to 430 mm. (17 in.) of homogeneous steel armor.

Shilleleigh is controlled in flight by an unusual infrared COMMAND GUIDANCE system, with semi-automatic command to line-of-sight (SACLOS). An infrared flare in the base of the missile is detected by a sensor in the gunner's sight. A FIRE CONTROL computer then calculates the angle between the missile's course and the gunner-target line of sight, to generate steering commands in order to bring the missile onto that line. These commands are transmitted to the missile through a coded infrared beacon, which is received by a detector in the base of the missile. All the gunner need do is keep the target centered in his sight until impact.

While technically ingenious, Shilleleigh was not successful in practice. The need to maintain a line of sight to the target throughout the engagement prevents the vehicle from shifting its position, a vulnerability magnified by the tendency of the missile to drop several feet after launch, until "gathered" by the command guidance beam; that prevents the use of more sheltered "hull-down" firing positions. In addition, the gathering phase imposes a minimum engagement range of 1000 m., while the maximum effective range of the gun-fired 152-mm. HEAT round is only 800 m., leaving a "dead zone" which happens to coincide with the ranges at which most tank battles occur. Moreover, the recoil of the gun when firing HEAT rounds frequently disables the fire control computer on which the missile depends for guidance, and the IR command beacon to the missile can be seriously attenuated by fog, rain, and smoke. Finally, the missile proved to be a serious fire hazard, its rocket motor often "cooking off" after the vehicle was hit. Most Shilleleigh-armed vehicles were withdrawn in 1979, but about 50 Sheridans remain with the U.S. Army's 82nd Airborne Division for want of anything better. The Soviet army recently introduced its own cannon-launched ATGM, the AT-8 SONGSTER, fired from the 125-mm. gun of the T-64 and T-80 tanks.

Specifications **Length:** 45 in. (1143 mm.). **Diameter:** 6 in. (152 mm.). **Span:** 11.5 in. (292 mm.). **Weight, launch:** 59 lb. (26.8 kg.). **Speed:** 1100 m./sec. (2460 mph). **Range:** 4000 m.

SHIPWRECK (SS-N-19): NATO code name for a Soviet ship- and submarine-launched ANTI-SHIP MISSILE introduced in 1981. A further development of the SS-N-3 SHADDOCK and SS-N-12 SANDBOX, the SS-N-19 is a large, fast, and powerful missile intended mainly for attacks on U.S. CARRIER BATTLE GROUPS. It is the primary armament of KIROV-class battle cruisers and OSCAR-class guided-missile submarines. In the former, Shipwreck is housed in a 20-round VERTICAL LAUNCH SYSTEM (VLS); and in the latter, in 12 pairs of fixed launch tubes built into the outer hull. In contrast to its predecessors, Shipwreck can be launched underwater, greatly reducing the vulnerability of the associated submarine.

The missile is believed to be an airplane-configured missile with sharply swept, high-mounted wings (which fold for storage), a small swept tailplane, and a small vertical rudder. It is believed to be powered by two strap-on solid rocket boosters and a large turbojet fed from a ventral air intake. Its high speed eliminates the need for midcourse guidance updates via a DATA LINK (as in both Shaddock and Sandbox). Shipwreck can rely instead on INERTIAL GUIDANCE through midcourse, with active radar homing in the terminal phase. It may also have a HOME-ON-JAM capability to exploit enemy electronic countermeasures. Payload options include 2200-lb. (1000-kg.) high-explosive and 360-KT nuclear warheads.

Specifications Length: 30 ft. (9.15 m.). Diameter: 3.5 ft. (1.07 m.). Weight, launch: 9977 lb. (4535 kg.). Speed: Mach 2.5 (1600 mph/2560 kph). Range: 330 mi. (528 km.).

SHIRANE: A class of two Japanese helicopter-carrying DESTROYERS (DDHs) commissioned in 1980–81. Enlarged and improved developments of the HARUNA class, the Shiranes are intended to form the nucleus of two ANTI-SUBMARINE WARFARE (ASW) escort groups of YAMAGUMO- and MINEGUMO-class hunter-killer destroyers (DDK).

Flush-decked, the Shiranes combine a standard destroyer superstructure forward with a large landing deck and hangar aft. Most radar, communication, and electronic warfare antennas are mounted on two large "macks" (combination mast and stack). A PRAIRIE MASKER bubble generator is fitted to reduce radiated noise, while the landing deck has a Canadian "Bear Claw" haul-down device to allow safer helicopter landings in high seas. Two sets of fin stabilizers improve seakeeping.

Primary armament consists of 3 large SH-3D SEA KING ASW helicopters, supplemented by 2 5-in. 54-caliber dual-purpose guns and an ASROC launcher on the bow, 2 sets of Mk.32 triple tubes for MK.46 ASW homing torpedoes amidships, a PHALANX 20-mm. radar-controlled anti-missile gun between the macks, and an 8-round NATO SEA SPARROW short-range surface-to-air missile launcher on top of the hangar. Eight HARPOON anti-ship missiles in individual launch canisters are to be added in the early 1990s. The Shiranes are also equipped with 2 satellite communication terminals and an NTDS data link. See also DESTROYERS, JAPAN.

Specifications Length: 521 ft. (158.8 m.). Beam: 57.4 ft. (17.5 m.). Draft: 17.4 ft. (5.3 m.). Displacement: 5200 tons standard/6800 tons full load. Powerplant: twin-shaft steam: 2 oil-fired boilers, 2 sets of geared turbines, 70,000 shp. Speed: 32 kt. Crew: 370. Sensors: 1 OPS-12 3-dimensional air-search radar, 1 OPS-28 surface-search radar, 1 OPS-22 navigation radar, 1 WM-25 Sea Sparrow guidance radar, 2 Type 72 and 2 VPS-2 fire control radars, 1 OQS-101 hull-mounted medium-frequency sonar, 1 SQR-18A TACTASS towed array sonar, 1 SQS-35(J) variable depth sonar. Electronic warfare equipment: 1 NOLQ electronic countermeasures array, 1 OLR-9B electronic signal monitoring array, 2 Mk.36 SRBOC chaff launchers.

SHOCK: The psychological effect of enemy attack, or close combat in any form. Before the machine-gun era, shock was exemplified by the effect of the cavalry charge, rarely a physical effect, because one side or the other would usually break and run before any actual physical contact.

With modern weapons, physical contact is even less likely, but shock remains a major factor in battle. Its psychological effects are generated by sudden attacks, by the sheer proximity of the enemy, and his closure, especially at speed; tanks advancing and firing en masse, with their overwhelming noise and menacing appearance, can shock all but the most disciplined troops into flight, and less classic forms of attack can do so even more easily.

Modern, long-range weapons can achieve shock by fire alone because of their sudden effect; concentrations of direct and indirect fire against a single target area ("time-on-target" fires) are meant specifically to maximize shock. Intense "hurricane" barrages are also meant to paralyze the victims by shock, to render them temporarily defenseless against an immediate ground attack.

The exploitation of shock effects is an important aspect of MANEUVER warfare.

SHRIKE: U.S. AGM-45 ANTI-RADIATION MISSILE (ARM). Developed by the Naval Weapons Center at China Lake in the early 1960s, Shrike entered service in 1963 and was much used in the Vietnam War by U.S. Navy E-6B and EA-6A IN-TRUDERS, and Air Force F-105F/G WILD WEASELS. Initial results were disappointing, but several incremental upgrades alleviated most of the problems. By 1988, more than 18,000 Shrikes had been built in at least 18 different models. Though superseded by the larger AGM-88 HARM, Shrike remains in service with the U.S., Israel, and Great Britain. It can be carried by a variety of tactical aircraft, including the A-4 SKYHAWK, A-7 CORSAIR, F-4 PHANTOM, IAI KFIR, and EF-111 RAVEN.

Shrike has four cruciform wings at midbody for steering as well as lift, and four smaller, fixed tail fins. The nose houses an antenna and a monopulse crystal receiver tuned to the frequency of a specific enemy radar; a different receiver must be used for each type of enemy radar (most Shrike improvements have involved the introduction of new receivers to deal with additional emitters). Immediately behind the seeker and the guidance electronics there is a 145-lb. (65.9-kg.) blast-fragmentation warhead fitted with a proximity fuze. The missile is powered by a Rockwell Mk.39, Aerojet Mk.53, or improved Mk.78 dual-thrust solid-rocket engine.

To launch Shrike, the aircraft must first acquire the target, either visually or on its RADAR WARNING RECEIVERS, then turn towards the target to allow the missile's seeker to lock onto its emissions. Once lock-on is achieved, the missile may be released to execute its attack. A major drawback of Shrike is its inability to continue homing on the target if the enemy radar shuts down even very briefly. This countertactic was used with some success in Vietnam and the Middle East, though of course the missile would then at least partially accomplish its mission, by forcing the enemy off the air, achieving "virtual attrition."

Specifications Length: 10 ft. (3.05 m.). **Diameter:** 8 in. (203 mm.). **Span:** 36 in. (915 mm.). **Weight, launch:** 390 lb. (177.3 kg.). **Speed:** Mach 2 (1300 mph/2080 kph). **Range:** 18–25 mi. (30–42 km.).

SICBM: MGM-134A "Small ICBM," a proposed U.S. intercontinental ballistic missile popularly known as "Midgetman." A relatively small, mobile missile intended to replace or supplement the much larger PEACEKEEPER (MX), the Midgetman's development was initiated in 1985, but has been hindered by political dissension and military controversy over its role, cost, and configuration. The logic of the program is that the U.S.-Soviet strategic balance can be stabilized only if both sides deploy only single-warhead missiles, logic that requires the elimination of all other ICBMs on both sides to achieve the Midgetman's purpose. It also fails to contend with the COUNTERFORCE capabilities of other strategic weapon types.

The original concept called for a single-warhead missile mounted on a high-mobility, wheeled transporter-erector-launcher (TEL). Normally based on U.S. military reservations, the missiles would be dispersed to remote launch sites during crises, to reduce their vulnerability to a Soviet counterforce strike (the Soviet SS-25 SICKLE is employed in a similar fashion). The original concept called for a three-stage solid-fuel missile some 50 ft. (15.24 m.) long, with only one REENTRY VEHICLE (RV) and a launch weight no greater than 30,000 lb. (13,608 kg.); but—predictably enough—the U.S. Air Force first proposed a larger version with an overall length of 38 ft. (11.58 m.), a diameter of 3.5 ft. (1.07 m.), and a launch weight of 35,000 lb. (15,876 kg.), and then an even larger version with three MULTIPLE INDEPENDENTLY TARGETED REENTRY VEHICLES (MIRVs) and a complex PENAID package to counter possible Soviet ballistic missile defenses. This missile would be some 50.2 ft. (15.3 m.) long, with a launch weight of 38,000 lb. (17,272 kg.); with its three warheads, it would also undermine the "stability" logic of the entire program.

All proposed designs would have wound KEVLAR casings to reduce weight. The multiple-warhead version would also have a liquid-fuel POST-BOOST VEHICLE (PBV) to allow each RV to be directed against a different target. All proposed versions have a specified range of some 6200 mi. (10,000 km.). Any SICBM would no doubt rely on an advanced form of INERTIAL GUIDANCE, with ring-laser gyroscopes. Accuracy would depend considerably on whether the launch site had been presurveyed, but the median error radius (CEP) would be on the order of 150–250 m. The first flight test of the single-warhead version took place in mid-1989, but was not successful.

The prototype TEL is a tractor-trailer with 8x8 drive. The cab is hardened to withstand blast pressure and has its own closed environmental system with an NBC filter system. All necessary communications, fire control, and power-generating equipment would be mounted on-board. The trailer has hydropneumatic suspension, to allow it to "squat" on the ground when deployed, thus reducing its vulnerability to blast. Sponsons on the sides of the trailer fold out to create a smooth, aerodynamic profile, further reducing blast vulnerability. The top of the trailer folds back, and the missile is erected to the vertical for launch. See also ICBMS, UNITED STATES.

SICKLE (SS-25): NATO code name for the Soviet RS-12M *Topol* "light," mobile ICBM. Introduced in the early 1980s, Sickle is a three-stage, solid-fuel missile based on the earlier SS-16 SINNER, and roughly comparable to the U.S. MINUTEMAN III. The initial Mod 1 version was deployed in 60 fixed, hardened silos at Yoshkar Ola (replacing the SS-16), but the later Mod 2 is deployed on self-propelled transporter-erector-launchers (TELs) derived from the MAZ 548/7910 8x8 truck. While normally based at centralized control and maintenance complexes, in wartime these missiles would be dispersed to remote launch sites to enhance their survivability: mobile missiles are extremely difficult to locate and track with surveillance satellites. For the same reason, they pose a major problem for the verification of ARMS CONTROL agreements. When the SS-25 first appeared, it was claimed by the U.S. to be a violation of the SALT II Treaty (which permitted only one new ICBM for each side), because the Soviet Union had previously announced that the SS-24 SCALPEL was its one new ICBM. The U.S.S.R. responded by claiming that the SS-25 is not a new missile, but simply an upgrade of the old SS-13 SAVAGE; in the light of Sickle's vast improvement in range, payload, and accuracy, this claim was not at all credible. Nonetheless, no action was taken by the U.S., because the treaty was unratified. By 1989 more than 125 Sickles had been deployed, with more on the way.

The first stage is believed to have four gimballed nozzles for attitude control and steering, while the second and third stages have a single gimballed nozzle. The missile casing is believed to be fabricated of wound glassfiber composites to reduce weight. Sickle has a solid-fuel POST-BOOST VEHICLE (PBV) with an assessed throw weight of 2200 lb. (1000 kg.). The standard payload is a single 550-kT REENTRY VEHICLE (RV), but there is sufficient volume for at least three 150-kT MULTIPLE INDEPENDENTLY TARGETED REENTRY VEHICLES (MIRVs).

It is controlled by INERTIAL GUIDANCE and employs General Energy Management Steering (GEMS), whereby the missile makes excursions above and below its nominal trajectory, in order to allow each stage to burn to exhaustion. That eliminates the need for unreliable and inaccurate thrust-termination ports, as used on the unsuccessful SS-13. Quite accurate, the SS-25 can attack semi-hardened targets.

Sickle employs the COLD LAUNCH technique, whereby the missile is housed in a launcher/storage canister, which is either lowered into the silo or mounted on the TEL. In the latter case, the missile is stowed horizontally for travel and raised vertically off the rear of the vehicle for launch. In both cases, a cold gas generator ejects the missile from its canister to a height of some 150 ft. (50 m.) before the main engines ignite. See also ICBMS, SOVIET UNION; RVSN.

Specifications **Length:** 59 ft. (18 m.). **Diameter:** 5.58 ft. (1.7 m.). **Weight, launch:** 81,550 lb. (37,070 kg.). **Range:** 6200 mi. (10,000 km.). **CEP:** (silo) 250 m.; (mobile) 350 m.

SIDEARM: U.S. AGM-122, an anti-radiation modification of AIM-9C SIDEWINDER air-to-air missiles. Equipped with a broad-bond passive seeker to detect radio frequency emissions, Sidearm homes automatically on hostile radar antennae. The missile is carried by attack aircraft as well as AH-1 COBRA and AH-64 APACHE attack helicopters. Specifications are similar to those of the AIM-9. See also ANTI-RADIATION MISSILE.

SIDE-LOOKING AERIAL RADAR (SLAR): A form of ground-mapping or surveillance RADAR which is mounted on the fuselage sides, so that the carrying aircraft can observe areas of interest while flying parallel to them. SLAR is thus ideally suited for standoff surveillance, and is frequently mounted on reconnaissance aircraft (e.g., the Lockheed U-2 and SR-71). Some SLARs, such as the advanced J-STARS, employ SYNTHETIC APERTURE techniques to generate high-resolution radar imagery.

SIDEWINDER: The U.S. AIM-9 short-range, INFRARED-HOMING air-to-air missile (AAM). Easily the most successful and widely deployed weapon of its type, with total production now exceeding 120,000, Sidewinder is considered by many to be the ideal guided missile: simple, cheap, effective, compatible with many different launch platforms (it can be fitted to almost any combat aircraft), and suitable for incremental improvement. Its basic configuration and many of its unique design features have been copied in the French Matra M.550 MAGIC, the Israeli SHAFRIR and PYTHON, the Soviet AA-2 ATOLL, and the Chinese PL-2, among other missiles. Sidewinder is produced under license in Japan, and by a European consortium. A ground-based version, the MIM-72 CHAPARRAL, is in service with the U.S. Army and Marine Corps and the armed forces of Egypt, Israel, Morocco, Taiwan, and Tunisia.

Developed by the U.S. Naval Weapon Center at China Lake, the design incorporated a number of off-the-shelf components, including the motor of the ZUNI 5-in. (127-mm.) rocket; in a manner all too unusual, simplicity was sought in all aspects. Thus, e.g., the missile body was made of aluminum tubing, and rather than using heavy, complex gyroscopes, the design team invented the "rolleron," a slipstream-driven gyro-wheel (one of which is mounted on each tail fin). The complete missile has fewer than 24 moving parts. Flight tests of the prototype AIM-9A began in 1953, and the first production model (AIM-9B) entered service with the U.S. Navy in 1956.

The AIM-9B had four cruciform control canards at the nose, and four fixed tail fins—a configuration followed in all subsequent versions (with some minor variations in length, weight, and span). It was armed with a 10-lb. (4.54-kg.) high-explosive fragmentation warhead fitted with a radar proximity (VT) fuze, and powered by a single-pulse solid rocket motor. Its uncooled lead-sulfide (PbS) seeker had a 25° field of view, was relatively insensitive, and could only be used for shots from dead astern, at close range, at high altitude, and in good visibility. Although it had a kill probability (P_K) of 70 percent in tests, under combat conditions the seeker tended to lock onto the sun, chimneys, or even trucks with overheated engines; in Vietnam, the AIM-9B had a P_k of some 10–12 percent.

The engagement sequence was quite simple and applies to all models. When the operator acquires the target in his gunsight and selects a missile, the missile's seeker head scans until it too acquires a target, as indicated by an audio "growl" in the pilot's headphones. When the missile locks on to the target, the growl changes to a high-pitched "tone." The pilot can then launch the missile, which homes autonomously on the thermal (INFRARED) emissions of the target aircraft's tailpipe. More than 80,000 AIM-9Bs were produced by Philco-Ford and General Electric, and 15,000 more in Europe; a considerable number of these missiles are still in service. The European version had an improved seeker with carbon dioxide (CO_2) cooling and solid-state electronics, which improved accuracy and reliability to some extent.

In 1962, the AIM-9B was supplemented by the AIM-9C and AIM-9D. The former relied on SEMI-ACTIVE RADAR HOMING (SARH) and was intended to provide a head-on engagement capability for the F-8 Crusader. It was not very successful, and only about 1000 were produced; many were converted in the 1980s to AGM-122A SIDEARM anti-radiation missiles. The AIM-9D was an IR missile with a PbS seeker of increased sensitivity cooled by nitrogen gas, and an expanded 40° field of view. Both models had a new engine which extended the maximum range, larger control surfaces for greater maneuverability, and a new 22.4-lb. (10.2-kg.) CONTINUOUS ROD WARHEAD with IR and VT fuzes. Although only 1000 AIM-9Ds were built, their design was the basis of many subsequent models.

The AIM-9E was a modification of some 5000 U.S. Air Force -9Bs with a thermo-electrically cooled PbS seeker with a 40° field of view and improved electronics, to bring performance up to AIM-9D standards.

The AIM-9G was an improved version of the -9D, with an off-boresight lock-on capability. Only 2120 were built by Raytheon before they were superseded by the AIM-9H, with all solid-state electronics and double-delta canards for enhanced maneuverability. Some 3000 of these were built by Ford for the U.S. Navy.

The AIM-9J was a major upgrade of some 14,-000 AIM-9Bs, with new solid-state electronics, the seeker of the AIM-9E, and double-delta canards. The AIM-9N is similarly a rebuilt AIM-9E. The AIM-9P is a further improvement of the AIM-9J, with a new engine, new fuzes, and improved reliability. About 13,000 have been converted by Ford, mainly for export.

The AIM-9L, introduced in 1977, was the first of the "third generation" Sidewinders. It has an argon gas–cooled Indium-Tin (InSb) seeker with all-aspect capability, long-chord double-delta canards for even better maneuverability, a new prefragmented continuous-rod warhead, and a LASER-diode proximity fuze. Some 16,000 were delivered by 1985, and 3500 more were built under license in Europe and Japan. Used in combat by the Israeli air force in the 1982 Lebanon War, and by British forces in the Falklands, the AIM-9L has demonstrated a P_k of roughly 60 percent, and proved resistant to most INFRARED COUNTERMEASURES (IRCM).

Essentially an improved AIM-9L, the current AIM-9M seeker head obtains improved resistance to IRCM through two-frequency filtering. The 25-lb. (11.4-kg.) continuous rod warhead is mounted behind the seeker and guidance electronics. It is powered by a Thiokol TX-683 low-smoke solid-fuel rocket, for reduced visibility.

Because of delays in the Anglo-U.S. Advanced Short-Range Air-to-Air Missile (ASRAAM) program, a new Sidewinder (AIM-9R) is now in development. It is reported to have a new, uncooled, charge-coupled device (CCH) IMAGING INFRARED (IIR) seeker with twice the lock-on range of the AIM-9M seeker, and almost total immunity to existing IRCM. Production of the AIM-9R is to begin in the early 1990s.

Specifications Length: (B) 9.28 ft. (2.82 m.); (L/M) 9.35 ft. (2.85 m.). **Diameter:** 5 in. (127 mm.). **Span:** (B) 22 in. (558 mm.); (L/M) 24.8 in. (630 mm.). **Weight, launch:** (B) 155 lb. (70.45 kg.); (L/M) 190 lb. (86.36 kg.). **Speed:** Mach 2.5 (1625 mph/2600 kph). **Range:** (B) 2 mi. (3.2 km.); (C/D/L/M) 11 mi. (17.6 km.). **Operators:** Arg, Bel, Bra, Can, Chi, Den, Gre, Iran, Isr, It, Jap, Ku, Malay, Mor, Neth, Nor, NZ, Pak, Phi, Por, ROK, S Ar, Sing, Sp, Swe, Tai, Tun, Tur, UK, US.

SIERRA: NATO code name for a class of three Soviet nuclear-powered attack SUBMARINES (SSNs) completed in 1983 and 1989; several more are under construction. The Sierras are one of the *three* new Soviet SSN classes introduced in the 1980s (along with the AKULA and MIKE classes). The exact role of the Sierras is not clear to Western analysts, but they may be successors to the high-speed ALFA class, with improved sensors and better acoustical silencing.

The Sierras have a modified teardrop hullform, with a bulbous bow, an oval midsection, and tapered stern. Like all Soviet nuclear submarines, they have a double-hull configuration with ballast tanks between the outer casing and pressure hull (which is fabricated of super-strong titanium). A long, low sail (conning tower) is located amidships. Control surfaces consist of retractable bow diving planes, and a cruciform rudder and stern plane arrangement mounted ahead of the propeller. A large, egg-shaped pod mounted on the upper rudder probably houses a towed array sonar. The outer hull is covered with CLUSTER GUARD anechoic tiles to reduce radiated noise and the effective range of hostile active sonar. These submarines are reported to be very quiet (perhaps equal to early units of the Los Angeles class) because of their internal soundproofing, Cluster Guard tiles, and improved propeller design (the latter due to illegal technology transfers from Japan and Norway).

Armament consists of 8 bow-mounted torpedo tubes, 6 of 21-in. (533-mm.) and two of 26-in. (550-mm.) diameter. The former can accommodate a variety of acoustic and wire-guided homing TORPEDOES, plus the SS-N-15 STARFISH anti-submarine missile; the latter can launch the Type 65 long-range wake-homing torpedo, the SS-N-16 STALLION anti-submarine missile, and the SS-N-21 SAMPSON long-range CRUISE MISSILE. The basic load is 24 weapons of all types; MINES can replace missiles and torpedoes on a 2-for-1 basis. The Sierras also carry a Pert Spring satellite communication terminal and a VLF towed antenna. See also SUBMARINES, SOVIET UNION.

Specifications Length: 360.9 ft. (110 m.). **Beam:** 40.7 ft. (12.4 m.). **Displacement:** 6765 tons surfaced/8060 tons submerged. **Powerplant:** single-shaft nuclear: 2 pressurized-water reactors, 2 sets of geared steam turbines, 45,000 shp. **Speed:** 20 kt. surfaced/ 33 kt. submerged. **Max. operating depth:** 1312 ft. (400 m.). **Collapse depth:** 3280 ft. (1000 m.). **Crew:** 70. **Sensors:** 1 bow-mounted low-frequency sonar, 1 low-frequency active sonar, 1 low-frequency towed array sonar, 1 "Snoop Pair" surface-search radar, 1 "Park Lamp" radio direction finder (D/F), 1 "Rim Hat" electronic signal monitoring array, 2 periscopes.

SIGINT: Signals Intelligence, the monitoring, collection, and analysis of electromagnetic signals, normally subdivided into communications intelligence (COMINT) and electronic intelligence (ELINT). The former covers the monitoring of radio and telephone message traffic, the decryption of coded messages, and the analysis of their contents. ELINT, on the other hand, mainly covers the detection and analysis of radar, telemetry, and other noncommunications signals, to discover performance parameters and develop countermeasures.

Strategic SIGINT is concerned with the overall deployments, capabilities, and intentions of actual or potential enemies, while tactical SIGINT focuses on the location and classification of enemy weapons, sensors, and formations, in order to support combat operations.

Signals are collected by passive receivers and recording devices, collectively known as ELECTRONIC SIGNAL MONITORING (ESM) equipment. ESM arrays have become common on combat aircraft and warships, while the SIGINT units of ground forces often have vehicle-mounted receiver arrays. Strategic SIGINT is often performed clandestinely from modified aircraft and specialized intelligence vessels, as well as by satellites. In the U.S., SIGINT is the responsibility of the NATIONAL SECURITY AGENCY (NSA); the Main Intelligence Directorate (GRU) performs a similar function for the U.S.S.R., while in Britain SIGINT is performed by the Government Communications Headquarters (GCHQ).

Because military operations have become so dependent on both radio and radar, SIGINT has become vitally important; in reaction, advanced countries are willing to go to extraordinary lengths to protect their codes and operating frequencies. See also INTELLIGENCE.

SIGNATURE: The characteristic pattern of detectable emissions and RADAR or SONAR echoes generated by ships, aircraft, vehicles or weapons. Such emissions include thermal (INFRARED), radio-frequency, and acoustical, as well as visual, signatures. Over time, the signatures of many weapons, platforms, and systems can be cataloged for identification and TARGET ACQUISITION.

SILEX (SS-N-14): NATO code name for a Soviet ship-launched anti-submarine missile introduced in 1969. Roughly analogous to the Australian IKARA and French MALAFON, Silex is the primary medium-range ANTI-SUBMARINE WARFARE (ASW) weapon of the KRIVAK frigates, UDALOY destroyers, KARA and KRESTA cruisers, and the battle cruiser KIROV. Silex is launched from fixed quadruple canisters, two of which are mounted on each ship, except for the Krivaks, which have a single, 4-round trainable launcher. Only the Kirov carries reload missiles.

Silex has four broad-chord swept wings and four smaller tail fins, and is powered by a solid-fuel rocket engine. It relies on a combination of INERTIAL GUIDANCE with radio COMMAND GUIDANCE updates: the missile is first aimed at the approximate target position, and in-flight course corrections are then transmitted to the missile via "Head Light" or "Eye Bowl" guidance radars. Most ships have two "Eye Bowls," allowing two missiles to be guided simultaneously. The Kresta IIs, however, rely on the "Head Light" guidance radar, also used with the SA-N-3 GOBLET surface-to-air missile.

Silex has two payload options: a 17.7-in. (450-mm.) acoustic homing TORPEDO with a 220-lb. (100-kg.) explosive warhead; or a 2.5-KT nuclear DEPTH CHARGE. Both are released from the missile by remote command, and employ a braking parachute to slow water entry. The nuclear version may also have a secondary anti-ship capability.

Specifications Length: 24.9 ft. (7.6 m.). Diameter: 1.8 ft. (550 mm.). Span: 3.6 ft. (1.1 m.). Weight, launch: 2,205 lb. (1000 kg.). Speed: Mach .92 (690 mph/1100 kph). Range: 35 mi. (55 km.).

SILKWORM: Chinese version of the Soviet SS-N-2c STYX anti-ship missile, produced in both ship- and shore-launched models. The missile achieved some notoriety when deployed by Iranian forces near the Straits of Hormuz in 1987, prompting the U.S. to begin tanker escort operations.

SILO: An underground storage and launch facility for BALLISTIC MISSILES, intended to provide protection from nuclear attack. A silo is a reinforced concrete tube, closed at the bottom and buried vertically in the earth so that the open upper end is flush with the surface. The silo opening is covered by a massive steel or concrete sliding door designed to resist the blast overpressure, heat, and intense radiation generated by nuclear explosions. The cover and its supporting structure (the "headworks") transmit the vertical load of a blast to the supports of the cover rather than to the silo proper, thereby minimizing the danger of a structural failure. The cover is often shaped like a plow, and is opened by powerful engines, to allow it to be forced open even if buried by debris.

In order to protect it from damage caused by shock waves from nearby nuclear groundbursts, the missile is placed on a sprung platform and held in place by shock absorbers ("isolators"). The upper part of the silo is surrounded by a multi-level structure housing monitoring and test equipment, from which the missile can be checked periodically.

Missiles are generally stored in unattended silos; their status is monitored by remote sensors and test equipment. Launch commands are transmitted to the missile by land lines from an underground control facility (which usually controls several silos), or by radio signal from an airborne command post.

Missiles are launched from their silos by one of two techniques. In a "hot launch," the missile's main engines are ignited within the silo, and the missile then flies out under its own power. The silo must therefore be somewhat wider than the missile to permit the venting of exhaust gases, and the silo must be completely refurbished after each launch (a process which may take several weeks). In COLD LAUNCH, by contrast, the missile is ejected from the silo by compressed gasses, and the main engine ignites only at some distance above the ground, to protect the silo from exhaust damage. Cold-launched missiles are loaded into their silos in combination transport/launch canisters that contain all necessary electrical and data connections. Cold launch has two advantages: first, the missile can be enlarged to fill the entire volume of the silo; second, the silo can be reloaded in a matter of hours. The latest ICBMS all use the cold launch technique.

Silos are inherently resistant to many nuclear weapon effects. The overhead door and headworks protect them from thermal and radiation effects. The main threat, therefore, is blast overpressure, transmitted through the ground to the entire silo structure. The resistance of silos is therefore usually measured in pounds per square inch (PSI) of peak overpressure. Most are now hardened to 1000–2000 psi, but some designs now under consideration have theoretical resistance levels of 3000–5000 psi. See also COUNTERFORCE; NUCLEAR WEAPON.

SINCGARS: Single Channel Ground and Airborne Radio System, a lightweight, jam-resistant UHF/FM tactical radio, now entering service with the U.S. Army. Developed in the late 1970s to replace all existing army tactical radios, SINCGARS uses frequency-hopping techniques to evade enemy electronic countermeasures; it also incorporates a voice scrambler and a data encryption device to enhance communications security. A contract for some 44,000 PRC-119 SINCGARS backpack sets was awarded to AT&T in 1983, with final delivery scheduled for 1992. Follow-on contracts could bring total production to more than 400,000 radios, including those made under license in Israel by Tadiran, which participated in their development. The program was hindered by reliability problems, which have reportedly been resolved.

SINNER (SS-16): NATO code name for the Soviet RS-14 ICBM. First detected in the early 1970s, the SS-16 was only the second Soviet solid-fuel ICBM, built as the successor of the SS-13 SAVAGE. Sinner became the object of much controversy, because the U.S.S.R. denied its deployment, while the U.S. claimed it was potentially a mobile missile, and hence a violation of the SALT II Treaty. It was also believed at one time that considerable numbers of Sinners had been stockpiled to provide the Soviet Union with a rapid reload capability for its ballistic missile silos.

It is now generally agreed that some 60 SS-16s were in fact deployed in fixed silos at Yoshkar Ola, to replace the SS-13. Those SS-16s in turn were replaced by the SS-25 SICKLE in the early 1980s. Though not built in great numbers, the SS-16 was important as the first successful Soviet solid-fuel

ICBM, whose development advanced the technology employed by the later SS-24 SCALPEL and SS-25 Sickle. In addition, its first two stages were developed into the SS-20 SABER IRBM, while its third stage was used in the SS-23 SPIDER short-range ballistic missile.

The first stage of the three-stage Sinner has four gimballed nozzles for attitude control and steering, while the upper stages have a single gimballed nozzle. All three stages are fabricated of rolled steel. The payload is a single 600-kT REENTRY VEHICLE (RV), which may be mounted on a POST-BOOST VEHICLE (PBV). Sinner relies on navigating INERTIAL GUIDANCE, in contrast to the SS-13, which employed the traditional Soviet preprogrammed "fly-by-wire" technique (though the latter is inherently unsuitable for solid-fuel missiles). The SS-16 uses General Energy Management Steering (GEMS), whereby the missile makes excursions above and below its nominal trajectory to allow each stage to burn to exhaustion, thus eliminating the need for unreliable and inaccurate thrust-termination ports (as used on the SS-13). As a result, Sinner is much more accurate than Savage.

Sinner is believed to employ the COLD LAUNCH technique, whereby the missile is housed in a launcher/storage canister lowered into the silo. A cold gas generator ejects the missile from the canister to a height of 150 ft. (45 m.) before the main engine ignites. The empty canister can then be removed, and the silo reloaded in 12 to 24 hours. See also ICBMS, SOVIET UNION; RVSN.

Specifications Length: 67.25 ft. (20.5 m.). **Diameter:** 5.87 ft. (1.8 m.). **Weight, launch:** 75,000 lb. (34,090 kg.). **Range:** 6200 mi. (10,000 km.). **CEP:** 480 m.

SINS: Ship's Inertial Navigation System, a U.S. precision navigation device of ballistic missile SUBMARINES, which is used to provide launch position data for missile guidance. Employing INERTIAL NAVIGATION techniques, SINS determines the submarine's position by continually measuring course changes and the distance traveled from a known point, by using a complex arrangement of gyroscopes and accelerometers. Because gyros "drift" over time, SINS must be periodically updated by a celestial, radio, or satellite position "fix." The most modern types are accurate to within 50–100 m.

SIOP: Single Integrated Operations Plan, U.S. contingency plans for strategic nuclear war. The SIOP provides the president and the NATIONAL COMMAND AUTHORITIES with a variety of attack options, each with its own targets, timing, tactics, and force requirements. The SIOP is continuously updated by the Joint Strategic Target Planning Staff, collocated with the Strategic Air Command Headquarters at Omaha, Nebraska.

SIREN (SS-N-7): NATO code name for a Soviet submarine-launched ANTI-SHIP MISSILE. Introduced in 1968 aboard the CHARLIE I nuclear-powered guided-missile submarines (SSGNs), Siren was the first Soviet anti-ship missile which could be launched from underwater.

An airplane-configured missile, Siren has mid-mounted delta wings (which probably fold for storage) and cruciform tail fins. Powered by a solid-fuel rocket engine, it relies on active radar homing throughout its flight; an ANTI-RADIATION variant has also been reported. It is armed with either an 1100-lb. (500-kg.) high-explosive or a 200-kT nuclear warhead. The missile is launched from fixed tubes mounted between the Charlie's pressure hull and the outer casing. After launch, it rises to the surface, where the rocket engine ignites. Siren then performs a "pop-up" maneuver to acquire the target, before diving to surface-skimming height (100 ft./30 m.) for the final approach.

Siren's vulnerability to ELECTRONIC COUNTERMEASURES is not known, but its surface-skimming trajectory and high speed would make it difficult to detect, and would allow little time in which to respond. The U.S. Navy considers this type of attack the most difficult to counter; in fact, the introduction of Siren prompted the development of the PHALANX radar-controlled anti-missile gun system.

Specifications **Length:** 23 ft. (7 m.). **Diameter:** 1.8 ft. (550 mm.). **Span:** 9 ft. (2.75 m.). **Weight, launch:** 6390 lb. (2900 kg.). **Speed:** Mach 0.95 (650 mph/1040 kph). **Range:** 34 mi. (55 km.).

SJOORMAN: A class of five Swedish diesel-electric attack SUBMARINES (SSs) commissioned between 1967 and 1969. Far more advanced successors to the small DRAKEN-class coastal submarines, they incorporated a number of advanced features for their time. The Sjoormans have a modified teardrop hullform (round bow, cylindrical center section, and tapered stern) in a single-hull configuration with a long, streamlined sail (conning tower) forward, and a flat deck casing covering most of the hull behind the sail. The forward diving planes are mounted on the sail, while the stern planes have a computer-controlled X-configuration (as in all later Swedish submarines and in the Dutch ZEELEEUW class) which provides excellent maneuverability, especially at low speeds. A re-

tractable snorkel allows diesel operations from shallow submergence.

Armament consists of 6 bow-mounted torpedo tubes, 4 of 21-in. (533-mm.) and 2 of 15.75-in. (400-mm.) diameter. The former can launch Type 61B anti-ship homing TORPEDOES, while the latter are for short Type 42 anti-submarine torpedoes. The basic load is 8 Type 61Bs and 4 Type 42s; MINES can replace the former on a 2-for-1 basis. All sensors are linked to an IBS-A17 torpedo FIRE CONTROL computer. See also SUBMARINES, SWEDEN.

Specifications **Length:** 167.3 ft. (51 m.). **Beam:** 20 ft. (6.1 m.). **Displacement:** 1125 tons surfaced/1400 tons submerged. **Powerplant:** 4 Hedemara-Pielstick V12A2 diesels, 2100 hp., 1 1500-hp. ASEA electric motor. **Speed:** 5 kt. surfaced/20 kt. submerged. **Endurance:** 21 days. **Max. operating depth:** 492 ft. (150 m.). **Collapse depth:** 982 ft. (300 m.). **Crew:** 24. **Sensors:** 1 Krupp-Atlas CSU3-2 bow-mounted low-frequency active/passive sonar, 1 Terma surface-search radar, 1 electronic signal monitoring array, 2 periscopes.

SKEET: A SELF-FORGING FRAGMENT (SFF) anti-tank submunition under development by Avco for the U.S. Air Force. Initiated in the early 1970s as part of the ASSAULT BREAKER program, Skeet is a small cylinder, only 3.75 in. (95.25 mm.) long and 5.25 in. (133.35 mm.) in diameter, which houses a concave tungsten disk backed by plastic explosives. When the charge is detonated by Skeet's INFRARED (IR) proximity fuze, the force of the explosion converts the disk into a penetrator bolt which is projected along the original axis of the disk at a velocity of several thousand m./sec. Each submunition now costs roughly $1250.

Two versions have been developed: the CBU-97/B Sensor-Fuzed Weapon (SFW) and the BLU-101/B Extended-Range Anti-Armor Mine (ERAM). The SFF consists of a BLU-108/B Skeet Delivery Vehicle (SDV) housing four Skeet submunitions. Each SDV is 31 in. (787.4 mm.) long and 5.25 in. (133.35 mm.) in diameter, with a launch weight of 55 lb. (25 kg.). Ten SDVs are packed into a SUU-64/B Tactical Munition Dispenser (TMD), to form a CBU-97 CLUSTER BOMB. The assembled cluster bomb is 92 in. (2.34 m.) long and 15.6 in. (396 mm.) in diameter, with a launch weight of 850 lb. (387 kg.).

After it is released from the aircraft, the SUU-64/B breaks apart at a preselected altitude, scattering the ten BLU-108 SDVs over a 1000-by-1000-ft. (305-by-305-m.) area. Braking parachutes then deploy from the tails of the SDVs, to orient them nose-downward. At a height of 75 ft. (determined by a laser altimeter), the braking chute is jettisoned and a retrorocket in the nose of the SDV arrests its descent, while a pair of spin jets simultaneously cause it to rotate at several hundred revolutions per minute. The Skeet submunitions are then extended on articulated arms and released. Centrifugal force cause them to fly outward to cover a wide area with their IR proximity fuzes. If the Skeet passes over a vehicle, the fuze detonates the warhead, causing the SFF to penetrate through the target's thin top armor. If no target is detected, the submunition explodes on contact with the ground, acting as an anti-personnel grenade. One attack aircraft carrying 12 CBU-97/B SFWs can release 480 Skeet submunitions, sufficient to cover one square mile. The air force has a current requirement for 14,000 SFWs at a cost of some $125,000 each. The system should become operational by the early 1990s.

The BLU-101/B ERAM, by contrast, is an air-delivered off-route MINE, consisting of a disk-shaped housing with a trainable turret holding two Skeet submunitions. The BLU-101 is 9 in. (229 mm.) high and 14 in. (356 mm.) in diameter, with a loaded weight of 88 lb. (40 kg.) and a unit cost of $1717. Nine ERAM are packaged in a SUU-65/B TMD to form a CBU-92/B Wide Area Anti-Armor Mine (WAAM). WAAM is 94 in. (2.39 m.) long and 15.6 in. (396 mm.) in diameter, and has a launch weight of 850 lb. (387 kg.).

WAAMs are meant to be released in the path of oncoming enemy forces. The cluster bomb splits open at a preselected altitude, releasing the nine ERAMs, which then descend to the ground on parachutes. On landing, the ERAMs deploy three acoustic sensors which can detect the engine noise of moving vehicles. Should a vehicle approach within 500 ft. (150 m.) of an ERAM, the sensors trains the turret towards it and fires a Skeet like a clay pigeon. The Skeet's IR sensor scans the area immediately below, and detonates the SFF warhead as it passes over the target. If no target is detected, the submunition detonates on contact with the ground.

Further development of WAAM has been suspended by the air force, but the U.S. Army is testing a hand-emplaced variant of ERAM for its Wide Area Mine (WAM) program. The army also has a self-forging fragment submunition in development, SADARM.

SKIFF (SS-N-23): NATO code name for a Soviet submarine-launched ballistic missile (SLBM)

introduced in 1986 as the primary armament of the DELTA IV nuclear-powered ballistic-missile submarine (SSBN). Developed in the early 1980s from the SS-N-18 STINGRAY of the delta IIIs, Skiff has greater range and payload, and considerably better accuracy. It may be retrofitted to the Delta IIIs, but is too large for the tubes of earlier Deltas.

A three-stage, liquid-fuel missile, like its predecessor, Skiff employs a hypergolic, bipropellant mix of undimensional dimethyl hydrazine (UDMH) and nitrogen tetroxide (NTO), despite the danger of keeping such volatile compounds in an enclosed submarine. The first stage has 4 gimballed main engines for attitude control and steering, the second stage has 2 gimballed main engines, and the third stage has a single fixed main engine and 2 gimballed vernier engines.

Armed with ten 100-kT MULTIPLE INDEPENDENTLY TARGETED REENTRY VEHICLES (MIRVs) on a liquid-fuel POST-BOOST VEHICLE (PBV), Skiff relies on INERTIAL GUIDANCE with stellar update ("stellar-inertial"). After launch, a star sensor in the nose locks onto one or more celestial objects to obtain a precise position fix, thereby compensating for launch position errors caused by gyro drift in the submarine's INERTIAL NAVIGATION unit. It is believed that Skiff employs the traditional Soviet "fly-by-wire" technique whereby the missile is constrained to a preprogrammed pitch-velocity profile by thrust-magnitude control. Though fairly accurate, Skiff's small warheads make it unsuitable for attacks against hardened targets. See also SLBMS, SOVIET UNION.

Specifications **Length:** 46.3 ft. (14.2 m.). **Diameter:** 6.7 ft. (2 m.). **Weight, launch:** 88,000 lb. (40,000 kg.). **Range:** 5190 mi. (8305 km.). **CEP:** 640 m.

SKIPJACK: A class of six U.S. nuclear-powered attack SUBMARINES (SSNs) commissioned between 1959 and 1961. Only three remain in service: USS *Scorpion* (SSN-589) was lost off the Azores in 1968, while *Scamp* (SSN-588) and *Snook* (SSN-592) were retired in 1986. The survivors are to be converted to training boats. The *Skipjack* embodied a revolution in submarine design, by combining nuclear power with the hydrodynamically superior teardrop hullform pioneered by the experimental diesel-electric submarine *Albacore*. At the time of their introduction, the Skipjacks were the fastest submarines in the world, and were also capable of radical, airplanelike maneuvers. Many of the design features developed for the Skipjacks have become standard on subsequent nuclear submarines, both U.S. and foreign. Two Skipjack hulls were lengthened while under construction to be converted to the *George Washington* configuration, making them the first U.S. ballistic missile submarines.

The hull has a perfect teardrop form, with a round bow and maximum beam well forward, tapering gradually to a single propeller at the stern. This form minimizes submerged drag, and, unlike the long, cigar-shaped hulls of earlier submarines, is stable at high speeds. The Skipjacks were also the first U.S. submarines with a single-hull configuration; i.e., the pressure hull itself forms the outer skin of the vessel. Ballast tanks are located either inside the pressure hull or outside it at the extreme bow and stern. A tall, streamlined sail with a small, raised spine extending from its rear base is located near the bow. The control surfaces introduced what has become the standard U.S. configuration: large "fairwater" diving planes mounted on the sail, and a cruciform rudder and stern-plane arrangement mounted just ahead of the propeller. The Skipjacks have the same reactor as the later PERMIT and STURGEON classes. As compared to later U.S. submarines, the Skipjacks are relatively noisy, because their hull was too small for elaborate silencing techniques.

Armament consists of six bow-mounted 21-in. (533-mm.) torpedo tubes, which currently can accommodate only MK.48 wire-guided homing torpedoes; a total of 24 are normally carried. Sensors have been upgraded several times, but remain obsolescent. All sensors were originally linked to a Mk.113 analog FIRE CONTROL unit, subsequently replaced by a digital Mk.117. See also SUBMARINES, UNITED STATES.

Specifications **Length:** 252 ft. (76.6 m.). **Beam:** 31.5 ft. (9.6 m.). **Displacement:** 3075 tons surfaced/3500 tons submerged. **Powerplant:** single-shaft nuclear: 1 S5W pressurized-water reactor, 1 set of geared turbines, 15,000 shp. **Speed:** 20 kt. surfaced/33–36 kt. submerged. **Max. operating depth:** 985 ft. (303 m.). **Collapse depth:** 1640 ft. (500 m.). **Crew:** 114. **Sensors:** 1 BQS-4 medium-frequency active/passive sonar, 1 BQR-2C passive sonar, 1 BPS-12 surface-search radar, 1 electronic signal monitoring array, 2 periscopes.

SKIPPER: U.S. AGM-123A air-to-surface missile, developed by the Naval Surface Weapons Center at China Lake as a low-cost standoff weapon for attacks on ships and fixed, hardened targets. Essentially a 1000-lb. (454-kg.) PAVEWAY II laser-guided glide bomb fitted with the solid-fuel rocket engine of the AGM-45 SHRIKE anti-radiation missile, Skipper began flight tests in 1984 and entered service in the following year.

The nose section consists of the laser seeker, guidance unit, and canard control fins of the Paveway II; the midsection is a Mk.83 low-drag general-purpose (LDGP) bomb; and the tail section consists of the pop-out fins of the Paveway II and the Shrike rocket motor; the latter extends range considerably, especially for low-altitude attacks. The engagement sequence is identical to that of a standard Paveway II.

Specifications **Length:** 14.08 ft. (4.3 m.). **Diameter:** 14 in. (356 mm.). **Span:** 5.25 ft. (1.6 m.). **Weight, launch:** 1283 lb. (582 kg.). **Range:** 10–15 mi. (16–24 km.).

SKORYY: A class of 72 Soviet destroyers completed between 1948 and 1954. The single largest class of Soviet surface warships ever built, all are now being scrapped. See also DESTROYERS, SOVIET UNION.

SKOT: Czech-produced version of the Soviet BTR-60 wheeled ARMORED PERSONNEL CARRIER.

SKYFLASH: British radar-guided, medium-range air-to-air missile, essentially an improved version of the U.S. AIM-7E SPARROW. Skyflash was developed from 1969 by fitting a more accurate monopulse SEMI-ACTIVE RADAR HOMING (SARH) guidance unit into the existing Sparrow airframe, to increase the missile's effectiveness against maneuvering low-altitude targets. Flight tests began in 1975, and the first production missiles were delivered by British Aerospace in 1978. They now arm Royal Air Force PHANTOMS and TORNADO interceptors, and Swedish Air Force JA.37 VIGGEN fighters.

Skyflash has the same configuration and external dimensions as the AIM-7E2, but is some 31 lb. (14 kg.) lighter at 425 lb. (193.2 kg.). Its new solid-state electronics are more reliable and require less prelaunch warmup. The AIM-7E's 66-lb. (30-kg.) CONTINUOUS ROD WARHEAD has been retained, but has been fitted with an advanced EMI radar proximity fuze. The AIM-7E's Aerojet Mk.52 solid-rocket engine has also been retained, and this limits maximum range to 31 mi. (50 km.); if required, the more powerful AIM-7F engine could be retrofitted, extending the range to 62 mi. (100 km.).

The monopulse guidance unit has proven quite accurate, at least in tests: more than half of all missiles fired actually hit the target, while the miss distance for the remainder was considerably less than for other radar-guided missiles (actual combat effectiveness is another matter). This performance prompted the U.S. to develop its own AIM-7M monopulse version of Sparrow.

Plans for an ACTIVE HOMING Skyflash Mk.2 were scuttled by an agreement with the U.S. for the RAF to deploy the AIM-120 AMRAAM. Nonetheless, an export version called Skyflash 90 is being developed as a low-cost alternative to AMRAAM.

SKYHAWK: McDonnell Douglas A-4 carrier-based ATTACK AIRCRAFT, in service with the U.S. Navy and Marine Corps, as well as in Argentina, Australia, Indonesia, Israel, Kuwait, Malaysia, New Zealand, and Singapore. Developed to a U.S. Navy specification for a jet-powered successor to the piston-engined A-1 Skyraider, the Skyhawk was designed by the legendary Ed Heinemann to be the smallest, lightest, and simplest aircraft capable of meeting the requirement. The first prototype flew on 22 June 1954, and the first A-4As entered service in 1956. One of the most successful of post-1945 combat aircraft, it remained in production for 26 years: the last of 2960 Skyhawks was completed on 27 February 1979.

The initial 142 A-4As, with a maximum payload of 7000 lb. (3182 kg.), were quickly superseded by 542 A-4Bs, with a more powerful engine, provisions for AGM-12 BULLPUP air-to-ground missiles, a computer navigation unit, an AERIAL REFUELING probe, and redundant hydraulic controls. The subsequent A-4C had an extended nose with a TERRAIN AVOIDANCE RADAR, an advanced autopilot, and a Low Altitude Bombing System (LABS); 648 were built between 1959 and 1964.

The much more capable A-4E of 1965 introduced a more fuel-efficient Pratt and Whitney J52 engine, for much greater range, and additional weapon pylons for a maximum payload of 8200 lb. (3719 kg.); 497 were built through 1967. The similar A-4F has an uprated engine, a "zero-zero" ejection seat, and additional AVIONICS in a distinctive "camelback" dorsal fairing. Some 139 were built, and most A-4Es were later brought up to F standard. The TA-4F and TA-4J are two-seat trainer variants. Much used during the Vietnam War, the A-4E/F equipped most U.S. Navy and Marine Corps light attack squadrons, and even managed to score two aerial victories over North Vietnamese MIGs. A few remain in navy service as TOP GUN aggressor aircraft, and some 37 were sold to Israel in 1973–74.

Israel became the first major export customer in 1967, when it ordered 90 A-4H and 10 TA-4H trainers. Otherwise similar to the A-4F, the H had a broad-chord, square-tipped vertical stabilizer, a braking parachute, and two 30-mm. DEFA cannon in place of the standard 20-mm. guns. New Zealand ordered 10 of the similar A-4K, and recently also acquired 10 ex-Australian A-4Gs (similar to the A-4F), while Kuwait purchased 30 A-4KUs in the early 1980s (many of which escaped to Saudi Arabia after the Iraqi invasion of 2 August 1990).

In 1970, the marine corps adopted the A-4M Skyhawk II, with more powerful engines (for a 50 percent improvement in climb rate), the broad vertical stabilizer of the A-4H, a raised cockpit for improved visibility, an Angle-Rate Bombing System (ARBS), and a 9100-lb. (4100-kg.) weapon load. Israel purchased 117 of the similar A-4Ns, with an APG-53A terrain avoidance radar, an ASN-41 INERTIAL NAVIGATION unit, an Elliott 546 Head-Up Display (HUD), and Israeli-made defensive avionics. Of some 267 single-seat and 27 two-seat Skyhawks delivered to Israel, more than 50 have been lost in combat or in accidents and 28 A-4E/Fs were sold to Indonesia between 1979 and 1985. Of the survivors, some 100 are in service, with the rest in storage; all have been brought up to A-4N standard and also fitted with extended tailpipes as protection against infrared-homing surface-to-air missiles.

Many earlier Skyhawks have been modernized (and, in some cases, completely rebuilt) for export customers. Fifty A-4Bs and 25 A-4Cs refurbished for the Argentine air force were redesignated A-4Ps, while 16 A-4Bs upgraded for the Argentine navy were designated A-4Q. Both were used in the 1982 Falklands War, when some 30 were shot down. Grumman converted 34 A-4Cs to A-4PTMs for Malaysia, which also has 6 TA-4PTM trainers with two separate cockpits. Singapore has 40 A-4S conversions of the A-4C, with uprated engines and 30-mm. ADEN cannon. Most of these were further modified in 1982 to A-4S-1 standard, with new avionics and reinforced weapon pylons. Singapore is presently reengining its Skyhawks with turbofan engines for greatly improved power loading and fuel efficiency, while New Zealand is modernizing its ex-Australian A-4Gs with a new cockpit (including multi-functional video displays in place of instrument dials), new avionics, and a Ferranti Red Fox sea surveillance radar. The U.S. Marine Corps has itself converted some two dozen TA-4Fs to OA-4M forward air control (FAC) aircraft. With additional modifications, Skyhawks could remain in service till the end of the century.

The A-4M's sloping nose houses the LASER

seeker and drift sensors associated with the ARBS, together with an APN-153(V) DOPPLER navigation radar. The pilot sits high over the nose under a bulged clamshell canopy which provides an excellent forward and side view (though rearward visibility is obscured by the camelback "hump"). Avionics include a HUD, a TACAN receiver, several RADAR WARNING RECEIVERS, an ACTIVE JAMMING unit, and a CHAFF dispenser. Two 20-mm. Mk.12 cannons with 200 rounds per gun are mounted in the wing roots. Israeli Skyhawks have two 30-mm. DEFA 553 cannons with 150 rounds per gun, while the Singaporean A-4S has two similar 30-mm. Adens with 125 rounds per gun. Immediately behind the cockpit is a 240-gal. (908-lit.) self-sealing fuel tank. The rear fuselage houses the J52 engine, which is fed by semi-circular cheek inlets behind the cockpit.

The low-mounted, moderately swept, rounded-delta wing is so short that it does not need a heavy wing-folding mechanism for carrier handling. It has a very strong 3-spar structure which forms an integral, self-sealing 560-gal. (2121-lit.) fuel tank. Full-span leading-edge slats provide excellent low-speed handling for arrested carrier landings, while large outboard ailerons give the Skyhawk an outstanding roll rate. The inboard flaps are split to act as speed brakes on landing, supplementing paddle-type air brakes on the sides of the fuselage. The main landing gear, housed in fairings under the wing, retract forward, so that the slipstream can blow down the gear in case of a hydraulics failure. The tailplane, mounted at the base of the vertical stabilizer, has the same planform as the wing, while the rudder has distinctive external stiffeners.

The A-4M has one fuselage pylon rated at 3575 lb. (1622 kg.), two inboard wing pylons, each rated at 2240 lb. (1016 kg.), and two outboard wing pylons, rated at 1000 lb. (454 kg.) each. The centerline and inboard pylons are plumbed for 400-gal. (1514-lit.) drop tanks, or 250-gal. (1135-litre) drogue-equipped, D-704 "Buddy" tanks for inflight refueling. Ordnance carried includes LDGP, NAPALM and CLUSTER BOMBS, PAVEWAY laser-guided bombs, AGM-62 WALLEYE TV-guided glide bombs, AGM-12 BULLPUP and AGM-65 MAVERICK air-to-ground missiles, AGM-45 SHRIKE and AGM-88 HARM anti-radiation missiles, and GABRIEL anti-ship missiles (on Israeli Skyhawks) as well as 2.75-in. (70-mm.) and 5-in. (127-mm.) rocket pods, and (for self-defense) AIM-9 SIDEWINDER, Rafael SHA-

FRIR and PYTHON, or Matra MAGIC infrared-homing air-to-air missiles.

Specifications **Length:** (M/N) 40.3 ft. (12.27 m.); (TA) 42.6 ft. (12.98 m.). **Span:** 27.5 ft. (8.38 m.). **Powerplant:** (A) Wright J65-W-2 turbojet, 7400 lb. (3493 kg.) of thrust; (B) J65-W-16A, 8500 lb. (3865 kg.) of thrust; (E) Pratt and Whitney J52-P-6A turbojet, 8500 lb. (3865 kg.) of thrust; (F) J52-P-8A, 9300 lb. (4218 kg.) of thrust; (M/N) J52-PW-408, 11,200 lb. (5080 kg.) of thrust; (S) J65-W-20, 8100 lb. (3674 kg.) of thrust; (S-1) General Electric F404 turbofan, 10,600 lb. (4818 kg.) of thrust. **Fuel:** 4434 lb. (2011 kg.). **Weight, empty:** (M/N) 10,465 lb. (4747 kg.). **Weight, normal loaded:** (M/N) 24,500 lb. (11,113 kg.). **Weight, max. Takeoff:** (M/N) 27,420 lb. (12,437 kg.). **Speed, max.:** (M/N) 685 mph (1150 kph) clean/645 mph (1083 kph) w/payload. **Initial climb:** (M/N) 8440 ft./min. (2573 m./min.). **Service ceiling:** (M/N) 42,250 ft. (12,881 m.). **Combat radius:** (M/N) 340 mi. (547 km.) hi-lo-hi. **Range, max.:** (M/N) 920 mi. (1480 km.).

SKYWARRIOR: Douglas A-3 twin-engine, carrier-based ATTACK AIRCRAFT of 1950s vintage; a few remain in service with the U.S. Navy as tankers and electronic warfare aircraft.

SLAR: See SIDE-LOOKING AERIAL RADAR.

SLAVA: A class of four large Soviet guided-missile CRUISERS, the first of which was completed in 1982. Designated "rocket cruisers" (*Raketnyy Kreysera*, RKR) by the Soviet navy, and apparently originally designed as a "low risk" insurance policy against the failure of the KIROV-class nuclear BATTLE CRUISER program, the Slavas are extremely powerful vessels, optimized for ANTI-SURFACE WARFARE (ASUW) against U.S. CARRIER BATTLE GROUPS but which also have considerable ANTI-AIR WARFARE (AAW) capabilities.

The Slavas are flush-decked, with a graceful shear and sharply raked bow, a long, low superstructure concentrated forward, and the pyramidal mast typical of Soviet cruisers. Armament consists of a 130-mm. twin DUAL PURPOSE gun at the bow; 8 twin launchers for SS-N-12 SANDBOX anti-ship missiles abreast of the forward superstructure, firing forward at a fixed elevation of 8°; 6 ADG-6-30 30-mm. radar-controlled guns for anti-missile defense (2 forward and 4 amidships); 2 12-barrel RBU-6000 anti-submarine rocket launchers ahead of the bridge; 8 8-round vertical launchers for SA-N-6 GRUMBLE long-range surface-to-air missiles (SAMs) aft; and 2 twin launchers for SA-N-4 GECKO short-range SAMs on the fantail. Two sets of quintuple 21-in. (533-mm.) tubes amidships behind shutters in the hull are probably intended for anti-submarine TORPEDOES, but could have a secondary anti-ship capability. A Ka-25 HORMONE B helicopter, operated from a landing deck and hangar aft, provides OVER-THE-HORIZON TARGETING (OTH-T) for the Sandbox missiles. The Slavas also have two "Punch Bowl" satellite communication terminals to receive OTH-T data. See also CRUISERS, SOVIET UNION.

Specifications   Length: 607.9 ft. (185.33 m.). Beam: 65.6 ft. (20 m.). Draft: 21 ft. (6.4 m.). Displacement: 7375 tons standard/10,200 tons full load. Powerplant: twin-shaft COGOG: 2 12,000-shp. gas turbines (cruise), 4 30,000-shp. gas turbines (boost). Speed: 32 kt. Range: 2000 n.mi. at 30 kt./8800 n.mi. at 15 kt. Crew: 720. Sensors: 1 "Top Pair" and 1 "Top Steer" 3-dimensional air-search radar, 1 "Top Dome" SA-N-6 guidance radar, 2 "Pop Group" SA-N-4 guidance radars, 1 "Front Door" SA-N-12 guidance radar, 3 "Palm Frond" navigation radars, 1 "Kite Screech" 130-mm. fire control radar, 3 "Bass Tilt" 30-mm. fire control radars, 4 "Plinth Group" electro-optical gun directors, 1 bow-mounted low-frequency sonar, 1 variable depth sonar. Electronic warfare equipment: 1 "Side Globe" electronic signal monitoring array, 4 "Rum Tub" and several "Bell" series electronic countermeasures arrays, 1 "High Pole B" IFF, two chaff launchers.

SLBM: Submarine-Launched Ballistic Missile, a BALLISTIC MISSILE launched from a vertical tube normally built into a specialized ballistic-missile SUBMARINE (SSB). The concept was first developed in Germany late in the Second World War, and some experiments were conducted by the U.S. Navy in the late 1940s, but the first operational SLBMs were deployed by the Soviet navy in the mid-1950s. Though extremely crude (the earliest SS-N-4 Sark could be launched only from the surface), they did generate considerable political leverage as the only Soviet nuclear weapons (bombers aside) capable of striking the continental U.S. before the advent of practical ICBMs.

The deployment of Soviet SLBMs in the context of the MISSILE GAP controversy forced the U.S. into a crash program which resulted in the POLARIS missile and the George Washington class of nuclear-powered ballistic-missile submarines (SSBNs), which were deployed in 1960. The Polaris system was so advanced and so successful that it set the standard for all subsequent SLBMs. A small, two-stage solid-fuel missile, Polaris was stowed in 16 vertical tubes on each submarine. Launch position data was fed to the missile's INERTIAL GUIDANCE unit from a Ship's Inertial Navigation system (SINS). The missiles could be launched from shallow submergence while hovering or moving at very low speed by using steam or compressed air to eject the missile with sufficient force to break the surface before ignition of the main engine.

All subsequent U.S., French, and British SLBMs emulated Polaris in employing solid fuel, while incorporating better accuracy, longer range, multiple warheads, and other improvements. The U.S.S.R., on the other hand, chose to rely mainly on liquid-fuel missiles, which are bulkier and present a greater risk of explosion, but which could employ technology derived from the Soviet land-based missile program. Only in the late 1970s were solid-fuel SLBMs deployed in quantity in the Soviet fleet, and these still complete with new liquid-fuel missiles.

The earliest SLBMs were inherently less accurate than their land-based counterparts, because the SINS could not achieve the precision of a surveyed site on land. The Polaris A-1 had a median error radius (CEP) of some 2000 m., while early Soviet SLBMs were even less accurate. Thus they were suitable only for attacks on cities and other large area targets, and were armed either with a single large warhead, or several smaller MULTIPLE REENTRY VEHICLES (MRVs). Improvements in guidance technology, especially the introduction of "stellar-inertial" guidance (in which a sensor on the missile updates its position with a "fix" on one or more celestial objects), have reduced CEPs for the latest SLBMs to less than 250 m. In addition, these missiles can now carry up to ten MULTIPLE INDEPENDENTLY TARGETED REENTRY VEHICLES (MIRVs) capable of destroying hardened point targets. But SLBMs are still not well suited for coordinated COUNTERFORCE attacks, because of the difficulty of communicating with submerged submarines, and need to launch SLBMs sequentially (which precludes a simultaneous mass attack).

The main advantage of SLBMs over land-based missiles is in their survivability: their submarines can be very quiet and very difficult to detect. Thus SLBMs are suited for secure SECOND STRIKE forces meant to deter counterforce attacks. See also separate SLBM listings for BRITAIN; CHINA; SOVIET UNION; UNITED STATES; for France, see MSBS.

SLBMS, BRITAIN: After the 1962 collapse of the U.S. Skybolt air-launched missile program left the RAF bomber force without an effective stand-off weapon (at a time when Soviet air defenses were overestimated), the U.S. agreed to supply POLARIS missiles and the associated technology instead. This resulted in the construction of four RESOLUTION-class nuclear-powered ballistic-missile submarines (SSBNs) in 1967–68, and the acquisition of 102 UGM-27C Polaris A-3 missiles by the Royal Navy (plus 31 later, to compensate for attrition). Initially armed with three British-designed 200-kT MULTIPLE REENTRY VEHICLES (MRVs), during the 1970s the missiles were rearmed with the complex and expensive CHEVALINE payload or three 60-kT MRVs and a penetration aid (PENAID) package (Soviet ballistic missile defenses around Moscow are consequential for a small force). In the early 1980s, the solid rocket engines of all surviving Polaris missiles were repacked to extend their service into the 1990s.

Polaris is to be replaced in the late 1990s by the TRIDENT II missile, with eight British-made MULTIPLE INDEPENDENTLY TARGETED REENTRY VEHICLES (MIRVs). Some 100 missiles may be acquired for up to four new Vanguard-class SSBNs. See also SUBMARINES, BRITAIN.

SLBMS, CHINA: The Chinese SLBM program began in the early 1960s, on the basis of Soviet technology, but Soviet assistance was soon stopped. It was not until 1982 that the first test launch of a Chinese SLBM took place, from a Golf-class ballistic-missile submarine (SSB) built from old, Soviet-supplied blueprints. Designated CSS-N-2 by the U.S., this two-stage solid-fuel missile is some 32.8 ft. (10 m.) long and 4.92 ft. (1.5 m.) in diameter, with a launch weight of 30,864 lb. (14,-030 kg.). Armed with a single 1-MT REENTRY VEHICLE (RV), it has a maximum range of 1678 mi. (2685 km.). Accuracy is poor, with the median error radius (CEP) on the order of 2800 m. Only a few were produced between 1975 and 1984.

The current CSS-N-3, also a two-stage solid-fuel missile, is deployed on the XIA-class nuclear-powered ballistic-missile submarines (SSBNs). It is to be superseded in turn by the CSS-N-4 (Chinese designation JL-1 "Great Wave"), which is reported to be some 42 ft. (12.8 m.) long and 7.55 ft. (2.3 m.) in diameter, with a launch weight of 44,-100 lb. (20,045 kg.). Armed with a 1-MT RV, the CSS-N-4 should have a maximum range of some 1988 mi. (3180 km.). See also SUBMARINES, CHINA.

SLBMS, FRANCE: See MSBS.

SLBMS, SOVIET UNION: The Soviet SLBM program began by the early 1950s, and relied heavily on captured German technology. In 1955, the Soviet navy successfully launched a modified SS-1 SCUD tactical ballistic missile from a surfaced Zulu-class diesel-electric submarine. This experiment was followed by the SS-N-4 Sark, a two-stage liquid-fuel missile with a range of some 350 mi. (560 km.). Initially deployed in 1956 aboard five Zulu V ballistic-missile submarines (SSBs), Sarks were deployed from 1958 aboard the purpose-built Golf-class SSBs and Hotel-class nuclear-powered ballistic-missile submarines (SSBNs). An extremely crude missile which could be launched only from the surface, Sark was armed with a 1-MT REENTRY VEHICLE (RV), and had a median error radius (CEP) of some 5000 m. Each Golf and Hotel submarine carried three Sarks in tubes running from the keel to the top of the sail (conning tower), while the Zulu Vs carried only two.

From 1964, Sark was replaced by the SS-N-5 Serb, also a two-stage, liquid-fuel missile, but capable of underwater launch. Though smaller than Sark, it had a range of 990 mi. (1585 km.); the CEP was assessed at 3000 m., and the missile was armed with a single 800-kT reentry vehicle. Sark and Serb accounted for the bulk of the Soviet SLBM force until the late 1960s. In the meanwhile, the U.S. Navy had introduced the solid-fuel POLARIS missile, 16 of which were carried by SSBNs. Greatly superior to the Soviet missiles in range, accuracy, and reliability, Polaris was superseded in 1970 by the even more powerful Poseidon, with ten MULTIPLE INDEPENDENTLY TARGETED REENTRY VEHICLES (MIRVs), and deployed on the improved LAFAYETTE-class SSBNs.

The next Soviet SSBN, the YANKEE, was obviously an imitation of the Lafayette, but its missile, the SS-N-6 SAWFLY, retained liquid-fuel propulsion, despite the resulting bulk and volatility. This allowed the U.S.S.R. to achieve some commonality with its land-based missile program, but the main reason for choosing liquid fuel was probably the inability of Soviet guidance technology to accommodate acontinuous-thrusting solid-fuel missiles. Armed with a single 700-kT RV, Sawfly initially had a range of 1490 mi. (2385 km.) and a CEP of 800–1000 m. The Mod 2 variant of 1972 had a range of 1800 mi. (2880 km.) and a 600-kT RV, while the Mod 3 of 1974 carried two 350-kT MULTIPLE REENTRY VEHICLES (MRVs).

In 1972, the Soviet navy launched the first of the DELTA-class SSBNs, an enlarged derivative of the Yankee. It was armed with the SS-N-8, a large, two-stage, liquid-fuel missile with considerably better range than Sawfly. Armed with a single 800-kT warhead, the SS-N-8 Mod 1 had a range of 4845 mi. (7755 km.) and a CEP of 1410 m., achieved through stellar-inertial guidance. The Mod 2 of 1977 was similar, but had a range of 5655 mi. (9050 km.).

In 1976, the U.S.S.R. introduced its first solid-fuel SLBM, the SS-N-17 SNIPE, 12 of which were deployed on a modified Yankee II. A three-stage missile with a range of 2350 mi. (3760 km.), Snipe was also the first Soviet SLBM equipped with a POST-BOOST VEHICLE (PBV), to deliver an 800-kT reentry vehicle with a CEP of some 1500 m. Not deployed in large numbers, the Snipe was apparently intended as a testbed for technology used in later missiles.

In 1978, the Soviet navy commissioned the first Delta III submarine, armed with 16 SS-N-18 STINGRAY missiles. A reversion to the two-stage, liquid-fuel configuration, Stingray was the first Soviet SLBM with MIRV capability. Armed with three 200-kT MIRVs, the initial Mod 1 had a range of 4040 mi. (6465 km.) and a CEP of 1410 m. The Mod 2, with a range of 4970 mi. (7955 km.) and CEP of 1550 m., is armed with a single 450-kT reentry vehicle, while the Mod 3, similar to the Mod 1, has seven 100-kT MIRVS.

In 1980 the first of the huge TYPHOON-class SSBNs entered service. The largest submarines in the world, the Typhoons are armed with 20 SS-N-20 STURGEONS, very large, three-stage solid-fuel missiles with a range of 5160 mi. (8255 km.). Armed with nine 500-kT MIRVS, and fitted with an improved stellar-inertial guidance for a CEP of only 600 m., Sturgeon is the first Soviet SLBM sufficiently accurate to attack some hardened targets.

The latest Soviet SLBM is the SS-N-23 SKIFF, introduced in 1986 on the Delta IV SSBNs. A two-stage, liquid-fuel missile armed with ten 100-kT MIRVS, Skiff has a range of 5190 mi. (8305 km.) and a CEP of less than 900 m. See also SUBMARINES, SOVIET UNION.

SLBMS, UNITED STATES: U.S. SLBM development lagged until the appearance of the first Soviet ballistic missile submarines in 1956 prompted the initiation of a crash program, which resulted in the POLARIS missile and the George Washington nuclear-powered ballistic-missile submarines (SSBNs) of 1960; they served as the basis for all subsequent U.S. SLBM systems. A two-stage, solid-fuel missile, 16 of which were carried by each submarine, the initial Polaris A-1 was seen as an interim weapon; it had a range of 1380 mi. (2208 km.), a single 500-kT REENTRY VEHICLE (RV), and a median error radius (CEP) of some 2000 m. The A-1 was superseded in 1962 by the Polaris A-2, with a range of 1727 mi. (2764 km.), first deployed on Ethan Allen—class SSBNs. The definitive A-3 of 1964, first deployed on LAFAYETTE SSBNs, had a range of 2880 mi. (4610 km.) and a CEP of some 930 m. Originally armed with a single 1-MT RV, it was later refitted with three 200-kT MULTIPLE REENTRY VEHICLES (MRVs). Polaris was retired from U.S. service in the early 1980s, when the last Ethan Allens were withdrawn from the SSBN mission. Polaris A-3s were also supplied to the Royal Navy for its RESOLUTION-class SSBNs, armed with three British-made 200-kT MRVs (see SLBMS, BRITAIN).

Polaris was superseded from 1970 by the UGM-73 POSEIDON C-3, retrofitted into the Lafayette class (whose launch tubes had originally been sized for the larger missile). Poseidon is a three-stage solid-fuel missile with much greater range, a much heavier payload, and better accuracy than the Polaris A-3. The first SLBM equipped with a POST-BOOST VEHICLE (PBV), it is armed with ten 40-kT MULTIPLE INDEPENDENTLY TARGETED REENTRY VEHICLES (MIRVs). Maximum range is 3230 mi. (5170 km.), and the CEP some 550 m. Poseidon will continue in service until the retirement of the last Lafayettes in the early 1990s.

Even as Poseidon was entering service, the U.S. Navy was already developing the TRIDENT SLBM, with enough range to strike Soviet territory from U.S. coastal waters. The OHIO-class SSBNs were designed specifically for Trident, but would not be ready until the early 1980s, while Trident was too large to fit the Lafayettes' tubes. To bridge the gap, a smaller interim version, the Trident I (C-4), was built to be retrofitted into the last 19 Lafayettes, and was later deployed on the first Ohios as well, because delivery of the definitive Trident II (D-5) missile was delayed until 1990. See also SUBMARINES, UNITED STATES.

SLEP: Service Life Extension Program, U.S. Navy major overhaul and modernization programs for AIRCRAFT CARRIERS, meant to extend their service lives by some 15–20 years. "SLEPed" aircraft carriers are thus expected to remain in service for up to 45 years.

SLEP involves hull repairs and the complete renovation of all engine machinery, catapults, elevators, and arresting gear; the installation of the latest sensors, electronics, and modifications to defensive armament; and the rearrangement of aviation fuel and munitions stowage, and of aircraft maintenance facilities. The entire process requires from 24 to 36 months, and costs some $500 million. Carriers of the FORRESTAL and KITTY HAWK classes have been rotating through SLEP since 1980. The nuclear-powered ENTERPRISE underwent a three-year SLEP from 1979 to 1982. See also AIRCRAFT CARRIERS, UNITED STATES.

SLOC: See Sea Lines of Communication.

SLQ-32: U.S. shipboard ELECTRONIC COUNTERMEASURES (ECM) system. Developed on "design-to-cost" principles with modular "building blocks," three variants are now in service. The simplest SLQ-32(V)1 version, intended for frigates, amphibious warfare ships, and auxiliaries, operates in the H/I/J-bands to provide threat warning, identification, and the bearing of incoming radar-guided missiles and of their launch platforms. The (V)2 version, intended for guided-missile frigates, guided-missile destroyers, and the SPRUANCE-class anti-submarine destroyers, is similar to the (V)1, but covers more frequency bands. The (V)3, for cruisers, battleships, command ships, and fast replenishment vessels, combines the features of the other variants with a quick-reaction ACTIVE JAMMING capability specifically designed to counter "pop-up" submarine-launched missiles, and missiles launched from fast attack craft hiding in shore "clutter." U.S. aircraft carriers do not carry the SLQ-32, but instead have the more comprehensive SLQ-29.

The SLQ-32 is designed to operate in conjunction with the Mk.33/34 RBOC and Mk.36 SRBOC chaff and flare launchers, which can be activated automatically by the ECM system (or manually from the COMBAT INFORMATION CENTER).

The SLQ-32 has suffered reliability and maintainability problems. In addition, questions have been raised about its ability to identify the full range of potential threats (the unit on the USS *Stark* never detected the Exocet missiles which hit the vessel), and about its ability to cope with high-angle missile attacks. More generally, the SLQ-32 exemplified the tendancy to favor shipbuilding over the provision of adequate electronic warfare equipment for each ship.

SMALL ARM: A weapon that can be carried and operated by one man; or, more broadly, a light infantry weapon as opposed to armor and artillery. Currently, this broader definition includes pistols, SUBMACHINE GUNS, ASSAULT RIFLES, RIFLES, rifle-caliber LIGHT MACHINE GUNS, GRENADE LAUNCHERS,

light anti-tank rocket launchers (LAWS), MORTARS of 82 mm. or less, and most recoilless weapons.

SMG: See SUBMACHINE GUN.

SNAKEYE: U.S. RETARDED BOMB kit, developed during the Vietnam War to facilitate low-altitude bombing. Much used and widely copied, Snakeye consists of a spring-loaded fin assembly, which can be attached to the tail of a standard Mk.80-series low-drag general-purpose (LDGP) bomb. When activated, the fins open into a cruciform air brake, rapidly decelerating the bomb and thus allowing its release much closer to the target (while also allowing the aircraft to escape from the bomb's fragmentation envelope).

SNIPE (SS-N-17): NATO code name for a Soviet submarine-launched ballistic missile (SLBM) developed from the late 1960s, the first Soviet SLBM with solid-fuel propulsion. After flight tests in 1975–76, it entered service in 1977 aboard a single YANKEE II nuclear-powered ballistic missile submarine (SSBN) with 12 missile tubes. Though technically successful, Snipe was never mass-produced; it was apparently meant serve as a testbed for technologies incorporated into the later SS-N-20 STURGEON.

The three-stage Snipe's first stage has either one large or four small gimballed nozzles for attitude control and steering, while the second and third stages each have a single gimballed nozzle. The payload is a single 800-kT REENTRY VEHICLE (RV), mounted on a liquid-fuel POST-BOOST VEHICLE (PBV). Snipe was the first Soviet SLBM with a PBV, and though only one large RV is carried, at least two smaller MULTIPLE INDEPENDENTLY TARGETED REENTRY VEHICLES (MIRVs) could be carried instead. It relies on navigating INERTIAL GUIDANCE with General Energy Management Steering (GEMS), whereby the missile makes excursions above and below its nominal trajectory, in order to allow each stage to burn to exhaustion, thus eliminating the need for unreliable and inaccurate thrust-termination ports. Given its relatively poor accuracy, it could be used effectively only against cities and other area targets. See also SLBMS, SOVIET UNION.

Specifications Length: 34.75 ft. (10.6 m.). **Diameter:** 5.4 ft. (1.65 m.). **Weight, launch:** 44,000 lb. (20,000 kg.). **Range:** 2350 mi. (3900 km.). **CEP:** 1410 m.

SNIPER: A rifleman specially trained and equipped to kill individuals by long-range, aimed fire. Although the material effects of sniper fire are slight, the psychological effect can be very great indeed. By targeting officers or scouts, snipers can throw entire units into confusion, and the knowledge that a sniper is active in an area can bring attacks to a standstill while he is hunted out. Snipers can also attack vehicle commanders and heavy-weapon crews, to deprive enemy infantry of their supporting fire at critical moments.

Armies differ in their emphasis on snipers. The Soviet army has always valued them, and the duels between Soviet and German snipers in the Second World War are legendary. Today, one sniper is organic to each Soviet Motorized Rifle and airborne squad; they are selected from the best marksmen and provided with special training as well as specialized rifles. The U.S. Marine Corps, deeply affected by its campaign in the Pacific, also values highly trained snipers, but the U.S. Army does not. The Israeli army neglected this specialty until the 1968–70 War of Attrition, but has since cultivated the skill.

Armies that have specialized snipers all follow the same training program. Of course, a great deal of attention is given to marksmanship, but much of the training is devoted to camouflage, fieldcraft, and target selection. Snipers commonly work in two-man teams, one to select firing positions and designate targets, the other to do the actual shooting.

Snipers are normally armed with high-power RIFLES firing full-size 7.62 to 7.92-mm. ammunition. Bolt-action rifles are preferred, but some semiautomatic weapons are also in service. Many sniper rifles are derived from civilian hunting rifles, or are "accuratized" versions of standard military rifles, but a number of purpose-built sniper weapons are also in service, e.g., the Soviet SVD DRAGUNOV, the French FR-F1, the Walther WA2000, and the Beretta Sniper. Regardless of origin, all sniper rifles are equipped with telescopic sights (4x to 8x are most common), and many can also be fitted with INFRARED or IMAGE-INTENSIFIER night sights. Given adequate visibility, a well-trained sniper can easily hit a man-sized target at 600–800 m., and kills out to 1200 m. have been documented.

"SOFT KILL":   The neutralization of a target by means other than physical attack ("HARD KILL"). The term commonly defines the intended effect of ELECTRONIC COUNTERMEASURES, as e.g., when ACTIVE JAMMING blinds a radar or diverts a guided missile from its target. The term is also applied to the effects of certain DIRECTED ENERGY WEAPONS, such as PARTICLE BEAMS, which could disrupt electronic circuitry at the molecular level without causing externally visible damage.

SOFT RECOIL: Techniques applied to large-caliber, high-velocity guns to minimize their recoil, thereby allowing them to be mounted on relatively light vehicles. For example, the standard British/U.S. 105-mm. rifled tank gun generates a peak recoil of some 37.5 tons and requires a carrying vehicle with a minimum weight of 30–35 tons, but a soft recoil variant of the same gun has a peak recoil of only 15 tons, and can be mounted on vehicles weighing as little as 15 tons. The development of soft recoil techniques has thus caused a renaissance of the LIGHT TANK, because they allow the firepower of a MAIN BATTLE TANK to be placed on a light armored vehicle.

Soft recoil techniques include the use of carefully designed muzzle brakes, combined with modified recoil mechanisms with much longer strokes. In the former, multiple baffles divert a portion of the propellant gases rearward, thereby countering some of the recoil, while the latter allow the gun to recoil further, over a longer period, thus distributing the recoil load and reducing peak forces by half. The effect on gun ballistics is minimal.

SOFT TARGET: See TARGET, SOFT.

SOMAN: Lethal NERVE AGENT (U.S. designation GD) developed in Germany during the Second World War. A colorless, odorless percutaneous (skin-permeable) liquid, Soman, like other nerve agents, acts by binding permanently with acetochlorinesterase (AChE), a body enzyme essential for the proper functioning of the central nervous system. Symptoms of Soman intoxication include involuntary contractions of the large muscles, convulsions, froth at the mouth, and death from respiratory collapse within 15 minutes. Soman is extremely potent, with an LD-50 (the dose at which 50 percent of all exposed personnel will die) of only 25 milligrams per min. per cubic m. (mg./min./m.^{3}) inhaled, or 10,000 mg./min./m.^{3} percutaneously. Because Soman is skin-permeable, troops must wear complete defensive suits in addition to respirator masks to be protected.

Troops can be "hardened" against Soman through a prophylactic regime of synergistic drugs. One set of drugs, administered before exposure, *temporarily* binds with some AChE, to form a protected pool of the enzyme. After exposure, a second set of drugs is administered to release the protected enzyme. The only other form of first aid consists of massive doses of adrenaline or atropine administered within minutes of exposure.

Soman has a relatively low boiling point, and will dissipate within a few hours, but a variant, "thickened Soman" (GD[T]), is far more persistent and requires up to six days to dissipate.

Both GD and GD(T) can be delivered by artillery shells, mortar bombs, rocket or missile warhead, aerial bombs, or aerosol dispenser. GD(T) forms a significant portion of the Soviet chemical arsenal. See also CHEMICAL WARFARE.

SONAR: Sound Navigation and Ranging, a communications and position-finding device used in underwater navigation, target detection, and weapon control (the British acronym ASDIC is synonymous). The primary sensor for submarine and ANTI-SUBMARINE WARFARE (ASW), sonar equips all submarines, most surface warships, and many anti-submarine helicopters; in the form of expendable SONOBUOYS, it is also used by fixed-wing anti-submarine patrol aircraft. Small sonar sets guide acoustic homing TORPEDOES and trigger some naval MINES.

Sonars are classified as passive or active. Passive sonars, or hydrophones, are simply underwater microphones which receive all sounds, natural and man-made, from the water around them. Submarines and other targets can be detected by engine and propeller noises, and by other radiated sounds such as water flows. Developed during the First World War, passive sonar remains the most important type, albeit in vastly more sophisticated forms. Whereas early hydrophones received and amplified sounds only in the audible frequency range, modern passive sonars can cover a very broad spectrum from very low to very high frequencies. Further, while early hydrophones depended exclusively on the trained ear of the operator to distinguish target sounds from background noise (and the sea and its creatures can be *very* noisy), modern passive sonars are assisted by powerful, computer-driven signal processors, which not only suppress the background, but can also compare target noises with known ship and submarine acoustical "signatures" in their memory.

The basic type of passive sonar for both ships and submarines is the circular or spherical "array" with several hundred individual hydrophones distributed on a cylindrical or spherical frame to provide 360° coverage. Usually mounted near the bow, each hydrophone in the array covers only a relatively narrow sector. By comparing the relative strengths of the signals received by each hydrophone in the array, it is possible to determine the target's bearing within one or two degrees.

Passive ranging sonar, such as the U.S. BQQ-4 PUFFS, can determine target range in addition to bearing. Two or three smaller hydrophone arrays at the bow and stern with a known baseline can be used to compare the relative target bearing from each, to then determine the target's position by triangulation. Flank-mounted CONFORMAL ARRAYS perform the same function more effectively, because their hydrophones are embedded in the sides of the ship or submarine, so that drag and flow noises are greatly reduced.

All ship and submarine sonars, both active and passive, have a blind zone dead astern, because of propeller, machinery, and wake noises. This gap can be covered by a passive TOWED ARRAY, essentially a series of hydrophones attached to a buoyant cable, which can be towed at a considerable distance behind the ship or submarine. Because such arrays operate well away from the towing platform's engine and flow noises, detection ranges are greatly enhanced. Moreover, when used in conjunction with a bow sonar, a towed array can provide a very long baseline for highly accurate triangulation. Thus towed arrays are a major ASW innovation that achieves much more than covering the blind zone.

Sonar intercept arrays are passive sonars meant to receive and analyze the signals of hostile active sonars, for ranging and target classification.

Active sonar, developed in Britain between the world wars, operates on the echo-ranging principle, and is thus analogous to RADAR: sound waves transmitted through the water are reflected by objects in their path, and received by hydrophones for classification. Although the speed of sound in water fluctuates somewhat, it is roughly 1500 m./sec., so that the time interval between transmission and echo reception (if there is an echo) can be used to derive range. The basic elements of active sonar are a transducer (a piezoelectric oscillator to produce a sound wave), a hydrophone to receive the target echo, and a display to present target data in audio and video format. In many systems, hydrophones and transducers are combined in "active/passive" arrays.

The earliest active sonars were narrow-beam, high-frequency, manually scanned "searchlight" types. The operator would point the transducer in the desired direction with a servo-device, press a trigger to generate a sonic pulse ("ping"), and then wait for a return echo over a period equivalent to the sonar's maximum range. If no target was detected, he would shift the transducer a few degrees to repeat the process. Because the sonar beam itself was only a few degrees wide, it was relatively easy for a submarine to escape detection by radical maneuvers. Moreover, as the transducer was fixed in elevation, blind zones would develop above and below the beam as the range closed. Late in World war II, the U.S. and Britain both introduced sonars which could be directed vertically as well as horizontally. Postwar active sonars had automatic electro-mechanical scanning, while the latest types are electronically scanned. When used in conjunction with spherical or circular hydrophone arrays, modern active sonar can provide near-instantaneous coverage, except for a narrow arc dead astern.

The performance of active sonars depends on their operating frequency. High-frequency sonars have good angular resolution and allow accurate range-finding, but their signals are rapidly attenuated, limiting their maximum range. Low-frequency sonars can have ranges of up to 100 mi. (160 km.), but have relatively poor resolutions. Navies operating mainly in coastal waters generally favor medium-frequency active/passive sonars, sometimes backed up by high-frequency active FIRE CONTROL sonars. The U.S. Navy, which operates mainly in deep waters, prefers low-frequency active/passive arrays.

Transducer size also depends on frequency. In surface ships, early high-frequency sonars were housed in retractable "domes" under the keel amidships. Larger medium-frequency arrays are generally housed in fixed, hull-mounted domes (often made of rubber), while the largest low-frequency arrays are housed in bulged bow domes. The bow position is preferable because it is farthest from machinery and propeller noises, but can have an adverse effect on ship handling (though with careful design bow domes can actually enhance hull dynamics). In submarines, smaller arrays are mounted in the bow, above or below the torpedo tubes, but low-frequency spherical arrays can be large enough to fill the entire bow, displacing the tubes to amidships.

While active sonar can provide accurate range and bearing, its audible ping can reveal its presence, warning the potential target. Thus passive sonar is normally preferred, with active sonar reserved for last-minute fire control.

Both active and passive sonars are affected by the highly variable behavior of sound in water. Unlike radar waves, which always travel at the constant speed of light, the velocity of sound in water varies with depth, temperature, and salinity. Moreover, sound waves can be refracted or reflected by temperature gradients (THERMOCLINES) under which submarines can escape detection. VARIABLE DEPTH SONARS can be lowered below the thermoclines, to relay data to the surface by cable.

Underwater telephones are a secondary application of sonar; voice messages can be transmitted for short distances by a transducer, for reception by a suitable hydrophone array.

SONGSTER (AT-8): NATO code name for the Soviet *Kobra* cannon-launched ANTI-TANK GUIDED MISSILE (ATGM). First reported in the late 1970s, Songster can be launched through the 2A46 125-mm. smoothbore gun of the T-64 and T-80 main battle tanks (MBTs); intended as a very long-range weapon for specially-designated "sniper" tanks, it could also be used to engage enemy ATGM vehicles and attack helicopters. Because of their size, nor more than four or five missiles could be carried by each tank.

Songster is armed with a shaped-charge (HEAT) warhead capable of penetrating 25.6 in. (650 mm.) of homogeneous steel armor. Expelled from the gun barrel by a special cartridge, Songster leaves the muzzle at a velocity of 150 m./sec. (335 mph), and then accelerates to maximum velocity. Songster is controlled by UHF radio COMMAND GUIDANCE with semi-automatic command to line-of-sight (SACLOS); in action, the gunner tracks the target in the main gunsight and launches the missile, while the tank's FIRE CONTROL computer determines the angle between the missile's course and the gunner-target line of sight, and generates steering commands to bring the missile onto that line. Its main drawback is the need to track the target throughout the engagement, resulting in a relatively low engagement rate (compared with ordinary tank rounds), and forcing the tank to remain at least partially exposed throughout the engagement. The narrow-band UHF guidance link is relatively immune to ACTIVE JAMMING, but can be detected by RADAR WARNING RECEIVERS on the target.

Specifications Length: 3.9 ft. (1.2 m.). Diameter: 4.9 in. (125 mm.). Weight, launch: 55 lb. (25 kg.). Speed: 500 m./sec. (1118 mph). Range envelope: 100–4000 m.

SONOBUOY: A remotely monitored, expendable SONAR float dropped from helicopters and fixed-wing aircraft to detect and track submerged submarines. Developed in 1943, sonobuoys are widely employed from land- and carrier-based patrol aircraft.

A sonobuoy consists of a buoyant float, a battery, a sonar unit, and a radio transmitter to relay sonar data. The sonar unit itself can often be lowered to a considerable depth at the end of a cable deployed from the float, thus enabling the sonar reach below any sound-reflecting THERMOCLINES.

The simplest sonobuoys have passive, omnidirectional sonars. Dropped in a linear pattern across the suspected path of a submarine, they act as detectors and can also provide a rough position fix, once the strengths of the signals received by each buoy in the line are compared. Typical of this type is the British SSQ-904 Jezebel. More advanced passive buoys, such as the U.S. SQQ-954 DIFAR, can also provide target bearing; they are often dropped in the vicinity of submarines first detected by omnidirectional buoys. Two or more directional buoys can pinpoint a submarine's location by triangulation.

Active sonobuoys have transducers, which generate an echo-ranging "ping." They can provide very accurate range, bearing, and positional data, but they can also be detected by the target submarine, hence they are most often dropped for last-minute target acquisition just before an attack.

Simple, lightweight omnidirectional buoys are most often deployed from automated dispensers built into the bellies of aircraft; larger and more complex directional and active buoys are usually dropped manually through a chute. See also ANTI-SUBMARINE WARFARE.

SOSUS: Sound Surveillance System, a U.S. submarine-detection network of hydrophone arrays emplaced on the seabed in coastal waters and across narrow water (CHOKEPOINTS) frequently traversed by Soviet submarines. Each SOSUS array is linked by cable to one of several processing and tracking centers ashore which monitor the movements of Soviet submarines. Because sound carries for a considerable distance underwater, each SOSUS array can detect submarines up to several hundred miles away, and the exactly correlated data of two or more arrays can determine submarine positions by triangulation. Powerful computers process incoming data to suppress background noise, and scan "libraries" of known submarine acoustical signatures to identify detected submarines.

SOSUS is a general surveillance and early-warning system; it is not accurate enough for actual target acquisition or weapon guidance, but it can serve to direct aircraft and other anti-submarine forces towards likely submarine contacts, in a manner analogous to the use of surveillance radar in air defense. A similar hydrophone network has been deployed by the U.S.S.R. See also ANTI-SUBMARINE WARFARE; SONAR.

SOUTHAG: Southern Army Group, a NATO ground command subordinate to SHAPE and responsible for all Allied ground forces deployed in and around the Mediterranean and the Balkans.

SOVREMENNY: A class of eight Soviet guided-missile DESTROYERS, the first of which was completed in 1980; more are under construction. Though designated *Eskadrenny Minonosets* (EM, literally "Fleet Minelayers"), the Sovremennys are large, general-purpose escorts which clearly complement the UDALOY-class ANTI-SUBMARINE WARFARE (ASW) destroyers; they represent a major advance over earlier Soviet destroyers in size, seaworthiness, endurance, armament, and sensors. Unlike previous Soviet destroyers, designed for short-term "sea-denial" missions, the Sovremennys are optimized for sustained "sea control" and "power projection" operations in Third World settings.

The Sovremennys have a broad transom stern and a raised forecastle, the latter a departure from previous Soviet designs. A long, low superstructure runs the length of the ship, with the bridge and a large stack concentrated amidships. The power-plant is believed to be both automated and pressure-fired, allowing the ships to accelerate from 10 to 32 kt. in less than two minutes. The main radar antennas are mounted on a solid, pyramidal mast (as in Soviet cruisers). Armament consists of 1 twin 130-mm DUAL PURPOSE gun and 1 single-arm launcher for SA-N-7 GADFLY surface-to-air missiles at each end of the ship; a fixed, forward-facing quadruple launcher for SS-N-22 SUNBURN anti-ship missiles on each side of the bridge; 2 twin 21-in. (533-mm.) tubes amidships, for ASW homing TORPEDOES; 2 ADG-6-30 30-mm. radar-controlled guns for anti-missile defense abreast of the stack; and 2 RBU-1000 ASW rocket launchers aft. Up to 100 MINES can be carried on rails running from amidships to the fantail. A Ka-25 HORMONE B helicopter, operated from a landing deck and hangar amidships, can provide OVER-THE-HORIZON TARGETING (OTH-T) for the SS-N-22 missiles. See also DESTROYERS, SOVIET UNION.

Specifications Length: 511.8 ft. (156 m.). **Beam:** 56.8 ft. (17.5 m.). **Draft:** 21.3 ft. (6.5 m.). **Displacement:** 6300 tons standard/7900 tons full load. **Powerplant:** twin-shaft steam: 4 oil-fired boilers, 2 sets of geared turbines, 110,000 shp. **Speed:** 34 kt. **Range:** 2400 n.mi. at 32 kt./6500 n.mi. at 20 kt. **Crew:** 320. **Sensors:** 1 "Top Steer" 3-dimensional air-search radar, 3 "Palm Frond" navigation radars, 6 "Front Dome" SA-N-7 guidance radars, 1 "Kite Screech" 130-mm. fire control radar, 2 "Bass Tilt" 30-mm. fire control radars, 1 "Band Stand" SS-N-22 guidance radar, 1 hull-mounted medium-frequency sonar. **Electronic warfare equipment:** 2 "Bell Squat," 2 "Bell Shroud," and 2 "Shot Rock" electronic counter-measures arrays, 1 "High Pole B" IFF, 2 twin chaff launchers.

SOYUZ: Soviet manned spacecraft, first flown in 1967, and now used as an expendable shuttle vehicle, first with the SALYUT and later with the *Mir* space stations. An incremental development of the original Soviet *Vostok* spacecraft, Soyuz can carry three men into low earth orbit for periods of up to one week. Though exceedingly crude by U.S. standards, Soyuz has achieved a high degree of reliability and economy through serial production.

Boosted into orbit by the A-2 booster (a modified version of the SS-6 Sapwood ICBM first used to place *Vostok-1* into orbit in 1961), Soyuz is 26 ft. (7.94 m.) long, and has a maximum diameter of 8.92 ft. (2.72 m.) and an all-up weight of 14,994 lb. (6815.5 kg.). The vehicle is divided into three sections: an orbital module up front, an ascent/descent module in the middle, and a propulsion/equipment module in the rear. Spherical in shape, the orbital module provides storage and work space, and has an air lock at each end, one for access to the ascent/descent module, the other for docking with other spacecraft, including the *Mir* space station.

The bell-shaped ascent/descent module has side-by-side seating for three and contains all flight controls. Its blunt rear end has an ablative heat shield to protect the crew during reentry, and a large parachute to slow the capsule for a dry landing. The propulsion and equipment module houses retrorockets, attitude-control thrusters, batteries and fuel cells, oxygen bottles, and communications equipment. During takeoff, the crew is strapped into the ascent/descent module. Once in orbit, the cosmonauts normally move into the orbital module until docking with a space station. For reentry, the spacecraft undocks from the station and maneuvers into the proper orbit for landing; the crew returns to the ascent/descent module, the orbital

module is jettisoned, and retrorockets are fired. The propulsion/equipment module is then jettisoned as well, and the crew returns to earth in the ascent/descent module.

In contrast to U.S. spacecraft, the crew has only limited control over the Soyuz: almost all critical functions, including docking, are either automated or controlled from the ground. This has allowed Soyuz to serve as the basis for the *Prognoz* (Progress) unmanned resupply vehicle. Essentially a Soyuz with the ascent/descent module replaced by a cargo pod, *Prognoz* is used to replenish the food, water and oxygen of Soviet space stations, allowing them to be manned for years at a time.

Though the U.S.S.R. has also developed a reusable space shuttle *(Buran)*, it is likely that Soyuz will remain in service for routine transfers between *Mir* and earth. See also SPACE, MILITARY USES OF; SPACE TRANSPORTATION SYSTEM.

SP: Self-propelled. See ARTILLERY.

SP-70: Multi-national 155-mm. self-propelled GUN-HOWITZER project. Intended as a replacement for the U.S. M109, the SP-70 program was initiated in 1973 with a "Memorandum of Understanding" signed by Britain, Italy, and West Germany (the team leader). The requirement called for a new chassis, a fully enclosed, NBC-protected turret, and an automatic loader to provide a burst rate of fire of 12–15 rds./min. By 1985, only 15 prototypes had been completed, but these proved so unreliable that the entire program was canceled in 1986, after an expenditure of more than $500 million. Each of the participants is now developing its own SP howitzer.

SPACE-BASED INTERCEPTOR: A proposed BALLISTIC MISSILE DEFENSE weapon being developed under the STRATEGIC DEFENSE INITIATIVE. Intended for orbital deployment, the SBI consists of a "kinetic kill vehicle" (KKV) the size and weight of a 1-gal. paint can, which is mounted on a two-stage rocket booster. Several SBIs would be housed in a satellite "carrier vehicle" (CV), several hundred of which would be placed into overlapping orbits to ensure global coverage. When enemy missiles are detected, a FIRE CONTROL system on the CV would compute the launch time and course for the interceptor. The booster would accelerate the KKV towards the computed interception point, and then separate. The KKV would then home on the target by using a combination of INFRARED, millimeter wave (MMW), and ultraviolet sensors, and a series of "puff jet" maneuvering thrusters.

SPACED ARMOR: A form of ARMOR protection in which two armor plates are separated by a space which may either be void or filled with a fluid or nonmetallic material. On tanks and other armored vehicles, spaced armor can consist of a thin outer plate to detonate shaped-charge (HEAT) warheads prematurely, ahead of the main (inner) plate. But this method protects only against small HEAT warheads; larger HEAT warheads have optimal standoff distances of several feet, and their detonation by the outer plate can actually enhance their penetration. On the other hand, if the space is filled with energy-absorbing materials (e.g., glass), the method can be more effective.

On armored warships (e.g., BATTLESHIPS, all-gun CRUISERS, and some larger AIRCRAFT CARRIERS), the opposite arrangement predominates: a thick outer layer is intended to resist penetration, while a thin inner layer is designed to resist the splinters of shells that do penetrate and explode in the void between the layers. See also CHOBHAM ARMOR.

SPACE, MILITARY USES OF: Commonly defined as beginning at an altitude of 50 mi. (80 km.) above the earth, space first became a theater of war when German V-2 BALLISTIC MISSILES passed through the ionosphere in 1944; despite repeated exhortations to maintain the "purity" of space, it has been the scene of intense military activity ever since then.

To date, most of these activities have been in support of earth-based operations, in the form of early warning, attack assessment, surveillance, reconnaissance, communications, navigation, meteorology, and geodesy satellite operations (see SATELLITES, MILITARY, for details).

For surveillance/reconnaissance and communications, satellites have become essential adjuncts of earth-based activities. Hence counterreconnaissance capabilities have also been developed in the form of ANTI-SATELLITE (ASAT) weapons. Only the U.S.S.R. has an operational ASAT system, a coorbital "killer satellite" which can destroy other satellites in low-earth orbit (LEO); the U.S. has tested but not deployed a direct-ascent ASAT in the form of an air-launched missile. The next step would be the deployment of permanently orbiting "defensive" satellites (DSATS), equipped with SPACE-BASED INTERCEPTOR missiles (SBIs) to attack enemy satellites, and defend one's own. SBIs could also form the backbone of a space-based BALLISTIC MISSILE DEFENSE (BMD) system, together with early-warning and space-tracking satellites. If rival DSATs were deployed, one of their major tasks would be destruction of opposing DSATs. Thus the evolution of war in space would parallel the development of war in the atmosphere, with DSATs in the role of fighters.

Manned spacecraft would provide added flexibility and the capability to conduct real-time surveillance and analysis (as already demonstrated by the military use of Soviet SALYUT space stations). In addition, a manned spacecraft could provide facilities for the maintenance and repair of unmanned satellites, and also possibly for their COMMAND AND CONTROL, but in space combat the action would be too fast for humans to intervene, except to override automated systems. Moreover, any manned platform would be a high-value target and thus require complex defenses. For further aspects of warfare in space, see FRACTIONAL ORBIT BOMBARDMENT SYSTEM; ORBITAL BOMBARDMENT SYSTEM.

SPACE TRANSPORTATION SYSTEM: The U.S. Space Shuttle, a manned, reusable spacecraft used to boost passengers and payloads into low earth orbit (LEO). The Space Shuttle program began in 1969 in succession to NASA's Project Apollo. Originally intended as part of a wider system with a permanently orbiting space station, the shuttle was developed independently after the space station program was canceled in the early 1970s. Initial designs called for a fully reusable vehicle, with a winged "fly-back" booster and a smaller winged orbiter, but funding cuts resulted in a more modest, partially reusable system. To spread the (escalating) costs of the program, the U.S. Air Force (USAF) was induced to use the shuttle as its primary launch vehicle, but to meet military payload requirements (notably the very large KH-12 reconnaissance satellite), the shuttle had to be redesigned on a larger scale. Captive and gliding tests of the prototype shuttle *Enterprise* began in 1977, and the first fully functional shuttle, *Columbia*, made its initial orbital flight in 1981. *Columbia* was followed by *Challenger*, *Discovery*, and *Atlantis*, but a proposed fifth shuttle was canceled. After 24 successful shuttle flights, *Challenger* was lost to a booster failure in 1986, as a result of which the entire shuttle fleet was grounded until September 1988. That forced USAF to reconsider its reliance on the shuttle, and in 1989 it was decided to revert to the use of expendable rockets. A new shuttle has been built to replace *Challenger*, and construction of an additional shuttle has been recommended as insurance against another catastrophic accident.

The Space Transportation System consists of the

Shuttle Orbiter (the shuttle proper), an expendable external tank, and two solid rocket boosters. The Shuttle Orbiter is a large "lifting body" vehicle with a length of 122.2 ft. (37.24 m.) and a wingspan of 78.06 ft. (23.79 m.). It has a large, boxy fuselage with a pressurized crew compartment in the nose, a large cargo hold amidships, and three 470,000-lb. (213,152-kg.) thrust shuttle main engines plus two 6000-lb. (2720-kg.) thrust orbital maneuvering system (OMS) pods in the tail. The crew compartment has two levels with accommodations for up to seven astronauts. The cargo bay is 60 ft. (18.3 m.) long and 15 ft. (4.6 m.) wide, and has a payload capacity of some 60,000 lb. (27,272 kg.). Payloads are released in space through two large clamshell doors; with their own auxiliary boosters, satellites can reach beyond LEO. An articulated remote manipulator arm can be used to retrieve objects from space for stowage in the payload bay, where they can be examined, repaired, or returned to earth.

The Shuttle Orbiter's low-mounted wing has a compound-delta planform, and the trailing edges are fitted with elevons for aerodynamic pitch and roll control; a tall vertical stabilizer has a rudder/airbrake for aerodynamic yaw and speed control. The outer surfaces of the orbiter are covered with refractory tiles for thermal shielding during reentry, instead of the usual ablative heat shield. The orbiter has an empty weight of some 165,000 lb. (75,000 kg.) and a maximum landing weight of 188,000 lb. (84,260 kg.).

The External Tank (ET) is a lightweight shell containing cryogenically cooled liquid hydrogen/liquid oxygen propellants for the Shuttle Main Engines (SMEs). Covered with orange foam insulation, the ET is 154.2 ft. (47 m.) long and 27.56 ft. (8.38 m.) in diameter, with a launch weight of 1,638,873 lb. (743,253 kg.). The two Solid Rocket Boosters (SRBs) are each 149.16 ft. (45.5 m.) long and 12.14 ft. (3.7 m.) in diameter, with a launch weight of 1,293,246 lb. (586,506 kg.). Generating some 2,650,000 lb. (1,201,815 kg.) of thrust each, the SRBs are built in segments and assembled at the launch site. A failure of a seal between two segments was responsible for the loss of *Challenger;* those seals have since been redesigned. Completely assembled, the Space Transportation System has a maximum liftoff weight of some 4,500,000 lb. (2,045,454 kg.).

The shuttle is launched from Kennedy Space Flight Center at Cape Canaveral, Florida. The air force built a second launch site at Vandenberg AFB, California, specifically for military missions, but this site was mothballed after the loss of *Challenger.*

The three main engines and both SRBs are ignited at liftoff. The SRBs burn for roughly 120 seconds and are then jettisoned at an altitude of some 28 mi. (45 km.); after descending by parachute, they are recovered and reconditioned for future use. The SMEs then propel the shuttle into orbit, at which point the ET is also jettisoned to burn up in the atmosphere. The two OMS engines are used for minor orbital adjustments and as retrorockets for reentry. Attitude control is provided by additional small thrusters at the nose and tail. After reentry, the shuttle normally glides to a landing at Edwards AFB, California; alternative landing sites are available at White Sands, New Mexico; Rota, Spain; and Kennedy Space Flight Center.

The Soviet Union began flight-testing its own shuttle *(Buran)* in 1988. Externally it is quite similar to the U.S. shuttle-orbiter, but *Buran* differs fundamentally in that the main engines are mounted on the very powerful *Energia* booster and not on the reusable shuttle. See also SATELLITES, MILITARY; SPACE, MILITARY USES OF.

SPANDREL (AT-5): NATO code name for the Soviet 9M 113 *Konkurs* ANTI-TANK GUIDED MISSILE (ATGM) introduced in the early 1970s. Intended mainly for vehicles rather than use on foot, Spandrel has replaced the AT-3 SAGGER on BRDM-2 armored cars, and on some BMP infantry fighting vehicles. Spandrel has been exported only to Czechoslovakia, East Germany, and Poland.

The Spandrel system consists of the missile, housed in a sealed launcher/storage canister, and a sight/tracker unit with a guidance computer. Similar in many respects to the Euromissile HOT, the Spandrel is believed to incorporate illegally transferred Western technology. Powered by a dual-impulse solid rocket engine, Spandrel is armed with an 8.8-lb. (4-kg.) shaped-charge (HEAT) warhead capable of penetrating up to 29 in. (750 mm.) of homogeneous steel armor.

The missile is controlled by WIRE GUIDANCE with semi-automatic command to line-of-sight (SACLOS). In action, the operator first centers the target in his sight, and then launches the missile. The guidance computer determines the angle between the missile's course and the operator-target line of sight, and generates steering commands to bring the missile onto that line. The U.S. Army estimates that Spandrel has an 89–93 percent chance of hitting a stationary tank, but combat experience suggests a much lower hit probability. The drawbacks of Spandrel, common to most ATGMs, include a long flight time, the exposure of the operator to enemy counterfire, interruption of the line of sight by smoke and terrain masking, and degradation of its HEAT warhead by REACTIVE ARMOR.

Specifications   **Length:** 3.28 ft. (1 m.). **Diameter:** 5.1 in. (130 mm.). **Weight, launch:** 26.45 lb. (12 kg.). **Speed:** 185 m./sec. (414 mph). **Range envelope:** 100–4000 m. **Flight time:** 22 sec.

SPANKER (SS-17):   NATO code name for the Soviet RS-16 "lightweight" ICBM. Introduced in 1975 as an intended replacement for the SS-11 SEGO, the SS-17 was developed in competition with the less innovative but larger SS-19 STILETTO. Spanker was deemed the less effective of the two, and only 150 were eventually deployed in hardened silos at Kostroma and Yedrovo in the northwestern U.S.S.R. The SS-17's advanced features, including its navigating inertial guidance and COLD LAUNCH technique, were used on subsequent Soviet ICBMs.

A two-stage missile, like most other Soviet liquid-fuel missiles, Spanker burns a hypergolic bipropellant mix of undimensional dimethyl hydrazine (UDMH) and nitrogen tetroxide (NTO). The first stage has two gimballed main engines for steering and attitude control, while the second stage has a single fixed engine and two gimballed vernier engines. Spanker was built in three distinct versions. The original Mod 1 had four 750-kT MULTIPLE INDEPENDENTLY TARGETED REENTRY VEHICLES (MIRVs) mounted on a solid-fuel POST-BOOST VEHICLE (PBV). The Mod 2 of 1977, with a single 6-MT REENTRY VEHICLE (RV), was apparently intended for attacks on hardened command bunkers, while the Mod 3 of 1980 carries four 450-kT MIRVs. The RV release sequence is unusual, in that the RVs are not spin-stabilized, and the PBV makes a lateral separation maneuver with its small auxiliary rockets. This method is inherently less accurate than the more conventional "spin-up and back-away" sequence of other Soviet ICBMs.

Spanker's navigating INERTIAL GUIDANCE also differs from the traditional Soviet "fly-by-wire" technique. In the latter, the missile is constrained to a preprogrammed pitch-velocity profile by thrust-magnitude control. In the navigating inertial technique, by contrast, the guidance unit continuously computes the instantaneous velocity vector needed to hit the target, and steers the missile accordingly. While both methods work equally

well for liquid-fuel missiles, the navigating inertial technique is essential for accurate guidance of solid-fuel missiles, and in fact the guidance unit developed for the SS-17 was used as the basis for those of the solid-fuel SS-16 SINNER, SS-24 SCALPEL, and SS-25 SICKLE.

Spanker is housed in a launcher/storage canister which is lowered into the silo as a complete unit. At launch, a cold-gas generator ejects the missile from the silo to a height of some 150 ft. (45 m.) before the main engines ignite. The empty canister can then be removed, and the silo reloaded. See also ICBMS, SOVIET UNION; RVSN.

Specifications   **Length:** 78.75 ft. (24 m.). **Diameter:** 8.2 ft. (2.5 m.). **Weight, launch:** 143,000 lb. (65,000 kg.). **Range:** ($^1/_3$) 6200 mi. (10,000 km.); (2) 6850 mi. (10,960 km.). **CEP:** (1) 440 m.; (2) 425 m.; (3) 350 m.

SPARROW: The Raytheon AIM-7 radar-guided, medium-range air-to-air missile (AAM), first developed for the U.S. Navy in the 1950s, and still the standard medium AAM for many Western air forces. A naval surface-to-air variant (RIM-7) has been developed into the NATO Sea Sparrow (see POINT DEFENSE MISSILE SYSTEM), while the Sparrow airframe has also served as the basis for the British SKYFLASH and Italian ASPIDE missiles. A land-based derivative (Sparrowhawk) is in development.

The initial AIM-7A Sparrow I entered service with the U.S. Navy in 1956. Intended as an anti-bomber weapon, it relied on BEAM-RIDING guidance and had a maximum range of roughly 2 mi. (3.2 km.); some 2000 were produced between 1953 and 1957. A radar ACTIVE HOMING variant, the AIM-7B Sparrow II, was developed in 1955, but not placed in production. Instead, in 1958 the navy opted in 1958 for the SEMI-ACTIVE RADAR HOMING (SARH) AIM-7C Sparrow III, which has formed the basis of all subsequent versions.

The AIM-7C, some 2000 of which were built, formed the primary armament of the navy's F-3 Demon and F-4B PHANTOM carrier-based interceptors. It was superseded in 1960 by the AIM-7D, with a prepacked liquid-fuel rocket engine, and this version was also adopted by the U.S. Air Force for its F-4C Phantoms; some 7500 were produced. A shipboard version, the RIM-7C Sea Sparrow, armed the navy's Basic Point Defense Missile System (BPDMS).

The AIM-7D was superseded in turn by the AIM-7E, which reverted to a solid-fuel rocket with a higher maximum speed and a slightly increased range. It also introduced a more effective continuous-rod warhead in place of the earlier blast-fragmentation warhead. Some 25,000 AIM-7Es were produced by 1977. Many were expended during the Vietnam War, but their combat performance was very disappointing. Essentially designed to destroy large, unmaneuverable bombers, the Sparrow performed poorly against small, nimble fighters. Both operational and technical constraints usually precluded its use for beyond-visual range (BVR) intercepts, while pilots found it difficult to obtain good firing positions at shorter ranges. Throughout the Vietnam War, the kill probability (P_k) for Sparrows was 3–5 percent. In response, Raytheon introduced the AIM-7E2, with a shorter minimum range and greater maneuverability; effectiveness improved modestly. Nevertheless the AIM-7E is still in service with the F-104S STAR-FIGHTER, while the shipboard RIM-7E arms the NATO Sea Sparrow system.

The RIM-7F, introduced in 1977, has all solid-state electronics for greater reliability, a larger warhead, and a more powerful rocket engine that doubles the maximum range. A new seeker head, meant to give better performance at low altitudes, is compatible with the pulse-Doppler radars of the F-15 EAGLE and FA-18 HORNET. Used by the Israeli air force in the 1982 Lebanon War, AIM-7Fs achieved a P_k of 15–20 percent, and were more useful to break up enemy formations than to actually destroy aircraft. Some 3000 were produced through 1982.

The current production version, the AIM-7M, has an improved seeker with digital monopulse homing. Current plans call for the production of some 2000 missiles, after which Sparrow will be superseded by the AIM-120 AMRAAM.

The AIM-7M has an ogival nose and cylindrical body, with four cruciform wings at midbody for steering and lift, and four smaller, fixed tail fins. A nose radar-receiver antenna detects target echoes from the monopulse illumination radar of the launch aircraft, while a secondary reference antenna in the the tail receives the unreflected radar signal. Comparison of the two signals allows the missile to determine the relative velocity of the target by measuring the DOPPLER shift. The guidance electronics are located immediately behind the radar receiver, ahead of the 88-lb. (40-kg.) Mk.71 CONTINUOUS ROD WARHEAD, which is fitted with both contact and radar-proximity (VT) fuzes. The missile is powered by a Hercules or Aerojet Mk.58 solid-fuel rocket engine. Minimum range,

constrained by guidance and fuzing, is roughly 1 mi. (1.6 km.). One unfortunate feature of the Sparrow's engine is its prominent smoke trail, which can reveal the launch aircraft's position and warn the intended target.

To launch Sparrow, the fighter must first acquire the target on its radar, and then lock onto it; tracking proceeds with the radar in the illumination mode. Once lock-on is confirmed, the missile is launched and begins receiving radar echoes from the target. The launch aircraft must continue to illuminate the target throughout the engagement, and since nose radars have scans limited to some 60° left and right of boresight, this means that the fighter must continue flying towards the target, nullifying some of the Sparrow's range advantage. Moreover, while illuminating the target, the launch aircraft cannot engage in radical evasive maneuvers. In spite of its defects, Sparrow remains useful against bombers, and to break up formations of fighters.

Specifications   **Length:** 12.1 ft. (3.69 m.). **Diameter:** 8 in. (203 mm.). **Span:** 40 in. (1.02 m.). **Weight, launch:** (D) 440 lb. (200 kg.); (E) 452 lb. (205 kg.); (F/M) 503 lb. (229 kg.). **Speed:** Mach 4 (2600 mph/4160 kph). **Range:** (C) 25 mi. (40 km.); (E) 28 mi. (45 km.); (F/M) 62 mi. (100 km.). **Operators:** FRG, Gre, Iran, Isr, It, Jap, ROK, Sp, Tur, UK, US.

SPARTAN: 1. The U.S. XLIM-49A ANTI-BAL-LISTIC MISSILE (ABM) interceptor missile, associated with the abortive SENTINEL and SAFE-GUARD ABM systems. Developed from 1959 as part of the earlier Nike X program, Spartan was meant to function as a long-range, exoatmospheric interceptor, to complement the high-speed, short-range SPRINT endoatmospheric interceptor. Flight tests began in 1968, and the missile was ready for deployment by 1970. It was duly deployed at the Safeguard ABM site at Grand Forks, North Dakota, and declared operational on 1 October 1975, but the entire system was deactivated the following day, in accordance with U.S. policy after ratifying the 1972 ABM TREATY. The missiles were stockpiled and finally dismantled in 1983. Much of the research and development for Spartan, however, was been applied in later ABM development.

2. Armored personnel carrier variant of the British SCORPION tracked armored reconnaissance vehicle.

SPARVIERO: A class of seven Italian missile-armed HYDROFOILS, commissioned between 1974 and 1984. In the early 1970s, the Italian navy ordered eight U.S-designed PEGASUS-class hydrofoils, but canceled the order when costs became prohibitive, in favor of the smaller, simpler Sparvieros. Though short-ranged and expensive to operate, they are well suited for the restricted waters of the Adriatic and central Mediterranean.

Similar in configuration to the Pegasus (both were derived from the Boeing Tucumcari design), their conventional, aluminum V-bottom hulls are equipped with retractable, fully submerged foils in a "canard" configuration, with one small foil at the bow and two larger foils at the stern. A small deckhouse is located right aft. Armament consists of a single OTO-MELARA 76.2-mm. "Compact" DUAL PURPOSE gun on the bow, and two canister launchers for OTOMAT Mk.II anti-ship missiles on the fantail. See also FAST ATTACK CRAFT, ITALY.

Specifications Length: 75.4 ft. (23 m.). **Beam:** 22.9 ft. (7 m.). **Draft:** 5.2 ft. (1.6 m.). **Displacement:** 62.5 tons full load. **Powerplant:** CODOG: 1 180-hp. GM 6V-53N diesel driving a single propeller (hull-borne), 1 5044-shp. Rolls Royce Proteus gas turbine driving a waterjet (on foils). **Fuel:** 11 tons. **Speed:** 8 kt. (hull-borne)/50 kt. (on foils). **Range:** 1050 n.mi. at 8 kt. (hull-borne)/400 n.mi. at 45 kt. (on foils). **Crew:** 10. **Sensors:** 1 SPS-701 surface-search radar, 1 RTN-10X fire control radar. **electronic warfare equipment:** radar warning receivers.

SPEARFISH: British submarine-launched, wire-guided acoustical homing TORPEDO, successor to the Mk.24 TIGERFISH. Spearfish has been developed by Marconi to a 1975 Royal Navy requirement for a heavy anti-ship/anti-submarine torpedo with greater speed, range, and reliability than Tigerfish, capable of catching and sinking the latest Soviet nuclear submarines, most of whose double hulls can withstand smaller torpedo warheads. The first trials were carried out in 1983, and the Royal Navy accepted the first production models in 1987 for in-service trials.

Spearfish has a classic torpedo configuration, with a round nose, cylindrical centerbody, and tapered tail with four cruciform fins. The nose houses an advanced active/passive SONAR and guidance computer originally developed for Marconi's lightweight STINGRAY torpedo. A 550-lb. (250-kg.) SHAPED-CHARGE warhead mounted immediately behind the guidance unit should be capable of penetrating even the largest Soviet submarines. The centerbody contains solid monopropellant (OTTO) fuel for a Sundstrand 21TP01 gas turbine engine in the tail, which drives a shrouded pump-

jet, rather than a traditional propeller. A low-speed, multi-blade turbine mounted between two sets of stator blades inside an annular duct, a pumpjet is considerably quieter than a propeller, especially at high speeds. The tail unit also houses an inertialess reel for the guidance wires.

For long-range attacks, the torpedo would initially move at low speed, to search passively with its own sonar or by using data relayed through the guidance wires from the submarine's FIRE CONTROL unit. If the target is acquired by the torpedo's sonar, the wires can be severed while the torpedo shifts to its high-speed mode, using active sonar in the terminal phase of the engagement. Alternatively, the torpedo can be launched without wire guidance, to search and home autonomously. Its guidance computer can be programmed to run a variety of search patterns and is designed to achieve physical contact with the target to maximize the effect of the shaped-charge warhead. If it performs to specification (and torpedoes are notorious for erratic performance), Spearfish will finally provide British submarines with a weapon which can fully exploit their excellent sensors and fire controls.

Specifications Length: 27.9 ft. (8.5 m.). **Diameter:** 21 in. (533 mm.). **Weight, launch:** 4400 lb. (2000 kg.). **Speed:** 24/65 kt. **Range:** 22.7 n.mi. (high)/14.25 n.mi. (low).

SPECIAL FORCES: 1. A common designation for elite units in many armies.

2. The "Green Berets," a SPECIAL OPERATIONS force of the U.S. Army. When first established, the all-volunteer Special Forces were meant to operate behind enemy lines (e.g., in Eastern Europe) to wage GUERRILLA warfare and assist local irregulars within the context of "regular" warfare, along the lines of the partisans of World War II. The first such unit, the 10th Special Forces Group (SFG), raised in 1952 at Fort Bragg, North Carolina, was followed in 1953, 1957, and 1961 by the 77th, 1st, and 5th SFGs, respectively.

When the Special Forces were sent to Vietnam, however, their mission changed from insurgency to counterinsurgency, and their units raised and trained tribal groups to oppose the Vietcong and North Vietnamese forces. A significant portion of their activity was devoted to "Civil-Military Affairs" and the "Hearts-and-Minds" program, which included the establishment of village schools and medical clinics.

The Special Forces had captured the imagination of President Kennedy, who ordered a massive expansion, with a concomitant loss in personnel quality. When the U.S. withdrew from Indochina in 1973, the Special Forces concept was discredited by association, and there was a substantial reduction in its establishment. Special operations forces were emphasized once more from the early 1980s, and the Special Forces were restored to a strength of eight SFGs, including two in the Army Reserve and one in the National Guard. In 1984, they were grouped into two Special Forces battalions under the army's First Special Operations Command.

A Special Forces Group consists of some 250 men organized into "A-," "B-," and "C-" teams. The basic tactical unit is the 12-man A-Team, consisting of a captain, a lieutenant, and 10 noncommissioned officers (NCOs), each of whom has a technical specialty (operations, heavy weapons, communications, engineering, or medical). Each A-Team is in theory capable of organizing and leading a 650-man guerrilla force, but they can also conduct independent patrols and raids. Each B-Team, commanded by a major, consists of 3 A-Teams plus a headquarters group of 5 officers and 18 NCOs. A Special Forces Group, commanded by a lieutenant colonel, consists of administrative overhead and an operational C-Team of 3 B-Teams.

Special Forces personnel are selected, parachute-qualified volunteers, and each must possess at least two of the recognized special skills. All selectees are subjected to rigorous physical training followed by additional training in foreign languages, transcultural communications, first aid, the use of foreign weapons, unconventional warfare, etc.

SPECIAL OPERATIONS: U.S. term for both commando operations—unorthodox, relatively low-cost, high-payoff surprise combat actions in enemy-controlled territory—and also insurgency and counterinsurgency. Specific modes of the former include clandestine SCOUTING, RAIDS and sabotage, TERRORISM and counterterrorism, PSYCHOLOGICAL WARFARE, and assassination. The latter are in fact radically different, and require radically different skills, notably transcultural communications. This terminological confusion reflects a substantive confusion of purposes within the U.S. military.

Commando operations are normally conducted by elite forces with specific training and equipment, e.g., the U.S. RANGERS, SEALS, and DELTA FORCE; the British SAS and SBS; the Israeli SAYARET MATKAL; and the Soviet SPETSNAZ. The U.S. SPECIAL FORCES ("Green Berets") are unique in being specialized for cooperation with indigenous forces for both insurgency and counterinsurgency.

SPECIAL OPERATIONS COMMAND, U.S. JOINT (JSOC):   A joint command of the U.S. armed forces, responsible for coordination of certain U.S. special operations forces (the army's DELTA FORCE and 160th Aviation Group, the navy's SEAL Team 6, and the air force's 2nd Air Division) for counterterrorism and hostage rescue. Formed at Fort Bragg, North Carolina, in 1981, the command is also responsible for training, deployment, and doctrinal development for the counterterrorism and hostage-rescue mission.

The U.S. Joint Special Operations Command is thus distinct from the U.S. Army's First Special Operations Command (SOCOM), formed in 1983 (also at Fort Bragg). SOCOM controls all active army SPECIAL OPERATIONS forces in the U.S. (except those under JSOC), including four SPECIAL FORCES Groups, the RANGER Regiment, an aviation group, a PSYCHOLOGICAL WARFARE group, a civil affairs battalion, and an INTELLIGENCE battalion.

SPECIFIED COMMAND: U.S. term for a command with broad and continuing missions, composed of forces from a single military service, as opposed to UNIFIED COMMANDS composed of two or more services. At present, there are two specified commands: the U.S. Army Forces Command (FORSCOM), and the U.S. Air Force STRATEGIC AIR COMMAND (SAC).

SPETSNAZ: *Spetsial'noye   Naznacheniye* (lit., "Troops of Special Designation"), the COMMANDO forces of the Soviet Union. Special units for long-range, clandestine SCOUTING, RECONNAISSANCE, sabotage, and GUERRILLA operations have been part of the Soviet armed forces since 1918. They played a significant role in the Second World War, when NKVD Osnaz (*Voiska Osnogo Naznacheniye*, or "Special Purpose Troops") organized, armed, and led partisan bands behind German lines. Such units never disappeared from the Soviet order of battle, but the current Spetsnaz were probably formed in the mid-1950s, as an adjunct to high-speed operations on the nuclear battlefield.

Spetsnaz did not come to Western attention until the 1968 invasion of Czechoslovakia, and it was not until the late 1970s that the total extent of its threat to Western security was realized.

The primary mission of Spetsnaz is *razvedka*, or "special reconnaissance," a term which includes not simply information-gathering, but also sabotage, raiding, diversionary actions, assassination, and other disruptive behind-the-lines actions. In wartime, their highest priority would be the identification and destruction of enemy "weapons of mass destruction" (nuclear and chemical stockpiles and their delivery systems), either by direct attack or by providing targeting data for long-range weapons. Of equal importance would be the disruption of enemy command centers and communications links, and the killing of key enemy leaders. Spetsnaz would spread confusion in rear areas by attacking supply units and spreading disinformation (with troops disguised in enemy uniforms). Diversionary raids would also be mounted to hide Soviet intentions (see MASKIROVKA).

In contingency operations, Spetsnaz would secure airports and other transit facilities to open the way for follow-on forces, and carry out sabotage and assassinations to disrupt local resistance (as in the 1979 invasion of Afghanistan). In Afghanistan, they were also pressed into service for counterinsurgency operations against the Mujahideen, and were generally held to be the most effective Soviet forces. Elsewhere, Spetsnaz have carried out penetrations of sensitive area by naval, air, and ground infiltration.

The Soviet Union maintains some 30,000 Spetsnaz, a huge number for elite troops of this kind, and indicative of the importance placed on such operations. Spetsnaz activities are coordinated by the Second (Intelligence) Department of the Soviet General Staff, which also controls directly a Long-Range Reconnaissance Regiment responsible for strategic missions on behalf of the GRU (Main Intelligence Directorate). Each Theater of Military Operations (TVD) has an attached Spetsnaz regiment of 600–700 men. Each regiment comprises a headquarters, a signals company and support units, and six or seven Spetsnaz anti-VIP companies, intended for the elimination of enemy political and military leaders.

Each FRONT, MILITARY DISTRICT, and Group of Forces Abroad has a Spetsnaz "brigade" of up to 1000 men, controlled by the Second (intelligence) Directorate of the Front, District, or Group. Each Spetsnaz brigade consists of a headquarters, a signals company and support units, an anti-VIP company, and three or four 360-man Spetsnaz battalions. Each battalion in turn is organized into a headquarters, a signal platoon, and three 115-man Spetsnaz companies, each of three parachute platoons (30 men) plus support elements. Each Army also has an attached Spetsnaz company, controlled by the Second (intelligence) Department of the Army headquarters.

The Soviet navy has its own Spetsnaz. Each of the four main fleets has a Naval Spetsnaz brigade controlled by the Second (intelligence) Department of the Fleet HQ. Each brigade consists of a headquarters, a signals company and support units, an anti-VIP company, a midget submarine group (see MINISUBS), a parachute battalion, and two or three diver battalions.

Spetsnaz enlisted personnel are selected from the annual conscript intake. Recruits are carefully screened for physical fitness, intelligence, and political reliability. Officers are selected from line units by Spetsnaz and GRU "talent scouts," and are also subjected to extensive physical and psychological screening. Training is intense, comprehensive, and (by Western standards) quite brutal. Spetsnaz are trained in foreign languages, foreign weapons and equipment, demolitions and booby traps, unarmed combat, silent killing, parachuting (including HALO), survival, and evasion. During annual exercises, Spetsnaz units are often pitted against KGB security troops in mock raids and reconnaissance missions. The casualty rates of these exercises would be considered prohibitive by Western armies, but they produce tough, intelligent, and resourceful troops.

In addition to standard Soviet small arms, Spetsnaz are also equipped with mines and demolition charges, silenced pistols, and exotic weapons such as a spring-loaded knife for silent killing at ranges out to 10 m.

A unique aspect of Spetsnaz is their enrollment of world-class athletes for the covert penetration of potential wartime operating areas. Most Soviet sport parachutists, divers, cross-country skiers, and marksmen have an Army affiliation and in fact belong to Spetsnaz; they are generally assigned to the elite anti-VIP companies. Covert operations are supported by specific Spetsnaz agent networks of foreign nationals who supply Spetsnaz with safe houses, access to special equipment, and tactical intelligence.

In order to maintain their cover, Spetsnaz units are always collocated with line formations (usually airborne, air-mobile, or air assault units), and wear their uniforms. Within the Soviet army, they are known as *Raydoviki* (Raiders), *Okhotniki* (Hunters), or *Vysotniki* (High-Altitude Troops), while the term Spetsnaz is never used.

The KGB maintains its own special operations troops (reportedly still called Osnaz), to carry out high-level assassinations and sabotage in both peace and war.

SPIDER (SS-23): NATO code name for the Soviet OTR-23 short-range (tactical) BALLISTIC MISSILE. Introduced in the early 1980s as a replacement for the SS-1c SCUD, Spider is intended as a delivery system for nuclear, chemical, and cluster warheads at the Army and FRONT levels. As compared with its predecessor, Spider has greater range, greatly improved accuracy, and a much faster reaction time. By the late 1980s, the Soviet army had deployed some 78 launchers and 167 missiles in 6 brigades of 12 to 18 launchers. One brigade was to be attached to each Army, and 2 or more were to be attached to each Front, but the SS-23 is now being withdrawn under the terms of the 1987 INF TREATY.

A single-stage solid-fuel missile, believed to be derived from the third stage of the SS-16 SINNER ICBM, Spider's casing is fabricated of wound fiberglass to reduce weight. The engine has two gimballed nozzles for steering and attitude control, supplemented by four small fins for aerodynamic steering. Payload options include 200-kT nuclear and 2200-lb. (1000-kg.) high-explosive, cluster, or chemical warheads. Spider's INERTIAL GUIDANCE unit is much more accurate than that of the Scud, greatly reducing the median error radius (CEP) when fired from presurveyed sites. Unlike Scud, the SS-23 is thus suitable for nuclear attacks on hardened targets and nonnuclear attacks on larger targets such as airfields.

The SS-23 is launched from a self-propelled transporter-erector-launcher (TEL) based on the MAZ-543 8 × 8 truck. The missile is stowed horizontally in an environmentally controlled compartment of the TEL for traveling, and is raised vertically off the rear of the vehicle for launching.

Specifications Length: 24.66 ft. (5.51 m.). Diameter: 3.02 ft. (0.92 m.). Weight, launch: 10,987 lb. (4995 kg.). Range envelope: 50–311 mi. (80–500 km.). CEP: 280 m.

SPIGOT (AT-4): NATO code name for the Soviet *Faggot* ANTI-TANK GUIDED MISSILE (ATGM). Introduced in the mid-1970s to replace the AT-3 SAGGER, Spigot has become the standard Soviet infantry ATGM, and has also been fitted to BMP and BMD infantry fighting vehicles, as well as to some BRDM armored cars outfitted as tank destroyers. Spigot is also in service in other Warsaw Pact armies, and has been exported to Finland, Iraq, Libya, Syria, and the Polisario guerrilla movement.

The Spigot system consists of the missile in a sealed launcher/storage canister, and a sight/tracker unit. In performance and general configuration, it is quite similar to the Euromissile MILAN, and may in fact be based on "transferred" technology. Powered by a dual-impulse solid rocket engine, Spigot is armed with a 6.6-lb. (3-kg.) shaped-charge (HEAT) warhead capable of penetrating 23.5 in. (600 mm.) of homogeneous steel armor. The sight/tracker unit incorporates a periscopic optical sight (as on Milan), which allows the operator to remain prone. The launcher and sight unit are both mounted on a light tripod or pintle mount.

Spigot is controlled by WIRE GUIDANCE with semi-automatic command to line-of-sight (SACLOS). The operator must keep the target centered in his sight throughout the engagement. As the missile flies downrange, the guidance computer measures the angle between the missile's course and the line of sight to the target, and generates steering commands to bring the missile onto that line. The U.S. Army estimates that Spigot has an 89–92 percent chance of hitting a stationary tank at ranges between 300 and 2000 m., but experience in the 1982 Lebanon War indicates that the actual hit probability is much lower.

The main drawbacks of Spigot, common to most ATGMs, include its relatively long time of flight, the exposure of the operator to counterfire, the interruption of the line of sight by smoke or terrain masking, and the degradation of its small HEAT warhead by REACTIVE ARMOR or modern composite (CHOBHAM) armor.

Specifications Length: (canister) 3.93 ft. (1.2 m.); (missile) 3.21 ft. (970 mm.). Diameter: (canister) 5.1 in. (130 mm.); (missile) 4.7 in. (120 mm.). Weight: (system) 88.5 lb. (40.25 kg.); (missile) 15.4 lb. (7 kg.). Speed: 185 m./sec. (414 mph). Range envelope: 70–2000 m. Flight time: 11 sec.

SPIRAL (AT-6): NATO code name for a large Soviet ANTI-TANK GUIDED MISSILE (ATGM) introduced in 1978 as the main armament of the Mi-24 HIND D/E and Mi-8 HIP E combat helicopters. The AT-6 system consists of the missile in a sealed launcher/storage canister mounted on twin or triple launch racks, and a sight/tracker unit (operated by the weapon system officer in Hind).

Roughly analogous to the U.S. AGM-114 HELLFIRE, and powered by a dual-impulse solid rocket motor, Spiral is armed with a 22-lb. (10-kg.) shaped-charge (HEAT) warhead capable of penetrating 31.5 in. (800 mm.) of homogeneous steel armor. Spiral relies on jam-resistant radio COM-

MAND GUIDANCE with semi-automatic command to line-of-sight (SACLOS). The operator acquires the target and centers it in his sight. The guidance computer then measures the angle between the missile's course and the line of sight to the target, generating steering commands to bring the missile onto that line. The shortcomings of Spiral, common to most ATGMs, include a relatively long time of flight at maximum range, the exposure of the launch platform to counterfire, the interruption of the line of sight by smoke or terrain masking, and the degradation of its HEAT warhead by REACTIVE ARMOR or modern composite (CHOBHAM) armor; Spiral's radio command link also remains vulnerable to ACTIVE JAMMING and other ELECTRONIC COUNTERMEASURES.

Specifications   Length: 5 ft. (1.53 m.). **Diameter:** 5.5 in. (140 mm.). **Weight, launch:** 70.5 lb. (32 kg.). **Speed:** 450 m./sec. (1006 mph). **Range:** 5000–7000 m. **Flight time:** 20–25 sec.

SPOT JAMMING: A form of ACTIVE JAMMING in which disruptive signals are broadcast on a narrow frequency band to mask hostile radars and radios operating on that frequency. Spot jamming is more power-efficient than BARRAGE JAMMING (which masks entire wave bands), but requires more precise information on enemy operating frequencies. Spot jamming can be countered by "frequency-agile" transmitters which shift automatically between a number of frequencies over a broad wave band. Multi-spot jammers use broadband frequency scanners to detect these shifts and adjust the jamming frequency accordingly. See also ELECTRONIC COUNTERMEASURES.

SPOTTING: The process of determining, by visual or electronic observation, the deviation of ARTILLERY or naval gunfire from an imaginary line running from observer to target (the "O-T Line"), and range errors, to provide data for the adjustment of fire. With DIRECT FIRE, the observer is usually collocated with, or close to the weapon. For INDIRECT FIRE, spotting is provided by FORWARD OBSERVERS, by FORWARD AIR CONTROLLERS, or, since quite recently, by REMOTELY PILOTED VEHICLES; the latter advantageously replace manned aircraft for spotting in heavily defended areas.

SPRINT: U.S. ANTI-BALLISTIC MISSILE (ABM) interceptor missile, associated with the abortive SAFEGUARD ABM system. The short-range, hypervelocity, endoatmospheric complement to the long-range, exoatmospheric SPARTAN, Sprint was developed from 1959 as part of the Nike X program. Its design was finalized by 1965, and flight

tests began in that same year. Sprint was ready by 1971, and was indeed deployed at the Safeguard ABM site at Grand Forks, North Dakota, but only for one day: declared operational on 1 October 1975, the site was deactivated the following day. Placed in storage, the missiles were finally dismantled in 1983, but much of their technology was later applied to the LOADS and HEDI missiles.

SPRUANCE: A class of 31 U.S. ANTI-SUBMARINE WARFARE (ASW) DESTROYERS commissioned between 1975 and 1983. Developed in the 1960s to replace for aging World War II destroyers, the much larger Spruances form the "high" end of a "high-low mix" with the PERRY-class guided-missile FRIGATES: the Spruances would escort high-speed carrier battle groups, while the Perrys would escort merchant convoys. The Spruances were designed to accept modular armament and sensors, to be tailored either for ASW or ANTI-AIR WARFARE (AAW). Because of budget limitations, the U.S. Navy did not procure the latter version, but four were ordered by the Shah of Iran and later purchased by the U.S. as the KIDD class. The Spruance hull and powerplant have proven to be extremely adaptable because of built-in growth margins. Notably, after the cancellation of the nuclear-powered "Strike Cruiser," the U.S. Navy selected the Spruance hull for its TICONDEROGA-class cruisers. Large growth margins have also allowed the Spruances to accommodate substantial modifications; they are universally regarded as the most powerful ASW escorts in service.

The Spruances are flush-decked, except for a short, recessed quarterdeck, and have a large, blocky superstructure amidships with two stacks offset to port and starboard. The machinery is highly automated and normally operated from a single remote control station. In keeping with the ASW role, the powerplant is extensively soundproofed, and a PRAIRIE MASKER bubble generator installed under the hull below the engine rooms reduces radiated noise which could interfere with sonar performance.

Armament as completed consisted of a 5-in. 54-caliber DUAL PURPOSE gun at each end of the ship; an 8-round ASROC launcher (with 16 reloads) forward of the bridge; 2 quadruple launchers for HARPOON anti-ship missiles amidships; a PHALANX 20-mm. radar-controlled gun for anti-missile defense at each end of the superstructure; 2 sets of Mk.32 triple tubes for MK.46 ASW homing torpedoes, launched through shutters in the sides of the hull; and an 8-round NATO SEA SPARROW short-range surface-to-air missile launcher aft. The primary ASW weapons, however, are 2 SH-60B SEAHAWK (LAMPS III) helicopters, operated from a flight deck and hangar amidships. RAST (Recovery Assist and Secure Traverse) haul-down gear allows safer flight operations in heavy seas. The ASROC launcher is being replaced by a 61-round Mk.41 VERTICAL LAUNCH SYSTEM (VLS), which can accommodate ASROC, Harpoon, and TOMAHAWK cruise missiles (STANDARD MR-2 surface-to-air missiles could be launched under remote guidance by an accompanying AEGIS ship). All Spruances also have three WSC-3 satellite communications terminals and an NTDS data link. See also CRUISERS, UNITED STATES; DESTROYERS, UNITED STATES.

Specifications **Length:** 563.2 ft. (171.7 m.). **Beam:** 55.1 ft. (16.8 m.). **Draft:** 29 ft. (8.85 m.). **Displacement:** 5770 tons standard/7810 tons full load. **Powerplant:** 4 20,000-shp. General Electric LM2500 gas turbine engines, 80,000 shp., 2 shafts. **Fuel:** 1650 tons. **Speed:** 33 kt. **Range:** 3300 n.mi. at 30 kt./6000 n.mi. at 20 kt./8000 n.mi. at 17 kt. **Crew:** 296. **Sensors:** 1 SPS-55 surface-search radar, 1 SPS-40 air-search radar, 1 SPQ-9 A track-while-scan fire control radar, 1 SPG-60 missile guidance radar, 1 SQS-53 bow-mounted low-frequency sonar, 1 SQR-19 TACTASS passive towed array sonar. **Fire controls:** 1 Mk.116 ASW fire control unit, 1 Mk.91 missile fire control unit, 1 Mk.86 gun fire control unit (ships with Tomahawk also have SWG-3 fire controls). **Electronic warfare equipment:** 1 SLQ-32 (V) 2 electronic countermeasures array, 4 Mk.36 SRBOC chaff launchers, 1 Mk.25 Nixie torpedo countermeasures sled.

SQUAD: The smallest subunit of infantry, of 6–15 men under the command of a NONCOMMISSIONED OFFICER (NCO). Three or four squads and (often) a heavy weapons squad make up a PLATOON.

SQUADRON: A common designation for air force units subordinate to GROUPS or WINGS. Fighter/attack squadrons can have more than 20 aircraft (though only about a dozen in British, French, and Soviet-style air forces), while bomber, tanker, and transport squadrons can number fewer than 6.

SR-71: U.S. strategic reconnaissance aircraft. See BLACKBIRD.

SRAM: Boeing AGM-69 Short-Range Attack Missile, a nuclear standoff missile which arms B-52G/H STRATOFORTRESS, FB-111, and B-1 bombers of the U.S. STRATEGIC AIR COMMAND (SAC). SCRAM is a lightweight, high-performance missile with a range of more than 100 mi. (160 km.), developed from 1966 for the suppression of Soviet air defense sites during deep penetration (nuclear) bombing missions. Flight tests began in 1969, and the first of a total of 1500 SRAMs were produced in 1972; 1200 remain in service.

Very compact, SCRAM has a long, slender nose, a cylindrical body, and a tapered tail. Relying mainly on body lift, it has only three tail fins for steering. (Its frontal radar cross section is said to be equivalent to that of a .50-caliber/12.7-mm. bullet.) The payload is a 200-KT W69 nuclear warhead fitted with barometric and contact fuzes. SCRAM is powered by a dual-impulse (booster/sustainer) solid-rocket engine (the original Lockheed engines are now being replaced by improved Thiokol models with a longer shelf life). Controlled by a highly accurate Singer-Kearfott KT-76 INERTIAL GUIDANCE unit with radar altimeter, it can one of fly four basic flight profiles: semi-ballistic, terrain-following, "pop-up," and combined inertial/terrain following.

B-52s can carry up to 20 SRAMs, 8 on an internal rotary launcher and 12 more on wing pylons. The FB-111 can carry 6, 2 in the weapon bay and 4 on wing pylons. B-1s can carry up to 38 SRAMs (24 on three rotary launchers and 14 on eight external hardpoints).

An improved version, the AGM-69B, with a new engine, guidance system, and warhead, was proposed in 1977, but was not placed in production. Instead, SRAM will be replaced in the mid-1990s by a new missile, the AGM-131A SRAM II, now in development. Intended for the B-1B and B-2 bombers, SRAM II is powered by an advanced fuel-pulse solid-rocket engine. Armed with a 300-kT W80 warhead, SRAM II has an advanced inertial guidance unit for greater accuracy. Total production may exceed 1600 missiles.

Specifications **Length:** (69) 15.38 ft. (4.7 m.); (131) 14 ft. (4.27 m.). **Diameter:** (69) 17.5 in. (444.5 mm.); (131) 16 in. (406 mm.). **Span:** 15 in. (381 mm.). **Weight, launch:** (69) 2230 lb. (1013.63 kg.); (131) 1800 lb. (818 kg.). **Speed:** (69) Mach 2.8 (1820 mph/2912 kph); (131) Mach 4.2 (2750 mph/4365 kph). **Range:** (69) 35–105 mi. (56–168 km.); (131) 50–200 mi. (80–320 km.). **CEP:** (69) 100 m.; (131) 75 m.

SRBOC: Super Rapid-Blooming Offboard Chaff, the U.S. Mk.36 shipboard countermeasures launcher, successor to the Mk.33/34 RBOC aboard U.S. and other Western warships. SRBOC is built in two versions. The Mk.36 Mod 1, intended for frigates and small combatants, comprises two 6-barrel mortars; the Mk.36 Mod 2, for destroyers and larger vessels, has four 6-barrel mortars. Both can fire the Mk.183 CHAFF cartridge to dispense a cloud of foil strips at a height of up to 800 ft. (244 m.) to deceive missile guidance radars. Other cartridges in development include the Torch INFRARED COUNTERMEASURES package, which deploys flares to decoy INFRARED HOMING missiles. SRBOC may be fired automatically by the ship's electronic countermeasures system, or manually from the COMBAT INFORMATION CENTER (CIC). SRBOC is to be replaced in the 1990s by the NATO Sea Gnat.

SS-: NATO prefix designator for Soviet land-based BALLISTIC MISSILES, notably ICBMS, IRBMS, MRBMS, and SRBMS with nuclear warheads. Models currently or recently in service include: SS-1c SCUD; SS-4 SANDAL; SS-11 SEGO; SS-12 SCALEBOARD; SS-13 SAVAGE; SS-16 SINNER; SS-17 SPANKER; SS-18 SATAN; SS-19 STILETTO; SS-20 SABER; SS-21 SCARAB; SS-23 SPIDER; SS-24 SCALPEL; and SS-25 SICKLE.

Land-based missiles no longer in service include the SS-5 Skean, SS-6 Sapwood, SS-7 Saddler, SS-8 Sasin, SS-9 Scarp, and SS-10 Scrag; the SS-5, SS-6 and SS-9 are still used as space launch vehicles. The designation SS-22 was originally assigned to a modified Scaleboard later designated SS-12b. See, more generally, IRBMS, SOVIET UNION; ICBMS, SOVIET UNION; RVSN.

SS-N-: NATO prefix designator for all Soviet ship- and submarine-launched surface-to-surface missiles, including submarine-launched ballistic missiles (SLBMs), ship- and submarine-launched long-range cruise missiles (SLCMs), ship- and submarine-launched anti-ship missiles, and ship- and submarine-launched anti-submarine missiles. Models currently or recently in service include: SS-N-2 STYX anti-ship missile; SS-N-3 SHADDOCK anti-ship missile; SS-N-6 SAWFLY SLBM; SS-N-7 SIREN anti-ship missile; SS-N-8 SLBM; SS-N-9 STARBRIGHT anti-ship missile; SS-N-12 SANDBOX anti-ship missile; SS-N-14 SILEX anti-submarine missile; SS-N-15 STARFISH anti-submarine missile; SS-N-16 STALLION anti-submarine missile; SS-N-17 SNIPE SLBM; SS-N-18 STINGRAY SLBM; SS-N-19 SHIPWRECK anti-ship missile; SS-N-20 STURGEON SLBM; SS-N-21 SAMPSON SLCM; SS-N-22 SUNBURN anti-ship missile; SS-N-23 SKIFF SLBM; and SS-N-24 SLCM.

SS-N-8: NATO designation for a Soviet submarine-launched ballistic missile (SLBM), now the primary armament of DELTA I (12 missiles) and II

(16 missiles) nuclear-powered ballistic-missile submarines (SSBNs). Developed in the mid-1960s, the SS-N-8 began flight tests in 1969, followed by sea trials aboard converted GOLF- and HOTEL-class submarines. Since its operational debut in 1972 it has been upgraded several times.

Like its predecessor, the SS-N-6 SAWFLY, the SS-N-8 is a two-stage, liquid-fuel missile, which burns a storable, hypergolic bipropellant mix of undimensional dimethyl hydrazine (UDMH) and nitrogen tetroxide (NTO); despite the apparent danger of such volatile fuels in an enclosed submarine, no accidents are known. The first stage appears to have four main engines with some form of thrust-vector control. The second stage has a single main engine and two gimballed verniers for steering and attitude control.

Armed with a single 800-kT REENTRY VEHICLE, the original SS-N-8 Mod 1 was superseded in 1977 by the longer-ranged Mod 2. Both versions are controlled by INERTIAL GUIDANCE with stellar update ("stellar-inertial") to improve accuracy. After launch, a star sensor in the nose locks onto one or more celestial objects to obtain a position fix, in order to correct launch point errors in the submarine's INERTIAL NAVIGATION. It is believed that the SS-N-8 employs the traditional Soviet "fly-by-wire" technique, whereby the missile is constrained to a preprogrammed velocity-pitch profile through thrust magnitude control. Because of its relatively poor accuracy, the SS-N-8 could be used effectively only against cities and other area targets. See also SLBMS, SOVIET UNION.

Specifications Length: 42.3 ft. (12.9 m.). Diameter: 5.45 ft. (1.67 m.). Weight, launch: 66,000 lb. (30,000 kg.). Range: (1) 4345 mi. (7750 km.); (2) 5655 mi. (9050 km.). CEP: (1) 1410 m.; (2) 1550 m.

SSN-21: U.S. nuclear-powered attack submarine. See SEA WOLF.

SS-N-24: NATO designation for a large Soviet submarine-launched CRUISE MISSILE (SLCM), introduced in the early 1980s. Believed to be an airplane-configured missile with sharply swept wings and tail surfaces, the SS-N-24 is probably controlled by INERTIAL GUIDANCE and armed with a nuclear warhead in the 500-kT class. It is currently in limited service aboard a converted YANKEE-class ballistic-missile submarine; a specialized guided-missile submarine may be designed around this missile in the future. See also SUBMARINES, SOVIET UNION.

Specifications Length: 38.4 ft. (11.7 m.). Diameter: 3.3 ft. (1 m.). Weight, launch: 16,000 lb.

(7272 kg.). **Speed:** Mach 3.1 (2000 mph/3200 kph). **Ceiling:** 72,000 ft. (21,950 m.). **Range:** 2400 mi. (3850 km.).

STABALLOY: An alloy consisting primarily of depleted uranium—tailings of U-238 (a very low-radiation isotope of uranium) left as a by-product of uranium processing for powerplant and weapon applications. Staballoy penetrators for ARMOR PIERCING ammunition, such as APDS and APFSDS, have been favored by the U.S. Army because of their very high density and pyrophoric effects on impact; more recently, Staballoy has been added to composite (CHOBHAM) armors.

STALLION (SS-N-16): NATO code name for a Soviet submarine-launched anti-submarine missile introduced in 1972. Essentially an enlarged SS-N-15 STARFISH armed with a 17.7-in. (450-mm.) homing TORPEDO instead of a nuclear DEPTH CHARGE, Stallion is analogous to the U.S. SEA LANCE ASW-SOW.

The SS-N-16 can be launched from the outsized 650-mm. torpedo tubes of recent Soviet submarines (AKULA, CHARLIE II, MIKE, OSCAR, SIERRA, and VICTOR II/III). Powered by a solid-fuel rocket engine, Stallion relies on inertial guidance to reach target areas at the range and bearing determined by the submarine's active SONAR. After ejection from the torpedo tube, the missile rises to the surface, where the main engine ignites. It then flies a ballistic trajectory towards the estimated target position, where the torpedo separates, to enter the water with a braking parachute. Once in the water, the torpedo executes an autonomous search to acquire and attack the target. See also ANTI-SUBMARINE WARFARE.

Specifications **Length:** 21.3 ft. (6.5 m.). **Diameter:** 25.6 in. (650 mm.). **Weight, launch:** 4740 lb. (2155 kg.). **Speed:** Mach 1.5 (1050 mph/1680 kph). **Range:** 34.2 mi. (55 km.).

STANAVFORLANT: Standing Naval Force Atlantic, a NATO naval force of several destroyers offered by various members of the alliance, under the direct command of SACLANT, the NATO chief for the Atlantic. The stated purpose of STANAVFORLANT is to study the problems and prospects of multinational naval forces. See also ACLANT.

STANDARD ARM: U.S. AGM-78 air-launched ANTI-RADIATION MISSILE, a derivative of the RIM-66 STANDARD surface-to-air missile, developed from 1966 as a more powerful supplement to the AGM-45 SHRIKE; more than 3000 were delivered through 1976. Now phasing out in favor of the

AGM-88 HARM, small numbers of Standard ARM remain in service with U.S. Navy A-6E INTRUDERS and Air Force F-4G WILD WEASELS. Israel has converted some into ground-launched defense-suppression missiles ("Purple Fist").

With the same airframe as the RIM-66, Standard ARM has an ogival nose and a cylindrical body, with four narrow-span body strakes for lift and four tail fins for steering. The passive radar seeker in the nose was initially the same as the Shrike's, but later versions introduced an improved seeker which could scan a broad waveband and lock onto any one of a variety of hostile radars. Moreover, the guidance unit has an autopilot that can direct the missile to its target, even if the hostile radar shuts down—a capability lacking in Shrike. Powered by an Aerojet Mk.27 Mod 4 dual-impulse solid-fuel rocket, Standard ARM has 219-lb. (99.5-kg.) high-explosive fragmentation warhead fitted with both proximity and impact fuzes.

Specifications **Length:** 15 ft. (4.57 m.). **Diameter:** 13.5 in. (343 mm.). **Span:** 42.9 in. (1090 mm.). **Weight, launch:** 1800 lb. (818.2 kg.). **Speed:** Mach 2.5 (1625 mph/2600 kph). **Range:** 15 mi. (24 km.).

STANDARD MISSILE: Raytheon RIM-66/RIM-67 series of shipboard surface-to-air missiles, in service with the U.S. Navy and the navies of Australia, France, Italy, Japan, the Netherlands, Spain, and West Germany. Developed in the early 1960s as replacements for both the RIM-2 TERRIER and RIM-24 TARTAR, the Standard family entered service in 1968, the RIM-66 Standard MR (Medium Range) as the successor to Tartar, and the RIM-67 Standard ER (Extended Range) as successor to Terrier.

The basic RIM-66A (Standard Missile-1 MR) was similar in configuration to Tartar, with four broad-chord, narrow-span body strakes for lift, and four cruciform tail fins for steering. Armed with a 215-lb. (97.75-kg.) high-explosive fragmentation warhead fitted with both impact and PROXIMITY FUZES, the missile also had a secondary anti-ship role. Powered by a dual-pulse solid-rocket motor, it relied on SEMI-ACTIVE RADAR HOMING (SARH) guidance, with illumination provided by SPG-51 and SPG-55 shipboard radars. It was soon superseded by the RIM-66B, with a more powerful engine, longer range, and a higher ceiling. The current RIM-66C (Standard Missile-2 MR) version is intended for use with the AEGIS air defense system. Externally similar to the RIM-66B, the RIM-66C has solid-state electronics and a digital autopilot for midcourse guidance, which extend its maximum range and ceiling. Standard MR can be launched from Mk.11 (twin-arm) and Mk.13 (single-arm) launchers (see Tartar for details), as well as by the later twin-arm Mk.26 (with two 18-round vertical magazines) and the Mk.41 VERTICAL LAUNCH SYSTEM (VLS).

The long-range RIM-67A (Standard Missile-1 ER), similar in configuration to Terrier, is essentially a RIM-66 with an added tandem solid-fuel booster; the latter is discarded after burnout, and the main stage therefore has a *single*-pulse solid-rocket motor. The warhead and guidance unit are unchanged. In 1980 the RIM-67A was superseded by the RIM-67B (Standard Missile-2 ER). Though it cannot be launched from Aegis ships, it has the -2MR's autopilot and guidance unit, which allows it to be guided by Aegis. Externally identical to the RIM-67A, it is considerably heavier but has more powerful engines to extend its maximum range and ceiling. Standard ER can be launched from Mk.10 twin-arm launchers, with twin horizontal magazines (see Terrier for details).

A nuclear version was developed as a replacement for the RIM-2D(N) nuclear Terrier, but not deployed; armed with a low-yield W81 warhead, it was meant for use against massed attacks. Now in development, the Standard Missile-3 is an extended-range variant with a short, high-power tandem booster, to fit the Mk.41 VLS.

Specifications Length: (66A) 14.67 ft. (4.47 m.); (66B/C) 15.5 ft. (4.73 m.); (67A/B) 26.16 ft. (7.98 m.). **Diameter:** 13.5 in. (343 mm.). **Span:** 36 in. (915 mm.). **Weight, launch:** (66A) 1276 lb. (580 kg.); (66B) 1342 lb. (610 kg.); (66C) 1553 lb. (705.9 kg.); (67A) 2873 lb. (1306 kg.); (67B) 1372 lb. (1442 kg.). **Speed:** (66A/B/C) Mach 2 (1300 mph/2080 kph); (67A/B) Mach 2.5 (1525 mph/2600 kph). **Range:** (66A) 28.75 mi. (46 km.); (66B) 41.6 mi. (66.6 km.); (66C) 40 mi. (64 km.); (67A) 46 mi. (74 km.); (67B) 93 mi. (149 km.). **Height envelope:** (66A) 150–50,000 ft. (46–15,250 m.); (66B) 62,500 ft. (19.055 m.) max.; (66C) 80,000 ft. (23,390 m.) max.; (67A) 150–80,000 ft. (46–23,390 m.); (67B) 100,000 ft. (30,488 m.) max.

STANDOFF MISSILE: A missile (often with a nuclear warhead) that can be launched from a bomber or an attack aircraft at a considerable distance from the target, which the aircraft thus need not overfly. The concept was developed in the 1950s to extend the operational life of bombers in the face of improving air defenses. Since the 1980s they have also been deployed to reduce risks to tactical aircraft. Specific examples include the U.S. SRAM and ALCM, and the Soviet KITCHEN, KIPPER, AND KENT.

S-TANK: Swedish *Stridsvagn* (Strv-103) turretless MAIN BATTLE TANK (MBT). In the early 1950s, the Swedish army examined many alternative configurations for a new tank before choosing a most original design featuring a rigidly mounted main gun laid by turning and elevating the entire vehicle, an automatic loader, a three-man crew, and a composite diesel/gas turbine (CODAG) powerplant. A turretless configuration offers several advantages: simplicity, economy, ease of production, suitability for automatic loading, a reduced silhouette and frontal area, and the option of distributing the weight normally allocated to the turret for additional hull armor. On the other hand, very sophisticated steering and suspension are needed for precision traverse and elevation to point the gun. Even then, the vehicle cannot respond quickly to flanking threats, and cannot fire on the move.

Tests began in 1961, development was rapid, and production of some 300 vehicles was completed by 1971. Though tested by several other armies, including the British, West German, and U.S., the S-Tank has not been exported, nor has the turretless design been emulated.

The all-welded steel hull has a well-sloped frontal glacis plate with horizontal ribs to deflect armor-piercing shot. Maximum armor thickness is on the order of 90–100 mm. (but equivalent to more than twice that amount because of the slope of the glacis). The hull is divided into a driving/engine compartment up front, a fighting compartment in the middle, and a magazine in the rear. The S-Tank has a unique twin-engine powerplant with one diesel and one gas turbine, acting through a common transmission. The diesel is used for economical cruising, and supplies power to the hydraulic system to turn and elevate the vehicle for gun-laying. The turbine is a "sprint" engine, also used to start the diesel in extremely cold weather, and as an emergency backup. Engine change is a complex procedure requiring the removal of the glacis, an operation requiring several hours. The initial version was not amphibious, but the improved Strv-103B has an erectable flotation screen, and can cross still water at 3.6 mph (5.75 kph) propelled by track action.

The driver sits up front on the left, with the engines mounted transversely to his right. Because he also serves as the gunner, the driver is provided with an OPS-1 combined periscope and binocular sight with a 100° field of view and magnification settings of 1x, 6x, 10x, and 16x, as well as an observation periscope. The radio operator sits behind the driver, facing to the rear; he has two observation periscopes and controls for steering the tank in reverse. The commander sits on the right behind the radio operator, under a trainable cupola with four periscopes and an OPS-1 sight. The cupola has only 208° traverse, but is stabilized in azimuth, while the OPS-1 is stabilized in elevation from −11° to +16°. In combat, the commander would first acquire the target in his sight, then turn the entire vehicle with a tiller control (overriding the driver) to bring the gun into line, and then select the appropriate ammunition from the automatic loader. Though the commander can fire the main gun, that task is usually left to the driver/gunner so that the commander can observe the fall of shot, and acquire additional targets.

The main armament is a British-designed L74 105-mm. rifled gun, a lengthened (and more powerful) version of the famous L7 which arms most Western tanks. Fitted with twin vertical breech blocks, the L.74 is connected to a 50-round automatic loader in the rear of the tank, which allows a maximum rate of fire of 15 rds./min. (twice that of manually loaded guns). Ammunition type is selected with a push button. Empty shell casings are automatically ejected, and the magazine can be reloaded through rear doors in less than ten minutes. The radio operator can load the gun manually in emergencies.

The L74 can fire APDS rounds with a muzzle velocity of some 1500 m./sec. to an effective range of 2000 m., more modern APFSDS rounds with a muzzle velocity of 1600 m./sec. and an effective range of 3000 m., and HESH with a muzzle velocity of some 1000 m./sec. out to ranges of up to 5000 m.; HESH is also suitable against troops and soft targets as well as armored vehicles. A typical ammunition load consists of 25 APFSDS or APDS, 20 HESH, and 5 smoke rounds.

The gun is trained by traversing the entire vehicle with a sensitive "external crossbar" steering mechanism, and elevated by changing the pitch of the vehicle through its adjustable hydropneumatic suspension. Elevation limits are −10° and +12°, and the suspension is locked before firing to provide a more stable platform; that, of course, prevents firing on the move.

Secondary armament consists of two 7.62-mm. machine guns fixed in the left side of the hull,

aimed and fired in the same manner as the main gun. A third 7.62-mm. machine gun is pintle-mounted outside the commander's cupola. Eight smoke grenade launchers are fitted on each side of the hull, and some vehicles also have two flare launchers on the roof. A build-in bulldozer blade, for clearing obstacles and improving firing positions, slides down for use from under the bow.

From 1983, the Swedish army initiated an S-Tank improvement program, retrofitting existing vehicles with a new transmission, a LASER rangefinder and FIRE CONTROL computer, a muzzle reference system (to compensate for barrel droop), and additional external fuel stowage. In addition, the Rolls Royce diesel is being replaced by a 300-hp. Detroit Diesel 6V53T which increases the cruising speed. Deliveries of the modified tanks (Strv-103C) began in 1986 and should be completed by late 1990.

Specifications **Length:** 23 ft. (7.01 m.). **Width:** 11.16 ft. (3.4 m.). **Height:** 7.1 ft. (2.16 m.). **Weight, Combat:** 39 tons. **Powerplant:** 1 240-hp. Rolls Royce K60 diesel engine, 1 490-hp. Boeing 553 gas turbine. **Hp./wt. ratio:** 18.4 hp./ton. **Fuel:** 250 gal. (1100 lit.). **Speed, road:** 30 mph (48 kph). **Range, max.:** (diesels only) 235 mi. (376 km.).

STARBRIGHT (SS-N-9): NATO code name for a Soviet ship- and submarine-launched ANTI-SHIP MISSILE. Introduced in 1969, it was first deployed on the CHARLIE II nuclear guided-missile submarines (SSGNs), and later on the single PAPA SSGN and the NANUCHKA corvettes.

Believed to be an incremental development of the SS-N-7 SIREN, Starbright is an airplane-configured missile with midmounted delta wings (which probably fold for storage) and a small delta tailplane. Powered by a solid-fuel rocket engine, Starbright relies on inertial guidance through midcourse, with active radar (and possibly passive INFRARED) homing in the terminal phase. An anti-radiation option has also been reported. Midcourse guidance updates can be provided by the launch platform, or by another platform for OVER-THE-HORIZON TARGETING (OTH-T). Payloads include a 1100-lb. (500 kg.) high-explosive and a 20-kT nuclear warhead.

Starbright is launched from fixed tubes in submarines and from canisters on surface ships. When launched from submergence, it rises to the surface before the engine ignites. Both ship- and submarine-launched versions are surface-skimmers, flying at less than 200 ft. (60 m.) during the final approach to the target. Little is known about its vulnerability to ELECTRONIC COUNTERMEASURES, but its low-altitude approach and relatively high speed would make it difficult to detect and limits response times; the U.S. Navy considers "pop-up" attacks from submarine-launched surface-skimming missiles the most difficult of threats.

Specifications Length: 28.9 ft. (8.8 m.). Diameter: 1.8 ft. (550 mm.). Span: 8.2 ft. (2.5 m.). Weight, launch: 7275 lb. (3300 kg.). Speed: Mach 0.9 (600 mph/960 kph). Range: 68 mi. (110 km.).

STARFIGHTER: Lockheed F-104 FIGHTER/ATTACK AIRCRAFT, little used by the U.S. Air Force, but once the backbone of NATO air forces, and still in service (albeit in dwindling numbers) with Canada, Denmark, Greece, Italy, Taiwan, Turkey, and West Germany.

Developed from 1952 by the legendary Clarence "Kelly" Johnson, the Starfighter reflected the demands of Korean War pilots for a simple, lightweight fighter with greater speed, acceleration, and climb rate, even at the expense of turning performance. First flown in 1954, the prototype was dubbed "the manned missile" because of its Mach 2 performance, sleek, needle-nose fuselage, and diminutive wingspan (which greatly limited its maneuverability). The initial F-104A entered service in 1958 with USAF's Air Defense Command, and set numerous speed and climb records. But by then USAF's operational doctrine stressed long-range nuclear delivery and air defense with heavy, missile-armed interceptors, and the F-104A was quickly relegated to the Air National Guard (though 12 were sold to Jordan and 12 to Pakistan). The follow-on F-104C, with an AERIAL REFUELING probe and limited ground attack capability, entered service in 1958. Only 70 were built, serving with the TACTICAL AIR COMMAND through 1965.

In 1959, the West German *Luftwaffe* issued a requirement for a modern, multi-role fighter to replace its aging F-84 Thunderjets and F-86 SABRES, and Lockheed redesigned the Starfighter for the requirement. Through aggressive marketing, this F-104G, the similar RF-104G RECONNAISSANCE fighter, and the two-seat F-104F trainer were eventually sold to West Germany, Belgium, Denmark, Greece, Italy, Norway, Turkey, and Spain. Of 1487 delivered, some 440 were built by Lockheed, with the rest built under license in Germany, Belgium, the Netherlands, and Italy. Mitsubishi built 207 similar F-104Js for the Japan Air Self-Defense Force, while Canadair built 200 CF-104Gs for the Canadian air force.

Though it had the most advanced ground attack avionics of its time, with inertial navigation, multimode radar, and computerized weapon delivery, the F-104G became notorious for its very high accident rate. A demanding aircraft to fly, the Starfighter was simply too "hot" for many inexperienced NATO pilots. The *Luftwaffe*, the single largest operator with more than 900 Starfighters, had lost more than 200 by 1980. As pilots gained experience, the accident rate declined, but the Starfighter's evil reputation was partly responsible for its loss of the 1970 International Fighter contract to the F-5E TIGER.

From 1968, Fiat introduced the F-104S INTERCEPTOR. With uprated engines, improved avionics, and a new air-to-air radar compatible with ASPIDE and AIM-7 SPARROW radar-guided missiles, some 165 were delivered to the Italian air force, and 40 to the Turkish air force (which later acquired some additional ex-Italian aircraft)

The slender, area-ruled fuselage has a needle-nose radome, which in the F-104G houses an Autonetics NASARR (North American Search and Ranging Radar) with air-to-ground, ground-mapping, and TERRAIN-AVOIDANCE modes, linked to an automatic weapon-delivery computer. The F-104S has a more advanced R21G/H radar optimized for air-to-air missions, with a moving-target indicator (MTI) and improved ELECTRONIC COUNTER-COUNTERMEASURES (ECCM). The pilot sits in the nose under a hinged canopy which provides an excellent view forward and to the sides, but rearward visibility is obstructed by the fairing of the cockpit into the fuselage. The original F-104A had an impractical downward-firing ejection seat, replaced in later models by an upward-firing "zero-zero" seat. AVIONICS were comprehensive by 1960s standards: an optical reflector gunsight, a "head-down" radar display, a Litton LN-3 INERTIAL NAVIGATION unit (the first on a fighter aircraft), a Honeywell stick-steering autopilot, and several RADAR WARNING RECEIVERS.

In the F-104G/S, a 20-mm. M61 VULCAN 6-barrel cannon is mounted below the cockpit on the left, with a 725-round drum magazine immediately behind the cockpit. (In the RF-104G, the gun is replaced by a variety of cameras and other sensors.) Behind the cockpit, the fuselage houses an avionics bay, self-sealing tanks for fuel, and the engine, a General Electric J79 afterburning turbojet fed by cheek inlets behind the cockpit (the F-104F trainer has a second seat in place of the forward fuel tank).

The midmounted wings are notable for their short span, marked anhedral, and extreme thinness—only 4 in. (104 mm.) at the roots. The leading edges are so sharp that covers must be fitted on the ground to protect maintenance crews. The trailing edges have outboard ailerons, while the inboard flaps are "blown" by engine bleed air to provide reasonable low-speed handling. Even so, with a landing speed of some 175 mph (292 kph), the Starfighter is very hot indeed. The tail surfaces have a T-configuration, with the slab tailplane on top of a rather short fin (airflow over the tail can be obstructed at high angles of attack, leading to a "stable stall" from which recovery is impossible).

The F-104A had only 2 wing-tip pylons rated at 1000 lb. (454 kg.) each, for AIM-9 SIDEWINDER air-to-air missiles or 142-gal. (645-lit.) tip tanks. Later, 2 fuselage pylons were added to allow the simultaneous carriage of missiles and tip tanks. In addition to these, the F-104G/S has a centerline pylon rated at 2000 lb. (907 kg.), 2 inboard wing pylons rated at 1000 lb. (454 kg.) each, and 2 outboard pylons rated at 500 lb. (227 kg.) each, but total payload capacity is 4000 lb. (1818 kg.). Ordnance loads for a typical ground attack mission could include two Sidewinders on the wingtips, a drop tank on the centerline, and free-fall bombs, rocket pods, or air-to-surface missiles on the wing pylons. For maritime strike, West German *Marineflieger* aircraft carry a pair of KORMORAN anti-ship missiles. For air superiority missions, the F-104G can carry up to four Sidewinders, while the F-104S can add a pair of Sparrows or Aspides on the outboard wing pylons. RF-104s can carry a multi-sensor pod on the centerline, in addition to a pair of Sidewinders for self-defense.

Specifications (G/S): Length: 54.75 ft. (16.69 m.). **Span:** 21.9 ft. (6.68 m.). **Powerplant:** (F/G) 1 General Electric J79-GE-11A afterburning turbojet, 15,800 lb. (7167 kg.) of thrust; (S) 1 J79-GE-19, 17,900 lb. (8136 kg.) lb of thrust. **Fuel:** (F) 583 gal. (2650 lit.); (G/S) 746 gal. (3992 lit.). **Weight, empty:** (G) 14,082 lb. (6387 kg.); (S) 14,-900 lb. (6758 kg.). **Weight, max. takeoff:** (G) 28,779 lb. (13,045 kg.); (S) 31,000 lb. (14,068 kg.). **Speed, max.:** 910 mph (1520 kph) at sea level/Mach 2.2 (1450 mph/2234 kph) at 36,000 ft. (11,000 m.). **Initial climb:** 50,000 ft./min. (15,239 m./min.). **Service ceiling:** 58,000 ft. (17,677 m.). **Combat radius:** 300 mi. (501 km.) hi-lo-hi. **Range, max.:** 1815 mi. (2920 km.).

STARFISH (SS-N-15): NATO code name for a Soviet submarine-launched anti-submarine missile introduced in 1972. Similar to the U.S. SUBROC, the SS-N-15 provides Soviet submarines with a long-range weapon which can be launched from standard 21-in. (533-mm.) torpedo tubes. The SS-N-15 is deployed aboard AKULA, ALFA, MIKE, SIERRA, TANGO, and VICTOR class submarines.

Powered by a solid-fuel rocket engine, Starfish relies on INERTIAL GUIDANCE with target range and bearing data supplied by the submarine's active SONAR. The payload is believed to be a 15-KT nuclear DEPTH CHARGE. Upon ejection from the torpedo tube, the SS-N-15 rises to the surface before the main engine ignites. It then flies a ballistic trajectory to the estimated target position, where the warhead separates from the missile and deploys a braking parachute to retard entry into the water; it is detonated at a preselected depth by a hydrostatic fuze. A large nuclear warhead is needed because of the "dead time" between launch and water entry, during which the target could move beyond the range of a conventional warhead. See also ANTI-SUBMARINE WARFARE.

Specifications Length: 21.3 ft. (6.5 m.). **Diameter:** 21 in. (533 mm.). **Weight, launch:** 4189 lb. (1905 kg.). **Speed:** Mach 1.5 (1050 mph/1680 kph). **Range:** 23 mi. (37 km.).

STARLIFTER: Lockheed C-141 long-range transport aircraft, in service with the MILITARY AIRLIFT COMMAND (MAC) of the U.S. Air Force. Developed in response to a requirement for a high-speed intercontinental transport, the Starlifter entered service in 1965, was much used during the Vietnam War, and remains the workhorse of MAC's intra-theater routes. Of 284 Starlifters built between 1964 and 1970, some 266 remain in service, including 80 assigned to the Air Force Reserve and Air National Guard.

As built, the C-141A had a large navigational/weather radar in the nose, and an airline-type flight deck for the pilot, copilot, flight engineer, and navigator. Behind the cockpit, the unobstructed cargo hold has a power-operated clam-shell door and loading ramp at the rear. It could originally accommodate 154 passengers, 123 paratroops, or 80 stretcher cases. For long flights, a special pallet with a galley and toilet facilities could be installed in the front of the hold. Cargo capability was 10 standard pallets with a total weight of 70,195 lb. (31,840 kg.). Several Starlifters were modified to transport Minuteman missiles in their storage canisters, a total weight of some 86,-207 lb. (39,103 kg.). The high-mounted wing is moderately swept, with a high aspect ratio for economical cruising. A very tall T-tail configuration allows ample clearance for the ramp-door in the cargo hold.

Operations on the long trans-Pacific route revealed both the need for AERIAL REFUELING capability and that the Starlifter often ran out of space in the cargo hold long before reaching its maximum takeoff weight. Between 1979 and 1982, Lockheed modified 277 Starlifters to C-141Bs, installing a 13.33-ft. (4.06-m.) fuselage plug ahead of the wing, and a similar 10-ft. (3.05-m.) plug behind the wing. The cargo hold can now hold 13 standard pallets for a maximum cargo load of 89,152 lb. (45,040 kg.). Other changes include an aerial refueling receptacle installed over the cockpit, and modified wing-root fairings to reduce drag and increase fuel efficiency. Additional rebuilding extended the airframe life by some 30,000 flight hours. In 1987, further modifications were made to the wing center section to extend fatigue life by another 15,000 hours. AVIONIC improvements include an advanced autopilot, an all-weather landing system, new cockpit displays, and a satellite communications terminal.

Specifications **Length:** (A) 145 ft. (44.2 m.); (B) 168.33 ft. (51.3 m.). **Span:** 159 ft. (48.74 m.). **Cargo hold:** (A) 70 × 10.25 × 9.06 ft. (21.34 × 3.12 × 2.77 m.); (B) 93.33 ft. (28.45 m.). **Powerplant:** 4 Pratt and Whitney TF33-P-7 turbofans, 21,000 lb. (9525 kg.) of thrust each. **Weight, empty:** (B) 149,848 lb. (67,970 kg.). **Weight, max. takeoff:** (B) 344,900 lb. (156,444 kg.). **Speed, max.:** 570 mph (920 kph) at 25,000 ft. (7620 m.). **Speed, cruising:** 560 mph (900 kph). **Service ceiling:** 41,000 ft. (12,500 m.). **Range, max.:** (max. fuel) 5000 mi. (8500 km.); (max. payload) 4000 mi. (6450 km.).

START: Strategic Arms Reduction Talks, ARMS CONTROL negotiations between the U.S. and the Soviet Union, initiated in 1981 after the U.S. Senate's refusal to ratify the SALT II Treaty. The objective of START, in contrast to SALT, is the reduction of strategic nuclear arsenals, rather than their limitation. An early U.S. negotiating position was centered on the "build-down" concept: for each new weapon deployed, two would have to be destroyed or dismantled. The Soviet negotiating position has also included innovative proposals; the most spectacular called for an immediate 50 percent reduction in all weapons, followed by the complete abolition of all ICBMs. The negotiations were long stalemated by the Soviet demand for prohibitions on the STRATEGIC DEFENSE INITIATIVE,

but meetings continue in Geneva. See also DISARMAMENT.

STEALTH: Popular term for "low observables," i.e., technologies, materials, and designs intended to reduce the radar, infrared, acoustic, and visual signatures of aircraft, missiles, and ships. This reduces (ideally to zero) the range at which they can be detected by RADAR, SONAR, and INFRARED (IR) sensors. Stealth-type research began in World War II, but it was only in the 1970s that stealth became a principal design criterion for aircraft, because of the increasing lethality of radar- and IR-guided weapons. Stealth features were incorporated into some 1950s British aircraft, and into the Lockheed U-2 and SR-71 BLACKBIRD strategic reconnaissance aircraft of the 1960s, as well as the B-1B of the 1970s; but the first true "stealth aircraft" was the Lockheed F-117A NIGHTHAWK, introduced from 1981. The Northrop B-2 "Stealth Bomber," currently in flight testing, was also designed with stealth characteristics paramount, and future U.S. aircraft, including the Advanced Technology Fighter (ATF) and the Advanced Technology Aircraft (ATA) will have a high number of stealth features "built in." Warships are now also being designed for stealth, to reduced their vulnerability to anti-ship missiles; some missiles, such as the U.S. Advanced Cruise Missile and SRAM II, also incorporate stealth technology to impede enemy detection.

Given the predominance of radar in surveillance, tracking, and weapon-direction systems, the goal of most stealth development has been the reduction of RADAR CROSS SECTION (RCS), either through the use of RADAR ABSORBANT MATERIALS (RAM) or by designs that eliminate signature-amplifying "corner reflectors"—right-angle junctions of two surfaces. Examples include the elimination of the vertical stabilizer (in the B-2), the use of "butterfly" tail surfaces (in the F-117A) and rounded contours and blended wing-fuselage lines (in the B-1B). Frontal area must be minimized, engine intakes fitted with baffles to prevent the reflection of radar waves by rotating turbine blades (which generate a large and distinctive echo), and weapons carried internally to eliminate reflections from angular control fins. Radar waves can also be reflected by an aircraft's own radar antenna, hence stealth requires electronically scanned CONFORMAL ARRAYS, or planar arrays tilted away from the vertical. The effects of these techniques on RCS can be remarkable: the B-52 STRATOFORTRESS of 1956 has an RCS of some 1000 m.²; its successor the B-1B, with some stealth features, has an RCS of 10 m.²; and the B-2 "Stealth Bomber" has an RCS of only 0.001 m.² (about the size of a hummingbird).

Infrared signature reduction can be achieved inaircraft by "cool," nonafterburning turbofan engines, shrouded exhaust nozzles, and exhaust suppressors which blend hot engine gases with cool air. Speeds must be kept relatively low to minimize the heat generated by atmospheric friction (though special IR-absorptive paint is available).

Suppressing RCS and IR signatures is futile unless other electromagnetic emissions are also minimized. Ships and aircraft normally rely on radars and radios that are easily detected by RADAR WARNING RECEIVERS. The alternative is to rely on passive sensors (such as FORWARD-LOOKING INFRARED [FLIR], THERMAL IMAGING, or LOW-LIGHT TELEVISION [LLTV]), or on hard-to-detect active sensors (such as laser radar, LADAR). Equally, instead of relying on ground-mapping or DOPPLER radar for navigation, only nonemitting INERTIAL NAVIGATION or satellite-based global positioning systems can be used.

Stealth technology has several problems. One is the cost; combining disparate approaches to make a true stealth vehicle compounds costs, as the state of the art of several technologies is passed. In addition, there is a "performance opportunity cost" of stealth—the loss of speed, payload, etc. Finally, the possibility exists that countermeasures, such as more sensitive infrared sensors and multi-static or space-based radars, could negate much of the stealth advantage.

STERLING: British 9-mm SUBMACHINE GUN, introduced in 1953 as the successor to the famous Sten gun of World War II, still in service with the British army and exported to Ghana, India, Libya, Malaysia, Nigeria, Tunisia, and the UAE.

The basic L2A3 Sterling has a tubular body with a pistol grip, and a perforated cooling sleeve around the barrel; a metal stock folded under the body increases overall length to 27.16 in. (690 mm.) when extended. Sterling is fed from a 34-round box magazine inserted into the left side of the receiver. To reduce costs, most parts are stamped steel. A simple, blowback-operated weapon, the Sterling fires standard 9-mm. Parabellum rounds. A fairly low cyclic rate makes the weapon relatively reliable.

A silenced version, the L34A1, has a cylindrical suppressor in place of the perforated sleeve. The silencer reduces muzzle velocity, degrading performance below L2A3 levels. The L34A1 is issued to the British SAS and the SBS. The Australian army F1 submachine gun is basically an L2A3 optimized for jungle warfare, with the magazine feed moved from the side to the top of the receiver. The Canadian C1 submachine gun is essentially a modified Sterling with a 30-round magazine.

Specifications **Length OA:** 19 in. (483 mm.). **Length, barrel:** 7.8 in. (198 mm.). **Weight, empty:** (L2A3) 6 lb. (2.72 kg.); (L34A1) 7.92 lb. (3.6 kg.). **Muzzle velocity:** (L2A3) 390 m./sec.; (L34A1) 310 m./sec. **Cyclic rate:** 550 rds./min. **Effective range:** 100 m.

STEYR AUG: *Armee Universal Gewehr*, or Universal Army Rifle, an Austrian 5.56-mm. ASSAULT RIFLE, introduced in 1978 and now in service with the armies of Australia, Austria, Ireland, Morocco, New Zealand, Oman, and Saudi Arabia.

Designated *Sturmgewehr* (Stgw) 77 in the Austrian army, the AUG has a "bullpup" configuration, with receiver located behind the trigger; this allows the maximization of barrel length (for range and accuracy) within the shortest overall length. The stock, pistol grip, and trigger guard are all injection-molded plastic castings, as is a fold-down foregrip. A 1.5× optical sight is fixed over the barrel and doubles as a carrying handle; a variety of telescopic and night sights can also be fitted. To further reduce weight, the receiver is made of cast aluminum. The barrel is fitted with a flash suppressor compatible with a bayonet and various rifle grenades, and has a chromed bore to facilitate cleaning.

A gas-operated weapon, the AUG fires standard NATO 5.56- × 45-mm rounds from 30-round box magazines molded of clear plastic, so that the operator can easily determine his ammunition status. The AUG has selectable semi-automatic or automatic fire. Variants include a SUBMACHINEGUN, a carbine, and a heavy-barreled squad automatic weapon (SAW); the latter has a bipod and 42-round box magazines.

Specifications **Length OA:** (AUG) 31.1 in. (790 mm.); (SMG) 24.7 in. (626 mm.); (carbine) 27.1 in. (690 mm.); (SAW) 35.4 in. (900 mm.). **Length, barrel:** (AUG) 20 in. (508 mm.); (SMG) 13.8 in. (350 mm.); (carbine) 16 in. (407 mm.); (SAW) 24.4 in. (621 mm.). **Weight, loaded:** 9.02 lb. (4.2 kg.). **Muzzle velocity:** 965 m./sec. **Cyclic rate:** 650 rds./min. **Effective range:** 350 m.

STILETTO (SS-19): NATO code name for the Soviet RS-18 ICBM. Introduced in 1975 as a replacement for the SS-11 SEGO, Stiletto had been developed (as a conservative fallback) along with the technically more advanced SS-17 SPANKER. Stiletto proved to be superior, and 360 were eventually deployed in former SS-11 silos at four sites in the western U.S.S.R.: Derazhnaya, Kozelsk, Pervomaisk, and Tatishchevo. The Stiletto was considered particularly threatening by U.S. analysts because its combination of warhead yield and accuracy were more than sufficient for a counterforce attack on MINUTEMAN missile silos (along with the heavier SS-18 SATAN).

Although classed as a "light" ICBM under the SALT II definition, the Stiletto is actually comparable in size to the "heavy" U.S. PEACEKEEPER. A two-stage, liquid-fuel missile, like other Soviet liquid-fuel ICBMs, it burns a storable hypergolic fuel mix of undimensional dimethyl hydrazine (UDMH) and nitrogen tetroxide (NTO). The first stage has two gimballed main engines for steering and attitude control, while the second stage has a single fixed main engine and two gimballed vernier engines. Unlike other Soviet ICBMs of its generation, Stiletto is "hot-launched"; the main engines ignite within the silo, the missile flies out under its own power, and the exhaust gases are vented upward between the missile and the silo walls (making the silo unusable for further launches).

Stiletto has been built in three versions: the Mod 1 with a liquid-fuel POST-BOOST VEHICLE (PBV) with six 550-kT MULTIPLE INDEPENDENTLY TARGETED REENTRY VEHICLES (MIRVs); the Mod 2 of 1978 with a single 10-MT REENTRY VEHICLE (RV) for attacks on hardened command bunkers; and the more accurate Mod 3 of 1980, with six 450-kT MIRVs.

The Mod 3 employs the same composite INERTIAL GUIDANCE as the SS-18 Mod 4. The first and second stages employ the traditional Soviet "fly-by-wire" technique, whereby the missile is constrained to a preprogrammed pitch-velocity profile by thrust-magnitude control. The PBV, however, relies on navigating inertial guidance, which can compensate for errors accumulated from the boost phase. Because the PBV has a constant-thrust engine, it must release its RVs while oriented on the "range-insensitive direction" (RID). Excursions up the RID change the time of flight but do not substantially alter the point of impact. But this prevents Stiletto from directing two RVs at the same target simultaneously, which would lead to FRATRICIDE effects. Thus a "two-on-one cross-targeting" technique would have to be employed: two

different missiles can each direct an RV against the same target with the same time of arrival; fly-by-wire guidance can provide the precise time-on-target control needed to implement this tactic. The Mod 3 has a kill probability (P_k) of 65 percent against a current Minuteman silo, rising to 85 percent with two-on-one cross-targeting. See also ICBMS, SOVIET UNION; RVSN.

Specifications Length: 88.5 ft. (27 mm.). Diameter: 9 ft. (2.75 m.). Weight, launch: 172,000 lb. (78,181 kg.). Range: (1) 6000 mi. (9600 km.); (2/3) 6200 mi. (10,000 km.). CEP: (1) 390 m.; (2) 260 m.; (3) 240 m.

STINGER: The FIM-92A man-portable (shoulder-fired) INFRARED-HOMING surface-to-air missile (SAM), developed by General Dynamics from the mid-1960s as a replacement for the unsatisfactory REDEYE. Technical problems delayed production until 1981, when 1444 missiles were delivered to the U.S. Army. Since then, more than 43,700 missiles have been supplied to U.S. forces, and the missile has been widely exported. Stingers have also been supplied to the Afghan Mujahideen, who used them with great success against Soviet helicopters and attack aircraft. Stinger was also used by Chad against Libyan forces, and by Britain in the 1982 Falklands War. Licensed production is under way in both Europe and Japan.

Outwardly similar to Redeye, the Stinger system consists of the missile in its sealed launcher/storage canister, a gripstock/trigger assembly, and an optical sight. The missile has a blunt nose housing an infrared seeker, a cylindrical body, and a tapered tail, with four pop-out control canards and four folding tail fins. Powered by a two-stage solid-fuel rocket motor, Stinger is armed with a 6.6-lb. (3-kg.) high-explosive fragmentation warhead fitted with contact and proximity fuzes. Stinger's cryogenically cooled IR seeker is capable of ALL-ASPECT engagement and is resistant to flares and other simple INFRARED COUNTERMEASURES (IRCM). An improved version, Stinger-POST (Passive Optical Seeker Technology) has a "rosette-scan seeker" with dual-frequency (IR/UV) detectors, to provide better performance at very low altitudes, and greater resistance to IRCM. Stinger-POST also has a programmable microprocessor to permit future upgrades without costly hardware changes.

The engagement sequence is similar to that of other shoulder-fired SAMs. The missile, in its canister, is first attached to the a launcher unit consisting of the gripstock, a battery, a coolant bottle (for the IR seeker), an IFF interrogator, and the optical sight. The operator places the launcher on his shoulder and acquires the target in his sight. If IFF interrogation indicates that the target is hostile, the operator must wait until the seeker achieves lock-on, as indicated by an audio tone, and then pulls the trigger. The small initial stage ejects the missile from the launch tube, before the main engine ignites at a safe distance from the operator. The canards and tail fins then pop out, and the missile homes autonomously to the target with PROPORTIONAL NAVIGATION to steer a collision course.

In addition to its basic shoulder-fired version, Stinger has been incorporated into the Boeing "Avenger" air defense vehicle, under the U.S. Army FAADS program. Avenger consists of four Stingers on a pedestal launcher, mounted on a 4 × 4 HMMWV ("Hummer") light truck. Similar pedestal mounts have been proposed for small warships, and a helicopter-launched versions has also been developed.

Specifications **Length:** 5 ft. (1.53 m.). **Diameter:** 2.75 in. (70 mm.). **Span:** 3.16 in. **Weight, launch:** (missile) 22.3 lb. (10.1 kg.); (system) 33.3 lb. (15.15 kg.). **Speed:** Mach 2 (1300 mph/2080 kph). **Range envelope:** 660–16,400 ft. (200–5000 m.). **Ceiling:** 15,750 ft. (4800 m.). **Operators:** Chad, FRG, It, Jap, Neth, S Ar, Tur, UK, US.

STINGRAY: 1. (SS-N-18). NATO code name for the Soviet RSM-50 submarine-launched ballistic missile (SLBM). Developed in the early 1970s as the successor to the SS-N-8, Stingrays underwent flight tests in 1975 and have been deployed since 1978 aboard DELTA III nuclear-powered ballistic-missile submarines (SSBNs), each of which has 16 missile tubes. Stingray was the first Soviet SLBM with MULTIPLE INDEPENDENTLY TARGETED REENTRY VEHICLES (MIRVs).

Somewhat larger than its predecessor, Stingray is a two-stage, liquid-fuel missile that burns a storable, hypergolic mix of undimensional dimethyl hydrazine (UDMH) and nitrogen tetroxide (NTO), despite the hazard of keeping such volatile compounds in an enclosed submarine. The first stage has four gimballed main engines for steering and attitude control, while the second stage appears to have a single fixed engine and two gimballed vernier engines. The initial SS-N-18 Mod 1 had three 200-kT MIRVs mounted on a liquid-fuel POST-BOOST VEHICLE (PBV), while the Mod 2 of 1979 had a single 450-kT REENTRY VEHICLE (RV). The Mod 3, also introduced in 1979, has seven 100-kT MIRVs and improved guidance.

All versions rely on INERTIAL GUIDANCE with stellar update ("stellar-inertial") to enhance accuracy. After launch, a star sensor in the nose locks onto one or more celestial objects, to obtain a position fix with which launch position errors caused by gyro "drift" in the submarine can be corrected. The SS-N-18 is believed to employ the traditional Soviet "fly-by-wire" technique, whereby the missile is constrained to a preprogrammed pitch-velocity profile through thrust-magnitude control. Because of its relatively poor accuracy, the SS-N-18 could be used effectively only against cities and other area targets. See also SLBMS, SOVIET UNION.

Specifications **Length:** 44.6 ft. (13.6 m.). **Diameter:** 5.85 ft. (1.8 m.). **Weight, launch:** 74,950 lb. (34,068 kg.). **Range:** ($1/3$) 4040 mi. (6465 km.); (2) 4970 mi. (7952 km.). **CEP:** (1) 1410 m.; (2) 1550 m.; (3) 1370 m.

2. British lightweight anti-submarine homing TORPEDO, developed by Marconi to replace for the U.S. MK.46 in the Royal Navy; also in service with the navies of Egypt and Thailand. Stingray was the first British torpedo developed entirely by private industry, and this may account for its relatively trouble-free gestation. Designed for launch from ships, helicopters, and fixed-wing aircraft alike, Stingray can be used in both deep and shallow waters with equal facility. A few were actually aboard British warships during the 1982 Falklands War, though none were used in combat. Officially, operational service with the Royal Navy and the RAF did not begin until 1983.

Stingray is compatible with all tubes and bomb racks used for the Mk.46. It has a conventional configuration, with an ogival nose, a cylindrical centerbody, and a tapered tail with four cruciform fins. The nose houses an active/passive multimode, multi-beam SONAR and a digital guidance computer (also in the Mk.24 Mod 2 TIGERFISH and Marconi SPEARFISH torpedoes), which can be programmed to counter evasive maneuvers and torpedo countermeasures. Stingray is armed with an 88-lb. (40-kg.) SHAPED-CHARGE warhead, to penetrate Soviet double-hulled submarines. The centerbody houses a seawater-activated battery for the electrically driven pumpjet propulsor; the latter, a low-speed, multi-blade turbine mounted between two sets of stator blades inside an annular duct, is considerably quieter than a conventional propellor, especially at higher speeds.

In action, Stingray's sonar would first operate in the passive mode while searching for the target. If the target is not detected immediately, the torpedo can execute a number of preprogrammed search patterns. After target acquisition, the guidance computer determines the proper intercept course, and can vary the sonar mode in response to detected evasive maneuvers and torpedo countermeasures.

Specifications Length: 8.52 ft. (2.6 m.). Diameter: 12.75 in. (324 mm.). Weight, launch: 585.2 lb. (266 kg.). Speed: 45 kt. Range: 6.9 n.mi. Max. operating depth: 2625 ft. (800 m.).

STOL: Short Takeoff and Landing, an aircraft attribute generally defined as the ability to take off and land in less than 1500 ft. (450 m.). Aircraft with this capability usually have wings equipped with high-lift devices, including full span leading-edge slats and full-span Fowler flaps.

STRATEGIC: That which pertains to the highest levels of national decision making, i.e., overall war planning and direction. Loosely used as a synonym for long-range (as in "strategic" transport), and less accurately for nuclear weapons aimed at the enemy homeland, as in "strategic nuclear weapons." See also OPERATIONAL; TACTICAL.

STRATEGIC AIR COMMAND (SAC): The operational command of the U.S. Air Force (USAF) responsible for all BOMBERS and land-based intercontinental ballistic missiles (ICBMS) and their ancillaries. Established in 1946 as a heavy-bomber strike force, SAC soon became the largest command in USAF. SAC Headquarters at Offutt Air Force Base (AFB) near Omaha, Nebraska, are housed in a hardened underground complex with elaborately redundant COMMAND AND CONTROL facilities. In addition, the SAC fleet of EC-135 "Looking Glass" and E-4B NATIONAL EMERGENCY AIRBORNE COMMAND POST aircraft could maintain control even if the main complex is destroyed. SAC is divided into two numbered AIR FORCES (8th and 15th), the 544th Strategic Reconnaissance Wing (SRW), and the 1st Combat Evaluation Group.

STRATEGIC AIR WAR: A war waged by nonnuclear bombing focused against the enemy's war-making capacity (typically industrial facilities, transport infrastructure, and energy sources), rather than his more elusive combat forces. Between great powers, this mode of war may be obsolete (prenuclear), but it has been used against lesser powers since 1945 as an alternative to nuclear attack. Small powers typically import their war materials, so that bombing targets are mostly

transport facilities (especially "nodes" and "interfaces" such as harbors), rather than factories.

STRATEGIC DEFENSE INITIATIVE: Official term for the post-1983 U.S. BALLISTIC MISSILE DEFENSE (BMD) research program popularly known as "Star Wars." A number of low-key BMD research programs survived the signing of the 1972 ABM TREATY, but in March 1983, President Reagan, influenced by the Defense Technologies Study Team (DTST) or "Fletcher Reports," proclaimed the goal of harnessing advanced technology to develop a comprehensive, nonnuclear defensive "shield" against (ballistic) missiles. Only in April 1984 was the Strategic Defense Initiative Office (SDIO) established to oversee the effort. While it has conducted high-level feasibility studies and analyses, SDIO primarily coordinates the BMD activities of the various services and national laboratories.

SDI is divided into four main functional areas: Surveillance, Acquisition, Tracking and Kill Assessment (SATKA); DIRECTED ENERGY WEAPONS (DEW); Kinetic Energy Weapons (KEW); BATTLE MANAGEMENT and Command, Control, and Communications (BM/C^3); and Survivability and Critical Technologies.

SATKA covers the development of space- and earth-based sensors; DEW covers all "beam" weapons, including LASERS and PARTICLE BEAMS; KEW covers interceptor missiles and electromagnetic launchers (rail guns); BM/C^3 covers the computers, software, and communications required to integrate the different elements of a system; and the Survivability and Critical Technologies division is responsible for identifying potential Soviet countermeasures.

Since its inception, SDI has been very controversial. Critics have emphasized its technical difficulties, potentially exorbitant costs, and the negative impacts on arms control, notably on the ABM Treaty. Advocates have stressed the considerable technical progress quickly achieved since 1983, relative costs (e.g., as compared to strategic offensive weapons), and the desirability of replacing ASSURED DESTRUCTION with a more defensive policy. As the controversy continues, funding for SDI has remained stable at some $3 billion to $4 billion per year—not much more than BMD outlays prior to SDI. Initially, SDIO focused on very exotic technologies (e.g., space-based lasers), but it later shifted to more conventional, near-term solutions, specifically a "layered defense" of kinetic energy SPACE-BASED INTERCEPTORS, and long-range ERIS and short-range HEDI missiles. These technologies have matured to the point where a decision for deployment could be made by the early 1990s. See also ANTI-BALLISTIC MISSILE.

STRATEGIC ROCKET FORCES, SOVIET:
See RVSN.

STRATEGY: The wisdom of war, which differs from other wisdom because the logic of war differs from normal logic. As against the linear/ formal logic of everyday life, the logic of war, and more broadly of conflict, is paradoxical ("if you want peace, prepare for war"), and dialectical (action yields not only a result, but also a reaction that modifies, and may utterly reverse, that result). In the realm of conflict, action cannot therefore proceed straightforwardly, but instead will normally reach a culminating point of achievement, which evokes corresponding adversary reactions so that decline ensues in the absence of additional effort (hence the transient effectiveness of weapons effective enough to have evoked countermeasures; the inability of victorious forces to continue winning by unaltered means; the rising resistance to further expansion encountered by expanding states, etc.). If action is nevertheless straightforwardly pursued, its results will normally tend to evolve towards their opposites (victory yields defeat, war yields peace, etc.). At each extreme, the paradoxical logic of strategy results in the coincidence of opposites, as in the case of nuclear weapons, for example, which are made useless in most cases by their excessive destructive efficiency, so that they are both the most powerful *and* least powerful of weapons (even daggers were more powerful for U.S. troops in Vietnam and Soviet troops in Afghanistan). In the realm of conflict, therefore, common-sense logic is not a trusty guide but a delusive snare, and the same is true of its derivatives, such as all normal criteria of efficiency. Many of the crimes and follies of mankind have resulted from the millenially obstinate attempt to pursue success in the realm of conflict by linear-logical methods, whose results must be paradoxical (thus the relentlessly peaceful attract war on themselves, the relentlessly expansionist are diminished, relentless invaders overthrow their own strength, disarmers cause arms races, and arms racers cause disarmament).

Other, more conventional definitions include:

"A science, an art, or a plan (subject to revision) governing the raising, arming, and utilization of the military forces of a nation (or coalition) to the end that its interests will be effectively promoted or secured against enemies, actual, potential, or merely presumed" (King, ed., *Lexicon of Military Terms* [1960], p. 14).

A modern U.S. definition of official military origin is more inclusive: "The art and science of developing and using political, economic, psychological, and military forces as necessary during peace and war, to afford the maximum support to policies, in order to increase the probabilities and favorable consequences of victory and lessen the chances of defeat" (U.S. Joint Chiefs of Staff, *Dictionary of United States Military Terms for Joint Usage* (1964), p. 135).

Even broader is the standard definition of strategy from Webster's *Third New International Dictionary:* "The science and art of employing political, economic, psychological, and military force of a nation or group of nations to afford the maximum support to adopted policies in peace and war."

The definition found in the official *Soviet Military Strategy,* attributed to Marshal V. D. Sokolovskiy, reveals both Marxist and bureaucratic preoccupations, and differentiates between descriptive and prescriptive meanings:

Military strategy is a system of scientific knowledge dealing with the laws of war as an armed conflict in the name of definite class interests. Strategy—on the basis of experience, military and political conditions, economic and moral potential of the country, new means of combat, and the views and potential of the probable enemy—studies the conditions and the nature of future war, the methods for its preparation and conduct, the services of the armed forces and the foundations of their strategic utilization, as well as the foundations for the material and technical support and leadership of the war and the armed forces. At the same time, this is the area of the practical activity of the higher military and political leadership, of the supreme command, and of the higher headquarters, that pertains to the art of preparing a country and the armed forces for war and conducting the war (Scott, ed., *Soviet Military Strategy* [1975], p. 11).

Finally, there is Gen. Andre Beaufre's succinct definition: *"l'art de la dialectique des volontes employant la force pour resoudre leur conflict"* (the art of the dialectics of wills that use force to resolve their conflict) (*Introduction à la strategie* [1963], p. 16).

As for the word itself, *strategy* is derived from the Greek *strategos* (general) via the Italian *strategia,* but it is not a Greek usage; *strategike epis-*

time (general's knowledge) and *strategon sophia* (general's wisdom) are the Greek equivalents.

STRATOFORTRESS: Boeing B-52 long-range heavy BOMBER, in service with the United States Air Force STRATEGIC AIR COMMAND (SAC). The mainstay of SAC's bomber force for more than 35 years, the B-52 was developed from 1948 as a turboprop replacement for the piston-engined B-50 (itself a derivative of the World War II B-29 Superfortress). With the introduction of the first fuel-efficient turbojet engines, the design was re-cast as a swept-wing jet with no fewer than eight engines in four underwing nacelles. The prototype flew in 1952, followed in 1954 by three B-52As employed mainly as technology testbeds. The first operational aircraft were 50 B-52Bs delivered in 1955–56; 27 of these were later converted to RB-52Bs for strategic RECONNAISSANCE. Thirty-five B-52Cs produced in 1956 were followed in that same year by 170 B-52Ds, with improved navigation/attack systems and electronic countermeasures. They were followed by in turn by 100 B-52Es and 89 similar B-52Fs in 1957–58.

Designed for high-altitude bombing with free-fall nuclear weapons, the B-52 became much more vulnerable with the introduction of the Soviet V-75 *Dvina* (NATO code name SA-2 GUIDELINE) surface-to-air missile. In response, Boeing developed the B-52G, optimized for low-level penetration. Externally distinguished by its short vertical stabilizer, the B-52G also introduced a "wet wing" with increased fuel, a remote-controlled, radar-directed gun turret in the tail, and pylons for the external carriage of Hound Dog supersonic cruise missiles. The most numerous variant, 193 B-52Gs were built between 1958 and 1961. They were followed in 1961–62 by the final version, 102 B-52Hs with more efficient turbofan engines and a 20-mm. tail gun, for a grand total of 744 Stratofortresses delivered between 1952 and 1962.

That the B-52 is still operational is a tribute to the soundness and flexibility of the original design. Designed as a nuclear bomber, the Stratofortress was nevertheless much used in Southeast Asia to deliver conventional bombs. From 1965, most B-52Ds received a "Big Belly II" modification, to accommodate no fewer than 105 M117 750-lb. (340-kg.) bombs, for a total payload of 78,750 lb. (35,795 kg.). From 1972, all B-52G/H were modified to launch AGM-69A SRAMs (Short-Range Attack Missiles) for standoff attacks against air defense sites. From 1982, 98 B-52Gs were further modified as CRUISE MISSILE carriers, for up to 20

AGM-86B ALCMS; 69 non-ALCM B-52Gs have been given a secondary maritime support role, while all surviving B-52Hs have been modified to carry a larger conventional payload. SAC now operates some 165 B-52Gs and 96 B-52Hs in 12 bomb wings of 13–30 aircraft each; earlier models were retired in the early 1980s. The B-52H now shares the penetration-bombing mission with the B-1B, but if the B-2 "Stealth Bomber" enters service in the mid-1990s, the H will replace the B-52G as a cruise missile carrier, with the latter relegated to nonnuclear missions. Because of numerous modification programs the B-52 has remained an effective weapon, and will remain in the U.S. inventory through the year 2000.

The rectangular, slab-sided fuselage has a prominent nose radome housing an APQ-156 multimode RADAR with TERRAIN-FOLLOWING and SYNTHETIC APERTURE capabilities. In the chin position below the radar are two fairings for an ASQ-151 Electro-optical Viewing System (EVS). The starboard fairing houses a Hughes AAQ-6 Forward-Looking Infrared (FLIR) sensor, while the port fairing houses a Westinghouse AVQ-22 low-light television (LLTV). Images from both sensors can be displayed on cockpit video terminals, together with relevant flight data, allowing the pilot or copilot to navigate at low altitudes at night or in bad weather without using the radar.

The B-52G/H has a two-level cockpit, with the pilot and copilot on the upper level, the gunner and defensive electronics operator behind them facing to the rear, and the navigator and bombardier on the lower level, facing forward. The B-52 has a comprehensive AVIONICS suite, including normal blind-flying instruments and radar, a Raytheon ASQ-38 digital weapon-delivery computer, a Honeywell ASN-131 INERTIAL NAVIGATION unit with Terrain Comparison and Matching (TERCOM) update, a NAVSTAR Global Positioning System receiver, DOPPLER navigation radar, an AFSATCOM satellite DATA LINK, and ELF and VLF antennas to maintain communications with SAC headquarters under nuclear-war conditions. The current Phase VI defensive electronics include an ALQ-122 SNOE (Smart Noise Operating Equipment) BARRAGE JAMMING transmitter, ALQ-155(V) and ALQ-172 DECEPTION JAMMERS, an ALR-46 digital RADAR WARNING RECEIVER, an ALE-24 bulk CHAFF dispenser, an ASG-15 tail gun FIRE CONTROL radar, and a Westinghouse AGS-153 pulse-Doppler tail-warning radar.

Behind the cockpit, the fuselage houses unique

bicycle landing gear, the bomb bay, and a full eight fuel cells; an AERIAL REFUELING receptacle is located over the cockpit. In the extreme tail, the B-52G has a remote-controlled turret with four .50-caliber (12.7-mm.) machine guns, while the B-52H has a VULCAN 6-barrel 20-mm. cannon. All B-52s have a braking parachute at the base of the vertical stabilizer.

The high-mounted wing, swept at 35°, has a very high aspect ratio for fuel economy in long-range cruising. B-52G cruise missile carriers have wing-root strakes as identifiers for arms control treaty verification, but these also reduce drag. Fitted with huge Fowler flaps, spoilers for roll control, and retractable outrigger landing gear, the wings house integral fuel tanks; there is also a fixed 700-gal. (2650-lit.) external tank under each outboard section. The eight engines are mounted in pairs in underwing nacelles. The nacelles and the external tanks also act as inertia dampers to limit wing flexing. In the B-52G and earlier models, the engines are J57 turbojets with water injection for boost on takeoff. The B-52H has lighter, more fuel efficient TF33 turbofans, which do not need a heavy water-injection system; they are also considerably quieter, reducing crew fatigue.

The B-52 can carry a wide range of payloads. On nuclear penetration missions, the B-52H can carry up to 20 SRAMs (8 in an internal rotary launcher, and 12 on pylons under the inboard wing sections), and several free-fall nuclear bombs. Alternatively, 12 SRAMs can be carried externally for defense suppression, with up to 12 free-fall weapons in the bomb bay. As a cruise missile carrier, the B-52G can carry 8 ALCMs internally in the rotary launcher, and 12 more on the wing pylons. For conventional bombing, modified B-52Hs can carry up to 84 Mk.82 500-lb. (227-kg.) high-explosive bombs internally, and 24 more on wing pylons, for a total payload of 54,000 lb. (24,545 kg.); unmodified aircraft can carry only 56 bombs internally (28,000 lb./12,701 kg.), in addition to 24 bombs externally, for a total payload of 40,000 lb. (18,181 kg.). In the maritime support role, the B-52G can carry 12 AGM-84 HARPOON anti-ship missiles externally, or up to 74 MINES with both internal and external loading.

Specifications Length: 160.83 ft. (49.04 m.). Span: 185 ft. (56.39 m.). Powerplant: (G) 8 Pratt and Whitney J57-P-43WB turbojets, 13,750 lb. (6237 kg.) of thrust each w/water injection; (H) 8 Pratt and Whitney TF33-P-3 turbofans, 17,000 lb. (7711 kg.) of thrust each. Fuel: 308,548 lb. (139,-955 kg.). Weight, empty: 158,737–172,066 lb. (72,-002–78,048 kg.). Weight, max. takeoff: 488,000 lb. (231,353 kg.). Speed, max.: 595 mph (958 kph) at 36,000 ft. (10,980 m.). Speed, cruising: 509 mph (819 kph). Speed, penetration: 405–420 mph (652–676 kph) at 500 ft. (152 m.). Range, max.: (G) 7500 mi. (12,070 km.); (H) 10,000 mi. (16,093 km.). Endurance: several days w/aerial refueling.

STRATOTANKER: Boeing KC-135 AERIAL REFUELING tanker, in service with the U.S. Air Force (USAF) and the French *Armee de l'Air*. Developed to a 1954 USAF requirement for a jet-propelled tanker/transport able to operate in conjunction with the B-52 STRATOFORTRESS bomber, the KC-135 resembles the Boeing 707 airliner, but with a narrower fuselage and no cabin windows. The first prototypes flew in 1956, and the first KC-135s entered service in 1957. More than 732 were built between 1956 and 1966, of which some 590 remain operational, including about 30 KC-135Bs with turbofan engines, and 12 KC-135Fs built for France. In addition, USAF acquired some 45 C-135 "Stratolifter" transports, without aerial refueling equipment. Over the years, many KC-135s have been converted for specialized roles, including the EC-135A/L for long-range communication relay, the EC-135B/N for range instrumentation, the EC-135H/J/K airborne command posts, 13 EC-355C "Looking Glass" launch-control aircraft (for MINUTEMAN ICBMs), the EC-135N for spacecraft tracking, the RC-135 M/S for electronic intelligence (ELINT), the RC-135 U/V/W with SIDE-LOOKING AERIAL RADAR (SLAR), the WC-135B for weather reconnaissance, and 56 KC-135Q tankers specialized to refuel SR-71 BLACKBIRD and U-2 high-altitude strategic reconnaissance aircraft with low-volatility JP-7 fuel. There have also been a number of "one-off" research aircraft, such as the NKC-135A Airborne Laser Laboratory.

The almost cylindrical fuselage has a navigational/weather RADAR in the nose, and a conventional, airline-style flight deck for the pilot, copilot, flight engineer, and radio operator. AVIONICS include a wide range of navigation and communications systems, including a TACAN beacon to facilitate airborne rendezvous. A galley and bunks for relief crews are located behind the cockpit. The rear fuselage has two levels—an upper cabin for up to 80 passengers or 16,000 lb. (7273 kg.) of cargo, and a lower hold with fuel tanks and the aerial refueling equipment. In the tanker role, the KC-135 can transfer up to 32,100 gal. (118,105 lit.) or 224,700 lb. (102,136 kg.) of fuel. U.S. Stratotankers employ the "flying boom" technique, whereby a rigid, telescoping boom housed under the rear fuselage is inserted into a receptacle in the upper fuselage of the receiving aircraft. A ventral window allows the boom operator to "fly" the boom into the receptacle by using a joystick to manipulate two winglets at the end of the boom. French KC-135Fs also have booms, but these terminate in a "hose-and-drogue" assembly. RECONNAISSANCE and ELECTRONIC WARFARE conversions, which do not have refueling equipment, are equipped with a variety of radomes and blade antennas.

From the mid-1980s, 134 KC-135As were reengined with TF33 turbofans taken from surplus Boeing 707s, and redesignated KC-135Es. From 1984, 375 other KC-135As were reengined with more fuel-efficient SNECMA high-bypass turbofans. These aircraft, designated KC-135Rs, were also extensively modernized with new avionics, reinforced landing gear, and an enlarged tailplane. Most KC-135s will also be subjected to a Life Extension Structural Modification program to extend their service lives through the year 2020.

Specifications **Length:** 136.35 ft. (41.53 m.). **Span:** 130.83 ft. (39.88 m.). **Powerplant:** (A) 4 Pratt and Whitney J57-P-59W turbojets, 13,500 lb. (6124 kg.) of thrust each w/water injection; (B/E) 4 Pratt and Whitney TF33-P-5 low-bypass turbofans, 18,000 lb. (8165 kg.) of thrust each; (R) 4 SNECMA F108-CF-100 high-bypass turbofans, 22,000 lb. (9979 kg.) of thrust each. **Fuel:** (A) 189,-702 lb. (86,047 kg.). **Weight, empty:** 98,466 lb. (44,663 kg.). **Weight, max. takeoff:** (A) 297,000 lb. (135,000 kg.); (E) 300,000 lb. (136,078 kg.); (R) 322,500 lb. (146,590 kg.). **Speed, max.:** 600 mph (966 kph). **Speed, cruising:** 530 mph (853 kph) at 40,000 ft. (12,190 m.). **Intial climb:** 1290 ft./min. (393 m./min.). **Service ceiling:** 50,000 ft. (15,244 m.). **Mission radius:** 1150 mi. (1850 km.). **Range, max.:** 9200 mi. (15,364 km.).

STRIDWAGEN: See S-TANK.

STRIKE EAGLE: Strike/interdiction variant of the F-15 EAGLE fighter.

STRIKER: British FV102 TANK DESTROYER, a variant of the SCORPION CVR(T) armed with SWING-FIRE anti-tank guided missiles.

STS: SPACE TRANSPORTATION SYSTEM, official designation of the U.S. space shuttle.

STURGEON: 1 (SS-N-20). NATO code name for the Soviet RSM-52 submarine-launched ballistic missile (SLBM), introduced in 1983 as the principal armament of the TYPHOON nuclear-powered ballistic-missile submarines (SSBNs). A three-stage solid-fuel missile developed from 1973, Sturgeon is equivalent in range and payload to the U.S. TRIDENT II missile, and can hit targets anywhere in the U.S. from protected "bastions" within the Arctic Circle, close to Soviet shores.

The first stage has either one large or four small nozzles, gimballed for steering and attitude control, and the second- and third-stage nozzles may be recessed to reduce length. All three stages are fabricated from wound glassfiber composites to reduce weight. Its liquid-fuel POST-BOOST VEHICLE (PBV) contains nine 500-kT MULTIPLE INDEPENDENTLY TARGETED REENTRY VEHICLES (MIRVs).

Sturgeon relies on INERTIAL GUIDANCE with stellar update ("stellar-inertial"). Shortly after launch, a star sensor in the nose locks onto one or more celestial objects to obtain a position fix, with which launch position errors caused by gyro drift in the submarine's inertial navigation unit can be corrected. Sturgeon employs navigating inertial guidance with General Energy Management Steering (GEMS), whereby the missile makes excursions above and below its nominal trajectory in order to allow each stage to burn to exhaustion, eliminating the need for heavy, unreliable, and inaccurate thrust-termination ports. Combined with its relatively high-yield warheads, Sturgeon's accuracy is sufficient to attack some hardened targets. See also SLBMS, SOVIET UNION.

Specifications **Length:** 9.2 ft. (15 m.). **Diameter:** 6.6 ft. (2 m.). **Weight, launch:** 132,000 lb. (60,000 kg.). **Range:** 5160 mi. (8260 km.). **CEP:** 600 m.

2. A class of 37 U.S. nuclear-powered attack SUBMARINES (SSNs) commissioned between 1967 and 1975. The largest class of nuclear submarines before the LOS ANGELES class, the Sturgeons are enlarged, incrementally improved developments of the prior PERMIT class, intended mainly for anti-submarine warfare with only a secondary anti-ship capability.

Sturgeons were built in two subclasses. All retain the modified teardrop hullform and single-hull configuration of the Permits; the pressure hull is fabricated of HY-80 steel. A tall, streamlined sail is located well forward, and in many Sturgeons a large extension is added to its rear, to house a "Bustle" VLF towed antenna. Control surfaces are arranged in standard U.S. fashion, with forward "fairwater" planes mounted on the sail, and cruciform rudders and stern planes ahead of the propeller. All machinery is individually soundproofed and the entire powerplant is mounted on a resilient sound-isolation raft; the Sturgeons are still among the quietest nuclear submarines in service.

Armament consists of 4 21-in. (533-mm.) torpedo tubes amidships, for MK.48 wire-guided homing torpedoes, UGM-84 HARPOON anti-ship missiles, UGM-109 TOMAHAWK cruise missiles, and SEA LANCE nuclear anti-submarine missiles. The basic load in the late 1980s consisted of 11 Mk.48s, 4 Harpoon, and 8 Tomahawk. All sensors are linked to a Mk.113 analog FIRE CONTROL unit (now being replaced by a digital Mk.117, compatible with Harpoon and Tomahawk). The Sturgeons also have a WSC-3 satellite communications terminal.

Because of their excellent acoustical silencing, several Sturgeons were modified in the late 1960s for covert intelligence-gathering missions (codenamed "Holystone") off the coast of the Soviet Union. One Sturgeon, USS *Cavalla* (SSN-684), has been converted into a troop transport, with SWIMMER DELIVERY VEHICLES (SDVs) housed in a cylindrical "dry deck shelter" placed behind the sail. Three others, *Pintado, Hawkbill,* and *William H. Bates,* have been modified to carry Deep Submergence Rescue Vehicles (DSRVs) for underwater rescue and salvage.

USS *Narwhal* (SSN-671) is an enlarged Sturgeon built in 1967 to evaluate the S5G natural-circulation reactor, which relies on convection-cooling at low speeds to minimize the use of noisy pumps. USS *Glennard P. Lipscomb,* launched in 1973, is another experimental variant, built to evaluate turbo-electric drive. See also SUBMARINES, UNITED STATES.

Specifications **Length:** (first 28) 292 ft. (89 m.); (last 9) 302 ft. (92.07 m.). **Beam:** 31.67 ft. (9.7 m.). **Displacement:** (first 28) 4250 tons surfaced/4780 tons submerged; (last 9) 4460 tons surfaced/4690 tons submerged. **Powerplant:** single-shaft nuclear: 1 S5W pressurized-water reactor, 2 geared turbines, 15,000 shp. **Speed:** 18 kt. surfaced/26 kt. submerged. **Max. operating depth:** 1315 ft. (400 m.). **Collapse depth:** 1970 ft. (600 m.). **Crew:** 121–141. **Sensors:** 1 BQQ-5 digital sonar system, 1 BQR-7 bow-mounted passive low-frequency spherical sonar, 1 BQS-13 low-frequency active sonar, 1 BQS-15 passive towed array, 1 BQS-14 underice sonar, 1 BPS-15 surface-search radar, 1 WLQ-4 electronic signal monitoring array, 2 periscopes.

STYX (SS-N-2): NATO code name for a Soviet ship-launched ANTI-SHIP MISSILE introduced in 1958. Initially developed to provide anti-ship

capabilities to small coastal defense vessels, Styx eventually became the most common of Soviet anti-ship missiles, deployed on KILDIN and KASHIN destroyers, as well as KOMAR and OSA missile boats, MATKA hydrofoils, TARANTUL corvettes, and export versions of the NANUCHKA corvettes. Styx achieved notoriety by sinking the Israeli destroyer *Elat* in 1968. A ground-launched variant, the SS-C-3, is also in service. China produces a modified copy (Silkworm) in both ship- and ground-launched variants.

Styx has been produced in three versions: the original SS-N-2a, first deployed on the Komar class, and later on the Osa I missile boats; the SS-N-2b, with folding wings and a more compact launcher, introduced in 1964 for the Osa II, Kildins, and Kashins; and the current SS-N-2c (initially designated SS-N-11), introduced in 1967. With improved range and guidance, the latter has replaced the -2b throughout the Soviet fleet, and has been widely exported.

A bulky, airplane-configured missile, Styx has midmounted cropped-delta wings that fold up at the roots for storage, unlike the small, swept tail surfaces. Powered by a ventrally mounted strap-on booster rocket and a sustaining turbojet, it flies a semi-ballistic path and then executes a diving attack on the target. All versions are armed with an 1100-lb. (500-kg.) high-explosive warhead. The SS-N-2a/b is controlled by an autopilot during the first minutes of flight, and then by radar ACTIVE HOMING activated at a range of some 5 mi. from the estimated target position. The -2c supplements active radar with INFRARED HOMING in the terminal phase.

The missile does not have a DATA LINK for over-the-horizon targeting, so target acquisition is limited to the radar horizon of the launch vessel. In all versions, the radar seeker can be easily jammed or decoyed, while the IR seeker of the -2c can be spoofed by flares and other INFRARED COUNTERMEASURES. In addition, its low speed and lofted trajectory make Styx vulnerable to interception or evasion. None of the 50-odd Styx launched against Israeli missile boats in the 1973 Yom Kippur War hit their targets. Though obsolescent against modern warships, Styx is cheap and still effective in Third World contexts; with its relatively large warhead it is also well suited for attacks on merchant vessels lacking ELECTRONIC COUNTERMEASURES.

Specifications **Length:** 20.67 ft. (6.3 m.). **Diameter:** 2.45 ft. (746 mm.). **Span:** 9 ft. (2.75 m.). **Weight, launch:** 6614 lb. (3006 kg.). **Speed:** Mach

0.9 (600 mph/960 kph). **Range:** (a/b) 28.6 mi. (46 km.); (c) 46 mi. (74 km.).

SU-: Designation of aircraft developed by the Soviet Sukhoi design bureau, known chiefly for its attack aircraft and interceptors. Models in service include: Su-7/17/20/22 FITTER; Su-15 FLAGON; Su-24 FENCER; Su-25 FROGFOOT; and Su-27 FLANKER.

SUBMACHINE GUN (SMG): A simple, lightweight automatic weapon which fires pistol cartridges. First introduced during World War I (in the face of opposition motivated by their low power and unsuitability for parade drills), SMGs were popularized by Chicago gangsters during the 1920s and much used in World War II.

Almost all SMGs operate by direct blowback action: when the cartridge is fired, expansion gases propel the bullet forward and concurrently push back the heavy bolt against a powerful spring; as the bolt moves back, it first extracts and ejects the spent cartridge, and then—as it retreats over the magazine lip—it allows the next round to be pushed up by the magazine spring and seated in the breach. If the trigger remains squeezed, the spring behind the bolt pushes it forward to seat the round and then detonate it, to repeat the process.

The pistol ammunition used by SMGs is low-powered, and the further loss of power in the firing mechanism means that the muzzle velocity is low, so that maximum accurate ranges are on the order of 50–100 m. World War II SMGs (e.g., the British Sten, Soviet PPS, and U.S. M3 "Grease Gun") tended to be very crude, firing whenever the trigger was pulled—and sometimes when it was not. Some postwar SMGs, such as the Israeli UZI, have safe triggers, good finishes, and refinements such as roll-around bolts, which reduce overall length while improving balance; with practice, they can be accurate out to 150 m. for single shots. Commandos, armored troops, and security forces all tend to use SMGs, but regular infantry forces have generally eliminated them since the advent of AS-SAULT RIFLES. In the Israeli army Uzis are issued mainly to women, but many Third World armies treat them as elite weapons. See, more generally, SMALL ARM.

SUBMARINE: A naval vessel designed for underwater operations. Originally intended for coastal defense, submarines emerged as important offensive weapons during World War I, and played a crucial role in World War II, when the U.S. submarines fought a decisive campaign against Japanese shipping, and German U-boats nearly succeeded in defeating Britain by blockade.

Since 1945, the submarine has become the dominant type of naval vessel, not only for the attack of naval and merchant vessels, but also for land attack and ANTI-SUBMARINE WARFARE. Ballistic missile submarines are a major element of U.S., Soviet, and French nuclear forces, predominate in the British force, and also arm China.

OPERATING PRINCIPLES

The first practical submarine, built by John Holland for the U.S. Navy in 1906, established the basic configuration and operating principles of most submarines to this day. The main technical requirements in submarine design are to surface and submerge, to control attitude (pitch) when submerged, to resist the pressure of the sea, and to provide an airless means of underwater propulsion.

Diving and surfacing are achieved by large, floodable "main ballast tanks" with valves at the bottom and vents at the top. When flooded, these tanks increase the weight of the submarine beyond that of the volume of water it displaces, causing it to sink. The tanks are emptied by closing the vents and "blowing" the water out with compressed air from storage flasks, causing the submarine to surface.

Attitude and depth control are maintained by "trim tanks" and diving planes. Trim tanks are smaller ballast tanks located fore and aft of the submarine's center of gravity. By varying the amount of water in each, the submarine can be "trimmed" with the bow up, down, or level. In addition, the submarine can be balanced with them, so that it weighs precisely what it displaces ("neutral buoyancy") and neither rises nor sinks. When the submarine is moving, attitude and depth can also be controlled by inclining the diving planes, small winglets at the bow and stern analogous to elevators on aircraft. Directional control is maintained by rudders, as in surface vessels.

Seawater exerts a pressure of 44.45 psi (3.13 kg./cm.²) for every 100-ft. (30.5-m.) increase in depth. To prevent collapse, submarines must have reinforced "pressure hulls," the strength of which determines how deep they can dive. The absolute limit, beyond which the pressure hull will "implode," is called the "collapse depth"; to provide a safety margin, submarines are normally limited to a "maximum operating depth," which amounts to 66–75 percent of their collapse depth. The ideal form for pressure hulls is the sphere, but this is impractical for ergonomic and hydrodynamic reasons; hence, submarine pressure hulls are cylindrical.

Several different hull configurations are compatible with hydrodynamic efficiency. In the "single-hull" configuration, the pressure hull also forms the outer skin (casing), except for streamlined bow and stern sections. Ballast and fuel tanks are either contained within the pressure hull or in the bow and stern extensions. In the "double-hull" configuration, the pressure hull is enclosed by a light outer casing, with ballast and fuel tanks contained between the two; the outer casing must be free-flooding, because it cannot resist sea pressure. In the "partial double hull" configuration, there is a double hull amidships, with a single hull at each end. The "saddle tank" configuration is essentially a single hull with the main ballast tanks attached as external "blisters."

Each configuration has its advantages. The single hull, favored by the U.S., Britain, France, and West Germany, maximizes internal volume for a given set of external dimensions, while the double hull, favored by the U.S.S.R., provides greater reserve buoyancy and allows multiple pressure-hull arrangements (as in the Soviet TYPHOON and Dutch DOLFIJN). Partial double hull and saddle tank configurations appear to be obsolete.

To provide an elevated navigation bridge when surfaced, and sometimes a control station when submerged, submarines normally have a raised "conning tower." In designs through World War II, this frequently contained an auxiliary pressure hull (the conning tower compartment) with ship handling facilities and fire controls. In modern submarines this compartment has been eliminated and all control facilities moved to the main hull. In place of the conning tower, a "sail" or "fin" now serves mainly to house the periscopes, radio antennas, and other mast-mounted sensors.

Since Holland's day, most nonnuclear submarines have relied on battery-powered electric motors for submerged propulsion, and on internal combustion engines (first gasoline, later diesel) while on the surface. Such "diesel-electric" submarines remain the most numerous type because of the great cost of nuclear propulsion. There are several types of diesel-electric propulsion. In the direct drive configuration, the diesels and the electric motors are connected to a common propeller shaft through a clutch-and-gearbox arrangement. More commonly, however, the diesels are connected to electrical generators to power the electric motors, which drive the propeller shaft. In both arrangements, diesel power is also used to recharge the batteries.

Their limited battery capacity is the Achilles heel of diesel-electric submarines. Most can operate submerged at high speed for only a few hours (though at very low speeds the most modern types can remain submerged for more than a week). Once the batteries are discharged, the submarine must surface to recharge them. Modern diesel-electric submarines have retractable SNORKEL tubes, which can supply air to the diesels while the submarine is submerged down to 40–60 ft. While surfaced or even snorkeling, submarines are extremely vulnerable, especially to aircraft and other submarines.

Because of their slow submerged speeds (less than 9 kt.) and short underwater endurance, submarines through World War II were actually "submersibles," surface vessels which could submerge briefly to approach targets or evade pursuit. Accordingly, they had boatlike hulls optimized for high surface speed, and many carried heavy deck guns which further reduced underwater performance. Most attacks were made at night, on the surface, and at relatively short ranges. But the general use of radar by 1943 made such attacks too hazardous. In response, the German navy first fitted its U-boats with snorkels (a Dutch invention of the 1930s) to reduce surface time, and then in 1944, it introduced the revolutionary Type XXI U-boat, the first submarine intended primarily for submerged operations. With a streamlined, cigar-shaped double hull, a streamlined conning tower, a retractable snorkel, and very light gun armament, the Type XXI was the archetype of many postwar diesel-electric designs, including the U.S. TANG, the British PORPOISE and OBERON, and the Soviet QUEBEC, WHISKEY, and ZULU.

With the introduction of nuclear propulsion on USS Nautilus in 1957, submarines at last acquired a powerplant completely independent of the surface, capable of high sustained underwater speed and unlimited submerged endurance. The earliest nuclear submarines (e.g, Nautilus and the Soviet NOVEMBER class) retained the cigar-shaped hullform, but this is unstable at speeds greater than 20 kt. In 1953, the U.S. Navy built the experimental USS Albacore, with a "body of revolution" or "teardrop" hullform, which was radially symmetrical and stable at all speeds. Moreover, teardrop hulls have more internal volume than conventional hulls of equal length, allow multi-deck internal layouts, and minimize submerged drag. The U.S. mated the teardrop hull with nuclear power in the SKIPJACK class of 1959, to achieve a speed of more than 33 kt. submerged. Modified by the addition of a cylindrical center section, the teardrop hull has been universally adopted for nuclear submarines, and now for the latest diesel-electric submarines as well.

WEAPONS

The original and still predominant submarine weapon is the TORPEDO. These were originally unguided "straight-running" weapons, but since World War II acoustical homing and wire-guided torpedoes have become standard. Most submarines carry 16–24 torpedoes, launched from 4 to 10 bow or amidships tubes. Some older submarines also have 2–4 stern tubes, now used mainly to release torpedo countermeasures or lightweight anti-submarine torpedoes.

During the 1970s, torpedoes were supplemented by anti-ship missiles launched through standard torpedo tubes, such as the U.S. TOMAHAWK and HARPOON, the French EXOCET, and the Soviet SS-N-21 SAMPSON. The Soviet Union, however, deployed larger missiles much earlier, aboard specialized guided-missile submarines. For long-range anti-submarine warfare, U.S. submarines also have SUBROC and SEA LANCE missiles; the Soviet equivalents are the SS-N-15 STARFISH and the SS-N-16 STALLION. Bottom-laid influence MINES may also be released from torpedo tubes, replacing torpedoes on a 2-for-1 basis. The U.S. also has a Submarine-Launched Mobile Mine (SLMM), essentially a torpedo body which can deliver a mine into an enemy harbor. The most powerful of all submarine weapons, however, are undoubtedly the submarine-launched ballistic missiles (SLBMS) and submarine-launched cruise missiles (SLCMs) deployed for "strategic" nuclear attack.

SENSORS

More than any surface warship, submarines are dependent on their remote sensors. The earliest and still one of the most important of sensors is the periscope—essentially a hollow tube with prismatic optics at each end. Most have several magni-

fication settings and contain coincidence or stadiametric rangefinders. Modern periscopes are quite sophisticated, and may incorporate low-light television (LLTV) or IMAGING INFRARED (IIR) sensors, as well as a videotape recorder. Most submarines have two periscopes: a large-aperture "search" periscope, and a small-aperture "attack" scope, which is harder to detect but has a narrower field of view.

The primary submarine sensor, however, is SONAR. The latest submarines have integrated systems which usually include a large, bow-mounted, low-frequency passive array, an active attack sonar, a flank-mounted CONFORMAL ARRAY sonar, a passive TOWED ARRAY, an underice sonar, and a sonar intercept array (to locate, range, and classify enemy active sonars). With modern passive sonars and sophisticated signal processing equipment, submarines can detect targets at ranges in excess of 62 mi. (100 km.), depending on water conditions (see THERMOCLINE), whereas active sonar is effective at barely half that range. Moreover, active sonar reveals the presence (and often the precise location) of the submarine; hence much emphasis is placed on passive detection and, conversely, on acoustical silencing. Techniques for the latter include the soundproofing of individual pieces of machinery, the mounting of the entire powerplant on a resilient "sound isolation raft," precision propeller design to avoid cavitation, and ANECHOIC tiles or other sound-absorbent coatings for the hull.

Other sensors, usually mounted on retractable masts in the sail, include surface-search RADAR, radio direction-finders (D/F), and ELECTRONIC SIGNAL MONITORING (ESM) arrays.

COMMUNICATIONS

The HF, VHF, and UHF radio waves normally used for military communications can penetrate seawater only to a depth of 40–50 ft. (12–15 m.), while modern submarines spend most of their time at much greater depths. To contact submarines cruising deep, Very Low Frequency (VLF) or Extremely Low Frequency (ELF) radio waves are now employed. The former requires the use of a trailing antenna or an antenna buoy towed behind the submarine, while the latter requires a very long trailing wire antenna. Both frequencies have low data rates, hence messages must be very short and precoded. Often, VLF or ELF messages are used only as "alarm bells," to tell the submarine crew to rise to periscope depth, so that a longer message can be transmitted on a higher frequency. The lack of speed and reliability in submarine communications is the major weakness of ballistic-missile submarines, and would generally preclude their use in COUNTERFORCE strikes. Experiments are now being conducted with blue-green LASERS (which can penetrate seawater) for underwater communications, but the results are inconclusive. Underwater telephones, used for submarine-to-submarine communications, combine active sonar transducers with hydrophones; they can exchange voice messages over short distances.

For specific trends in submarine development, see separate SUBMARINES entries for BRITAIN; CHINA; FEDERAL REPUBLIC OF GERMANY; FRANCE; ITALY; JAPAN; NETHERLANDS; SOVIET UNION; SWEDEN; UNITED STATES. See also ANTI-SUBMARINE WARFARE.

SUBMARINES, BRITAIN:

DIESEL-ELECTRIC

After the Royal Navy acquired several Type XXI U-boats at the end of World War II, some British wartime "T"- and "A"-class submarines were streamlined and equipped with new sensors in overhauls similar to the U.S. GUPPY program; several remained in service through the 1970s. The first operational postwar British submarines were the eight 2410-ton Porpoise-class diesel-electric attack submarines commissioned between 1956 and 1959. Combining features of the Type XXI and the U.S. Tang class, they were fairly fast, and were extremely quiet, making them excellent ANTI-SUBMARINE WARFARE (ASW) platforms. All eight were retired by 1984 to fund new construction.

The subsequent OBERON class was nearly identical, but had a glass-reinforced plastic sail and upper deck, and better electronics. Thirteen were built for the Royal Navy between 1960 and 1967, 6 for Australia, 3 for Brazil, 3 for Canada, and 2 for Chile.

The 2400-ton UPHOLDERS (Type 2400) are the first British diesel-electric submarines in more than 20 years. During the 1960s, the Royal Navy had adopted an "all-nuclear" policy for submarines, but high costs and a lack of qualified shipyards forced a change by the late 1970s. Still quite expensive, four of these state-of-the-art vessels are scheduled for completion by 1993, with eight more planned.

NUCLEAR-POWERED ATTACK

British research on nuclear propulsion for submarines began in 1946, but it was 1965 before an indigenous British reactor could enter service. In the interim, the Royal Navy purchased an S5W reactor from the U.S. to power its first nuclear attack submarine, HMS *Dreadnaught,* which served until 1981. The five 4900-ton VALIANT SSNs, commissioned between 1966 and 1970, were the first all-British nuclear submarines; essentially enlarged and refined Dreadnaughts with improved sensors, fire controls, and acoustical silencing, all are to remain in service through the 1990s.

The six 4500-ton SWIFTSURES commissioned between 1971 and 1979 were of a completely new design, with a very broad, short stern cone to maximize internal volume as well as much better sensors and acoustical silencing. All are to remain in service through the end of the century. The latest 5300-ton TRAFALGAR class combines the Swiftsure hullform with a new, very quiet "natural circulation" reactor, an innovative pumpjet propulsor, and an ANECHOIC coating to reduce radiated noise and impede hostile active sonars. Among the quietest of nuclear submarines, seven were commissioned between 1983 and 1991; a follow-on class (SSN 20) is now in advanced development.

NUCLEAR-POWERED BALLISTIC-MISSILE

The British ballistic-missile submarine program was started in 1962, after the cancellation of the U.S. Skybolt STANDOFF MISSILE left the RAF V-bomber force without a suitable nuclear weapon. The U.S. then agreed to help the Royal Navy build its own nuclear-powered ballistic-missile submarines (SSBNs) for U.S. POLARIS A3 missiles (with British nuclear warheads). The result, the four 8400-ton RESOLUTION-class SSBNs commissioned between 1966 and 1980, now comprise the British nuclear force. They will be replaced in the late 1990s by four 15,850-ton Vanguard-class SSBNs, each armed with 16 TRIDENT D-5 long-range SLBMs.

SUBMARINES, CHINA: The Chinese People's Liberation Army's navy operates a large number of submarines, mainly for coastal defense. Soviet models built in China are now being supplemented by several indigenous designs. Chinese submariners are said to be competent and aggressive, but are hindered by obsolete equipment.

China received several Soviet WHISKEY-class diesel-electric attack submarines by the early 1960s, and copied a dozen more from 1964, of which about 15 remain in service. The U.S.S.R. also supplied a few ROMEO-class attack submarines, and that design was placed in mass production, with more than 100 built between 1960 and 1982. Seven have been transferred to North Korea (which is now building its own copies), and 4 were sold to Egypt in 1982–83. Some Chinese Romeos may yet be refitted with Western sensors and fire controls.

The Romeo was developed into the E5SE (code-named "Ming" class), three of which have been completed since 1975. The Ming, in turn, was later developed into the E5SG guided-missile submarine with six (surface-launched) C801 Yingi ANTI-SHIP MISSILES. Roughly equivalent to the Soviet JULIETT class, the first E5SG was launched in 1986.

The Chinese nuclear-powered-submarine program was initiated in the early 1960s, and the first of four HAN-class attack submarines was laid down in 1965. Construction was severely delayed by the Cultural Revolution, and the Hans were finally commissioned between 1974 and 1988. These 4500-ton submarines have a modern teardrop hull-form and a twin-shaft powerplant based on a pressurized-water reactor with turbo-electric drive.

The Soviet Union also provided China with the plans of the diesel-powered GOLF ballistic-missile submarine before the breakdown of military cooperation. One was completed in 1964, but the development of a suitable submarine-launched ballistic missile proceeded more slowly, and the first launch of a Chinese SLBM did not occur until 1982. By that time, the Chinese navy had launched the first of four 700-ton XIA-class nuclear-powered ballistic-missile submarines, armed with 12 CSS-N-3 missiles. See also SLBMS, CHINA.

SUBMARINES, FEDERAL REPUBLIC OF GERMANY: Germany had been the world leader in submarine technology, but from 1945 to 1954, all developmental work ceased. After the establishment of the Federal Republic of Germany, the Western powers agreed to let the new German navy construct coastal submarines of up to 350 tons, for operations in the Baltic. To train crews while the new submarines were being built, West Germany raised and refurbished one Type XXI and two smaller Type XXIII U-boats scuttled in 1945.

The first postwar German submarines were the 450-ton TYPE 205s, 14 of which were completed between 1961 and 1968 by ILK shipyards. Six remain in service; 2 Improved Type 205s were also built for Denmark in 1968–69. The Improved Type 205 was further developed for the Norwegian navy as the TYPE 207 (Kobben class). Fifteen were built between 1964 and 1967, of which 11 remain in service. The Type 205 was followed by the 600-ton TYPE 206. Eighteen were built for the German navy between 1971 and 1974; all remain in service. Three modified (Gal-class) variants were built under license in Britain for the Israeli navy.

After completing the Types 205 and 206, IKL developed designs for export. The most successful so far has been the TYPE 209 medium-range, oceangoing submarine. More than 40 have been sold in five distinct versions of 1290, 1390, 1440, and 1850 tons displacement, as well as a 600-ton coastal variant. Type 209s are now in service with Argentina, Columbia, Ecuador, Greece, Indonesia, Peru, Turkey, and Venezuela.

During the mid-1970s, Thyssen-Henschel shipyards entered the submarine design field in competition with IKL, but with a focus on much larger vessels, e.g., their 3364-ton TR-1700 long-range submarine. Six were ordered by Argentina, and 2 were delivered in 1984–85, before funding was cut off. In 1982, Thyssen was awarded a contract to built 6 new 1300-ton TYPE 210 (Ula-class) submarines to replace Norway's Type 207s. Thyssen is also to build up to 12 1450-ton TYPE 211s, to replace West Germany's remaining Type 205s during the early 1990s. A planned Type 212 is to have closed-cycle fuel cells, instead of a diesel-electric powerplant.

SUBMARINES, FRANCE:

DIESEL-ELECTRIC

The first French postwar submarines, of the Narval class, were essentially refined versions of the German Type XXI U-boat. Six of these 1910-ton attack submarines, commissioned between 1954 and 1958, were finally retired in the mid-1980s. The subsequent four Aretheuse-class coastal submarines, commissioned between 1955 and 1958, displaced only 669 tons and proved too small to be effective; all were retired by the early 1980s.

The 1043-ton DAPHNE class, 25 of which were completed between 1958 and 1967, are enlarged derivatives of the Aretheuse. Eleven are in service with the French navy, and 10 were built for export to South Africa, Pakistan, and Portugal (Spain built 4 under license). Two Daphnes were lost in 1970–71 because of faulty snorkels.

The Daphnes were followed by the larger 1725-ton AGOSTA class; 4 were commissioned in 1977–78, and 2 more were sold to Pakistan; Spain also built 4 under license. It is the announced intention of the French navy to acquire only nuclear submarines in the future.

NUCLEAR-POWERED BALLISTIC-MISSILE

The French nuclear submarine program was specifically meant to create an all-French ballistic missile force. Unlike Britain, France received no U.S. technical assistance, yet managed to produce both nuclear-powered ballistic-missile submarines (SSBNs) and their missiles, fairly quickly and fairly cheaply. The 9000-ton REDOUTABLE SSBNs, six of which were completed between 1967 and 1985, loosley resemble the U.S. LAFAYETTES, but have turbo-electric drive and a natural circulation reactor. The last of this class (Inflexible) was built to a slightly modified design and is sometimes listed as a separate class. The Redoutables will be replaced by the 14,200-ton Triomphant class from the mid-1990s.

NUCLEAR-POWERED ATTACK

A nuclear attack submarine (SSN) program did not begin until 1974, because of the higher priority assigned to the SSBNs. The resulting 2670-ton RUBIS class, four of which were commissioned between 1983 and 1988, are the smallest operational SSNs in the world. They are to be supplemented in the 1990s by the Amethyste SSNs, which are slightly longer with a hydrodynamically refined hull. See also MSBS for details of French submarine-launched ballistic missiles.

SUBMARINES, ITALY: After World War II, the Italian navy operated 3 of its surviving wartime boats and 5 U.S. GUPPY types, all now retired. The first postwar Italian design was the TOTI class,

4 582-ton coastal anti-submarine "hunter-killer" submarines commissioned in 1967–68. The Totis were followed by 6 oceangoing, 1631-ton SAURO attack submarines, commissioned between 1976 and 1988; the last 2, built to a modified design, are sometimes known as the Pelosi class. Italy still operates one U.S. TANG-class diesel-electric attack submarine, acquired in 1974.

SUBMARINES, JAPAN: The Japanese Maritime Self-Defense Force (JMSDF) was first equipped with a number of ex-U.S. Fleet and GUPPY submarines. The first postwar Japanese design was the Oyashio class of small coastal submarines commissioned from 1959; these were followed by four improved Hayashio-class submarines, commissioned in 1961–62. All were retired in the early 1970s. They were followed by the 1650-ton Ooshio class of five submarines commissioned between 1963 and 1967. Based on the U.S. TANG design, the Ooshios were retired by 1986.

The 7 1850-ton UZUSHIOS commissioned between 1970 and 1977 have a teardrop hullform based on the U.S. BARBEL class. The subsequent 2200-ton YUUSHIO class of 11 submarines, commissioned between 1980 and 1990, are improved Uzushios. Both the Yuushios and Uzushios have excellent SONAR and FIRE CONTROLS. In every respect except submerged endurance they are comparable to nuclear submarines.

SUBMARINES, NETHERLANDS: The first post–World War II Dutch submarines were 4 DOLFIJN-class boats commissioned between 1959 and 1965. With a displacement of 1826 tons, the Dolfijns have a unique double-hull configuration with three separate pressure hulls. To reduce cost and complexity, the Dutch navy reverted to a more conventional layout for 2 subsequent 2640-ton ZWAARDVIS-class submarines commissioned in 1970–71, whose design was derived from the U.S. BARBEL class; 2 were purchased by Taiwan in 1987. The Zwaardvis have been supplemented by 2 new ZEELEEUW (originally Walrus) class boats commissioned in 1989–90; 2 more are scheduled for completion in 1992–93, and a final 2 are on order.

SUBMARINES, SOVIET UNION: For the Soviet navy, the submarine is the decisive naval weapon, with surface ships relegated to secondary and supporting roles. It has the world's largest and most varied submarine fleet, and vast resources have been dedicated to the improvement of submarine design and of underwater sensors and weapons. Once markedly inferior to Western designs in such vital areas as acoustical silencing and sensors, recently Soviet submarines have substantially narrowed the qualitative gap.

DIESEL-ELECTRIC ATTACK

In World War II, the Soviet navy had a large but generally ineffectual submarine force. After the 1945 capture of several German Type XXI U-boats, and components for several more, the Soviet navy built three new submarines during the 1950s, all derived from the Type XXI: the 740-ton Quebec class, for coastal defense; the 1350-ton, medium-range WHISKEY class; and the 2300-ton, long-range Zulu class.

The 25 Quebecs built between 1950 and 1955 proved too small to be useful, and were discarded in the 1970s. By contrast, the Whiskeys were very successful, and more than 236 were built between 1951 and 1957. A number were exported to Albania, Bulgaria, Egypt, Poland, Indonesia, and China (which also produced a copy). Many Whiskeys were converted for special roles, including the Whiskey "Canvas Bag" radar PICKET, and the "Twin Cylinder" and "Long Bin" conversions with SS-N-3 SHADDOCK anti-ship missiles. Some basic Whiskeys remain in Soviet service. The Zulu class of some 35 boats completed between 1951 and 1957 was an enlarged version of the Whiskey. From 1955 to 1957, at least 7 were converted into makeshift ballistic-missile submarines with two SS-N-4 Sarks each. Only three "Zulu IVs" remain in service as research vessels.

In the mid-1950s the Whiskey was superseded by the similar but refined ROMEO class, with a displacement of 1800 tons. Only 20 were built between 1958 and 1962, because a change in Soviet doctrine shifted the mission of offshore defense to bombers and guided-missile submarines. All Soviet Romeos have been sold or placed in reserve; several were transferred to China, which produced some 100 copies between 1960 and 1982. Chinese Romeos in turn have been exported, and the type is still being built in North Korea.

For its part, the Zulu was superseded by the 2400-ton FOXTROT class, some 62 of which were built between 1958 and 1983. Many Soviet Foxtrots remain active, and 17 more, built specifically for export, have been sold to Cuba, India, and

Libya. The Foxtrot hull also served as the basis for the JULIETT-class SSG, the Golf-class SSB, and the India-class minisub carriers. In line with the Soviet navy's belief in the continued value of diesel-electric submarines, the Foxtrots were followed by the much improved 3900-ton TANGO class, 22 of which were built between 1972 and 1982.

In 1982, the Soviet navy introduced the totally new, 2900-ton KILO class, the first Soviet diesel submarines with a teardrop hullform, and capable of 20 kt. submerged for short bursts. More than 14 are in Soviet service, and some 15 have been exported to Algeria, India, Romania, and Poland.

NUCLEAR-POWERED, ATTACK

The Soviet nuclear submarine program began in the early 1950s, and the first 14 NOVEMBER-class attack submarines (SSNs) were completed between 1958 and 1962. One has been lost at sea, while another was scrapped after a nuclear accident. Despite their shortcomings, the survivors still remain in service. The November also served as the basis for the ECHO-class guided-missile submarines (SSGNs), and the Hotel-class ballistic-missile submarines (SSBNs). The Novembers were followed by the VICTOR class, at least 44 of which have been completed (in three subtypes) since 1967.

In 1979 the Soviet navy introduced the revolutionary ALFA class, a rather small, 3680-ton attack submarine whose 45-knot submerged speed and 4000-ft. (1220-m.) maximum depth shocked Western navies. Six were completed between 1979 and 1983 with minor variation, clearly as testbeds for advanced technologies incorporated into later designs.

The Soviet navy has recently produced three separate SSN classes: the AKULA, MIKE, and SIERRA. The 10,000-ton Akulas, possibly successors to the Victor III, are almost as quiet as the best Western SSNs; at least three have been completed since 1984. The 6400-ton Mike was believed to be an enlarged Alfa follow-on because of its titanium pressure hull and liquid metal reactor. Only one was completed in 1986; it was lost at sea in April 1989. The 7550-ton Sierras are apparently anti-ship successors to the Victor class. Capable of 34–36 kt. submerged, the Sierras are almost as quiet as the U.S. LOS ANGELES SSNs, in part because of propeller milling technology acquired covertly from Japan and Norway. Two have been completed since 1983.

GUIDED-MISSILE

In the late 1950s, the Soviet navy determined that the best way of attacking U.S. carrier battle groups was to use long-range anti-ship missiles; consequently it developed specialized guided-missile submarines (SSGs) for the purpose. The first Whiskey "Twin Cylinder" and "Long Bin" conversions, with two and four SS-N-3 Shaddock missiles, respectively, were extremely crude, and Shaddock could be launched only from the surface; all have been retired. These were followed by more refined Shaddock launchers, the nuclear-powered Echo-class SSGNs and the Juliett-class SSGs (the former for open-ocean operations, the latter for in-shore defense). The Echoes, derived from the Novembers, were built in two subclasses. The five 5500-ton Echo Is, built between 1960 and 1962 and converted to SSNs between 1970 and 1974, were followed by 29 6000-ton Echo IIs built between 1962 and 1967. Fourteen 3750-ton Julietts were built between 1961 and 1968, and all remain in service.

The subsequent CHARLIE-class SSGNs, apparently derived from the Victor class, marked a significant advance. The original Charlie I, displacing 5000 tons, is armed with eight SS-N-7 SIREN missiles—the first anti-ship missiles capable of submerged launch. Of the 11 Charlie Is built between 1968 and 1972, 1 was lost at sea in 1983, while another was transferred to India in 1988. The 6 5500-ton Charlie IIs completed between 1973 and 1982 are longer and carry eight improved SS-N-9 STARBRIGHT missiles, but are otherwise similar to the Charlie I.

Completed in 1970, the single 8000-ton PAPA SSGN is now regarded as a testbed for components of the subsequent 14,500-ton OSCAR class, four of which have been completed since 1982. The huge Oscar is capable of 35 kt. submerged, has a maximum operating depth of 1640 ft. (500 m.), and is very quiet.

BALLISTIC-MISSILE

The Soviet ballistic-missile submarine (SSB) program began in the early 1950s, and by 1955 the U.S.S.R. succeeded in launching an SS-1a SCUD short-range ballistic missile from a surfaced Zulu submarine. From 1956 to 1958, five Zulus were converted to Zulu V SSBs, each with two SS-N-4 Sark missiles. Extremely crude, the Zulu Vs were mainly of propaganda value, but probably provided useful data for the design of the Golf-class SSBs (23 built from 1958 to 1962) and the Hotel-class SSBNs (eight built from 1959 to 1962). Both types had three launch tubes in an extended sail.

Compared with the U.S. POLARIS, the Zulu V, Golf, and Hotel were all extremely primitive, and it was not until 1967 that the Soviet navy was able to acquire a comparable capability in its YANKEE-class SSBNs, 34 of which were completed between 1967 and 1974. Displacing 9300 tons and capable of 27 kt. submerged, the Yankees were obviously imitative of the U.S. LAFAYETTE class. To comply with SALT Treaty limitations, at least 14 Yankees have since been converted to other roles. Some are now used as attack submarines (after removal of their missile tubes), while others have been converted into SSGNs armed with SS-N-21 SAMPSON and SS-N-24 cruise missiles.

The Yankees were followed in 1974 by the considerably larger and more sophisticated DELTA class. The original 11,750-ton Delta I was armed with 12 SS-N-8 missiles; 18 were completed between 1972 and 1977. They were followed in 1974–75 by 4 12,750-ton Delta IIs, armed with 16 missiles. The 13,250-ton Delta III has a raised whaleback housing for 16 SS-N-18 STINGRAY missiles with up to 7 MULTIPLE INDEPENDENTLY TARGETED REENTRY VEHICLE warheads each; 14 were completed between 1975 and 1985. The last variant is the 13,550-ton Delta IV, of which at least 4 have been completed since 1985. It has an even larger whaleback housing for 16 SS-N-23 SKIFF missiles, each with 10 MIRVs.

In 1981, the Soviet navy commissioned the first of the TYPHOON-class SSBNs, at 30,000 tons the largest submarines in the world; four have been completed to date. Both Typhoon and Delta IV are believed to be intended for deployment within defended ocean "bastions" under the arctic ice near Soviet home waters; they would launch their missiles from open "ponds" (*polynas*) in the ice. See also SLBMS, SOVIET.

SUBMARINES, SWEDEN: The first post-1945 Swedish submarines of the Hajen class were closely based on the German Type XXI U-boat of 1944–45. The six built between 1956 and 1960 have since been retired. They were followed by six derived 1110-ton Draken-class submarines commissioned in 1960–61; now obsolescent, they will be retired during the 1990s.

The Drakens were followed in turn by five 1400-

ton SJOORMAN-class submarines commissioned in 1967–68. The Sjoormans have a modified teardrop hullform and also introduced the innovative, computer-controlled X-configuration stern plane found on all subsequent Swedish submarines, as well as the Dutch ZEELEEUW class. The subsequent 1125-ton NAKEN class is generally similar, but even more automated (the standard complement is only 18 men). Three were commissioned in 1980–81. The latest Swedish submarines are the four 1140-ton VASTERGOTLANDS commissioned between 1987 and 1990. The Swedish navy is now investigating closed-cycle propulsion systems.

SUBMARINES, UNITED STATES:

DIESEL-ELECTRIC

The U.S. Navy ended World War II with a force of some 200 modern "Fleet" submarines of the similar 2400-ton Gato, Balao, and Tench classes. Trials with captured German Type XXI U-boats revealed that they were greatly superior in submerged speed and endurance, operating depth, acoustical silencing, and sonar. Because it was then impossible to fund new construction, the U.S. Navy initiated the GUPPY (Greater Underwater Propulsion) program, modifying 41 Balaos and Tenches between 1948 and 1954 by streamlining their outer hulls, installing snorkels and more powerful batteries, and fitting improved sonar and fire controls. With several equipment upgrades, they continued to serve until the mid-1970s, and were also widely exported. Nineteen remain in service with the navies of Brazil, Greece, Peru, Taiwan, Turkey, and Venezuela. Other World War II fleet submarines were converted for various specialized roles, including radar PICKETS and COMMANDO carriers.

In 1951 the navy introduced its first postwar submarines with the 2700-ton TANG class, 6 of which were commissioned in 1951–52. Combining the best features of the Type XXI and the Fleet submarines, the Tangs were optimized for submerged operations, but were plagued by engine problems. Four were sold to Italy and Turkey in the early 1970s, and the remaining 2 were laid up in mothballs. The Tangs were followed in 1956 by the 2388-ton Darter, an improved version with more reliable engines. Only 1 was built, but 2 other Darter hulls were converted into Greyback-class guided-missile submarines as a MISSILE GAP expedient. Commissioned in 1957, they had a large

bow hangar for a surface-launched Regulus I cruise missile with a nuclear warhead; both were retired in the early 1970s.

In 1958 the navy launched the experimental submarine *Albacore*, whose revolutionary "teardrop" hullform allowed aircraftlike maneuvers and submerged speeds in excess of 30 kt. *Albacore* was used as a testbed for a variety of sensor, propulsion, and control surface schemes, many of which were incorporated into later U.S. submarines. The next operational U.S. submarines, the three 2894-ton BARBELS, had teardrop hulls based on the *Albacore* design. Commissioned in 1958–59, they were the last diesel-electric submarines built by the U.S., but were also the basis for the Dutch ZWAARDVIS and Japanese UZUSHIO and YUUSHIO classes.

U.S. nuclear submarine development efforts were initiated in 1949 under the direction of Capt. (later Adm.) Hyman Rickover, the dominant force in U.S. submarine development for 35 years through his control of the entire nuclear propulsion program.

NUCLEAR-POWERED ATTACK

The first U.S. nuclear submarine, the 4040-ton *Nautilus*, commissioned in 1954, made history with a submerged transit to the North Pole in 1958. *Nautilus* was followed in 1957 by the generally similar *Sea Wolf*, which tested a liquid-sodium reactor; this proved unreliable, and all subsequent U.S. nuclear submarines have been powered by pressurized-water reactors (see NUCLEAR PROPULSION). *Nautilus* and *Sea Wolf* were experimental designs, though capable of combat operations, and had conventional hullforms based on the Tang class. The first operational U.S. nuclear attack submarines (SSNs), the four 2547-ton Skates commissioned in 1957–58, served until the early 1980s. The Skate design was the basis of the nuclear guided-missile submarine (SSGN) *Halibut*, a nuclear version of the Greybacks armed with the Regulus II CRUISE MISSILE. At the same time, the U.S. commissioned the radar picket submarine *Triton*. The largest submarine built till then, Triton resembled a stretched Skate and had two separate nuclear reactors. Both *Halibut* and *Triton* were retired during the 1970s.

The next U.S. submarines, the SKIPJACK class, combined nuclear power with the teardrop hullform of the *Albacore*. Commissioned between

1958 and 1960, the six 2523-ton Skipjacks were the fastest submarines yet built, with a maximum submerged speed of more than 33 kt. They introduced the single-hull configuration (with no outer casing), single-screw propulsion, and sail-mounted ("fairwater") diving planes, which became standard on subsequent U.S. submarines. Three remain in service; one, USS *Scorpion*, was lost off the Azores in 1968, and two others were retired in 1987. The Skipjacks were followed by *Tullibee*, a small (2640-ton) experimental submarine intended as a specialized anti-submarine "hunter-killer." *Tullibee*'s advanced features included a large, bow-mounted sonar array that displaced the torpedo tubes amidships, and turbo-electric drive. But with a top speed of only 20 kt. submerged, it was considered too slow, as well as too small to accommodate enough weapons and sensors, and the design was not repeated.

The U.S. Navy opted instead for a larger, general-purpose attack submarine (SSN) design, the PERMIT class (originally the Thresher class, after the lead boat, lost on diving trials in 1963). At 4300 tons, the Permits were significantly larger than the Skipjacks, and the additional displacement was used for additional sensors (including a bow sonar array), and better acoustical silencing, thus beginning a trend towards larger, quieter submarines. With the same powerplant as the Skipjacks, the larger Permits had a maximum speed of only 27 kt., but fabricated of super-strong HY-80 steel, their hulls have a collapse depth of 1970 ft. (500 m.). The Permits were the first submarines capable of launching the SUBROC anti-submarine missile in addition to torpedoes, and were later modified to launch HARPOON anti-ship missiles as well. Fourteen Permits were commissioned between 1961 and 1966; the 13 survivors are scheduled for retirement in the 1990s.

The subsequent STURGEON class, 37 of which were commissioned between 1966 and 1974, were enlarged and refined Permits. Displacement increased to 4640 tons to accommodate more silencing and additional electronics; because they retained the Skipjack's reactor, their maximum submerged speed was reduced to some 26 kt. All remain in service. The 5350-ton *Narwhal*, commissioned in 1967, is a modified Sturgeon with an S5G natural-circulation reactor, which minimizes the use of noisy cooling pumps by relying on convection cooling alone at low speeds. USS *Glennard P. Lipscomb*, commissioned in 1973, is another experimental submarine intended to test improved tech-

niques of acoustical silencing. Specialized features include turbo-electric drive, the individual soundproofing of machinery, and the mounting of the entire powerplant on a resilient sound isolation raft. Weapons and sensors are similar to those of the Sturgeon class.

The next attack submarines class was the LOS ANGELES, some 38 of which have been commissioned to date, with a further 24 on order; the Los Angeles are the single largest class of SSNs in the world. Designed specifically to recover the speed lost since the Skipjacks, the Los Angeles displace 6927 tons, but with their 30,000-shp. powerplant they have a maximum speed of 31 kt. submerged. Their acoustical silencing continued to be enhanced, and the later versions are thought to be the quietest nuclear attack submarines in the world. Later versions also have 12 vertical launch tubes in the bow for TOMAHAWK cruise missiles.

The SEA WOLF (SSN-21) class, now in development, is intended to match the speed and depth of the latest Soviet submarines while being even quieter than the Los Angeles class, and without the latter's defects—notably its limited armament and lack of underice capability. The Sea Wolfs are to displace some 9150 tons, with a maximum submerged speed of 35 kt., and a maximum operating depth of more than 1800 ft. (550 m.). They are to be armed with eight 26-in. (650-mm.) tubes for a basic load of 36 weapons, including the MK.48 ADCAP torpedo, Harpoon, Tomahawk, and the SEA LANCE ASW-SOW. The Sea Wolf design has been criticized for its cost (more than $1 billion each), size, and technological conservatism. The first is scheduled for completion by 1996.

NUCLEAR-POWERED BALLISTIC-MISSILE

The U.S. ballistic-missile submarine program was initiated in the mid-1950s when a crash program led to development of the POLARIS submarine-launched ballistic missile (SLBM). The first U.S. nuclear-powered ballistic-missile submarines (SSBNs) of the George Washington class were derived from the Skipjacks; in fact, the first two George Washingtons were actually Skipjack hulls modified while still under construction by the insertion of a 130-ft. (39.6-m.) section behind the sail for 16 missile launch tubes. This configuration became standard for subsequent U.S. SSBNs, and was also copied by the British, French, Soviet, and Chinese navies. Commissioned in 1960–61, the five George Washingtons displaced 6888 tons and had a maximum submerged speed of 27 kt. They were followed by five 7880-ton Ethan Allen—class SSBNs commissioned between 1960 and 1962. Generally similar to the George Washingtons, the Ethan Allens' additional displacement was devoted mainly to better sensors and acoustical silencing. Limited to the Polaris missile by the diameter of their launch tubes, the George Washingtons and Ethan Allens were converted to attack submarines in the early 1980s, and were all retired by 1988.

Continuing a process of incremental development, the Ethan Allens were followed by 31 8250-ton LAFAYETTE-class SSBNs commissioned between 1962 and 1966 with improved sensors, enhanced acoustical silencing, and enlarged missile tubes. Initially armed with Polaris, they were refitted with the 10-warhead POSEIDON missile during the 1970s. The last 12 submarines (the "Franklin" class) have been rearmed with the heavier TRIDENT I missile, but their missile tubes are too small for the later Trident II. Earlier Lafayettes are now being retired to conform to SALT limits on SLBM launchers.

The latest U.S. ballistic-missile submarines, of the OHIO class, were designed during the 1970s specifically for the large Trident II missile. At 18,-700 tons, they are second in size only to the Soviet TYPHOON class. The Ohios are presently armed with the Trident I, but will be rearmed with Trident II in the 1990s. Capable of 24 kt. submerged, the Ohios have much acoustical silencing and are considered the quietest of all nuclear-powered submarines in service. Eight have been commissioned since 1981, with seven more under construction and an additional five on order. See also SLBMS, UNITED STATES.

SUBROC: UUM-44A Submarine Rocket, a U.S. submarine-launched anti-submarine missile. Development began in 1958 when it was realized that sonar detection ranges (40+ mi./60+ km.) greatly exceeded practical torpedo ranges. SUBROC was intended to provide attack submarines with a long-range standoff weapon which could minimize the "dead time" between launch and impact. Introduced in 1965, SUBROC was deployed aboard the PERMIT class and all subsequent U.S nuclear-powered attack submarines.

SUBROC is launched from standard 21-in. torpedo tubes. It consists of a 5-kT W58 nuclear

DEPTH CHARGE delivered by an inertially guided solid-fuel rocket. Steering is provided by jet deflectors in the rocket nozzle. In action, target heading and range provided by the submarine's sonar are fed to a Mk.113 FIRE CONTROL computer, which determines the desired impact point, launch time, and time of flight. When these data are transferred to the missile's guidance unit, the missile is ejected from the torpedo tube. At a safe distance from the submarine, the rocket motor ignites, propelling the missile to the surface. It then turns to the programmed target heading and flies a parabolic trajectory to the target area. At a preprogrammed time before impact, the depth charge separates from the rocket, flying ballistically to the impact point. Minor course corrections can still be made with two small fins at the base of the depth charge. Shortly before impact, a braking parachute deploys from the tail, to slow the charge for water entry. The depth charge is then detonated at a preset depth by a hydrostatic fuze.

A nuclear depth charge was deemed necessary to compensate for inherent errors in fire control and missile guidance, and for the residual dead time (some two minutes) during which the target could deviate from its predicted course. On the other hand, as a nuclear weapon, SUBROC's utility was limited to conditions of GENERAL WAR. As a result, SUBROC is now being replaced by the Sea Lance Anti-Submarine Stand-Off Weapon (ASW-SOW), which is similar but more accurate, and armed with a Mk.50 BARRACUDA homing torpedo. See also ANTI-SUBMARINE WARFARE.

Specifications Length: 22 ft. (6.71 m.). Diameter: 21 in. (533 mm.). Weight, launch: 4000 lb. (1814 kg.). Speed: Mach 1.5 (1150 mph/1830 kph). Range: 35 mi. (56 km.).

SUBVERSION: Activities meant to achieve hidden but effective control over a population or group nominally under the control of the overtly established authority. Subversion is the principal technique of REVOLUTIONARY WAR. In Vietnam, for example, the agents of subversion (the Vietcong) extracted food, conscripts, and information from populations theoretically governed by the Republic of Vietnam. When this level of subversion is achieved, the constituted government remains in apparent control (because its military and security forces "hold" the territory), but the population is actually administered by the agents of subversion.

The tools of subversion are PROPAGANDA and TERRORISM. The former conditions the population to accept the covert administration, while the latter is used to intimidate or eliminate those who resist or could lead popular resistance (including entire "classes"); its main purpose is to reduce the population into isolated individuals, unwilling to risk their own lives to oppose the process. Once the population is thus atomized, propaganda plus the mere threat of terrorism are sufficient to control it, and the agents of subversion can then covertly collect food, conscripts, and information by simple request, instead of at gunpoint. Those resources are then used to support the GUERRILLA, the other arm of revolutionary war.

Subversion usually requires a covert network of "civilians" operating in their home communities, and also a separate group of covert agents. The latter may be part of the guerrilla forces, but they have a liaison role between the guerrillas and the "civilians." The latter carry out propaganda action, act as informers for the executioners, and collect combat intelligence for the guerrillas. This enables terrorism to be aimed against carefully selected targets (e.g., the honest government official rather than than the corrupt one, who is useful for propaganda); at the same time, surviving police authorities cannot find the executioners, who do not belong to the locality and who only come into it briefly, to act. The more subversion, the more food, information, and conscripts can be supplied to the guerrillas, who can then further reduce the government presence, which in turn facilitates further subversion. The population may prefer to withhold its crops and sons from the guerrillas, but it cannot deny them unless it is effectively protected. Only local security and counter-terrorism efforts can protect the population, not air power or the maneuver of conventional forces.

SUFFREN: A class of two French guided-missile DESTROYERS, commissioned in 1967 and 1970. Originally classified as light cruisers, the Suffrens are intended as ANTI-AIR WARFARE (AAW) escorts for the CLEMENCEAU-class aircraft carriers; they were the first French warships specifically designed to carry surface-to-air missiles. The Suffrens have a blocky superstructure concentrated forward, surmounted by a large, mushroom-shaped radome and a large "mack" (combination mast and stack); there is also a small, detached deckhouse aft. Their relatively low freeboard is typical of French warships and not a handicap for Mediterranean operations. Three pairs of gyro-controlled fin stabilizers improve seakeeping.

Armament consists of 2 single 100-mm. DUAL PURPOSE guns forward, a twin-arm MASCURA sur-

face-to-air missile launcher (with 48 missiles) aft, 4 EXOCET MM.38 anti-ship missile canisters and a MALAFON anti-submarine missile launcher (with 13 missiles) amidships, 4 fixed 21-in. (533-mm.) tubes for L5 anti-submarine homing TORPEDOES, and 4 manually operated 20-mm. ANTI-AIRCRAFT guns. These ships have no effective close-in defense against aircraft or surface-skimming missiles. See also DESTROYERS, FRANCE.

Specifications Length: 517 ft. (157.6 m.). **Beam:** 57 ft. (17.37 m.). **Draft:** 23.75 ft. (7.25 m.). **Displacement:** 5090 tons standard/6090 tons full load. **Powerplant:** twin-shaft steam: 4 oil-fired boilers, 2 sets of geared turbines, 72,500 shp. **Speed:** 32 kt. **Range:** 2000 n.mi. at 30 kt./2400 n.mi. at 29 kt./5100 n.mi. at 18 kt. **Crew:** 355. **Sensors:** 1 DRB123 air-search and target designation radar (under the large radome), 1 DRBV50 surface-search radar, 2 DRBR51 Mascura guidance radars, 1 DRBC33 fire control radar, 1 DRBN32 navigation radar, 1 Piranha III electro-optical/laser tracking unit, 1 DUBV33 hull-mounted medium-frequency sonar, 1 DUBV43 variable depth sonar. **Electronic warfare equipment:** 1 ARBB32 electronic countermeasures array, 2 Sagaie and 2 Dagaie chaff rocket launchers.

SUMNER: A class of 70 U.S. DESTROYERS commissioned in 1944–45, no longer in U.S. service but still in service elsewhere. Essentially a more heavily armed development of the FLETCHER class, the Sumners are flush-decked, with twin stacks and a relatively small superstructure forward. As built, they had 6 5-in. 38-caliber DUAL PURPOSE guns in 3 twin mounts (2 forward, 1 aft), 12 40-mm. BOFORS GUNS in 2 twin and 2 quad mounts, 11 20-mm. ANTI-AIRCRAFT guns, 10 21-in. torpedo tubes in 2 quintuple mounts, and 2 DEPTH CHARGE racks. After 1945, the Sumners (and the slightly larger Gearings) were retained in active service, but by the late 1950s it was apparent that these ships were fast approaching obsolescence, and there was no possibility of replacement by equal numbers of new ships. As a substitute, the U.S. Navy instituted the Fleet Rehabilitation and Modernization (FRAM) program, a major overhaul intended to extend the service lives of these ships by some ten years. Most of the Gearings received the full "FRAM I" upgrade, while 33 of the smaller Sumners received the lesser FRAM II modification, which included a thorough overhaul of the hull and machinery, revised armament, and modest improvements to the sensors.

Under FRAM II, the 40-mm. guns, torpedo tubes, and depth charges were removed, to be replaced by an enclosed bridge; two HEDGEHOG depth charges mortars; a DASH (Drone Anti-Submarine Helicopter) landing pad, hangar, and control facilities; two sets of Mk.32 triple tubes for lightweight anti-submarine homing TORPEDOES; and an SQS-36 VARIABLE DEPTH SONAR. In addition, the original SQS-4 hull-mounted SONAR was upgraded, and new ELECTRONIC COUNTERMEASURES installed. Current operators have further modernized the sensors, and some have also installed EXOCET or HARPOON anti-ship missiles. See also DESTROYERS, UNITED STATES.

Specifications Length: 376.5 ft. (114.76 m.). **Beam:** 40.83 ft. (12.45 m.). **Draft:** 14.16 ft. (4.32 m.). **Displacement:** 2610 tons standard/3218 tons full load. **Powerplant:** twin-shaft steam: 4 oil-fired boilers, 2 sets of geared turbines, 60,000 shp. **Fuel:** 650 tons. **Speed:** 32 kt. **Range:** 800 n.mi. at 32 kt./3300 n.mi. at 20 kt. **Crew:** 270–80. **Sensors:** see text. **Electronic warfare equipment:** see text.

SUNBURN (SS-N-22): NATO code name for a Soviet ship-launched ANTI-SHIP MISSILE. Introduced in 1981, Sunburn was at first believed to be an incremental development of the SS-N-9 STARBRIGHT, but now appears to be a completely new missile with a different configuration. Sunburn is deployed in quadruple canisters on SOVREMENNY-class destroyers, and in twin canisters on TARANTUL-class corvettes.

Sunburn has a conical nose cone and a cylindrical body, with four small, midmounted cruciform wings. Probably powered by a dual-pulse solid-fuel rocket engine, the missile relies on a combination of INERTIAL GUIDANCE and radar ACTIVE HOMING (in the terminal phase); an ANTI-RADIATION version has also been reported. A surface-skimming missile, Sunburn's combination of high speed and low altitude would make it very difficult to detect, intercept, or evade. Payload options include 1100-lb. (500-kg.) high explosive and 200-kT nuclear warheads.

Specifications Length: 30 ft. (9.15 m.). **Diameter:** 2 ft. (610 mm.). **Weight, launch:** 6000 lb. (2727 kg.). **Speed:** Mach 2.5 (1750 mph/2800 kph). **Range:** 60–93 mi. (95–150 km.).

SUPER ETENDARD: See ETENDARD.

SUPER FRELON: Sud-Aviation (Aerospatiale) SA.321 heavy-lift HELICOPTER, in service with the French navy and the air forces of China, Iraq, Israel, Libya, South Africa, and Zaire. Developed from 1959 with technical assistance from Si-

korsky and Fiat, the prototype flew in 1963, and the first of 24 SA.321Gs entered service with the French *Aeronavale* in 1966 for troop transport and ANTI-SUBMARINE WARFARE (ASW). The largest helicopters ever built in Western Europe, 99 Super Frelons were produced between 1962 and 1980 in several military and civilian variants, including the SA.321H, a simplified, nonamphibious transport; the SA.321K, a nonamphibious assault helicopter for the Israeli air force (12 delivered, 8 still in service); the SA.321L, a similar transport for Libya and South Africa; and the SA.321M, an amphibious ASW/search-and-rescue helicopter for Libya.

The Super Frelon has been used in combat by Israel as a transport and electronic warfare platform. Iraq armed its Super Frelons with EXOCET AM.39 anti-ship missiles for use against tankers and other "naval targets" during the Iran-Iraq War.

The fuselage has a boat-shaped, watertight hull, and ends in a high-mounted boom for the five-bladed, anti-torque tail-rotor. The helicopter has fixed tricycle landing gear; on amphibious versions, the main gear are attached to float-sponsons, while most nonamphibious versions have simple outrigger landing gear. In all versions, the pilot and copilot sit side-by-side in the large, glazed nose. The SA.321G has a Heracles I/II search RADAR in a nose-mounted thimble-radome, a MAGNETIC ANOMALY DETECTOR (MAD) "bird," and an HS.12 dipping SONAR. Its other AVIONICS include a Crouzet Nadir Mk.1 DOPPLER navigation radar, surveillance radars in the float-sponsons, and a variety of RADAR WARNING RECEIVERS. Behind the cockpit, the main cabin has accommodations for 27–37 troops, or up to 11,023 lb. (5000 kg.) of cargo. Access is provided by large sliding side doors, and by a power-operated rear clamshell ramp-doors. Most versions have a rescue hoist by the right side door, and a belly hook for slinging cargo. The engines are mounted over the main cabin, with one to starboard and two to port. The complex transmission was developed by Fiat, while the main rotors and rotor hub were derived from the Sikorsky S-61 SEA KING. Most Super Frelons are unarmed, though the SA.321.G and SA.321L can carry up to four lightweight homing torpedoes for ASW, or two Exocets for anti-shipping missions.

Specifications Length: 63.9 ft. (19.4 m.). **Rotor diameter:** 62 ft. (18.9 m.). **Powerplant:** 3 1550-shp. Turbomeca IIIC6/E6 turboshafts; (Israeli) 3 1870-shp. General Electric T58-GE-16s.

Weight, empty: 4,775 lb. (6702 kg.); (G) 15,130 lb. (6863 kg.). **Weight, max. takeoff:** 28,660 lb. (13,000 kg.). **Speed, max.:** 171 mph (275 kph). **Speed, cruising:** 155 mph (250 kph). **Initial climb:** 1312 ft./min. (400 m./min.). **Service ceiling:** 10,325 ft. (3150 m.). **Range, max.:** 633 mi. (920 km.). **Endurance:** 4 hr.

SUPER SABRE: North American F-100 FIGHTER/ATTACK AIRCRAFT, once the backbone of the U.S. Air Force TACTICAL AIR COMMAND, and widely exported. Though obsolescent, it continues to serve with the air forces of Taiwan and Turkey.

SUPPLY: See LOGISTICS.

SUPPORT: The assisting of infantry and armored units with firepower, intelligence, engineering, or logistics; units providing such assistance are said to be "in support." See also DIRECT SUPPORT; GENERAL SUPPORT.

SUPPRESSION: Activities meant to neutralize or inhibit the action of enemy forces, weapons, or sensors—as opposed to outright destruction. Typically, extreme accuracy is not as important for suppression as a high volume of fire; thus multiple rocket launchers, which can saturate a wide area with a single salvo, are highly valued as suppressive weapons, despite their relatively poor accuracy, which limits the actual destruction they inflict. Similarly, infantry tactics may emphasize high-volume automatic fire rather than deliberate aimed fire, to obtain a greater suppressive effect for maneuver (even if thereby inflicting less attrition).

Against surface-to-air missiles and other radar-directed anti-aircraft weapons, temporary suppression by the use of ELECTRONIC COUNTERMEASURES (which jam or mask enemy radars, thereby preventing effective weapon direction) is the alternative to destruction, usually more difficult to achieve.

In some cases, the mere presence of certain forces can achieve suppression through deterrence, as when air defense radars shut down in the presence of aircraft armed with ANTI-RADIATION MISSILES.

SURFACE-TO-AIR MISSILE: See ANTI-AIRCRAFT (weapons) and MISSILE, GUIDED.

SURTASS: Surveillance Towed Array Sonar System, a U.S. submarine detection system intended to supplement the SOSUS network of seabed hydrophones. SURTASS consists of the UQQ-2 towed hydrophone array, which is some 3 mi. (4.8 km.) long, and can be towed at 3 kt. by specialized "T-AGOS" ships. Data from the UQQ-2 is relayed via a WSC-6 satellite communications terminal to a shore station for processing. Extremely sensitive, SURTASS can detect submarines at ranges out to several hundred miles. Current plans call for a force of 18–24 T-AGOS ships, to conduct 60- to 90-day patrols in deep ocean areas not covered by SOSUS. See also TOWED ARRAY SONAR.

SURVEILLANCE: The systematic observation of a given area for activities of any kind, as opposed to more focused SCOUTING or RECONNAISSANCE. Surveillance may be undertaken visually or electronically, by aircraft, ground forces, ships, or satellites. A key purpose of routine surveillance is to update information on patterns of normal activity in order to detect departures from those norms, which may provide early warning of major enemy initiatives.

SURVEILLANCE SATELLITE: See SATELLITES, MILITARY.

SVERDLOV: A class of 14 Soviet light CRUISERS completed between 1951 and 1955. The last all-gun cruisers built in the U.S.S.R., the Sverdlovs are still in service for amphibious gunfire support and as training vessels. Two have been converted to command cruisers (*Korabl'Upravleniye*, KU) and now serve as fleet flagships on overseas station. One, the *Dzerzinskiy*, was converted to a guided-missile cruiser in 1961 and laid up in 1972; another, the *Admiral Nakhimov*, used as a test platform for the SS-N-1 SCRUBBER anti-ship missile, was scrapped in 1961. Finally, the *Ordzhonikize*, transferred to Indonesia in 1962, was scrapped in 1972.

SWIFTSURE: A class of six British nuclear-powered attack SUBMARINES (SSNs) commissioned between 1973 and 1981. Successors to the VALIANT class, the Swiftsures have a revised hull and powerplant, and are significantly quieter. The hull has a modified teardrop form, with a blunt bow and cylindrical center section which ends in a broad, conical stern. This arrangement maximizes internal volume and is reported to have some hydrodynamic advantages. Like all British nuclear submarines, the Swiftsures have a single-hull configuration in which ballast tanks are concentrated outside the pressure hull at the bow and stern. A tall, streamlined sail is located amidships. The control surfaces consist of retractable bow diving planes, with cruciform rudders and stern planes just ahead of the propeller. The Swiftsures' PWR1 is a natural circulation reactor which relies on convection cooling alone at low speeds, thereby minimizing the use of noisy pumps (which are still needed at higher speeds). All machinery is mounted on resilient sound isolation rafts, making the Swiftsure class among the quietest nuclear submarines in service.

Armament consists of 5 21-in. (533-mm.) tubes amidships, 4 on the sides angled outward, and 1 angled downward beneath the hull. An automatic loader can reload all 5 tubes in only 18 seconds. The primary weapon is now the Mk.24 Mod 2 TIGERFISH wire-guided homing TORPEDO, supplemented by the UGM-84 HARPOON anti-ship missile, but Tigerfish is to be superseded in the early 1990s by the SPEARFISH torpedo. The basic load is currently 20 Tigerfish and 4 Harpoon; "Stonefish" bottom-laid MINES can replace these weapons on a 2-for-1 basis. The SONAR suite was originally identical to that of the Valiant class, but was upgraded in 1985. All sensors are linked to a DCB automated torpedo and missile FIRE CONTROL system based on two Ferranti 1600B digital computers. See also SUBMARINES, BRITAIN.

Specifications **Length:** 272 ft. (82.9 m.). **Beam:** 32.33 ft. (9.8 m.). **Displacement:** 4200 tons surfaced/4900 tons submerged. **Powerplant:** single-shaft nuclear: 1 PWR1 pressurized-water reactor, two sets of geared turbines, 15,000 shp. **Speed:** 20 kt. surfaced/30 kt. submerged. **Max. operating depth:** 1315 ft. (400 m.). **Collapse depth:** 1970 ft. (600 m.). **Crew:** 130. **Sensors:** 1 Type 2020 chin-mounted low-frequency active/passive sonar, 1 Type 2007 flank-mounted conformal array sonar, 1 Type 2019 passive/active range and intercept sonar (PARIS), 1 Type 2026 passive towed array sonar, 1 Type 1007 surface-search radar, 1 electronic signal monitoring array, 2 periscopes.

SWIMMER DELIVERY VEHICLE (SDV): A small submersible for the transport of combat swimmers from an offshore mother ship to their target. Linear descendants of the "Pig" and "Chariot" human torpedoes used by the Italian and British navies in World War II, SDVs of various kinds are employed by naval special operations forces, including the U.S. Navy's SEALS, the British SBS, and Soviet naval SPETSNAZ, and Israeli naval commandos.

SDVs are typically some 20–30 ft. (6–9 m.) long and 2–3 ft. (600–915 mm.) in diameter, with a cylindrical, torpedolike body. The combat swimmers ride in open cockpits, breathing through their own personal apparatus. Powered by electric motors, SDVs can maintain speeds of 3–4 kt. for several hours, thereby greatly extending the swimmers' radius of action. In addition, SDVs can accommodate weapons, mines, and other equipment. SDVs are usually carried on the decks of submarines, but surface vessels (frequently disguised as fishing boats) can also be used as mother ships.

The U.S. Navy is believed to have some 15 SDVs. The smallest, a converted Mk.37 torpedo, can carry a two-man team, while the largest can carry up to six divers; the latter are carried in "dry deck shelters" fitted to some STURGEON nuclear-powered attack submarines.

SWINGFIRE: British ANTI-TANK GUIDED MISSILE (ATGM) developed from 1961 as a replacement for towed anti-tank guns and the 120-mm. Wombat RECOILLESS rifle. Swingfire entered service with the Royal Artillery in 1969, and remains the standard British ATGM. It is also in service with Belgium, Egypt, Kenya, and Sudan.

Intended as a vehicle-mounted weapon, Swingfire is fitted to the FV.438 variant of the TROJAN armored personnel carrier, and to the Striker tank-destroyer variant of the SCORPION armored reconnaissance vehicle. In both cases, the launchers consists of an armored box which folds flush with the roof for traveling, and which can be elevated to 45° for launch. The FV.438 has a twin launcher and carries 14 reloads, while the Striker has a quintuple launcher and 5 reloads. A portable version called Beeswing has also been developed, and THERMAL IMAGING night sights have been fitted to both FV.438 and Striker.

Four folding tail fins are used exclusively for stabilization; steering is accomplished by thrust-vector control with a gimballed rocket nozzle. Swingfire's 15.4-lb. (7-kg.) shaped-charge (HEAT) warhead is capable of penetrating 31.5 in. (800 mm.) of homogeneous steel armor. The missile is powered by a dual-impulse solid rocket; because TVC steering is employed, the main engine burns over the missile's entire trajectory. Swingfire relies on WIRE GUIDANCE with a unique combination of semi-automatic (SACLOS) and manual (MCLOS) command to line-of-sight. After launch, the missile is automatically "gathered" onto the operator-target line. Thereafter, the missile is steered manually with a joystick. This technique eliminates the most difficult phase of MCLOS and yet allows the operator to be positioned up to 100 m. from the missile launcher. Tactically, that means the missile can be launched from a hull-down position, minimizing the risk of enemy counterfire.

The main drawbacks of Swingfire, common to most ATGMs, include its relatively long flight time, the exposure of the operator and/or launch vehicle to counterfire, interruption of the line of sight by smoke and terrain masking, and the degradation of its HEAT warhead by REACTIVE and composite (CHOBHAM) armor.

Specifications Length: 3.5 ft. (1.07 m.). Diameter: 6.7 in. (170 mm.). **Span:** 14.7 in. (373 mm.). **Weight, launch:** 60 lb. (27 kg.). **Speed:** 186 m./sec. (417 mph). **Range envelope:** 150–4000 m. Flight time: 22 sec.

SYNTHETIC APERTURE RADAR: A signal processing technique usually applied in SIDE-LOOKING AERIAL RADAR (SLAR) and other ground-mapping radars. A synthetic aperture radar takes a number of sequential "snapshots" of the same area, while the platform moves a considerable distance. These snapshots are then processed and collated to produce photolike, high-resolution radar imagery.

In a variation called "Inverse Synthetic Aperture Radar" (ISAR), the radar remains stationary while the target moves. See also J-STARS; RADAR.

T

T-10: Soviet heavy TANK of 1950s vintage, still used by some Third World armies.

T-34: Soviet medium TANK of World War II vintage, still in service with some Third World armies.

T-54/55: Family of Soviet MAIN BATTLE TANKS (MBTs), in service with Soviet reserve formations and more than 40 Warsaw Pact and Third World armies. The T-54 was developed from the T-44, an interim model combining a new, low-profile hull with the turret and 85-mm. gun of the T-34/85. Rushed into production in 1944, the T-44 was not a success. In 1947, a 100-mm. gun mounted in a new dome-shaped turret was mated to a modified T-44 hull to create the prototype T-54, which entered production in 1949. It became the mainstay of Soviet and Warsaw Pact forces until the early 1960s, when it was superseded by the similar but upgraded T-55. The latter became the standard Soviet MBT until the early 1970s, when it was replaced in turn by the T-62. The T-54 and T-55 were produced in many different subtypes in factories in the Soviet Union, Czechoslovakia, and Poland. By the time production ended in 1981, more than 80,000 had been delivered, not including several thousand built (without license) in China as the T-59 (total U.S. production of all MBTs from 1945 to 1980 was 28,000). The T-54/55 has been used in combat in the Middle East, Africa, Latin America, Afghanistan, and India. Though inferior in several respects to its Western contemporaries, the T-54/55 was considerably cheaper and usually available in much greater numbers.

The T-54 is of conventional design, but is notable for its small dimensions as compared to Western MBTs. The all-welded steel hull is divided into a crew compartment forward and an engine compartment in the rear. Armor protection varies between 100 mm. on the frontal glacis and 70 mm. on the upper sides. The glacis is sloped at 58°, doubling the effective thickness of the plate. The turret, a circular, flattened dome cast as a single piece, offers excellent all-around protection, has a maximum armor thickness of 203 mm., and features a prominent roof ventilator. The driver is provided with two observation periscopes, one of which can be replaced by an INFRARED night vision device. The remaining crewmen are in the turret, with the gunner on the left, the commander above and behind him, and the loader on the right. In contrast to Western MBTs, the T-54 has no turret basket (i.e., a floor which rotates with the turret); instead, crew seats are attached directly to the turret. The loader must therefore retrieve rounds from ammunition racks while avoiding the breech of the gun as the turret rotates. His position on the right side of the turret also forces him to lift shells into the breech and ram them home with his left hand. The commander has a rotating cupola with two observation periscopes and a TPK-1 stadiametric gun sight for target acquisition, while the gunner has a TSh 2-22 stadiametric sight with magnification settings of 3.5x and 7x. The commander acquires the target, and the gunner lays and fires the gun. Both the commander's and gunner's sights can be replaced with infrared night sights. As compared to Western MBTs, such FIRE CONTROLS are rudimentary.

The main armament, a D-10T 100-mm. rifled gun (derived from a dual-purpose naval gun), has a muzzle velocity of 1415 m./sec. firing APDS, and 900 m./sec. firing HE, HEAT, and APHE. Effective range varies between 900 m. for HEAT and 2800 m. for APDS, but lack of a fire control computer limits practical engagement ranges to some 1000 m. The APDS and HEAT rounds can destroy 1960s-vintage tanks, but are only marginally effective against late-model tanks with advanced armor. The gun has a top-loading breech and must be elevated to +10° for loading, reducing the maximum rate of fire to 3–4 rds./min. (as compared to 6–8 rds. for Western MBTs). Elevation limits of only −4° and +17°, caused by the low roofline of the turret, force the T-54 to expose more of its hull when firing at targets below it than Western tanks, a serious tactical shortcoming. A total of 34 rounds of 100-mm. ammunition are carried, with three ready rounds in the turret and the remainder on the floor or beside the driver. Not all the ammunition is accessible with the turret traversed.

Secondary armament consists of a remotely operated, bow-mounted 7.62-mm. DT machine gun operated by the driver; a second, coaxial DT; and a pintle-mounted 12.7-mm. DSHK heavy machine gun by the loader's hatch. Two infrared searchlights are carried, one to the right of the main gun, the other over the commander's hatch. COLLECTIVE FILTRATION for NBC defense, standard on later models, is generally retrofitted on earlier tanks. A modified Christie suspension with five large bogie wheels and no return rollers gives good mobility in snow and mud, but is prone to track-throwing at high speeds. Streams up to 13 ft. (4 m.) deep can be forded with an erectable snorkel to provide air to the engine and crew. The T-54 can generate its own smoke screen by injecting diesel oil into the engine exhaust pipe.

The original T-54 was often upgraded: the T-54A, introduced in the mid-1950s, has a D-10TG gun, stabilized in elevation, and an NBC system; the T-54B, introduced in 1957–58, has a D-10T2S gun, stabilized in elevation and traverse; and the T-54C, introduced in the late 1950s, has a flush hatch for the loader and no DShK.

Eventually the T-54 incorporated so many changes as to warrant a new designation, T-55. Externally quite similar, the T-55 can best be distinguished by its D-10T2S gun with a muzzle-mounted fume extractor; the elimination of the turret roof ventilator and the bow machine gun are less visible. The more substantive internal differences include the provision of a turret basket and a more powerful 580-hp. V-2-55 engine; in addition, 9 more rounds of main gun ammunition (for a total of 43) and increased internal fuel take the place of the bow machine gun. Eliminated in early versions, the DShK machine gun was restored in the T-55A, introduced in 1963. The last Soviet version, the T-55A(M), has an anti-radiation lining of lead-impregnated foam for the crew spaces. Remaining T-55s in the Soviet army are being further upgraded with new fire controls, possibly including a laser rangefinder; T-55s may also be equipped with add-on REACTIVE ARMOR against HEAT rounds.

Israel captured several hundred T-54/55s in 1967 and 1973. About 200 were rebuilt, with new diesel engines, L7A1/M68 105-mm. guns, and new fire controls. Designated M-1967, these tanks were used (often against Egyptian T-55s) in the 1973 Yom Kippur War. (Israel now offers similar upgrade kits to T-54/55 operators.) The T-54/55 chassis has also been used for a number of specialized vehicles, including the IT-122 and IT-130 ASSAULT GUNS, the ZSU-57-2 anti-aircraft tank, several different armored recovery vehicles, two different bridgelayers, a combat engineer vehicle, and a flamethrower tank.

Though it has a low silhouette and good cross-country mobility, the T-54/55 has a number of serious operational shortcomings. Armor protection is light; the vehicle is cramped, badly laid out, and fatiguing to operate. With driver, gunner, and commander in line on the left side, a single hit can kill all three. This may not matter, because the close proximity of fuel and ammunition within the hull often results in catastrophic explosions when the armor is penetrated. The engine, transmission, and tracks all wear out quickly (having an average life of 125 hr., roughly half that of Western tanks); this reflects the Soviet preference for building tanks in sufficient quality to replace entire units in combat, rather than attempting field repairs. Finally, as noted, fire controls are exceedingly primitive.

Specifications Length: 21.16 ft. (6.58 m.). Width: 10.8 ft. (3.29 m.). Height: 7.85 ft. (2.39 m.). Weight, combat: 36 tons. Powerplant: 520-hp. V-2-54 12-cylinder diesel. Hp./wt. ratio: 14.4 hp./ton. Fuel: 128 gal. (563 lit.) + 84 gal. (396 lit.) in fender panniers + 88 gal. (387 lit.) in 2 rear-mounted drums. **Speed:** 31 mph (51 kph) road/20 mph (33 kph) cross-country. **Range, max.:** 310 mi. (518 km.) road/185 mi. (309 km.) cross-country.

T-59: Chinese MAIN BATTLE TANK (MBT), essentially an unlicensed copy of the Soviet T-54, a number of which were supplied to China in the early 1950s. In service from 1958–59, about 12,000 were built before production terminated in the early 1980s. **Operators:** Alb, Con, Iran, Iraq, Kam, N Kor, Pak, PRC, Sud, Viet, Zai, Zam, Zim.

T-62: A Soviet MAIN BATTLE TANK (MBT) developed from the T-54/55 series. Designed in the late 1950s, the T-62 entered production in 1961, but was not seen publicly until 1965. It remained the standard Soviet MBT from the mid-1960s until the late 1970s, when it was superseded by the T-64 and T-72. More than 20,000 T-62s were built by the Soviet Union and Czechoslovakia before production ended in 1970; the vehicle is still being built in small numbers by North Korea. Surprisingly, the T-62 was not supplied in large numbers to non-Soviet Warsaw Pact armies, possibly due to the speed with which the T-72 was developed, prompting those forces to forego the T-62 in favor of the more advanced tank. The T-62 has been in combat in the Middle East, the Iran Iraq War, Afghanistan, and Angola. Marginally inferior to 1960s-vintage Western tanks, the T-62 is no match for the latest MBTs (e.g., the M1 ABRAMS, LEOPARD 2, CHALLENGER, and MERKAVA).

The T-62 is externally quite similar to the T-54/55, with a low-slung hull and flat, circular turret. The all-welded steel hull is divided into a crew compartment forward and engine compartment in the rear. Armor protection is 102 mm. on the frontal glacis and 79 mm. on the upper sides, but the glacis is sloped at 60°, effectively doubling the protection of the plate. The turret, cast as a single piece with a welded roof, has a maximum thickness of 242 mm.; its flattened dome shape provides excellent all-around protection. The layout of the T-62 is also similar to that of the T-55. The driver has two observation periscopes, one of which can be replaced by a TVN-2 INFRARED night vision scope. The remaining three crewmen sit in the turret, with the gunner on the left, the commander above and behind him, and the loader on the right. The commander has a rotating cupola with four observation periscopes and a TKN-3 stadiametric gun sight with an integral infrared night sight, used for target designation. The gunner has a TSh2B-41u stadiametric sight with magnification

settings of 3.5x and 7x, and a TPN1-41-11 infrared night sight. The commander acquires the target and slews the turret to the proper bearing; the gunner then lays and fires the gun. These FIRE CONTROLS are crude by Western standards.

Main armament, a U-5TS 115-mm. smoothbore, high-velocity gun, can be distinguished from the 100-mm. D-10T of the T-54/55 by its greater length, and by the fume extractor located at mid-barrel. The U-5TS fires armor-piercing fin-stabilized discarding-sabot (APFSDS), fin-stabilized HEAT, and fin-stabilized HE ammunition. Muzzle velocity is 1680 m./sec. for APFSDS; 900 m./sec. for HEAT; and 750 m./sec. for He. Maximum effective range is more than 2000 m., but rudimentary fire controls impose a practical limit of about 1000 m. The gun is stabilized in both azimuth and elevation, theoretically allowing fire on the move, but this is precluded by the inadequate fire controls. The U-5TS has a top-loading breech, and is automatically elevated to +3.5° for loading; in addition, the turret cannot be traversed while the gun is being loaded. These factors limit the maximum rate of fire to 3–4 rds./min., roughly half that of Western tanks. After firing, empty shell cases are automatically ejected through a hatch in the rear of the turret. Elevation limits of −3° and +17°, caused by the low roofline of the turret, force the T-62 to expose more of its hull when firing at targets below it than Western tanks, a serious tactical shortcoming inherited from the T-54/55. A total of 40 main gun rounds are carried, with 2 ready rounds in the turret, 16 to the right of the gunner, and 20 in the rear of the crew compartment. Secondary armament consists of a coaxial 7.62-mm. PKT machine gun, and a pintle-mounted 12.7-mm. DSHK heavy machine gun by the loader's hatch. Two white-light/infrared searchlights are fitted, one to the right of the main gun, the other over the commander's hatch. The T-62 has a positive overpressure COLLECTIVE FILTRATION unit for NBC defense. In addition, the turret and crew compartment have an anti-radiation lining of lead-impregnated foam. A modified Christie suspension with five large bogie wheels and no return rollers provides good mobility in mud and snow, but is prone to track-throwing at high speeds. The T-62 can ford streams up to 13 ft. (4 m.) deep with an erectable snorkel to provide air to the engine and crew, but the practice is hazardous and unpopular with crews. Like the T-54/55, the T-62 can generate its own smoke screen by injecting diesel oil into the engine exhaust pipes.

The T-62 has been built in two other variants: the T-62A, with a revised turret which has a revolving cupola for the loader; and the T-62M, with a "live track" with return rollers, similar to that of the T-72. In Afghanistan, T-62s have been observed with laser rangefinders, add-on REACTIVE ARMOR to defeat HEAT rounds, and armored side skirts. Vehicles developed from the T-62 chassis, include the SU-130 ASSAULT GUN (a few of which may still be in service), the T-62K command tank, a flamethrower tank, and an armored recovery vehicle.

The T-62 has all the virtues and vices of the T-54/55. On the one hand, it is fairly easy to maintain, and has good mobility, a low silhouette, and an excellent gun. On the other hand, the vehicle is a cramped "ergonomic slum" (e.g., the loader must insert and ram a 45- to 50-lb. round into the gun breech using his left hand), leading to rapid crew fatigue. Armor is rather thin, and the proximity of fuel and ammunition within the hull can lead to catastrophic explosions when the armor is penetrated. In addition, track, engine, and transmission life are only half that of Western tanks (reflecting the Soviet preference for replacing, rather than repairing, major items of equipment).

Specifications Length: 21.75 ft. (6.63 m.). **Width:** 10.7 ft. (3.26 m.). **Height:** 7.85 ft. (2.39 m.). **Weight, combat:** 40 tons. **Powerplant:** 580-hp. V-2-62 12-cylinder diesel. **Hp./wt. ratio:** 14.5 hp./ ton. **Fuel:** 148 gal. (650 lit.) + 63 gal. (277 lit.) in fender panniers + 88 gal. (387 lit.) in 2 rear-mounted drums. **Speed:** 30 mph (50 kph) road/20 mph (33 kph) cross-country. **Range, max.:** 310 mi. (518 km.) road/186 mi. (310 km.) cross-country. **Operators:** Afg, Ang, Alg, Bul, Cze, DDR, Egy, Iran, Iraq, N Kor, Lib, Mon, Rom, Syr, Viet, Yem, Yug.

T-64: Soviet MAIN BATTLE TANK (MBT) developed in the late 1960s as replacement for the T-62; together with the T-72, it is the first truly new Soviet tank design since the T-34. Of the two new tanks, the T-64 was the more innovative design—too innovative, perhaps, and not entirely successful: production was halted in 1981, after some 5000 were delivered, a low figure by Soviet standards.

The T-64 and T-72 are externally similar and use many of the same components, though the T-64's turret shape and suspension are quite different. The all-welded steel hull is divided into a driver's compartment forward, a fighting compartment in the middle, and an engine compartment in the rear. The well-sloped frontal glacis may be fabricated of composite armor (similar to CHOBHAM ARMOR); if so, though roughly 200 mm. thick it provides ballistic protection equivalent to 500-600 mm. of steel. Side armor is 50–80 mm. thick. The steel turret, circular and very flat, is cast as a single piece with a maximum thickness of 280 mm. In contrast to most other MBTs (with the exception of the T-72 and T-80), the T-64 has a three-man crew: driver, commander, and gunner, with the traditional loader replaced by a mechanical system. The driver has a single wide-angle periscope with an integral INFRARED night vision device. The gunner and commander sit in the turret, on the left and right, respectively. The commander has a rotating cupola with two rear-facing observation periscopes and an optical sight with integral night vision, used mainly for target acquisition; the commander may also have a stadiametric rangefinder. The gunner has a TPD-2 day sight and a TPN-1-49-23 night sight, both connected to a laser rangefinder.

Main armament, a 125-mm. 2A46 high-velocity, smoothbore gun (which also arms the T-72 and T-80), can fire three types of fin-stabilized ammunition: APFSDS "arrow," HEAT, and HE rounds. The ammunition is semi-fixed (i.e., the projectile and propellant charge are loaded separately); the casing is combustible, except for a brass base stub—a very advanced feature. Muzzle velocity is 1615 m./sec. for APFSDS, and about 900 m./sec. for HEAT and HE. Maximum effective range is 2100 m. for APFSDS and 4000 m. for the other rounds. The 2A46 may be the most powerful tank gun in the world: its APFSDS round can penetrate 300 mm. or armor at 1000 m., while the HEAT round can penetrate 475 mm. at any range. The carousel-type auto-loader on the floor of the turret has two separate clips for different types of ammunition. In theory, it allows a rate of fire of 8 rds./min., but is not entirely reliable, exposing the gunner to injury. Elevation limits of −5° and +18°, while better than in earlier Soviet tanks, are still smaller than those of Western MBTs; thus the T-64 must expose more of its hull when firing at targets below it, a serious tactical disadvantage. The gun is fully stabilized to allow fire on the move, a tactic made possible by the installation of a fire control computer for the first time on a Soviet MBT. A total of 39 rounds of main gun ammunition are carried (normally 12 APFSDS, 6 HEAT and 21 HE). Secondary armament comprises a coaxial 7.62-mm. PKT machine gun, and a remotely controlled 12.7-

mm. heavy machine gun mounted over the commander's hatch. Two white light/infrared searchlights are fitted, one to the right of the main gun, the other by the commander's hatch. The T-64 has a positive overpressure COLLECTIVE FILTRATION system for NBC defense; in addition, the turret and crew compartments have an anti-radiation lining of lead-impregnated foam.

Powered by an entirely new engine, the T-64 also has a "live" track with hydro-mechanical suspension, six small, stamped road wheels, and four return rollers (as opposed to the Christie suspension used on all earlier Soviet MBTs). Apparently, the T-64's suspension was troublesome, since it was not repeated on the T-80. The T-64 can ford streams up to 18 ft. (5.5 m.) deep with two erectable snorkels to provide air to the engine and crew.

T-64 subtypes include the T-64A, with smoke grenade launchers mounted on the turret, a gunsight with an enlarged aperture, and armored side skirts; the T-64B, with a modified gun which can also launch the AT-8 *Kobra* (NATO code name SONGSTER) anti-tank guided missile (all T-64As are being brought up to T-64B standard); and the T-64K, a specialized command vehicle. All T-64s can be fitted with add-on REACTIVE ARMOR for further protection against HEAT rounds.

Specifications **Length:** 21 ft. (6.4 m.). **Width:** 11.1 ft. (3.38 m.). **Height:** 7.5 ft. (2.28 m.). **Weight, combat:** 38 tons. **Powerplant:** 750-hp. 5-cylinder opposed piston multi-fuel diesel. **Hp./ wt. ratio:** 19.7 hp./ton **Fuel:** 150 gal. (660 lit.) + 86 gal. (378 lit.) in 3 fender panniers + 88 gal. (387 lit.) in 2 rear-mounted drums. **Speed, road:** 43 mph (72 kph). **Range, max.:** 300 mi. (500 km.).

T-72: Late-model Soviet MAIN BATTLE TANK (MBT). The T-72 and the T-64 were designed in the late 1960s to replace the T-62; together they represent the most significant development in Soviet tank design since the T-34 of 1940, though the T-72 is both the cheaper and more conservative of the two. Now in production in the Soviet Union, Czechoslovakia, Poland, India, and Yugoslavia, more than 15,000 T-72s have been delivered to date; Soviet production continues mainly for export while the T-80 is being acquired for the Soviet army.

The welded steel hull is divided into a driver's compartment forward, a fighting compartment in the middle, and an engine compartment in the rear. The well-sloped frontal glacis plate is believed to be fabricated of an advanced composite armor (similar to CHOBHAM armor), with a maxi-

mum thickness of 200 mm., but providing ballistic protection equivalent to 500–600 mm. of homogenous steel armor. The side armor, probably steel, has a thickness of 50–80 mm. The elliptical steel turret, cast as a single piece, has a maximum thickness of 280 mm. Spring-mounted armored skirts protect the upper sides and tracks. In contrast to earlier MBTs, the T-72 has a three-man crew: driver, commander, and gunner. The driver has a single wide-angle periscope with integral INFRARED night vision. The gunner and commander sit in the turret, on the left and right, respectively, while the loader has been replaced by a mechanical system. The commander has a rotating cupola with two observation periscopes and an optical gun sight with integral infrared night vision (used mainly for target acquisition), and a backup stadiametric rangefinder. The gunner has a TPD-2 day sight and a TPN1-49-23 IR night sight, both linked to a laser rangefinder mounted in front of the gunner's hatch.

Main armament, a 125-mm. 2A46 high-velocity smoothbore gun (which also arms the T-64 and T-80), fires three types of fin-stabilized ammunition: APFSDS "arrow" rounds, HEAT, and HE. The ammunition is semi-fixed (i.e., the projectile and propellent charge are loaded separately), with a combustible casing except for a brass base stub. Muzzle velocity is is 1615 m./sec. for APFSDS, and about 900 m./sec. for HEAT and HE. Maximum effective range is 2100 m. for APFSDS and 4000 m. for the other rounds. The 2A46 may be the most powerful tank gun in the world: its APFSDS round can penetrate 300 mm. of armor at 1000 m., while the HEAT round can penetrate 475 mm. at any range. The carousel-type auto-loader on the turret floor has two separate clips for different types of ammunition. In theory, it allows a rate of fire of eight rds./min., but is not entirely reliable, and has sometimes injured the gunner. Elevation limits of −5° and +18° are better than in earlier Soviet tanks, but still less than those of Western MBTs; thus the T-72 must expose more of its hull when firing at targets below it, a serious tactical disadvantage. The gun is fully stabilized to allow fire on the move, a tactic made possible by a fire control computer. A total of 39 rounds of ammunition are carried (normally 12 APFSDS, 6 HEAT, and 21 HE). Secondary armament consists of a coaxial 7.62-mm. PKT machine gun, and a pintle-mounted 12.7-mm. heavy machine gun mounted over the commander's hatch. Later T-72s have four smoke grenade launchers mounted on either side of the turret.

Two white infrared searchlights are fitted, one to the left of the main gun, the other by the commander's hatch. The T-72 has a positive overpressure COLLECTIVE FILTRATION system for NBC defense; in addition, the turret and crew compartments of Soviet T-72s have an anti-radiation lining of lead-impregnated foam. The T-72 has a "live" track with torsion bar suspension, six small road wheels, and four return rollers (a similar suspension was tested on the T-62M). It can ford streams up to 18 ft. (5.5 m.) deep with an erectable snorkel to provide air to the engine and crew. The T-72, like the T-62, can generate its own smoke screen by injecting diesel oil into the engine exhaust pipes. A dozer blade folds down from the bow, enabling the tank to dig its own firing position or clear obstacles.

The T-72 has been produced in several distinct subtypes: the T-72 M1980/1, with fabric armor over the side storage boxes and suspension; the M1981/2, with additional armor over the sides and rear deck; the M1981/3, with increased frontal armor; and the M1984, with increased armor on the turret roof as well. All T-72s can also be fitted with add-on REACTIVE ARMOR for further protection against HEAT rounds. Specialized vehicles based on the T-72 chassis include the T-72K command tank and the BREM armored recovery vehicle.

The T-72 has been used in combat by Syria and Iraq. In the 1982 Lebanon War, Israeli tanks apparently had no great difficulty in destroying them (but only with advanced, Israeli-made 105-mm. APDFSDS rounds with "long-rod" penetrators). The T-72 is currently assessed as superior to 1960s vintage Western MBTs, but inferior to the latest Western tanks (the M1 ABRAMS, LEOPARD 2, CHALLENGER and MERKAVA III) especially in protection, crew comfort, and fire control. Its advantages include a very low silhouette and a very powerful gun, but the crew spaces are still cramped and badly laid out, and the concentration of main-gun ammunition in the (open) auto-loader ensures a catastrophic explosion if the tank is penetrated.

Specifications Length: 22.8 ft. (6.95 m.). **Width:** 11.8 ft. (3.6 m.). **Height:** 7.8 ft. (2.38 m.). **Weight, combat:** 41 tons. **Powerplant:** 780-hp. W-46 12-cylinder multi-fuel diesel. **Hp./wt. ratio:** 19 hp./ton. **Fuel:** 150 gal. (660 lit.) + 86 gal. (378 lit.) in 3 fender panniers + 88 gal. (387 lit.) in 2 rear-mounted drums. **Speed, road:** 37.5 mph (62 kph). **Range, max.:** 290 mi. (485 km.). **Operators:** Alg, Ang, Cuba, Fin, Ind, Iraq, Lib, Syr, USSR and other Warsaw Pact forces, Yug.

T-80: The latest Soviet MAIN BATTLE TANK (MBT), issued exclusively to front-line Soviet forces. Introduced at the end of the 1970s, the T-80 combines many features of the earlier T-64 and T-72, adding improvements to the armor, powerplant, and fire controls. More than 10,000 have been produced to date, with more rolling off the assembly line at a rate of 2000–2500 per year. Externally similar to both the T-64 and T-72, the T-80 weighs almost 20 tons less than current Western MBTs. The all-welded steel hull appears to be derived from the T-64, but is about 3 ft. longer. The well-sloped frontal glacis is fabricated from advanced composite armor. The flat, circular turret is also similar to that of the T-64, but may also be cast from composite armor rather than steel, greatly increasing protection. Armored side skirts protect the hull and suspension, while add-on REACTIVE ARMOR increases protection against HEAT rounds. The internal layout of the T-80 is similar to the T-64 and T-72: the driver up front, and the gunner and commander in the turret on the left and right, respectively. Periscopes, gun sights, and fire controls are identical to those of late model T-64s and T-72s. The main armament is the same 2A46 125-mm. smoothbore gun used by the T-64B, capable of firing APFSDS, HEAT, and HE rounds, plus the laser-guided AT-8 *Kobra* (NATO code name SONGSTER) anti-tank guided missile. The automatic loader is an improved version of the carousel device of the T-64 and T-72. The main gun is fully stabilized and can be fired accurately on the move because of computerized fire controls. A total of 39 rounds are carried. Secondary armament consists of a coaxial 7.62-mm. PKT machine gun and a remotely operated 12.7-mm. heavy machine gun on the gunner's cupola. Six smoke grenade launchers are mounted on either side of the turret.

Two INFRARED searchlights are fitted, one to the right of the main gun, the other over the commander's hatch. The T-80 has a positive overpressure COLLECTIVE FILTRATION system for NBC defense. In addition, the turret and crew compartment have an anti-radiation lining of lead-impregnated foam. Powered by a gas turbine engine, the T-80 has a "live" track using a torsion bar suspension derived from that of the T-72. The T-80 can ford streams up to 18 ft. (5.5 m.) deep with an erectable snorkel. A dozer blade folds down from the bow, enabling the tank to improve its own firing position and clear obstacles.

The advantages of the T-80 include a low silhouette, good mobility, an excellent gun, and better armor protection than its predecessors. On the other hand, its internal layout is cramped, while fire controls and, above all, protection are probably inferior to those of late-model Western tanks (which have almost 20 tons of additional armor), but superior numbers may compensate for many deficiencies. There are reports of a new Soviet tank under development, with a 130-mm. gun mounted semi-externally.

Specifications Length: 24.25 ft. (7.4 m.). Width: 11.16 ft. (3.4 m.). Height: 7.25 ft. (2.2 m.). Weight, combat: 43 tons. Powerplant: 980-shp. gas turbine. Hp./wt. ratio: 23 hp./ton. Fuel: 280 gal. (1232 lit.). Speed, road: 47 mph (78 kph). Range, max.: 285 mi. (475 km.).

TABLE OF ORGANIZATION AND EQUIPMENT (TO&E): Authorized levels of personnel and major equipment in formations and their subunits. TO&Es represent nominal or "paper" strengths, seldom achieved or long maintained by forces in combat. Uncovering the TO&Es of opposing forces is essential for the compilation of ORDER OF BATTLE intelligence.

TABUN: Ethyl NN-dimethyl-phosphoramidocyanidate (U.S. code name GA), the first antichlorinesterase or NERVE AGENT (see CHEMICAL WARFARE). Developed in Germany in 1937, Tabun is an odorless, colorless, skin-permeable liquid which disrupts the central nervous system by binding permanently with the nerve-connecting enzyme acetylchlorinesterase (AChE). Symptoms of exposure include involuntary muscle contractions, convulsions, and eventually death from respiratory collapse. Tabun, a relatively persistent agent, is also highly lethal: the LD-50, or average dose needed to kill half of all those exposed, is only 400 milligrams per min. per cubic m. inhaled, or 1000 milligrams percutaneously (i.e., through the skin); victims are incapacitated within 10 minutes and die within 15 minutes. Protection against Tabun requires not only a mask, but a complete set of protective clothing, which seriously degrades troop performance. Troops can be given some advance protection ("hardened") by a prophylactic regimen of synergistic drugs which *temporarily* bind a portion of the body's AChE, forming a protected "reservoir" of the enzyme. After exposure, a second series of drugs is administered to unbind the AChE in the reservoir, neutralizing the effects

of the nerve agent. The only other form of first aid consists of large doses of epinephrine or adrenaline injected within seconds of exposure. Tabun is included in the chemical arsenals of the United States, the Soviet Union, and other countries; it is known to have been used by the Iraqis against Iranian troops and Kurdish civilians. The agent can be delivered in artillery shells, aerial bombs, or rocket warheads, or by aerosol dispensers. Other nerve agents include SARIN (GB), SOMAN (GD), and the extremely potent VX.

TACAMO: "Take Charge And Move Out," a U.S. Navy airborne relay system for communications between the NATIONAL COMMAND AUTHORITIES (NCA) and submerged ballistic-missile submarines (SSBNs). Developed from the early 1960s as an emergency backup for ground transmitter sites, TACAMO consists of specially modified EC-130Q variants of the HERCULES transport aircraft fitted with a USC-13 communications suite. The heart of the system is a VLF (Very Low Frequency) transmitter with two 35,000-ft. (10,670-m.) trailing wire antennas paid out from the tail of the aircraft. The TACAMO aircraft, orbiting over the ocean at 30,000 ft. (9146 m.) so that the antenna wires hang nearly vertical, receive messages from navy shore bases or U.S. Air Force airborne command posts (see NATIONAL EMERGENCY AIRBORNE COMMAND POST) for retransmission in VLF to submerged SSBNs.

The EC-130Q is now being replaced by the more capable E-6A HERMES, derived from the Boeing 707-320 airliner. With a higher speed and ceiling, and long range, the E-6A will be able to stay on station longer, and will be equipped with improved communications equipment, including a MILSTAR satellite communications terminal and a solid-state VLF transmitter.

TACAN: Tactical Air Navigation, an airborne ultra-high frequency (UHF) radio navigation aid which receives continuous bearing and slant-range readings from TACAN ground stations.

TACHIKAZE: A class of three Japanese guided-missile DESTROYERS (DDGs) commissioned between 1976 and 1983. Larger, more powerful successors to Japan's first DDG, the *Amatsukaze*, they reflect the Japanese Maritime Self-Defense Force's increasing focus on the threat of Soviet Naval Aviation (AV-MF) to Japan's sea lines of communication (SLOCs).

The Tachikazes are flush-decked, with a large, boxy superstructure forward, and a separate, lower deckhouse aft. Like most Japanese warships, they are excellent sea boats and heavily armed for their size. Although designed primarily for ANTI-AIR WARFARE (AAW), the Tachikazes have considerable ANTI-SUBMARINE WARFARE (ASW) and ANTI-SURFACE WARFARE (ASUW) capabilities. The main AAW weapon is a stern-mounted Mk.13 single-arm launcher supplied by a magazine with 40 STANDARD-1MR surface-to-air missiles. It is supplemented by 2 5-in. 54-caliber DUAL PURPOSE guns, one on the bow and the other on the after deck-house firing over the missile launcher. A PHALANX radar-controlled 20-mm. gun was added in 1987–88 to enhance anti-missile defenses. Because of the missile launcher aft, the Tachikazes have no helicopter facilities. ASW weapons consist of an 8-round ASROC pepperbox launcher with 8 reloads, mounted behind the forward 5-in. gun, and 2 sets of Mk.32 triple tubes for lightweight homing TORPEDOES. Anti-surface warfare capability is provided two quadruple launchers for HARPOON anti-ship missiles, backed up by the 2 5-in. guns. Each ship also has two SATCOM satellite communications terminals.

The Tachikazes have been succeeded by the larger HATAKAZE class, with gas turbine propulsion, improved fire controls, and rearranged armament. See, more generally, DESTROYERS, JAPAN.

Specifications **Length:** 469 ft. (143 m.). **Beam:** 46.75 ft. (14.25 m.). **Draft:** 15 ft. (4.5 m.). **Displacement:** 3850 tons standard/4800 tons full load. **Powerplant:** twin-shaft steam; 4 oil-fired boilers, 2 sets of geared turbines, 70,000 shp. **Speed:** 32 kt. **Crew:** 277. **Sensors:** 1 SPS-52D 3-dimensional air-search radar, 1 OPS-11B surface-search radar, 2 Type 72 fire control radars, 2 SPG-51C missile guidance radars, 1 OQS-3/4 hull-mounted medium-frequency sonar. **Electronic warfare equipment:** 1 OLT-3 electronic signal monitoring array, 4 Mk.36 SRBOC chaff launchers.

TACIT RAINBOW: Code name for the AGM-136A air-to-surface ANTI-RADIATION MISSILE (ARM) developed by Northrop for the U.S. Air Force. In contrast to most other ARMs, which are short-range, high-speed missiles, Tacit Rainbow is a CRUISE MISSILE: it can be launched at a considerable distance from its target to loiter for up to 30 minutes over enemy AIR DEFENSE positions; if a radar is activated in its vicinity, the missile will dive to attack it. Tacit Rainbow can therefore suppress radar-controlled air defenses for an extended period simply by its presence in the battle area; in effect, it amounts to an unmanned WILD WEASEL aircraft.

Derived from the BQM-74 Chukar drone, Tacit Rainbow has an airplane configuration; the wings fold into the body for carriage, extending only after launch. Tacit Rainbow is armed with a 30-lb. (13.63-kg.) high-explosive fragmentation warhead, and powered by a Williams International J400-404 turbojet generating 240 lb. (109 kg.) of thrust. The guidance system sends the missile in the general direction of the target; after a present run, it begins orbiting until a radar emitter is detected. An on-board computer stores the location of detected emitters, so that they may be attacked for up to several minutes after they have shut down. Tacit Rainbow will enter service in the early 1990s if the program survives the usual cost escalation, and the usual preference for funding manned aircraft instead of munitions.

Specifications **Length:** 8.33 ft. (2.54 m.). **Diameter:** 14 in. (355 mm.). **Span:** 5.78 ft. (1.76 m.). **Weight, launch:** 450 lb. (204.5 kg.). **Speed:** 600 mph (1000 kph). **Range:** 300 mi. (500 km.).

TACTASS: Tactical Towed Array Sonar System, an advanced U.S. low-frequency SONAR deployed from cruisers, destroyers, and frigates. The array is towed by cable at a considerable distance behind the ship, hence away from its propeller, machinery, and hull-flow noises, which can mask target sounds. Unlike earlier towed arrays, TACTASS can be used at relatively high speeds and in rough seas. Modular construction facilitates the repair or replacement of its hydrophone elements aboard ship.

TACTASS has been developed in two variants. One, designated SQR-19, is intended for TICONDEROGA-class cruisers, SPRUANCE-, KIDD-, and BURKE-class destroyers, and PERRY-class frigates, and became operational in 1983. The array has 16 separate receiver modules (8 VLF, 4 LF, 2 Mf, and 2 HF), and is towed at the end of a 5200-ft. (1585-m.) cable. The other model, designated SQR-18A(V), designed for the KNOX-class frigates, is available in two versions: the SQR-18A(V)1, for frigates already equipped with VARIABLE DEPTH SONAR (VDS), is attached to the VDS transducer, or "fish"; and the the SQR-18A(V)2, for frigates without VDS. The array, 730 ft. (222.5 m.) long and consisting of 32 acoustically isolated hydrophones, is attached to a 5000-ft. (1525-m.) cable. The SQR-18A(V) became operational in 1980, and has been exported to several allied navies.

TACTICAL: 1. The lowest of the command levels (strategic, operational, and tactical), generally applied to ground formations of BRIGADE size or smaller, air units on the SQUADRON level, and naval forces at the squadron, detachment, or individual ship level.

2. An adjective loosely applied to shorter-range weapons, meant for use on the battlefield itself; e.g., tactical nuclear forces. Also used as a euphemism for unimportant, as in "tactical withdrawal." See also TACTICS.

TACTICAL AIR COMMAND (TAC): A major operational command of the U.S. Air Force (USAF), responsible for all FIGHTER, INTERCEPTOR, ATTACK, tactical reconnaissance, and SPECIAL OPERATIONS squadrons based in the continental United States (CONUS), i.e., not assigned to an overseas command such as USAFE, PACAF, or the Alaskan Air Command. TAC also controls a number of ELECTRONIC WARFARE and tanker squadrons to support its combat forces. TAC is organized into three numbered air forces (the 1st, 9th, and 12th), the 28th Air Division, the Fighter Weapons Center, and the Tactical Air Warfare Center. First Air Force, formerly known as ADTAC, is responsible for the air defense of CONUS, under the joint U.S.-Canadian NORAD.

TACTICAL AIRCRAFT: Generally, shorter-ranged COMBAT AIRCRAFT meant for AIR SUPERIORITY, AIR DEFENSE, CLOSE AIR SUPPORT, INTERDICTION, and transport, including FIGHTERS, INTERCEPTORS, ATTACK AIRCRAFT, and shorter-range transports. The term is sometimes applied to certain types of specialized aircraft (some of *long* range), including ELECTRONIC WARFARE (EW), AERIAL REFUELING, and airborne warning and control system (AWACS) aircraft.

TACTICAL NUCLEAR FORCES: Also called "battlefield nuclear forces"; the term usually (but not always) describes low-yield NUCLEAR WARHEADS (under 100 KT) delivered by short-range weapons, and intended for battlefield use. Western tactical NUCLEAR WEAPONS usually have yields of less than 20 kT, with the trend being towards ever smaller weapons; many have yields of less than 5 kT, and subkiloton artillery shells are also in service. Current weapons of this category include NUCLEAR ARTILLERY, short-range BALLISTIC MISSILES (SRBMs), and TACTICAL AIRCRAFT, as well as surface-to-air missile warheads, depth charges, and torpedoes.

TACTICS: The craft of using armed forces in battle. Tactics therefore define specific uses of military forces, within the limitations set by the strategic and operational context. Thus, for example, the *strategic* response to an air defense threat may be the deployment of a system of interceptors, missiles, and radars; the *operational*-level decision would determine the precise mix of forces needed to implement the chosen strategy, and the relationships among them; and *tactics* would define the detailed methods to be used by each element of the system against a particular threat. More conventionally, if country A decides to invade country B at the level of national ("grand") strategy, the choice of invasion axis comes under "theater strategy," the overall scheme of the invasion is defined by an operational-level decision, and the specifics of conduct in battle are the tactics.

Traditionally, strategic decisions have been made at the highest political levels, while tactical decisions have been made by military officers exclusively, with the operational level often contested (in the classic example, during the Vietnam War, individual bombing targets were selected by the president of the United States).

Since World War I, infantry tactics have been dominated by the difficulty of advancing over ground swept by rifle, machine-gun, and artillery fire. Once the near-impossibility of frontal attacks against well-entrenched forces armed with modern weapons was established, there was a threefold response: dispersion, suppression, and concealment.

Dispersion forces the enemy to dilute his firepower over a wider area, reducing its effectiveness. At the most detailed level, dispersed forces are less likely to suffer multiple casualties from a single weapon (a dominant consideration with nuclear weapons). Further, dispersed forces are more difficult to detect; hence they may be able to avoid the effects of firepower entirely. Defensive forces, in turn, disperse to cover the longer frontage of offensive forces, to dilute the effects of offensive weapons (particularly artillery), and to avoid detection by the attacking force. Dispersion, for both offense and defense, has been made possible only by the advent of reliable radio COMMUNICATIONS, which enable commanders to control scattered forces.

SUPPRESSION is the use of firepower to neutralize enemy firepower, either by destroying weapons or forcing their operators to seek cover.

Concealment is partly derived from dispersal, but can be enhanced by CAMOUFLAGE and active deception. The objective is to circumvent enemy firepower by avoiding detection entirely.

These factors are combined in the two basic methods that are the building blocks of all modern ground tactics: FIRE AND MOVEMENT and INFILTRATION. In fire and movement, attacking forces are divided into "fire" and "movement" elements, which alternately advance and direct suppressive fire against known or suspected enemy positions. The effect of the "fire" element is twofold: it distracts the enemy's attention away from the "movement" element, and its suppressive fire can neutralize enemy firepower. By moving forward in bounds, the fire and movement elements can gradually close on enemy positions.

Infiltration tactics exploit concealment: using terrain, weather, or man-made cover (e.g., smoke), an attacking force attempts to go around or between defensive positions, either to cut them off from their rear, or to attack them from unexpected directions.

Both fire and movement and infiltration can also be applied defensively. In the former, the defender seeks to halt or delay the attacker by firepower, and then to counterattack with his "movement" reserves. Examples of defensive infiltration include ACTIVE DEFENSE, whereby concealed/dispersed mobile groups counterattack the flanks an attacking force; but the classic use of concealment is the ambush.

For other ground tactics (and operational methods), see ARMORED WARFARE; DEFENSE-IN-DEPTH; GUERRILLA WARFARE; HEDGEHOG; INDIRECT APPROACH; MARCHING FIRE; OVERWATCH; SPECIAL OPERATIONS. For naval tactics, see ANTI-AIR WARFARE; ANTI-SUBMARINE WARFARE; ANTI-SURFACE WARFARE. For aerial tactics, see AIR COMBAT MANEUVERING; BOMBING TECHNIQUES.

TAKATSUKI: A class of four Japanese DESTROYERS commissioned between 1967 and 1970 as successors to the smaller, diesel-powered "hunter-killer" destroyers of the YAMAGUMO and MINEGUMO classes. Much larger, faster, and more powerful vessels, the Takasukis are flush-decked, with sharply raked bows and a graceful shear line.

Optimized for ANTI-SUBMARINE WARFARE (ASW), as completed they had very limited ANTI-AIR WARFARE (AAW) and ANTI-SURFACE WARFARE (ASUW) capabilities. The original weapons suite consisted of 2 5-in. 54-caliber DUAL PURPOSE guns, one on the bow and the other on the superstructure aft; an 8-round ASROC pepperbox launcher behind the forward gun; a Bofors 375-mm. 4-barrel ASW rocket launcher mounted ahead of the forward gun; 2 sets of Mk.32 triple tubes for MK.46 lightweight homing TORPEDOES amidships; and a landing deck and hangar for 2 remotely piloted DASH (Drone Anti-Submarine Helicopters). From 1981–82, when the ships were overhauled and modernized, the DASH system and aft 5-in. gun were removed to make room for 2 quadruple launchers for HARPOON anti-ship missiles, an 8-round NATO SEA SPARROW surface-to-air missile launcher, and a PHALANX radar-controlled gun for anti-missile defense, giving the Takatsukis much more balanced armament. The Takasukis were superseded by the larger HATSUYUKI class. See, more generally, DESTROYERS, JAPAN.

Specifications Length: 446 ft. (136 m.). Beam: 44 ft. (13.41 m.). Draft: 14.5 ft. (4.42 m.). Displacement: 3200 tons standard/4500 tons full load. Powerplant: twin-shaft steam: 4 oil-fired boilers, 2 sets of geared turbines, 60,000 shp. Fuel: 900 tons. Speed: 32 kt. Range: 7000 n.mi. at 20 kts. Crew: 270. Sensors: 1 OPS-11B air-search radar, 1 OPS-17 surface-search radar, 2 Mk.56 fire control radars, 1 CPS-2 fire control radar, 1 OQS-3 hull-mounted medium-frequency sonar, 1 SQS-35(J) variable depth sonar, 1 SQR-18A(V)1 TACTASS towed passive array. **Electronic warfare equipment:** 1 NOLR electronic countermeasures array, 2 Mk.36 SRBOC chaff launchers.

TALOS: The RIM-8 shipboard surface-to-air missile (SAM) in service with the U.S. Navy from 1957 to the early 1980s; a few remain in service as target drones.

TAM: *Tanque Argentino Mediano*, the lightest MAIN BATTLE TANK (MBT) now in service, designed by the German Thyssen-Henschel firm for the Argentine army, using many automotive components of the MARDER infantry fighting vehicle. The first prototype was delivered in 1976; since then, about 200 have been built, with some exported to Panama and Peru.

The all-welded steel hull is almost identical to the Marder's, but has been reinforced to cope with the heavier armor and the recoil stresses of a large-caliber gun. Protection, very light for an MBT, is only capable of resisting armor-piercing rounds up to 20 mm. The interior is divided into a driver/engine compartment in front, and a fighting compartment in the rear. The driver is provided with three observation periscopes with night vision optics. The other three crewmen sit in the all-welded turret, mounted at the rear of the vehicle, with the gunner on the right, the commander above and behind him, and the loader on the left. The commander has eight observation periscopes and a nonstabilized TRP-2A panoramic gunsight (identical to that of early model LEOPARD 1s) and an optical coincidence rangefinder. The gunner has a Zeiss TZF optical sight mounted in a trainable periscope, while the loader has a single, trainable observation scope. There is an escape hatch at the rear of the vehicle.

The main armament, a British-designed L7A 105-mm. rifled gun, fires APFSDS, APDS, HEAT, HESH, and smoke rounds. Fifty rounds are carried: 20 in the turret and 30 in the hull. Secondary armament includes a coaxial 7.62-mm. FN-MAG machine gun, a pintle-mounted FN-MAG on the turret roof, and four smoke grenade launchers mounted on each side of the hull. COLLECTIVE FILTRATION for NBC defense and an automatic fire extinguisher are standard. The TAM can ford streams up to 13 ft. (4 m.) deep using an erectable snorkel.

Other vehicles developed from the TAM chassis (mostly single prototypes) include the VCI ARMORED PERSONNEL CARRIER, the VCTP INFANTRY FIGHTING VEHICLE (similar to Marder), the Dragon 30-mm. anti-aircraft tank, a 155-mm. self-propelled howitzer, a support tank with a Bofors 57-mm. automatic cannon, and the TH 301, an improved TAM designed for the Thai army, with better fire controls, passive night-vision sights, a fully stabilized Rheinmetall 105-mm. gun, and an up-rated engine. Nominally a main battle tank, the TAM, with its heavy armament, high speed, and thin armor, is actually more of a TANK DESTROYER; it could fight competently handled MBTs only defensively, in favorable terrain.

Specifications **Length:** 22.2 ft. (6.77 m.). **Width:** 10.67 ft. (3.25 m.). **Height:** 7.9 ft. (2.4 m.). **Combat:** 30.5 tons. **Powerplant:** 720-hp. MTU MB833 Ka500 6-cylinder diesel. Hp./wt. ratio: 23.27 hp./ton. **Fuel:** 165 gal. (726 lit.). **Speed, road:** 46.6 mph (78 kph). **Range, max.:** 340 mi. (577 km.).

TANG: A class of six U.S. diesel-electric attack SUBMARINES commissioned in 1951–52, no longer with the U.S. Navy, but still in service with the navies of Italy and Turkey. At the end of the Second World War, the examination of captured German Type XXI U-boats revealed that U.S. submarines were inferior in submerged speed and endurance, operating depth, and acoustical silencing. The U.S. Navy modified 52 fleet submarines under the GUPPY (Greater Underwater Propulsion) program by streamlining, installing a snorkel and more powerful batteries, and modernizing sensors

and electronics, but the Guppies were still limited by their origins as surface-oriented fleet boats. The Tangs, by contrast, were designed from the keep up to incorporate wartime innovations (mostly German) to maximize underwater performance.

Their streamlined, cigar-shaped hulls show the influence of the Type XXI design. The Tangs have a double-hull configuration, with fuel and ballast stored between the outer casing and the pressure hull; the latter is fabricated of STS (Special Tensile Strength) steel, for greater diving capabilities than the Guppies. Fabricated from glass-reinforced plastic (GRP), the sail is merely an elevated navigating bridge (for use on the surface), and a streamlined fairing for the periscopes and other sensors; the traditional conning tower was eliminated because it offered too much frontal area for high underwater speed (and because it had been discovered during the war that the navigating and fire control functions of the conning tower were better exercised from the control room in the main hull). The control surfaces are conventional, consisting of retractable bow diving planes, a single rudder, and stern planes mounted ahead of the propellers. Originally, the Tangs had four revolutionary "pancake" diesel engines, which were very compact but also very unreliable, and which had to be replaced by more conventional engines (the hull was lengthened by 20 ft. in the process). The boats have a retractable snorkel for diesel operations at shallow submergence, while the battery is a variant of the standard GUPPY unit.

Armament consists of 8 21-in. (533-mm.) torpedo tubes, 6 in the bow and 2 in the stern. The forward tubes are full length, and can launch a variety of free-running and acoustical homing TORPEDOES; the short stern tubes are used for anti-submarine torpedoes and TORPEDO COUNTERMEASURES. Eighteen torpedoes are carried forward, and 6 aft. All sensors are linked to a Mk.106 analog torpedo FIRE CONTROL system.

Two Tangs (*Trigger* and *Harder*) were transferred to Italy in 1973–74; 2 more (*Tang* and *Gudgeon*) were transferred to Turkey in 1974 and 1983. The remaining 2 (*Wahoo* and *Trout*) were to be transferred to Iran in 1980; when this sale was canceled, the boats were laid up in reserve. The Tang class served as the basis for the similar DARTER class and the Greyback guided-missile submarines. See, more generally, SUBMARINES, UNITED STATES.

Specifications **Length:** 287 ft. (87.5 m.). **Beam:** 27.33 ft. (8.33 m.). **Displacement:** 2050 tons surfaced/2700 tons submerged. **Powerplant:** twin-shaft diesel-electric: 3 15,000-hp. Fairbanks-Morse diesels, 2 2800-hp. Westinghouse electric motors. **Speed:** 15.5 kt. surfaced/16 kt. submerged. **Range:** 7600 n.mi. at 15 kt. (snorkeling)/170 n.mi. at 9 kt. (batteries). **Max. operating depth:** 600 ft. (183 m.). **Collapse depth:** 984 ft. (300 m.). **Crew:** 85. **Sensors:** 1 BSQ-4 medium-frequency active/passive sonar, 1 BQG-4 PUFFS passive-ranging sonar, 1 BPS-12 surface-search radar, 1 electronic signal monitoring array, 2 periscopes.

TANGO: NATO code name for a class of 18 Soviet diesel-electric attack SUBMARINES completed between 1972 and 1982. The Tangos are incremental improvements of the long-range FOXTROT class, with a slightly larger and more hydrodynamically refined hull, improved propulsion, and far more effective weapons and sensors. The Tangos have a streamlined, cigar-shaped hull similar to the Foxtrots', and bearing a strong resemblance to their common ancestors, the Soviet Zulu class and German Type XXI. Like them, the Tango has a double-hull configuration with fuel and ballast tanks located between the pressure hull and the outer casing, but the latter has fewer free-flooding holes, thereby reducing underwater drag and flow noises. A tall, streamlined sail (conning tower) is located amidships; an extension of its aft end houses a snorkel mast and diesel exhaust vent. The outer hull is covered with a rubberized ANE-CHOIC coating to reduce radiated noise and impede active sonar; this has a tendency to come loose at high speeds, and is not as effective as the CLUSTER GUARD tiles on Soviet nuclear submarines. The control surfaces are conventional, consisting of retractable bow planes, a single rudder, and stern planes mounted behind the propellers. Although the Tangos have a snorkel, a very powerful battery allows a fully submerged endurance of more than seven days at low speeds.

Armament consists of 8 21-in. (533-mm.) torpedo tubes, 6 in the bow and 2 in the stern, for a variety of free-running or acoustical homing TORPEDOES and the SS-N-15 STARFISH anti-submarine missile. A total of 18 weapons are carried, which can be replaced by MINES a 2-for-1 basis. The SONAR suite is much more capable than that in the Foxtrots, and all sensors are linked to an automatic torpedo FIRE CONTROL system.

Although now apparently superseded by the teardrop-hull KILO class, the Tangos are very quiet and effective. Their primary mission appears to be ANTI-SUBMARINE WARFARE conducted from ambush

positions in sectors such as the GIUK GAP. See, more generally, SUBMARINES, SOVIET UNION.

Specifications **Length:** 301.8 ft. (92 m.). **Beam:** 29.5 ft. (9 m.). **Displacement:** 3100 tons surfaced/3900 tons submerged. **Powerplant:** triple-shaft diesel-electric: 3 diesel generators, 3 electric motors, 6000 hp. **Speed:** 16 kt. surfaced/15.5 kt. submerged. **Max. operating depth:** 984 ft. (300 m.). **Collapse depth:** 1640 ft. (500 m.). **Crew:** 62. **Sensors:** 1 bow-mounted low-frequency passive sonar, 1 active medium-frequency fire control sonar, 1 "Snoop Tray" surface-search radar, 1 "Brick Group" electronic signal monitoring array, 2 periscopes.

TANK: A fully tracked, heavily armed AR-MORED FIGHTING VEHICLE. The concept is quite ancient; aside from Roman precursors, Leonardo da Vinci designed a "tank" in 1484. There were earlier proposals to combine the internal combustion engine with track-laying and armor, but it was not until 1915 that the first practical tank was developed by the British Admiralty (at the instigation of the First Lord, Winston Churchill), as a means of traversing the shell-pocked, wire-strewn, machine-gun-swept expanses of No Man's Land between the trench lines of the Western Front. The first tanks, essentially mobile armored boxes armed with machine guns and quick-firing naval cannon, were rhomboidal in shape (the tracks running completely around them), with all armament mounted in sponsons on the sides of the chassis (or "hull"). Although slow, short-ranged, and notoriously unreliable, these tanks were successful in crossing trench lines, and their potential ability to do more was soon recognized by many theorists, including J. F. C. Fuller and B. H. Liddell-Hart in Britain and de Gaulle in France; and by practical soldiers such as Guderian in Germany, Tukachevsky in the Soviet Union, and Patton in the United States.

By the 1930s, the tank acquired the general configuration it still has today, with the main armament moved from side-mounted sponsons into a rotating armored turret mounted on top of the hull (allowing the tank to engage targets in any direction); and with a driver (and formerly also a machine-gunner) seated in the bow, the vehicle commander, gunner, and loader in a central fighting compartment in or under the turret, and the engine mounted in the rear.

In Britain, France, and the Soviet Union, pre–World War II tank designs diverged into two distinct types: "infantry" (I) tanks, direct successors of the original World War I vehicles, slow, heavily armored for their time, and intended to operate in small units in direct support of infantry; and "cavalry" or "cruiser" (C) tanks, which were lighter, faster, and designed for pursuit and exploitation. In other words, the tank was subordinated to the infantry in one version, and configured as a new form of light cavalry in the other.

The German army, prohibited by the Treaty of Versailles from developing or fielding tanks, experimented and cogitated instead; the outcome was the COMBINED ARMS *Panzer* division, intended to fight *independently* in the depths of enemy territory, i.e., to fight in a new way, instead of enhancing traditional roles. Accordingly, it was tanks combining speed, reasonable armor, and heavy firepower that were needed, not distinct "I" or "C" tanks with only two of these attributes.

During World War II, tank designs fell into three categories: light, medium, and heavy. LIGHT TANKS were meant mainly for reconnaissance; medium tanks were the main combat type; and heavy tanks were meant as long-range anti-tank vehicles, as well as to spearhead high-priority attacks against strongly defended positions. In general, heavy tanks carried larger guns and had more armor than mediums, but were slower and less mobile across country.

By the end of World War II, the evolution of tank design combined the medium and heavy types into the MAIN BATTLE TANK (MBT), with the armor and firepower of heavy tanks and the mobility of mediums; the Soviet T-34 was the precursor of this concept, and the German Panther its best application.

Since World War II, MBT main guns have increased in caliber, from 76.2–90 mm. in 1945 to 105–125 mm., while gun performance has also improved due to better metallurgy and propellant technology, and new types of ARMOR-PIERCING ammunition, chiefly APDS and, more recently, APFSDS. Firepower has been further enhanced by advances in FIRE CONTROL. In World War II, gunners still had to estimate range visually in most tanks. By the early 1960s, optical (stereoscopic or coincidence) rangefinders as well as ranging machine guns were available. By the early 1980s, most MBTs had laser rangefinders connected with ballistic computers which also assimilate ammunition type, relative target motion, and meteorological data into the fire control solution, giving a high probability of a first-round hit out to ranges of 2000 m. and more. Moreover, main guns are now stabilized in both elevation and traverse, allowing them to be fired accurately while the tank is moving.

Armor protection has also improved. Aside from using thicker and better steel, other methods of improving protection include sloping armor away from the vertical (to increase its effective thickness and present a glancing surface), emphasizing protection in the frontal arcs at the expense of the sides and rear, lowering the vehicle silhouette, and using other materials in combination with steel. By the 1980s, two new developments dramatically increased the level of protection, without a concomitant weight increase: CHOBHAM-type composite armor, with several times the effectiveness of steel; and so-called REACTIVE ARMOR (actually special explosive charges) capable of defeating shaped-charge (HEAT) warheads.

Mobility improvements since 1945 include the universal adoption of diesels in place of gasoline engines (improving fuel economy while reducing the risks of fire and explosion), increases in engine output through supercharging and turbocharging, and the introduction of gas turbines. Current 1500-hp. engines (vs. 500 in World War II) allow vehicles weighing 45 to 60 tons to achieve speeds of 45 mph (75 kph) or more on roads. Horsepower-to-weight ratios, a useful index of relative vehicle mobility, have improved from 10 hp. per ton or less in World War II, to 15–20 hp. per ton in the 1960s, to 20–25 hp. per ton today (the latest MBTs, such as the M1 ABRAMS and LEOPARD 2, are faster and more mobile than the light tanks of World War II). Other automotive improvements, just as important in combination, include automatic transmissions, stronger suspensions, and more durable tracks, all of which make modern tanks more reliable than their predecessors, though still far less reliable mechanically than any commercial vehicle.

Costs have risen more than proportionately: whereas typical World War II medium tanks cost roughly $50,000 or less, current top-quality MBTs cost more than $1.5 million. Few countries can afford to design and build tanks, and fewer still can build them in large numbers. New design concepts, intended to reduce size, weight, and above all cost, include externally mounted, remotely operated guns; automatic loaders (standard only on Soviet tanks); and even completely unmanned, remotely piloted vehicles. In spite of many revolutionary projects, the basic configuration of MBTs has hardly evolved since World War II; see MERKAVA and S-TANK for exceptions.

For a survey of alternative methods of integrating tanks into ground forces, see ARMORED FORCES; ARMORED WARFARE; BLITZKRIEG. For methods of countering tanks, see ANTI-TANK.

TANK DESTROYER: A specialized ARMORED FIGHTING VEHICLE (AFV) intended to engage TANKS and other AFVs defensively from behind terrain cover, to offset light protection or the lack of all-round traverse.

Two separate types evolved during World War II: fast, lightly armored, turreted vehicles mounting a large, high-velocity gun; and turretless vehicles. Although it looked like a tank, the turretted tank destroyer (developed by the U.S. Army) had to rely on stealth and speed for survival; because of its light armor protection, it could not engage in stand-up tank duels. The Swedish IKV.91 and French AMX-13 are postwar examples.

German and Soviet tank destroyers, by contrast, were heavily armored turretless AFVs with a powerful but limited traverse gun in a frontal casemate. This type of tank destroyer could engage in tank duels, and could be produced at lower cost than tanks, but its limited gun traverse was tactically restrictive on the offensive. The Soviet Union and West Germany both continued to produce such tank destroyers after World War II, notably the Soviet SU-100, ISU-152, and the light, airborne ASU-85; and the German JAGDPANZER *Kanone* (JPzK). But Soviet tank destroyers now remain in service only with reserve units, and West German JPzKs may be converted to missile armament.

More unusual postwar tank destroyers include the Ontos, an 8-ton tracked vehicle mounting six 106-mm. RECOILLESS RIFLES, once in service with the U.S. Marine Corps; French heavy ARMORED CARS (including the Panhard AML and EBR, and the current AMX-10RC), armed with 90-mm. guns firing shaped-charge (HEAT) ammunition; the Italian Centauro heavy armored car, with a full-power 105-mm. gun; and ANTI-TANK GUIDED MISSILES (ATGMs) mounted on light armored vehicles (the Soviet BRDM family, the U.S. M901 IMPROVED TOW VEHICLE, the French Panhard VCR/AT, and the British STRIKER). West Germany has taken a different approach by mounting ATGMs on the well-armored *Jadgpanzer* chassis (*Jagdpanzer Rakete*).

The recent development of SOFT RECOIL allows high-velocity 105-mm. guns to be mounted on a chassis weighing less than 20 tons. This has led to a revival of the LIGHT TANK in the tank destroyer role. A number of such vehicles, wheeled and tracked, are now being evaluated as low-cost sub-

stitutes for main battle tanks, or for use by air-landed forces.

TARANTUL: NATO code name for a class of some 38 Soviet guided-missile CORVETTES completed from 1979. Intended as replacements for the OSA-class missile boats, the Tarantuls are much larger, better armed, and more seaworthy, reflecting a new Soviet preference for coastal defense at greater distances from Soviet shores. At least 16 are now in the Soviet navy, and 3 early models have been transferred to Poland and 1 to East Germany.

With the same hull and machinery as the PAUK-class anti-submarine corvettes, the Tarantuls are flush-decked, with a tall, blocky superstructure amidships. There are three distinct subclasses: the Tarantul I and II differ only in their sensors, but the Tarantul III has a much more powerful armament. In the Tarantul I/II, the main armament consists of 4 SS-N-2c STYX anti-ship missiles in fixed twin launched canisters mounted on either side of the bridge, while the Tarantul III has 4 new, supersonic, surface-skimming SS-N-22 SUNBURN missiles. All have a 76.2-mm. DUAL PURPOSE gun on the bow, a quadruple launcher for SS-N-5 GRAIL or SS-N-8 GREMLIN short-range surface-to-air missiles on the stern, and 2 30-mm. radar-controlled guns for anti-missile defense on the superstructure amidships. See also FAST ATTACK CRAFT, SOVIET UNION.

Specifications Length: 185.33 ft. (56.5 m.). Beam: 34.45 ft. (10.5 m.). Draft: 8.25 ft. (2.5 m.). Displacement: 480 tons standard/580 tons full load. Powerplant: (I/II) twin-shaft COGOG: 2 3000-shp. gas turbines (cruise), 2 12,000-shp. NK-12M gas turbines (sprint); (III) twin-shaft CODOG: 1 5000-hp. M504 diesel, 2 NK-12M turbines. Fuel: 50 tons. Speed: 36 kt. Range: 400 n.mi. at 36 kt./2000 n.mi. at 20 kt. Crew: 40. Sensors: 1 "Kivach-3" surface-search radar, 1 "Plank Shave" Styx target acquisition radar, 1 "Bass Tilt" 30-mm. fire control radar; (II) 1 "Light Bulb" SS-N-22 guidance data link, 1 "Bank Stand" 76.2-mm. fire control radar, 1 30-mm. electro-optical director. Electronic warfare equipment: 4 radar warning receivers, 2 16-round chaff launchers.

TARAWA: A class of five U.S. amphibious assault ships (LHAS) commissioned between 1976 and 1980. Intended to combine the capabilities of a helicopter carrier (LPH) and a dock landing ship (LPD) in a single hull, the Tarawas were the largest amphibious warfare ships in service until the advent of the similar WASP class in 1989.

Resembling small aircraft carriers, the Tarawas have full-length flight decks 32 ft. (40.2 m.) wide, with large island superstructures to starboard housing the usual ship-handling and flight control facilities, as well as extensive COMMAND AND CONTROL facilities for an amphibious force command group. Two elevators, one on the deck edge and the other at the fantail, connect the flight deck with the hangar, which is 820 ft. (250 m.) × 78 ft. (23.78 m.) × 20 ft. (6.09 m.); there are also five cargo elevators between the flight deck and the holds. The Tarawas can operate up to 35 transport and attack HELICOPTERS, and have on occasion embarked detachments of U.S. Marine Corps AV-8 HARRIER V/STOL attack aircraft.

A floodable well deck, 268 ft. (81.7 m.) × 78 ft. (23.78 m.), is built into the stern and accessed through large transom doors. It can accommodate up to 4 tank landing craft (LCTS), or 2 LCTs and 2 medium landing craft (LCMS), or 17 LCMs, or up to 45 LVTP-7 amphibious assault vehicles, but only 1 LCAC hovercraft, because of the internal arrangements of the well. There is also a vehicle deck below the hangar for up to 31 additional LVTP-7s, accommodations for up to 1900 fully equipped troops, and hospital beds for up to 300 casualties. Defensive armament consists of 2 8-round pepperbox launchers for SEA SPARROW short-range surface-to-air missiles, 3 5-in. 54-caliber DUAL PURPOSE guns, a 20-mm. PHALANX radar-controlled gun for anti-missile defense, and 6 manually operated 20-mm. anti-aircraft guns.

Specifications Length: 833 ft. (237.1 m.). **Beam:** 106 ft. (32.2 m.). **Draft:** 26 ft. (7.9 m.). **Displacement:** 25,120 tons standard/39,400 tons full load. **Powerplant:** twin-shaft steam: 2 oil-fired boilers, 2 sets of geared turbines, 70,000 shp. **Fuel:** 5900 tons. **Speed:** 24 kt. **Range:** 10,000 n.mi. at 20 kt. **Crew:** 940. **Sensors:** 1 SPS-10 and SPS-53 surface-search radar, 1 SPS-40 2-dimensional air-search radar, 1 SPS-52 3-dimensional air-search radar, 1 SPG-60 and SPQ-9A fire control radar. **Electronic warfare equipment:** one SLQ-32(V3) electronic countermeasures array, 4 Mk.36 SRBOC chaff launchers.

TARGET: 1. A geographic point, area, or installation designated for capture or destruction; or the aim point for a projectile weapon.

2. In INTELLIGENCE usage, a nation, area, installation, group, or individual against which intelligence activities are directed.

3. In sensor-design terminology, any discrete object which either radiates or reradiates detectable energy; alternatively, the object of sensor search or surveillance.

TARGET ACQUISITION: Also, or simply "acquisition"; the detection, classification, and tracking of an object with sufficient accuracy to direct weapons against it. Target acquisition can be performed visually, or by RADAR, SONAR, INFRARED, ELECTRO-OPTICAL, or optical sensors. In most cases, target acquisition must be performed before a weapon is fired or launched, but some PRECISION-GUIDED MUNITIONS are capable of autonomous search and target acquisition, and can thus be launched in the general direction of targets without prior acquisition. See also FIRE AND FORGET; FIRE CONTROL; OVER-THE-HORIZON TARGETING.

TARGET, AREA: A TARGET covering a significant physical expanse, such as a city, a factory complex, an air base, a large body of troops, or a naval anchorage. In general, such targets are "soft," i.e., not reinforced to resist blast effects.

TARGET, HARD: A TARGET reinforced to resist blast and/or other weapon effects, such as a missile SILO, a TANK, or an armored warship. See also HARD, HARDNESS, HARDENING; HARDPOINT.

TARGET, POINT: A TARGET consisting of one small, localized object, such as a building, a bunker, a COMBAT VESSEL, a COMBAT AIRCRAFT, or a missile SILO, which may or may not be HARDENED to resist weapon effects. See also HARDPOINT.

TARGET, SOFT: Any TARGET not reinforced, or HARDENED, to resist weapon effects. It may consist of a single point, or extend over a considerable area.

TARTAR: The RIM-short to medium-range shipboard surface-to-air missile (SAM), no longer with the U.S. Navy, but still serving in several allied navies. Tartar, the last of the three missiles developed under Project Bumblebee in the early 1950s (the others being TALOS and TERRIER), was designed as a light, compact weapon for use on DESTROYERS. In essence, it is a SEMI-ACTIVE RADAR HOMING, tail-controlled Terrier, without the latter's first (booster) stage. The missile has four short-span, broad-chord body strakes for lift, and four folding tail fins for control. The narrow strakes allow the missile to be stowed vertically and loaded automatically, which, combined with relatively short length, permits its installation on relatively small ships. Tartar is armed with a 275-lb. (125-kg.) CONTINUOUS ROD high-explosive warhead (with a lethal envelope of approximately 65 ft./19.8 m.), triggered by radar proximity (VT) fuzing.

The initial RIM-24A of 1961 was intended as a DUAL PURPOSE weapon with secondary anti-ship capability, but was seldom tested in that capacity. Like all missiles of its era, Tartar was unreliable. The Product Improved Tartar (RIM-24B), which entered service in 1963, had an improved radar seeker (developed for the Terrier HT-3), which increased nominal lethality to 50 percent, and a new motor for longer maximum range. But reliability continued to be unacceptable, and the missile was scheduled for replacement by Typhon, a Tartar-sized missile offering much longer range and better performance against surface-skimming targets. Typhon was canceled during the Vietnam War, and Tartar was eventually replaced by the STANDARD-MR missile, which has the same configuration and dimensions, but completely new propulsion and guidance.

Tartar can be launched from several fairly compact automatic launchers, the oldest of which, the Mk.11 fitted on older ADAMS-class destroyers, is a twin-arm launcher mounted over two rotary magazines, each holding 21 rounds. For loading, the launcher arms are aligned vertically over two hatches through which the missiles are hydraulically transferred to the launcher rails, for a nominal rate of fire (ROF) of 2 rounds every 18 seconds. The Mk.11 was unreliable, and was superseded by the widely deployed Mk.13, a single-arm launcher with a 40-round rotary magazine and ROF of 6–8 rds./min. Later Mk.13s can also launch Standard-MR and HARPOON anti-ship missiles. The Mk.22 launcher, essentially a scaled-down Mk.13 with a 16-round magazine, arms the BROOKE-class guided-missile FRIGATES.

Specifications Length: 15.5 ft. (4.72 m.). Diameter: 13.5 in. (343 mm.). Span: 24 in. (610 mm.). Weight, launch: 1310 lb. (595.5 kg.). Speed: Mach 1.85. Range envelope: (A) 1.15–9.2 mi. (1.92–15.36 km.); (B, max.) 18.4 mi. (30.72 km.). Height envelope: 50–50,000 ft. (15.25–15,244 m.).

TASM: Tomahawk Anti-Ship Missile, a variant of the General Dynamics BGM-109 Tomahawk CRUISE MISSILE designed to attack warships at sea. For details, see TOMAHAWK.

TASS: Towed Array Sonar System, the U.S. SQR-15 passive hydrophone array, deployed by GARCIA-class frigates. The SQR-15 became operational in 1973, and also equipped the first six SPRUANCE-class destroyers until replaced by the more advanced SQR-19 TACTASS.

TBILISI: NATO code name for a Soviet large-deck, nuclear-powered AIRCRAFT CARRIER (a.k.a. *Brezhnev*, a.k.a. *Kremlin*) now undergoing landing trials in the Black Sea. The *Tbilisi* was spotted by reconnaissance satellites in 1979, leading analysts to conclude that construction had begun sometime in 1977–78. Because of its size, the ship was constructed in two halves, each of which was launched separately, and then joined together in a drydock. The finished hull was launched on 5 December 1985; five days later, the keel was laid for a second ship of the class. The *Tbilisi* is expected to complete flight deck training in 1991–92, and become fully operational in 1993–94. The second carrier was launched in 1988, for completion in 1993, and a third carrier Ulyanovsk (of a modified design displacing some 10,000 tons more) was laid down in 1989. Because the Soviet Union has only one assembly dock of sufficient size, the maximum rate of production for these ships is estimated at only one every 4–5 years (four Tbilisi could be in service by 2000). With the four KIEV-class V/STOL carriers and two MOSKVA-class helicopter carriers, the Tbilisi could provide useful air support to Soviet submarine operations, but may be most effective for gunboat diplomacy in Third World contingencies.

Various specifications for the Tbilisi were released over time, reflecting additional intelligence and more refined analysis, but there is also some evidence that the Soviet navy modified the design at least once during construction, probably to take advantage of technical developments in the aircraft to be carried. In contrast to the earlier Kiev V/STOL carriers (which have heavy, guided-missile CRUISER armament forward), the Tbilisi has a full-length flight deck, angled landing deck cantilevered to port, and a large island superstructure offset to starboard. The ship has a ski-jump ramp at the bow for short takeoffs, but a catapult and arresting gear may still be fitted to the angled deck to permit the operation of conventional takeoff and landing (CTOL) aircraft. The ship has three elevators: two on the starboard deck edge, fore and aft of the island, and one inboard, amidships. In contrast to the Kievs, the *Tbilisi* does not carry heavy anti-surface, anti-air, or anti-submarine weapons. Armament appears to be limited to several vertical launchers for SA-N-9 short-range surface-to-air missiles, and as many as eight ADG6-30 6-barrel 30-mm. radar-controlled guns for anti-missile defense (this "last-ditch-only" defensive armament parallels the U.S. Navy's concept, which relies mainly on the air group and escort vessels for protection).

Initially believed to have Combined Nuclear and Steam (CONAS) propulsion as in the KIROV battle cruisers, the ship is now assessed to have conventional oil-fired boilers. The Ulyanovsk, however, does appear to have either a nuclear or a CONAS powerplant.

The composition of the Tbilisi's air group has not yet been determined. In the short term, it could operate a mixed group of 40–60 aircraft, including up to 30 Yak-36 FORGER and improved Yak-41 V/STOL fighter/attack aircraft, and 20 Ka-27 HELIX helicopters. The V/STOL fighters may be supplemented or replaced in the mid-1990s by a navalized version of the (CTOL) Su-27 FLANKER (a very powerful aircraft in the class of the U.S. F-14 TOMCAT or F-15 EAGLE), which has been observed practicing ski-jump takeoffs and arrested landings on a simulated flight deck. Specialized AIRBORNE EARLY WARNING (AEW) and fixed-wing anti-submarine aircraft have not yet been developed to take full advantage of the carrier's potential. See also AIRCRAFT CARRIERS, SOVIET UNION.

Specifications **Length:** 985 ft. (300.3 m.). **Beam:** 120 ft. (36.58 m.). **Draft:** unknown. **Displacement:** 64,000–70,000 tons full load. **Powerplant:** 4-shaft steam, 8 oil-fired boilers, 4 sets of geared turbines, 200,000 shp. **Fuel:** 10,500 tons. **Speed:** 32–34 kt. (superheat). **Sensors:** 1 Sky Watch phased array radar, 1 "Top Plate" air-and-surface-search radar, 2 Strut Pair 2-dimensional air search radars, 1 Plate Steer 3-dimensional air search radar, 3 Palm Frond navigation radars, 3 Cross Sword Fire Control radars, 1 TACAN-type navigational beacon, 1 small hull-mounted sonar. **Electronic warfare equipment:** 1 electronic signal monitoring array, several chaff launchers.

TECHINT: Technical Intelligence: 1. Information generated by technical means, including signals intelligence (SIGINT) and its branches, electronic intelligence (ELINT) and communications intelligence (COMINT) aerial photography; satellite imagery; etc.; as opposed to HUMINT, or human intelligence collected by people.

2. Information on scientific and technological developments by actual or potential adversaries, particularly those with military applications. See also INTELLIGENCE; NATIONAL TECHNICAL MEANS; SATELLITES, MILITARY.

TERCOM: See TERRAIN COMPARISON AND MATCHING GUIDANCE.

TERMINAL DEFENSE: Weapons and tactics intended to defeat hostile missiles during the last, TERMINAL PHASE of their trajectories.

TERMINAL GUIDANCE: Missile guidance applied between the end of the midcourse phase and impact.

TERMINAL IMAGING RADAR (TIR): A proposed PHASED ARRAY RADAR intended for the terminal layer of BALLISTIC MISSILE DEFENSE (BMD) systems being developed under the STRATEGIC DEFENSE INITIATIVE. TIR is designed to discriminate incoming REENTRY VEHICLES from decoys and other PENAIDS by using digital signal processing to create two-dimensional DOPPLER "images" of detected objects. The TIR would also be capable of tracking incoming RVs with sufficient accuracy to provide TARGET ACQUISITION data for weapons such as ERIS or HEDI.

Because radar is vulnerable to nuclear BLACKOUT, efforts are being made to reduce the size and cost of TIR, to allow many such radars to be deployed, so that if one radar is neutralized, its sector can be scanned by others waiting in standby mode. The size and weight of TIR could be reduced to the point where it could be deployed on railroad cars. Mobility and dispersal would provide an additional degree of survivability, by preventing the enemy from targeting the radars, and by permitting some degree of reconfiguration to cover gaps resulting from technical failures or enemy attack. The TIR program has experienced considerable technical difficulties.

TERMINALLY GUIDED SUBMISSILE (TGSM): A "smart" anti-armor submunition under development by LTV for the U.S. Army. Initiated in the early 1970s as a means of increasing the effectiveness of conventionally armed LANCE tactical missiles, TGSM was further developed as part of the ASSAULT BREAKER system.

The TGSM is a small, autonomously homing glide bomb dispensed from a BALLISTIC MISSILE at a present range and height from the target area. It is approximately 35 in. (889 mm.) long, 5.9 in. (150 mm.) in diameter, and weights 35.2 lb. (16 kg.). Its cylindrical body has a blunt nose section housing either an IMAGING INFRARED or passive millimeter wave (MMW) radar sensor, and a small guidance computer. Lift is provided by four wraparound, pop-out wings, and steering by four smaller, pop-out tail fins. It is armed with a relatively small shaped-charge (HEAT) warhead to attack the thin top armor of tanks and other armored fighting vehicles. Up to nine TGSMs can be carried in the warhead section of a Lance-size ballistic missile.

In action, the missile would be fired in the general direction of a concentration of enemy armored vehicles; once the missile was past the apogee of its trajectory, a linear charge would rupture the warhead section, dispersing the TGSMs. A para-balloon would be deployed from the tail of each submission, to decelerate and stabilize it; next, the wings and tail fins would pop out, and the parabal-lon would be jettisoned. The sensor then would then begin scanning the scene below; when an appropriate target signature was detected, the guidance computer would lock on and steer the TGSM towards it, unless the computer deter-mined that the target was outside the kinematic limits of the control system, in which case it would break lock and resume scanning.

TGSMs have been test-fired from Lance mis-siles and the LTV T-22 Assault Breaker missile since 1974. It is now being developed for the MUL-TIPLE LAUNCH ROCKET SYSTEM (MLRS) and the Army Advanced Tactical Missile System (ATACMS). See also ANTI-TANK; DEEP STRIKE; IM-PROVED CONVENTIONAL MUNITIONS.

TERMINAL PHASE: 1. The final portion of a ballistic-missile trajectory, from reentry into the atmosphere to impact, which begins when aerody-namic effects introduce significant perturbations into the nominal ballistic (free-fall) trajectory.

2. The portion of a missile trajectory during which TERMINAL GUIDANCE is in operation.

TERRAIN-AVOIDANCE RADAR: An air-borne RADAR that scans ahead of the flight path to detect high terrain or other obstructions, and which alerts the pilot to its presence, extent, and location. Unlike TERRAIN-FOLLOWING RADARS, it is not linked to a navigation/control system in order to automatically guide the aircraft over or around the obstruction.

TERRAIN COMPARISON AND MATCH-ING GUIDANCE (TERCOM): A form of (long-range) missile guidance in which ground-mapping data received by an on-board radar is compared to a radar "map" stored in the guidance computer, to update the missile's INERTIAL GUIDANCE unit. Ca-pable of unparalleled accuracy (on the order of 10 ft. over 1500 mi.), TERCOM is employed on the U.S. ALCM and TOMAHAWK cruise missiles, and per-haps the Soviet AS-15 KENT and SS-N-21 SAMPSON.

TERRAIN-FOLLOWING RADAR (TER): An airborne radar that scans the area ahead of the flight path to detect high terrain or other obstruc-tions, and passes the data to an automatic pilot, which directs the aircraft over or around the ob-struction. TRF systems are essential for high-speed CONTOUR FLYING. TFR can be jammed,

spoofed, or (more easily) detected by enemy RADAR WARNING RECEIVERS—hence, passive devices such as LANTIRN (which has an IMAGING INFRARED sen-sor) are regarded as more reliable. See also TER-RAIN-AVOIDANCE RADAR.

TERRIER: The U.S. RIM-2 shipboard me-dium-range surface-to-air missile, still serving in small numbers with the Italian navy. Terrier, the closely related TARTAR, and the long-range TALOS were all developed in the late 1940s under Project Bumblebee, initiated late in World War II in re-sponse to the novel kamikaze and guided bomb threats. Terrier actually originated as a Talos Su-personic Test Vehicle in 1949, but was rushed to completion before Talos as an interim weapon for World War II CRUISERS converted to carry guided missiles. Terrier is smaller than Talos and has a much more compact launcher, allowing its installa-tion on ships displacing as little as 5000 tons (a lightweight version was installed experimentally on the 2800-ton GEARING-class destroyer *Gyatt*). Flight tests began in 1950, and the missile entered service with the U.S. Navy in 1954.

Early versions of the two-stage, solid-fuel rocket-propelled missile had four movable cruci-form wings and four small fixed tail fins on the second (sustainer) stage, plus four larger fixed fins on the first (booster) stage; such versions were designated "W," for wing-controlled. In later models the wings were replaced by four fixed, nar-row-chord body strakes, with control provided by four movable tail fins; these missiles are desig-nated "T" for tail-controlled. Two alternative guidance modes are provided: BEAM RIDING ("B") or SEMI-ACTIVE RADAR HOMING ("H"). On launch, the booster burns for 2.5 seconds before dropping away; the sustainer then burns for about 20 sec-onds. The initial production version, the RIM-2A/B (Terrier BW-0/1), was armed with a 218-lb. (99-kg.) continuous-rod, high-explosive warhead with a radar proximity–fuzed lethal radius of 65 ft. (19.81 m.). Later models had a 275-lb. (125-kg.) warhead.

The RIM-2C (BT-3), the first "wingless" model, entered service in 1956. It had improved guidance and a more powerful engine. The RIM-2D (BY-3A) was similar, but also had a secondary anti-ship capability. The RIM-2D(N), armed with a 1-kT W45 nuclear warhead (beam riding and command detonation were required under the U.S. POSITIVE CONTROL doctrine for nuclear weapons) was re-tained by the U.S. Navy into the 1980s in the ab-sence of a modern replacement. The RIM-2E

(HT-3), the first radar-homing version, was introduced in 1955 and had a much better low-altitude engagement capability. The RIM-2F, also introduced in 1955, was similar to the E model, but had an improved sustainer motor. The last Terrier was delivered in 1966, after some 8000 had been built.

Terrier proved to be quite effective, shooting down several MiGs off the coast of North Vietnam; but like all missiles of its era, it was mechanically unreliable. In the early 1960s it was subjected to an improvement scheme (the "3-T Get Well Program"), which eliminated some (but not all) of its defects. It had been planned to replace both Terrier and the smaller Tartar in the late 1960s with a new Typhon missile, but though it was technically successful, Typhon was canceled for financial reasons. Terrier was replaced by the closely related but more capable STANDARD-ER (Extended Range), originally meant as a stopgap measure after the cancellation of Typhon, but still in service today.

Terrier is fired from the twin-arm Mk.10 launcher, with the missiles stowed horizontally in several rotary magazines behind and below the launcher. For loading, a missile is raised to a transfer room immediately behind the launcher, where the wings and booster fins are attached manually. The missile is then transferred by overhead rail through blast-proof doors onto the launcher. Launchers have either two or three 20-round rotary magazines, depending upon the model and the size of the ship. The Mk.10 Mod 1 can also launch ASROC anti-submarine missiles, eliminating the need for a separate launcher. The Mk.10 can also accommodate Standard-ER with minor adjustments.

Specifications Length: 26.16–27.1 ft. (7.98–8.26 m.). **Diameter:** 13.5 in. (343 mm.). **Span:** 46 in. (1.17 m.). **Weight, launch:** 2900–3090 lb. (1318–1404 kg.). **Speed:** (A/B) Mach 1.8; (C/D) Mach 3. **Range:** (A/B) 13.2 mi. (22 km.); (C/D) 19 mi. (31.75 km.); (F) 53 mi. (88 km.). **Height envelope:** 5000–50,000 ft. (1524–15,240 m.); (C/D) 75,000 ft. (22,865 m.) max.

TERRITORIAL ARMY, BRITAIN: Reserve forces of the British army, intended for the rapid reinforcement of the British Army of the Rhine (BAOR), for rear area security, and for home defense. The Territorial Army (TA), first established in 1908 and important in both world wars, was allowed to languish for lack of funds in the 1960s and '70s, but has since been revitalized, and today numbers roughly 85,000 men. In wartime, the TA

would compose about 30 percent of BAOR, and almost all home-defense troops. It is currently organized into 35 infantry battalions (21 motorized, 14 "straight leg"), 3 parachute battalions, 3 security battalions, 2 armored reconnaissance regiments, 2 artillery battalions, 6 engineer battalions, 4 air defense regiments, and 2 small SAS "regiments."

Territorial units have the same organization as the active forces, though equipment is usually older. Personnel are all volunteer part-timers with the exception of a small cadre. Training is conducted during an annual two-week call-up, and at 12 one-day training sessions; many TA units meet more frequently. Only a small proportion of TA soldiers have prior military experience, and turnover is high, factors which reduce effectiveness (some TA formations, however, are very good indeed, notably the SAS, paratroops, and artillery). The TA is completely integrated into the British army: on mobilization, TA battalions are attached to divisions and brigades as "round-out" forces. They participate regularly in BAOR exercises.

The TA is backed up by a 218,000-man Reserve, made up of former military personnel liable to recall in emergencies. They receive very little refresher training, but could provide casualty replacements.

TERRITORIAL ARMY, FEDERAL REPUBLIC OF GERMANY: Reserve forces of the West German army (Heer), intended for rapid reinforcement of the active army. These are the only West German troops not under NATO control, and are not counted against the 500,000-man limit placed on West German forces. The Territorial Army is manned by conscripts recently discharged from active service, who can be recalled instantly by the minister of defense without prior notification to the Bundestag. It is organized into three Territorial Commands (Terrkdo): Terrkdo Nord, in the NORTAG area; Terrkdo Sud, in the CENTAG area; and a separate Terrkdo for Schleswig-Holstein. Terrkdo Nord and Sud are in turn divided into two regional commands (Wehrbereichskommando, WBK), each of which has several subregional commands. There are static and mobile units. The static forces, consisting of 150 independent home defense companies and 300 independent security platoons, would mobilize in the area they are assigned to defend; their missions include rear area security, traffic control, and civil defense. Mobile troops are organized into 12 home defense brigades (Heimatshutzbrigaden, HSBs), 2 of which are assigned to each WBK. These are high-quality mechanized formations, each consisting of one or two *Panzer* battalions, one or two motorized JAGER battalions, and a field artillery battalion. The HSBs, well equipped with tanks and other weapons passed down from the active forces, are intended to cover gaps in the NATO frontage, and to intercept Soviet deep penetrations, airborne DESANTS, and OPERATIONAL MANEUVER GROUPS.

TERRORISM: The use of violence against civilians by covert or clandestine organizations for political purposes. Terrorist targets are often selected for their symbolic significance, and are frequently attacked in a manner calculated to cause maximum shock. By bombings, shootings, kidnappings, hijackings, and assassinations, terrorists seek to lower public morale, reduce confidence in official authorities and institutions, obtain concessions, and force governments into acts of repression which they hope will lead to popular revolt.

Terrorists fall into several distinct categories: alienated individuals; ethnic, religious, or nationalist factions; ideological groupings; and covert agents of other states. Terrorism exists on both the extreme right and left of the political spectrum, but the beliefs of some terrorist groups are incoherent. The defining characteristic of "left wing" terrorism is its support by communist or revolutionary regimes; such support includes funding, training, the supply of equipment and intelligence, and the provision of safe havens. States recently identified as supporting terrorism include Afghanistan, Iran, Libya, North Korea, and Yemen. The Soviet Union and other Warsaw Pact states (notably Bulgaria and Czechoslovakia) have provided support to Palestinian, Armenian, and Turkish terrorist organizations to destabilize NATO allies: terrorism can be a very cost-effective form of warfare. In the Third World, terrorism has been widely used as an instrument of REVOLUTIONARY WAR.

In response to the increase in long-range terrorism after 1968, Western states now share intelligence and expertise, and also strive to deny political sanctuary to terrorist organizations. Almost every major state now has at least one specialized counterterrorist and hostage rescue unit, the best-known being Israel's SAYARET MATKAL, West Germany's GSG-9, the British SAS, and the U.S. DELTA FORCE.

TESEO: Italian navy designation for the OTO-MAT Mk.II anti-ship missile.

TEST BAN TREATIES: A series of ARMS CONTROL agreements limiting the testing of nuclear warhead or devices. The first such agreement, officially the Treaty Banning Nuclear Weapons Tests in the Atmosphere, in Outer Space, and Underwater, was signed initially by the United States, the Soviet Union, and Great Britain in 1963; subsequently, more than 100 states became signatories, though two nuclear powers, France and the People's Republic of China, have steadfastly refused to sign. The terms of the treaty prohibit the detonation of nuclear devices in the atmosphere, underwater, or in outer space; underground testing is still permitted, but only if no residue or fallout escapes into the atmosphere. The treaty also prohibits the atmospheric testing of nuclear warheads by proxies which are not signatories.

The 1974 Threshold Test Ban Treaty, signed by the U.S. and U.S.S.R., restricts the yield of underground test explosions to 150 kilotons (KT). A third treaty, governing the conduct of nuclear tests for "peaceful" purposes, was signed in 1976.

Although a large number of underground tests have been conducted by both the U.S. and U.S.S.R., the treaties have been a severe constraint on the further development of NUCLEAR WEAPONS by the two countries. Underground testing is less reliable as a source of data on certain nuclear effects (e.g., prompt radiation and ELECTROMAGNETIC PULSE), while the threshold test ban impedes the testing of full-scale "strategic" nuclear weapons (the 150-kT limit is adequate for fission-fusion weapons, but not higher-yield fission-fusion-fission devices). In addition, the treaties impede the development of so-called "directed energy" nuclear weapons, such as X-RAY LASERS for BALLISTIC MISSILE DEFENSE. Underground testing is also significantly more expensive than atmospheric testing, and can (in the United States, at least) evoke political resistance in the areas in which tests are conducted.

Any signatory power can withdraw from the test ban treaties within three months of announcing its intention to do so; this "escape" clause was inserted to protect the parties against technological breakthroughs by a rival power, but the political costs of resuming atmospheric testing would be very high for any signatory.

Attempts to negotiate a Comprehensive Test Ban Treaty (CTB), outlawing all nuclear testing, even underground, have not succeeded to date. Opposition is motivated by doubts about verification of compliance, and by the desire to further explore nuclear weapon effects, to develop lower-yield, more reliable weapons, and to test current nuclear weapons in order to enhance the credibility of their use.

TFR: See TERRAIN-FOLLOWING RADAR.

TGSM: See TERMINALLY GUIDED SUBMISSILE.

THEATER OF OPERATIONS: Sometimes called "area of operations"; a large, geographically distinct region within which military OPERATIONS can interact, even if they are not closely coordinated. For example, during World War II, Western Europe, the Mediterranean, Eastern Europe, the Middle East, the South Pacific, the Southwest Pacific, the Central Pacific, and China-Burma-India were all individual theaters of operations, each under a single theater command. See also TVD.

THERMAL IMAGING (DEVICE): A passive INFRARED sensor which converts detected heat into video images; such devices are used mostly for night surveillance and in gun sights on TANKS and other ARMORED FIGHTING VEHICLES, aircraft, and long-range crew-served weapons (e.g., ANTI-TANK GUIDED MISSILES).

THERMOCLINE: A well-defined temperature gradient (or "boundary layer") in the ocean, a phenomenon of particular importance in ANTI-SUBMARINE WARFARE (ASW) because of its effect on SONAR performance. Generally, the temperature of seawater decreases as depth increases, but sometimes temperature changes can be quite sudden due to submarine currents. Conversely, currents upwelling from the seabed can cause temperature inversions, cold layers above warmer water. Cold water is less dense than warm water, hence sound travels more slowly through it (e.g., the speed of sound through water increases from 4700 to 5300 ft./sec. [1433 to 1615 m./sec.] as the temperature rises from 30° to 85° F); as a result, sound waves passing through different temperature layers tend to be reflected or refracted.

As a rule, sound passing through warm water tends to be reflected when it hits a well-defined cold layer, or thermocline. In addition, sound is refracted (bent) away from warmer water towards colder water. If the temperature increases with depth, sound directed downward will be refracted upward; if temperature decreases with depth, sound directed downward will be bent further in that direction. A cold layer trapped between two warmer layers can form a "sound channel," through which sound waves can travel for consid-

erable distances, but which is isolated from sounds above or below it; such channels are generally found at depths below 300 ft. (91 m.).

Submarines can exploit such temperature effects on sonar performance by hiding below thermoclines. Active sonar signals are reflected off thermoclines, effectively hiding the submarine from surface forces. Conversely, sound radiated by the submarine is trapped by the underside of the thermocline, shielding it from passive sonar. That is why surface vessels and helicopters use VARIABLE DEPTH SONAR (VDS) to lower active or passive sonars below the thermocline. In addition, VDS lowered into sound channels can be used for very long-range detection. Submarines hunting other submarines can avoid thermoclines by moving above or below them as required. By traveling in sound channels, submarines can passively detect targets at ranges greatly exceeding 100 n.mi., but may themselves be detected by sonar within the sound channel.

Other factors affecting sound propagation through water are salinity and density. The former is relatively unimportant at normal submarine operating depths (500–1500 ft./152–457 m.); the latter is a linear function, the speed of sound increasing with depth (everything else being equal). Temperature and pressure effects frequently combine to create "convergence zones." As the sound waves travels deeper, pressure increases at its leading edge, bending the wave back towards the surface; but as the wave approaches the surface, it is refracted downward once more. A convergence zone is the area covered by the sound wave as it nears the surface again after its initial transmission. Sound may be retransmitted several times in this manner, creating two or three convergence zones: in general, the first such zone occurs at 10–40 n.mi. from the sound source, and additional convergence zones occur at roughly equal intervals thereafter. Convergence zones can be exploited for long-range detection by low-frequency sonars on surface vessels.

Temperature gradients can be detected by "bathythermographs," which record water temperature at various depths. It is also possible to detect sound channels and predict the location of convergence zones with bathythermographs. Naval forces employ both expendable and retrievable models. See also OCEANOGRAPHY.

THRESHOLD: Recognizable demarcation lines in the level of violence in armed conflicts between states. The value of the concept lies in its application to nuclear warfare. Thresholds are implicit in the policy of Flexible Response, including: (1) the use of low-yield NUCLEAR WEAPONS on the battlefield; (2) the use of nuclear weapons of any size on targets outside of the enemy's homeland; (3) recognizably selective nuclear strikes within the enemy homeland; (4) MASSIVE RETALIATION, or "spasm," a term used by Herman Kahn to describe a situation wherein a state about to die simply launches all the nuclear weapons at its disposal at the adversary. See also LIMITED WAR; TOTAL WAR.

THROW WEIGHT: The total mass a missile can deliver over a stated range in a given trajectory. Throw weight depends on the thrust of the missile's engines, the range, and the chosen trajectory. The term is generally applied to BALLISTIC MISSILES, and includes the mass of the POST-BOOST VEHICLE (PBV), of all REENTRY VEHICLES (RVs) and penetration aids (PENAIDS), of the payload shroud, and of any ancillary equipment.

THUNDERBOLT II: The Fairchild A-10A CLOSE AIR SUPPORT (CAS) aircraft in service with the U.S. Air Force (USAF). The A-10 was developed in response to a 1967 USAF specification for a specialized CAS aircraft to replace the piston-engined A-1 Skyraider, which called for a simple, rugged aircraft with excellent endurance, good maneuverability, and the ability to destroy small hard targets, mainly tanks. Maximum emphasis was placed on low cost, easy maintainability, and short takeoff and landing (STOL) performance. Two competing prototypes, the Northrop A-9 and the Fairchild A-10, both began flight tests in 1972; after a "flyoff," USAF selected the latter. The first production aircraft were delivered in 1976, and a total of 713 were built through 1984.

One of the ugliest aircraft ever built (unofficially known as the Warthog), the A-10 is literally designed around the long General Electric GAU-8/A Avenger 7-barrel, 30-mm. Gatling gun and its huge magazine. As large as a compact car, with a loaded weight of 4191 lb. (1905 kg.), the GAU-8 can fire milk-bottle-size depleted uranium (STABALLOY) armor-piercing incendiary (API) and high-explosive incendiary (HEI) ammunition at a rate of either 2100 or 4200 rds./min. A total of 1174 rounds are carried in the drum magazine behind the gun. Easily the most powerful aircraft gun in service, the GAU-8 can tear open the thin top and side armor of most tanks with ease. Immediately over the gun barrel, ahead of the cockpit, is an AERIAL REFUELING receptacle for a standard USAF

flying boom. The cockpit is a titanium armor "bathtub" which can resist 23-mm. cannon fire; the pilot sits in a raised ACES-II zero-zero ejection seat under a clear bubble canopy, providing excellent all-round visibility, especially downward and to the rear. AVIONICS are fairly basic, inasmuch as no radar is included: a wide-angle HUD, VHF and UHF radios (including links to army networks), a TACAN receiver, an INERTIAL NAVIGATION system and map display, an instrument landing system, an IFF transponder, a PAVE PENNY laser ranger/target designator mounted on the right side of the nose, an ITEK ALR-46(V) RADAR HOMING AND WARNING RECEIVER, an ALR-69 RADAR WARNING RECEIVER, and four internal ALE-40 CHAFF/flare dispensers. The space behind the cockpit houses two self-sealing fuel tanks with a capacity of 6795 lb. (3088 kg.).

The low-mounted wings are straight and have a very deep cross section to provide lift at low speeds. The inboard wing sections have small leading-edge slats, while the trailing edge has large Fowler flaps inboard, and even larger split ailerons outboard, which double as dive brakes. The main landing gear retract forward, and can be lowered by gravity and airflow if the hydraulic system fails. Even when fully retracted the main tires remain exposed to reduce damage during belly landings. Two 343-gal. (1510-lit.) fuel tanks are housed in the inboard wing sections. The tail section features a large horizontal stabilizer with twin vertical stabilizers as endplates. The engines are housed in widely separated nacelles over the fuselage between the wings and the tailplane. This unique configuration masks infrared emissions to seekers on the ground, and minimizes the probability of a single hit destroying both engines. In addition, the nacelles themselves are partially armored.

The A-10 was built to take considerable punishment and still return to base: to minimize vulnerability to ground fire and facilitate field maintenance, all control runs are triply redundant and widely separated, fuel lines run through (not around) the self-sealing, foam-filled tanks, and vital components such as oil and hydraulic reservoirs are protected by armor. To simplify maintenance, many components such as the tail fins, ailerons, and engines are interchangeable for either side, while the fuselage has no complex curves, expediting field repairs.

The A-10 has 8 wing and 3 fuselage pylons, with a maximum payload capacity of 16,000 lb. (7272 kg.), which would never be carried operationally.

Payloads include 600-gal. (2640-lit.) drop tanks, free-fall or retarded LDGP bombs, NAPALM and CLUSTER BOMBS, rocket and gun pods, PAVEWAY laser guided bombs, HOBOS electro-optical glide bombs, and AGM-65 MAVERICK air-to-ground missiles. On a typical anti-tank mission, the A-10 could carry 4 Mk.20 ROCKEYE 500-lb. anti-tank cluster bombs and 6 Mavericks, with an ALQ-119 or -131 jamming pod and an ALE-37 chaff/flare dispenser pod on the outboard pylons. Since 1986, the aircraft have also been wired to carry up to 2 AIM-9 SIDEWINDER air-to-air missiles for self-defense.

To survive on the modern battlefield, A-10 pilots must resort to very low-level flying, making maximum use of terrain masking, and popping up only long enough to acquire and attack targets; these techniques place a premium on pilot skill and are extremely fatiguing (pilots must sometimes be lifted out of the cockpit after sorties). USAF and the U.S. Army have devised special coordinated tactics for A-10s and attack helicopters: the helicopters would attack enemy air defense weapons with ANTI-TANK GUIDED MISSILES, clearing the way for A-10s to attack tanks and other targets with Mavericks, Rockeyes, and the GAU-8.

Despite its ungainly appearance, large size, and modest speed, the A-10 is very agile, with a phenomenal low-speed turn rate, which allows A-10 pilots to spot camouflaged targets (a prerequisite for CAS), and may enable them to evade much faster high-performance fighters. In addition, the A-10 is one of the few fixed-wing aircraft which can successfully attack helicopters.

On the other hand, its low engine power and acceleration make the aircraft very vulnerable to anti-aircraft weapons because of its slow "getaway" speed, and also degrades maneuverability and STOL performance in hot and high conditions. Further, the avionics suite is totally inadequate for flying conditions on the NATO Central Front (80 percent cloud cover 60 percent of the time). A proposed a two-seat variant, with a FLIR/LLTV sensor and a ground mapping/TERRAIN-FOLLOWING RADAR, failed to attract USAF's interest (a few two-seaters, designated OA-10A, were acquired for use as FORWARD AIR CONTROL aircraft, but they have only the basic avionics). USAF originally intended to equip the A-10 with LANTIRN night-vision/target-designation pods, but this plan was scrapped after price escalation reduced the number of pods available. There is now no program to upgrade or modernize the A-10, which the USAF plans to replace in the mid-1990s with a much

higher-performance variant of the F-16 FALCON or A-7 CORSAIR only marginally suited to the task. In the meantime, the Soviet Union has introduced its own heavily armored CAS aircraft, the Su-25 FROG-FOOT.

Specifications Length: 53.33 ft. (16.26 m.). Span: 57.5 ft. (17.53 m.). Powerplant: 2 General Electric TF34-GE-100 high-bypass turbofans, 9065 lb. (41290 kg.) of thrust each. Fuel: 11,220 lb. (5100 kg.). Weight, empty: 19,856 lb. (9025 kg.). Weight, normal loaded: 38,136 lb. (17335 kg.). Weight, max. takeoff: 46,786 lb. (21,266 kg.). Speed, max.: 423 mph (706 kph) w/bomb load. Speed, cruising: 345 mph (576.15 kph). Attack speed: 299 mph (500 kph). Initial climb: 3090 ft./min. (942 m/min.). Combat radius: (CAS, 1.8-hr. loiter) 288 mi. (481 km.); (Interdiction, lo-lo-lo) 620 mi. (1035 km.). Range, max.: 2542 mi. (4245 km.).

TICONDEROGA (CG-47): A class of 27 U.S. guided-missile CRUISERS (CGs) equipped with the AEGIS air defense system. Developed from the SPRUANCE-class destroyer, the first of these ships was commissioned in 1983; since then, an additional 12 have been commissioned, 3 are under construction, and 11 are on order. Aegis is designed specifically to defend CARRIER BATTLE GROUPS. The U.S. Navy originally planned to install it on a nuclear-powered guided-missile cruiser based on the VIRGINIA class; when Congress refused to fund the project due to its huge cost, the navy selected the Spruance as the smallest hull capable of accommodating Aegis with a reasonable weapon load.

The Ticonderogas thus have the same hull and machinery as the Spruances, but the installation of Aegis has increased the fully loaded displacement from 7810 tons to 9600 tons, using up most of the available growth margin in the hull. Despite this, the Ticonderogas have essentially the same range, speed, and maneuverability as the much lighter Spruances. The only modification to the hull is a raised bulwark at the bow, to compensate for slightly reduced freeboard.

The most noticeable external change is the tall, angular superstructure for the Aegis SPY-1A/B PHASED ARRAY radar. Two arrays are mounted forward, pointing ahead and to starboard, while the aft deckhouse supports arrays facing astern and to port. Other radars include an SPS-49(V)6 two-dimensional air-search radar, an SPG-55 or SPG-67 surface-search radar, an SPQ-9A gun fire control radar, and four SPG-62 missile guidance radars.

The Ticonderogas retain the SQS-53A/B bow-mounted, low-frequency active/passive sonar and SQR-19 TACTASS passive towed array of the Spruances. ELECTRONIC WARFARE systems include two SLQ-32(V)3 ELECTRONIC COUNTERMEASURE arrays and two Mk.36 SRBOC chaff launchers (not nearly enough). In addition, the ships are provided with two SLQ-25 NIXIE torpedo countermeasures sleds.

The first five Ticonderogas were built with the same weapons as the KIDD class destroyers (an ANTI-AIR WARFARE variant of the Spruance): 2 5-in. 54-caliber DUAL PURPOSE guns, 2 PHALANX radar-controlled guns for anti-missile defense, 2 quadruple HARPOON anti-ship missile launchers, 2 sets of Mk.36 triple tubes for lightweight homing TORPEDOES, and 2 twin-arm Mk.26 missile launchers for STANDARD-MR surface-to-air missiles, as well as ASROC anti-submarine missiles. The Ticonderogas also retain the Spruances' landing deck and hangar for two SH-60B SEAHAWK anti-submarine helicopters.

Beginning with the sixth ship, USS *Bunker Hill* (CG-52), the Mk.26 launchers have been replaced by two 61-round Mk.41 VERTICAL LAUNCH SYSTEMS (VLS), capable of firing Standard, ASROC, and TOMAHAWK cruise missiles. Early units of the class will be retrofitted with the Mk.41 during major overhauls. With the Mk.41s, each capable of firing one round per second, the Ticonderogas can take full advantage of the capabilities of Aegis, but at more than $900 million each, their cost is high.

TIGER: The Northrop F-5 series of twin-engine light jet fighters, in service with the air forces of more than 30 countries. The F-5 originated in a 1954 U.S. Air Force (USAF) requirement for a simple, rugged jet fighter for export under the Military Assistance Program (MAP). In response, Northrop developed a design designated N.156, from which a supersonic advanced trainer variant, the T-38 Talon, was ordered in 1956. The fighter variant (N.156C) was selected in 1962 as the FX or "Freedom Fighter," with the designation F-5A. The first aircraft entered service in 1964; in October 1965, 18 were sent to South Vietnam under Project "Skoshi Tiger"; these aircraft were eventually transferred to the Vietnamese air force (which later acquired more through MAP), and the F-5A was widely exported. A total of 1199 were built between 1965 and 1972, including 320 built under license in Canada, Spain, and the Netherlands. Many users have since modified their aircraft with specialized equipment.

The F-5A incorporated many advanced aerodynamic features for its time. The fuselage has a very pronounced area rule ("coke bottle") profile to reduce trans-sonic drag. The nose houses a simple ranging RADAR, behind which are mounted two M39 20-mm. cannons with 200 rounds per gun. The cockpit has a clear one-piece windscreen and a clamshell canopy faired into a prominent dorsal spine which obstructs rearward visibility. AVIONICS are fairly simple: a radar-ranging optical gunsight, a UHF radio, an IFF transponder, and a TACAN receiver. The pilot has a rocket-powered ejection seat.

The straight, tapered, and low-mounted wings are equipped with simple slotted flaps inboard, outboard ailerons, and full-span leading edge flaps. The wide track tricycle landing gear retracts under the cockpit and into the inboard wing sections. The F-5A has one fuselage and six wing pylons, including two wingtip pylons for AIM-9 SIDEWINDER air-to-air missiles or fuel tanks. The centerline pylon usually carries a 229-gal. drop tank, while the remaining wing pylons can accommodate a variety of ordnance, including LDGP bombs, rocket pods, NAPALM and CLUSTER BOMBS, and the AGM-12 BULLPUP air-to-surface missile, up to a maximum payload of 6200 lb. (2828 kg.).

An operational trainer version, the F-5B, has a second seat behind the pilot, displacing some internal fuel. Range is slightly reduced, but otherwise performance is practically unaffected. The RF-5A is a tactical reconnaissance variant with forward, downward, and oblique cameras in the nose. The F-5G, modified for Norway, has extended nose gear, a drag chute, and automatic maneuvering flaps for safer landings on icy runways. The Canadian CF-5A and Dutch NF-5A also have automatic leading-edge flaps and better avionics, including a RADAR WARNING RECEIVER (RWR) and internal flare/CHAFF dispensers. Many F-5As have also been fitted with removable AERIAL REFUELING probes.

The F-5A and its variants proved to be economical and effective. In air combat they perform well because they are small, difficult to spot, and highly maneuverable. Shortcomings include short range, basic avionics, and relatively low speed with a payload. In 1972, Northrop introduced the more powerful and better-equipped F-5E Tiger II, lengthened and widened slightly to increase internal fuel capacity. A much more comprehensive avionics suite includes full blind-flying instruments, an angle-of-attack (AOA) sensor, an air data computer, an instrument landing system, an INERTIAL

NAVIGATION unit, a CRT display screen, and an Emerson APQ-153 or -159 multi-mode radar for use against air and ground targets. ELECTRONIC COUNTERMEASURES include two ALE-40 chaff/flare dispensers, an ITEK ALR-46 programmable RWR, and, in some cases, an internal ACTIVE JAMMING transmitter. The wingspan was increased, and small leading edge root extensions (LERX) were added, increasing lift by 38 percent for only a 4 percent increase in wing area. The F-5E also incorporates the automatic maneuvering flaps and leading-edge flaps of the CF-5A and NF-5A, which greatly improve sustained turn rates. The landing gear were strengthened for rough field operations.

The F-5E has 18 percent more power, the maximum payload has been increased to 7000 lb. (3182 kg.), and now includes the AGM-65 MAVERICK electro-optical air-to-ground missile. The F-5F is a two-seat version equivalent to the F-5B. The RF-5E Tigereye is a reconnaissance variant with a modular multi-sensor package in the nose, displacing one M39 cannon and its ammunition drum. More than 1500 F-5E/Fs were produced between 1972 and 1987, including licensed production in Taiwan and South Korea. They were sold or given to many U.S. allies. The USAF and the U.S. Navy use small numbers of F-5E/Fs in "Aggressor Squadrons" for dissimilar air combat training at the TOP GUN and RED FLAG schools (the Tiger II simulates quite closely the size and performance of the Soviet MiG-21 FISHBED). In training exercises the F-5E has been able to hold its own against larger and more capable aircraft such as the F-14 TOMCAT and F-15 EAGLE, if only under rules which preclude the use of beyond visual range (BVR) missiles, and when flown by highly experienced pilots.

In 1980, Northrop introduced a further development of the F-5, the F-20 TIGERSHARK, which offered Mach 2 speed, a power-to-weight ratio in excess of unity, and state-of-the-art avionics. But the F-20 program was canceled after it failed to attract foreign buyers (because of the USAF's own preference for the F-16 FALCON).

Specifications Length: (A/B) 47.16 ft. (14.38 m.); (E/F) 48.16 ft. (14.68 m.). Span: (A/B) 25.3 ft. (7.71 m.); (E/F) 26.67 ft. (8.13 m.). Powerplant: (A) 2 General Electric J85-GE-13 afterburning turbojets, 4080 lb. (1855 kg.) of thrust each; (E/F) 2 J85-GE-21As, 5000 lb. (2273 kg.) of thrust each. Fuel: (A/B) 3040 lb. (1382 kg.); (E/F) 3800 lb. (1727 kg.). Weight, empty: (A) 8085 lb. (3675 kg.); (E/F) 9683 lb. (4401 kg.). Weight, max. takeoff: (A) 20,677 lb. (9398 kg.); (E/F) 24,676 lb. (11,216 kg.).

Speed, max.: (A) Mach 1.4 (924 mph/1543 kph) at 36,000 ft. (11,000 m.); (E/F) Mach 1.63 (1077 mph/1800 kph) at 36,000 ft. (11,000 m.). Initial climb: (A) 28,700 ft./min. (8750 m./min.); (E/F) 34,500 ft./min. (10,518 m./min.). Service ceiling: 50,500 ft. (15,396 m.). Combat radius: (A) 195 mi. (325 km.) at sea level; (E/F) 138 mi. (230 km.) at sea level/w max. wpns. Range, max.: (A) 1565 mi. (2614 km.); (E/F) 1779 mi. (2970 km.). Operators: (A/B) Can, Gre, Iran, Lib, Mor, Neth, Nor, Phi, ROK, Sp, Tai, Thai, Tur; (E/F) Bah, Bra, Chi, Eth, Indo, Iran, Jor, Ken, Malay, Mex, Mor, N Ye, Phi, ROK, S Ar, Su, Swi, Tai, Thai, Tun, US.

TIGER CAT: A land-based version of the British SEA CAT naval surface-to-air missile (SAM), consisting of three missiles mounted on a lightweight trailer/launcher, controlled by an electro-optical or radar COMMAND GUIDANCE unit mounted on a second trailer. OPERATORS: Arg, Ind, Iran, Jor, Qat, S Af, UK (RAF).

TIGERFISH: The British Mk.24 submarine-launched, WIRE-GUIDED acoustical homing TORPEDO, built by Marconi for the Royal Navy and the Brazilian navy. Tigerfish originated in 1959 as "Project Onager." In 1961, the Royal Navy drafted a "staff target" for a new torpedo to counter future Soviet submarines, but it was 1964 before a Staff Requirement was issued to start research and development at the Admiralty Underwater Weapons Establishment (AUWE). In the late 1960s, the AUWE subcontracted Marconi to develop the acoustical homing system. Because of numerous technical problems, the initial Mk.24 Mod 0 did not enter service until 1974, more than seven years after the planned in-service date. Even then, the Mod 0 suffered from so many technical problems that in 1977 the Admiralty turned the entire program over to Marconi, which in 1978 developed an improved version, the Mod 1. Neither version was particularly reliable, and each had severe tactical limits: the Mod 0 could be used only against submarines, while the Mod 1 was effective only against surface targets. In 1986–87, Marconi introduced the dual-role Mod 2, with a longer range and (at last) fair reliability.

Electrically powered, the Mk.24 is controlled by wire guidance during the initial phase of its run; when the target is acquired by the torpedo's own active/passive sonar, the wire is cut and the torpedo homes autonomously, first by passive homing, but switching to active sonar for the terminal phase. The Mod 2 has the sonar and guidance computer of the Marconi STINGRAY lightweight

ASW torpedo. All versions are armed with a 295-lb. (234-kg.) high-explosive warhead (rather small for a full-size torpedo). Because of its relatively low speed, short range, and small warhead, the Tigerfish is considered only marginally effective against the latest generation of Soviet nuclear submarines; it is to be replaced by the Marconi SPEARFISH in the mid-1990s.

Specifications **Length:** 21.2 ft. (6.46 m.). **Diameter:** 21 in. (533 mm.). **Weight:** 3410 lb. (1550 kg.). **Speed:** 24–35 kt. **Range:** (Mod 0/1) 8–13 n.mi.; (Mod 2) 11.2–18 n.mi.

TIGERSHARK: The Northrop F-20 lightweight fighter, a greatly improved derivative of the F-5 TIGER series. The F-20 was developed by Northrop specifically for the Foreign Military Sales (FMS) market, in part as a replacement for the company's earlier F-5As and F-5Es. As compared with the F-5E, the Tigershark was faster, more maneuverable, could carry a heavier payload over a longer range, and could accommodate a wider variety of weapons, including the AIM-7F SPARROW and AIM-120 AMRAAM radar-guided air-to-air missiles. It was also reported to be significantly cheaper than its principal competitor, the General Dynamics F-16 FALCON. Despite that, the Tigershark failed to attract any buyers because FMS recipients were unwilling to buy an aircraft not in service with the U.S. Air Force (USAF), while USAF for its part promoted FMS sales of the F-16 to increase the production run and thus reduce its own procurement costs. In 1985–86, USAF staged a fly-off between the F-20 and the F-16, to choose a fighter for the Air National Guard. There was no clear-cut winner, so the competition was decided on price; when the contract went to the F-16, this decision effectively ended the F-20 program after only three aircraft had been built (Northrops's loss reportedly exceeded $1 billion).

While retaining the basic configuration of the F-5E (the aircraft was originally designated F-5G), the Tigershark has a completely redesigned nose and tail. The fuselage has a broad "sharknose" radome housing an advanced APG-67 multimode pulse-Doppler RADAR, with search, track, and snapshot air-to-air modes; and ground mapping, Doppler beam-sharpened mapping, ranging, and freeze-frame air-to-ground modes. The radome also generates lift at high angles of attack (AOA), thereby enhancing maneuverability. The completely revised cockpit includes a HUD, two multimode CRT displays, and Hands-On-Throttle-And-Stick (HOTAS) controls. Other AVIONICS include UHF and VHF radio, IFF, TACAN, RADAR WARNING RECEIVERS, and ACTIVE JAMMING transmitters.

The F-20 has 70 percent more power than the F-5E. To accommodate a larger engine, the aft fuselage was redesigned, and a dorsal ram-air cooling duct was added at the base of the vertical stabilizer. All avionics, the powerplant, and auxiliary equipment are modular for easy replacement (the F-20 required only 5.6 maintenance man-hours/flight hour, less than any other fighter). The wings have been reinforced to withstand stresses of up to 9 G. Large Leading Edge Root Extensions (LERX) enhance instantaneous and sustained turn rates, especially at high AOA. Like the F-5E, the Tigershark has automatic maneuvering flaps and full span leading-edge flaps, both controlled by the air data computer. With a combat thrust-to-weight ratio of 1 to 1, the F-20 has outstanding maneuverability and acceleration; the initial rate of climb is among the best in the world.

Surprisingly, the F-20 retains the two M39 20-mm. cannons used on the F-5E, even though more advanced guns are available. The aircraft has one centerline and six wing pylons, including two wingtip pylons for AIM-9 SIDEWINDER air-to-air missiles. The two outboard wing pylons are wired for Sparrow or AMRAAM, which are compatible with the APG-67 radar. Other possible ordnance includes drop tanks, free-fall and LASER-GUIDED BOMBS, NAPALM, CLUSTER BOMBS, rocket and gun pods, and the AGM-65 MAVERICK air-to-ground missile, up to a maximum payload of 8300 lb. (3773 kg.).

Specifications **Length:** 46.6 ft. (14.2 m.). **Span:** 26.75 ft. (8.15 m.). **Powerplant:** 1 General Electric F404-GE-100A afterburning turbofan, 17,000 lb. (7727 kg.) of thrust. **Weight, empty:** 11,810 lb. (5368 kg.). **Weight, normal loaded:** 18,000 lb. (8182 kg.). **Weight, max. takeoff:** 27,500 lb. (12,500 kg.). **Speed, max.:** Mach 2 (1323 mph/2210 kph) at 36,000 ft. (11,000 m.). **Initial climb:** 52,800 ft./min. (16,098 m./min.). **Service ceiling:** 54,700 ft. (16,677 m.). **Combat radius:** (hi-lo-hi w/ext. fuel and 2500-lb. [1136-kg.] bomb load) 633 mi. (1057 km.); (combat air patrol w/97-min. loiter) 345 mi. (576 km.).

TIME-ON-TARGET: 1. A method of firing ARTILLERY in which all guns involved are fired so as to strike the target simultaneously, maximizing the material and morale effect of the fire. See also INDIRECT FIRE.

2. The length of time during which aircraft are scheduled to attack (or photograph) a target.

3. The time at which the warhead of a BALLISTIC MISSILE detonates on (or over) the target. Strict control of time-on-target is necessary to prevent FRATRICIDE. See also COUNTERFORCE.

TIR: See TERMINAL IMAGING RADAR.

TITAN: An obsolete U.S. land-based intercontinental ballistic missile (ICBM), still in service as a launch vehicles for military satellites. See, more generally, ICBMS, UNITED STATES.

TLAM: Tomahawk Land Attack Missile, a variant of the General Dynamics BGM-109 Tomahawk CRUISE MISSILE designed to attack fixed targets on land. See TOMAHAWK for details.

TNT: Tri-Nitro Toluene, a high explosive developed in Germany during the 1890s, widely used for DEMOLITIONS as well as in weapons of all types. See EXPLOSIVES, MILITARY.

TOMAHAWK: The General Dynamics AGM/BGM-109 long-range CRUISE MISSILE, in service with the U.S. Air Force (USAF) and Navy. Tomahawk originated in a 1972 navy requirement for a compact strategic missile sized to fit standard 21-in. (533-mm.) diameter submarine torpedo tubes. This Submarine-launched Cruise Missile (SLCM), with a nuclear warhead and a range of 1600 n.mi., was meant to supplement submarine-launched ballistic missiles (SLBMS) in the strategic attack role. At the time, some saw SLCM only as a "bargaining chip" in ARMS CONTROL negotiations with the Soviet Union, but SLCM was soon merged with another requirement for a long-range, conventionally armed tactical missile for anti-ship missions. Competitive designs were submitted by GD and LTV, and the GED proposal was accepted in 1977; at the same time, the missile was also recommended for development in a ground-launched version. After delays caused by arms control considerations, it was decided in late 1980 to deploy tactical version of Tomahawk (now designated Tomahawk anti-ship missile, or TASM), which first entered service aboard surface ships in 1982, and aboard LOS ANGELES–class nuclear-powered attack submarines in 1983. In 1984 USAF also deployed the BGM-109G Ground Launched Cruise Missile (GLCM) version as part of NATO's LONG-RANGE INTERMEDIATE NUCLEAR FORCES (LRINF). A total of 464 GLCMs were planned; all are now being withdrawn and destroyed under the terms of the 1987 INF TREATY. Also in 1984, the navy began to deploy conventionally armed Tomahawk Land Attack Missiles (TLAM-C) as long-range, precision-guided weapons against fixed targets such as submarine bases

and airfields. It was not until late 1987 that the original nuclear-armed version (TLAM-N) was finally deployed aboard some surface ships and submarines. In addition, air-launched, nonnuclear versions (AGM-109) will equip navy A-6E INTRUDERS and USAF B-52G STRATOFORTRESSES for long-range anti-ship (TSAM) and airfield attack (TLAM-C) missions. Current plans call for the procurement of some 600 TSAMs, 760 TLAM-Ns, and more than 3200 TLAM-Cs. In 1983, McDonnell-Douglas was selected as a second production source for the missile.

Tomahawk has a torpedo-shaped body, mid-mounted pop-out wings, and four fold-out, cruciform tail fins. The missile body is modular in construction, consisting of a guidance section in the nose, a warhead section forward of the wings, a fuel tank aft of the wings, and an engine compartment in the tail. All components from the wings aft are common to all versions; with different guidance and warhead sections, the missile can be configured as a TLAM-N, TLAM-C, or TSAM.

TLAM-N (BGM-109A), has TERRAIN COMPARISON AND MATCHING (TERCOM) guidance, in which data from a ground-mapping radar is compared to a terrain map to update an INERTIAL GUIDANCE unit. This provides outstanding accuracy (a CEP of less than 15 m. over 1500 n.mi.), allowing the missile to be used against hard targets with its 200-KT W80 nuclear warhead. The BGM-109G GLCM is essentially identical to TLAM-N. TLAM-C is produced in several different subvariants, all of which rely on TERCOM guidance in conjunction with either active radar or optical scene-matching TERMINAL GUIDANCE. The BGM-109C has the 1000-lb. (454-kg.) high-explosive warhead of the AGM-12 BULLPUP missile; the BGM-109D has a payload of anti- personnel/anti-material (APAM) submunitions; and the BGM-109F is armed with runway-cratering submunitions. Both variants were used successfully against Iraqi targets in the 1991 Persion Gulf conflict. Air-launched versions of TLAM-C include the AGM-109C and J, both with 1000-lb. unitary warheads; the AGM-109I, with a secondary anti-ship capability; and the AGM-109H, with runway-attack submunitions. Both TASM versions have 1000-lb. (454-kg.) conventional warheads, but rely on a combination of inertial guidance and active radar terminal guidance (the latter derived from the HARPOON anti-ship missile). The BGM-109B has the Bullpup warhead, while the BGM-109E has a magnesium-based incendiary warhead.

All versions of the missile are powered by a Williams International F107-WR-400 turbofan rated at 600 lb. (273 kg.) of thrust, and fed by a pop-out air intake under the tail section. Ship-, ground-, and submarine-launched versions also have a tandem solid-fuel booster which adds 2 ft. (610 mm.) to overall length and about 550 lb. (250 kg.) to the gross weight. Because the nuclear warhead is relatively light and compact, additional fuel can be carried in the warhead section ahead of the wings.

All versions cruise at medium altitude, before descending to a height of 50–100 ft. (15–30 m.) for the final run to the target. Combined with its small size, this flight profile makes the missile difficult to detect both visually and by radar. Moreover, TASM can be programmed to fly doglegs to hide the position of the launch platform, and can also perform an autonomous search if the target is not at its programmed location.

Ship-based Tomahawks are launched from either the 4-round Mk.141 Armored Box Launcher (ABL), or from the Mk.41 VERTICAL LAUNCH SYSTEM (VLS). Submarine-launched versions can be launched either from standard torpedo tubes or from vertical launch tubes installed in the bow of later units of the Los Angeles class. The after part of submarine-launched missiles is encased in a waterproof capsule, which breaks away when they emerge from the water. Air-launched versions are compatible with standard weapon pylons. Finally, the GLCM was launched from a four-round, towed transporter/erector/launcher (TEL). See also ALCM; ANTI-SHIP MISSILE.

Specifications **Length:** 18.16 ft. (5.53 m.). **Diameter:** 21 in. (533 mm.). **Span:** 8.67 ft. (2.64 m.). **Weight, launch:** 2650 lb. (1205 kg.). **Speed:** 550 mph (919 kph). **Range:** (nuclear) 1725 mi. (2880 km.); (HE) 287–345 mi. (333–576 km.).

TOMCAT: The Grumman F-14 two-seat, twin-engine carrier-based FIGHTER/INTERCEPTOR aircraft. The Tomcat, at present the U.S. Navy's standard interceptor, originated in a 1969 replacement proposal for the F-4 PHANTOM (after the cancellation of the abortive F-111B). To speed development, the F-14 incorporated the radar, missile system, engines, and variable geometry (swing-wing) technology of the F-111B, but it also included features which reflected the lessons of the Vietnam War. The result is a very powerful interceptor which also has AIR COMBAT MANEUVERING (ACM) capabilities not normally required by pure (anti-bomber) interceptors.

The navy ordered 12 prototypes in 1970; development was rapid, and the first F-14As entered squadron service in October 1972. Intended as an interim version, the F-14A was to be superseded by the definitive F-14B with more powerful F401 engines. But Congress refused to fund development of the F401 engine, and the F-14A, relatively underpowered as it was, remained in production until 1987. The navy acquired more than 600 between 1972 and 1987; Iran also bought 80 in the mid-1970s (only a handful of these remain in service, the remainder having been either destroyed or cannibalized). It was not until 1984 that the navy finally obtained funding to upgrade the Tomcat with General Electric F110 engines (comparable in power to the F401), and a new, all-digital avionics suite. This new F-14D version began entering service in 1990. In the meantime, the F-14A Plus, with the new engines but still with the original avionics, entered service in 1988. The navy originally planned to buy some 450 F-14Ds, but the program was canceled in 1990 to make funds available for a naval version of the U.S. Air Force Advanced Technology Fighter (ATF).

Very large for a fighter, the F-14 has a fuselage consisting of two widely separated engine pods with a central cockpit and avionics pod in between. All three pods are aerodynamically faired together to create a very broad fuselage which generates more than one-third of total lift, and which contributes to the aircraft's maneuverability. The engines were widely separated to ensure a straight flow of air from the intakes to the engines, and to minimize the possibility of both engines being damaged by a single hit (a hard-learned lesson of Vietnam).

Because its primary mission is long-range air defense, the Tomcat is designed around its RADAR, the very large, liquid-cooled AWG-9, the first fighter radar with coherent pulse-Doppler signal processing for LOOK-DOWN/SHOOT-DOWN capability, and also the first with multiple TRACK-WHILE-SCAN capability. Although designed in the early 1960s, the AWG-9 remains the most powerful fighter radar in service anywhere: it has a detection range of more than 172 mi. (288 km.), and can track up to 20 targets simultaneously, automatically prioritize them by the immediacy of the threat, and direct missiles against 6 targets concurrently. The long range of the AWG-9 creates a problem of identifying aircraft as friend or foe in conditions short of total war. As a partial solution, a Northrop TCS (Television Camera Set) ELECTRO-OPTICAL telescope is slaved to the radar for visual identification, but its range is only 15 mi. (25 km.) for fighter-sized target and 70 mi. (117 km.) for bomber-sized ones. This constrains the use of beyond-visual-range (BVR) missiles in airspace shared by both friendly and enemy aircraft. The F-14D has a new programmable signal processor for the AWG-9, which improves its performance against small, surface-skimming targets such as ANTI-SHIP MISSILES. In addition, solid-state components increase reliability and facilitate maintenance. An infrared search-and-track scanner (IRST) supplements the TCS for passive target acquisition.

Behind the radar, on the lower left side of the nose, is an M61 VULCAN 6-barrel 20-mm. Gatling gun, with 675 rounds of ammunition; a retractable aerial refueling probe is mounted on the right. The pilot and the Radar Intercept Officer (RIO) are seated in a tandem cockpit with a large clamshell canopy designed for good all-around visibility. The pilot has both a HUD and a head-down display, while the RIO has several multi-function display screens and an AVA-12 vertical situation display (the F-14D has a new digital cockpit with multi-functional display screens in place of analog instruments). Other AVIONICS include a digital weapon and flight data computer; a ring-laser gyro INERTIAL NAVIGATION system; a JTIDS Class-2 DATA LINK terminal; an ALR-67 RADAR HOMING AND WARNING RECEIVER; an ALR-45 RADAR WARNING RECEIVER; an ALE-39 CHAFF/flare dispenser; an ACTIVE JAMMING system; a TACAN receiver; and an IFF transponder. The space behind the cockpit houses an avionics bay and three fuel cells with a total capacity of 1781 gal. (12,200 lb.). The landing gear, which retracts beneath the cockpit and into the wing roots, is stressed for carrier landings and has an integral catapult tow bar. An arrester hook is mounted under the tail between the engines.

The high-mounted wing is swept automatically between 20° and 68° by the flight data computer, which matches sweep to airspeed and angle of attack (AOA); the wings can be "overswept" for deck stowage, eliminating the need for a wing-folding system. The sweep mechanism is hydraulically actuated, and requires less than two seconds to move between the minimum and maximum settings. With its variable geometry, the Tomcat can operate efficiently over a wide range of flight conditions, from supersonic cruise, to high-g ACM, to low-speed carrier deck landings. The wing has large fixed inboard sections ("gloves"), which are faired into the upper surface of the fuselage; two small winglets ("vanes") recessed into the gloves are extended automatically at high sweep setting to act as canards, improving maneuverability and stability at high AOA. The outboard (pivoting) wing sections have full-span flaps and full-span leading-edge slats for low-speed stability (both controlled automatically by the flight data computer), and spoilers for lateral control. The outer wing panels each house 2000-lb. (909-kg.) integral fuel cells. The horizontal stabilizers are mounted slightly below the level of the wings, so that at full sweep the Tomcat has a delta configuration; the stabilizers move differentially at high sweep settings for lateral control. Coordination between the stabilizer and spoilers is also controlled by the flight data computer. The Tomcat has twin vertical stabilizers, canted slightly outboard to provide directional stability at high AOA.

The TF30 engines of the F-14A are significantly less powerful than the intended F401 engine. Developed for the F-111 interdiction aircraft, they are not optimized for the rapid throttle cycling and disturbed airflows of air combat maneuvers: over the years, compressor stalls and flameouts have resulted in the loss of several aircraft. The F-14A Plus and D have 28 percent more power, which gives a combat thrust-to-weight ratio of 1.05 to 1 (comparable to the latest fighters), with a corresponding improvement in acceleration and agility. In addition, the F110 is designed specifically for ACM and can function reliably in spite of rapid variations in throttle setting and AOA.

The primary armament is the large AIM-54 PHOENIX active radar-homing air-to-air missile, which has a maximum range of 143 mi. (238 km.). The Tomcat can carry up to 6 Phoenix, 4 on semi-recessed body pallets and 2 on pylons beneath the wing gloves. Alternatively, the aircraft can carry up to 6 medium-range AIM-7 SPARROW or AIM-120 AMRAAM missiles (or a mix of all three). Two AIM-9 SIDEWINDER short-range infrared homing missiles can be carried concurrently on rails mounted on the sides of the glove pylons. Combined with the M61 cannon, its various missiles enable the Tomcat to engage targets from a distance of 100 mi. (167 km.) down to dogfight ranges, a capability unmatched by any other fighter. The Tomcat also has one fuselage hardpoint and two more hardpoints under the engine intakes, either for two 222-gal. drop tanks, or a TARPS multi-sensor reconnaissance pod. Up to 14,500 lb. (6590 kg.) of air-to-ground ordnance could be carried, but the Tomcat does not have specialized air-to-ground avionics, and its crews are not trained for this role.

Specifications Length: 61.16 ft. (18.65 m.). **Span:** (68°) 38.16 ft. (11.63 m.); (20°) 64.1 ft. (19.54 m.). **Powerplant:** (A) 2 Pratt and Whitney TF30-PW-412 afterburning turbofans, 20,900 lb. (9500 kg.) of thrust each; (A Plus/D) 2 F110-GE-400 afterburning turbofans, 29,000 lb. (13,182 kg.) of thrust each. **Fuel:** 16,200 lb. (7364 kg.). **Weight, empty:** 37,500 lb. (17,045 kg.). **Weight, normal loaded:** 55,000 lb. (25,000 kg.). **Weight, max. takeoff:** 72,000 lb. (32,727 kg.). **Speed, max.:** Mach 2.34 (1564 mph/2612 kph) at 36,000 ft. (11,000 m.)/Mach 1.02 (910 mph/1520 kph) at sea level. **Initial climb:** (A) 30,000 ft./min. (9146 m./min.); (A Plus/D) 50,000 ft./min. (15,243 m./min.). **Service ceiling:** 56,000 ft. (17,073 m.) **Combat radius:** 800 mi. (1336 km.). **Range, max.:** 2300 mi. (3841 km.).

TOPAS: Czech-produced version of the Soviet BTR-50 tracked ARMORED PERSONNEL CARRIER.

TOP GUN: Unofficial name of the U.S. Navy's Post-Graduate Course in Fighter Weapons, Tactics, and Doctrine, held at Naval Air Station Miramar, near San Diego, California. Top Gun was initiated after air combat over North Vietnam revealed that navy fighter pilots were inadequately trained in AIR COMBAT MANEUVERING. The Top Gun program provides realistic training against special "Aggressor" squadrons flying F-5 TIGERS, A-4 SKYHAWKS, IAI KFIRS, and F-16N FALCONS to simulate Soviet fighter aircraft ("Dissimilar Air Combat Training"). Aggressor pilots are supposed to replicate Soviet fighter doctrine and tactics, so that trainees can learn effective countertactics. Training missions are conducted over a fully instrumented range, whose sophisticated sensors and computers record the action for playback in post-flight debriefing. Every year, the best pilots of each navy fighter squadron are sent to Top Gun for advanced training; after graduation, they return to their squadrons as training officers. Aggressor pilots, for their part, are periodically rotated back to the fleet to pass on their experience. The efficacy of Top Gun is beyond doubt: prior to 1968, the navy's air-to-air kill ratio over Vietnam was barely 2 to 1; in 1972–73 (after Top Gun), it jumped to more than 8 to 1. The U.S. Air Force established a similar program, RED FLAG, in 1972.

TORNADO: A twin-engine, two-seat, long-range INTERDICTION aircraft (with an INTERCEPTOR variant) produced by the multi-national Panavia consortium (British Aerospace, Messerschmidt-Bolkow-Blohm, and Aeritalia), initially for the Royal Air Force, West German *Luftwaffe* and *Marineflieger* (Naval Aviation), and the Italian *Aeronautica Militare,* and subsequently for Saudi Arabia, Oman, and Jordan. The program originated in a 1969 "Multi-Role Combat Aircraft" (MRCA) proposal for a supersonic, all-weather interdiction aircraft with interceptor potential, intended to replace the Lockheed F-104 STARFIGHTER fighter-bombers, British Vulcan and CANBERRA bombers, and the BUCCANEER maritime strike aircraft. The MRCA concept resembled that of the U.S. F-111, but Panavia had the earlier project as an object lesson.

The 1970 design featured a tandem cockpit, twin engines, and a shoulder-mounted variable geometry ("swing") wing. Development proceeded rapidly (for a multi-national project) on the attack version, designated Tornado IDS (Interdictor/Strike). The first prototype began flight tests in 1974, and the first production IDS was delivered to a Tri-National Training Unit in 1980. IDS orders to date total 937 aircraft, including 125 combat-capable, dual-control trainers: 220 for the RAF (designated Tornado GR.1), 228 for the *Luftwaffe,* 96 for the *Marineflieger,* 100 for the *Aeronautica Militare,* 72 for Saudi Arabia, and 8 for Jordan. Except for the *Marineflieger* aircraft, which are meant for maritime strike, the primary mission of the Tornado IDS is deep INTERDICTION, the attack of heavily defended fixed targets such as bridges and airfields by high-speed, low-level penetration at night and in bad weather. In that role, the aircraft has been an unqualified success, repeatedly winning the U.S. Strategic Air Command's international bombing competition.

Considering the difficulty of its mission, its large payload, and its elaborate ELECTRONIC COUNTERMEASURES (ECM), the Tornado IDS is a remarkably compact aircraft compared to the F-111. The nose houses a Texas Instruments radar system which combines TERRAIN-FOLLOWING RADAR (TFR) and a ground-mapping radar (GMR) capabilities. The TFR can fly the aircraft automatically at heights of less than 200 ft. (61 m.) and speeds in excess of 600 mph (1000 kph). The GMR is the principal attack sensor, with target identification, ranging, lock-on, and navigation modes; it also has a secondary air-to-air capability for self-defense. Immediately behind the radar is a Ferranti LASER rangefinder/target designator, designed to work in conjunction with LASER-GUIDED BOMBS and missiles.

The pilot and WEAPON SYSTEM OPERATOR (WSO) sit under a single clamshell canopy in Martin-Baker Mk.10 zero-zero ejection seats. The pilot has a head-up display linked to the TFR and laser designator, plus a head-down display and a digital map display. The WSO has several multi-function displays to control the radar, communication, navigation, and ECM systems. The aircraft has triplex digital FLY-BY-WIRE flight controls with a manual backup. Other AVIONICS include a Litief Spirit-3 digital flight data computer and autopilot; a Ferranti digital INERTIAL NAVIGATION system; an automatic instrument landing system; VHF, UHF, and HF radios; a TACAN receiver; an IFF transponder; a radar altimeter; and a Decca DOPPLER navigation system. British GR.1s have a modular RADAR WARNING RECEIVER (RWR) and an ARI.23246 modular ACTIVE JAMMING unit. German and Italian aircraft carry several RADAR HOMING AND WARNING RECEIVERS (RHWRs) and the EL/73 DECEPTION JAMMER. Two IKWA/Mauser 27-mm. automatic cannons are mounted below the pilot's seat. A retractable AERIAL REFUELING probe can be added on the right side of the cockpit. The space behind the cockpit contains several avionics bays and two bag-type fuel cells. The landing gear, which retracts under the nose and air intakes, has very large tires and is reinforced for rough-field operations.

The wings can be swept manually between 28° and 66°. Full-span double-slotted flaps, full-span slats on the outboard (pivoting) sections, and leading-edge Kruger flaps on the fixed inboard sections ("gloves") enhance short takeoff and landing performance. The outboard wing sections contain integral fuel tanks. The horizontal stabilizers can move differentially for lateral and pitch control; with wings extended, they are supplemented by large wing spoilers. The vertical stabilizer is very tall for directional stability at low speeds and high angles of attack (AOA); it also contains an overflow (expansion) fuel tank. Somewhat underpowered, the IDS has a thrust-to-weight ratio is only 1-to-2 at maximum weight, making the Tornado a very fast airplane at low level but rather lacking in acceleration; as a result, time to 30,000 ft. (9146 m.) is almost two minutes.

The Tornado IDS has 5 fuselage and 4 wing pylons, with a total payload capacity of 18,000 lb. (8182 kg.). The fuselage pylons (2 tandem pairs and a centerline hardpoint) are all rated at 2000 lb. (909 kg.). The 4 wing pylons pivot as the wing changes sweep; the inner pair are rated at 3000 lb. (1364 kg.) each, while the outer pair are rated at 1000 lb. (454 kg.). The fuselage pylons are mainly for air-to-ground ordnance, while the wing pylons are reserved for 330-gal. (1452-lit.) drop tanks inboard, and missiles or ECM pods outboard. The inboard pylons also have side-mounted launch rails for two AIM-9 SIDEWINDER air-to-air missiles, to be carried at all times for self-defense. Planned ordnance loads vary with the user: the RAF favors the JP.233 submunitions dispenser for airfield attack (used against Iraqi runways in 1991), 1000-lb. (454-kg.) free-fall and retarded bombs, 1000-lb. (454-kg.) PAVEWAY laser-guided bombs, NAPALM canisters, BL.755 CLUSTER BOMBS, rocket pods, ALARM anti-radiation missiles, and SEA EAGLE anti-ship missiles; the *Luftwaffe* prefers a variety of free-fall bombs, plus the MBB MW-1 submunitions dispenser, and AGM-65 MAVERICK missiles. The *Marineflieger* emphasizes free-fall bombs and the KORMORAN anti-ship missile, while the *Aeronautica Militare* mostly relies on free-fall, Maverick, and Kormoran.

Britain developed the Tornado ADV (Air Defense Variant), or F.2 in its RAF designation, from the early 1970s to replace aging BAe Lightnings and F-4K/M PHANTOMS, in their primary mission of defending British airspace over the North Sea and GIUK GAP. Despite more than 80 percent commonality with the Tornado GR.1, development of the F.2 was protracted due to funding limits and technical problems with the radar. Full-scale development was authorized in 1976, the first flight tests followed in 1979, and the first production aircraft was delivered in 1984. Funding cuts slowed production, delaying initial squadron service from 1986 to 1988. The RAF plans to acquire a total of 162 F.2s by 1990. Saudi Arabia and Oman have placed orders for 24 and 8 ADVs, respectively.

The F.2 has the basic airframe, wings, and engines of the IDS version, but avionics and weapons are different, and the fuselage has been lengthened by a 21.25-in. (539-mm.) midfuselage extension to accommodate two tandem pairs of medium-range air-to-air missiles in semi-recessed belly wells. This extension also accommodates an additional 1360 lb. (618 kg.) of fuel and more avionics. The slightly longer nose radome houses a Marconi/Ferranti "Foxhunter" pulse-Doppler, frequency-modulated interrupted continuous-wave (FMICW) radar with a detection range of 120 mi. (200 km.) and LOOK-DOWN/SHOOT-DOWN and multiple TRACK-WHILE-SCAN capabilities (to track between 12 and 20 targets simultaneously).

The radar incorporates advanced ELECTRONIC COUNTER-COUNTERMEASURES (ECCM) to resist enemy jamming, a Threat Evaluation Display (TED), and an integrated IFF interrogator. Because of its extra avionics, including a programmable RHWR, the F.2 carries only one Mauser cannon. The wings are identical to the IDS, except for a 2° increase in the sweep of the wing gloves, to 68°. But the F.2 does not have automatic sweep control (as in the F-14·TOMCAT), which would greatly enhance its maneuverability. The engines are up-rated, giving the F.2 somewhat better acceleration than the IDS. The principal armament now consists of four AIM-7F SPARROW or similar SKYFLASH semi-active radar homing missiles (supplemented by two Sidewinders and the cannon), pending the availability of the active homing AIM-120 AMRAAM. Although intended as an interceptor, the F.2 is surprisingly agile (easily outturning the Phantom but not more modern fighters), and very controllable at low speeds and high AOA.

Germany is currently developing a new version, the Tornado ECR (Electronic Combat and Reconnaissance), which will combine the roles of jamming platform and WILD WEASEL aircraft against enemy air defenses. The ECR will resemble the IDS, but will have new RB.199 Mk.105 engines and new avionics. To date, Germany and Italy have ordered about 30 ECRs, with deliveries scheduled to begin in late 1989.

Within its design limitations (neither version was meant for air combat maneuvering), the Tornado is a powerful if extremely expensive aircraft.

Specifications Length: (IDS) 54.8 ft. (16.7 m.); (F.2) 59.25 ft. (18.06 m.). **Span:** (28°) 45.6 ft. (13.9 m.); (66°) 28.2 ft. (8.6 m.). **Powerplant:** (IDS) 2 Turbo-Union RB.199 afterburning turbofans, 15,000 lb. (6818 kg.) of thrust each; (F.2) 2 RB.199 Mk.104s, 16,000 lb. (7273 kg.) of thrust each. **Fuel:** (IDS) 14,000 lb. (6364 kg.); (F.2) 15,360 lb. (6982 kg.). **Weight, empty:** (IDS) 31,065 lb. (14,120 kg.); (F.2) 31,500 lb. (14,318 kg.). **Weight, normal loaded:** (IDS) 45,000 lb. (20,454 kg.); (F.2) 47,500 lb. (21,590 kg.). **Weight, max. takeoff:** (IDS) 60,000 lb. (27,272 kg.); (F.2) 61,700 lb. (27,986 kg.). **Speed, max.:** (IDS) Mach 2 (1452 mph/2425 kph) at 36,000 ft. (11,000 m.)/Mach 1.2 (920 mph/1536 kph) at sea level; (F.2) Mach 2.27 (1500 mph/2505 kph) at 36,000 ft. (11,000 m.)/Mach 1.2 (920 mph/1536 kph) at sea level. **Initial climb:** (F.2) 50,000 ft./min. (15,243 m./min.). **Service ceiling:** (F.2) 50,-000 ft. (15,243 m.). **Combat radius:** (IDS, hi-lo-hi w/8000-lb. [3636-kg.] bomb load) 863 mi. (1396 km.); (F.2, 2.33-hr. loiter) 375 mi. (626 km.). **Range, max.:** (IDS) 2420 mi. (4041 km.).

TORPEDO: A self-propelled underwater weapon intended for attacks against surface vessels and SUBMARINES. Torpedoes may be launched from aircraft, surface ships, and submarines; or they may be delivered as the payload of surface-to-underwater missiles such as ASROC, IKARA, MALAFON, SS-N-14 SILEX, SS-N-15 STARFISH, and SS-N-16 STALLION. The term *Torpedo* originally described what is now called a naval MINE; the first *self-propelled* torpedo was invented by Robert Whitehead in 1866. Although modern torpedoes are infinitely more capable than Whitehead's, their basic configuration and operating principles have remained unchanged.

A torpedo consists of four basic elements: a streamlined body, a propulsion unit, a guidance and control system, and a warhead. All torpedoes have cylindrical steel bodies with either a blunt, rounded, or ogival nose (invariably housing the warhead and any homing system), a cylindrical center section (housing the power supply for the propulsion unit), and a tapered tail section (with the propulsion unit), terminating in four cruciform fins and one or more propellers. Torpedoes come in several sizes, and are usually categorized by diameter and weight: the most common are the 21-in. (533-mm.) "heavy" and the 12.75-in. (324-mm.) "light," but the French navy also has torpedoes of 21.7-in. (550-mm.) diameter, while the Soviet navy has torpedoes of both 21-in. and 26-in. (650-mm.) diameter. The overall length of torpedoes is defined by the space needed to fit the specified warhead, propulsion system, and guidance system, on the one hand; and by the length of available torpedo tubes, aircraft weapon bays, or external pylons on the other. Heavy torpedoes (21-in. and up) are generally 20 to 22 ft. (6.09 to 6.7 m.) long and weigh between 3100 and 3500 lb. (1409 and 1590 kg.), though Soviet 26-in. torpedoes may reach 30 ft. (9.15 m.) and weight 4000 to 4500 lb. (1818 to 2045 kg.). Shorter versions of the 21-in. torpedo, meant for ANTI-SUBMARINE WARFARE (ASW), carry lighter warheads and have shorter ranges. Small-diameter, lightweight torpedoes, exclusively for ASW, are 8 to 10 ft. (2.44 to 3.05 m.) long and weight between 200 and 800 lb. (90 and 364 kg.).

The original Whitehead torpedo was powered by a piston engine driven by compressed air released from a pressure vessel. Of very short range and low speed, the compressed-air torpedo was soon superseded by the steam (or, more properly, air-steam) torpedo. In this, compressed air, water, and alcohol from separate flasks in the torpedo midsection are mixed and ignited in a combustion chamber, producing high-pressure steam and exhaust gases which drive a turbine or piston engine, and thus the propeller. Exhaust gases, consisting of water vapor, carbon dioxide, and nitrogen, are vented into the sea, where the water vapor and carbon dioxide are absorbed, but the nitrogen rises to the surface, forming a highly visible wake, or "track," which can reveal the position of the torpedo and sometimes even its launch platform. In spite of this, the air-steam torpedo remained standard issue in most navies well into World War II. Typically, they have maximum speeds of 30 to 45 kt., and ranges of 10,000 to 12,000 yd. (9146 to 10,975 m.). The Soviet Union continues even now to rely heavily on air-steam propulsion in its heavy torpedoes.

As an alternative to air-steam propulsion, the Japanese developed the so-called "oxygen" (steam) torpedo in the 1930s, with HYDROGEN PEROXIDE (H_2O_2) in place of compressed air. The peroxide passed through a catalyst, decomposing into oxygen and water, which were then mixed with alcohol and burned to generate steam for a turbine engine. Because no nitrogen was released, the torpedo did not generate a visible wake, and because H_2O_2 generates more power per pound of fuel carried, oxygen torpedoes were faster and had longer range than air-steam torpedoes of equivalent size. On the other hand, peroxide fuel is highly unstable, and presents a considerable explosive hazard. Although Japanese oxygen torpedoes had outstanding performance (43,500 yd./39,786 m. at 36 kt.), and although the Japanese experienced no particular problems with peroxide fuel, this method was not adopted on a large scale by other navies, except for the Swedish navy (the few U.S. Mk.16 peroxide or "Navol" torpedoes were quickly discarded).

By 1941–42, the German navy acquired electrically powered torpedoes on a large scale. In these, instead of alcohol, water, and air flasks, a bank of lead-acid batteries supplied power to an electric motor. Electric torpedoes leave no visible wake, and are considerably quieter than steam torpedoes. On the other hand, they are also slower and have shorter ranges. In addition, battery output (hence speed) can be affected by variations in ambient water temperature, complicating the fire control problem. Nonetheless, the advantages outweighed the shortcomings, and electric torpedoes were soon copied by the United States, Britain, and Japan. Electric torpedoes are still one of the principal types, especially short-range ASW models. Advances in battery technology (e.g., solid-state nickel-cadmium batteries) have improved the speed, range, and reliability of electric torpedoes. Models currently in service include the British Mk.24 TIGERFISH heavy and STINGRAY light torpedoes; the Swedish Types 42 and 61; and most Soviet lightweight torpedoes.

After World War II, the United States developed solid monopropellant ("OTTO") torpedoes. OTTO is essentially a solid or liquid rocket propellant containing its own oxidizer, which is burned to generate high-velocity gases that drive a piston or turbine engine. OTTO propulsion has a much higher specific energy than either steam or electrical propulsion, and compares favorably with peroxide-powered oxygen torpedoes. OTTO-fuel torpedoes now in service include the U.S. MK.46 lightweight and MK.48 heavyweight and the British SPEARFISH heavy torpedoes.

The latest torpedo propulsion technique, still in development, is the Stored Chemical Energy Propulsion System (SCEPS). It relies on the chemical decomposition of a solid compound into two liquids to generate heat for steam. The liquid residues are stored within the torpedo; because they need not be vented against the pressure of the sea, SCEPS torpedoes can function down to their collapse depth. SCEPS, expected to give performance comparable to OTTO engines, is already used in the U.S. Mk.50 BARRACUDA Advanced Lightweight Torpedo (AWLT), and is the preferred technique for any follow-on to the Mk.48.

The earliest and simplest form of torpedo, the straight or "free" runner, maintains a preset course and depth until it either hits a target or exhaust its fuel. The course is maintained by a gyroscope mechanically connected to the rudders, while depth is maintained using a hydrostatic sensor connected to the horizontal control surfaces. By altering the angle at which the gyro is initially set, a torpedo may be launched as much as 90° from the target azimuth. The main drawback of the free runner is the need for very precise prelaunch FIRE CONTROL. Because the target is moving during the time the torpedo to travels from the launch platform to the target, the torpedo must be aimed a considerable distance ahead of the target's position at launch; and, of course, the target can alter speed or course during the "dead time" between launch and impact. Thus it is standard procedure to fire several free-running torpedoes in a fan-shaped "spread," to compensate for errors in the fire control solution or post-launch variations in target course or speed. The uncertainties associated with dead time increase with range; thus it is common practice to fire free runners from relatively short ranges (under 3000 yd./2743 m.), except against large area targets such as a CONVOY.

In 1943, the Germans introduced the "pattern running" torpedo to reduce their dependency on short-range precision attacks. After a preset straight run, this type of torpedo executes a preset circular, box, or spiral pattern. Thus, if the torpedo missed the target on its straight run, it could still have a chance of a hit during the pattern. If fired at a large group of ships in a convoy, the pattern obviated the need for accurate fire control by covering an area large enough to ensure a high hit probability.

Also in 1943, the German, British and the U.S. navies all introduced the acoustical homing torpedo. The German and U.S. versions were submarine-launched (GNAT and "Cutie," respectively), and mainly for use against escort vessels. The Britain and the U.S. also developed an air-launched ASW torpedo. All relied on passive homing: hydrophones in the nose of the torpedo tracked the sound generated by the propellers of surface ships or submarines, and could be decoyed by relatively simple TORPEDO COUNTERMEASURES, notably noisemaker sleds towed by ships. Later passive homing torpedoes incorporated signal processors to discriminate between decoys and targets. By the early 1950s, active SONAR homing, which is not as vulnerable to simple countermeasures, was introduced. Most acoustical torpedoes now in service combine passive homing with active homing for terminal guidance, as well as pattern running programs to search for targets. WIRE GUIDANCE, an adjunct to acoustical homing, is used on the Mk.48, Tigerfish, and Spearfish, among others. Steering commands are transmitted from the launch platform's fire control system to the torpedo through two thin wires paid out from spools in its tail section, so as to take advantage of the superior sonar and computers of the launch platform. In the Mk.48, data can also be transmitted from the torpedo's own sonar to the launch platform (in a manner analogous to TRACK-VIA-MISSILE guidance). During the terminal phase, the wires are severed, and the torpedo homes autonomously towards the target.

Soviet torpedoes capable of homing on the wake turbulence generated by surface ships and submarines have been reported. Wake turbulence is extensive and persistent, allowing attacks at very long range. The Soviet Type 65 (26-in.) wake homer has a reported range of 50 n.mi. at 50 kt., or 100 n.mi. at 30 kt.

Most torpedoes are armed with high-explosive warheads weighing between 800 and 2000 lb. (364 and 909 kg.) for heavy torpedoes, or 75–200 lb. (34–91 kg.) for lightweight ASW torpedoes. The most common EXPLOSIVES are TNT, Torpex, HBX, H-6, and PBX. They are detonated by either contact or PROXIMITY FUZES. Contact-fuzed torpedoes explode against the side of the target hull, leaving a large hole, but if the vessel has watertight subdivisions, one or even two such hits may not suffice to sink it. Proximity-fuzed torpedoes, however, can be set to explode *under* the target to break its keel and ensure the ship's destruction. The most common proximity fuze is the magnetic exploder, which exploits the magnetic field generated by ship hulls to close an electrical firing circuit; but this method is relatively unreliable due to anomalies in the earth's magnetic field. U.S., British, and German torpedoes all had magnetic exploders at the start of World War II, but these were removed due to the high rate of duds and premature explosions. The Soviet navy, faced with the need to attack very large U.S. warships, reportedly relies on some form of bottom attack fuze. Homing torpedoes can also be fitted with acoustical proximity fuzes activated by their active sonar systems.

Recently, it has been suggested that the warheads of Western lightweight ASW torpedoes such as the Mk.46 and Stingray are too small to penetrate the pressure hulls of larger Soviet submarines. As a result, the U.S. and Britain are developing new "directed energy" warheads employing the shaped charge (HEAT) or SELF-FORGING FRAGMENT principles.

Both the U.S. and the Soviet Union have also developed nuclear-armed heavy torpedoes. The American weapon, the Mk.45 ASTOR (Atomic Submarine Torpedo), was equipped with a 5 to 10-KT warhead and, in compliance with U.S. positive control requirements, employed wire guidance and command detonation. It was extremely unpopular with U.S. submarines, and was withdrawn from service in 1976 when the Mk.48 became available in large numbers. The Soviet navy has nuclear torpedoes with 15-kT warheads for use against American carrier battle groups.

Surface-ship and submarine torpedoes are generally launched from tubes (though some surface-ship torpedoes can be dropped over the side from simple racks). In surface-ship tubes, whether fixed or trainable, torpedoes are ejected by gas from a small propellant cartridge inserted into the breech of the tube. Heavy torpedoes were once the primary ANTI-SURFACE WARFARE (ASUW) weapons of DESTROYERS and FRIGATES, but after World War II, Western navies generally abandoned the heavy 21-in. torpedo in favor of lightweight ASW types, launching from fixed or trainable double and triple mounts such as the U.S. Mk.32. Exceptions to this trend include the navies of Sweden, Finland, and West Germany; operating as they do in the confined and shallow Baltic Sea, they retain heavy anti-surface torpedoes aboard their FAST ATTACK CRAFT as a supplement to ANTI-SHIP MISSILES. The Soviet navy, in contrast, retains 21-in. tubes aboard most of its larger surface combatants, including the KIEV-class aircraft carriers. They are intended primarily for ASW, but are capable of ASUW as well. Tubes are generally mounted in groups of four or five.

Submarine torpedo tubes have pressure-tight doors at each end. When the outer door is closed, the inner door can be opened to load or remove a torpedo from the tube. Launch was originally accomplished by compressed air, but as submarines began operating at greater depth, hydraulically operated "positive impulse" ejection was adopted to overcome the increased water pressure. All these methods produce a large acoustical signature which can reveal the presence of the submarine. If the torpedo is smaller than the diameter of the tube, on the other hand, it can swim out under its own power. This method is favored for lightweight ASW torpedoes, and some navies which use 21-in. torpedoes deliberately equip their submarines with 22-in. (560-mm.) tubes, to rely on swim-out exclusively. Its principal drawback is the inability to ensure safe separation if the submarine is moving at high speed.

Aircraft-, helicopter-, or missile-delivered torpedoes are simply dropped into the water. To prevent damage, especially to the homing system or guidance package, most have a braking parachute attached to the tail, which separates once the torpedo enters the water.

TORPEDO COUNTERMEASURES: Weapons, other equipment, and tactics meant to divert, evade, or otherwise neutralize TORPEDOES, particularly homing torpedoes. For surface vessels, sheer

speed and aggressive ANTI-SUBMARINE WARFARE (ASW) patrols (to prevent a submarine from achieving a favorable firing position) are the most effective countermeasures. Against torpedoes themselves, a number of different tactics can be used. First (depending on the range and bearing at which the torpedo is detected), the ship can turn either towards or away from the torpedo, to reduce its target area. If the torpedo is a straight or pattern runner, this tactic minimizes the chances of a hit.

Against acoustic homing torpedoes, a variety of active countermeasures can be employed; the simplest are sled-mounted decoys towed some distance behind the ship. These simulate machinery and propeller noises, to attract the torpedo. The first such device (Fanfare) was developed by the British in World War II as a counter to German homing torpedoes. The Germans developed a counter-countermeasure: they programmed the torpedo to execute a circling maneuver before hitting a detected target; the torpedo could thus come out of the turn ahead of the decoy, to home on the true target. The British responded by towing two Fanfares in tandem; the rear decoy would trigger the torpedo circling maneuver, so that the torpedo would come out of its turn to home on the second decoy (this type of countermeasure is still in use today, the current U.S. version being the SLQ-25 NIXIE). But as each ship can carry only a few decoy sleds, a salvo of several torpedoes can overwhelm the countermeasure.

Submarines rely on stealth rather than speed to avoid engagement, and instead of towed decoys, they rely on expendable noisemakers and self-propelled decoys. The former are small, lightweight devices, whose function is analogous to that of the CHAFF or flares used against radar- or infrared-guided missiles; i.e., they present to the homing torpedo a more prominent source of noise than the submarine itself. If, however, the torpedo has active sonar, noisemakers are ineffective. Self-propelled decoys are more sophisticated. Launched from torpedo tubes, they follow a preset course away from the submarine. Equipped with noisemakers to simulate the acoustical signature of the submarine, they can also contain sophisticated deception jammers. The current U.S. model is MOSS (Mobile Submarine Simulator).

If decoys fail to divert the torpedo, both surface ships and submarines can still employ on-board DECEPTION JAMMING—to transmit spurious signals on the torpedo's sonar frequency so as to create a false target position. If active jamming fails, submarines can also resort to radical changes in speed, heading, or depth, but such desperate, last-minute efforts have only a small chance of success in most cases. Surface vessels have one additional countermeasure: anti-torpedo weapons. These are essentially depth charges launched in the path of the torpedo. The turbulence generated by the explosion may be sufficient to detonate the torpedo, disrupt its homing mechanism, or throw it off its course. Soviet RBU-series rocket launchers are said to have a specific anti-torpedo as well as an anti-submarine capability.

TORPEX (TPX): A high explosive introduced by the U.S. Navy in 1943 as a replacement for TNT in TORPEDO warheads. Approximately twice as powerful as TNT, Torpex has itself been superseded by HBX, an even more powerful explosive. See also EXPLOSIVES, MILITARY.

TOSS BOMBING: A low-altitude bombing technique whereby the aircraft pulls up sharply while still some distance from the target, to release its bomb (or bombs) at an upward angle so as to impart the velocity and elevation required to reach the desired impact point. The aircraft usually reverses course after release by completing a half-loop (Immelman) maneuver. Toss bombing (like loft bombing and over-the-shoulder delivery) is especially suitable to deliver nuclear bombs safely at low altitudes, but is too inaccurate for conventional bombing without sophisticated navigation/attack equipment. See, more generally, BOMBING TECHNIQUES.

TOT: See TIME-ON-TARGET.

TOTAL WAR: A theoretical concept, implying the use of all available resources and weapons in war, and the elimination of all distinctions between military and civilian targets. Even Hitler's Germany refrained from using all its available weapons (e.g., NERVE AGENTS) and refrained from some success-maximizing measures, such as the execution of unproductive prisoners of war.

The term is propagandistic and literary; strategic discourse uses CENTRAL WAR for direct combat between the nuclear superpowers, and GENERAL WAR for combat between them which includes nuclear weapons.

TOTI: A class of four Italian diesel-electric attack SUBMARINES commissioned in 1967–68. The first submarines built by Italy after World War II, the Totis are small and optimized for anti-submarine warfare in the shallow waters of the Mediterranean. The Totis have a modified teardrop hull, with a rather straight bow and a tapered stern in a single-hull configuration. A tall, streamlined sail (conning tower) is located amidships. Control surfaces are of standard pattern, with retractable bow planes, and a cruciform arrangement of rudders and stern planes ahead of the propeller. A retractable snorkel allows diesel operation from shallow submergence. Armament consists of four 21-in. (533-mm.) tubes in the bow for six acoustical homing TORPEDOES. All sensor data is processed by a computerized IPD-60/64 FIRE CONTROL system.

Although small and short-ranged, the extremely maneuverable and very quiet Totis would be ideal "hunter-killer" submarines in chokepoints such as the Straits of Gibraltar or Messina. They would also be more effective in shallow, inshore waters than larger submarines. See, more generally, SUBMARINES, ITALY.

Specifications Length: 151.5 ft. (46.19 m.). Beam: 15.5 ft. (4.72 m.). Displacement: 524 tons surfaced/582 tons submerged. Powerplant: single-shaft diesel-electric: 2 2200-hp. Fiat MB820 diesel-generators, 1 2200-hp. electric motor. Speed: 14 kt. surfaced/15 kt. submerged. Range: 7500 n.mi. at 4.5 kt. (snorkel); 15 kt. for 1 hr. (batteries). Max. operating depth: 591 ft. (180 m.). Collapse depth: 964 ft. (300 m.). Crew: 26. Sensors: 1 Velox medium-frequency passive ranging sonar, 1 JP-64 active sonar, 1 3RM20/SMG surface-search radar, 1 electronic signal monitoring array, 1 periscope.

TOURVILLE: A class of three French DESTROYERS commissioned between 1974 and 1977. Successors to the unsuccessful *Aconit*, the Tourvilles are intended primarily for ANTI-SUBMARINE WARFARE (ASW), but also have considerable ANTI-AIR WARFARE (AAW) and ANTI-SURFACE WARFARE (ASUW) capabilities.

The Tourvilles are flush-decked (except for a short well deck at the stern), and have relatively low freeboard, a characteristic of French ships, which are meant mainly for the Mediterranean. Two sets of gyro-controlled fin stabilizers make them good gun platforms even in rough seas. But their hulls were apparently too lightly constructed, and subsequently they had to be reinforced with additional steel strakes at the upper deck level.

Armament consists of 2 100-mm. DUAL PURPOSE guns, one mounted on the foredeck, the other on the superstructure ahead of the bridge; 2 20-mm. ANTI-AIRCRAFT guns immediately aft of the bridge; 6 single canisters for EXOCET MM.38 anti-ship missiles mounted amidships; a MALAFON ASW missile launcher with 13 missiles, also amidships; an

8-round Naval CROTALE surface-to-air missile launcher with 26 missiles, mounted on the aft superstructure; 2 launch racks for L5 lightweight ASW TORPEDOES; and 2 Westland LYNX WG.13 ASW helicopters, operated from a landing deck and hangar aft. The Tourvilles were the first French destroyers designed from the outset with helicopter facilities. When first commissioned, the ships had a third 100-mm. gun mount in place of the naval Crotale, which was not yet ready; the missile launcher was retrofitted in its place between 1979 and 1981. See more generally, DESTROYERS, FRANCE.

Specifications Length: 501 ft. (152.75 m.). **Beam:** 50 ft. (15.25 m.). **Draft:** 21 ft. (6.4 m.). **Displacement:** 4800 tons standard/5800 tons full load. **Powerplant:** twin-shaft steam: 4 oil-fired boilers, 2 sets of geared turbines, 54,000 shp. **Speed:** 32 kt. **Range:** 1900 n.mi. at 30 kt./4000 n.mi. at 18 kt. **Crew:** 282. **Sensors:** 2 DRBN 32 navigation radars, 1 DRBV 26 air-search radar, 1 DRBV 51B surface-search radar, 1 DRBC 32C fire control radar, 1 DUBV 23 bow-mounted low-frequency active/passive sonar, 1 DUBV 43 low-frequency towed array sonar. **Electronic warfare equipment:** 1 ARBR 16 electronic signal monitoring array, 1 ARBB 32 active jamming system, 2 Corvus chaff rocket launchers.

TOW: Tube-launched, Optically tracked, Wire-guided, the Hughes BGM/MGM-71 ANTI-TANK GUIDED MISSILE (ATGM) developed for the U.S. Army and now also in service with the armies of 36 other countries. Designed in the mid-1960s as a replacement for the 106-mm. RECOILLESS RIFLE in company and battalion ANTI-TANK units, TOW entered service in 1970, and has since seen combat in Vietnam (1972), the Middle East (1973 and 1982), and Chad (1986–88). TOW has proven to be reliable and easy to operate, which, along with its relatively low price ($8000–$10,000 each), accounts for its great success (more than 350,000 produced to date).

The basic BGM-71A was intended for ground operation, either from a tripod or a variety of vehicle mounts, ranging from a simple, hand-operated jeep mount to the complex, power-operated Emerson turret of the M901 ITV (IMPROVED TOW VEHICLE). In its simplest configuration, the weapon consists of four elements: the tripod, the missile, a launch tube, and a tracker/sight unit. The missile comes packaged in a sealed container, which in action is opened and attached to the rear end of the launch tube; this secures all required electrical connections. The tracker unit is attached to the left side of the launcher, and incorporates the trigger mechanism. A TAS-4 THERMAL IMAGING sight can also be attached to the tracker unit for operations at night or in poor visibility. The entire weapon weighs about 175 lb. (79.5 kg.) ready to fire, which means that it is only marginally man-portable.

To fire the missile, the operator acquires the target in the crosshairs of the sight and pulls the trigger, activating a gas generator which blows the missile out of the launch tube; at a safe distance, the main engine ignites. TOW has Semi-Automatic Command to Line-of-Sight (SACLOS) guidance: the tracker detects the infrared emissions of a flare mounted in the base of the missile, and automatically generates steering commands to bring the missile onto the line of sight between the operator and the target. These commands are transmitted to the missile over two thin copper wires which are paid out from an inertialess reel inside the missile. All the operator needs to do is keep the target in the crosshairs of his sight until missile impact.

The TOW missile has been produced in several versions with progressively heavier warheads and other improvements to increase lethality against advanced forms of armor. The missile is controlled in flight by four pop-out cruciform wings and stabilized by four pop-out tail fins. It is powered by a two-pulse solid-rocket motor with two exhaust nozzles mounted on opposite sides of the missile between the wings. The motor burns for several seconds, boosting the missile to a maximum speed; thereafter it coasts, decelerating slowly out to its maximum range of 3000 m. (the limit of the control wire), at which point speed is down to about 250 mph (417 kph). Time of flight at maximum range is about 12–15 seconds. The BGM-71A has a 5-in. (127-mm.) shaped-charge (HEAT) warhead weighing 7.7 lb. (3.5 kg.), which can penetrate up to 23.6 in. (600 mm.) of homogeneous steel armor.

The BGM-71B Improved TOW (I-TOW), introduced in 1976, has a 27-in. (686-mm.) telescoping fuze probe in the nose, which increases the stand-off distance when it detonates to improve penetration. The third version, the BGM-71C Extended Range TOW, is identical to I-TOW but has an improved motor and longer control wires. The current version, the BGM-71D TOW-2, introduced in the early 1980s, has the motor of the BGM-71C and a new 6-in. (152-mm.), 13-lb. (5.9-kg.) HEAT warhead with fuze probe, and is capable of defeating 31.5 in. (800 mm.) of steel armor.

The MGM-71 is an air-launched version which

can be mounted on a number of different helicopters including the UH-1 HUEY, the AH-1 COBRA, the UH-60 BLACKHAWK, the BO-105, the Westland LYNX, and the MD-500 DEFENDER. Identical to the equivalent ground-launched versions, the missile in its launch tube is mounted on outrigger pylons, while the tracker and sight unit are usually mounted in the nose, on the cabin roof, or above the rotor mast of the helicopter.

TOW's operational drawbacks include a large backblast at launch, which prevents it from being fired from enclosed spaces, and which can reveal the launch position to the enemy; and the need to keep the target in sight throughout the missile's flight, during which the operator is vulnerable to enemy counterfire. Placing the operator under armor, as in the M901 ITV or M2/3 BRADLEY fighting vehicle, mitigates the risk (but exposes the vehicle), while helicopters in particular must remain exposed as they hover for ten seconds or more waiting for the missile to strike. This problem, common to all wire-guided ATGMs, limits the value of TOW, especially in its original role as a man-portable, tripod-mounted weapon.

Recent advances in armor technology, including composite CHOBHAM ARMOR and explosive REACTIVE ARMOR, have reduced the effectiveness of most HEAT warheads; TOW-2 is now the smallest missile still deemed effective against modern MAIN BATTLE TANKS. Two new TOW variants are being developed specifically to counter reactive armor: TOW-2A, with a precursor charge in the fuze probe (to initiate the reactive armor before the main warhead is detonated); and TOW-2B, with a downward-firing SELF-FORGING FRAGMENT warhead meant to penetrate the thin top armor of most tanks.

Specifications   Length: 3.85 ft. (1.16 m.). Diameter: 6 in. (152 mm.). **Span:** 13.5 in. (343 mm.). **Weight, launch:** (A) 49.6 lb. (22.5 kg.); (D) 61.95 lb. (28.15 kg.). **Speed, max.:** 623 mph (1002 kph). **Range, min.:** 65 m. **Range, max.:** (A/B) 3000 m.; (C/D) 3750 m. **Operators:** Bah, Can, Den, Egy, Fin, FRG, Gre, Iran, Isr, It, Jor, Ku, Leb, Mor, Neth, N Ye, Nor, Oman, Pak, Por, ROK, S Ar, Swe, Tai, Thai, Tur, UAE, UK, US.

TOWED ARRAY SONAR: A passive linear array of hydrophones, which is towed behind surface ships or submarines at the end of a long cable, in order to isolate the sensitive receivers from propeller, engine, and hull-flow noises which would otherwise interfere with the detection of targets. Moreover, being at a considerable distance behind the towing vessel, the array can be used in conjunction with bow-mounted sonars to passively locate and range targets by triangulation. The most elaborate types incorporate their own cable winches and deployment equipment, but some navies prefer clip-on types which are attached to a vessel as it leaves on patrol. This allows the optimum use of available arrays, but limits the maneuvering freedom of towing vessels, because most towed arrays have rather low speed and turning envelopes. Specific towed arrays include SURTASS, TACTASS, and TASS. See also SONAR.

TR-1:   An advanced version of the Lockheed U-2 strategic reconnaissance aircraft, in service with the U. S. Air Force.

TR-1700 (A.K.A. SANTA CRUZ):   A class of German-designed diesel-electric attack SUBMARINES built by Thyssen Nordseewerke for the Argentine navy. Ordered in 1977 as replacements for Argentina's old GUPPY-class submarines, the first 3 TR-1700s were commissioned in 1984, 1985 and 1986, but construction of the remaining 3 has been halted by a lack of funds.

The TR-1700s have a modified teardrop hull, with a broad, blunt bow (to ensure smooth water flow around the torpedo tubes), a cylindrical center section, and a tapered stern. The upper hull is covered by a wide, flat deck casing. A tall, streamlined sail (conning tower) is located amidships. The TR-1700s have a single-hull configuration, fabricated of HTS (High Tensile Strength) steel. The control surfaces consists of "fairwater" diving planes on the sail, and a cruciform arrangement of rudders and stern planes ahead of the propeller. A retractable snorkel allows diesel operation from shallow submergence. The electric motor is powered by eight 120-cell high-capacity, fast-charging VARTA batteries for exceptional submerged endurance.

Armament consists of 6 21-in. tubes for several types of free-running and acoustical homing TORPEDOES, using the swim-out launch technique; 16 reloads are carried, for a total of 22 torpedoes. All sensors are linked to an HSA SINBADS digital FIRE CONTROL system and a SAGEM automated plotting table. See, more generally, SUBMARINES, FEDERAL REPUBLIC OF GERMANY.

Specifications   Length: 216.5 ft. (66 m.). **Beam:** 23.9 ft. (7.28 m.). **Displacement:** 2150 tons surfaced/3364 tons submerged. **Powerplant:** single-shaft diesel-electric: 4 1475-hp. MTU 16V652 MB80 diesel generators, 1 8900-hp. Siemens electric motor. **Speed:** 13 kt. surfaced/25 kt. submerged. **Fuel:** 314 tons. **Range:** 8000 n.mi. **Patrol endurance:** 70 days. **Max. operation depth:** 964 ft. (300 m.). **Collapse depth:** 1640 ft. (500 m.). **Crew:** 29. **Sensors:** 1 Krupp-Atlas CSU-3-4 integrated active/passive low-frequency sonar, 1 French DUUX 3 passive ranging sonar, 1 SMA surface-search radar, 1 electronic signal monitoring array, 2 periscopes.

TRACKER: The Grumman S-2 twin-engine, carrier-based ANTI-SUBMARINE WARFARE (ASW) patrol aircraft. A few still serve (mostly from shore bases) with Argentina, Brazil, Canada, Peru, Thailand, Turkey, Uruguay, and Venezuela.

TRACK-VIA-MISSILE (TVM): A method of missile guidance whereby data from the missile's active or semi-active radar are relayed to a ship platform or ground launcher via a DATA LINK, for processing by FIRE CONTROL computers into guidance commands, which are then transmitted back to the missile. A primitive form of TVM is applied in some versions of the Soviet SS-N-3 SHADDOCK anti-ship missile: the radar "picture" generated by the missile's own tracking radar is transmitted back to the launch platform, where it is compared to target acquisition data supplied by on-board sensors or OVER-THE-HORIZON TARGETING; the data are then updated and transmitted back to the missile. More sophisticated, fully automated TVM guidance is applied in the Soviet SA-10 GRUMBLE and American PATRIOT surface-to-air missiles.

TVM guidance takes advantage of the more powerful data processing capabilities of the launch platforms in order to reduce vulnerability to jamming, spoofing, or other ELECTRONIC COUNTERMEASURES, while on the other hand, sensors mounted on the missiles greatly extend the radar horizon, facilitating long-range attack.

TRACK-WHILE-SCAN (TWS): The ability of a RADAR or SONAR to lock onto and track a specific target while simultaneously searching for additional targets. A system with TWS capability can therefore direct weapons against a specific target while displaying the positions of all other objects within its field of view. The use of digital computers for electronic beam steering (as in PHASED ARRAY radar or sonar), combined with digital signal processing, enable many modern sensors to track more than one object simultaneously, a capability called multiple track-while-scan. The AWG-9 radar of the F-14 TOMCAT fighter, for instance, can track up to 20 objects at one time, while guiding missiles against 6. The SPY-1 phased array radar of the AEGIS naval air defense system can track several dozen targets while simultaneously directing missiles against as many as 18.

TRAFALGAR: A class of 7 British nuclear-powered attack SUBMARINES, the first 4 of which were commissioned between 1983 and 1988; the remaining 3 will be completed in the early 1990s. The Trafalgars are incremental developments of the prior SWIFTSURE class, incorporating substantial improvements in sensors, FIRE CONTROLS, and acoustical silencing.

The Trafalgars have a modified teardrop hull-form, with a blunt bow, cylindrical center section, and a stern section that terminates in a 45° cone. This configuration maximizes internal volume while reducing overall length. In the single-hull configuration, the ballast tanks are outside the pressure hull at the bow and stern. A tall, streamlined sail (conning tower) is located slightly forward of amidships. The control surfaces are arranged in the standard British pattern, with fold-up bow diving planes, and cruciform rudders and stern planes. The hull is covered with ANECHOIC tiles (similar to Soviet CLUSTER GUARD tiles), to reduce radiated noise and impede enemy active sonars. Armament consists of 5 21-in. (533-mm.) torpedo tubes amidships for a total of 20 TIGERFISH or SPEARFISH wire-guided acoustical homing TORPEDOES, and 5 UGM-84 HARPOON anti-ship missiles; Stonefish bottom-laid mines can replace torpedoes on a 2-for-1 basis. All sensor data are fed to a DCB Action Information Organization and Fire Control System.

The Trafalgars' PWR.1 is a natural circulation reactor; at low speed, convection currents within the reactor are sufficient to supply an adequate flow of coolant through the primary loop, thereby reducing the use of noisy pumps. All but the first boat have pumpjets instead of conventional propellers. In a pumpjet, a single, multi-blade rotor revolves at low speed between stator vanes inside a duct, reducing cavitation noise, especially at high speeds. All machinery is mounted on resilient sound-isolation rafts to reduce radiated noise. Combined with the anechoic tiles, these features make the Trafalgars among the quietest nuclear submarines in service. See more generally, SUBMARINES, BRITAIN.

Specifications **Length:** 280 ft. (85.36 m.). **Beam:** 33 ft. (10.06 m.). **Displacement:** 4800 tons surfaced/5300 tons submerged. **Powerplant:** single-shaft nuclear: 1 PWR.1 pressurized-water reactor, 2 sets of geared turbines, 15,000 shp. **Speed:** 20 kt. surfaced/30 kt. submerged. **Max. operating depth: 1315 ft. (400 m.). **Collapse depth:** 2000 ft. (610 m.). **Crew:** 98. **Sensors:** 1 Type 2020 bow-mounted low-frequency passive sonar, 1 Type 2007 passive flank-mounted conformal array sonar, 1 Type 2026 active attack sonar, 1 Type 2026 passive towed array sonar, 1 Type 2046 sonar intercept array, 1 Type 1007 surface-search radar, 1 electronic signal monitoring array, 2 periscopes.

TRAINING, MILITARY: Although details vary from country to country, and among the different armed services of each, the principles of military training are now more or less standardized. Upon induction, recruits undergo several weeks of "basic" training, which has several different purposes: physical conditioning, the breaking down of individual identity to create a group identity, the instilling of military discipline, introductory weapons training, and (sometimes) the reduction of societal inhibitions against killing. After basic training, personnel are separated into specialty tracks for more advanced individual training. In ground forces, troops assigned to combat arms are posted to units within which they undergo team training on the building block principle: the individual learns to function as part of a squad, the squad as part of a platoon, the platoon as part of a company, etc. To maintain proficiency, individuals and units must also undergo periodic "refresher" training; retraining may also be needed when new equipment or tactics are introduced.

Elite troops, usually recruited from fully trained soldiers in "line" units, are subjected to more rigorous training, designed to test the candidate's endurance and fortitude as well as to teach specialized skills. NONCOMMISSIONED OFFICERS (NCOs) are normally trained as such in small unit leadership courses.

Officers can be selected from the ranks, or educated in special academies, or trained in (short) officer candidate programs, sometimes in conjunction with civilian higher education (as in the U.S. ROTC). New officers normally undergo basic and advanced individual training before being posted to specialized technical or command schools. As officers advance in rank, they may be sent to staff colleges for instruction in military theory, doctrine, command skills, and, above all, military administration.

Some countries also have some form of part-time preinduction military training for teenagers (e.g., DOSAAF in the Soviet Union), which prepares them for basic training.

TRANSALL: The C.160 twin turboprop tactical transport aircraft built by the Franco-German Transport Allianz consortium (Nord-Aviation, Hamburger Flugzeugbau, and Vereintige Flugtechnische Werke), originally for the French *Armee de l'Air* and the West German *Luftwaffe*. The Transall project began in 1959 as a replacement for the French piston-engined Noratlas. Three prototypes were constructed, one by each member of the consortium, and flight tests began in 1963. The first production aircraft were delivered in 1967, and a total of 169 were completed by 1972. The production line was reopened in 1978–79 to build an additional 25 aircraft for France. South Africa also purchased 9 modified aircraft (C.160Z), while Turkey acquired 20 ex-*Luftwaffe* Transalls (C-160T); none of those aircraft remain in service.

The Transall has a conventional layout. A weather radar is mounted in the nose, and the pilot, copilot, and engineer sit in an airline-style flight deck just ahead of the cargo hold. The tail has a fold-down ramp-door, and the high-mounted wing provides an unobstructed cargo deck for a payload of up to 35,275 lb. (14,670 kg.); alternatively, fold-up seats can accommodate up to 99 paratroops. The landing gear are designed for short, rough field landings, with twin nose wheels beneath the cockpit, and two quadruple main wheels mounted in sponsons alongside the fuselage. The wing has large-span Fowler flaps for short takeoff and landing.

While range is sufficient for the *Luftwaffe*, it is completely inadequate for the *Armee de l'Air*, which has long-range commitments. For the 1978 intervention in Kolwezi, French paratroops had to be dropped from borrowed U.S. C-130 HERCULES transports after their transit to Zaire by Air France commercial jets. The 25 C.160s built in the late 1970s have additional internal fuel and an AERIAL REFUELING probe; these modifications may be retrofitted to older French Transalls. See also AIRBORNE FORCES, FRANCE; REPUBLIC OF GERMANY.

Specifications Length: 106.3 ft. (32.4 m.). **Span:** 131.25 ft. (40 m.). **Powerplant:** 2 6100-hp. Rolls Royce Tyne RTy.20 turboprops. **Weight, empty:** 63,405 lb. (28,820 kg.). **Weight, max. takeoff:** 112,435 lb. (51,106 kg.). **Speed, max.:** 368 mph (615 kph). **Range:** (w/17,637-lb. [8016-kg.] payload) 2983 mi. (4981 km.).

TRIAD, STRATEGIC: A term referring to the entire spectrum of U.S. long-range nuclear DELIVERY SYSTEMS—intercontinental ballistic missiles

(ICBMS), submarine-launched ballistic missiles (SLBMS); and manned BOMBERS—and connoting the deterrence-enhancing synergy of their complementary attributes. Land-based ICBMs, extremely accurate and with very secure COMMAND AND CONTROL links, are most effective against hardened point targets (e.g., enemy missile silos), but if themselves based in fixed silos, they are also the most vulnerable. SLBMs are currently secure (due to their concealment underwater), but lack the accuracy and reliable command and control required for counterforce attacks. Bombers are relatively slow and may be intercepted by AIR DEFENSES, but only they can be placed on (airborne) alert during crises, reducing their vulnerability and reinforcing the credibility of nuclear threats. They can also locate and strike mobile targets, and unlike missiles, they can be recalled. In addition, the very fact that bombers can be intercepted means (paradoxically) that potential adversaries will divert resources from offensive forces to maintain air defenses.

The "triad" was the result of an ad hoc decision made in the late 1950s, but it does ensure that a Soviet technological breakthrough in one particular area (e.g., ANTI-SUBMARINE WARFARE) will not neutralize the entire U.S. retaliatory force. Air-launched cruise missiles still rely on bomber platforms, but naval cruise missiles with nuclear warheads and strategic reach have added a fourth mode of attack to the triad.

TRIDENT: A U.S. submarine-launched ballistic missile (SLBM), successor to the POSEIDON C-3, and currently in service aboard the LAFAYETTE- and OHIO-class nuclear-powered ballistic-missile submarines (SSBNs). The Trident program emerged from a 1966 study of future U.S. strategic forces (STRAT-X), which called for an "Undersea Long-range Missile System" (ULMS). Two alternative designs were proposed: a missile with a range of 6000 n.mi., which was too large for existing SSBN launch tubes, and a smaller missile, with a range of only 4500 n.mi., but which could fit in the tubes of the Lafayette class. Only the larger missile could reach all potential targets in the Soviet Union from launch positions within U.S. territorial waters, complicating Soviet anti-submarine operations, but new submarines to carry it could not be ready until the late 1970s at the earliest. Thus it was decided to build the smaller missile, later designated UGM-96A Trident I (C-4), as an interim weapon until new SSBNs of the Ohio class would become available, when it would be super-

seded by the larger UGM-133 Trident II (D-5) missile. The decision to proceed was made in 1972 in the wake of the 1972 SALT I accords, and the first contract for the Trident I was issued to Lockheed in 1974. The missile was first test-fired in 1977, and entered service in 1979, to equip 12 submarines of the Lafayette class and the first of the Ohio class (the latter have volume-reducing inserts in their missile tubes to accommodate the C-4).

The Trident I is a three-stage, solid-fuel missile similar to the earlier Poseidon, with greater range and better accuracy, but roughly the same throw weight. All three stages are fabricated from spun KEVLAR fiber, which is lighter and stronger than the spun glassfiber of the Poseidon. The missile has a very blunt nose, to maximize internal volume within the dimensional constraints of the Lafayettes' launch tubes. To improve aerodynamic performance at supersonic speeds, a telescoping "aerospike" extends from the nose at liftoff. Essentially a long pole with a small, flat disc at the end, the aerospike generates a supersonic shockwave ahead of the missile, achieving the effect of greater length to improve ballistic and drag coefficients.

The first and second stages have a conventional configuration, and use cold gas injected into the exhaust (at the nozzle exit) for the thrust vector control of steering and attitude. The third-stage motor has a reduced diameter to fit inside an annular POST-BOOST VEHICLE (PBV), and burns storable liquid fuel (probably undimensional dimethyl hydrazine and nitrogen tetroxide) in a throttleable engine which can be shut down and restarted. The PBV carries eight Mk.4 REENTRY VEHICLES (RVs), each armed with a 100-KT (W76) thermonuclear warhead. The PBV can be programmed to direct RVs against widely separated targets, hence Trident has MULTIPLE INDEPENDENTLY TARGETED REENTRY VEHICLE (MIRV) capability. The missile has implicit (navigating) INERTIAL GUIDANCE with stellar update (a "stellar-inertial" system). Trident does not have thrust termination blowout ports for velocity control (as in earlier U.S. SLBMs) but instead uses the General Energy Management Steering (GEMS) technique to burn off unexpended fuel in shorter-range shots.

The Trident II D-5, operational since 1990 aboard Ohio-class submarines, has the same general configuration as the C-4, but is longer and heavier, for greater range; the CEP has been further reduced by improvements in the guidance unit. The D-5 is armed with up to seven Mk.5

MIRVs, each containing a 475-kT (W76) warhead. In the future, it could carry a smaller number of Mk.500 MANEUVERABLE REENTRY VEHICLES (MaRVs); penetration aids (PENAIDS) are already provided. The combination of low CEP and a high warhead yield gives the Trident II a "hard target" attack capability, but it is not a useful "first strike" weapon because the command and control limitations of submarines would impede the mass coordinated launches required for a disarming counterforce attack.

The Trident I has also been purchased by the British Royal Navy for the new Vanguard-class SSBNs, which are to succeed the POLARIS-armed RESOLUTION class. That choice has been criticized within the Royal Navy because it is absorbing funds which some believe should be spent on attack submarines and surface vessels. See also BALLISTIC MISSILE; SLBMS, UNITED STATES; SUBMARINES, UNITED STATES.

Specifications Length: (I) 34.1 ft. (10.39 m.); (II) 45.8 ft. (13.96 m.). **Diameter:** (I) 6.16 ft. (1.87 m.); (II) 6.9 ft. (2.1 m.). **Weight, launch:** (I) 70,000 lb. (31,818 kg.); (II) 126,000 lb. (57,272 kg.). **Range:** (I) 4370 mi. (7298 km.); (II) 6900 mi. (11,-523 km.). **CEP:** (I) 250–500 m.; (II) 250 m.

"TRIGGER": A strategic theory meant to justify small, "independent" national nuclear forces. It argues that a nuclear force too small in itself to deter attack by a superpower could do so by provoking the release of another superpower's nuclear forces. While the usual "equalizer" argument for an independent nuclear force is that it frees its possessor from the need for an alliance with a superpower (on the claim that any nuclear strike, no matter how small, is "unacceptable"), the trigger theory runs rather the other way: its aim is to ensure that an alliance would become operative, by preventing the superpower from "decoupling" from its smaller ally. Beyond the cataclysmic presumption that a general war would result more or less automatically from any nuclear strike against either superpower, a more subtle version has also been proposed: a small power allied to superpower A could protect a vital interest from superpower B by threatening it with a COUNTERFORCE nuclear strike, which might eliminate enough of B's nuclear strength to make B vulnerable to attack by A. At present, however, the "independent forces"—the British POLARIS and the French FORCE DE DISSUASION—are too small to make a significant dent in the nuclear strength of either superpower. For a more sinister use of small nuclear forces, see CATALYTIC WAR.

TRIPWIRE: A military force, too small in itself to resist successfully but nonetheless deployed against a potential enemy, in order to trigger the intervention of more powerful forces. The term occurs in the continuing debate on European security, and specifically refers to the idea of replacing most U.S. forces in Europe with a token screen of U.S. troops, on the presumption that an attack by Soviet forces would still trigger a U.S. nuclear response. That proposal has been ridiculed by suggesting that a single U.S. soldier might suffice, by interposing himself in the path of the Soviet army. In practice, the tripwire concept is an application of the old policy of MASSIVE RETALIATION, designed to achieve security at minimal cost.

Another form of tripwire was the implicit strategy of the United Nations Emergency Force which patrolled the Israeli-Egyptian armistice lines until May 1967; it was envisaged as sufficiently strong to arrest the conflict until the intervention of one or more of the Great Powers.

TROJAN: The FV.432 tracked ARMORED PERSONNEL CARRIER (APC), built by GKN for the British army as a replacement for the SARACEN 6 × 6 wheeled APC; a total of 3000 were delivered between 1963 and 1971. Trojan is strictly a "battle taxi," meant to deliver troops to the battlefield, where they are to dismount and fight on foot (in contrast to INFANTRY FIGHTING VEHICLES, designed for mounted combat). Similar to the U.S. M113 APC, the Trojan has a boxy, all-welded hull, but is fabricated of steel (rather than aluminum), and weighs considerably more. Armor thickness ranges between 6 and 12 mm., sufficient for protection against small arms and shell splinters.

Internally, the Trojan is divided into a frontal engine compartment and a rear crew/troop compartment. The driver and a commander/gunner sit on the right side of the vehicle, with the driver up front; both have observation periscopes with night vision devices. Up to ten fully equipped infantrymen can be carried, with access through a large rear door and a roof hatch. A 7.62-mm. machine gun is mounted by the commander's hatch for self-defense, and three smoke grenade launchers are fitted on each side of the hull. Uniquely for Western APCs of its generation, the Trojan has positive overpressure COLLECTIVE FILTRATION for NBC defense. Unlike the M113, the Trojan is not amphibious. When initially delivered, it was fitted with an erectable flotation screen, but this proved to be too clumsy and was later removed.

A number of specialized vehicles have been developed from the Trojan, including a fire support vehicle armed with the 30-mm. RARDEN turret of the FOX armored car, an armored ambulance, an armored command vehicle, an 81-mm. MORTAR carrier, a minelayer, an engineer vehicle with two Giant Viper mine-clearing rockets, an artillery spotting vehicle with a COUNTERBATTERY radar, and a signals vehicle. The ABBOT self-propelled 105-mm. gun is also based on the chassis and drive train of the FV.432. Trojan equips mechanized battalions in the British Army of the Rhine, supplemented from 1987 by the MICV-80 WARRIOR.

Specifications Length: 17.25 ft. (5.26 m.). **Width:** 9.16 ft. (2.79 m.). **Height:** 7.5 ft. (2.28 m.). **Weight, combat:** 15.25 tons. **Powerplant:** 240-hp. Rolls Royce K60 No.4 Mk.4F 6-cylinder diesel. **Hp/wt. ratio:** 15.7 hp./ton. **Fuel:** 120 gal. (528 lit.). **Speed, road:** 32 mph (54 kph). **Range, max.:** 300 mi. (500 km.).

TROMP: A class of two Dutch guided-missile FRIGATES commissioned in 1976–77 as replacements for the cruisers *de Ruyter* and *de Zeven Provincien*. Though classified only as frigates, the Tromps are the largest surface warships of the Dutch navy, and are outfitted as flagships for ANTI-SUBMARINE WARFARE (ASW) groups (of KORTENAER- and HEEMSKERCK-class frigates). The Tromps are flush-decked with high freeboard, and are equipped with gyro-controlled fin-stabilizers which, combined with their broad beam, makes them excellent sea boats for North Sea conditions.

As general purpose escorts, the Tromps have a balanced armament. Their primary ANTI-AIR WARFARE (AAW) weapon is a Mk.13 single-arm launcher with 40 STANDARD-MR surface-to-air missiles (SAMs) on the after superstructure. Secondary AAW weapons include an 8-round NATO SEA SPARROW short-range SAM launcher (with 52 reload missiles) ahead of the bridge, and a twin 120-mm. DUAL PURPOSE gun mount on the foredeck. For ANTI-SURFACE WARFARE, the ships have two quadruple launchers for HARPOON anti-ship missiles amidships, backed up by the 120-mm. guns. The principal ASW weapon is a Westland LYNX WG.13 helicopter, operated from a landing deck and hangar on the fantail. For short-range ASW, there are two sets of Mk.32 triple tubes for lightweight homing TORPEDOES. The *Tromp* will undergo a midlife overhaul to extend its service life into the mid-

1990s; its sister ship *de Ruyter* will be retained only until 1992. See also FRIGATES, NETHERLANDS.

Specifications Length: 453 ft. (138 m.). **Beam:** 49 ft. (14.93 m.). **Draft:** 21.5 ft. (6.55 m.). **Displacement:** 4300 tons standard/5400 tons full load. **Powerplant:** twin-shaft COGOG: 2 4100-shp. Rolls Royce Tyne RM1C gas turbines (cruise), 2 25,000-shp. Rolls Royce Olympus TM3B turbines (sprint). **Fuel:** 600 tons. **Speed:** 30 kt. **Range:** 5000 n.mi. at 18 kt. **Crew:** 306. **Sensors:** 1 SPS-01D 3-dimensional air-search radar (in a large, mushroom-shaped dome), 2 Decca 1226 navigational/surface search radars, 1 WM-25 fire control radar, 2 SPG-51C missile guidance radars, 1 CWE 610 hull-mounted medium-frequency sonar, 1 Type 162 bottom-mapping sonar. **Electronic warfare equipment:** 1 Ramses electronic signal monitoring array, 4 Mk.36 SRBOC chaff launchers.

TRUXTON: A U.S. nuclear-powered guided-missile CRUISER (CGN-35). Commissioned in 1967, the *Truxton* is a nuclear-powered version of the BELKNAP class, with an enlarged, modified hull, and rearranged topsides; the additional displacement is due mainly to the shielding for the nuclear reactors. The Truxton has the same weapon and sensor suite as the Belknaps, but the positions of the 5-in. DUAL PURPOSE gun and Mk.10 TERRIER launcher have been transposed—the *Truxton* carries the gun forward and the missile launcher aft. In addition, its radar and communications antennas are mounted on two large lattice masts, rather than on "macks" (combination mast and exhaust stacks). See, more generally, CRUISERS, UNITED STATES.

Specifications Length: 564 ft. (171.95 m.). **Beam:** 58 ft. (17.68 m.). **Draft:** 31 ft. (9.45 m.). **Displacement:** 8200 tons standard/9200 tons full load. **Powerplant:** twin-shaft nuclear; 2 D2G pressurized-water reactors, 2 sets of geared turbines, 60,000 shp. **Speed:** 32 kt. **Crew:** 591. **Sensors:** see BELKNAP. **Electronic warfare equipment:** see BELKNAP.

TRY-ADD: NATO code name for a family of Soviet tracking and guidance RADARS associated with the GALOSH ANTI-BALLISTIC MISSILE (ABM) system. Try-Add consists of three radars of two different types, one large and two small. The large radar, Try-Add A, tracks incoming ballistic missile REENTRY VEHICLES (RVs) after initial data is supplied by CAT HOUSE or DOG HOUSE phased-array early warning radars. Try-Add A data is in turn supplied to a fire control system which transmits missile guidance commands via the two smaller

Try-Add B radars. Each Galosh site has 1 or 2 Try-Add As, and 2 or 4 Try-Add Bs. Each Try-Add system is believed to be capable of tracking two incoming RVs simultaneously with each Try-Add A, and of directing one Galosh missile with each Try-Add B. See also BALLISTIC MISSILE DEFENSE; COMMAND GUIDANCE.

TTMTC: See TYURATAM MISSILE TEST CENTER.

TUPOLEV: A Soviet aircraft design bureau (named after its founder, A. N. Tupolev), known primarily for its large BOMBER, patrol, long-range INTERCEPTOR, and AIRBORNE EARLY WARNING aircraft. Tupolevs currently in service include: Tu-16 BADGER; Tu-20 BEAR; Tu-22 BLINDER; Tu-26 BACKFIRE; Tu-28 FIDDLER; Tu-126 MOSS; and Tu-160 BLACKJACK.

TURYA: NATO code name for a class of some 50 Soviet hydrofoil torpedo boats derived from the OSA-class fast missile boat; the first was completed in 1971. The Turyas have the hull and machinery of the Osa, but with fixed, surface-piercing foils added to increase speed by some 5 kt. Armed with acoustical homing torpedoes, which have some anti-surface capabilities but are mainly for ANTI-SUBMARINE WARFARE (ASW), the Turyas are meant for fast-reaction inshore ASW, usually in concert with helicopters, fixed-wing aircraft, or larger (but slower) patrol ships.

Main armament is four 21-in. (533-mm.) TORPEDOES launched from fixed tubes mounted amidships. Secondary armament consists of a twin 25-mm. ANTI-AIRCRAFT gun forward, and a twin 57-mm. DUAL PURPOSE gun aft. Sensors include a dipping SONAR similar to that of the Kamov Ka-25 HORMONE ASW helicopter, a "Pat Hand" surface-search RADAR, a "Muff Cat" FIRE CONTROL radar, a "High Pole" IFF, and a "Square Head" IFF interrogator.

A total of 31 Turyas were delivered to the Soviet navy; an additional 17 supplied to Cuba, Ethiopia, Kampuchea, and Vietnam do not have sonar, and are apparently meant for surface attack. The similar MATKA class, introduced in 1978, is armed with anti-ship missiles instead of torpedoes. See, more generally, FAST ATTACK CRAFT, SOVIET UNION.

TVD: *Teatr voyennykh deystviy*, "theater of military operations," a Soviet term that describes a large regional military command higher than a FRONT, and responsible directly to the Soviet High Command. TVDs serve as the focus of planning and control for major military actions on frontages of up to several thousand kilometers, and a corresponding depth.

Western intelligence tentatively identified at least 8 separate TVDs during the 1980s. Five are primarily on land, while 3, the Atlantic, Arctic, and Pacific TVDs, cover oceanic zones. The land TVDs are designated Northwestern, Western, Southwestern, Southern, and Far East; the first 3 cover most NATO territory. The Northwestern TVD is responsible for Scandinavia and the Baltic Sea; at this time its forces include 12 DIVISIONS, 1400 TANKS, 3100 other ARMORED FIGHTING VEHICLES (AFVs), 2000 ARTILLERY pieces, and 180 TACTICAL AIRCRAFT.

The Western TVD, by far the most important, is responsible for all of Western Europe. The Southwestern TVD covers the Balkans, the Levant, and North Africa; it also controls the Soviet Black Sea Fleet. The Southern TVD covers most of Western Asia, including Iran, Afghanistan, Pakistan, and the Indian subcontinent. The Far East TVD covers all of Eastern Asia, and is probably the second most important command.

The Oceanic TVDs correspond to Soviet fleets. The Arctic TVD controls the Northern Fleet, and covers the Arctic Sea, North Sea, and GIUK GAP. Its primary mission is to support Northwestern TVD operations against Scandinavia, by interdicting reinforcement from the United States, and protecting Soviet ballistic missile submarines operating in "bastions" close to Soviet home waters. The Atlantic TVD controls the Baltic Fleet. Its primary missions are to support the operations of the Western TVD in the Baltic and the mid-Atlantic, and to interdict NATO convoys. The Pacific TVD controls the Pacific Fleet, and supports the operations of the Far East TVD by operations throughout the Pacific Ocean.

In wartime, each TVD controls a number of Fronts corresponding roughly to peacetime Military Districts. Each Front, in turn, would control several ARMIES, each composed of four to six divisions. In addition to the forces assigned to TVDs, the Soviet Union maintains a strategic RESERVE retained under the direct control of the Defense Committee or the Supreme High Command (*Verkhovnoye glavnokomandovaniye*, or VGK). These forces can be assigned to any TVD as required, and also have an internal security role.

TVM: See TRACK-VIA-MISSILE.

TWO-MAN RULE: A mechanical procedure designed to prohibit any one person from firing NUCLEAR WEAPONS, by requiring the active cooperation of at least two individuals—each of whom can prevent release. See also FAILSAFE; PERMISSIVE ACTION LINK; POSITIVE CONTROL.

TWS: See TRACK-WHILE-SCAN.

TYPE 205: A class of 14 West German diesel-electric attack SUBMARINES commissioned between 1961 and 1969. The first German submarines built after World War II, the Type 205s are small, coastal boats optimized for Baltic operations; they bear a considerable resemblance to the wartime Type XXIII U-boat.

The Type 205s have a modified teardrop hull with a single-hull configuration and a relatively large sail (conning tower) amidships, surrounded by a short, raised deck casing. The first six 205s were constructed of a special nonmagnetic steel, which, however, proved highly vulnerable to corrosion (causing the first two to be scrapped); the remaining 205s were constructed of ordinary steel. The control surfaces are of a unique design: each forward diving plane has a different angle of incidence, one to generate a bow-up, the other a bow-down attitude; they are extended differentially to achieve the desired angle. Four cruciform tail fins support the stern planes and the single large rudder. A retractable snorkel allows diesel operations from shallow submergence. Armament consists of eight bow-mounted tubes for 21-in. torpedoes launching with the "swim-out" technique. No reloads are carried, and the tubes can be loaded only through the bow caps in drydock. Sensors are linked to an HSA Mk.8 automated FIRE CONTROL set.

Only six Type 205s remain in service; the remainder were laid up or scrapped between 1968 and 1982. The survivors will be replaced by the new TYPE 211 between 1990 and 1993. Two slightly modified Improved Type 205s were built for Denmark in 1968–69; these will be replaced by six TYPE 210s in the mid-1990s. Fifteen similar TYPE 207s (Kobben class) were built for Norway between 1964 and 1967. See, more generally, SUBMARINES, FEDERAL REPUBLIC OF GERMANY.

Specifications Length: 144 ft. (43.9 m.). **Beam:** 14.9 ft. (4.54 m.). **Displacement:** 419 tons surfaced/450 tons submerged. **Powerplant:** single-shaft diesel-electric: 2 600-hp. MTU 12V493AZ diesel generators, 1 2300-hp. electric motor. **Speed:** 10 kt. surfaced/17 kt. submerged. **Max. operating depth:** 500 ft. (152 m.). **Collapse depth:** 750 ft. (228 m.). **Crew:** 22. **Sensors:** 1 SRS-M1H bow-mounted medium-frequency passive sonar, 1 GHG AN5039A1 high-frequency active attack sonar, 1 Thomson-CSF surface-search radar, 1 electronic signal monitoring array, 2 periscopes.

TYPE 206: A class of 21 diesel-electric SUBMARINES designed by the West German IKL shipyard (18 for West Germany and 3 for Israel), commissioned between 1971 and 1976. The Type 206 is an enlarged and improved version of the TYPE 205 coastal submarine, with more advanced sensors, fire controls, and armament, and are virtually identical in layout and external configuration to the 205s. A retractable snorkel allows diesel operation from shallow submergence, and a high-capacity, fast-charging silver-zinc Varta battery reduces recharge periods. Armament consists of eight bow-mounted 21-in. (533-mm.) torpedo tubes, capable of launching free-running or acoustical homing TORPEDOES with the swim-out method. No reloads are carried; as in the 205s, the tubes can be reloaded only though the bow caps while in drydock. Up to 16 bottom-laid MINES can be loaded in the tubes as an alternative payload, and an additional 24 mines can be carried in external containers. All sensors are linked to an HSA WM-08 torpedo FIRE CONTROL system.

The three Israeli (Gal-class) submarines were built under license by Vickers in Britain in 1975–76. Also known as the IKL-540 class, they are slightly shorter and broader than the standard Type 206, and are equipped with British electronics and fire controls. From 1988, 12 of the German boats will be modernized with new fire controls and torpedoes, as the Type 206A. See, more generally, SUBMARINES, FEDERAL REPUBLIC OF GERMANY.

Specifications Length: 159.4 ft. (48.6 m.); (Gal) 146.75 ft. (44.74 m.). **Beam:** 15.1 ft. (4.6 m.); (Gal) 15.4 ft. (4.7 m.). **Displacement:** 450 tons surfaced/500 tons submerged; (Gal) 420 tons surfaced/680 tons submerged. **Powerplant:** single-shaft diesel-electric: 2 640-hp. MTU 12V493AZ diesel generators, 1 1800-hp. electric motor. **Speed:** 10 kt. surfaced/18 kt. submerged. **Range:** 4500 n.mi. at 5 kt. (snorkel)/200 n.mi. at 5 kt. (batteries). **Max. operating depth:** 500 ft. (152 m.). **Collapse depth:** 750 ft. (228 m.). **Crew:** 22. **Sensors:** 1 AN 5039A1 bow-mounted low-frequency passive sonar, 1 AN 410A4 high-frequency active sonar, 1 French DUUX 2 active fire control sonar, 1 Thomson-CSF Calypso surface-search radar, 1 electronic signal monitoring array, 2 periscopes.

TYPE 207: Also called the Kobben class, 15 German-designed diesel-electric attack submarines built by IKL for the Norwegian navy between 1964 and 1967, with half the cost paid by the U.S. The Type 207 is essentially a slightly enlarged version of the TYPE 205 coastal submarine, with the pressure hull fabricated of HTS (High Tensile Strength) steel for increased operating depth. One boat, the *Svenner*, used as an officer training vessel, is slightly longer. The hullform, external configuration, and internal layout are all similar to the Type 205.

Armament consists of eight bow-mounted 21-in. (533-mm.) tubes, for either the Type 61 anti-ship wire-guided acoustical homing TORPEDO, or the 19-in. (483-mm.) NT-37C anti-submarine homing torpedo. As in the 205s, no reloads are carried, and the tubes can be loaded only through the bow caps while in drydock. Up to 16 MINES can be carried as an alternative load.

Only 14 Type 207s remain in service, the oldest boat having been retired in 1982. Six of the remaining boats will be replaced by the TYPE 210, or Ula class, in the mid-1990s. The survivors will be overhauled, modernized, and sold to Denmark. See, more generally, SUBMARINES, FEDERAL REPUBLIC OF GERMANY.

Specifications Length: 148.9 ft. (45.4 m.); *(Svenner)* 152.2 ft. (46.4 m.). **Beam:** 15.1 ft. (4.6 m.). **Displacement:** 370 tons surfaced/482 tons submerged; *(Svenner)* 485 tons submerged. **Powerplant:** single-shaft diesel-electric: 2 600-hp. MTU 12V493AZ diesel generators, 1 2300-hp. electric motor. **Speed:** 13.5 kt. surfaced/17 kt. submerged. **Max. operating depth:** 623 ft. (190 m.). **Collapse depth:** 934 ft. (285 m.). **Crew:** 17; *(Svenner)* 18. **Sensors:** 1 KAE bow-mounted medium-frequency passive sonar, 1 KAE high-frequency active sonar, 1 Thomson-CSF Calypso surface-search radar, 1 electronic signal monitoring array, 2 periscopes.

TYPE 209: A class of more than 40 German-designed diesel-electric attack SUBMARINES built by IKL Shipyards for export. The Type 209 is a medium-range, oceangoing submarine which has been built in six different subclasses to meet customer requirements, with differences in length, displacement, and equipment.

The basic Type 209/1100 has a modified teardrop hull with a very broad, blunt bow (to ensure smooth water flow around the torpedo tubes), a cylindrical center section, and a tapered stern. The boat has a single-hull configuration with a flat deck casing on the bow, and a streamlined sail (conning tower) amidships. The pressure hull is fabricated from HTS (High Tensile Strength) steel. The control surfaces repeat the otherwise unique design developed by IKL for the Type 205s: the two forward diving planes have different angles of incidence, one to generate a bow-up, the other a bow-down attitude; they are extended differentially to achieve the desired angle. The rudders and stern planes have a conventional cruciform configuration. A retractable snorkel allows diesel operation at shallow submergence. The electric motor is driven by high-capacity, quick-charging Hagen/VARTA batteries, which occupy 20 percent of total internal volume. Armament consists of eight bow-mounted 21-in. torpedo tubes for a variety of acoustical and wire-guided TORPEDOES, using the swim-out technique. Unlike earlier IKL submarines, the 209s carry six reload torpedoes and have automated loading equipment. MINES can replace torpedoes on a 2-for-1 basis. All sensors are linked to an automated FIRE CONTROL system, an HSA 8/24 in early boats, or an HSA SINBADS or a Ferranti KAFS in later models.

The 209/1100 was bought by Greece, which received 4 boats in 1971–72. Other variants include the 209/1200, also bought by Greece (4 boats in 1979–80), as well as by Argentina (2 boats in 1974), Colombia (2 in 1975), Peru (6 between 1975 and 1983), and Turkey (12 boats, including 9 built under license, from 1975 to 1988); the 209/1300, bought by Ecuador (2 boats in 1976–77), Indonesia (2 in 1981, plus 4 on order), and Venezuela (2 in 1976–77); the 209/1400, bought by Chile (2 boats in 1984) and Brazil (5 between 1984 and 1988); and the 209/1500, bought by India (4 boats in 1986, with more on order). These variants are lengthened or shortened by adding or removing prefabricated modules from the center section of the hull. The additional length is used mainly to store additional fuel and add battery cells to extend range and patrol endurance. Performance for all types is essentially the same as for the 209/1100. The Type 209s are quiet, maneuverable, and economical. Although IKL has since introduced newer designs, the 209 continues to attract orders, and will long remain in service. See, more generally, SUBMARINES, FEDERAL REPUBLIC OF GERMANY.

Specifications Length: (1100) 178.5 ft. (54.42 m.); (1200) 184 ft. (56.09 m.); (1300) 195.25 ft. (59.52 m.); (1400) 200 ft. (60.97 m.); (1500) 211 ft. (64.33 m.). **Beam:** 20.33 ft. (6.2 m.). **Displacement:** 1105 tons surfaced/1230 tons submerged; (1200) 1180 tons surfaced/1290 tons submerged; (1300) 1260 tons surfaced/1390 tons submerged; (1400) 1320 tons surfaced/1440 tons submerged; (1500) 1660 tons surfaced/1850 tons submerged.

Powerplant: single-shaft diesel-electric: 4 600-hp. MTU 12V493TY60 diesel generators, 1 Siemens 2400-hp. electric motor. **Speed:** 11 kt. surfaced/22 kt. submerged (burst)/12 kt. (snorkel). **Endurance:** 50 days. **Max. operating depth:** 984 ft. (300 m.). **Collapse depth:** 1640 ft. (500 m.). **Crew:** 30–33. **Sensors:** 1 Krupp-Atlas CSU-3-2 low-frequency active sonar, 1 PRS-3-4 passive conformal array sonar (or 1 CSU-3-4 active/passive sonar), 1 Thomson-CSF Calypso surface-search radar, 1 electronic signal monitoring array, 2 periscopes. **Operators:** Arg, Bra, Chi, Col, Ecu, Gre, Ind, Indo, Peru, Tur, Ven.

TYPE 210:   Also called the Ula class, six German-designed diesel-electric attack SUBMARINES being built by Thyssen Nordseewerke for the Norwegian navy. The Type 210s are replacements for the IKL TYPE 207, or Kobben class; the first was commissioned in 1989. Much larger and more capable than the 207s, they have a modified teardrop hull, with a broad, blunt bow (to ensure smooth water flow over the torpedo tubes), a cylindrical center section, and a tapered stern; a flat deck casing extends over most of the upper surface of the hull. The boats have a single-hull configuration with a pressure hull fabricated of HTS (High Tensile Strength) steel. A tall, streamlined sail (conning tower) is located amidships. The forward diving planes may be mounted either on the hull or on the sail, while the rudders and stern planes have a cruciform configuration. A retractable snorkel allows diesel operation at shallow submergence. Armament consists of eight bow-mounted 21-in. torpedo tubes for a variety of acoustical and wire-guided TORPEDOES. The torpedo room has automatic loading equipment and holds six reloads. MINES can replace torpedoes on a 2-for-1 basis. All sensors are linked to an MSI 9OU digital FIRE CONTROL system.

The Type 210 is very similar to the TYPE 211 proposed by Thyssen for the West German navy; the latter displaced 410 tons more, but had the same armament, systems, and performance. See, more generally, SUBMARINES, FEDERAL REPUBLIC OF GERMANY.

Specifications   **Length:** 193.6 ft. (59 m.). **Beam:** 17.7 ft. (5.4 m.). **Displacement:** 1040 tons surfaced/1300 tons submerged. **Powerplant:** single-shaft diesel-electric: 2 MTU diesel generators, 2520 hp., 1 electric motor. **Speed:** 11 kt. surfaced/23 kt. submerged. **Endurance:** 40 days. **Max. operating depth:** 820 ft. (250 m.). **Collapse depth:** 1640 ft. (500 m.). **Crew:** 18–20. **Sensors:** 1 KAE CSU83 low-frequency active/passive sonar, 1 Type 2007 surface-search radar, 1 electronic signal monitoring array, 2 periscopes.

TYPE 212:   A class of 12 diesel-electric SUBMARINES now being built by Thyssen Nordseewerke for the West German navy. A medium-range oceangoing submarine intended as a replacement for the much smaller TYPE 205 coastal submarines, the first 7 212s will replace the remaining 205s on a 1-for-1 basis between 1991 and 1994. The remaining 5 are to be completed between 1994 and 1996. The type 212 program was initiated in 1988 after the larger Type 211 (a variant of the TYPE 210 built for Norway) proved too expensive to build.

In the late 1990s, the West German navy plans to acquire a derivative Type 212, with closed-cycle fuel cell propulsion. See, more generally, SUBMARINES, FEDERAL REPUBLIC OF GERMANY.

TYPHOON: NATO code name for a class of 6 Soviet nuclear-powered ballistic-missile SUBMARINES (SSBNs), completed between 1981 and 1989.

The Typhoons are the largest submarines ever built—nearly twice the size of the next largest, the U.S. OHIO-class SSBNs. They have a unique double-hull configuration. The outer casing is twice as broad as it is deep, with a bulbous bow and a short, tapered stern. A very long sail is located aft, with 20 missile tubes arranged in two rows of 10 forward. It is believed that the Typhoon actually consists of two separate pressure hulls arranged side-by-side, each containing one row of missile tubes, crew accommodations, and machinery. Judging by their size, these pressure hulls may be sections of the DELTA-class SSBNs, or developments thereof. In addition, a semi-cylindrical casing at the base of the sail may indicate yet a third (short) pressure hull built over the two main hulls, housing all (or most) of the ship and weapon controls. The outer hull is notably clean, with few protuberances. The numerous free-flooding holes appear to be covered with flush-fitting hatches to minimize flow noises. All outer surfaces are covered with CLUSTER GUARD anechoic tiles to reduce radiated noise and the effective range of hostile active sonars. There seems to be much empty space between the outer and pressure hulls, which is only partially taken up by ballast and fresh water tanks, auxiliary batteries, and ancillary equipment; this free-flooding space may serve as anti-torpedo protection by detonating warheads some distance from the pressure hull. The standoff distance is probably great enough to defeat most Western lightweight torpedoes (e.g., the U.S. Mk.46), and at least partially neutralize heavier ones (for this reason the next generation of ASW torpedoes, the BARRACUDA, SPEARFISH, and STINGRAY, will have "directed energy" warheads based on the shaped-charge principle). The Typhoon's control surfaces, conventional in every respect but size, consist of retractable bow diving planes, and cruciform rudders and stern planes (the former on a massive vertical fin). A flat body flap extends aft from the stern; its purpose has not been determined.

Primary armament consists of 20 SS-N-20 STURGEON SLBMs. Sturgeon has a range of 5160 mi. (8617 km.), carries nine 500-kT MULTIPLE INDEPENDENTLY TARGETED REENTRY VEHICLES MIRVs, and has a CEP of approximately 500 m. As noted, the missiles are mounted in the forward part of the hull, reversing standard SSBN design practice. It is not known why this configuration was adopted, but it has been speculated that it is easier to compensate for weight changes and maintain trim during multiple missile launches if the missiles are forward of the submarine's center of gravity (on several occasions the Typhoon has fired no fewer than four missiles simultaneously, assisted, no doubt, by an automatic hover control system).

Secondary armament consists of 8 bow-mounted torpedo tubes arranged in 2 horizontal rows of 4 (4 tubes in each pressure hull), with a total of 24 weapons, divided among acoustical homing and nuclear anti-submarine TORPEDOES, and SS-N-15 STARFISH anti-submarine missiles.

Sensors are optimized for long-range passive detection. The Typhoon also has a comprehensive set of communications equipment, including VHF, UHF, and HF aerials; VLF antenna buoys; a "Pert Spring" satellite communications terminal; and an ELF trailing wire antenna. They undoubtedly have one or more INERTIAL NAVIGATION systems similar to the U.S. SINS, with provision for celestial and satellite position updates.

Some credit the Typhoon with a maximum speed of more than 30 kt., but a more realistic assessment based on hullform and displacement is only 24 kt. The Typhoons are relatively quiet, although not as quiet as Western SSBNs, the principal source of noise being cavitation caused by their two relatively small, and therefore high-speed,

propellers (the single, large, low-speed screw of Western SSBNs is much quieter).

Given its size, the Typhoon should be capable of long-range patrol throughout the world's oceans. On the other hand, although difficult to detect passively, its immense bulk is easy to find with active sonar, while its low speed and lack of maneuverability would hinder evasion. Typhoons are therefore more likely to remain in protected "bastions"—waters close to the Soviet Union, especially under the Arctic ice pack. Once there, if defended by Soviet attack submarines and ASW forces, the Typhoon's size would not be serious handicap; indeed, it would be useful for breaking through the ice. A large submarine with many missiles also makes good economic and military sense under a bastion method of operations: the number of expensive platforms can be reduced, and the ratio of attack submarines to SSBNs kept high. See, more generally, SUBMARINES, SOVIET UNION.

Specifications **Length:** 558 ft. (170.12 m.). **Beam:** 75 ft. (22.87 m.). **Displacement:** 26,000 tons surfaced/33,000 tons submerged. **Powerplant:** twin-shaft nuclear: 4 pressurized-water reactors (2 in each pressure hull), 4 sets of geared turbines, 70,000–80,000 shp. **Speed:** 20 kt. surfaced/24 kt. submerged. **Max. operating depth:** 1300 ft. (400 m.). **Collapse depth:** 2000 ft. (610 m.). **Crew:** 150. **Sensors:** 1 large bow-mounted low-frequency passive sonar, 1 medium-frequency active fire control sonar, 1 passive flank-mounted conformal array sonar, (1?) passive towed array sonar, 1 "Snoop Tray" surface-search radar, 1 "Rim Hat" electronic signal monitoring array, 2 periscopes.

TYURATAM MISSILE TEST CENTER (TTMTC): The principal Soviet spaceport, site of all Soviet manned spaceflight operations (including the new Soviet space shuttle), a test center for intercontinental ballistic missiles (ICBMS), and an operational center for the Soviet coorbital ANTISATELLITE (ASAT) system. TTMTC is located at 48°48′, 65°50′ E, in Soviet Central Asia, near the town of Tyuratam, east of the Aral Sea. It is sometimes called the Baikonur Cosmodrome in Soviet sources, even though the town of Baikonur is several hundred miles away. TTMTC is one of three Soviet spaceports; the other two, the PLESETSK MISSILE TEST CENTER and the KASPUTIN YAR MISSILE TEST CENTER, are used exclusively for military operations. See also SPACE, MILITARY USES OF.

U

U-2: U.S. high-altitude strategic RECONNAIS-
SANCE aircraft. Until the early 1950s, the U.S. attempted to collect photographic INTELLIGENCE on
Soviet military activities using modified B-29,
B-50, and B-36 bombers for reconnaissance, but
these were increasingly vulnerable to interception
by Soviet jet fighters. In 1954, the U.S. Air Force
and the Central Intelligence Agency (CIA) asked
Lockheed designer Clarence L. "Kelly" Johnson to
develop a specialized aircraft that could penetrate
deep into the Soviet Union and return safely. Johnson designed the aircraft as a jet-propelled sail-
plane, whose long, high-aspect-ratio wing allowed
flight at altitudes in excess of 70,000 ft., well above
the ceiling of contemporary Soviet INTERCEPTORS.
Development proceeded rapidly and in great secrecy. The first prototypes flew in 1955, and operational flights were under way by 1957. Under the
cover designation U (Utility Aircraft)-2, these machines were publicly described as atmospheric research platforms for NASA.

The initial U-2A had a very conventional, albeit
lightly constructed, airframe. The pilot sat in a
small unpressurized cockpit in the nose, and had to
wear a full pressure suit. The canopy was
equipped with an opaque glare shield to protect
the pilot from the intense ultraviolet radiation present at high altitudes. Behind the cockpit, a ventral
bay housed several high-resolution cameras. The
U-2 had bicycle landing gear, with two small
"pogo" wheels attached to each wingtip to stabilize the fuel-heavy aircraft during takeoff; these
were jettisoned once the aircraft was airborne.

The rest of the fuselage contained fuel tanks and
the engine. The midmounted, unswept wings were
equipped with conventional flaps and ailerons, and
housed four additional fuel tanks. The U-2A was
powered by a single Pratt and Whitney J57-PW-
13A turbojet; fed by two small cheek inlets behind
the cockpit, the engine burned a special low-
volatility fuel which would not evaporate at high
altitude. Typical mission endurance was 5.5 hours.
The CIA ordered a total of 48 single-seat and 5
two-seat training aircraft. Based in Japan and Germany, U-2As began regular overflights of the Soviet Union and China in 1957 from a variety of
staging fields.

The U-2A was soon superseded by the U-2B,
with a more powerful J75 engine, additional fuel,
and more equipment. The U-2B could also be fitted with two 605-lb. (275-kg.) wing-mounted slip-
per tanks, increasing mission endurance to 6.5
hours. The subsequent U-2C was essentially identical except for larger engine intakes to improve
the efficiency of the J75 engine, and a dorsal spine
for additional avionics. In addition to cameras, the
U-2C could carry air-sampling sensors to detect
atmospheric nuclear tests. (The U-2D was a 2-seat
trainer version of the U-2C, five of which were
built; the U-2CT was a 2-seat conversion of a standard C model; while the U-2E, F, and G were all
conversions of U-2B/C models actually intended
for atmospheric research.)

In May 1960, a U-2B flown by CIA employee
Francis Gary Powers was shot down by an SA-2
GUIDELINE surface-to-air missile over the Soviet

city of Sverdlovsk. Powers was captured, provoking an international incident and ending U-2 overflights of the Soviet Union. Thereafter, U-2 operations were transferred to the air force, and though several other U-2s were shot down over China and Cuba, the aircraft remained in service, providing critical intelligence unavailable from satellites at the time, notably on the progress (or lack thereof) of Soviet intercontinental ballistic missile (ICBM) programs, the installation of Soviet ballistic missiles in Cuba in 1962, and the great famine in China of 1962–63. U-2s were also used extensively in Southeast Asia between 1962 and 1968. A number of U-2Bs, operated by Taiwan in the early 1960s, continued overflights of the Chinese mainland for several years, with some ten aircraft lost in the process.

The U-2 production line closed in 1963 after some 100 aircraft had been built. But following attrition from accidents (the U-2 is a very difficult aircraft to fly) and enemy action, the line was reopened in 1968 to produce a much larger variant, the U-2R, intended primarily for standoff reconnaissance (the overflight mission had been taken over by the Mach 3 Lockheed SR-71 BLACKBIRD). The U-2R has an extended nose that houses additional sensors and avionics. The landing gear are closer together, and the tail section has been raised to leave space in the fuselage for additional fuel and equipment, allowing the elimination of the dorsal spine. The wingspan has been extended, increasing the wing area to 1000 sq. ft. from the 565 sq. ft. of the U-2A/B/C. Large, fixed pods under the inboard wing sections house palletized sensors, ELECTRONIC COUNTERMEASURES (ECM), and 714 lb. (325 kg.) of fuel. The powerplant is unchanged, but the larger wing increases the operational ceiling, and typical mission ensurance is 7.5 hours. Some 14 U-2Rs were built between 1968 and 1975; most are still in service.

The last variant of the U-2 is the TR-1, essentially a refinement of the U-2R optimized for standoff SURVEILLANCE and ELECTRONIC WARFARE. It has an Advanced SYNTHETIC APERTURE RADAR System (ASARS) based on the UPD-X SIDE-LOOKING AERIAL RADAR (SLAR), with a range of some 30 mi.; new pod-mounted avionics; and very elaborate ECM and ELECTRONIC SIGNAL MONITORING (ESM), cameras, and infrared line scanners. A digital DATA LINK can transmit radar and photogrammetric imagery down to ground stations in real time. The TR-1 has the same dimensions as the U-2R, but can carry a much greater load of internal fuel, for a mission time of 12 hours. The first TR-1A flew in August 1981; the air force now has 33, plus 2 TR-1B two-seat trainers, and 1 ER-2 research variant for NASA. Ten TR-1As are earmarked as platforms for the Precision Location Strike System (PLSS), a long-range ESM system meant to provide real-time target acquisition data on enemy radars and other emitters. The TR-1 was also proposed as a platform for the Joint Surveillance and Target Acquisition Radar System (J-STARS), an advanced synthetic aperture radar designed to detect and identify enemy vehicles and equipment far behind the front lines.

The U-2R and TR-1 are currently in service with the 9th Strategic Reconnaissance Wing of the STRATEGIC AIR COMMAND at Beale Air Force Base, California, and with the 17th Reconnaissance Wing at Alconbury, England. A number of U-2B/Cs are still used as civilian research aircraft.

Specifications Length: (A) 49.6 ft. (15.12 m.); (B/C/R/TR) 63 ft. (19.2 m.). **Span:** (A/B/C) 80 ft. (24.4 m.); (B/C/R/TR) 103 ft. (31.39 m.). **Powerplant:** (A) 1 Pratt and Whitney J57-PW-13A turbojet, 11,200 lb. (5091 kg.) of thrust; (B/C/R/TR) 1 Pratt and Whitney J75-PW-13 turbojet, 17,000 lb. (7727.27 kg.) of thrust. **Weight, empty:** (A) 9920 lb. (4509 kg.); (B/C) 11,700 lb. (5318 kg.); (R) 14,990 lb. (6814 kg.); (TR) 16,000 lb. (7272 kg.). **Weight, max. takeoff:** (A) 14,800 lb. (6727.27 kg.); (B/C) 16,000 lb. (7272.75 kg.) clean/17,270 lb. (7850 kg.) w/tanks; (R) 29,000 lb. (13182 kg.); (TR) 40,000 lb. (18.182 kg.). **Speed, max.:** (A) 494 mph (790.4 kph); (B/C) 528 mph (845 kph); (R) 510 mph (816 kph); (TR) 494 mph (790.5 kph). **Service ceiling:** (A) 70,000 ft. (21,342 m.); (B/C) 85,000 ft. (25,915 m.); (R/TR) 90,000 ft. (27,440 m.). **Range, max.:** (A) 2200 mi. (3520 km.); (B/C) 3000 mi. (4800 km.); (R/TR) 4000 mi. (6400 km.).

UDALOY: NATO code name for a class of large Soviet DESTROYERS; 11 have been completed since 1978, and more are under construction. Classified by the Soviet navy as Large Anti-Submarine Ships (*Bol'shoy protivolodochnyy korabl'*, BPK), the Udaloys are specialized ANTI-SUBMARINE WARFARE (ASW) vessels, broadly similar to the ASW version of the U.S. SPRUANCE class; like the Spruances, they are intended as long-range escorts for mixed BATTLE GROUPS built around TBILISI- or KIEV-class aircraft carriers, or KIROV-class battle cruisers. Accordingly, their design emphasizes endurance and sustainability, in sharp contrast to earlier Soviet destroyers. The Udaloys are powerful vessels, larger than earlier Soviet ships rated as CRUISers. The hull has a wide waterplane area and high freeboard, making the Udaloys excellent sea boats, and providing a great deal of internal volume for fuel, weapons, and sensors.

The Udaloys carry a very heavy ASW armament, but have only limited ANTI-AIR WARFARE (AAW) and ANTI-SURFACE WARFARE (ASUW) capabilities. ASW weapons comprise 2 quadruple launch canisters for SS-N-14 SILEX ASW missiles alongside the bridge; 2 sets of quadruple 21-in. (533-mm.) tubes for acoustical homing TORPEDOES amidships; and 2 12-barrel RBU-6000 ASW rocket launchers on the aft deckhouse. The Udaloys also have 2 Kamov Ka-27 HELIX ASW helicopters, operated from a large landing deck over the fantail. The Udaloys are the first Soviet destroyers with more than one helicopter; strangely, each helicopter is provided with its own separate hangar.

For defense against air attack, the Udaloys have eight octuple vertical launchers for SS-N-9 short-range surface-to-air missiles; 4 30-mm. ADG6-30 radar-controlled guns amidships; and 2 100-mm. DUAL PURPOSE guns, both forward. For ASUW, the ships have their 100-mm. guns, while the SS-N-14 is believed to have a secondary anti-ship capability. MINES can be laid from rails on the fantail under the flight deck. See also DESTROYERS, SOVIET UNION.

Specifications Length: 531.4 ft. (162 m.). **Beam:** 63.3 ft. (19.3 m.). **Draft:** 20.3 ft. (6.2 m.). **Displacement:** 6500 tons standard/7900 tons full load. **Powerplant:** twin-shaft COGOG: 4 30,000-shp. gas turbines. **Speed:** 35 kt. **Range:** 2500 n.mi. at 32 kt./5000 n.mi. at 20 kt./10,500 n.mi. at 14 kt. **Crew:** 300. **Sensors:** 1 large bow-mounted low-frequency active/passive sonar, 1 variable depth sonar, 2 "Strut Pair" or 1 "Top Plate" surveillance radar, 2 "Eye Bowl" SS-N-14 guidance radars, 2 "Cross Swords" SA-N-9 guidance radars, 1 "Kite Screech" 100-mm. fire control radar, 2 "Bass Tilt" 30-mm. fire control radars, 1 "Palm Frond" navigational radar. **Electronic warfare equipment:** 2 "Bell Shroud" and 2 "Bell Pot" active jamming units, 2 "Salt Pot" IFF units, 2 twin chaff launchers.

UDT: Underwater Demolition Team, official term for U.S. Navy combat diver units. UDTs were first formed in World War II to perform beach RECONNAISSANCE, obstacle DEMOLITION, and sabotage missions as precursors to AMPHIBIOUS ASSAULT operations. Additional postwar duties have included the recovery of spacecraft for NASA.

UDTs are manned entirely by volunteers, who undergo a rigorous 24-week training program of physical conditioning, scuba diving, open-ocean swimming, demolitions, scouting, survival, escape and evasion, land navigation, parachute training, and submarine egress.

The U.S. Navy has four UDTs, each of 15 officers and 111 enlisted men. Two UDTs are controlled by the Naval Special Warfare Group on each coast, based at San Diego, California, and Norfolk, Virginia. UDTs are the primary source of recruits for Navy SEAL teams.

UH-1: U.S. utility helicopter. See HUEY.

UH-60: U.S. utility helicopter. See BLACK-HAWK.

UHF: Ultra-High Frequency, radio signals in the 150–400 MHz frequency band, used mainly for short-range COMMUNICATIONS, tactical DATA LINKS, missile COMMAND-GUIDANCE channels, and surveillance RADARS.

UKADGE: United Kingdom Air Defense Ground Environment, an integrated warning and control network designed to protect Great Britain from air attack. Successor to the manual network established in the Second World War, the UKADGE perimeter extends for 1000 n.mi. (1852 km.) from Ireland in the west, the Faeroes in the north, and France in the East. RADAR stations around the perimeter are linked to three Sector Operations and Control Centers (SOCs), which process target data and control several interceptor and surface-to-air missile units. The three SOCs are linked in turn to the Air Defense Operations Center (ADOC) at High Wycombe, which is responsible for coordinating the actions of the SOCs. UKADGE is also linked to the NATO Air Defense Ground Environment (NADGE) through the ADOC. To ensure the survival of the system, there is a standby ADOC at Bently Priory, and each SOC is capable of independent operations (albeit at reduced efficiency).

ULA: A class of 14 Norwegian diesel-electric attack SUBMARINES (SS) commissioned between 1964 and 1967, essentially identical to the West German TYPE 207. See also SUBMARINES, FEDERAL REPUBLIC OF GERMANY.

UNCONVENTIONAL WARFARE: A generic term for all military and quasi-military operations other than ordinary combat between regular forces. Specific forms include REVOLUTIONARY WAR and its constituents, SUBVERSION and GUERRILLA; commando raids and other special operations; terrorism; and counterterrorism. Nuclear, chemical, and biological warfare are not generally included, but are not CONVENTIONAL WAR either.

UNDERWAY REPLENISHMENT: A naval supply technique whereby individual warships or battle groups are rearmed, refueled, and reprovisioned at sea. Perfected by the U.S. Navy during World War II, underway replenishment requires the establishment of a fleet logistics train (defined in the U.S. Navy as an Underway Replenishment Group), with one or more tankers (U.S. designation AOR), provisions ships (AK), munitions ships (AE), and stores ships (AFS), or their multi-purpose equivalents (AOE). Oil is transferred through hoses extended from booms between two ships sailing parallel in close formation (bow-stern hose connections, an easier but more fragile technique, is used by less skilled navies). Other matériel is transferred on hoist-lines between ships.

Underway replenishment groups generally follow well behind the battle group, to avoid areas of intense enemy activity, and are usually guarded by an escort group of several DESTROYERS or FRIGATES.

Underway replenishment cycles vary: destroyers and frigates must refuel after 3 to 5 days of high-speed cruising, and even nuclear-powered AIRCRAFT CARRIERS have only enough aviation fuel and munitions to support 3–4 days of intensive air operations. When a force commander determines the need for underway replenishment, he establishes a rendezvous between the replenishment group and the battle group, when possible out of range of enemy forces.

In addition to logistic ships, aircraft can also be used for underway replenishment. Aircraft carriers can be supplied by fixed-wing Carrier Onboard Delivery (COD) flights, while smaller ships depend on HELICOPTERS for Vertical Replenishment (VERTREP). In both cases, only high-value, low-volume critical items such as missiles, engine parts, and electronic components can be supplied by air.

The U.S. Navy's Fast Combat Support Ships (AOEs), which carry a mixed cargo of fuel, food, munitions, and general supplies, displace more than 50,000 tons, and have on-board helicopters (and landing decks) for VERTREP. Unlike the traditional flotilla of replenishment ships, AOEs are fast enough to accompany CARRIER BATTLE GROUPS, thereby eliminating the need to withdraw from the combat area to replenish; moreover, AOEs are more economically efficient. But by placing all immediate logistic support into one large ship, a rather vulnerable, high-value target is created; the

loss of an AOE could cripple the battle group's ability to remain on station.

UNIFIED COMMAND: U.S. term for a command with broad and continuing missions, with operational control over forces from two or more military services. At present, U.S. unified commands include Space Command, SPECIAL OPERATIONS COMMAND, Transportation Command, Pacific Command, Atlantic Command, European Command, CENTRAL COMMAND, and Southern Command.

UPHOLDER: A class of four British diesel-electric attack SUBMARINES (SSs), the first of which was commissioned in 1989. Also known as the Type 2400s, the Upholders are intended to replace the aging Porpoise and OBERON classes, the last of which was commissioned in 1967. At that time the Royal Navy had announced its intention to build only nuclear-powered submarines, but the escalating costs of nuclear propulsion, and the fact that the only nuclear-certified shipyard in Britain (Vickers Barrow) would be fully committed throughout the 1990s to build TRIDENT ballistic-missile submarines (SSBNs), led to a change of policy, which was additionally motivated by the discovery by all navies (other than the USN) that small, conventionally powered submarines can be actually superior to SSNs in some roles, notably ultra-quiet operations in shallow and confined waters.

The Upholders have a cylindrical hullform and single-hull configuration derived from late-model British nuclear submarines, with fuel and ballast tanks concentrated outside the pressure hull at the bow and stern. The pressure hull is fabricated of high tensile strength NQ-1 steel, while the streamlined sail amidships is constructed of glass-reinforced plastic (GRP) to reduce weight. Control surfaces are arranged in standard British fashion, with the forward diving planes on the bow, and cruciform rudders and stern planes ahead of the propeller. A snorkel allows diesel operation from shallow submergence, while the two 240-cell chloride lead-acid batteries which provide power to the electric motor when fully submerged provide 50 percent greater endurance than the Oberon's batteries; an exceptionally high speed of 19 kt. can be sustained while snorkeling. All machinery is mounted on resilient sound-isolation rafts to reduce radiated noise. Patrol endurance is 49 days.

Armament consists of six 21-in. (533-mm.) torpedo tubes in the bow, equipped with a positive impulse system which permits weapons to be launched at any depth. At present, the standard load is 18 Mk.24 TIGERFISH or SPEARFISH wire-guided acoustical homing TORPEDOES. HARPOON anti-ship missiles can be substituted for torpedoes, while Marconi Stonefish MINES can replace other weapons on a 2-for-1 basis.

Sonars are equivalent to those of late-model British nuclear submarines, and all sensors are linked a DCC Action Information Organization and torpedo FIRE CONTROL system. Derived from the DCA and DCB systems employed on British SSNs, the DCC consists of two FM 1600E digital computers with three work stations, and can simultaneously track up to 35 contacts and guide 4 torpedoes against 4 separate targets.

If the first four Upholders are successful, the Royal Navy plans to acquire six more. A 3000-ton stretched variant for longer-range deployments may be built from the fifth boat onward. The Upholders are quite expensive; they are, in effect, conventionally powered SSNs, and that may inhibit completion of the entire program. See also SUBMARINES, BRITAIN.

Specifications Length: 230.6 ft. (70.3 m.). Beam: 25 ft. (7.6 m.). Displacement: 2125 tons surfaced/2362 tons submerged. Powerplant: 2 2035-hp. Paxman Valenta 16-cylinder diesel-generators, 1 4070-hp. GEC electric motor. Speed: 12 kt. surfaced/20 kt. submerged. Fuel: 200 tons. Range: 10,000 n.mi. Max. operating depth: 984 ft. (300 m.). Collapse depth: 1640 ft. (500 m.). Crew: 44. Sensors: 1 Type 2040 bow-mounted low frequency active/passive array, 1 Type 2007 flank-mounted conformal array, 1 Type 2019 sonar intercept system, 1 Type 2026 passive towed array, 1 Type 1007 surface-search radar, 1 Porpoise electronic signal monitoring array, 2 periscopes.

URBAN WARFARE: Known in the U.S. Army as Military Operations in Urban Terrain (MOUT), urban warfare defines military operations conducted in terrain dominated by man-made constructions. Most urban areas can be converted into formidable defended zones, with Stalingrad in 1942–43 and Breslau in 1945 the classic examples. Cities can constitute significant barriers to movement per se, insofar as they contain important road and rail junctions in densely built-up central zones. Even small towns can seriously impede an offensive, if they are fortified and contain artillery to interdict adjacent zones; a number can form the nucleus of a DEFENSE-IN-DEPTH.

The defensive value of urban terrain depends on the predominant type and density of construc-

tion. Wood-frame dwellings and industrial sheds have little resistance to fire, and offer concealment only; but brick, stone, and, above all, reinforced-concrete structures are easily converted into outright fortifications. High-rise buildings are particularly hard to reduce, if well built and competently defended.

Urban terrain acts like a sponge, absorbing large numbers of troops for small gains; hence, attackers will normally attempt to bypass or encircle a city, rather than fight through it. But in highly developed settings, urban sprawl and strip development may preclude bypassing, and overextend encircling forces. Soviet doctrine offers two options: the hasty attack ("accelerated assault"), meant to catch the defender by surprise, before defensive preparations can be completed; or the methodical and costly "deliberate" attack. Since hasty attacks can end in a bloody repulse, deliberate attack is the more usual method.

The tactics of artillery or aerial bombardment will not normally suffice to break resistance, and if a surrender is not obtained quickly they will be counterproductive, because road-blocking rubble is an even greater obstacle than intact structures. Tactical nuclear weapons may break the defense, but the attacker must then pass through contaminated terrain, the road and rail net may be irrevocably damaged, and there is of course the risk of retaliation and escalation.

Fields of observation and fire are short; mechanized forces are confined to streets with little scope for maneuver; fighting takes place in three dimensions, at street level, on upper floors and rooftops, and in basements and sewers; and radio communications are impaired by the density of the buildings. In combination, these tactical factors devalue firepower relative to combat in open terrain. Specifically, tanks and mechanized infantry are distinctly subordinate to dismounted INFANTRY, and indirect ARTILLERY fire is generally ineffective.

The defense is based on mutually supporting strongpoints, i.e., fortifying buildings by loopholing walls, blocking street-level access, reinforcing basements, and barricading doors and windows. Entire city blocks may be converted into defensive islands, with the individual buildings connected by openings in basement walls; buildings between two such islands may be demolished to block avenues of attack and create killing zones with clear fields of fire. To defeat step-by-step conquest, the defenders can bring up reinforcements through sewers, utility tunnels, subways, and interbasement connections. This allows the defender to move more securely than attackers forced to advance at street level. Artillery and armor, nearly useless in their traditional roles, can be dug in to provide heavy direct firepower for the defense. The most useful weapons for both sides are often the lightest: SUBMACHINE GUNS, small MORTARS, and GRENADES. SNIPERS, of only nuisance value in open terrain, can actually disrupt attacks by picking off unit commanders and heavy-weapon crews. Armored vehicles are particularly susceptible to overhead attacks with grenades, BAZOOKA-type weapons, or simply Molotov cocktails (gasoline-filled bottles with burning wicks).

The attack is led by infantry on foot, with other arms in support. The attack must generally proceed in two phases: the first to penetrate between defended blocks and isolate them, and the second to reduce each block in detail. Attacking forces are normally organized into combined-arms assault groups, typically a reinforced rifle company (150–200 men) plus 3–5 tanks; several may be required to reduce one block, or even a single, very large building. The only feasible tactics are costly: after suppressive fire is directed against the selected target and any supporting positions, an assault team (30–50 men) attempts to breach doors or walls to force an entry; once inside, they attempt to clear the building, room to room, from ground floor to rooftop. Once the building is secured, the team, reinforced by the rest of the assault group, clears the basement and attempts to break into other buildings, to repeat the process until the entire "island" is isolated from supply and reinforcement. But defending forces may be able to infiltrate behind the attackers, and isolate them in turn. Under such circumstances, an attacker must have a large numerical and firepower advantage to have a reasonable possibility of success. Casualties and matériel losses can be extremely heavy (historically, on the order of 50–75 percent for assault echelons).

ENGINEERS and SAPPERS are the most valuable support forces in urban warfare. On the defense, they prepare fortifications and obstacles, open subterranean routes, lay MINES, and set booby traps. On the attack, engineers form an integral part of every assault force, to breach walls with satchel charges, locate mines, and disarm booby traps. They may have special equipment, such as flamethrowers, invaluable for clearing basements and tunnels.

Although West Germany is more than 65 percent urbanized, NATO has never adopted an explicit policy of defending cities tenaciously, because to defend a city is, in effect, to destroy it, with concomitant casualties among civilians.

USAFE: United States Air Forces in Europe, a major overseas command of the U.S. Air Force. USAFE consists of three numbered air forces (3rd, 16th, and 17th) and a tactical intelligence wing. USAFE is an administrative organization; in war, its forces would be assigned to one of NATO's Allied Tactical Air Forces (ATAFS), under the overall control of Allied Air Forces, Central Europe (AAFCE). See also AIR FORCE, UNITED STATES.

UZI: Ubiquitous Israeli SUBMACHINE GUN, in service with many armed forces, and also much used by bodyguards, the U.S. Secret Service, internal security forces, guerrillas, drug traffickers, and terrorists. Developed in the late 1940s by Major Uziel Gal on the basis of a Czech design, the Uzi is a rugged, compact, and reliable, if somewhat heavy, weapon, manufactured largely of stamped metal parts to reduce costs. The Uzi introduced several revolutionary features which have since been widely copied. The magazine housing is built into the pistol grip, ensuring a reliable fit to the receiver mechanism (a major problem in other submachine guns) and making reloading much easier in the dark. Muzzle rise in automatic fire is minimized by the wraparound bolt, machined with a recess to accept the barrel, which extends into the receiver; that also reduces overall length while providing adequate barrel length for accuracy. Finally, a rear safety "trigger" in the pistol grip must be depressed fully to engage the trigger itself, thereby reducing accidents.

Early versions had a fixed wooden stock, but that was soon replaced by a folding metal stock which reduces length by some 9 in. (22.8 cm.) when closed. The Uzi, like most submachine guns, fires 9-mm. Parabellum ammunition, is designed for close combat at shorter ranges, and normally has only simple open sights. The Uzi operates on the blowback principle, with selectable semi- or full automatic fire. Spring-loaded box magazines are available in 25-, 32- and 40-round versions.

Still the most accurate weapon of its type, the Uzi has proven itself in thousands of firefights around the world, and has come to symbolize both the Israeli army and the Israeli arms industry. It is still in large-scale production by Israel Military Industries and Fabrique Nationale (FN) of Belgium.

Specifications Length OA: 26.37 in. (670 mm.). **Length, barrel:** 10.24 in. (260 mm.). **Weight, loaded:** 7.7 lb. (3.5 kg.). **Muzzle velocity:** 390 m./sec. **Cyclic rate:** 600 rds./min. **Effective range:** 200 m. **Operators:** Isr, Bel, Iran, Neth, FRG, many African and Latin American armies.

UZUSHIO: A class of seven Japanese diesel-electric attack SUBMARINES (SSs) commissioned between 1970 and 1977. Derived from the U.S. BARBEL class, the Uzushios have a teardrop-shaped hull, but unlike the Barbels, the Uzushios have a double-hull configuration, with fuel and ballast tanks installed between the pressure hull and outer casing. Control surfaces resemble those of U.S. nuclear submarines, with "fairwater" diving planes mounted on a tall, streamlined sail, and cruciform rudders and stern planes ahead of the propeller. The Uzushios served as the basis for the slightly larger, and greatly improved, YUUSHIO class.

The Uzushios are armed with six 21-in. (533-mm.) torpedo tubes amidships, and carry up to 18 homing TORPEDOES, notably the old Mk.37 (Japan refused the more modern Mk.48 unless it could build it under license, while the U.S. refused to sell the most advanced ADCAP version). See also SUBMARINES, JAPAN.

Specifications Length: 236.2 ft. (72 m.). **Beam:** 29.5 ft. (9 m.). **Displacement:** 1850 tons surfaced/3600 tons submerged. **Powerplant:** single-shaft diesel-electric: 2 1750-hp. Kawasaki-MAN diesel-generators, 1 7200-hp. electric motor. **Speed:** 12 kt. surfaced/20 kt. submerged. **Max. operating depth:** 650 ft. (198 m.). **Collapse depth:** 910 ft. (277 m.) **Crew:** 80. **Sensors:** 1 ZQQ-4 bow-mounted low-frequency passive spherical sonar, 1 SQS-36J active attack sonar, 1 ZPS-6 surveillance radar, 1 electronic signal monitoring array, 2 periscopes.

V

V-22: U.S. tilt-rotor V/STOL aircraft. See OS-PREY.

VAB: *Vehicule de l'Avant Blindee* ("Armored Front-Line Vehicle"), Renault 4 × 4 wheeled AR-MORED PERSONNEL CARRIER (APC) introduced in 1976, and now serving with French, Moroccan, and Lebanese forces.

The VAB's all-welded steel hull is divided into a crew compartment in front, an engine compartment in the center, and a troop compartment in the rear. Maximum armor thickness is only 12 mm., adequate for protection only against small arms and shell fragments.

The driver and the commander both have large armored glass windows equipped with steel shutters; active/passive INFRARED night vision devices are provided on many VABs. Up to ten fully equipped infantrymen can be accommodated on bench seats in the troop compartment, which has three armored windows on each side for use as firing ports for rifles and machine guns. Access is via 2 large rear doors, 2 side doors, and 3 roof hatches. VABs in French service are equipped with a COLLECTIVE FILTRATION system for NBC defense. Fully amphibious with minor preparations, French VABs are propelled through the water by tire motion only; most export models are equipped with a water jet.

Tactically, the VAB can only be employed as a "battle taxi" for transport to the battle area; in most cases, troops would fight dismounted, though firing ports are provided as insurance against surprise engagements.

The standard troop carrier version (VTT) is armed only with a pintle-mounted 7.62-mm. machine gun over the commander's hatch. Variants include the VAB ECH armored repair vehicle; the VCAC Mephisto TANK DESTROYER with HOT anti-tank guided missiles; the VAB PC armored command vehicle (ACV); the VAB *Sanitaire* armored ambulance; and VTM tractor for 120-mm. mortars. The VAB VIB, used by the French air force for base security, is armed with a turret-mounted 20-mm. cannon and a coaxial 7.62-mm. machine gun. A 6 × 6 APC and the VBC-90 armored car are derivatives intended for export.

Specifications **Length:** 19.6 ft. (5.98 m.). **Width:** 6.75 ft. (2.49 m.). **Height:** 8.16 ft. (2.06 m.) **Weight, combat:** 13 tons. **Powerplant:** 235-hp. MAN 6-cylinder diesel. **Speed, road:** 57 mph (92 kph). **Range, max.:** 621 mi. (1000 km.).

VALIANT: A class of five British nuclear-powered attack SUBMARINES (SSNs) commissioned between 1967 and 1970 (sometimes known as the Churchill class). The first British nuclear submarines with a completely indigenous propulsion system, the Valiants have a modified teardrop hull-form and a single-hull configuration with ballast tanks located at the extreme bow and stern. A tall, streamlined sail (conning tower) is located amidships, while the control surfaces are arranged in the standard British fashion: retractable bow diving planes, and cruciform rudders and stern planes mounted ahead of the propeller. A 112-cell battery can provide emergency power.

Armament consists of six 21-in. (533-mm.) tubes,

for a variety of unguided and homing TORPEDOES, including the Mk.24 TIGERFISH and its replacement, SPEARFISH. In the 1980s the Valiants were also modified to launch HARPOON anti-ship missiles through the torpedo tubes. At present, the basic load consists of 26 torpedoes and 8 Harpoons; Stonefish mines can replace other weapons on a 2-for-1 basis. All sensors are linked to a DCA Action Information Organization and FIRE CONTROL unit in the attack center.

Like their contemporaries of the U.S. SKIPJACK class, the Valiants have relatively little acoustical insulation. Unlike the Skipjacks, however, the Valiants have their machinery mounted on sound-isolation "rafts" inside the pressure hull. But these rafts, which float on spring mounts, must be locked rigidly in place at high speed, thereby negating their effect. In 1982, HMS *Conqueror* of this class sank the Argentine CRUISER *General Belgrano* with a salvo of old, guided Mk.8 steam-powered torpedoes. See also SUBMARINES, BRITAIN.

Specifications **Length:** 285 ft. (86.9 m.). **Beam:** 33.25 ft. (10.1 m.). **Displacement:** 4200 tons surfaced/4900 tons submerged. **Powerplant:** PWR-1 pressurized-water reactor, 2 geared turbines, 15,000 shp. **Speed:** 20 kt. surfaced/29 kt. submerged. **Max. operating depth:** 985 ft. (300 m.). **Collapse depth:** 1640 ft. (500 m.). **Crew:** 126. **Sensors:** 1 Type 2001 low-frequency active/passive sonar (being replaced a Type 2020 electronically scanned array), 1 Type 2007 passive flank-mounted conformal array, 1 Type 197 intercept array, 1 Type 2026 clip-on passive towed array, 1 Type 1007 surface-search radar, 1 electronic signal monitoring array, 2 periscopes.

VAN SPEIJK: A class of six Dutch FRIGATES based on the British LEANDER class.

VARIABLE DEPTH SONAR (VDS): An active/passive SONAR whose array can be lowered from surface ships to selected depths below temperature and salinity gradients opaque to sonar at shallow depths. A VDS consists of a transducer and a receiver housed in a torpedo-shaped capsule ("paravane" or "fish"), which can be lowered by cable over the stern of the ship. Control surfaces allow towing at any selected depth down to the cable limit. Electrical power and data transmission lines from the VDS to the ship's fire control system are incorporated into the cable. Dipping sonar is a form of VDS used from HELICOPTERS: the helicopter hovers at low altitude to lower the sonar by cable to the required depth. See also ANTI-SUBMARINE WARFARE; THERMOCLINE.

VASTERGOTLAND: A class of four Swedish diesel-electric attack SUBMARINES, the first of which was commissioned in 1987. Replacements for the aging DRAKEN class, the Vastergotlands are incremental developments of the NACKEN class, with greater underwater endurance and better electronics.

The Vastergotlands have a modified teardrop hullform in a single-hull configuration, with only 7 percent reserve buoyancy—more than most, they are true submarines rather than submersible boats. Control surfaces follow standard Swedish practice, with "fairwater" diving planes mounted on the streamlined sail, and X-configuration stern planes acting differentially as both diving planes and rudders under automatic control. A retractable snorkel allows diesel operation from shallow submergence, while three 84-cell Tudor steel-lead storage batteries provide underwater power.

Armament comprises 9 torpedo tubes mounted in the bow: six of 21-in. (533-mm.) diameter, and 3 of 15.75-in. (400-mm.) diameter. The large tubes can launch heavy, wire-guided anti-ship/anti-submarine TORPEDOES, while the smaller tubes are for light anti-submarine torpedoes. The basic load consists of 12 Type 61 heavy and 6 Type 42 light torpedoes; 2 MINES can be substituted for each heavy torpedo. In the future these boats may be retrofitted with RBS-17 ANTI-SHIP MISSILES fired from vertical launch tubes in the rear half of the sail. All sensors are linked to an Ericcson IPS-17 FIRE CONTROL computer. See also SUBMARINES, SWEDEN.

Specifications **Length:** 159.1 ft. (48.5 m.). **Beam:** 20 ft. (6.1 m.). **Displacement:** 1070 tons surfaced/1140 tons submerged. **Powerplant:** single-shaft diesel-electric: 2 1080-hp. Hedemora V12A/15U6 diesels, 2 Jeumont-Schneider 760-kilowatt generators, 1 1800-hp. ASEA electric motor. **Speed:** 12 kt. surfaced/20 kt. submerged. **Max. operating depth:** 984 ft. (300 m.). **Collapse depth:** 1640 ft. (500 m.). **Crew:** 17. **Sensors:** 1 KAE CSU-83 low-frequency active/passive sonar, 1 PEAB/Terma surface-search radar, 1 Argo electronic signal monitoring array, 2 periscopes.

VDS: See VARIABLE DEPTH SONAR.

VDV: *Vozdushno-desantnyye voyska*, Airborne Troops. See AIRBORNE FORCES, SOVIET UNION.

VEINTICINCO DE MAYO: Argentine AIRCRAFT CARRIER, originally the British light carrier HMS *Venerable*, laid down in 1942 and commissioned in 1945. Purchased by the Netherlands in 1948 and recommissioned as the *Karel Doorman*, it was extensively rebuilt between 1955 and 1958, with an angled flight deck, a steam catapult, a mirror landing system, a new anti-aircraft battery, and modern radar; new boilers were installed during a 1967 overhaul. The ship was purchased by Argentina in 1968, and superficially overhauled in 1969. In 1980 the flight deck was enlarged to provide additional deck parking.

The flight deck is 695 ft. (211.9 m.) long and 138.5 ft. (42.2 m.) wide, with a 540-ft. (164.6-m.) angled landing deck cantilevered to port, a steam catapult forward, and two deck-edge elevators to the hangar deck. Defensive armament consists of nine 40-mm. BOFORS GUNS. The *Veinticinco de Mayo* has a British CAAIS data display system compatible with the ADAWS 4 DATA LINKS of the two SHEFFIELD-class destroyers in the Argentine fleet.

The original Argentine air group consisted of 6 to 9 A-4Q SKYHAWK attack aircraft, 5 S-2A TRACKER anti-submarine warfare (ASW) aircraft, and 4 SH-3D SEA KING ASW helicopters. In 1982 the replacement of the Skyhawks with Dassault Super ETENDARD attack aircraft armed with EXOCET anti-ship missiles had begun, but only 2 were ready in time for the Falklands War. The air group now consists of 11 Super Etendards, 5 Trackers, and 4 Sea Kings.

The *Veinticinco de Mayo* has had persistent boiler problems; this had a significant impact during the Falklands War, because the ship could not achieve the speed required to launch its Skyhawks, and spent most of the war in Argentine home waters. Despite its age and mechanical infirmity, the carrier will probably continue in service during the 1990s. Since 1987, the *Veinticinco de Mayo* has been out of commission for a major overhaul which may include the replacement of the steam plant with diesel engines. See also AIRCRAFT CARRIERS, BRITAIN.

Specifications Length: 693.25 ft. (211.3 m.). **Beam:** 80 ft. (24.4 m.). **Draft:** 25 ft. (7.6 m.). **Displacement:** 15,892 tons standard/19,896 tons full load. **Powerplant:** twin-shaft steam: 4 oil-fired boilers, 2 sets of geared turbines, 40,000 shp. **Fuel:** 3200 tons. **Speed:** 24.5 kt. (18 kt. sustained). **Range:** 12,000 n.mi. at 14 kt. **Crew:** 1509. **Sensors:** 2 LW-01 air-search radars, 1 SGR height-finding radar, 1 ZW-01 surface-search radar, 1 DA-02 fire control radar.

VELA: U.S. military SURVEILLANCE satellites, designed to detect the detonation of NUCLEAR WEAPONS on the earth's surface, in the atmosphere,

and out to 100 million mi. (167 million km.) in space. Vela had visible-light and INFRARED sensors to detect the characteristic flash of nuclear explosions; they could not be used to detect underground detonations. The Vela program was initiated in 1959, originally to investigate the use of satellites to verify a nuclear test ban. The first two satellites were launched in October 1963, and placed into 60,000-mi. (96,000-km.) circular orbits 180° apart to provide coverage of the entire planet. The initial Velas could detect only surface or space detonation; two additional satellites of the same type were launched in 1965. They were followed in 1969 by two improved Velas which could also detect atmospheric bursts. The last of the series were Velas 11 and 12, launched in 1969–70.

After 1970 the nuclear-detection surveillance mission was gradually taken over by the Integrated Missile Early Warning System (IMEWS) satellites, later renamed DEFENSE SUPPORT PROGRAM (DSP) satellites, which have infrared sensors to detect enemy ballistic-missile launches as well as nuclear detonations. See also SATELLITES, MILITARY; SPACE, MILITARY USES OF.

VERTICAL ENVELOPMENT: An operational technique whereby enemy forces are to be cut off and surrounded by the insertion of AIRBORNE or AIR-MOBILE forces into their rear, while ground forces attack their front. The concept was devised by Col. J. F. C. Fuller in his "Plan 1919"; it was actually pioneered by German airborne forces in 1940–41, and attempted on a much larger scale by Allied forces in Sicily (1943) and in Holland (1944). In practice, no vertical envelopment has ever been achieved, and given the development of highly effective AIR DEFENSES, it is increasingly unlikely that it ever will be. The phrase is now used to describe almost any air insertion of combat troops. See also AIR ASSAULT.

VERTICAL LAUNCH SYSTEM (VLS): A type of missile launcher, primarily for warships, in which the missiles are stowed and launched vertically through hatches, rather than from a trainable rail, as in traditional launchers. This most useful simplification is made possible by advances in missile design: after ascending high enough to clear the ship's superstructure, VLS-suitable missiles are programmed or commanded to pitch over towards their target. In U.S. vertical launch systems, typified by the Mk.41 VLS of the TICONDEROGA cruisers and BURKE destroyers, the launcher consists of several 8-cell modules. Each cell is an individual storage and launch canister with its own hatch. In the Soviet VLS for the SA-N-6 GRUMBLE surface-to-air missile, the missiles are housed in a revolving magazine, which rotates each missile in turn under the single launch port.

Though the U.S. version is the simplest, both types are much simpler and thus more reliable than rail-type launchers, with their hoists, elevation, and traverse gear. The U.S. Mk.41 VLS offers superior tactical flexibility compared to the Soviet system, because it can accommodate a mixed load of missiles for instant launch; e.g., STANDARD surface-to-air missiles, Vertical Launch ASROC anti-submarine missiles, HARPOON anti-ship missiles, and TOMAHAWK cruise missiles. In addition, the Mk.41 can fire at a much higher rate (one missile per second). The Soviet VLS, on the other hand, is more economical of deck space. The success of both U.S. and Soviet VLS designs suggests that future warships will incorporate them.

VERTREP: Vertical Replenishment, U.S. Navy term for the UNDERWAY REPLENISHMENT of warships by helicopters from supply ships or shore depots. VERTREP is used mainly for the delivery of lightweight, low-volume, critical items such as replacement electronic components, crew mail, and new videotapes.

VHF: Very High Frequency, radio signals in the 30–150 MHz frequency band, used mainly for short-range tactical COMMUNICATIONS, IFF, TACAN, and long-range surveillance RADARS.

VHSIC: Very High-Speed Integrated Circuits, a U.S. government program for a new electronics technology. VHSIC is meant to allow picosécond (1 trillionth of a second) electronic-switching delays, vs. the nanosecond (1 billionth of a second) delays already achieved. Potential military applications include BATTLE MANAGEMENT; command, control, and communications (c^3); weapon guidance; and artificial intelligence (AI) systems. The development of VHSIC has both been accelerated by government funding and impeded by government micromanagement: key producers have refused to participate in the program. See also VLSI.

VICKERS MBT: British-designed MAIN BATTLE TANK (MBT) intended for export, now in service with the armies of India, Kenya, Kuwait, and Nigeria. Designed in the late 1950s, the Vickers tank has the same 105-mm. L7A1 gun as the CENTURION MBT, and the engine, transmission, steering, and (rudimentary) fire controls of the CHIEFTAIN MBT. The first prototype was completed in

1963, and production of the initial Mk.1 began in 1964.

Armor protection is limited to reduce weight: the frontal glacis and turret face have only 80 mm. of homogeneous steel plate, while the sides, rear, and top of the tank have armor varying between 60 and 25 mm. Superficially similar to the Centurion, this tank has an all-welded steel hull divided into a driver's compartment forward, a fighting compartment in the middle, and an engine compartment in the rear. The four-man crew consists of a driver, a commander, a gunner, and a loader. The driver is provided with a wide-angle periscope which can be replaced by a passive INFRARED (IR) periscope for night driving. Twenty-five rounds of main gun ammunition are stowed on the driver's left.

The fighting compartment is surmounted by an all-welded turret, in which the commander sits on the right side, above and behind the gunner, with the loader opposite the gunner on the left. The commander has a single-piece hatch with a $10 \times$ optical sight and six periscopes providing 360° observation. The gunner has a sighting telescope with a ballistic recticle, while the loader's hatch has a single observation periscope.

The 105-mm. L7A1 rifled gun can fire APDS, APFSDS, HEAT, and HESH, and is effective against most other tanks out to a range of 2000 m. Electrically controlled with manual backup, the gun has elevation limits of $+20°$ and $-7°$, and is fully stabilized to allow firing on the move, although accuracy is then seriously degraded given the rudimentary fire controls, which consist of a 12.7-mm. ranging machine gun ballistically matched to the main gun. The gunner fires tracer bullets at the target; when the machine gun hits, the main gun is fired. This method is accurate out to some 1800 m. (if visibility is good). Nineteen rounds of main gun ammunition are stored in the fighting compartment, for a total of 44 rounds.

Secondary armament consists of a coaxial 7.62-mm. MACHINE GUN mounted on the left side of the 105-mm. gun and a pintle-mounted 7.62-mm. machine gun by the commander's hatch. Six smoke grenade launchers are attached to each side of the turret. As with other British tanks of its generation, the Vickers Mk.1 has a low power-to-weight ratio and mediocre off-road mobility.

The Mk.1 was adopted by the Indian army in 1964. Coproduction began in 1965, serving as the basis for the indigenously produced VIJAYANTA

MBT. Kuwait operates a total of 70 Mk.1s, delivered between 1970 and 1972.

In 1975, Vickers introduced the Mk.3, with a redesigned turret face for improved ballistic protection, a new engine, and better fire controls. The commander's position is provided with a Pilkington PE Condor day/night sight, while the gunner has a Vickers L23 periscopic sight incorporating a LASER rangefinder, which is linked to a Marconi SFC 600 FIRE CONTROL computer that allows the gun to be accurately fired on the move. Ammunition stowage has been increased to 50 rounds. Seventy-six Mk.3s were delivered to Kenya between 1977 and 1982, and 72 were delivered to Nigeria from 1981.

In addition to the basic models, Vickers developed a number of specialized vehicles based on the MBT chassis, including an armored bridgelayer (AVLB), an armored recovery vehicle, a 155-mm. self-propelled gun mount, and an anti-aircraft tank.

Specifications Length: 25.33 ft. (7.72 m.). **Width:** 10.3 ft. (3.15 m.). **Height:** 8.6 ft. (2.62 m.). **Weight, combat:** 38 tons. **Powerplant:** (Mk.1) 650-hp. Leyland 6-cylinder water-cooled diesel or 800-hp. Rolls Royce CV12TCA 12-cylinder turbocharged diesel; (Mk.3) 720-hp. General Motors 12V-71T turbocharged diesel. **Fuel:** 250 gal. (1000 lit.). **Speed, road:** 30 mph (49 kph). **Range, max.:** (Mk.1) 276 mi. (480 km.); (Mk.3) 360 mi. (600 km.).

VICTOR: NATO code name for a class of 41 Soviet nuclear-powered attack SUBMARINES (SSNs), the first of which was completed in 1968. The Victors have been built in three distinct subclasses. Sixteen Victor Is, completed between 1968 and 1972, were designed primarily to detect and destroy U.S. ballistic-missile submarines (SSBNs) near U.S. ports. Along with the similar CHARLIE-class guided-missile submarines (SSGNs), the Victor Is were the first Soviet submarines to feature a teardrop-shaped hull optimized for underwater speed and maneuverability. The Victor I has the standard Soviet double-hull configuration, with ballast tanks installed between the pressure hull and the outer casing. It has a small, streamlined sail (conning tower) amidships, retractable bow diving planes, and cruciform stern planes and rudders.

Two small auxiliary thrusters facilitate maneuvering, and also allow low-speed, silent propulsion. Although quieter than earlier Soviet submarines, the Victor Is are relatively noisy compared to the contemporary U.S. PERMIT-class SSNs, partly because of poor sound isolation of the machinery, and partly because of the large number of free-flooding holes in the outer hull.

Armed with six 21-in. (533-mm.) torpedo tubes mounted in the lower half of the bow, the Victor I can carry a variety of free-running and acoustical homing TORPEDOES, plus the SS-N-15 STARFISH anti-submarine missile. The basic load is normally 8 21-in. anti-ship torpedoes, 10 15.75-in. (400-mm.) anti-submarine torpedoes, and 2 Starfish; alternatively, up to 36 mines can be carried.

The seven Victor IIs completed between 1972 and 1978 were designed primarily for the defense of Soviet SSBNs operating in "sanctuaries" close to Soviet home waters; their primary targets are U.S. and NATO SSNs hunting Soviet SSBNs, rather than Western SSBNs. The Victor II has a 16-foot (4.88-m.) section added to the hull forward of the sail to accommodate additional weapons and fire-control equipment. The most important change to the Victor II was the addition of two 26-in. (650-mm.) torpedo tubes for the SS-N-16 STALLION ASW missile, an enlarged version of Starfish. Sensors are similar to those of the Victor I, with the addition of a VLF communications buoy and an ELF trailing antenna.

The Victor IIIs, still in production with 22 completed since 1978, have improved sensors and acoustical silencing; armament is identical to the Victor II's. The most visible change is the addition of a large, teardrop-shaped pod on top of the upper rudder; this is believed to house a passive TOWED ARRAY SONAR, to enhance the Victor III's ability to detect quiet U.S. submarines at long range. The Victor III has two counterrotating four-bladed propellers which reduce cavitation noise; U.S. analysts claim it is equal in silencing to the U.S. STURGEON class. Crew size has been increased, probably to man the additional sensors and fire control equipment. The Victor III is now being supplemented by the more refined AKULA class. See also SUBMARINES, SOVIET UNION.

Specifications Length: (I) 308.1 ft. (93.9 m.); (II) 328 ft. (100 m.); (III) 341.16 ft. (104 m.). **Beam:** 32.83 ft. (10 m.). **Displacement:** (I) 4300 tons surfaced/5300 tons submerged; (II) 4700 tons surfaced/5700 tons submerged; (III) 5000 tons surfaced/6300 tons submerged. **Powerplant:** single-shaft nuclear: 1 pressurized-water reactor, 2 geared turbines, 30,000 shp. **Speed:** 20 kt. surfaced/32 kt. submerged. **Max. operating depth:** 1315 ft. (400 m.). **Collapse depth:** 1970 ft. (600 m.).

Crew: (I/II) 80; (III) 85. **Sensors:** 1 large bow-mounted low-frequency active/passive sonar, 1 medium-frequency active fire control sonar, 1 "Snoop Tray" surveillance radar, "Brick Split" and "Brick Pulp" electronic signal monitoring arrays, 1 "Park Lamp" radio direction-finder (D/F), 2 periscopes.

VIGGEN: The Swedish Saab 37 multi-role fighter/attack/reconnaissance aircraft. Designed in the late 1960s to replace the Saab 35 DRAKEN, the Viggen had to satisfy Swedish air force requirements for battlefield INTERDICTION, maritime attack, AIR SUPERIORITY, anti-bomber interception, and tactical RECONNAISSANCE capabilities, combined with short takeoff and landing (STOL) performance sufficient to operate from highways and dispersal airfields. The resulting design had many innovative features, including a double-delta planform with large canards ahead of the wing, an afterburning turbofan engine with thrust reversers, and tandem-bogie main landing gear with powerful brakes stressed for no-flare landings. The first prototype flew in 1967, and the first AJ.37 attack variant entered service in 1971.

The AJ.37 has an area-ruled ("coke-bottle") fuselage with an LM Erickson multi-mode search and attack RADAR in the nose. Other AVIONICS include a DOPPLER navigation system, a radio altimeter, and an air data computer, all feeding information to a Saab CK-37 miniaturized digital computer, which displays the processed data on a Marconi Avionics Head-Up Display (HUD). The pilot sits in a conventional cockpit immediately behind the radome; surprisingly, for a 1970s fighter, rearward visibility is obscured by a prominent fuselage spine. Behind the cockpit, the fuselage houses additional avionics, fuel cells, and the engine, which is fed by two cheek inlets.

The most distinctive characteristic of the Viggen is its double-delta planform. The low-mounted delta main wing extends from midfuselage to the extreme tail, with large Fowler flaps, and elevons for both for roll and pitch control. The main landing gear are housed in the inboard wing sections, opening outward to provide a wide, stable track. The canard foreplanes, mounted higher than the main wing to avoid interference drag and generate lift-enhancing vortexes, have a square-tipped, cropped delta planform with a span of approximately 20 ft. (6.1 m.). The canards are fixed, but have slotted trailing-edge flaps to enhance low-speed performance. Canards and wings have a combined area of 495 sq. ft. (46 m.²), for a typical

wing load of 70–81 lb. per sq. ft. (369 kg./m.²), resulting in excellent sustained-turning performance. The AJ.37 is powered by a Volvo Flygmotor RM8A afterburning turbofan (a license-built derivative of the Pratt and Whitney JT-8D-22); the exhaust nozzle is fitted with a thrust reverser, allowing the Viggen to land in a space of less than 1650 ft. (503 m.), even on icy surfaces (takeoff distance is even shorter).

The AJ.37 has no fixed armament, but can carry up to 13,200 lb. (6000 kg.) of ordnance on two wing, two inlet, and one centerline pylon, including MAVERICK and other air-to-surface missiles, Bofors 135-mm. rocket pods, free-fall bombs, a 30-mm. ADEN gun pod, and FALCON air-to-air missiles. Some 120 AJ-37s were produced between 1967 and 1974.

From 1975 Saab produced the SH-37 RECONNAISSANCE variant. Optimized for maritime surveillance, with a nose-mounted surface surveillance radar, a pod-mounted long-range search radar, an FFV Red Baron INFRARED sensor pod, and various ELECTRONIC COUNTERMEASURES (ECM) pods, and a pair of SIDEWINDER air-to-air missiles for self-defense, its external dimensions and performance are similar to those of the AJ.37.

The SF.37, introduced in 1977, is a specialized photo-reconnaissance aircraft equipped with a total of nine cameras for forward, oblique, vertical, and panoramic views. Other sensors include an infrared line scanner and a photoflash dispenser for night photography. Performance is similar to the AJ.37. Some 30 reconnaissance version have been delivered to date.

Introduced in 1978, the last Viggen variant is the JA.37 fighter/interceptor, an extensively redesigned aircraft with new avionics and armament and a new engine. Its avionics include an LM Erickson PS-46 pulse-Doppler multi-mode air search and tracking radar, and a Singer Kearfott INERTIAL NAVIGATION platform linked to a Singer central digital computer. The cockpit is totally redesigned, with three multi-functional video displays in place of instrument dials, a U.S. Smith Industries HUD, and a Hands-on-Throttle-and-Stick (HOTAS) configuration for all essential controls.

The JA.37 is armed with a fixed belly-mounted 30-mm. Oerlikon KCA cannon, and can carry either four SKYFLASH semi-active radar homing (SARH) or four Sidewinder air-to-air missiles. Its more powerful engine enhances both sustained turn performance and rate of climb. A total of 149

JA.37s have been delivered to the Swedish air force. Some 30 two-seat trainers have also been produced. The Viggen will probably continue in service beyond the year 2000; the new Saab 39 GRIPEN is now in development.

Specifications Length: (AJ) 50.67 ft. (15.45 m.); (JA) 51.1 ft. (15.58 m.). Span: 34.75 ft. (10.6 m.). Powerplant: (AJ) RM8A afterburning turbofan, 26,015 lb (11,800 kg.) of thrust; (JA) RM8B afterburning turbofan, 28,153 lb. (12770 kg.) of thrust. Fuel: 9750 lb. (4432 kg.). Weight, empty: 26,015 lb. (11,800 kg.). Weight, max. takeoff: 45,-085 lb. (20,450 kg.). Speed, max.: Mach 2 (1320 mph/2112 ,ph) at 40,000 ft./Mach 1.1 (835 mph/1335 kph) at sea level. Initial climb: 40,000 ft./min. (12,200 m./min.). Service ceiling: 60,000 ft. (18,300 m.). Combat radius: (hi-lo-hi) 620 mi. (1000 km.)/(lo-lo-lo) 310 mi. (500 km.).

VIJAYANTA: Indian-produced derivative of the VICKERS MBT. After licensed production began in 1965, progressive modifications made the Vijayanta significantly different. Due to the use of Indian materials and production techniques, combat weight increased from 37 to 41 tons, with some gain in protection.

From 1981, the Vijayantas were retrofitted with Marconi digital FIRE CONTROLS and British-made LASER rangefinders, bringing the tanks close to Vickers Mk.3 standard. From 1983, the Vijayantas were reengined with Leyland L60 diesels. India has also developed a number of specialized vehicles based on the Vijayanta chassis, including a self-propelled version of the Soviet 130-mm. field gun. It is believed that India has produced between 1200 and 1500 Vijayantas of all types to date.

VIKING: Lockheed S-3 twin-engine carrier-based ANTI-SUBMARINE WARFARE (ASW) patrol aircraft, in service with the U.S. Navy. Developed from 1967 as a replacement for the piston-engine S-2 TRACKER, the prototype S-3 first flew in 1972, and the initial S-3A version entered service in 1974. A total of 179 were produced through 1978 to equip 13 10-aircraft squadrons (1 for each active aircraft carrier); some 160 were upgraded to S-3B standard in the late 1980s. A tanker variant (KS-3A) was not placed in production, but four preproduction aircraft were converted to US-3A transports for Carrier On-board Delivery (COD). Sixteen S-3Bs are to be converted to ES-3Bs for electronic intelligence (ELINT) collection, replacing the aging EA-3B SKYWARRIOR.

Of conventional monocoque construction, the S-3B has a capacious, slab-sided fuselage with a large nose radome housing a high-resolution APS-116 surface-search RADAR. The cockpit has two-by-two seating, with the pilot and copilot up front, and the ASW tactical coordinator (TACCO) and weapon system operator (WSO) in the rear. AVIONICS include an INERTIAL NAVIGATION unit, a DOPPLER navigation radar, a TACAN receiver, a UHF direction-finder (D/F), and an automatic carrier landing system (ACLS).

ASW equipment includes a forward-looking infrared (FLIR) sensor in a retractable ventral turret, wingtip-mounted ELECTRONIC SIGNAL MONITORING arrays, an ASQ-81 MAGNETIC ANOMALY DETECTOR in a retractable tail "stinger," and 60 SONOBUOYS in a ventral fuselage dispenser. All sensors are linked to an IBM digital computer tactical display system. A midfuselage weapons bay can accommodate four MK.46 or MK.50 lightweight homing TORPEDOES, or up to 2400 lb. (1090 kg.) of bombs, DEPTH CHARGES, or flares. A retractable AERIAL REFUELING probe is housed over the cockpit.

The moderately swept, shoulder-mounted wing, with single-slotted Fowler flaps and outboard ailerons, folds up for more compact deck parking, while the tall vertical stabilizer folds to the side to clear the 25-ft. (7.62-m.) ceilings of carrier hangar decks. There are two underwing pylons for additional torpedoes, AGM-84 HARPOON anti-ship missiles, or up to 3000 lb. (1364 kg.) of bombs, depth charges, rocket pods, MINES, or auxiliary fuel tanks. See also AIRCRAFT CARRIERS, UNITED STATES.

Specifications Length: 49.45 ft. (15.06 m.). Span: 68.67 ft. (20.93 m.). Powerplant: 2 General Electric TF34-GE-400A high-bypass turbofans, 9275 lb. (4207 kg.) of thrust each. Weight, empty: 26,650 lb. (12,088 kg.) Weight, max. takeoff: 52,-539 lb. (23,831 kg.). Speed, max.: 518 mph (834 kph) at sea level. Speed, cruising: 426 mph (686 kph). Speed, loiter: 184 mph (296 kph). Range, Patrol: 2300 mi. (3700 km.). Range, max.: 3450 mi. (5761 ,km.).

VIKRANT: Indian AIRCRAFT CARRIER, originally laid down in 1943 as the British light carrier HMS *Hercules*. Purchased incomplete by the Indian government 1959, it finally entered service in 1961. As completed, the ship had an enclosed bow and an angled landing deck cantilevered to port. The flight deck, 682 ft. (208 m.) long and 128 ft. (39 m.) wide, is equipped with a steam catapult, arresting gear, a mirror landing system, and two deck-edge elevators for access to the hangar deck.

Defensive armament consists of seven 40-mm. BO-FORS GUNS.

The air group was originally built around 6–9 Hawker Sea Hawk ATTACK AIRCRAFT of late 1940s design, plus up to 11 Breguet ALIZE anti-submarine warfare (ASW) aircraft and several helicopters. In 1983 India purchased 8 Sea HARRIER V/STOL fighters to replace the ancient Sea Hawks; a second batch of 8 was ordered in 1986. In 1982–83, *Vikrant* underwent an overhaul to make the ship compatible with the Sea Harrier, but the catapult and arresting gear have been retained to permit continued operation of the Alize. A ski-jump bow ramp for the Harriers was installed in 1987–88. By that time, India had acquired another British light fleet carrier, the VIRAAT (ex-HMS *Hermes*), already equipped for Sea Harrier operations.

During the 1965 and 1971 Indo-Pakistani Wars, the *Vikrant*'s air group dominated the Bay of Bengal, and the new Sea Harrier air group amounts to a substantial improvement in its strike and air superiority capabilities. Materially, *Vikrant* is said to be in remarkably good condition, and should continue to serve well until an India-designed carrier is built in the late 1990s. See also AIRCRAFT CARRIERS, INDIA.

Specifications **Length:** 700 ft. (213.4 m.). **Beam:** 80 ft. (24.4 m.). **Draft:** 24 ft. (7.3 m.). **Displacement:** 15,700 tons standard/19,500 tons full load. **Powerplant:** twin-shaft steam: 4 oil-fired boilers, two sets of geared turbines, 40,000 shp. **Fuel:** 3200 tons. **Speed:** 24 kt. **Range:** 12,000 n.mi. at 14 kt./6200 n.mi. at 23 kt. **Crew:** 1340. **Sensors:** 1 LW-05 air-search radar, 1 ZW-06 surface-search radar, LW-10 and LW-11 tactical search radars, 1 Type 963 air-traffic control radar.

VIRAAT: Indian AIRCRAFT CARRIER, originally the British light carrier HMS *Hermes*. Though laid down in 1944, the ship was not completed until 1958, and was finally commissioned in 1959. In 1971 the Royal Navy converted *Hermes* into a helicopter-equipped amphibious assault ship; in 1976, it was converted again into an ANTI-SUBMARINE WARFARE (ASW) carrier (CVS) with an air group of SEA KING ASW helicopters. In 1980 the ship was modified yet again to operate Sea HARRIER V/STOL fighters; a ski-jump ramp was installed in the bow to enhance their range (by eliminating the need for vertical takeoffs). *Hermes* was meant to serve only as an interim training vessel, pending the completion of the first INVINCIBLE-class V/STOL carrier in 1981, but it was hastily reactivated for the 1982 Falklands War, during which

it operated an air group of 12 Sea Harriers and 12 Sea Kings. *Hermes* was laid up in 1984 and sold to India in 1986, undergoing a major overhaul before being recommissioned as the *Viraat* in late 1987. The *Viraat* reinforces the status of India as a major regional power. Together with the smaller VIKRANT, the *Viraat* will continue to serve into the 1990s.

The flight deck is 735 ft. (224 m.) long and 160 ft. (48.8 m.) wide, with an angled landing deck cantilevered 6.5° to port. The "ski-jump" is 148.25 ft. (45.2 m.) long, 44.95 ft. (13.7 m.) wide, and 16.07 ft. (4.9 m.) high, with a maximum elevation of 12°. The ship has no catapults or arresting gear, and cannot operate conventional fixed-wing aircraft. Defensive armament consists of two quadruple SEA CAT short-range surface-to-air missile launchers. The air group now consists of six Sea Harriers and nine Sea Kings. See also AIRCRAFT CARRIERS, INDIA.

Specifications **Length:** 744 ft. (226.9 m.). **Beam:** 90 ft. (27.4 m.). **Draft:** 28.5 ft. (8.7 m.). **Displacement:** 23,900 tons standard/28,700 tons full load. **Powerplant:** twin-shaft steam: 4 oil-fired boilers, 2 sets of geared turbines, 76,000 shp. **Fuel:** 4200 tons. **Speed:** 28 kt. **Range:** 6500 n.mi. at 14 kt. **Crew:** 1170. **Sensors:** 1 Type 965 air-search radar, 1 Type 992 surface-search radar, 1 Type 1006 navigation radar, 2 GWS22 Sea Cat guidance radars, 1 Type 184 hull-mounted sonar. **Electronic warfare equipment:** two radar warning receivers, 2 Knebworth Corvus chaff rocket launchers.

VIRGINIA: A class of four U.S. nuclear-powered guided-missile CRUISERS commissioned between 1972 and 1980. Successors to the CALIFORNIA class, the Virginias are ANTI-AIR WARFARE (AAW) escorts for CARRIER BATTLE GROUPS, with secondary ANTI-SUBMARINE WARFARE (ASW) and ANTI-SURFACE WARFARE (ASUW) capabilities. Flush-decked, with sharply raked bows, they have a blocky superstructure amidships, and two large, pyramidal masts.

The weapon and sensor suites are derived from the AAW variant of the SPRUANCE-class destroyers (i.e., the KIDD class). Primary armament consists of a Mk.26 twin-arm missile launcher at each end of the ship. The forward Mk.26 Mod 0 launcher, with a magazine capacity of 24 missiles, is mainly for ASROC ASW rockets; the aft Mk.26 Mod 1 launcher, with a capacity of 44 missiles, is normally loaded with STANDARD MR surface-to-air missiles and some HARPOON anti-ship missiles. Other armament consists of a 5-in. 54-caliber DUAL PURPOSE gun at the bow and stern, 2 PHALANX 20-mm. radar-controlled guns for anti-missile defense, 2 quadruple launch canisters for HARPOON, and 2 sets of Mk.32 triple tubes for lightweight ASW TORPEDOES. As completed, these ships had a landing deck and hangar for two ASW helicopters on the fantail, but the hangar proved unsatisfactory, leaking heavily in rough seas. Hangar and landing deck were both replaced by three armored box launchers, each with four TOMAHAWK cruise missiles. Two satellite communications terminals and a NAVAL TACTICAL DATA SYSTEM (NTDS) data link enhance the ability of the Virginias to serve as flagships.

Originally, the navy planned to build 11 Virginias; a modified version, the CGN-42 "Strike Cruiser," was proposed in 1979, but even during the Reagan defense buildup the cost of such nuclear-powered escorts was deemed excessive. Very capable ships, though they lack AEGIS radar capability, the Virginias will continue in service well into the next century; likely upgrades include the replacement of the Mk.26 launchers with two Mk.41 VERTICAL LAUNCH SYSTEMS, to increase both the number and variety of missiles carried. See also CRUISERS, UNITED STATES.

Specifications Length: 581.7 ft. (177.3 m.). Beam: 63 ft. (19.2 m.). Draft: 31.2 ft. (9.5 m.). Displacement: 10,400 tons standard/11,300 tons full load. Powerplant: twin-shaft nuclear: 2 D2G pressurized-water reactors, 2 sets of geared turbines, 100,000 shp. Speed: 32+ kt. Crew: 473. Sensors: 1 SPS-48A 3-dimensional air-search radar, 1 SPS-40B 2-dimensional air-search radar, 1 SPS-55 surface-search radar, 1 SPQ-9A gun fire control radar, 1 LN-66 navigational radar, 1 SPG-60D gun and missile control radar, 2 SPG-51D missile guidance radars, 1 bow-mounted SQS-53A low-frequency active/passive sonar. Fire controls: 1 Mk.13 weapons control system, 1 Mk.86 gun control system, 1 Mk.74 missile control system, 1 Mk.114 ASW control system. Electronic warfare equipment: two SLQ-32(V)3 electronic countermeasure arrays, 4 Mk.36 SRBOC chaff launchers.

VITTORIO VENETO: Italian air-capable CRUISER (CGH) commissioned in 1969. *Vittorio Veneto*, an enlarged and improved version of the prior ANDREA DORIA class, has a large flight deck aft, while retaining a traditional cruiser superstructure and weapons forward. Intended as the nucleus and flagship of a DESTROYER/FRIGATE escort group, it combines powerful ANTI-SUBMARINE WARFARE (ASW), ANTI-AIR WARFARE (AAW), and ANTI-SURFACE WARFARE (ASUW) armament.

The flight deck, 131.2 ft. (40 m.) long and 60.7 ft. (18.5 m.) wide, is raised two levels above the main deck to leave room for a 90-by-50-ft. (27.5-by-15.3-m.) hangar below it; the two are connected by a large centerline elevator. The ship can operate either nine Agusta Bell AB.212 HUEY or four SH-3D SEA KING ASW helicopters.

Other SAW armament consists of 2 sets of Mk.32 triple tubes for lightweight ASW TORPEDOES, and ASROC ASW rockets launched from a twin-arm Mk.10 dual-purpose launcher, primarily used for STANDARD ER surface-to-air missiles; the magazine capacity of 60 missiles is usually allocated to 40 Standard and 20 ASROC. Secondary armament consists of 8 OTO-MELARA 76.2-mm. DUAL PURPOSE guns and 6 twin Breda 40-mm. ANTI-AIRCRAFT guns. The 40-mm. guns are linked to a Dardo FIRE CONTROL system which allows them to be used effectively against missiles. In 1984 the ship was also equipped with 4 Teseo launch canisters for OTOMAT Mk.2 anti-ship missiles.

With the advent of the GARIBALDI-class aircraft carrier, no further ships of the Vittorio Veneto class will be constructed. A powerful ship especially in Mediterranean conditions, *Vittorio Veneto* will no doubt continue to receive periodic upgrades to maintain its effectiveness. See also CRUISERS, ITALY.

Specifications Length: 589.25 ft. (179.6 m.). Beam: 63.6 ft. (19.4 m.). Draft: 19.7 ft. (6 m.). Displacement: 7500 tons standard/8850 tons full load. Powerplant: twin-shaft steam: 4 oil-fired boilers, 2 sets of geared turbines, 73,000 shp. Fuel: 1200 tons. Speed: 32 kt. Range: 6000 n.mi. at 20 kt./3000 n.mi. at 28 kt. Crew: 557. Sensors: 1 SPS-40 long-range search radar, 1 SPS-52C air-search radar, 1 SPS-702 surface-search radar, 3 RM-7 navigation radars, 2 SPG-55C missile-guidance radars, 4 RTN-10X Argo 76.2-mm. fire control radars, 2 RTN-20X 40-mm. fire control radars, 1 SQS-23 bow-mounted low-frequency active/passive sonar. Electronic warfare equipment: 1 British-made Abbeyhill electronic signal monitoring array, 2 SCLAR chaff rocket launchers.

VLF: Very Low Frequency, a radio wave band between 3 and 30 KHz, mostly used for submarine communications; though its data rate is low, it can be received at depths down to 50 ft. (15 m.). Two types of VLF antennas for SUBMARINES are known to be in use: a trailing wire (up to 1500 ft./457.3 m.) long, and a loop mounted on a tethered buoy.

Long-range VLF transmission requires large and powerful shore stations. It is believed that the U.S. has 7 such facilities, the Soviet navy more than 26, NATO 2, and Great Britain 1. Communications at depths greater than 50 ft. require transmissions in the Extra Low Frequency (ELF) wave band. See also COMMUNICATIONS.

VLS: See VERTICAL LAUNCH SYSTEM.

VLSI: Very Large Scale Integration, an emerging electronics technology which will increase the amount of information stored on a single microprocessor by at least one order of magnitude. The size, power consumption, and cost of computers will decrease (and reliability increase), as individual microchips replace larger multi-chip modules. The most promising military application of VLSI is in the development of small, reliable computers for *autonomous*, precision-guided "brilliant munitions." See also VHSIC.

V/STOL: Vertical/Short Takeoff and Landing. The only fixed-wing aircraft that can take off and land either vertically or after a very short rollout (to save fuel) are the British/U.S. AV-8 HARRIER and Soviet Yak-36 FORGER. Tilt rotor aircraft such as the V-22 OSPREY combine features of both helicopters and fixed wing types. See also COMBAT AIRCRAFT.

VT: Variable Time (fuze), U.S. designation for a radar PROXIMITY FUZE. Invented in 1943–44, the VT fuze is designed to detonate a projectile in flight within a set distance of any material capable of reflecting electromagnetic radiation—not only metal objects such as ships or aircraft, but also the ground—to achieve airburst effects. A VT fuze is essentially a very small RADAR, small enough to fit into projectiles down to 40-mm. cannon shells. It consists of a radio transmitter and a receiver, a firing condenser, a squib battery, an electrical detonator, and some form of safety device. When the projectile's safety mechanism is released, a vial of electrolyte is broken to activate the battery, energizing the condenser and starting the radio transmission. If radio waves from the transmitter strike a target, they are reflected back to the projectile and picked up by the receiver. The amplitude of this return signal is inversely proportional to the range from projectile to target; as the range closes the signal amplitude increases. When the projectile comes within the preset range and the signal reaches the amplitude required to close the firing circuit, the condenser is discharged into the electrical detonator, which in turn sets off the explosive charge in the projectile.

A Controlled Variable Time (CVT) fuze allows a choice of fuze-activation delays, e.g., to fire artillery safely over the heads of friendly forces.

VT and CVT fuzes are built into most ANTI-AIRCRAFT shells as well as surface-to-air and air-to-air missiles. They are also fitted into some ARTILLERY shells. Against troops and other exposed soft targets, the fuze can be set to burst at the height deemed optimal to inflict maximum splinter damage.

VT fuzes are vulnerable to ELECTRONIC COUNTERMEASURES called fuze jammers (a form of DECEPTION JAMMER), which repeat the radar signal from the fuze at a higher amplitude, in order to activate the firing circuit prematurely.

VTA: *Voenno-transportnaya aviatsiya*, Military Transport Aviation, the branch of the Soviet Air Force (VVS) responsible for both strategic and tactical airlift. The VTA also maintains close coordination with the Airborne Troops (VDV) command. In wartime, operational control of *tactical* airlift would normally be assigned to the Air Armies attached to each Theater of Military Operations (TVD), which would in turn allocate aircraft to the various FRONTS; *strategic* airlift would normally remain under the direct control of the Supreme High Command (VGK). See also AIRBORNE FORCES, SOVIET UNION.

VTOL: Vertical Takeoff and Landing. The capability is inherent in HELICOPTERS and tilt-rotor aircraft such as the V-22 OSPREY, and is also present in two fixed-wing aircraft, the British/U.S. HARRIER and the Soviet FORGER, both of which, however, are operated in a STOL (short takeoff and landing) mode whenever possible, to save fuel. See also COMBAT AIRCRAFT.

VULCAN: The General Electric M61 20-mm. CANNON, widely used by U.S. and other armed forces as an aircraft, ground, and naval weapon. Developed from 1950 as a fighter weapon, the Vulcan was the first of modern cannon to use the "rotary-barrel" or "Gatling" action, whereby several barrels (six, in the case of the Vulcan), mounted rigidly about a common axis, are fed from a single breech; each barrel is rotated successively into line with the breech for firing, allowing very high cyclic rates to be achieved without barrel overheating. The prototype Vulcan was test-fired in 1953, and entered service in 1958 as the main armament of the F-104 STARFIGHTER; still the standard U.S. aircraft gun, it has armed the F-105 Thunderchief, the F-106 Delta Dart, the F-111, the F-4E PHANTOM, the A-7D/E CORSAIR, the F-14

TOMCAT, the F-15 EAGLE, the F-16 FALCON, and the FA-18 HORNET, as well as the tail turrets of the B-58 Hustler and B-52H STRATOFORTRESS bombers.

The basic aircraft version is 73.8 in. (1.875 m.) long, and has an empty weight of 265 lb. (120 kg.). Driven by a 35-hp. electric motor, the Vulcan has selectable rates of fire of 6000 rounds per minute for aerial combat and 4000 rounds per minute for ground attack. Connected by a linkless feed chute to drum magazines ranging in size from 650 to 2084 rounds, the Vulcan can fire standard 20-mm. HE, HEI, and AP rounds at a muzzle velocity of 3400 ft. (1036 m.) per second.

GE also developed two similar podded versions (SUU-16 and SUU-23), both 199 in. (5.055 m.) long and 22 in. (560 mm.) in diameter. The SUU-16 houses an M61A1 cannon with 1200 rounds of ammunition for a loaded weight of 1716 lb. (780 kg.); driven by a ram-air turbine, the gun cannot be fired reliably at speeds below 304 mph (650 kph). The SUU-23 has the self-powered GAU-4 version of the Vulcan, which allows firing at any speed; with 1200 rounds of ammunition, it is slightly heavier at 1730 lb. (785 kg.).

The M197 is a three-barreled, lightweight version of the Vulcan that arms the chin turrets of AH-1 COBRA attack helicopters. With a length of 74.5 in. (1.89 m.) and empty weight of 145 lb. (66 kg.), it has selectable rates of fire from 400 to 3000 rounds per minute, and an effective ground attack range of some 1200 m.

A six-barreled variant is employed in the U.S. Navy's PHALANX radar-controlled anti-missile gun system, and an optically directed Sea Vulcan mount has been developed for FAST ATTACK CRAFT.

The U.S. Army's Vulcan Air Defense System consists of a six-barreled M168 Vulcan cannon in a self-contained turret with a simple range-only radar, an optical gunsight, and a drum magazine. Two versions are in service: the M163, mounted on a modified M113 armored personnel carrier, and the M167, on a light, two-wheeled trailer. In both cases, the gun has selectable rates of fire of 1000 and 300 rounds per minute, with burst-fire settings of 10, 30, 60, and 100 rounds. The self-propelled M163 has an 1100-round magazine with 1000 additional rounds in the vehicle, while the towed M167 only has a 500-round magazine.

The electrically powered turret has a traverse rate of 60° per second and an elevation rate of 45° per second, with elevation limits of −5° and +80°. The effective range is 1.36 mi. (2.2 km.) and the effective ceiling is 3937 ft. (1200 m.). The ef-fectiveness of the system is limited by its short range and lack of an on-mount surveillance radar.

The M163 is in service with U.S. Army "heavy" divisions, and the armies of Ecuador, Israel, Morocco, North Yemen, Portugal, South Korea, and Tunisia. The towed M167 is in service with U.S. Army "light" divisions, and the armies of Belgium, Ecuador, Israel, Jordan, Morocco, North Yemen, Saudi Arabia, Somalia, South Korea, and Sudan.

VVS: *Voyenno-vozdushnyye sily,* (Soviet) Air Forces. There are actually three separate Soviet air services: the RVSN (*Raketnyye voyska strategicheskogo naznacheniya,* or Strategic Rocket Forces), responsible for long-range nuclear missiles, space launches, and satellite SURVEILLANCE; the PVO (*Protivovozdushnaya Oborona,* or Air Defense Forces), responsible for aerospace defense, with INTERCEPTOR aircraft, strategic surface-to-air missiles, ANTI-BALLISTIC MISSILES, and ANTI-SATELLITE (ASAT) systems; and the VVS or Air Force proper.

The VVS is organized into three major commands: the ADD (*Aviatsiya dal'nego deystviya,* or Long-Range Aviation); the FA (*Frontovaya aviatsiya,* or Frontal Aviation); and the VTA (*Voyenno-tranportnaya Aviatsiya,* or Military Transport Aviation). The VVS now has more than 12,000 combat aircraft and 400,000 men (and is thus considerably larger than the U.S. Air Force).

VX: O-ethyl S-2-disoprylaminoethylmethylphosphonothiolate, an extremely lethal antichloruinesterase, or NERVE AGENT, developed in Great Britain during the 1950s. Like all nerve agents, it acts by binding permanently with the nerve-connecting enzyme acetochlorinesterase (AChE), thereby disrupting the entire functioning of the central nervous system. Symptoms of exposure include involuntary spasms of the large muscles, convulsions, frothing at the mouth, and death from respiratory collapse within 15 minutes. Considerably more potent than the other known nerve agents, SARIN (GA), TABUN (GB), and SOMAN (GD), VX has a lethal dose (LD-50) of only 0.5 to 15 milligrams per cubic m. Like all nerve agents, VX is percutaneous (skin-permeable), and so can kill by contact as well as by inhalation; the lethal percutaneous dose is only 15 milligrams. Troops must therefore wear complete protective suits as well as respirator masks to avoid injury.

Troops can be "hardened" against exposure by a prophylactic regime of synergistic drugs. One set of drugs, administered before exposure, binds *tem-*

porarily with some AChE, forming a protected pool of the enzyme within the body. After exposure, a second set of drugs releases this pool, bypassing the effects of the agent. Because of the speed with which it acts, no other form of first aid is effective against VX.

With a high boiling point and the consistency of motor oil, VX is extremely persistent and can present a continuing hazard for several days, or even weeks, by lingering in the soil or on foliage, uniforms, and equipment. Decontamination is extremely difficult and hazardous. VX could be dispersed from spray canisters, aerial bombs, rockets, and artillery shells; it is a standard nerve agent in both the U.S. and Soviet chemical arsenals. See also CHEMICAL WARFARE.

W

WALLEYE: AGM-62 electro-optically guided GLIDE BOMB developed by the U.S. Navy in the 1960s as a standoff precision weapon against heavily defended targets. Walleye is still in service with U.S. and Israeli forces in several versions.

The original AGM-62A Walleye I consists of an ELECTRO-OPTICAL sensor (a television camera) in the nose, a guidance and control unit, an 825-lb. (374-kg.) SHAPED CHARGE warhead, four cruciform delta wings with movable control surfaces, and a windmill-driven electrical generator in the tail. A total of 4531 were produced by Martin Marietta between 1966 and 1970, including a few nuclear-armed variants for the U.S. Air Force. Walleye can be carried by most U.S. tactical aircraft at speeds up to 1200 mph (1920 kph); range varies with the release speed and altitude. In action, the operator (usually the pilot) must first acquire the target with the weapon's electro-optical sensor, whose field of view is displayed on a cockpit video terminal. The pilot then places a cursor over the target, locking the weapon onto a contrast point that the sensor can recognize. After lock-on, the weapon can be released to home autonomously to the target. The median error radius (CEP) varies between 10 and 15 ft. (3 and 4.5 m.).

Walleye I was much used in Vietnam, where it was discovered that the 825-lb. warhead was too small to destroy targets such as major bridges. Hughes therefore developed the Walleye II with a 2000-lb. (909-kg.) warhead derived from the Mk.84 LDGP bomb, and enlarged wings to extend maximum range.

"Lock-on before launch" (LOBL) requirement dictated a direct line of sight to the target before release, reducing Walleye's usefulness in poor visibility and exposing the launching aircraft to enemy air defenses. An extended range (ER) Walleye with larger wings only made dissatisfaction with LOBL more acute. In 1974, the navy introduced the Walleye Extended Range Data Link (ERDL), which can lock on after launch (LOAL). In action, the weapon is released in the general direction of the target, transmitting its video images back to the aircraft through a DATA LINK in the tail. When the pilot acquires the target on his video terminal, he can lock the weapon onto a sharp contrast as before, or he can guide the weapon manually with a joystick, transmitting steering commands through the data link. It is also possible to release a Walleye ERDL from one aircraft and guide it from another. Since 1978, some 1400 Walleye I and 2400 Walleye II have been converted to ERDLs.

To allow the employment of Walleye at night or in poor weather, an IMAGING INFRARED (IIR) seeker head identical to that of the Air Force's GBU-15 glide bomb and MAVERICK AGM-65 missile was provided in 1977. Though out of production, Walleye is still the subject of modification and upgrade programs.

Specifications **Length:** (I) 135 in. (3.44 m.); (II) 159 in. (3.04 m.). **Diameter:** (I) 12.5 in. (317 mm.); (II) 18 in. (457 mm.). **Span:** (I) 45.5 in. (1.15 m.); (II) 51 in. (1.13 m.). **Weight, launch:** (I) 1100 lb. (499 kg.); (II) 2400 lb. (1089 kg.). **Range, max.:** (I) 16 mi. (26 km.); (II) 35 mi. (56 km.).

WALRUS: Original designation of the Dutch ZEELEEUW class of diesel-electric attack SUBMARINES, changed after a dockyard fire delayed completion of the name boat. See also SUBMARINES, NETHERLANDS.

WAR: A form of international relations in which organized violence is used in addition to other instruments of policy. To protect what they have, or acquire more, states employ formal diplomacy, the modulation of trade, PROPAGANDA, POLITICAL WARFARE, ECONOMIC WARFARE, the threat of war, and, finally, war itself. Nonstate entities which lack some of these instruments may still be capable of waging war. War and the threat of war are thus ordinary instruments of power, in the absence of supranational regulation and enforcement.

According to Karl von Clausewitz, "War . . . is an act of violence intended to compel our opponent to fulfill our will," moreover, "war is not merely a political act, but also a real political instrument, a contribution to political commerce." That classical definition implies limits ("The result of war is never absolute . . .") that nuclear war could easily transcend; indeed, any political objective implies limits.

Total destruction may, however, be the objective of idealistic war, that is, a "crusade" or JIHAD fought against opponents whose very existence is seen as intolerable to the deity (or to stated ideals), with whom therefore no compromise is possible, the only remaining choices being either unconditional surrender or annihiliation.

No war between nuclear powers can be both unrestrained and a political phenomenon, although the threat of nuclear war remains a useful instrument of policy. Nuclear threats, however, must be either partial or conditional to have political meaning, as in ASSURED DESTRUCTION, MASSIVE RETALIATION, or LOCAL WAR.

This classical view of war as a normal manifestation of political life in a world of nation-states is disputed. Three rival conceptions of war have some currency.

1. In Marxist-Leninist doctrine, capitalist classes in control of the state are held to use war in their perpetual struggle for more labor, raw materials, and markets. While their own working class is deceived (or forced) into fighting for a supposedly national cause, the goal is never national but exclusively capitalistic, and often motivated by the urge to exploit foreign populations by "imperialist" wars. In "diversionary" wars, on the other hand, the aim is to stifle revolutionary sentiment by diverting anger into manipulated nationalist quarrels. War between "capitalist" and "socialist" states is a direct extension of the class struggle, wherein both sides use the resources of the respective states to further their different class interests. This explanation of the causes of war thus implies the impossibility of wars between socialist states: because war is class struggle, there can be no war between states controlled by the same universal "working class." In Marxism-Leninism, therefore, war is only a temporary phenomenon, caused by the residual existence of waning capitalist systems.

The central assumption of Marxist-Leninist doctrine, that in "socialist" countries the state is controlled by the "working class," has of course been contradicted by experience. Classes do not rule, do not conduct diplomacy or make war; the political leadership which does rule has its own (nonclass) interests and can engage in war as any other leadership, with "socialist" countries as well.

2. According to the nonviolent resistance school of thought, war is not a necessary instrument of policy. No political leadership should engage in war, and therefore none should maintain armed forces. National defense can instead be assured by nonviolent resistance against an aggressor, to nullify the expected benefits of aggression and induce withdrawal. Direct propaganda and demonstrations to demoralize the invading troops and passive resistance by bureaucrats (to incapacitate the administration of the regime imposed by an army of occupation) were some of the means employed by the Czechs against the Soviet-led Warsaw Pact invasion of 1968; but the attempt failed. Nonviolent resistance against the British in India and in what was then Palestine included the nonpayment of taxes and the formation of human barricades to prevent troop movements; those attempts were successful. No account is taken of aerial bombing, naval blockades, and other remote forms of military force in nonviolent resistance theory, which implies the presence of enemy troops on the ground. It assumes in any case that the invader can always be resisted nonviolently, but that is true only if his values or self-image inhibit the use of terror to enforce obedience; against determined foes who freely employ violence and other means of compulsion, nonviolence fails. Thus this alternative to war has a very limited scope: the forces

against which it works are also less likely to engage in aggression in the first place.

Pacifism is not subject to this limitation: its rejection of the use of violence is absolute and unconditional.

3. In the "peace research and conflict resolution" view, war is an (avoidable) event, rather than an instrument of policy. Wars are seen as the outcome of political phenomena and structural peculiarities which "generate" conflict, rather than as the result of deliberate decisions.

Because wars are "events," peace researchers seek to uncover their causes, on the assumption that once they are known, the illness can be cured. Thus conceptual models of international systems have been developed under varied behavioral parameters, to discover those which do not generate war. In this activity, which resembles economic model building, the existence of a warless equilibrium is assumed. The ultimate aim is to create a "balancing" institution (rather like a more powerful UN) which would act to undo the causes of conflict. Neither empirical research nor model building has yet isolated the "causes" of war; and no credible institutional framework for conflict resolution (as opposed to "conflict avoidance") has been suggested.

As long as humans are organized into sovereign entities, as long as those entities are vehicles of power accumulation, and as long as war remains a tool of power accumulation, there will be parties willing to initiate war, forcing others to engage in it as well. In the present era, however, the utility of war appears to have declined overall, because while its costs and risks have increased, the benefits of invasion and occupation have declined.

WAR GAMES: Also described as "conflict simulations," war games attempt to synthesize the phenomena of war in peacetime as a training aid, to test new weapons, strategies, operational methods, and tactics, and generally develop military knowledge. War games have been "played" for thousands of years (chess being a very ancient and very abstract example), but it was in Prussia (and, later, Imperial Germany) that their use was formalized. Such *Kriegspiels* ranged from map or sand-table exercises to full-scale military maneuvers. The basic elements of the *Kriegspiel*—its free play and reliance on umpires—are now emulated worldwide.

Since World War II, a new form of war game has been developed from the techniques of operations research. Very detailed simulations, such war games attempt to predict combat outcome and the performance of specific weapons by quantitative analysis. Only in the case of nuclear-warfare simulations are the results persuasive. Since the 1960s, computers have been increasingly important in the development and conduct of war games, to automate much of the bookkeeping involved and replace human umpires. Complex algorithms with near-real-time implementation are now in use. War games cannot, however, transcend their subjective assumptions; hence their model of "reality" diverges unpredictably from reality itself.

WARSAW PACT: The nominally multilateral military alliance of the Soviet Union, Albania, Bulgaria, Czechoslovakia, East Germany, Hungary, Poland, and Romania, formally established in May 1955 to supplement prior bilateral treaties, and in de facto dissolution by 1990. In February 1991, the USSR announced its intention to disolve the military aspect of the Pact in April 1991; however, it is expected to maintain bi-lateral defense treaties with several Pact countries. The Warsaw Treaty Organization has a (military) High Command and a parallel Political Consultative Committee consisting of party secretaries, heads of government, and the foreign and defense ministers of each member state. Its offices are in Moscow, and both the commander-in-chief and chief of staff have always been Soviet officers. In the Joint Secretariat and Permanent Commission, the senior posts are also filled by Soviet personnel.

Major pact commands include the Northern Group of Forces, with headquarters in Legnica, Poland, the Southern Group of Forces with headquarters in Tokol, Hungary, and the Group of Soviet Forces in Germany (GSFG), with headquarters in Wunsdorf. Most Soviet forces in Eastern Europe, however, are now being withdrawn, so the continued existence of these commands is doubtful. NUCLEAR WEAPONS are invariably kept under Soviet control, though nuclear-capable missiles have been supplied to other pact members.

Albania did not participate in any pact activities after 1961, and Romania was only a nominal member after 1968.

The Warsaw Pact was comparable to NATO only in narrow military terms; while NATO is a voluntary alliance that functions multi-nationally, the Warsaw Pact was a Soviet instrument of control. Hence NATO's cohesion was naturally resilient, albeit threat-dependent, while the pact's cohesion derived largely from the effectiveness of Soviet compulsion.

WARRIOR: British MICV-80 INFANTRY FIGHTING VEHICLE (IFV), now in service with the mechanized infantry units of armored divisions in the British Army of the Rhine. Developed from the early 1970s by GKN Sankey as a faster, better-armed, and better-protected replacement for the FV.432 TROJAN armored personnel carrier, the Warrior encountered serious technical and financial problems which delayed completion of the first prototype until 1979. After rigorous trials, Warrior was accepted by the British army in 1984, but the first production vehicles were not delivered until 1987. A total of 1048 Warriors have been ordered, of which some 400 were delivered through 1990.

The all-welded aluminum hull, armored against shell splinters and small arms up to 12.7 mm., is divided into a driver's compartment in the front, a turreted fighting compartment in the middle, and a passenger compartment in the rear. The driver has a single wide-angle observation periscope, which can be replaced by a passive night-vision scope. The commander and gunner, in the welded aluminum turret, are provided with passive night sights.

The main armament is a 30-mm. RARDEN cannon, which can penetrate all light armored vehicles, as well as the side and rear armor of older main battle tanks. Secondary armament is a coaxial 7.62-mm. CHAIN GUN. Four smoke grenade launchers are attached to each side of the turret, and Warrior is also equipped with a positive overpressure COLLECTIVE FILTRATION unit for NBC defense.

The passenger compartment in the rear can seat seven fully equipped infantrymen, with access provided by a roof hatch and two large rear doors. In contrast to other IFVs (e.g., the U.S. BRADLEY, West German MARDER, and Soviet BMP), Warrior does not have firing ports in the passenger compartment; this reflects the British army's insistence that the infantry fight dismounted, with the Warrior used strictly for fire support from covered positions, rather than for mounted combat. In contrast to most other IFVs, Warrior is not amphibious. Proposed variants include a command vehicle, a mortar carrier, a recovery vehicle, and an artillery command post.

Specifications **Length:** 20.85 ft. (6.34 m.). **Width:** 9.95 ft. (3.03 m.). **Height:** 9 ft. (2.73 m.). **Weight, combat:** 22.5 tons. **Powerplant:** 550-hp. Rolls Royce CV8 TCA 8-cylinder diesel. **Speed, road:** 46.6 mph (75 kph). **Range, max.:** 310 mi. (500 km.).

WASP: A class of five U.S. amphibious assault ships (LHDS), the first of which was commissioned in 1989. Developed from the similar TARAWA-class LHAS, the Wasps are the largest amphibious-warfare ships in service, and differ from their predecessors mainly in having expanded aviation facilities at the expense of some vehicle parking space.

Resembling small aircraft carriers, the Wasps have a full-length flight decks with an island superstructure offset to starboard, housing the usual ship-handling and flight control facilities, as well as COMMAND AND CONTROL facilities for an amphibious landing command group. One deck-edge and one fantail elevator connect the flight deck with the hangar, which is larger than in the Tarawas. There are also six (vs. five) cargo elevators between the flight deck and the holds. The Wasps can operate up to 40 (vs. 35) attack and transport HELICOPTERS, and have improved support facilities for U.S. Marine Corps AV-8 HARRIER V/STOL attack aircraft.

As in the Tarawas, the Wasps have a floodable docking well in the stern, but this has been modified to hold up to 3 (vs. 1) LCAC hovercraft, or up to 12 medium landing craft (LCMS). There is also a vehicle deck below the hangar (though smaller than in the Tarawas), accommodations for up to 1900 troops, and hospital beds for up to 600 (vs. 300) casualties. Defensive armament consists of 2 8-round pepperbox launchers for SEA SPARROW short-range surface-to-air missiles, 3 20-mm. PHALANX radar-controlled guns for anti-missile defense, and 8 manually operated .50-caliber (12.7-mm.) machine guns.

Specifications **Length:** 844 ft. (257 m.). **Beam:** 106 ft. (32.3 m.). **Draft:** 26.67 ft. (8 m.). **Displacement:** 25,800 tons standard/40,530 tons full load. **Powerplant:** twin-shaft steam: 2 oil-fired boilers, 2 sets of geared turbines, 71,000 shp. **Fuel:** 6200 tons. **Speed:** 24 kt. **Range:** 9500 n.mi. at 20 kt. **Crew:** 1081. **Sensors:** 1 SPS-64 surface-search radar, 1 SPS-49 2-dimensional air-search radar, 1 SPS-52 3-dimensional air-search radar, 2 Mk.91 Sea Sparrow guidance radars. **Electronic warfare equipment:** 1 SLQ-32(V)3 electronic countermeasures array, 3 Mk.36 SRBOC chaff launchers, 1 SLQ-25 Nixie torpedo countermeasures sled.

WEAPON: A man-made object (or an *objet trouvé*) intended as a means of killing or incapacitating humans, or destroying their artifacts. *Strategic* weapons are those capable of striking at the "homeland" of an opponent, his industrial and population centers. *Tactical* weapons, on the other hand, are presumed to be of use only against battlefield manifestations of the enemy. The former are generally long-range and the latter short-range weapons, but the distinction lacks meaning outside a specific geographic context (in some cases, pistols may be "strategic"). *Conventional* is an unfortunate adjective for all but nuclear, chemical, or biological weapons. *Defensive* is a far more useful adjective: one's own weapons are defensive, while those of an opponent are invariably offensive. "Weapons of mass destruction" is the preferred Soviet term for nuclear, chemical, and biological weapons, which are known in the West as NBC (Nuclear, Chemical, Biological) or CBR (Chemical, Biological, Radiological). NUCLEAR WEAPONS are either fission (A-bomb) or fusion (H-bomb) devices (the latter are also known as "thermonuclear"), but fusion bombs include a fission "trigger." *Special weapons* is a euphemism for nuclear and chemical weapons.

WEAPON SYSTEM: An integrated system consisting of a WEAPON, a vehicle or static "platform" to carry and launch it, a FIRE CONTROL or weapon-direction device, and often some maintenance and support equipment. Thus a MAIN BATTLE TANK is a weapon system formed by a heavy gun (the weapon), a self-propelled armored chassis (the platform), some form of fire control, and various tools and test devices to maintain those components. Most surface-to-air missiles, on the other hand, are "subsystems," i.e., parts of weapon systems which include the missile itself, the launcher, and the fire controls, as well as any sensors not built into the missile (e.g., radars).

WEAPON SYSTEM OPERATOR: A nonpiloting aircraft crewman in multi-place COMBAT AIRCRAFT responsible for the operation of weapons and sensors. Radar intercept officer (RIO) and bombardier-navigator are equivalent. Larger aircraft, such as BOMBERS or patrol aircraft, may have two or more weapons system operators; e.g., one for offensive systems and one for defensive ELECTRONIC WARFARE systems.

WELDED WING: An AIR COMBAT MANEUVERING (ACM) formation of two fighters, one of which is the designated leader, and the other the overwatching "wingman." The leader alone normally engages the enemy, while the wingman covers him. Two such pairs flying together form a "Finger Four." Developed by the German *Luftwaffe* during the Spanish Civil War, these became the standard air combat formations for all air forces in World War II. Since the 1960s, both have been largely superseded in first-class air forces by LOOSE DEUCE and FLUID FOUR formations, in which leader and wingman both attack while providing mutual support. These formations are more flexible and more demanding; hence, Welded Wing is still employed by air forces short of skilled pilots.

WESSEX: A variant of the Sikorsky S-58 HELICOPTER built under license by the British Westland company. Although its basic design dates to the early 1950s, the Wessex is still used in large numbers by Britain's Royal Air Force and Fleet Air Arm, and the air forces of Australia, Brunei, Ghana, and Iraq in a variety of roles, including ANTI-SUBMARINE WARFARE (ASW), search and rescue (SAR), casualty evacuation (CASEVAC), assault transport, and logistic support.

The initial version (HAS.1), introduced in 1960 as an ASW helicopter for the Royal Navy, carried a variety of SONOBUOYS, DEPTH CHARGES, and homing TORPEDOES. Most were later upgraded to an HAS.3 standard, with improved ASW sensors, a dorsal surface-search RADAR, and improved AVIONICS.

The Royal Air Force HC.2, a transport with a more powerful engine, could carry either 16 troops, 7 stretcher cases, or a 4000-lb. (1818-kg.) slung load. In the RAF, the HC.2 was replaced by the HU.5, now also used by the Royal Marines as an assault helicopter.

Both the HU.5 and HAS.3 can be fitted with a variety of MACHINE GUNS, rocket pods, and guided missiles. During the 1982 Falklands War, an HAS.3 crippled the Argentine submarine *Santa Fe* with a SEA SKUA missile and machine-gun fire, while HU.5s were much used as assault and transport helicopters, and for liaison and CASEVAC.

The Australian Mk.31B version is similar to the HAS.1; other foreign versions are similar to the HC.2. Although rugged and reliable, the Wessex lacks the performance of more modern machines, and is being replaced by Westland SEA KING and LYNX helicopters in British service.

Specifications (HU.5) **Length:** 48.4 ft. (14.75 m.). **Rotor diameter:** 56 ft. (17.07 m.). **Powerplant:** 1 1550-shp. Rolls Royce Coupled Gnome twin turboshaft. **Weight, empty:** 8657 lb. (3935 kg.). **Weight, max. takeoff:** 13,500 lb. (6136 kg.). **Speed, max.:** 132 mph (212 kph). **Speed, cruising:** 121 mph (197 kph). **Service ceiling:** 14,100 ft. (4300 m.). **Range, max.:** 390 mi. (625 km.).

WETEYE: The Mk.116 air-delivered chemical bomb, developed by the U.S. Navy in 1963. Derived from the Mk.83 low-drag, general-purpose (LDGP) bomb, Weteye is normally filled with sarin (GB), a nonpersistent nerve agent. In operation, a small burster charge set off by a mechanical or vt fuze expels the agent through ports in the weapon casing. Weteye remains the most important U.S. chemical weapon for long-range delivery, but it is to be replaced by the BLU-80/B bigeye, a 500-lb. binary weapon. See also chemical warfare.

Specifications Length: 90 in. (2.29 m.). Diameter: 14 in. (356 mm.). Weight: 500 lb. (227 kg.).

WHISKEY: NATO code name for the largest class of Soviet postwar diesel-electric attack submarines (SS); some 240 were completed by the Soviet Union between 1949 and 1957, and an additional 21 were built in China between 1960 and 1964. These medium-range submarines have many features derived from the German Type XXI U-Boat of 1944, 5 of which were captured by Soviet forces in 1945.

The Whiskeys have a conventional, cigar-shaped hullform and a double-hull configuration, with fuel and ballast tanks installed between the pressure hull and the outer casing. The latter has numerous free-flooding holes which cause much flow noise at high speeds. A tall, streamlined sail (conning tower) is located amidships. Control surfaces include retractable bow diving planes, a single rudder, and stern planes behind the propellers. A retractable snorkel allows diesel operation from shallow submergence.

Armament consists of 6 21-in. (533-mm.) tubes (4 forward, 2 aft), for a total of 12 free-running and acoustical homing torpedoes (both conventional and nuclear). mines can be substituted for torpedoes on a 2-for-1 basis. Sensors and fire controls are of course primitive by modern standards.

In 1958, five boats were converted into guided-missile submarines (SSGs), designated "Whiskey Twin Cylinder" by NATO. They were armed with two SS-N-3 shaddock anti-ship missiles in erectable container/launchers on the upper deck. Seven "Whiskey Long Bin" conversions followed; lengthened to 275.6 ft. (84 m.) and displacing 1800 tons submerged, they were armed with four Shaddocks housed in tubes faired into the sail. Both types had to surface to launch their missiles, making them vulnerable to detection. All were retired by the mid-1980s, having served as precursors to the echo- and juliett-class guided-missile submarines. Several other Whiskeys were converted to "Whiskey Canvas Bag" radar picket submarines

(SSRs), with a large surveillance radar in the sail; all have been retired.

The Whiskey class was superseded by the improved romeo class, 18 of which were completed between 1958 and 1961. In spite of their age, a few Whiskeys remain in Soviet service, mainly for coastal defense, minelaying, training, and special operations (one achieved notoriety by running aground inside a Swedish naval base in 1981). See also submarines, soviet union.

Specifications Length: 249.3 ft. (76 m.). Beam: 21.3 ft. (6.5 m.). Displacement: 1050 tons surfaced/1350 tons submerged. Powerplant: twin-shaft diesel-electric: 2 2000-hp. Type 37D diesel-generators, 2 1250-hp. electric motors. Speed: 17 kt. surfaced/13 kt. submerged. Range: 6000 n.mi. at 5 kt. Patrol endurance: 45 days. Max. operating depth: 656 ft. (200 m.). Collapse depth: 1115 ft. (340 m.). Crew: 50. Sensors: 1 Tamir 5 medium-frequency active sonar, 1 small passive sonar, 1 mast-mounted "Snoop Plate" surveillance radar, 2 periscopes. Operators: Alb (3), Bul (2), Egy (3), Indo (1), N Kor (4) PRC (20?), USSR (30?).

WIG: See wing-in-ground-effect.

WILD WEASEL: U.S. Air Force term for aircraft and the tactics optimized for the suppression of enemy air defenses (sead). The Wild Weasel program was a Vietnam-era response to the proliferation of North Vietnamese radar-controlled anti-aircraft guns and surface-to-air missiles. The first Wild Weasel aircraft was the two-seat F-100F super sabre jet fighter, equipped for the role with radar homing and warning receivers (RHWRs) and cluster bombs to locate and destroy enemy fire control radars. The Super Sabre was quickly replaced by the faster, more capable F-105F, the two-seat version of the Thunderchief strike fighter, with additional avionics, and launchers for shrike and standard ARM anti-radiation missiles (ARMs). This was followed in turn by the F-105G, a more extensive conversion of the F-105F with built-in RHWRs, active jamming equipment, and other electronic countermeasures. In the early 1980s, the F-105s were replaced by (116) F-4G conversions of the F-4E phantom multi-role fighter. These now have more advanced avionics and weapons, including the AGM-88 harm missile.

In combat, one Weasel aircraft equipped with ARMs can be paired with a standard fighter armed with cluster bombs. Preceding the main strike force, they must fly nonevasively to allow enemy tracking, fire control, and missile guidance radars to track and even lock onto them; once detected by the Wild Weasel's RHAWs, enemy radars are attacked either with ARMs or simply with cluster bombs. In Vietnam, the mere presence of Wild Weasels could suppress air defenses by inducing radar crews to shut down—an effective if counterproductive response, now partially negated by HARM's backup inertial guidance. Wild Weasel missions are inherently hazardous; even with radical evasive maneuvers, CONTOUR FLYING, and ample electronic countermeasures, combat losses could be heavy.

"WINDOW OF VULNERABILITY": A finite period within a more prolonged confrontation, during which one side achieves some key military advantage that makes the other more vulnerable to attack. This situation is then inherently destabilizing in two ways: the superior power may be confident of its ability to defeat the adversary, and it may calculate that its advantage is fleeting; if so, it may be tempted into aggression. The inferior power, on the other hand, if aware of its vulnerability and uncertain of the adversary's intentions, may choose to risk a PREEMPTIVE STRIKE to eliminate the perceived threat. The term was widely used in the U.S. during the late 1970s with specific reference to the growing threat of Soviet strategic nuclear weapons; in spite of rhetorical abuse, it usefully draws attention to the importance of "phasing effects."

WING-IN-GROUND-EFFECT (WIG): A designation for an air vehicle resembling a conventional aircraft that exploits the "ground effect" air cushion formed between the wing and the ground when flying at very low altitudes, to lift much heavier loads than would be possible relying on wings alone.

When an aircraft flies at a height equal to half of its wingspan or less, a cushion of air is trapped between the lower surface of the wing and the surface of the earth; this cushion reduces drag by up to 70 percent, with a concomitant reduction in fuel consumption and increase in range and payload—an increase on the order of 500 percent. (At higher altitudes, WIG vehicles can maneuver like aircraft, but lose the benefits of the ground-effect cushion.)

WIG vehicles could thus carry substantial payloads over very long distances (for, e.g., amphibious assault, inter-theater transport, or fast attack), offering the speed of aircraft with the payload efficiency of ships. The Soviet Union has been the leading exponent of WIG technology, having built two experimental prototypes on the Black Sea. One is apparently a troop transport; the other (NATO designation "Utka") is armed with four SS-N-22 SUNBURN anti-ship missiles. See also HOVERCRAFT.

WIRE GUIDANCE: A form of COMMAND GUIDANCE in which steering signals are passed from operator to weapon through thin wires paid out from spools in the weapon. Wire guidance is widely used in ANTI-TANK GUIDED MISSILES (ATGMs) and TORPEDOES. In the former, the two basic modes are manual command to line-of-sight (MCLOS) and semi-automatic command to line-of-sight (SACLOS). In the latter, signals from platform to torpedo can be complemented by signals from the torpedo's on-board sonar transmitted back through the wires to the launch platform's FIRE CONTROL system. Fiberoptics are now being investigated as a replacement for metal wires. They are lighter, have greater resistance to electromagnetic interference, and offer much higher data rates, but are also more fragile. In the U.S. Army's Fiberoptic Guided Missile (FOG-M), video signals from the TV camera in the missile's nose are passed back to the operator, allowing indirect fire and target lock-on after launch.

WSO: See WEAPON SYSTEM OPERATOR.

XIA: Western code name for a class of two Chinese nuclear-powered ballistic-missile submarines (SSBNs), the first of which was completed in 1984–85. At least two more are believed to be under construction. Little is known about these vessels; the configuration imitates the U.S. LA-FAYETTE class, with a cylindrical hull, bulbous bow, and two rows of six missiles housed vertically in tubes under a "whaleback" casing behind the sail.

The Xia's 12 CSS-N-3 submarine-launched ballistic missiles (SLBMS), two-stage, solid-fuel missiles with one nuclear warhead each, have an estimated range of 1675 mi. (2795 km.). Six to eight torpedo tubes are present as secondary armament. See also SUBMARINES, CHINA; SLBMS, CHINA.

Specifications Length: 393 ft. (120 m.). Beam: 32.8 ft. (10 m.). Displacement: 7000 tons submerged. Powerplant: single-shaft nuclear. Speed: 20 kt. submerged.

X-RAY KILL: Neutralization of a weapon in outer space by X-ray radiation. Much of the energy released by the detonation of NUCLEAR WEAPONS is in the form of X-ray radiation. Endoatmospherically, X-rays interact with atmospheric molecules and are transformed into thermal energy. Exoatmospherically, X-rays are the principal effect of nuclear explosions; ANTI-BALLISTIC MISSILE (ABM) interceptors such as the proposed U.S. SPARTAN and SPRINT and the deployed Soviet GALOSH rely on X-rays as the kill mechanism against enemy REENTRY VEHICLES (RVs). High-energy (or "hard") X-rays hitting the surface of an RV heat shield would cause very rapid heating, resulting in explosive vaporization, which would in turn generate shock waves damaging to their internal structure. Secondary effects include molecular alterations in the solid-state electronics of warhead fuzing mechanisms, resulting in their failure to detonate as intended. Lower-energy ("soft") X-rays would also cause surface heating in RVs, but too slowly for shock effects; it is possible, however, that asymmetrical heating would sufficiently distort the ballistic shape of RVs to cause aerodynamic instability, resulting in their breakup during reentry. Shielding against X-ray effects is technically possible, but only with a large reduction in payload, because only very dense materials could suffice for the purpose.

X-RAY LASER: A proposed type of high-energy LASER which would generate a coherent, monochromatic beam of electromagnetic energy in the X-ray frequency band (.0001 to 1000-angstrom wavelengths). An X-ray laser would consist of a special alloy rod which would act as the lasing medium when "pumped" (or "excited") by an extremely powerful energy source. To date, the only practical source of such energy would be a low-yield (subkiloton) nuclear device: if it were detonated in close proximity to the lasting rod, large numbers of photons would be generated, to be projected from the rod in a narrow beam. Although the duration of the beam would be very short (the rod would be vaporized after a few milliseconds), it would have enough energy to destroy almost any target. X-rays can penetrate all but the most dense materials, so that shielding would be impossible if weight is a constraint. Obviously, the X-ray laser could only be a one-shot weapon; to be cost-effective, many rods would have to be mounted on the nuclear device, each already aimed at an individual target. Because X-rays are absorbed rapidly by atmospheric molecules, X-ray lasers could be effective only as space-based weapons, as proposed for both ANTI-SATELLITE use and BALLISTIC MISSILE DEFENSE. In the U.S., X-ray lasers are being investigated by the STRATEGIC DEFENSE INITIATIVE Office under the Excalibur program.

XRL: See X-RAY LASER.

Y

YAK-28: Soviet light bomber/interceptor. See BREWER.

YAK-36: Soviet naval V/STOL fighter. See FORGER.

YAMAGUMO: A class of six Japanese hunter-killer DESTROYERS (DDKs), commissioned between 1966 and 1978. Specialized ANTI-SUB-MARINE WARFARE (ASW) vessels, the Yamagumos are almost identical to the earlier MINEGUMO class, but were completed with an 8-round ASROC ASW rocket launcher instead of a DASH (Drone Anti-Submarine Helicopter) hangar and landing deck. Other armament includes 2 twin 3-in. 50-caliber anti-aircraft guns, a 4-barrel Bofors 375-mm. DEPTH CHARGE mortar, and 2 sets of Mk.32 triple tubes for lightweight ASW TORPEDOES. The 4 3-in. guns are being replaced by 2 OTO-MELARA 76.2-mm. DUAL PURPOSE guns. See also DESTROYERS, JAPAN.

Specifications Length: 377 ft. (114.9 m.). Beam: 38.75 ft. (11.8 m.). Draft: 13 ft. (4 m.). Displacement: 2100 tons standard/2700 tons full load. Powerplant: 6 Mitsubishi 12UEV 30/40N diesels, 26,500 hp. total, twin shafts. Speed: 27 kt. Range: 7000 n.mi. at 20 kt. Crew: 210. Sensors: 1 OPS-11 surface-search radar, 1 OPS-17 air-search radar, 2 GFCS-1 fire control radars, 1 bow-mounted SQS-23 low-frequency active/passive sonar, 1 SQS-35(J) variable depth sonar. **Electronic warfare equipment:** 1 NORL-1B or NORL-5 radar warning receiver.

YANKEE: NATO code name for a class of 34 Soviet nuclear-powered ballistic-missile submarines (SSBNs) completed between 1967 and 1976. The configuration of the Yankee, the first *efficient* Soviet SSBN, was copied from the U.S. George Washington class, with a bulbous bow, cylindrical hull, and two rows of eight missile tubes housed under a "whaleback" casing behind the sail. The Yankees differ from their U.S. model in their double-hull configuration, twin-shaft propulsion, and lack of hydrodynamic refinement.

The outer hull has numerous free-flooding holes (which generate considerable flow noise at high speeds), and a streamlined sail located forward. Control surfaces are unusual for a Soviet submarine, with sail-mounted "fairwater" diving planes, and cruciform rudders and stern planes ahead of the propellers. Their small, high-speed propellers make the Yankees considerably noisier than contemporary Western SSBNs, which have single, very large, low-speed propellers. Lack of effective sound insulation further contributes to high noise levels. These submarines are equipped with a VLF towed communications buoy, and an ELF trailing wire antenna.

Primary armament consists of 16 SS-N-6 SAWFLY submarine-launched ballistic missiles (SLBMS), whose initial version (SS-N-6 Mod 1), was 42.75 ft. (13 m.) long and 6 ft. (1.83 m.) in diameter, with a range of some 1500 mi. (2400 km.). These missiles were replaced from 1973 by the Mod 2, with a range of 1850 mi. (2960 km.); the latter in turn was superseded from 1978 by the Mod 3, with a range of 2000 mi. (3200 km.) and a payload of three MULTIPLE REENTRY VEHICLES (MRVs). Sawfly is a two-stage, liquid-fuel missile which burns a corrosive and volatile hydrazine–nitrogen tetroxide mixture. One Yankee was lost in the Atlantic in October 1986 when a Sawfly exploded in its tube after a fuel leak.

One submarine completed in 1977 (*Yankee II*) serves as a testbed for 12 SS-N-17 snipe SLBMs but is otherwise identical to the Yankee I (Snipe, the first Soviet solid-fuel SLBM, served as a developmental prototype for the larger, more capable SS-N-20 missile of the typhoon-class SSBNs). Secondary armament on all Yankees consists of 6 21-in. (533-mm.) torpedo tubes, for a total of 18 torpedoes.

By the early 1980s, the Yankees were being replaced as SSBNs by the deltas and Typhoons. Only 18 Yankees still remain in their original role; to remain within the limits of the salt ii Treaty, the Soviet navy removed the missile compartments of 14 Yankees, converting them to Yankee III attack submarines (SSNs). In 1982, however, some were modified again as cruise missile carriers ("Yankee Notch"), with a "wasp-waist" containing horizontal launch tubes for SS-N-21 sampson cruise missiles; it is believed that up to 40 missiles could be carried. One other Yankee has been converted to serve as an experimental launch platform for the very large, supersonic ss-n-24 Scorpion SLCM. These changes add considerably to Soviet nuclear delivery capabilities, and will probably extend the service life of the Yankee class by many years. See also submarines, soviet union; slbms, soviet union.

Specifications Length: (SSBN) 426.5 ft. (130 m.); (SSN) 329.6 ft. (100.5 m.); (Notch) 459.2 ft. (140 m.). Beam: 38.1 ft. (11.6 m.). Displacement: (SSBN) 7700 tons surfaced/9300 tons submerged; (SSN) 5600 tons submerged; (Notch) 11,500 tons submerged. Powerplant: twin-shaft nuclear: 2 pressurized-water reactors, 2 sets of geared turbines, 50,000 shp. Speed: 20 kt. surfaced/27 kt. submerged. Max. operating depth: 1315 ft. (400 m.). Collapse depth: 1970 ft. (600 m.). Crew: 120. Sensors: 1 low-frequency active/passive sonar, 1 active medium-frequency fire control sonar, 1 "Snoop Tray" surface-search radar, 1 "Brick Group" electronic signal monitoring array, 2 periscopes.

"YELLOW RAIN": A Soviet-developed chemical warfare agent, thus called because of the dense yellow clouds associated with its release. Yellow rain was first reported in the late 1970 by Hmong tribesmen fleeing Laos and Cambodia. After 1979, it was also reported by Afghan guerrillas. Samples of clothing, soil, and vegetation brought to U.S. government officials revealed abnormally high concentrations of mycotoxins—natural poisons produced by several species of fungi, particularly fusaria, whose specific mycotoxin is tricothecene.

Fusaria and related fungi can accidentally infect stored grain and have been known to cause fatalities, most notably by rye ergot poisoning. That induced some U.S. scientists to dispute official U.S. allegations of Soviet (and Soviet client-state) use of mycotoxins as weapons, but abundant evidence supports that finding, including the presence of tricothecenes in the absence of fusaria fungi, finds of tricothecenes produced by fungi alien to the areas in question, heavy, localized concentrations of tricothecenes after Soviet and Vietnamese bombing attacks (shown on film), defector reports, and autopsy evidence.

The symptoms associated with Yellow Rain are particularly gruesome: dermal necrosis (blackening and putrefaction of the skin), nausea and vomiting, shock, nervous disorders, massive internal hemorrhaging, blood poisoning, and, eventually, death from respiratory collapse.

Mycotoxins are very easy to manufacture in large quantities by standard fermentation processes, and can be effective in relatively small doses. Unlike blister agents or nerve agents, however, they cannot penetrate skin; hence, they can be reliably negated by gas masks and filters, which makes them especially suitable for use against backward opponents such as the Hmong. On the other hand, mycotoxins cannot be detected by current Western chemical warning systems, and this could make them effective as surprise weapons. See also biological warfare.

YUBARI: A class of two Japanese frigates (FFs) commissioned in 1983–84. Enlarged versions of the Ishikari class (which proved to be too cramped), the Yubaris are intended as general-purpose escorts with emphasis on anti-surface warfare (ASUW). Armament includes 1 OTO-Melara 76.2-mm. dual purpose gun, 8 harpoon anti-ship missiles, 1 Bofors 4-barrel, 375-mm. depth charge mortar, 2 sets of Mk.32 triple tubes for lightweight anti-submarine warfare (ASW) torpedoes, and one phalanx radar-controlled gun for anti-missile defense. See also frigates, japan.

Specifications Length: 295.75 ft. (90.16 m.). Beam: 35.1 ft. (10.7 m.). Draft: 11.3 ft. (3.45 m.). Displacement: 1400 tons standard/1690 tons full load. **Powerplant:** twin-shaft CODOG: 1 28390-shp. Kawasaki-Olympus TM-3B gas turbine (surge), 1 5000-hp. Mitsubishi 6DRV 35/44 diesel (cruise). **Speed:** 25 kt. **Crew:** 98. **Sensors:** 1 OPS-28 surveillance radar, 1 GFCS-1 fire control radar, one OQS-1 medium-frequency sonar. **Electronic warfare equipment:** 1 NOLQ-6 radar warning receiver, 1 OLT-3 active jamming unit, 2 Mk.36 SRBOC chaff launchers.

YUUSHIO: A class of 11 Japanese diesel-electric attack SUBMARINES (SSs), commissioned between 1980 and 1990. Improved versions of the UZUSHIO class, they have a teardrop hullform based on that of the U.S. BARBEL class, but unlike the Barbels, both the Yuushios and Uzushios have a double-hull configuration, with fuel and ballast tanks between the pressure hull and the outer casing. The pressure hull is fabricated from super-strong NS-90 steel, allowing a 50 percent increase in depth capability over the Uzushios. Control surfaces are similar to those on contemporary U.S. nuclear submarines, with "fairwater" planes mounted on the tall, streamlined sail, and cruciform rudders and stern planes mounted ahead of the propeller. A computerized system automatically controls both course and depth. A retractable snorkel allows diesel operation from shallow submergence. A successor submarine, equipped with an advanced Hitachi weapon-control system, is under development.

Armed with six 21-in. (533-mm.) torpedo tubes amidships, the Yuushios have a basic load of 18 TORPEDOES, notably old U.S. Mk.37 anti-ship and Mk.36 (short) anti-submarine wire-guided homing torpedoes (Japan refused to buy the modern Mk.48 when license production was not allowed, and was denied the latest ADCAP version). HARPOON anti-ship missiles, for launch through the torpedo tubes, can replace torpedoes on a 1-for-1 basis. See also SUBMARINES, JAPAN.

Specifications **Length:** 50 ft. (76.2 m.). **Beam:** 32.5 ft. (9.9 m.). **Displacement:** 2200 tons surfaced/2730 tons submerged. **Powerplant:** single-shaft diesel-electric: 2 1700-hp. Kawasaki-MAN V8/V24-30 AMTL diesel-generators, 1 7200-hp. Fuji electric motor. **Speed:** 12 kt. surfaced/20 kt. submerged. **Range:** 9000 n.mi. at 9 kt. **Max. operating depth:** 984 ft. (300 m.). **Collapse depth:** 1640 ft. (500 m.). **Crew:** 80. **Sensors:** 1 bow-mounted ZQQ-4 (Japanese version of the U.S. BQS-4) low-frequency passive sonar, 1 SQS-36(J) active attack array, 1 ZPS-6 surface-search radar, several electronic signal monitoring arrays, 2 periscopes.

Z

ZAMPOLIT: *Zamestitel Po Politcheskim Delom*, Deputy Commander for Political Affairs. Every Soviet ground and air force unit, from regiment down to company level, and every naval vessel has a *Zampolit* appointed by the Main Political Directorate of the Armed Forces, which is the military arm of the Communist party. In higher formations, the Main Political Directorate is represented by Political Departments, headed by party officials.

The *Zampolit* is responsible for the political orthodoxy of his unit, as opposed to military command. His functions include political indoctrination, the assessment of the political reliability of officers and men, and the upkeep of unit MORALE. As a representative of the party, the *Zampolit* is independent of the military chain of command. Theoretically he should not interfere in purely military matters, but as the de facto coequal of the military commander he may question or even override the latter's decisions, despite a lack of formal military training. The office of *Zampolit* is itself indicative of the Communist party's distrust of the armed forces, as potential threats to its power.

ZEELEEUW: A class of 6 Dutch diesel-electric attack SUBMARINES (SSs), originally known as the Walrus class. The first was scheduled for completion in 1986, but was seriously delayed by a shipyard fire. The first 2 Zeeleeuws were commissioned in 1989–90, and the remaining 4 are scheduled for completion at two-year intervals. Ordered in 1979 as replacements for the DOLFIJN

class, the Zeeleeuws were originally planned as incremental improvements on the previous ZWAARDVIS class, but changing operational requirements, notably a 50 percent increase in the required operating depth, led to a major redesign.

Like the Zwaardvis, the Zeeleeuws have teardrop-shaped hulls derived from the U.S. BARBEL class. In their partial double-hull configuration, the pressure hull consists of a large cylinder amidships, with smaller cylinders at each end covered by a streamlined outer casing with fuel and ballast tanks in between. The pressure hull is constructed of high-tensile-strength French MAREL steel. These boats have a tall, streamlined sail amidships which carry the forward diving planes. The stern planes have the unusual, computer-controlled X-configuration pioneered by Swedish submarines. A retractable snorkel allows diesel operation from shallow submergence, while three 140-cell storage batteries provide long submerged endurance. Special attention has been given to acoustical silencing, and the diesels are mounted on a resilient sound-isolation raft, to reduce radiated noise.

The Zeeleeuws have four 21-in. (533-mm.) torpedo tubes equipped with a positive discharge "water slug" ejection mechanism, which allows torpedoes to be launched at any depth. The basic load of 24 weapons will include Northrop NT-37D and MK.48 homing torpedoes and HARPOON antiship missiles; mines can replace other weapons on a 2-for-1 basis. All sensors are linked to a SEWACO VIII Action Information Center and torpedo fire control unit. See also SUBMARINES, NETH-ERLANDS.

Specifications **Length:** 222.1 ft. (67.7 m.). **Beam:** 27.6 ft. (8.4 m.). **Displacement:** 2390 tons surfaced/2740 tons submerged. **Powerplant:** single-shaft diesel-electric: 3 SEMT-Pielstick 12PA14V200 diesel-generators, 3950 hp. 1 Helec electric motor, 5430 hp. **Speed:** 13 kt. surfaced/21 kt. submerged (burst). **Fuel:** 300 tons. **Range:** 10,000 n.mi. at 9 kt. **Patrol endurance:** 60 days. **Max. operating depth:** 1148 ft. (350 m.). **Collapse depth:** 2297 ft. (700 meters.). **Crew:** 50. **Sensors:** 1 bow-mounted Octopus active/passive low-frequency circular sonar array, 1 British Type 2026 passive towed array, 1 Passive/Active Range and Intercept Sonar (PARIS), 1 Decca Type 1001 surface-search radar, 1 electronic signal monitoring array, 2 periscopes.

ZERO OPTION: ARMS CONTROL jargon for the complete elimination of a particular class of weapon. The term was first used during the negotiations that led to the INF TREATY, when the U.S. proposed, and eventually obtained, the elimination of all U.S. and Soviet intermediate-range ballistic missiles (IRBMS).

ZPU-4: Soviet quadruple ANTI-AIRCRAFT gun, which consists of four KPV 14.5-mm. machine guns mounted on a towed, four-wheel carriage, with 360° manual traverse, and elevation limits of −19° to +90°. Each gun is fed by a 150-round belt stored in a drum magazine. The cyclic rate of fire is 600 rounds per minute per barrel, but the practical rate of reloading allows only 150 rounds per minute per barrel. The ZPU-4 has a maximum effective range of 1400 m. against aircraft and 2000 m. against ground targets, but its mechanically computing optical gunsight functions only in good visibility. The ZPU-4 weighs 3700 lb. (1682 kg.) in firing order, and requires a crew of five men.

In the Soviet army the ZPU-4 has been replaced by the 23-mm. ZU-23-2, but it is still in service with other Soviet-supplied forces and in China. During the Vietnam War, the ZPU-4's small size and mobility made it quite effective in an anti-aircraft ambush role.

ZSU-: *Zenitnaya Samokhodnaya Ustanovka* (Zenith Self-Propelled Mount), the Soviet term for self-propelled ANTI-AIRCRAFT guns. See ZSU-23-4; ZSU-30; ZSU-57-2.

ZSU-23-4: A Soviet self-propelled battlefield AIR DEFENSE system, also known as *Shilka*. Developed in the early 1960s, the ZSU-23-4 entered service in 1966 as a radar-directed, all-weather replacement for the optically directed ZSU-57-2. Its four 23-mm. AZP-23 water-cooled CANNONS are enclosed in a power-operated turret mounted on a lightly armored tracked chassis derived from the PT-76 light tank. Hull and turret are protected by 9–15 mm of armor—adequate protection against small arms and splinters. The ZSU-23-4 has been produced in larger numbers than any other self-propelled anti-aircraft gun. The Soviet army alone has more than 2000, and large numbers have been exported.

The AZP-23 has a cyclic rate of fire of 800–1000 rounds per minute, but barrel heating and ammunition reload rates limit the practical rate of fire to 200 rounds per minute. In the quadruple ZSU-23-4, the cyclic rate is thus 3200–4000 rounds per minute, and the practical rate 800 rounds per minute; 2000 rounds of ammunition are stored on board in 40 50-round box magazines. The gun can fire a mix of HEI, API, and fragmentation rounds, with a maximum muzzle velocity of 970 m./sec. The turret has 360° traverse at a rate of 45° per second, with elevation limits of −7° and +80°.

The "Gun Dish" RADAR at the rear of the turret can acquire targets at ranges up to 20 km. (12 mi.), and track them from 8 km. (4.8 mi.). It has a moving target indicator (MTI) to suppress ground clutter, CHAFF, and NOISE JAMMING but is relatively ineffective against targets below 60 m. (200 ft.). In action, radar tracking data are fed to an analog FIRE CONTROL computer which provides azimuth and elevation cues to the gunner. Both the guns and the radar are fully stabilized to permit firing on the move, but this degrades accuracy by up to 50 percent. Backup optical sights can be used in two modes: electro-optical, in which Gun Dish provides range only, with tracking performed optically; and fully optical, obviously usable only in good visibility, but also faster and more effective than radar against targets below 200 ft. Under radar direction, the guns can engage aircraft at a maximum range of 3500 m. (2.1 mi.) at heights between 60 and 4875 m. (200 and 16,000 ft.), although they are most effective below 1525 m. (5000 ft.). The U.S. Army credits the *Shilka* with a 53 percent hit probability against a maneuvering aircraft at a range of 1000 m. (0.6 mi.) with a 40-round burst; or 30 percent at 1500 m. (1 mi.). Against ground targets, the guns can be fired using optical sights out to an effective range of 2000 m. (1.2 mi.); they are capable of penetrating up to 20 mm. of armor at 1000 m., and thus are quite effective against light armored vehicles.

The *Shilka* is the standard anti-aircraft gun in Soviet tank and Motorized Rifle REGIMENTS, with one PLATOON of four *Shilkas* attached to every regimental air defense COMPANY; in combat, ZSU-23-4s would follow closely behind the lead combat echelons. In addition, one *Shilka* platoon is attached to each Army-level SA-4 GANEF surface-to-air missile battalion to provide close-in defense.

Shilkas were in limited service with the North Vietnamese army from 1972, but their true baptism of fire was in the 1973 Yom Kippur War, during which Egyptian and Syrian forces employed them in large numbers. They proved highly effective, destroying 31 to 47 of the 103 Israeli aircraft lost in combat, and damaging many more. The absence of effective ELECTRONIC COUNTERMEASURES (ECM) and the sheer density of fire over relatively small battle areas were compounded by the synergy between the *Shilka* and the SA-6 GAINFUL—to avoid the latter, Israeli aircraft flew low, into the *Shilka*'s lethal envelope. Once the initial surprise wore off, however, the *Shilka* revealed serious operational shortcomings, including its extreme vulnerability to direct fire compounded by a distinctively high silhouette, limited cross-country mobility, gun and radar cooling problems, and, above all, the vulnerability of Gun Dish to DECEPTION JAMMING and the lack of an on-board surveillance radar. An improved post-1973 variant, the ZSU-23-4M, has a modified Gun Dish with a wider frequency range, more jam resistance, and the ability to rotate and search independently of turret orientation. The *Shilka* is now being replaced by the new ZSU-30.

Specifications **Length:** 21.4 ft. (6.54 m.). **Width:** 9.67 ft. (2.95 m.). **Height:** 7.33 ft. (2.25 m.) w/radar folded/12.5 ft. (3.8 m.) w/radar erected. **Weight, combat:** 20.5 tons. **Powerplant:** 280-hp. V-6R 6-cylinder in-line diesel. **Fuel:** 57 gal. (250 lit.). **Speed, road:** 27 mph (44 kph). **Range, max.:** 162 mi. (260 km.). **Operators:** Afg, Alg, Ang, Bul, Cuba, Cze, DDR, Egy, Eth, Hun, Ind. Iran, Iraq, Jor, Kam, Lib, Moz, N. Kor, Nig, N Ye, Peru, Pol, Som, S Ye, Syr, USSR, Viet, Yug.

ZSU-30: Soviet self-propelled battlefield AIR DEFENSE system introduced in the late 1980s as a replacement for the ZSU-23-4 *Shilka*. The ZSU-30 reportedly consists of two 30-mm. ANTI-AIRCRAFT guns in an armored turret mounted on a T-72 tank chassis. Two launch tubes for SA-15 short-range surface-to-air missiles provide additional firepower. It is said to resemble the West German GEPARD, including its separate surveillance and tracking radars. The ZSU-30 could also rectify automotive deficiencies of the ZSU-23-4, including its inadequate armor and relatively low mobility. The 30-mm. gun, if based on the Neudelman-Richter NR-30 aircraft cannon, should have much greater striking power than the *Shilka*'s 23-mm. gun.

ZSU-57-2: Soviet self-propelled AIR DEFENSE system, consisting of two 57-mm. S-60 ANTI-AIRCRAFT guns in a lightly armored, open-top turret mounted on a T-54 tank chassis. The ZSU-57-2 was the standard regimental anti-aircraft gun from the mid-1950s until it began to be replaced by the ZSU-23-4 *Shilka* from 1966. The S-60 gun has a cyclic rate of fire of 105–120 rounds per minute, limited by ammunition reloading to 70 rounds per minute; the ZSU-57-2 thus has a total rate of fire of 140 rounds per minute. The S-60 can fire a mix of HE, HEI, APC, and API rounds, some fitted with radar proximity (VT) fuzes. Muzzle velocity is 1000 m./sec. (328 ft./sec.), and the maximum range against aircraft is 4000 m. (2.4 mi.) with a ceiling of 1540 m. (5050 ft.). In practice, the mechanically computing optical sights limit effective ranges much more restrictively and require good visibility. The guns are highly effective against ground targets out to 2000 m. (1.2 mi.), with armor penetration of up to 106 mm. at 1000 m.

The hull, engine, drive train, and performance are essentially identical to those of the T-54 tank. The turret has a maximum traverse rate of 30° per second, with elevation limits of −5° and +85° at 20° per second. The crew of six consists of a driver, commander, gunner, and three ammunition handlers. The ZSU-57-2 lacks radar, and, moreover, the open-topped turret exposes the crew to enemy fire.

Specifications **Length:** 20.5 ft. (6.22 m.). **Width:** 10.75 ft. (3.27 m.). **Height:** 9 ft. (2.75 m.). **Weight, combat:** 28 tons. **Powerplant:** 520-hp. V-12 diesel. **Fuel:** 275 gal. (1212 lit.). **Speed, road:** 30 mph (48 kph). **Range, max.:** 370 mi. (595 km.). **Operators:** Bul, DDR, Egy, Fin, Hun, Iran, Iraq, N Kor, Pol, Rom, Syr, USSR (reserves), Viet, Yug.

ZU-23-2: Soviet towed, twin 23-mm. ANTI-AIRCRAFT gun. The same AZP-23 gun arms the self-propelled ZSU-23-4 *Shilka*, but the ZU-23-2's guns are air-cooled rather than water-cooled; hence, they are equipped with quick-change barrels to avoid overheating. The cyclic rate of fire is 1000 rounds per minute per barrel, but overheating limits the practical rate to 200 rounds per minute per barrel. Effective gun ranges are unchanged, but system ranges are defined by the lack of radar, and the ZU-23-2's mechanically computing optical gun sight of course requires good visibility. The wheels of its two-wheeled carriage fold down in firing position, with the weapon resting on jacks. The carriage has manually operated 360° traverse, with elevation limits of −10° and +90°. It weighs 1850 lb. (840 kg.) in firing order and requires a crew of five.

Operators: Ang, Egy, Fin, Iran, Iraq, Lib, Moz, Pak, Soviet and Warsaw Pact airborne forces.

ZULU: NATO designation for a class of 28 Soviet diesel-electric attack SUBMARINES (SSs) completed between 1951 and 1955; only 3 remain in service as research vessels. See SUBMARINES, SOVIET UNION.

ZUNI: U.S. 5-in. (127-mm.) unguided, folding-fin aerial rocket (FFAR), developed by the U.S. Navy in 1957 as a successor to the fixed-fin 5-in. High Velocity Aerial Rocket (HVAR) of World War II. Both of the two alternative shaped-charge (HEAT) and general-purpose (HE) warheads can be fitted with either impact or radar proximity (VT) fuzes. Zuni was originally intended for an air-to-air as well as air-to-ground role, but has seldom been used as such (though one MIG-17 FRESCO was shot down with Zunis in the Vietnam War). The rocket is quite accurate, and extremely effective against hardened targets such as bunkers. Zunis are usually employed with a four-round, reusable LAU-10/A pod, which weighs 533 lb. (242 kg.) loaded, and can be carried by most tactical aircraft. The Zuni's rocket motor was also used to power the original AIM-9A SIDEWINDER air-to-air missile.

Specifications **Length:** 110 in. (2.79 m.). **Diameter:** 5 in. (127 mm.). **Weight, launch:** 107 lb. (48.65 kg.). **Speed, max.:** 750 m./sec. **Effective range:** 3000 m.

ZWAARDVIS: A class of two Dutch diesel-electric attack SUBMARINES (SSs), commissioned in 1972. The Zwaardvis have a teardrop-shaped hull derived from the U.S. BARBEL class, with a partial double-hull configuration; fuel and ballast tanks are installed between the pressure hull and the outer casing at the bow and stern. Control surfaces consist of sail-mounted "fairwater" diving planes, and cruciform stern rudders and stern planes ahead of the propeller. A retractable snorkel allows diesel operation from shallow submergence, while two 140-cell batteries provide fully submerged power. Special attention was paid to acoustical silencing, and the diesel-generators are mounted on resilient sound-isolation rafts. The Zwaardvis have now been supplemented by the improved ZEELEEUW class.

The Zwaardvis are armed with six bow-mounted 21-in. (533-mm.) torpedo tubes. At present, the basic load consists of 20 U.S. Mk.37 or MK.48 homing torpedoes, but HARPOON anti-ship missiles may replace some torpedoes in the future. All sensors are linked to an HSA M8 Torpedo FIRE CONTROL computer, which allows the simultaneous launch of two torpedoes.

Taiwan purchased two slightly modified Zwaardvis, commissioned in 1987 as the Hai Lung class. The main differences are the sensors, which include an HSA "SIASS" integrated sonar system, a Decca surface-search radar, and an HSA "RAPIDS" ELECTRONIC SIGNAL MONITORING array. All sensors are linked to a highly automated SINBADS-M fire control unit which can track up to eight targets simultaneously. Other changes include two 196-cell batteries for greater underwater endurance, and a Sperry Mk.29 Mod 2A INERTIAL NAVIGATION unit. Orders for four more submarines were canceled by the Netherlands after protests from Beijing. See also SUBMARINES, NETHERLANDS.

Specifications **Length:** 216.5 ft. (66 m.). **Beam:** 27.6 ft. (8.4 m.). **Displacement:** 2350 tons surfaced/2640 tons submerged. **Powerplant:** single-shaft diesel-electric: 3 1400-hp. Werkspoor RUB 215X12 diesel-generators, 1 5100-hp. electric motor. **Speed:** 13 kt. surfaced/20 kt. submerged. **Range:** 10,000 n.mi. at 9 kt. **Max. operating depth:** 984 ft. (300 m.). **Collapse depth:** 1640 ft. (500 m.). **Crew:** 67. **Sensors:** 1 Thomson-CSF active/passive low-frequency sonar, 1 British Type 2026 passive towed array sonar, 1 Decca Type 1001 surface-search radar, 2 periscopes.

GLOSSARY

Anhedral: For aircraft, the downward slope of the wings as viewed from the front.

AOA: Angle of Attack. The angle between an aircraft's velocity vector and the chord line of the wings. If AOA exceeds certain limits, the wing stalls and the aircraft could enter a spin.

CEP: Circular Error, Probable. See entry.

Dihedral: For aircraft, the upward slope of the wings as viewed from the front.

Hi-hi-hi: For military aircraft, a flight profile in which the entire mission is flown at high altitude.

Hi-lo-hi: For military aircraft, a flight profile in which the first phrase of the mission is flown at high altitude (for fuel efficiency), after which the aircraft decends to low altitude to approach and attack the target (to avoid defenses), and then returns to base at high altitude.

Lo-lo-lo: For military aircraft, a flight profile in which the entire mission is flown at low altitude.

OA: Overall (as in length)

Recce: Reconnaissance

Seakeeping: For naval vessels, a combination of seaworthiness, stability, dryness, and ruggedness that makes a ship a superior weapon and sensor platform.

Specific Impulse (I_{sp}): For rockets and missiles, a measure of the specific energy of propellants. Specifically, the time in seconds that one unit of fuel can lift one unit of mass one foot (or one meter in metric terms).

MAJOR WEAPONS

AIRCRAFT

NAME	DESIGNATION	NATIONALITY
Attack		
Alpha Jet		Fr/FRG
AMX		Br./It.
ATA	A-12	US
Buccaneer	S.2	UK
Corsair II	A-7	US
Dragonfly	A-37	US
Etendard		Fr.
Fencer	Su-24	USSR
Fitter	Su-7/17/22	USSR
Flogger	MiG-27	USSR
Forger	Yak-38	USSR
Frogfoot	Su-25	USSR
Harrier	AV-8	US/UK
Hawk		UK
Hunter		UK
Intruder	A-6	US
Jaguar		Fr./UK
Lavi		Israel
	F-111	US

NAME	DESIGNATION	NATIONALITY
Skyhawk	A-4	US
Skyraider	A-1	US
Skywarrior	A-3	US
Thunderbolt	A-10	US
Tornado	IDS	UK/FRG/It.
Vigilante	A-5	US

Bomber

NAME	DESIGNATION	NATIONALITY
Backfire	Tu-26	USSR
Badger	Tu-16	USSR
Beagle	Il-28	USSR
Bear	Tu-95	USSR
Bison	Mya-4	USSR
Blackjack	Tu-160	USSR
Blinder	Tu-22	USSR
Brewer	Yak-28I	USSR
Canberra		UK/US
Mirage IV		Fr.
na	B-2	US
na	FB-111	US
	B-1B	US
Stratofortress	B-52	US

Fighter

NAME	DESIGNATION	NATIONALITY
ATF	YF-22/23	US
Crusader	F-8	US
Delta Dart	F-106	US
Draken	J.35	Sw
Eagle	F-15	US
Falcon	F-16	US
Farmer	MiG-19	USSR
Firebar	Yak-28P	USSR
Fishbed	MiG-21	USSR
Flagon	Su-15	USSR
Flanker	Su-27	USSR
Flogger	MiG-23	USSR
Foxbat	MiG-25	USSR
Foxhound	MiG-31	USSR
Fresco	MiG-17	USSR
Fulcrum	MiG-29	USSR
Gripen	JAS.39	Sw
Hornet	FA-18	US
Kfir		Israel
Lightning		UK
Mirage 2000		Fr
Mirage F.1		Fr
MirageIII/5		Fr.
Nighthawk	F-117	US
Phantom	F-4	US
Sabre	F-86	US
Starfighter	F-104	US
Super Sabre	F-100	US
Tiger	F-5	US
Tigershark	F-20	US

NAME	DESIGNATION	NATIONALITY
Tomcat	F-14	US
Tornado	ADV	UK/FRG/It.
Viggen	JA.37	Sw.

Anti-Submarine Warfare

NAME	DESIGNATION	NATIONALITY
Alize		Fr.
Atlantic		Fr.
Mail	Be-12	USSR
May	Il-38	USSR
Neptune	P-2	US
Nimrod		UK
Orion	P-3	US
Tracker	S-2	US
Viking	S-3	US

Reconnaissance

NAME	DESIGNATION	NATIONALITY
Blackbird	SR-71	US
	TR-1	US
	U-2	US
Vigilante	RA-5	US

Airborne Early Warning/Electronic Warfare

NAME	DESIGNATION	NATIONALITY
Hawkeye	E-2	US
Hermes	E-6	US
J-Stars	E-8	US
Mainstay	Il-76	USSR
Moss	Tu-	USSR
NEACP	E-4	US
Prowler	EA-6B	US
Raven	EF-111A	US
Sentry	E-3	US

Transport/Tanker

NAME	DESIGNATION	NATIONALITY
Candid	Il-76	USSR
Cock	An-22	USSR
Colt	An-2	USSR
Condor	An-124	USSR
Coot	Il-18	USSR
Crate	Il-14	USSR
Cub	An-12	USSR
Extender	KC-10	US
Galaxy	C-5	US
Hercules	C-130	US
Midas	Il-78	USSR
	C-17	US
Osprey	V-22	US
Starlifter	C-141	US
Stratotanker	KC-135	US
Transall	C.160	Fr/FRG
Victor	na	UK

NAME	DESIGNATION	NATIONALITY

HELICOPTERS

NAME	DESIGNATION	NATIONALITY
Alouette		Fr.
Apache	AH-64	US
Blackhawk	UH-60	US
	BO.105	FRG
	BO.105	FRG
Chinook	CH-47	US
Cobra	AH-1	US
Defender	500.MD	US
Gazelle		Fr.
Halo	Mil-26	USSR
Havoc	Mil-28	USSR
Haze	Mil-17	USSR
Helix	Ka-27	USSR
Hind	Mil-24	USSR
Hip	Mil-8	USSR
Hockum		USSR
Hook	Mil-6	USSR
Hoplite	Mil-2	USSR
Hormone	Ka-25	USSR
Huey	UH-1	US
Kiowa	OH-58	US
Lynx		UK
Puma		Fr.
Sea Hawk	SH-60	US
Sea King	SH-3	US
Sea Knight	CH-46	US
Sea Sprite	SH-2	US
Sea Stallion	H-53	US
Super Frelon	SA.321	Fr.
Wessex		UK

WARSHIPS

CLASS	NATIONALITY

Aircraft Carriers

CLASS	NATIONALITY
Enterprise	US
Forrestal	US
Garibalidi	It.
Invincible	UK
Kiev	USSR
Kitty Hawk	US
Midway	US
Minas Gerais	Br.
Moskva	USSR
Nimitz	US
Principe de Asturias	Sp.
Tbilisi	USSR
Veinticento de Mayo	Ar.
Vikrant	In.
Viraat	In.

CLASS	NATIONALITY

Battleships

CLASS	NATIONALITY
Iowa	US

Battle Cruisers

CLASS	NATIONALITY
Kirov	USSR

Cruisers

CLASS	NATIONALITY
Andrea Doria	It.
Bainbridge	US
Belknap	US
California	US
Colbert	Fr.
Kara	USSR
Kresta	USSR
Kynda	USSR
Leahy	US
Long Beach	US
Slava	USSR
Sverdlov	USSR
Ticonderoga	US
Truxton	US
Virginia	US
Vittorio Veneto	It.

Destroyers

CLASS	NATIONALITY
Adams	US
Animoso	It.
Asagiri	Jap.
Audace	It.
Bristol	UK
Burke	US
Cassard	Fr.
Coontz	US
County	UK
Fletcher	US
Gearing	US
George Leygues	Fr.
Hamburg	FRG
Haruna	Jap.
Hatakaze	Jap.
Hatsuyuki	Jap.
Iroquois	Can.
Kanin	USSR
Kashin	USSR
Kidd	US
Kildin	USSR
Kotlin	USSR
Luda	PRC
Lutjens	FRG
Minegumo	Jap.
Perth	Aus.
Sheffield	UK
Shirane	Jap.
Skoryy	USSR

CLASS	NATIONALITY
Sovremenyy	USSR
Spruance	US
Suffren	Fr.
Sumner	US
Tachikaze	Jap.
Takatsuki	Jap.
Tromp	Ne.
Udaloy	USSR
Yamagumo	Jap.

Frigates

CLASS	NATIONALITY
Amazon	UK
Annapolis	Can.
Baleres	Sp.
Bremen	FRG
Broadsword	UK
Bronstein	US
Brooke	US
Chikugo	Jap.
Commandante Riviere	Fr.
d'Estinne d'Orves	Fr.
Dealey	US
Descubierta	Sp.
Garcia	US
Halifax	Can.
Heemskerck	Ne.
Joao Coutinho	Port.
Karel Doorman	Ne.
Knox	US
Koln	FRG
Kortenaer	Ne.
Krivak	USSR
Leander	UK
Lupo	It.
Maestrale	It.
MEKO	FRG
Niteroi	Br.
Perry	US
Restigouche	Can.
Riga	USSR
River	Aus.
Saint Laurent	Can.
van Speijk	Ne.
Yubari	Jap.

Corvettes

CLASS	NATIONALITY
Grisha	USSR
Mirka	USSR
Nanuchka	USSR
Pauk	USSR
Petya	USSR
Poti	USSR
Tarantul	USSR

CLASS	NATIONALITY

Fast Attack Craft

CLASS	NATIONALITY
Combattante	Fr.
Komar	USSR
Matka	USSR
Osa	USSR
Pegasus	US
Reshef	Is.
Sa'ar	Is.
Sparviero	It.
Turya	USSR

Amphibious Assault

CLASS	NATIONALITY
Alligator	USSR
Ivan Rogov	USSR
Tarawa	US
Wasp	US
Polnocny	USSR

Hovercraft

CLASS	NATIONALITY
LCAC	US
Aist	USSR
Pomornik	USSR

SUBMARINES

CLASS	TYPE	NATIONALITY
Agosta	SS	Fr.
Akula	SSN	USSR
Alfa	SSN	USSR
Barbel	SS	US
Charlie	SSGN	USSR
Daphne	SS	Fr.
Delta	SSBN	USSR
Dolfijn	SS	Ne.
Draken	SS	Sw.
Echo	SSGN	USSR
Ethan Allen	SSBN	US
Foxtrot	SS	USSR
Franklin	SSBN	US
George Washington	SSBN	US
Golf	SSB	USSR
Guppy	SS	US
Han	SSN	PRC
Hotel	SSBN	USSR
India	SSR	USSR
Inflexible	SSBN	Fr.
Juliett	SSG	USSR
Kilo	SS	USSR
Lafayette	SSBN	US
Los Angeles	SSN	US
Mike	SSN	USSR
Nacken	SS	Sw.
November	SSN	USSR

CLASS	TYPE	NATIONALITY
Oberon	SS	UK
Ohio	SSBN	US
Oscar	SSGN	USSR
Papa	SSGN	USSR
Permit	SSN	US
Porpoise	SS	UK
Quebec	SS	USSR
Redoutable	SSBN	Fr.
Resolution	SSBN	UK
Romeo	SS	USSR
Rubis	SSN	Fr.
Sauro	SS	It.
Sea Wolf	SSN	US
Sierra	SSN	USSR
Sjoorman	SS	Sw.
Skipjack	SSN	US
Sturgeon	SSN	US
Swiftsure	SSN	UK
Tang	SS	US
Tango	SS	USSR
Toti	SS	It.
TR-1700	SS	FRG
Trafalgar	SSN	UK
Tullibee	SSN	US
Type 205	SS	FRG
Type 206	SS	FRG
Type 207	SS	FRG
Type 209	SS	FRG
Type 210	SS	FRG
Type 211	SS	FRG
Typhoon	SSBN	USSR
Upholder	SS	UK
Uzushio	SS	Jap.
Valiant	SSN	UK
Vastergotland	SS	Sw.
Whiskey	SS	USSR
Xia	SSBN	PRC
Yankee	SSBN	USSR
Yuushio	SS	Jap.
Zeeleeuw	SS	Ne.
Zulu	SS	USSR
Zwaardvis	SS	Ne.

MISSILES

NAME	DESIGNATION	NATIONALITY

Anti-Tank Guided Missiles

NAME	DESIGNATION	NATIONALITY
Bill	Rbs.56	Sw.
Cobra		FRG
Dragon	M47	US
Hellfire	AGM-114	US
HOT		FR/FRG
Mamba		FRG

NAME	DESIGNATION	NATIONALITY
Milan		FR/FRG
Sagger	AT-3	USSR
Saxhorn	AT-7	USSR
Shillelagh	MGM-51	US
Snapper	AT-1	USSR
Songster	AT-8	USSR
Spandrel	AT-5	USSR
Spigot	AT-4	USSR
Spiral	AT-6	USSR
Swatter	AT-2	USSR
Swingfire		UK
TOW	BGM-71	US

Air-To-Air

NAME	DESIGNATION	NATIONALITY
Acrid	AA-6	USSR
Acrid	AA-7	USSR
Alamo	AA-10	USSR
Alkalai	AA-1	USSR
Amos	AA-9	USSR
AMRAAM	AIM-120	US
Anab	AA-3	USSR
Aphid	AA-8	USSR
Archer	AA-11	USSR
Ash	AA-5	USSR
Aspide		It.
ASRAAM	AIM-132	UK
Atoll	AA-2	USSR
Falcon	AIM-4	US
Genie	AIR-2	US
Magic	R.550	Fr.
Matra	R.530	Fr.
Phoenix	AIM-54	US
Python		Is.
Shafrir		Is.
Sidewinder	AIM-9	US
Sky Flash		UK
Sparrow	AIM-7	US

Air-To-Ground

NAME	DESIGNATION	NATIONALITY
	AGM-130	US
ALARM		UK
ALCM	AGM-86	
ASMP		UK
Bullpup	AGM-12	US
	GBU-15	US
HARM	AGM-88	US
HOBOS	GBU-8	US
Kangaroo	AS-3	USSR
Karen	AS-10	USSR
Kedge	AS-14	USSR
Kegler	AS-12	USSR
Kelt	AS-5	USSR
Kennel	AS-1	USSR
Kent	AS-15	USSR

NAME	DESIGNATION	NATIONALITY	NAME	DESIGNATION	NATIONALITY
Kerry	AS-7	USSR	Gainful	SA-6	USSR
Kilter	AS-11	USSR	Gammon	SA-5	USSR
Kingfish	AS-6	USSR	Ganef	SA-4	USSR
Kipper	AS-2	USSR	Gaskin	SA-9	USSR
Kitchen	AS-4	USSR	Gecko	SA-8/SA-N-4	USSR
Kyle	AS-9	USSR	Giant	SA-12B	USSR
Martel		Fr/UK	Gladiator	SA-12A	USSR
Maverick	AGM-65	US	Goa	SA-3/SA-N-1	USSR
PAVEWAY	GBU-12/16/24	US	Goblet	SA-N-3	USSR
Shrike	AGM-45	US	Gopher	SA-13	USSR
Sidearm	AGM-122	US	Grail	SA-7/SA-N-5	USSR
Skipper	AGM-123	US	Gremlin	SA-14/SA-N-8	USSR
SRAM	AGM-69	US	Grumble	SA-10/SA-N-6	USSR
Standard ARM	AGM-78	US	Guideline	SA-2/SA-N-2	USSR
Tacit Rainbow	AGM-136	US	HAWK	MIM-23	US
Tomahawk	AGM-109	US	Javelin		UK
Walleye	AGM-62	US	Mascura		Fr.
			Mistral		Fr.

Anti-Ship

			Nike Hercules	MIM-14	US
Exocet	MM.38/40	Fr.	Patriot	MIM-104	US
Gabriel		Is.	RAM	RIM-11	US
Harpoon	AGM-84	US	Rapier		UK
Kangaroo	AS-3	USSR	Redeye	FIM-43	US
Kelt	AS-5	USSR	Roland		Fr/FRG
Kent	AS-15	USSR	Sea Cat		UK
Kingfish	AS-6	USSR	Sea Dart		UK
Kipper	AS-2	USSR	Sea Sparrow	RIM-7	US
Kitchen	AS-4	USSR	Sea Wolf		UK
Kormoran		FRG	Standard	RIM 66/67	US
Martel		Fr/UK	Stinger	FIM-92	US
Otomat		It.	Talos	RIM-8	US
Penguin	AGM-119	Nor/US	Tartar	RIM-24	US
Sampson	SS-N-21	USSR	Terrier	RIM-2	US
Sandbox	SS-N-12	USSR	Tigercat		UK
Scrubber	SS-N-1	USSR			
Sea Eagle		UK			
Sea Skua		UK			

Anti-Submarine

Shaddock	SS-N-3	USSR	ASROC	RUR-5A	US
Shipwreck	SS-N-19	USSR		FRAS-1	USSR
Siren	SS-N-7	USSR	Hedghog		US
	SS-N-24	USSR	Ikara		Aus.
Starbright	SS-N-9	USSR	Limbo		UK
Styx	SS-N-2	USSR	Malafon		Fr.
Sunburn	SS-N-22	USSR		RBU	USSR
Tomahawk	BGM-109	US	Sea Lance		US
			Silex	SS-N-14	USSR

Surface-To-Air

			Stallion	SS-N-16	USSR
ADATS		Swiss/US	Starfish	SS-N-15	USSR
Aspide		It.	SUBROC	UUM-44	US
Bloodhound		UK			
Blowpipe		UK			

Intercontinental Ballistic Missiles (ICBMs)

Chaparral	MIM-72	US			
Crotale		Fr.	Minuteman	LGM-30	US
FOG-M		US	Peacekeeper	LGM-118	US
Gadfly	SA-11/SA-N-7	USSR	Satan	SS-18	USSR

NAME	DESIGNATION	NATIONALITY
Savage	SS-13	USSR
Scalpel	SS-24	USSR
Sego	SS-11	USSR
SICBM		
Sickle	SS-25	USSR
Sinner	SS-16	USSR
Spanker	SS-17	USSR
Stilletto	SS-19	USSR
Titan	LGM-25	US

Submarine-Launched Ballistic Missiles (SLBMs)

NAME	DESIGNATION	NATIONALITY
MSBS		Fr.
Polaris	UGM-27	US
Poseidon	UGM-73	US
Sark	SS-N-4	USSR
Sawfly	SS-N-6	USSR
Serb	SS-N-5	USSR
Skiff	SS-N-23	USSR
Snipe	SS-N-17	USSR
	SS-N-8	USSR
Stingray	SS-N-18	USSR
Sturgeon	SS-N-20	USSR
Trident	UGM-96	US

Intermediate-Range Ballistic Missiles (IRBMs)

NAME	DESIGNATION	NATIONALITY
Saber	SS-20	USSR
SSBS	S-2	Fr.
SSBS	S-3	Fr.

Medium-Range Ballistic Missiles (MRBMs)

NAME	DESIGNATION	NATIONALITY
Al Husayn		Iraq
Condor		Arg/Egp.
Pershing	MGM-31	US
Sandal	SS-4	USSR

Short-Range Ballistic Missiles (SRBMs)

NAME	DESIGNATION	NATIONALITY
Hades		Fr.
Lance	MGM-52	US
Pluton		Fr.
Scaleboard	SS-12	USSR
Scarab	SS-21	USSR
Scud	SS-1	USSR
Spider	SS-23	USSR

TORPEDOES

NAME	DESIGNATION	NATIONALITY
	Mk.46	US
	Mk.48	US
Barracuda	Mk.50	US
Tigerfish	Mk.24	UK
Stingray		UK
Spearfish		UK

ARMORED FIGHTING VEHICLES

Main Battle Tanks

NAME	DESIGNATION	NATIONALITY
Abrams	M1	US
	AMX-30	Fr.
Centurion		UK
Challenger		UK
Chieftain		UK
Leopard 1		FRG
Leopard 2		FRG
Merkava		Is.
Patton	M47	US
Patton	M48	US
Patton	M60	US
S-Tank	Strv.103	Swe.
Sherman	M4	US
	T-10	USSR
	T-34/85	USSR
	T-55/55	USSR
	T-62	USSR
	T-64	USSR
	T-72	USSR
	T-80	USSR
TAM		Arg.
Vickers		UK
Vijayanta		Ind.

Light Tanks/Armored Reconnaissance Vehicles/Tank Destroyers

NAME	DESIGNATION	NATIONALITY
	AMX-13	Fr.
	ASU-57	USSR
	ASU-85	USSR
FISTV		US
	Ikv.91	Swe.
Improved TOW Vehicle	M901	US
Jagdpanzer		FRG
Piranha		Swiss
	PT-76	USSR
Scorpion	CVR(T)	UK
Sheridan	M551	US

Armored Cars

NAME	DESIGNATION	NATIONALITY
	AML	Fr.
	AMX-10RC	Fr.
	BRDM	USSR
	BTR-40	USSR
Commando		US
	EBR	Fr.
	ERC	Fr.
Ferret		UK
Fox		UK
Luchs		FRG
Saladin		UK

NAME	DESIGNATION	NATIONALITY
Armored Personnel Carriers/Infanty Fighting Vehicles		
	AMX-10P	Fr.
	BMD	USSR
	BMP	USSR
Bradley	M2/M3	US
	BTR-50	USSR
	BTR-60	USSR
	BTR-70	USSR
	BTR-80	USSR
LAV		US
	M113	US
Marder		FRG
	MT-LB	USSR
Saracen		UK
Saxon		UK
Trojan	FV.432	UK
VAB		Fr.
Warrior	MICV-80	UK

NAME	DESIGNATION	NATIONALITY
Air Defense Vehicles		
DIVAD		US
Gepard		FRG
Shilka	ZSU-23-4	USSR
Vulcan	M163	US
	ZSU-30	USSR
	ZSU-57-2	USSR

NAME	DESIGNATION	NATIONALITY
Self-Propelled Artillery		
Abbot	FV.438	UK
LARS		FRG
	M107	US
	M108	US
	M109	US
	M110	US
MLRS		US
	SP-70	FRG/It./UK

TREATIES AND ALLIANCES

ABM Treaty
Antarctic Treaty
ANZUS
CAFE
Geneva
Conventions
INF Treaty
MBFR
NATO
Non-Proliferation
Treaty
Outer Space Treaty
Rio Treaty
SALT I
SALT II
START
Test Ban Treaty
Warsaw Pact